Topological Fixed Point Principles for Boundary Value Problems

Topological Fixed Point Theory and Its Applications

VOLUME 1

Topological Fixed Point Principles for Boundary Value Problems

by

Jan Andres

Department of Mathematical Analysis,
Palacky University, Olomouc, Czech Republic

and

Lech Górniewicz

Juliusz Schauder Center for Nonlinear Studies,
Nicholas Copernicus University, Torun, Poland

KLUWER ACADEMIC PUBLISHERS
DORDRECHT / BOSTON / LONDON

A C.I.P. Catalogue record for this book is available from the Library of Congress.

Published by Kluwer Academic Publishers,
P.O. Box 17, 3300 AA Dordrecht, The Netherlands.

Sold and distributed in North, Central and South America
by Kluwer Academic Publishers,
101 Philip Drive, Norwell, MA 02061, U.S.A.

In all other countries, sold and distributed
by Kluwer Academic Publishers,
P.O. Box 322, 3300 AH Dordrecht, The Netherlands.

to Helena and Maria

Chapter II

General principles... **127**

Chapter III

Application to differential equations and inclusions....... **233**

Table of Contents

Appendices

Appendices

References

PREFACE

Our book is devoted to the topological fixed point theory both for single-valued and multivalued mappings in locally convex spaces, including its application to boundary value problems for ordinary differential equations (inclusions) and to (multivalued) dynamical systems. It is the first monograph dealing with the topological fixed point theory in non-metric spaces.

Although the theoretical material was tendentially selected with respect to applications, we wished to have a self-consistent text (see the scheme below). Therefore, we supplied three appendices concerning almost-periodic and derivo-periodic single-valued (multivalued) functions and (multivalued) fractals. The last topic which is quite new can be also regarded as a contribution to the fixed point theory in hyperspaces. Nevertheless, the reader is assumed to be at least partly familiar in some related sections with the notions like the Bochner integral, the Aumann multivalued integral, the Arzelà–Ascoli lemma, the Gronwall inequality, the Brouwer degree, the Leray–Schauder degree, the topological (covering) dimension, the elemens of homological algebra, ... Otherwise, one can use the recommended literature.

Hence, in Chapter I, the topological and analytical background is built. Then, in Chapter II (and partly already in Chapter I), topological principles necessary for applications are developed, namely:

- the fixed point index theory (resp. the topological degree theory),
- the Lefschetz and the Nielsen theories both in absolute and relative cases,
- periodic point theorems,
- topological essentiality,
- continuation-type theorems.

All the above topics are related to various classes of mappings including compact absorbing contractions and condensing maps. Besides the (more powerful) homological approach, the approximation techniques are alternatively employed as well.

In Chapter III, boundary value problems for differential equations and inclusions are investigated by means of the results in Chapter II. The following problems are mainly considered:

- existence of solutions,
- topological structure of solution sets,
- topological dimension of solution sets,
- multiplicity results,
- (subharmonic) periodic and almost-periodic solutions,

xi

– Ważewski-type results.

Equations (inclusions) are considered on compact or non-compact intervals. The particular attention is, however, paid to asymptotic boundary value problems, because they require special techniques for the representing Hammerstein operators in Fréchet spaces. Moreover, they are necessary for studying almost-periodic problems which are systematically treated here in various metrics. In the applications, an important role is also played by the Poincaré operators along the trajectories of solutions. Since the Poincaré operator associated with a given boundary value problem is shown to be admissible in the sense of the second author, the obtained fixed point results can be applied here, too.

Every chapter concludes by the section with many remarks and comments. It gives to the reader the possibility of further studies and historical information. The added bibliography is very (maybe too) large, but we wanted to put there everything connected with our goal.

Since in some branches of mathematics (from our point of view: especially topology and differential equations) different notations and terminology are used, those in our book are appropriate to the discussed subject. For example, we use:

- grad or ∇ for the gradient of a mapping,
- cl A or $\overline{A}$ for the closure of the set A,
- fr A or ∂A for the boundary of the set A.

Similarly, we do not distinguish between a multivalued map or a set-valued map or a multimap or a multioperator. On the other hand, the Hutchinson–Barnsley map and the Hutchinson–Barnsley operator defined in Appendix 3 are not synonyma. The notion of an attactor can have different meaning according to the context.

Although the majority of presented results is our own, we believe that the monograph can be useful for a wider auditorium, i.e. for post-graduate students and researchers working in the following fields of mathematics:

- topological fixed point theory,
- topological methods in nonlinear analysis,
- differential and functional equations and inclusions,
- boundary value and asymptotic problems,
- evolution problems,
- dynamical systems,
- systems with variable structure (including dry friction problems),
- nonlinear oscillations,
- almost-periodic and derivo-periodic problems,
- method of Liapunov functions,
- optimal control,
- chaos and fractals.

We would also like to stimulate the interest of mathematical economists, population dynamics experts, theoretical physicists exploring the topological dynamics, etc.

We are indebted to many coauthors of the quoted papers and to several younger colleagues who read and commented the manuscript. The latter group includes

T. Fürst, K. Leśniak, K. Pastor, D. Rozpłoch-Nowakowska, R. Skiba, ...

The partial support of grants (J14/98: 153100011 of the Council of Czech Government, 201-00-0768 of the Grant Agency of Czech Republic, 2 P03A 024 16 of the Polish KBN and M/2/2001 of the Nicolaus Copernicus University) is also highly appreciated.

The authors wish to express their gratitude to M. Czerniak who prepared the electronic version of the text.

Jan Andres, Lech Górniewicz

Olomouc–Toruń, December, 2002

Scheme for the relationship of single sections

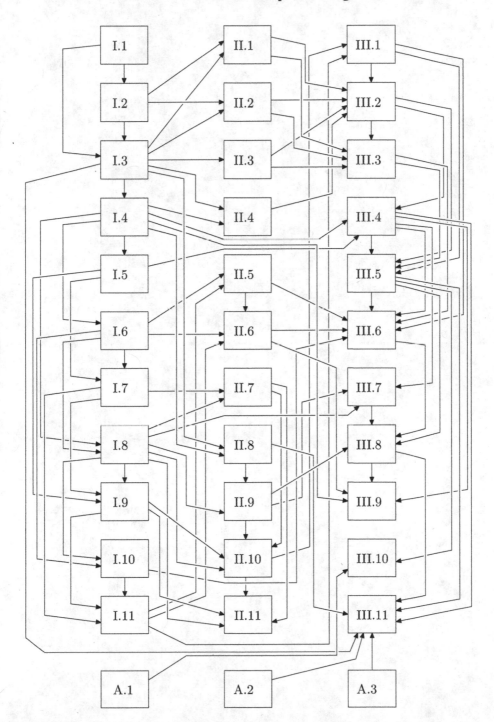

CHAPTER I

THEORETICAL BACKGROUND

I.1. Structure of locally convex spaces

At first, we recall some basic properties of topological vector spaces and, in particular, locally convex spaces; for more details, see e.g. [Kö-M], [Lu-M], [RoRo-M], [Ru-M], [Sch-M].

(1.1) DEFINITION. By a *topological vector space* we understand, as usual, a vector space, say E, over a field of scalars $\mathbb{K}$ ($\mathbb{K} = \mathbb{R}$ or $\mathbb{K} = \mathbb{C}$), endowed with the Hausdorff (T_2)-topology in which the operations

+ (addition): $E \times E \to E$, $\forall x \in E \ \forall y \in E \ +(x, y) = x + y$,
· (multiplication): $\mathbb{K} \times E \to E$, $\forall t \in \mathbb{K} \ \forall x \in E \ \cdot(t, x) = tx$

are continuous.

It is well-known that every topological vector space E is a *Tikhonov space* ($T_{3\frac{1}{2}}$-space), i.e., a totally regular T_1-space, namely the T_1-space such that, for every $x \in E$ and for every closed $F \subset E$ with $x \notin F$, there exists a continuous function $f : E \to [0, 1]$ such that $f(x) = 0$ and $f(y) = 1$, for every $y \in F$.

Moreover, if A is a compact subspace of a Tikhonov space X, then for every compact set $B \subset X \setminus A$ there exists a continuous function $f : X \to [0, 1]$ such that $f(x) = 0$, for every $x \in A$, and $f(x) = 1$, for every $x \in B$.

In the entire text, all topological vector spaces will be real (i.e. over $\mathbb{R}$). Furthermore, if E is a topological vector space and $A \subset E$, $B \subset E$, $x \in E$, $t \in \mathbb{R}$, then we adopt the notation:

$$x + A := \{x + a \mid a \in A\},$$
$$x - A := \{x - a \mid a \in A\},$$
$$A + B := \{a + b \mid a \in A, \ b \in B\},$$
$$tA := \{ta \mid a \in A\}.$$

(1.2) DEFINITION. Let E be a topological vector space. We say that a subset $A \subset E$:

(1.2.1) *absorbs* $B \subset E$ if there exists a number $\lambda_0 > 0$ such that $B \subset \lambda A$, for every $|\lambda| \geq \lambda_0$,

(1.2.2) is *absorbing* if A absorbs every finite subset of E, namely if for every $x \in E$ there exists a number $\lambda_0 > 0$ such that $x \in \lambda A$, for every $|\lambda| \geq \lambda_0$,

(1.2.3) is *convex* if, for every $t \in [0,1]$ and every $x, y \in A$, we have $tx + (1-t)y \in A$,

(1.2.4) is *balanced* if, for every λ with $|\lambda| \leq 1$, we have $\lambda A \subset A$,

(1.2.5) is *absolutely convex* if it is both convex and balanced, i.e. if (1.2.3) as well as (1.2.4) hold,

(1.2.6) is *star-shaped* w.r.t zero if, for every $x \in A$ and every $t \in [0,1]$, we have $tx \in A$.

Observe that if $A \subset E$ is balanced, then $A = -A$, i.e. A is *symmetric*. The same is obviously true for an absolutely convex set which is at the same time also star-shaped w.r.t. zero.

(1.3) DEFINITION. A function $p : E \to \mathbb{R}$ is a *seminorm* on E if

(1.3.1) $$\forall x \in E \quad p(x) \geq 0,$$

(1.3.2) $$\forall x \in E \quad \forall t \in \mathbb{R} \ p(tx) = |t|p(x),$$

(1.3.3) $$\forall x \in X \quad \forall y \in X \ p(x+y) \leq p(x) + p(y).$$

If moreover $p(x) = 0$ implies that $x = 0$, then p is the *norm* (on E).

A family P of seminorms on E is called *separating* if

(1.3.4) $$\forall x \in E \setminus \{0\} \ \exists p \in P \quad p(x) \neq 0.$$

Obviously, if p is a seminorm on E, then $p(0) = 0$; moreover, we have:

(1.3.5) $\quad \forall x \in E \quad \forall y \in E \ |p(x) - p(y)| \leq p(x - y),$

$$\text{and} \quad \{x \mid p(x) = 0\} \text{ is a subspace of } E.$$

Furthermore, E is *normable* if there exists a seminorm p on E such that

$$\forall x \in E \setminus \{0\} \quad p(x) \neq 0.$$

For an absorbing set $A \subset E$, the *Minkowski functional* on A is defined as follows

$$p_A : E \to \mathbb{R}, \quad p_A(x) := \inf\{t > 0 \mid x \in tA\}, \quad \text{for every } x \in E.$$

It is easy to verify the following:

(1.4) PROPOSITION. *Let p be a seminorm on E and $A = \{x \in E \mid p(x) < 1\}$. Then A is absorbing, absolutely convex and $p = p_A$.*

If $A \subset E$ is absorbing and absolutely convex, then the Minkowski functional p_A is a seminorm on E.

Now, let E be a topological vector space, $a \in E$, $t \in \mathbb{R} \setminus \{0\}$, and define the *translation operator* $T_a : E \to E$ as well as the *multiplication operator* $M_t : E \to E$ by the formulas

$$T_a(x) := a + x, \quad \text{for every } x \in E,$$
$$M_t(x) := tx, \qquad \text{for every } x \in E.$$

Then T_a, M_t are easily checked to be homeomorphisms of E onto E. Moreover, one can see that any subset of E is open if and only if the same is true for each of its translates. Thus, the Hausdorff topology $\mathcal{T}$ in E is completely determined by any local base. For our purposes, a local base can always be fixed as the one at 0 (if U is such a base, then $U_p = \{p + V \mid V \in U\}$ is the base at $p \in E$).

If $x \in E$, then the neighbourhood U_x of x is a set $U \in \mathcal{T}$ ($\mathcal{T}$ denotes the Hausdorff topology) with $x \in U$. Recall that a local base is a collection $\mathcal{B}$ of neighbourhoods of 0 such that each neighbourhood of 0 contains an element of $\mathcal{B}$. The open sets of E are then the unions of translates of elements of $\mathcal{B}$.

From the continuity of multiplication in E immediately follows that each neighbourhood of 0 is absorbing and contains some balanced neighbourhood of 0. Therefore, *every topological vector space has a local base consisting of balanced sets.*

(1.5) DEFINITION. Let $(E, \mathcal{T})$ be a topological vector space. We say that $A \subset E$ is *bounded* if for every neighbourhood U of 0 in E there exists $s > 0$ such that

$$\forall t \in (s, \infty) \quad A \subset tU.$$

(1.6) DEFINITION. A metric d in E is said to be *invariant* if

$$\forall x \in E \; \forall y \in E \; \forall z \in E \quad d(x + z, y + z) = d(x, y).$$

If E is a metric space with a metric d, then $A \subset E$ is called *d-bounded* if there is $s > 0$ such that

$$\forall x \in A \; \forall y \in A \quad d(x, y) \le s.$$

(1.7) REMARK. For a topological vector space E with a compatible metric d, i.e., a metric which generates the same topology on E, the bounded sets and the d-bounded sets need not be the same, even if d is invariant.

For instance, if E is a normed space and d is the metric induced by its norm, then these two notions coincide. However, they do not coincide for the metric d^* defined by

$$\forall x \in E \; \forall y \in E \quad d^*(x,y) = \frac{\|x - y\|}{1 + \|x - y\|},$$

which is invariant and induces the same topology on E as d.

The notion of boundedness can be reformulated in terms of sequences as follows:

(1.8) PROPOSITION. *For $A \subset E$, the following conditions are equivalent:*

(1.8.1) *A is bounded;*

(1.8.2) *If $\{x_n\}$ is a sequence in A and $\{\alpha_n\}$ is a sequence of reals such that $\lim_{n \to \infty} \alpha_n = 0$, then $\lim_{n \to \infty} \alpha_n x_n = 0$.*

(1.9) DEFINITION. We say that a topological vector space E is *locally bounded* if 0 has a bounded neighbourhood. E is said to be *locally compact* if 0 has a neighbourhood whose closure is compact.

Obviously, every locally compact space is locally bounded.

(1.10) PROPOSITION. *For a topological vector space E, the following is true:*

(1.10.1) *If E is locally bounded, then it has a countable local base;*

(1.10.2) *E is metrizable if and only if it has a countable local base;*

(1.10.3) *E has finite dimension if and only if it is locally compact;*

(1.10.4) *If $\dim E = n \in \mathbb{N}$, then every (algebraic) isomorphism of E onto $\mathbb{R}^n$ is a homeomorphism.*

Now, we can proceed to an important class of locally convex spaces.

(1.11) DEFINITION. A topological vector space E is said to be *locally convex* if it has a local base whose elements are convex.

(1.12) PROPOSITION. *Every locally convex space has a local base $\mathcal{B}$ consisting of absolutely convex sets. If $\mathcal{B}$ is such a base and $U \in \mathcal{B}$, then $P_\mathcal{B} := \{p_U \mid U \in \mathcal{B}\}$ represents a separating family of continuous seminorms on E, where p_U denotes the Minkowski functional on $U \in \mathcal{B}$.*

(1.13) PROPOSITION. *Let E be a topological vector space and P be a separating family of seminorms on E. Associate to every pair $(p,n) \in P \times \mathbb{N}$ the set $U(p,n) = \{x \mid p(x) < \frac{1}{n}\}$ and denote by $\mathcal{B}$ the collection of all finite intersections of the sets $U(p,n)$. Then $\mathcal{B}$ consists of absorbing absolutely convex sets.*

Furthermore, E can be endowed with a topology $\mathcal{T}$ making E a locally convex space with the following properties:

(1.13.1) *$\mathcal{B}$ is a local base in E;*

(1.13.2) *Every $p \in P$ is continuous;*

(1.13.3) *$A \subset E$ is bounded $\Leftrightarrow \forall p \in P \; \exists \alpha_p \in \mathbb{R} \; \forall x \in A \; p(x) < \alpha_p$.*

(1.14) DEFINITION. We say that the topology $\mathcal{T}$ in (1.13) is *induced by given system P of seminorms* in a topological vector space E.

If E is a locally convex space with a topology, say $\mathcal{T}_1$, then it has a local base $\mathcal{B}$ whose elements are absolutely convex. Thus, the related system $P_\mathcal{B}$ of Minkowski's functionals of sets in $\mathcal{B}$ represents, according to (1.12), a separating family of seminorms in E. If $\mathcal{T}$ is the topology in E induced by $P_\mathcal{B}$, then $\mathcal{T} = \mathcal{T}_1$.

Indeed, since every $p \in P_\mathcal{B}$ is, according to (1.12), $\mathcal{T}_1$-continuous, every set $U(p,n)$ is open in $\mathcal{T}_1$ and so we can conclude that $\mathcal{T} \subset \mathcal{T}_1$. For the reverse inclusion, fix an arbitrary $U \in \mathcal{B}$. Since U is absorbing and absolutely convex, the Minkowski functional p_U is, according to (1.4), a seminorm on E. Thus, in view of (1.4), $U = \{x \mid p_U(x) < 1\} = U(p_U, 1) \in \mathcal{T}$, and subsequently $\mathcal{T}_1 \subset \mathcal{T}$, as claimed.

For $p_1, p_2 : E \to \mathbb{R}$, define $\max\{p_1, p_2\} : E \to \mathbb{R}$ as follows:

$$\forall x \in E \quad \max\{p_1, p_2\}(x) = \max\{p_1(x), p_2(x)\}.$$

(1.15) PROPOSTION. *Let P_1, P be separating families of seminorms in a topological vector space E and denote*

$$U(p, n) := \left\{ x \;\middle|\; p(x) < \frac{1}{n} \right\}.$$

Assume that P is the smallest system of seminorms such that

(1.15.1) $P_1 \subset P,$

(1.15.2) $\forall p_1 \in P \; \forall p_2 \in P \quad \max\{p_1, p_2\} \in P.$

Then P induces the same topology $\mathcal{T}$ in E as P_1. Moreover, the system $\mathcal{B} = \{U(p, n) \mid p \in P, \; n \in \mathbb{N}\}$ represents a local base in $\mathcal{T}$.

PROOF. Let P_1, P induce topologies $\mathcal{T}_1$, $\mathcal{T}$, respectively. It follows from (1.13.1) that $\mathcal{B}$ is a local base in $\mathcal{T}$. Since $\{U(p, n) \mid p \in P_1, \; n \in \mathbb{N}\} \subset \{U(p, n) \mid p \in P, \; n \in \mathbb{N}\}$, we can conclude that $\mathcal{T}_1 \subset \mathcal{T}$. Since, for every $(p, n) \in P \times \mathbb{N}$, there exist $n \in \mathbb{N}$ and $p_1, \ldots p_r \in P_1$ such that $\bigcap_{j=1}^{n} U(p_j, r) = U(p, n)$, we can also conclude that $\mathcal{T} \subset \mathcal{T}_1$. $\square$

Evidently, $\max\{p_1, p_2\}$ is a seminorm if so are p_1 and p_2. Thus, the existence of P in (1.15) is ensured for every separating family P_1 of seminorms. Note that P can be considered as a partially ordered set, provided we define, for $p_1, p_2 \in P$,

$$p_1 \leq p_2 \Leftrightarrow \forall x \in E \quad p_1(x) \leq p_2(x).$$

Assume that a topological vector space E endowed with a topology $\mathcal{T}$ has a countable local base. Then, according to (1.10.2), E is metrizable. If E is only locally convex, then there exists a countable separating family $P = \{p_j \mid j = 1, 2, \dots\}$ of continuous seminorms in E and the related metric $d : E \times E \to \mathbb{R}$ can be directly defined in terms of seminorms in the following way:

$$(1.16) \qquad \forall x \in E \; \forall y \in E \quad d(x,y) = \sum_{j=1}^{\infty} \frac{1}{2^j} \frac{p_j(x-y)}{1 + p_j(x-y)}.$$

This d in (1.16) is an invariant metric in E which induces the primary topology $\mathcal{T}$ in E.

(1.17) REMARK. Let us note that, for $r > 0$, the d-ball $\{x \mid d(x,0) < r\}$ need not be convex. On the other hand, for every metrizable locally convex space, we can construct in a more complicated way an equivalent metric d^* whose open balls are convex sets.

(1.18) PROPOSITION. *Let E be a locally convex space whose topology is induced by a countable family $P = \{p_j \mid j = 1, 2, \dots\}$ of seminorms. If $\{x_n\}$ is a sequence in E, then*

$$\lim_{n \to \infty} x_n = x \Leftrightarrow (\forall j \in \mathbb{N} \quad \lim_{n \to \infty} p_j(x_n - x) = 0)$$
$$\Leftrightarrow (\forall j \in \mathbb{N} \; \forall \varepsilon > 0 \; \exists n_\varepsilon \in \mathbb{N} \; \forall n \in \mathbb{N} \; (n > n_\varepsilon \Rightarrow p_j(x_n - x) < \varepsilon)).$$

PROOF. Since E has a vector topology, it is enough to show that

$$\lim_{n \to \infty} x_n = 0 \Leftrightarrow \lim_{n \to \infty} p_j(x_n) = 0, \quad \text{for all } j \in \mathbb{N}.$$

So, let, $\lim_{n \to \infty} x_n = 0$, at first. Fix $j \in \mathbb{N}$ and take $\varepsilon > 0$. Since $\{x \mid p_j(x) < \varepsilon\}$ is an open neighbourhood of 0, there exists $n_\varepsilon \in \mathbb{N}$ such that

$$\forall n \in \mathbb{N} \; (n > n_\varepsilon \Rightarrow x_n \in \{x \mid p_j(x) < \varepsilon\}).$$

Therefore,

$$\forall n \in \mathbb{N} \; (n > n_\varepsilon \Rightarrow p_j(x_n) < \varepsilon)$$

which already implies that $\lim_{n \to \infty} p_j(x_n) = 0$.

Now, assume that $\lim_{n \to \infty} p_j(x_n) = 0$, for every $j \in \mathbb{N}$. Let U be a neighbourhood of 0. Obviously, there exist positive integers $j^*, i_1, i_2, \dots, i_{j^*}$ and positive reals $\varepsilon_1, \varepsilon_2, \dots, \varepsilon_{j^*}$ such that

$$\bigcap_{k=1}^{j^*} \{x \mid p_{i_k}(x) < \varepsilon_k\} \subset U.$$

Defining $\varepsilon = \min\{\varepsilon_k \mid k = 1, \ldots, j^*\}$, we obtain, by the hypothesis, that

$$\forall n \in \mathbb{N}\ (n > n_\varepsilon \Rightarrow x_n \in \bigcap_{k=1}^{j^*}\{x \mid p_{i_k}(x) < \varepsilon_k\} \subset U).$$

Since U was arbitrary, we can conclude that $\lim_{n\to\infty} x_n = 0$. $\qquad\square$

(1.19) REMARK. Under the assumptions of (1.18), it can be proved quite analogously that $\{x_n\}$ is a Cauchy sequence if and only if

$$\forall j \in \mathbb{N}\ \forall \varepsilon > 0\ \exists n_0 \in \mathbb{N}\ \forall m \in \mathbb{N}\ \forall n \in \mathbb{N}\quad ((m > n_0, n > n_0) \Rightarrow p_j(x_m - x_n) < \varepsilon).$$

(1.20) DEFINITION. A locally convex space which is metrizable and complete is called a *Fréchet space*.

If, in a locally convex space E, a norm p is defined, then E with the topology given by the one element family $P = \{p\}$ is the metrizable (Hausdorff) space called a *normed space*.

It follows from the well-known Kolmogorov theorem that if a locally convex space E is locally bounded (see (1.9)), then it is normable. This yields that, in a metrizable, but not normable space E, the balls $B(0, \varepsilon) = \{x \in E \mid d(x, 0) < \varepsilon\}$ are not bounded.

In fact, the following statement is true for any topological vector (Hausdorff) space:

(1.21) PROPOSITION. *A topological vector space is normable if and only if it is locally bounded and locally convex.*

(1.22) REMARK. If U is a bounded absolutely convex neighbourhood of 0, then the Minkowski functional p_U can be taken as the related norm.

Obviously, every Banach space is Fréchet. Perhaps the most popular examples of Fréchet spaces are however:

$C(J, \mathbb{R}^n)$,

i.e. the one of continuous functions $x : J \to \mathbb{R}^n$, where $J \subset \mathbb{R}$ is an arbitrary interval, with the topology of a uniform convergence on compact subintervals of J, and

$C^{(k)}(J, \mathbb{R}^n)$, $k \in \mathbb{N}$,

i.e. the one of continuously differentiable functions of the kth order $x : J \to \mathbb{R}^n$, where $J \subset \mathbb{R}$ is an arbitrary interval, with the topology of a uniform convergence on compact subintervals of J of x as well as their jth derivatives $x^{(j)} : J \to \mathbb{R}^n$, where $j = 1, \ldots, k$.

To be more precise, denote by $\emptyset \neq J \subset \mathbb{R}$ and $\mathcal{K} \subset \mathbb{R}$ an arbitrary interval (e.g. open) and a compact interval with nonempty interior, respectively. Furthermore, for $k \in \mathbb{N}$, set

$$C^{(k)}(J, \mathbb{R}^n) := \{q \in C(J, \mathbb{R}^n) \mid \forall j = 1, \ldots, k \quad q^{(j)} \in C(J, \mathbb{R}^n)\},$$
$$C^{(k)}(\mathcal{K}, \mathbb{R}^n) := \{q \in C(\mathcal{K}, \mathbb{R}^n) \mid \forall j = 1, \ldots, k \quad \exists w \in C(\mathcal{K}, \mathbb{R}^n) \; w|_{\mathrm{int}\mathcal{K}} = q^{(j)}\}.$$

For $q \in C^{(k)}(\mathcal{K}, \mathbb{R}^n)$ and $j \in \{1, \ldots k\}$, we reserve the symbol $q^{(j)}$ for a continuous extension of the jth derivative of q on $\mathcal{K}$. The following pairs of real vector spaces with the associated norms are well-known to be Banach:

$$(C(\mathcal{K}, \mathbb{R}^n), \|\cdot\|_1), \quad \text{where for all } q \in C(\mathcal{K}, \mathbb{R}^n) \quad \|q\|_1 := \max_{t \in \mathcal{K}} |q(t)|,$$

$$(C^{(k)}(\mathcal{K}, \mathbb{R}^n), \|\cdot\|_2), \quad \text{where for all } q \in C^{(k)}(\mathcal{K}, \mathbb{R}^n) \quad \|q\|_2 := \sum_{j=0}^{k} \max_{t \in \mathcal{K}} |q^{(j)}(t)|.$$

Let $\{\mathcal{K}_i\}$ be a sequence of compact subintervals of J. Assume that

$$(1.23) \qquad\qquad \bigcup_{i=1}^{\infty} \mathcal{K}_i = J,$$

$$(1.24) \qquad\qquad \forall i \in \mathbb{N} \quad \mathcal{K}_i \subset \mathcal{K}_{i+1},$$

and define, for every $i \in \mathbb{N}$, the function $p_i : C(J, \mathbb{R}^n) \to \mathbb{R}$ by the formula

$$p_i(q) := \max_{t \in \mathcal{K}_i} |q(t)|.$$

Furthermore, for $i \in \mathbb{N}$ and $k \in \mathbb{N}$, define

$$p_i^* : C^{(k)}(J, \mathbb{R}^n) \to \mathbb{R}, \quad \forall q \in C^{(k)}(J, \mathbb{R}^n) \quad p_i^*(q) = \max_{t \in \mathcal{K}_i} |q(t)| + \max_{t \in \mathcal{K}_i} |q^{(k)}(t)|.$$

These functions are obviously seminorms and (1.24) implies

$$\forall i \in \mathbb{N} \; \forall q \in C(J, \mathbb{R}^n) \qquad p_i(q) \leq p_{i+1}(q),$$
$$\forall i \in \mathbb{N} \; \forall q \in C^{(k)}(J, \mathbb{R}^n) \quad p_i^*(q) \leq p_{i+1}^*(q).$$

Denoting $P := \{p_i \mid i \in \mathbb{N}\}$, $P^* := \{p_i^* \mid i \in \mathbb{N}\}$, we obtain from (1.23) that both systems are countable and separating families in $C(J, \mathbb{R}^n)$ and $C^{(k)}(J, \mathbb{R}^n)$, respectively. Moreover, they are closed under max (see (1.15.2)). Therefore, in view of (1.13) and (1.15), systems

$$\{U(p_i, n) \mid i \in \mathbb{N}, \; n \in \mathbb{N}\}, \quad \{U(p_i^*, n) \mid i \in \mathbb{N}, \; n \in \mathbb{N}\},$$

where

$$\forall i \in \mathbb{N} \ \forall n \in \mathbb{N} \quad U(p_i, n) = \left\{ q \in C(J, \mathbb{R}^n) \, | \, p_i(q) < \frac{1}{n} \right\},$$

$$\forall i \in \mathbb{N} \ \forall n \in \mathbb{N} \quad U(p_i^*, n) = \left\{ q \in C^{(k)}(J, \mathbb{R}^n) \, | \, p_i^*(q) < \frac{1}{n} \right\},$$

form local bases for some topologies $\mathcal{T}$, $\mathcal{T}^*$ in $C(J, \mathbb{R}^n)$ and $C^{(k)}(J, \mathbb{R}^n)$, respectively.

Endowing $C(J, \mathbb{R}^n)$ and $C^{(k)}(J, \mathbb{R}^n)$ with these topologies, they become (in view of the above theory) Fréchet.

As pointed out in (1.16), the related topologies can be generated by the metrics (see (1.23), (1.24))

$$(1.25) \qquad d(x, y) = \sum_{j=1}^{\infty} \frac{1}{2^j} \cdot \frac{p_j(x - y)}{1 + p_j(x - y)},$$

where $p_j(q) = \max_{t \in \mathcal{K}_j} |q(t)|$, and

$$(1.26) \qquad d(x, y) = \sum_{j=1}^{\infty} \frac{1}{2^j} \cdot \frac{p_j^*(x - y)}{1 + p_j^*(x - y)},$$

where $p_j^*(q) = p_j(q) + p_j(q^{(k)})$, respectively.

Note that $C^{(k)}(J, \mathbb{R})$ can be embedded into a closed subset of $C(J, \mathbb{R}^{k+1})$ via the map $x \to (x^{(0)}, \ldots, x^{(k)})$.

Because of applications, it will be convenient to recall basic properties of $C(J, \mathbb{R}^n)$ and $C^{(k)}(J, \mathbb{R}^n)$, $k \in \mathbb{N}$. Let $q \in C(J, \mathbb{R}^n)$ $[q \in C^{(k)}(J, \mathbb{R}^n)]$ and $\{q_i\}$ be in $C(J, \mathbb{R}^n)$ [in $C^{(k)}(J, \mathbb{R}^n)$]. Then, according to (1.18), the following conditions are equivalent:

$(1.27) \quad \lim\limits_{i \to \infty} q_i = q,$

$(1.28) \quad \lim\limits_{i \to \infty} (q_i|_{\mathcal{K}} - q|_{\mathcal{K}}) = 0$ in $(C(\mathcal{K}, \mathbb{R}^n), \| \cdot \|_1)$ [in $(C^{(k)}(\mathcal{K}, \mathbb{R}^n), \| \cdot \|_2)$],

for every compact subinterval $\mathcal{K}$ such that $\emptyset \neq \text{int}\mathcal{K} \subset J$. As pointed out, these spaces are endowed with the topologies of the uniform convergence on compact subintervals of J.

Furthermore, according to (1.13.3), $S \subset C(\mathcal{K}, \mathbb{R}^n)$ $[S \subset C^{(k)}(\mathcal{K}, \mathbb{R}^n)]$ is bounded if and only if there exists $\varphi \in C(\mathcal{K}, \mathbb{R})$ such that

$$\forall q \in S \ \forall t \in J \quad |q(t)| < \varphi(t) \quad [\forall q \in S \ \forall t \in J \ \forall j = 0, \ldots, k \ |q^{(j)}(t)| < \varphi(t)].$$

Since both spaces are metrizable (see (1.25), (1.26)), the following conditions are equivalent:

(1.29) S is closed,

(1.30) $\lim\limits_{i\to\infty} q_i = q \Rightarrow q \in S$,   for every $q \in C(J, \mathbb{R}^n)$ $[q \in C^{(k)}(J, \mathbb{R}^n)]$

 and every sequence $\{q_i\}$ in S.

At last, it follows from the well-known Arzelà–Ascoli lemma that $S \subset C(J, \mathbb{R}^n)$ $[S \subset C^{(k)}(J, \mathbb{R}^n)]$ is relatively compact if and only if it is bounded and functions in S [their kth order derivatives] are equicontinuous in every $t \in J$, i.e.

$$\forall t \in J \; \forall \varepsilon > 0 \; \exists \delta > 0 \; \forall t^* \in J \; \forall q \in S \quad |t - t^*| < \delta \Rightarrow |q(t) - q(t^*)| < \varepsilon$$

$$[\forall t \in J \; \forall \varepsilon > 0 \; \exists \delta > 0 \; \forall t^* \in J \; \forall q \in S \; \forall j = 0, \ldots, k$$

$$|t - t^*| < \delta \Rightarrow |q^{(j)}(t) - q^{(j)}(t^*)| < \varepsilon].$$

Below, we give (only briefly) two further examples of Fréchet spaces which are suitable for applications.

(1.31) EXAMPLE. The space $L^1_{\mathrm{loc}}(J, \mathbb{R})$, where $J \subset \mathbb{R}$ is an arbitrary interval, of locally Lebesgue-integrable functions $x : J \to \mathbb{R}^n$, i.e. integrable on every compact subinterval of J, is a Fréchet space. The related family of seminorms $\{p_{\mathcal{K}_j}\}$ is of the form $p_{\mathcal{K}_j}(q) = \int_{\mathcal{K}_j} |q| \, dt$, where $\{\mathcal{K}_j\}$ is a sequence of compact subintervals of J such that $\bigcup_{j=1}^{\infty} \mathcal{K}_j = J$ and $\mathcal{K}_j \subset \mathcal{K}_{j+1}$, for all $j \in \mathbb{N}$. In particular, for a compact $J \subset \mathbb{R}$, $L^1(J, \mathbb{R}^n)$ becomes Banach.

(1.32) EXAMPLE. The space $H^{k,1}_{\mathrm{loc}}(J, \mathbb{R}^n)$, where $J \subset \mathbb{R}$ is an arbitrary interval, of continuously differentiable (up to the $(k-1)$th order) functions $x : J \to \mathbb{R}^n$ with locally absolutely continuous $(k-1)$th derivatives, is a Fréchet space. The related family of seminorms $\{p_{\mathcal{K}_j}\}$ is of the form

$$p_{\mathcal{K}_j}(q) = \sum_{i=1}^{k-1} \max_{t \in \mathcal{K}_j} |x^{(i)}(t)| + \int_{\mathcal{K}_j} |x^{(k)}(t)| \, dt,$$

where $\{\mathcal{K}_j\}$ is a sequence of compact subintervals of J such that $\bigcup_{j=1}^{\infty} \mathcal{K}_j = J$ and $\mathcal{K}_j \subset \mathcal{K}_{j+1}$, for all $j \in \mathbb{N}$. In particular, $\mathrm{AC}_{\mathrm{loc}}(J, \mathbb{R}^n) = H^{1,1}_{\mathrm{loc}}(J, \mathbb{R}^n)$ is a Fréchet space and, for a compact $J \subset \mathbb{R}$, $H^{k,1}(J, \mathbb{R}^n)$ and $\mathrm{AC}(J, \mathbb{R}^n)$ become Banach.

The following statement will be important for us in the sequel.

(1.33) THEOREM (Mazur) (cf. [Mu-M, Theorem 21.4] or [Au-M, p. 271] or [De3-M]). *If E is a normed space and the sequence $\{x_k\} \subset E$ is weakly convergent to $x \in$*

E, then there exists a sequence of linear combinations $y_m = \sum_{k=1}^{m} a_{m_k} x_k$, where $a_{m_k} \geq 0$, for $k = 1, 2, \ldots, m$, and $\sum_{k=1}^{m} a_{m_k} = 1$, which is strongly convergent.

We conclude this section by a generalization of Schauder's approximation theorem. Let X be a (Hausdorff) topological vector space. For $Y \subset X$, denote by the symbol $\operatorname{Cov}_X(Y)$ the set of all open in X, coverings of Y ($\operatorname{Cov}(X) = \operatorname{Cov}_X(X)$). The elements of $\operatorname{Cov}_X(Y)$ will be denoted by Greek letters α, β, γ. Hence,

$$\alpha \in \operatorname{Cov}_X(Y) \Leftrightarrow \alpha = \{U_t\}_{t \in T} \quad \text{and} \quad Y \subset \bigcup_{t \in T} U_t,$$

where every set U_t is open in X.

(1.34) DEFINITION. For $\alpha \in \operatorname{Cov}(X)$, we say that continuous mappings $f, g : Y \to X$ are α-close if, for every $x \in Y$, there exists an open set $U_x \in \alpha$ such that $f(x), g(x) \in U_x$.

(1.35) PROPOSITION. Let Y be a compact subset of a topological vector space X and $i : Y \to X$, $i(x) = x$, be the inclusion map. Then the following two statements are equivalent:

(1.35.1) for every $\alpha \in \operatorname{Cov}(X)$, there exists a continuous map $g : Y \to X$ which is α-close to i,

(1.35.2) for every open neighbourhood V of the zero point in X, there exists a continuous map $g : Y \to X$ such that $(x - g(x)) \in V$, for every $x \in Y$.

PROOF. Obviously (1.35.1) implies (1.35.2). Conversely, let $\alpha = \{U_t\}_{t \in T} \in \operatorname{Cov}(X)$. For every $x \in X$, there exists an open neighbourhood U_x of the point zero in X such that the covering $\beta = \{x + U_x\}_{x \in X}$ is a refinement of α. For every $x \in X$, we choose V_x to be an open neighbourhood of the zero point in X such that $V_x + V_x \subset U_x$. Then $\beta_1 = \{x + V_x\}_{x \in Y}$ is again a refinement of α. Let $\{x_1 + V_{x_1}, x_2 + V_{x_2}, \ldots, x_k + V_{x_k}\}$ be a finite refinement of β_1 and let $V = \bigcap_{i=1}^{k} V_{x_i}$. Observe that if $g : Y \to X$ is a map such that $(x - g(x)) \in V$, for every $x \in Y$, then g is α-closed to $i : Y \to X$ and the proof is completed. $\qquad \square$

In the frame of normed (vector) spaces, one speaks about ε-close maps, i.e. about particular class of α-close maps, where α denotes the covering of a normed space by the balls with centers in arbitrary points and radius $\varepsilon > 0$. The related result concerning the approximation of mappings is due to J. P. Schauder.

(1.36) SCHAUDER'S APPROXIMATION THEOREM. Let E be a normed (vector) space and K its compact subset. Then, for every $\varepsilon > 0$, there exist a finite subset $N = \{c_1, \ldots, c_n\}$ of E and a mapping $F_\varepsilon : K \to \operatorname{conv} N$ such that

(1.36.1) $\qquad \qquad \|F_\varepsilon(x) - x\| < \varepsilon, \quad \text{for every } x \in E,$

(1.36.2) $\qquad \qquad F_\varepsilon(E) \subset \operatorname{conv} N,$

where conv N *denotes the convex hull of* N.

The generalization of (1.36) for locally convex spaces leads to the following result.

(1.37) THEOREM. *Let E be a locally convex space, K a compact subset of E and C a convex subset of E with $K \subset C$. Then given an open neighbourhood U of 0 (the zero element of E), there exists a continuous mapping $\pi_U : K \to C$, with*

(1.37.1) $$\pi_U(K) \subset L$$

(1.37.2) $$(\pi_U(x) - x) \in U, \quad \text{for } x \in K,$$

where L is a finite dimensional subspace of E.

PROOF. Without any loss of generality, assume that U is convex and balanced. Let

$$|x|_U := \inf\{\alpha > 0 \mid x \in \alpha U\}$$

be the Minkowski functional associated with U. Obviously, $x \mapsto |x|_U$ is a continuous seminorm on E and

$$U = \{x \mid x \in E \quad \text{and} \quad |x|_U < 1\}.$$

Since K is compact, there exists a finite set $\{a_i, \ldots, a_n\} \subset K$ such that

$$K \subset \bigcup_{i=1}^{n} U(a_i),$$

where $U(a) := U + a$, for $a \in E$. Define the function μ_i, $i = 1, \ldots, n$, by

$$\mu_i(x) := \max\{0, 1 - |x - a_i|_U\}, \quad \text{for } x \in E.$$

Since $| \cdot |_U$ is a continuous function on E, we have that $\mu_i(\cdot)$, $i = 1, \ldots, n$, is also a continuous function on E. In addition, for $i = 1, \ldots, n$,

$$0 \leq \mu_i(x) \leq 1, \quad \text{for } x \in E,$$

with

$$\mu_i(x) = 0, \text{ when } x \notin U(a_i), \quad \text{and} \quad \mu_i(x) > 0, \text{ otherwise.}$$

Let

$$\pi_U(x) := \frac{\sum_{i=1}^{n} \mu_i(x) a_i}{\sum_{i=1}^{n} \mu_i(x)}, \quad \text{for } x \in K.$$

Notice that π_U is well-defined, because if $x \in K$, then $x \in U(a_i)$, for some $i \in \{1, \ldots, n\}$, and so $\sum_{i=1}^{n} \mu_i(x) \neq 0$. Observe also that π_U is a continuous function on K. Its values obviously belong to the linear subspace L, generated by $\{a_i \mid i = 1, \ldots, n\}$. In addition, since $K \subset C$ and C is convex, we have that

$$\pi_U(x) \in C, \quad \text{for each } x \in K.$$

Therefore,

$$\pi_U(x) \in L \cap C, \quad \text{for each } x \in K.$$

Furthermore, notice that

$$\pi_U(x) - x := \frac{\sum_{i=1}^n \mu_i(x)(a_i - x)}{\sum_{i=1}^n \mu_i(x)}, \quad \text{for } x \in K,$$

and so

$$|\pi_U(x) - x|_U = \frac{\sum_{i=1}^n \mu_i(x)|a_i - x|_U}{\sum_{i=1}^n \mu_i(x)} < 1, \quad \text{for } x \in K,$$

because, for any $i = 1, \ldots, n$, either $\mu_i(x) = 0$ and $|a_i - x|_U \geq 1$ or $\mu_i(x) > 0$ and $|a_i - x|_U < 1$. This immediately yields $\pi_U(x) - x \in U$, for $x \in K$. $\qquad \square$

Now, from (1.37) and the classical Brouwer fixed point theorem, we get:

(1.38) THEOREM (Schauder–Tikhonov). *Let E be a locally convex space, C a convex subset of E and $F : C \to E$ a continuous mapping such that*

$$F(C) \subset K \subset C$$

with K compact. Then F has at least one fixed point.

PROOF. Let U be an open, convex, balanced neighbourhood of 0 (the zero element of E) and π_U be as in Theorem (1.37). Define a function F_U by

$$F_U(x) := \pi_U F(x), \quad \text{for } x \in C.$$

Since π_U takes values in the space L (defined in Theorem (1.37)), we shall restrict our considerations to this space. From Theorem (1.37), we have that

(1.38.1) $$F_U(L \cap C) \subset \pi_U(K) \subset L \cap C,$$

because if $x \in L \cap C$, then $F(x) \in K$, and so

$$F_U(x) = \pi_U(F(x)) \subset L \cap C.$$

Let K^* denote the convex hull of the compact set $\pi_U(K)$ in L. Note that K^* is compact. In addition, (1.38.1) and

$$\pi_U(A) \subset K^* \subset L \cap C$$

imply that

(1.38.2) $$F_U(K^*) \subset K^*.$$

Appling the Brouwer fixed point theorem, we can deduce that there exists $x \in K^*$ with $x = F_U(x)$. That is, x satisfies

(1.38.3) $$x - F(x) \in U,$$

because $x = F_U(x)$ is equivalent to $x = \pi_U(F(x))$, and therefore, by Theorem (1.37), we have that
$$\pi_U(F(x)) - F(x) \in U.$$

We have shown:

(1.38.4) $$\begin{cases} \text{to any open neighbourhood } U \text{ of } 0, \text{ there exists} \\ \text{at least one } x \in K^* \subset C \text{ such that (1.38.3) holds.} \end{cases}$$

Suppose now that $x \neq F(x)$, for all $x \in C$. The continuity of F and the fact that E is Hausdorff guarantee that there exist two open neighbourhoods V_x and W_x of 0 with the properties[1]

(1.38.5) $$F(C \cap V_x(x)) \subset W_x(F(x))$$

and

(1.38.6) $$V_x(x) \cap W_x(F(x)) = \emptyset.$$

Choose U_s to be another open neighbourhood of 0 such that

(1.38.7) $$2U_x \subset V_x \cap W_x.$$

Since K is compact, there exists a finite set $\{a_i \mid i = 1, \ldots, n\} \subset K$ with
$$K \subset \bigcup_{i=1}^{n} U_{a_i}(a_i).$$

We claim that, for any $x \in C$, there exists $j \in \{1, \ldots, n\}$ such that

(1.38.8) $$x - F(x) \subset U_{a_j}$$

---

[1]For simplicity, we let $W_x(F(x)) = F(x) + W_x$.

cannot hold. Fix $x \in C$. Since $y = F(x) \in K$, there exists $j \in \{1, \ldots, n\}$ with $y \in U_{a_j}(a_j)$. In addition, we have

$$(1.38.9) \qquad\qquad U_{a_j}(y) \subset V_{a_j}(a_j).$$

To see this, notice that

$$y = u + a_j, \quad \text{for some } u \in U_{a_j}.$$

Therefore, if $z \in U_{a_j}(y)$, then there exists $w \in U_{a_j}$ with

$$z = w + y = w + u + a_j,$$

and consequently

$$z \in 2U_{a_j} + a_j \subset V_{a_j}(a_j),$$

according to (1.38.7).

Suppose that (1.38.8) is not true. Then, for any $x \in C$ we have that $x \in U_{a_j}(y)$ with $y = F(x)$, and so from (1.38.9), we see that $x \in V_{a_j}(a_j)$. Now, (1.38.5) guarantees that

$$y = F(x) \in W_{a_j}(F(a_j)).$$

However, $y \in W_{a_j}(F(a_j))$ and (1.38.6) imply

$$y \notin V_{a_j}(a_j),$$

which contradicts to (1.38.9). Therefore, (1.38.8) cannot be true.

Choose U such that

$$U \subset \bigcap_{i=1}^{n} U_{a_i}.$$

From what we deduced above, it follows that

$$x - F(x) \notin U, \quad \text{for all } x \in C.$$

Consequently, there exists $x \in C$ with $x = F(x)$. $\qquad\qquad\square$

(1.39) COROLLARY. *Let C be a convex subset of a locally convex space E. Suppose that $F : C \to C$ is a continuous, compact map. Then F has at least one fixed point in C.*

(1.40) REMARK. Topological vector spaces with properties (1.37.1), (1.37.2) are called *admissible in the sense of V. Klee* They represent a very large class of spaces. Besides locally convex spaces, playing a fundamental role for our purposes, they contain the space of measurable functions on a compact interval, the $L^p(0,1)$-space, where $0 < p < 1$, Hardy's spaces H^r, and many others. For more details, see [Kle], [Rob].

I.2. ANR-spaces and AR-spaces

We also need some facts from the geometric topology. For the proofs, we recommend [Go5-M]. For more details concerning the theory of retracts, see [Brs-M], [Hu2-M]. If we deal with ANR-spaces or AR-spaces, then we assume that the related spaces are metric.

We say that a space X *possesses an extension property* (written $X \in ES$) if, for every space Y, every closed $B \subset Y$, and every continuous map $f : B \to X$, there exists a continuous extension $\tilde{f} : Y \to X$ of f onto Y, i.e., $\tilde{f}(x) = f(x)$, for each $x \in B$. Similarly, X *possesses a neighbourhood extension property* (written $X \in NES$) if, for every space Y, every closed $B \subset Y$, and every $f : B \to X$ there exists an open neighbourhood U of B in Y and an extension $\tilde{f} : U \to X$ of f onto U.

Of course, every ES-space is NES. Before we formulate more properties of these spaces, we need the notion of a retract. Recall that a subset $A \subset X$ is called a *retract of* X if there exists a (continuous) retraction $r : X \to A$, i.e., $r(x) = x$, for every $x \in A$. Observe that A is a retract of X if and only if the identity map id_A over A possesses a continuous extension onto X. It is also easy to see that if A is a retract of X, then A is a closed subset of X. Similarly, we say that A is a *neighbourhood retract* of X if there exists an open subset $U \subset X$ such that $A \subset U$ and A is a retract of U.

Below we collect some simple but important properties of ES and NES-spaces.

(2.1) PROPERTIES.

(2.1.1) *If X is homeomorphic to Z and $X \in ES$ ($X \in NES$), then $Z \in ES$ ($Z \in NES$),*

(2.1.2) *if $X \in ES$ ($X \in NES$) and A is a retract of X, then $A \in ES$ ($A \in NES$),*

(2.1.3) *if $X \in NES$ and V is an open subset of X, then $V \in NES$,*

The proof of (2.1) is self-evident and, therefore, is left to the reader.

In a vector space E, we again understand by the convex hull, $\mathrm{conv}(A)$, of a subset $A \subset E$ the set of all points $y \in E$ of the form:

$$y = \sum_{i=1}^{n} t_i a_i,$$

where a_i are in A, and the coefficients t_i are greater or equal to zero ($t_i \geq 0$) and their sum is equal to 1. It is easy to see that $\mathrm{conv}(A)$ is equal to the intersection of all the convex subsets of E which contain A.

It is well-known that the theorem of Tietze asserts that each real continuous function defined on a closed subset of a metric space X can be (continuously) extended onto X. The generalization of this theorem proved by J. Dugundji (cf. [DG-M]) shows that, for the range space, we can even take any locally convex space. More precisely, we have the following:

(2.2) DUGUNDJI'S EXTENSION THEOREM. *If E is locally convex, then $E \in ES$. Moreover, for every closed subset B of a paracompact space Y and for every map $f : B \to E$, there exists a continuous extension $\tilde{f} : Y \to E$ such that:*

(2.2.1) $$\tilde{f}(Y) \subset \mathrm{conv}(f(B)).$$

As an immediate consequence of the above theorem, we get

(2.3) COROLLARY. *Let C be a convex subset of a normed space E. Then $C \in ES$.*

(2.4) COROLLARY. *Let $S^n = \{x \in R^{n+1} \mid \|x\| = 1\}$ be the unit sphere in R^{n+1}. Then $S^n \in NES$.*

Now, we are going to express extension spaces (neighbourhood extension spaces) in terms of absolute retracts (absolute neighbourhood retracts).

Before doing it, we prove an important embedding theorem ([ArEe]).

(2.5) THEOREM (Arens–Eells's Embedding Theorem). *Let X be a metric space. Then there exists a normed space E and an isometry $\Theta : X \to E$ on X into E such that $\Theta(X)$ is a closed subset of E.*

Now, following K. Borsuk ([Brs-M]), we introduce the notion of absolute retracts (AR-spaces) and the notion of absolute neighbourhood retracts (ANR-spaces).

It is useful to use the notion of an r-map. A mapping $r : Z \to T$ is called an *r-map* if there exists a map $s : T \to Z$ such that $r \circ s = \mathrm{id}_T$, i.e., $(r \circ s)(t) = t$, for every $t \in T$.

We shall also use the notion of an embedding. Namely, by an *embedding* of a space X into Y we understand any homeomorphism $h : X \to Y$ from X to Y such that $h(X)$ is a closed subset of Y.

Now, we are able to formulate the following

(2.6) DEFINITION. We say that $X \in AR$ ($X \in ANR$) if, for any space Y and for any embedding $h : X \to Y$, the set $h(X)$ is a retract of Y ($h(X)$ is a neighbourhood retract of Y).

In view of (2.5), we obtain

(2.7) PROPOSITION.
(2.7.1) $X \in AR$ if and only if X is an r-image of some normed space E,
(2.7.2) $X \in ANR$ if and only if X is an r-image of some open subset U of a normed space E.

PROOF. For the proof of (2.7.1), it is sufficient to show that if there exists an r-map $r : E \to X$ from a normed space onto X, then $X \in AR$. Let Y be an arbitrary space and $h : X \to Y$ be an embedding. We have to prove that $h(X)$ is a retract of Y. Let us denote $h(X)$ by B. So B is a closed subset of Y. We define $f : h(X) \to E$ by putting:

$$f = s \circ h^{-1}, \quad \text{where } r \circ s = \mathrm{id}_X.$$

Since $E \in ES$, we have the extension $\widetilde{f} : Y \to E$ of f onto Y. Then the map $\varrho : Y \to h(X)$ given by $\varrho = r \circ \widetilde{f}$ is the needed retraction, and the proof of (2.7.1) is complete. The proof of (2.7.2) is strictly analogous and, therefore, we leave it to the reader. $\square$

If, in particular, a space X is homeomorphic to a neighbourhood retract $Y \subset \mathbb{R}^n$ in $\mathbb{R}^n$, then we speak about a *Euclidean neighbourhood retract* (ENR). Obviously, $ENR \subset ANR$. For more details abour ENR-spaces, see e.g. [Do-M].

Now, we shall prove the main result of this section.

(2.8) THEOREM. *In the class of metric spaces, we have:*

(2.8.1) $$X \in ES \iff X \in AR,$$

(2.8.2) $$X \in NES \iff X \in ANR.$$

PROOF. Since the proof of (2.8.2) is analogous to the proof of (2.8.1), we will restrict our considerations to the proof of (2.8.1) only.

At first, assume that $X \in ES$. To prove that $X \in AR$, let $h : X \to Y$ be an embedding. We let

$$B = h(X).$$

Then B is a closed subset of Y. We consider the map $f : B \to X$ defined by $f = h^{-1}$. Since $X \in ES$, there exists an extension $\widetilde{f} : Y \to X$ of f onto Y. Then the map $r : Y \to h(X)$ defined as follows:

$$r = h \circ \widetilde{f}$$

is a retraction from Y onto $h(X)$ what proves that $X \in AR$.

Now, assume on the contrary that $X \in AR$. We would like to prove that $X \in ES$. Let B be a closed subset of Y and let $f : B \to X$ be a mapping. For the proof, it is sufficient to define the extension $\widetilde{f} : B \to X$ of f onto Y.

Since $X \in AR$, in view of (2.7.1), there exists a normed space E and an r-map $r : E \to X$ (i.e., there exists $s : X \to E$ such that $r \circ s = \mathrm{id}_X$). We define $f_1 : B \to E$ by the formula: $f_1 = s \circ f$.

Since $E \in ES$, we obtain an extension

$$\widetilde{f_1} : Y \to E$$

of f_1 onto Y. Then the map $\widetilde{f} : Y \to X$ given by:

$$\widetilde{f} = r \circ f_1$$

is an extension of f onto Y, and the proof of (2.8.1) is complete. $\square$

We suggest the reader to prove (2.8.2).

In view of (2.8), we see that all properties of ES and NES-spaces remain valid for AR and ANR-spaces, respectively. In particular, an r-image of AR-space (ANR-space) is AR-space (ANR-space) again.

The notion of a homotopy plays an important role in the geometric topology. In what follows, by $[0,1]$ we shall denote the closed unit interval in $\mathbb{R}$.

Consider two maps $f, g : X \to Y$. We say that f is *homotopic to* g (written $f \sim g$) if there exists a mapping $h : X \times [0,1] \to Y$ such that:

$$h(x,0) = f(x) \quad \text{and} \quad h(x,1) = g(x), \qquad \text{for every } x \in X.$$

In what follows, the mapping h is called the homotopy joining f and g.

The notion of a homotopy can be reinterpreted in terms of the extension property. Namely, let us consider a closed subset $(X \times \{0\}) \cup (X \times \{1\})$ of $X \times [0,1]$ and a map $\overline{f} : X \times \{0\} \cup X \times \{1\} \to Y$ defined as follows:

$$\overline{f}(x,t) = \begin{cases} f(x), & \text{for } t = 0, \\ g(x), & \text{for } t = 1. \end{cases}$$

One says that $f \sim g$ if $\overline{f}$ possesses an extension $\widetilde{f}$ over $X \times [0,1]$.

Then, of course, $\widetilde{f}$ is a homotopy joining f and g. We have:

(2.9) PROPOSITION. *If $Y \in AR$, then any two mappings $f, g : X \to Y$ are homotopic.*

For given X and Y, we shall denote by $C(X,Y)$ the set of all (continuous) mappings from X to Y. We have:

(2.10) PROPOSITION. *The relation '$\sim$' is an equivalence relation in the set of continuous maps $C(X,Y)$.*

We define:

$$[X, Y] = C(X, Y)/_{\sim}.$$

Then $[X, Y]$ is called the set of all homotopy classes under the homotopical equivalence.

(2.11) COROLLARY. *If $Y \in ES$, then $[X, Y]$ is a singleton.*

Corollary (2.11) immediately follows from (2.9).

Two spaces X and Y are said to be *homotopically equivalent* (written $X \sim Y$) if there are two maps:

$$f : X \to Y \quad \text{and} \quad g : Y \to X$$

such that

(2.12) $$g \circ f \sim \mathrm{id}_X$$

and

(2.13) $$f \circ g \sim \mathrm{id}_Y.$$

Of course, if two spaces X and Y are homeomorphic, then they are homotopically equivalent.

The following notion is extremely important in our considerations.

(2.14) DEFINITION. A space X is called *contractible* if it is homotopically equivalent to the one-point space $\{p\}$, i.e., $X \sim \{p\}$.

One can easily see that the space X is contractible if and only if there exists a point $x_0 \in X$ such that:

$$\mathrm{id}_X \sim g,$$

where $g : X \to X$ is defined by: $g(x) = x_0$, for every $x \in X$. Moreover, the above consideration does not depend on the choice of the point $x_0 \in X$, because every two one-point spaces are homeomorphic, and so homotopically equivalent.

From this we deduce

(2.15) PROPOSITION. *If $X \in AR$, then X is a contractible space.*

Observe that the converse to (2.15) is not true. Namely, consider the so called comb space $C \subset \mathbb{R}^2$, i.e.,

$$C = \left\{ (x, y) \in \mathbb{R}^2 \,\middle|\, \left(x = 0, \frac{1}{2}, \frac{1}{3}, \frac{1}{4}, \dots, \text{ and } 0 \le y \le 1 \right) \text{ or} \right.$$

$$\left. (x \in [0, 1] \text{ and } y = 0) \right\}.$$

Evidently, C is a contractible space, but id_C cannot be extended over $\mathbb{R}^2$, so C is not an AR-space.

As we already know, ANR-spaces need not be contractible (compare S^n or a non contractible polyhedron). We are going to explain what type of contractibility is implied by ANR-spaces.

A space X is said to be *locally contractible at a point* $x_0 \in X$ if, for each $\varepsilon > 0$, there exists $\delta > 0$ ($\delta < \varepsilon$) and a homotopy $h : [0,1] \times B(x_0, \delta) \to B(x_0, \varepsilon)$ such that:

$$h(x, 0) = x_0 \quad \text{and} \quad h(x, 1) = x, \qquad \text{for every } x \in B(x_0, \delta);$$

in other words, the ball $B(x_0, \delta) = B_{x_0}^{\delta}$ is contractible in $B(x_0, \varepsilon) = B_{x_0}^{\varepsilon}$.

It is evident that the local contractibility at a point x_0 implies the local arcwise connectivity at this point. A space X is said to be locally contractible if it is locally contractible at each of its points. For the sake of brevity, we shall write $X \in LC$ if X is a locally contractible space. We see that every open subset of a locally contractible space is itself locally contractible.

Now, let us observe that open subsets of normed spaces are locally contractible, because the open balls are convex. Moreover, it is easy to see that every r-image of a locally contractible space is locally contractible. Summing up the above, we obtain:

(2.16) PROPOSITION. *If* $X \in$ ANR, *then* $X \in LC$.

Observe that the comb space $C \subset \mathbb{R}^2$ is not locally contractible, so $C \notin$ ANR.

In the case of compact metric spaces, it is useful to consider uniformly locally contractible spaces (ULC-spaces). Namely, a compact metric space (A, d) is said to be a ULC-space if, for every $\varepsilon > 0$, there is $\delta > 0$ and a map:

$$g : [0,1] \times \{(a,b) \in A \times A \mid d(a,b) < \delta\} \to A$$

such that

$$g(0, a, b) = a, \quad g(1, a, b) = b, \quad g(t, a, a) = a$$

and

$$\text{diam}\{g(a, b, t) \mid t \in [0,1]\} < \varepsilon,$$

where

$$\text{diam}\{g(a, b, t) \mid t \in [0,1]\}$$
$$= \sup\{d(c, d) \mid c = g(a, b, t_1),\ d = g(a, b, t_2), t_1, t_2 \in [0,1]\}.$$

A compact space (A, d) is called k-ULC, $k \geq 1$, if, for every $\varepsilon > 0$,there exists $\delta > 0$ such that any map $\overline{g} : S^k \to A$, with $\text{diam}(f(S^k)) < \delta$, is homotopic to a constant map by a homotopy $h : S^k \times [0,1] \to A$ such that $\text{diam}(h(S^k \times [0,1])) < \varepsilon$.

Of course, every ULC-space is an LC-space. In view of (2.16) and compactness of A, we conclude:

(2.17) PROPOSITION. *If A is a compact ANR-space, then A is a ULC-space.*

(2.18) DEFINITION. A compact nonempty metric space is called an R_δ-set if, there exists a decreasing sequence $\{A_n\}$ of compact absolute retracts such that

$$A = \bigcap_{n \geq 1} A_n.$$

Note that any intersection of a decreasing sequence of R_δ-sets is R_δ. Observe that A is not an AR-space, in general. Furthermore, A need not be contractible, in general.

(2.19) EXAMPLE. We shall construct an R_δ-space which is not contractible. Let $f : (0, (1/\pi)] \to \mathbb{R}$ be a function defined as follows:

$$f(x) = \sin \frac{1}{x}.$$

Let $B = \{(x,y) \in \mathbb{R}^2 \mid y = f(x),\ x \in (0,(1/\pi)]\}$, $C = \{(x,y) \in \mathbb{R}^2 \mid x = 0 \text{ and } -1 \leq y \leq 1\}$, and $A = B \cup C$. We have:

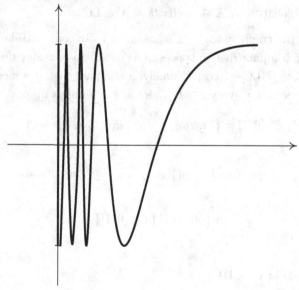

Figure 1

Of course, A is not contractible (in fact, it is not locally contractible!). We let:

$$A_n = \left[0, \frac{1}{n\pi}\right] \times [-1,1] \cup B_n,$$

where

$$B_n = \left\{ (x,y) \in \mathbb{R}^2 \mid \frac{1}{n\pi} \le x \le \frac{1}{\pi} \text{ and } y = f(x) \right\}, \quad \text{i.e.,}$$

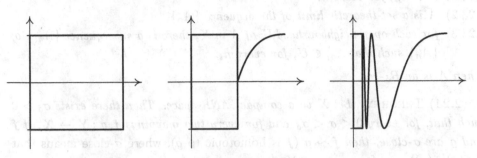

Figure 2

then $A = \bigcap_{n \ge 1} A_n$. The fact that A_n is an AR-space for every n follows.

(2.20) DEFINITION. Let A be a compact subset of X. We say that A is an ∞-*proximally connected* subset of X if, for every $\varepsilon > 0$, there exists $\varepsilon > \delta > 0$ such that, for every $n = 0, 1, 2, \dots$ and for every map $g : \partial \Delta^{n+1} \to O_\delta(A)$, there is a map $\widetilde{g} : \Delta^{n+1} \to O_\varepsilon(A)$ such that $g(x) = \widetilde{g}(x)$, for every $x \in \partial \Delta^{n+1}$, where Δ^{n+1} is an $(n+1)$-dimensional standard simplex, $\partial \Delta^{n+1}$ stands for the boundary of Δ^{n+1} and $O_\varepsilon(A) = \{ y \in X \mid \text{dist}(y, A) < \varepsilon \}$ is an ε-hull of A in X.

It is easy to see that in the above definition, we can replace Δ^{n+1} by the unit ball K^{n+1} in R^{n+1} and $\partial \Delta^{n+1}$ by the unit sphere S^n.

We can give the following characterization theorem proved by D. M. Hyman in 1969 (see [Hy]).

(2.21) THEOREM. *Let $X \in$ ANR and $A \subset X$ be a compact nonempty subset. Then the following statements are equivalent:*

(2.21.1) *A is an R_δ-set,*

(2.21.2) *A is an intersection of a decreasing sequence $\{A_n\}$ of compact contractible spaces,*

(2.21.3) *A is ∞-proximally connected,*

(2.21.4) *for every $\varepsilon > 0$, the set A is contractible in $O_\varepsilon(A)$.*

Observe, that as an immediate consequence of (2.21.2), we obtain.

(2.22) COROLLARY. *An intersection of a decreasing sequence of R_δ-sets is again an R_δ-set.*

We shall make use of the following

(2.23) PROPOSITION (cf. [BrGu]). *Let $\{A_n\}$ be a sequence of compact AR-s contained in X, and let A be a subset of X such that the following conditions hold:*

(2.23.1) *$A \subset A_n$, for every n,*

(2.23.2) *A is a set-theoretic limit of the sequence $\{A_n\}$,*

(2.23.3) *for each open neighbourhood U of A in X, there is a subsequence $\{A_{n_i}\}$ of $\{A_n\}$ such that $A_{n_i} \subset U$, for every n_i.*

Then A is an R_δ-set.

(2.24) THEOREM. *Let X be a compact ANR-space. Then there exists $\sigma_\delta > 0$ such that, for every $0 < \sigma < \sigma_\delta$ and for every two mappings $f, g : Y \to X$, if f and g are σ-close, then $f \sim g$ (f is homotopic to g), where σ-close means that $d(f(x), g(x)) < \sigma$, for every $x \in Y$.*

(2.25) DEFINITION. Let X be a space and $K \subset X$ be a closed (not necessarily compact) subset of X. The set K is ∞-*proximally connected in X* (written $K \in PC_X^\infty$) if, for every open neighbourhood U of K in X, there exists an open neighbourhood $V \subset U$ of K in X, such that for every $n = 0, 1, 2, \ldots$ and for every map $g : \partial\Delta^{n+1} \to V$, there exists a mapping $\tilde{g} : \Delta^{n+1} \to U$ such that $\tilde{g}(x) = g(x)$, for every $x \in \partial\Delta$.

Since, for any open neighbourhood W of a compact $K \subset X$, there is $\varepsilon > 0$ such that $O_\varepsilon(K) \subset W$, we see that Definition (2.25) is equivalent to Definition (2.20) formulated above. Therefore, we shall use the notation $K \in m\,PC_X^\infty$, when U and V are replaced by $O_\varepsilon(K)$ and $O_\delta(K)$.

Obviously, we have:

(2.26) If K is compact, then $K \in m\,PC_X^\infty$ if and only if $K \in PC_X^\infty$.

Generally, $PC_X^\infty \not\subset m\,PC_X^\infty$ and $m\,PC_X^\infty \not\subset PC_X^\infty$.

(2.27) EXAMPLE. Consider the set $K \subset \mathbb{R}^2$ defined as follows:

$$K = \{(x,y) \in \mathbb{R}^2 \mid y = 0 \text{ and } x \geq 1\} \cup \{(x,y) \in \mathbb{R}^2 \mid x = 1 \text{ and } 0 \leq y \leq 1\}$$
$$\cup \left\{(x,y) \in \mathbb{R}^2 \mid x \geq 1 \text{ and } y = \frac{1}{x}\right\}.$$

Then the set K is homeomorphic to $\mathbb{R}$. Hence, $K \in PC_{\mathbb{R}^2}^\infty$. Moreover, $K \notin m\,PC_{\mathbb{R}^2}^\infty$, because, for every $\varepsilon > 0$, the set $O_\varepsilon(K)$ is homotopically equivalent to S^1.

(2.28) EXAMPLE. Consider $K \subset \mathbb{R}^2$ defined as follows:

$$K = \{(x,y) \in \mathbb{R}^2 \mid x = 0 \text{ and } y \geq 1\} \cup \left\{(x,y) \in \mathbb{R}^2 \,\middle|\, x > 0 \text{ and } y = \frac{1}{x}\right\}.$$

Then, for every $\varepsilon > 0$, the set $O_\varepsilon(K)$ is contractible, and subsequently $K \in$ $m PC_{\mathbb{R}^2}^\infty$. Since there is an open neighbourhood U of K in $\mathbb{R}^2$ such that $U = U_1 \cup U_2$ and $U_1 \cap U_2 = \emptyset$, we conclude that $K \notin PC_{\mathbb{R}^2}^\infty$.

Observe that Theorem (2.21) is not true for non-compact K. On the other hand, we are able to prove the following:

(2.29) THEOREM. *Let K be a closed contractible subset of X and $X \in$ ANR. Then $X \in PC_X^\infty$.*

In what follows, we need some additional topological notions. A metric space (X, d) is C^n (i.e., *n-connected*) if, for every $k \leq n$, every continuous map from the k-sphere S^k into X is null homotopic (i.e. homotopic to a constant map), namely, every continuous map $f : S^k \to X$ has a continuous extension over the closed ball B^{n+1}, where S^n and B^{n+1} stand for the unit sphere and the unit ball in the Euclidean $(n+1)$-space $\mathbb{R}^{n+1}$, respectively.

A space X is C^∞ (i.e., *infinitely connected*), if it is C^n, for every n. A collection $\mathcal{E} \subset 2^X$ is equi-LC^n if, for every $y \in \bigcup\{B \mid B \in E\}$, every neighbourhood V of y in X contains a neighbourhood W of y in X such that, for all $B \in E$ and $k \leq n$, every map from S^k into $W \cap B$ is null homotopic over $V \cap E$ (i.e., a homotopy taking values in $V \cap E$).

I.3. Multivalued mappings and their selections

In this section, we shall survey the most important properties of multivalued mappings which we use in the sequel. There are several monographs devoted to multivalued mappings; see e.g. [Be-M], [BrGMO1-M]–[BrGMO5-M], [CV-M], [Go5-M], [HP1-M], [LR-M], [ReSe-M], [Sr-M].

In what follows, we assume that all topological spaces are the Tikhonov $T_{3\frac{1}{2}}$-spaces.

Let X and Y be two spaces and assume that, for every point $x \in X$, a nonempty closed (sometimes we will assume only that $\varphi(x) \neq \emptyset$) subset $\varphi(x)$ of Y is given; in this case, we say that φ is a *multivalued mapping* from X to Y and we write $\varphi : X \multimap Y$. In what follows, the symbol $\varphi : X \to Y$ is reserved for single-valued mappings, i.e., $\varphi(x)$ is a point of Y.

Let $\varphi : X \multimap Y$ be a multivalued map. We associate with φ the *graph* Γ_φ of φ by putting:

$$\Gamma_\varphi = \{(x, y) \in X \times Y \mid y \in \varphi(x)\}$$

and two natural projections $p_\varphi : \Gamma_\varphi \to X$, $q_\varphi : \Gamma_\varphi \to Y$ defined as follows: $p_\varphi(x, y) = x$ and $q_\varphi(x, y) = y$, for every $(x, y) \in \Gamma_\varphi$.

The point-to-set mapping $\varphi : X \multimap Y$ extends to a set-to-set mapping by putting:

$$\varphi(A) = \bigcup_{x \in A} \varphi(x), \quad \text{for } A \subset X.$$

Then $\varphi(A)$ is called the *image* of A under φ. If $\varphi : X \multimap Y$ and $\psi : Y \multimap Z$ are two maps, then the *composition* $\psi \circ \varphi : X \multimap Z$ of φ and ψ is defined by:

$$(\psi \circ \varphi)(x) = \bigcup \{\psi(y) \mid y \in \varphi(x)\}, \quad \text{for every } x \in X.$$

If $X \subset Y$ and $\varphi : X \multimap Y$, then a point $x \in X$ is called a *fixed point* of φ if $x \in \varphi(x)$. We let:

$$\text{Fix}(\varphi) = \{x \in X \mid x \in \varphi(x)\}.$$

For $\varphi : X \multimap Y$ and any subset $B \subset Y$, we define the *small counter image* $\varphi^{-1}(B)$ and the *large counter image* $\varphi_+^{-1}(B)$ of B under φ as follows:

$$\varphi^{-1}(B) = \{x \in X \mid \varphi(x) \subset B\},$$
$$\varphi_+^{-1}(B) = \{x \in X \mid \varphi(x) \cap B \neq \emptyset\}.$$

If $\varphi : X \multimap Y$ and $A \subset X$, then by $\varphi|_A : A \multimap Y$ we denote the restriction of φ to A. If, moreover, $\varphi(A) \subset B$, then the map $\widetilde{\varphi} : A \multimap B$, $\widetilde{\varphi}(x) = \varphi(x)$, for every $x \in A$, is the restriction of φ to the pair (A, B).

Below, we will summarize the properties of an image and a counter image.

(3.1) PROPOSITION. *Let* $\varphi : X \multimap Y$ *be a multivalued map,* $A \subset X$ *and* $B \subset Y$, $B_j \subset Y$, $j \in J$. *Then we have:*

(3.1.1) $\varphi^{-1}(\varphi(A)) \supset A,$

(3.1.2) $\varphi(\varphi^{-1}(B)) \subset B,$

(3.1.3) $X \setminus \varphi^{-1}(B) \supset \varphi^{-1}(Y \setminus B),$

(3.1.4) $\varphi^{-1}\left(\bigcup_{j \in J} B_j\right) \supset \bigcup_{j \in J} \varphi^{-1}(B_j),$

(3.1.5) $\varphi^{-1}\left(\bigcap_{j \in J} B_j\right) = \bigcap_{j \in J} \varphi^{-1}(B_j),$

(3.1.6) $\varphi_+^{-1}(\varphi(A)) \supset A,$

(3.1.7) $\varphi(\varphi_+^{-1}(B)) \supset B \cap \varphi(X),$

(3.1.8) $X \setminus \varphi_+^{-1}(B) = \varphi^{-1}(Y \setminus B),$

(3.1.9) $\varphi_+^{-1}\left(\bigcup_{j \in J} B_j\right) = \bigcup_{j \in J} \varphi_+^{-1}(B_j),$

(3.1.10) $\varphi_+^{-1}\left(\bigcap_{j \in J} B_j\right) \subset \bigcap_{j \in J} \varphi_+^{-1}(B_j).$

The proof of (3.1) is straightforward and we leave it to the reader.

For given two maps $\varphi, \psi : X \multimap Y$, we let:

$$\varphi \cup \psi : X \multimap Y \quad \text{and} \quad \varphi \cap \psi : X \multimap Y$$

as follows:

$$(\varphi \cup \psi)(x) = \varphi(x) \cup \psi(x) \quad \text{and} \quad (\varphi \cap \psi)(x) = \varphi(x) \cap \psi(x),$$

for every $x \in X$.

Of course, the map $\varphi \cap \psi$ is defined, provided that $\varphi(x) \cap \psi(x) \neq \emptyset$, for every $x \in X$.

By an easy observation, we obtain:

(3.2) PROPOSITION. *Let* $\varphi, \psi : X \multimap Y$ *be such that* $\varphi \cap \psi$ *is defined and let* $B \subset Y$. *Then we have:*

(3.2.1) $$(\varphi \cup \psi)^{-1}(B) = \varphi^{-1}(B) \cap \psi^{-1}(B),$$

(3.2.2) $$(\varphi \cap \psi)^{-1}(B) \supset \varphi^{-1}(B) \cup \psi^{-1}(B),$$

(3.2.3) $$(\varphi \cup \psi)_+^{-1}(B) = \varphi_+^{-1}(B) \cup \psi_+^{-1}(B),$$

(3.2.4) $$(\varphi \cap \psi)_+^{-1}(B) \subset \varphi_+^{-1}(B) \cap \psi_+^{-1}(B).$$

If we have two maps $\varphi : X \multimap Y$ and $\psi : Y \multimap Z$, then, for any $B \subset Z$, we obtain:

(3.3) PROPOSITION.

(3.3.1) $$(\psi \circ \varphi)^{-1}(B) = \varphi^{-1}(\psi^{-1}(B)),$$

(3.3.2) $$(\psi \circ \varphi)_+^{-1}(B) = \varphi_+^{-1}(\psi_+^{-1}(B)).$$

Finally, let us consider two maps:

$$\varphi : X \multimap Y \quad \text{and} \quad \psi : X \multimap Z.$$

Then we can define the Cartesian product $\varphi \times \psi : X \multimap Y \times Z$ of φ and ψ by putting:

$$(\varphi \times \psi)(x) = \varphi(x) \times \psi(x), \quad \text{for every } x \in X.$$

By an easy observation, we obtain:

(3.4) PROPOSITION. *Letting* $B \subset Y$ *and* $D \subset Z$, *we have:*

(3.4.1) $$(\varphi \times \psi)^{-1}(B \times D) = \varphi^{-1}(B) \cap \psi^{-1}(D),$$

(3.4.2) $$(\varphi \times \psi)_+^{-1}(B \times D) = \varphi_+^{-1}(B) \cap \psi_+^{-1}(D).$$

To make the notion of a multivalued map more transparent, below we shall present a number of examples (comp. [Go5-M]).

(3.5) EXAMPLES.

(3.5.1) Let $\varphi : [0,1] \multimap [0,1]$ be the map defined as follows:

$$\varphi(x) = \begin{cases} 1 & x < 1/2, \\ \{0,1\} & x = 1/2, \\ 0 & x > 1/2. \end{cases}$$

(3.5.2) Let $\varphi : [0,1] \multimap [0,1]$ be given:

$$\varphi(x) = \begin{cases} 1 & x < 1/2, \\ [0,1] & x = 1/2, \\ 0 & x > 1/2. \end{cases}$$

(3.5.3) Let $\varphi : [0,1] \multimap [0,1]$ be defined as follows:

$$\varphi(x) = [x,1].$$

(3.5.4) We let $\varphi : [0,1] \multimap [0,1]$ as follows:

$$\varphi(x) = \begin{cases} [0,1/2] & x \neq 1/2, \\ [0,1] & x = 1/2. \end{cases}$$

(3.5.5) Let $\varphi : [0,1] \multimap [0,1]$ be given:

$$\varphi(x) = \begin{cases} [0,1] & x \neq 1/2, \\ [0,1/2] & x = 1/2. \end{cases}$$

(3.5.6) Let $\varphi : [0,\pi] \multimap \mathbb{R}$ be defined:

$$\varphi(x) = \begin{cases} [\operatorname{tg} x, 1 + \operatorname{tg} x] & x \neq \pi/2, \\ \{0\} & x = \pi/2. \end{cases}$$

(3.5.7) Let $\varphi : \mathbb{R}^+ = [0,+\infty) \multimap \mathbb{R}$ be defined: $\varphi(x) = [e^{-x},1]$.

(3.5.8) Let $\varphi : \mathbb{R}^2 \multimap \mathbb{R}^2$ be defined:

$$\varphi(x,y) = \{(x + z_1, y + z_2) \in \mathbb{R}^2 \mid z_1, z_2 > 0 \text{ and } z_1 \cdot z_2 = 1\}.$$

(3.5.9) Let $K^2 = \{(x,y) \in \mathbb{R}^2 \mid |(x,y)| \leq 1\}, S^1 = \{(x,y) \in \mathbb{R}^2 \mid |(x,y)| = 1\}$. We define a map $\varphi : K^2 \to K^2$ by putting:

$$\varphi(x,y) = \{(x,y) \in K^2 \mid |(x,y)| = \rho(x,y)\} \cup \{(x,y) \in S^1 \mid |(x,y)| \geq \rho(x,y)\},$$

where $\rho(x,y) = 1 - |(x,y)| + |(x,y)|^2$.

Let us make the geometrical illustrations of the above mappings:

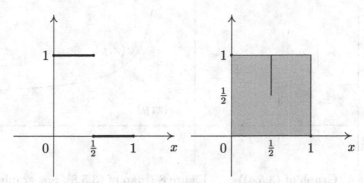

Figure 1 (Graph of (3.5.1))   Figure 5 (Graph of (3.5.5))

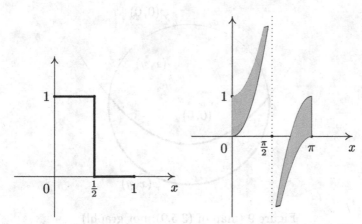

Figure 2 (Graph of (3.5.2))   Figure 6 (Graph of (3.5.6))

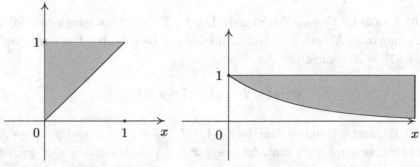

Figure 3 (Graph of (3.5.3))   Figure 7 (Graph of (3.5.7))

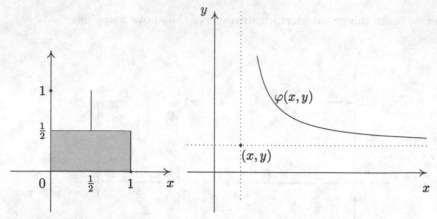

Figure 4 (Graph of (3.5.4))   Figure 8 (map of (3.5.8), not graph!)

For every $(x, y) \in K^2$, the set $\varphi(x, y)$ is homeomorphic to S^1.

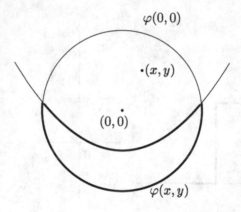

Figure 9 (map of (3.5.9), not graph!)

Let us also present some more general examples stimulating our consideration of multivalued maps.

(3.6) EXAMPLE (Inverse functions). Let $f : X \to Y$ be a (single-valued) continuous map from X onto Y. Then its inverse can be considered as a multivalued map $\varphi_f : Y \multimap X$ defined by:

$$\varphi_f(y) = f^{-1}(y), \quad \text{for } y \in Y.$$

(3.7) EXAMPLE (Implicit functions). Let $f : X \times Y \to Z$ and $g : X \to Z$ be two continuous maps such that, for every $x \in X$, there exists $y \in Y$ such that $f(x, y) = g(x)$.

The implicit function (defined by f and g) is a multivalued map $\varphi : X \multimap Y$ defined as follows:

$$\varphi(x) = \{y \in Y \mid f(x,y) = g(x)\}.$$

(3.8) EXAMPLE. Let $f : X \times Y \to R$ be a continuous map. Assume that there is $r > 0$ such that for every $x \in X$ there exists $y \in Y$ such that $f(x,y) \le r$. Then we let $\varphi_r : X \multimap Y$, $\varphi_r(x) = \{y \in Y \mid f(x,y) \le r\}$.

(3.9) EXAMPLE (Multivalued dynamical systems). Dynamical systems determined by autonomous ordinary differential equations without the uniqueness property are multivalued maps.

(3.10) EXAMPLE (Metric projection). Let A be a compact subset of a metric space (X, d). Then, for every $x \in X$, there exists $a \in A$ such that

$$d(a, x) = \text{dist}(x, A).$$

We define the metric projection $P : X \multimap A$ by putting:

$$P(x) = \{a \in A \mid d(a, x) = \text{dist}(x, A)\}, \quad x \in X.$$

Note that the metric retraction is a special case of a metric projection.

(3.11) EXAMPLE (Control problems). Consider the following control problem:

$$(3.11.1) \qquad \begin{cases} \dot{x}(t) = f(t, x(t), u(t)), \\ x(0) = x_0, \end{cases}$$

controlled by parameters $u(t)$ (the controls), where $f : [0, a] \times \mathbb{R}^n \times \mathbb{R}^m \to \mathbb{R}^n$.

In order to solve (3.11.1), we define a multivalued map $F : [0, a] \times \mathbb{R}^n \multimap \mathbb{R}^n$ as follows:

$$F(t, x) = \{f(t, x, u)\}_{u \in U}.$$

Then solutions of (3.11.1) are those of the following differential inclusions:

$$(3.11.2) \qquad \begin{cases} \dot{x}(t) \in F(t, x(t)), \\ x(0) = x_0. \end{cases}$$

Thus, any control problem (3.11.1) can be transformed, by means of multivalued maps, into problem (3.11.2).

Note that many other examples come from the game theory, mathematical economics, convex analysis and nonlinear analysis.

Let $\varphi : X \multimap Y$ be a multivalued map and $f : X \to Y$ be a single-valued map. We say that f is a *selection* of φ (written $f \subset \varphi$) if $f(x) \in \varphi(x)$, for every $x \in X$.

The problem of existence of good selections for multivalued mappings is very important in the fixed point theory.

The concept of upper semicontinuity is related to the notion of the small counter image of open sets. On the other hand, the concept of lower semicontinuity is related to the large counter image of open sets. Note that the concept of lower semicontinuity will be treated by ourselves later.

(3.12) DEFINITION. A multivalued map $\varphi : X \multimap Y$ is called *upper semicontinuous* (u.s.c.) if for every open $U \subset Y$ the set $\varphi^{-1}(U)$ is open in X.

In terms of closed sets, we can reformulate (3.12) as follows:

(3.13) PROPOSITION. *A multivalued map $\varphi : X \multimap Y$ is u.s.c. iff for every closed set $A \subset Y$ the set $\varphi_+^{-1}(A)$ is a closed subset of X.*

The proof of (3.13) is an immediate consequence of (3.1.3) and (3.1.8).

(3.14) REMARK. Observe that multivalued mappings considered in examples (3.5.1)–(3.5.4), (3.5.7) and (3.5.9) are u.s.c.

(3.15) PROPOSITION. *If $\varphi : X \multimap Y$ is u.s.c., then the graph Γ_φ is a closed subset of $X \times Y$.*

PROOF. We should prove that $X \times Y \setminus \Gamma_\varphi$ is open, i.e., $y \notin \varphi(x)$, for any $(x, y) \in X \times Y \setminus \Gamma_\varphi$. Hence, we choose an open neighbourhood V_y of y in Y and $V_{\varphi(x)}$ of $\varphi(x)$ in Y such that $V_y \cap V_{\varphi(x)} = \emptyset$ (we consider $T_{3\frac{1}{2}}$-spaces).

Let $U_x = \varphi^{-1}(V_{\varphi(x)})$. Then U_x is an open neighbourhood of x in X. Consequently, the set $U_x \times V_y$ is an open neighbourhood of (x, y) in $X \times Y$. We can observe that $U_x \times V_y \cap \Gamma_\varphi = \emptyset$. In fact, if $(x', y') \in U_x \times U_y$, then $\varphi(x') \subset V_{\varphi(x)}$, but $V_{\varphi(x)} \cap V_y = \emptyset$. Thus, $y' \notin V_{\varphi(x)}$ and $(x', y') \notin \Gamma_\varphi$. The proof of (3.15) is completed. $\square$

Observe that the map $f : \mathbb{R} \to \mathbb{R}$ defined as follows:

$$f(x) = \begin{cases} 1/x & x \neq 0, \\ 0 & x = 0 \end{cases}$$

has a closed graph, but it is not u.s.c., i.e., continuous.

In general, if $f : X \to Y$ is a continuous map from X onto Y, then the inverse map $\varphi_f : Y \multimap X$ considered in (3.6) has a closed graph, but is not necessarily u.s.c. However, we have:

(3.16) PROPOSITION. *Assume $\varphi : X \multimap Y$ is a multivalued map such that $\varphi(X) \subset K$ and the graph Γ_φ of φ is closed, where K is a compact set. Then φ is u.s.c.*

PROOF. Assume on the contrary that φ is not u.s.c. at a point $x \in X$. Let us denote by $\{W_\alpha\}$ the basis of an open topology in X at the point x. By the hypothesis, there exists an open neighbourhood $V_{\varphi(x)}$ of $\varphi(x)$ in Y such that, for every α, we have $\varphi(W_\alpha) \not\subset V_{\varphi(x)}$.

Now, we take a point $x_\alpha \in W_\alpha$ and $y_\alpha \in \varphi(x_\alpha)$ such that $y_\alpha \notin V_{\varphi(x)}$. Since the generalized sequence $\{y_\alpha\}$ is contained in the compact set K, we can assume, without any loss of generality, that $\lim_\alpha y_\alpha = y$. Since $\lim_\alpha x_\alpha = x$ and Γ_φ is closed, we get that $(x, y) \in \Gamma_\varphi$. Conversely, $\{y_\alpha\} \subset Y \setminus V_{\varphi(x)}$ and $Y \setminus V_{\varphi(x)}$ is closed. Consequently, $y \in Y \setminus V_{\varphi(x)}$, and we get a contradiction.   □

Taking into account the inverse map mentioned above, we have the following

(3.17) PROPOSITION. *If $f : X \to Y$ is a closed continuous map from X onto Y, then the inverse map $\varphi_f : Y \multimap X$ (defined in (3.6)) is u.s.c. In fact, we have:*

$$(\varphi_f)_+^{-1}(A) = f(A),$$

for every closed subset $A \subset X$.

The proof of (3.17) is an immediate consequence of (3.2.1).

(3.18) PROPOSITION. *Assume that $\varphi, \psi : X \multimap Y$ are two u.s.c. mappings and let $Y \in T_4$ be a normal space. Then:*

(3.18.1) *the map $\varphi \cup \psi : X \multimap Y$ is u.s.c.,*

(3.18.2) *the map $\varphi \cap \psi : X \multimap Y$ is u.s.c., provided it is well-defined.*

PROOF OF (3.18.2). Let $x \in X$ and V be an open neighbourhood of $\varphi(x) \cap \psi(x)$ in Y. Then $\varphi(x) \setminus V$ and $\psi(x) \setminus V$ are closed subsets of Y such that $(\varphi(x) \setminus V) \cap (\psi(x) \setminus V) = \emptyset$. Let W_1 and W_2 be open neighbourhoods of $(\varphi(x) \setminus V)$ and $(\psi(x) \setminus V)$, respectively, such that $W_1 \cap W_2 = \emptyset$ (because of normality). Since φ is u.s.c., we can choose an open neighbourhood U_1 of x in X such that:

$$\varphi(U_1) \subset V \cup W_1$$

and an open neighourhood U_2 of x in X such that:

$$\psi(U_2) \subset U \cup W_2.$$

Let $U = U_1 \cap U_2$. Then $(\varphi \cap \psi)(U) \subset V$, and the proof is completed.   □

(3.19) PROPOSITION. *Let $\varphi : X \multimap Y$ and $\psi : X \multimap Z$ be two u.s.c. mappings. Then the map $\varphi \times \psi : X \multimap Y \times Z$ is u.s.c. as well.*

Note that (3.19) follows immediately from (3.4.1).

Now, we will restrict our considerations to multivalued maps with compact values.

(3.20) PROPOSITION. *Let $\varphi : X \multimap Y$ be an u.s.c. map with compact values and let A be a compact subset of X. Then $\varphi(A)$ is compact.*

PROOF. Let $\{V_t\}$ be an open covering of $\varphi(A)$. Since $\varphi(x)$ is compact, for every $x \in X$, we infer that there exists a finite number of sets V_t such that $\varphi(x) \subset W_x$, where W_x is the union of the sets V_t, for every $x \in A$. This implies that the family $\{W_x\}_{x \in A}$ is an open covering of $\varphi(A)$. Let $U_x = \varphi^{-1}(W_x)$, for each $x \in A$. Then $\{U_x\}_{x \in A}$ is an open covering of A in X. Since A is compact, there exists a finite subcovering $U_{x_1}, \ldots, U_{x_n}$ of this covering. Consequently, the sets $W_{x_1}, \ldots, W_{x_n}$ cover $\varphi(A)$, and since every W_{x_i} is a finite union of sets in $\{V_t\}$, we obtain a finite subcovering $V_{t_1}, \ldots, V_{t_k}$ of the covering $\{V_t\}$. This completes the proof. $\square$

Now, from (3.20) and (3.3.1), we obtain:

(3.21) PROPOSITION. *If $\varphi : X \multimap Y$ and $\psi : Y \multimap Z$ are two u.s.c. mappings with compact values, then the composition $\psi \circ \varphi : X \multimap Z$ of φ and ψ is an u.s.c. map with compact values.*

Let us observe that the upper semicontinuity for mappings with compact values (in metric spaces) can be reformulated in the Cauchy sense as follows:

(3.22) PROPOSITION. *Let $\varphi : X \multimap Y$ be a multivalued map with compact values in metric spaces. Then φ is u.s.c. if and only if*

(3.22.1) *for all $x \in X$ and for all $\varepsilon > 0$, there exists $\delta > 0$ such that $\varphi(B(x, \delta)) \subset O_\varepsilon(\varphi(x))$.*

It is easy to see that if φ is u.s.c., then (3.22.1) holds true. Conversely, if U is an open neighbourhood of $\varphi(x)$ in Y, then, in view of compactness of $\varphi(x)$, we can choose an $\varepsilon > 0$ such that $O_\varepsilon(\varphi(x)) \subset U$, and our assertion follows from (3.22.1).

Let E be a Banach space and let $\varphi : X \multimap E$ be an u.s.c. map with compact values. We define the map

$$\overline{\mathrm{conv}}\,\varphi : X \multimap E$$

by putting:

$$\overline{\mathrm{conv}}\,\varphi(x) = \overline{\mathrm{conv}}(\varphi(x)),$$

where $\overline{\mathrm{conv}}(\varphi(x))$ denotes the closed convex hull of $\varphi(x)$ in E. Since $\varphi(x)$ is compact, according to the Mazur Theorem (1.33), the set $\overline{\mathrm{conv}}(\varphi(x))$ is compact, too. Recall that $\overline{\text{conv}}(\varphi(x))$ is the intersection of all convex closed subsets of E containing $\varphi(x)$.

We can state the following

(3.23) PROPOSITION. *If* $\varphi : X \multimap E$ *is an u.s.c. map with compact values and* X *is a metric space, then* $\overline{\text{conv}}\,\varphi : X \multimap E$ *is also u.s.c. with compact values.*

PROOF. In the proof, we shall use (3.22). Let $\varepsilon > 0$ and $0 < \varepsilon_1 < \varepsilon$. Assume that $x_0 \in X$. Since φ is u.s.c., there exists $\delta > 0$ such that $\varphi(B(x_0, \delta)) \subset O_{\varepsilon_1}(\varphi(x_0))$.

Consequently, $\varphi(B(x_0, \delta)) \subset O_{\varepsilon_1}(\overline{\text{conv}}\,\varphi(x_0))$. Since $O_{\varepsilon_1}(\overline{\text{conv}}\,\varphi(x_0))$ is convex, we deduce that

$$\text{conv}\,\varphi(B(x_0, \delta)) \subset O_{\varepsilon_1}(\overline{\text{conv}}\,\varphi(x_0)),$$

and so

$$\overline{\text{conv}}\,\varphi(B(x_0, \delta)) \subset \text{cl}(O_{\varepsilon_1}(\overline{\text{conv}}\,\varphi(x_0))) \subset O_\varepsilon(\overline{\text{conv}}\,\varphi(x_0)),$$

where $\text{conv}\,\varphi(x) = \text{conv}(\varphi(x))$ is the convex hull of $\varphi(x)$.

Thus, we have proved (3.23). □

Using the large counter image, instead of a small one, we get:

(3.24) DEFINITION. Let $\varphi : X \multimap Y$ be a multivalued map. If, for every open $U \subset Y$, the set $\varphi_+^{-1}(U)$ is open in X, then φ is called a *lower semicontinuous* (*l.s.c.*) *map*.

Note that, for $\varphi = f : X \to Y$, the notion of upper semicontinuity coincides with the lower semicontinuity which means nothing else than the continuity of f.

In what follows, we also say that a multivalued map $\varphi : X \multimap Y$ is *continuous* if it is both u.s.c. and l.s.c.

(3.25) REMARK. Note that the maps presented in: (3.5.3), (3.5.5), (3.5.7), (3.5.8) and (3.5.9) are l.s.c. The maps (3.5.3), (3.5.7) and (3.5.9) are continuous. The maps (3.5.1), (3.5.2) and (3.5.4) are u.s.c., but not l.s.c. The maps (3.5.5) and (3.5.8) are l.s.c., but not u.s.c.

In terms of a small counter image, we can define the lower semicontinuity as follows:

(3.26) PROPOSITION. *A map* $\varphi : X \multimap Y$ *is l.s.c. if and only if, for every closed* $A \subset Y$, *the set* $\varphi^{-1}(A)$ *is a closed subset of* X.

Proposition (3.26) is an immediate consequence of (3.1.8) and (3.24).

The following two propositions are straightforward (see (3.2.3) and (3.3.2)).

(3.27) PROPOSITION.

(3.27.1) *If $\varphi, \psi : X \multimap Y$ are two l.s.c. mappings, then $\varphi \cup \psi : X \multimap Y$ is l.s.c. as well.*

(3.27.2) *If $\varphi : X \multimap Y$ and $\psi : Y \multimap Z$ are two l.s.c. maps, then the composition $\psi \circ \varphi : X \multimap Z$ of φ and ψ is l.s.c. as well, provided, for every $x \in X$, the set $\psi(\varphi(x))$ is closed.*

We would like to emphasize that the intersection of two l.s.c. mappings need not be l.s.c.

(3.28) EXAMPLE. Consider two multivalued mappings $\varphi, \psi : [0, \pi] \multimap \mathbb{R}^2$ defined as follows:

$$\varphi(t) = \{(x, y) \in \mathbb{R}^2 \mid y \geq 0 \text{ and } x^2 + y^2 \leq 1\},$$

for every $t \in [0, \pi]$;

$$\psi(t) = \{(x, y) \in \mathbb{R}^2 \mid x = \lambda \cos t, \; y = \lambda \sin t, \; \lambda \in [-1, 1]\}.$$

Then φ is a constant map and so even continuous, ψ is a l.s.c. map, but $\varphi \cap \psi$ is no longer l.s.c. (to see it, put $t = 0$ or $t = \pi$).

One can prove the following:

(3.29) PROPOSITION. *Let $\varphi : X \multimap Y$ be a l.s.c. map, $f : X \to Y$ and $\lambda : X \to (0, \infty)$ be continuous and $\psi : X \multimap Y$ be defined as follows:*

$$\psi(x) = B(f(x), \lambda(x)).$$

Assume, furthermore, that, for every $x \in X$, we have:

$$\varphi(x) \cap B(f(x), \lambda(x)) \neq \emptyset.$$

Then $\varphi \cap \operatorname{cl} \psi : X \multimap Y$ is a l.s.c. map.

PROOF. Let $x_0 \in X$ and V be an open set of Y such that $V \cap (\varphi \cap \operatorname{cl} \psi)(x_0) \neq \emptyset$. Let $y_0 \in V \cap (\varphi(x_0) \cap B(f(x_0), \lambda(x_0)))$ and let V_{y_0} be an open neighbourhood of y_0 in Y such that $V_{y_0} \subset V \cap B(f(x_0), \lambda(x_0))$.

The continuity of f and λ implies that there is an open neighbourhood U_{x_0} of x_0 in X such that:

$$V_{y_0} \subset \psi(x), \quad \text{for every } x \in U_{x_0}.$$

Consequently, since φ is l.s.c., we can choose an open neighbourhood W_{x_0} of x_0 in X such that $\varphi(x) \cap V_{y_0} \neq \emptyset$, for every $x \in W_{x_0}$. Let $U = U_{x_0} \cap W_{x_0}$. Then, we get that $(\varphi \cap \operatorname{cl} \psi)(x) \cap V_{y_0} \neq \emptyset$, for every $x \in U$, and the proof is completed. $\square$

For further formulations of (3.29), we recommend [BrGMO1-M] (see also the references in [BrGMO2-M]–[BrGMO5-M]).

The most famous continuous selection theorem is the following result due to E. A. Michael (see [Mi1]–[Mi3], for more details).

(3.30) THEOREM (E. A. Michael). *Let X be a paracompact space, E a Banach space and $\varphi : X \multimap E$ a l.s.c. map with closed convex values. Then there exists $f : X \to E$, a continuous selection of φ ($f \subset \varphi$).*

PROOF. (i) Let us begin with proving the following claim: given any convex (not necessarily closed) valued l.s.c. map $\Phi : X \multimap E$ and every $\varepsilon > 0$, there exists a continuous $g : X \to E$ such that $\text{dist}(g(x), \Phi(x)) \leq \varepsilon$, i.e., $g(x) \in \text{cl}\, O_\varepsilon(\Phi(x))$, for every $x \in X$.

In fact, for every $x \in X$, let $y_x \in \Phi(x)$ and let $\delta_x > 0$ be such that $B(y_x, \varepsilon) \cap \Phi(x') \neq \emptyset$, for x' in $B(x, \delta_x)$. Since X is paracompact, there exists a locally finite refinement $\{U_x\}_{x \in X}$ of $\{B(x, \delta_x)\}_{x \in X}$. Let $\{L_x\}_{x \in X}$ be an associated partition of unity. The mapping $g : X \to E$ defined as follows:

$$g(u) = \sum_{x \in X} L_x(u) \cdot y_x$$

is continuous, because it is locally a finite sum of continuous functions. Fix $x \in X$. Whenever $L_x(u) > 0$, $u \in B(x, \delta_x)$. Hence, $y_x \in O_\varepsilon(\Phi(u))$. Since the latter set is convex, any convex combination of such y's belongs to it.

(ii) Furthermore, we claim that we can define a sequence $\{f_n\}$ of continuous mappings from X to E with the following properties:

(a) $\qquad\qquad \text{dist}(f_n(u), \varphi(u)) \leq \dfrac{1}{2^n}, \qquad n = 1, 2, \dots, \; u \in X,$

(b) $\qquad\qquad \|f_n(u) - f_{n-1}(u)\| \leq \dfrac{1}{2^{n-2}}, \qquad n = 2, \dots, \; u \in X.$

For $n = 1$, it is enough to take in the claim of part (i), $\Phi = \varphi$ and $\varepsilon = 1/2$.

Assume we have defined mappings f_n satisfying (a) up to $n = k$. We shall define f_{k+1} satisfying (a) and (b) as follows.

Consider the set $\Phi(u) = B(f_k(u), 1/2^n) \cap \varphi(u)$. By (a), it is not empty, and it is a convex set. By (3.29), the map Φ is l.s.c., so by the claim (i), there exists a continuous g such that

$$\text{dist}(g(x), \Phi(x)) < \frac{1}{2^{n+1}}.$$

Set $f_{k+1}(u) = g(u)$. Then $\text{dist}(f_{k+1}(u), \varphi(u)) < 1/2^{k+1}$, proving (a). Also

$$f_{k+1}(u) \in O_{1/2^{k+1}}(\Phi(u)) \subset B\left(f_k(u), \frac{1}{2^k} + \frac{1}{2^{k+1}}\right),$$

i.e.,

$$\|f_{k+1}(u) - f_n(u)\| \leq \frac{1}{2^{k-1}}$$

proving (b).

(iii) Since the series $\sum(1/2^n)$ converges, $\{f_n\}$ is a Cauchy sequence uniformly converging to a continuous f. Since the values of φ are closed, by (a) of part (ii), f is a selection of φ. The proof is completed.   □

Some applications of the Michael selection theorem to differential inclusions will be shown later. Now, we explain the connection between the continuous selection property and the extension property. Namely, we would like to obtain from (3.30) the following version of the Dugundji extension theorem.

(3.31) COROLLARY. *If E is a Banach space, then, for any paracompact space X, a closed subset $A \subset X$ and a continuous map $f : A \to E$, there exists a continuous extension $\widetilde{f} : X \to E$ of f onto X, i.e., $\widetilde{f}(x) = f(x)$, for every $x \in A$.*

PROOF. Let A be a closed subset of X and let $f : A \to E$ be a continuous map. We define $\varphi : X \multimap E$ as follows:

$$\varphi(x) = \begin{cases} f(x) & x \in A, \\ \overline{\mathrm{conv}}(f(A)) & x \notin A. \end{cases}$$

Then φ is l.s.c. with convex, closed values. So, it has a selection $\widetilde{f} \subset \varphi$. It is evident that $\widetilde{f}$ is an extension of f.   □

The Michael selection theorem is not true for u.s.c. mappings (cf. (3.5.1) and (3.5.2)), but under some natural assumptions u.s.c. mappings are σ-selectionable.

(3.32) DEFINITION. We say that a map $\varphi : X \multimap Y$ is *σ-selectionable* if there exists a decreasing sequence of compact valued u.s.c. maps $\varphi_n : X \multimap Y$ satisfing:

(3.32.1) for all $n \geq 0$, φ_n has a continuous selection,
(3.32.2) for all $x \in X$, $\varphi(x) = \bigcap_n \varphi_n(x)$.

We can state the following:

(3.33) THEOREM. *Let $\varphi : X \multimap E$ be an u.s.c. map with compact convex values from a metric space X to a Banach space E. If $\mathrm{cl}(\varphi(X))$ is a compact set, then φ is σ-selectionable. Actually, there exists a sequence of u.s.c. mappings φ_n from X to $\overline{\mathrm{conv}}(\varphi(X))$, which approximate φ in the sense that, for all $x \in X$, we have:*

$$(3.33.1) \quad \begin{cases} \text{for all } n \geq 0, \ \varphi(x) \subset \cdots \subset \varphi_{n+1}(x) \subset \varphi_n(x) \subset \cdots \subset \varphi_0(x), \\ \text{for every } \varepsilon > 0, \ \text{there exists } n_0 = n_0(\varepsilon, x) \\ \quad \text{such that, for all } n \leq n_0, \ \varphi_n(x) \subset \overline{0_\varepsilon(\varphi(x))}, \end{cases}$$

and, moreover, the maps φ_n can be written in the following form:

$$(3.33.2) \qquad \text{for all } x \in X, \quad \varphi_n(x) = \sum_{i \in I_n} L_i^n(x) \cdot C_i^n,$$

where the subsets C_i^n are compact and convex and where the functions L_i^n form a locally Lipschitzian finite partition of unity.

(3.34) REMARK. Note that (3.33.1) implies

$$\varphi(x) = \bigcap_{n \geq 0} \varphi_n(x).$$

If X is compact, then the sets I_n, which appear in formula (3.33.2), are finite.

PROOF OF (3.33). Let $K = \overline{\mathrm{conv}}\varphi(X)$. Then K is a compact convex subset of E. We fix $\varrho > 0$. Let us cover X by the open balls $\{B(x, \varrho)\}_{x \in X}$. Let $\{\Omega_i^{(0)}\}_{i \in I(0)}$ be a locally finite refinement of $\{B(x, \varrho)\}_{x \in X}$. For any $i \in I(0)$, there exists $x_i^{(0)} \in X$ with $\Omega_i^{(0)} \subset B(x_i^{(0)}, \varrho)$. We then define, for any $i \in I(0)$, the set $C_i^{(0)} = \overline{\mathrm{conv}}\varphi(B(x_i^{(0)}, 2\varrho))$, which is a nonempty, closed, convex subset of K.

Now, we can associate a locally Lipschitzian partition of unity $\{L_i^{(0)}\}_{i \in I(0)}$ to the open covering $\{\Omega_i^{(0)}\}_{i \in I(0)}$ (see Theorem 0.1.1 in [AuC-M]).

We define the map $\varphi_{(0)} : X \multimap E$ by putting:

$$\varphi_0(x) = \sum_{i \in I(0)} L_i^{(0)}(x) C_i^{(0)}.$$

Thus, the map $f_0 : X \to E$ given by:

$$f_0(x) = \sum_{i \in I(0)} L_i^{(0)}(x) \cdot y_i^{(0)}$$

is a locally Lipschitzian selection of φ_0, where $y_i^{(0)} \in C_i^{(0)}$ is fixed, for every $i \in I(0)$. In order to define φ_1, we do the same as before with the open covering $\{B(x, \varrho/3)\}_{x \in X}$. Thus, we consider its locally finite refinement $\{\Omega_i^{(1)}\}_{i \in I(1)}$ and the associated locally Lipschitzian partition of unity $\{L_i^{(1)}\}_{i \in I(1)}$.

As before, we set for all $i \in I(1)$:

$$C_i^{(1)} = \overline{\mathrm{conv}}\,\varphi\left(B\left(x_i^{(1)}, \frac{2\varrho}{3}\right)\right) \subset K$$

and define $\varphi_1 : X \multimap E$ by putting:

$$\varphi_1(x) = \sum_{i \in I(1)} L_i^{(1)}(x) \cdot C_i^{(1)}.$$

The map φ_1 satisfies the same properties as φ_0.

Now, we shall prove that $\varphi_1(x) \subset \varphi_0(x)$, for every $x \in X$. Let us fix $x \in X$. Then we define:

$$I_{(0)}^x = \{i \in I_{(0)} \mid x \in B(x_i^{(0)}, \varrho)\},$$

$$I_{(1)}^x = \left\{i \in I_{(1)} \mid x \in B\left(x_i^{(1)}, \frac{\varrho}{3}\right)\right\}.$$

Let $i_{(0)} \in I_{(0)}^x$ and $i_{(1)} \in I_{(1)}^x$ be given. If $y \in B(x_{i(1)}^{(1)}, 2\varrho/3)$, we obtain:

$$d(y, x_{i(1)}^{(1)}) < \frac{2\varrho}{3} \quad \text{with} \quad d(x, x_{i(0)}^{(0)}) < \varrho \quad \text{and} \quad d(x, x_{i(1)}^{(1)}) < \frac{\varrho}{3}.$$

Thus, we have:

$$d(y, x_{i(0)}^{(0)}) \leq \frac{2\varrho}{3} + \frac{\varrho}{3} + \varrho = 2\varrho.$$

Subsequently, $B(x_{i(1)}^{(1)}, 2\varrho/3) \subset B(x_{i(0)}^{(0)}, 2\varrho)$, for all $i_{(0)} \in I_{(0)}^x$ and all $i_{(1)} \in I_{(1)}^x$. This leads to $C_{i(1)}^{(1)} \subset C_{i(0)}^{(0)}$ for such indices.

For all $i_{(1)} \in I_{(1)}^x$, we have:

$$C_{i(1)}^{(1)} \subset \sum_{i \in I_{(0)}^x} L_i^{(0)}(x)C_i^{(0)} = \sum_{i \in I_{(0)}} L_i^{(0)}(x)C_i^{(0)} = \varphi_0(x).$$

This becomes true by convexity arguments, and since $\{L_i^{(0)}\}_{i \in I_{(0)}}$ is a locally finite partition of unity associated with $\{\Omega_i^{(0)}\}_{i \in I(0)}$, and thus also with $\{B(x_i^{(0)}, \varrho)\}_{i \in I(0)}$. In particular, this says that $L_i^{(0)}(x) = 0$, whenever $i \in I_{(0)}^x$.

Consequently, by the same reasons, we get:

$$\varphi_1(x) = \sum_{i \in I(1)} L_i^{(1)}(x) \cdot C_i^{(1)} = \sum_{i \in I_{(1)}^x} L_i^{(1)}(x)C_i^{(1)} \subset \varphi_0(x).$$

Moreover, it is easy to see that $\varphi(x) \subset \varphi_1(x)$, for every $x \in X$.

Now, let us define $\varrho_n = (1/3) \cdot \varrho$, for any $n = 1, 2, \ldots$ Then we can construct by induction a sequence of multivalued maps $\varphi_n : X \to E$, each of them being u.s.c. nonempty, convex, compact valued and satisfying the first part of (3.33.1) (as φ_0 for $\varrho_0 = \varrho$ and φ_1 for $\varrho_1 = \varrho/3$).

To conclude the proof, we have to show that, for every $\varepsilon > 0$, there exists $n_0 = n_0(\varepsilon, x)$ such that $\varphi_n(x) \subset O_\varepsilon(\varphi(x))$, for all $n \geq n_0$.

Let $x \in X$ be given. Since φ is u.s.c., for any $\varepsilon > 0$, there exists $\eta = \eta(\varepsilon, x)$ such that $d(y, x) \leq \eta$ implies $\varphi(y) \subset O_\varepsilon(\varphi(x))$. Then, there obviously exists $n_0 = n_0(\varepsilon, x)$ such that, for $n \geq n_0$, we have $\varrho_n \leq \eta/3$.

Let us define, as before, $I_{(n)}^x = \{i \in I_{(n)} \mid x \in B(x^{(n)}, \varrho_n)\}$. By the same reasons as for φ_0 and φ_1, we can write:

$$\varphi_n(x) = \sum_{i \in I_{(n)}^x} L_i^{(n)}(x) \cdot C_i^{(n)},$$

where $C_i^{(n)} = \overline{\mathrm{conv}}\,\varphi(B(x_i^{(n)}, 2\varrho_n)) \subset K$. Thus, for all $y \in B(x_i^{(n)}, 2\varrho_n)$ with $i \in I_{(n)}^x$, we have:

$$d(y, x) \leq d(y, x_i^{(n)}) + d(x_i^{(n)}, x)$$
$$\leq 2\varrho_n + \varrho_n = 3\varrho_n < \eta, \quad \text{taking } n > n_0.$$

Thus, for all $n \leq n_0$, we have $\varphi(y) \subset O_\varepsilon(\varphi(x))$, for all $y \in B(x_i^{(n)}, 2\varrho_n)$ with $i \in I_{(n)}^x$.

But since $\overline{O_\varepsilon(\varphi(x))}$ is closed and convex, we obtain: $C_i^{(n)} \subset \overline{O_\varepsilon(\varphi(x))}$, and by convexity, we infer:

$$\varphi_n(x) \subset \overline{O_\varepsilon(\varphi(x))},$$

for all $n \geq n_0$. Therefore, the proof of (3.33) is completed. $\square$

(3.35) REMARKS.

(3.35.1) If X is a compact space, then φ is automatically compact.

(3.35.2) If $E = \mathbb{R}^n$, then any bounded u.s.c. map with convex compact values satisfies assumptions of (3.33).

(3.36) REMARK (see Definition (3.32)). Besides σ-selectionable mappings, we can define, for example, Lipschitz σ-selectionable (L-σ-selectionable) mappings or locally Lipschitz σ-selectionable mappings (LL-σ-selectionable) or Carathéodory σ-selectionable (Ca-σ-selectionable) mappings, when in (3.32.1) we ask φ_n to have a Lipschitz selection or a locally Lipschitz selection or a Carathéodory selection, for every $n \geq 0$, respectively.

The following result, due to A. Lasota and J. A. Yorke (see [Go2-M]), is useful for showing that a map is LL-σ-selectionable.

(3.37) THEOREM (Lasota–Yorke Approximation Theorem). *Let E be a normed space, X a metric space and $f : X \to E$ be a continuous single-valued map. Then, for each $\varepsilon > 0$, there is a locally Lipschitz (single-valued) map $f_\varepsilon : X \to E$ such that:*

$$\|f(x) - f_\varepsilon(x)\| < \varepsilon, \quad \textit{for every } x \in X.$$

PROOF. Let $V_\varepsilon(x) = \{y \in X \mid \|f(y) - f(x)\| < (\varepsilon/2)\}$. Then $\alpha = \{V_\varepsilon(x)\}_{x \in X}$ is an open covering of the metric space X. Since X is paracompact, there exists a locally finite subcovering $\beta = \{W_\lambda\}_{\lambda \in \Lambda}$ of α. For every $\lambda \in \Lambda$, we obtain

$$\mu_\lambda : X \to \mathbb{R}$$

by putting

$$\mu_\lambda(x) = \begin{cases} 0, & \text{for } x \notin W_\lambda, \\ \inf\{d(x,y) \mid y \in \partial W_\lambda\}, & \text{for } x \in W_\lambda. \end{cases}$$

Thus, μ_λ is a Lipschitz map with the constant 1. Since β is locally finite, we deduce that, for every $\lambda \in \Lambda$, the map:

$$\eta_\lambda : X \to [0,1],$$

defined as follows

$$\eta_\lambda(x) = \frac{\mu_\lambda(x)}{\sum_{\varrho \in \Lambda} \mu_\varrho(x)},$$

is a locally Lipschitz map.

We let $f_\varepsilon : X \to E$ by putting

$$f_\varepsilon(x) = \sum_{\lambda \in \Lambda} \eta_\lambda(x) \cdot f(a_\lambda),$$

where $a_\lambda \in W_\lambda$ is an arbitrary but fixed element.

We get:

$$\|f_\varepsilon(x) - f(x)\| = \left\| \sum_{\lambda \in \Lambda} \eta_\lambda(x) \cdot f(a_\lambda) - \sum_{\lambda \in \Lambda} \eta_\lambda(x) \cdot f(x) \right\|$$

$$\leq \sum_{\lambda \in \Lambda} \eta_\lambda(x) \|f(a_\lambda) - f(x)\| < 1 \cdot \varepsilon = \varepsilon,$$

and the proof is completed. $\square$

The Michael selection theorem is not true for l.s.c. mappings with arbitrary compact values. For example, see (3.5.9). In 1988, A. Bressan ([Bre1], [Bre2]) observed that there can exist a directionally continuous selection.

In the following, a set $\Gamma \subset \mathbb{R}^m$ will be called a *cone* if Γ is a nonempty, closed, convex subset of $\mathbb{R}^m$ such that

(i) $x \in \Gamma, \ \lambda \geq 0 \Rightarrow \lambda x \in \Gamma,$

(ii) $\Gamma \cap (-\Gamma) = \{0\}.$

We now introduce the basic related concept.

(3.38) DEFINITION. Let Γ be a cone in $\mathbb{R}^m$ and let Y be a metric space. A map $f : \mathbb{R}^m \to Y$ is *Γ-continuous at a point* $\overline{x} \in \mathbb{R}^m$ if, for every $\varepsilon > 0$, there exists $\delta > 0$ such that $d(f(x), f(\overline{x})) < \varepsilon$, for all $x \in B(\overline{x}, \delta) \cap (\overline{x} + \Gamma)$. We say that f is *Γ-continuous on A* if it is Γ-continuous, at every point $\overline{x} \in A$.

In the above setting, some preliminary results concerning directional continuity are now listed.

(3.39) PROPOSITION.

(a) f *is Γ-continuous at $\overline{x}$ if and only if $\lim_{n \to \infty} f(x_n) = f(\overline{x})$, for every sequence x_n tending to $\overline{x}$ such that $(x_n - \overline{x}) \in \Gamma$, for all $n \geq 1$.*

(b) *If f is Γ-continuous at $\overline{x}$, then f is also Γ'-continuous at $\overline{x}$, for every cone $\Gamma' \subset \Gamma$.*

(c) *If $(f_n)_{n \geq 1}$ is a sequence of Γ-continuous functions which converges uniformly to f, then f is Γ-continuous.*

All proofs are straightforward. Particularly interesing is the case, where $\Gamma = \mathbb{R}^m_+$ is the cone of all points in $\mathbb{R}^m$ with non-negative coordinates (with respect to the canonical basis). In fact, a large class of $\mathbb{R}^m_+$-continuous functions can be constructed. For every integer k, define the partition $\mathcal{P}_k$ of $\mathbb{R}^m$ into half-open cubes with the side length 2^{-k}:

(3.39.1) $\mathcal{P}_k = \{Q^k_\eta \mid \eta = (\eta_1, \ldots, \eta_m) \in \mathbb{Z}^m\}$, where:

(3.39.2) $Q^k_\eta = \{x = (x_1, \ldots, x_m) \mid 2^{-k}(\eta_i - 1) \leq x_i < 2^{-k}\eta_i, i = 1, \ldots, m\}$.

In this setting, we have:

(3.40) PROPOSITION.

(a) *Let k be an arbitrary integer. If the restriction of f to each cube $Q^k_\eta \in \mathcal{P}_k$ is $\mathbb{R}^m_+$-continuous, then f is $\mathbb{R}^m_+$-continuous on $\mathbb{R}^m$.*

(b) *If, for some k, f is constant, on every cube $Q^k_\eta \in \mathcal{P}_k$, then f is $\mathbb{R}^m_+$-continuous.*

PROOF. Let $\mathcal{T}^+$ be the topology on $\mathbb{R}^m$, generated by the sets

$$A_{x,\varepsilon} = \{y \in \mathbb{R}^m \mid (y - x) \in \mathbb{R}^m_+, |y - x| < \varepsilon\},$$

with $x \in \mathbb{R}^m$, $\varepsilon > 0$. Saying that a map f is $\mathbb{R}^m_+$-continuous simply means that f is continuous with respect to the topology $\mathcal{T}^+$. Since all cubes Q^k_η are closed-open sets in $\mathcal{T}^+$, assertions (a) and (b) follow. $\qquad \square$

The following result provides a useful tool for reducing a problem concerning an arbitrary cone Γ to the special case, where $\Gamma = \mathbb{R}^m_+$.

(3.41) PROPOSITION.

(a) *Let Γ be a cone in $\mathbb{R}^m$, Y a metric space. Let L be an invertible linear operator on $\mathbb{R}^m$. Then a map $f : \mathbb{R}^m \to Y$ is Γ-continuous at a point $\overline{x}$ if and only if the composed map $f \circ L^{-1}$ is $L(\Gamma)$-continuous at $L(\overline{x})$.*

(b) *For every cone $\Gamma \subset \mathbb{R}^m$, there exists an invertible linear operator ψ on $\mathbb{R}^m$ such that $\psi(\Gamma) \subset \mathbb{R}^m_+$.*

To prove (a), assume that $f \circ L^{-1}$ is $L(\Gamma)$-continuous at $L(\overline{x})$. Take any sequence $x_n \to \overline{x}$ such that $(x_n - \overline{x}) \in \Gamma$, for all $n \geq 1$. Then $L(x_n) \to L(\overline{x})$ and $(L(x_n) - L(\overline{x})) \in L(\Gamma)$. Hence,

$$d(f(x_n), f(\overline{x})) = d(f \circ L^{-1}(L(x_n)), f \circ L^{-1}(L(\overline{x}))) \to 0,$$

showing that f is Γ-continuous at $\overline{x}$. The converse is obtained by replacing L with L^{-1}.

To prove (b), let Γ be given and consider the positive dual cone

$$\Gamma^+ = \{y \in \mathbb{R}^m \mid \langle y, x \rangle \geq 0, \quad \text{for all } x \in \Gamma\}.$$

Since Γ^+ has a nonempty interior (see [Del-M]), there exists a unit vector $w_1 \in \mathrm{int}(\Gamma^+)$ and some $\varepsilon > 0$ such that

(3.41.1) $\langle w_1, x \rangle \geq \varepsilon |x|$,   for all $x \in \Gamma$.

Choose $m - 1$ unit vectors $w_2, \ldots, w_m$ such that $\{w_1, w_2, \ldots, w_m\}$ is an orthonormal basis, for $\mathbb{R}^m$, and let $\{e_1, \ldots, e_m\}$ be the canonical basis. Define the invertible operators L_1, L_2 on $\mathbb{R}^m$ by setting:

$$L_1(w_1) = w_1 + \varepsilon^{-1}(w_2 + w_3 + \cdots + w_m),$$
$$L_1(w_i) = w_i, \qquad\qquad\qquad\qquad i = 2, \ldots, m,$$
$$L_2(w_j) = e_j, \qquad\qquad\qquad\qquad j = 1, \ldots, m.$$

The transformation $L = L_2 \circ L_1$ then satisfies our requirement. Indeed, if $u = \sum_{j=1}^{m} \lambda_j w_j \in \Gamma$, (3.41.1) implies

(3.41.2) $$\lambda_1 \geq \varepsilon \left(\sum_{j=1}^{m} \lambda_j^2 \right)^{1/2} \geq \varepsilon |\lambda_i|,$$

for all $i = 1, \ldots, m$. Let $L_1(u) = \sum_{i=1}^{m} \mu_i w_i$. Then $\mu_1 = \lambda_1 \geq 0$ and $\mu_i = \lambda_i + \varepsilon^{-1} \lambda_i \geq 0$, for $2 \leq i \leq m$, because of (3.41.2). Therefore, $L(u) = \sum_{i=1}^{m} \mu_i e_i \in \mathbb{R}_+^m$.

Now, we are able to prove the main related result.

(3.42) THEOREM. *Let $\varphi : \mathbb{R}^m \multimap Y$ be a l.s.c. map with nonempty closed values and Y be a complete (metric) space. Then, for every cone $\Gamma \subset \mathbb{R}^m$, the mapping φ admits a Γ-continuous selection.*

PROOF. The proof will be given at first in the special case $\Gamma = \mathbb{R}_+^m$, and then extended to an arbitrary cone Γ. We begin by constructing a sequence of approximate selections $\{f_n\}_{n \geq 1}$ on the half-open unit cube $Q = \{(x_1, \ldots, x_m) \in \mathbb{R}^m \mid 0 \leq x_i < 1, i = 1, \ldots, m\}$. Each f_n will have the following properties:

(i)$_n$ There exists an integer $h = h(n)$ such that f_n is constant on every cube $Q_\eta^{h(n)} \subset Q$ of the partition $\mathcal{P}_{h(n)}$, defined in (3.39.1); say, $f(x) = y_\eta^n \in Y$, for all $x \in Q_\eta^{h(n)}$, (ii)$_n$ $d(y_\eta^n, \varphi(x)) < 2^{-n}$, for all $x \in \overline{Q_\eta^{h(n)}}$,

(iii)$_n$ $d(f_n(x), f_{n-1}(x)) < 2^{-n+1}$, for all $x \in Q (n \geq 2)$.

To define f_1, choose a finite set of points $a_1, \ldots, a_k \in \overline{Q}$, elements $y_i \in \varphi(a_i)$ and open neighborhoods $V_1, \ldots, V_k$ such that

(a) $$a_i \in V_i,$$

(b) $$\bigcup_{i=1}^k V_i \supset \overline{Q},$$

(c) $$d(y_i, \varphi(x)) < 2^{-1}, \quad \text{for all } x \in V_i.$$

All of these can be done, because φ is lower semicontinuous on the compact set $\overline{Q}$. Let λ be a Lebesgue number for the covering $\{V_i\}$ of $\overline{Q}$ and choose an integer $h = h(1)$ so large that the closure of every cube $Q_\eta^h \subset \overline{Q}$ is entirely contained in some V_i. This is certainly the case if $\sqrt{m} \cdot 2^{-h} < \lambda$. For each $Q_\eta^h \subset Q$, choose a V_i such that $\overline{Q_\eta^h} \subset V_i$ and define $f_1(x) = y_i$, for all $x \in Q_\eta^h$. Clearly (i)$_1$ and (ii)$_1$ hold. Let now f_n be defined and satisfy (i)$_n$ – (iii)$_n$. We shall construct f_{n+1} separately on each cube of the partition $\mathcal{P}_{h(n)}$. Fix $\sigma \in \mathbb{Z}^m$ such that $Q_\sigma^{h(n)} \subset Q$. By (i)$_n$, $f_n(x) = y_\sigma^n$ is constant on $Q_\sigma^{h(n)}$. Choose a finite set of points $a_1, \ldots, a_k \in \overline{Q_\sigma^{h(n)}}$ elements $y_i \in F(a_i)$ and open neighborhoods $V_1, \ldots, V_k$ such that

(a) $$a_i \in V_i,$$

(b) $$\bigcup_{i=1}^k V_i \supset \overline{Q_\sigma^{h(n)}},$$

(c) $$d(y_i, F(x)) < 2^{-n-1}, \quad \text{for all } x \in V_i,$$

(d) $$d(y_i, y_\sigma^n) < 2^{-n}.$$

Notice that all of these can be done, because φ is lower semicontinuous, $\overline{Q_\sigma^{h(n)}}$ is compact and (ii)$_n$ holds. Choose an integer $h(\sigma)$ so large that every closed cube $\overline{Q_\eta^{h(n)}} \subset \overline{Q_\sigma^{h(n)}}$ is entirely contained in some V_i. For every $Q_\eta^{h(\sigma)}$, select a V_i for which $\overline{Q_\eta^{h(\sigma)}} \subset V_i$ and define $f_{n+1}(x) = y_i$, for all $x \in Q_\eta^{h(\sigma)}$. Repeating this construction on every cube $Q_\sigma^{h(n)} \subset Q$, we obtain an approximate selection f_{n+1} defined on the whole cube Q. Setting $h(n+1) = \max\{h(\sigma) \mid \sigma \in \mathbb{Z}^m, Q_\sigma^{h(n)} \subset Q\}$, all conditions (i)$_{n+1}$ – (iii)$_{n+1}$ are satisfied.

By induction, we can now assume that a sequence $\{f_n\}_{n\geq 1}$, satisfying (i)–(iii) has been constructed. By (iii) and the completeness of Y, the sequence $\{f_n\}$ has a uniform limit $f : Q \to Y$. Property (i) together with Propositions (3.40)(b) and (3.39)(a) imply that f is $\mathbb{R}_+^m$-continuous. Moreover, $f(x) \in \varphi(x)$, for all $x \in Q$, because of (ii) and of the closure of $\varphi(x)$. Therefore, f is an $\mathbb{R}_+^m$-continuous selection of φ on Q. Repeating the same construction on every cube Q_η^0, $\eta \in \mathbb{Z}^m$, and recalling Proposition (3.40)(a), we obtain an $\mathbb{R}_+^m$-continuous selection of φ defined on the whole space $\mathbb{R}^m$.

Consider now the case, where $\Gamma \subset \mathbb{R}^m$ is an arbitrary cone. Using Proposition (3.41)(b), let L be an invertible linear operator such that $L(\Gamma) \subset \mathbb{R}_+^m$. Construct an $\mathbb{R}_+^m$-continuous selection g of the lower semicontinuous map $\varphi \circ L^{-1}$ and set $f = g \circ L$. This yields

$$f(x) = g(L(x)) \in \varphi \circ L^{-1}(L(x)) = \varphi(x),$$

for every $x \in \mathbb{R}^m$. Hence, f is a selection of φ. Since $L(\Gamma) \subset \mathbb{R}_+^m$, Propositions (3.39)(b) and (3.41)(a) imply that g is $L(\Gamma)$-continuous and f is Γ-continuous. The proof of (3.42) is completed. $\qquad\qquad\qquad\qquad\qquad\qquad\qquad\qquad$ $\square$

Theorem (3.42) has important applications to the existence results for differential inclusions with l.s.c. right-hand sides.

Note that Theorem (3.42) is not a generalization of the Michael selection theorem. There exists another result connected with the Michael selection theorem. Namely, in 1983 A. Fryszkowski [Fry2] (comp. also [FryRz], [Fry3]) proved the selection theorem for l.s.c. mappings with closed decomposable values (for the definition, see (3.71) below). A generalization of Fryszkowski's theorem for separable spaces has been proved by A. Bressan and G. Colombo in 1988 (see Theorem (3.81)).

Finally, we would like to point out that in the following part, we will present the Kuratowski–Ryll–Nardzewski selection theorem [KRN] (cf. (3.49)) frequently used in the theory of differential inclusions.

Apart from semicontinuous multivalued mappings, multivalued measurable mappings will be of the great importance in the sequel. Throughout this section, we assume that Y is a separable metric space, and $(\Omega, \mathcal{U}, \mu)$ is a measurable space, i.e., a set Ω equipped with σ-algebra $\mathcal{U}$ of subsets and a countably additive measure μ on $\mathcal{U}$. A typical example is when Ω is a bounded domain in the Euclidean space $\mathbb{R}^k$, equipped with the Lebesgue measure.

(3.43) DEFINITION. A multivalued map $\varphi : \Omega \multimap Y$ with closed values is called *measurable* if $\varphi^{-1}(V) \in \mathcal{U}$, for each open $V \subset Y$.

(3.44) DEFINITION. A multivalued map $\varphi : \Omega \multimap Y$ with closed values is called *weakly measurable* if $\varphi^{-1}(A) \in \mathcal{U}$, for each closed $A \subset Y$.

Another way of defining measurability is by requiring the measurability of the graph Γ_φ of φ in the product $\Omega \times Y$, equipped with the minimal σ-algebra $\mathcal{U} \otimes \mathcal{B}(Y)$

generated by the sets $A \times B$ with $A \in \mathcal{U}$ and $B \in \mathcal{B}(Y)$, where $\mathcal{B}(Y)$ denotes the family of all Borel subsets of Y.

For our convenience, we collect some relations between these definitions in the following

(3.45) PROPOSITION. *Assume that $\varphi, \psi : \Omega \multimap Y$ are two multivalued mappings. Then the following hold true:*

(3.45.1) *φ is measurable if and only if $\varphi_+^{-1}(A) \in \mathcal{U}$, for each closed $A \subset Y$,*

(3.45.2) *φ is weakly measurable if and only if $\varphi_+^{-1}(V) \in \mathcal{U}$, for each open $V \subset Y$,*

(3.45.3) *if φ is measurable, then φ is also weakly measurable,*

(3.45.4) *if φ has compact values, measurability and weak measurability of φ are equivalent,*

(3.45.5) *φ is weakly measurable if and only if the distance function $f_y : \Omega \to R$, $f_y(x) = \mathrm{dist}(y, \varphi(x))$ is measurable, for all $y \in Y$,*

(3.45.6) *if φ is weakly measurable, then the graph Γ_φ of φ is product measurable,*

(3.45.7) *if φ and ψ are measurable, then so is $\varphi \cup \psi$,*

(3.45.8) *if φ and ψ are measurable, then so is $\varphi \cap \psi$,*

(3.45.9) *if φ and ψ are measurable, then so is $\varphi \times \psi$.*

The proof of (3.45) is straightfoward and, therefore, we left it to the reader.

Let us note that the composition of two measurable multivalued mappings need not be measurable.

(3.46) EXAMPLE. Let $\Omega = [0,1]$ be equipped with the Lebesgue measure and let $f : \Omega \to \mathbb{R}$ be a strictly increasing Cantor function which is obviously measurable. It is well-known that one can find a measurable set $\mathcal{D} \subset \mathbb{R}$ such that $f^{-1}(\mathcal{D})$ is not measurable. If we define

$$\varphi : \Omega \multimap \mathbb{R} \quad \text{and} \quad \psi : \mathbb{R} \multimap \mathbb{R}$$

by

$$\varphi(t) = \{f(t)\} \quad t \in \Omega,$$

$$\psi(u) = \begin{cases} \{1\} & \text{if } n \in \mathcal{D}, \\ \{0\} & \text{if } n \notin \mathcal{D}, \end{cases}$$

then both φ and ψ are measurable, but $\psi \circ \varphi$ is not.

Now, we collect the results and the counter-examples obtained so far for the presentation of semicontinuity or measurability properties in the following table:

φ, ψ	u.s.c.	l.s.c.	measurable
$\varphi \cup \psi$	yes	yes	yes
$\varphi \cap \psi$	yes	no	yes
$\varphi \times \psi$	yes*	yes*	yes
$\varphi \circ \psi$	yes	yes	no

* if φ and ψ have compact values.

A famous relation between measurability and continuity of single-valued functions is established by Luzin's theorem which states, roughly speaking, that $f : \Omega \to Y$ is measurable if and only if f is continuous up to subsets of Ω of an arbitrarily small measure. It is not surprising that this result has an analogue for multi-valued mappings (for details, see [ADTZ-M]). Below we shall sketch Luzin's-type multivalued results.

(3.47) DEFINITION. We say that a multivalued map $\varphi : \Omega \multimap Y$ with closed values has the *Luzin property* if, given $\delta > 0$, one can find a closed subset $\Omega_\delta \subset \Omega$ such that $\mu(\Omega \setminus \Omega_\delta) \leq \delta$ and the restriction $\varphi|_{\Omega_\delta}$ of φ to Ω_δ is continuous (of course, we have assumed that Y is a metric space).

We have:

(3.48) THEOREM. *A multivalued map* $\varphi : \Omega \multimap Y$ *is measurable if and only if* φ *has the Luzin property.*

In what follows, we shall use the following Kuratowski–Ryll–Nardzewski selection theorem (see [KRN] and cf. [AuF-M], [AuC-M], [Ki-M], [Sr-M]).

(3.49) THEOREM (Kuratowki–Ryll–Nardzewski). *Let* Y *be a separable complete space. Then every measurable* $\varphi : \Omega \multimap Y$ *has a (single-valued) measurable selection.*

PROOF. Without any loss of generality, we can change the metric of Y by an equivalent metric, preserving completeness and separability, so that Y becomes a bounded (say, with the diameter M) complete metric space. Now, let us divide the proof into two steps.

(a) Let C be a countable dense subset of Y. Set $\varepsilon_0 = M$, $\varepsilon_i = M/2^i$. We claim that we can define a sequence of mappings $s_m : \Omega \to C$ such that:

(i) $\qquad\qquad\qquad s_m$ is measurable,

(ii) $\qquad\qquad\qquad s_m(x) \in O_{\varepsilon_m}(\varphi(x))$,

(iii) $\qquad\qquad\qquad s_m(x) \in B(s_{m-1}(x), \varepsilon_{m-1}), \quad m > 0$.

Arrange the points of C into a sequence $\{c_j\}_{j=0,1,2,\ldots}$ and define s_0 by putting:

$$s_0(x) = c_0, \quad \text{for every } x \in \Omega.$$

Then (i) and (ii) are clearly satisfied.

Assume we have defined functions s_m satisfying (i) and (ii) up to $m = p - 1$, and define s_p, satisfying (i), (ii) and (iii), as follows.

Set

$$A_j = \varphi_+^{-1}(B(c_j, \varepsilon_p)) \cap s_{p-1}^{-1}(B(c_j, \varepsilon_{p-1}))$$

and

$$E_0 = A_0, \quad E_j = A_j \setminus (E_0 \cup \ldots \cup E_{j-1}).$$

We claim that

$$\Omega = \bigcup_{j=0}^{\infty} E_j.$$

Obviously, E_j, $j = 0, 1, \ldots$, is measurable (comp. (3.45)). Let $x \in \Omega$ and consider $s_{p-1}(x)$ and $\varphi(x)$. By (ii), $s_{p-1}(x) \in O_{\varepsilon_{p-1}}(\varphi(x))$; by the density of C, there is a c_j such that at once $s_{p-1}(x) \in B(c_j, \varepsilon_{p-1})$ and $\varphi(x) \cap B(c_j, \varepsilon_{p-1}) \neq \emptyset$, i.e., $x \in A_j$. Finally, either $x \in E$, or it is in some E_i, $i < j$. In either case, $x \in \bigcup_{j=0}^{\infty} E_j$. We define $s_p : \Omega \to C$ by putting:

$$s_p(x) = c_j, \quad \text{whenever } x \in E_j.$$

Then s_p satisfies (i)–(iii).

Condition (iii) implies that $\{s_m(x)\}$ is a Cauchy sequence, for every $x \in \Omega$. We let $s : \Omega \to Y$ as follows:

$$s(x) = \lim_{m \to \infty} s_m(x), \quad x \in \Omega.$$

Since φ has closed values, by (ii), we deduce that $s(x) \in \varphi(x)$, for every $x \in \Omega$.

It remains to show that s is measurable. This is equivalent to show that counter images of closed sets are measurable. Let K be a closed subset of Y. Then each set $s_m^{-1}(O_{\varepsilon_m}(K))$ is measurable. We shall complete the proof by showing that $s^{-1}(K) = \bigcap s_m^{-1}(O_{\varepsilon_m}(K))$. In fact, if $x \in s^{-1}(K)$, then $s(x) \in K$ and since $d(s_m(x), s(x)) < \varepsilon_m$, $s_m \in O_{\varepsilon_m}(K)$, for every m. On the other hand, if $x \in s_m^{-1}(O_{\varepsilon_m}(K))$, for all m, then $s_m(x) \in O_{\varepsilon_m}(K)$ and since $\{s_m(x)\}$ converges to $s(x)$ and K is closed, we get $s(x) \in K$. The proof of (3.49) is completed. □

Now, we shall be concerned with multivalued mappings which are defined on the topological product of some measurable set with the Euclidean space $\mathbb{R}^n$. We are particulary interested in Carathéodory multivalued mappings and the Scorza–Dragoni multivalued mappings. Besides their fundamental importance in all fields of multivalued analysis, such multivalued mappings are useful for differential inclusions.

Let $\Omega = [0, a]$ be equipped with the Lebesgue measure and $Y = \mathbb{R}^n$.

(3.50) DEFINITION. A map $\varphi : [0, a] \times \mathbb{R}^n \multimap \mathbb{R}^n$ with nonempty compact values is called u-Carathéodory (resp., l-Carathéodory; resp., Carathéodory) if it satisfies:

(3.50.1) $t \multimap \varphi(t, x)$ is measurable, for every $x \in \mathbb{R}^n$,

(3.50.2) $x \multimap \varphi(t, x)$ is u.s.c. (resp., l.s.c.; resp., continuous), for almost all $t \in [0, a]$,

(3.50.3) $|y| \leq \mu(t)(1 + |x|)$, for every $(t, x) \in [0, a] \times \mathbb{R}^n$, $y \in \varphi(t, x)$, where $\mu : [0, a] \to [0, +\infty)$ is an integrable function.

As before, by $\mathcal{U} \otimes \mathcal{B}(\mathbb{R}^n)$, we denote the minimal σ-algebra generated by the Lebesgue measurable sets $A \in \mathcal{U}$ and the Borel subsets of $\mathbb{R}^n$, and so the term "product-measurable" means measurability with respect to $\mathcal{U} \otimes \mathcal{B}(\mathbb{R}^n)$.

(3.51) PROPOSITION. Let $\varphi : [0, a] \times \mathbb{R}^n \multimap \mathbb{R}^m$ be a Carathéodory multivalued map. Then φ is product-measurable.

PROOF. Consider the countable dense subset $Q^n \subset \mathbb{R}^n$ of rationals. For closed $A \subset \mathbb{R}^n$, $a \in Q^n$ and k, the set

$$G_k(A, a) = \{t \in [0, a] \mid \varphi(t, a) \cap O_{1/k}(A) \neq \emptyset\} \times B\left(a, \frac{1}{k}\right)$$

belongs to $\mathcal{U} \otimes \mathcal{B}(\mathbb{R}^n)$. Since φ is l.s.c. in the second variable, we have:

$$\varphi_+^{-1}(A) \leq \bigcap_{k=1}^{\infty} \bigcup_{a \in Q^n} G_k(A, a),$$

while the u.s.c. of φ implies the reverse inclusion. The proof is completed. □

The following example shows that an l-Carathéodory multivalued map need not be product measurable.

(3.52) EXAMPLE. Let $\varphi : [0,1] \times \mathbb{R} \multimap \mathbb{R}$ be defined as follows:

$$\varphi(t,u) = \begin{cases} \{0\}, & \text{for } u = 0, \\ [0,1], & \text{otherwise.} \end{cases}$$

Then φ is l-Carathéodory (not u-Carathéodory), but not product measurable.

An analogous example can be constructed for u-Carathéodory mappings.

Let $\varphi : [0,a] \times \mathbb{R}^n \multimap \mathbb{R}^n$ be a fixed multivalued map. We are interested in the existence of Carathéodory selections, i.e., Carathéodory functions $f : [0,a] \times \mathbb{R}^n \to \mathbb{R}^n$ such that $f(t,u) \in \varphi(t,u)$, for almost all $t \in [0,a]$ and all $u \in \mathbb{R}^n$. It is evident that, in the case when φ is u-Carathéodory, this selection problem does not have a solution in general (the reason is exactly the same as in Michael's selection principle). For l-Carathéodory multivalued maps φ, however, this is an interesting problem.

We are now going to study this problem. We use the following notation:

$$C(\mathbb{R}^n, \mathbb{R}^n) = \{f : \mathbb{R}^n \to \mathbb{R}^n \mid f \text{ is continuous}\}.$$

We shall understand that $C(\mathbb{R}^n, \mathbb{R}^n)$ is equipped with the topology of uniform convergence on compact subsets of $\mathbb{R}^n$. This topology is metrizable (see [BPe-M], for example). Moreover, as usually, by $L^1([0,a], \mathbb{R}^n)$ we shall denote the Banach space of Lebesgue integrable functions.

There are, essentially, two ways to deal with the above selection problem. Let $\varphi : [0,a] \times \mathbb{R}^n \multimap \mathbb{R}^n$ be a l-Carathéodory mapping. We can show that the multivalued map:

$$\Phi : [0,a] \multimap C(\mathbb{R}^n, \mathbb{R}^n),$$

$$\Phi(t) = \{u \in C(\mathbb{R}^n, \mathbb{R}^n) \mid u(x) \in \varphi(t,u(x)) \text{ and } u \text{ is continuous}\}$$

is measurable. Then, if we assume that φ has convex values, in view of the Michael selection theorem (3.30), we obtain that $\Phi(t) \neq \emptyset$, for every t. Moreover, let us observe that every measurable selection of Φ will then give rise to a Carathéodory selection of φ.

On the other hand, we can show that the multivalued map:

$$\Psi : \mathbb{R}^n \multimap L_1([0,a], \mathbb{R}^n),$$

$$\Psi(x) = \{u \in L_1([0,a], \mathbb{R}^n) \mid u(t) \in \varphi(t,u(t)), \text{ for almost all } t \in [0,a]\}$$

is a l.s.c. mapping.

Consequently, continuous selections of Ψ will then give rise to Carathéodory selections of φ.

Hence, our problem can be solved by using the Michael and the Kuratowski–Ryll–Nardzewski selection theorems.

Let us formulate, only for informative purposes, the following result due to A. Cellina ([Ce1]).

(3.53) THEOREM. *Let $\varphi : [0,a] \times \mathbb{R}^n \multimap \mathbb{R}^n$ be a multivalued map with compact convex values. If $\varphi(\cdot, x)$ is u.s.c., for all $x \in \mathbb{R}^n$, and $\varphi(t, \cdot)$ is l.s.c., for all $t \in [0,a]$, then φ has a Carathéodory selection.*

We conclude this part by introducing mappings having the Scorza–Dragoni property.

(3.54) DEFINITION. We say that a multivalued map $\varphi : [0,a] \times \mathbb{R}^n \multimap \mathbb{R}^n$ with closed values has the *u-Scorza–Dragoni property* (resp., *l-Scorza–Dragoni property*; resp., *Scorza–Dragoni property*) if, given $\delta > 0$, one can find a closed subset $A_\delta \subset [0,a]$ such that the measure $\mu([0,a] \setminus A_\delta) \leq \delta$ and the restriction $\widetilde{\varphi}$ of φ to $A_\delta \times \mathbb{R}^n$ is u.s.c. (resp. l.s.c.; resp. continuous).

Let us observe that the Scorza Dragoni property plays the same role for multivalued mappings of two variables as the Luzin property for multivalued mappings of one variable.

There is a close connection between Carathéodory multivalued mappings and multivalued mapping having the Scorza Dragoni property.

(3.55) PROPOSITION. *Let $\varphi : [0,a] \times \mathbb{R}^m \multimap \mathbb{R}^n$ be a multivalued map with closed values satisfing (3.50.3). Then we have:*

(3.55.1) *φ is Carathéodory if and only if φ has the Scorza–Dragoni property,*

(3.55.2) *if φ has the u-Scorza–Dragoni property, then φ is u-Carathéodory,*

(3.55.3) *if φ has the l-Scorza–Dragoni property, then φ is l-Carathéodory,*

(3.55.4) *if φ is product-measurable l-Carathéodory, then φ has the l-Scorza–Dragoni property.*

Assume, furthermore, that φ satisfies the Filippov condition, i.e., for every open $U, V \subset \mathbb{R}^n$, the set:

$$\{t \in [0,a] \mid \varphi(t, U) \subset V\}$$

is Lebesgue measurable. Then:

(3.55.5) *φ is an u-Carathéodory multivalued map if and only if φ has the u-Scorza–Dragoni property.*

Proposition (3.55) is taken from [ADTZ-M]. All proofs are rather technical and need sometimes long calculations.

Therefore, we shall present below only two examples showing that *l*-Carathéodory (*u*-Carathéodory) map need not have the *l*-Scorza–Dragoni (*u*-Scorza–Dragoni) property.

(3.56) EXAMPLE. Let $\varphi : [0,1] \times \mathbb{R} \multimap \mathbb{R}$ be the map defined as follows:

$$\varphi(t,u) = \begin{cases} \{0\} & \text{if } u = t \text{ and } t \in [0,1] \setminus A, \\ \{1\} & \text{if } u = t \text{ and } t \in A, \\ [0,1] & \text{otherwise,} \end{cases}$$

where A is a non-measurable subset of $[0,1]$. Then, obviously, φ is l-Carathéodory, but does not have the l-Scorza–Dragoni property.

(3.57) EXAMPLE. Let $\varphi : [0,1] \times \mathbb{R} \multimap \mathbb{R}$ be defined as follows:

$$\varphi(t,u) = \begin{cases} [0,1] & \text{if } t = u \text{ and } t \in A, \\ \{0\} & \text{otherwise,} \end{cases}$$

where A is a non-measurable subset of $[0,1]$. It is not hard to see that φ is u-Carathéodory, but does not have the u-Scorza–Dragoni property.

Until the end of this part, X is a metric separable space and Ω a complete measurable space. We also assume that $\varphi : \Omega \times X \multimap X$ is a product-measurable multivalued mapping with compact values.

At first, we state:

(3.58) PROPOSITION. *If $\varphi : \Omega \times X \multimap X$ is product-measurable, then the function $f : \Omega \times X \to [0,\infty)$ defined by the formula:*

$$f(\omega, x) = \text{dist}(x, \varphi(\omega, x))$$

is also product measurable.

PROOF. We have:

$$\{(\omega, x) \in \Omega \times X \mid f(\omega, x) < r\} = \{(\omega, x) \in \Omega \times X \mid \varphi(\omega, x) \cap O_r(\{x\}) \neq 0\}.$$

Therefore, our assertion follows directly from the assumption that φ is measurable. $\qquad \square$

(3.59) THEOREM (R. J. Aumann). *If $\varphi : \Omega \multimap X$ is a multivalued map with compact values such that the graph Γ_φ of φ is measurable, then φ possesses a measurable selection.*

The proof of (3.59) is not in the scope of our book. Therefore, for the proof, we recommend [Hi2] (comp. also [CV-M], [Ki-M], [Ox-M]).

A map $\varphi : [a,b] \times \mathbb{R}^n \multimap \mathbb{R}^n$ is said to be *integrably bounded* if there exists an integrable function $\mu \in L^1([a,b])$ such that $|y| \leq \mu(t)$, for every $x \in \mathbb{R}^n$, $t \in [a,b]$ and $y \in \varphi(t,x)$.

We say that φ has (at most) a *linear growth* if there exists an integrable function $\mu \in L^1([a, b])$ such that

$$|y| \leq \mu(t)(1 + |x|),$$

for every $x \in \mathbb{R}^n$, $t \in [a, b]$ and $y \in \varphi(t, x)$.

The following Scorza Dragoni type result describes possible regularizations of Carathéodory maps.

(3.60) THEOREM. *Let X be a compact subset of $\mathbb{R}^n$ and $\varphi : [0, a] \times X \multimap \mathbb{R}^n$ be a nonempty compact convex valued Carathéodory map. Then there exists an u-Scorza Dragoni $\psi : [0, a] \times X \multimap \mathbb{R}^n$ with nonempty compact convex values such that:*

(3.60.1) *$\psi(t, x) \subset \varphi(t, x)$, for every $(t, x) \in [0, a] \times X$,*

(3.60.2) *if $\Delta \subset [0, a]$ is measurable, $u : \Delta \to \mathbb{R}^n$ and $v : \Delta \to X$ are measurable maps and $u(t) \in \varphi(t, v(t))$, for almost all $t \in \Delta$, then $u(t) \in \psi(t, v(t))$, for almost all $t \in \Delta$.*

Now, we prove:

(3.61) THEOREM. *Let E, E_1 be two separable Banach spaces and $\varphi : [a, b] \times E \multimap E_1$ be an u-Scorza–Dragoni map with compact convex values, then φ is Ca-σ-selectionable. The maps $\varphi_k : [a, b] \times E \multimap E_1$ (see Definition (3.32)) are u-Scorza–Dragoni, and we have*

$$\varphi_k(t, e) \subset \left(\bigcup_{x \in E} \varphi(t, x) \right).$$

Moreover, if φ is integrably bounded, then φ is mLL-σ-selectionable.

PROOF. Consider the family $\{B(y, r_k)\}_{y \in E}$, where $r_k = (1/3)^k$, $k = 1, 2, \ldots$. Using well-known Stone's theorem, for every $k = 1, 2, \ldots$, we get localy finite subcovering $\{U_i^k\}_{i \in I_k}$ of $\{B(y, r_k)\}_{y \in E}$. For every $i \in I_k$, $k = 1, 2, \ldots$, we fix the center $y_i^k \in E$ such that $U_i^k \subset B(y_i^k, r_k)$. Now, let $\eta_i^k : E \to [0, 1]$ be a locally Lipschitz partition of unity subordinated to $\{U_i^k\}_{i \in I_k}$.

Define

$$\psi_i^k : [0, a] \multimap E \quad \text{and} \quad f_i^k : [0, a] \to E$$

as follows:

$$\psi_i^k(t) = \overline{\text{conv}}\left(\bigcup_{y \in B(y_i^k, 2r_k)} \varphi(t, y) \right),$$

and let f_i^k be a measurable selection of ψ_i^k which exists in view of the Kuratowski–Ryll–Nardzewski theorem (3.49).

Finally, we define

$$\varphi_k : [a,b] \times E \multimap E_1, \quad f_k : [a,b] \times E \to E_1$$

as follows:

$$\varphi_k(t,z) = \sum_{i \in I_k} \eta_i^k(z) \cdot \psi_i^k(t),$$

$$f_k(t,z) = \sum_{i \in I_k} \eta_i^k(z) \cdot f_i^k(t).$$

Then $f_k \subset \varphi_k$. Fix $t \in [a,b]$. If $\varphi(t, \cdot)$ is u.s.c., then $\varphi(t,z) = \bigcap_{k=1}^{\infty} \varphi_k(t,z)$ and $\varphi_{k+1}(t,z) \subset \varphi_k(t,z)$, for every $z \in E$. By the assumptions on φ, the map $\varphi(t, \cdot)$ is u.s.c., for almost all $t \in [0,a]$, and the first part of (3.61) follows. The remaining part is an immediate consequence of the first one, and the theorem is proved. $\square$

In what follows, for a metric space (Y,d), we shall denote by $B(Y)$ $(C(Y))$ the family of all nonempty closed bounded (compact) subsets of Y. We shall consider $B(Y)$ as a metric space with the Hausdorff metric $d_H{}^2$ and $C(Y)$ as a metric space with the Borsuk metric $d_C{}^3$ (for more details, see Part II.20 in [Go5-M]). We will consider mappings of the type

$$F : X \to B(Y) \quad \text{or} \quad F : X \to C(Y).$$

Any such a map can be reinterpreted as a multivalued map $\varphi : X \multimap Y$ with closed bounded and nonempty values or, respectively, with compact nonempty values defined as follows:

$$\varphi(x) = F(x), \quad \text{for every } x \in X.$$

For simplicity, we shall use only one notion F, in the place of φ. We hope it will not cause any confusion.

Observe, that for $F : X \to B(Y)$, we have the notion of continuity with respect to the metric given in X and d_H in $B(Y)$, but, for $F : X \to C(Y)$, we can also speak about the continuity of F with respect to the metric given in X and d_C in $C(Y)$.

As a first observation, we get the following

---

[2] Recall that $d_H(A,B) = \inf\{r > 0 \mid A \subset O_r(B) \text{ and } B \subset O_r(A)\}$.

[3] We let $d_C(A,B) = \inf\{r > 0 \mid \exists f : A \to B \text{ and } g : B \to A \text{ such that } d(y,g(y)) \leq r \text{ and } d(x,f(x)) \leq r, \text{ for every } x \in A \text{ and } y \in B\}$.

(3.62) PROPOSITION. *If $F : X \to C(Y)$ is continuous with respect to d_C, then F is continuous with respect to d_H.*

In fact, our claim follows from the following inequality obtained in Part II.20 in [Go5-M]):

$$d_C(F(x), F(y)) \geq d_H(F(x), F(y)),$$

for every $x, y \in X$.

(3.63) REMARK. Note that the continuity with respect to d_H does not imply the one with respect to d_C.

For example, the mapping φ considered in (3.5.1) is d_H-continuous, but not d_C continuous.

In what follows, we shall also say that d_H-continuous maps are Hausdorff continuous and d_C-continuous maps are Borsuk continuous. We prove the following:

(3.64) THEOREM. *A mapping $F : X \to C(Y)$ is Hausdorff continuous if and only if it is both u.s.c. and l.s.c.*

PROOF. Assume that F is d_H-continuous and let U be an open subset of Y. At first, we shall prove that the set:

$$F^{-1}(U) = \{x \in X \mid F(x) \subset U\}$$

is open.

Let $x_0 \in F^{-1}(U)$. Then $F(x_0) \subset U$. Since $F(x_0)$ is compact, there exists $\varepsilon > 0$ such that $O_\varepsilon(F(x_0)) \subset U$. Since F is d_H-continuous, we can find $\delta > 0$ such that, for every $x \in B(x_0, \delta)$, we have $d_H(F(x_0), F(x)) < \varepsilon$. It implies that

$$F(x) \subset O_\varepsilon(F(x_0)) \subset U.$$

So, $B(x_0, \delta) \subset F^{-1}(U)$ and $F^{-1}(U)$ is open. Now, we would like to show that the set:

$$F_+^{-1}(U) = \{x \in U \mid F(x) \cap U \neq \emptyset\}$$

is open.

Let $x_0 \in F_+^{-1}(U)$. So $F(x_0) \cap U \neq 0$. Let $y_0 \in F(x_0) \cap U$. We take $\varepsilon > 0$ such that $B(y_0, \varepsilon) \subset U$. We can find $\delta > 0$ such that, for every $x \in B(x_0, \delta)$, we have:

$$d_H(F(x_0), F(x)) < \frac{\varepsilon}{2}.$$

We claim that $F(x) \cap B(y_0, \varepsilon) \neq \emptyset$. Assume, on the contrary, that $F(x) \cap B(y_0, \varepsilon) = \emptyset$. At the same time, we have:

$$F(x_0) \subset O_{\varepsilon/2}(F(x)).$$

Therefore, $y_0 \in O_{\varepsilon/2}(F(x))$, and there exists $z_0 \in F(x)$ such that $d(y_0, z_0) < \varepsilon/2$. It implies $z_0 \in B(y_0, \varepsilon)$, and we obtain a contradiction.

Now, assume that F is both u.s.c. and l.s.c. and let $\varepsilon > 0$, $x_0 \in X$.

We let $U = O_\varepsilon(F(x_0))$. Then the sets $F^{-1}(U)$ and $F_+^{-1}(U)$ are open and $x_0 \in F^{-1}(U) \cap F_+^{-1}(U)$. Put $V = F^{-1}(U) \cap F_+^{-1}(U)$. Then V is an open neighbourhood of x_0 such that:

$$F(x) \subset O_\varepsilon(F(x_0)), \quad \text{for every } x \in V.$$

We are looking for $\delta > 0$ such that $B(x_0, \delta) \subset V$ and $F(x_0) \subset O_\varepsilon(F(x))$, for every $x \in B(x_0, \delta)$. For this, we cover the compact set by n open balls $B(y_i, \varepsilon)$, $i = 1, \ldots, n$. Then $F(x_0) = \bigcup_{i=1}^n B(y_i, \varepsilon) \subset O_{\varepsilon/2}(F(x_0))$ and since F is l.s.c., there are open balls $B(x_0, \delta_i) \subset V$ such that

$$F(x) \cap B\left(y_i, \frac{\varepsilon}{2}\right) \neq \emptyset, \quad \text{for every } x \in B(x_0, \delta_i).$$

Let $\delta = \min\{\delta_1, \ldots, \delta_n\}$. Then $B(x_0, \delta) \subset V$ and any $y \in F(x_0)$ belongs to $B(y_i, \varepsilon/2)$, for some i. Furthermore, we know that, for any $x \in B(x_0, \delta)$, $F(x) \cap B(y_i, \varepsilon/2) \neq \emptyset$, for every $i = 1, \ldots, n$. Thus, for every $x \in B(x_0, \delta)$ and $y \in F(x_0)$, there exists $i = 1, \ldots, n$ such that:

$$\text{dist}(y, F(x)) \leq d(y, y_i) + \text{dist}(y_i, F(x_i)) < \frac{\varepsilon}{2} + \frac{\varepsilon}{2} = \varepsilon.$$

Therefore, for every $x \in B(x_0, \delta)$, we obtain:

$$F(x_0) \subset O_\varepsilon(F(x))$$

and the proof is completed. $\square$

Note that if F is only u.s.c. or l.s.c., then F is not Hausdorff continuous, in general. For example, the mappings defined in (3.5.1) or (3.5.2) are u.s.c., but not d_H-continuous. Below we present an example of a l.s.c. map which is not d_H-continuous.

(3.65) EXAMPLE. Let $X = Y = [0, 1]$. Let $F : [0, 1] \to C([0, 1])$ be defined as follows:

$$F(x) = \begin{cases} [0, 1] & x \neq 0, \\ \{0\} & x = 0. \end{cases}$$

Then F is l.s.c. with compact values, but

$$d_H(F(0), F(x)) = 1, \quad \text{for every } x \neq 0.$$

So, F is not Hausdorff continuous.

We can give:

(3.66) THEOREM. *Let $F : X \to B(X)$ be Hausdorff continuous, then F is l.s.c.*

The proof of (3.66) is strictly analogous to the respective part of the proof of (3.64). Note that, under assumptions of (3.66), the map F need not be u.s.c.

(3.67) EXAMPLE. Let $X = \mathbb{R}$ be the Euclidean space of real numbers and let $Y = \mathbb{R}^2$ be equipped with the bounded metric d defined as follows:

$$d(x, y) = \frac{|x - y|}{1 + |x - y|}.$$

We consider the mapping $F : \mathbb{R} \to B(\mathbb{R}^2) \subset 2^{\mathbb{R}^2} \setminus \{\emptyset\}$ defined as follows:

$$F(t) = \{(t, y) \mid y \in \mathbb{R}\}, \quad \text{for every } t \in \mathbb{R}.$$

Thus, we have:

$$d_H(F(t), F(t')) \leq 2|t - t'|,$$

and F is so Hausdorff continuous.

Let $U = \{(x, y) \in \mathbb{R}^2 \mid |y| < 1/x \text{ and } x = 0\}$. Then U is an open subset of $\mathbb{R}^2$, but

$$F^{-1}(U) = \{0\}$$

is not open in $\mathbb{R}$. Consequently, F is not u.s.c.

A multivalued map $F : X \to B(X)$ is called a *contraction* if there exists $k \in [0, 1)$ such that:

$$d_H(F(x), F(y)) \leq k \cdot d(x, y), \quad \text{for every } x, y \in X.$$

We have the following result of H. Covitz and S. B. Nadler, Jr. (see [CN]).

(3.68) THEOREM. *Let (X, d) be a complete metric space and $F : X \to B(X)$ be a contraction map. Then there exists $x \in X$ such that $x \in F(x)$.*

PROOF. Assume that:

$$d_H(F(x), F(y)) \leq k \cdot d(x, y), \quad \text{for every } x, y \in X,$$

where $k \in [0, 1)$. We let $\lambda = l^{-1} \cdot k$, where $l \in [k, 1]$. Letting $x \in X$, we have:

$$\overline{B(x, \lambda \operatorname{dist}(x, F(x)))} \cap F(x) \neq \emptyset,$$

and so we can select $x_1 \in F(x)$ such that:

$$d(x, x_1) \leq \lambda \operatorname{dist}(x, F(x)).$$

Then, for such $x_1 \in X$, select $x_2 \in F(x_1)$ such that

$$d(x_1, x_2) \le \lambda \operatorname{dist}(x_1, F(x_1)).$$

Repeating this procedure, we can find a sequence $\{x_n\} \subset X$ such that:

$$d(x_n, x_{n+1}) \le \lambda \operatorname{dist}(x_n, F(x_n)).$$

Hence, it follows:

$$d(x_n, x_{n+1}) \le \lambda \operatorname{dist}(x_n, F(x_n)) \le \lambda \, d_H(F(x_{n-1}), F(x_n))$$

$$\le \lambda d(x_{n-1}, x_n) \le \lambda^n \operatorname{dist}(x, F(x)).$$

So, it is easy to verify that $\{x_n\}$ is a Cauchy sequence. We let $u = \lim_n x_n$. Then we get:

$$\{x_n\} \to u$$

and

$$x_{n+1} \in F(x_n), \quad \text{for every } n = 1, 2, \dots$$

Since F is u.s.c., we can deduce that the graph Γ_F of F is closed, and consequently we obtain:

$$u \in F(u).$$

The proof of (3.68) is completed. $\square$

There are many extensions of Theorem (3.68). We recommend [GoeKi-M], [Ki-M], [KiSi-M] and [We-M], for details. Below, we shall concentrate our considerations on the topological structure of the set of fixed points of contraction mappings. At first, observe that a multivalued contraction can possess not necessarily a unique fixed point.

(3.69) REMARKS.

(3.69.1) Note that if $F : X \to B(X)$ is a contradiction with compact values, then Fix(F) is compact (comp. [AG1]).

(3.69.2) Example.

Let $F : \mathbb{R} \to B(\mathbb{R})$ be a map defined as follows:

$$F(x) = [0, 1], \quad \text{for every } x \in \mathbb{R}.$$

Then F as a constant map is a contraction. Obviously, we have:

$$\text{Fix}(F) = \{x \in \mathbb{R} \mid x \in F(x)\} = [0, 1].$$

Observe that, for $A \in B(\mathbb{R})$, we can construct a multivalued contraction such that Fix$(F) = A$ (for example, a compact map).

Since, unlike in the single-valued case, Fix(F) of a contraction F may have many elements, it is interesting to look for topological properties of it. In this framework, the following two results are well-known:

(3.70) THEOREM ([Ri1]). *Let E be a Banach space and let X be a nonempty, convex, closed subset of E. Suppose $F : X \to B(X)$ is a contraction with convex values. Then the set* Fix(F) *is an absolute retract.*

Let $(T, \mathcal{F}, \mu)$ be a finite, positive, nonatomic measure space and let $(E, \| \cdot \|)$ be a Banach space. We denote by $L^1(T, E)$ the Banach space of all (equivalence classes of) μ-measurable functions $u : T \to E$ such that the function $t \to \|u(t)\|$ is μ-integrable, equipped with the norm

$$\|u\|_{L^1(T,E)} = \int_T \|u(t)\| d\mu.$$

We always assume that the space $L^1(T, E)$ is separable. Now, we introduce

(3.71) DEFINITION. A nonempty set $K \subset L^1(T, E)$ is said to be *decomposable* if for every $u_1, u_2 \in K$ and every μ-measurable subset A of T, one has

$$(\chi_A \cdot u_1 + (1 - \chi_A) \cdot u_2) \in K,$$

where χ_A denotes the characteristic function of $A \subset T$.

Some basic facts about decomposable sets in $L^1(T, E)$ are collected in the following:

(3.72) REMARKS.

(3.72.1) It can be easily seen that every decomposable subset of $L^1(T, E)$ is contractible and, consequently, infinitely connected,

(3.72.2) Any closed decomposable subset of $L^1(T, E)$ is an absolute retract,

(3.72.3) A simple calculation shows that the open (or closed) unit ball of $L^1(T, E)$ is not decomposable.

For more details concerning the notion of decomposability, we recommend: [Fry2], [Fry3], [Ol3].

(3.73) THEOREM ([BCF]). *If $X = L^1(T, E)$, for some measure space T, and $F : X \to B(X)$ is a contraction with decomposable values, then* Fix(F) *is an absolute retract.*

Below, we establish a result (see [GMS] or [GM]) which unifies and extends to a larger class of multivalued contractions defined on arbitrary complete absolute retracts both Theorems (3.70) and (3.73).

In the proof of the main related result, the following proposition will play an important role.

(3.74) PROPOSITION. *Let (X, d) be a metric space and let $\Phi : X \multimap X$ be a Lipschitzian multivalued function. Set $f(x) = d(x, \Phi(x))$, for every $x \in X$. Then the function $f : X \to [0, +\infty)$ is Lipschitzian.*

PROOF. Let $L \geq 0$ be such that $d_H(\Phi(x'), \Phi(x'')) \leq Ld(x', x'')$, for all $x', x'' \in X$. Pick $x', x'' \in X$ and choose $\varepsilon > 0$. According to the definition of f, there exists $z' \in \Phi(x')$ satisfying

$$-f(x') < -d(x', z') + \varepsilon.$$

Applying the inequality $d(z', \Phi(x'')) \leq Ld(x', x'')$, we can find $z'' \in \Phi(x'')$ such that

$$d(z', z'') < Ld(x', x'') + \varepsilon.$$

Therefore,

$$f(x'') - f(x') < d(x'', \Phi(x'')) - d(x', z') + \varepsilon$$
$$\leq d(x'', z'') - d(x', z') + \varepsilon < (L+1)d(x', x'') + 2\varepsilon.$$

Since ε is arbitrary, we actually have

$$f(x'') - f(x') \leq (L+1)d(x', x'')$$

and, interchanging x' with x'',

$$f(x') - f(x'') \leq (L+1)d(x', x'').$$

This completes the proof. $\square$

We now recall the notion of the Michael family of subsets of a metric space ([GMS, Definition 1.4]). In what follows, by $\mathcal{M}$ we denote the class of all metric spaces.

(3.75) DEFINITION. Let $X \in \mathcal{M}$ and let $M(X)$ be a family of closed subsets of X, satisfying the following conditions:

(1) $X \in M(X)$, $\{x\} \in M(X)$, for all $x \in X$, and if $\{A_i\}_{i \in I}$ is any subfamily of $M(X)$, then $\bigcap_{i \in I} A_i \in M(X)$.

(2) For every $k \in \mathbb{N}$ and every $x_1, x_2, \ldots, x_k \in X$, the set

$$A(x_1, x_2, \ldots, x_k) = \bigcup \{A \mid A \in M(X), x_1, x_2, \ldots, x_k \in A\}$$

is infinitely connected.

(3) To each $\varepsilon > 0$ there corresponds $\delta > 0$ such that, for any $A \in M(X)$, any $k \in \mathbb{N}$, and any $x_1, x_2, \ldots, x_k \in O_\delta(A)$, one has $A(x_1, x_2, \ldots, x_k) \subset O_\varepsilon(A)$.

(4) $A \cap B(x, r) \in M(X)$, for all $A \in M(X)$, $x \in X$, and $r > 0$.

Then we say that $M(X)$ is a *Michael family* of subsets of X.

Let us note (cf. [Bi1], [GMS]) that, in the Michael selection theorem (3.30), the notion of convexity can be replaced by a Michael family.

Namely, we obtain:

(3.76) PROPOSITION. *Let $X, Y \in \mathcal{M}$ and let $\Phi : X \multimap Y$ be a lower semicontinuous multivalued function. If Y is complete and there exists a Michael family $M(Y)$ of subsets of Y such that $\Phi(x) \in M(Y)$, for each $x \in X$, then, for any nonempty closed set $X_0 \subset X$, every continuous selection f_0 from $\Phi|_{X_0}$ admits a continuous extension f over X such that $f(x) \in \Phi(x)$, for all $x \in X$.*

The preceding result gains in interest if we realize that significant classes of sets are examples of Michael families.

(3.77) EXAMPLES.

(3.77.1) Let X be a convex subset of a normed space and let $M(X)$ be the class of all sets $A \subset X$ such that $A = \emptyset$ or A is closed and convex in X. Then $M(X)$ is a Michael family of subsets of X.

(3.77.2) Let $X \in \mathcal{M}$ and let $M(X)$ be the family of all simplicially convex closed subsets of X (in the sense of [Bi1]) or closed convex sets with respect to an abstract convex structure (see [Wie-M]). Then $M(X)$ is a Michael family of subsets of X.

In (3.77.2), we only indicated some non-typical examples of Michael's families. The next definition is crucial in what follows.

(3.78) DEFINITION. Let $X \in \mathcal{M}$, let $\Phi : X \multimap X$ be a lower semicontinuous multivalued function, and let $\mathcal{D} \subset \mathcal{M}$. We say that Φ has the *selection property with respect to $\mathcal{D}$*, when, for any $Y \in \mathcal{D}$, any pair of continuous functions $f : Y \to X$ and $h : Y \to (0, \infty)$ such that

$$\Psi(y) = \overline{\Phi(f(y)) \cap B(f(y), h(y))} \neq \emptyset, \quad y \in Y,$$

and, for any nonempty closed set $Y_0 \subset Y$, every continuous selection g_0 from $\Psi|_{Y_0}$ admits a continuous extension g over Y fulfilling $g(y) \in \Psi(y)$, for all $y \in Y$. If $\mathcal{D} = \mathcal{M}$, then we say that φ has the *selection property* (in symbols, $\Phi \in SP(X)$).

The above notion has some meaningful features, as it is pointed out below.

(3.79) EXAMPLE. Let $X \in \mathcal{M}$ and let $\Phi : X \multimap X$ be a l.s.c. mapping. If X is complete and there exists a Michael family $M(X)$ of subsets such that $\Phi(x) \in M(X)$, for all $x \in X$, then $\Phi \in SP(X)$ (see (3.76)).

Now, we establish the following result:

(3.80) THEOREM. *Let X be a nonempty closed subset of $L^1(T, E)$ and let $\varphi :$ $X \multimap X$ be a lower semicontinuous map with decomposable values. Then φ has the selection property with respect to the family $\mathcal{D}$ of all separable metric spaces.*

PROOF. Throughout this proof, we write 0 to denote the zero vector of $L^1(T, E)$ with $\| \cdot \|_{L^1(T,E)}$. Pick $Y \in \mathcal{D}$ and a pair of continuous functions $f : Y \to X$, $h : Y \to (0, \infty)$ such that $\psi(y) = \mathrm{cl}(\varphi(f(y)) \cap B(f(y), h(y))) \neq \emptyset$, for all $y \in Y$. If Y_0 is a nonempty closed subset of Y and g_0 denotes a continuous selection of $\psi|_{Y_0}$, then the function $k_0 : Y_0 \to L^1(T, E)$ defined by

$$k_0(y) = h(y)^{-1}[g_0(y) - f(y)], \quad y \in Y_0,$$

is a continuous selection of $\eta|_{Y_0}$, where

$$\eta(y) = \mathrm{cl}(h(y)^{-1}[\varphi(f(y)) - f(y)] \cap B(0, 1)), \quad y \in Y.$$

Evidently, the proof will be completed as soon as we show that k_0 admits a continuous extension k over Y, with the property $k(y) \in \eta(y)$, for every $y \in Y$.

At first, we define

$$\xi(y) = \begin{cases} \{k_0(y)\} & \text{if } y \in Y_0, \\ h(y)^{-1}[\varphi(f(y)) - f(y)] & \text{if } y \in Y \setminus Y_0. \end{cases}$$

It is a simple matter to see that the multivalued map: $\xi : Y \to L^1(T, E)$ is lower semicontinuous and with decomposable values. Hence, by Theorem 3 in [BC1], for any $y \in Y$ and any $u \in \xi(y) \cap B(0, 1)$, there exists a continuous selection $k_{y,u} : Y \to L^1(T, E)$ from ξ such that $k_{y,u}(y) = u$. Let

$$V_{y,u} = \{z \in Y \mid \|k_{y,u}(z)\|_1 < 2^{-1}(1 + \|u\|_1)\}.$$

The family of sets $\{V_{y,u} \mid y \in Y, u \in \xi(y) \cap B(0, 1)\}$ is an open covering of the separable metric space Y. So, it has a countable neighbourhood finite refinement $\{V_n \mid n \in \mathbb{N}\}$. For each $n \in \mathbb{N}$, choose $y_n \in Y$ and $u_n \in \xi(y_n) \cap B(0, 1)$ such that $V_n \subset V_{y_n, u_n}$, and define $k_n = k_{y_n, u_n}$. Let $\{p_n\}$ be a continuous partition of unity subordinated to the covering $\{V_n\}$ and let $\{h_n\}$ be a sequence of continuous functions from Y into $[0, 1]$, fulfilling the conditions $h_n(y) = 1$ on $\mathrm{supp}\, p_n$, $\mathrm{supp}\, h_n \subset V_n$, $n \in \mathbb{N}$. We now set, for any $y \in Y$,

$$\varphi_n(y)(t) = \|k_n(y)(t)\|, \quad t \in [0, a] \text{ and } n \in \mathbb{N},$$
$$l(y) = \frac{1}{2}\left[1 - \sum_{n=0}^{\infty} \frac{1 + \|u_n\|_1}{2} p_n(y)\right]^{-1} \sum_{n=1}^{\infty} h_n(y).$$

Since $u_n \in B(0,1)$ and the above summations are locally finite, the function l is well-defined, positive, and continuous. Therefore, there exists a continuous function $r : Y \to (0, \infty)$ and a family $\{A_{r,\lambda} \mid r > 0,\ \lambda \in [0,1]\}$ of measurable subsets of T satisfying (comp. Lemma 2 in [BC1]):

(a) $A_{r,\lambda_1} \subset A_{r,\lambda_2}$ if $\lambda_1 \leq \lambda_2$,

(b) $\mu(A_{r_1,\lambda_1} \Delta A_{r_2,\lambda_2}) \leq |\lambda_1 - \lambda_2| + 2|r_1 - r_2|$ and $\mu(A_{r,\lambda}) = \lambda \mu(T)$,

(c)   for each $y \in Y$, $\lambda \in [0,1]$, and $n \in \mathbb{N}$, if $h_n(y) = 1$, then

$$\left| \int_{A_{r(y),\lambda}} \varphi_n(y)(t)\,d\mu - \lambda \int_T \varphi_n(y)(t)\,d\mu \right| < \frac{1}{4l(y)}.$$

Finally, let us define, for $y \in Y$ and $n \in N$, $\lambda_0(y) = 0$, $\lambda_n(y) = \sum_{m \leq n} p_m(y)$, $\chi_{y,n} = \chi_{A_{r(y),\lambda_n(y)} \setminus A_{r(y),\lambda_{n-1}(y)}}$,

$$k(y) = \sum_{n=1}^{\infty} \chi_{y,n} \cdot k_n(y).$$

Keeping in mind condition (b), it is a simple matter to see that the function $k : Y \to L^1(T, E)$ is continuous. Furthermore, for any $y \in Y$, one has $k(y) \in \xi(y)$, because $\xi(y)$ is decomposable. Thus, to complete the proof, we only need to show that $\|k(y)\|_1 < 1$ at all points of Y. Fix$(y) \in Y$ and observe that if $I(y) = \{n \in \mathbb{N} \mid p_n(y) > 0\}$, then $1 \leq \sharp I(y) \leq \sum_{n=1}^{\infty} h_n(y)$. From (a)–(c), we deduce

$$\int_T \|k(y)(t)\| d\mu \leq \sum_{n \in I(y)} \int_{A_{r(y),\lambda_n(y)} \setminus A_{r(y),\lambda_{n-1}(y)}} \varphi_n(y)(t)\,d\mu$$

$$= \sum_{n \in I(y)} \left[\int_{A_{r(y),\lambda_n(y)}} \varphi_n(y)(t)\,d\mu - \lambda_n(y) \int_T \varphi_n(y)(t)\,d\mu \right.$$

$$\left. - \int_{A_{r(y),\lambda_{n-1}(y)}} \varphi_n(y)(t)\,d\mu + \lambda_{n-1}(y) \int_T \varphi_n(y)(t)\,d\mu + p_n(y) \int_T \varphi_n(y)(t)\,d\mu \right]$$

$$< \frac{\sharp I(y)}{2l(y)} + \sum_{n=1}^{\infty} \frac{1 + \|u_n\|_1}{2} p_n(y) \leq \frac{1}{2l(y)} \sum_{n=1}^{\infty} h_n(y) + \sum_{n=1}^{\infty} \frac{1 + \|u_n\|_1}{2} p_n(y).$$

Hence, by the definition of l, $\|k(y)\|_1 < 1$ as required. $\square$

The following generalizattion of the Fryszkowski selection theorem [Fry2] was proved in a similar way in [BC1].

(3.81) THEOREM (Fryszkowski-like). *If X is a separable metric space and $\varphi : X \multimap L^1(T, E)$ is a l.s.c. mapping with decomposable (closed) values, then φ has a continuous selection.*

Now, we are in a position to prove the main related result.

(3.82) THEOREM. *Let X be a complete absolute retract and $\Phi : X \multimap X$ be a multivalued contraction such that $\Phi \in SP(X)$. Then* Fix(Φ) *is a complete* AR-*space.*

PROOF. Since Fix(Φ) is nonempty and closed in X, we only have to show that if $Y \in \mathcal{M}$, Y^* is a nonempty closed subset of Y, and $f^* : Y^* \to$ Fix(Φ) is a continuous function, then there exists a continuous extension $f : Y \to$ Fix(Φ) of f^* over Y.

Let d be a metric in X, let $L \in (0,1)$ be such that $d_H(\Phi(x'),\Phi(x'')) \leq Ld(x',x'')$, for all $x',x'' \in X$, and let $M \in (1,L^{-1})$. The assumption $X \in AR$ yields a continuous function $f_0 : Y \to X$ fulfilling $f_0(y) = f^*(y)$ in Y^*. We claim that there is a sequence $\{f_n\}$ of continuous functions from Y into X with the following properties:

(i) $f_n|_{Y^*} = f^*$, for every $n \in \mathbb{N}$,

(ii) $f_n(y) \in \Phi(f_{n-1}(y))$, for all $y \in Y$, $n \in \mathbb{N}$,

(iii) $d(f_n(y), f_{n-1}(y)) \leq L^{n-1} d(f_1(y), f_0(y)) + M^{1-n}$, for every $y \in Y$, $n \in \mathbb{N}$.

To see this, we proceed by induction on n. From Proposition (3.74), it follows that the function $h_0 : Y \to (0,\infty)$ defined by

$$h_0(y) = \text{dist}(f_0(y), \Phi(f_0(y))) + 1, \quad y \in Y,$$

is continuous; moreover, one clearly has $\Phi(f_0(y)) \cap B(f_0(y), h_0(y)) \neq \emptyset$, for all $y \in Y$. Keeping in mind that $\Phi \in SP(X)$, we obtain a continuous function $f_1 : Y \to X$ satisfying $f_1(y) = f^*(y)$ in Y^* and $f_1(y) \in \Phi(f_0(y))$ in Y. Hence, conditions (i), (ii), and (iii) are true for f_1. Suppose now we have constructed p continuous functions $f_1, f_2, \ldots, f_p$ from Y into X in such a way that (i), (ii), and (iii), hold whenever $n = 1, 2, \ldots, p$. Since Φ is Lipschitzian with the constant L, (ii) and (iii) can be applied if $n = p$, and $LM < 1$, for every $y \in Y$. We achieve

$$\text{dist}(f_p(y), \Phi(f_p(y))) \leq d_H(\Phi(f_{p-1}(y)), \Phi(f_p(y))) \leq Ld(f_{p-1}(y), f_p(y))$$
$$\leq L^p d(f_1(y), f_0(y)) + LM^{1-p} < L^p d(f_1(y), f_0(y)) + M^{-p},$$

so that

$$\Phi(f_p(y)) \cap B(f_p(y), L^p d(f_1(y), f_0(y)) + M^{-p}) \neq \emptyset.$$

Because of the assumption $\Phi \in SP(X)$, this procedure yields a continuous function $f_{p+1} : Y \to X$ with the properties:

$$f_{p+1}|_{Y^*} = f^*, \quad f_{p+1}(y) \in \Phi(f_p(y)), \quad \text{for every } y \in Y,$$
$$\text{dist}(f_{p+1}(y), f_p(y)) \leq L^p d(f_1(y), f_0(y)) + M^{-p}, \quad \text{for all } y \in Y.$$

Thus, the existence of the sequence $\{f_n\}$ is established.

Furthermore, we define, for any $a > 0$, $Y_a = \{y \in Y \mid d(f_1(y), f_0(y)) < a\}$. Obviously, the family of sets $\{Y_a \mid a > 0\}$ is an open covering of Y. Moreover, due to (iii) and the completeness of X, the sequence $\{f_n\}$ converges uniformly on each Y_a. Let $f : Y \to X$ be the point-wise limit of $\{f_n\}$. It is easy to see that the function f is continuous. Moreover, due to (i), one has $f|_{Y^*} = f^*$. Finally, the range of f is a subset of $\mathrm{Fix}(\Phi)$, because, by (ii), $f(y) \in \Phi(f(y))$, for all $y \in Y$. This completes the proof. $\qquad\qquad\qquad\qquad\qquad\qquad\qquad\qquad\qquad\qquad\qquad\square$

The same arguments used to prove Theorem (3.82) actually produce the following more general result.

(3.83) THEOREM. *Let $\mathcal{D} \subset \mathcal{M}$, let X be a complete absolute retract, and let $\Phi : X \multimap X$ be a multivalued contraction having the selection property with respect to $\mathcal{D}$. Then, for any $Y \in \mathcal{D}$ and any nonempty closed set $Y_0 \subset Y$, every continuous function $f_0 : Y_0 \to \mathrm{Fix}(\Phi)$ admits a continuous extension over Y.*

At last, note that (3.70) and (3.73) are special cases of (3.83).

I.4. Admissible mappings

To present the class of admissible mappings, we need some information concerning the homology theory. We recall some basic facts concerning the Čech homology functor with compact carriers (for details, see [Go1-M] or [Go5-M]).

By a *pair of spaces* (X, X_0), we understand a pair consisting of a topological space X and of its subset X_0. A pair of the form $(X, \emptyset)$ will be identified with the space X. Let (X, X_0), (Y, Y_0) be two pairs; if $X \subset Y$ and $X_0 \subset Y_0$, then the pair (X, X_0) is a subpair of (Y, Y_0), and we indicate this by writing $(X, X_0) \subset (Y, Y_0)$. A pair (X, X_0) is called *compact* if X is a compact space and X_0 is a closed subset of X. By a *map* $f : (X, X_0) \to (Y, Y_0)$ we understand a continuous map $f : X \to Y$ satisfying the condition $f(X_0) \subset Y_0$.

The category of all pairs and maps will be denoted by $\mathcal{E}$. By $\widetilde{\mathcal{E}}$ will be denoted the subcategory of $\mathcal{E}$ consisting of all compact pairs and maps of such pairs. For maps of pairs, we can also consider the notion of homotopy. Namely, two maps $f, g : (X, X_0) \to (Y, Y_0)$ are said to be *homotopic* (written $f \sim g$) if there is a map $h : (X \times [0, 1], X_0 \times [0, 1]) \to (Y, Y_0)$ such that $h(x, 0) = f(x)$ and $h(x, 1) = g(x)$, for every $x \in X$. Let us observe that if (X, X_0) is a pair in $\widetilde{\mathcal{E}}$, then $(X \times [0, 1], X_0 \times [0, 1])$ is in $\widetilde{\mathcal{E}}$, too.

Below we recall some basic facts concerning the Čech homology functor. For details, we recommend [Do-M], [ES-M], [Sp-M].

By H_*, we denote the *Čech homology functor* with the coefficients in the field of rational numbers $\mathbb{Q}$ (or in a group G if necessary) from the category $\tilde{\mathcal{E}}$ ($\mathcal{E}$) to the category $\mathcal{A}$ of graded vector spaces over $\mathbb{Q}$ and linear maps of degree zero.

Thus, for a pair (X, X_0),

$$H_*(X, X_0) = \{H_q(X, X_0)\}$$

is a graded vector space and, for $f : (X, X_0) \to (Y, Y_0)$, we have $H_*(f)$ to be the induced linear map:

$$H_*(f) = f_* = \{f_{*q}\} : H_*(X, X_0) \to H_*(Y, Y_0),$$

where $f_{*q} : H_q(X, X_0) \to H_q(Y, Y_0)$.

We have assumed, as well-known, that the functor H_* satisfies all of the Eilenberg–Steenrod axioms (see [ES-M]), provided we consider the rational coefficients case.

Now, let (X, X_0) be an arbitrary pair in $\mathcal{E}$. We shall denote by $\mathcal{M} = \{(A_\alpha, A_{0\alpha})\}$ the directed set of all compact pairs such that $(A_\alpha, A_{0\alpha}) \subset (X, X_0)$, for each α, with the natural quasi-order relation $\leq$ defined by the condition

$$(A_\alpha, A_{0\alpha}) \leq (A_\beta, A_{0\beta}), \quad \text{whenever} \quad (A_\alpha, A_{0\alpha}) \subset (A_\beta, A_{0\beta}).$$

If $(A_\alpha, A_{0\alpha}) \leq (A_\beta, A_{0\beta})$, then we shall denote by $i_{\alpha\beta} : (A_\alpha, A_{0\alpha}) \to (A_\beta, A_{0\beta})$ the inclusion map. For each pair $(A_\alpha, A_{0\alpha})$, we consider the graded vector space $H_*(A_\alpha, A_{0\alpha})$, together with the linear map $i_{\alpha\beta*}$ given for $(A_\alpha, A_{0\alpha}) \leq (A_\beta, A_{0\beta})$. Then the family $\{H_*(A_\alpha, A_{0\alpha}), i_{\alpha\beta*}\}$ is a directed system in the category $\mathcal{A}$ over $\mathcal{M}$. We define a graded vector space

$$H(X, X_0) = \varinjlim_{\alpha} \{H_*(A_\alpha, A_{0\alpha}), i_{\alpha\beta*}\},$$

where the symbol $\varinjlim_{\alpha}$ stands for the direct limit of the system of vector spaces $\{H_*(A_\alpha, A_{0\alpha}), i_{\alpha\beta*}\}$, for more details, see [Do-M], [ES-M], [Sp-M].

It is easy to see that

$$H(X, X_0) = \{H_q(X, X_0)\},$$

where

$$H_q(X, X_0) = \varinjlim_{\alpha} \{H_q(A_\alpha, A_{0\alpha}), i_{\alpha\beta*}\}, \quad \text{for each } q.$$

Let $f : (X, X_0) \to (Y, Y_0)$ be a map. Consider the directed sets $\mathcal{M} = \{(A_\alpha, A_{0\alpha})\}$ and $\mathcal{N} = \{(B_\gamma, B_{0\gamma})\}$, for (X, X_0) and (Y, Y_0), respectively. We define $F : \mathcal{M} \to \mathcal{N}$ by the formula

$$F((A_\alpha, A_{0\alpha})) = (f(A_\alpha), f(A_{0\alpha})), \quad \text{for each } (A_\alpha, A_{0\alpha}) \in \mathcal{M}.$$

We observe that if $(A_\alpha, A_{0\alpha}) \le (A_\beta, A_{0\beta})$,then

$$F((A_\alpha, A_{0\alpha})) \le F((A_\beta, A_{0\beta})).$$

For each α, by $f_\alpha : (A_\alpha, A_{0\alpha}) \to (f(A_\alpha, A_{0\alpha}))$, we denote a map given by $f_\alpha(x) = f(x)$, for every $x \in A_\alpha$. Then the map F and the family $\{f_{\alpha*}\}$ is a map of directed systems $\{H_*(A_\alpha, A_{0\alpha}), i_{\alpha\beta*}\}$ and $\{H_*(B_\beta, B_{0\beta}), i_{\delta\gamma*}\}$. We define the induced linear map $H(f)$, for f, by putting

$$H(f) = f_* = \varinjlim_\alpha \{f_{*\alpha}\}.$$

Thus, we have $f_{*q} = \varinjlim_\alpha \{f_{\alpha*q}\}$, for every q.

From the functoriality of $\varinjlim$, we deduce that $H : \mathcal{E} \to \mathcal{A}$ is a covariant functor. The functor H is said to be the *Čech homology functor with compact carriers*.

We note that if (X, X_0) is a compact pair, then the family consisting of the single pair (X, X_0) is a cofinal subset of $\mathcal{M} = \{(A_\alpha, A_{0\alpha})\}$, for (X, X_0), and hence, we obtain $H_*(X, X_0) = H(X, X_0)$. Similarly, if $f : (X, X_0) \to (Y, Y_0)$ is a map of compact pairs, then $H_*(f) = H(f)$.

The following properties of H clearly follow from the Eilenberg–Steenrod axioms for H_* and some simple properties of $\varinjlim$.

(4.1) PROPERTY. *If $f, g : (X, X_0) \to (Y, Y_0)$ are homotopic maps, then the induced linear maps are equal, that is, $f_* = g_*$.*

A space X is called *acyclic* if $X \ne \emptyset$ and

$$H_q(X) = \begin{cases} 0 & q > 0, \\ \mathbb{Q} & q = 0. \end{cases}$$

A space X is called *contractible* if there exists a homotopy $h : X \times [0, 1] \to X$ such that

$$h(x, 0) = x \quad \text{and} \quad h(x, 1) = x_0, \quad \text{for every } x \in X.$$

Let us recall that compact space X is called an *R_δ-set* if there exists a decreasing sequence of compact contractible metric spaces $\{X_n\}$ such that:

$$X = \bigcap_n X_n.$$

We have:

(4.2) PROPOSITION. *If X is contractible or X is an R_δ-set, then X is acyclic.*

A space X is called of a *finite type* (with respect to the functor H_*) if $H_n(X)$ is a vector space of a finite dimension, for every $n \geq 0$, and $H_n(X) = 0$, for almost all n.

Evidently, any acyclic space or, in particular, any R_δ-space is of a finite type. We have (see [Go5-M]):

(4.3) PROPOSITION. *Assume X is a compact metric space of a finite type. Then there exists $\varepsilon > 0$ such that, if $f, g : Y \to X$ are ε-close, then $f_* = g_*$.*

Let $p : Y \to X$ be a continuous surjection; p is called *proper* if, for every compact $K \subset X$, the set $p^{-1}(K)$ is compact; p is called *closed* if, for every closed $A \subset Y$, the set $p(A)$ is closed; p is called *perfect* if p is closed and, for every $x \in X$, the set $p^{-1}(x)$ is compact.

It is well-known that (comp. [Eng1-M])

(4.4) PROPOSITION. *If $p : Y \to X$ is perfect, then p is proper.*

Let us observe the following:

(4.5) PROPOSITION. *If $p : Y \to X$ is proper and Y is a metric space, then p is perfect.*

We also need the notion of a Vietoris map.

(4.6) DEFINITION. A continuous map $p : Y \to X$ is called a *Vietoris map* if p is perfect and, for every $x \in X$, the set $p^{-1}(x)$ is acyclic; a map $p : (Y, Y_0) \to (X, X_0)$ is called a Vietoris map if $p_Y : Y \to X$, $p_Y(y) = p(y)$ and $p_{Y_0} : Y_0 \to X_0$, $p_{Y_0}(y) = p(y)$ are Vietoris maps and $p^{-1}(X_0) = Y_0$.

In what follows, we shall reserve the symbol $p : (Y, Y_0) \Rightarrow (X, X_0)$ for Vietoris maps.

Let us consider the following diagram:

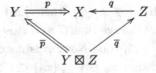

where $Y \boxtimes Z = \{(y, z) \in Y \times Z \mid p(y) = q(z)\}$, $\overline{p}(y, z) = y$ and $\overline{q}(y, z) = z$. Then $\overline{p}$ is called the *fibre product* of p and q. Note, that $\overline{p}$ is a Vietoris map, too.

The following property is straightforward:

(4.7) PROPOSITION.

(4.7.1) *If $p : Y \Rightarrow X$ and $\bar{p} : X \Rightarrow Z$ are Vietoris maps, then the composition:*

$$\bar{p} \circ p : Y \Rightarrow Z$$

is a Vietoris map, too.

(4.7.2) *If $p : Y \Rightarrow X$ is Vietoris map, then, for every $A \subset X$, the map $\widetilde{p} : p^{-1}(A) \Rightarrow A$, $\widetilde{p}(y) = p(y)$ is a Vietoris map, too.*

The following theorem is a generalization of the theorem presented in 1927 by L. Vietoris [Vie] (comp. [Go5-M] or [Go1-M]).

(4.8) THEOREM (Vietoris Mapping Theorem). *If $p : (Y, Y_0) \Rightarrow (X, X_0)$ is a Vietoris map, then $p_* : H_*(Y, Y_0) \to H_*(X, X_0)$ is an isomorphism.*

Now, we are able to define the class of admissible multivalued maps as defined for the first time in [Go1-M].

(4.9) DEFINITION. A multivalued map $\varphi : X \multimap Y$ is called *admissible* if there exists a diagram:

$$X \overset{p}{\Longleftarrow} \Gamma \overset{q}{\longrightarrow} Y,$$

in which p is Vietoris and q continuous such that, for every $x \in X$, we have:

$$\varphi(x) = q(p^{-1}(x)).$$

Since p is closed, we get that any admissible map is u.s.c.

An u.s.c. map $\varphi : X \to Y$ is called *acyclic* if $\varphi(x)$ is a compact acyclic set, for every $x \in X$.

We state:

(4.10) PROPOSITION. *Any acyclic map is admissible.*

PROOF. Assume that $\varphi : X \multimap Y$ is an acyclic map. Let Γ_φ be the graph of φ and $p_\varphi : \Gamma_\varphi \to X$, $p_\varphi(x, y) = x$, $q_\varphi : \Gamma_\varphi \to Y$, $q_\varphi(x, y) = y$ be natural projections. Since $p_\varphi^{-1}(x)$ is homeomorphic to $\varphi(x)$, we get that $p_\varphi^{-1}(x)$ is compact acyclic. Therefore, for the proof it is sufficient to show that p_φ is closed. Let C be a closed subset of Γ_φ. If $x_0 \notin p_\varphi(C)$, then $\{x_0\} \times \varphi(x_0) \cap C = \emptyset$, and so $\{x_0\} \times \varphi(x_0) \subset (X \times Y) \setminus C$. Observe that C as a closed subset of a closed subset Γ_φ of $X \times Y$ is closed in $X \times Y$, too. Now, φ is u.s.c. and, therefore, there are two open sets V and U such that $x_0 \in V$, $\varphi(x_0) \subset U$, $V \times U \subset X \times Y \setminus C$ and $V = \varphi^{-1}(U)$. Thus, $V \cap p_\varphi(C) = \emptyset$, which proves our proposition. $\qquad\square$

The following example shows that the composition of two acyclic mappings need not be acyclic.

(4.11) EXAMPLE. Let S^1 denote the unit sphere in $\mathbb{R}^2$. Let $\varphi : S^1 \multimap S^1$ be defined as follows:

$$\varphi(x) = \{y \in S^1 \mid |x - y| \leq \sqrt{3}\}.$$

Obviously φ is acyclic, but $\varphi(\varphi(x)) = S^1$, for every x, and consequently $\varphi \circ \varphi$ is not acyclic.

However, we have:

(4.12) PROPOSITION. *Let $\varphi : X \multimap Y$ and $\psi : Y \multimap Z$ be admissible mappings, then the composition $\psi \circ \varphi : X \multimap Z$ of φ and ψ is also admissible. In particular, the composition of two acyclic mappings is admissible.*

For the proof of (4.12), it is sufficient to consider the following diagram:

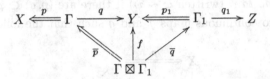

in which the pair (p, q) determines φ, (p_1, q_1) determines ψ, $f(y, z) = y$, $\overline{q}(y, z) = z$ and $\overline{p}$ is the fibre product of p_1 and q. Then $p \circ \overline{p}$ is a Vietoris map and, therefore, we get:

$$(\psi \circ \varphi)(x) = (q_1 \circ \overline{q})((p \circ \overline{p})^{-1}(x)).$$

In what follows, the pair (p, q) determining an admissible map φ will be called a *selected pair* for φ (written $(p, q) \subset \varphi$).

(4.13) DEFINITION. Let $\varphi : X \multimap Y$ be an admissible map and (p, q) its selected pair. We define the induced linear map:

$$q_* \circ p_*^{-1} : H_*(X) \to H(Y)$$

and the induced set $\{\varphi_*\}$ of linear mappings by putting:

$$\{\varphi_*\} = \{q_* \circ p_*^{-1} \mid (p, q) \subset \varphi\}.$$

It easily follows from the example (4.11) that $\{\varphi_*\}$ can be even an infinite set. Nevertheless, we have

(4.14) PROPOSITION. *If $\varphi : X \multimap Y$ is an acyclic map, then $\{\varphi_*\} = \varphi_*$ is a singleton.*

PROOF. Consider the graph Γ_φ of φ and natural projections: $p_\varphi : \Gamma_\varphi \Rightarrow X$, $q_\varphi : \Gamma_\varphi \to Y$. Assume, furthermore, that $(p, q) \subset \varphi$. Then the following diagram is commutative:

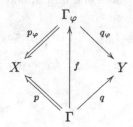

where $f(u) = (p(u), q(u))$.

Now, the commutativity of the above diagram implies that $q_* \circ p_*^{-1} = (q_\varphi)_* \circ (p_\varphi)_*^{-1}$, and the proof is completed.   □

(4.15) DEFINITION. Let $\varphi, \psi : X \multimap Y$ be two admissible maps. We shall say that φ is *homotopic to* ψ (written $\varphi \sim \psi$) if there are $(p, q) \subset \varphi$, $(p_1, q_1) \subset \psi$, a pair $(\overline{p}, \overline{q})$, and two continuous maps f and g such that the following diagram is commutative:

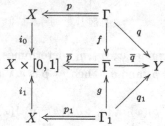

where $i_0(x) = (x, 0)$, $i_1(x) = (x, 1)$.

Since $(i_0)_*$ and $(i_1)_*$ are equal isomorphisms by applying to the above diagram the functor H_*, we obtain:

(4.16) PROPOSITION. *If* $\varphi \sim \psi$, *then* $\{\varphi_*\} \cap \{\psi_*\} \neq \emptyset$.

I.5. Special classes of admissible mappings

In this section, we shall assume that all multivalued mappings are admissible and all topological spaces are regular.

(5.1) DEFINITION. A multivalued map $\varphi : X \multimap Y$ is called *compact* if the set $\varphi(X) = \bigcup_{x \in X} \varphi(x)$ is contained in a compact subset of Y.

Observe that if $(p, q) \subset \varphi$, then φ is compact if and only if (a single-valued map) q is compact. We let:

$$\mathbb{K}(X, Y) = \{\varphi : X \multimap Y \mid \varphi \text{ is compact}\};$$

if $X = Y$, then we write $\mathbb{K}(X)$ to be equal $\mathbb{K}(X, X)$.

(5.2) DEFINITION. A map $\varphi : X \multimap Y$ is called *locally compact* if, for every $x \in X$, there exists an open neighbourhood U_x of x in X such that the restriction: $\varphi|_{U_x} : U_x \multimap Y$ of φ to U_x is a compact map.

Observe that if X is a locally compact space, then any admissible map $\varphi : X \multimap Y$ is locally compact. Obviously, any compact map is locally compact. We let:

$$\mathbb{K}_{\mathrm{loc}}(X,Y) = \{\varphi : X \multimap Y \mid \varphi \text{ is locally compact}\}$$

and $\mathbb{K}_{\mathrm{loc}}(X) = \mathbb{K}_{\mathrm{loc}}(X,X)$. We have:

$$\mathbb{K}(X,Y) \subset \mathbb{K}_{\mathrm{loc}}(X,Y).$$

(5.3) DEFINITION. A map $\varphi : X \multimap X$ is called *eventually compact* if there exists n such that the nth iteration $\varphi^n : X \multimap X$ of φ is a compact map.

We let

$$\mathbb{E}(X) = \{\varphi : X \multimap X \mid \varphi \text{ is eventually compact}\}.$$

We have:

$$\mathbb{K}(X) \subset \mathbb{E}(X).$$

(5.4) DEFINITION. A map $\varphi : X \multimap X$ is called *asymptotically compact* if

(5.4.1) for every $x \in X$, the orbit $\bigcup_{n=1}^{\infty} \varphi^n(x)$ is contained in a compact subset of X, and

(5.4.2) the *center*, sometimes also called the *core*, $C_\varphi = \bigcap_{n=1}^{\infty} \varphi^n(X)$ of φ is nonempty, contained in a compact subset of X.

We let

$$\mathbb{ASC}(X) = \{\varphi : X \multimap X \mid \varphi \text{ is asymptotically compact}\}.$$

(5.5) DEFINITION. A map $\varphi : X \multimap X$ is called a *compact attraction* if there exists a compact set $K \subset X$ such that, for every open neighbourhood U of K in X and, for every $x \in X$, there exists $n = n_x$ such that $\varphi^m(x) \subset U$, for every $m \geq n$; then K is called the *attractor* of φ.

We let

$$\mathbb{CA}(X) = \{\varphi : X \multimap X \mid \varphi \text{ is a compact attraction}\}.$$

(5.6) DEFINITION. A map $\varphi : X \multimap X$ is called a *compact absorbing contraction* if there exists an open set $U \subset X$ such that:

(5.6.1) $\varphi(U) \subset U$ and the map $\tilde{\varphi} : U \multimap U$, $\tilde{\varphi}(x) = \varphi(x)$, for every $x \in U$, is compact, and

(5.6.1) for every $x \in X$, there exists $n = n_x$ such that $\varphi^n(x) \subset U$.

We let

$$\mathbb{C}AC(X) = \{\varphi : X \multimap X \mid \varphi \text{ is a compact absorbing contraction}\}.$$

Evidently, we have:

(5.7)
$$\mathbb{K}(X) \subset \mathbb{C}AC(X) \subset \mathbb{C}A(X).$$

At first, we prove:

(5.8) PROPOSITION. $\mathbb{E}C(X) \subset \mathbb{A}SC(X)$.

PROOF. Let $\varphi \in \mathbb{E}C(X)$ and assume that $K = \varphi^{n_0}(X)$ is a compact set. Then, for every $x \in X$, we have:

$$\bigcup_{i=1}^{\infty} \varphi^i(x) \subset \{x\} \cup \{\varphi(x)\} \cup \ldots \cup \{\varphi^{n_0-1}(x)\} \cup \varphi^{n_0}(X).$$

Thus, the set $\{x\} \cup \{\varphi(x)\} \cup \ldots \cup \{\varphi^{n_0-1}(x)\} \cup \overline{\varphi^{n_0}(X)}$ is compact, i.e. every orbit is a relatively compact set. Moreover, we have:

$$\forall i \geq 0 \; \varphi^{n_0+i}(K) = \varphi^{n_0}(\varphi^i(K)) \subset \varphi^{n_0}(X) \subset K.$$

Consequently,

$$\emptyset \neq \bigcap_{i=0}^{\infty} \varphi^{n_0+i}(K) \subset \bigcap_{i=0}^{\infty} \varphi^{n_0+i}(X) = \bigcap_{i=0}^{\infty} \varphi^i(X)$$

which implies that the core C_φ of φ is nonempty and relatively compact. So the proof of (5.8) is completed. $\square$

(5.9) PROPOSITION. $\mathbb{A}SC(X) \subset \mathbb{C}A(X)$.

PROOF. Let $\varphi \in \mathbb{A}SC(X)$. Then the set $\overline{C_\varphi}$ is nonempty compact. It is enough to show that $\overline{C_\varphi}$ is an attractor of φ. Let V be an open neighbourhood of $\overline{C_\varphi}$ in X and let $x \in X$. We put $L = \overline{\bigcup_{i=1}^{\infty} \varphi^i(x)}$. Then, for $0 \leq j \leq n$, and for arbitrary n we get:

$$\varphi^{2n}(x) \subset \bigcup_{i=0}^{\infty} \varphi^{i+j}(x) = \varphi^j \left(\bigcup_{i=0}^{\infty} \varphi^i(x) \right) \subset \varphi^j(L),$$

and so:

$$\varphi^{2n}(x) \subset \bigcap_{j=0}^{n} \varphi^j(L).$$

Therefore, it is enough to show that there is a natural number n_x such that:

$$\bigcap_{j=0}^{n_x} \varphi^j(L) \subset V.$$

In fact, we obtain:

$$\bigcap_{n=0}^{\infty} A_n \subset \bigcap_{n=0}^{\infty} \varphi^n(X) = C_\varphi \subset V,$$

where $A_n = \bigcap_{i=0}^n \varphi^i(L)$. Since A_n is a decreasing sequence of compact sets, this implies that there are natural numbers $n_1 < n_2 < \ldots < n_k$ such that

$$A_{n_1} \cap A_{n_2} \cap \ldots \cap A_{n_k} \subset V,$$

but

$$A_{n_1} \cap A_{n_2} \cap \ldots \cap A_{n_k} = A_{n_k},$$

and so $n_x = n_k$ is the required natural number. The proof of (5.9) is completed.$\square$

Summing up the above, we get:

(5.10)
$$
\begin{array}{ccccc}
\mathbb{E}C(X) & \subset & \mathbb{A}SC(X) & \subset & \mathbb{C}A(X) \\
\cup & & & & \cup \\
\mathbb{K}(X) & & \subset & & \mathbb{C}AC(X)
\end{array}
$$

In what follows, we need the following notations:

$$
\begin{aligned}
\mathbb{E}C_0(X) &= \mathbb{E}C(X) \cap \mathbb{K}_{loc}(X), \\
\mathbb{A}SC_0(X) &= \mathbb{A}SC(X) \cap \mathbb{K}_{loc}(X), \\
\mathbb{C}A_0(X) &= \mathbb{C}A(X) \cap \mathbb{K}_{loc}(X).
\end{aligned}
$$

We start by the following:

(5.11) LEMMA. $\mathbb{E}C_0(X) \subset \mathbb{C}AC(X)$.

PROOF. Let $\varphi \in \mathbb{E}C_0(X)$, $M = \varphi^{n_0}(X)$ be compact and let $K = \bigcup_{i=0}^{n_0-1} \varphi^i(M)$. Then K is compact and

$$\varphi(K) = \bigcup_{i=1}^{n_0} \varphi^i(M) \subset K \cup M = K.$$

Since φ is locally compact, there exists an open neighbourhood W of K in X such that $L = \overline{\varphi(W)}$ is compact. Now, we define open sets $V_0, \ldots, V_{n_0}$ such that $L \cap \overline{\varphi(V_i)} \subset V_{i-1}$ and $K \cup \varphi^{n_0-i}(L) \subset V_i$, $i = 1, \ldots n_0$. Namely, we put $V_0 = W$. If $V_0, \ldots, V_i$ are needed sets then $(K \cup \varphi^{n_0-i}(L)) \cap (L \setminus V_i) = \emptyset$. Therefore, there is an open set V such that $K \cup \varphi^{n-i}(L) \subset V \subset \overline{V} \subset V_i \cup (X \setminus L)$. We let $V_{i+1} = \varphi^{-1}(V)$. We obtain

$$\varphi(K \cup \varphi^{n_0-(i+1)}(L)) = \varphi(K) \cup \varphi(\varphi^{n_0-(i+1)}(L)) \subset K \cup \varphi^{n_0-i}(L) \subset V,$$

and so

$$K \cup \varphi^{n_0-(i+1)}(L) \subset V_{i+1}.$$

Moreover, we have

$$\varphi(V_{i+1}) \subset \overline{V} \subset V_i \cup (X \setminus L),$$

so

$$L \cap \overline{\varphi(V_{i+1})} \subset V_i.$$

Letting $U = V_0 \cap \ldots \cap V_{n_0}$, we have $M \subset K \subset U$ and

$$\varphi(U) \subset \varphi(V_0) \cap \ldots \cap \varphi(V_{n_0}) \subset L \cap \overline{\varphi(V_1)} \cap \ldots \cap \overline{\varphi(V_{n_0})}.$$

Consequently, we obtain

$$\overline{\varphi(U)} \subset (L \cap \overline{\varphi(V_1)}) \cap \ldots \cap (L \cap \overline{\varphi(V_{n_0})}) \cap L \subset V_0 \cap \ldots \cap V_{n_0} = U,$$

and $\overline{\varphi(U)}$ is compact.

Since $M \subset U$, for every $x \in X$, we get $\varphi^{n_0}(x) \subset U$, and the proof is completed. $\square$

The following example shows that $\mathbb{E}C(X) \not\subset \mathbb{C}AC(X)$.

(5.12) EXAMPLE. Let $C = \{\{x_n\} \subset \mathbb{R} \mid \{x_n\} \text{ is a bounded sequence}\}$ be the space of bounded sequences with the usual supremum norm. We define $F : C \to C$ as follows

$$F(\{x_n\}) = \{0, x_1, 0, x_3, 0, x_5, 0, \ldots\}.$$

Then $F^2 = 0$, so $F \in \mathbb{E}C(C)$, but $F \notin \mathbb{C}AC(C)$.

Now, we prove

(5.13) PROPOSITION. $\mathbb{C}A_0(X) \subset \mathbb{C}AC(X)$.

PROOF. Let $\varphi \in \mathbb{C}A_0(X)$ and let K be a compact attractor for φ. Since φ is locally compact, there exists an open neighbourhood W of K in X such that $L = \overline{\varphi(W)}$ is a compact set. We have

$$L \subset X \subset \bigcup_{i=0}^{\infty} \varphi^{-i}(W).$$

It implies that $\{\varphi^{-i}(W)\}_{i=1,2,\ldots}$ is an open covering of L in X. Therefore, there is a finite subcovering

$$\varphi^{-i_1}(W),\ldots,\varphi^{-i_j}(W).$$

Let $n = \max\{i_1,\ldots,i_j\}$ and $V = \bigcup_{i=0}^{n} \varphi^{-i}(W)$. Then we have

$$L \subset V \quad \text{and} \quad W \subset V \quad \text{and} \quad \varphi^{-i}(W) \subset \varphi^{-i}(V).$$

Consequently,

$$X \subset \bigcup_{i=0}^{\infty} \varphi^{-i}(W) \subset \bigcup_{i=0}^{\infty} \varphi^{-i}(V).$$

We get

$$\varphi(V) = \bigcup_{i=0}^{n} \varphi^{-i+1}(W) = \varphi(W) \cup \bigcup_{i=1}^{n-1} \varphi^{-i}(W) \subset \varphi(W) \cup V \subset L \cup V = V$$

and, moreover, we have

$$\varphi^{n+1}(V) \subset \bigcup_{i=0}^{n} \varphi^{n-i+1}(W) = \bigcup_{i=0}^{n} \varphi^{i}(\varphi(W)) \subset \bigcup_{i=0}^{n} \varphi^{i}(L).$$

Consequently, the set $\bigcup_{i=0}^{n} \varphi^{i}(L)$ is a compact subset of V, and so we have shown that $\widetilde{\varphi} \in \mathbb{E}C(X)$, where $\widetilde{\varphi} = \varphi|_V$. In view of (5.11), we infer $\widetilde{\varphi} \in \mathbb{C}AC(V)$, but it immediately implies that $\varphi \in \mathbb{C}AC(X)$. It fact, if U is an open set of V such that

$$V \subset \bigcup_{i=0}^{\infty} \widetilde{\varphi}^{-i}(U) \subset \bigcup_{i=0}^{\infty} \varphi^{-i}(U,)$$

then

$$X \subset \bigcup_{i=0}^{\infty} \varphi^{-i}(V) \subset \bigcup_{i=0}^{\infty} \varphi^{-i}(U),$$

and $\overline{\widetilde{\varphi}(U)} = \overline{\varphi(U)}$ is a compact subset of V. The proof is completed. $\square$

Summing up the above, we get

(5.14) $\mathbb{K}(X) \subset \mathbb{E}C_0(X) \subset \mathbb{A}SC_0(X) \subset \mathbb{C}A_0(X) \subset \mathbb{C}AC(X).$

Finally, we show that all the above inclusions are proper.

EXAMPLE.

(5.14.1) Let $F : C \to C$ be defined as follows

$$F(\{x_n\}) = \{x_2, x_3, x_4, \dots\}.$$

Then $F \in \mathbb{C}A(C)$ with the compact attractor $K = \{0\}$, but $F \notin \mathbb{C}AC(C)$.

(5.14.2) Let $f : \mathbb{R} \to \mathbb{R}$ be defined as follows

$$f(x) = \begin{cases} x - 1, & x > 1, \\ 0, & -1 \le x \le 1, \\ x + 1, & x < -1, \end{cases}$$

and let $F : C \to C$ be given by the formula

$$F(\{x_n\}) = \{f(x_1), f(x_2), \dots\}.$$

Then $F \in \mathbb{C}AC(C) \cap \mathbb{C}A(C)$ with the compact attractor $K = \{0\}$, but $F \notin \mathbb{C}A_0(C)$.

(5.14.3) Let $f : [0, \infty) \to [0, \infty)$ be defined

$$f(x) = \begin{cases} \dfrac{1}{x}, & x < 1, \\ 1, & x \ge 1. \end{cases}$$

Then $f^2(x) = \{1\}$, for every $x \in [0, \infty)$. Therefore, $f \in \mathbb{E}C_0([0, \infty))$, and $f \notin \mathbb{K}([0, \infty))$.

(5.14.4) Let $A = \{(x, y) \in \mathbb{R}^2 \mid (x < 0 \vee y < 0) \wedge x^2 + y^2 \le 1\} \cup \{(x, y) \in \mathbb{R}^2 \mid (x \ge 0 \wedge y \ge 0) \wedge (y \le 1 \wedge x \le 1) \vee y > 1 \wedge 0 < x \le 1)\}$. We define $f : A \to A$ by putting $f(x, y) = \left(\frac{1}{2}x, \frac{1}{2}y\right)$. Then $f \in \mathbb{E}C(A)$ and $f \in \mathbb{A}SC_0(X)$.

(5.14.5) Let $f : \mathbb{R} \to \mathbb{R}$ be defined $f(x) = \frac{x}{2}$. Evidently, $f \in \mathbb{C}A_0(\mathbb{R})$ with the compact attractor $K = \{0\}$, but $f \notin \mathbb{A}SC(\mathbb{R})$, because $C_f = \mathbb{R}$.

Now, we shall discuss the notion of the measure of noncompactness in locally convex spaces. This notion is necessary to define the class of admissible condensing and μ-set contraction mappings.

Let $\mathbb{E}$ be a locally convex space. The family of all bounded nonempty subsets of $\mathbb{E}$ is denoted by $B(\mathbb{E})$.

(5.15) DEFINITION. Let S be a set of indices. A family of nonnegative functions $\mu = \{\mu_s\}_{s \in S} : B(\mathbb{E}) \to \mathbb{R}^S$ is called a *measure of noncompactness* (MNC) on $\mathbb{E}$ if the following conditions are satisfied:

(μ_1) $\overline{X}$ is compact $\Leftrightarrow \forall s \in S \ \mu_s(X) = 0$,

(μ_2) $\mu_s(\overline{\mathrm{conv}}(X)) = \mu_s(X)$,

(μ_3) $\mu_s(X \cup Y) = \max\{\mu_s(X), \mu_s(Y)\}$,

(μ_4) $\mu_s(t \cdot X) = |t|\mu_s(X)$, $t \in \mathbb{R}$,

(μ_5) $\mu_s(X + Y) \le \mu_s(X) + \mu_s(Y)$, where $\mathbb{R}^S$ denotes the Cartesian product $\prod_{s \in S} \mathbb{R}_s$, $\mathbb{R}_s = \mathbb{R}$ is the field of real numbers.

(5.16) PROPOSITION. *Any measure of noncompactness $\mu : B(\mathbb{E}) \to \mathbb{R}^S$ has the following properties:*

(μ_6) $X \subset Y \Rightarrow \mu_s(X) \leq \mu_s(Y)$, *for all* $s \in S$,

(μ_7) K *is compact* $\Rightarrow \mu_s(X + K) = \mu_s(X \cup K) = \mu_s(X)$, *for all* $s \in S$,

(μ_8) *If sequences* $\{x_i\}$, $\{y_i\} \subset \mathbb{E}$ *are bounded and* $\overline{\{x_i - y_i\}}$ *is compact, then*
$$\mu_s(\{x_i\}) = \mu_s(\{y_i\}), \text{ for all } s \in S,$$

(μ_9) *If* $X_1 \supset X_2 \supset \dots$, $\emptyset \neq X_i = \overline{X_i} \in B(\mathbb{E})$ *and* $\mu_s(X_i) \to 0$ *as* $i \to \infty$, *for all* $s \in S$, *then* $X = \bigcap_{i=1}^{\infty} X_i$ *is a nonempty compact set.*

PROOF. (μ_6) is immediate from (μ_3).

(μ_7): By (μ_3) and (μ_1), $\mu_s(X \cup K) = \max\{\mu_s(X), \mu_s(K)\} = \mu_s(X)$. By ($\mu_5$), $\mu_s(X + K) \leq \mu_s(X) + \mu_s(K) = \mu_s(X)$. Since $X \subset X + K + (-K)$, the reverse inequality follows from (μ_6).

(μ_8): Clearly, $\{x_i\} \subset \{x_i - y_i\} + \{y_i\}$ and $\{y_i\} \subset \{x_i\} + \{y_i - x_i\}$. Since the sets $\{x_i - y_i\}$ and $\{y_i - x_i\}$ are relatively compact, the conclusion follows from (μ_6) and (μ_7).

(μ_9): Let $x_i \in X_i$, $i = 1, 2, \dots$ Since any finite set is compact, for any $s \in S$ and any $N = 1, 2, \dots$, it holds:

$$\mu_s(\{x_i\}) = \max\{\mu_s(\{x_i \mid i \leq N\}), \mu_s(\{x_i \mid i > N\})\}$$
$$= \mu_s(\{x_i \mid i > N\}) \leq \mu_s(X_{N+1}) \to 0 \text{ as } N \to \infty.$$

It follows that $\mu_s(\{x_i\}) = 0$, for all $s \in S$. Hence, $\overline{\{x_i\}}$ is compact. Thus, $\{x_i\}$ has a cluster point $x_0 \in \mathbb{E}$. Then $x_0 \in \overline{X_i} = X_i$, for all i, so $x_0 \in X$. It follows that X is a nonempty set with $\mu_s(X) = 0$, for all $s \in S$, and, since X is closed, it must be compact. $\square$

(5.17) PROPOSITION (Restriction to a subspace). *Let $\mathbb{E}_1$ be a closed subspace of $\mathbb{E}$ and let $\mu : B(\mathbb{E}) \to \mathbb{R}^S$ be a measure of noncompactness on $\mathbb{E}$. Then the restriction of μ to $B(\mathbb{E}_1)$:*

$$\mu|_{\mathbb{E}_1} : B(\mathbb{E}_1) \to \mathbb{R}^S$$

is a measure of noncompactness on $\mathbb{E}_1$.

PROOF. The proof is trivial. $\square$

(5.18) PROPOSITION (Product of measures). *Let $\mathbb{E} = \prod_{j \in J} \mathbb{E}_j$ be a topological product of spaces with measures of noncompactness $\mu^{(j)} : B(\mathbb{E}_j) \to \mathbb{R}^{S^{(j)}}$. Define*

$$\mu_{s,j}(X) = \mu_s^{(j)}(\pi_j(X)), \quad s \in S^{(j)}, \ X \in B(\mathbb{E}).$$

Then $\{\mu_{s,j}\}_{s \in S^{(j)}, j \in J}$ is a measure of noncompactness on $\mathbb{E}$, called the product of $\{\mu^{(j)}\}_{j \in J}$. ($\pi_j : \mathbb{E} \to \mathbb{E}_j$ denotes the canonical projection).

PROOF. Let X be a bounded subset of $\mathbb{E}$. Then $\pi_j(X)$ is bounded in $\mathbb{E}_j$. Hence, the functions $\mu_{s,j}$ are well-defined. We only prove that the axioms (μ_1) and (μ_2) are satisfied. The proofs of the remaining ones are trivial.

(μ_1): If $\overline{X}$ is compact, then $\pi_j(\overline{X})$ is compact, and so $\pi_j(X)$ is relatively compact in $\mathbb{E}_j$. Thus, $\mu_s^{(j)}(\pi_j(X)) = 0$, for all $s \in S^{(j)}$, $j \in J$.

Conversely, let $\mu_s^{(j)}(\pi_j(X)) = 0$, for all $s \in S^{(j)}$, $j \in J$. Then $\overline{\pi_j(X)}$ is compact, for all $j \in J$, and the set $\prod_{j \in J} \overline{\pi_j(X)}$ is a compact set containing X.

(μ_2): Note that $\mathrm{conv}(\pi_j(X)) = \pi_j(\mathrm{conv}(X))$ and, since π_j is continuous, $\pi_j(\overline{X}) = \overline{\pi_j(X)}$, $j \in J$. Consequently,

$$\mu_s^{(j)}(\pi_j(\overline{\mathrm{conv}}(X))) = \mu_s^{(j)}(\pi_j(\overline{\mathrm{conv}}(X)))$$
$$= \mu_s^{(j)}(\overline{\mathrm{conv}}(\pi_j(X))) = \mu_s^{(j)}(\pi_j(X)),$$

for any $j \in J$, $s \in S^{(j)}$. $\qquad\square$

(5.19) COROLLARY (Extension to $\mathbb{E} \times \mathbb{R}^n$). *Let $\mu : B(\mathbb{E}) \to \mathbb{R}^S$ be a measure of noncompactness on $\mathbb{E}$. Then μ extends to a measure of noncompactness $\tilde{\mu} : B(\mathbb{E} \times \mathbb{R}^n) \to \mathbb{R}^S$ by the formula:*

$$\mu_s(X) = \mu_s(\pi(X)), \quad s \in S, \ X \in B(\mathbb{E} \times \mathbb{R}^n),$$

where $\pi : \mathbb{E} \times \mathbb{R}^n \to \mathbb{E}$ is the canonical, projection. In what follows, the same letter will be used for μ and its extension to $B(\mathbb{E} \times \mathbb{R}^n)$.

PROOF. Note that $\tilde{\mu} = \mu \times \mathcal{O}$, the product measure defined in (5.18), where $\mathcal{O}$ is the trivial measure of noncompactness on $\mathbb{R}^n$. $\qquad\square$

(5.20) THEOREM (Limit measure). *Let $\mathbb{E}$ be a strict inductive limit of a sequence of complete spaces $\mathbb{E}_0 \subset \mathbb{E}_1 \subset \ldots$ with measures of noncompactness $\mu^{(n)} : B(\mathbb{E}_n) \to \mathbb{R}^{S^{(n)}}$. Define $\mu_{(s,n)}(X) = \mu_s^{(n)}(X \cap \mathbb{E}_n)$. Then $\lim_{n \to \infty} \mu^{(n)} = \{\mu_{(s,n)}\}_{s \in S^{(n)}, n=0,1,\ldots}$ is a measure of noncompactness on $\mathbb{E}$, called the limit of $\{\mu^{(n)}\}$.*

PROOF. The proof is straightforward. $\qquad\square$

(5.21) REMARK. Let $\mu : B(\mathbb{E}) \to \mathbb{R}^{S \times \{1,\ldots,n\}}$ be a measure of noncompactness. Then any of the following functions from $B(\mathbb{E})$ to $\mathbb{R}^S$ is also a measure of noncompactness:

$$X \to \max\{\mu_{s,i}(X) \mid i = 1, \ldots, n\},$$

$$X \to \sum_{i=1}^n \mu_{s,i}(X),$$

$$X \to \left\{ \sum_{i=1}^n (\mu_{s,i}(X))^p \right\}^{1/p}, \quad 1 < p < \infty,$$

$s \in S, X \in B(\mathbb{E})$.

This conclusion follows from the properties of the corresponding norms on the product $\mathbb{R}^n$ of $\mathbb{R}$'s.

(5.22) REMARK. Let $\mu : B(\mathbb{E}) \to \mathbb{R}^n$ be a measure of noncompactness. Then the function:

$$\widetilde{\mu} : B(\mathbb{E}) \to \mathbb{R}, \quad \widetilde{\mu}(X) = \sum_{n=1}^{\infty} 2^{-n} \frac{\mu_n(X)}{1 + \mu_n(X)},$$

satisfies the axioms $(\mu_{1,2,3,5})$, but not (μ_4). Instead, it is symmetric: $\widetilde{\mu}(-X) = \widetilde{\mu}(X)$. Such a function is called nonhomogeneous, symmetric measure of noncompactness. Examples of such measures are related to Fréchet spaces.

It must be stressed that not every space $\mathbb{E}$ is given a measure of noncompactness. In the examples below, the set S will be related to a base of continuous seminorms on $\mathbb{E}$.

In the first two examples below, $\mathbb{E}$ is a locally convex quasi-complete space with the base S of continuous seminorms. We use the notation:

$$\mathrm{diam}_\rho(X) = \sup\{\rho(x - y) \mid x, y \in X\}, \quad \rho \in S, \ X \in B(\mathbb{E}),$$
$$B_\rho = \{x \in \mathbb{E} \mid \rho(x) < 1\}, \quad \rho \in S.$$

(5.23) THEOREM. *The Kuratowski measure of noncompactness* $\alpha : B(\mathbb{E}) \to \mathbb{R}^S$ *on a quasi-complete space* $\mathbb{E}$ *is given by the formula:*

$$\alpha_\rho(X) = \inf\{t > 0 \mid \ \textit{there exists a finite covering } X_1 \cup \ldots \cup X_n = X$$
$$\textit{with } \mathrm{diam}_\rho(X_i) \leq t, \ i = 1, 2, \ldots\},$$

$\rho \in S, X \in B(\mathbb{E})$. α *satisfies the axioms* (μ_1) *to* (μ_5).

PROOF. (μ_1): Since $\mathbb{E}$ is quasi-complete, a set $X \subset \mathbb{E}$ is relatively compact $\Leftrightarrow X$ is totally bounded $\Leftrightarrow X$ is totally bounded in each ρ-semimetric $\Leftrightarrow$ for all $\rho \in S$, $\alpha_\rho(X) = 0$ (comp. [RoRo-M], [Sch-M]).

(μ_2): α obviously satisfies (μ_6). Hence, it is enough to show the inequality $\alpha_\rho(\overline{\mathrm{conv}}(X)) \leq \alpha_\rho(X)$.

Let $X = X_1 \cup \ldots \cup X_n$, $\mathrm{diam}_\rho(X_i) \leq t$. Then $\overline{X} = \overline{X}_1 \cup \ldots \cup \overline{X}_n$, $\mathrm{diam}_\rho(\overline{X}_i) \leq t$, and so $\alpha_\rho(\overline{X}) \leq \alpha_\rho(X)$. It remains to prove that $\alpha_\rho(\mathrm{conv}(X)) \leq \alpha_\rho(X)$. It is easy to show that $\mathrm{conv}(X)$ is the set of convex combinations of points $x_i \in \mathrm{conv}(X_i)$, $i = 1, \ldots, n$, and that $\mathrm{diam}_\rho(\mathrm{conv}(X_i)) = \mathrm{diam}_\rho(X_i) \leq t$. There is a covering $\{A_1, \ldots, A_m\}$ of the set $\{(a_1, \ldots, a_n) \in [0, 1]^n \mid \sum a_i = 1\}$ such that the set

$$Y_j = \left\{ \sum_{i=1}^{n} a_i x_i \ \middle| \ x_i \in \mathrm{conv}(X_i), \ a_i \in A_j \right\}, \quad j = 1, \ldots, m,$$

has the ρ-diameter less than $t + \varepsilon$. Obviously, $\mathrm{conv}(X) = Y_1 \cup \ldots \cup Y_m$. Letting $\varepsilon \to 0$, $t \to \alpha_\rho(X)$, we obtain the conclusion.

(μ_3): Let $X = X_1 \cup \ldots \cup X_n$, $Y = Y_1 \cup \ldots \cup Y_m$, where $\alpha_\rho(X_i) \leq t$ and $\alpha_\rho(X_j) \leq t$, for all i, j. Then $X \cup Y = X_1 \cup \ldots \cup X_n \cup Y_1 \cup \ldots \cup Y_m$ and letting $t \to \max\{\alpha_\rho(X), \alpha_\rho(Y)\}$, we obtain $\alpha_\rho(X \cup Y) \leq \max\{\alpha_\rho(X), \alpha_\rho(Y)\}$. The reverse inequality is obvious.

(μ_4): We note that

$$X = X_1 \cup \ldots \cup X_n \Leftrightarrow t \cdot X = t \cdot X_1 \cup \ldots \cup t \cdot X_n,$$

and

$$\mathrm{diam}_\rho(t \cdot X_i) = t \cdot \mathrm{diam}_\rho(X_i), \quad i = 1, \ldots, n.$$

The conclusion follows.

(μ_5): If $X = X_1 \cup \ldots \cup X_n$ and $Y = Y_1 \cup \ldots \cup Y_m$, then

$$X + Y \subset \bigcup_{i=1}^{n} \bigcup_{j=1}^{m} X_i + Y_j$$

and

$$\mathrm{diam}_\rho(X_i + Y_j) \leq \mathrm{diam}_\rho(X_i) + \mathrm{diam}_\rho(Y_j).$$

The conclusion follows. $\qquad\qquad\qquad\qquad\qquad\qquad\qquad\qquad\qquad\quad\square$

(5.24) THEOREM. *The Hausdorff ρ-ball measure of noncompactness $\gamma : B(\mathbb{E}) \to \mathbb{R}^S$ on a quasi-complete space $\mathbb{E}$ is defined by the formula:*

$$\gamma(X)(\rho) = \inf\{t > 0 \mid \text{ there exists a finite net } N \subset \mathbb{E} \text{ with } X \subset N + t \cdot \overline{B}_\rho\},$$

$\rho \in S$, $X \in B(\mathbb{E})$. *γ satisfies the axioms (μ_1) to (μ_5).*

PROOF. We only prove (μ_1), because the proofs of the remaining axioms are the same as in (5.23). We again note that a set $X \subset \mathbb{E}$ is relatively compact $\Leftrightarrow X$ is totally bounded $\Leftrightarrow \forall \varepsilon > 0 \ \forall \rho \in S$ exists a finite set $N \subset X$ with $X \subset N + \varepsilon \cdot \overline{B}_\rho \Rightarrow \forall \rho \in S$, $\gamma(X)(\rho) = 0$. For the reverse last implication, let $X \subset \{x_1, \ldots, x_n\} + \varepsilon \cdot \overline{B}_\rho/2$, $x_i \in \mathbb{E}$, $i = 1, \ldots, n$. For any x_i, let us choose $y_i \in X$ with $y_i \in x_i + \varepsilon \cdot \overline{B}_\rho/2$. Then $x_i + \varepsilon \cdot \overline{B}_\rho/2 \subset y_i + \varepsilon \cdot \overline{B}_\rho$ and, consequently,

$$X \subset \{y_1, \ldots, y_n\} + \varepsilon \cdot B_\rho, \quad y_i \in X, \ i = 1, \ldots, n. \qquad\qquad \square$$

Let X be a locally convex quasi-complete space, $\mathbb{F}$ a Banach space and $\mathbb{E} = C(X, \mathbb{F})$ the space of continuous $\mathbb{F}$-valued functions on X. Let S be a family of subsets of X such that any compact K in X is contained in some $A \in S$. We consider the subspace $\mathbb{E}_S \subset \mathbb{E}$ of functions which are bounded on sets $A \in S$ with the topology τ_S of uniform convergence on elements of S. In particular, if $S = \mathbb{K}$ is a family of compact sets satisfying the above condition, then $\mathbb{E}_K = \mathbb{E}$ and τ_K is the compact-open topology.

$\mathbb{E}_S$ is a complete locally convex space with the base of seminorms

$$\rho_A(f) = \sup_{x \in A} \|f(x)\|, \quad A \in S, \ f \in \mathbb{E}_S,$$

and α_ρ in (5.23) are measures of noncompactness on it.

Now, let $\mu : B(\mathbb{F}) \to \mathbb{R}$ be any measure of noncompactness on a Banach space and $\mathbb{F}$ and X be locally compact. Define

$$\mu_K^*(F) = \sup_{x \in K} \mu(F(x)), \quad F \in B(\mathbb{E}), \ K \in \mathbb{K},$$

where $F(x) = \{u(x) \mid u \in \mathbb{F}\}$, and

$$e_K(F) = \sup_{x \in K} \limsup_{y \to x} \sup_{F \in F} \|f(x) - f(y)\|, \quad F \in B(\mathbb{E}), \ K \in \mathbb{K}.$$

(5.25) THEOREM. *The measure of nonequicontinuity (on $C(X, \mathbb{F})$)*

$$e_{\mu, K} : B(C(X, \mathbb{F})) \to \mathbb{R}^K$$

is defined by the formula:

$$e_{\mu, K}(F) = e_K(F) + \mu_K^*(F), \quad F \in B(C(X, \mathbb{F})), \ K \in \mathbb{K};$$

$e_{\mu, K}$ *is a measure of noncompactness on $(\mathbb{E}, \tau_K)$.*

PROOF. If F is bounded, then $F(K)$ is bounded, for any $K \in \mathbb{K}$, and the expressions $\mu_K(F)$, $e_K(F)$ are finite.

(μ_1): Since $\mu_K^*(F) = 0$, for all $K \in \mathbb{K} \Leftrightarrow \overline{F(x)}$ is compact, for all $x \in X$, and $e_K(F) = 0$, for all $K \in \mathbb{K} \Leftrightarrow F$ is equicontinuous, the conclusion follows from the Arzelà–Ascoli lemma.

(μ_2): Note that (μ_6) is satisfied trivially. Hence, it is enough to verify the inequalities

$$\mu_K^*(\overline{\text{conv}}(F)) \le \mu_K^*(F),$$
$$e_K(\text{conv}(F)) \le e_K(F),$$
$$e_K(\overline{F}) \le e_K(F), \quad F \in B(\mathbb{E}), \ K \in \mathbb{K}.$$

The first one follows from the inclusion $\overline{\mathrm{conv}}(F(X)) \subset \mathrm{conv}(\overline{F(x)})$, $x \in X$, which is easy to prove. To show the second one, let $g = \sum_{j=1}^{m} \lambda_j f_j$ be a convex combination of $f_1, \ldots, f_n \in F$. Then, for any $x, y \in K$,

$$\|g(x) - g(y)\| \leq \sum_{j=1}^{m} \lambda_j \|f_j(x) - f_j(y)\|$$

$$\leq \sum_{j=1}^{m} \lambda_j \cdot \sup_{f \in F} \|f(x) - f(y)\|$$

$$= \sup_{f \in F} \|f(x) - f(y)\|.$$

By passing to upper limits, we reach the conclusion. The last inequality is immediate. $\qquad\square$

In what follows, E is a locally convex space equiped with a measure of noncompactness μ. Let X, Y be two subsets of E.

(5.26) DEFINITION. An admissible map $\varphi : X \multimap Y$ is called

(5.26.1) a *k-set contraction* if, for every $s \in S$, there exists a constant $0 \leq k_s < 1$ such that, for every bounded $A \subset X$, we have

$$\mu_s(\varphi(A)) \leq k_s \mu_s(A),$$

(5.26.2) *condensing* if, for any $s \in S$ and any bounded $A \subset X$, we have

$$\mu_s(A) \neq 0 \Rightarrow \mu_s(\varphi(A)) < \mu_s(A),$$
$$\mu_s(A) = 0 \Rightarrow \mu_s(\varphi(A)) = 0.$$

We let

$$\mathbb{C}(X, Y) = \{\varphi : X \multimap Y \mid \varphi \text{ is condensing}\},$$
$$\mathbb{C}_k(X, Y) = \{\varphi : X \multimap Y \mid \varphi \text{ is a } k\text{-set contraction}\}$$

and we also put

$$\mathbb{C}(X) = \mathbb{C}(X, X), \quad \mathbb{C}_k(X) = \mathbb{C}_k(X, X).$$

Evidently, we have

(5.27) $$\mathbb{K}(X, Y) \subset \mathbb{C}_k(X, Y) \subset \mathbb{C}(X, Y).$$

If E is a Banach space, then the set of indices S can be taken as a singleton.

(5.28) PROPOSITION. *Let $X \subset Y$ and Y be a bounded subset of E. If $\varphi \in \mathbb{C}(X,Y)$, then the set $\mathrm{Fix}(\varphi)$ is a relatively compact subset of E. If $\varphi \in \mathbb{C}(Y,Y)$ and U is a closed subset of E, then $\mathrm{Fix}(\varphi)$ is compact.*

PROOF. Observe that $\mathrm{Fix}(\varphi) \subset \varphi(\mathrm{Fix}(\varphi))$. For every $s \in S$, we have

$$\mu_s(\varphi(\mathrm{Fix}(\varphi))) < \mu_s(\mathrm{Fix}(\varphi)),$$

and $\mathrm{Fix}(\varphi) \subset \varphi(\mathrm{Fix}(\varphi))$ implies

$$\mu_s(\mathrm{Fix}(\varphi)) \leq \mu_s(\varphi(\mathrm{Fix}(\varphi))).$$

So, $\mu_s(\mathrm{Fix}(\varphi))$ must be equal zero, and the proof is completed. $\square$

We can give

(5.29) THEOREM. *If Y is a closed bounded subset of E and $\varphi \in \mathbb{C}(Y)$, then $\varphi \in \mathbb{C}A(Y)$.*

For the proof of (5.29), we need the following lemma.

(5.30) LEMMA. *Let Y_n, $n = 1, 2, \ldots$, be subsets of a bounded set Y in a locally convex space E. Let $\mathcal{A}$ denote the family of all sets A which are representable in the form:*

$$(5.30.1) \qquad A = \bigcup_{n=1}^{\infty} A_n,$$

where A_n is a finite subset of Y_n, for all n. Then, for every $s \in S$, there exists $A_s^* \in \mathcal{A}$ such that:

$$(5.30.2) \qquad \mu_s(A_s^*) = \sup\{\mu_s(A) \mid A \in \mathcal{A}\}.$$

PROOF. We let $r_s = \sup\{\mu_s(A) \mid A \in \mathcal{A}\}$, for every $s \in S$. We fix some $s \in S$. Let $A^k \in \mathcal{A}$, $k = 1, 2, \ldots$, be such that:

$$\lim_{k \to \infty} \mu_s(A^k) = r_s.$$

Since $A^k \in \mathcal{A}$, we can set

$$A^k = \bigcup_{n=1}^{\infty} A_n^k,$$

where A_n^k are finite sets satisfying assumptions of our lemma.

Consider the set $\widetilde{A}^k = \bigcup_{n=k}^{\infty} A_n^k$. Since $\widetilde{A}^k$ differs from A^k only by finitely many terms, we have:

$$\mu_s(\widetilde{A}^k) = \mu_s(A^k) \quad \text{and} \quad \lim_{k \to \infty} \mu_s(\widetilde{A}^k) = r_s.$$

We put: $A_s^* = \bigcup_{k=1}^{\infty} \widetilde{A}^k$. Because of $A_s^* = \bigcup_{k=1}^{\infty} \bigcup_{n=k}^{\infty} A_n^k = \bigcup_{n=1}^{\infty} \bigcup_{k=1}^{n} A_n^k$ and $\left(\bigcup_{k=1}^{n} A_n^k \right) \subset Y_n$, we obtain that $A_s^* \in \mathcal{A}$.

Now, we claim that $A^* \in \mathcal{A}$, established above. On the other hand,

$$\mu_s(A_s^*) \geq \mu_s(\widetilde{A}^k), \quad \text{for any } k, \text{ because } \widetilde{A}^k \subset A^*.$$

Consequently, we get

$$\mu_s(A_s^*) \geq \lim_{k \to \infty} \mu_s(\widetilde{A}^k),$$

for every $s \in S$ and the proof is complete. $\qquad \square$

PROOF OF THEOREM (5.29). Assume $\varphi : Y \multimap Y$ is a condensing map. Letting

$$Y_1 = \varphi(Y), \ldots, Y_n = \varphi(Y_{n-1}), \ldots, \quad n = 1, 2, \ldots,$$

we have

$$\overline{Y}_1 \supset \overline{Y}_2 \supset \ldots \supset \overline{Y}_n \supset \ldots$$

Let $M = \bigcap_{n=1}^{\infty} \overline{Y}_n$. Obviously, M is an attractor of φ, if M is compact and nonempty. Since $\{\overline{Y}_n\}$ is a decreasing sequence, it is enough to show that

$$\lim_{n \to \infty} \mu_s(Y_n) = 0, \quad \text{for every } s \in S.$$

Applying Lemma (5.30), there exists a set A_s^* for which (5.30.2) holds true, i.e.

$$\mu_s(A_s^*) = r_s = \sup\{\mu_s(A) \mid A \in \mathcal{A}\}.$$

Let the representation (5.30.1) of A_s^* be $A_s^* = \bigcup_{n=1}^{\infty} A_n^*$.

Futhermore, for each $x \in A_n^*$, pick an element $y \in \varphi(Y_{n-2})$ such that $x \in \varphi(y)$, and then denote the set of all elements y constructed in this manner by B_{n-1}. Now, set $B = \bigcup_{n=1}^{\infty} B_n$. Clearly, $B \in \mathcal{A}$ and so:

$$\mu_s(B) \leq \mu_s(A_s^*).$$

Moreover,

$$A^* = \bigcup_{n=2}^{\infty} A_n^* \subset \varphi(B),$$

and subsequently

$$\mu_s(\varphi(B)) \geq \mu_s(A^*) \geq \mu_s(B).$$

Since φ is condensing, it follows from the above inequality that $\mu_s(B) = 0$, for arbitrary $s \in S$. Therefore, the set $\overline{B}$ is compact. But then $\varphi(\overline{B})$ is also compact and, consequently, we get that $\overline{A^*}$ is compact.

Now, suppose that $\lim_{n\to\infty} \mu_s(Y_n) > 0$, for some $s \in S$. This means that there are a positive ε_0 and an infinite increasing sequence n_k, $k = 1, 2, \ldots$, such that

$$\mu_s(Y_{n_k}) > \varepsilon_0, \quad \text{for any } k = 1, 2, \ldots$$

For every $k = 1, 2, \ldots$, take the point $y_k \in Y_{n_k}$ such that the pseudonorm $|y_k - y_m|_s \geq \varepsilon_0$, for every $k \neq m$. Consider the set $A = \overline{\{y_k\}_k}$. Then we get $\mu_s(A) \geq \varepsilon_0$, but A is a set belonging to $\mathcal{A}$ (considered in Lemma (5.30)), so $\mu_s(A) \leq \mu_s(A_s^*) = 0$, a contradiction. $\qquad \square$

Finally, we would like to note that for many fixed point applications an abstract measure of noncompactness is useful. Definition (5.15) in an abstract form can be formulated as follows:

(5.31) DEFINITION (Abstract Measure of Noncompactness in Locally Convex Spaces). Let E be a quasi-complete locally convex space. Let A be a lattice with a minimal element, which will be denoted by 0, too. Then a map $\mu : 2^E \to A$ is called an *abstract measure of noncompactness* on E if the following conditions hold, for any $X, Y \in 2^E$:

(5.31.1) (regularity) $\mu(X) = 0$ iff $\overline{X}$ is compact,

(5.31.2) (convex hull property) $\mu(\overline{\text{conv}}(X)) = \mu(X)$,

(5.31.3) (nonsingularity) $\mu(X \cup Y) = \max\{\mu(X), \mu(Y)\}$,

(5.31.4) (monotonicity) if $X \subset Y$, then $\mu(X) \leq \mu(Y)$.

An admissible map $\varphi : X \multimap Y$ is called *condensing* (with respect to μ) if $D \subset X$ and $\mu(D) \leq \mu(\varphi(D))$, then D is relatively compact, i.e., $\mu(D) = 0$. Analogously φ is called a *k-set contraction* with (respect to μ) if there exists $0 \leq k \leq 1$ such that, for any $D \subset X$, we have:

$$\mu(\varphi(D)) \leq k\mu(D).$$

We keep the notations: $\mathbb{C}(X, Y)$ and $\mathbb{C}_k(X, Y)$ respectively for condensing and k-set contraction mappings (with respect to μ) from X to Y.

In the case, when E is a Fréchet space, the abstract measure of noncompactness is a function $\beta : 2^E \to \mathbb{R}$ which satisfies (5.31.1)–(5.31.4) and, additionaly,

(5.31.5) if $Y_1 \supset \ldots \supset Y_n \supset \ldots$ is a decreasing sequence of closed subsets of E and $\lim_n \beta(Y_n) = 0$, then $\bigcap_{n=1}^{\infty} Y_n$ is compact and nonempty.

I.6. Lefschetz fixed point theorem for admissible mappings

At first, we recall some algebraic preliminaries (for details, see [Do-M], [DG-M], [Go1-M], [Go5-M], [Sp-M]).

In what follows, all the vector spaces are taken over $\mathbb{Q}$. Let $f : E \to E$ be an endomorphism of a finite-dimensional vector space E. If $v_1, \dots, v_n$ is a basis for E, then we can write

$$f(v_i) = \sum_{j=1}^{n} a_{ij} v_j, \quad \text{for all } i = 1, \dots, n.$$

The matrix $[a_{ij}]$ is called the matrix of f (with respect to the basis $v_1, \dots, v_n$). Let $A = [a_{ij}]$ be an $(n \times n)$-matrix; then the trace of A is defined as $\sum_{i=1}^{n} a_{ii}$. If $f : E \to E$ is an endomorphism of a finite-dimensional vector space E, then the trace of f, written $\mathrm{tr}(f)$, is the trace of the matrix of f with respect to some basis for E. If E is a trivial vector space then, by definition, $\mathrm{tr}(f) = 0$. It is a standard result that the definition of the trace of an endomorphism is independent of the choice of the basis for E.

We recall the following two basic properties of the trace:

(6.1) PROPERTY. *Assume that, in the category of finite-dimensional vector spaces, the following diagram commutes*

$$
\begin{array}{ccc}
E' & \xrightarrow{\ f\ } & E'' \\
{\scriptstyle f'}\big\uparrow & \searrow{\scriptstyle g} & \big\uparrow{\scriptstyle f''} \\
E' & \xrightarrow{\ f\ } & E''.
\end{array}
$$

Then $\mathrm{tr}(f') = \mathrm{tr}(f'')$; *in other words,* $\mathrm{tr}(gf) = \mathrm{tr}(fg)$.

(6.2) PROPERTY. *Given a commutative diagram of finite-dimensional vector spaces with exact rows*

$$
\begin{array}{ccccccccc}
0 & \longrightarrow & E' & \longrightarrow & E & \longrightarrow & E'' & \longrightarrow & 0 \\
 & & {\scriptstyle f'}\big\downarrow & & {\scriptstyle f}\big\downarrow & & {\scriptstyle f''}\big\downarrow & & \\
0 & \longrightarrow & E' & \longrightarrow & E & \longrightarrow & E'' & \longrightarrow & 0
\end{array}
$$

we have $\mathrm{tr}(f) = \mathrm{tr}(f') + \mathrm{tr}(f'')$.

Let $E = \{E_q\}$ be a graded vector space of a finite type, i.e. $E_q = 0$, for almost all q and $\dim E_q < \infty$, for every q. If $f = \{f_q\}$ is an endomorphism of degree zero of such a graded vector space, then the (*ordinary*) *Lefschetz number* $\lambda(f)$ *of* f *is defined by*

$$\lambda(f) = \sum_{q} (-1)^q \mathrm{tr}(f_q).$$

Let $f : E \to E$ be an endomorphism of an arbitrary vector space E. Denote by $f^{(n)} : E \to E$ the nth iterate of f and observe that the kernels

$$\ker f \subset \ker f^{(2)} \subset \ldots \subset \ker f^{(n)} \subset \ldots$$

form an increasing sequence of subspaces of E. Let us now put

$$N(f) = \bigcup_n \ker f^{(n)} \quad \text{and} \quad \widetilde{E} = E/N(f).$$

Clearly, f maps $N(f)$ into itself and, therefore, induces the endomorphism

$$\widetilde{f} : \widetilde{E} \to \widetilde{E}$$

on the factor space $\widetilde{E} = E/N(f)$.

(6.3) PROPERTY. We have $f^{-1}(N(f)) = N(f)$; consequently, the kernel of the induced map $\widetilde{f} : \widetilde{E} \to \widetilde{E}$ is trivial, i.e., $\widetilde{f}$ is a monomorphism.

PROOF. If $v \in f^{-1}(N(f))$, then $f(v) \in N(f)$. This implies that, for some n, we have $f^{(n)}(f(v)) = 0 = f^{(n+1)}(v)$ and $v \in N(f)$. Conversely, if $v \in N(f)$, then $f^{(n)}(v) = 0$, for some n. Thus, $f^{(n)}(f(v)) = 0$, and so $f(v) \in N(f)$, i.e., $v \in f^{-1}(N(f))$. $\square$

Let $f : E \to E$ be an endomorphism of a vector space E. Assume that $\dim \widetilde{E} < \infty$. In this case, we define the generalized trace $\mathrm{Tr}(f)$ of f by putting $\mathrm{Tr}(f) = \mathrm{tr}(\widetilde{f})$.

(6.4) PROPERTY. Let $f : E \to E$ be an endomorphism. If $\dim E < \infty$, then $\mathrm{Tr}(f) = \mathrm{tr}(f)$.

PROOF. We have the commutative diagram with exact rows

$$
\begin{array}{ccccccccc}
0 & \longrightarrow & N(f) & \longrightarrow & E & \longrightarrow & E/N(f) & \longrightarrow & 0 \\
& & \downarrow{\overline{f}} & & \downarrow{f} & & \downarrow{\widetilde{f}} & & \\
0 & \longrightarrow & N(f) & \longrightarrow & E & \longrightarrow & E/N(f) & \longrightarrow & 0
\end{array}
$$

in which $\widetilde{f}$ is induced by f.

Applying (6.4) to the above diagram, we obtain

(6.4.1) $\qquad \mathrm{tr}(f) = \mathrm{tr}(\overline{f}) + \mathrm{tr}(\widetilde{f}), \quad \text{where } \mathrm{tr}(\widetilde{f}) = \mathrm{Tr}(f).$

We prove that $\mathrm{tr}(\overline{f}) = 0$. Since $\dim E < \infty$, we can assume that $N(f) = \ker(f^{(n)})$, for some $n \geq 1$. Now, consider the commutative diagram

$$
\begin{array}{ccccccccc}
\ker(f) & \longrightarrow & \ker(f^{(2)}) & \longrightarrow & \cdots \longrightarrow & \ker(f^{(n-1)}) & \longrightarrow & \ker(f^{(n)}) \\
\scriptstyle{0 = \overline{f}_1}\downarrow & \swarrow & \scriptstyle{\overline{f}_2}\downarrow & & & \scriptstyle{\overline{f}_{n-1}}\downarrow & \swarrow & \scriptstyle{\overline{f}_n = \overline{f}}\downarrow \\
\ker(f) & \longrightarrow & \ker(f^{(2)}) & \longrightarrow & \cdots \longrightarrow & \ker(f^{(n-1)}) & \longrightarrow & \ker(f^{(n)})
\end{array}
$$

where the maps $\overline{f}^i$, $f^{(i)}$, $i = 1, \ldots, n$ are given by f (observe that if $v \in \ker(f^{(i)})$, then $f(v) \in \ker(f^{(i-1)})$, for every $i > 1$).

Then, from (6.3), we infer

$$\operatorname{tr}(\overline{f}) = \operatorname{tr}(\overline{f}_{n-1}) = \ldots = \operatorname{tr}(\overline{f}_2) = \operatorname{tr}(\overline{f}_1) = 0.$$

Finally, from (6.4.1), we obtain $\operatorname{Tr}(f) = \operatorname{tr}(\widetilde{f}) = \operatorname{tr}(f)$, and the proof of (6.4) is completed. □

Let $f = \{f_q\}$ be an endomorphism of degree zero of a graded vector space $E = \{E_q\}$. We say that f is a *Leray endomorphism* if the graded vector space $\widetilde{E} = \{\widetilde{E}_q\}$ is of finite type. For such an f, we define the (*generalized*) *Lefschetz number* $\Lambda(f)$ of f by putting

$$\Lambda(f) = \sum_q (-1)^q \operatorname{Tr}(f_q).$$

It is immediate from (6.4) that

(6.5) PROPERTY. *Let* $f : E \to E$ *be an endomorphism of degree zero, i.e.,* $f = \{f_q\}$ *and* $f_q : E_q \to E_q$ *is a linear map. If* E *is a graded vector space of finite type, then* $\Lambda(f) = \lambda(f)$.

The following property of the Leray endomorphism is of an importance:

(6.6) PROPERTY. *Assume that, in the category* $\mathcal{A}$ *of graded vector spaces over* $\mathbb{Q}$, *the following diagram commutes:*

$$
\begin{array}{ccc}
E' & \xrightarrow{\ f\ } & E'' \\
{\scriptstyle f'}\big\uparrow & \nwarrow{\scriptstyle g} & \big\uparrow{\scriptstyle f''} \\
E' & \xrightarrow{\ f\ } & E''.
\end{array}
$$

Then if either f' *or* f'' *is a Leray endomorphism, then the other is a Leray endomorphism as well, and in that case* $\Lambda(f') = \Lambda(f'')$.

PROOF. By the hypothesis, we have, for each q, the following commutative diagram in the category of vector spaces:

$$
\begin{array}{ccc}
E'_q & \xrightarrow{\ f_q\ } & E''_q \\
{\scriptstyle f'_q}\big\uparrow & \nwarrow{\scriptstyle g_q} & \big\uparrow{\scriptstyle f''_q} \\
E'_q & \xrightarrow{\ f_q\ } & E''_q.
\end{array}
$$

For the proof, it is sufficient to show that if either $\mathrm{Tr}(f_q')$ or $\mathrm{Tr}(f_q'')$ is defined, then so is the other trace, and in that case $\mathrm{Tr}(f_q') = \mathrm{Tr}(f_q'')$. We observe that the commutativity of the above diagram implies that the following diagram commutes:

$$
\begin{array}{ccc}
E_q'/N(f_q') & \xrightarrow{\;\tilde{f}_q\;} & E_q''/N(f_q'') \\
\widetilde{f_q'}\Big\uparrow & \overset{\tilde{g}_q}{\nwarrow} & \Big\uparrow\widetilde{f_q''} \\
E_q'/N(f_q') & \xrightarrow{\;\tilde{f}_q\;} & E_q''/N(f_q'').
\end{array}
$$

Since $\tilde{f}_q$, and $\tilde{g}_q$ are monomorphisms, the commutativity of the above diagram implies that $\dim(E_q'/N(f_q')) < \infty$ if and only if $\dim(E_q''/N(f_q'')) < \infty$, and hence we conclude that $\mathrm{Tr}(f_q')$ is defined if and only if $\mathrm{Tr}(f_q'')$ is defined. Moreover, from (6.1), we deduce that $\mathrm{Tr}(f_q') = \mathrm{Tr}(f_q'')$, provided $\mathrm{Tr}(f_q'')$ is defined. The proof of (6.6) is completed. $\qquad\qquad\square$

Assume that the following diagram

$$
\begin{array}{ccc}
E' & \xrightarrow{\;f\;} & E'' \\
f'\Big\uparrow & & \Big\uparrow f'' \\
E' & \xrightarrow[\sim]{\;f\;} & E''.
\end{array}
$$

is commutative. Then we obtain the following commutative diagram:

$$
\begin{array}{ccc}
E' & \xrightarrow{\;f\;} & E'' \\
f'\Big\uparrow & \overset{f'\circ f^{-1}}{\nwarrow} & \Big\uparrow f'' \\
E' & \xrightarrow{\;f\;} & E''.
\end{array}
$$

Therefore, from (6.6), we obtain:

(6.6.1) PROPERTY. *Assume that, in the category $\mathcal{A}$ of graded vector spaces over $\mathbb{Q}$, the following diagram is commutative:*

$$
\begin{array}{ccc}
E' & \xrightarrow{\;f\;} & E'' \\
f'\Big\uparrow & & \Big\uparrow f'' \\
E' & \xrightarrow{\;f\;} & E''.
\end{array}
$$

and f is an isomorphism. Then the conclusion of (6.6) holds true.

(6.7) PROPERTY. *Let*

$$\cdots \longrightarrow E'_q \longrightarrow E_q \longrightarrow E''_q \longrightarrow E'_{q-1} \longrightarrow \cdots$$

$$\Big\downarrow f'_q \qquad \Big\downarrow f_q \qquad \Big\downarrow f''_q \qquad \Big\downarrow f'_{q-1}$$

$$\cdots \longrightarrow E'_q \longrightarrow E_q \longrightarrow E''_q \longrightarrow E'_{q-1} \longrightarrow \cdots$$

be a commutative diagram of vector spaces in which the rows are exact. If two of the following endomorphisms $f = \{f_q\}$, $f' = \{f'_q\}$, $f'' = \{f''_q\}$ are the Leray endomorphisms, then so is the third, and, moreover, in that case we have:

$$\Lambda(f'') + \Lambda(f') = \Lambda(f).$$

PROOF. This immediately follows from (6.2). □

Among the above properties of the Leray endomorphisms, we also point out some information about weakly nilpotent endomorphisms.

(6.8) DEFINITION. A linear map $f : E \to E$ of a vector space E into itself is called *weakly nilpotent* if, for every $x \in E$, there exists n_x such that $f^{n_x}(x) = 0$.

Observe that if $f : E \to E$ is weakly nilpotent, then $N(f) = E$, and so, we have:

(6.9) PROPERTY. *If $f : E \to E$ is weakly nilpotent, then $\mathrm{Tr}(f)$ is well-defined and $\mathrm{Tr}(f) = 0$.*

Assume that $E = \{E_q\}$ is a graded vector space and $f = \{f_q\} : E \to E$ is an endomorphism. We say that f is *weakly nilpotent* if f_q is weakly nilpotent, for every q.

From (6.9), we deduce:

(6.10) PROPERTY. *Any weakly nilpotent endomorphism $f : E \to E$ is a Leray endomorphism, and $\Lambda(f) = 0$.*

Now, the Lefschetz number will be defined for admissible mappings. Namely, let $\varphi : X \multimap X$ be an admissible map and $(p, q) \subset \varphi$ be a selected pair of φ. Then the induced homomorphism:

$$q_* \circ p_*^{-1} : H_*(X) \to H_*(X)$$

is an endomorphism of the graded vector space $H_*(X)$ into itself. So, we can define the Lefschetz number $\Lambda(p, q)$ of the pair (p, q) by putting:

$$\Lambda(p, q) = \Lambda(q_* \circ p_*^{-1}),$$

provided the Lefschetz number $\Lambda(q_* \circ p_*^{-1})$ is well-defined.

It allows us to define the Lefschetz set Λ of φ as follows:

$$\Lambda(\varphi) = \{\Lambda(p,q) \mid (p,q) \subset \varphi\}.$$

In what follows, we say that the Lefschetz set $\Lambda(\varphi)$ of φ is *well-defined* if, for every $(p,q) \subset \varphi$, the Lefschetz number $\Lambda(p,q)$ of (p,q) is defined.

Moreover, from the homotopy property, we get:

(6.11) PROPERTY.

(6.11.1) *If* $\varphi, \psi : X \multimap X$ *are homotopic* $(\varphi \sim \psi)$, *then:* $\Lambda(\varphi) \cap \Lambda(\psi) \neq \emptyset$.

(6.11.2) *If* $\varphi : X \multimap X$ *is an admissible map and X is acyclic, then the Lefschetz set $\Lambda(\varphi)$ is well-defined and* $\Lambda(\varphi) = \{1\}$.

Now, let $\varphi : (X, A) \multimap (X, A)$ be an admissible map, i.e., $\varphi_X : X \multimap X$, where $\varphi_X(x) = \varphi(x)$, for all $x \in X$, and a contraction $\varphi_A : A \multimap A$ being a restriction of φ_X to the pair (A, A), be admissible. Then (6.7) can be reformulated as follows:

(6.12) PROPERTY. *For every* $(p,q) \subset \varphi$, *there are* $(p_1, q_1) \subset \varphi_X$ *and* $(p_2, q_2) \subset \varphi_A$ *such that*

(i) *if two of the following Lefschetz numbers* $\Lambda(p,q)$, $\Lambda(p_1, q_1)$, $\Lambda(p_2, q_2)$ *are defined, then so is the third,*

(ii) $\Lambda(p,q) = \Lambda(p_1, q_1) - \Lambda(p_2, q_2)$, *provided two of them are defined.*

Finally, Property (6.10) can be reformulated as follows:

(6.13) PROPERTY. *Assume that* $\varphi : (X, A) \multimap (X, A)$ *is an admissible map such that A absorbs compact sets, i.e., for every compact $K \subset X$, there exists* $n = n(K)$ *such that* $\varphi^n(K) \subset A$.

Then the Lefschetz set $\Lambda(\varphi)$ *of φ is well-defined, and* $\Lambda(\varphi) = \{0\}$.

It is useful to formulate

(6.14) COINCIDENCE THEOREM (see (12.8) in [Go5-M]). *Let U be an open subset of a finite dimensional normed space E. Consider the following diagram:*

$$U \xleftarrow{p} \Gamma \xrightarrow{q} U$$

in which q is a compact map.

Then the Lefschetz number $\Lambda(p,q)$ *of the pair* (p,q), *given by the formula*

$$\Lambda(p,q) = \Lambda(q_* \circ p_*^{-1}),$$

is well-defined, and $\Lambda(p,q) \neq 0$ *implies that* $p(y) = q(y)$.

Theorem (6.14) can be reformulated in terms of multivalued mappings as follows.

Let $U \subset E$ be the same as in (6.11) and let $\varphi : U \multimap U$ be a compact, admissible map, i.e., $\varphi \in \mathbb{K}(U)$. We let

$$\Lambda(\varphi) = \{\Lambda(p,q) \mid (p,q) \subset \varphi\},$$

where $\Lambda(p,q) = \Lambda(q_* p_*^{-1})$. Then we have:

(6.15) THEOREM.

(6.15.1) *The set* $\Lambda(\varphi)$ *is well-defined, i.e., for every* $(p,q) \subset \varphi$, *the generalized Lefschetz number* $\Lambda(p,q)$ *of the pair* (p,q) *is well-defined, and*

(6.15.2) $\Lambda(\varphi) \neq \{0\}$ *implies that the set* $\mathrm{Fix}(\varphi) = \{x \in U \mid x \in \varphi(x)\}$ *is nonempty.*

At first, we shall generalize (6.14) to the case, when $\varphi \in \mathbb{K}(U)$ and U is an open subset of a locally convex space E. To do it, we recall some additional notions and facts.

Two maps $\varphi, \psi : X \multimap Y$ are called α-*close*, $\alpha \in \mathrm{Cov}(Y)$, if, for every $x \in X$, there exists $U_x \in \alpha$ such that $\varphi(x) \cap U_x \neq \emptyset$ and $\psi(x) \cap U_x \neq \emptyset$.

Assume that $X \subset Y$ and $\alpha \in \mathrm{Cov}(Y)$. A point $x \in X$ is called an α-*fixed point* of $\varphi : X \multimap Y$ if there exists $U_x \in \alpha$ such that $x \in U_x$ and $\varphi(x) \cap U_x \neq \emptyset$.

We prove:

(6.16) LEMMA. *Let* $\varphi \in \mathbb{K}(X)$ *and* $D \subset \mathrm{Cov}(X)$ *be a cofinal family of open coverings. If, for every* $\alpha \in D$, *there exists an* α-*fixed point of* φ, *then* $\mathrm{Fix}(\varphi) = \{x \in X \mid x \in \varphi(x)\}$ *is a nonempty set.*

PROOF. Assume, on the contrary, that $\mathrm{Fix}(\varphi) = \emptyset$. Then $x \notin \varphi(x)$, for every $x \in X$. Since we consider X to be a $T_{3\frac{1}{2}}$-space, there are U_x and V_x such that $x \in U_x$, $\varphi(x) \subset V_x$ and $U_x \cap V_x = \emptyset$. Observe that the set $\{x \in X \mid \varphi(x) \subset V_x\}$ is an open neighbourhood of x (we have assumed that φ is u.s.c.).

Let $W_x = U_x \cap \{x \in X \mid \varphi(x) \subset V_x\}$. Then $\beta = \{W_x\}_{x \in X}$ is an open covering of X. Since D is cofinal in $\mathrm{Cov}(X)$, there exists $\gamma = \{O_x\} \in D$ such that $\gamma \geq \beta$. Then φ has no γ-fixed point, and we obtain a contradiction. $\square$

(6.17) LEMMA. *Let* U *be an open subset of a locally convex space* E. *Then, for every* $\alpha \in \mathrm{Cov}(U)$, *there exists a refinement* $\beta \in \mathrm{Cov}(U)$ *such that, for any space* X *and any two single-valued (continuous) maps* $f, g : X \to U$, *if* f *and* g *are* β-*close, then* f *and* g *are* α-*stationary homotopic, i.e. there is a homotopy* $h : X \times [0,1] \to U$ *joining* f *and* g *and such that, for every* $x \in U$, *there is*

$V \in \alpha$ *for which we have* $h(x,t) \in V$, *for every* $t \in [0,1]$, *and* $h(x,t) = $ *constants,*
whenever $f(x) = g(x)$.

PROOF. Let $z \in U$. Then there exists $W \in \alpha$ and an open neighbourhood W_z
of the zero point in E such that $z + W_z \subset W$. Let V_z be an open neighbourhood
of the zero point in E such that $V_z + V_z \subset W_z$. Finally, let U_z be an open
neighbourhood of $0 \in E$ such that $z + U_z \subset V_z$. We let:

$$\beta = \{z + U_z\}_{z \in U}$$

and assume that $f, g : X \to U$ are β-close. It implies that, for every $x \in X$, there
exists $z \in U$ such that: $f(x), g(x) \in (z + U_z) \in \beta$.

We define the homotopy $h : X \times [0,1] \to U$ by putting:

$$h(t,x) = (1 - t)f(x) + tg(x).$$

Then $h(t,x) \in [z + ((1-t)U_z + tU_z)] \subset z + V_z + V_z \subset z + W_z$, and the proof is
completed. $\square$

(6.18) LEMMA. *Let* U *be an open subset of a locally convex space* E *and let*
$\varphi \in \mathbb{K}(U)$. *Then:*
(6.18.1) *the set* $\Lambda(\varphi)$ *is well-defined, and*
(6.18.2) $\Lambda(\varphi) \neq \{0\}$ *implies* $\text{Fix}(\varphi) \neq \emptyset$.

PROOF. Let $(p, q) \subset \varphi$ and assume that: $U \xleftarrow{p} \Gamma \xrightarrow{q} U$, where p is a Vietoris
map and q is compact.

Thus, there exists a compact $K \subset U$ such that $q(\Gamma) \subset K$. Let $\alpha \in \text{Cov}_U(K)$.
We choose $\beta \in \text{Cov}_U(K)$, according to the Lemma (6.16). Now, applying the
approximation theorem (1.37), we obtain a finite dimensional map $\pi_\beta : K \to U$
which is β-close to the inclusion map $i : K \to U$. Then the map $q_\beta = \pi_\beta \circ q : \Gamma \to U$
satisfies the following conditions:

  (i) q_β is compact,
 (ii) q and q_β are β-close,
(iii) q and q_β are stationary α-homotopic.
(iv) $\overline{q_\beta(\Gamma)} \subset U_\beta$, where $U_\beta = U \cap E^n$ and E^n is a finite dimensional subspace
      of E chosen according to the approximation theorem.

Summing up the above, we get the following commutative diagram:

$$
\begin{array}{ccc}
H_*(U_\beta) & \xrightarrow{\ j_*\ } & H_*(U) \\
{\scriptstyle (\overline{q}_\beta)_* (p_\beta)_*^{-1}} \Big\uparrow & {\scriptstyle (q'_\beta)_* p_*^{-1}} \nwarrow & \Big\uparrow {\scriptstyle q_{\beta_*} p_*^{-1}} \\
H_*(U_\beta) & \xrightarrow{\ j_*\ } & H_*(U),
\end{array}
$$

in which $j : U_\beta \to U$ is the inclusion map and $\bar{q}_\beta$, p_β, q'_β are the respective restrictions of q_β and p.

Now, from the commutativity of the above diagram and (6.11), we get

$$\Lambda((\bar{q}_\beta)_*(p_\beta)_*^{-1}) = \Lambda(q_* p_*^{-1})$$

which proves (6.18.1).

Finally, if we assume that

$$\Lambda(q_* p_*^{-1}) \neq 0, \quad \text{for some } (p,q) \subset \varphi, \text{ then } \Lambda((\bar{q}_\beta)_*(p_\beta)_*^{-1}) \neq 0$$

and, in view of (6.18.1), we get that φ has an α-fixed point. Consequently, our claim follows from (6.16), and the proof is completed.   □

As the last lemma we state:

(6.19) LEMMA. *Let U be the same as in (6.18) and assume that $\varphi \in \mathbb{C}AC(U)$. Then the Lefschetz set $\Lambda(\varphi)$ of φ is well-defined and $\Lambda(\varphi) \neq \{0\}$ implies that* $\mathrm{Fix}(\varphi) \neq \emptyset$.

PROOF. Lef V be chosen according to the definition of compact absorbing contractions. Since V absorbs a point and φ is u.s.c., it follows that V absorbs compact sets.

We consider the following map of pairs:

$$\widetilde{\varphi} : (U, V) \multimap (U, V).$$

Let $\varphi|_V : V \to V$, $\varphi|_V(x) = \varphi(x)$, for every $x \in V$. Then $\varphi|_V \in \mathbb{K}(V)$, so from (6.18) and (6.12) it follows that the set $\Lambda(\varphi)$ is well-defined.

Now, if we assume that $\Lambda(\varphi) \neq \{0\}$ and $\Lambda(p,q) \neq 0$, for some $(p,q) \subset \varphi$, then we have the diagram:

$$U \xleftarrow{p} \Gamma \xrightarrow{q} U,$$

and let $\Gamma_1 = p^{-1}(V)$, $p_1 : \Gamma_1 \Rightarrow V$, $p_1(y) = p(y)$, $q_1 : \Gamma_1 \to V$, $q_1(y) = q(y)$, for every $y \in \Gamma_1$.

Consequently, in view of (6.7.1), we obtain $\Lambda(p,q) = \Lambda(p_1,q_1) \neq 0$. Hence, our lemma follows from (6.18).   □

(6.20) THE LEFSCHETZ FIXED POINT THEOREM. *Let X be a retract of an open subset U of a locally convex space E. Assume, furthermore, that $\varphi \in \mathbb{C}AC(X)$. Then:*

(6.20.1) *the Lefschetz set $\Lambda(\varphi)$ of φ is well-defined,*

(6.20.2) *if $\Lambda(\varphi) \neq \{0\}$, then $\mathrm{Fix}(\varphi) \neq \emptyset$.*

PROOF. Let $r : U \to X$ be the retraction map and $i : X \to U$ the inclusion map. We have the following commutative diagram:

$$
\begin{array}{ccc}
X & \xrightarrow{\ i\ } & U \\
\varphi \downarrow & \searrow {\scriptstyle \varphi \circ r} & \downarrow {\scriptstyle i \circ \varphi \circ r} \\
X & \xrightarrow{\ i\ } & U.
\end{array}
$$

Consequently, by using (6.6.1) and (6.19), we obtain our theorem (observe that $i \circ \varphi \circ r : U \multimap U$ is a compact absorbing map); the proof is completed. $\square$

Several fixed point results can be obtained as a corollary from (6.20).

(6.21) COROLLARY. *Let X be the same as in (6.20) and $\varphi : X \multimap X$ be a map such that one of the following possibilities occurs:*

(6.21.1) $\varphi \in \mathbb{K}(X)$,

(6.21.2) $\varphi \in \mathrm{EC}_0(X)$,

(6.21.3) $\varphi \in \mathrm{ASC}_0(X)$,

(6.21.4) $\varphi \in \mathbb{CA}_0(X)$.

Then the Lefschetz set $\Lambda(\varphi)$ of φ is well-defined, and $\Lambda(\varphi) \neq \{0\}$ implies

$$\mathrm{Fix}(\varphi) \neq \emptyset.$$

(6.22) COROLLARY. *Let X be an acyclic retract of an open subset U in a locally convex space. If $\varphi \in \mathbb{CAC}(X)$ or, in particular, $\varphi \in \mathbb{K}(X)$ or $\varphi \in \mathrm{EC}_0(X)$ or $\varphi \in \mathrm{ASC}_0(X)$ or $\varphi \in \mathrm{EA}_0(X)$, then φ has a fixed point.*

I.7. Lefschetz fixed point theorem for condensing mappings

In the preceding section, the Lefschetz fixed point theorem for admissible mappings which are compact or, more generally, $\mathbb{CAC}$ was discussed. So, the problem of admissible condensing mappings is natural. We have:

(7.1) THEOREM. *Let U be an open subset of a Fréchet space E and let $\varphi : U \multimap U$ be an admissible condensing mapping with a compact attractor. Then:*

(7.1.1) *the Lefschetz set $\Lambda(\varphi)$ of φ is well-defined, and*
(7.1.2) *of $\Lambda(\varphi) \neq \{0\}$, then $\mathrm{Fix}(\varphi)) \neq \emptyset$.*

In the case, when E is a Banach space, Theorem (7.1) was proved by:

  (i) R. Nussbaum for single-valued mappings (see [Nu1]);
 (ii) G. Fournier and D. Violette for compositions of acyclic mappings (see [FV2], [FV3]).

Our formulation is slightly more general, but the proof of G. Fournier and D. Violette can be repeated up to some natural changes. To do it, we must define the fixed point index for compactifying mappings in Fréchet spaces and to prove several lemmas concerning condensing mappings (comp. (6.1)–(6.5) in [FV2], [FV3]).

Since everything can be done in an easy but technically complicated way, the proof of (7.1) is omitted here.

(7.2) DEFINITION. A closed bounded subset X of a Fréchet space E is called a *special neighbourhood retract* (written, $X \in \text{SNR}(E)$) if there exists an open set U of E such that $X \subset U$ and a (continuous) retraction $r : U \to X$ such that:

(7.2.1) $\mu(r(A)) \leq \mu(A)$, for every $A \subset U$, where μ is an abstract measure of noncompactness.

The main result of this short section is the following:

(7.3) THEOREM. *Assume that $X \in \text{SNR}(E)$ and $\varphi : X \multimap X$ is a condensing admissible map. Then:*

(7.3.1) *the Lefschetz set $\Lambda(\varphi)$ of φ is well-defined,*
(7.3.2) *if $\Lambda(\varphi) \neq \{0\}$, then $\text{Fix}(\varphi) \neq \emptyset$.*

PROOF. Let $r : U \to X$ be a retraction map, chosen according to the Definition (7.2).

Then the map $\psi = i \circ \varphi \circ r$ is an admissible condensing map, where $i : X \to U$ is the inclusion map. In view of (5.29), we can assume that φ has a compact attractor A.

Now, it follows from (7.2.1) that A is a compact attractor for ψ and, moreover, we get that ψ is an admissible condensing map. Consequently, ψ satisfies all assumptions of (7.1). Furthermore, we have the following commutative diagram of admissible mappings:

$$\begin{array}{ccc} X & \xrightarrow{\ i\ } & U \\ {\scriptstyle\varphi}\big\downarrow & {\scriptstyle\varphi \text{ or } \psi} & \big\downarrow{\scriptstyle\psi} \\ X & \xrightarrow{\ i\ } & U. \end{array}$$

Thus, the Lefschetz set $\Lambda(\varphi)$ of φ is well-defined iff $\Lambda(\psi)$ of ψ is so.

It implies that, for a selected pair $(p,q) \subset \varphi$ and the induced selected pair $(\overline{p}, \overline{q})$ of ψ, the Lefschetz numbers are equal.

Consequently, (7.3) follows from (7.1).   □

(7.4) COROLLARY. *Let $X \in \text{SNR}(E)$ and $\varphi \in \mathbb{C}_k(X)$. Then*

(7.4.1) *the Lefschetz set $\Lambda(\varphi)$ of φ is well-defined,*
(7.4.2) *if $\Lambda(\varphi) \neq \{0\}$, then $\text{Fix}(\varphi) \neq \emptyset$.*

Corollary (7.4) can be obtained from (7.3), because every k-set contraction mapping is condensing.

(7.5) COROLLARY. *Let $X \in \mathrm{SNR}(E)$, for $U = E$ (see Definition (7.2)). If $\varphi \in \mathbb{C}(X)$ or $\varphi \in \mathbb{C}_k(X)$, then φ has a fixed point.*

In fact, in the considered case, we obtain that $\Lambda(\varphi) \neq \{0\}$, and (7.5) follows from (7.3).

Finally, we would like to formulate the following open problem:

(7.6) OPEN PROBLEM. *Is it possible to prove (7.1) for addmissible condensing mappings with compact attractors on open sets in locally convex quasi-complete spaces?*

In our opinion, the answer to (7.6) is positive, but it requires to develope the adequate fixed point index theory for compactifying admissible mappings.

I.8. Fixed point index and topological degree for admissible maps in locally convex spaces

The aim of this section is to define the fixed point index for compact admissible mappings on retracts of open sets in locally convex spaces.

At first, following [Go1-M] and [Go5-M], we recall the definition and properties of the fixed point index for admissible maps on open sets in the Euclidean space $\mathbb{R}^n$.

Consider the diagram:

$$X \xleftarrow{\;p\;} \Gamma \xrightarrow{\;q\;} Y.$$

The above diagram induces a multivalued mapping $\varphi(p, q) : X \multimap Y$ defined by the formula:

$$\varphi(p, q)(x) = q(p^{-1}(x)), \quad \text{for every } x \in X.$$

In what follows, we shall identify the map $\varphi(p, q)$ with the pair (p, q). Of course, $\varphi(p, q)$ is admissible, but we keep only one selected pair (p, q) of this map. Moreover, $\varphi(p, q)$ is compact if and only if q is compact.

For a multivalued map $(p, q) : X \multimap Y$, we denote the set of coincidence points and fixed points as:

$$C(p, q) = \{z \in \Gamma \mid p(z) = q(z)\},$$
$$\mathrm{Fix}(p, q) = \{x \in X \mid x \in q(p^{-1}(x))\}.$$

Evidently,

$$p(C(p, q)) = \mathrm{Fix}(p, q)$$

and so:

$$\mathrm{Fix}(p, q) \neq \emptyset \Leftrightarrow C(p, q) \neq \emptyset.$$

By S^n, we shall denote the unit sphere in $\mathbb{R}^{n+1}$, i.e., $S^n = \{x \in \mathbb{R}^{n+1} \mid |x| = 1\}$. It is well-known that the one point compactification of $\mathbb{R}^n$ is S^n, i.e., $S^n = \mathbb{R}^n \cup \{\infty\}$.

Let U be an open subset of $\mathbb{R}^n$ and assume that $(p,q) : U \multimap \mathbb{R}^n$ is an admissible map such that $\operatorname{Fix}(p,q)$ is compact. We have the following diagram:

$$(8.1) \qquad\qquad U \xleftarrow{\ p\ } \Gamma \xrightarrow{\ q\ } \mathbb{R}^n$$

and its induced one:

$$(8.2) \qquad (U, U \setminus \operatorname{Fix}(p,q)) \xleftarrow{\ \bar{p}\ } (\Gamma, p^{-1}(U \setminus \operatorname{Fix}(p,q)) \xrightarrow{\ \bar{q}\ } (\mathbb{R}^n, \mathbb{R}^n \setminus \{0\}),$$

where $\bar{p}(z) = p(z)$ and $\bar{q}(z) = p(z) - q(z)$, for every $z \in \Gamma$. Observe that $\bar{p}$ and $\bar{q}$ are well-defined and $\bar{p}$ is a Vietoris map.

Now, we can extend the diagram (8.2) to the following one:

$$(8.3) \quad S^n \xrightarrow{\ i\ } (S^n, S^n \setminus \operatorname{Fix}(p,q)) \xleftarrow{\ j\ } (U, U \setminus \operatorname{Fix}(p,q)) \xleftarrow{\ \bar{p}\ }$$
$$(\Gamma, p^{-1}(U \setminus \operatorname{Fix}(p,q)) \xrightarrow{\ \bar{q}\ } (\mathbb{R}^n, \mathbb{R}^n \setminus \{0\}),$$

where i, j are the respective inclusions. We would like to apply the Čech homology functor with compact carriers and coefficients in integers $\mathbb{Z}$.

Note that, by the excision axiom, j_* is an isomorphism. Moreover,

$$H_n(S^n) = H_n(\mathbb{R}^n, \mathbb{R}^n \setminus \{0\}) = \mathbb{Z}.$$

By applying the functor H_n to (8.2), we can define the *fixed point index* $\operatorname{ind}(p,q)$ of (p,q) by putting (comp. [Go1-M] or [Go5-M]):

$$(8.4) \qquad\quad \operatorname{ind}(p,q) = ((\bar{q}_{*n} \circ (\bar{p}_{*n})^{-1} \circ (j_{*n})^{-1} \circ i_{*n}))(1) \in \mathbb{Z}.$$

It is useful to consider the homology class

$$\mathcal{O}_{\operatorname{Fix}(p,q)} = ((j_{*n})^{-1} \circ i_{*n})(1) \in H_n(U, U \setminus \operatorname{Fix}(p,q))$$

called the *fundamental class* of $\operatorname{Fix}(p,q)$ in U. Using the notion of the fundamental class, one can define:

$$(8.5) \qquad\qquad \operatorname{ind}(p,q) = (\bar{q}_{*n} \circ (\bar{p}_{*n})^{-1})(\mathcal{O}_{\operatorname{Fix}(p,q)}).$$

Below we shall collect the most important properties of the fixed point index (comp. [Go1-M] or [Go5-M]).

(8.6) Proposition. *Assume that* $(p,q) : U \multimap \mathbb{R}^n$ *is a multivalued map and* $\mathrm{Fix}(p,q)$ *is compact.*

(8.6.1) (*Existence*). *If* $\mathrm{ind}(p,q) \neq 0$, *then* $\mathrm{Fix}(p,q) \neq \emptyset$.

(8.6.2) (*Localization*). *If* V *is an open subset of* $\mathbb{R}^n$ *such that* $\mathrm{Fix}(p,q) \subset V \subset U$, *then*

$$\mathrm{ind}(p,q) = \mathrm{ind}(p_1,q_1),$$

where

$$V \overset{p_1}{\Longleftarrow} p^{-1}(V) \overset{q_1}{\longrightarrow} \mathbb{R}^n, \quad p_1(z) = p(z), q_1(z) = q(z),$$

i.e., $(p_1,q_1) : V \multimap \mathbb{R}^n$ *is the restriction of* (p,q).

(8.6.3) (*Additivity*). *Assume that* $U = U_1 \cup U_2$, *where* U_1, U_2 *are open in* $\mathbb{R}^n$. *Assume, furthermore, that* $(p_1,q_1) : U_1 \multimap \mathbb{R}^n$, $(p_2,q_2) : U_2 \multimap \mathbb{R}^n$ *are respective restrictions of* (p,q), $\mathrm{Fix}(p_1,q_1)$, $\mathrm{Fix}(p_2,q_2)$ *are compact and* $\mathrm{Fix}(p_1,q_1) \cap \mathrm{Fix}(p_2,q_2) = \emptyset$, *then*

$$\mathrm{ind}(p,q) = \mathrm{ind}(p_1,q_1) + \mathrm{ind}(p_2,q_2).$$

(8.6.4) (*Homotopy*). *If* (p_1,q_1), $(p_2,q_2) : U \multimap \mathbb{R}^n$ *are homotopic and the joining homotopy of* (p_1,q_1) *with* (p_2,q_2) *has a compact set of fixed points, then*

$$\mathrm{ind}(p_1,q_1) = \mathrm{ind}(p_2,q_2).$$

(8.6.5) (*Contraction*). *Assume that* $q(p^{-1}(U)) \subset \mathbb{R}^{n-k}$ *and let* $\widetilde{U} = U \cap \mathbb{R}^{n-k}$, $(p_1,q_1) : \widetilde{U} \multimap \mathbb{R}^{n-k}$ *be the respective contraction of* (p,q). *Then:*

$$\mathrm{ind}(p,q) = \mathrm{ind}(p_1,q_1).$$

(8.6.6) (*Multiplicity*). *Let* $U \subset \mathbb{R}^n$, $U' \subset \mathbb{R}^{n'}$ *be open sets and* $(p,q) : U \multimap \mathbb{R}^n$, $(p',q') : U' \multimap \mathbb{R}^{n'}$ *be two maps such that* $\mathrm{Fix}(p,q)$ *and* $\mathrm{Fix}(p',q')$ *are compact sets of fixed points and*

$$\mathrm{ind}(p \times p', q \times q') = \mathrm{ind}(p,q) \cdot \mathrm{ind}(p',q').$$

(8.6.7) (*Normalization*). *Assume that* $(p,q) : U \multimap U$ *is compact. Then*

$$\mathrm{ind}(p,\widehat{q}) = \Lambda(p,q),$$

where $\widetilde{q} : \Gamma \to \mathbb{R}^n$, $\widetilde{q}(z) = q(z)$, *for every* $z \in \Gamma$.

Instead of calculating the fixed points of a map (p, q), we can calculate the points for which zero is in its image. Such points are usually called *equilibrium points*. A good tool to do it is the topological degree. We shall explain below what the topological degree is and what is its connection with the fixed point index.

Let U be an open subset of $\mathbb{R}^n$ and let $(p, q) : \overline{U} \multimap \mathbb{R}^n$ be a map such that:

$$(8.7) \qquad 0 \notin (x - \varphi(p, q))(x), \quad \text{for every } x \in \partial U.$$

Assume, furthermore, that (p, q) is a compact map and the map $\Phi(p, q) : \overline{U} \multimap \mathbb{R}^n$ is defined by the formula:

$$\Phi(p, q)(x) = x - \varphi(p, q)(x), \quad \text{for every } x \in \overline{U}.$$

Then $\Phi(p, q)$ is called the *vector field* associated with $\varphi(p, q)$, and Φ is called a *compact vector field*, provided $\varphi(p, q)$ is compact.

(8.8) DEFINITION. Let $\Phi(p, q) : \overline{U} \multimap \mathbb{R}^n$ be a compact vector field such that:

$$\{x \in \overline{U}, \ 0 \in \Phi(p, q)(x)\} \cap \partial U = \emptyset.$$

We define the *topological degree* $\deg(\Phi(p, q); U)$ of $\Phi(p, q)$ with respect to U by the formula:

$$\deg(\Phi(p, q), U) = \text{ind}(p, q).$$

Let us observe that, under the above assumptions, the set $\text{Fix}(p, q)$ is compact.

Now, the properties of the topological degree analogous to those presented in (8.6) can be formulated as follows.

(8.9) PROPOSITION. *Let $\Phi(p, q) : \overline{U} \multimap \mathbb{R}^n$ be a compact vector field such that:*

$$\{x \in \overline{U} \mid 0 \in \Phi(p, q)(x)\} \cap \partial U = \emptyset.$$

(8.9.1) (*Existence*). *If $\deg(\Phi(p, q); U) \neq 0$, then there is $x \in U$ such that $0 \in \Phi(p, q)(x)$.*

(8.9.2) (*Localization*). *If $\{x \in \overline{U} \mid \Phi(p, q)(x)\} \subset V \subset U$, then $\deg(\Phi(p, q), U) = \deg(\Phi(p, q), V)$.*

(8.9.3) (Additivity) *Let U, U_1, U_2 be open subset of $\mathbb{R}^n$ such that $U_1 \cup U_2 \subset U$, $U_1 \cap U_2 = \emptyset$ and $0 \notin \phi(p, q)(\overline{U} \setminus (U_1 \cup U_2))$, then*

$$\deg(\phi(p, q), U) = \deg(\phi(p, q), U_1) + \deg(\phi(p, q), U_2).$$

(8.9.4) (Homotopy)[4] *If $\phi(p, q) \sim \psi(p', q')$, then*

$$\deg(\varphi(p, q), U) = \deg(\psi(p', q'), U).$$

----

[4]Here, the homotopy is understood as a homotopy of compact vector fields with no zero, for every $x \in \partial U$ and every $t \in [0, 1]$.

(8.9.5) (Multiplicity) *Assume that $U \subset \mathbb{R}^n$ and $V \subset \mathbb{R}^m$ are open and $\phi_1 : \overline{U} \multimap$* $\mathbb{R}^n$, $\phi_2 : \overline{V} \multimap \mathbb{R}^m$ *are admissible compact vector fields with no zero on the boundaries ∂U and ∂V, respectively. Then*

$$\deg(\phi \times \psi, U \times V) = \deg(\phi, U) \cdot \deg(\psi, V).$$

(8.9.6) (Contraction) *Let $\phi(p,q) : \overline{U} \multimap \mathbb{R}^n$ and assume that $\phi(p,q)(\overline{U}) \subset \mathbb{R}^{n-k}$, for some $k = 1, 2, \ldots$ Then*

$$\deg(\phi(p,q), U) = \deg(\phi(p,q)|_{U \cap \mathbb{R}^{n-k}}, U \cap \mathbb{R}^{n-k}).$$

Now, we are going to define the topological degree and the fixed point index for admissible compact vector fields (admissible mappings) in locally convex spaces. We shall start by the topological degree. As above, (p,q) is a compact multivalued map defined on $\overline{U}$ with values in E, where U is an open subset of a locally convex space E. We shall put:

$$\phi = \phi(p,q) : \overline{U} \multimap E$$

to be a compact vector field defined as follows:

$$\phi(x) = x - q(p^{-1}(x)) = \{x - y \mid y \in q(p^{-1}(x))\}.$$

We shall also assume that:

(8.10) $0 \notin \phi(x)$,   for every $x \in \partial U$.

We prove the following

(8.11) PROPOSITION. *If $\phi : \overline{U} \multimap E$ is a compact vector field, then ϕ is a closed map, i.e., for every closed $A \subset \overline{U}$, the set $\phi(A)$ is closed in E.*

PROOF. Assume that A is a closed subset of $\overline{U}$ and $y \in \overline{\phi(A)}$. It is enough to show that $y \in \phi(A)$. Since $\phi = \phi(p,q)$ and q is compact, there is a compact set $K \subset E$ such that $qp^{-1}(\overline{U}) \subset K$. Since $y \in \overline{\phi(A)}$, it follows that every open neighbourhood V of the zero point in E, we have:

$$(y + V) \cap \phi(A) \neq \emptyset.$$

Let $y_V \in (y + V) \cap \phi(A)$ and let $x_V \in A$ be such that $y_V \in (x_V - q(p^{-1}(x_V)))$. Let us consider the generalized sequence $\{x_V - y_V\}_V$, where V is an neighbourhood of the zero point in E. Since $\{x_V - y_V\} \subset K$, we can assume that it converges to a point $x \in K$. Now, it is easy to observe that $\{x_V\}$ converges to $x + y$. Since A

is closed, we get $(x + y) \in A$ and the u.s.c. of ϕ implies that the graph Γ_ϕ of ϕ is closed. Consequently $y \in \phi(x + y)$, and so $y \in \phi(A)$. The proof is completed. $\square$

We let

$$\mathcal{A}(U, \partial U) = \{\phi = \phi(p, q) \mid \phi : \overline{U} \to E, \ \phi \text{ satisfies } (8.10)\}.$$

We would like to define a function:

$$\text{Deg} : \mathcal{A}(U, \partial U) \to \mathbb{Z},$$

which satisfies the analogous properties to (8.9.1)–(8.9.6), where $\mathbb{Z}$ denotes the set of integers.

Let $\phi = \phi(p, q) \in \mathcal{A}(U, \partial U)$. Since $\phi(\partial U)$ is a closed subset of E and $0 \notin \phi(\partial U)$, there exists an open neighbourhood V_0 of the zero point in E such that $V_0 \cap \phi(\partial U) = \emptyset$. Let $V = V_0/2$ and $K = \overline{q(p^{-1}(\overline{U}))}$. Since K is compact, there exists a finite set $\{y_1, \ldots, y_k\} \subset K$ such that $K \subset \bigcup_{i=1}^{k}\{y_i + V\}$. We let $\alpha = \{y_i + V\}_{i=1,\ldots,k}$. Now, let β be a refinement of α such that any two β-close mappings are α-stationary homotopic. By using the approximation Theorem (1.37), we get a β-approximation $\pi_\beta : K \to E$ of the inclusion $i : K \to E$ such that $\pi_\beta(K) \subset E^{n=n(\beta)} \subset E$. We let $q_\beta = \pi_\beta \circ q$ and let $\phi_\beta = \phi(p, q_\beta) : \overline{U} \to E$.

By a standard calculation (cf. [GGa-M]) it is easy to see that:

$$(8.12) \qquad\qquad \phi_\beta \in \mathcal{A}(U, \partial U).$$

Now, we let $\phi_{\beta, n(\beta)}$ to be the contraction of ϕ_β to the pair $(\overline{U} \cap E^{n(\beta)}, E^{n(\beta)})$. Then we have

$$(8.13) \qquad\qquad \phi_{\beta, n(\beta)} \in (U \cap E^{n(\beta)}, E^n(\beta)).$$

Consequently, the topological degree $\deg(\phi_{\beta, n(\beta)}, U \cap E^{n(\beta)})$ is well-defined.

(8.14) DEFINITION.

$$\text{Deg}(\phi, U) = \deg(\phi_{\beta, n(\beta)}, U \cap E^{n(\beta)}).$$

Using the properties (8.9.1)–(8.9.6), it is easy to see that Definition (8.14) is correct (for details see [GGa-M]).

We would like to add that properties (8.9.1)–(8.9.6) can be reformulated for the topological degree Deg defined above. We left the respective formulations to the reader.

Now, we shall show that the topological degree $\mathrm{Deg} : \mathcal{A}(U, \partial U) \to \mathbb{Z}$ will help us to define the fixed point index for compact admissible mappings in locally convex spaces.

Let $\varphi \in \mathbb{K}(U, E)$, where U is an open subset in a locally convex space E. Moreover, we shall assume:

(8.15) $\mathrm{Fix}(\varphi)$ is a compact subset of U.

We put

$$\mathbb{K}_0(U, E) = \{\varphi \in \mathbb{K}(U) \mid (8.15) \text{ is satisfied}\}.$$

At first, we are going to define the fixed point index ind on $\mathbb{K}_0(U, E)$ as a function:

$$\mathrm{ind} : \mathbb{K}_0(U, E) \to \mathbb{Z}.$$

To get $\mathrm{ind}(\varphi)$ as a singleton, we will assume that $(p, q) \subset \varphi$ is fixed, i.e., $\varphi = \varphi(p, q) = q \circ p^{-1}$.

Let $\varphi = \varphi(p, q) \in \mathbb{K}_0(U, E)$. Then there is an open subset U_0 of E such that:

$$\mathrm{Fix}(\varphi) \subset U_0 \subset \overline{U}_0 \subset U.$$

Observe that then we have

$$\phi = \phi(p, q) \in \mathcal{A}(U_0, \partial U_0).$$

Consequently, we let:

(8.16) $\mathrm{ind}(\varphi) = \mathrm{Deg}(\phi, U_0).$

The additivity of the topological degree Deg implies that (8.16) does not depend on the choice of U_0.

Assume that X is a retract of an open $U \subset E$ and $r : U \to X$ is a retraction map. Assume, furthermore, that: $\varphi = \varphi(p, q) : V \multimap X$ is a compact map with $\mathrm{Fix}(\varphi)$ being compact and V to be an open subset of X. Then we have the diagram

$$r^{-1}(V) \xrightarrow{r} V \xrightarrow{\varphi} X \xrightarrow{i} E.$$

It is easy to see that

$$\varphi \circ r \in \mathbb{K}_0(r^{-1}(V), E).$$

We define

(8.17) $\mathrm{ind}(\varphi) = \mathrm{ind}(\varphi \circ r).$

Note that the above definition depends on the choice of the retraction map.

Nevertheless, we have:

(8.18) THEOREM. *Assume that X is a neighbouhood retract of E and $r : U \to X$ is a given retraction map.*

(8.18.1) (Additivity) *Let V_1, V_2, V be open subsets of X such that $V_1 \cap V_2 = \emptyset$, $V_1 \cup V_2 \subset V$ and let $\varphi \in \mathbb{K}_0(V, X)$, then*

$$\mathrm{ind}(\varphi) = \mathrm{ind}(\varphi_1) + \mathrm{ind}(\varphi_2),$$

where φ_i, $i = 1, 2$ are restrictions of φ to V_1 and V_2, respectively.

(8.18.2) (Existence) *If $\mathrm{ind}(\varphi) \neq 0$, then $\mathrm{Fix}(\varphi) \neq \emptyset$.*

(8.18.3) (Homotopy) *If $\varphi(p, q) \sim \varphi(p_1, q_1)$ with the homotopy having the compact fixed point set, then*

$$\mathrm{ind}(\varphi(p, q)) = \mathrm{ind}(\varphi(p_1, q_1)).$$

(8.18.4) (Multiplicity) *Assume that $V_i \subset E_i$, $i = 1, 2$ are neighbourhood retracts of E_1 and E_2, respectively, and r_i, $i = 1, 2$ are fixed retractions. Assume $\varphi_i \in \mathbb{K}(V_i, E_i)$, $i = 1, 2$ and $\varphi = \varphi_i \times \varphi_2$. Then*

$$\mathrm{ind}(\varphi) = \mathrm{ind}(\varphi_1) \cdot \mathrm{ind}(\varphi_2).$$

(8.18.5) (Normalization) *If $\varphi = \varphi(p, q) \in \mathbb{K}_0(X, X)$, then*

$$\mathrm{ind}(\varphi) = \Lambda(p, q).$$

The proof of (8.18) is straighforward and follows from the respective properties of the topological degree Deg.

I.9. Noncompact case

In this section, we shall generalize the fixed point index defined in the last section to the classes of compact absorbing contractions and condensing mappings.

At first, we assume that X is a neighbourhood retract of a locally convex space E and

$$\varphi = \varphi(p, q) \in \mathbb{CAC}(X).$$

For given open $V \subset X$, we let

$$\mathbb{CAC}_V(X) = \{\varphi = \varphi(p, q) \in \mathbb{CAC}(X) \mid \mathrm{Fix}(\varphi) \cap \partial V = \emptyset\},$$

where ∂V denotes the boundary of V in X.

Let $\varphi = \varphi(p, q) \in \mathbb{CAC}_V(X)$. We choose $W \subset X$, according to the Definition (5.6). Evidently, $\mathrm{Fix}(\varphi) \subset W$, and obviously W is a neighbourhood retract of E. We let $V_1 = W \cap V$. Then $\varphi(V_1) \subset W$ and $\widetilde{\varphi}(p,q) : V_1 \to W$ is a compact map with a compact set of fixed points, where $\widetilde{\varphi}(p,q)$ is the respective restriction of $\varphi(p,q)$.

We define:

$$(9.1) \qquad \mathrm{ind}(\varphi(p,q), V) = \mathrm{ind}(\widetilde{\varphi}(p,q)).$$

The correctness of the above definition follows from the additivity property of the fixed point index for compact admissible mappings.

Let us note that Theorem (8.18) can be also formulated for $\mathbb{C}AC$-mappings. We left it to the reader. Below, we shall prove the normalization property, only.

(9.2) THEOREM (Normalization Property). *Let* $\varphi = \varphi(p,q) \in \mathbb{C}AC(X)$, *where* X *is a neighbourhood retract of a locally convex space* E. *Then*

$$\mathrm{ind}(\varphi(p,q), X) = \Lambda((p,q)).$$

PROOF. Let $W \subset X$ be chosen according to Definition (5.6). In view of (9.1), we have:

$$\mathrm{ind}(\varphi(p,q), X) = \mathrm{ind}(\widetilde{\varphi}(p,q)),$$

where $\widetilde{\varphi}(p,q) : W \multimap W$ is a compact map. Moreover, in view of (6.20), $\Lambda(\widetilde{\varphi}(p,q))$ is well-defined and by means of (6.11), we deduce

$$\Lambda(\varphi(p,q)) = \Lambda(\widetilde{\varphi}(p,q)).$$

Then from the normalization property of the index, for compact maps, we get:

$$\mathrm{ind}(\widetilde{\varphi}(p,q)) = \Lambda(\widetilde{\varphi}(p,q)),$$

and so:

$$\mathrm{ind}(\varphi(p,q), X) = \Lambda(\varphi(p,q)) = \Lambda(p,q),$$

which completes the proof. $\qquad\qquad\square$

I.10. Nielsen number

As in the last two sections, by a multivalued map we shall understand an admissible map represented by a fixed pair (p,q) of the form

$$X \xleftarrow{p} \Gamma \xrightarrow{q} Y.$$

For the clarity of our explanation, we shall present below some necessary well-known notions.

At first, we give an example which demonstrates that the multivalued setting is not a direct extension of the single-valued one.

(10.1) EXAMPLE. Let us consider the unit circle $S^1 = \mathbb{R}/\mathbb{Z}$ with the correspondence

$$\mathbb{R}/\mathbb{Z} \ni [t] \leftrightarrow e^{2\pi t i} \in S^1,$$

define a family of maps $p_\varepsilon : S^1 \to S^1$, $0 < \varepsilon \leq 1/2$,

$$p_\varepsilon[t] = \begin{cases} [t/2\varepsilon], & \text{for } 0 \leq t \leq \varepsilon, \\ [1/2], & \text{for } \varepsilon \leq t \leq 1-\varepsilon, \\ [(1-t)/2\varepsilon + 1], & \text{for } 1-\varepsilon \leq t \leq 1, \end{cases}$$

and put $q[t] = [kt]$, for a fixed $k \in \mathbb{Z}$.

Let us note that $p_\varepsilon^{-1}[y]$ is one point, for $[y] \neq [1/2]$, while $p_\varepsilon^{-1}[1/2] = \{[t] \mid \varepsilon \leq t \leq 1-\varepsilon\}$ is an arc. Thus, any counterimage is contractible. Let us fix a number ε_0 satisfying $0 < \varepsilon_0 \leq 1/2k$. Then $\{(p_\varepsilon, q)\}$, where ε runs through the interval $[\varepsilon_0, 1/2]$, is a homotopy between the multivalued maps $(p_{1/2}, q)$ and (p_ε, q). But since we can observe that $p_{1/2} = \mathrm{id}_{S^1}$, $(p_{1/2}, q)$ corresponds to the single-valued map $q[t] = [kt]$. It is known (see e.g. [KTs-M]) that the Nielsen number $N(q) = |k-1|$, by which we know that any single-valued map homotopic to q has at least $|k-1|$ fixed points. On the other hand, we will show that $[x] \in qp_{\varepsilon_0}^{-1}[x]$ only for $[x] = [0]$ or $[x] = [1/2]$. (In fact, for $0 < x < 1/2$, $p_\varepsilon^{-1}[x] = [2\varepsilon_0 x]$, so $qp_{\varepsilon_0}^{-1}[x] = [2k\varepsilon_0 x]$, but the assumption $0 < 2k\varepsilon_0 < 1$ implies $0 < 2k\varepsilon_0 x < x < 1/2$ which gives $[2k\varepsilon_0 x] \neq [x]$). If $1/2 < x < 1$, then

$$p_{\varepsilon_0}^{-1}[x] = [1 - 2\varepsilon_0(1-x)],$$

so

$$qp_{\varepsilon_0}^{-1}[x] = [k - 2k\varepsilon_0(1-x)] = [1 - 2k\varepsilon_0(1-x)].$$

It follows from the assumption $0 < 2k\varepsilon_0 < 1$ that $0 < 1 - 2k\varepsilon_0(1-x) < 1$, and subsequently $[x] = [1 - 2k\varepsilon_0(1-x)]$ if and only if $x = 1 - 2k\varepsilon_0(1-x)$. The last equality, however, yields $2k\varepsilon_0(1-x) = 1-x$, i.e., $2k\varepsilon_0 = 1$, after dividing by $x - 1 \neq 0$, which contradicts the assumption $0 < 2k\varepsilon_0 < 1$.

Thus, a multivalued homotopy of a single-valued map with the Nielsen number $N(q) = |k-1|$ gives the fixed point set

$$\mathrm{Fix}(p, q) = \{x \in X \mid x \in qp^{-1}(x)\}$$

consisting of only two elements!

The above example seems to suggest that the Nielsen fixed point theory fails to be extended to the multivalued case. On the other hand, we can observe that, in this example, the restriction $q : p_{\varepsilon_0}^{-1}[1/2] \to S^1$ covers the point $[1/2]$ k-times. This observation encourages us to estimate the *number of coincidences*

$C(p, q) = \{z \in \Gamma \mid p(z) = q(z)\}$ of the pair (p, q) rather than the *number of fixed points* $\text{Fix}(qp^{-1})$ of the map qp^{-1}.

Let $X \xleftarrow{p_0} \Gamma \xrightarrow{q_0} Y$ and $X \xleftarrow{p_1} \Gamma \xrightarrow{q_1} Y$ be two maps. We say that (p_0, q_0) is *homotopic* to (p_1, q_1) (written $(p_0, q_0) \sim (p_1, q_1)$) if there exists a multivalued map $X \times I \xleftarrow{p} \overline{\Gamma} \xrightarrow{q} Y$ such that the following diagram is commutative:

$$
\begin{array}{ccc}
X & \xleftarrow{p_i} \Gamma \xrightarrow{q_i} & Y \\
{\scriptstyle k_i}\downarrow & {\scriptstyle f_i}\downarrow \ \nearrow {\scriptstyle q} & \\
X \times I & \xleftarrow{p} \overline{\Gamma} &
\end{array}
$$

for $k_i(x) = (x, i)$, $i = 0, 1$, and for some $f_i : \Gamma \to \overline{\Gamma}$, $i = 0, 1$, i.e., $k_0 p_0 = p f_0$, $q_0 = q f_0$, $k_1 p_1 = p f_1$ and $q_1 = q f_1$.

If $(p_0, q_0) \sim (p_1, q_1)$ and $h : Y \to Z$ is a continuous map, then we write $(p_0, hq_0) \sim (p, hq)$. We say that a multivalued map $X \xleftarrow{p} \Gamma \xrightarrow{q} Y$ represents a *single-valued map* $\rho : X \to Y$ if $q = p\rho$. Now, we assume that $X = Y$ and we are going to estimate the cardinality of the coincidence set $C(p, q)$. We begin by defining a Nielsen-type relation on $C(p, q)$. This definition requires the following conditions on $X \xleftarrow{p} \Gamma \xrightarrow{q} Y$:

(10.2) X, Y are paracompact, connected, locally contractible topological spaces (observe that then they admit universal coverings),

(10.3) $p : \Gamma \Longrightarrow X$ is a Vietoris map,

(10.4) for any $x \in X$, the restriction $q_1 = q|_{p^{-1}(x)} : p^{-1}(x) \to Y$ admits a lift $\widetilde{q_1}$ to the universal covering space $(p_Y : \widetilde{Y} \to Y)$:

$$
\begin{array}{ccc}
 & & \widetilde{Y} \\
 & {\scriptstyle \widetilde{q_1}}\nearrow & \Vert {\scriptstyle p_Y} \\
p^{-1}(x) & \xrightarrow{q_1} & Y
\end{array}
$$

Consider a single-valued map $\rho : X \to Y$ between two spaces admitting universal coverings $p_X : \widetilde{X} \Longrightarrow X$ and $p_Y : \widetilde{X} \Longrightarrow Y$. Let $\theta_X = \{\alpha : \widetilde{X} \to \widetilde{X} \mid p_X \alpha = p_X\}$ be the group of natural transformations of the covering p_X. Then the map ρ admits a lift $\widetilde{\rho} : \widetilde{X} \to \widetilde{Y}$. We can define a homomorphism $\widetilde{\rho}_! : \theta_X \to \theta_Y$ by the equality

$$\widetilde{q}(\alpha \cdot \widetilde{x}) = \widetilde{q}_!(\alpha)\widetilde{q}(\widetilde{x}) \quad (\alpha \in \theta_X, \ \widetilde{x} \in \widetilde{X}).$$

It is well-known (see for example [Sp-M]) that there is an isomorphism between the fundamental group $\pi_1(X)$ and θ_X which may be described as follows. We fix points $x_0 \in X$, $\widetilde{x} \in \widetilde{X}$ and a loop $\omega : I \to X$ based at x_0. Let $\widetilde{\omega}$ denote the unique lift of ω starting from $\tilde{x}_0$. We subordinate to $[\omega] \in \pi_1(X, x_0)$ the unique transformation from θ_X sending $\tilde{\omega}(0)$ to $\tilde{\omega}(1)$. Then the homomorphism $\tilde{\rho}_! : \theta_X \to \theta_Y$ corresponds to the induced homomorphism between the fundamental groups $\rho_\# : \pi_1(X, x_0) \to \pi_1(Y, \rho(x_0))$.

We will show that, under the assumptions (10.2)–(10.4), a multivalued map (p, q) admits a lift to a multivalued map between the universal coverings. These lifts will split the coincidence set $C(p, q)$ into Nielsen classes. Besides that, we will also show that the pair (p, q) induces a homomorphism $\theta_X \to \theta_Y$ giving the Reidemeister set in this situation.

We start with the following lemma.

(10.5) LEMMA. *Suppose we are given Y, a paracompact locally contractible space, Γ a topological space, $\Gamma_0 \subset \Gamma$ a compact subspace, $q : \Gamma \to Y$, $\tilde{q}_0 : \Gamma_0 \to Y$ continuous maps for which the diagram*

$$
\begin{array}{ccc}
\Gamma_0 & \xrightarrow{\tilde{q}_0} & \tilde{Y} \\
\downarrow{\scriptstyle i} & & \downarrow{\scriptstyle p_Y} \\
\Gamma & \xrightarrow{q} & Y
\end{array}
$$

commutes (here, $p_Y : \tilde{Y} \to Y$ denotes the universal covering). In other words, $\tilde{q}_0$ is a partial lift of q. Then $\tilde{q}_0$ admits an extension to a lift onto an open neighbourhood of Γ_0 in Γ.

PROOF. Let us fix a covering $\{W_i\}$ of the space Y consisting of open connected sets satisfying: if $\mathrm{cl}\, W_i \cap \mathrm{cl}\, E_j \neq \emptyset$, then $\mathrm{cl}\, W_i \cup \mathrm{cl}\, W_j$ is contained in a contractible subset of Y.

Let $\{\widetilde{W}_j\}$ denote the covering consisting of connected components of the covering $\{p_Y^{-1} W_i\}$. We notice that the restriction of p_Y to any of sets $\widetilde{W}_j$ is a homeomorphism.

Let $\{U_i\}$ be a finite covering of Γ_0 such that $\{\mathrm{cl}\, U_i\}$ is subcovering of $\{\tilde{q}_0^{-1} \widetilde{W}_i\}$. For any U_i, we fix an open subset $V_i \subset \Gamma$ satisfying $V_i \cap \Gamma_0 = U_i$.

We can assume that $\mathrm{cl}\, V_i$ is disjoint with $\bigcup \{\mathrm{cl}\, U_j \mid \mathrm{cl}\, U_j \cap \mathrm{cl}\, U_i = \emptyset\}$ (notice that the sets $\mathrm{cl}\, U_i$ and $F_i = \bigcup \{\mathrm{cl}\, U_j \mid \mathrm{cl}\, U_j \cap \mathrm{cl}\, U_i = \emptyset\}$ are disjoint and closed. Hence, there exists an open subset $S \subset \Gamma$ satisfying $\mathrm{cl}\, U_i \subset S \subset \mathrm{cl} S \subset \Gamma - F_i$, and so we can put $V_i := V_i \cap S$).

Let $V_i' = V_i - \bigcup \{\mathrm{cl}\, U_j \mid \mathrm{cl}\, U_j \cap \mathrm{cl}\, U_i = \emptyset\}$. Then

(1) $V_i' = V_i \cap \Gamma_0 = U_i$,

(2) if $V_i' \cap V_i \neq \emptyset$, then $\mathrm{cl}\, U_i \cap \mathrm{cl}\, U_j \neq \emptyset$.

For any V_i', we fix $\widetilde{W}_{\alpha(i)}$ satisfying $\tilde{q}_0(\mathrm{cl}\, U_i) \subset \widetilde{W}_{\alpha(i)}$ and we put $V_i'' = V_i' \cap q^{-1}(\widetilde{W}_{\alpha(i)})$. The covering $\{V_i'\}$ also satisfies the above conditions (1) and (2). For any i, we denote by $\varphi_i : \widetilde{W}_{\alpha(i)} \to \widetilde{W}_\alpha$ the homomorphism inverse to the projection p_Y.

Now, we can define an extension of the lift $\widetilde{q}_0$ onto the neighbourhood $\bigcup V_i'$. We define the map $\widetilde{q}_i : V_i'' \to \widetilde{Y}$ by the formula $\widetilde{q}_i = \varphi q(x) \in \widetilde{W}_{\alpha(i)} \subset \widetilde{Y}$.

It remains to show that the maps $\widetilde{q}_i$ and $\widetilde{q}_j$ are consistent. Let $V_i'' \cap V_j'' \neq \emptyset$. Then there is a point $x \in \operatorname{cl} U_i \cap \operatorname{cl} U_j$ which implies $\widetilde{q}_0(x) \in \widetilde{W}_{\alpha(i)} \cap \widetilde{W}_{\alpha(j)}$. Let $S \subset Y$ be a contractible set containing $\pi_Y \widetilde{W}_{\alpha(i)} \cup \widetilde{W}_{\alpha(j)}$ and let $\widetilde{S}$ be the component of $p_Y^{-1}(S)$ containing $\widetilde{Q}_0(x)$. Then $\widetilde{W}_{\alpha(i)} \cup \widetilde{W}_{\alpha(j)} \subset \widetilde{S}$, and so the values of the sections $\widetilde{q}_i$, $\widetilde{q}_j$ are contained in $\widetilde{S}$ which implies that they must be consistent. $\quad\square$

(10.6) LEMMA. *Suppose we are given a multivalued map $X \xleftarrow{p} \Gamma \xrightarrow{q} Y$ satisfying (10.2), where X is simply-connected. Then there exists a map $\widetilde{q} : \Gamma \to \widetilde{Y}$ making the diagram*

$$
\begin{array}{ccc}
 & & \widetilde{Y} \\
 & \overset{\widetilde{q}}{\nearrow} & \big\uparrow {\scriptstyle p_X} \\
\Gamma & \xrightarrow{\quad q \quad} & Y
\end{array}
$$

commutative.

PROOF. (a) Let $X = [0, 1]$. Then Γ is compact. Let us fix $t_0 \in [0, 1]$. By Lemma (10.5), there exists an open set U, $\Gamma_{t_0} = p^{-1}(t_0) \subset U \subset \Gamma$ and a lift $\widetilde{q} : U \to \widetilde{V}$. Since Γ is compact, there exists $\varepsilon \geq 0$ satisfying $p^{-1}[t_0 - \varepsilon, t_0 + \varepsilon] \subset U$. Thus, for any to $t_0 \in [0, 1]$, any lift $\widetilde{q} : \Gamma_{t_0} \to \widetilde{Y}$ extends onto $p^{-1}[t_0 - \varepsilon, t_0 + \varepsilon]$, for an $\varepsilon > 0$. Now, if we fix a sufficiently fine division $0 = t_0 < t_1 < \ldots < t_n = 1$, then we can extend any lift from Γ_0 onto $p^{-1}[t_0, t_1]$, and subsequently onto the whole $\Gamma = p^{-1}[0, 1]$.

(b) $X = [0, 1]$. The proof is similar.

(c) X is an arbitrary simply-connected space satisfying (10.2). Fix a point $x_0 \in X$ and a lift $\widetilde{q}_{x_0} : \Gamma_{x_0} \to \widetilde{Y}$. Let $x_1 \in X$ be another point. We choose a path $\omega : I \to X$ satisfying $\omega(i) = \chi_i$, for $i = 0, 1$. Now, we can apply (a) to the induced fibering $\omega^* = \{(z, t) \in \Gamma \times I \mid p(z) = \omega(t)\}$ and the obtained lift $\omega : \Gamma \to \widetilde{Y}$ defines, for $t = 1$, a lift on $\Gamma_{x_1} = \omega_{x_1}^* = \{(z, t) \in \omega^* \mid t = 1\}$. If we take another path ω' from x_0 to x_1, then there is a homotopy $H : I \times I \to Y$ joining these two paths, because Y is simply-connected. Now, (b) shows that both obtained lifts coincide. $\quad\square$

Consider again a multivalued map $X \xleftarrow{p} \Gamma \xrightarrow{q} Y$ satisfying (10.2). Define

$$
\widetilde{\Gamma} = \{(\widetilde{x}, z) \in \widetilde{X} \times \Gamma \mid p_X(\widetilde{x}) = p(z)\}
$$

(a pullback). This gives the diagram

$$
\begin{array}{ccccc}
\widetilde{X} & \xleftarrow{\ \widetilde{p}\ } & \widetilde{\Gamma} & \xrightarrow{\ \widetilde{q}\ } & \widetilde{Y} \\
{\scriptstyle p_X}\downarrow & & {\scriptstyle p_\Gamma}\downarrow & & \downarrow{\scriptstyle p_Y} \\
X & \xleftarrow{\ p\ } & \Gamma & \xrightarrow{\ q\ } & Y
\end{array}
$$

where $\widetilde{p}(\widetilde{x}, z) = \widetilde{x}$ and $p_\Gamma(\widetilde{x}, z) = z$. Notice that the restrictions of $\widetilde{p}$ are homeomorphic on fibres.

Now, we can apply Lemma (10.5) to the multivalued map $\widetilde{X}\xLeftarrow{\widetilde{p}}\widetilde{\Gamma}\xrightarrow{qp_\Gamma}Y$, and so we get a lift $\widetilde{q} : \widetilde{\Gamma} \to \widetilde{Y}$ such that the diagram

$$
\begin{array}{ccccc}
\widetilde{X} & \xLeftarrow{\ \widetilde{p}\ } & \widetilde{\Gamma} & \xrightarrow{\ \widetilde{q}\ } & \widetilde{Y} \\
{\scriptstyle p_X}\downarrow & & {\scriptstyle p_\Gamma}\downarrow & & \downarrow{\scriptstyle p_Y} \\
X & \xLeftarrow{\ p\ } & \Gamma & \xrightarrow{\ q\ } & Y
\end{array}
$$

is commutative. Let us note that the lift $\widetilde{p}$ is given by the above formula, but $\widetilde{q}$ is not precised. We fix such a $\widetilde{q}$.

Observe that $p : \Gamma \Longrightarrow X$ and the lift $\widetilde{p}$ induce a homomorphism $\widetilde{p}^! : \theta_X \to \theta_\Gamma$ by the formula $\widetilde{p}^!(\alpha)(\widetilde{x}, z) = (\alpha\widetilde{x}, z)$. It is easy to check that the homomorphism $\widetilde{p}^!$ is an isomorphism (any natural transformation of $\widetilde{\Gamma}$ is of the form $\alpha \cdot (\widetilde{x}, z) = (\alpha\widetilde{x}, z)$) and that $\widetilde{p}^!$ is inverse to $\widetilde{p}_!$. Recall that the lift $\widetilde{q}$ defines a homomorphism $\widetilde{q}_! : \theta_\Gamma \to \theta_Y$ by the equality $\widetilde{q}(\lambda) = \widetilde{q}_!(\lambda)\widetilde{q}$.

In the sequel, we will consider the composition $\widetilde{q}_!\widetilde{p}^! : \theta_X \to \theta_Y$.

(10.7) LEMMA. *Let a multivalued map (p, q) satisfying (10.2) represent a single-valued map ρ, i.e. $q = \rho p$. Let $\widetilde{\rho}$ be the lift of ρ which satisfies $\widetilde{q} = \widetilde{\rho}\widetilde{p}$. Then $\widetilde{\rho}_!\widetilde{p}^! = \rho_!$.*

PROOF. $\widetilde{\rho}_!\widetilde{p}^! = (\widetilde{\rho p})_!\widetilde{p}^! = \widetilde{\rho}_!\widetilde{p}_!\widetilde{p}^! = \rho_!$.   □

Now, we are in a position to define the Nielsen classes. Consider a multivalued self-map $X\xLeftarrow{p}\Gamma\xrightarrow{q}X$ satisfying (10.2). By the above consideration, we have a commutative diagram

$$
\begin{array}{ccc}
\widetilde{\Gamma} & \xrightarrow{\ \widetilde{p},\widetilde{q}\ } & \widetilde{X} \\
{\scriptstyle p_\Gamma}\downarrow & & \downarrow{\scriptstyle p_X} \\
\Gamma & \xrightarrow{\ p,q\ } & X
\end{array}
$$

Following the single-valued case (see [BJ-M]), we can prove

(10.8) LEMMA.

(i) $C(p,q) = \bigcup_{\alpha \in \theta_X} p_\Gamma C(\widetilde{p}, \alpha\widetilde{q})$,

(ii) if $p_\Gamma C(\widetilde{p}, \alpha\widetilde{q}) \cap p_\Gamma C(\widetilde{p}, \beta\widetilde{q})$ is not empty, then there exists a $\gamma \in \theta_X$ such that $\beta = \gamma \circ \alpha \circ (\widetilde{q}_! \widetilde{p}^! \gamma)^{-1}$,

(iii) the sets $p_\Gamma C(\widetilde{p}, \alpha\widetilde{q})$ are either disjoint or equal.

PROOF. (i) Let $p(z) = q(z)$ and $\widetilde{z} \in p_\Gamma^{-1}(z)$. Then $\widetilde{p}(\widetilde{z})$, $\widetilde{q}(\widetilde{(z)}) \in p_X^{-1}(p(z))$. Thus, there exists $\alpha \in \theta_X$ such that $\widetilde{p}(\widetilde{z}) = \alpha\widetilde{q}(\widetilde{z})$, which implies $\widetilde{z} \in C(\widetilde{p}, \alpha\widetilde{q})$.

(ii) Let $z \in p_\Gamma C(\widetilde{p}, \alpha\widetilde{q}) \cap p_\Gamma C(\widetilde{p}, \beta\widetilde{q})$. Then there exist, $\widetilde{x}$, $\widetilde{x}' \in \widetilde{X}$ such that $(\widetilde{x}, z) \in C(\widetilde{x}, \alpha\widetilde{q})$, $(\widetilde{x}', z) \in C(\widetilde{x}, \beta\widetilde{q})$ and $\widetilde{x} = \alpha\widetilde{q}(\widetilde{x}, z)$, $\widetilde{x}' = \beta\widetilde{q}(\widetilde{x}', z)$. On the other hand, $\widetilde{p}_X \widetilde{x}' = pz$ implies $\widetilde{x}' = \gamma\widetilde{x}$, for a $\gamma \in \theta_X$. Thus,

$$\gamma\widetilde{x} = \widetilde{x}' = \beta\widetilde{q}(\widetilde{x}', z) = \beta\widetilde{q}(\gamma\widetilde{x}, z) = \beta(\widetilde{q}_! \widetilde{p}^! \gamma)\widetilde{q}(\widetilde{x}, z) = \beta(\widetilde{q}_! p^! \gamma)\alpha^{-1}(\widetilde{x}),$$

which implies $\gamma = \beta(\widetilde{q}_! p^! \gamma)\alpha^{-1}$ and $\beta = \gamma\alpha(\widetilde{q}_! p^! \gamma)^{-1}$.

(iii) It remains to prove that

$$p_\Gamma C(\widetilde{p}, \alpha\widetilde{q}) = p_\Gamma C(\widetilde{p}, \gamma\alpha(\widetilde{q}_! \widetilde{p}^! \gamma)^{-1}\widetilde{q}).$$

Let $(\widetilde{x}, z) \in C(\widetilde{p}, \gamma\alpha(\widetilde{q}_! p^! \gamma)^{-1}\widetilde{q})$. Then $\widetilde{x} = \gamma\alpha\widetilde{q}_! (\widetilde{p}^! \gamma)^{-1})\widetilde{q}(\widetilde{x}, z)$, $\widetilde{x} = \gamma\alpha\widetilde{q}(\gamma^{-1}\widetilde{x}, z)$. Hence, $\gamma^{-1}\widetilde{x} = \alpha\widetilde{q}(\gamma^{-1}\widetilde{x}, z)$ and $\widetilde{p}(\gamma^{-1}\widetilde{x}, z) = \alpha\widetilde{q}(\gamma^{-1}\widetilde{x}, z)$. Thus, $p_!(\gamma^{-1}) \cdot (\widetilde{x}, z) = (\gamma^{-1}\widetilde{x}, z) \in C(\widetilde{p}, \alpha\widetilde{q})$, which implies $\widetilde{p}_\Gamma(\widetilde{x}, z) = \widetilde{p}_\Gamma(p^!(\gamma)(\widetilde{x}, z)) \in p_\Gamma C(\widetilde{x}, \alpha\widetilde{q})$. $\square$

Define an action of θ_X on itself by the formula

$$\gamma \circ \alpha = \gamma\alpha(\widetilde{q}_! p^! \gamma).$$

The quotient set will be called the *set of Reidemeister classes* and will be denoted by $R(p,q)$. The above lemma defines an injection

$$\text{Set of Nielsen classes} \to R(p,q),$$

given by $A \to [\alpha] \in R(p,q)$, where $\alpha \in \theta_X$ satisfies $A = p_\Gamma(C(\widetilde{p}, \alpha\widetilde{q}))$.

Now, we are going to prove that our definition does not depend on $\widetilde{q}$.

Let us recall that the homomorphism $\widetilde{q}_! : \theta_\Gamma \to \theta_Y$ is defined by the relation $\widetilde{q}\alpha = \widetilde{q}_!(\alpha)\widetilde{q}$, for $\alpha \in \theta_\Gamma$. If $\widetilde{q}' = \gamma\widetilde{q}$ is another lift of q ($\widetilde{\gamma} \in \theta_\Gamma$), then the induced homomorphism $\widetilde{q}'_! : \theta_\Gamma \to \theta_Y$ is defined by the relation $\widetilde{q}'\alpha = \widetilde{q}'_!(\alpha)\widetilde{q}'$.

(10.9) LEMMA. If $\widetilde{q}' = \gamma \cdot \widetilde{q}$ is another lift of q, then $\gamma \cdot \widetilde{q}_!(\alpha) \cdot \gamma^{-1} = \widetilde{q}'_!(\alpha)$, for all $\alpha \in \theta_\Gamma$.

PROOF. The equalities $\widetilde{q}' = \gamma \cdot \widetilde{q}$ and $\widetilde{q}'(\alpha\widetilde{u}) = \widetilde{q}'_!(\alpha)\widetilde{q}'(\widetilde{u})$ imply $\gamma \cdot \widetilde{q}(\alpha\widetilde{u}) = \widetilde{q}'_!(\alpha) \cdot \gamma \cdot \widetilde{q}(\widetilde{u})$, by which $\gamma \cdot \widetilde{q}_!(\alpha) \cdot \widetilde{q}(\widetilde{u}) = \widetilde{q}'_!(\alpha) \cdot \gamma \cdot \widetilde{q}'(\widetilde{u})$. Thus, $\gamma \cdot \widetilde{q}_!(\alpha) = \widetilde{q}'_!(\alpha) \cdot \gamma$ and finally $\widetilde{q}'_!(\alpha) \cdot \widetilde{q}(\widetilde{u}) = \gamma \cdot \widetilde{q}_!(\alpha) \cdot \gamma^{-1}$. $\square$

(10.10) THEOREM. *Let us fix two lifts $\widetilde{q}$ and $\widetilde{q}'$. Let $\gamma \in \theta_X$ denote the unique transformation satisfying $\widetilde{q} = \gamma \cdot \widetilde{q}'$. Then $\alpha, \beta \in \theta_X$ are in the Reidemeister relation with respect to $\widetilde{q}$ if and only if so are $\alpha \cdot \gamma^{-1}, \beta \cdot \gamma^{-1}$ with respect to $\widetilde{q}'$.*

PROOF. Suppose that $\beta = \delta \cdot \alpha \cdot \widetilde{q}_! \widetilde{p}^!(\delta^{-1})$. Then $\beta \cdot \gamma^{-1} = \delta \cdot \alpha \cdot \widetilde{q}_! \widetilde{p}^!(\delta^{-1}) \gamma^{-1} = \delta \cdot \alpha \cdot \gamma^{-1}(\gamma \cdot \widetilde{q}_! \widetilde{p}^!(\delta^{-1}) \gamma^{-1}) = \delta(\alpha \cdot \gamma^{-1}) \widetilde{q}_! \widetilde{p}^!(\delta^{-1}) \gamma^{-1}) \delta \cdot \alpha \cdot \gamma^{-1} \cdot \gamma \cdot \widetilde{q}_! \widetilde{p}^!(\delta^{-1})$. Then

$$\delta \cdot \alpha \cdot \gamma^{-1} \cdot \gamma \cdot \widetilde{q}_! \widetilde{p}^!(\delta^{-1}) = \delta(\alpha \cdot \gamma^{-1}) \widetilde{q}'_! \widetilde{p}^!(\delta^{-1}).$$

$\square$

The above consideration shows that the Reidemeister sets obtained by different lifts of q are canonically isomorphic. That is why we write $R(p, q)$ omitting tildes.

(10.11) THEOREM. *If $X \times \mathcal{I} \xleftarrow{p} \Gamma \xrightarrow{q} Y$ is a homotopy satisfying (10.2)–(10.4), then the homomorphism*

$$\widetilde{q}_{t!} \widetilde{p}_t^! : \theta_X \to \theta_Y.$$

does not depend on $t \in [0,1]$, where the lifts used in the definitions of these homomorphisms are restrictions of some fixed lifts p, q of the given homotopy.

PROOF. The commutative diagram of maps

$$
\begin{array}{ccccc}
X & \xLeftarrow{p_t} & \Gamma_t & \xrightarrow{q_t} & Y \\
{\scriptstyle i_{X,t}}\downarrow & & {\scriptstyle i_{\Gamma,t}}\downarrow & & \downarrow{\scriptstyle \mathrm{id}} \\
X \times \mathcal{I} & \xLeftarrow{p} & \Gamma & \xrightarrow{q} & Y
\end{array}
$$

where $i_{X,t}(x) = (x, t)$, $i_{\Gamma,t}(\widetilde{x}, z) = (\widetilde{x}, t, z)$, induces the commutative diagram of homomorphisms

$$
\begin{array}{ccccc}
\theta_X & \xrightarrow{p_t^!} & \theta_{\Gamma_t} & \xrightarrow{q_{t!}} & \theta_Y \\
{\scriptstyle (i_{X,t})^!}\uparrow & & {\scriptstyle (i_{\Gamma,t})^!}\uparrow & & \uparrow{\scriptstyle \mathrm{id}} \\
\theta_{X \times \mathcal{I}} & \xrightarrow{p^!} & \theta_\Gamma & \xrightarrow{q_!} & \theta_Y
\end{array}
$$

and it remains to notice that $(i_{x,t})^! : \theta_{X \times \mathcal{I}} \to \theta_X$ is an isomorphism. $\square$

(10.12) REMARK. *If (p, q) represents a single-valued map $\rho : X \to Y (q = \rho p)$, then $\widetilde{q}_! \widetilde{p}^!$ equals $\widetilde{\rho}_!$ (here the chosen lifts satisfy $\widetilde{q} = \widetilde{\rho} \widetilde{p}$).*

Indeed. Let us fix a point $(\widetilde{x}, z) \in \widetilde{\Gamma}$, and $\alpha \in \theta_X$. Then

$$\widetilde{q}(\alpha \widetilde{x}, z) = \widetilde{\rho} \widetilde{p}(\alpha \widetilde{x}, z) = \widetilde{\rho}(\alpha \widetilde{x}) = \widetilde{\rho}_!(\alpha) \widetilde{\rho}(\widetilde{x}) = \widetilde{\rho}_!(\alpha) \widetilde{\rho} \widetilde{p}(\widetilde{x}, z) = \widetilde{\rho}_!(\alpha) \widetilde{q}(\widetilde{x}, z).$$

On the other hand, $\tilde{q}(\alpha\tilde{x}, z) = \tilde{q}(\tilde{p}_!(\alpha)(\tilde{x}, z)) = \tilde{q}_!\tilde{p}_!(\alpha) \cdot \tilde{q}(\tilde{x}, z)$. Since the natural transformations $\tilde{p}_!(\alpha)$, $\tilde{q}_!\tilde{p}_!(\alpha) \in \theta_Y$ coincide at the point $\tilde{q}(\tilde{x}, z) \in \tilde{Y}$, they are equal.

Below we shall define the Nielsen relation modulo a subgroup.

Let us point out that the above theory can be modified onto the relative case. Consider again a multivalued pair (p, q) satisfying (10.2)–(10.4). Let $H \subset \theta_X$, $H' \subset \theta_Y$ be normal subgroups. Then the action of H on $\tilde{X}$ gives the quotient space $\tilde{X}_H$ and the map $p_{XH} : \tilde{X}_H \to X$ is also a covering. Similarly, we get $p_{YH'} : \tilde{Y}'_H \to Y$. On the other hand, the action of H on $\tilde{\Gamma}$ given by $h \circ (\tilde{x}, z) = (h\tilde{x}, z)$ determines the quotient space $\tilde{\Gamma}_H$ with the natural map $\tilde{p}_H : \tilde{\Gamma}_H \to \tilde{X}_H$ induced by $\tilde{p}$. Assume that $\tilde{q}_!\tilde{p}^!(H) \subset H'$. Observe that this condition does not depend on the choice of the lifts $\tilde{p}$, $\tilde{q}$, because the subgroups H, H' are the normal divisors. Thus, $\tilde{q} : \tilde{\Gamma} \to \tilde{Y}$ induces a map $\tilde{q}_H : \tilde{\Gamma}_H \to \tilde{Y}_{H'}$ and the diagram

$$\begin{array}{ccccc} \tilde{X}_H & \xleftarrow{\tilde{p}_H} & \tilde{\Gamma}_H & \xrightarrow{\tilde{q}_H} & \tilde{Y}_H \\ {\scriptstyle p_{XH}}\downarrow & & {\scriptstyle p_{\Gamma H}}\downarrow & & \downarrow{\scriptstyle p_{YH'}} \\ X & \xleftarrow{p} & \Gamma & \xrightarrow{q} & Y \end{array}$$

commutes. Now, we can get the homomorphisms $\tilde{q}_{H!}\tilde{p}^!_H : \theta_{XH} \to \theta_{YH'}$, where θ_{XH}, $\theta_{YH'}$ denote the groups of natural transformations of $\tilde{X}_H$ and $\tilde{Y}_{H'}$, respectively.

Assuming $X = Y$ and $H = H'$. We can give

(10.13) LEMMA.

(10.13.1) $C(p, q) = \bigcup_{\alpha \in \theta_{XH}} p_{\Gamma H} C(\tilde{p}_H, \alpha\tilde{q}_H)$,

(10.13.2) if $p_{\Gamma H} C(\tilde{p}_H, \alpha\tilde{q}_H) \cap p_{\Gamma H} C(\tilde{p}_H, \beta\tilde{q}_H)$ is not empty, then there exists a $\gamma \in \theta_{XH}$ such that $\beta = \gamma \circ \alpha \circ (\tilde{q}_{H!}\tilde{p}^!_H\gamma)^{-1}$,

(10.13.3) the sets $p_{\Gamma H} C(\tilde{p}_H, \alpha\tilde{q}_H)$ are either disjoint or equal.

Hence, we get the splitting of $C(p, q)$ into the H-Nielsen classes and the natural injection from the set of H-Nielsen classes into the set of Reidemeister classes modulo H, namely, $R_H(p, q)$.

Now, we would like to exhibit the classes which do not disappear under any (admissible) homotopy. For this, we need however (besides (10.2)–(10.4)) the following two assumptions on the pair $X \xleftarrow{p} \Gamma \xrightarrow{q} Y$.

(10.14) Let X be a connected retract of an open set in a paracompact locally convex space, p is a Vietoris map and $\mathrm{cl}(q(\Gamma)) \subset X$ is compact, i.e. q is a compact map.

(10.15) There exists a normal subgroup $H \subset \theta_X$ of a finite index satisfying $\tilde{q}_!\tilde{p}^!(H) \subset H$.

(10.16) DEFINITION. We call a pair (p,q) N-*admissible* if it satisfies (10.2)–(10.4), (10.14) and (10.15).

(10.17) REMARK. The pairs satisfying (10.14) are called admissible.

Let us recall that, under the assumption (10.14), the Lefschetz number $\Lambda(p,q) \in \mathbb{Q}$ is defined. This is a homotopy invariant (with respect to the homotopies satisfying (10.14)) and $\Lambda(p,q) \neq 0$ implies $C(p,q) \neq \emptyset$ (comp. Chapter I.6).

The assumption (10.15) gives rise to the commutative diagram

$$
\begin{array}{ccccc}
\widetilde{X}_H & \xleftarrow{\ \widetilde{p}_H\ } & \widetilde{\Gamma}_H & \xrightarrow{\ \widetilde{q}_H\ } & \widetilde{Y}_H \\
{\scriptstyle p_{XH}}\downarrow & & {\scriptstyle p_{\Gamma H}}\downarrow & & \downarrow{\scriptstyle p_{YH'}} \\
X & \xleftarrow{\ p\ } & \Gamma & \xrightarrow{\ q\ } & Y
\end{array}
$$

where the coverings p_{XH}, $p_{\Gamma H}$, p_{YH} are finite, because the subgroup $H \in \theta_{XH}$ has a finite index. Now, we can observe that the pair $(\widetilde{p}_H, \alpha \widetilde{q}_H)$, for any $\alpha \in \theta_{XH}$, also satisfies (10.14) ($\widetilde{p}^{-1}(\widetilde{x}) = p^{-1}(x)$, $\mathrm{cl}(\alpha \widetilde{q}_H(\widetilde{\Gamma}_H)) \subset p_{XH}^{-1}(\mathrm{cl}\, q(\Gamma))$ and the last set is compact, because the covering p_{XH} is finite).

Let $A = p_{\Gamma H} C(\widetilde{p}, \alpha \widetilde{q})$ be a Nielsen class of an N-admissible pair (p,q). We say that (the N-Nielsen class) A is *essential* if $\Lambda(\widetilde{p}, \alpha \widetilde{q}) \neq 0$. The following lemma explains that this definition it correct, i.e. does not depend on the choice of α.

(10.18) LEMMA. *If* $p_{\Gamma H} C(\widetilde{p}, \alpha \widetilde{q}) = p_{\Gamma H} C(\widetilde{p}, \alpha' \widetilde{q}) \neq \emptyset$, *for some* $\alpha, \alpha' \in \theta_{XH}$, *then* $\Lambda(\widetilde{p}, \alpha \widetilde{q}) = \Lambda(\widetilde{p}, \alpha' \widetilde{q})$.

PROOF. Since α, α' represent the same element in $R_H(p,q)$, there exists $\gamma \in \theta_{XH}$ such that $\alpha' = \gamma \circ \alpha \circ \widetilde{q}_! \widetilde{p}^!(\gamma^{-1})$. Thus,

$$
\begin{aligned}
\Lambda(\widetilde{p}, \alpha' \widetilde{q}) &= \mathrm{Tr}((\widetilde{p}^*)^{-1}(\alpha' \cdot \widetilde{q})^*) = \mathrm{Tr}((\widetilde{p}^*)^{-1}(\gamma \circ \alpha \circ (\widetilde{q}_! \widetilde{p}^!(\gamma^{-1})) \circ \widetilde{q})^*) \\
&= \mathrm{Tr}((\widetilde{p}^*)^{-1}(\gamma \circ \alpha \circ \widetilde{q}_!(\widetilde{p}^!(\gamma^{-1})))^*) = \mathrm{Tr}((\widetilde{p}^*)^{-1}((\widetilde{p}^!(\gamma^{-1})))^* \circ (\alpha \widetilde{q})^* \circ \gamma^*) \\
&= \mathrm{Tr}((\widetilde{p}^* \widetilde{p}^!(\gamma))^*)^{-1} \circ (\alpha \widetilde{q})^* \circ \gamma^*) = \mathrm{Tr}((\widetilde{\gamma}^*)^{-1}(\widetilde{p}^*)^{-1} \circ (\alpha \widetilde{q})^* \circ \gamma^*) \\
&= \mathrm{Tr}((\widetilde{p}^*)^{-1} \circ (\alpha \widetilde{q})^*) = \Lambda(\widetilde{p}, \alpha \widetilde{q}).
\end{aligned}
$$

$\square$

(10.19) DEFINITION. Let (p,q) be an N-admissible multivalued map (for a subgroup $H \subset \theta_X$). We define the Nielsen number modulo H as the number of essential classes in θ_{XH}. We denote this number by $N_H(p,q)$.

(10.20) REMARK. Observe that the above method allows us to define only essential classes (and the Nielsen number) modulo a subgroup of a finite index in $\theta_X = \pi_1 X$. The problem how to get similar notions in an arbitrary case we leave open.

The following theorem is an easy consequence of the homotopy invariance of the Lefschetz number.

(10.21) THEOREM. $N_H(p,q)$ is a homotopy invariant (with respect to N-admissible homotopies) $X \times [0,1] \overset{p}{\Longleftarrow} \Gamma \overset{q}{\longrightarrow} X$. Moreover, (p,q) has at least $N_H(p,q)$ coincidences.

The following theorem shows that the above definition is consistent with the classical Nielsen number for single-valued maps.

(10.22) THEOREM. If an N-admissible map (p,q) is N-admissibly homotopic to a pair (p',q'), representing a single-valued map p (i.e. $q' = \rho p'$), then (p,q) has at least $N_H(\rho)$ coincidences (here H denotes also the subgroup of $\pi_1 X$ corresponding to the given $H \subset \theta_X$ in (10.4)).

PROOF. Consider a covering space $p : \widetilde{X}_H \to X$, corresponding to H. So, ρ admits a lift $\widetilde{\rho} : \widetilde{X}_H \to \widetilde{X}_H$ and in the diagram

$$
\begin{array}{ccccc}
\widetilde{X}_H & \overset{\widetilde{p}'}{\longleftarrow} & \widetilde{\Gamma}_H & \overset{\widetilde{q}'}{\longrightarrow} & \widetilde{Y}_H \\
{\scriptstyle p_{XH}}\downarrow & & {\scriptstyle p_{\Gamma H}}\downarrow & & \downarrow{\scriptstyle p_{YH'}} \\
X & \overset{p'}{\longleftarrow} & \Gamma & \overset{q'}{\longrightarrow} & Y
\end{array}
$$

we can put $\widetilde{q}' = \widetilde{\rho}\widetilde{p}'$. Thus, a homotopy between (p',q') and (p,q) lifts onto the coverings and we get lifts $(\widetilde{p}, \widetilde{q})$. Since

$$\widetilde{q}_! \widetilde{p}^! = (\widetilde{q}')_! (\widetilde{p}')^! = (\widetilde{\rho}\widetilde{p}')_! \widetilde{p}'^! = \widetilde{\rho},$$

there is a natural bijection between the Reidemeister sets $R(\rho)$ and $R(p,q)$. It remains to show that the essential classes correspond to the essential classes in the both Reidemeister sets. Consider a class $[\alpha] \in R_H(\rho)$. This class is essential if and only if the index of $p_H(\mathrm{Fix}(\alpha\widetilde{\rho}))$ is non-zero. But $\mathrm{ind}(\alpha\widetilde{\rho}) = \Lambda(\alpha\widetilde{\rho})$ is a non-zero multiplicity of $\mathrm{ind}(p_{XH}(\mathrm{Fix}(\alpha\widetilde{\rho})))$, i.e. it is also non-zero. Thus,

$$0 \neq \Lambda(\alpha\widetilde{\rho}) = \Lambda(\widetilde{p}', \alpha\widetilde{\rho}\widetilde{p}') = \Lambda(\widetilde{p}', \alpha\widetilde{q}') = \Lambda(\widetilde{p}, \alpha\widetilde{q}),$$

and $[\alpha] \in R_H(p,q)$ is also essential. $\qquad\square$

Although in the general case the theory, presented in the previous sections, requires special assumptions on the considered pair (p,q), we shall see that in the case of multivalued self-maps on a torus it is enough to assume that this pair satisfies only (10.9), i.e. it is admissible. We will do this by showing that in the case of any pair satisfying (10.9), it is homotopic to a pair representing a single-valued map.

(10.23) LEMMA. *For any compact space X, if $\widetilde{H}^1(X;\mathbb{Q}) = 0$, then $\widetilde{H}^1(X;\mathbb{Z}) = 0$.*

PROOF. Recall that $\widetilde{H}^k(X;\mathbb{Q}) = \lim_{\rightarrow} H^k(N(\alpha);\mathbb{Q})$, where $N(\alpha)$ denotes the nerve of a covering α. Since X is compact, we can consider only finite coverings. So, by the Universal Coefficient Formula (see [Sp-M, Theorem 5.5.10]), the natural homomorphism

$$H^k(X_\alpha;\mathbb{Z}) \otimes \mathbb{Q} \rightarrow H^k(X_\alpha;\mathbb{Q})$$

is an injection. Since $\widetilde{H}^k(X;\mathbb{Q}) = \lim_{\rightarrow} H^k(N(\alpha);\mathbb{Q})$ and the direct limit functor is exact, the homomorphism $\widetilde{H}^q(X;\mathbb{Z}) \otimes \mathbb{Q} \rightarrow \widetilde{H}^k(X;\mathbb{Q})$ is also mono. Thus, $\widetilde{H}^q(X;\mathbb{Z}) \otimes \mathbb{Q} = 0$, which implies that any element in $\widetilde{H}^q(X;\mathbb{Z})$ is a torsion. By another Universal Coefficient Formula (see [Sp-M, Theorem 5.5.3]),

$$H^1(X_\alpha;\mathbb{Z}) = \mathrm{Hom}(H_1(X_\alpha;\mathbb{Z},\mathbb{Z}) \otimes \mathrm{Ext}(H_0(X_\alpha;\mathbb{Z}),\mathbb{Z}).$$

Since $H_0(X_\alpha;\mathbb{Z})$ is free, Ext $= 0$. Now, $H^1(X_\alpha;\mathbb{Z}) = \mathrm{Hom}(H_1(H_\alpha;\mathbb{Z}),\mathbb{Z})$ is torsion free and $\widetilde{H}^1(X;\mathbb{Z})$, as the direct limit of torsion free groups, is also torsion free. Therefore, $\widetilde{H}^1(X;\mathbb{Z})$ must be zero. $\square$

(10.24) THEOREM. *Any multivalued self-map (p,q) on the torus satisfying (10.9) is admissibly homotopic to a pair representing a single-valued map.*

PROOF. At first, we prove that (p,q) satisfies (10.4), i.e. that the restriction $q : p^{-1}(x) \rightarrow \mathbb{T}^n$ admits a lift to the universal cover $\mathbb{R}^n \rightarrow \mathbb{T}^n$ (for any $x \in \mathbb{T}^n$). It is enough to show that any such restriction is contractible. On the other hand, since any map into the n-torus $\mathbb{T}^n = S^1 \times \ldots \times S^1$ splits into n-maps into the circle, it is enough to show that any map from $p^{-1}(x)$ to S^1 is contractible. By the well-known Hopf theorem (see e.g. [Sp-M]),

$$[p^{-1}(x), S^1] = \widetilde{H}^1(p^{-1}(x);\mathbb{Z}).$$

On the other hand, Lemma (10.23) implies that $\widetilde{H}^1(p^{-1}(x);\mathbb{Z}) = 0$. Thus, any restriction $q : p^{-1}(x) \rightarrow S^1$ is contractible. $\square$

Now, we prove that $X \xleftarrow{p} \Gamma \xrightarrow{q} \mathbb{T}^n$ is homotopic to a pair representing a single-valued map. By the above consideration, there is a commutative diagram

$$
\begin{array}{ccccc}
\widetilde{X} & \xleftarrow{\;\widetilde{p}\;} & \widetilde{\Gamma} & \xrightarrow{\;\widetilde{q}\;} & \mathbb{R}^n \\
{\scriptstyle p_X}\downarrow & & {\scriptstyle p_\Gamma}\downarrow & & {\scriptstyle p_T}\downarrow \\
X & \xleftarrow{\;p\;} & \Gamma & \xrightarrow{\;q\;} & \mathbb{T}^n
\end{array}
$$

which gives rise to the induced homomorphism $\widetilde{q}_! \widetilde{p}^! : \theta_X \to \theta_{\mathbb{T}^n}$. Since the torus is a $K(\pi, 1)$ space and there is an isomorphism between the groups θ_X and $\pi_1 X$, there exists a single-valued map $\rho : X \to \mathbb{T}^n$ such that the induced homomorphism $\widetilde{\rho} : \theta_X \to \theta_{\mathbb{T}^n}$ coincides with $\widetilde{q}_! \widetilde{p}^!$. We will show that (p, q) is homotopic to $(p, \rho p)$. Define a homotopy $\widetilde{q}_t : \widetilde{\Gamma} \to \mathbb{R}^n$ by putting $\widetilde{q}_t(\widetilde{x}, z) = (1 - t)\widetilde{q}(\widetilde{x}, z) + t\widetilde{\rho}(\widetilde{x})$.

The equalities

$$\widetilde{q}_t \widetilde{p}^!(\gamma)(\widetilde{x}, z) = (1 - t)\widetilde{q}(\widetilde{x}, z) + t\widetilde{\rho}(\widetilde{x}) = (1 - t)\widetilde{q}(p^!(\gamma)(\widetilde{x}, z)) + t\widetilde{\rho}\widetilde{p}(\widetilde{p}^!(\gamma)(\widetilde{x}, z))$$
$$= (1 - t)\widetilde{q}_! p^!(\gamma)\widetilde{q}(\widetilde{x}, z) + t\widetilde{\rho}(\widetilde{p}_!(\widetilde{p}^!(\gamma)))\widetilde{p}(\widetilde{x}, z)$$
$$= (1 - t)\widetilde{q}_! p^!(\gamma)\widetilde{q}(\widetilde{x}, z) + t\widetilde{\rho}(\gamma)\widetilde{p}(\widetilde{x}, z)$$
$$= \widetilde{\rho}_!(\gamma)[(1 - t)\widetilde{q}(\widetilde{x}, z) + t\widetilde{\rho}(\widetilde{x})] = \widetilde{\rho}_!(\gamma)\widetilde{q}_t(\widetilde{x}, z)$$

verify that

$$\widetilde{q}_t \widetilde{p}^!(\gamma)(\widetilde{x}, z) = \widetilde{\rho}_!(\gamma)\widetilde{q}_t(\widetilde{x}, z).$$

Since any natural transformation on $\widetilde{\Gamma}$ is of the form, $\widetilde{p}^!(\gamma)$ for some $\gamma \in \theta_X$, $\widetilde{q}_t(\gamma)$ induces a homotopy $q_t : \Gamma \to \mathbb{T}^n$ for which the diagram

$$
\begin{array}{ccccc}
\widetilde{X} & \xleftarrow{\;\widetilde{p}\;} & \widetilde{\Gamma} & \xrightarrow{\;\widetilde{q}_t\;} & R E^n \\
{\scriptstyle p_X}\downarrow & & {\scriptstyle p_\Gamma}\downarrow & & \downarrow{\scriptstyle p_T} \\
X & \xleftarrow{\;p\;} & \Gamma & \xrightarrow{\;q_t\;} & \mathbb{T}^n
\end{array}
$$

commutes and the obtained homotopy satisfies (10.2) ($\widetilde{p}$ does not vary). For $t = 1$, we get $\widetilde{q}_1(\widetilde{x}, z) = \rho(\widetilde{x}) = \rho\widetilde{p}(\widetilde{x}, z)$ which implies $q_1(z) = \rho p(z)$.

(10.25) THEOREM. *Let* $\mathbb{T}^n \xleftarrow{p} \Gamma \xrightarrow{q} \mathbb{T}^n$ *be such that* p *is a Vietoris map. Let* $\rho : \mathbb{T}^n \to \mathbb{T}^n$ *be a single-valued map representing a multivalued map homotopic to* (p, q) *(according to Theorem (10.24), such a map always exists). Then* (p, q) *has at least* $N(\rho)$ *coincidences.*

PROOF. Let us recall (cf. [BBPT]) that $N(\rho) = |\Lambda(\rho)| = |\det(I - A)|$, where A is an integer $(n \times n)$-matrix representing the induced homotopy homomorphism $\rho_\# : \pi_1 \mathbb{T}^n \to \pi_1 \mathbb{T}^n$. Moreover, if $\det(I - A) \neq 0$, then $\operatorname{card}(\pi_1(\mathbb{T}^n)/\operatorname{Im}(\rho_\#)) = |\det(I - A)|$.

The case $N(\rho) = 0$ is obvious. Assume that $N(\rho) \neq 0$. By Theorem (10.24), it is enough to find a subgroup (of a finite index) $H \subset \pi_1 \mathbb{T}^n = \mathbb{Z}^n$ satisfying (a) $\rho_\#(H) \subset H$ and (b) $N_H(\rho) = N(\rho)$. We define $H = \{z - \rho_\#(x) \mid x \in \pi_1 X\}$. Then (a) is clear. Recall that, for any endomorphism of an abelian group $\rho : G \to G$, the Reidemeister set is the quotient group $R(\rho) = G/(\operatorname{Im}(\operatorname{id} - \rho)$. In our case, $H = \operatorname{Im}(\operatorname{id} - \rho)$ and the natural map $G \to G/H$ induces the bijection between

$R(\rho) = G/H$ and $R_H(\rho) = (G/H)/(H/H)$. Thus, we get the bijection $R_H(\rho) = R(\rho)$. Finally, we notice that all the Nielsen classes of ρ have the same index ($= \text{sign}(\det(I - A))$). Thus, all involving classes in $R_H(\rho)$ in $R(\rho)$ are essential, which proves $N_H(\rho) = N(\rho)$. $\square$

I.11. Nielsen number: noncompact case

This section can be regarded as a summary of Chapters I.9 and I.10. Namely, the Nielsen number is defined here for CAC-mappings.

It has been shown in (6.20) in Chapter I.6 that, for any multivalued CAC-map on a retract of an open subset of a locally convex space, the Lefschetz number $\Lambda(p, q) \in \mathbb{Z}$ is defined and $\Lambda(p, q) \neq 0$ implies the existence of a coincidence point $z \in \Gamma$ $(p(z) = q(z))$ of the pair (p, q).

On the other hand, we have constructed in Chapter I.10 the Nielsen number $N(p, q)$ for a class of compact multivalued self-maps on, in particular, a connected retract of an open set in a paracompact locally convex space. $N(p, q)$ is a non-negative integer, a homotopy invariant and $\#C(p, q) \geq N(p, q)$.

In this section, we generalize this construction: we drop out the compactness assumption imposed on (p, q) by replacing it to be a CAC.

As in the single-valued case, the definition of a Nielsen number is done in two stages: at first, $C(p, q)$ is split into disjoint classes (Nielsen classes) and then we define essential classes.

Fix a universal covering[5] $p_X : \widetilde{X} \to X$. We define $\widetilde{\Gamma} = \{(\widetilde{x}, z) \in \widetilde{X} \times \Gamma \mid p_X(\widetilde{x}) \supset p(z)\}$ (pullback) and the map $\widetilde{p} : \widetilde{\Gamma} \to \widetilde{X}$ by $\widetilde{p}(\widetilde{x}, z) = \widetilde{x}$.

PROPERTY A. *For any $x \in X$, the restriction $q_1 = q|_{p^{-1}(x)} : p^{-1}(x) \to X$ admits a lift $\widetilde{q}_1$, making the diagram*

$$
\begin{array}{ccc}
 & & \widetilde{X} \\
 & \overset{\widetilde{q}_1}{\nearrow} & \big\downarrow p_X \\
p^{-1}(x) & \xrightarrow{\ q_1\ } & X
\end{array}
$$

commutative.

(11.1) REMARK. Note that a sufficient condition for guaranteeing Property A is, for example, that $p^{-1}(x)$ is an ∞-proximally connected set, for every $x \in X$ (see [KM]). It is well-known (see (2.21)) that, on ANR-spaces, any ∞-proximally connected compact (nonempty) subset is an R_δ-set and vice versa.

---

[5]In what follows, we shall assume that $\widetilde{X}$ is a retract of an open set in a locally convex space. For metric ANR-s, this assumption is satisfied automatically (see: the proof of Lemma 2.5 in [AGJ2])

(11.2) LEMMA. *If (p,q) satisfies (CAC + A), then there is a lift $\widetilde{q} : \Gamma \to \widetilde{X}$ making the diagram*

$$
\begin{array}{ccccc}
\widetilde{X} & \xleftarrow{\;\widetilde{q}\;} & \widetilde{\Gamma} & \xrightarrow{\;\widetilde{p}\;} & \widetilde{X} \\
{\scriptstyle p_X}\downarrow & & {\scriptstyle p_\Gamma}\downarrow & & \downarrow{\scriptstyle p_X} \\
X & \xleftarrow{\;p\;} & \Gamma & \xrightarrow{\;q\;} & X
\end{array}
$$

commutative.

PROOF. Notice that the assumptions (10.2)–(10.4) in Chapter I.10 are satisfied. Let

$$\theta_X = \{\alpha : \widetilde{X} \to \widetilde{X} \mid p_X \alpha = p_X\}$$

denote the group of covering transformations of the covering $\widetilde{X}$. Similarly, we define θ_Γ.

The lifts $\widetilde{p}$, $\widetilde{q}$ define homomorphisms:

$$\widetilde{p}^! : \theta_X \to \theta_\Gamma \quad \text{by the formula } p^!(\alpha)(\widetilde{x}, z) = (\alpha \widetilde{x}, z)$$

and

$$\widetilde{q}_! : \theta_\Gamma \to \theta_X \quad \text{by the equality } \widetilde{q} \cdot \alpha = \widetilde{q}_!(\alpha) \cdot \widetilde{q}.$$

Let us recall that θ_X is isomorphic with $\pi_1(X)$ and if (p,q) represents a single-valued map (i.e. $qp^{-1}(x) = \eta(x)$, for a single-valued map $\eta : X \to X$), then $\widetilde{q}_!\widetilde{p}^! : \theta_X \to \theta_X$ is equal to the homomorphism $\widetilde{\eta}_! : \theta_X \to \theta_X$ given by $\widetilde{\eta} \cdot \alpha = \widetilde{\eta}_!(\alpha) \cdot \widetilde{\eta}$, where $\widetilde{\eta}$ is given by the formula $\widetilde{\eta}(\widetilde{x}) = \widetilde{q}\widetilde{p}^{-1}(x)$. However, the homomorphism $\widetilde{\eta}_!$ corresponds to the induced map $\eta_\# : \pi_1(X) \to \pi_1(X)$.

Thus, the composition $\widetilde{q}_!\widetilde{p}^! : \theta_X \to \theta_X$ can be considered as a generalization of the induced homotopy homomorphism. □

PROPERTY B. *There is a normal subgroup $H \subset \theta_X$ of a finite index (θ_X/H-finite), invariant under the homomorphism $q_!p^!$ ($q_!p^!(H) \subset H$).*

(11.3) REMARK. In particular, if X is a connected space such that the fundamental group $\pi_1(X)$ of X is abelian and finitely generated, then X satisfies Property B (see [Sp-M]). Observe also that if (p,q) is admissibly homotopic to a single-valued map f, then Property B holds true (see Chapter I.10).

Let us note that (CAC + A + B) makes the diagram

$$
\begin{array}{ccccc}
\widetilde{X}_H & \xleftarrow{\;\widetilde{q}_H\;} & \widetilde{\Gamma}_H & \xrightarrow{\;\widetilde{p}_H\;} & \widetilde{X}_H \\
{\scriptstyle p_{XH}}\downarrow & & {\scriptstyle p_{\Gamma H}}\downarrow & & \downarrow{\scriptstyle p_{XH}} \\
X & \xleftarrow{\;p\;} & \Gamma & \xrightarrow{\;q\;} & X
\end{array}
$$

commutative, where $p_{XH} : \widetilde{X}_H \to X$ is a covering corresponding to the normal subgroup $H\Delta\theta_X \sim \pi_1 X$ and $\widetilde{\Gamma}_H$ is a pullback. As above, we can define homomorphisms $\widetilde{p}_H^! : \theta_{XH} \to \theta_{\Gamma H}$, $\widetilde{q}_{H!} : \theta_{\Gamma H} \to \theta_{XH}$, where $\theta_{XH} = \{\alpha \mid \widetilde{X}_H \to \widetilde{X}_H \mid p_{XH}\alpha = p_{XH}\}$.

(11.4) LEMMA (cf. Lemma (10.8)). *We have:*

(11.4.1) $C(p,q) = \bigcup_{\alpha \in \theta_{XH}} p_{\Gamma H} C(\widetilde{p}_H, \alpha\widetilde{q}_H)$,

(11.4.2) *if* $p_{\Gamma H} C(\widetilde{p}_H, \alpha\widetilde{q}_H) \cap p_{\Gamma H} C(\widetilde{p}_H, \beta\widetilde{q}_H)$ *is not empty, then there exists a* $\gamma \in \theta_{XH}$ *such that* $\beta = \gamma \circ \alpha \circ (\widetilde{q}_{H!}\widetilde{p}_H^!\gamma)^{-1}$,

(11.4.3) *the sets* $p_{\Gamma H} C(\widetilde{p}_H, \alpha\widetilde{q}_H)$ *are either disjoint or equal.*

Thus, $C(p,q)$ *splits into disjoint subsets* $p_{\Gamma H} C(\widetilde{p}_H, \alpha \circ \widetilde{q}_H)$ *called Nielsen classes modulo a subgroup* H.

Now, we shall define essential classes. We consider the diagram

$$
\begin{array}{ccccc}
\widetilde{X}_H & \xleftarrow{\alpha\widetilde{q}_H} & \widetilde{\Gamma}_H & \xrightarrow{\widetilde{p}_H} & \widetilde{X}_H \\
\downarrow{\scriptstyle p_{XH}} & & \downarrow{\scriptstyle p_{\Gamma H}} & & \downarrow{\scriptstyle p_{XH}} \\
X & \xleftarrow{p} & \Gamma & \xrightarrow{q} & X
\end{array}
$$

(11.5) LEMMA. *The multivalued map* $(\widetilde{p}_H, \widetilde{q}_H)$ *is a* $\mathbb{C}AC$.

PROOF. Since $\widetilde{p}_{\Gamma H}$ is a homeomorphism between $\widetilde{p}_H^{-1}(x)$ and $\widetilde{p}^{-1}(px)$, $\widetilde{p}_H$ is Vietoris. If $U \subset X$ satisfies the definition of the $\mathbb{C}AC$ for (p,q), then $\widetilde{U} = p_{XH}^{-1}(U)$ satisfies the same for $(\widetilde{p}_H, \alpha\widetilde{q}_H)$. To see the last relation, we note that

$$
\mathrm{cl}\widetilde{\varphi}(\widetilde{U}) \subset \mathrm{cl}(p_{XH}^{-1}(\varphi(U))) \subset \mathrm{cl}(p_{XH}^{-1}(\mathrm{cl}(\varphi(U)))).
$$

Since $\mathrm{cl}(\varphi(U))$ is compact and covering p_{XH} is finite, $p_{XH}^{-1}(\mathrm{cl}(\varphi(U)))$ is also compact. Thus, so is $\mathrm{cl}\varphi\widetilde{U})$. $\qquad\square$

(11.6) DEFINITION. A Nielsen class mod H of the form $p_{\Gamma H} C(\widetilde{p}_H, \alpha\widetilde{q}_H)$ is called *essential* if $\Lambda(\widetilde{p}_H, \alpha\widetilde{q}_H) \neq 0$.

By Lemma 10.18 in Chapter I.10, this definition is correct, i.e., if

$$
p_{\Gamma H} C(\widetilde{p}_H, \alpha\widetilde{q}_H) = p_{\Gamma H} C(\widetilde{p}_H, \beta\widetilde{q}_H),
$$

then

$$
\Lambda(\widetilde{p}_H, \alpha\widetilde{q}_H) = \Lambda(\widetilde{p}_H, \beta\widetilde{q}_H).
$$

(11.7) DEFINITION. The number of essential classes of (p,q) mod a subgroup H is called the *H-Nielsen number* and is denoted by $N_H(p,q)$.

Now, we can give two main theorems of this section.

(11.8) THEOREM. *A multivalued map (p, q) satisfying* (CAC + A + B) *has at least $N_H(p, q)$ coincidence points.*

PROOF. We show that each essential H-Nielsen class is nonempty. Consider an essential class $p_\Gamma H C(\widetilde{p}_H, \alpha \widetilde{q}_H)$. Then $\Lambda(\widetilde{p}_H, \alpha \widetilde{q}_H) \neq 0$ implies a point $\widetilde{z} \in C(\widetilde{p}_H, \alpha \widetilde{q}_H)$, by which $p_\Gamma H C(\widetilde{p}_H, \alpha \widetilde{q}_H)$ is nonempty as required. $\square$

(11.9) THEOREM. *$N_H(p, q)$ is a homotopy invariant (with respect to homotopies satisfying* (CAC + A + B)).

PROOF. Let the map (p_t, q_t) be such a homotopy. It is enough to show that the class $p_\Gamma H C(\widetilde{p}_{0H}, \alpha \widetilde{q}_{0H})$ is essential if and only if the same is true for the class $p_\Gamma H C(\widetilde{p}_{1H}, \alpha \widetilde{q}_{1H})$. However, this is implied by the equality of Lefschetz numbers

$$\Lambda(\widetilde{p}_{0H}, \alpha \widetilde{q}_{0H}) = \Lambda(\widetilde{p}_{1H}, \alpha \widetilde{q}_{1H}).$$ $\square$

(11.10) REMARK. We would like to point out that any metric ANR-space is a retract of an open subset of a normed space. Therefore, the obtained results (in Chapters I.6–I.11) contain as a particular case the Nielsen, the Lefschetz and the degree theories for metric spaces.

I.12. Remarks and comments

I.1.

The collection of the basic facts and properties of locally convex spaces was determined by the scope of our book. For more details, see e.g. [Kö-M], [Lu-M], [RoRo-M], [Ru-M], [Sch-M].

The generalization of the Schauder approximation theorem in (1.37) allows us to give a modern proof of the Schauder–Tikhonov fixed point theorem (1.38). A more general version of the Schauder approximation theorem, related to the notion of admissibility in the sense of V. Klee ([Kle]), can be found e.g. in [RN-M].

I.2.

More details concerning the geometric topological objects, namely ES-spaces, NES-spaces, AR-spaces, ANR-spaces, R_δ-sets, ∞-proximally connected sets, etc., can be found in [Brs-M], [DG-M], [Go5-M], [Hu2-M], [Clp], [Str1], [Su], [Woj].

I.3.

There is a vast literature about various aspects of multivalued mappings and their selections, see e.g. [ADTZ-M], [AuF-M], [Be-M], [BrGMO1-M]–[BrGMO5-M], [CV-M], [Go5-M], [HP1-M], [ReSe-M], [Sr-M], and the references therein.

For more details about multivalued mappings, see also [Au-M], [AuC-M], [Cu-M], [De3-M], [Dz-M], [GG-M], [Go1-M], [Go4-M], [Kry2-M], [LR-M], [Ma-M], [Pet-M], [Pr-M], [We-M], [ACZ1], [ACZ4], [AuC], [Ap], [AF2], [BGK], [Beb], [Bee1]–[Bee5], [Beg], [Be2], [Bi1]–[Bi3], [BeGa], [Bon], [Bry], [BG2], [BrFH], [BCF], [BC1], [C1], [CFRS], [DBM1], [DBM4], [DBM5], [Dan1], [Dan2], [Dar], [Dav], [DoSh], [FM], [Fry1], [Fry2], [FryGo], [Ge4], [Go11], [Go14], [GKr1], [GL], [GSc], [Gra], [GJ1], [Hal1], [Her], [Hi1], [Hi2], [HPV], [HPVV], [HV2], [Hov], [Ja2], [Kry1], [Kry2], [Kry4], [Kry8], [Kuc1]–[Kuc3], [KN], [KRN], [McCl2], [Mi1]–[Mi3], [Neu], [Ol3], [Rhe], [Ri4], [Ri5], [Rbn1], [Rz2], [Srv], [St3], [St4], [Su], [Vie].

As concerns the multivalued integration, it is sufficient to be restricted, in the entire text, to the simplest related concept due to R. J. Aumann [Aum], i.e. the set of integrals of single-valued selections. For other concepts of a multivalued integration, see e.g. [ArBy], [BDP], [Byr], [Deb], [DBLa1], [DBLa2], [HiaUm], [Hu3], [JK2], [MarSa1], [MarSa2], [Ol4], [Pap8], [Sam1]–[Sam3], [Sil]. Various concepts of a set-valued differentiation are discussed and employed in Appendix 2; for more details, see e.g. [AuCe-M], [AuFr-M], [BaJa], [DBl3], [FV1], [Her], [HP1-M], [LVH1], [LVH2], [MaVi1], [Sil]. The standard references for the Bochner integrals for maps acting in Banach spaces are [DS-M], [Ysd-M].

I.4.

The notion of admissible mappings was introduced for the first time in 1975 by the second author. Admissible maps in metric spaces are treated in detail in [Go1-M], [Go5-M]. In topological spaces, Vietoris maps must be assumed to be still perfect in order to have a proper definition of admissible maps (see 4.9)), cf. also [Kry1-M], [Kry2-M], [RN-M].

The homological techniques were applied for the first time to the fixed point problems in 1946 by S. Eilenberg and D. Montgomery (cf. [EM], where the Lefschetz-type fixed point theorem was developed for acyclic mappings). Then this approach was followed by the second author. The main tool to do it was the Čech homology functor with compact carriers which is a generalization of the „ordinary" Čech homology functor. For more details about the Čech homology functor, see e.g. [Do-M], [ES-M], [Sp-M]. In particular, for the Čech homology functor with compact carriers, playing an essential sole in our investigations, see [Go1-M], [Go5-M]. For further aspects of the homological methods, see [BrGMO3-M], [Cu-M], [Da-M], [Dz-M], [Go1-M]–[Go5-M], [LR-M], [Ma-M], [We-M], [RN-M], [Bow1], [Bow2], [Bry], [BG2], [BGoPr], [Cal], [FG1], [Go2]–[Go14], [GGr1]–[GGr5], [GR], [Ja1]–[Ja5], [JP], [Kry1-M], [Kry2-M], [Ne1]–[Ne3], [Pat], [Po1]–[Po3], [Sko1]–[Sko4], [Ski1].

I.5.

Noncompact single-valued maps were considered in the fixed point theory by
J. Leray [Ler], B. O'Neil [Ne3], A. Dold [Do], R. S. Palais (unpublished), W. P.
Petryshyn [Pet1]–[Pet3], R. Nussbaum [Nu1]–[Nu6], [Nu-M] and many others.

The CAC-maps (compact absorbing contractions) were defined and systematically studied by the second author and G. Fournier, both in the single-valued
and multivalued cases, see [Fo2]–[Fo4], [FG2], and cf. also [GR], [RN-M]. For
more details concerning condensing maps and k-set contractions, based on the
notion of measure of noncompactness, see [AKPRS-M], [BrGMO1-M]–[BrGMO5-
M], [BaGo-M], [HP1-M], [KOZ-M], [CAl], [Go], [Hei], [MTY], [KZS], [Sad1], [Sad2],
[Vae2].

I.6.

The history of the Lefschetz fixed point theorem starts in 1923, when S. Lefschetz formulated it for compact manifolds. In 1928, H. Hopf gave its new proof for
self-maps of polyhedra. In 1967, A. Granas extended it to arbitrary ANR-spaces.
His proof was based on the fact that all compact ANR's are homotopically equivalent with polyhedra and that the case of noncompact ANR's can be reduced,
by using the generalized trace theory introduced by J. Leray, to the compact
case. More details and the related references can be found e.g. in [Br1-M], [DG-
M]; for multivalued versions, see e.g. [Go1-M], [Go5-M]. For some further related
references, see e.g. [Da-M], [Dz-M], [Gr4-M], [BoGl], [FG1], [FG2], [Fo2]–[Fo4],
[Go2]–[Go5], [Go17], [GR], [Gr1], [Gr3]–[Gr6], [Pei], [Po1], [Sav], [Ski1], [Sko3],
[Sr3], [Sr4].

I.7.

The results presented in this section, dealing with the Lefschetz fixed point
theorem for condensing maps on subsets of Fréchet spaces are formally new. In
Banach spaces, the similar results were obtained, in the single-valued case, in [Nu1]
and [FV2], [FV3].

Although a more general approach was indicated, again in the single-valued
case, in [Go17], the related problem in locally convex spaces remains open.

I.8.

Although the topological degree and the fixed point index was defined on many
various levels of abstraction, see e.g. [BrGM03-M], [Br2-M], [Bad-M], [DG-M], [Dz-
M], [FMMN-M], [GaM-M], [Go2-M], [Go4-M], [Go5-M], [Gr1-M], [Gr4-M], [KOZ-
M], [KrwWu-M], [Kry1-M], [Kry2-M], [Ll-M], [Ma-M], [Maw-M], [Ni-M], [Nu-M],
[Pr-M], [Rot-M], and the references therein, our presentation is new (cf. also [GGa-
M]).

For more details about the topological degree and the fixed index, see also

[Br1-M], [Cro-M], [FrGr-M], [Fuc-M], [FZ-M], [GD-M], [GGa-M], [GaM-M], [GG-M], [Go1-M], [Gr2-M], [Gr3-M], [Ks1-M], [Ks2-M], [KsZa-M], [LR-M], [Ptr-M], [RN-M], [Swr-M], [AndBa], [AGG1], [AG2], [AndVae], [BiG1], [BiG2], [BGoPl], [BKr1], [Bi2], [Bou1]–[Bou3], [Bro2], [Brw6], [Cal], [CL1], [Do], [DGP1]–[DGP3], [DBM3], [FiPe], [FiPe1], [FP1], [FP2], [FG3], [FPr2], [FV2], [FV3], [GaKr2], [GaPe1], [GaPe], [GGK1]–[GGK3], [Gr4], [Gr6], [Kry3], [Ku1], [Ku2], [LaR1], [LaR3], [MasNi], [Maw2], [Maw4], [Mlj1], [Nag], [Nu1]–[Nu6], [PetFi], [Se4], [SeS], [Sko1], [Sko2], [SY], [Th], [Vae2], [Vae3], [Vae5], [Wil], [Zel],...

I.9.

For more references concerning the information about the fixed point index for noncompact maps, see e.g. [AndBa], [FV2], [FV3], [FMMN-M], [KOZ-M], [Nu1]–[Nu6], [PetFi], [Vae2], [Vae3], [Vae5], [Zel].

I.10.

The Nielsen theory was originated by the Danish mathematician J. Nielsen in 1927, who studied with this respect self-maps of compact surfaces. Later on, finite polyhedra were systematically treated by F. Wecken in 1941–42, who also gave an alternative definition of the Nielsen relations. The crucial step in this development was an observation that the Nielsen number can be expressed by means of the generalized Lefschetz number for arbitrary (i.e. also infinite-dimensional) ANRs. More details and the related references can be found e.g. in [BJ-M], [Br1-M], [KTs-M], [McC-M].

The Nielsen number for multivalued maps was defined in [Dz-M], [Je2], [KM], [Sch2], [Sch3], but none of theses definitions was suitable for applications to differential inclusions. Our presentation is based on the papers [And6], [AGJ1], where however only metric spaces were considered. For further aspects of the Nielsen theory, see e.g. [Fel-M], [Go5-M], [And6], [And26], [AG2], [AGJ1]–[AGJ4], [Br8], [Brw1], [Brw2],[Brw6], [BBrS1], [BBrS2], [BBS], [BBPT], [BrwGS], [BrwS1], [BrwS2], [DbrKu], [Gua2], [Hal3], [Hea], [HeKe], [HePY], [HeSY], [HeY], [Je3], [JeMa], [Ji2], [McCo2], [NOW], [Sch5], [Sch7], [Wng1]–[Wng3].

I.11.

The first paper about the Nielsen number on noncompact sets, but for compact maps, was published in 1969 by R. F. Brown [Brw2]. Then U. K. Scholz [Scho] considered for the same goal even not necessarily compact maps in 1974. Our presentation is based on the papers [AGJ1], and especially [AGJ2]. It is an open problem how to define the Nielsen number for condensing maps.

CHAPTER II

GENERAL PRINCIPLES

II.1. Topological structure of fixed point sets: Aronszajn–Browder–Gupta-type results

In this section, we shall present current results concerning the Browder–Gupta-type theorems for single-valued and multivalued mappings.

(1.1) DEFINITION. Let X, Y be two metric spaces. A continuous map $f : X \to Y$ is called *proper at the point* $y \in Y$ if there exists $\varepsilon > 0$ such that, for any compact subset K of the open ball $B(y, \varepsilon) = B_y^\varepsilon$, the set $f^{-1}(K)$ is compact.

Recall that $f : X \to Y$ is called *proper* if, for any compact $K \subset Y$, the set $f^{-1}(K)$ is compact. Obviously, any proper map $f : X \to Y$ is proper at every point $y \in Y$.

We shall formulate the Browder–Gupta theorem (see [BrGu]) in the following form:

(1.2) THEOREM (comp. [Go5-M], [Go15]). *Let E be a Banach space and let $f : X \to E$ be a continuous map such that the following conditions are satisfied:*

(1.2.1) f *is proper at* $0 \in E$,

(1.2.2) *for any* $\varepsilon > 0$, *there exists a continuous map* $f_\varepsilon : X \to E$ *for which we have:*

   (i) $\|f(x) - f_\varepsilon(x)\| < \varepsilon$, *for every* $x \in X$,

   (ii) *the map* $\tilde{f}_\varepsilon : f_\varepsilon^{-1}(B(0, \varepsilon)) \to B(0, \varepsilon)$, $\tilde{f}_\varepsilon(x) = f_\varepsilon(x)$, *for every* $x \in f_\varepsilon^{-1}(B(0, \varepsilon))$, *is a homeomorphism.*

Then the set $f^{-1}(\{0\})$ *is an* R_δ-*set.*

PROOF. At first, we prove that $f^{-1}(\{0\})$ is nonempty. We take, for every $\varepsilon = 1/n$, $n = 1, 2, \ldots$ a map $f_n : X \to E$ which satisfies (1.2.2). In view of (1.2.2)(i), for every n, we can find a point $x_n \in X$ such that $f_n(x_n) = 0$. It follows that:

$$\|f(x_n)\| = \|f(x_n) - f_n(x_n)\| < \frac{1}{n}.$$

So, the sequence $\{f(x_n)\}$ converges to the point $0 \in E$. Since f is proper at $0 \in E$, we can assume without loss of generality that $\lim_n x_n = x$. Now, from the continuity of f it follows that $f(x) = 0$, and consequently $f^{-1}(\{0\}) \neq \emptyset$.

Furthermore, let us assume that $S = f^{-1}(\{0\})$. Evidently, S is compact (comp. (1.2.1)) and nonempty. For every $\varepsilon = 1/n$, $n = 1, 2, \ldots$, let $A_n = f_n(S)$, where f_n is chosen according to (1.2.2). Then from (1.2.2)(i) it follows that $A_n \subset B(0, 1/n)$ and, moreover, A_n is compact, for every n. We let

$$C_n = \overline{\text{conv}}(A_n)$$

to be the closed convex hull of A_n. Using Mazur's Theorem (1.33) in Chapter I.1, we infer that C_n is compact. Thus, C_n is a compact and convex subset of $B(0, 1/n)$, and so it is an absolute retract.

Consequently, we deduce that

$$D_n = f_n^{-1}(C_n)$$

is a compact absolute retract.

Observe that $S \subset D_n$, for every n, and S is a set theoretic limit of the sequence $\{D_n\}$. Thus, applying Proposition (2.23) from Chapter I.2, we obtain that S is an R_δ-set. The proof is completed. $\qquad\square$

(1.3) REMARKS.

(1.3.1) Let us note that Theorem (1.2) has exactly the same proof if we replace the Banach space E by an arbitrary Fréchet space E and open balls $B(0, \varepsilon)$ by convex symmetric open neighbourhoods of $0 \in E$.

(1.3.2) Let $F : X \to E$ be a compact map and assume that, for every $\varepsilon > 0$, there exists a compact ε-approximation F_ε of F. Then it folows from (1.2) that $\text{Fix}(F)$ is an R_δ-set.

(1.3.3) In (1.3.2), F and F_ε can be replaced by k-set contractions or by condensing mappings, respectively.

For more details and references, see [AGG3], [Go15].

Now, we are going to formulate the multivalued version of the Browder–Gupta Theorem.

We say that a multivalued map $\varphi : X \multimap Y$ is *proper* if the set

$$\varphi_+^{-1}(K) = \{x \in X \mid \varphi(x) \cap K \neq \emptyset\}$$

is compact, for every compact $K \subset Y$.

Now, we are in a position to prove a multivalued generalization of Theorem (1.2).

(1.4) THEOREM. *Let X be a metric space, E a Fréchet space, $\{U_k\}$ a base of open convex symmetric neighbourhoods of the origin in E, and let $\varphi : X \multimap E$ be an u.s.c. proper map with compact values. Assume that there is a sequence of compact convex valued u.s.c. proper maps $\varphi_k : X \multimap E$ such that*

(i) *$\varphi_k(x) \subset \varphi(O_{1/k}(x)) + U_k$, for every $x \in X$,*

(ii) *if $0 \in \varphi(x)$, then $\varphi_k(x) \cap \overline{U}_k \neq \emptyset$,*

(iii) *for every $k \geq 1$ and every $u \in E$ with $u \in U_k$, the inclusion $u \in \varphi_k(x)$ has an acyclic set of solutions.*

Then the set $S = \varphi^{-1}(0)$ is compact and acyclic.

PROOF. We show that S is nonempty. Observe that, for every $k \geq 1$, we can find $x_k \in X$ such that $0 \in \varphi_k(x_k)$. Assumption (i) implies that there are $z_k \in O_{1/k}(x_k)$, $y_k \in \varphi_k(z_k)$ and $u_k \in U_k$ such that $0 = y_k + u_k$. Thus, $y_k \to 0$. Consider the compact set $K = \{y_k\} \cup \{0\}$. Since φ is proper, the set $\varphi_+^{-1}(K)$ is compact. Moreover, $\{z_k\} \subset \varphi_+^{-1}(K)$. Thus, we can assume, without any loss of generality, that $\{z_k\}$ converges to some point $x \in X$. By the upper semicontinuity of φ, we have $0 \in \varphi(x)$, which implies that $S \neq \emptyset$.

Since φ is proper, the set S is compact. We show that it is acyclic. By assumption (ii), the set $A_k = \varphi_{k+}^{-1}(\overline{U}_k)$ is nonempty. Consider the map $\psi_k : A_k \multimap \overline{U}_k$, $\psi_k(x) = \varphi_k(x) \cap \overline{U}_k$. Since $\overline{U}_k$ is contractible and ψ_k is an u.s.c. convex valued surjection such that $\psi_{k+}^{-1}(y)$ is acyclic, for every $y \in \overline{U}_k$. It implies that A_k is acyclic.

Now, we show that, for every open neighbourhood U of S in X, there exists $k \geq 1$ such that $A_k \subset U$. Indeed, assume on the contrary that there is an open neighbourhood U of S in X such that $A_k \not\subset U$, for every $k \geq 1$. It means that there are $x_k \in A_k$ with $x_k \notin U$ and, consequently, there are $y_k \in \varphi_k(x_k)$ such that $y_k \in \overline{U}_k$. Assumption (i) implies that there are $z_k \in O_{1/k}(x_k)$, $v_k \in \varphi_k(z_k)$ and $u_k \in U_k$ such that $y_k = v_k + u_k$. Therefore, $v_k = y_k - u_k \in 2U_k$ which implies that $v_k \to 0$. Consider the compact set $K_0 = \{v_k\} \cup \{0\}$. Since φ is proper, we can assume that $\{z_k\}$ and, consequently, $\{x_k\}$ converges to some point $x \in X$. Thus, $x \in S$. On the other hand, $x \notin U$, a contradiction.

Finally, observe that for every open neighbourhood U of S there exists an acyclic compact Z such that $S \subset Z \subset U$. Thus, Z is acyclic in the cohomological sense (see: [Sp-M], [Ga3, Lemma 2.10]), and consequently in the sense of the Čech homology functor. The proof is completed. $\qquad\square$

(1.5) REMARK. It is easy to see that in the above result we can assume that X is a subset of a Fréchet space. Then, instead of ε-neighbourhoods, we can consider the sets $x + V_k$, where $\{V_k\}$ is the base of open convex symmetric neighbourhoods of the origin (for more details, see [AGG3]).

As a consequence of Theorem (1.4) and properties of a topological degree of u.s.c. compact convex valued maps, one can obtain the following theorem (comp. [Cza1], [Ga3]):

(1.6) THEOREM. *Let Ω be an open subset of a Fréchet space E, $\{U_k\}$ the base of open convex symmetric neighbourhoods of the origin in E, and $\Phi : \overline{\Omega} \multimap E$ a compact u.s.c. map with compact convex values. Suppose that $x \notin \Phi(x)$, for every $x \in \partial\Omega$, and $\mathrm{Deg}\,(j - \Phi, \Omega, 0) \neq 0$, where $j : \overline{\Omega} \to E$ is the inclusion. Assume that there exists a sequence $\{\Phi_k : \Omega \multimap E\}$ of compact u.s.c. maps with compact convex values such that*

(i) *$\Phi_k(x) \subset \Phi(x + U_k) + U_k$, for every $x \in \overline{\Omega}$,*

(ii) *if $x \in \Phi(x)$, then $x \in \Phi_k(x) + U_k$,*

(iii) *for every $u \in \overline{U_k}$, the set S_u^k of all solutions to the inclusion $x - \Phi_k(x) \ni u$ is acyclic or empty, for every $k > 0$.*

Then the fixed point set $\mathrm{Fix}(\Phi)$ of Φ is compact and acyclic.

PROOF. Define the maps φ, $\varphi_k : \overline{\Omega} \multimap E$, $\varphi = j - \Phi$, $\varphi_k = j - \Phi_k$. One can check that φ, φ_k are proper maps. To apply Theorem (1.4), it is sufficient to show that, for sufficiently big k and for every $u \in \overline{U_k}$ the set S_u^k is nonempty.

For each $k \geq 1$, define the map $\Psi_k : \overline{\Omega} \multimap E$, $\Psi_k(x) = \Phi_k(x) + u$, for every $x \in \overline{\Omega}$. We prove that, for sufficiently big k, $\mathrm{Deg}\,(j - \Psi_k, \Omega, 0) \neq 0$ which implies, by the existence property of a degree, a nonemptiness of S_u^k.

Since φ is a closed map, we can find, for sufficiently big k, a neighbourhood U_k of the origin such that $\varphi(\partial\Omega) \cap \overline{U_k} = \emptyset$, i.e., for every closed $A \subset \overline{\Omega}$, the set $\varphi(A)$ is closed in E.

Consider the following homotopy $H_k : \overline{\Omega} \times [0,1] \multimap E$, $H(x,t) = (1-t)\Phi(x) + t\Psi_k(x)$. We show that

$$Z_k = \{x \in \partial\Omega \mid x \in H_k(x,t), \text{ for some } t \in [0,1]\} = \emptyset,$$

for sufficiently big k. Suppose, on the contrary, that there are a subsequence of $\{H_k\}$ (we denote it also by $\{H_k\}$), points $x_k \in \partial\Omega$, and numbers $t_k \in [0,1]$ such that $x_k \in H_k(x_k, t_k)$, that is $x_k = (1 - t_k)y_k + t_k s_k + t_k u$, for some $y_k \in \Phi(x_k)$ and $s_k \in \Phi(x_k)$. Assumption (i) implies that there are $z_k \in x_k + U_k$ and $v_k \in \Phi(z_k)$ such that $s_k \in v_k + U_k$. By the compactness of Φ, we can assume that $y_k \to y$ and $v_k \to v$. Therefore, $s_k \to v$. Moreover, we can assume that $t_k \to t \in [0,1]$. This implies that $x_k \to x_0 = (1-t)y + tv + tu$ or, equivalently, that $0 = (1-t)(x_0 - y) + t(x_0 - v) - tu$. But by the upper semicontinuity of φ, we obtain that $x_0 - y \in \varphi(x_0)$ and $x_0 - v \in \varphi(x_0)$. Since φ is convex valued, $0 \in (1-t)\varphi(x_0) + t\varphi(x_0) - tu \subset \varphi(x_0) - tu$. This implies that $\varphi(x_0) \cap \overline{U_k} \neq \emptyset$, a contradiction, and the proof is completed. $\qquad\square$

Finally, we will show that in Theorem (1.4) one can weaken the assumption on a regularity of maps instead of strengthening a connection between φ and φ_k.

(1.7) THEOREM. *Let X be a metric space, E a Fréchet space, $\{U_k\}$ a base of open convex symmetric neighbourhoods of the origin in E, and let $\varphi : X \multimap E$ be an u.s.c. proper map with compact values. Assume that there is a sequence of compact valued u.s.c. proper maps $\varphi_k : X \multimap E$ such that*

(i) $\varphi_k(x) \subset \varphi(O_{1/k}(x)) + U_k$, *for every $x \in X$,*
(ii) $\varphi(x) \subset \varphi_k(x)$, *for every $x \in X$,*
(iii) *for every $k \geq 1$ the set $S_k = \varphi_{k+}^{-1}(0)$ is acyclic.*

Then the set $S = \varphi^{-1}(0)$ is compact and acyclic.

PROOF. At first, one can easily prove that S is nonempty and compact (cf. the proof of Theorem (1.4)).

Consider, for every $n \geq 1$, the following open sets $W_n = O_{2/n}(S)$ and $V_n = O_{1/n}(S)$. We show that there is $k_0 = k_0(n) \geq 1$ such that $1/k_0 < 1/n$ and $\varphi(x) \cap U_k = \emptyset$, for every $k \geq k_0$ and $x \in X \setminus V_n$.

Suppose on the contrary, that there are sequences $k_1 < k_2 < \ldots$ and $\{x_{k_i}\} \subset X \setminus V_n$ such that $\varphi(x_{k_i}) \cap U_{k_i} \neq \emptyset$. This implies that there is a sequence $\{y_{k_i}\} \subset E$ such that $y_{k_i} \in \varphi(x_{k_i}) \cap U_{k_i}$. Thus, $y_{k_i} \to 0$. Consider the set $K = \{y_{k_i}\} \cup \{0\}$ and take its preimage $\varphi_+^{-1}(K)$. Since $\{x_{k_i}\} \subset \varphi_+^{-1}(K)$ and φ is proper, we can assume that $x_{k_i} \to x$, for some $x \in X \setminus V_n$. On the other hand, since φ is u.s.c., it follows that $0 \in \varphi(x)$, which means that $x \in S$, a contradiction.

Notice that $S_k \subset W_n$, for every $k \geq k_0$. Indeed, suppose that there exists $x \notin W_n$ such that $x \in S_k$, for some $k \geq k_0$. Then $0 \in \varphi_k(x) \subset \varphi(O_{1/k}(x)) + U_k$, which implies that there exist $z_k \in N_{1/k}(x)$ and $y_k \in \varphi(z_k)$ such that $y_k \in U_k$. But $z_k \in O_{1/k}(X \setminus W_n) \subset X \setminus V_n$, and so, $\varphi(z_k) \cap U_k = \emptyset$, a contradiction.

Thus, we have $S_k \subset W_n$. Assumption (ii) implies that $S \subset S_k$, for each $k \geq 1$, and similarly as in the proof of (1.4), we obtain that S is acyclic. $\square$

II.2. Topological structure of fixed point sets: inverse limit method

Let us recall that an *inverse system* of topological spaces is a family $S = \{X_\alpha, \pi_\alpha^\beta, \Sigma\}$, where Σ is a set directed by the relation $\leq$, X_α is a topological space, for every $\alpha \in \Sigma$ (we assume that all topological spaces are Hausdorff) and $\pi_\alpha^\beta : X_\beta \to X_\alpha$ is a continuous mapping, for each two elements $\alpha, \beta \in \Sigma$, such that $\alpha \leq \beta$. Moreover, for each $\alpha \leq \beta \leq \gamma$, the following conditions should hold:

$$\pi_\alpha^\beta = \mathrm{id}_{X_\alpha} \quad \text{and} \quad \pi_\alpha^\beta \pi_\beta^\gamma = \pi_\alpha^\gamma.$$

A subspace of the product $\prod_{\alpha \in \Sigma} X_\alpha$ is called a *limit of the inverse system S* and it is denoted by $\lim_{\leftarrow} S$ or $\lim_{\leftarrow} \{X_\alpha, \pi_\alpha^\beta, \Sigma\}$ if

$$\lim_{\leftarrow} S = \left\{ (x_\alpha) \in \prod_{\alpha \in \Sigma} X_\alpha \;\middle|\; \pi_\alpha^\beta(x_\beta) = x_\alpha, \text{ for all } \alpha \le \beta \right\}.$$

An element of $\lim_{\leftarrow} S$ is called a *thread* or a *fibre* of the system S. One can see that if we denote by $\pi_\alpha : \lim_{\leftarrow} S \to X_\alpha$ a restriction of the projection $p_\alpha : \prod_{\alpha \in \Sigma} X_\alpha \to X_\alpha$ onto the αth axis, then we obtain $\pi_\alpha = \pi_\alpha^\beta \pi_\beta$, for each $\alpha \le \beta$.

Now, we summarize some useful properties of limits of inverse systems.

(2.1) PROPOSITION (see [Eng1-M]). *Let $S = \{X_\alpha, \pi_\alpha^\beta, \Sigma\}$ be an inverse system.*

(2.1.1) *The limit $\lim_{\leftarrow} S$ is a closed subset of $\prod_{\alpha \in \Sigma} X_\alpha$.*

(2.1.2) *If, for every $\alpha \in \Sigma$, X_α is*

    (i) *compact, then $\lim_{\leftarrow} S$ is compact,*

    (ii) *compact and nonempty, then $\lim_{\leftarrow} S$ is compact and nonempty,*

    (iii) *a continuum, then $\lim_{\leftarrow} S$ is a continuum,*

    (iv) *acyclic and $\lim_{\leftarrow} S$ is nonempty, then $\lim_{\leftarrow} S$ is acyclic,*

    (v) *metrizable and Σ is countable, then $\lim_{\leftarrow} S$ is metrizable,*

    (vi) *R_δ-set and $\Sigma = \mathbb{N}$ is the directed set of natural numbers, then then $\lim_{\leftarrow} S$ is an R_δ-set.*

(2.2) REMARK. It is easy to see that there are inverse systems such that X_α are nonempty but noncompact and their limits are empty.

The following example shows that a limit of an inverse system of AR-spaces need not be an AR-space.

(2.3) EXAMPLE. Consider a family $\{X_n\}_{n=1}^\infty$ of subsets of $\mathbb{R}^2$ defined as follows:

$$X_n = \left(\left[0, \frac{1}{n\pi}\right] \times [-1,1] \right) \cup \left\{ (x,y) \;\middle|\; y = \sin\frac{1}{x} \text{ and } \frac{1}{n\pi} < x \le 1 \right\}.$$

One can see that, for each $m, n \ge 1$ such that $m \ge n$, we have $X_m \subset X_n$.

Define the maps $\pi_n^m : X_m \to X_n$, $\pi_n^m(x) = x$. Therefore, $S = \{X_n, \pi_n^m, \mathbb{N}\}$ is an inverse system of compact AR-spaces. It is evident that $\lim_{\leftarrow} S$ is homeomorphic to the intersection of all X_n. On the other hand,

$$X = \bigcap_{n=1}^\infty X_n = \{(0,y) \mid y \in [-1,1]\} \cup \left\{ (x,y) \;\middle|\; y = \sin\frac{1}{x} \text{ and } 0 < x \le 1 \right\}$$

and $X \notin$ AR-space, for instance, X is not locally connected.

Let us give important examples of inverse systems.

(2.4) EXAMPLE. Let, for every $m \in \mathbb{N}$, $C_m = C([0,m], \mathbb{R}^n)$ be a Banach space of all continuous functions of the closed interval $[0,m]$ into $\mathbb{R}^n$ and $C = C([0,\infty), \mathbb{R}^n)$ be an analogous Fréchet space of continuous functions.

Consider the maps $\pi_m^p : C_p \to C_m$, $\pi_m^p(x) = x|_{[0,m]}$. It is easy to see that C is isometrically homeomorphic to a limit of the inverse system $\{C_m, \pi_m^p, \mathbb{N}\}$. The maps $\pi_m : C \to C_m$, $\pi_m(x) = x|_{[0,m]}$ correspond to suitable projections.

(2.5) REMARK. By the same manner as above we can show that Fréchet spaces $C(J, \mathbb{R}^n)$, where J is an arbitrary interval, $L^1_{\text{loc}}(J, \mathbb{R}^n)$ of all locally integrable functions, $\text{AC}_{\text{loc}}(J, \mathbb{R}^n)$ of all locally absolutely continuous functions and $C^k(J, \mathbb{R}^n)$ of all continuously differentiable functions up to the order k can be considered, as limits of suitable inverse systems.

More generally, every Fréchet space is a limit of some inverse system of Banach spaces.

Now, we introduce the notion of multivalued maps of inverse systems. Suppose that two systems $S = \{X_\alpha, \pi_\alpha^\beta, \Sigma\}$ and $S' = \{Y_{\alpha'}, \pi_{\alpha'}^{\beta'}, \Sigma'\}$ are given.

(2.6) DEFINITION. By a *multivalued map of the system S into the system S'* we mean a family $\{\sigma, \varphi_{\sigma(\alpha')}\}$ consisting of a monotone function $\sigma : \Sigma' \to S$, that is $\sigma(\alpha') \leq \sigma(\beta')$, for $\alpha' \leq \beta'$, and of multivalued maps $\varphi_{\sigma(\alpha')} : X_{\sigma(\alpha')} \multimap T_{\alpha'}$ with nonempty values, defined for every $\alpha' \in \Sigma'$ and such that

(2.6.1)
$$\pi_{\alpha'}^{\beta'} \varphi_{\sigma(\beta')} \subset \varphi_{\sigma(\alpha')} \pi_{\sigma(\alpha')}^{\sigma(\beta')},$$

for each $\alpha' \leq \beta'$.

A map of systems $\{\sigma, \varphi_{\sigma(\alpha')}\}$ induces a *limit map* $\varphi : \lim_\leftarrow S \multimap \lim_\leftarrow S'$ defined as follows:

$$\varphi(x) = \prod_{\alpha' \in \Sigma} \varphi_{\sigma(\alpha')}(x_{\sigma(\alpha')}).$$

In other words, a limit map is a map such that

(2.6.2)
$$\pi_{\alpha'} \varphi = \varphi_{\sigma(\alpha')} \pi_{\sigma(\alpha')},$$

for every $\alpha' \in \Sigma'$.

Since a topology of a limit of an inverse system is the one generated by the base consisting of all sets of the form $\pi_\alpha(U_\alpha)$, where α runs over an arbitrary set cofinal in Σ and U_α are open subsets of the space X_α, it is easy to prove the following continuity property for limit maps:

(2.7) PROPOSITION. *Let $S = \{X_\alpha, \pi_\alpha^\beta, \Sigma\}$ and $S' = \{Y_{\alpha'}, \pi_{\alpha'}^{\beta'}, \Sigma'\}$ be two inverse systems and $\varphi : \lim_\leftarrow S \multimap \lim_\leftarrow S'$ be a limit map induced by the map $\{\sigma, \varphi_{\sigma(\alpha')}\}$. If, for every $\alpha' \in \Sigma'$, $\varphi_{\sigma(\alpha')}$ is*

(i) *u.s.c., then φ is u.s.c.,*

(ii) *l.s.c., then φ is l.s.c.,*

(iii) *continuous, then φ is continuous.*

PROOF. It is sufficient to show that preimages of sets of the form $\pi_{\alpha'}^{-1}(U_{\alpha'})$ are open in $\lim_{\leftarrow} S$. Notice that

$$\varphi^{-1}(\pi_{\alpha'}^{-1}(U_{\alpha'})) = \{x \in \lim_{\leftarrow} S \mid \varphi(x) \subset \pi_{\alpha'}^{-1}(U_{\alpha'})\}$$
$$= \{x \in \lim_{\leftarrow} S \mid \pi_{\alpha'}\varphi(x) \subset U_{\alpha'}\}$$
$$= \{x \in \lim_{\leftarrow} S \mid \varphi_{\sigma(\alpha')}\pi_{\sigma(\alpha')}(x) \subset U_{\alpha'}\}.$$

Now, if $\varphi_{\sigma(\alpha')}$ is u.s.c., then the composition $\varphi_{\sigma(\alpha')}\pi_{\sigma(\alpha')}$ is u.s.c. and the above set is open in $\lim_{\leftarrow} S$. It implies that φ is u.s.c. and the proof of (i) is complete.

Similarly, notice that

$$\varphi_+^{-1}(\pi_{\alpha'}^{-1}(U_{\alpha'})) = \{x \in \lim_{\leftarrow} S \mid \varphi(x) \cap \pi_{\alpha'}^{-1}(U_{\alpha'}) \neq \emptyset\}$$
$$= \{x \in \lim_{\leftarrow} S \mid \exists y \in \lim_{\leftarrow} S' y \in \varphi(x) \text{ and } \pi_{\alpha'}(y) \in U_{\alpha'}\}$$
$$= \{x \in \lim_{\leftarrow} S \mid \pi_{\alpha'}\varphi(x) \cap U_{\alpha'} \neq \emptyset\}$$
$$= \{x \in \lim_{\leftarrow} S \mid \varphi_{\sigma(\alpha')}\pi_{\sigma(\alpha')}(x) \cap U_{\alpha'} \neq \emptyset\}.$$

Therefore, a lower semicontinuity of $\varphi_{\sigma(\alpha')}$ implies a lower semicontinuity of φ.

The statement (iii) is an immediate consequence of (i) and (ii). $\quad\square$

Now, we are able to formulate and prove the main result.

(2.8) THEOREM. *Let $S = \{X_\alpha, \pi_\alpha^\beta, \Sigma\}$ be an inverse system and $\varphi : \lim_{\leftarrow} S \multimap \lim_{\leftarrow} S$ be a limit map induced by a map $\{\mathrm{id}, \varphi_\alpha\}$, where $\varphi_\alpha : X_\alpha \multimap X_\alpha$. If fixed point sets of φ_α are acyclic and the fixed point set of φ is nonempty, then it is acyclic, too.*

PROOF. Denote by $\mathcal{F}_\alpha$ the fixed point set of φ_α, for every $\alpha \in \Sigma$, and by $\mathcal{F}$ the fixed point set of φ. We will show that $\pi_\alpha^\beta(\mathcal{F}_\beta) \subset \mathcal{F}_\alpha$.

Let $x_\beta \in \mathcal{F}_\beta$. Then $x_\beta \in \varphi_\beta(x_\beta)$ and $\pi_\alpha^\beta(x_\beta) \in \pi_\alpha^\beta\varphi_\beta(x_\beta) \subset \varphi_\alpha\pi_\alpha^\beta(x_\beta)$, which implies that $\pi_\alpha^\beta(x_\beta) \in \mathcal{F}_\alpha$.

Similarly, we show that $\pi_\alpha(\mathcal{F}) \subset \mathcal{F}_\alpha$. Denote by $\overline{\pi}_\alpha^\beta : \mathcal{F}_\beta \to \mathcal{F}_\alpha$ the restriction of π_α^β. One can see that $\overline{S} = \{\mathcal{F}_\alpha, \overline{\pi}_\alpha^\beta, \pm\}$ is an inverse system.

By Proposition (2.1), the set $\mathcal{F}$ is acyclic and the proof is complete. $\quad\square$

(2.9) COROLLARY. *Let $S = \{X_\alpha, \pi_\alpha^\beta, \Sigma\}$ be an inverse system and $\varphi : \lim_{\leftarrow} S \multimap \lim_{\leftarrow} S$ be a limit map induced by a map $\{\mathrm{id}, \varphi_\alpha\}$, where $\varphi_\alpha : X_\alpha \multimap X_\alpha$ is closed valued, for every $\alpha \in \Sigma$. Assume that all the sets X_α are complete AR-spaces and* all φ_α are contractions, i.e. they are Lipschitz with a constant $0 \leq k < 1$, and have the selection property.

If the fixed point set $\mathcal{F}$ of φ is nonempty, then it is acyclic. If all φ_α are compact valued, then $\mathcal{F}$ is nonempty, compact and acyclic.

PROOF. By Theorem (3.82) in Chapter I.3, all the fixed point sets $\mathcal{F}_\alpha$ of φ_α are AR-spaces. Thus, they are acyclic. Therefore, the first statement is a consequence of Proposition (2.1.2)(iv).

If, moreover, all φ_α have compact values, then condition (3.69.1) from Chapter I.3 implies a compactness of $\mathcal{F}_\alpha$. By Proposition (2.1.2)(ii), we get the second statement. $\square$

Moreover, we get:

(2.10) COROLLARY. If all X_α are Fréchet spaces and all φ_α are contractions with convex, closed [resp. compact] values, then the fixed point set of the limit map φ is acyclic (if it is nonempty) [resp. nonempty, compact and acyclic].

Let us mention one information about the structure of the fixed point set of a limit map in a special case of functional spaces.

(2.11) PROPOSITION. Let $\{\mathrm{id}, \varphi_m\}$ be a map of the inverse system $\{C_m, \pi_m^p, \mathbb{N}\}$ considered in Example (2.4). If all the fixed point sets $\mathcal{F}_m$ of φ_m are convex [resp. convex, compact], then the fixed point set $\mathcal{F}$ of the induced limit map is convex (possibly, empty) [resp. nonempty, convex, compact].

PROOF. The statement is a consequence of a linearity of the maps π_m^p.

In fact, let $x, y \in \mathcal{F}$. We want to show that $tx + (1-t)y \in \mathcal{F}$, for every $t \in [0, 1]$. But $(tx + (1-t)y|_{[0,m]}) = tx|_{[0,m]} + (1-t)y|_{[0,m]}$ and, by a linearity of π_m^p, we have

$$\pi_m^p(tx|_{[0,p]} + (1-t)y)|_{[0,p]} = tx|_{[0,m]} + (1-t)y|_{[0,m]},$$

which completes the proof. $\square$

We shall complete this section by considering the following example and then by the formulations of obtained results in function spaces.

(2.12) EXAMPLE. Consider the map $f : C([0, \infty), \mathbb{R}) \to C([0, \infty), \mathbb{R})$, $f(x) = x/2$. This map is a contraction (with $1/2$ as a constant of contractivity) with respect to each seminorm p_m.

Suppose that there is k, $0 \leq k < 1$, such that

$$d(f(x), f(y)) \leq k d(x, y), \quad \text{for any } x, y \in C([0, \infty), \mathbb{R}).$$

Take L, $\max\{1/2, k\} < L < 1$. We show that there are functions $x, y \in C([0, \infty), \mathbb{R})$ such that $d(f(x), f(y)) \geq Ld(x, y)$.

Indeed, let $y \equiv 0$ and $x \equiv 2L/(1-L)$. Then $p_m(x - y) = 2L/(1-L)$, for every $m \geq 1$. One can easily check that

$$L\frac{p_m(x-y)}{1 + p_m(x-y)} = \frac{2L^2}{1+L},$$

and

$$\frac{p_m(f(x) - f(y))}{1 + p_m(f(x) - f(y))} = \frac{p_m(x-y)/2}{1 + p_m(x-y)/2} = L > \frac{2L^2}{1+L}.$$

Hence,

$$Ld(x, y) = L \sum_{m=1}^{\infty} \frac{1}{2^m} \frac{p_m(x-y)}{1 + p_m(x-y)}$$

$$\leq \sum_{m=1}^{\infty} \frac{p_m(f(x) - f(y))}{1 + p_m(f(x) - f(y))} = f(f(x), f(y)).$$

This implies that f is not a contraction with respect to the metric in $C([0, \infty), \mathbb{R})$.

Now, we formulate the results obtained in this section in a special case of function spaces.

(2.13) COROLLARY. *Let* $\varphi_m : C([0, m], \mathbb{R}^n) \multimap C([0, m], \mathbb{R}^n)$ *(respectively,* $\varphi_m : L^1([0, m], \mathbb{R}^n) \multimap L^1([0, m], \mathbb{R}^n))$, $m \geq 1$, *be compact valued contractions having the selection property and such that* $\varphi_p(x)|_{[0,m]} = \varphi_m(x|_{[0,m]})$, *for every* $x \in C([0, p], \mathbb{R}^n)$ *(respectively,* $L^1([0, p], \mathbb{R}^n))$, $p \geq m$. *Define* $\varphi : C([0, \infty), \mathbb{R}^n) \multimap C([0, \infty), \mathbb{R}^n)$ *(respectively,* $\varphi_m : L^1([0, \infty), \mathbb{R}^n) \multimap L^1([0, \infty), \mathbb{R}^n))$, $\varphi(x)|_{[0,m]} = \varphi_m(x|_{[0,m]})$, *for every* $x \in C([0, \infty), \mathbb{R}^n)$ *(respectively,* $L^1([0, \infty), \mathbb{R}^n))$. *Then* $\text{Fix}(\varphi)$ *is* R_δ.

(2.14) REMARK. From the above corollary, one can infer that if each φ_m has convex values (or decomposable – in the case of L^1-spaces), then $\text{Fix}(\varphi)$ is R_δ.

The inverse system approach described above gives us an easy way to study the topological structure of solution sets of differential problems on noncompact intervals. Namely, a suitable operator with solutions as fixed points can be often considered as a limit map induced by maps of Banach spaces of functions defined on compact intervals (see [AGG2], [AGG3]).

II.3. Topological dimension of fixed point sets

In this section, we would like to study the topological (covering) dimension $\dim \text{Fix}(F)$ of the set of all fixed points for some multivalued contractions. For the definitions of topological dimensions and more details, see e.g. [AlPa-M], [Eng2-M], [HW-M]. The notation is the same as in Chapter I.3 (see (3.68)–(3.73)).

The following result is due to Z. Dzedzej and B. Gelman ([DzGe]).

(3.1) THEOREM. *Let E be a Banach space and $F : E \to B(E)$ be a contraction with convex values and a constant $\alpha < 1/2$. Assume, furthermore, that the topological dimension $\dim F(x)$ of $F(x)$ is greater or equal to n, for some n and every $x \in E$. If $\mathrm{Fix}(F)$ is compact, then $\dim \mathrm{Fix}(F) \geq n$.*

(3.2) OPEN PROBLEM. Is it possible to prove (3.1), for a complete AR-space $E = X$ and $F : X \to B(X)$ with values belonging to a Michael family $M(X)$?

Following D. Miklaszewski, we would like to discuss some generalizations of (3.1) (comp. also [AG1]).

(3.3) THEOREM. *Let X be a retract of a Banach space E, and $F : X \to B(X)$ be a compact continuous multivalued map with values being such elements of the Michael family $\mathcal{M}(X)$ that $F(x) \setminus \{x\} \in C^{k-2}$, for every $x \in \mathrm{Fix}(F)$. Then the set $\mathrm{Fix}(F)$ has the dimension greater or equal to k.*

PROOF. Suppose on the contrary that $\dim(\mathrm{Fix}(F)) < k$. Let us consider the maps $\psi : \mathrm{Fix}(F) \to BC(E)$ and $\varphi : \mathrm{Fix}(F) \to E \setminus \{0\}$ defined by the formulas: $\psi(x) = F(x) - x = \{y - x \mid y \in F(x)\}$ and $\varphi(x) = \psi(x) \setminus \{0\} = (F(x) \setminus \{x\}) - x$. We are going to prove that the family $\{\varphi(x) \mid x \in \mathrm{Fix}(F)\}$ is equi-LC^∞. Let $y \in \varphi(x_0)$ and r be a positive number such that $0 \notin B_E(y, 3r)$. Suppose that the set $B_E(y, r) \cap \varphi(x)$ is non-empty, for a fixed point x of F. Then $B_E(y, r) \cap \varphi(x) = [(B_E(y + x, r) \cap F(x)) - x]$. Let $z \in B_E(y + x, r) \cap F(x)$. It is easy to show that $B_E(y + x, r) \cap F(x) \subset B_E(y + x, 3r) \cap F(x)$. But the second set of these three sets being in the Michael family $M(X)$ is C^∞ as well as its translation, so the inclusion of $B_E(y, r) \cap \varphi(x)$ into the set $B_E(y, 3r) \cap \varphi(x)$ is homotopically trivial, and the family $\{\varphi(x) \mid x \in \mathrm{Fix}(F)\}$ is equi-LC^∞. It follows from Proposition (3.76) in Chapter I.3 that φ has a selection f. Then the map $g : \mathrm{Fix}(F) \to X$ defined by the formula: $g(x) = f(x) + x$ is a selection of F. We conclude that, in view of Proposition I.(3.76) again, there exists a selection h of F being an extension of g. But h has a fixed point $x' \in \mathrm{Fix}(F)$, $h(x') = g(x') = f(x') + x' = x'$, $f(x') = 0 \in \varphi(x)$, which is a contradiction.   □

In the case when $\dim X < \infty$, by analogous considerations as in the proof of (3.3), we obtain:

(3.4) THEOREM. *Let X be a retract of a Banach space E and $F : X \to B(X)$ be a continuous (i.e. both l.s.c. and u.s.c.) map such that $\overline{F(X)} = \bigcup\{F(x) \mid x \in X\}$ is a compact set. Assume that the values of F satisfy the following conditions:*

  (i) *$F(x) \setminus \{x\}$ is C^{k-2}, for every $x \in \mathrm{Fix}(F)$,*
  (ii) *$F(x)$ is C^k, for every $x \in X$,*
  (iii) *$\{F(x) \mid x \in \mathrm{Fix}(F)\}$ is equi-LC^{k-2} in E,*
  (iv) *$\{F(x) \mid x \in X\}$ is equi-LC^k in X.*

Then $\dim(\mathrm{Fix}(F)) \geq k$.

The proof of (3.4) is quite analogous to that of (3.3). Finally, note that one can show an example of a continuous (i.e. both l.s.c. and u.s.c.) map with contractible values of the local dimension 2 such that (iii) and (iv) are satisfied, but the dimension of the set of fixed points equals 1.

II.4. Topological essentiality

Topological essentiality plays a similar role as the topological degree. In fact, one can interpretate it as the topological degree mod 2. Since it has less properties, it is possible to define the topological essentiality for a more general class of mappings. Below, we shall follow [RN-M] (see also [AG1], [GS]).

In this section by E, E_1, we shall denote locally convex spaces and by U an open subset of E.

We let

$$\mathcal{A}(\overline{U}, E_1) = \{\varphi : \overline{U} \multimap E_1 \mid \varphi \text{ is admissible}\},$$
$$\chi(\varphi) = \{x \in \overline{U} \mid 0 \in \varphi(x)\}.$$

If, additionally, V is an open subset of U, then we put:

$$\mathcal{A}_V^0(\overline{U}, E_1) = \{\varphi \in \mathcal{A}(\overline{U}, E_1) \mid \varphi \text{ is compact and}$$
$$\varphi(x) = \{0\}, \text{ for every } x \in \overline{U} \setminus V\},$$
$$\mathcal{A}_V(\overline{U}, E_1) = \{\varphi \in \mathcal{A}(\overline{U}, E_1) \mid \chi(\varphi) \subset V\}.$$

(4.1) DEFINITION. An admissible map $\varphi \in \mathcal{A}(\overline{U}, E_1)$ is *essential* with respect to $\mathcal{A}_V^0(\overline{U}, E_1)$ if, for every $\psi \in \mathcal{A}_V^0(\overline{U}, E_1)$, there exists $x \in V$ such that $\varphi(x) \cap \psi(x) \neq \emptyset$.

Let

$$\mathcal{A}_V^{\mathrm{ess}}(\overline{U}, E_1) = \{\varphi \in \mathcal{A}(\overline{U}, E_1) \mid \varphi \text{ is essential with respect to } \mathcal{A}_V^0(\overline{U}, E_1)\}.$$

(4.2) REMARK. Observe that if $V = U$, then $\overline{U} \setminus U = \overline{U} \setminus V = \partial U$ is the boundary of U. Observe also that if $V_1 \subset V_2$, then $\overline{U} \setminus V_2 \subset \overline{U} \setminus V_1$, and consequently:

$$\mathcal{A}_{V_1}(\overline{U}, E_1) \subset \mathcal{A}_{V_2}(\overline{U}, E_1).$$

Now, we start to study properties of the topological essentiality.

(4.3) (Existence). *If* $\varphi \in \mathcal{A}_V^{\mathrm{ess}}(\overline{U}, E_1)$, *then* $\chi(\varphi) \neq \emptyset$.

For the proof, it is sufficient to observe that a map $\psi : \overline{U} \to E_1$, $\psi(x) = \{0\}$, belongs to $\mathcal{A}_V^0(\overline{U}, E_1)$, for every $x \in \overline{U}$.

(4.4) (Compact perturbation). *If $\varphi \in \mathcal{A}_V^{\text{ess}}(\overline{U}, E_1)$ and $\eta \in \mathcal{A}_V^0(\overline{U}, E_1)$, then $(\varphi + \eta) \in \mathcal{A}_V^{\text{ess}}(\overline{U}, E_1)$.*

PROOF. At first, observe that $\chi(\varphi + \eta) \subset V$. In fact, if we assume that $0 \in (\varphi(x) + \eta(x))$ and $x \in \overline{U} \setminus V$, then $\eta(x) = 0$, and so $x \in \chi(\varphi)$, a contradiction.

Now, assume that $\psi \in \mathcal{A}_V^0(\overline{U}, E_1)$. Then $(\psi - \eta) \in \mathcal{A}_V^0(\overline{U}, E_1)$. Since φ is essential, we get a point $x \in V$ such that:

$$\varphi(x) \cap (\psi - \eta)(x) \neq \emptyset.$$

It implies that $\psi(x) \cap (\varphi + \eta)(x) \neq 0$, and the proof is completed. $\qquad \square$

(4.5) (Coincidence). *Let $\varphi \in \mathcal{A}_V^{\text{ess}}(\overline{U}, E_1)$ and $\varphi : \overline{U} \multimap E_1$ be a compact admissible map. Assume that:*

$$A = \{x \in \overline{U} \mid \exists t \in [0, 1] \text{ such that } \varphi(x) \cap (t\eta)(x) \neq \emptyset\}$$

is a subset of V. Assume, furthermore, that A is compact or E is a normal space. Then there exists $x \in V$ such that $\varphi(x) \cap \eta(x) \neq \emptyset$.

PROOF. Since φ is essential, there exists $x \in V$ such that $0 \in \varphi(x)$, and so $0 \in 0 \cdot \eta(x) = \{0\}$. Thus, $A \neq \emptyset$. Obviously, A is a closed subset of E. In both cases, i.e. A is compact or E is normal[6], we can find an Urysohn function $s : \overline{U} \to [0, 1]$ such that $s(x) = 1$, for every $x \in A$ and $s(x) = 0$, for every $x \in \overline{U} \setminus V$. We let $\psi : \overline{U} \to E_1$, $\psi(x) = s(x) \cdot \eta(x)$. Then $\psi \in \mathcal{A}_V^0(\overline{U}, E)$ and from the essentiality of φ, we obtain a point $x_0 \in V$ such that

$$\varphi(x_0) \cap \psi(x_0) \neq \emptyset.$$

Then $x_0 \in A$, and consequently $s(x_0) = 1$. Therefore, $\varphi(x_0) \cap \eta(x_0) \neq \emptyset$, and the proof is completed. $\qquad \square$

(4.6) REMARK. The assumption that A is compact or E is a normal topological space can be replaced by assuming that φ is proper (see Chapter II.1). This remark remains true in (4.9) and (4.11) below.

(4.7) (Normalization). *The inclusion $i : \overline{U} \to E$, $i(x) = x$ is essential with respect to $\mathcal{A}_V^0(\overline{U}, E)$ if and only if $0 \in V$.*

PROOF. For the proof, it is sufficient to show that $0 \in V$ implies essentiality of i. Let $\psi \in \mathcal{A}_V^0(\overline{U}, E)$ and let $\widetilde{\psi} : E \to E$ be defined as follows:

$$\widetilde{\psi}(x) = \begin{cases} \psi(x) & x \in \overline{U}, \\ \{0\} & x \in E \setminus \overline{U}. \end{cases}$$

It is easy to see that $\widetilde{\psi}$ is compact and admissible. Then, from I.(6.20), we deduce that there is $\widetilde{x} \in E$ such that $\widetilde{x} \in \widetilde{\psi}(\widetilde{x})$. Since $0 \in V$, it implies that $\widetilde{x} \in V$. Therefore, $\widetilde{x} \in \psi(\widetilde{x})$ and $i(\widetilde{x}) = \widetilde{x}$, and so i is essential. The proof is completed. $\square$

---

[6] Observe that every locally convex space is always a $T_{3\frac{1}{2}}$-topological space (cf. Chapter I.1).

(4.8) (Localization). *Assume that* $\varphi \in \mathcal{A}_V^{\mathrm{ess}}(\overline{U}, E_1)$ *and* W *is an open subset of* U *such that* $\chi(\varphi) \subset W$. *Then the restriction* $\varphi|_{\overline{W}}$ *of* φ *to* $\overline{W}$ *is an element in* $\mathcal{A}_{V \cap W}(\overline{W}, E_1)$ *and* $\varphi|_{\overline{W}}$ *is essential with respect to* $\mathcal{A}_{V \cap W}^0(\overline{W}, E_1)$.

PROOF. It follows from the existence property that $\chi(\varphi) \neq \emptyset$. From the above assumptions, we also deduce that $\chi(\varphi) \subset V \cap W$.

Let $\psi \in \mathcal{A}_{V \cap W}^0(\overline{W}, E_1)$. We define $\widetilde{\psi} \in \mathcal{A}_V^0(\overline{U}, E_1)$ by putting:

$$\widetilde{\psi}(x) = \begin{cases} \psi(x) & x \in W, \\ \{0\} & x \in \overline{U} \setminus W. \end{cases}$$

Then there exists $x \in V$ such that

$$\varphi(x) \cap \widetilde{\psi}(x) \neq \emptyset.$$

Now, it is easy to see that $x \in W$. Consequently, we get $\varphi|_{\overline{W}} \cap \psi(x) \neq \emptyset$, and the proof is completed. □

(4.9) (Homotopy). *Let* $\varphi \in \mathcal{A}_V^{\mathrm{ess}}(\overline{U}, E_1)$ *and* $\tau : \overline{U} \times [0,1] \multimap E_1$ *be an admissible compact map such that* $\tau(x, 0) = \{0\}$, *for every* $x \in \overline{U} \setminus V$. *Assume, furthermore, that the set:*

$$B_\psi = \{x \in \overline{U} \mid \exists t \in [0,1] \; \varphi(x) \cap (\psi(x) + \tau(x, t)) \neq \emptyset\}$$

is compact, for every $\psi \in \mathcal{A}_V^0(\overline{U}, E_1)$, *provided* E *is not a normal space.*

If, for $\psi(x) = \{0\}$ *and for every* $x \in \overline{U}$, *the set* B_ψ *is contained in* V, *then* $(\varphi - \tau(\cdot, 1)) \in \mathcal{A}_V^{\mathrm{ess}}(\overline{U}, E_1)$.

PROOF. At first, let us observe that

$$\chi(\varphi - \tau(\cdot, t)) \subset V, \quad \text{for every } t \in [0,1].$$

Let $\psi \in \mathcal{A}_V^0(\overline{U}, E_1)$. Since $(\psi + \tau(\cdot, 1)) \in \mathcal{A}_V^0(\overline{U}, E_1)$, from the existence property, we get that $B_\psi \neq \emptyset$.

Now, by a simple calculation, we obtain that $B_\psi \cap (\overline{U} \setminus V) = \emptyset$. So, we can consider the Urysohn function

$$s : \overline{U} \to [0,1]$$

such that $s(x) = 1$, for every $x \in B_\psi$ and $s(x) = 0$, for every $x \in \overline{U} \setminus V$. Then the admissible map:

$$\eta(x) = \psi(x) + \tau(x, s(x))$$

belongs to $\mathcal{A}_V^0(\overline{U}, E_1)$.

Consequently, the essentiality of φ implies that $\varphi(x) \cap \eta(x) \neq \emptyset$, for some $x \in V$. It follows that $x \in B_\psi$, and so, we get

$$(\varphi(x) - \tau(x, 1)) \cap \psi(x) \neq \emptyset.$$

The proof is completed.   □

Let us show how to get some topological consequences of the topological essentiality.

(4.10) THEOREM (Invariance of Domain). *Let* $\varphi \in \mathcal{A}_V^{ess}(\overline{U}, E_1)$ *be a closed and proper map. Assume that* Ω *is a connected component of the set* $E_1 \setminus \varphi(\overline{U} \setminus V)$ *which contains* $0 \in E_1$. *Then* $\Omega \subset \varphi(V)$.

PROOF. Assume on the contrary that there is $\omega \in \Omega$ such that $\omega \notin \varphi(V)$. Since φ is closed, the set $E_1 \setminus \varphi(\overline{U} \setminus V)$ is a closed subset of a locally convex space E_1, and consequently Ω is an open subset of E_1. Moreover, it is arc connected.

Let $\alpha : [0, 1] \to \Omega$ be an arc such that $\alpha(0) = 0$ and $\alpha(1) = \omega$. Then $K = \alpha([0, 1])$ is a compact subset of Ω containing both $0 \in E_1$ and ω. Consider the set:

$$\widetilde{K} = \{x \in \overline{U} \mid \varphi(x) \cap K \neq \emptyset \text{ and } \varphi(x) \subset \varphi(\overline{U} \setminus V)\}.$$

Since φ is proper, we get that $\widetilde{K}$ is compact and $\widetilde{K} \cap \overline{U} \setminus V = \emptyset$. Let

$$s : \overline{U} \to [0, 1]$$

be an Urysohn function such that $s(x) = 1$, for every $x \in \widetilde{K}$ and $s(x) = 0$, for every $x \in \overline{U} \setminus V$. Finally, we define an admissible compact map $\psi : \overline{U} \multimap E_1$ by putting:

$$\psi(x) = \alpha(s(x)), \quad \text{for every } x \in \overline{U}$$

(in fact, ψ is a compact single-valued map). Observe that $\psi \in \mathcal{A}_V^0(\overline{U}, E_1)$ is by φ essential, so there is a point $x \in V$ such that $\varphi(x) \cap \psi(x) \neq \emptyset$, i.e., $\psi(x) = \alpha(s(x)) = \alpha(1) = \omega \in \varphi(x)$ and $x \in V$, but it is a contradiction. The proof is completed.   □

(4.11) THEOREM (Generalized Nonlinear Alternative). *Let* $\varphi \in \mathcal{A}_V^{ess}(\overline{U}, E_1)$ *and* $\psi : \overline{U} \multimap E_1$ *be a compact admissible map. In the case when* E *is not a normal space, we assume that the set*

$$\{x \in \overline{U} \mid \exists t \in [0, 1] \text{such that } \varphi(x)(t\psi(x) + \eta(x)) \neq \emptyset\}$$

is compact, for every $\eta \in \mathcal{A}_V^{ess}(\overline{U}, E_1)$.

If, for every $x \in \overline{U} \setminus V$, *we have* $\varphi(x) \cap \psi(x) = \emptyset$, *then:*

(4.11.1) $\varphi(x) \cap \psi(x) \neq \emptyset$, *for some* $x \in V$, *or*

(4.11.2) *there exists* $\lambda \in (0, 1)$ *and there exists* $x \in \overline{U} \setminus V$ *such that* $\varphi(x) \cap \lambda \psi(x) \neq \emptyset$.

PROOF. Assume that, for every $\lambda \in (0, 1)$ and for every $x \in \overline{U} \setminus V$, we have:

$$\varphi(x) \cap \lambda\psi(x) = \emptyset.$$

Moreover, we have:

$$\varphi(x) \cap \psi(x) = \emptyset, \quad \text{for every } x \in \overline{U} \setminus V.$$

Consider a homotopy:

$$\tau(x, t) = t \cdot \psi(x).$$

Then all assumptions of (4.9) are satisfied. Thus, $\varphi - \tau(\,\cdot\,, 1) = \varphi - \psi$ is essential and, from the existence property, we get a point $x \in V$ such that $0 \in (\varphi - \psi)(x)$. Hence, $\varphi(x) \cap \psi(x) \neq \emptyset$, and the proof is completed. $\square$

As a direct consequence of (4.11) and (4.8), we obtain

(4.12) COROLLARY (Nonlinear Alternative). *Assume that $\psi : \overline{U} \multimap E$ is a compact admissible map and $0 \in U$. Assume, furthermore, that in the case when E is not the normal space, the set*

$$\{x \in \overline{U} \mid \exists t \in [0, 1] \; \mathrm{Fix}(\eta + t\psi) \neq \emptyset\}$$

is compact, for every $\eta \in \mathcal{A}_V^0(\overline{U}, E)$.
   Then:
(4.12.1) $\mathrm{Fix}(\psi(x)) \neq \emptyset$ *or*
(4.12.2) $\exists \lambda \in (0, 1) \; \exists x \in \overline{U} \setminus U$ *such that $x \in (\lambda\psi(x))$.*

We shall complete this section by showing simple examples of essential mappings.

(4.13) EXAMPLES.

(4.13.1) Let U be an open connected subset of E and V be an open subset of U. We take $E_1 = \mathbb{R}$. Assume, furthermore, that there are two different points $x_1, x_2 \in \overline{U} \setminus V$ such that $u > 0$, for every $u \in \varphi(x_1)$, and $u < 0$, for every $u \in \varphi(x_2)$, where $\varphi \in \mathcal{A}_V(\overline{U}, \mathbb{R})$. It is easy to see that $\varphi \in \mathcal{A}_V^{\mathrm{ess}}(\overline{U}, \mathbb{R})$.

(4.13.2) Let $L : E \to E_1$ be a continuous linear isomorphism and U be an open subset of E and V be an open subset of U containing the zero point $0 \in E$. Then the restriction $L|_{\overline{U}}$ of L to U is essential, i.e., $L|_{\overline{U}} \in \mathcal{A}_V^{\mathrm{ess}}(\overline{U}, E_1)$.

(4.13.3) Let U be an open subset of E and let $p : \overline{U} \Rightarrow E_1$ be a Vietoris map such that $\chi(p) \subset V$, where V is an open subset of U. Then $p \in \mathcal{A}_V^{\mathrm{ess}}(\overline{U}, E_1)$.

II.5. Relative theories of Lefschetz and Nielsen

In this section, by X and A we shall denote retracts of some open subsets of a locally convex space E. We shall also assume that $A \subset X$.

Similarly as in Chapter I, by a multivalued mapping, we shall understand a pair (p, q), where:

$$X \xleftarrow{p} \Gamma \xrightarrow{q} Y,$$

but we shall also consider multivalued maps of pairs of spaces, i.e., the diagrams of the following type:

$$(X, A) \xleftarrow{p} (\Gamma, \Gamma_0) \xrightarrow{q} (Y, B)$$

(comp. Definition (4.6) in Chapter I.4), and, as before, we shall use the following notation

$$(p, q) : (X, A) \multimap (Y, B)$$

or

$$\varphi = \varphi(p, q) : (X, A) \multimap (Y, B).$$

For such a pair, we let:

$$(p, q)_X : X \multimap Y$$

and

$$(p, q)_A : A \multimap B,$$

where $(p, q)_X = (p_\Gamma, q_\Gamma)$ and $(p, q)_A = (p_{\Gamma_0}, q_{\Gamma_0})$.

For a multivalued map:

$$(p, q) : (X, A) \multimap (X, A),$$

we define the *generalized Lefschetz number* $\Lambda((p, q))$ by putting

$$\Lambda((p, q)) = \Lambda(q_* \circ p_*^{-1}),$$

(comp. (6.14) in Chapter I.6.)

(5.1) DEFINITION. A multivalued map $(p, q) : (X, A) \multimap (X, A)$ is called *compact (compact absorbing contraction)* if both $(p, q)_X$ and $(p, q)_A$ are compact (compact absorbing contractions).

As a direct consequence of (6.7) in Chapter I.6, we get:

(5.2) PROPOSITION. *Let* $(p, q) : (X, A) \multimap (X, A)$. *If for any two maps of:* (p, q), $(p, q)_X$, $(p, q)_A$, *the generalized Lefschetz number is well-defined, then so is the third, and in that cases we have:*

$$\Lambda((p, q)) = \Lambda((p, q)_X) - \Lambda((p, q)_A).$$

Now, we are able to formulate the relative version of the Lefschetz fixed point theorem:

(5.3) THEOREM (Relative Version of the Lefschetz Fixed Point Theorem). *If* $(p,q):(X,A) \multimap (X,A)$ *is a compact absorbing contradiction, then:*

(5.3.1) *the Lefschetz number* $\Lambda((p,q))$ *of* (p,q) *is well-defined, and*

(5.3.2) $\Lambda((p,q)) \neq 0$ *implies that there exists a point* $x \in \mathrm{Fix}((p,q))$ *such that* $x \in \overline{X \setminus A}$.

PROOF. First of all, observe that from (6.7) and (6.18) in Chapter I.6 it follows that $\Lambda((p,q)_X)$ and $\Lambda((p,q)_A)$ are well-defined and so

$$(1) \qquad \Lambda((p,q)) = \Lambda((p,q)_X) - \Lambda((p,q)_A)$$

is well-defined, too.

Now, assume that $\Lambda((p,q)) \neq 0$ and $\mathrm{Fix}((p,q)) \cap \overline{X \setminus A} = \emptyset$. It implies that $\mathrm{Fix}((p,q)) \subset X \setminus \overline{X \setminus A} = \mathrm{int}_X A$. Using the fixed point index defined in Chapter I.8, we get:

$$(2) \qquad \mathrm{ind}((p,q)_X, \mathrm{int}_X A) = \mathrm{ind}((p,q)_X, X) = \Lambda((p,q)_X)$$

and

$$(3) \qquad \mathrm{ind}((p,q)_A, \mathrm{int}_X A) = \mathrm{ind}((p,q)_A, A) = \Lambda((p,q)_A).$$

Moreover, since $\mathrm{Fix}(p,q) \subset \mathrm{int}_X A \subset A$, we get

$$(4) \qquad \mathrm{ind}((p,q)_X, \mathrm{int}_X A) = \mathrm{ind}((p,q)_A, \mathrm{int}_X A).$$

Consequently, from (2)–(4), we obtain

$$(5) \qquad \Lambda((p,q)_X) = \Lambda((p,q)_A)$$

and, in view of (1), we have:

$$\Lambda((p,q)) = \Lambda((p,q)_X) - \Lambda((p,q)_A) = 0,$$

but it is a contradiction. The proof is complete. $\qquad \square$

Observe that from (5.3) we get not only the existence of a fixed point, but also its localization.

As a simple corollary of (5.3), we obtain:

(5.4) COROLLARY. *Let $(p,q) : (X,A) \multimap (X,A)$ be a compact map. Then:*

(5.4.1) *the Lefschetz number $\Lambda((p,q))$ of (p,q) is well-defined, and*

(5.4.2) *$\Lambda((p,q)) \neq \emptyset$ implies that there exists $x \in \text{Fix}(p,q)$ such that $x \in \overline{X \setminus A}$.*

Now, we are going to present the relative Nielsen theory.

We shall also assume that X (which is a retract of an open set in a locally convex space E) admits a universal covering $p_X : \tilde{X} \to X$ (comp. [AGJ1], [AGJ2]).

Moreover, we consider all admissible maps $(p,q) : X \multimap X$ for which the following conditions are satisfied:

(A) For any $x \in X$, the restriction $q|_{p^{-1}(x)} : p^{-1}(x) \to X$ admits a lift $\tilde{q}$ to the universal covering space, i.e., $p_X \circ \tilde{q} = q$.

(B) There exists a normal subgroup $H \subset \pi_1(X)$ of a finite index such that $\tilde{q}_! \tilde{p}^!(H) \subset H$. Here, $\tilde{q}_! \tilde{p}^!$ denotes a homomorphism of the fundamental group $\pi_1(X)$ induced by $(p,q) : X \multimap X$ (comp. Chapter I.10); if (p,q) represents a single-valued map, i.e., $q(p^{-1}(x))$ is a singleton, for every $x \in X$, then $\tilde{q}_! \tilde{p}^!$ coincides with the induced homomorphism $f_\# : \pi_1(X) \to \pi_1(X)$.

Finally, in what follows, we shall assume that $A \subset X$ is, moreover, closed and connected.

Assume that $(p,q) : X \multimap X$ is a compact absorbing contraction (CAC) map of the form

$$X \xleftarrow{p} \Gamma \xrightarrow{q} X.$$

Denote $\Gamma_A = p^{-1}(A)$ and consider the restriction $A \xleftarrow{p|} \Gamma_A \xrightarrow{q|} A$. If $U \subset X$ is an open subset in the definition of a CAC-mapping, for (p,q), then $U \cap A$ can be associated to the CAC-mapping of $(p_|, q_|)$. Let $H_0 = i_\#^{-1}(H) \subset \pi_1 A$. Since the induced homomorphism $i_\# : (\pi_1 A)/H_0 \to (\pi_1 X)/H$ is mono, H_0 is also a normal subgroup of a finite order. Hence, $A \xleftarrow{p|} \Gamma_A \xrightarrow{q|} A$ (where $p_|, q_|$ denote the natural restrictions) also satisfies the assumptions (A) and (B).

Let us note that the diagram

$$
\begin{array}{ccc}
A & \xleftarrow{p|} \Gamma_A \xrightarrow{q|} & A \\
\Big\downarrow{i} & \Big\downarrow{i_\Gamma} & \Big\downarrow{i} \\
X & \xleftarrow{p} \Gamma \xrightarrow{q} & X
\end{array}
$$

where the vertical lines are natural inclusions, is commutative.

Let $p_X : \tilde{X} \to X, p_A : \tilde{A} \to A$ be fixed coverings corresponding to the subgroups H, H_0, respectively. In view of the results in Chapter I.10, there exist lifts $\tilde{q}, \tilde{q}_A$

making the diagrams

$$\widetilde{X} \xleftarrow{\ \widetilde{p}\ } \widetilde{\Gamma} \xrightarrow{\ \widetilde{q}\ } \widetilde{X}_H \qquad\qquad \widetilde{A} \xleftarrow{\ \widetilde{p}_A\ } \widetilde{\Gamma}_A \xrightarrow{\ \widetilde{q}_A\ } \widetilde{A}$$

$$p_X \downarrow \qquad p_\Gamma \downarrow \qquad p_X \downarrow \qquad\qquad p_A \downarrow \qquad p_{\Gamma_A} \downarrow \qquad p_A \downarrow$$

$$X \xleftarrow{\ p\ } \Gamma \xrightarrow{\ q\ } X \qquad\qquad A \xleftarrow{\ p_|\ } \Gamma_A \xrightarrow{\ q_|\ } A$$

commutative, where

$$\widetilde{\Gamma} = \{(\widetilde{x}, z) \in \widetilde{X} \times \Gamma;\ p_X(\widetilde{x}) = p(z)\}, \qquad \widetilde{p}(\widetilde{x}, z) = \widetilde{x}, \qquad p_\Gamma(\widetilde{x}, z) = z,$$

$$\widetilde{\Gamma}_A = \{(\widetilde{a}, z) \in \widetilde{A} \times \Gamma_A;\ p_A(\widetilde{a}) = p_A(z)\}, \quad \widetilde{p}_A(\widetilde{a}, z) = \widetilde{a}, \quad p_{\Gamma_A}(\widetilde{a}, z) = z.$$

Let $O_X = \{\alpha : \widetilde{X} \to \widetilde{X};\ p_X\alpha = p_X\}$, $O_A = \{\alpha : \widetilde{A} \to \widetilde{A};\ p_A\alpha = p_A\}$ denote the groups of the covering transformations. Recall that $O_X \sim \pi_1 X$, $O_A \sim \pi_1 A$.

(5.5) LEMMA. *There exist maps $\widetilde{i} : \widetilde{A} \to \widetilde{X}$, $\widetilde{i}_\Gamma : \widetilde{\Gamma}_A \to \widetilde{\Gamma}$ making the following diagrams commutative.*

$$\widetilde{A} \xrightarrow{\ \widetilde{i}\ } \widetilde{X} \qquad\qquad \widetilde{\Gamma}_A \xrightarrow{\ \widetilde{i}_\Gamma\ } \widetilde{\Gamma}$$

$$p_A \downarrow \qquad p_X \downarrow \qquad\qquad p_{\Gamma_A} \downarrow \qquad p_\Gamma \downarrow$$

$$A \xrightarrow{\ i\ } X \qquad\qquad \Gamma_A \xrightarrow{\ i_\Gamma\ } \Gamma$$

Let us fix such maps $\widetilde{i}$, $\widetilde{i}_\Gamma$. Then, for any $\alpha_0 \in O_A$, there is exactly one $\alpha \in O_A$ making the following diagram

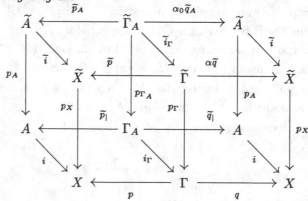

commutative.

PROOF. Since $(ip_A)_\#(\pi_1\widetilde{A}) = i_\#(H_0) \subset H = \operatorname{im}(p_X)_\#$, there exists a lift $\widetilde{i}$ making the first diagram commutative. We define $\widetilde{i}_\Gamma : \widetilde{\Gamma}_A \to \widetilde{\Gamma}$ by putting $\widetilde{i}_\Gamma(\widetilde{a}, z) = (\widetilde{i}(\widetilde{a}), i_\Gamma(z))$, and the second diagram commutes.

Now, we consider the big diagram. The commutativity of all squares, but the deck ones, follow from the already discussed diagrams. It remains to check the commutativity of the deck squares. One can easily check that $\widetilde{p}\,i_\Gamma = \widetilde{i}\,\widetilde{p}_A$. Then we can examine the right deck square. At first, we notice that $p_X(\widetilde{q}\,i_\Gamma) = p_X(\widetilde{i}\alpha_0 q_A)$. Thus, $\widetilde{q}\,i_\Gamma$, $\widetilde{i}\alpha_0 q_A$ are lifts of the same map $qi_\Gamma = iq_|$, by which there is exactly one $\alpha \in O_X$ such that $\alpha \widetilde{q}\,i_\Gamma = \widetilde{i}\alpha_0 q_A$. $\square$

(5.6) REMARK. We can replace the lifts $\widetilde{q}$, $\widetilde{q}_A$ by those for which the above big diagram commutes (i.e. it commutes for α_0 and α being identities). Then the mapping $O_A \ni \alpha_0 \to \alpha \in O_X$ defined in the Lemma (5.5) coincides with the homomorphism $\pi_1 A \to \pi_1 X$, induced by the inclusion $A \to X$. So, we denote it by $i_\#$.

Let us recall that a pair (p,q) satisfying (A) induces a homomorphism $\widetilde{q}_!\widetilde{p}^! :$ $O_X \to O_X$, corresponding, in the single-valued case ($\rho = qp^{-1}$), to $\rho : \pi_1 X \to \pi_1 X$. We define the action of O_X on itself $\gamma \circ \alpha = \gamma\alpha\widetilde{q}_!\widetilde{p}^!\gamma^{-1}$. By an analogy with the single-valued case, we define the quotient set, the *set of Reidemeister classes*, and we denote it by $\mathcal{R}_H(p,q)$. We define the *Nielsen class* corresponding to a class $[\alpha] \in \mathcal{R}(p,q)$, as $p_\Gamma(C(\widetilde{p},\alpha\widetilde{q}))$. This splits $C(p,q)$ into disjoint Nielsen classes and defines the natural injection $\eta : \mathcal{N}_H(p,q) \to \mathcal{R}_H(p,q)$. If $\Lambda(\widetilde{p},\alpha\widetilde{q}) \neq 0$, then $C(\widetilde{p},\alpha\widetilde{q}) \neq \emptyset$, and the Nielsen class corresponding to $[\alpha] \in \mathcal{R}_H(p,q)$ is called *essential*. We define the *H-Nielsen number* as the number of essential classes and denote it by $N(p,q)$.

(5.7) LEMMA. *The homomorphism $i_\#$ in Remark (5.6) induces a map of the Reidemeister sets $\mathcal{R}(i) : \mathcal{R}_{H_0}(p_|,q_|) \to \mathcal{R}_H(p,q)$.*

PROOF. It is enough to show the commutativity of the diagram

$$
\begin{array}{ccc}
O_A \times O_A & \xrightarrow{\;\;\circ\;\;} & O_A \\
{\scriptstyle i_\# \times i_\#}\big\downarrow & & \big\downarrow{\scriptstyle i_\#} \\
O_X \times O_X & \xrightarrow{\;\;\circ\;\;} & O_X
\end{array}
$$

where the horizontal lines are given by the Reidemeister action

$$(\gamma, \alpha) \to \gamma \circ \alpha = \gamma\alpha\widetilde{q}_!\widetilde{p}^!\gamma^{-1}.$$

In fact, $i_\#$ is a homomorphism which implies $i_\#(\gamma \circ \alpha) = i_\#(\gamma\alpha\widetilde{q}_!\widetilde{p}^!\gamma^{-1}) = i_\#(\gamma) \cdot i_\#\alpha \cdot i_\#(\widetilde{q}_!\widetilde{p}^!\gamma^{-1}) = i_\#(\gamma) \cdot i_\#\alpha \cdot (\widetilde{q}_!\widetilde{p}^!i_\#\gamma^{-1}) = i_\#\gamma \circ i_\#\alpha.$ $\square$

(5.8) LEMMA. *The following diagram commutes*

$$
\begin{array}{ccc}
\mathcal{N}_{H_0}(p_|,q_|) & \xrightarrow{\;\mathcal{N}(i)\;} & \mathcal{N}_H(p,q) \\
{\scriptstyle \eta}\big\downarrow & & \big\downarrow{\scriptstyle \eta} \\
\mathcal{R}_{H_0}(p_|,q_|) & \xrightarrow{\;\mathcal{R}(i)\;} & \mathcal{R}_H(p,q)
\end{array}
$$

Let $S_H(p, q; A) \subset \mathcal{R}_H(p, q)$ denote the set of essential Reidemeister classes which contain no essential class from $\mathcal{R}_{H_0}(p_|, q_|)$.

(5.9) THEOREM. *Under the assumptions* (CAC), (A) *and* (B), *the pair* (p, q) *has at least* $N_H(p, q) + (\#S_H(p, q; A))$ *coincidences.*

PROOF. We choose a point $z_1, \ldots, z_k \in C(f_|, g_|)$, from each essential class of $f_|, g_|$, and a point $w_1, \ldots, w_l \in C(f, g)$ from each one in $S_H(p, q; A)$. It remains to show that $i(z) \neq w$, for any $z = z_1, \ldots, z_k$, $w = w_1, \ldots, w_l$. Suppose the contrary. Then, by Lemma (5.8), the essential Nielsen class of $(p_|, q_|)$ containing z is involved in the class of (p, q) containing w. Since the last class belongs to $S_H(p, q; A)$, we get a contradiction. $\qquad\square$

(5.10) REMARK. Since any essential Reidemeister class corresponds always to a nonempty Nielsen class, in the definition of $S_H(p, q; A)$, the name of Reidemeister can be replaced by Nielsen: $S_H(p, q; A)$ becomes the set of Nielsen classes (from $\mathcal{N}_H(p, q)$) which contains no essential Nielsen class (from $\mathcal{N}_{H_0}(p_|, q_|)$).

The following theorem gives a lower bound for the number of coincidences of the pair (p, q) lying outside Γ_A. Let $SN_H(p, q; A)$ be the cardinality of the set of essential classes in $\mathcal{R}_H(p, q) \setminus \operatorname{im} \mathcal{R}(i)$.

Finally, we have:

(5.11) THEOREM. *Under the assumptions* (CAC), (A), (B), *the pair* (p, q) *has at least* $SN_H(p, q; A)$ *coincidences in* $\Gamma \setminus \Gamma_A$.

For the proof, it is sufficient to observe that each essential class from $\mathcal{R}_H(p, q) \setminus \operatorname{im} \mathcal{R}(i)$ is nonempty and disjoint from Γ_A.

II.6. Periodic point principles

Let $E = \{E_q\}$ be a graded vector space over $\mathbb{Q}$ (comp. Chapter I.6) and let $f : E \to E$ be a Leray endomorphism. For, such f we have defined the generalized generalized Lefschetz number $\Lambda(f)$. Now, we shall define, for such f, the *generalized Euler characteristic* $\chi(f)$ by putting:

$$\chi(f) = \sum_q (-1)^q \dim(\widetilde{E}_q),$$

where the graded vector space $\widetilde{E} = \{\widetilde{E}_q\}$ is constructed in Chapter I.6.

The following proposition is a direct consequence of our definitions:

(6.1) PROPOSITION. *Let $f : E \to E$ be an endomorphism and $f^n : E \to E$ denote its n-iterate, i.e., $f^n = \underbrace{f \circ \ldots \circ f}_{n\text{-times}}$. We have:*

(6.1.1) *f is a Leray endomorphism if and only if f^n is a Leray endomorphism and, in this case, we obtain:*

$$\chi(f) = \chi(f^n).$$

Let $\mathbb{Q}\{x\}$ denote the integral domain of all formal power series:

$$S = a_0 + a_1 x + a_2 x^2 + \ldots = \sum_{n=0}^{\infty} a_n x^n$$

with coefficients $a_n \in \mathbb{Q}$. Then $\mathbb{Q}\{x\}$ contains the polynomial ring $\mathbb{Q}[x]$.

(6.2) DEFINITION. *Let $f : E \to E$ be a Leray endomorphism. The Lefschetz power series $L(f)$ of f is an element of $\mathbb{Q}\{x\}$ defined by:*

$$L(f) = \chi(f) + \sum_{n=0}^{\infty} \Lambda(f^n) x^n.$$

It is well-known (see [DG-M] or [Bow2], [AGJ4]) that

$$L(f) = u \cdot v^{-1},$$

where u and v are relatively prime polynomials with $\deg u < \deg v$ ($u \neq 0$). We let

(6.3) $$P(f) = \deg v, \quad \text{where } L(f) = u \cdot v^{-1}.$$

The above can be summarized as follows:

(6.4). *Let $f : E \to E$ be a Leray endomorphism. Then we have:*

(6.4.1) $\chi(f) \neq 0$ *implies* $P(f) \neq 0$,

(6.4.2) $P(f) \neq 0 \Leftrightarrow \Lambda(f^n) \neq 0$, *for some n,*

(6.4.3) $P(f) \neq 0$, *then for any natural m one of the numbers:*

$$\Lambda(f^{m+1}), \Lambda(f^{m+2}), \ldots, \Lambda(f^{m+P(f)})$$

is different from zero.

We recall that an admissible map $(p, q) : X \multimap X$ $((p, q) : (X, A) \multimap (X, A))$ is called a Lefschetz map if $(p, q)_* : H(X) \longrightarrow H(X)$ $((p, q)_* : H(X, A) \longrightarrow H(X, A))$ is a Leray endomorphism (see Chapter I.6, for details). We shall consider, for the sake of simplicity, admissible maps of the form $(p, q) : X \multimap X$, but all formulations remain the true for pairs of spaces. We let:

$$\Lambda(p, q) = \Lambda((p, q)_*),$$
$$\chi(p, q) = \chi((p, q)_*),$$
$$L(p, q) = L((p, q)_*),$$
$$P(p, q) = P((p, q)_*),$$

whenever (p, q) is a Lefschetz map.

Then Proposition (6.1) can be reformulated as follows:

(6.5) PROPOSITION. *An admissible map* $(p, q) : X \multimap X$ *is a Lefschetz map if and only if p is any iterate φ^n of φ. In this case, we have:*

$$\chi(p, q) = \chi((p, q)^n),$$

where $(p, q)^n = \underbrace{(p, q) \circ \ldots \circ (p, q)}_{n\text{-times}}.$

Finally, the preceding discussion can be summarized as follows (cf. Proposition (6.4)).

(6.6) THEOREM. *Let* $(p, q) : X \multimap X$ *be a Lefschetz map. We have:*

(6.6.1) $\chi(p, q) \neq 0$ *implies* $P(p, q) \neq 0$;

(6.6.2) $P(p, q) \neq 0$ *if and only if* $\Lambda((p, q)^n) \neq 0$, *for some natural n*;

(6.6.3) $P(p, q) = k \neq 0$ *implies that, for any natural m, at least one of the coefficients*

$$\Lambda((p, q)^{m+1}), \Lambda((p, q)^{m+2}), \ldots, \Lambda((p, q)^{m+k})$$

of the series $L(p, q)$ must be different from zero.

For a multivalued map $(p, q) : (X, A) \multimap (X, A)$, a point $x \in X$ is called *n-periodic* if $x \in \text{Fix}((p, q)^n)$.

According to the Chapter II.5, by X and A we shall denote retracts of some open sets in a locally convex space E. We shall also assume that $A \subset X$.

We are able to prove:

(6.7) THEOREM (Periodic Point Theorem). *Let* $(p, q) : X \multimap X$ $((p, q) : (X, A) \multimap (X, A))$ *be a $\mathbb{C}AC$-map.*

If $\chi(p,q) \neq 0$ or $P(p,q) \neq 0$, then (p,q) has an n-periodic point (an n-periodic point in $\overline{X \setminus A}$), for some $m + 1 \leq n \leq m + P(p,q)$ and arbitrary $m \geq 0$.

PROOF. It follows from (6.18) in Chapter I.6 (resp. (5.3) in Chapter II) that $\Lambda(p,q)$ is well-defined. In view of (6.6), it is sufficient to assume that $P(p,q) \neq 0$. Then, using (6.18) in Chapter I.6 (resp. (5.3) in Chapter II.5), we deduce that $\text{Fix}((p,q)^n) \neq \emptyset$ (and $\text{Fix}((p,q)^n) \cap \overline{X \setminus A} \neq \emptyset$). Observe that every point $x \in \text{Fix}((p,q)^n)$ is n-periodic for (p,q). The proof is completed. $\square$

Consider a map:

$$X \xleftarrow{p} \Gamma \xrightarrow{q} X.$$

A sequence of points $(z_1, \ldots, z_k)$, satisfying $z_i \in \Gamma$, $i = 1, \ldots, k$, $q(z_i) = z_{i+1}$, $i = 1, \ldots, k-1$ and $q(z_k) = z_1$, will be called a *k-periodic orbit of coincidences*, for (p,q).

Let us note that, for $(p,q) = (\text{id}_X, f)$, a k-periodic orbit of coincidences equals the orbit of periodic points for f. In what follows, we shall consider periodic orbits of coincidences with the fixed first element $(z_1, \ldots, z_k)$, unless otherwise stated. Then $(z_2, z_3, \ldots, z_k, z_1)$ is considered as another periodic orbit. We say that two orbits $(z_1, \ldots, z_k)$, $(z_1', \ldots, z_k')$ are *cyclically equal* if

$$(z_1', \ldots, z_k') = (z_l, \ldots, z_k; z_1, \ldots, z_{l-1}), \quad \text{for an } l = 1, \ldots, k.$$

Otherwise, we call them *cyclically different*. Let us also note that, in the single-valued case, a k-periodic point x determines the whole orbit $\{x, fx, f^2 x, \ldots, f^{k-1} x\}$, but for a multivalued map $X \xleftarrow{p} \Gamma \xrightarrow{q} X$, there can be distinct orbits starting from a given z_1 (the second element z_2 satisfies only $z_2 \in q^{-1}(pz_1)$, so it need not be uniquely determined).

Let us denote $\Gamma_k = \{(z_1, \ldots, z_k); z_i \in \Gamma, qz_i = pz_{i+1}, i = 1, \ldots, k-1\}$. We define the maps $p_k, q_k : \Gamma_k \to X$ by the formulae $p_k(z_1, \ldots, z_k) = p(z_1)$, $q_k(z_1, \ldots, z_k) = q(z_k)$.

(6.8) REMARK. A sequence of points $(z_1, \ldots, z_k) \in \Gamma_k$ is an orbit of coincidences if and only if $(z_1, \ldots, z_k) \in C(p_k, q_k)$.

Thus, the study of k-periodic orbits of coincidences reduces to the coincidences of the pair $X \xleftarrow{p_k} \Gamma_k \xrightarrow{q_k} X$. Here, we shall try to find an estimation of the number of k-orbits of coincidences of the pair (p,q). However, to lift the multivalued map $X \xleftarrow{p} \Gamma \xrightarrow{q} X$ to covering spaces, we need the following assumption. Let X be compact space which admits a universal covering $p_X : \widetilde{X} \to X$.

(A) $p : \Gamma \Rightarrow X$ is a Vietoris map and there exists a map $\widetilde{q}$ making the following commutative diagram:

$$\begin{array}{ccccc}
\widetilde{X} & \xleftarrow{\ \widetilde{p}\ } & \widetilde{\Gamma} & \xrightarrow{\ \widetilde{q}\ } & \widetilde{X} \\
{\scriptstyle p_X}\downarrow & & {\scriptstyle p_\Gamma}\downarrow & & \downarrow{\scriptstyle p_X} \\
X & \xleftarrow{\ p\ } & \Gamma & \xrightarrow{\ q\ } & X
\end{array}$$

($\widetilde{\Gamma} = \{(\widetilde{x}, z) \in \widetilde{X} \times \Gamma;\ p_X(\widetilde{x}) = p(z)\}$ is the pullback with natural projections $\widetilde{p}(\widetilde{x}, z) = \widetilde{x}$, $p_\Gamma(\widetilde{x}, z) = z$).

We assume that (p, q) satisfies (A) and we fix such a lift $\widetilde{q}$.

(6.9) LEMMA. *If (p, q) satisfies* (A), *then so does* (p_k, q_k).

PROOF. We prove that $p_k : \Gamma_k \Longrightarrow X$ is Vietoris. In fact, we note that p_k equals the composition $\Gamma_k \xrightarrow{r_k} \Gamma_{k-1} \xRightarrow{r_{k-1}} \ldots \xRightarrow{r_2} \Gamma_1 \xRightarrow{p} X$, where $r_i(z_1, \ldots, z_i) = (z_1, \ldots, z_{i-1})$. Now, each r_i is Vietoris, because

$$r_i^{-1}(z_1^0, \ldots, z_{i-1}^0) = \{(z_1^0, \ldots, z_{i-1}^0, z_i);\ p(z_i) = q(z_{i-1}^0)\} \sim p^{-1}(q(z_{i-1}^0))$$

is acyclic. Thus, p_k is Vietoris as the composition of Vietoris maps.

It remains to prove that (p_k, q_k) admits a lift to covering spaces. Let us denote $\widetilde{\Gamma}_k = \{(\widetilde{x}, (z_1, \ldots, z_k)) \in \widetilde{X} \times \Gamma_k;\ p_X(\widetilde{x}) = p_k(z_1, \ldots, z_k)\}$ the standard pullback making the diagram

$$\begin{array}{ccc}
\widetilde{X} & \xleftarrow{\ \widetilde{p}_k\ } & \widetilde{\Gamma}_k \\
{\scriptstyle p_X}\downarrow & & \downarrow{\scriptstyle p_{\Gamma_k}} \\
X & \xleftarrow{\ p_k\ } & \Gamma_k
\end{array}$$

commutative with $\widetilde{p}_k(\widetilde{x}, (z_1, \ldots, z_k)) = \widetilde{x}$, $p_{\Gamma_k}(\widetilde{x}, (z_1, \ldots, z_k)) = (z_1, \ldots, z_k)$.

Now, it is possible to give a formula of the required lift $\widetilde{q}_k$. However, it will be more convenient to replace the above pullback by another (isomorphic) diagram and then to define an appropriate lift (denoted by $\overline{q}_k$).

We show that the above diagram is isomorphic to the diagram

$$\begin{array}{ccc}
\widetilde{X} & \xleftarrow{\ \overline{p}_k\ } & \overline{\Gamma}_k \\
{\scriptstyle p_X}\downarrow & & \downarrow{\scriptstyle p_{\overline{\Gamma}_k}} \\
X & \xleftarrow{\ p_k\ } & \Gamma_k
\end{array}$$

where $\overline{\Gamma}_k = \{(\overline{z}_1, \ldots, \overline{z}_k);\ \overline{z}_i \in \widetilde{\Gamma},\ (i = 1, \ldots, k)\ \widetilde{q}\overline{z}_i = \widetilde{p}\overline{z}_{i+1},\ (i = 1, \ldots, k-1)\}$, $p_{\overline{\Gamma}_k}(\overline{z}_1, \ldots, \overline{z}_k) = (p_\Gamma(\overline{z}_1), \ldots, p_\Gamma(\overline{z}_k))$ and $\overline{p}_k(\overline{z}_1, \ldots, \overline{z}_k) = \widetilde{p}\overline{z}_1$.

More precisely, we show that there exists a bijective map $\varphi : \widetilde{\Gamma}_k \to \widetilde{\Gamma}_k$ making the following diagram commutative.

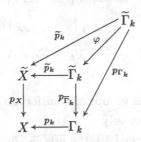

This will follow from the next lemma. Thus, in the sequel, we can consider, instead of the first diagram, the second one which turns out to be more convenient in calculations. $\qquad\square$

(6.10) LEMMA. *For any point* $(\widetilde{x}, (z_1, \dots, z_k)) \in \widetilde{\Gamma}_k$, *there exists a unique point* $(\overline{z}_1, \dots, \overline{z}_k) \in \overline{\Gamma}_k$ *satisfying* $\widetilde{p}(\overline{z}_1) = \widetilde{x}$ *and* $\mathrm{pr}(\overline{z}_i) = z_i$ $(i = 1, \dots, k)$. *The map* $\varphi : \widetilde{\Gamma}_k \longrightarrow \overline{\Gamma}_k$ *given by this correspondence* $(\varphi(\widetilde{x}; z_1, \dots, z_k)) = (\overline{z}_1, \dots, \overline{z}_k))$ *is a bijection for which the second diagram commutes. The opposite map* $\psi : \overline{\Gamma}_k \to \widetilde{\Gamma}_k$ *is given by the formula* $\psi(\overline{z}_1, \dots, \overline{z}_k) = (\widetilde{p}\overline{z}_1; \mathrm{pr}\,\overline{z}_1, \dots, \mathrm{pr}\,\overline{z}_k)$.

PROOF. We define $(\overline{z}_1, \dots, \overline{z}_k) \in \overline{\Gamma}_k$ by induction. Let $\overline{z}_1 = (\widetilde{x}, z_1)$. Suppose that $\overline{z}_i$ is already defined. We put $\overline{z}_{i+1} = (\widetilde{q}(\overline{z}_i), z_{i+1})$. Then, we check:

(1) $\overline{z}_{i+1} \in \widetilde{\Gamma}$.

This follows from $p_X(\widetilde{q}(\overline{z}_i)) = q\,\mathrm{pr}(\overline{z}_i) = q(z_i) = p(z_{i+1})$.

(2) $(\overline{z}_1, \dots, \overline{z}_k) \in \overline{\Gamma}_k$.

This follows from the equalities $\widetilde{p}(\overline{z}_{i+1}) = \widetilde{p}(\widetilde{q}(\overline{z}_i), z_{i+1}) = \widetilde{q}(\overline{z}_i)$.

The sequence is unique, because $\mathrm{pr}(\overline{z}_i) = z_i$ is determined by the given sequence $(z_1, \dots, z_k)$ and $\widetilde{p}(\overline{z}_i) = \widetilde{q}(\overline{z}_{i-1})$ is determined by the preceding element $\overline{z}_{i-1}$.

One easily checks that the map ψ is well-defined and $\varphi\psi$ and $\psi\varphi$ are the identities. It remains to check that the diagram commutes, i.e. $p_{\overline{\Gamma}_k}\varphi = \mathrm{pr}_k$ and $\overline{p}_k\varphi = \widetilde{p}_k$.

But $p_{\overline{\Gamma}_k}\varphi(\widetilde{x}, (z_1, \dots, z_k)) = p_{\overline{\Gamma}_k}(\overline{z}_1, \dots, \overline{z}_k) = (z_1, \dots, z_k) = \mathrm{pr}_k(\widetilde{x}, (z_1, \dots, z_k))$ and, moreover,

$$\overline{p}_k\varphi(\widetilde{x}, (z_1, \dots, z_k)) = \overline{p}_k(\overline{z}_1, \dots, \overline{z}_k) = \widetilde{p}(\overline{z}_1) = \widetilde{x} = \widetilde{p}_k(\widetilde{x}, (z_1, \dots, z_k)). \qquad\square$$

PROOF OF LEMMA (6.9) (CONTINUED). By the above, we can consider (in the pullback diagram) $\overline{\Gamma}_k$, instead of $\widetilde{\Gamma}_k$. Now, the map $\overline{q}_k : \overline{\Gamma}_k \longrightarrow \widetilde{X}$ given by the formula $\bar{q}_k(\bar{z}_1, \ldots, \bar{z}_k) = \tilde{q}(\bar{z}_k)$ makes the following diagram commutative:

$$
\begin{array}{ccccc}
\widetilde{X} & \xleftarrow{\;\overline{p}_k\;} & \overline{\Gamma}_k & \xrightarrow{\;\overline{q}_k\;} & \widetilde{X} \\
{\scriptstyle p_X}\downarrow & & {\scriptstyle p_{\overline{\Gamma}_k}}\downarrow & & {\scriptstyle p_X}\downarrow \\
X & \xleftarrow{\;p_k\;} & \Gamma_k & \xrightarrow{\;q_k\;} & X
\end{array}
$$

$\square$

Let $\mathcal{O}_X, \mathcal{O}_\Gamma, \mathcal{O}_{\overline{\Gamma}_k}$ denote the groups of transformations of coverings: $\widetilde{X} \longrightarrow X$, $\overline{\Gamma} \longrightarrow \Gamma$ and $\overline{\Gamma}_k \longrightarrow \Gamma_k$, respectively.

By the arguments in Chapter I.10, the lifts $\overline{p}_k, \overline{q}_k$ define homomorphisms $\overline{p}_k^!$: $\mathcal{O}_X \longrightarrow \mathcal{O}_{\Gamma_k}$ and $\overline{q}_{k!} : \mathcal{O}_{\overline{\Gamma}_k} \longrightarrow \mathcal{O}_X$ by the commutative diagrams:

$$
\begin{array}{ccc}
\widetilde{X} \xleftarrow{\;\overline{p}_k\;} \overline{\Gamma}_k & \qquad & \overline{\Gamma}_k \xrightarrow{\;\overline{q}_k\;} \widetilde{X} \\
{\scriptstyle\alpha}\downarrow \quad {\scriptstyle\overline{p}_k^!(\alpha)}\downarrow & & {\scriptstyle\alpha}\downarrow \quad {\scriptstyle\overline{q}_{k!}(\alpha)}\downarrow \\
\widetilde{X} \xleftarrow{\;\overline{p}_k\;} \overline{\Gamma}_k & & \overline{\Gamma}_k \xrightarrow{\;\overline{q}_k\;} \widetilde{X}
\end{array}
$$

Hence, we can define the action of $\mathcal{O}_X$ on itself by the formula $\alpha \circ \beta = \alpha \cdot \beta \cdot [(\overline{q}_k)_!(\overline{p}_k)^!(\alpha^{-1})]$. The quotient set will be called the *set of Reidemeister classes* and will be denoted by $\mathcal{R}(\overline{p}_k, \overline{q}_k)$ (cf. Chapter I.10). Notice that the lifts $\overline{p}_k, \overline{q}_k$, are fixed, because the lifts $\tilde{p}_k, \tilde{q}_k$ were fixed. The composition $(\overline{q}_k)_!(\overline{p}_k)^!$ is a generalization of the homomorphism $\rho_\# : \mathcal{O}_X \longrightarrow \mathcal{O}_X$ induced by a (single-valued) map $\rho : X \to X$ (or, equivalently, to the homomorphism of the fundamental group $\rho_\# : \pi_1 X \longrightarrow \pi_1 X$). In fact, if (p, q) represents a single-valued map ρ (i.e. $\rho(x) = qp^{-1}(x)$), then $\rho_\# = q_! p^!$ (see Chapters I.10 and II.5).

(6.11) LEMMA. *Let $\alpha \in \mathcal{O}_X$ and let $(\bar{z}_1, \ldots, \bar{z}_k) \in \overline{\Gamma}_k$, $\bar{z}_i = (\tilde{x}_i, z_i)$. Let us denote $\overline{p}_k^!(\alpha)(\bar{z}_1, \ldots, \bar{z}_k) = (\bar{z}_1', \ldots, \bar{z}_k')$, where $\bar{z}_i' = (\tilde{x}_i', z_i)$. Then*

$$
\tilde{x}_i' = [\tilde{q}_! \tilde{p}^!]^{i-1}(\alpha)(\tilde{x}_i).
$$

Conversely, if $(\bar{z}_1, \ldots, \bar{z}_k)$, $(\bar{z}_1', \ldots, \bar{z}_k') \in \overline{\Gamma}_k$ satisfy the above equality, then

$$
\overline{p}_k^!(\alpha)(\bar{z}_1, \ldots, \bar{z}_k) = (\bar{z}_1', \ldots, \bar{z}_k').
$$

PROOF. It is enough to show that the map $\overline{\Gamma}_k \ni (\bar{z}_1, \ldots, \bar{z}_k) \longrightarrow (\bar{z}_1', \ldots, \bar{z}_k') \in \overline{\Gamma}_k$ is

(1) well-defined, i.e. $(\bar{z}_1', \ldots, \bar{z}_k') \in \overline{\Gamma}_k$;

(2) a natural transformation of the covering space;

and (3) the above diagram on the left commutes.

Ad (1). We check that $\widetilde{q}(\overline{z}_i') = \widetilde{p}(\overline{z}_{i+1}')$. We have:

$$\widetilde{q}(\overline{z}_i') = \widetilde{q}(([\widetilde{q}_!\widetilde{p}^!]^{i-1}(\alpha))(\widetilde{x}_i), z_i) = \widetilde{q}(\widetilde{p}^!([\widetilde{q}_!\widetilde{p}^!]^{i-1}(\alpha))(\widetilde{x}_i, z_i))$$
$$= (\widetilde{q}_!\widetilde{p}^!([\widetilde{q}_!\widetilde{p}^!]^{i-1}(\alpha))\widetilde{q}(\widetilde{x}_i, z_i)) = ([\widetilde{q}_!\widetilde{p}^!]^i(\alpha))(\widetilde{q}(\overline{z}_i)) = ([\widetilde{q}_!\widetilde{p}^!]^i(\alpha))\widetilde{p}(\overline{z}_{i+1})$$
$$= ([\widetilde{q}_!\widetilde{p}^!]^i(\alpha))(\widetilde{x}_{i+1}) = \widetilde{x}_{i+1}' = \widetilde{p}(\overline{z}_{i+1}').$$

Ad (2). This follows from $p_\Gamma \overline{z}_i = z_i = p_\Gamma \overline{z}_i'$.

Ad (3).

$$\overline{p}_k(\overline{p}_k^!(\alpha))(\overline{z}_1, \dots, \overline{z}_k) = \overline{p}_k(\overline{z}_1', \dots, \overline{z}_k') = \widetilde{p}(\overline{z}_1') = \widetilde{x}_1' = \alpha\widetilde{x}_1$$
$$= \alpha\widetilde{p}(\overline{z}_1) = \alpha\overline{p}_k(\overline{z}_1, \dots, \overline{z}_k). \qquad \square$$

(6.12) COROLLARY. $(\overline{q}_k)_!(\overline{p}_k)^!(\alpha) = [\overline{q}_!\overline{p}^!]^k(\alpha)$.

PROOF. Let us notice that

$$\overline{q}_k(\overline{p}_k^!(\alpha)(\overline{z}_1, \dots, \overline{z}_k)) = \overline{q}_k(\overline{z}_1', \dots, \overline{z}_k') = \widetilde{q}(\overline{z}_k')$$
$$= \widetilde{q}([\widetilde{q}_!\widetilde{p}^!]^{k-1}(\alpha)(\widetilde{x}_k), z_k) = \widetilde{q}(\widetilde{p}_!([\widetilde{q}_!\widetilde{p}^!]^{k-1}(\alpha))(\widetilde{x}_k, z_k))$$
$$= \widetilde{q}_!(\widetilde{p}_!([\widetilde{q}_!\widetilde{p}^!]^{k-1}(\alpha))\widetilde{q}(\widetilde{x}_k, z_k))$$
$$= [\widetilde{q}_!\widetilde{p}^!]^k(\alpha)\overline{q}_k(\overline{z}_1, \dots, \overline{z}_k).$$

On the other hand, by the commutativity of the right diagram,

$$\overline{q}_k(\overline{p}_k^!(\alpha)(\overline{z}_1, \dots, \overline{z}_k)) = \overline{q}_{k!}\overline{p}_k^!(\alpha)\overline{q}_k(\overline{z}_1, \dots, \overline{z}_k),$$

which proves the equality

$$(\overline{q}_k)_!(\overline{p}_k)^!(\alpha) = [\widetilde{q}_!\widetilde{p}^!]^k(\alpha). \qquad \square$$

(6.13) COROLLARY. $(\overline{q}_k)_!(\overline{p}_k)^! = [(\widetilde{q}_l)_!(\widetilde{p}_l)^!]^{k/l}$ for $l|k$.

(6.14) LEMMA. Let $j_{k,l} : C(p_l, q_l) \longrightarrow C(p_k, q_k)$, $i_{k,l} : \mathcal{O}_X \to \mathcal{O}_X$ be given by the formula

$$j_{kl}(z) = \underbrace{(z, \dots, z)}_{(k/l)\text{-times}},$$

$$i_{kl}(\alpha) = \alpha(\overline{q}_!\overline{p}^!)^l(\alpha)(\overline{q}_!\overline{p}^!)^{2l}(\alpha)\dots(\overline{q}_!\overline{p}^!)^{(r-1)l}(\alpha).$$

Then $j_{kl}(p_{\overline{\Gamma}_l}(C(\overline{p}_l, \alpha\overline{q}_l))) \subset p_{\overline{\Gamma}_k}(C(\overline{p}_k, i_{kl}(\alpha)\overline{q}_k))$.

PROOF. Let $\overline{z} = (\overline{z}_1, \dots, \overline{z}_l) \in C(\overline{p}_l, \alpha\overline{q}_l)$ and let $z = p_{\Gamma_l}(\overline{z})$. We define an element $\widehat{z} \in (\widetilde{\Gamma})^k$ for which we show that

(1) $\widehat{z} \in \overline{\Gamma}_k$;
(2) $\widehat{z} \in C(\overline{p}_k, i_{kl}(\alpha)\overline{q}_k)$;
(3) $p_{\overline{\Gamma}_k}(\widehat{z}) = j_{kl}(p_{\overline{\Gamma}_l}(\overline{z}))$.

We define $\widehat{z} = (p^\mathsf{l}(\gamma_1)\overline{z}_1, \ldots, p^\mathsf{l}(\gamma_k)\overline{z}_k)$, where we put $\overline{z}_j = \overline{z}_i$, for $i \leq l$ congruent to j modulo l. Moreover, $\gamma_i \in \mathcal{O}_X$ are defined inductively as follows.

Let us denote, for the sake of simplicity, $\rho = \widetilde{q}_!\widetilde{p}^\mathsf{l}$. We put $\gamma_1 = \mathrm{id}$ and, for $i > 1$,

if i does not divide l, then we put $\gamma_{i+1} = \rho(\gamma_i)$ and,

if i divides l, then we put $\gamma_{i+1} = \rho(\gamma_i)\alpha^{-1}$.

Thus, the sequence $\gamma_1, \ldots, \gamma_k$ is of the form

$$\mathrm{id}, \mathrm{id}, \ldots, \mathrm{id};$$
$$\alpha^{-1}, \rho(\alpha^{-1}), \ldots, \rho^{l-1}(\alpha^{-1});$$
$$\rho^l(\alpha^{-1})\alpha^{-1}, \rho^{l+1}(\alpha^{-1})\rho(\alpha^{-1}), \ldots, \rho^{2l-1}(\alpha^{-1})\rho^{l-1}(\alpha^{-1});$$
$$\cdots\cdots\cdots\cdots\cdots\cdots\cdots\cdots\cdots\cdots\cdots\cdots\cdots\cdots\cdots\cdots\cdots\cdots\cdot$$
$$\rho^{(r-2)l}(\alpha^{-1})\rho^{(r-3)l}(\alpha^{-1})\ldots\rho^l(\alpha^{-1})\alpha^{-1}, \ldots,$$
$$\rho^{(r-1)l-1}(\alpha^{-1})\rho^{(r-2)l-1}(\alpha^{-1})\ldots\rho^{l-1}(\alpha^{-1})$$

(in each line, there are l elements).

Ad (1). To show that $\widehat{z} \in \overline{\Gamma}_k$, we check that $\widetilde{q}(\widetilde{p}^\mathsf{l}(\gamma_i)\overline{z}_i) = \widetilde{p}(\widetilde{p}^\mathsf{l}(\gamma_{i+1})\overline{z}_{i+1})$. Let us assume that i is not a multiplicity of l. Then

$$\widetilde{q}(\widetilde{p}^\mathsf{l}(\gamma_i)\overline{z}_i) = \widetilde{q}_!\widetilde{p}^\mathsf{l}(\gamma_i)\widetilde{q}(\overline{z}_i) = \widetilde{q}_!\widetilde{p}^\mathsf{l}(\gamma_i)\widetilde{p}(\overline{z}_{i+1})$$
$$= \gamma_{i+1}\widetilde{p}(\overline{z}_{i+1}) = \widetilde{p}(\widetilde{p}^\mathsf{l}(\gamma_{i+1})\overline{z}_{i+1}).$$

Now, we suppose that $i \geq l$ is a multiplicity of l. Then

$$\widetilde{q}(\widetilde{p}^\mathsf{l}(\gamma_i)\overline{z}_i) = \widetilde{q}_!\widetilde{p}^\mathsf{l}(\gamma_i)\widetilde{q}(\overline{z}_i) = \widetilde{q}_!\widetilde{p}^\mathsf{l}(\gamma_i)\alpha^{-1}\widetilde{p}(\overline{z}_{i+1})$$
$$= \gamma_{i+1}\widetilde{p}(\overline{z}_{i+1}) = \widetilde{p}(\widetilde{p}^\mathsf{l}(\gamma_{i+1})\overline{z}_{i+1}).$$

Ad (2). Now,

$$\widetilde{p}^\mathsf{l}(\gamma_k) = \widetilde{p}^\mathsf{l}((\widetilde{q}_!\widetilde{p}^\mathsf{l})^{(r-1)l-1}(\alpha^{-1})(\widetilde{q}_!\widetilde{p}^\mathsf{l})^{(r-2)l-1}(\alpha^{-1})\ldots(\widetilde{q}_!\widetilde{p}^\mathsf{l})^{l-1}(\alpha^{-1}))$$

(see the form of γ_k given above).

Moreover, $\overline{p}_k(\widehat{z}) = \widetilde{p}(\overline{z}_1)$ and

$$\overline{q}_k(\widehat{z}) = \widetilde{q}(\widetilde{p}^\mathsf{l}(\gamma_k)\overline{z}_k) = \widetilde{q}_!\widetilde{p}^\mathsf{l}(\gamma_k)\widetilde{q}(\overline{z}_k)$$
$$= \widetilde{q}_!(\widetilde{p}^\mathsf{l}[(\widetilde{q}_!\widetilde{p}^\mathsf{l})^{(r-1)l-1}(\alpha^{-1})\ldots(\widetilde{q}_!\widetilde{p}^\mathsf{l})^{l-1}(\alpha^{-1})]\widetilde{q}(\overline{z}_k)$$
$$= (\widetilde{q}_!\widetilde{p}^\mathsf{l})^{(r-1)l}(\alpha^{-1})\ldots(\widetilde{q}_!\widetilde{p}^\mathsf{l})^l(\alpha^{-1})\alpha^{-1}\widetilde{p}(\overline{z}_1).$$

Thus,

$$\alpha(\widetilde{q}_!\widetilde{p}^\mathsf{l})^l(\alpha)\ldots(\widetilde{q}_!\widetilde{p}^\mathsf{l})^{(r-1)l}(\alpha)\overline{q}_k(\widehat{z}) = \widetilde{p}(\overline{z}_1),$$

and subsequently $i_{kl}(\alpha)\bar{q}_k(\hat{z}) = \bar{p}_k(\hat{z})$.

Ad (3).

$$p_{\Gamma_k}(\hat{z}) = p_{\Gamma_k}(\bar{p}^!(\gamma_1)\bar{z}_1, \ldots, \bar{p}^!(\bar{\gamma}_k)\bar{z}_k) = (z_1, \ldots, z_l; \ldots; z_1, \ldots, z_l)$$
$$= j_{kl}(z_1, \ldots, z_l) = j_{kl}(z) = j_{kl}(p_{\overline{\Gamma}_l}(\bar{z})). \qquad \square$$

In this part, we show that, as in the single-valued case, the set of periodic coincidences splits into disjoint closed and open subsets (Nielsen classes).

Since periodic coincidences are exactly coincidences of the map $X \xLeftarrow{p_k} \Gamma_k \xrightarrow{q_k} X$, we can apply Lemma (10.8) in Chapter I.10 to this map (and the covering $\tilde{X} \xLeftarrow{\tilde{p}_k} \overline{\Gamma}_k \xrightarrow{\tilde{q}_k} \tilde{X}$).

(6.15) LEMMA. *Let the multivalued map* $X \xLeftarrow{p} \Gamma \xrightarrow{q} X$ *satisfy* (A). *Then*

(1) $C(p_k, q_k) = \bigcup_{\alpha \in \mathcal{O}_X} p_{\overline{\Gamma}_k}(C(\bar{p}_k, \alpha\bar{q}_k))$.

(2) *If* $p_{\overline{\Gamma}_k}(C(\bar{p}_k, \alpha\bar{q}_k)) \cap p_{\overline{\Gamma}_k}(C(\bar{p}_k, \alpha'\bar{q}_k)) \neq \emptyset$, *then* $[\alpha] = [\alpha'] \in \mathcal{R}(\bar{p}_k, \bar{q}_k)$.

(3) *If* $[\alpha] = [\alpha'] \in \mathcal{R}(\bar{p}_k, \bar{q}_k)$, *then* $p_{\overline{\Gamma}_k}(C(\bar{p}_k, \alpha\bar{q}_k)) = p_{\overline{\Gamma}_k}(C(\bar{p}_k, \alpha'\bar{q}_k))$.

Thus, $C(p_k, q_k) = \bigcup p_\Gamma(C(\bar{p}_k, \alpha\bar{q}_k))$ *is a disjoint sum, where the summation runs one* α *from each Reidemeister class in* $\mathcal{R}(\bar{p}_k, \bar{q}_k)$.

Thus, any Reidemeister class $[\alpha]$ determines a subset $p_{\overline{\Gamma}_k}(C(\bar{p}_k, \alpha\bar{q}_k))$ which will be called the *Nielsen class*. The set of all (nonempty) Nielsen classes will be denoted by $\mathcal{N}(p_k, q_k)$. Thus, we have a natural injection $\mathcal{N}(p_k, q_k) \to \mathcal{R}(p_k, q_k)$.

Let us notice that following the arguments in Chapter I.10, we can reformulate all the above, when replacing the universal covering $p_X : \tilde{X} \to X$ by any finite regular covering; i.e. for any normal subgroup $H \triangleleft \mathcal{O}_X$ of a finite index we take a covering $p_{XH} : \tilde{X}_H \to X$ determined by this group. This allows us to define the spaces $\tilde{\Gamma}_H$, $\tilde{\Gamma}_{kH}$, $\overline{\Gamma}_{kH}$ and the homomorphisms $\tilde{p}_H^!, \tilde{q}_{!H}, \bar{p}_H^!, \bar{q}_{!H}$. This generalization works once we assume the condition

(AH) $p : \Gamma \Longrightarrow X$ is a Vietoris map and there exists a map $\tilde{q}_H$ making the following diagram commutative

$$
\begin{array}{ccccc}
\tilde{X}_H & \xleftarrow{\tilde{p}_H} & \tilde{\Gamma}_H & \xrightarrow{\tilde{q}_H} & \tilde{X}_H \\
\downarrow{\scriptstyle p_{XH}} & & \downarrow{\scriptstyle p_{\Gamma H}} & & \downarrow{\scriptstyle p_{XH}} \\
X & \xleftarrow{p} & \Gamma & \xrightarrow{q} & X.
\end{array}
$$

Observe that (A) implies (AH) for any normal subgroup of a finite index $H \triangleleft \mathcal{O}_X$. In particular, Lemma (6.15) becomes

(6.16) LEMMA. *Let the multivalued map* $X \xleftarrow{p} \Gamma \xrightarrow{q} X$ *satisfy* (AH). *Then*

(1) $C(p_k, q_k) = \bigcup_{\alpha \in \mathcal{O}_{XH}} p_{\overline{\Gamma}_{kH}}(C(\overline{p}_{kH}, \alpha \overline{q}_{kH}))$.

(2) *If* $p_{\overline{\Gamma}_{kH}}(C(\overline{p}_{kH}, \alpha \overline{q}_{kH})) \cap p_{\overline{\Gamma}_{kH}}(C(\overline{p}_{kH}, \alpha' \overline{q}_{kH})) \neq \emptyset$, *then*

$$[\alpha] = [\alpha'] \in \mathcal{R}(\overline{p}_{kH}, \overline{q}_{kH}).$$

(2) *If* $[\alpha] = [\alpha'] \in \mathcal{R}(\overline{p}_{kH}, \overline{q}_{kH})$, *then* $p_{\overline{\Gamma}_k}(C(\overline{p}_k, \alpha \overline{q}_k)) = p_{\overline{\Gamma}_k}(C(\overline{p}_{kH}, \alpha' \overline{q}_{kH}))$. *Thus,* $C(p_k, q_k) = \bigcup p_\Gamma(C(\overline{p}_{kH}, \alpha \overline{q}_{kH}))$ *is a disjoint sum, where the summation runs one* α *from each Reidemeister class from* $\mathcal{R}(\overline{p}_{kH}, \overline{q}_{kH})$.

Since the spaces $\widetilde{\Gamma}$, $\overline{\Gamma}_k$ can be quite arbitrary, we are not able to follow the case of polyhedra or ANR's to define essential classes. We will follow the ideas presented in Chapter I.10. We fix a normal subgroup $H \lhd \mathcal{O}_X$ of a finite index satisfying $\widetilde{q}_! \widetilde{p}^!(H) \subset H$ and we say that an H-Nielsen class $p_\Gamma(C(\widetilde{p}, \alpha \widetilde{q}))$ is essential if $\Lambda(\widetilde{p}, \alpha \widetilde{q}) \neq 0$. The last (Lefschetz) number is defined, because the map is CAC as the finite covering of (p, q).

Observe that we have a natural map

$$\nu : \mathcal{R}(p_k, q_k) \to \mathcal{R}(p_k, q_k)$$

given by $\nu[\alpha] = [\widetilde{q}_! \widetilde{p}^!(\alpha)]$. One can check that this is well-defined and, moreover, $\nu^k[\alpha] = [\widetilde{q}_! \widetilde{p}^!(\alpha)] = [\alpha]$. Hence, we get an action of $\mathbb{Z}_k$ on $\mathcal{R}(p_k, q_k)$. We can talk about *orbits of Reidemeister classes*. On the other hand, we define a map $\nu' : C(p_k, q_k) \longrightarrow C(p_k, q_k)$ by $\nu'(z_1, \ldots, z_k) = (z_2, \ldots, z_n, z_1)$.

(6.17) LEMMA. *The following diagram commutes:*

$$
\begin{array}{ccc}
\mathcal{N}(p_k, q_k) & \xrightarrow{\nu'} & \mathcal{N}(p_k, q_k) \\
\mu \downarrow & & \mu \downarrow \\
\mathcal{R}(p_k, q_k) & \xrightarrow{\nu} & \mathcal{R}(p_k, q_k)
\end{array}
$$

where μ *denotes the natural inclusion.*

PROOF. Let $(z_1, \ldots, z_k) \in C(p_k, q_k)$ be of the form

$$(z_1, \ldots, z_k) = p_{\overline{\Gamma}_k}(\overline{z}_1, \ldots, \overline{z}_k), \quad \text{for an } (\overline{z}_1, \ldots, \overline{z}_k) \in C(\overline{p}_k, \alpha \overline{q}_k) \subset \overline{\Gamma}_k.$$

Then, $\nu\mu([z_1, \ldots, z_k]) = \nu(\alpha) = [\widetilde{q}_! \widetilde{p}^!(\alpha)]$.

On the other hand, $\nu'[z_1, \ldots, z_k] = [z_2, \ldots, z_k, z_1]$. Hence, it remains to show that $(z_2, \ldots, z_k, z_1) \in p_{\overline{\Gamma}_k}(C(\overline{p}_k, \widetilde{q}_! \widetilde{p}^!(\alpha) \overline{q}_k))$. But this is true since

$$(\overline{z}_2, \ldots, \overline{z}_k, \widetilde{p}^!(\alpha^{-1})\overline{z}_1) \in C(\overline{p}_k, (\widetilde{q}_! \widetilde{p}^!(\alpha))\overline{q}_k)$$

and

$$p_{\overline{\Gamma}_k}(\overline{z}_2, \dots, z_k, \widetilde{p}^!(\alpha^{-1})\overline{z}_1) = (z_2, \dots, z_k, z_1).$$

Since the last equality is evident, it remains to show the inclusion.

We notice that $\overline{p}_k(\overline{z}_2, \dots, \overline{z}_k, \widetilde{p}^!(\alpha^{-1})\overline{z}_1) = \widetilde{p}\overline{z}_2$, while

$$(\widetilde{q}_!\widetilde{p}^!(\alpha))\overline{q}_k(\overline{z}_2, \dots, \overline{z}_k, \widetilde{p}^!(\alpha^{-1})\overline{z}_1) = ((\widetilde{q}_!\widetilde{p}^!(\alpha))(\widetilde{q}\widetilde{p}^!(\alpha^{-1}))\overline{z}_1)$$
$$= ((\widetilde{q}_!\widetilde{p}^!(\alpha))(\widetilde{q}_!\widetilde{p}^!(\alpha^{-1}))\widetilde{q}\overline{z}_1) = \widetilde{q}\overline{z}_1.$$

Now, the both sides are equal, because $(z_1, \dots, z_k) \in \overline{\Gamma}_k$. $\qquad\square$

(6.18) COROLLARY. *The map* $j_{kl} : C(p_l, q_l) \to C(p_k, q_k)$ *sends the Nielsen class corresponding to* $[\alpha] \in \mathcal{R}(\widetilde{p}_l, \widetilde{q}_l)$ *to the Nielsen class corresponding to* $[i_{kl}(\alpha)] \in \mathcal{R}(\widetilde{p}_k, \widetilde{q}_k)$. *In other words, the following diagram commutes:*

$$
\begin{array}{ccc}
\mathcal{N}(p_l, q_l) & \xrightarrow{\ j_{kl}\ } & \mathcal{N}(p_k, q_k) \\
\downarrow & & \downarrow \\
\mathcal{R}(\overline{p}_l, \overline{q}_l) & \xrightarrow{\ i_{kl}\ } & \mathcal{R}(\overline{p}_k, \overline{q}_k)
\end{array}
$$

(6.19) COROLLARY. *If the Nielsen classes* A, $A' \subset C(p_k, q_k)$ *contain cyclically equal orbits* $(x_1, \dots, x_k) \in A$, $(x_{i+1}, \dots, x_k; x_1, \dots, x_i) \in A'$, *then* $\mu(A)$, $\mu(A')$ *belong to the same orbit of the Reidemeister classes.*

PROOF. Since $(x_{i+1}, \dots, x_k; x_1, \dots, x_i) = \nu'^i(x_1, \dots, x_k)$, we have $A' = \nu^i(A)$. Now, $\mu(A') = \mu\nu'^i(A) = \nu^i\mu(A)$. $\qquad\square$

An orbit of the Reidemeister classes will be called *essential* if

$$\Lambda(\overline{p}_k, \alpha\overline{q}_k) \neq 0,$$

for an α from this orbit. By the above lemma, the coincidence set is nonempty for the lift corresponding to *any* Reidemeister class from an essential orbit.

The Corollary (6.13) makes the following definitions consistent.

(6.20) DEFINITION. A k-orbit of coincidences $(z_1, \dots, z_k)$ is called *reducible* if

$$(z_1, \dots, z_k) = j_{kl}(z_1, \dots, z_l), \quad \text{for an } l < k \text{ dividing } k.$$

(6.21) DEFINITION. A Reidemeister class $[\alpha] \in \mathfrak{R}(\widetilde{p}_k, \widetilde{q}_k)$ is called *reducible* if it lies in the image of $i_{kl} : \mathcal{R}(\overline{p}_l, \widetilde{q}_l) \to \mathcal{R}(\overline{p}_k, \widetilde{q}_k)$, for an $l < k$ dividing k.

(6.22) DEFINITION. An orbit of Reidemeister classes in $\mathcal{R}(\overline{p}_k, \widetilde{q}_k)$ is called *reducible* if it contains a reducible element.

(6.23) LEMMA. *Let $[\alpha] \in \mathcal{R}(\overline{p}_k, \overline{q}_k)$ represent an essential irreducible orbit of Reidemeister classes. Then $p_{\overline{\Gamma}_k}(C(\overline{p}_k, \alpha \overline{q}_k) \neq \emptyset$. Moreover, if*

$$(z_1, \ldots, z_k) \in p_{\overline{\Gamma}_k}(C(\overline{p}_k, \alpha \overline{q}_k)),$$

then $(z_1, \ldots, z_k) \neq (z_1, \ldots, z_l; z_1, \ldots, z_l; z_1, \ldots, z_l)$, for any $l \mid k$, $l < k$.

PROOF. $p_{\overline{\Gamma}_k}(C(\overline{p}_k, \alpha \overline{q}_k)) \neq \emptyset$ follows from the essentiality of $[\alpha]$ and Lemma (6.11). Suppose that $(z_1, \ldots, z_k) = (z_1, \ldots, z_l; z_1, \ldots, z_l; z_1, \ldots, z_l)$. Then

$$(z_1, \ldots, z_l) \in C(p_l, q_l) \quad \text{and} \quad j_{kl}(z_1, \ldots, z_l) = (z_1, \ldots, z_k).$$

Assume that

$$(z_1, \ldots, z_l) \in p_{\overline{\Gamma}_k}(C(\overline{p}_l, \beta \overline{q}_l)).$$

Lemma (6.11) implies $(z_1, \ldots, z_k) \in p_{\overline{\Gamma}_k}(C(\overline{p}_k, j_{kl}(\beta)\overline{q}_l))$. Now,

$$(z_1, \ldots, z_k) \in p_{\overline{\Gamma}_k}(C(\overline{p}_k, \alpha \overline{q}_k)) \cap p_{\overline{\Gamma}_k}(C(\overline{p}_k, j_{kl}(\beta)\overline{q}_k))$$

which implies $[\alpha] = [i_{kl}(\beta)] = i_{kl}([\beta])$. Thus, $[\alpha]$ is reducible which contradicts to the assumption. $\qquad \square$

Let $S_k(\widetilde{p}, \widetilde{q})$ denote the number of irreducible and essential orbits in $\mathcal{R}(\overline{p}_k, \overline{q}_k)$.

(6.24) THEOREM. *The multivalued map (p, q) satisfying condition (A) and $\mathbb{C}AC$ has at least $S_k(\widetilde{p}, \widetilde{q})$ irreducible cyclically different orbits of coincidences.*

PROOF. We choose an orbit of points from each essential irreducible orbit of the Reidemeister classes. By Corollary (6.18), they are cyclically different. By Lemma (6.11), they are also irreducible. $\qquad \square$

(6.25) REMARK. Since the essentiality is a homotopy invariant and irreducibility is defined in terms of Reidemeister classes, $S_k(\widetilde{p}, \widetilde{q})$ is a homotopy invariant.

II.7. Fixed point index for condensing maps

In this section, E is a locally convex space, C a convex subset of E and U an open subset of C. In the first part of this section, we would like to define the fixed point index for mappings in $\mathbb{K}(U, C)$ (comp. Chapter I.8). Then, in the second part, we want to do this for mappings in $\mathbb{C}^0(U, C) = \{\varphi \in \mathbb{C}(U, C) \mid \mathrm{Fix}(\varphi)$ is a compact subset of $U\}$ (comp. Chapter I.5). Obviously, $\mathbb{K}_0(U, C) \subset \mathbb{C}^0(U, C)$. In fact, we have more, namely:

$$\mathbb{K}_0(U, C) \subset \mathbb{C}^0_p(U, C) \subset \mathbb{C}^0(U, C), \quad \text{for every } p \in [0, 1),$$

where

$$\mathbb{C}_p^0(U,C) = \{\varphi \in \mathbb{C}_p(U,C) \mid \mathrm{Fix}(\varphi) \text{ is a compact subset of } U\}.$$

In what follows, an admissible map $\varphi : U \multimap C$ will be identified with the pair (p,q), as in Chapters I.8, II.5 and II.6.

Note that, for such a pair (p,q), we are able to define a unique number as the fixed point index, Lefschetz number, etc.

Let $(p,q) \in \mathbb{K}_0(U,C)$. We denote $K = \overline{qp^{-1}(U)}$. By the hypothesis, K is a compact subset which contains the set $\mathrm{Fix}(p,q)$.

(7.1) PROPOSITION. *There exists $V \in \mathcal{N}_E(0)$ ($\mathcal{N}_E(0)$ denotes the basis of topology at $0 \in E$) such that, for every V-approximation $\pi_V : K \to C$ of the inclusion $i : K \to C$, i.e., a continuous map $\pi_V : K \to C$ such that $\pi_V(x) \in (x + V)$ and $\pi_V(K) \subset L$, where L is a finite dimensional subspace of E (see (1.37) in Chapter I.1), we have that $\mathrm{Fix}(p, \pi_V \circ q)$ is a compact subset of U.*

The proof of (7.1) is straightforward (comp. Chapter I.1).

Note that if (7.1) holds true in some $V \in \mathcal{N}_E(0)$ and $V_1 \in \mathcal{N}_E(0)$ is a subset of V, then (7.1) is also true for V_1.

Now, for given $(p,q) \in \mathbb{K}_0(U,C)$, we choose V and $\pi_V : K \to C$ such that $\pi_V(x) \in (x + V)$ and $\pi_V(K) \subset L$, where L is a finite dimensional subspace of E.

We let:

$$U^L = U \cap L \quad \text{and} \quad C^L = C \cap L$$

and

$$U^L \xleftarrow{p^L} p^{-1}(U^L) \xrightarrow{q^L} C^L, \quad \text{where } q^L = \pi_V \circ q.$$

Then the set $\mathrm{Fix}(p^L, q^L)$ is a compact subset of U^L and $(p^L, q^L) \in \mathbb{K}_0(U^L, C^L)$. Since C^L, as a convex subset of L, is an ANR-space, the fixed point index $\mathrm{ind}((p^L, q^L), U^L)$ is well-defined (comp. Chapter I.8 or [Go5-M]).

We define

(7.2) $$\mathrm{ind}((p,q),U) = \mathrm{ind}((p^L, q^L), U^L).$$

Now, it is a standard procedure (comp. Chapter I.8) to verify that (7.2) is correct.

Moreover, Theorem (8.18) in Chapter I.8 can be reformulated in the considered case. So, properties (8.18.1)–(8.18.5) hold true.

(7.3) REMARK. Observe that if $(p,q) \in \mathbb{K}_0(C,C)$, i.e., $U = C$, then from (8.18.5), we have

$$\mathrm{ind}((p,q),C) = \Lambda(p,q),$$

but convexity of C implies that $\Lambda(p,q) = 1$, for every $(p,q) : C \multimap C$. So, we have

$$\mathrm{ind}((p,q),C) = 1.$$

Consequently, from the existence property (8.18.2), we get that $\mathrm{Fix}(p,q) \neq \emptyset$. For single-valued mappings, we have proved it in (1.38), Chapter I.1.

(7.4) COROLLARY. *If* $(p,q) \in \mathbb{K}(C,C)$, *then* $\mathrm{Fix}(p,q) \neq \emptyset$.

Starting from now, we shall assume that E is a quasicomplete locally convex space, C is a convex subset of E, U is an open subset of C and $\mu : 2^E \to A$ is an abstract measure of noncompactness (see (5.31) in Chapter I.5).

(7.5) PROPOSITION. *If* $\varphi \in \mathbb{C}(U,C)$, *then* $\mathrm{Fix}(\varphi)$ *is relatively compact.*

For the proof, see (5.28) in Chapter I.5.

We shall also use the notion of a fundamental set introduced in 1972 by M. A. Krasnosielskiĭ, W. W. Strygin and P. P. Zabreiko [KSZ].

(7.6) DEFINITION. Let $(p,q) \in \mathbb{C}(U,C)$. A compact, convex, nonempty set $S \subset C$ is called a *fundamental set* if:

(7.6.1) $q(p^{-1}(U \cap S)) \subset S$,
(7.6.2) if $x \in \overline{\mathrm{conv}}(\varphi(x) \cup S)$, then $x \in S$.

A set S is called ω-*fundamental* for (p,q) if S satisfies all conditions of (7.6), but it is not necessarily compact.

Below, we collect some important properties of fundamental sets.

(7.7) PROPERTIES. *Assume* $(p,q) \in \mathbb{C}(U,C)$.

(7.7.1) *If* S *is a fundamental set for* (p,q), *then* $\mathrm{Fix}(p,q) \subset S$,
(7.7.2) *if* S *is an* ω-*fundamental set, for* (p,q), *and* $P \subset S$, *then the set* $S' = \overline{\mathrm{conv}}(qp^{-1}(S \cap U) \cup P)$ *is* ω-*fundamental, for* (p,q), *too,*
(7.7.3) *intersection of any family of fundamental sets* (ω-*fundamental sets*), *for* (p,q), *is also fundamental, for* (p,q),
(7.7.4) *if* S *is the intersection of all* ω-*fundamental sets, for* (p,q), *then we have:*

$$S = \overline{\mathrm{conv}}(qp^{-1}(S \cap U)),$$

(7.7.5) *for every compact* $K \subset C$, *there exists an* ω-*fundamental set such that*

$$S = \overline{\mathrm{conv}}(qp^{-1}(S \cap U) \cup K),$$

(7.7.6) *the family of all fundamental sets for* (p,q) *is nonempty.*

PROOF. Observe that (7.7.1)–(7.7.4) are obvious. To prove (7.7.5), we let:

$$\mathcal{M} = \{S \subset \overline{C} \mid S \text{ is fundamental, for } (p,q), \text{ and } K \subset S\}.$$

It is easy to see that the set

$$S_0 = \bigcap \{S \mid S \in \mathcal{M}\}$$

is fundamental for (p,q).

In view of (7.7.4), the set

$$S' = \overline{\text{conv}}(qp^{-1}(S_0 \cap U) \cup K)$$

is fundamental, too.

On the other hand, we have $S' \subset S_0$, because $K \subset S'$. It implies that $S' = S_0$. So our claim is proved, provided $\mathcal{M} \neq \emptyset$, but it is evident that $\overline{C} \in \mathcal{M}$. The proof of (7.7.5) is completed.

To prove (7.7.6), we let $K = \{y\}$, for an arbitrary $y \in C$. In view of (7.7.5), there exists an ω-fundamental set S such that:

$$S = \overline{\text{conv}}(qp^{-1}(S \cap U) \cup \{y\}).$$

Consequently, it is enough to prove that S is compact. Indeed, if we assume on the contrary, that S is not compact, then from the fact that (p,q) is condensing follows:

$$\mu(S) = \mu(\overline{\text{conv}}(qp^{-1}(S \cap U) \cup \{y\})) = \mu(qp^{-1}(S \cap U) \cup \{y\}) < \mu(S),$$

and we get a contradiction. The proof of (7.7.6), and subsequently of (7.7), is completed. □

Now, we are able to extend the fixed point index to the class $\mathbb{C}^0(U, C)$. In what follows, we shall additionally assume, that C is a closed convex subset of a locally convex space.

From the above assumption and (7.7), it follows, for any $(p,q) \in \mathbb{C}(U, C)$, that there exists a fundamental set S such that $S \subset C$.

Let $(p,q) \in \mathbb{C}^0(U, C)$ and $S = S(p,q)$ be a fundamental set for (p,q). Then $U_S = U \cap S$ is an open subset of S and the compact set $\text{Fix}(p,q)$ is contained in U_S.

For $(p,q) : U \xleftarrow{p} \Gamma \xrightarrow{q} C$, we consider

$$(p_S, q_S) : U_S \xleftarrow{p_S} p^{-1}(U_S) \xrightarrow{q_S} S,$$

where p_S, q_S are the respective restrictions of p and q.

Therefore, $(p_S, q_S) \in \mathbb{K}_0(U_S, S)$ and $\mathrm{ind}((p_S, q_S), U_S)$ is defined as in (7.2). We define

$$(7.8) \qquad \mathrm{ind}((p, q), U) = \mathrm{ind}((p_S, q_S), U_S).$$

It is straigtforward that definition (7.8) is correct.

Moreover, properties (8.18.1)–(8.18.4) in Chapter I.8 can be reformulated for our index defined in (7.8).

The problem of the normalization property is more complicated. At first, let us observe that we are able to obtain the following

(7.9) THEOREM. *If $(p, q) \in \mathbb{C}(C, C)$ (in particular, $(p, q) \in \mathbb{C}_\lambda(C, C)$, $\lambda \in [0, 1]$) with C to be a closed convex subset of a locally convex space E, then $\mathrm{Fix}(p, q) \neq \emptyset$.*

PROOF. Let $(p, q) : C \xleftarrow{p} \Gamma \xrightarrow{q} C$ and let S be a fundamental set for (p, q). We define:

$$(p_S, q_S) : S \xleftarrow{p_S} p^{-1}(S) \xrightarrow{q_S} C$$

(as in (7.8)).

Then $\mathrm{Fix}(p, q) = \mathrm{Fix}(p_S, q_S)$, but, in view of (6.20) in Chapter I.6, we get $\mathrm{Fix}(p_S, q_S) \neq \emptyset$, and the proof is completed. $\qquad\square$

Now, assume that U is an open subset of C and $(p, q) : U \multimap U$ is condensing (or, in particular, a λ-set contraction) and $\mathrm{Fix}(p, q)$ is compact.

(7.10) OPEN PROBLEM.

(7.10.1) *Is it true that the Lefschetz number $\Lambda((p, q))$ of (p, q) is well-defined?*

(7.10.2) *Assume that we know that $\Lambda((p, q))$ of (p, q) is well-defined. Is it true that $\Lambda((p, q)) \neq 0$ implies $\mathrm{Fix}(p, q) \neq \emptyset$?*

We shall return to the fixed point theory for condensing mappings in Chapters II.10 and II.11.

II.8. Approximation methods in the fixed point theory of multivalued mappings

Many results developed in Chapters I.6–I.11 and II.5–II.7 can be obtained by means of approximation methods, instead of homological methods. Let us add that approximation methods are analytical and simpler than homological, but the obtained results are unfortunately weaker. By an approximation, we mean the one on the graph of multivalued mappings by single-valued maps. In this section, we shall sketch all possible results. For details, we recommend [GGa-M] (see also: [Kry1-M], [Kry2-M] and [Go2-M]).

In what follows, we assume at least that X, Y are Hausdorff topological spaces and $\varphi : X \multimap Y$ is an u.s.c. map with closed values.

(8.1) DEFINITION. Let W be an open neighbourhood of Γ_φ in $X \times Y$ and $A \subset X$. A continuous map $f : A \to Y$ is called a W-*approximation* of φ (on the graph) if $\Gamma_f \subset W$.

By an *admissible covering* α of the graph Γ_φ of φ we shall understand the following family:

$$\alpha = \{U_x^\alpha \times W_x^\alpha \mid x \in U_x^\alpha \text{ and } \varphi(x) \subset W_x^\alpha, \text{ for every } x \in X, \text{ and}$$
$$U_x^\alpha \text{ is open in } X \text{ and } W_x^\alpha \text{ is open in } Y\}.$$

We let:

$$\mathcal{U}(\varphi) = \{\alpha \mid \alpha \text{ is an admissible covering of } \Gamma_\varphi\}$$

and

$$|\alpha| = \bigcup_{x \in X} (U_x^\alpha \times W_x^\alpha), \quad \alpha \in \mathcal{U}(\varphi).$$

Assume that $\alpha, \beta \in \mathcal{U}(\varphi)$. We shall say that β is a *subcovering* of α (written $\beta < \alpha$) if $U_x^\beta \subset U_x^\alpha$ and $W_x^\beta \subset W_x^\alpha$, for every $x \in X$.

If X and Y are metric spaces, then we define the diameter $\delta(\alpha)$ of $\alpha \in \mathcal{U}(\varphi)$ by putting

$$\delta(\alpha) = \max\{\sup_{x \in X} \delta(U_x^\alpha), \sup(\inf\{r > 0 \mid W_x^\alpha \subset O_r(\varphi(x))\})\}.$$

(8.2) DEFINITION. Let $\alpha \in \mathcal{U}(\varphi)$, $A \subset X$ and $\varphi : X \multimap Y$ be an u.s.c. map with closed values. A continuous map $f : A \to Y$ is called an α-*approximation* of φ if $\Gamma_f \subset |\alpha|$ (or, equivalently, $\forall x \in A \, \exists y \in X \, (x, f(x)) \in U_y^\alpha \times W_y^\alpha$).

We let

$$a(A, \varphi, \alpha) = \{f : A \to Y \mid f \text{ is an } \alpha\text{-approximation of } \varphi\}$$

and

$$a(\varphi, \alpha) = a(X, \varphi, \alpha).$$

Now, assume that E, E_1 are two locally convex topological spaces, $X \subset E$ and $Y \subset E_1$. Let us denote by $\mathcal{N}_E(0)$ (resp. $\mathcal{N}_{E_1}(0)$) the basis of topology at $0 \in E$ ($0 \in E_1$) consisting of all open neighbourhoods of 0 which are star-shaped w.r.t. 0 (λU) $\in \mathcal{N}_E(0)$ ($\mathcal{N}_{E_1}(0)$), for every $\lambda \neq 0$, and $U \in \mathcal{N}_E(0)$ ($\mathcal{N}_{E_1}(0)$) and $(U + V) \in \mathcal{N}_E(0)$ ($\mathcal{N}_{E_1}(0)$), for every $U, V \in \mathcal{N}_E(0) \times \mathcal{N}_{E_1}(0)$.

(8.3) DEFINITION. Let $(U, V) \in \mathcal{N}_E(0) \times \mathcal{N}_{E_1}(0)$, $A \subset X$ and $\varphi : X \multimap Y$ be an u.s.c. map with closed values. A continuous map $f : A \to Y$ is called a (U, V)-*approximation* of φ if:

$$f(x) \in ((\varphi(x + U) \cap X) + V) \cap Y, \quad \text{for every } x \in A.$$

We let:

$$a(A, \varphi, U, V) = \{f : A \to Y \mid f \text{ is a } (U, V)\text{-approximation of } \varphi\}$$

and

$$a(\varphi, U, V) = a(X, \varphi, U, V).$$

Below, we shall formulate the connections between (8.1)–(8.3).

(8.4) THEOREM. *Let* $X \subset E$, $Y \subset E_1$ *and* $\varphi : X \multimap Y$ *be an u.s.c. map with closed values. Consider the following conditions:*

(8.4.1) *for every open neighbourhood* W *of* Γ_φ *in* $X \times Y$, *there exists a* W-*approximation of* φ,

(8.4.2) *for every* $\alpha \in \mathcal{U}(\varphi)$, *there exists an* α-*approximation of* φ,

(8.4.3) *for every* $(U \times V) \in \mathcal{N}_E(0) \times \mathcal{N}_{E_1}(0)$, *there exists a* (U, V)-*approximation of* φ.

Then we have:

   (i) (8.4.1) $\Leftrightarrow$ (8.4.2),

  (ii) (8.4.1) $\Rightarrow$ (8.4.3),

 (iii) (8.4.3) $\Rightarrow$ (8.4.1), *whenever* X *is compact.*

The proof of (8.4) is straightforward and therefore it is left to the reader (see: [GGa-M, Theorem 2.2]). Note also that the following properties can be easily verified (see: [GGa-M, Theorem 2.4]).

(8.5) THEOREM. *Let* X, Y *and* φ *be the same as in* (8.4).

(8.5.1) *Assume that* X *is compact and let* Z *be a subset of a locally convex space. If* $\psi : Y \multimap Z$ *is an u.s.c. map with compact values, then, for every* $\alpha \in \mathcal{U}(\psi \circ \varphi)$, *there are* $\beta_1 \in \mathcal{U}(\varphi)$ *and* $\beta_2 \in \mathcal{U}(\psi)$ *such that, for every* $f_1 \in a(\varphi, \beta_1)$, $f_2 \in a(\psi, \beta_2)$, *we have*

$$f_2 \circ f_1 \in a(\psi \circ \varphi, \alpha).$$

(8.5.2) *Let* C *be a compact subset of* X *and* $y \in Y$. *If* $\varphi_+^{-1}(y) \cap C = \emptyset$, *then there exists* $\alpha_0 \in \mathcal{U}(\varphi)$ *such that* $f^{-1}(y) \cap C = \emptyset$, *for every* $f \in a(\varphi, \alpha_0)$.

(8.5.3) *Let C be a compact subset of X. Then, for every $\alpha \in \mathcal{U}(\varphi|_C)$, there exists $\beta \in \mathcal{U}(\varphi)$ such that $f|_C \in a(\varphi|_C, \alpha)$, whenever $f \in a(\varphi, \beta)$.*

(8.5.4) *Let X be compact and $\chi : X \times [0,1] \multimap Y$ be an u.s.c. map with compact values. Denote by $\chi_t : X \multimap Y$ the map defined by $\chi_t(x) = \chi(x,t)$, for every $t \in [0,1]$ and $x \in X$. Then, for every $t \in [0,1]$ and for every $\alpha \in \mathcal{U}(\chi_t)$, there exists $\beta \in \mathcal{U}(\chi)$ such that $h_t \in a(\chi_t, \alpha)$, whenever $h \in \mathcal{U}(\chi, \beta)$.*

(8.5.5) *Let Z and T be subsets of locally convex spaces E_1 and E_2, respectively. Assume, furthermore, that $\psi : Z \multimap Z$ is an u.s.c. map with closed values and define $\varphi \times \psi : X \times Z \multimap Y \times Z$ by putting $(\varphi \times \psi)((x,z)) = \varphi(x) \times \psi(z)$. Then, for every $\alpha \in \mathcal{U}(\varphi \times \psi)$, there are $\beta_1 \in \mathcal{U}(\varphi)$ and $\beta_2 \in \mathcal{U}(\psi)$ such that $f \times g \in a(\varphi \times \psi, \alpha)$, whenever $f \in a(\varphi, \beta_1)$ and $g \in a(\psi, \beta_2)$.*

Let $X \subset E$ and $Y \subset E_1$, C be a compact subset of X and $y \in Y$. We let:

$$AP_0(X,Y) = \{\varphi : X \multimap Y \mid \forall \alpha \in \mathcal{U}(\varphi) \text{ we have } a(\varphi, \alpha) \neq \emptyset\}$$

$$AP_0(X) = AP_0(X,X),$$

$$AP(X,Y) = \{\varphi \in AP_0(X,Y) \mid \forall \alpha \in \mathcal{U}(\varphi) \, \exists \beta \in \mathcal{U}(\varphi) \, f,g \in a(\varphi, \beta)$$
$$\Rightarrow \exists h : X \times [0,1] \to Y \; h(x,0) = f(x), \; h(x,1) = g(x)$$
$$\text{and } h_t \in a(\varphi, \alpha), \text{ for every } t\},$$

$$AP_C(X,Y;y) = \{\varphi \in AP(X,Y) \mid \varphi_+^{-1}(y) \cap C = \emptyset\}$$

$$J(X,Y) = \{\varphi : X \multimap Y \mid \varphi \text{ is u.s.c. with (see I.(2.25))}$$
$$\varphi(x) \in PC_Y^\infty, \text{ for every } x \in X\},$$

$$J(X) = J(X,X),$$

$$J_C(X,Y;y) = \{\varphi \in J(X,Y) \mid \varphi_+^{-1}(y) \cap C = \emptyset\}.$$

In what follows, we shall say that $AP_0(X,Y)$ is the class of *approximable mappings* and $AP(X,Y)$ is the class of *strongly approximable mappings*.

Observe also that if $f : X \to Y$ is a continuous map and $\psi \in J(Y,Z)$, then $(\psi \circ f) \in J(X,Z)$.

We also need the notion of homotopy in the set $AP_C(X,Y;y)$.

(8.6) DEFINITION. Let $\varphi, \psi \in AP_C(X,Y;y)$. We shall say that φ and ψ are homotopic (written $\varphi \sim_C \psi$) if there exists $\chi \in AP_0(X \times [0,1], Y)$ such that:

$$\chi(x,0) = \varphi(x), \; \chi(x,1) = \psi(x), \quad \text{for every } x \in X$$
$$\text{and } y \notin \chi(x,t), \quad \text{for every } x \in C \text{ and } t \in [0,1].$$

Note that "$\sim_C$" is an equivalence relation. We let

$$[AP_C(X,Y;y)] = AP_C(X,Y;y)/\sim_C$$

to be the set of all homotopy classes of $AP_C(X, Y; y)$ with respect to "$\sim_C$".

Now, if we denote by:

$$S_C(X, Y; y) = \{f : X \to Y \mid f \text{ is continuous and } f(x) \neq y, \text{ for every } x \in C\}$$

then we have the relation "$\sim_C$" with χ to be a single-valued map, and so we get:

$$[S_C(X, Y; y)] = S_C(X, Y; y) / \sim_C .$$

Let $\varphi \in AP_C(X, Y; y)$ and assume that C is closed in X. In view of (8.5.2), there exists $\alpha_0 \in \mathcal{U}(\varphi)$ such that, for every $f \in a(\varphi, \alpha_0)$, we have $f^{-1}(y) \cap C = \emptyset$. Since $\varphi \in AP(X, Y)$, there exists $\alpha \in \mathcal{U}(\varphi)$, $\alpha < \alpha_0$ such that, for every $f, g \in a(\varphi, \alpha)$, we have $f \sim_C g$. Consequently, we can define:

(8.7) $F : [AP_C(X, Y; y)] \to [S_C(X, Y; y)]$

by putting:

$$F([\varphi]) = [f],$$

where $f \in a(\varphi, \alpha)$ and α is chosen as above.

Correctness of (8.7) immediately follows from (8.5). Since

$$S_C(X, Y; y) \subset AP_C(X, Y; y),$$

we infer that F is onto.

We can give:

(8.8) THEOREM. *Let X, Y are subsets of locally convex spaces E, E_1, respectively, and C is a closed subset of the compact set X. Assume, furthermore that, for every covering $\lambda \in \mathrm{Cov}_{E_1}(Y)$ and for every Hausdorff topological space Z, there exists a subcovering $\lambda' \in \mathrm{Cov}_E(Y)$ such that any two λ'-close mappings $f, g : Z \to Y$ are λ-homotopic.*

Under the above assumptions, we have that the map:

$$F : [AP_C(X, Y; y)] \to [S_C(X, Y; y)]$$

is a bijection.

PROOF. For the proof, it is sufficient to show that F is an injection. We claim that this follows from the following:

(8.8.1) $\begin{cases} \text{for every } \varphi \in AP_C(X, Y; y), \text{ there exists} \\ \beta \in \mathcal{U}(\varphi) \text{ such that any } \beta\text{-approximation} \\ \text{of } \varphi \text{ belongs to } [\varphi]. \end{cases}$

In fact, assume that (8.8.1) holds true and let $\varphi, \psi \in AP_C(X, Y; y)$ be such that

$$F([\varphi]) = F([\psi]) = [f],$$

where $f \in a(\varphi, \alpha') \cap a(\psi, \alpha'')$ and $\alpha' \subset \beta(\varphi)$, $\alpha'' \subset \beta(\psi)$ and $\beta(\varphi)$, $\beta(\psi)$ comes from (8.8.1), for φ and ψ, respectively. Then it follows from (8.8.1) that $f \in [\varphi]$ and $f \in [\psi]$, and so $[\varphi] = [\psi]$. We shall complete it by proving (8.8.1). Let $\varphi \in AP_C(X, Y; y)$ and $\alpha \in \mathcal{U}(\varphi)$, $\alpha = \{U_z^\alpha \times W_z^\alpha \mid z \in X\}$ be such that $y \notin W_z^\alpha$, for every $z \in C$. For every $z \in X$, we choose $U_z \in \mathcal{N}_E(0)$ such that:

$$(z + U_z) \cap X \subset U_z^\alpha \quad \text{and} \quad \varphi((z + U_z) \cap X) \subset W_z^\alpha.$$

Since C and $\overline{X} \setminus C$ are compact, we can get:

$$X = \bigcup_{i=1}^m \{(x_i + U_{x_i}) \cap X \mid x_i \in C\} \cup \bigcup_{i=m+1}^k \{(x_i + U_{x_i}) \cap X \mid x_i \in X \setminus C\}$$

and

$$C \subset \bigcup_{i=1}^m \{(x_i + U_{x_i}) \cap X \mid x_i \in C\}.$$

We take $W \in \mathcal{N}_{E_1}(0)$ and $U' \in \mathcal{N}_E(0)$ such that $y \notin (\overline{W_{x_i}^\alpha} + W)$, for every $i = 1, \dots, m$ and $(C + U') \cap X \subset U$. Finally, let $U'' \in \mathcal{N}_E(0)$ be such that: $U'' + U'' \subset U'$.

We define $\alpha' \in \mathcal{U}(\varphi)$ by putting:

$$\alpha' \in \{U_z^{\alpha'} \times W_z^{\alpha'} \mid z \in X\},$$

where $U_z^{\alpha'} = z + (U'' \cap U_z)$ and $W_z^{\alpha'} \bigcap \{W_{x_i}^\alpha \mid z \in X_i + U_{x_i}\}$.

We consider an arbitrary $\beta \in \mathcal{U}(\varphi)$ such that $\beta \subset \alpha'$ and the following condition is satisfied:

if $g, f \in a(\varphi, \beta)$ and $f^{-1}(y) \cap C = \emptyset$, $g^{-1}(y) \cap C = \emptyset$, then $f \sim_C g$ and $h_t \in a(\varphi, \alpha')$, for every $t \in [0, 1]$.

We claim that such a β satisfies (8.8.1).

Indeed, let $f \in a(\varphi, \beta)$. We define $\chi : X \times [0, 1] \multimap Y$ by putting:

$$\chi(x, t) = \begin{cases} \varphi(x) & t \in [0, 1/3], \\ \mathrm{cl}(\bigcup \{W_z^\alpha \mid z \in (x + \overline{U''}) \cap X\}) & t \in [1/3, 2/3), \\ f(x) & t \in [2/3, 1]. \end{cases}$$

Then χ is a required homotopy (for details, see [GGa-M, Theorem 2.7]), and the proof is completed. $\qquad\square$

(8.9) REMARK. Note that if Y is a retract of an open subset U in some locally convex space E, then Y satisfies assumptions of (8.8). One can observe that, for any $\alpha \in \mathrm{Cov}_{E_1}(Y)$, there is a subcovering $\beta = \{W_t\}_{t \in T}$ such that W_t is convex, for every $t \in T$. This observation implies that our claim is true for $Y = W$ being open. Finally, observe that the required property is hereditary with respect to any retraction.

As a very simple application of the above results, we get:

(8.10) THEOREM. *Assume that X is a compact subset of a locally convex space E which possesses the fixed point property with respect to continuous (single-valued) mappings, i.e., any continuous $f : X \to X$ has a fixed point. Then, for any $\varphi \in AP(X)$, we have $\mathrm{Fix}(\varphi) \neq \emptyset$.*

PROOF. Let $\alpha \in \mathcal{U}(\varphi)$ and let $f_\alpha \in a(\varphi, \alpha)$. By the hypothesis, there exists $x_\alpha \in X$ such that $f_\alpha(x_\alpha) = x_\alpha$. It implies that, for any $V \in \mathcal{N}_E(0)$, we have $(x_\alpha + V) \cap \varphi(x_\alpha) \neq \emptyset$, and our theorem follows from (6.14) in Chapter I.6. $\square$

Now, we would like to show how large the class $AP(X, Y)$ is. Namely, we are going to prove the following:

(8.11) THEOREM. *If $X = P$ is a finite polyhedron, then $J(P, Y) \subset AP(P, Y)$.*

We shall prove (8.11) in two steps. At first, we prove:

(8.12) THEOREM. *Let P_0 be a subpolyhedron, of P and $\varphi \in J(P, Y)$. Then, for every $\alpha \in \mathcal{U}(\varphi)$, there exists $\beta \in \mathcal{U}(\varphi)$ such that any approximation $f_0 \in a(P_0, \varphi, \beta)$ can be extend to $f \in a(\varphi, \alpha)$.*

PROOF. Since P is a finite polyhedron, we can assume without any loss of generality that P is a compact subset of the Euclidean space $\mathbb{R}^q$ (with the Euclidean norm).

Let $\varphi \in \mathcal{U}(\varphi)$. We let $n = \dim P \geq n_0 = \dim P_0$. For every $z \in P$, we can find a positive real number $\eta_n(z) > 0$ such that the open ball $B(z, \eta_n(z))$ is a subset of U_z^α and $\varphi(B(z, \eta_n(z)) \subset W_z^\alpha$. Since P is compact, the covering $\{B(z, \eta_n(z))\}_{z \in P}$ possesses a finite subcovering $\{B(z_i^n, \eta_n(z_i^n))\}_{i=1,\dots,r(n)}$. Let $\theta_n > 0$ be the Lebesgue number of this subcovering and let $\eta_n = \min\{\eta_n(z_i^n) \mid i = 1, \dots, r(n)\}$.

We define

$$W_z^{\alpha_n} = \begin{cases} W_{z_i^n}^{\alpha_n} & z = z_i^n, \\ \bigcap\{W_{z_i^n}^{\alpha_n} \mid z \in B(z_i^n, \eta_n(z_i^n)) \cap W_z^\alpha\} & \text{if } i = 1, \dots, r(n) \text{ and } z \notin z_i^n, \end{cases}$$

and

$$\alpha_n\{U_z^{\alpha_n} \times W_z^{\alpha_n} \mid z \in P\}.$$

Assume that we have defined α_{k+1}, for every $k \leq j$, where $j = 0, \ldots, n-1$. It follows from the assumption $\varphi(z) \in PC_Y^\infty$ that there exists $\overline{\alpha_n} \in \mathcal{U}(\varphi)$ such that, for every $z \in P$, we have:

$$U_z^{\overline{\alpha_n}} \subset U_z^{\alpha_{n+1}}, \quad W_z^{\overline{\alpha_n}} \subset W_z^{\alpha_{n+1}},$$

and any continuous map $g : \partial \Delta^n \to W_z^{\overline{\alpha_n}}$ can be extended to a continuous map $g : \Delta^n \to W_z^{\alpha_{k+1}}$ (comp. (2.25) in Chapter I.2) and

$$B(z, \eta_k(z)) \subset U_z^{\overline{\alpha_k}} \quad \text{and} \quad \varphi(B(z, \eta_n(z))) \subset W_z^{\overline{\alpha_k}}.$$

Again, since P is compact, we get

$$P = \bigcup_{z \in P} B(z, \eta_{k(z)}/n) = \bigcup_{i=1}^{r(k)} B(z_i^k, \eta_{n_k(z_i^k)}/n).$$

We denote by θ_k the Lebesgue number of the last covering and let

$$\eta_k = \min\{\eta_{k(z_i^k)} \mid i = 1, \ldots, r(k)\}.$$

Then define

$$W_z^{\alpha_k} = \begin{cases} W_{z_i^k}^{\overline{\alpha_k}} & \text{if } z = z_i^k, \\ \bigcap\{W_{z_i^k}^{\overline{\alpha_k}} \mid z \in B(z_i^k, \eta_{k(z_i^k)}) \cap W_z^\alpha\} & \text{if } z \neq z_i^k \text{ and for } i = 1, \ldots, r(k), \end{cases}$$

and

$$\alpha_k \{U_z^{\alpha_k} \times W_z^{\alpha_k} \mid z \in P\}.$$

Consequently, by induction, we can finally define

$$\alpha_0 \{U_z^{\alpha_0} \times W_z^{\alpha_0} \mid z \in P\},$$

and then we let

$$\beta = \alpha_0 \quad \text{and} \quad f_0 : P_0 \to Y$$

be a β-approximation of φ.

We consider a triangulation (T, T_0) of (P, P_0) such that the diameter of T is less than $\min\{\theta_n, \ldots, \theta_0\}$. It implies that, for every simplex $s \in T$, there exists z_i^k such that $s \subset B(z_i^k, \eta_{k(z_i^k)}/4)$. By T^k, $k = 0, \ldots, n$ we shall denote the n_k-dimensional skeleton of T. Analoguosly, we keep T_0^k for T_0. Finally, $P^k = |T^k|$, $P_0^k = |T_0^k|$ is a simplicial complex determined by T^k (resp. T_0^k). Then $P^n = P$ and $P_0^{n_0} = P_0$.

We shall construct a sequence of continuous maps $\{f^k : P^k \to Y\}_{k=0,\ldots,n}$ such that

(i) $f^k|_{P_0^k} = f_0|_{P_0^k}$, for every $0 \leq k \leq n_0$,
(ii) $f^{k+1}|_{P^k} = f^k$, for every $0 \leq k \leq n-1$,
(iii) f^k is a β-approximation of φ, $k = 0, \ldots, n$, where $\beta_k = \{U_z^{\beta_k} \times W_z^{\beta_k} \mid z \in P\}$ is such that $U_z^{\beta_k} = B(z, \eta_k/n) \cap B(z, \eta_{k(z)})$ and $W_z^{\beta_k} = W_z^{\alpha_k}$.

Let $P^0 = \{x_1, \ldots, x_q, x_{q+1}, \ldots x_p\}$, where $x_i \in P_0^0$, for every $i \leq q$ and $x_i \notin P_0^0$, for $i \geq q+1$.

We define

$$f^0(x_i) = \begin{cases} f_0(x_i), & \text{for } 0 \leq i \leq q, \\ y_i, & \text{where } y_i \in \varphi(x_i), \text{ for } q+1 \leq i \leq p. \end{cases}$$

Obviously, we have defined $f^0, \ldots, f^k$ ($k \leq n-1$) satisfying (i)–(iii). We shall define f^{k+1}.

Let s be a $(k+1)$-dimensional simplex in T and ∂s be its boundary. By the hypothesis, there exists $z_i^k \in P$ such that $s \subset B(z_i^k, \eta_{k(z_i^k)}/4)$. Let $x \in \partial s$, then $x \in P^k$. We find a point $z \in P$ such that $(x, f^k(x)) \in U_z^{\beta_k} \times W_z^{\beta_k}$. Therefore, $\|x - z\|\eta_k/2$ and so:

$$\|z - z_i\| < \eta_{k(z_i^k)}/4 + \eta_{k(z_i^k)}/2 < \eta_k(z_i^k).$$

Consequently,

$$W_z^{\beta_k} = W_z^{\alpha_k} \subset W_{z_i^k}^{\alpha_k}$$

and

$$f^k(x) \in W_{z_i^k}^{\alpha_n}, \quad (x, f^k(x)) \in B(z_i^k, \eta_{k(z_i^k)}/4) \times W_{z_i^k}^{\alpha_k}.$$

If $s \in T_0^{k+1}$, then we let $f^{k+1}|_s = f_0|_s$. If $k+1 > n_0$ or $s \in T_0^{k+1}$, then there exists $f^{k+1} : s \to W_{z_i^k}^{\alpha_{k+1}}$ such that $f^{k+1}|_{\partial s} = f^k|_{\partial s}$.

Now, by a simple calculation, we can show that f^{k+1} is a β_{k+1}-approximation of φ. Therefore, the sequence $f^0, \ldots, f^{k+1}$ satisfies (i)–(iii).

Finally, it is easy to see that f^n, constructed in this way, is the required approximation, i.e., $f^n \in a(\varphi, \alpha)$ and f^n is a continuous extension of f_0. The proof (8.12) is completed. $\qquad \square$

(8.13) COROLLARY. $J(P, Y) \subset AP_0(P, Y)$.

In the second step, we prove:

(8.14) THEOREM. *If* $\varphi \in J(P, Y)$, *then, for every* $\alpha \in \mathcal{U}(\varphi)$, *there exists* $\beta \in \mathcal{U}(\varphi)$ *such that, for every two* $f, g \in a(\varphi, \beta)$, *there exists a continuous homotopy* $h : P \times [0, 1] \to Y$ *such that* $f = h_0$, $g = h_1$ *and* $h_t \in a(\varphi, \alpha)$, *for every* $t \in [0, 1]$.

PROOF. Consider the polyhedron $P' = P \times [0, 1]$ and its standard triangulation T'. Let us recall that, for every simplex $s \in P$ with vertices $\{x_0, \ldots, x_p\}$ (in some triangulation T of P), we get a family of simplices in T' of the following form:

$$x_0 \times \{0\}, \ldots, x_i \times \{0\}, x_i \times \{1\}, \ldots, x_p \times \{1\}, \quad i = 0, \ldots, p.$$

Let $P_0' = (P \times \{0\}) \cup \{P \times \{1\}\}$ and $\varphi' : P \multimap Y$ be defined as follows:

$$\varphi'(x, t) = \varphi(x), \quad \text{for every } (x, t) \in P \times [0, 1].$$

Consider a covering $\alpha' \in \mathcal{U}(\varphi)$, $\alpha = \{U_z^\alpha \times W_z^\alpha \mid z \in P\}$. We define $\alpha' = \{U_{(z,t)}^{\alpha'} \times W_{(z,t)}^{\alpha'} \mid (z, t) \in P'\} \in \mathcal{U}(\varphi)$ by putting:

$$U_{(z,t)}^{\alpha'} = U_z^\alpha \times [0, 1], \quad W_{(z,t)}^{\alpha'} = W_z^\alpha.$$

Since P' is a finite polyhedron, in view of (8.12), there exists $\beta' \in \mathcal{U}(\varphi')$ such that any $h_0 \in a(P_0', \varphi', \beta')$ can be extended to $h \in a(\varphi', \alpha')$. Moreover, let us observe that, for every $z \in P$, there exists $r(z) > 0$ such that

$$B((z, 0), r(z)) \subset U_{(z,0)}^{\beta'} \quad \text{and} \quad W_z^\beta = W_{(z,0)}^{\beta'} \cap W_{(z,1)}^{\beta'}$$

and

$$\beta = \{U_z^\beta \times W_z^\beta \mid z \in P\}.$$

For given $f, g \in a(\varphi, \beta)$, we define $h_0 : P_0' \to Y$ by the formula:

$$h_0(x, t) = \begin{cases} f(x) & t = 0, \\ g(x) & t = 1. \end{cases}$$

Then we have:

$$((x, t), h_0(x, t)) = \begin{cases} ((x, 0), f(x)) & t = 0, \\ ((x, 1), g(x)) & t = 1. \end{cases}$$

If $x \in P$, then there are $z, u \in P$ such that $(x, f(x)) \in U_z^\beta \times W_z^\beta$ and $(x, g(x)) \in U_u^\beta \times W_u^\beta$. Thus, $(z, 0) \in B((z, 0), r(z))$ and $f(x) \in W_{(z,0)}^{\beta'}$. Analogously,

$$((x, 1), g(x)) \in U_{(u,1)}^{\beta'} \times W_{(u,1)}^{\beta'}.$$

It implies that h_0 is a β'-approximation of φ'. Let $h : P' \to Y$ be a continuous extension of h_0 such that $h \in a(\varphi', \alpha')$ and let $(z, t) \in P'$ satisfy:

$$((x, t), h(x, t)) \in U_{(z,t)}^{\alpha'} \times W_{(z,t)}^{\alpha'}.$$

From the definition of α', we deduce that $x \in U_z^\alpha$ and $h_t(x) = h(x, t) \in W_z^\alpha$. Thus, $h_t \in a(\varphi, \alpha)$, for every $t \in [0, 1]$, and the proof of (8.14) is completed. $\square$

(8.15) REMARK. Summing up (8.12) and (8.14), we get (8.11).

Having the above results, many topological consequences can be obtained. In the next section, we shall show how to define the topological degree by means of our approximation approach.

II.9. Topological degree defined by means of approximation methods

We start with some definitions and notations. Let Ω be an open subset of a locally convex space E.

(9.1) DEFINITION. A compact map $\Phi : \overline{\Omega} \multimap E$ is called *decomposable*[7] (written $\Phi \in D(\overline{\Omega}, E)$) if there exists a subset T of a locally convex space E, a map $\gamma \in J(\overline{\Omega}, T)$ and a continuous map $g : T \to E$ such that $\Phi = g \circ \gamma$.

(9.2) REMARK. Note that if, moreover, γ has compact values, then Φ is an admissible map, i.e., $\Phi \in D(\overline{\Omega}, E)$ satisfies I.(4.9) (cf. Remark 1.18 in Part 1B in [Kry2-M]).

If $\Phi \in D(\overline{\Omega}, T)$ and $\Phi = g \circ \gamma$, then the diagram

$$D : \overline{\Omega} \xrightarrow{\gamma} T \xrightarrow{g} E$$

is called a *decomposition* of Φ.

We also let:

$$D_{\partial\Omega}(\overline{\Omega}, E) = \{\Phi \in D(\overline{\Omega}, E) \mid \mathrm{Fix}(\Phi) \cap \partial\Omega = \emptyset\}.$$

By similar arguments as in (8.10), we get:

(9.3) THEOREM. *If $\Phi \in D(E, E)$, then $\mathrm{Fix}(\Phi) \neq \emptyset$.*

We need the notion of a homotopy in $D_{\partial\Omega}(\overline{\Omega}, E)$.

Let $\Phi, \Psi \in D_{\partial\Omega}(\overline{\Omega}, E)$ be two maps with the following decompositions:

$$D_\Phi : \overline{\Omega} \xrightarrow{\gamma_1} T_1 \xrightarrow{g_1} E \quad \text{and} \quad D_\Psi : \overline{\Omega} \xrightarrow{\gamma_2} T_2 \xrightarrow{g_2} E.$$

We say that Φ and Ψ are *homotopic* in $D_{\partial\Omega}(\overline{\Omega}, E)$ (written $\Phi \sim_{\partial\Omega} \Psi$) if there exists a subset T of a locally convex space $\widetilde{E}$ and two maps $\chi \in J(\overline{\Omega} \times [0,1], T)$, $g : T \to E$ such that $H = g \circ \chi$ is compact and there are mappings $k_i : T_i \to T$, $i = 1, 2$ such that

$$k_1 \circ \gamma_1 = \chi \circ i_0, \quad k_2 \circ \gamma_2 = \chi, \quad g \circ k_i = g_1, \quad i = 1, 2.$$

Let $j : \overline{\Omega} \to E$ be the inclusion map. We shall consider the set of compact vector fields $FD(\overline{\Omega}, E)$ on $\overline{\Omega}$ defined as follows:

$$FD(\overline{\Omega}, E) = \{\varphi \mid \varphi = j - \Phi \text{ and } \Phi \in D(\overline{\Omega}, E)\}$$

---

[7]The notion of decomposable maps is not related to the notion of decomposable sets defined in (3.71) in the Chapter I.3.

and
$$FD_{\partial\Omega}(\overline{\Omega}, E) = \{\varphi \mid \varphi = j - \Phi \text{ and } \Phi \in D_{\partial\Omega}(\overline{\Omega}, E)\}.$$
Observe that if $\varphi \in FD_{\partial\Omega}(\overline{\Omega}, E)$, then $0 \notin \varphi(\partial\Omega)$.

If $\varphi = j - \Phi$, then we say that φ is a *vector field* associated with Φ.

Vector fields will be denoted by $\varphi, \psi, \ldots$ and their compact parts by capital $\Phi, \Psi, \ldots$

Two vector fields $\varphi, \psi \in FD_{\partial\Omega}(\overline{\Omega}, E)$ are *homotopic* if their compact parts are homotopic in $D_{\partial\Omega}(\overline{\Omega}, E)$, i.e., $\Phi \sim_{\partial\Omega} \Psi$.

Now, we shall define the topological degree $\mathrm{Deg} : D_{\partial\Omega}(\overline{\Omega}, E) \to \mathbb{Z}$. In Chapter I.8, we defined it by means of the Čech homology functor. Here, using the approximation method, we shall reduce the respective definition to the topological degree of single-valued mappings, i.e., to the Brouwer degree, whenever $E = \mathbb{R}^n$ is a n-dimensional Euclidean space and to the Leray–Schauder degree in an infinite dimensional case.

(9.4) The case $E = \mathbb{R}^n$.

Let $\Phi \in D_{\partial\Omega}(\overline{\Omega}, E)$, $\Phi = g \circ \gamma$. Then $\mathrm{Fix}(\Phi)$ is a compact subset of Ω. Consequently, we can take an open bounded set W such that $\mathrm{Fix}(\Phi) \subset W \subset \overline{W} \subset \Omega$. We can assume, without any loss of generality, that $\overline{W}$ is a finite polyhedron.

We consider $\Phi_W = g \circ (\gamma|_{\overline{W}})$ and $\varphi_W = j - \Phi_w$. In view of (8.5), there exists $\alpha_0 \in \mathcal{U}(\Phi_W)$ such that $\mathrm{Fix}(u) \cap \partial W = \emptyset$, for every $u \in a(\Phi_W, \alpha_0)$. There exists a covering $\alpha_1 \in \mathcal{U}(\gamma|_{\overline{W}})$ such that, for every $\alpha \in \mathcal{U}(\gamma|_{\overline{W}})$, we get
$$g \circ f \in a(\Phi_W, \alpha_0).$$
Finally, there exists a covering $\alpha \in \mathcal{U}(\gamma|_{\overline{W}})$ such that, all $f, k \in (\gamma|_{\overline{W}}, \alpha)$ are homotopic and the joining homotopy $h : \overline{W} \times [0, 1] \to T$ satisfies the following condition: $h_t \in a(\gamma|_{\overline{W}}, \alpha_1)$, for every $t \in [0, 1]$. Observe that then we can define a homotopy $h_W : \overline{W} \times [0, 1] \to E$, $h_W(x, t)g(h(x, t))$, for which we have
$$h_w(\cdot, 0) = g \circ f, \quad h_w(\cdot, 1) = g \circ k, \quad h_w(\cdot, t) \in a(\Phi_W, \alpha_0).$$
Consequently, $h_W(x, t) \neq x$, for every $(x, t) \in \partial W \times [0, 1]$.

Now, let $f \in a(\gamma|_{\overline{W}}, \alpha)$ be an arbitrary approximation. We define the topological degree $\mathrm{Deg}(\varphi, D, \Omega)$ of φ with respect to the decomposition $D : \Omega \xrightarrow{\gamma} T \xrightarrow{g} E$ by putting:

(9.4.1) $\deg(\varphi, D, \Omega) = \deg(\varphi_W, D_{\overline{W}}, W) = \deg(j - g \circ f, W)$,

where $\deg(j - g \circ f, W)$ is the Brouwer degree of the map $(j - g \circ f) : \overline{W} \to \mathbb{R}^n$.

Above, we proved that (9.4.1) does not depend on the choice of the approximation f. Using the localization (or more generally, additivity) property of the Brouwer degree, it is easy to see that (9.4.1) does not depend on the choice of W.

The following example shows us that (9.4.1) depends on the decompositions D.

(9.4.2) EXAMPLE. Let $E = \mathbb{R}^2$ be the field of complex numbers. Let

$$K^2 = \{z \in \mathbb{R}^2 \mid |z| \le 1\}$$

and

$$S^1 = \{z \in \mathbb{R}^2 \mid |z| = 1\}.$$

Let $\gamma : K^2 \multimap K^2$ be defined as follows:

$$\gamma(z) = \begin{cases} se^{i(r+t)} & t \in [0, 3\pi/2], \ z = se^{ir}, \\ 0 & z = 0. \end{cases}$$

Obviously, γ is u.s.c. with contractible values. Hence, $\gamma \in J(K^2, \mathbb{R}^2)$.

Let $p_1, p_2 : K^2 \to \mathbb{R}^2$ be defined by the formulas:

$$p_1(z) = z^2, \quad \text{for every } z \in K^2,$$
$$p_2(z) = z^3, \quad \text{for every } z \in K^2.$$

We have the diagrams:

$$D_1 : K^2 \overset{\overline{\gamma}}{\multimap} K^2 \times K^2 \overset{k_1}{\longrightarrow} \mathbb{R}^2,$$
$$D_2 : K^2 \overset{\overline{\gamma}}{\multimap} K^2 \times K^2 \overset{k_2}{\longrightarrow} \mathbb{R}^2,$$

where $\overline{\gamma}(z, w) = \{(z, u) \mid u \in \gamma(w)\}$, $k_1(u, w) = u - p_1(w)$, $k_2(u, w) = u - p_2(w)$.

Let $\Phi : K^2 \multimap \mathbb{R}^2$ be defined by:

$$\Phi(z) = \{z - v \mid v \in K^2\}.$$

Evidently, $\Phi = k_1 \circ \overline{\gamma} = k_2 \circ \overline{\gamma}$, but $\deg(\varphi, D_1, K^2) = 2$ and $\deg(\varphi, D_2, K^2) = 3$, where $\varphi = j - \Phi$.

Summing up the above, we can say that (9.4.1) is correct for a given pair (Φ, D), i.e., for a given map and its decomposition.

Let us also note that the topological degree defined in (9.4.1) has all properties formulated in (8.9) in Chapter I.8. For more details, see [GGa-M] or [Kry1-M] or [Go5-M].

(9.5) The case of an arbitrary locally convex space E.

Let $\Phi \in D_{\partial\Omega}(\overline{\Omega}, E)$, where E is an arbitrary locally convex space and let $D : \overline{\Omega} \overset{\overline{\gamma}}{\multimap} T \overset{g}{\longrightarrow} E$ be a decomposition of Φ. Obviously, $0 \in \varphi(\partial\Omega)$, whenever $\varphi = j - \Phi$.

We have proved in (8.11) in Chapter I.8 that, under our assumptions, φ is a closed map. Consequently, we deduce that $\varphi(\partial U)$ is a closed subset of E and $0 \notin \varphi(\partial U)$. Therefore, there is $V_0 \in \mathcal{N}(0)$ such that $\varphi(\partial U) \cap V_0 = \emptyset$.

Let $V = V_0/2$. Since the set $K = \overline{\Phi(\Omega)}$ is compact, we infer that:

$$K \subset \Theta = \bigcup_{i=1}^{n}(y_i + V), \quad y_i \in K,$$

and there exists a map $\pi^L : \Theta \to L$, where $L = \mathrm{span}\{y_1, \ldots, y_n\}$ is a subspace of E, spanned on $y_1, \ldots, y_n$, such that $\Omega \cap L \neq \emptyset$ and $(y - \pi^L(y)) \in V$, for every $y \in \Theta$ (see (1.37) in Chapter I.1).

Observe that if $x \in \partial\Omega$ and $z \in \pi^L(\Phi(x))$, then there is $y \in \Phi(x)$ such that $z = \pi^L(y)$. Hence, $(x - y) \notin V$, $(y - z) \in V$ and $x - z = (x - y) + (y - z) \neq 0$.

Since Ω is open and L is a closed subspace of E, we get:

$$\overline{\Omega \cap L} = \overline{\Omega} \cap L \quad \text{and} \quad \partial(\Omega \cap L) = \partial\Omega \cap L.$$

Let us introduce the following notation:

$$\Omega^L = \Omega \cap L, \quad \Phi^L = (\pi^L \circ g) \circ \gamma|_{\overline{\Omega}^L},$$

$$\varphi^L = j - \Phi^L \quad \text{and} \quad D^L : \overline{\Omega}^L \overset{\gamma|_{\overline{\Omega}^L}}{\multimap} T \overset{\pi^L \circ g}{\longrightarrow} L.$$

We define

(9.5.1) $$\mathrm{Deg}(\varphi, D, \Omega) = \deg(\varphi^L, D^L, \Omega^L),$$

where $\deg(\varphi^L, D^L, \Omega^L)$ is defined in (9.4.1).

Now, we shall show that (9.5.1) is correct. Assume that $V_0' \in \mathcal{N}(0)$ is an open neighbourhood of 0 such that

$$\varphi(\partial\Omega) \cap V_0' = \emptyset.$$

For the proof, it is sufficient to consider the cases when $V_0' \subset V_0$. Assume that $V_0' \subset V_0$. Let $V' = V_0'/2$ and

$$\Theta' = \bigcup_{i=n+1}^{m}(y_i + V') \supset K, \quad y_i \in K,$$

$$\pi^{L'} : \Theta' \to L' = \mathrm{span}\{y_{n+1}, \ldots, y_m\} \subset E$$

be such that $\Omega \cap L' \neq \emptyset$ and $(y - \pi^{L'}(y)) \in V'$, for every $y \in \Theta'$.

We define

$$\Theta_0 = \bigcup_{i=1}^{m}(y_i + V) \supset \Theta \cup \Theta'$$

and

$$\pi^G : \Theta_0 \to G = \text{span}\{y_1, \dots, y_n, \dots, y_m\}$$

such that $(y\pi^G(y)) \in V$, for every $y \in \Theta_0$. It is enough to show that

$$\deg(\varphi^L, D^L, \Omega^L) = \deg(\varphi^G, D^G, \Omega^G),$$

where φ^G, D^G, Ω^G are defined according to Θ_0 an G.

But the last equality follows from the localization and contraction (restriction) properties of the topological degree defined in (9.4).

(9.5.2) REMARK. Now, by standard arguments, we can obtain the following properties of the topological degree obtained in (9.5.1):

- additivity,
- existence,
- localization,
- homotopy,
- multiplicity,
- restriction,

(comp. Chapter I.8). For more details, we recommend [GGa-M] or [Kry2-M] or [Go5-M].

(9.5.3) REMARK. The class of admissible mappings for which we considered the topological degree theory in Chapter I.8 is larger than the class of decomposable mappings whose all components are compact valued considered in this section, in general.

We would like to point out one important observation. Namely, if Φ is admissible, then Φ has the following decomposition:

$$D : X \xrightarrow{p^{-1}} \Gamma \xrightarrow{q} Y,$$

i.e., $\Phi = q \circ p^{-1}$. Note that p^{-1} is defined by a Vietoris map $p : \Gamma \Rightarrow X$. Therefore, the values of $p^{-1}(x)$ are compact acyclic.

On the other hand, if we have a decomposable $\Phi : X \multimap Y$ with the decomposition:

$$D : X \xrightarrow{\gamma} \Gamma \xrightarrow{q} Y,$$

then γ is u.s.c. with closed (not necessarily compact) PC_T^∞ values. We also have that $\Phi = g \circ \gamma$. So, in the case of admissible mappings Φ, p^{-1} and q have compact values, but for decomposable mappings only Φ and g have compact values.

We shall complete this section by showing some topological consequences of (9.5.1).

(9.6) THEOREM. *If $\Phi \in D(E, E)$, with a decomposition D, then*

$$\text{Deg}(\varphi, D, E) = 1,$$

where $\varphi = j - \Phi$.

PROOF. Let D be of the form:

$$D : E \xrightarrow{\gamma} T \xrightarrow{g} E.$$

We define a homotopy $h : T \times [0, 1] \to E$, $h(y, t) = t \circ g(y)$ and the associated homotopy $H : E \times [0, 1] \multimap E$, $H = h \circ \overline{\gamma}$, where $\overline{\gamma}(x, t)\gamma(x)$, for every $(x, t) \in E \times [0, 1]$. Then H is a homotopy joining Φ with the constant map $c : E \to E$, $c(x) = 0$, for every $x \in E$. From the homotopy property, we get

$$\text{Deg}(\varphi, E, D) = \text{deg}(j - c \circ \gamma, E),$$

but $j - c \circ \gamma = \text{id}_E$, and so $\deg(j - c \circ \gamma, E) = \deg(\text{id}_E, E) = 1$. The proof of (9.6) is completed.   □

From (9.6), we get:

(9.7) COROLLARY. *If $\Phi \in D(E, E)$, then $\text{Fix}(\Phi) \neq \emptyset$.*

Corollary (9.7) immediately follows from (9.6) and the existence property of the topological degree.

By analogous arguments as in the proof of (9.6), we get:

(9.8) THEOREM. *Assume that $\Omega \in \mathcal{N}_E(0)$ and $\varphi \in FD_{\partial\Omega}(\overline{\Omega}, E)$ be a compact vector field satisfying the following condition:*
(9.8.1) $\lambda x \notin \varphi(x)$, *for every $x \in \partial\Omega$ and $\lambda > 0$.*
Then $\text{Deg}(\varphi, D, \Omega) = 1$.

Now, we are going to formulate a version of the theorem about antipodes in locally convex spaces. For the case of Banach spaces, see: [Go1-M], [Go5-M] and [GG1], [GG2].

(9.9) THEOREM. *Let $\Omega \in \mathcal{N}_E(0)$ an $\Phi \in J(\overline{\Omega}, E)$ satisfy the following condition:*

(9.9.1) $\varphi(x) \cap \lambda\varphi(-x) = \emptyset$,   *for every $x \in \partial\Omega$ and $\lambda > 0$*

(recall that Ω is open convex and symmetric). Then $\text{Deg}(\varphi, D, \Omega)$ is odd, where $D : \Omega \xrightarrow{\Phi} E \xrightarrow{\text{id}_E} E$.

PROOF. Consider the homotopy $H : \overline{\Omega} \times [0, 1] \multimap E$ defined as follows

$$H(x, t) = \frac{1}{1+t}[\Phi(x) - t\Phi(-x)], \quad \text{for every } (x, t) \in \overline{\Omega} \times [0, 1].$$

Let $\chi : \overline{\Omega} \times [0,1] \multimap E$,

$$\chi(x,t) = x - H(x,t).$$

Then we have:

$$\chi(x,t) = x - \frac{1}{1+t}[\Phi(x) - t\Phi(-x)]$$

$$= x - \frac{1}{1+t}[(x - \Phi(x)) - t(-x - \Phi(-x))]$$

$$= \frac{1}{1+t}[\varphi(x) - t\varphi(-x)].$$

We shall show that:

$$0 \notin \chi(x,t), \quad \text{for every } (x,t) \in \partial\Omega \times [0,1].$$

Assume, on the contrary, that there is $(x,t) \in \partial\Omega \times [0,1]$ such that $0 \in \chi(x,t)$. Then there are $y_1 \in \varphi(x)$ and $y_2 \in \varphi(-x)$ such that $0 = (y_1 - ty_2)/(1 + t)$, and so $y_1 = ty_2$. We have assumed that, for $t = 0$, $0 \notin \varphi(x)$. We conclude that $t > 0$ which contradics our assumption $\varphi(x) \cap \lambda\varphi(-x) = \emptyset$.

From the homotopy property of the topological degree, we get:

$$\text{Deg}(\varphi, D, \Omega) = \text{Deg}(\chi_1, D, \Omega).$$

Since H_1 and χ_1 are odd, we deduce that the set $K = \overline{H(\overline{\Omega} \times \{1\})}$ is symmetric. Let $V \in \mathcal{N}_E(0)$, according to the definition of the topological degree of χ_1. We get a finite covering:

$$K \subset \Theta = \bigcup_{i=1}^{n} \left(y_i + \frac{1}{2}V \right), \quad y_i \in K.$$

Assume that if $y_i \in \{y_1, \ldots, y_n\}$, then $(-y_i) \in \{y_1, \ldots, y_n\}$, too. Then Θ is a symmetric open set. Let $L = \text{span}\{y_1, \ldots, y_n\} \subset E$ and $\pi^L : \Theta \to L$ be such that $(\pi^L(y) - y) \in V/2$, for every $y \in \Theta$ (comp. (1.37) in Chapter I.1).

We define $\pi : \Theta \to L$ as follows:

$$\pi(y) = \frac{1}{2}(\pi^L(y) - \pi^L(-y)), \quad \text{for every } y \in \Theta.$$

Thus, we have:

$$y - \pi(y) = \frac{1}{2}y - \frac{1}{2}\pi^L(y) + \frac{1}{2}y + \frac{1}{2}\pi^L(-y)$$

$$= \left[\left(\frac{1}{2}y - \frac{1}{2}\pi^L(y) \right) - \left(\frac{1}{2}(-y) - \frac{1}{2}\pi^L(-y) \right) \right] \in \frac{1}{2}V + \frac{1}{2}V = V.$$

We let: $\Omega^L = \Omega \cap L$, $H^L : \overline{\Omega^L} \multimap L$, $H^L(x) = \pi(H(x,1))$. Then H^L is an odd mapping. Observe also that the set:

$$\text{Fix}(H^L) = \{x \in \overline{\Omega^L} \mid x \in H^L(x)\}$$

of fixed points of H^L is symmetric and compact. Consequently, the set $\overline{\text{conv}}(\text{Fix}(H^L))$ is a compact subset of Ω^L.

Let W be an open subset of $\overline{\text{conv}}(\text{Fix}(H^L))$ and let it be also symmetric and convex.

Let $\beta \in U(H_1|_{\overline{W}})$ be such that, for every $f \in a(H_1|_{\overline{W}}, \beta)$, we have:

$$\text{Deg}(\chi_1, D) = \deg(j - \pi \circ f, W).$$

Since the set $\overline{W}$ is compact, there are $U_\chi \in \mathcal{N}_L(0)$ and $V_\chi \in \mathcal{E}(0)$ such that $K + V_\chi \subset \Theta$ and, for every $f \in a(H_1|_{\overline{W}}, U_\chi, V_\chi)$, we have $f \in a(H_1|_{\overline{W}}, \beta)$.

Let $V_0 = V_\chi/2$. Since ϕ is u.s.c. for every $x \in \overline{W}$ there exists $U_x \in \mathcal{N}_L(0)$ such that $U_x \subset U_\chi$ and $\phi(x + U_x) \cap \overline{W}) \subset \phi(x) + V_0$ and $\phi(-x + U_x) \cap \overline{W}) \subset \phi(-x) + V_0$.

Let us consider a finite covering:

$$\overline{W} = \bigcap_{j=1}^{k} \left(x_j + \frac{1}{2}U_{x_j}\right) \cap \overline{W}.$$

We define:

$$U_0 = \bigcap_{j=1}^{k} \frac{1}{2}U_{x_j}.$$

Let $g \in a(\phi|_{\overline{W}}, U_0, V_0)$ and let $x \in \overline{W}$. Then there are $z_1 \in (x + U_0) \cap \overline{W}$, $z_2 \in (-x + U_0) \cap \overline{W}$ and $v_1 \in \phi(z_1)$, $v_2 \in \phi(z_2)$ such that

$$g(x) \in (v_1 + V_0), \quad g(-x) \in (v_2 + V_0).$$

Thus, we see that: $-z_2 \in (X + U_0) \cap \overline{W}$. Moreover, there exists j, $1 \leq j \leq k$, such that $x \in (x_j + U_{x_j}/2)$, and so $z_1, -z_2 \in x_j + U_j$. Consequently, $z_2 \in (-x_j + U_{x_j})$.

Therefore, we get

$$g(x) \in (v_1 + V_0) \subset (\phi(z_1) + V_0) \subset (\phi(x_j) + V_\chi)$$

and

$$g(-x) \in (v_2 + V_0) \subset (\phi(z_2) + V_0) \subset (\phi(-x_j) + V_\chi).$$

From the above, we obtain

$$\frac{1}{2}[g(x) - g(-x)] \subset \frac{1}{2}[\phi(x_j) - \phi(-x_j) + 2V_\chi] \subset (H_1(x_j) + V_\chi)$$
$$\subset [(H_1(x + U_\chi) \cap \overline{W}) + V_\chi].$$

Now, we define: $f : \overline{W} \to \Theta$ as follows:

$$f(x) = \frac{1}{2}(g(x) - g(-x)), \quad \text{for every } x \in \overline{W}.$$

Thus, $f \in a(H_1|_{\overline{W}}, U_\chi, V_\chi)$ and f is odd. So the map $(j - \pi \circ f)$ and the Borsuk Theorem about Antipodes, applied to the map $(j - \pi \circ f)$, imply our claim, and the proof is completed. $\qquad\square$

We shall prove one more application of our topological degree theory obtained by approximation methods.

(9.10) THEOREM. *Let $\Omega \in \mathcal{N}_E(0)$ and $\varphi \in FD_{\partial\Omega}(\overline{\Omega}, E)$ be such that*

$$\mathrm{Deg}(\varphi, D_\phi, \Omega) \neq 0.$$

Then, for every $b \in \partial\Omega$, there exists $\lambda_b > 0$ such that $(\lambda_b \cdot b) \in \varphi(\partial\Omega)$.

PROOF. Assume that ϕ is associated with φ and that it has the following decomposition

$$D_\phi : \overline{\Omega} \xrightarrow{\gamma} T \xrightarrow{q} E.$$

Assume, on the contrary, that there exists $b \in \partial\Omega$ such that $\lambda b \notin \varphi(\partial\Omega)$, for every $\lambda > 0$. So, if we consider the set $S = \{\lambda b \mid \lambda \geq 0\}$ then, we get $S \cap \varphi(\partial\Omega) = \emptyset$.

Now, by means of the Minkowski functional $P_\Omega : E \to \mathbb{R}$ defined for Ω by standard generalized sequences calculation (see Chapter I.1), we deduce that there exists $V \in \mathcal{N}_E(0)$ such that

$$(S + V) \cap \varphi(\partial\Omega) = \emptyset.$$

Furthermore, we consider an open set $\Theta \supset \overline{\phi(\overline{\Omega})}$, a finite dimensional subspace $L \subset E$ such that $b \in L$ and a map $\pi : \Theta \to L$ for which we use the following notations:

$$\Omega^L = \Omega \cap L, \quad \phi^L = (\pi \circ q) \circ \gamma|_{\overline{\Omega^L}}, \quad \varphi^L = j = \phi^L.$$

We get

$$S \cap \varphi^L(\partial\Omega^L) = \emptyset$$

and

(9.10.1) $\mathrm{Deg}(\varphi, D_\phi, \Omega) = \deg(\varphi^L, D^L, \Omega^L).$

Since $\phi^L(\overline{\Omega^L})$ is compact, there is $\lambda > 0$ such that $\phi^L(\overline{\Omega^L}) \subset \lambda\Omega^L$.

Let A be an open bounded set such that $A \in \mathcal{N}_L(0)$ and $\phi^L(\overline{\Omega^L}) \subset A/4$, $4(1 + \lambda)b \in A$.

Let $\Omega_0 = \Omega^L \cap A$. We have

$$\partial\Omega_0 = (\partial\Omega^L \cap \overline{A}) \cup (\partial A \cap \overline{\Omega^L}).$$

Therefore, $b \in \partial\Omega_0$. Observe that $b \in D$ and $b \in \partial\Omega^L$ and, moreover, $\Omega_0 \in \mathcal{N}_L(0)$ is a bounded set.

Now, by a simple calculation, we obtain

$$S \cap \varphi^L(\partial\Omega_0) = \emptyset.$$

So, from the localization property of the topological degree, we get

$$(9.10.2) \qquad \deg(\varphi^L, D^L, \Omega^L) = \deg(\varphi_0, D_0, \Omega_0),$$

where $\phi_0 = \pi \circ \phi|_{\overline{\Omega}_0}$, $\varphi_0 = j - \phi_0$ and D_0 is the induced decomposition.

Let $B = \{x \in \overline{\Omega}_0 \mid \exists s \in S \ s \in \varphi(x_0)\}$. Then $B \subset \Omega_0$ is a closed subset of L and since Ω_0 is a bounded, we conclude that B is compact. It implies that there exists an open set W and $\alpha \in U(V|_{\overline{W}})$ such that

$$B \subset W \subset \Omega_0$$

and, for every $f \in a(V|_{\overline{W}}, \alpha)$, we have

$$S \cap (j - \pi \circ g \circ f)(\partial W) = \emptyset.$$

Moreover, we have

$$(9.10.3) \qquad \deg(\varphi_0, D_0, \Omega_0) = \deg(j - \pi \circ g \circ f, W).$$

We define a homotopy $h : \overline{W} \times [0,1] \to L$ by putting

$$h(x,t) = t(j - (\pi \circ g \circ f))(x) - (1-t)b,$$

for every $x \in \overline{W}$ and $t \in [0,1]$. By a standard calculation, we see that $h(x,t) \neq 0$, for every $(x,t) \in \partial W \times [0,1]$. Consequently, from the homotopy property of the Brouwer topological degree, we obtain

$$(9.10.4) \qquad \deg(j - \pi \circ g \circ f, W) = \deg(h_0, W),$$

but h_0 is a constant map, and so $\deg(h_0, W) = 0$.

Finally, from (9.10.1)–(9.10.4), we infer that $\mathrm{Deg}(\varphi, D, \Omega) = 0$ which is a contradiction. The proof of (9.10) is completed. $\square$

(9.11) REMARK. We would like to point out the possibility to define the fixed point index by means of approximation methods. We can do it in a very similar way as in Chapter I.8. Since we have already defined the topological degree for decomposable mappings, we can proceed as in Chapter I.8, for obtaining the desired fixed point index for decomposable mappings (cf. [AGG1]). We will introduce this definition in Chapter III.7 (see Proposition (7.21)). The case of condensing decomposable mappings will be considered in Chapter II.10.

II.10. Continuation principles based on a fixed point index

Because of applications, especially to differential equations and inclusions, it will be very useful to formulate sufficiently general continuation principles.

In particular, for compact J-maps on open subsets of a retract of a locally convex space E into E, the definitions of a topological degree and a fixed point index were already given in Chapters I.8 and II.9.

Now, we will apply these results to formulating the appropriate continuation principles, at first.

Furthermore, we will replace compact maps by condensing maps. Although the fixed point index was already developed in Chapter II.7, we recall its construction in a more sophisticated way. Because of applications, we restrict our results only to Fréchet spaces.

We often need to study fixed points for maps defined on sufficiently fine sets (possibly with an empty interior), but with values out of them. Making use of the previous results, we are in position to make the following construction.

Assume that X is a retract of a Fréchet space E and D is an open subset of X. Let $\Phi \in J(D, E)$ be locally compact, Fix(Φ) be compact and let the following condition hold:

(A) $\forall x \in \text{Fix}(\Phi) \; \exists U_x \ni x, \; U_x$ is open in D such that $\Phi(U_x) \subset X$.

The class of locally compact J-maps from D to E with the compact fixed point set and satisfying (A) will be denoted by the symbol $J_A(D, E)$.[8] We say that $\Phi, \Psi \in J_A(D, E)$ are homotopic in $J_A(D, E)$ if there exists a homotopy $H \in J(D \times [0,1], E)$ such that $H(\,\cdot\,, 0) = \Phi$, $H(\,\cdot\,, 1) = \Psi$, for every $x \in D$, there is an open neighbourhood V_x of x in D such that $H|_{V_x \times [0,1]}$ is compact, and

(A_H) $\forall x \in D \; \forall t \in [0,1] \; [x \in H(x,t) \Rightarrow \exists U_x \ni x, \; U_x$ is open in D

$$H(U_x \times [0,1]) \subset X].$$

Note that the condition (A_H) is equivalent to the following one:

If $\{x_j\}_{j \geq 1} \subset D$ converges to $x \in H(x, t)$, for some $t \in [0,1]$, then $H(\{x_j\} \times [0,1]) \subset X$, for j sufficiently large.

Let $\Phi \in J_A(D, E)$. Then Fix(Φ) $\subset \bigcup \{U_x \mid x \in \text{Fix}(\Phi)\} \cap V =: D' \subset D$ and $\Phi(D') \subset X$, where V is a neighbourhood of the set Fix(Φ) such that $\Phi|_V$ is compact (by the compactness of Fix(Φ) and local compactness of Φ) and U_x is a neighbourhood of x as in (A).

---

[8]Note that this symbol differs from the symbol $J(D, E)$ considered in Chapters II.8 and II.9.

Define

(10.1) $$\operatorname{Ind}_A(\Phi, X, r, D) = \operatorname{ind}(\Phi|_{D'}, X, r, D'),$$

where ind on the right-hand side of (10.1) was defined in (8.17) in Chapter I.8 with the same retraction r and $\Phi = \varphi$. The localization property of ind defined in Chapter I.8 (cf. also Remark (9.11)) implies that the definition (10.1) is independent of the choice of D'.

In the following theorem, we give some properties of Ind_A which will be used in the proof of the continuation Theorem (10.3). The simple proof is omitted.

(10.2) THEOREM.
   (i) (Existence) If $\operatorname{Ind}_A(\Phi, X, r, D) \neq 0$, then $\operatorname{Fix}(\Phi) \neq \emptyset$.
   (ii) (Localization) If $D_1 \subset D$ are open subsets of a retract X of a space E, $\Phi \in J_A(D, E)$ is compact, and $\operatorname{Fix}(\Phi)$ is a compact subset of D_1, then

$$\operatorname{Ind}_A(\Phi, X, r, D) = \operatorname{Ind}_A(\Phi, X, r, D_1).$$

   (iii) (Homotopy) If H is a homotopy in $J_A(D, E)$, then

$$\operatorname{Ind}_A(H(\,\cdot\,, 0), X, r, D) = \operatorname{Ind}_A(H(\,\cdot\,, 1), X, r, D).$$

   (iv) (Normalization) If $\Phi \in J(X)$ is a compact map, then $\operatorname{Ind}_A(\Phi, X, r, X) = 1$.

Now, we can formulate the continuation principle which is a generalization of Theorem 2.1 in [FPr2] in the case of emptiness of a domain's interior.

(10.3) THEOREM (Continuation Principle). *Let X be a retract of a Fréchet space E, D be an open subset of X and H be a homotopy in $J_A(D, E)$ such that*
   (i) $H(\,\cdot\,, 0)(D) \subset X$,
   (ii) *there exists $H' \in J(X)$ such that $H'|_D = H(\,\cdot\,, 0)$, H' is compact and $\operatorname{Fix}(H') \cap (X \setminus D) = \emptyset$.*
Then there exists $x \in D$ such that $x \in H(x, 1)$.

PROOF. Applying the localization property, we obtain

$$\operatorname{Ind}_A(H(\,\cdot\,, 0), X, r, D) = \operatorname{Ind}_A(H(\,\cdot\,, 0), X, r, X).$$

By the normalization property, $\operatorname{Ind}_A(H(\,\cdot\,, 0), X, r, X) = 1$. Thus, by the homotopy property, $\operatorname{Ind}_A(H(\,\cdot\,, 0), X, r, D) = \operatorname{Ind}_A(H(\,\cdot\,, 1), X, r, D) = 1$, which implies that $H(\,\cdot\,, 1)$ has a fixed point.   □

(10.4) COROLLARY. *Let X be a retract of a Fréchet space E and H be a homotopy in $J_A(X, E)$ such that $H(x,0) \subset X$, for every $x \in X$, and $H(\cdot, 0)$ is compact. Then $H(\cdot, 1)$ has a fixed point.*

(10.5) COROLLARY. *Let X be a retract of a Fréchet space E, D be an open subset of X and H be a homotopy in $J_A(D, E)$. Assume that $H(x,0) = x_0$, for every $x \in D$. Then there exists $x \in D$ such that $x \in H(x, 1)$.*

PROOF. It is sufficient to define $H' \in J(X)$, $H'(x) = x_0$ and to use Theorem (10.3). □

The following result generalizes the well-known Ky Fan theorem in the case of the Fréchet spaces (see [Fan1]) and follows also from (8.18) in Chapter I.8.

(10.6) COROLLARY. *Let X be a retract of a Fréchet space E and $\Phi \in J(X)$ be compact. Then Φ has a fixed point.*

Some problems for differential equations motivate us to consider weaker than (A_H) condition on H. Unfortunately, we cannot use the fixed point index technique described above. However, applying Corollary (9.7) in Chapter II.9, we obtain the following result generalizing Theorem 1.1 in [FPr1], for the case of set-valued maps.

(10.7) THEOREM (Continuation Principle). *Let X be a closed convex subset of a Fréchet space E and let $H \in J(X \times [0,1], E)$ be compact. Assume that*

  (i) $H(x,0) \subset X$, *for every $x \in X$,*
  (ii) *for any $(x,t) \in \partial X \times [0,1)$ with $x \in H(x,t)$, there exist open neighbourhoods U_x of x in X and I_t of t in $[0,1)$ such that $H((U_x \cap \partial X) \times I_t) \subset X$.*

Then there exists a fixed point of $H(\cdot, 1)$.

The idea of the proof is taken from [FPr1]. We need the following special case of Theorem (1.37) in Chapter I.1.

(10.8) LEMMA. *Let X be a convex closed subset of a Fréchet space E and $K \subset E$ be a compact subset such that $K \cap X \neq \emptyset$. Then, for every $\varepsilon > 0$, there exists a map $\pi_\varepsilon : K \to E$ whose image is contained in a finite dimensional space and such that*

  (i) $\pi_\varepsilon(K \cap X) \subset X$,
  (ii) $d(\pi_\varepsilon(x), x) < \varepsilon$, *for all $x \in K$.*

PROOF OF THEOREM (10.7). At first, let us suppose that E is finite dimensional. We also assume that $H \in D(X \times [0,1], E)$.

Let $r : E \to X$ be a retraction which sends a point into the nearest point in X. Define

$$\mathcal{F}_\lambda = \{x \in E \mid x \in H(r(x), A), \text{ for some } \lambda \in [0,1]\},$$
$$\mathcal{F} = \{x \in E \mid x \in H(r(x), A)\}.$$

By Corollary (9.7), we obtain that $\mathcal{F}_\lambda \neq \emptyset$, for every $\lambda \in [0,1]$.

Notice that our assertion can be reformulated as follows: $(\mathcal{F}_1, X) \neq \emptyset$. Suppose, by contradiction, that $\mathcal{F}_1 \cap X = \emptyset$. Since $\mathcal{F}_1$ is compact, $\text{dist}(\mathcal{F}_\lambda, X) = 2$, for some $\varepsilon > 0$, and there is an open set $V \supset X$ such that $\mathcal{F}_1 \cap \overline{V} = \emptyset$. We prove that there exists $(y, \lambda) \in \partial V \times [0,1)$ such that $H(r(y), X) \ni y$. Suppose that it is not true.

By the upper semicontinuity of H, $\text{dist}(\partial V, \mathcal{F} \cap \overline{V}) > 0$. Define the map $\sigma : E \to [0,1]$ as follows

$$\sigma(x) = \max\left\{1 - \frac{\text{dist}(x, \mathcal{F} \cap \overline{V})}{\text{dist}(\partial V, \mathcal{F} \cap \overline{V})}, 0\right\}.$$

Obviously, σ is continuous, $\sigma(x) = 1$ in $\mathcal{F} \cap \overline{V}$ and $\sigma(x) = 0$ in $E \setminus V$.

Now, we have a decomposable map $\widehat{H} : E \multimap E$, $\widehat{H}(x) = H(r(x), \sigma(x))$. Thus, there exists a fixed point $y \in \widehat{H}(y)$, which means that $y \in H(r(y), \sigma(y))$. Notice that $y \notin E \setminus V$, because $\mathcal{F}_0 \subset X$. Therefore, $y \in \mathcal{F} \cap \overline{V}$, which implies that $\sigma(y) = 1$, and so, $y \in \mathcal{F} \cap \overline{V}$. A contradiction.

Thus, we found for V a pair $(y, \lambda) \in \partial V \times [0,1)$ such that $H(r(y), \lambda) \ni y$. Take a sequence of open neighbourhoods V_n of X defined by $V_n = \{x \in E \mid \text{dist}(x, X) < \varepsilon/n\}$. Then, for every $n \in \mathbb{N}$, we can find $y_n \in \partial V_n$, $\lambda_n \in [0,1)$ and $x_n \in X$ such that $H(r(y_n), \lambda_n) \ni y_n$ and $\|x_n - y_n\| < \varepsilon/n$. By the compactness of $[0,1]$ and $\overline{H(X \times [0,1])}$, we can assume that $\lambda_n \to \lambda \in [0,1]$ and $y_n \to y \in E$. Thus, $x_n \to y$ and $y \in X$, because X is closed. This implies that $r(y_n) \to r(y)$ and, since $r(y) = y$ and H is u.s.c., $y \in H(y, \lambda)$. However, by the hypothesis $(\mathcal{F}_1 \cap X = \emptyset)$, we have $\lambda < 1$. By (ii), we get that there are open neighbourhoods $U_y \subset E$ and $I_\lambda \subset [0,1)$ of y and λ, respectively, such that $H((U_y \cap \partial X) \times I_\lambda) \subset X$. Notice that $r(y_n) \in U_y$, $r(y_n) \in \partial X$ (by the assumption on r), $y_n \in H(r(y_n), \lambda_n)$ and $y_n \notin X$, which is a contradiction.

Now, let E be infinite dimensional. Since H has a closed graph, it is sufficient to show that $\inf\{d(x, y) \mid x \in X, y \in H(x, 1)\} = 0$.

Suppose that

(10.9) $\inf\{d(x, y) \mid x \in X, y \in H(x, 1)\} = \varepsilon > 0$.

It follows that $x \in H(x, 1)$ in ∂X. Thus, by (ii), we can find for every $(x, \lambda) \in \partial X \times [0,1]$, $x \in H(x, \lambda)$, an open neighbourhood $\Omega_{(x,\lambda)}$ in $\partial X \times [0,1]$ such that

$H(\Omega_{(x,\lambda)}) \subset X$. Define

$$\Omega = \bigcup\{\Omega_{(x,\lambda)} \mid (x,\lambda) \in \partial X \times [0,1], \; x \in H(x,\lambda)\}.$$

Then $H(\Omega) \subset X$. Note that, by the "closed graph" argument, we can assume that ε is such that $\{(x,\lambda) \in \partial X \times [0,1] \mid \mathrm{dist}(x, H(x,\lambda)) < \varepsilon\} \subset X$.

Denote $K = \overline{H(X \times [0,1])}$. We know that K is compact and $K \cap X \neq \emptyset$, because $H(\,\cdot\,,0) \in J(X)$ and X is a retract of E. By Lemma (10.8), there exists a map $\pi_\varepsilon : K \to E$ such that $\pi_\varepsilon(K \cap X) \subset X$, $\pi_\varepsilon(K) \subset L$ ($\dim L < \infty$) and $d(\pi_\varepsilon(x), x) < \varepsilon$, for all $x \in K$.

Let us define $H_\varepsilon := \pi_\varepsilon \circ H : (L \cap X) \times [0,1] \multimap L$. Obviously, $H_\varepsilon \in D(X' \times [0,1], L)$, where $X' = L \cap X$. Notice that $H_\varepsilon(X' \times \{0\}) = \pi_\varepsilon(H(X' \times \{0\})) \subset \pi_\varepsilon(X \cap K) \subset X'$. We denote by $\partial_L X'$ a boundary of X' in L. Then, for $x \in \partial_L X'$ such that $x \in H_\varepsilon(x,\lambda)$, for some $\lambda \in [0,1)$, we have $x \in \partial X$ and $x = \pi_\varepsilon(y)$, for some $y \in H(x,\lambda)$. Thus, $d(x,y) < \varepsilon$, and so, $(x,\lambda) \in \Omega$. But this implies that $H(\Omega'_{(x,\lambda)}) \subset X$, where $\Omega'_{(x,\lambda)}$ is some open neighbourhood of (x,λ) in $\partial_L X' \times [0,1)$ and, consequently, $H_\varepsilon(\Omega'_{(x,\lambda)}) \subset X'$.

The first part of the proof permits us to conclude that there exists a fixed point $x \in H_\varepsilon(x,1)$. By the property of π_ε, there is $y \in H(x,1)$ such that $d(x,y) < \varepsilon$. This, however, contradicts our assumption (10.9). $\square$

(10.10) REMARK. Note that the convexity of X in Theorem (10.7) is essential only in the infinite dimensional case. For the proof, we have to intersect X with a finite dimensional subspace L.

Now, we would like to replace compact J-maps by condensing J-maps. For the sake of simplicity, by a *J-map* $\varphi : X \multimap Y$ (written $\varphi \in J(X,Y)$), we understand here (cf. Theorem (2.21) in Chapter I.2) only the one, for which $\varphi(x)$ is an R_δ-set, for every $x \in X$.

Let X be an ANR, $U \subset X$ be open, $\Phi : \overline{U} \multimap X$ be a set-valued map with the following property: there is an ANR Y, a J-map $\varphi : U \multimap Y$ and a continuous map $f : Y \to X$ such that $\Phi = f \circ \varphi$. In this case, we say again that Φ is a *decomposable map* and we denote the decomposition by

(10.11) $D(\Phi) : \overline{U} \xrightarrow{\varphi} Y \xrightarrow{f} X.$

We also recall the notion of homotopy of a decomposable map (cf. Chapter II.9). Let $\Psi : \overline{U} \to X$ be a set-valued map with a decomposition

$$D(\Psi) : \overline{U} \xrightarrow{\psi} Y \xrightarrow{g} X.$$

Then we say that Φ and Ψ are *homotopic* if there exists a J-map $\eta : \overline{U} \times [0,1] \multimap Y$ such that $\eta(\,\cdot\,,0) = \varphi$, $\eta(\,\cdot\,,1) = \psi$, and a continuous map $h : Y \times [0,1] \to$

X such that $h(\cdot, 0) = f$, $h(\cdot 1) = g$. In this case, we consider the composed homotopy between Φ and Ψ given by $\chi(\cdot, t) := \bigcup_{y \in \eta(x,t)} h(y, t)$ (clearly $\chi(\cdot, 0) = \Phi$, $\chi(\cdot, 1) = \Psi$).

Let us note that the class of decomposable maps is essentially larger than the class of J-maps (see [Go2-M], for a relevant example and also Chapter II.9).

Now, assume that the fixed point set $\mathrm{Fix}(\Phi) := \{x \in U \mid x \in \Phi(x)\}$ of Φ has an empty intersection with the boundary ∂U of U and Φ is a compact map, i.e. the range of Φ is contained in a compact subset of X. For the map Φ (more precisely, for the decomposition $D(\Phi)$ of Φ), one can define an integer valued fixed point index as in Chapter I.8 (see (8.17)):

$$(10.12) \qquad \mathrm{ind}_X(D(\Phi), U).$$

The index given in (10.12) has the following properties (see e.g. [AndBa] and cf. Theorem (8.18) in Chapter I.8).

(10.13) PROPOSITION.

(i) (Existence) If $\mathrm{ind}_X(D(\Phi), U) \neq 0$, then $\emptyset \neq \mathrm{Fix}(\Phi) \subset U$.

(ii) (Additivity) Let $\mathrm{Fix}(\Phi) \subset U_1 \cup U_2$, where U_1, U_2 are open disjoint subsets of U. Then $\mathrm{ind}_X(D(\Phi), U) = \mathrm{ind}_X(D(\Phi)|_{\overline{U}_1}, U_1) + \mathrm{ind}_X(D(\Phi)|_{\overline{U}_2}, U_2)$, where $D(\Phi)|_{\overline{U}_i} : \overline{U}_i \xrightarrow{\varphi|_{\overline{U}_i}} Y \xrightarrow{f} X$, for $i = 1, 2$.

(iii) (Homotopy) Let $\Psi : \overline{U} \to X$ be a decomposable map homotopic to Φ. Assume that the composed homotopy χ between Φ and Ψ is compact and satisfies $x \notin \chi(x, t)$, for each $x \in \partial U$, $t \in [0, 1]$. Then

$$\mathrm{ind}_X(D(\Phi), W) = \mathrm{ind}_X(D(\Psi), W).$$

(iv) (Weak normalization) Assume that f in $D(\Phi)$ is a constant map, i.e. $f(x) = a \notin \partial U$, for each $x \in Y$. Then

$$\mathrm{ind}_X(D(\Phi), U) = \begin{cases} 1, & \text{for } a \in U, \\ 0, & \text{for } a \notin U. \end{cases}$$

(v) (Contraction) Let X' be an ANR such that $X \overset{j}{\subset} X'$ and let U' be an open subset of X'. Let $\varphi : \overline{U}' \multimap Y$ be a J-map and consider the decomposition $D(\Phi) : \overline{U}' \xrightarrow{\varphi} Y \xrightarrow{f} X$ of the map $\Phi := f \circ \varphi$. Assume that Φ is compact and $\mathrm{Fix}(j \circ \Phi) \cap \partial U' = \emptyset$. Then $\mathrm{ind}_{X'}(j \circ D, U') = \mathrm{ind}_X(D|_{\overline{U' \cap X}}, U' \cap X)$.

In the sequel, let E again denote a Fréchet space, i.e. a completely metrizable locally convex topological vector space.

In order to generalize the fixed point index theory for condensing set-valued maps, we recall the notion of a fundamental set (see (7.6) in Chapter II.7).

Again, let $\varphi : E \supset D \multimap E$ be a set-valued mapping. Then a closed convex subset T of E is said to be *fundamental*, for φ if

(a) $$\varphi(D \cap T) \subset T,$$

(b) $$x_0 \in \overline{\mathrm{conv}}(\varphi(x_0) \cup T) \quad \text{implies } x_0 \in T.$$

For a set-valued homotopy $\chi : D \times [0,1] \multimap E$, a closed convex set T is called *fundamental* if it is fundamental for $\chi(\,\cdot\,, t)$, for each $t \in [0,1]$.

Let us note that F and $\overline{\mathrm{conv}}(\varphi(D))$ are examples of fundamental sets.

The following properties of fundamental sets can be easily verified, for more details, see (7.7) in Chapter II.7.

(10.14) LEMMA. *The fixed point set* $\mathrm{Fix}(\varphi)$ *is included in every fundamental set of* φ.

(10.15) LEMMA. *If* T *is a fundamental set for a multivalued mapping* χ *and* $P \subset T$, *then the set*

$$\widetilde{T} = \overline{\mathrm{conv}}(\chi((D \cap T) \times [0,1]) \cup P)$$

is also fundamental.

(10.16) LEMMA. *If* $\{T_\tau\}$ *is an arbitrary system of fundamental sets of* φ, *then the set*

$$\widehat{T} = \bigcap_\tau T_\tau$$

is also fundamental.

Now, it is time to develop the fixed point index theory suitable for applications to boundary value problems in Banach spaces on unbounded domains. At first, we apply the technique presented in Chapter II.9 to get the fixed point index for decomposable mappings in Fréchet spaces.

Since we tendentially omit the compactness asumptions above, the index can be regarded in many situations as a rather general one. On the other hand, this generalization is not straightforward and, moreover, requires some unpleasant restrictions.

Let X be a closed and convex subset of a Fréchet space E, $U \subset X$ be (relatively) open and let $\Phi : \overline{U} \multimap X$ have a decomposition

(10.17) $$D(\Phi) : \overline{U} \xrightarrow{\varphi} Y \xrightarrow{f} X,$$

where Y is an ANR (e.g. a closed convex subset of a given Fréchet space), φ is a J-map and f is a continuous map. Assume that $\mathrm{Fix}(\Phi) \cap \partial U = \emptyset$ and let Φ have a compact fundamental set.

We will define a fixed point index for the decomposition $D(\Phi)$ of the map Φ.

Let $\mathrm{Ind}_X(D(\Phi), U) := 0$, whenever $\mathrm{Fix}(\Phi) = \emptyset$. Otherwise, take an arbitrary nonempty fundamental set T of Φ and assume that $T \subset X$ (otherwise, take $T \cap X$). Since T is an AR, we may choose a retraction $r : X \to T$. Using these objects, we obtain a decomposition

$$D(\Phi, r, T) : \overline{U} \cap T \xrightarrow{\varphi} Y \xrightarrow{r \circ f} X.$$

Now, define the fixed point index by

(10.18) $$\mathrm{Ind}_X(D(\Phi), U) := \mathrm{ind}_T(D(\Phi, r, T), U \cap T),$$

where ind_T on the right-hand side is given by (10.12).

Since, for each $x \in \overline{U} \cap T$, we have that $r \circ f \circ \varphi(x) = \Phi(x)$, it is clear that the map given by the decomposition $D(\Phi, r, T)$, has no fixed points on the relative boundary of $U \cap T$, and thus the index on the right-hand side of (10.18) is well-defined.

(10.19) PROPOSITION. *The definition* (10.18) *is independent of the chosen fundamental set.*

PROOF. Let T and T' be compact fundamental sets for Φ. Without any loss of generality, we assume $T \overset{j}{\subset} T'$. Let $r : X \to T$ and $r' : X \to T'$ be retractions and construct the decompositions $D(\Phi, r, T)$ and $D(\Phi, r', T')$. Consider also

$$D : \overline{U} \cap T' \xrightarrow{\varphi} Y \xrightarrow{j \circ r \circ f} T'.$$

Now, the continuous map $h : Y \times [0,1] \to T'$, $h(y,t) := r'((1-t)f(y) + \mathrm{tr}(f(y)))$ shows that $D(\Phi, r', T')$ and D are homotopic decompositions and we obtain the composed homotopy $\chi : \overline{U} \cap T' \times [0,1] \multimap T'$ given by

$$\chi(x,t) = \bigcup_{y \in \varphi(x)} r'((1-t)f(y) + \mathrm{tr}(f(y))).$$

In order to apply the homotopy property (Propositon (10.13)(iii)), we need to show that $x \notin \chi(x,t)$, for $x \in \partial(U \cap T')$, $t \in [0,1]$. So assume, on the contrary, that there is $x_0 \in \partial(U \cap T')$ and $t_0 \in [0,1]$ with $x_0 \in \chi(x_0, t_0)$. Then there is $y \in \varphi(x_0)$ such that $x_0 = r'((1-t)f(y) + \mathrm{tr}(f(y)))$. Since T' is a fundamental set, we have $(1-t)f(y) + \mathrm{tr}(f(y)) \in (1-t)\Phi(x_0) + \mathrm{tr}(\Phi(x_0)) \subset T'$ and, therefore, indeed

$x_0 = (1 - t)f(y) + \mathrm{tr}(f(y))$. But then $x_0 \in \overline{\mathrm{conv}}(\Phi(x_0) \cup T)$, and subsequently $x_0 \in \Phi(x_0)$, which contradicts $\mathrm{Fix}(\Phi) \cap \partial U = \emptyset$. We obtain that

$$\mathrm{Ind}_{T'}(D(\Phi, r', T'), U \cap T') = \mathrm{Ind}_{T'}(D, U \cap T').$$

Finally, an application of the contraction property (Propositon (10.13)(v)) shows that

$$\mathrm{Ind}_{T'}(D, U \cap T') = \mathrm{Ind}_T(D(\Phi, r, T), U \cap T),$$

which concludes the proof.   □

Let us also note the following: If $\Phi : \overline{U} \multimap X$ given with a decomposition $D(\Phi)$ (see (10.17)) is a compact map such that $\mathrm{Fix}(\Phi) \cap \partial U = \emptyset$, then Φ has a compact fundamental set $T := \overline{\mathrm{conv}}(\Phi(\overline{U}))$. By an application of the contraction property (Proposition (10.13)(v)), we see that

$$\mathrm{Ind}_X(D(\Phi), U) = \mathrm{Ind}_T(D(\Phi, r, T), U \cap T).$$

Thus, the index in Definition (10.18) is consistent with the one in (10.12).

Let us collect some properties of the index in (10.18), whose proofs follow easily from Proposition (10.13).

(10.20) THEOREM.

   (i) (Existence) *If* $\mathrm{Ind}_X(D(\Phi), U) \neq 0$, *then* $\emptyset \neq \mathrm{Fix}(\Phi) \subset U$.

  (ii) (Additivity) *Let* $\mathrm{Fix}(\Phi) \subset U_1 \cup U_2$, *where* U_1, U_2 *are open disjoint subsets of* U. *Then*

$$\mathrm{Ind}_X(D(\Phi), U) = \mathrm{Ind}_X(D(\Phi)|_{\overline{U}_1}, U_1) + \mathrm{Ind}_X(D(\Phi)|_{\overline{U}_2}, U_2).$$

 (iii) (Homotopy) *Let* $\Psi : \overline{U} \to X$ *be a decomposable map homotopic to* Φ. *Assume that the composed homotopy* $\chi : \overline{U} \times [0, 1] \multimap X$ *has a compact fundamental set* T. *Moreover,* $x \notin \chi(x, t)$, *for each* $x \in \partial U$, $t \in [0, 1]$. *Then*

$$\mathrm{Ind}_X(D(\Phi), U) = \mathrm{Ind}_X(D(\Psi), U).$$

 (iv) (Weak normalization) *Assume that* f *in* $D(\Phi)$ *is a constant map, i.e.* $f(x) = a \notin \partial U$, *for each* $x \in Y$. *Then*

$$\mathrm{Ind}_X(D(\Phi), U) = \begin{cases} 1, & \text{for } a \in U, \\ 0, & \text{for } a \notin U. \end{cases}$$

As the main example of mappings having compact fundamental sets, we consider maps which are condensing w.r.t. a measure of noncompactness. In what follows, we shall use the abstract measure of noncompactness, as considered in (5.31) in Chapter I.5.

Let $\varphi : F \supset D \multimap F$ be an u.s.c. set-valued map. Let us recall that φ is *condensing* if

$$\beta(\varphi(\Omega)) \geq \beta(\Omega),$$

implies that Ω is relatively compact, for every $\Omega \subset D$. Furthermore, we say that an u.s.c. set-valued homotopy $\chi : D \times [0,1] \multimap F$ is condensing if, for each $\Omega \subset D$, we have that $\Omega, \chi(\Omega \times [0,1]) \in \mathcal{M}$ and if

$$\beta(\chi(\Omega \times [0,1])) \geq \beta(\Omega),$$

then Ω is relatively compact.

The following lemma shows that condensing maps possess a nonempty compact fundamental set.

(10.21) LEMMA. *Let D be a closed subset of E. Let $\chi : D \times [0,1] \multimap E$ be condensing and assume that χ has compact values. Then χ has a compact fundamental set T_∞ containing an arbitrarily prescribed point $p \in E$.*

For the proof see (7.7.6) in Chapter II.7.

(10.22) REMARK. Let $\Phi : \overline{U} \multimap X$ be given with a decomposition $D(\Phi)$ (see (10.17)) such that $\text{Fix}(\Phi) \cap \partial U = \emptyset$. Assume that Φ is condensing. Then the above Lemma (10.21) shows that the index $\text{Ind}_X(D(\Phi), U)$ is defined and has all the properties given in Theorem (10.20) (cf. Chapter II.7).

(10.23) REMARK. More delicate situations concerning fundamental sets will be still discussed in detail in the next Chapter II.11.

(10.24) LEMMA (cf. (7.9) in Chapter II.7). *Let X be a closed, convex subset of a Fréchet space E and let $\Phi : X \multimap X$ be given with a decomposition*

$$D(\Phi) : X \overset{\varphi}{\multimap} Y \overset{f}{\longrightarrow} X.$$

Assume that Φ is condensing. Then

$$\text{Ind}_X(D(\Phi), X) = 1,$$

and so Φ has a fixed point.

In the applications of the fixed point theory, we often need to consider maps with values in the Fréchet space and not in the closed convex set. We will extend our theory to this case, similarly as above.

Again, let E be a Fréchet space and X be a closed and convex subset of E. Let $U \subset X$ be open and consider a map $\varphi \in J_A(U, F)$, where the symbol $J_A(U, F)$ is again reserved for J-maps from U to F satisfying condition (A) at the beginning of this chapter. The notion of homotopy in J_A will be understood analogously. Thus, $\mathrm{Fix}(\varphi)$ is compact and φ has a compact fundamental set T. Set $\mathrm{Ind}_X^A(\varphi, U) := 0$, whenever $\mathrm{Fix}(\varphi) = \emptyset$. Otherwise, let $x_1, \ldots, x_n \in \mathrm{Fix}(\varphi)$ such that $\mathrm{Fix}(\varphi) \subset \bigcup_{i=1}^n U_{x_i} =: V$, where U_{x_i} are neighbourhoods of x_i such that $\overline{U}_{x_i} \subset U$ and satisfy condition (A). Then $\varphi|_V : V \multimap X$ is a J-map with compact fundamental set T and satisfies $\mathrm{Fix}(\varphi) \cap \partial V = \emptyset$. Thus, we can define

$$(10.25) \qquad \mathrm{Ind}_X^A(\varphi, U) := \mathrm{Ind}_X(\varphi|_{\overline{V}}, V).$$

The independence of this definition of the chosen set V follows from the additivity property (Theorem (10.20)(ii)). Furthermore, if $\varphi : \overline{U} \multimap X$ has a compact fundamental set and $\mathrm{Fix}(\varphi) \cap \partial U = \emptyset$, then $\mathrm{Ind}_X^A(\varphi, U)$ is defined and

$$\mathrm{Ind}_X^A(\varphi, U) = \mathrm{Ind}_X(\varphi, U).$$

Hence, the index in (10.25) is consistent with the one in (10.18) (cf. Remark (10.22)). The following proposition follows easily from the above argumentation.

(10.26) PROPOSITION.

(i) (Existence) *If* $\mathrm{Ind}_X^A(\varphi, U) \neq 0$, *then* $\emptyset \neq \mathrm{Fix}(\varphi)$.

(ii) (Additivity) *Let* $\mathrm{Fix}(\varphi) \subset U_1 \cup U_2)$, *where* U_1, U_2 *are open disjoint subsets of* U. *Then*

$$\mathrm{Ind}_X^A(\varphi, U) = \mathrm{Ind}_X^A(\varphi|_{U_1}, U_1) + \mathrm{Ind}_X^A(\varphi|_{U_2}, U_2).$$

(iii) (Homotopy) *Let* $\psi : U \multimap F$ *be homotopic in* J_A *to the map* φ. *Assume that the homotopy* $\chi : U \times [0, 1] \multimap F$ *has a compact fundamental set and the set*

$$(10.27) \qquad \Sigma := \{ (x, t) \in U \times [0, 1] \mid x \in \chi(x, t) \}$$

is compact. Then

$$\mathrm{Ind}_X^A(\varphi, U) = \mathrm{Ind}_X^A(\psi, U).$$

(iv) (Weak normalization) *Assume that* $\varphi : U \to F$ *is a constant map* $\varphi(x) = a \in F$, *for all* $x \in U$. *Then*

$$\mathrm{Ind}_X^A(\varphi, U) = \begin{cases} 1, & \text{for } a \in U, \\ 0, & \text{for } a \notin U. \end{cases}$$

Let us now formulate a continuation principle which is convenient for our applications to differential inclusions in Banach spaces.

(10.28) THEOREM (Continuation Principle). *Let X be a closed, convex subset of a Fréchet space E, let $U \subset X$ be open and let $\chi : U \times [0,1] \multimap F$ be a homotopy in J_A such that Σ (see (10.27)) is compact. Let χ be condensing and assume that there is $\varphi \in J(X)$ condensing such that $\varphi|_U = \chi(\cdot, 0)$ and $\mathrm{Fix}(\varphi) \cap (X \setminus U) = \emptyset$. Then $\chi(\cdot, 1)$ has a fixed point.*

PROOF. The proof follows, in view of the existence property (see Proposition (10.26)(i)), from the following equations:

$$\mathrm{Ind}_X^A(\chi(\cdot, 1), U) = \mathrm{Ind}_X^A(\chi(\cdot, 0), U),$$

by the homotopy property (Proposition (10.26)(iii)),

$$\mathrm{Ind}_X^A(\chi(\cdot, 0), U) = \mathrm{Ind}_X^A(\varphi|_U, U) = \mathrm{Ind}_X^A(\varphi, X),$$

by the additivity property (Proposition (10.26)(ii)). Finally, we see that

$$\mathrm{Ind}_X^A(\varphi, X) = \mathrm{Ind}_X(\varphi, X) = 1,$$

from Lemma (10.24). □

(10.29) COROLLARY. *Let $\chi : X \times [0,1] \multimap F$ be a condensing homotopy in J_A such that $\chi(x, 0) \subset X$, for every $x \in X$. Then $\chi(\cdot, 1)$ has a fixed point.*

II.11. Continuation principles based on a coincidence index

We could see that in locally convex spaces, many obstructions for condensing maps occur. Nevertheless, a scheme for continuation principles can be indicated as well. To this aim, we develop a coincidence index for noncompact maps on nonconvex sets in a rather axiomatic way. This approach is local in the spirit of the idea of topological essentiality in Chapter II.4, but technically it is completely different. For more details, see [AndVae].

Hence, let X, Y be topological spaces. We consider a fixed map $F : X \to Y$ and a fixed family $\mathcal{T}$ of "admissible" triples such that F provides a coincidence index from $\mathcal{T}$ in the following sense.

Unless something else is said, by a *triple*, we always mean a triple of the form (φ, Ω, K) or (h, Ω, K), where $K \subset Y$ is closed (possibly empty), $\Omega \subset K \subset Y$, and $\varphi : F^{-1}(\overline{\Omega}) \to Y$ respectively $h : I \times F^{-1}(\overline{\Omega}) \to Y$ (with some nonempty set I).

Given such a triple and $M \subset Y$, we define the *coincidence point image*

$$\mathrm{Coin}(F, \varphi, M) := F(\{x \mid F(x) = \varphi(x) \in M\}) = \{y \in M \mid y \in \varphi(F^{-1}(y))\},$$

respectively

$$\mathrm{Coin}(F, h, M) := \bigcup_{t \in I} \mathrm{Coin}(F, h(t, \cdot), M) = \{F(x) \mid F(x) \in h(I \times \{x\}) \cap M\}.$$

We call a triple (φ, Ω, K), respectively (h, Ω, K) F-*admissible* if Ω is open in K and the condition $\mathrm{Coin}(F, \varphi, \partial\Omega) = \emptyset$, respectively $\mathrm{Coin}(F, h, \partial\Omega) = \emptyset$ holds, where $\partial\Omega$ is the boundary of Ω with respect to K.

We assume that the given family $\mathcal{T}$ consists of F-admissible triples of the form (φ, Ω, K) or (h, Ω, K), where φ is a continuous map (respectively h is a homotopy) with values in K. We denote by $\mathcal{A}$ the family of all sets K with the property that $(\varphi, \Omega, K) \in \mathcal{T}$, for some φ and Ω. By our general assumptions on triples, all sets in $\mathcal{A}$ are closed.

(11.1) DEFINITION. We say that F provides a *coincidence index* ind_F for $\mathcal{T}$ if there is a map ind_F from $\mathcal{T}$ into a ring with 1 (typically $\mathbb{Z}$, $\mathbb{Q}$, or $\mathbb{Z}_2$) such that, for any $(\varphi, \Omega, K) \in \mathcal{T}$, the following holds:

(11.1.1) (Coincidence point property) If $\mathrm{ind}_F(\varphi, \Omega, K) \neq 0$, then $\mathrm{Coin}(F, \varphi, \Omega) \neq \emptyset$.

(11.1.2) (Normalization) If $\varphi(x) \equiv c \in \Omega$, then $\mathrm{ind}_F(\varphi, \Omega, K) = 1$.

(11.1.3) (Homotopy invariance) If $(h, \Omega, K) \in \mathcal{T}$, then $(h(t, \cdot), \Omega, K) \in \mathcal{T}$, for any $t \in [0, 1]$, and $\mathrm{ind}_F(h(t, \cdot), \Omega, K)$ is independent of $t \in [0, 1]$.

(11.1.4) (Contraction) If $K_0 \in \mathcal{A}$ and $\varphi(F^{-1}(\overline{\Omega})) \subset K_0 \subset K$, then $(\varphi, \Omega \cap K_0, K_0) \in \mathcal{T}$ and

$$(11.2) \qquad \mathrm{ind}_F(\varphi, \Omega, K) = \mathrm{ind}_F(\varphi, \Omega \cap K_0, K_0).$$

We say that ind_F satisfies the excision property, if additionally:

(11.1.5) (Excision) If $\Omega_0 \subset \Omega$ is open in K with $\Omega_0 \supset \mathrm{Coin}(F, \varphi, \overline{\Omega})$, then $(\varphi, \Omega_0, K) \in \mathcal{T}$ and

$$\mathrm{ind}_F(\varphi, \Omega, K) = \mathrm{ind}_F(\varphi, \Omega_0, K).$$

We call ind_F *additive* if, moreover:

(11.1.6) (Additivity) If $\Omega_1, \Omega_2 \subset \Omega$ are disjoint and open in K with $\Omega_1 \cup \Omega_2 \supset \mathrm{Coin}(F, \varphi, \overline{\Omega})$, then $(\varphi, \Omega_i, K) \in \mathcal{T}$ and

$$\mathrm{ind}_F(\varphi, \Omega, K) = \mathrm{ind}_F(\varphi, \Omega_1, K) + \mathrm{ind}_F(\varphi, \Omega_2, K).$$

The excision property is a special case of the additivity with $\Omega_2 = \emptyset$. We separated the additivity from the other requirements, because indices with values in $\mathbb{Z}_2$ will usually never be additive. Values in $\mathbb{Z}_2$ are natural for indices which are defined by purely homotopic methods. However, such indices will satisfy the contraction and the excision properties only in exceptional cases, because homotopy theory does not satisfy the excision axiom [ES-M]. For this reason, we have also separated the excision property from the other axioms. However, the contraction property is essential for all considerations in this section.

To avoid trivial case distinctions, we assume throughout that $(\varphi, \emptyset, \emptyset), (h, \emptyset, \emptyset) \in \mathcal{T}$, and thus $\emptyset \in \mathcal{A}$, which is no loss of generality, in view of the coincident point property: we must have

$$\mathrm{ind}_F(\varphi, \emptyset, \emptyset) = 0.$$

(11.3) DEFINITION. Given a system $\mathcal{A}$ of subsets of Y and some $R \subset Y$, then we denote by $\mathcal{T}(F, R, \mathcal{A})$ the system of all F-admissible triples (φ, Ω, K) respectively (h, Ω, K) with $K \in \mathcal{A}$ and $\overline{\Omega} \subset R$ such that φ is continuous (respectively h is a homotopy) with values in K.

We say that F *provides a coincidence point index* on $(R, \mathcal{A})$ if it provides a coincidence point index for $\mathcal{T}(F, R, \mathcal{A})$.

(11.4) REMARK. If F provides a coincidence index on $(R, \mathcal{A})$ and $\mathcal{A}$ contains all sets of the form $\{y\}$ with $y \in R$, then the normalization and coincidence point property imply that the range of F contains R. In this case, it is no loss of generality to assume that $F(X) = R$ (because one may replace X by $F^{-1}(R)$ and consider the corresponding restriction of F without changing the property that F provides a coincidence index on $(R, \mathcal{A})$).

(11.5) EXAMPLE. If $X = Y$ is a metric space, then $F = \mathrm{id}$ provides a coincidence index on $(Y, \mathcal{A})$, where $\mathcal{A}$ is the family of all compact neighbourhood retracts which are contained in Y. This index is the well-known uniquely determined fixed point index (see e.g. [Br1-M], [De3-M], [EiFe-M], [Nu1]). Recall that this index is a generalization of the Leray–Schauder degree in the following sense. If $X = Y$ is a Banach space, Ω is open in X with closure $\overline{\Omega}$, and $\varphi : \overline{\Omega} \to X$ is continuous with relatively compact range and has no fixed points on the boundary of Ω (with respect to X), then

(11.6) $\deg(\mathrm{id} - \varphi, \Omega, 0) = \mathrm{ind}_F(\varphi, \Omega \cap K, K),$

for $K = \overline{\mathrm{conv}}(\varphi(\overline{\Omega}))$ (this follows e.g. from [Nu1] in view of the contraction property which in turn is a special case of the commutativity).

A more general class of maps F for which the existence of a coincidence index is known is the following.

(11.7) DEFINITION. Let X and Y be metric spaces and $F : X \to Y$ be continuous, surjective and proper (i.e. preimages of compact sets are compact). Then we say:

(11.7.1) F is a $\mathbb{Z}$-Vietoris map if the fibres $F^{-1}(\{y\})$ are acyclic with respect to the Čech cohomology with coefficients in $\mathbb{Z}$.

(11.7.2) F belongs to the class $\mathcal{V}_0(\mathbb{K})$ if the fibres $F^{-1}(\{y\})$ are acyclic with respect to the Čech homology with coefficients in a field $\mathbb{K}$.

(11.7.3) F belongs to $\mathcal{V}_n(\mathbb{K})$ $(n = 1, 2, \dots)$ if the fibres $F^{-1}(\{y\})$ consist of n components, each of which is acyclic with respect to the Čech homology with coefficients in a field $\mathbb{K}$, and if F^{-1} is also a lower semicontinuous map.

(11.7.4) F is a composed $\mathbb{K}$-Vietoris map if it can be written in the form $F = F_1 \circ \cdots \circ F_k$, where $F_k \in \mathcal{V}_{n_k}(\mathbb{K})$, for some k and $n_k \in \{0, 1, \dots\}$.

(11.8) PROPOSITION. *The class of $\mathbb{Z}$-Vietoris maps resp. composed $\mathbb{K}$-Vietoris maps is closed under compositions.*

PROOF. For composed $\mathbb{K}$-Vietoris maps, this is immediate from the definition. The fact that the class of $\mathbb{Z}$-Vietoris maps is closed under compositions follows from the Vietoris theorem on the invertibility of the map F^* induced on the Čech cohomology groups (for $\mathbb{Q}$ instead of $\mathbb{Z}$, the proof can e.g. be found in [Go5-M, Proposition (8.10)], it resembles the proof for $\mathbb{Z}$). □

Recall that a subset Z of some topological space Z_0 is called a *neighbourhood retract* if there is some neighbourhood $U \subset Z_0$ of Z and a *retraction* $r : U \to Z$, i.e. r is continuous with $r|_Z = \mathrm{id}$. A metric space Z is a (metric) ANR if for each metric space Z_0 in which Z is a closed subset, the set Z_0 is a neighbourhood retract of Z. If we assume the axiom of choice, then each neighbourhood retract of a locally convex metrizable space is an ANR by Dugundji's Extension Theorem (2.2) in Chapter I.2 (the (uncountable) axiom of choice is not required if Z is separable and complete [Vae2, Lemma 1.1]). Conversely, if Z is an ANR, then it is a closed subset of some normed space Z_0 by the Arens-Eells Embedding Theorem (2.5) in Chapter I.2, and thus a neighbourhood retract of Z_0.

(11.9) THEOREM. *Let X be a metric space and Z some (metric) ANR. Then there is a map which associates with any neighbourhood retract Y of Z, any $R \subset Y$ and any $\mathbb{Z}$-Vietoris map (respectively any composed $\mathbb{K}$-Vietoris map) $F : X \to R$ an additive index ind_F with values in $\mathbb{Z}$ (respectively $\mathbb{K}$) on $(R, \mathcal{A})$ where $\mathcal{A}$ is the family of all compact neighbourhood retracts of Y.*

For $\mathbb{Z}$-Vietoris maps, and if X has finite covering dimension, Theorem (11.9) has been proved in [Kry3] (see also [Go5-M, Theorem (47.8)]), but be aware of the fact that the assumption that X has finite dimension is mistakenly missing there). However, since we do not require a homotopic invariance for our index, essentially the same proof can be used to obtain the statement of Theorem (11.9) for $\mathbb{Z}$-Vietoris maps (for more details, see [Vae1]). Under mild additional assumptions on the fibres of F, the index of Theorem (11.9) is actually uniquely determined by its properties, see [Vae1]. However, without additional assumptions, we do not know whether the index is unique. In particular, we do not know whether the map in Theorem (11.9) may be chosen independent of Z for $\mathbb{Z}$-Vietoris maps, in general. The latter is true if F is a composed $\mathbb{K}$-Vietoris map: For such maps, Theorem (11.9) has been proved in a more general form in [Dz-M] (see also [Go5-M]).

We point out that the proofs of Theorem (11.9) are rather different for $\mathbb{Z}$-Vietoris maps and composed $\mathbb{K}$-Vietoris maps: For $\mathbb{Z}$-Vietoris maps, the essential tool is the Vietoris theorem (also used in the proof of Proposition (11.8)). This is the classical approach already employed in [EM]. In contrast, for composed $\mathbb{K}$-Vietoris maps, the construction is based on simplicial approximations, see [SeS].

Despite of the name, there is no direct connection between the coincidence index in Definition (11.1) (or of Theorem (11.9)) and the Mawhin coincidence degree [GaM-M], [Maw-M], [Maw2]–[Maw4], [Maw3] for Fredholm maps $F = L$ of index 0 (or, more general, of nonnegative index): Although Fredholm maps have acyclic (namely convex) fibres and are proper on bounded sets, they are not Vietoris maps, because they are not onto (recall in this connection also Remark (11.4)). One may of course just restrict Y to the range of L, but this would dramatically decrease the number of admissible homotopies: It is an essential property of the Mawhin degree that the homotopies are not assumed to take their values in the range of L. However, it is possible to combine the coincidence index from Theorem (11.9) with the ideas of the Mawhin coincidence degree to define a "generalized degree" for three maps L, F, φ which essentially "counts" the number of coincidences of the maps L and $\varphi \circ F^{-1}$. Here, L is a Fredholm operator of nonnegative index, and F a $\mathbb{Z}$-Vietoris map. This generalized degree was initiated by W. Kryszewski [GaKr2], [Kry2-M] (see also [Go5-M]). However, since the Mawhin coincidence degree (and even the generalized degree) satisfies the contraction property only for linear subspaces, it is much more difficult to deal with the noncompact situation. For this reason, we restrict ourselves to the coincidence index of Definition (11.1).

Our aim in the following is to extend the given index ind_F to a larger class of triples (φ, Ω, K). In particular, we want to get rid of the compactness assumption in Theorem (11.9). Also, by the way, we intend to drop the assumption that Ω is open and that φ is continuous on all of $F^{-1}(\overline{\Omega})$ and assumes its values in K.

(11.10) DEFINITION. Let (φ, Ω, K) be an arbitrary triple (in the sense explained earlier). We call a subset $K_0 \subset K$ a *retraction candidate* for (φ, Ω, K) (with respect to F) if $(\varphi, \Omega \cap K_0, K_0) \in \mathcal{T}$ and $\mathrm{Coin}(F, \varphi, \overline{\Omega}) \subset K_0$.

We point out that in case $\mathrm{Coin}(F, \varphi, \overline{\Omega}) = \emptyset$, the set $K_0 = \emptyset$ is a retraction candidate.

(11.11) EXAMPLE. If $\mathcal{T} = \mathcal{T}(F, R, \mathcal{A})$, then $K_0 \subset K$ is a retraction candidate for (φ, Ω, K) if and only if $K_0 \in \mathcal{A}$, the set $\Omega \cap K_0$ is open in K_0 and contains $\mathrm{Coin}(F, \varphi, \overline{\Omega})$, if $\overline{\Omega \cap K_0} \subset R$,

$$(11.12) \qquad\qquad \varphi(F^{-1}(\overline{\Omega \cap K_0})) \subset K_0,$$

and if the restriction of φ to $F^{-1}(\overline{\Omega \cap K_0})$ is continuous. (In case $\mathrm{Coin}(F, \varphi, \overline{\Omega}) = \emptyset$, the choice $K_0 = \emptyset$ is possible).

If (φ, Ω, K) has a retraction candidate K_0, one might be tempted to define the desired extension Ind of the given index ind_F by

$$(11.13) \qquad\qquad \mathrm{Ind}(\varphi, \Omega, K) := \mathrm{ind}_F(\varphi, \Omega \cap K_0, K_0).$$

However, if this would give a well-defined extension of ind_F, one would have in the case $(\varphi, \Omega, K) \in \mathcal{T}$ that the retraction equality (11.2) holds for each retraction candidate K_0. The following disappointing example shows that this is not true without additional assumptions, even for the usual fixed point index in $\mathbb{R}$.

(11.14) EXAMPLE. Let $X = Y = \mathbb{R}$, and $F = \mathrm{id}$ induce the classical fixed point index on $(Y, \mathcal{A})$, where $\mathcal{A}$ is the family of all compact intervals. For $\Omega := (-1, 1)$, define

$$\varphi(x) := \begin{cases} 2x & \text{if } x \in \Omega, \\ 2\,\mathrm{sgn}(x) & \text{if } |x| \geq 1. \end{cases}$$

For $K := [-2, 2]$, we have by (11.6) that

$$\mathrm{ind}_F(\varphi, \Omega, K) = \mathrm{ind}_F(\varphi, \Omega \cap K, K) = \deg(F - \varphi, \Omega) = \deg(-\mathrm{id}, \Omega) = -1.$$

The set $K_0 = \{0\}$ is a retraction candidate for (φ, Ω, K), but we have

$$\mathrm{ind}_F(\varphi, \Omega \cap K_0, K_0) = \mathrm{ind}_F(\varphi, \{0\}, \{0\}) = 1,$$

in view of the normalization property.

The essential idea in the following is to use (11.13) to define the extension ind_F for all triples (φ, Ω, K) which have a retraction candidate K_0 satisfying the additional property that the corresponding contraction equality (11.2) holds "without any additional assumptions". Unfortunately, the precise formulation of the latter is rather cumbersome.

In the following, we assume not only that F provides a coincidence index ind_F, for $\mathcal{T}$, but also that this index has an extension to a coincidence index ind_F^e for a slightly larger family $\mathcal{T}_{\cup}$.

(11.15) DEFINITION. By $\mathcal{T}_\cup$, we denote the union of $\mathcal{T}$ with all triples of the form (φ, Ω, K) which have the property that there are retraction candidates K_1, K_2 of (φ, Ω, K) with $K_1 \cap K_2 \neq \emptyset$ and $K = K_1 \cup K_2$.

(11.16) EXAMPLE. If $\mathcal{T} = \mathcal{T}(F, R, \mathcal{A})$, then $\mathcal{T}_\cup \subset \mathcal{T}(F, R, \mathcal{A}_\cup)$, where

$$\mathcal{A}_\cup = \{K_1 \cup K_2 \mid K_1, K_2 \in \mathcal{A}, \ K_1 \cap K_2 \neq \emptyset\} \cup \emptyset$$

(this can be verified straightforwardly, making use of the fact that all retraction candidates are closed). If the family $\mathcal{A}$ is closed under finite unions, we have of course $\mathcal{A}_\cup = \mathcal{A}$, and thus $\mathcal{T}_\cup = \mathcal{T}$. But even if $\mathcal{A}$ does not have this property, our requirements are reasonable: If F provides a coincidence index on $(R, \mathcal{A})$, for some $\mathcal{A}$, then the restriction of this index to a subfamily of $\mathcal{A}$ is still a coincidence index in the sense of Definition (11.1). Hence, replacing $\mathcal{A}$ by a sufficiently small subfamily, we can always arrange that F actually provides a coincidence index on $(R, \mathcal{A}_\cup)$ (and thus on $\mathcal{T}_\cup$).

For example, if F provides a coincidence index on $(R, \mathcal{A}_0)$, where $\mathcal{A}_0$ denotes the family of compact neighbourhood retracts of a locally convex metric vector space Y (e.g. if F is a Vietoris map with range R), then F provides a coincidence index on $(R, \mathcal{A}_\cup)$ (and thus for $\mathcal{T}_\cup$) if we let $\mathcal{A}$ be the family of all compact convex subsets of Y.

(11.17) DEFINITION. We say that a set K_0 is *pre-fundamental* for (φ, Ω, K) (on $\mathcal{T}$ with respect to ind_F^e) if K_0 is a retraction candidate for (φ, Ω, K), and if for any retraction candidate K_1 with $K_0 \cap K_1 \neq \emptyset$ the union $K_2 = K_0 \cup K_1$ ($\subset K$) satisfies

(11.18) $\mathrm{ind}_F^e(\varphi, \Omega \cap K_0, K_0) = \mathrm{ind}_F^e(\varphi, \Omega \cap K_2, K_2)$.

We say that K_0 is *fundamental*, for (φ, Ω, K) if K_0 is pre-fundamental, for any triple of the form (φ, Ω_0, K), where $\Omega_0 \subset \Omega$ is open in K with $\Omega_0 \supset \mathrm{Coin}(F, \varphi, \overline{\Omega})$.

We say that K_0 is *locally fundamental* for (φ, Ω, K), if there is some $\Omega_0 \subset \Omega$ which is open in K and with $\Omega_0 \supset \mathrm{Coin}(F, \varphi, \overline{\Omega})$ and such that K_0 is fundamental for (φ, Ω_0, K).

We define similar notions also for maps $h : [0, 1] \times F^{-1}(\overline{\Omega}) \to Y$:

(11.19) DEFINITION. We call K_0 *pre-fundamental* for (h, Ω, K) if $(h, \Omega, K_0) \in \mathcal{T}$ and if K_0 is pre-fundamental for each triple $(h(t, \cdot), \Omega, K)$ $(0 \leq t \leq 1)$.

We call K_0 *locally fundamental* for (h, Ω, K) if there is some $\Omega_0 \subset \Omega$ with $\Omega_0 \supset \mathrm{Coin}(F, h, \overline{\Omega})$ and $(h, \Omega_0 \cap K_0, K_0) \in \mathcal{T}$ such that K_0 is locally fundamental for each triple $(h(t, \cdot), \Omega_0, K)$ $(0 \leq t \leq 1)$.

In particular, in case $\mathcal{T} = \mathcal{T}(\mathcal{A}, R, F)$, K_0 is pre-fundamental (resp. locally fundamental) for (h, Ω, K_0) with a constant $h(t, \cdot) = \varphi$ if and only if K_0 is pre-fundamental (resp. locally fundamental) for (φ, Ω, K_0).

(11.20) PROPOSITION. *If K_0 is (locally) fundamental for (φ, Ω, K), then K_0 is (locally) fundamental for any triple of the form (φ, Ω_0, K), where $\Omega_0 \subset \Omega$ is open in K with $\Omega_0 \supset \mathrm{Coin}(F, \varphi, \overline{\Omega})$.*

PROOF. Assume at first that K_0 is fundamental for (φ, Ω, K). If $\Omega_1 \subset \Omega_0$ is open in K with $\Omega_1 \supset \mathrm{Coin}(F, \varphi, \overline{\Omega}_0)$, then also $\Omega_1 \supset \mathrm{Coin}(F, \varphi, \overline{\Omega})$, and so K_0 is pre-fundamental for (φ, Ω_1, K). But this means that K_0 is fundamental for (φ, Ω_0, K).

If K_0 is locally fundamental for (φ, Ω, K), then there is some $\Omega_1 \subset \Omega$ which is open in K with $\Omega_1 \supset \mathrm{Coin}(F, \varphi, \overline{\Omega})$ such that K_0 is fundamental for (φ, Ω_1, K). Then $\Omega_0 \cap \Omega_1 \supset \mathrm{Coin}(F, \varphi, \overline{\Omega}_0)$, and by what we proved before, K_0 is fundamental for $(\varphi, \Omega_1 \cap \Omega_0, K)$ and thus locally fundamental for (φ, Ω_0, K). $\square$

(11.21) PROPOSITION. *If $(\varphi, \Omega, K) \in \mathcal{T}$ resp. $(h, \Omega, K) \in \mathcal{T}$, then $K_0 := K$ is fundamental.*

PROOF. In Definition (11.17), we have $K_0 \subset K_2 \subset K$ and thus $K_2 = K_0$ which trivially implies (11.18). $\square$

In particular, each triple $(\varphi, \Omega, K) \in \mathcal{T}$ resp. $(h, \Omega, K) \in \mathcal{T}$ is $(\mathcal{T}, \mathrm{ind}_F^e)$-pre-admissible and locally $(\mathcal{T}, \mathrm{ind}_F^e)$-admissible in the following sense.

(11.22) DEFINITION. We say that the triple (φ, Ω, K) resp. (h, Ω, K) is $(\mathcal{T}, \mathrm{ind}_F^e)$-*pre-admissible* if it has a pre-fundamental set. We call it *locally $(\mathcal{T}, \mathrm{ind}_F^e)$-admissible* if it has a locally fundamental set.

We shall present the main extension theorem for the index. We note that all approaches we found in literature which provide some degree or fixed point index for noncompact maps use in their proofs implicitly a special case of this theorem. (The only exception that we know are the approaches from [FV2], [FV3], [Nu1] which however use implicitly a result very similar to the following theorem).

The main extension theorem for the index is the following.

(11.23) THEOREM. *Assume that F provides a coincidence index ind_F for $\mathcal{T}$ which has an extension to a coincidence index ind_F^e for $\mathcal{T}_\cup$. Assume also that ind_F satisfies the excision property.*

Then ind_F has a unique extension to an index Ind_F defined on all locally $(\mathcal{T}, \mathrm{ind}_F^e)$-admissible triples (φ, Ω, K) such that for each such triple the following holds:

(i) (Coincident point property) *If $\mathrm{Ind}_F(\varphi, \Omega, K) \neq 0$, then $\mathrm{Coin}(F, \varphi, \Omega) \neq \emptyset$.*

(ii) (Normalization) *If $\varphi(x) \equiv c \in \Omega$, for some $c \in F(X)$, then $\mathrm{Ind}_F(\varphi, \Omega, K) = 1$.*

(iii) (Homotopy invariance) *If (h, Ω, K) is locally $(\mathcal{T}, \mathrm{ind}_F^e)$-admissible, then* $\mathrm{Ind}_F(h(t, \cdot, \Omega, K)$ *is independent of* $t \in [0, 1]$.

(iv) (Fundamental contraction) *If $K_0 \in \mathcal{A}$ is fundamental, for (φ, Ω, K), then* $(\varphi, \Omega \cap K_0, K_0)$ *is locally $(\mathcal{T}, \mathrm{ind}_F^e)$-admissible, and we have*

$$(11.24) \qquad \mathrm{Ind}_F(\varphi, \Omega, K) = \mathrm{Ind}_F(\varphi, \Omega \cap K_0, K_0) \quad (= \mathrm{ind}_F(\varphi, \Omega \cap K_0, K_0))$$

The same holds if $K_0 \in \mathcal{A}$ is only locally fundamental with $(\varphi, \Omega \cap K_0, K_0) \in \mathcal{T}$ (in particular, with (11.12)).

(v) (Excision) *If $\Omega_0 \subset \Omega$ is open in K with $\Omega_0 \supset \mathrm{Coin}(F, \varphi, \overline{\Omega})$, then (φ, Ω_0, K) is locally $(\mathcal{T}, \mathrm{ind}_F^e)$-admissible, and*

$$\mathrm{Ind}_F(\varphi, \Omega, K) = \mathrm{Ind}_F(\varphi, \Omega_0, K).$$

If ind_F is additionally additive, we also have:

(vi) (Additivity) *If $\Omega_1, \Omega_2 \subset \Omega$ are disjoint and open with $\Omega_1 \cup \Omega_2 \supset \mathrm{Coin}(F, \varphi, \overline{\Omega})$, then (Φ, Ω_1, K) and (Φ, Ω_2, K) are locally $(\mathcal{T}, \mathrm{ind}_F^e)$-admissible, and*

$$(11.25) \qquad \mathrm{Ind}_F(\Phi, \Omega, K) = \mathrm{Ind}_F(\Phi, \Omega_1, K) + \mathrm{Ind}_F(\Phi, \Omega_2, K).$$

It should be noted that the excision property and the additivity are required only for ind_F, not for the extension ind_F^e. Note also that Ind_F is only an extension of ind_F and not necessarily an extension for ind_F^e (we do not even claim that the values of Ind_F and ind_F^e coincide for those triples for which both indices are defined).

There is an analogous result without the excision property.

(11.26) THEOREM. *Assume that F provides a coincidence index ind_F for $\mathcal{T}$ which can be extended to a coincidence index ind_F^e for $\mathcal{T}_\cup$. Then ind_F has a unique extension to an index Ind_F defined on all $(\mathcal{T}, \mathrm{ind}_F^e)$-pre-admissible triples (φ, Ω, K) such that for each such triple the following holds:*

(i) (Coincident point property) *If $\mathrm{Ind}_F(\varphi, \Omega, K) \neq 0$, then $\mathrm{Coin}(F, \varphi, \overline{\Omega}) \neq \emptyset$.*

(ii) (Normalization) *If $\varphi(x) \equiv c \in \Omega$, for some $c \in F(X)$, then $\mathrm{Ind}_F(\varphi, \Omega, K) = 1$.*

(iii) (Homotopy invariance) *If (h, Ω, K) is $(\mathcal{T}, \mathrm{ind}_F^e)$-pre-admissible, then $\mathrm{Ind}_F(h(t, \cdot), \Omega, K)$ is independent of $t \in [0, 1]$.*

(iv) (Fundamental contraction) *If $K_0 \neq \emptyset$ is pre-fundamental for (φ, Ω, K), then $(\varphi, \Omega \cap K_0, K_0)$ is $(\mathcal{T}, \mathrm{ind}_F^e)$-admissible, and (11.24) holds.*

PROOF THEOREM (11.23). For the uniqueness proof, we need only the fundamental contraction and the excision property. Let (φ, Ω, K) be locally $(\mathcal{T}, \mathrm{ind}_F^e)$-admissible with a locally fundamental set K_0. This means that we find some

$\Omega_0 \subset \Omega$ which is open in K and with $\Omega_0 \supset \mathrm{Coin}(F, \varphi, \overline{\Omega})$ such that K_0 is fundamental for (φ, Ω_0, K). If Ind_F is as in the theorem, we must have by the excision property that
$$\mathrm{Ind}_F(\varphi, \Omega, K) = \mathrm{Ind}_F(\varphi, \Omega_0, K),$$

and by the fundamental contraction property, we must also have

$$\mathrm{Ind}_F(\varphi, \Omega_0, K) = \mathrm{Ind}_F(\varphi, \Omega_0 \cap K_0, K_0).$$

Hence, the only way to define the index is to put

$$(11.27) \qquad \mathrm{Ind}_F(\varphi, \Omega, K) := \mathrm{Ind}_F(\varphi, \Omega_0 \cap K_0, K_0).$$

We now use (11.27) to define the extension Ind_F. We have to show that this definition is independent of the particular choice of Ω_0 and K_0. Let Ω_1 and K_1 be another choice for these sets. We have to show that

$$(11.28) \qquad \mathrm{Ind}_F(\varphi, \Omega_0 \cap K_0, K_0) = \mathrm{Ind}_F(\varphi, \Omega_1 \cap K_1, K_1)$$

holds, so that the right-hand side of (11.27) is indeed independent of the particular choice of Ω_0 and K_0. With $\Omega_2 = \Omega_0 \cap \Omega_1$, we have by the excision property of the given index that

$$\mathrm{ind}_F(\varphi, \Omega_i \cap K_i, K_i) = \mathrm{ind}_F(\varphi, \Omega_2 \cap K_i, K_i) \quad (i = 0, 1).$$

If the equation $F(x) = \varphi(x) \in \overline{\Omega}$ has no solution, then the coincident point property implies that both sides of (11.28) vanish, and we are done. Hence, we only have to consider the case that there is some solution x. Since K_0 and K_1 are retraction candidates, we have $F(x) \in K_0 \cap K_1$, in particular $K_0 \cap K_1 \neq \emptyset$. Put $K_2 = K_0 \cup K_1$. Since K_0 and K_1 are pre-fundamental for (φ, Ω_2, K), we have

$$\mathrm{ind}_F(\varphi, \Omega_2 \cap K_i, K_i) = \mathrm{ind}_F^e(\varphi, \Omega_2 \cap K_i, K_i) = \mathrm{ind}_F^e(\varphi, \Omega_2 \cap K_2, K_2) \quad (i = 0, 1).$$

Combining the above equalities, we now find (11.28).

If $(\varphi, \Omega, K) \in \mathcal{T}$, then we may choose $\Omega_0 = \Omega$ and $K = K_0$, in view of Proposition (11.21), and so (11.27) shows that Ind_F is indeed an extension of ind_F.

Let now K_0 be locally fundamental for (φ, Ω, K), and $\Omega_0 \subset \Omega$ open in K with $\Omega_0 \supset \mathrm{Coin}(F, \varphi, \overline{\Omega})$.

The fundamental contraction property follows from the fact that if additionally $(\varphi, \Omega \cap K_0, K_0) \in \mathcal{T}$, then the excision property of ind_F implies

$$\mathrm{ind}_F(\varphi, \Omega \cap K_0, K_0) = \mathrm{ind}_F(\varphi, \Omega_0 \cap K_0, K_0).$$

But the right-hand side of this equality is $\mathrm{Ind}_F(\varphi, \Omega, K)$ by (11.27), and the left-hand side is $\mathrm{Ind}_F(\varphi, \Omega \cap K_0, K_0)$, because Ind_F is an extension of ind_F.

Concerning the coincident point property, observe that if $0 \neq \mathrm{Ind}_F(\varphi, \Omega, K) = \mathrm{ind}_F(\varphi, \Omega_0 \cap K_0, K_0)$, then $\emptyset \neq \mathrm{Coin}(F, \varphi, \Omega_0 \cap K_0) \subset \mathrm{Coin}(F, \varphi, \Omega)$.

If $\varphi(x) \equiv c \in \Omega$ belongs to the range of F, then $c \in \mathrm{Coin}(F, \varphi, \overline{\Omega}) \subset K_0 \cap \Omega_0$. Hence, $\mathrm{Ind}_F(\varphi, \Omega, K) = \mathrm{ind}_F(\varphi, \Omega_0 \cap K_0, K_0) = 1$, by the normalization property of ind_F.

Now, we prove the additivity. The proof of the excision property is analogous (put $\Omega_2 = \emptyset$ in the following arguments). Let $\Omega_0 \subset \Omega$ be open in K with $\Omega_0 \supset \mathrm{Coin}(F, \varphi, \overline{\Omega})$, and let K_0 be fundamental for (φ, Ω_0, K). Then $\Omega_{i,0} := \Omega_i \cap \Omega_0$ $(i = 1, 2)$ is open in K with

$$\Omega_{i,0} \supset \mathrm{Coin}(F, \varphi, \overline{\Omega}_i) \quad (i = 1, 2),$$

because $\mathrm{Coin}(F, \varphi, \overline{\Omega}_i) = \mathrm{Coin}(F, \varphi, \overline{\Omega}) \cap \overline{\Omega}_i \subset (\Omega_1 \cup \Omega_2) \cap \overline{\Omega}_i = \Omega_i$, by the fact, that Ω_1 and Ω_2 are disjoint and open.

Since K_0 is fundamental for (φ, Ω_0, K), Definition (11.30) implies that K_0 is fundamental for $(\varphi, \Omega_{i,0}, K)$ $(i = 1, 2)$. Hence, (φ, Ω_i, K) is locally $(\mathcal{T}, \mathrm{ind}_F^e)$-admissible, and we have

$$\mathrm{Ind}_F(\varphi, \Omega_i, K) = \mathrm{ind}_F(\varphi, \Omega_{i,0} \cap K_0, K_0) \quad (i = 1, 2).$$

In view of (11.12), we have $\mathrm{Coin}(F, \varphi, \overline{\Omega_0 \cap K_0}) \subset K_0 \cap \mathrm{Coin}(F, \varphi, \overline{\Omega}) \subset K_0 \cap (\Omega_1 \cup \Omega_2) \cap \Omega_0 = (\Omega_{1,0} \cap K_0) \cup (\Omega_{2,0} \cap K_0)$. Hence, the additivity of ind_F (resp. the excision property if $\Omega_2 = \emptyset$) implies

$$\mathrm{ind}_F(\varphi, \Omega_{1,0} \cap K_0, K_0) + \mathrm{ind}_F(\varphi, \Omega_{2,0} \cap K_0, K_0)$$
$$= \mathrm{ind}_F(\varphi, \Omega_0 \cap K_0, K_0) = \mathrm{Ind}_F(\varphi, \Omega, K).$$

Combining the above equalities, the additivity follows.

To prove the homotopy invariance, let K_0 be locally fundamental for (h, Ω, K), i.e. there is some open $\Omega_0 \subset \Omega$ with $\Omega_0 \supset \mathrm{Coin}(F, h, \overline{\Omega})$ and $(h, \Omega_0 \cap K_0, K_0) \in \mathcal{T}$ such that K_0 is locally fundamental for each triple $(h(t, \cdot), \Omega_0, K)$ $(0 \leq t \leq 1)$. The excision and fundamental contraction properties imply

$$\mathrm{Ind}_F(h(t, \cdot), \Omega, K) = \mathrm{Ind}_F(h(t, \cdot), \Omega_0, K) = \mathrm{Ind}_F(h(t, \cdot), \Omega_0 \cap K_0, K_0)$$
$$= \mathrm{ind}_F(h(t, \cdot), \Omega_0 \cap K_0, K_0),$$

and the right-hand side is independent of $t \in [0, 1]$ by the homotopy invariance of ind_F. $\qquad \square$

PROOF OF THEOREM (11.26). The proof is analogous to the proof of Theorem (11.23). The main difference is that one has to put $\Omega_0 = \Omega$ in that proof. Then it suffices that K_0 is pre-fundamental for (φ, Ω, K), and also so the excision property of ind_F is not needed.   □

The reader who has some experience with related theorems for the fixed point index may be surprised that in Theorems (11.23) and (11.26) it is not necessary to assume, for the triple (φ, Ω, K), that K belongs to a certain class of "topologically nice" sets (like ANRs). However, the reason is that we do not claim that the index satisfies the contraction equality (11.24), for any set K_0 which contains $\varphi(F^{-1}(\overline{\Omega}))$. (The fundamental contraction property holds only for (locally) fundamental sets K_0 which by definition belong to $\mathcal{A}$).

Nevertheless, the fundamental contraction property contains the contraction property, as a special case, if we require $(\varphi, \Omega \cap K_0, K_0) \in \mathcal{T}$ (in particular $K_0 \in \mathcal{A}$). More precisely, the following holds.

(11.29) PROPOSITION. *Let F provide a coincidence index* ind_F^e *for* $\mathcal{T}_\cup$, *and* (φ, Ω, K) *be such that there is some $K_0 \subset K$ which contains $\varphi(F^{-1}(\overline{\Omega}))$ and $(\varphi, \Omega \cap K_0, K_0) \in \mathcal{T}$. Then K_0 is pre-fundamental for (φ, Ω, K) and even fundamental if ind_F satisfies the excision property. In particular, in the situation of Theorem (11.23) or (11.26), we have (11.24) (and both sides are defined).*

PROOF. Let $\Omega_0 = \Omega$ respectively $\Omega_0 \subset \Omega$ be open in K with $\Omega_0 \supset \mathrm{Coin}(F, \varphi, \overline{\Omega})$. The assumptions imply $(\varphi, \Omega_0 \cap K_0, K_0) \in \mathcal{T}$, and so K_0 is a retraction candidate for (φ, Ω_0, K). If K_1 is another retraction candidate, then the contraction property of ind_F^e implies for $K_2 := K_0 \cup K_1$ that

$$\mathrm{ind}_F^e(\varphi, \Omega_0 \cap K_0, K_0) = \mathrm{ind}_F^e(\varphi, \Omega_0 \cap K_2, K_2).$$

Hence, K_0 is pre-fundamental for (φ, Ω_0, K).   □

(φ, Ω, K)-deformation retracts.

For the rest of this discussion, we assume that we are given a family $\mathcal{A}$ of closed subsets of Y and some $R \subset Y$. We are interested in the case that F provides a coincidence index for $\mathcal{T} = \mathcal{T}(F, R, \mathcal{A})$.

We propose a general topological concept which allows us to prove that a given set K_0 is fundamental for some triple (φ, Ω, K).

(11.30) DEFINITION. Let a triple (φ, Ω, K) be given. Then we call a closed subset $K_0 \subset K$ with (11.12) a *neighbourhood (φ, Ω, K)-deformation retract* (with respect to F) if there is an open (in K) neighbourhood $\Omega_0 \subset K$ of $\Omega \cap K_0$ and a continuous map $\rho : [0,1] \times \varphi(F^{-1}(\overline{\Omega \cap \Omega_0})) \to K$ with the following properties:

(11.30.1) $\rho(0,y) = y$ and $\rho(1,y) \in K_0$, for all $y \in \varphi(F^{-1}(\overline{\Omega \cap \Omega_0}))$,

(11.30.2) $\rho(1,y) = y$ if $y \in \varphi(F^{-1}(\overline{\Omega \cap K_0})) \subset K_0$,

(11.30.3) $\rho(t,y) = y$ if $t \in [0,1]$

$\qquad\qquad$ and $y \in \varphi(F^{-1}(K_0 \cap \partial(\Omega \cap \Omega_0))) \subset \varphi(F^{-1}(K_0 \cap \partial\Omega))$,

(11.30.4) $F(x) = \rho(t, \varphi(x)) \Longrightarrow F(x) \in K_0$,

$\qquad\qquad$ for all $(t,x) \in [0,1] \times F^{-1}(\partial(\Omega \cap \Omega_0))$.

(All boundaries are understood with respect to K).

If the choice $\Omega_0 = \Omega$ is possible, we call K_0 a (φ, Ω, K)-*deformation retract*.

Before we discuss Definition (11.30), let us prove the following key property of (φ, Ω, K)-deformation retracts. Roughly speaking, the statement is that, for such sets, the problem of Example (11.14) cannot occur.

(11.31) PROPOSITION. *Let* F *provide a coincidence index* ind_F *for* $\mathcal{T} := \mathcal{T}(F, \mathcal{R}, \mathcal{A})$. *Let* $(\varphi, \Omega, K) \in \mathcal{T}$, *and* $K_0 \in \mathcal{A}$ *be a retraction candidate. If* K_0 *is a* (φ, Ω, K)-*deformation retract, then*

$$\mathrm{ind}_F(\varphi, \Omega, K) = \mathrm{ind}_F(\varphi, \Omega \cap K_0, K_0).$$

If ind_F *satisfies the excision property, it suffices that* $K_0 \in \mathcal{A}$ *is a neighbourhood* (φ, Ω, K)-*deformation retract.*

PROOF. Let ρ and Ω_0 be as in Definition (11.30). By the excision property, we have (since $K_0 \subset \Omega_0$ is a retraction candidate) that

$$\mathrm{ind}_F(\varphi, \Omega, K) = \mathrm{ind}_F(\varphi, \Omega \cap \Omega_0, K).$$

Put $D := F^{-1}(\overline{\Omega \cap \Omega_0})$, and define a homotopy $h : [0,1] \times D \to K$ by $h(t,x) := \rho(t, \varphi(x))$.

We claim that $(h, \Omega \cap \Omega_0, K)$ is F-admissible. Assume that $F(x) = h(t,x)$, for some $(t,x) \in [0,1] \times D$. We have to prove that $F(x) \notin \partial(\Omega \cap \Omega_0)$. Otherwise, we have, in view of $F(x) = \rho(t, \varphi(x))$, that $F(x) \in K_0$ (by the hypothesis). Hence, we actually have $x \in F^{-1}(K_0 \cap \partial(\Omega \cap \Omega_0))$, and so $\rho(t, \varphi(x)) = \varphi(x)$, which implies

(11.32) $\qquad\qquad F(x) = \varphi(x) \in \partial(\Omega \cap \Omega_0) \subset \overline{\Omega} \setminus (\Omega \cap \Omega_0)$.

Since (φ, Ω, K) is F-admissible, (11.32) implies $F(x) \in \Omega$, and so $F(x) \in \Omega \cap K_0 \subset \Omega_0$, i.e. $F(x) \in \Omega \cap \Omega_0$, a contradiction to (11.32).

This contradiction shows that $(h, \Omega, K) \in \mathcal{T}(F, R, \mathcal{A})$. Since $h(0, \cdot) = \varphi|_D$, we thus have

$$\mathrm{ind}_F(\varphi, \Omega \cap \Omega_0, K) = \mathrm{ind}_F(h(1, \cdot), \Omega \cap \Omega_0, K).$$

The crucial fact now is that we have $h(\{1\} \times D) \subset K_0$, and so the contraction property implies

$$\mathrm{ind}_F(h(1, \cdot), \Omega \cap \Omega_0, K) = \mathrm{ind}_F(h(1, \cdot), \Omega \cap \Omega_0 \cap K_0, K_0)$$
$$= \mathrm{ind}_F(h(1, \cdot), \Omega \cap K_0, K_0).$$

Since we have, for $x \in F^{-1}(\overline{\Omega \cap K_0})$, that $h(1, x) = \rho(1, \varphi(x)) = \varphi(x)$, the last index is actually $\mathrm{ind}_F(\varphi, \Omega \cap K_0, K_0)$, and the statement follows. □

Let now $\mathcal{T}_\cup$ correspond to $\mathcal{T} = \mathcal{T}(F, R, \mathcal{A})$, as above.

(11.33) COROLLARY. *Let F provide a coincidence index ind_F^e for $\mathcal{T}_\cup$, and (φ, Ω, K) have some retraction candidate $K_0 \in \mathcal{A}$. Assume that, for any set of the form $K_2 = K_0 \cup K_1$, where $K_1 \in \mathcal{A}$ is a retraction candidate with $K_0 \cap K_1 \neq \emptyset$, we have either (11.18), or the set K_0 is a $(\varphi, \Omega \cap K_2, K_2)$-deformation retract. Then K_0 is pre-fundamental, for (φ, Ω, K).*

If ind_F^e satisfies the excision property, it suffices that K_0 is a neighbourhood $(\varphi, \Omega \cap K_2, K_2)$-deformation retract.

Definition (11.30) is rather technical, but for applications the following test usually suffices (whence the name "(φ, Ω, K)-deformation retract"):

Let (φ, Ω, K) be F-admissible, and let $K_0 \subset K$ be closed. Assume, for a moment, that, instead of (11.12), we have the slightly more restrictive inclusion

$$(11.34) \qquad\qquad \varphi(F^{-1}(\overline{\Omega} \cap K_0)) \subset K_0.$$

If K_0 is a strong deformation retract of K, then the properties (11.30.1)–(11.30.3) are satisfied for the restriction ρ of the corresponding deformation. The crucial property (11.30.4) states that the deformation is admissible in the sense that it avoids certain coincidences outside K_0.

It turns out that (11.34) is slightly too restrictive for certain applications. We prove now that if we are only interested in neighbourhood (φ, Ω, K)-deformation retracts (i.e. if we assume the excision property of ind_F^e), we need only a deformation of a certain set containing $\varphi(F^{-1}(\overline{\Omega} \cap K_0))$ (and this deformation may still obtain values in K). Moreover, in this case the condition (11.34) may actually be dropped if e.g. Y is a metric space.

(11.35) PROPOSITION. *Let a triple (φ, Ω, K) be given, and $K_0 \subset K$ be closed with*

$$\varphi(F^{-1}(\overline{\Omega_0 \cap K_0})) \subset K_0.$$

Suppose, in addition, that K satisfies the T_5-separation axiom of Alexandroff–Hopf (e.g. that K is completely normal). Assume there is some open (in K) set $\Omega_0 \subset \Omega$ with $\Omega_0 \supset \Omega \cap K_0$ and a continuous map $\rho : [0,1] \times \varphi(F^{-1}(\overline{\Omega}_0)) \to K$ with the following properties:

(11.35.1) $\rho(0,y) = y$ *and* $\rho(1,y) \in K_0$, *for all* $y \in \varphi(F^{-1}(\overline{\Omega}_0))$,

(11.35.2) $\rho(t,y) = y$ *if* $t \in [0,1]$ *and* $y \in \varphi(F^{-1}(\overline{\Omega_0 \cap K_0})) \subset K_0$,

(11.35.3) $F(x) = \rho(t, \varphi(x)) \Longrightarrow F(x) \in K_0$,

$$\text{for all } (t,x) \in [0,1] \times F^{-1}(\overline{\Omega}_0).$$

Then K_0 is a neighbourhood (φ, Ω, K)-deformation retract.

PROOF. Put $A := \Omega_0 \cap K_0$. We claim that there is an open set $O \subset K$ with $O \supset A$ and $\overline{O} \cap K_0 \subset \overline{A}$. Note that we have then $\overline{(O \cap \Omega_0) \cap K_0} \supset \overline{A} \supset \overline{O} \cap K_0 \supset \overline{(\overline{O} \cap \Omega_0)} \cap K_0$ (and so, we have actually equality throughout). Hence, if we replace Ω_0 by $\Omega_0 \cap O$, the assumptions of the statement are still satisfied, but we have additionally

(11.36) $\varphi(F^{-1}(\partial(\Omega_0 \cap \Omega) \cap K_0)) \subset \varphi(F^{-1}(\overline{\Omega}_0 \cap K_0)) \subset \varphi(F^{-1}(\overline{\Omega_0 \cap K_0}))$.

Now, it is straightforward to verify that K_0 is indeed a neighbourhood (φ, Ω, K)-deformation retract.

To construct O, put $B := (\overline{\Omega}_0 \cap K_0) \setminus \overline{A}$, and let B_0 denote the closure of B in the space K_0. Since K_0 is closed in K, also B_0 is closed in K, i.e. $\overline{B} \subset B_0$. Moreover, since A is open in K_0 and disjoint from B, we have $A \cap B_0 = \emptyset$, and so $A \cap \overline{B} = \emptyset$; the relation $\overline{A} \cap B = \emptyset$ follows from the definition of B. Since K is a T_5 space, we find disjoint open sets $O_1, O_2 \subset K$ with $O_1 \supset A$ and $O_2 \supset B$. The set $O = O_1 \cap \Omega_0$ now has the required properties. Clearly, $O \supset A \cap \Omega_0 = A$. Since the set $\overline{O} \cap K_0$ is contained in $\overline{\Omega}_0 \cap K_0$ and disjoint from $O_2 \supset B$, the definition of B implies $\overline{O} \cap K_0 \subset \overline{A}$.   □

We point out that even if one may choose $\Omega_0 = \Omega$ in Proposition (11.35), it need not be the case that K_0 is a (φ, Ω, K)-deformation retract (the proposition implies only that it is a neighbourhood (φ, Ω, K)-deformation retract). The reason is that we had to pass in the proof to a subset Ω_0 of Ω which satisfies (11.36).

If we are interested in the application of Corollary (11.33) to find locally fundamental sets, we may essentially replace "completely normal" by "normal" in Proposition (11.35). The reason is that we may immediately pass to a smaller set Ω_0, in view of Definition (11.17). More precisely, we have the following result.

(11.37) THEOREM. *Let F provide a coincidence index ind_F^e for $\mathcal{T}_\cup$. Consider the triple (φ, Ω, K), where K is a T_4-space. Let $K_0 \subset K$ with $K_0 \in \mathcal{A}$ contain the set $C := \mathrm{Coin}(F, \varphi, \overline{\Omega})$, and let $\Omega_0 \subset \Omega$ be open in K with $\Omega_0 \supset C$, $\overline{\Omega_0} \cap K_0 \subset R$ such that the restriction $\varphi : F^{-1}(\overline{\Omega_0}) \to Y$ is continuous, and*

$$(11.38) \qquad\qquad \varphi(F^{-1}(\overline{\Omega_0 \cap K_0})) \subset K_0.$$

Suppose for any open $\Omega_1 \subset K$ with $\overline{\Omega_1} \subset \Omega_0$ and $\Omega_1 \supset C$ and for any retraction candidate $K_1 \in \mathcal{A}$ for (φ, Ω_1, K) with $K_0 \cap K_1 \neq \emptyset$, the set $K_2 := K_0 \cup K_1$ either satisfies

$$(11.39) \qquad\qquad \mathrm{ind}_F^e(\varphi, \Omega_1 \cap K_0, K_0) = \mathrm{ind}_F^e(\varphi, \Omega_1 \cap K_2, K_2),$$

or there is a continuous map $\rho : [0,1] \times \varphi(F^{-1}(\overline{\Omega_1 \cap K_2})) \to K_2$ with the following properties:

(11.39.1) $\rho(0, y) = y$ *and* $\rho(1, y) \in K_0$, *for all* $y \in \varphi(F^{-1}(\overline{\Omega_1 \cap K_2}))$.

(11.39.2) $\rho(t, y) = y$ *if* $t \in [0,1]$ *and* $y \in \varphi(F^{-1}((\overline{\Omega_1 \cap K_0}) \cup (K_0 \cap B))) \subset K_0$.

(11.39.3) $F(x) = \rho(t, \varphi(x)) \implies F(x) \in K_0$, *for all* $t \in [0,1]$ *and* $x \in F^{-1}(B) \subset F^{-1}(\overline{\Omega_1 \cap K_2})$.

Here we put $B := (\overline{\Omega_1 \cap K_2}) \setminus \Omega_1$ ($\subset K_2 \cap \partial\Omega_1$, when the boundary is understood with respect to K).

Assume, in addition, that either K is even a normal space and the restriction $F : \overline{F^{-1}(C)} \to K$ is continuous and closed, or that even $\overline{C} \subset \Omega_0$.

Then K_0 is locally fundamental for (φ, Ω, K), and in the situation of Theorem (11.23) (i.e. if ind_F satisfies the excision property), we have

$$(11.40) \qquad\qquad \mathrm{Ind}_F(\varphi, \Omega, K) = \mathrm{ind}_F(\varphi, \Omega_0 \cap K_0, K_0).$$

If not only ind_F satisfies the excision property, but even ind_F^e, we get a slight weakening of our assumptions.

(11.41) THEOREM. *Let ind_F^e satisfy the excision property. Then Theorem (11.37) holds with the following changes. In (11.39) and in the conditions involving ρ, one can replace Ω_1 by some appropriate open (in K) subset $\Omega_2 \subset \Omega_1$ with $\Omega_2 \supset C$ which may depend not only on Ω_1 but also on K_0, K_1 (and K_2).*

We prepare the proof by an elementary lemma which we prove even for homotopies h (the latter will be needed also in the sequel for applications).

(11.42) LEMMA. *Let $\Omega \subset K \subset Y$, I be a compact topological space, $F : X \to Y$, and $h : I \times F^{-1}(\overline{\Omega}) \to Y$. Put $C := \mathrm{Coin}(F, h, \overline{\Omega})$, and $D := \overline{F^{-1}(C)}$. If $M := h(I \times D) \cup F(D)$ is Hausdorff and if the restrictions $h : I \times D \to Y$ and*

$F : D \to Y$ are continuous, then $D_I := \{x \in D \mid F(x) \in h(I \times \{x\})\}$ is closed. In particular, if additionally $F : D \to Y$ is closed, then $C = F(D_I)$ is closed.

PROOF. For $J \subset I$, put $D_J := \{x \in D \mid F(x) \in h(J \times \{x\})\}$. To prove that D_I is closed, let some $x \in D$ with $x \notin D_I$ be given. We show that $x \notin \overline{D}_I$.

To this end, let $\mathcal{I}$ be the system of all open sets $J \subset I$ with $x \notin \overline{D}_J$. Then $\mathcal{I}$ is an open cover of I. Indeed, given some $t \in I$, we have $F(x) \neq h(t, x)$. Since M is Hausdorff, we find open neighbourhoods $O_1, O_2 \subset Y$ of $F(x)$ and $h(t, x)$, respectively, such that $O_1 \cap O_2 \cap M = \emptyset$. Since F and h are continuous, we find an open neighbourhood $J \subset I$ of t and a neighbourhood $N \subset D$ of x with $F(N) \subset O_1$ and $h(J \times N) \subset O_2$. Then N is disjoint from D_J, and so $x \notin \overline{D}_J$.

Since I is compact, we find finitely many $I_1, \ldots, I_n \in \mathcal{I}$ with $I_1 \cup \cdots \cup I_n = I$. Then $x \notin \overline{D}_{I_1} \cup \cdots \cup \overline{D}_{I_n} = \overline{D_{I_1} \cup \cdots \cup D_{I_n}} = \overline{D}_I$, as required.   □

PROOF OF THEOREM (11.37). Since any normal space is Hausdorff, Lemma (11.42) implies that we have $\overline{C} \subset \Omega_0$ in both cases. Since K is a T_4-space with $\overline{C} \subset \Omega_0$, we find some open $O \subset K$ with $C \subset O$ and $\overline{O} \subset \Omega_0$. We claim that K_0 is fundamental for (φ, O, K). Indeed, if $\Omega_1 \subset O$ is open in K such that (φ, Ω_1, K) is F-admissible and if $K_2 = K_0 \cup K_1$, where $K_1 \in \mathcal{A}$ is a retraction candidate for (φ, Ω_1, K) with $K_0 \cap K_1 \neq \emptyset$, we have to prove that (11.39) holds. However, either this holds by the hypothesis, or we have that K_0 is a $(\varphi, \Omega_1 \cap K_2, K_2)$-deformation retract (note that $B = (\overline{\Omega_1 \cap K_2}) \setminus (\Omega_1 \cap K_2)$ is the boundary of $\Omega_1 \cap K_2$ with respect to K_2), and so (11.39) holds by Proposition (11.31).

To prove (11.40), note that the excision property of Ind_F implies $\mathrm{Ind}_F(\varphi, \Omega, K) = \mathrm{Ind}_F(\varphi, \Omega_0, K)$, and so (11.40) follows, in view of the fundamental contraction property.   □

PROOF OF THEOREM (11.41). In view of the excision property of ind_F^e, the equality (11.39) is equivalent to

$$(11.43) \qquad \mathrm{ind}_F^e(\varphi, \Omega_2 \cap K_0, K_0) = \mathrm{ind}_F^e(\varphi, \Omega_2 \cap K_2, K_2).$$

Now, the proof is analogous to the proof of Theorem (11.37) with the difference that (11.39) is replaced everywhere by (11.43).   □

We point out that any Vietoris map $F : X \to Y$ is continuous and closed (on X, and thus on any closed subset). Indeed, any continuous proper map F in metric spaces is closed, because if $F(x_n) \to y$, then the preimage of $\{y, F(x_1), F(x_2), \ldots\}$ is compact; hence, $(x_n)_n$ has an accumulation point x, and $F(x) = y$.

Let us summarize some special cases of the previous results.

(11.44) COROLLARY. *Consider the situation in Theorem* (11.23), *and let a triple* (φ, Ω, K) *be given. Let* $K_0 \subset K$ *with* $K_0 \in \mathcal{A}$ *contain the set* $C = \mathrm{Coin}(F, \varphi, \overline{\Omega})$, *and let* $\Omega_0 \subset \Omega$ *be open in* K *with* $\Omega_0 \supset C$, $\overline{\Omega_0 \cap K_0} \subset R$, *such that the restriction* $\varphi : F^{-1}(\overline{\Omega}_0) \to Y$ *is continuous and* (11.38) *holds.*

Suppose for any open $\Omega_1 \subset K$ *with* $C \subset \Omega_1 \subset \Omega_0$ *and for any set* $K_2 = K_0 \cup K_1 \in \mathcal{A}_\cup$, *where* $K_1 \in \mathcal{A}$ *is a retraction candidate for* (φ, Ω_1, K) *with* $K_0 \cap K_1 \neq \emptyset$, *we find a continuous map* $\rho : [0,1] \times D \to K_2$ *with* $D := \varphi(F^{-1}(\overline{\Omega_1 \cap K_2})) \subset K_2$ *such that the following holds:*

(11.44.1) $\rho(0, y) = y$ *and* $\rho(1, y) \in K_0$, *for all* $y \in D$.

(11.44.2) $\rho(t, y) = y$ *for all* $(t, y) \in [0, 1] \times (K_0 \cap D)$.

(11.44.3) $F(x) = \rho(t, \varphi(x)) \implies F(x) \in K_0$, *for all* $(t, x) \in [0, 1] \times F^{-1}(\overline{\Omega_1 \cap K_2})$.

Suppose, in addition, that one of the following properties is satisfied:

(11.44.4) K *is a* T_4-*space, and* $\overline{C} \subset \Omega_0$.

(11.44.5) K *is a normal space, and the restriction* $F : \overline{F^{-1}(C)} \to K$ *is continuous and closed.*

(11.44.6) K *is a* T_5-*space, and* ind_F^e *satisfies the excision property.*

Then (φ, Ω, K) *is locally* $(\mathcal{T}(F, R, \mathcal{A}), \mathrm{ind}_F^e)$-*admissible with locally fundamental set* K_0, *and we have* (11.40).

PROOF. In the cases (11.44.4) and (11.44.5), the statement follows from Theorem (11.37), so it remains to consider the case (11.44.6). It has to be proved that (11.39) holds. Since ind_F^e satisfies the excision property, it suffices by Proposition (11.31) to show that that K_0 is a neighborhoood $(\varphi, \Omega_1 \cap K_2, K_2)$-deformation retract. But since K is a T_5-space, this follows from Proposition (11.35) (put there $\Omega_0 = \Omega$). The formula (11.40) follows as in the proof of Theorem (11.37). $\qquad \square$

We point out once more that if ind_F^e is (a restriction of) the index from Theorem (11.9), then even all of the three alternative additional assumptions are satisfied automatically.

(c_1, c_2)-fundamental sets.

In this subsection, we describe a way, how fundamental sets K_0 can be found (if they exist), under the assumption that Y admits an operation which has properties analogous to the passage to the convex hull.

Let $\mathcal{U}$ be a family of subsets of Y which satisfies $Y \in \mathcal{U}$ and $\mathcal{A} \subset \mathcal{U}$ and which, moreover, is closed under intersections. We can define a mapping c_1 which associates, to any $M \subset Y$, the smallest set $U \in \mathcal{U}$ which contains M, i.e. we put

$$c_1(M) := \bigcap \{U \in \mathcal{U} \mid U \supset M\}.$$

The family $\mathcal{U}$ is the range of the function c_1.

(11.45) EXAMPLE. Let Y be a convex resp. closed and convex subset of some topological vector space, and $\mathcal{U}$ be the system of all convex resp. closed and convex subsets of Y. Then $c_1(M) = \operatorname{conv}(M)$ resp. $c_1(M) = \overline{\operatorname{conv}}(M)$.

The crucial property of $\mathcal{U}$ is that it is closed under intersections. For this reason, the family $\mathcal{U}$ can serve as an abstract analogue of the family of convex sets for the "classical" coincidence index on convex sets. To obtain a satisfactory analogous theory, we need another ingredient, namely a substitution for "convex homotopies". To this end, we assume that we are given another function c_2 which associates to any set of the form $M = U \cup \{y\}$ with $U \in \mathcal{U}$ and $y \in Y$ a set $c_2(M) \subset Y$. We assume that c_2 is monotone in the sense that

$$c_2(U_0 \cup \{y\}) \subset c_2(U \cup \{y\}) \quad (U_0, U \in \mathcal{U}, \ y \in Y, \ U_0 \subset U).$$

(11.46) DEFINITION. We say that the pair (c_1, c_2) is *convex on* $\mathcal{A}$ if, for any $U \in \mathcal{A}$ and any $K_0 \in \mathcal{A}$ with $U_0 := U \cap K_0 \neq \emptyset$, the set U_0 is a strong deformation retract of U such that the corresponding deformation d satisfies in addition

$$d(\lambda, y) \in c_2(U_0 \cup \{y\}) \quad (\lambda \in [0, 1], \ y \in U).$$

"Convex homotopies" witness that the choice $c_2 = \operatorname{conv}$ is possible:

(11.47) PROPOSITION. *Let Y be a convex subset of a locally convex metric space. For $M \subset Y$, let $\overline{\operatorname{conv}}_Y(M)$ denote the smallest convex closed in Y subset of Y which contains M.*
Then the pairs

$$(\overline{\operatorname{conv}}_Y, \overline{\operatorname{conv}}_Y), \quad (\overline{\operatorname{conv}}_Y, \operatorname*{conv}_Y), \quad (\operatorname*{conv}_Y, \overline{\operatorname{conv}}_Y), \quad and \quad (\operatorname*{conv}_Y, \operatorname*{conv}_Y)$$

are convex on the system $\mathcal{A}$ of all compact convex subsets of Y.

PROOF. The only nontrivial property is that a function d as in Definition (11.46) exists. Thus, let $U_0 = U \cap K \neq \emptyset$ be given with $U, K \in \mathcal{A}$. Then U_0 is a compact and convex subset of a locally convex metric space, and consequently there is a retraction ρ from U onto U_0 (see e.g. [Vae5, Lemma 1.1]). Then the "convex homotopy" $d(\lambda, x) = \lambda \rho(x) + (1 - \lambda)x$ is the desired map. $\qquad \square$

(11.48) REMARK. The same proof shows that each of the four pairs (c_1, c_2) with $c_1, c_2 \in \{\operatorname{conv}_Y, \overline{\operatorname{conv}}_Y\}$ is convex on the system $\mathcal{A}$ of all separable complete convex subsets of Y. If we use Dugundji's Extension Theorem (2.2) in Chapter I.2, we find that these pairs are even convex on the system of all closed convex subsets of Y. Note, however, that the proof of Dugundji's extension theorem requires the (uncountable) axiom of choice which we do not want to assume.

For a case of a closed convex subset Y of a locally convex space, the following definition was introduced in [Vae3], for the $c_1 = \overline{\mathrm{conv}}$ and $c_2 = \mathrm{conv}$. In the classical situation of Example (11.5) and for $D = \overline{\Omega}$, this is the usual definition (7.6) of fundamental sets (which was first introduced in [KSZ]).

(11.49) DEFINITION. Let I be a nonempty set, $D \subset Y$, and $h : I \times F^{-1}(D) \to Y$. We call a set $U \in \mathcal{U}$ (c_1, c_2)-fundamental, for h on D (with respect to F) if the following holds:

(11.49.1) $h(I \times F^{-1}(D \cap U)) \subset U,$

(11.49.2) $F(x) \in D \cap c_2(\{h(t,x)\} \cup U) \Longrightarrow F(x) \in U,$
           for any $(t, x) \in I \times F^{-1}(D).$

Similarly, we call U (c_1, c_2)-fundamental, for $\varphi : D \to Y$ on D, if U is (c_1, c_2)-fundamental (on D), for the map $h : \{0\} \times D \to Y$, defined by $h(0, \cdot) := \varphi$.

The set Y is always (c_1, c_2)-fundamental. However, we are interested in (c_1, c_2)-fundamental sets which belong to $\mathcal{A}$. Since usually $\mathcal{A}$ consists only of compact sets, we are thus interested in (c_1, c_2)-fundamental sets which are as small as possible. There is an optimal choice which we get for $V = \emptyset$ in the following observation.

(11.50) PROPOSITION. *Let $h : I \times F^{-1}(D) \to Y$ with $I \neq \emptyset$ and $D \subset Y$.*

(11.50.1) *If U is (c_1, c_2)-fundamental on D, then U is also (c_1, c_2)-fundamental on any $D_0 \subset D$.*

(11.50.2) *The intersection of any nonempty family of (c_1, c_2)-fundamental sets is (c_1, c_2)-fundamental.*

(11.50.3) *If U if (c_1, c_2)-fundamental, for h on D, then for any $V \subset U$ the set*

$$U_0 = c_1(h(I \times F^{-1}(D \cap U)) \cup V)$$

is contained in U and (c_1, c_2)-fundamental.

(11.50.4) *For any $V \subset Y$, there is a smallest (c_1, c_2)-fundamental set $U_V \subset U$ which contains V. This set satisfies the equality*

(11.51) $U_V = c_1(h(I \times F^{-1}(D \cap U_V)) \cup V).$

PROOF. The first two properties are trivial. For statement (11.50.3), observe that by definition $U_0 \subset c_1(U \cup V) = c_1(U) = U$. Hence, $c_1(h(I \times F^{-1}(D \cap U_0)) \cup V) \subset c_1(h(I \times F^{-1}(D \cap U)) \cup V) = U_0$. Moreover, the inclusion $F(x) \in D \cap c_2(\{h(t,x)\} \cup U_0)$ implies in view of $U_0 \subset U$ that $F(x) \in U$, and so $x \in F^{-1}(D \cap U) \subset U_0$.

For the last statement, let $U = U_V$ be the intersection of all (c_1, c_2)-fundamental sets which contain V. Since the corresponding set U_0 considered above is also such a set, we must have $U_V \subset U_0$; on the other hand, $U_0 \subset U = U_V$, as observed before. $\qquad\square$

For most applications, Proposition (11.50) can be used to verify that there is a (c_1, c_2)-fundamental set $K_0 \in \mathcal{A}$. As an example, we give one corollary which is standard in the case $c_1 = \overline{\mathrm{conv}}$ (see e.g. [AKPRS-M]).

(11.52) DEFINITION. Let us call a function η, defined on the power set of Y with values in a partially ordered space, a *monotone c_1-measure of noncompactness*, if it satisfies the following:

(11.52.1) $\qquad\qquad U \subset V \subset Y$ implies $\eta(U) \le \eta(V)$,

(11.52.2) $\qquad\qquad \eta(c_1(U)) = \eta(U)$, for each $U \subset Y$.

We call a (multivalued) function $\Phi : D \multimap Y$ $(c_1, \mathcal{A})$-*condensing on* $D \subset Y$, if for any $M \in \mathcal{U} \setminus \mathcal{A}$, there is some monotone c_1-measure of noncompactness with $\eta(\Phi(M)) \not\ge \eta(M)$.

An example for a monotone $\overline{\mathrm{conv}}$-measure of noncompactness in a closed convex subset Y of a Banach space is the *Kuratowski measure of noncompactness*. In this case, if $\mathcal{A}$ contains all compact convex subsets of Y, the sum of a contraction and a compact operator is $(\overline{\mathrm{conv}}, \mathcal{A})$-condensing on closed bounded subsets.

(11.53) COROLLARY. *If $h(I \times F^{-1}(\cdot))$ is $(c_1, \mathcal{A})$-condensing on D, then there is some (c_1, c_2)-fundamental set $K_0 \subset c_1(h(I \times F^{-1}(D)))$ with $K_0 \in \mathcal{A}$.*

PROOF. Let K_0 be the smallest (c_1, c_2)-fundamental set for h. By (11.51), we have $K_0 = c_1(h(I \times F^{-1}(D \cap K_0)))$. In particular, $K_0 \in \mathcal{U}$, and we have, for each monotone c_1-measure of noncompactness, that $\eta(K_0) = \eta(h(I \times F^{-1}(D \cap K_0))) \le \eta(h(I \times F^{-1}(D \cap K_0)))$, and so $K_0 \in \mathcal{A}$. $\qquad\square$

It is suggested by the name that there is some relation between "(c_1, c_2)-fundamental" sets K_0 and "fundamental" sets in the sense of Definition (11.17). However, it is not true that any (c_1, c_2)-fundamental set K_0 is fundamental, because the latter requires $K_0 \subset K$ and $(\varphi, \Omega \cap K_0, K_0) \in \mathcal{T}$. But by the following theorem these necessary additional requirements are already sufficient.

(11.54) THEOREM. *Let $\mathcal{T} := \mathcal{T}(F, R, \mathcal{A})$, and let F provide a coincidence index ind_F^e on $\mathcal{T}_\cup$. Let (c_1, c_2) be convex. Consider a triple (φ, Ω, K), and suppose that $K_0 \subset K$ is (c_1, c_2)-fundamental for φ on $\overline{\Omega}$. If $K_0 \subset K$ and $(\varphi, \Omega \cap K_0, K_0) \in \mathcal{T}$, then K_0 is fundamental for (φ, Ω, K).*

An analogous result holds for triples (h, Ω, K).

PROOF. We show at first that K_0 is pre-fundamental for (φ, Ω, K). Since K_0 is (c_1, c_2)-fundamental for φ on $\overline{\Omega}$, we have (11.34). This implies (11.12), and since $F(x) = \varphi(x) \in \overline{\Omega}$ implies $F(x) \in c_2(K_0 \cup \{\varphi(x)\})$, and thus $F(x) \in K_0$, we have that K_0 is actually a retraction candidate. We have to prove that (11.18) holds, for any set $K_2 \in \mathcal{A}_\cup$ which has the form $K_2 = K_0 \cup U$ where $U \in \mathcal{A} \subset \mathcal{U}$ is a retraction candidate and for which $U_0 = U \cap K_0 \neq \emptyset$. Note that $U_0 \in \mathcal{U}$. Choose $d : [0,1] \times U \to U$ as in Definition (11.46). We define $\rho : [0,1] \times K_2 \to K_2$ by putting

$$\rho(\lambda, y) = \begin{cases} d(\lambda, y) & \text{if } y \in U, \\ y & \text{if } y \in K_0. \end{cases}$$

Note that ρ is well-defined, because for $y \in U \cap K_0 = U_0$, we have $d(\lambda, y) = y$. Since K_0 and U are closed in K_2, it follows that ρ is continuous (see e.g. [Hu2-M, Chapter I, Proposition 5.1]). By definition, ρ is a strong deformation retract of K_2 onto K_0. Moreover, the relation $F(x) = \rho(\lambda, \varphi(x)) \in \overline{\Omega \cap K_2}$ implies $F(x) \in K_0$. Indeed, in case $\varphi(x) \in K_0$, we have $F(x) = \rho(\lambda, \varphi(x)) = \varphi(x) \in K_0$, and in case $\varphi(x) \in U$, we have $F(x) = \rho(\lambda, \varphi(x)) = d(\lambda, \varphi(x)) \in c_2(U_0 \cup \{\varphi(x)\}) \subset c_2(K_0 \cup \{\varphi(x)\})$ which, in view of $F(x) \in \overline{\Omega \cap K_2} \subset \overline{\Omega}$, also implies $F(x) \in K_0$. Since $\varphi(F^{-1}(\partial \Omega_1 \cap K_0)) \subset K_0$, in view of (11.34), it follows that K_0 is a (φ, Ω_1, K_2)-deformation retract, and so Proposition (11.31) implies that (11.18) holds. Hence, K_0 is pre-fundamental for (φ, Ω, K).

To see that K_0 is even fundamental for (φ, Ω, K), we have to show that K_0 is pre-fundamental for (φ, Ω_0, K), for any open set $\Omega_0 \subset K$ with $\text{Coin}(F, h, \overline{\Omega}) \subset \Omega_0 \subset \Omega$. But since K_0 is (c_1, c_2)-fundamental, for φ on $\overline{\Omega}_0$ by Proposition (11.50), this follows from what we proved before.

For the statement about (h, Ω, K), observe that K_0 is (c_1, c_2)-fundamental, for any triple $(h(t, \cdot), \Omega, K)$ $(0 \leq t \leq 1)$, by what we just proved and that by assumption $(h, \Omega, K) \in \mathcal{T}$. □

For locally fundamental sets, it suffices that $K_0 \in \mathcal{A}$ is (c_1, c_2)-fundamental on Ω (not on $\overline{\Omega}$). In fact, the following holds.

(11.55) THEOREM. *Let $\mathcal{T} := \mathcal{T}(F, R, \mathcal{A})$, and let F provide a coincidence index ind_F^e on $\mathcal{T}_\cup$. Assume that (c_1, c_2) is convex on $\mathcal{A}$. Consider a triple (φ, Ω, K) where K is a T_4-space. Let $K_0 \in \mathcal{A}$ with $K_0 \subset K$ be (c_1, c_2)-fundamental, for φ on some open in K set $\Omega_0 \subset \Omega$ which contains $C := \text{Coin}(F, \varphi, \overline{\Omega})$ such that $\overline{\Omega_0 \cap K_0} \subset R$ and such that $\varphi : F^{-1}(\overline{\Omega}_0) \to Y$ is continuous.*

Assume in addition that either K is even a normal space and the restriction $F : \overline{F^{-1}(C)} \to K$ is continuous and closed, or that even $\overline{C} \subset \Omega_0$.

Then (φ, Ω, K) is locally $(\mathcal{T}, \text{ind}_F^e)$-admissible with locally fundamental set K_0, and in the situation of Theorem (11.23) (*i.e. if ind_F satisfies the excision property*), *we have* (11.40).

An analogous result holds for triples (h, Ω, K); here we have to assume that $h : [0,1] \times F^{-1}(\overline{\Omega}_0) \to Y$ is continuous.

PROOF. We verify that the assumptions of Theorem (11.37) are satisfied with the given sets K_0 and Ω_0. Since K_0 is (c_1, c_2)-fundamental on Ω_0, we have (11.38), and since $F(x) = \varphi(x) \in \overline{\Omega}$ implies $F(x) \in \Omega_0$ and $F(x) \in c_2(K_0 \cup \{\varphi(x)\})$. Hence, $F(x) \in K_0$ and finally we have that $C \subset K_0$.

Let $\Omega_1 \subset K$ be open with $\overline{\Omega}_1 \subset \Omega_0$ and $\Omega_1 \supset C$, and let $K_2 = K_0 \cup U$, where $U \in \mathcal{A}$ is a retraction candidate for (φ, Ω_1, K) with $U_0 = K_0 \cap U \neq \emptyset$. The same construction as in the proof of Theorem (11.54) provides a strong deformation retract $\rho : [0,1] \times K_2 \to K_2$ of K_2 onto K_1 which has the additional property that the relation $F(x) = \rho(\lambda, \varphi(x)) \in \overline{\Omega_1 \cap K_2}$ implies $F(x) \in K_0$. To see the latter, observe that we have in case $\varphi(x) \in K_0$ by construction $F(x) = \varphi(x) \in K_0$, and in case $\varphi(x) \in U$, the construction implies $F(x) \in c_2(K_0 \cup \{\varphi(x)\})$ which in view of $F(x) \in \overline{\Omega_1 \cap K_2} \subset \overline{\Omega}_1 \subset \Omega_0$ implies $F(x) \in K_0$.

For the statement about homotopies h, observe that we actually have $\overline{C} \subset \Omega_0$ in both cases. This follows from Lemma (11.42), since normal spaces are Hausdorff. Hence, by what we proved before, K_0 is locally (c_1, c_2)-fundamental, for any triple $(h(t, \cdot), \Omega, K)$ $(0 \leq t \leq 1)$. $\square$

Now, we can see that the coincidence index of Theorem (11.54) (resp. Theorem (11.55)) includes that of [Vae3]. For triples for which the index of [Vae3] is defined, it is straightforward to verify that Theorem (11.54) (resp. Theorem (11.55)) applies with the choice $c_1 := \overline{\text{conv}}$ and $c_2 := \text{conv}$.

If not all elements of $\mathcal{A}$ are connected, Definition (11.46) is rather restrictive, as we will see soon. However, if we assume that not only ind_F satisfies the excision property, but also ind_F^e, we can replace the assumption that (c_1, c_2) is convex in the previous results by a "local" assumption which is not so restrictive.

(11.56) DEFINITION. We call the pair (c_1, c_2) *convex for separated components on $\mathcal{A}$* if, for any $U \in \mathcal{A}$ and any $K_0 \in \mathcal{A}$ with $U_0 = U \cap K_0 \neq \emptyset$, there is some $\Omega \subset U$ which is open in U and contains U_0 and a continuous function $d : [0,1] \times \Omega \to U$ with the following properties:

(i) $d(0, y) = y$ $(y \in \Omega)$ and $d(\{1\} \times \Omega) \subset U_0$.
(ii) $d(\lambda, y) = y$, for all $(\lambda, y) \in [0,1] \times U_0$.
(iii) $d(\lambda, y) \in c_2(U_0 \cup \{y\})$, for all $(\lambda, y) \in [0,1] \times \Omega$.

(11.57) THEOREM. *Assume that, in the situation of Theorem (11.55), the restriction of the multivalued map F^{-1} to elements from $\mathcal{A}_\cup$ is upper semicontinuous and that the index ind_F^e satisfies the excision property. Then we can replace, in Theorem (11.55), the assumption that (c_1, c_2) is convex by the assumption that (c_1, c_2) is convex for separated components on $\mathcal{A}$.*

PROOF. The proof is analogous to the proof of Theorem (11.55) with the difference that Theorem (11.41) is applied. The required set Ω_2 and the function ρ are constructed as follows.

Let $\Omega_1 \subset K$ be open with $\overline{\Omega}_1 \subset \Omega_0$ and $\Omega_1 \supset C$, and let $K_2 = K_0 \cup U$, where $U \in \mathcal{A}$ is a retraction candidate for (φ, Ω_1, K) with $U_0 = K_0 \cap U \neq \emptyset$. Let $d : [0,1] \times \widetilde{\Omega} \to U$ be the function from Definition (11.56), where $\widetilde{\Omega} \subset U$ is open in U and contains U_0. We have $\widetilde{\Omega} = O_1 \cap U$, for some open $O_1 \subset K$. Note that

$$\varphi(F^{-1}(\overline{\Omega_1 \cap U_0})) \subset U_0.$$

Since $\varphi \circ F^{-1} : \overline{\Omega_1 \cap K_2} \to K_2$ is upper semicontinuous, we find some open $O_2 \subset K$ with $U_0 \subset O_2$ and

$$\varphi(F^{-1}(O_2 \cap (\overline{\Omega_1 \cap K_2}))) \subset O_1 \cap K_2.$$

Since K is a T_4-space, we find some open $O_3 \subset K$ with $U_0 \subset O_3$ and $\overline{O}_3 \subset O_1$. Putting $\Omega_2 = O_3 \cap \Omega_1$, we have that the set

$$D := \varphi(F^{-1}(\overline{\Omega_2 \cap K_2}))$$

is contained in $O_1 \cap K_2$. Hence, we can define $\rho : [0,1] \times D \to K_2$ by putting

$$\rho(\lambda, y) = \begin{cases} d(\lambda, y) & \text{if } y \in U \cap D \subset \widetilde{\Omega}, \\ y & \text{if } y \in K_0 \cap D. \end{cases}$$

For $y \in U \cap K_0 \cap D$, we have $y \in U_0$, and thus $d(\lambda, y) = y$. Consequently, ρ is well-defined and continuous.

We have $\rho(0, \cdot) = \mathrm{id}|_D$ and $\rho(\lambda, y) = y$, for $y \in K_0$ by construction. Also $\rho(\{1\} \times D) \subset K_0$, since for $y \in U \cap D$, we have $d(1, y) \in U_0 \subset K_0$. Finally, the relation $F(x) = \rho(\lambda, \varphi(x))$, for some $x \in F^{-1}(\overline{\Omega_2 \cap K_2})$, implies $F(x) \in K_0$. Indeed, this is trivial if $\varphi(x) \in K_0$. But if $\varphi(x) \in U$, then $F(x) = d(\lambda, \varphi(x)) \in c_2(U_0 \cup \{\varphi(x)\}) \subset \chi(K_0 \cup \{\varphi(x)\})$ implies, in view of $F(x) \in \overline{\Omega}_2 \subset \overline{\Omega}_1 \subset \Omega_0$, that $F(x) \in K_0$, because K_0 is (c_1, c_2)-fundamental on Ω_0. $\square$

We mentioned above that Definitions (11.46) and (11.56) are related with components. To describe this relation more precisely, the following notation is convenient. If B is a topological space and $A \subset B$, then $\mathrm{Comp}(A, B)$ denotes the union of all components of B which intersect A. In particular, if A is connected, then $\mathrm{Comp}(A, B)$ denotes the component of A in B.

(11.58) PROPOSITION. *The pair (c_1, c_2) is convex on $\mathcal{A}$ if and only if, for any $U \in \mathcal{A}$ and any $U_0 = U \cap K_0 \neq \emptyset$ with $K_0 \in \mathcal{A}$, we have $\mathrm{Comp}(U_0, U) = U$, and additionally the following holds. There is a function $V_0 \mapsto d_{V_0}$ which associates with any component V_0 of U_0 a homotopy $d_{V_0} : [0,1] \times V \to U$ with $V = \mathrm{Comp}(V_0, U)$ such that $V \cap U_0 \subset V_0$ and:*

(i) $d_{V_0}(0, y) = y$ $(y \in V)$ *and* $d_{V_0}(\{1\} \times V) \subset V_0$.

(ii) $d_{V_0}(\lambda, y) = y$, *for all* $(\lambda, y) \in [0,1] \times V_0$.

(iii) $d_{V_0}(\lambda, y) \in c_2(U_0 \cup \{y\})$, *for all* $(\lambda, y) \in [0,1] \times V$.

PROOF. Let the above properties be satisfied. Given $y \in U$, we find at most one component V_y of U_0 with $y \in \mathrm{Comp}(V_y, U)$. Indeed, if V_0 is a (possibly different) component of U_0 such that $V = \mathrm{Comp}(V_0, U)$ contains y, then V and $\mathrm{Comp}(V_y, U)$ are components of U which contain y, and so $V = \mathrm{Comp}(V_y, U)$. Hence, the assumption implies $\mathrm{Comp}(V_y, U) \cap U_0 \subset V_0$, in particular $V_y \subset V_0$. Since V_y and V_0 are components of U_0, this implies $V_y = V_0$.

Since $U = \mathrm{Comp}(U_0, U)$, we actually find, for each $y \in U$, some (unique) component V_y of U_0 with $y \in \mathrm{Comp}(V_y, U)$, and so we can define a function $d : [0,1] \times U \to U$ by putting $d(\lambda, y) = d_{(V_y, U)}(\lambda, y)$ (here, we used that $V_0 \mapsto d_{V_0}$ is a function). Since d is continuous on each component, it is continuous. The other properties of d are evident.

Conversely, let (c_1, c_2) be convex on $\mathcal{A}$. If V_0 is a component of U and $V = \mathrm{Comp}(V_0, U)$, then $d(\{1\} \times V)$ is a connected subset of U_0. Since this set contains the component V_0, it follows that $d(\{1\} \times V) = V_0$. This implies $V \cap U_0 = d(\{1\} \times (V \cap U_0)) \subset V_0$ and that the homotopy $d_{V_0} = d|_{[0,1] \times V}$ has the required properties. To prove $\mathrm{Comp}(U_0, U) = U$, we have to show that any component V_1 of U intersects U_0. But the set $d([0,1] \times V_1) \subset U$ is connected and intersects V_1, and thus is contained in V_1. Hence, V_1 contains $d(\{1\} \times V_1) \subset U_0$, and so intersects U_0. $\qquad\square$

(11.59) COROLLARY. *If the pair (c_1, c_2) is convex on $\mathcal{A}$, then for any $U \in \mathcal{A}$ and any $U_0 = U \cap K_0 \neq \emptyset$ with $K_0 \in \mathcal{A}$, the sets U and U_0 have the same number (in the sense of cardinality) of components.*

PROOF. Proposition (11.58) implies that, for any component V_0 of U_0, the set $V = \mathrm{Comp}(V_0, U)$ satisfies $V \cap U_0 \subset V_0$. Hence, the map $V_0 \mapsto V$ is an injection of the components of U_0 into the components of U. This map is onto, because the union over all its values is $\mathrm{Comp}(U_0, U) = U$ (Proposition (11.58)). $\qquad\square$

Corollary (11.59) means that (except of some pathological examples) there can only be a convex pair on $\mathcal{A}$ if the elements of $\mathcal{A}$ are connected.

If we are interested in pairs which are convex for separated components, we get of course a less restrictive condition. The following analogy to Proposition (11.58) gives in this case only a sufficient condition (which, however, is "almost" necessary).

(11.60) PROPOSITION. *Suppose, for any $U \in \mathcal{A}$ and any $U_0 = U \cap K_0 \neq \emptyset$ with $K_0 \in \mathcal{A}$, we find a neighbourhood $N \subset U$ of U_0 such that $\mathrm{Comp}(U_0, N)$ is a neighbourhood of U_0, and additionally the following holds. There is a function $V_0 \mapsto d_{V_0}$ which associates with any component V_0 of U_0 a homotopy $d_{V_0} : [0,1] \times V \to U$, where $V = \mathrm{Comp}(V_0, N)$ such that $V \cap U_0 \subset V_0$ and such that d_{V_0} satisfies the other three properties of Proposition (11.58).*

Then (c_1, c_2) is convex for separated components on $\mathcal{A}$.

Conversely, if (c_1, c_2) is convex for separated components on $\mathcal{A}$, then the above properties hold with the possible exception that $\mathrm{Comp}(U_0, N)$ is not a neighbourhood of U_0. However, $\mathrm{Comp}(U_0, U)$ is a neighbourhood of U_0.

PROOF. Let U, U_0 and N be as in the claim. Since $\mathrm{Comp}(U_0, N)$ is, by the hypothesis, a neighbourhood of U_0, we find some open $\Omega \subset U$ with $U_0 \subset \Omega \subset \mathrm{Comp}(U_0, N)$. The same argument as in the proof of Proposition (11.58) shows that, for each $y \in U$, there is at most one component V_y of U_0 with $y \in \mathrm{Comp}(V_y, N)$. For $y \in \Omega \subset \mathrm{Comp}(U_0, N)$, there is such a component by definition, and so $d(\lambda, y) = d_{V_y}(\lambda, y)$ defines a function with the required properties.

Conversely, let (c_1, c_2) be convex, for separated components on $\mathcal{A}$, and put $N = \Omega$. Let U and U_0 be as above. Given $y \in \Omega$, let $V_y = \mathrm{Comp}(\{y\}, U)$. Since $D_y = d([0,1] \times \mathrm{Comp}(\{y\}, \Omega)) \subset U$ is connected and $y \in D_y \cap V_y$, we have $D_y \subset V_y$. Observing that $d(\{1\} \times \Omega) \subset D_y \cap U_0$, we find $V_y \cap U_0 \neq \emptyset$, and so $\mathrm{Comp}(U_0, U)$ contains $V_y \ni y$. This proves that $\mathrm{Comp}(U_0, U) \supset \Omega$ is a neighbourhood of U_0.

For a given component V_0 of U_0, put $V = \mathrm{Comp}(V_0, N)$. Then $D = d(\{1\} \times V)$ is a connected subset of U_0 which contains the component V_0. Hence, $D = V_0$ which implies $V \cap U_0 = d(\{1\} \times (V \cap U_0)) \subset V_0$ and that the function $d_{V_0} = d|_{[0,1] \times V}$ has the required properties. $\square$

The essential advantage of Proposition (11.60) over the earlier Proposition (11.58) is that the crucial set $\mathrm{Comp}(U_0, N)$ on which the homotopies are defined may be much smaller than U which means that we may just "forget" many components of the larger set U. In particular, U_0 may have strictly less components than U_0 (the condition $V \cap U_0 \subset V_0$ implies that U_0 cannot have more components than U).

Concerning the reverse implication of Proposition (11.58), we show, in the following example, that even in a compact connected metric space U and a single-point set $U_0 \subset U$, for which $\mathrm{Comp}(U_0, U)$ is a neighbourhood of U_0, it need not be the case that $\mathrm{Comp}(U_0, N)$ is a neighbourhood of U_0, for any neighbourhood $N \subset U$ of U_0.

(11.61) EXAMPLE. Let $U \subset \mathbb{R}^2$ be the union of the line segments $[0,1] \times \{n^{-1}\}$ $(n \in \mathbb{N})$, $[0,1] \times \{0\}$, and $\{1\} \times [0,1]$. Then U is connected, and so we have for $U_0 = \{(0,0)\}$ that $\mathrm{Comp}(U_0, U) = U$. If $N \subset U$ is a neighbourhood of U_0 in U which is disjoint from $\{1\} \times [0,1]$, then $\mathrm{Comp}(U_0, N) \subset [0,1) \times \{0\}$ is not a neighbourhood of U_0.

An application for multivalued mappings.

In this subsection, we illustrate the previous results. At first, we consider a rather general situation.

Let Y be a neighbourhood retract of a locally convex metrizable space, and $\Omega \subset K \subset Y$ with K closed in Y. Let $\Phi_k(t, \cdot) = \Phi_{k,t}$ $(k = 1, \ldots, n)$ be multivalued maps such that the composition

$$\Phi(t, \cdot) := \Phi_{n,t} \circ \Phi_{n-1,t} \circ \cdots \circ \Phi_{1,t}$$

is defined on $[0,1] \times \overline{\Omega}$. We assume that each Φ_k takes values in (the power set of) a metric space X_k, where $X_n := Y$. We point out that $\Phi_{k,t}$ needs only be defined on a subset of X_{k-1} which can depend on t (but such that the composition Φ is defined on $[0,1] \times \overline{\Omega}$). Moreover, we require the following:

(1) Each Φ_k is upper semicontinuous with nonempty closed values $\Phi_k(t,x)$; for $k < n$, we assume additionally that the values are always compact.

(2) One of the following alternatives holds:

(a) We have, for each k, that whenever $\Phi_k(t,x)$ is compact, then it is acyclic with respect to the Čech cohomology with integer coefficients.

(b) There is a field $\mathbb{K}$ such that, for each k, one of the following alternatives holds:

(i) Whenever $\Phi_k(t,x)$ is compact, then it is acyclic with respect to the Čech homology with coefficients in $\mathbb{K}$.

(ii) Φ_k is also lower semicontinuous, and whenever $\Phi_k(t,x)$ is compact, then it consists of N_k components, each of which is acyclic with respect to the Čech homology with coefficients in $\mathbb{K}$.

We are interested in the fixed point sets

$$\mathrm{Fix}(\Phi(t, \cdot), A) := \{y \in A \mid y \in \Phi(t,y)\} \quad (t \in [0,1], \ A \subset Y)$$

and

$$\mathrm{Fix}(\Phi, A) := \bigcup_{t \in [0,1]} \mathrm{Fix}(\Phi(t, \cdot), A) = \{y \in A \mid y \in \Phi([0,1] \times \{y\})\} \quad (A \subset Y).$$

To reduce the question to a coincident point problem, we define the space

$$X := \{(t, x_0, \ldots, x_n) \mid t \in [0,1], \ x_0 \in \overline{\Omega}, \ x_k \in \Phi_k(t, x_{k-1})\}$$

and maps $F : X \to [0,1] \times \overline{\Omega}$ and $h : [0,1] \times X \to [0,1] \times Y$ by $F(t, x_0, \ldots, x_n) := (t, x_0)$ and $h(t, (\lambda, x_0, \ldots, x_n)) := (t, x_n)$. We also put

$$R := \{(t, x) \in [0,1] \times \overline{\Omega} \mid \Phi_n(t, y) \text{ is compact for each } y \in (\Phi_{n-1,t} \circ \cdots \circ \Phi_{1,t})(x)\}.$$

(in case $n = 1$, we put $(\Phi_{n-1,t} \circ \cdots \circ \Phi_{1,t})(x) := x$).

(11.62) PROPOSITION. *The map h is a homotopy which satisfies*

$$h(t, F^{-1}(\lambda, y)) = (t, \Phi(\lambda, y)) \quad \text{for } t, \lambda \in [0,1] \text{ and } y \in \overline{\Omega}.$$

In particular,

(11.63) $\mathrm{Coin}(F, h(t, \cdot), [0,1] \times M) = \{t\} \times \mathrm{Fix}(\Phi(t, \cdot), M) \quad (t \in [0,1], \ M \subset \overline{\Omega}).$

Moreover, the restriction $F : F^{-1}(R) \to R$ is a $\mathbb{Z}$-Vietoris map (respectively a composed $\mathbb{K}$-Vietoris map in case 2(b)).

PROOF. Only the last statement requires a proof.

Put $X_j := \{(t, x_0, \ldots, x_j) : (t, x_0) \in R, \ x_k \in \Phi_k(t, x_{k-1})\}$, and define $F_j : X_j \to X_{j-1}$ by $F_j(t, x_0, \ldots, x_{j-1}, x_j) := (t, x_0, \ldots, x_{j-1})$. Since upper semicontinuous maps with compact values send compact sets into compact sets, we obtain that each F_j is a $\mathbb{Z}$-Vietoris map (respectively a composed $\mathbb{K}$-Vietoris map). In view of Proposition (11.8), we conclude that also $F|_{F^{-1}(R)} = F_1 \circ \cdots \circ F_n$ is a $\mathbb{Z}$-Vietoris map (respectively a composed $\mathbb{K}$-Vietoris map). $\square$

In order to obtain hypotheses which can be easily verified, we restrict our attention now to the convex situation. More precisely, we assume now in addition that Y is a closed convex subset of a (not necessarily complete) locally convex metrizable space. We let $\mathcal{A}$ be the system of all compact convex subsets of $Z := [0,1] \times Y$. Theorem (11.9) implies, in view of Proposition (11.62), that F induces an additive coincidence index ind_F on $(R, \mathcal{A})$ and even an index ind_F^e on $(R, \mathcal{A}_\cup)$, where $\mathcal{A}_\cup$ is as in Example (11.16). Hence, putting $\mathcal{T} := \mathcal{T}(F, [0,1] \times R, \mathcal{A})$, we can define an index Ind as in Theorem (11.23).

(11.64) THEOREM (Scheme for Continuation Principles). *Assume that, in the above situation, there is some open in K set $\Omega_0 \subset \Omega$ with $\Omega_0 \supset \mathrm{Fix}(\Phi, \overline{\Omega})$ and some (possibly empty) $V \subset Y$ with compact $\overline{\mathrm{conv}}\, V$ and*

(11.65) $$\overline{\mathrm{conv}}(\Phi([0,1] \times \Omega_0) \cup V) \subset K.$$

Suppose that $\Phi_n(t, x)$ is separable, for each $t \in [0, 1]$ and each $x \in (\Phi_{t,n-1} \circ \cdots \circ$
$\Phi_{t,1})(\Omega_0)$, and that, for any countable $C \subset \Omega_0$, the relation

$$(11.66) \qquad \overline{C} = \Omega_0 \cap \text{conv}(\Phi([0, 1] \times C) \cup V)$$

implies that $\overline{\text{conv}}\,\Phi([0, 1] \times C)$ is compact. Then the triples $(h, [0, 1] \times \Omega, [0, 1] \times K)$
and $(h(t, \cdot), \{t\} \times \Omega, \{t\} \times K)$ $(0 \le t \le 1)$ are locally $(\mathcal{T}, \text{ind}_F^e)$-admissible, and

$$(11.67) \qquad \text{Ind}_F(h(0, \cdot), \{0\} \times \Omega, \{0\} \times K) = \text{Ind}_F(h(1, \cdot), \{1\} \times \Omega, \{1\} \times K).$$

PROOF. Let $\mathcal{U}$ be the set of all closed convex subsets of Z, i.e. we put $c_1 :=$
$\overline{\text{conv}}$. Putting $c_2 := \text{conv}$, Proposition (11.47) implies that (c_1, c_2) is a convex pair
on $\mathcal{A}$. By Proposition (11.50), we find a smallest (c_1, c_2)-fundamental set $U_V \in \mathcal{U}$
containing $[0, 1] \times V$ for h on $O_0 := [0, 1] \times \Omega_0$ which satisfies

$$U_V = \overline{\text{conv}}(h([0, 1] \times F^{-1}(O_0 \cap U_V)) \cup ([0, 1] \times V)).$$

By Proposition (11.62), this means

$$U_V = \overline{\text{conv}}([0, 1] \times (\Phi(O_0 \cap U_V) \cup V)) = [0, 1] \times \overline{\text{conv}}(\Phi(O_0 \cap U_V) \cup V).$$

Putting $U := \overline{\text{conv}}(\Phi(O_0 \cap U_V) \cup V)$, we thus have that $U_V = [0, 1] \times U$ and

$$(11.68) \qquad U = \overline{\text{conv}}(\Phi([0, 1] \times (\Omega_0 \cap U)) \cup V).$$

By (11.65), we have in particular $U \subset K$.

We claim that our hypothesis concerning (11.66) implies that U is compact. To
see this, note at first that $\Phi([0, 1] \times \{y\})$ is separable, for each $y \in \Omega_0$, because
uppersemicontinuous maps in metric spaces with separable values send separable
sets into separable sets (see e.g. [Vae2]). We now apply [Vae3, Corollary 4.2] with
$F = \text{id}$, $M = U$, and $N = \Omega_0$, observing that $\overline{\text{conv}}(\Phi([0, 1] \times (\Omega_0 \cap U)) \cup V) \setminus$
$\Omega_0 = U \setminus \Omega_0$ is closed, because U is a closed subset of K and Ω_0 is open in K.
Hence, to prove that the right-hand side of (11.68) is compact, it suffices by [Vae3,
Proposition 4.1], in view of (11.68), to show that, for any countable $C \subset \Omega_0 \cap U$
satisfying (11.66), the set $A_C := \overline{\text{conv}}(\Phi([0, 1] \times C) \cup V)$ is compact. Putting
$g(\lambda, y_1, y_2) = \lambda y_1 + (1 - \lambda)y_2$, we have

$$A_C \subset g([0, 1] \times \overline{\text{conv}}V \times \overline{\text{conv}}\Phi([0, 1] \times C)),$$

because the right-hand side is convex and compact (by hypothesis) and contains
$\Phi([0, 1] \times C) \cup V$. In particular, A_C is compact, as required. Hence, U is compact,
as claimed, i.e. $U_V = [0, 1] \times U \in \mathcal{A}$.

Since Φ_n assumes closed values, the compactness of (11.68) implies now that $[0,1] \times (U \cap \Omega_0) \subset R$. Note that $C := \mathrm{Coin}(F, h, [0,1] \times \overline{\Omega})$ is closed by Lemma (11.42), and that, in view of (11.63) and our hypothesis, we have $C = [0,1] \times \mathrm{Fix}(\Phi, \overline{\Omega}) \subset [0,1] \times \Omega_0$. Since K is normal, we thus find some open in K set $\Omega_1 \supset C$ with $\overline{\Omega}_1 \subset \Omega_0$. Then we have in particular $[0,1] \times (\overline{\Omega_1 \cap U}) \subset R$. Since $U_V = [0,1] \times U$ is (c_1, c_2)-fundamental for h on Ω_0, it is also (c_1, c_2)-fundamental for h on Ω_1. Moreover, in view of our particular form of F and h, it is easily verified that this implies also that $U_t := \{t\} \times U$ is (c_1, c_2)-fundamental for $h(t, \cdot)$ on Ω_1. Theorem (11.55) now implies, in view of $\overline{(\{t\} \times \Omega_1) \cap U_t} \subset R$ and $\overline{([0,1] \times \Omega_1) \cap U_V} \subset R$, that the triples $(h(t, \cdot), \{t\} \times \Omega, \{t\} \times K)$ and $(h, [0,1] \times \Omega, [0,1] \times K)$ are locally $(\mathcal{T}, \mathrm{ind}_F^e)$-admissible with locally fundamental set U_t respectively U_V, and that the relations

$$\mathrm{Ind}_F(h(t, \cdot), \{t\} \times \Omega, \{t\} \times K) = \mathrm{ind}_F(h(t, \cdot), \{t\} \times (\Omega_1 \cap U), \{t\} \times U)$$

and

$$\mathrm{Ind}_F(h(t, \cdot), [0,1] \times \Omega, [0,1] \times K) = \mathrm{ind}_F(h(t, \cdot), [0,1] \times (\Omega_1 \cap U), [0,1] \times U)$$

hold. Note that the left-hand side of the last equality is independent of t by the homotopy invariance of Ind_F. Moreover, in view of the contraction property of ind_F (and by our particular form of F and h), we have

$$\mathrm{ind}_F(h(t, \cdot), [0,1] \times (\Omega_1 \cap U), [0,1] \times U) = \mathrm{ind}_F(h(t, \cdot), \{t\} \times (\Omega_1 \cap U), \{t\} \times U).$$

Combining the above equalities, we find (11.67). □

(11.69) REMARK. Actually, we needed only a weaker compactness assumption: The compactness of $\overline{\mathrm{conv}}\Phi([0,1] \times C)$ is required only for countable sets $C \subset \Omega_0$ with (11.66) and the additional property that $C \subset U$, where U is as in the proof.

Some notes are in order: We call Theorem (11.64) a "scheme for continuation principles", because any condition on the function $\Phi_{k,0}$ which assures that the left-hand side of (11.67) is nonzero provides a continuation principle (we give an example below).

We point out that we need not assume for Theorem (11.64) that Φ assumes its values in K; the only requirement of this type is the "a priori estimate" (11.65) in a neighbourhood Ω_0 of the fixed points. This aspect is in the attitude of the "pushing assumption" (A) which was apparently firstly introduced in [FPr1], [FPr2] and further studied in [AGG1] and [AndBa] (see the foregoing Chapter II.10).

The set V in Theorem (11.64) was introduced only, because it slightly changes the meaning of (11.66), and it can be sometimes more convenient to prove the compactness of $\overline{\text{conv}}\Phi([0,1] \times C)$, when $V \neq \emptyset$. In most cases, however, one would probably choose $V = \emptyset$.

To understand a deeper meaning of Theorem (11.64), we point out the following observation (since we do not know whether the index of Theorem (11.9) is unique, in general, we have to be careful with the formulation).

(11.70) PROPOSITION. *The index map of Theorem* (11.9) *can be chosen such that*

$$\text{ind}_F(h(t, \cdot), \{t\} \times \Omega, \{t\} \times K)$$

depends (when it exists) only on the maps $\Phi_{1,t}, \ldots, \Phi_{n,t}$, *but not on* t. *In this case,* Ind_F *has the same property.*

PROOF. The last statement follows immediately from the fundamental contraction property. The first statement follows, in case of a $\mathbb{Z}$-Vietoris map, straightforwardly from the multiplicativity property of the index from [Kry3], and in case of composed $\mathbb{K}$-Vietoris maps e.g. by the homotopy invariance of the index from [Dz-M]. □

If we choose ind_F as in Proposition (11.70), we can define

$$\text{Ind}(h(t, \cdot), \{t\} \times \Omega, \{t\} \times K)$$

(if it exists) as a "fixed point index" of the decomposition $\Phi_{1,t}, \ldots, \Phi_{n,t}$ (on Ω with respect to K). Theorem (11.23) implies, in view of (11.63), that this index has the expected properties of additivity, normalization and the fixed point property. Theorem (11.64) can be considered as the homotopy invariance of this index. This index generalizes the index constructed in Chapter II.10. To see this, note that if Φ is such that the fixed point index in Chapter II.10 exists, then Theorem (11.64) applies with $V = \emptyset$ (more precisely, Remark (11.69) can be applied by observing that the set U constructed in the proof of Theorem (11.64) satisfies (11.68) and is also contained in the smallest set which is fundamental for Φ in the sense used in Chapter II.10, and thus compact). However, we will not make use of Proposition (11.70) in the following.

The method of measures of noncompactness can be used in any topological vector space to find nonempty, invariant, compact, convex sets [MTY]. However, in Fréchet spaces, these conditions are particularly handy:

(11.71) PROPOSITION. *Let* Y *be a closed convex subset of some Fréchet space,* $\Omega_0 \subset \Omega$, *and assume that* $\Phi_n(t, x)$ *is compact, for each* $t \in [0, 1]$ *and each* $x \in (\Phi_{t,n-1} \circ \cdots \circ \Phi_{t,1})(\overline{\Omega}_0)$. *Suppose that* Φ *is countably condensing on* Ω_0 *in the* *following sense: For any countable $C \subset \Omega_0$ with noncompact $\overline{C}$, there is some measure of noncompactness β (cf. Chapter I.5) with*

$$(11.72) \qquad\qquad \beta(\Phi([0,1] \times C)) \not\leq \beta(C).$$

Then the relation (11.66), for some $C \subset \Omega_0$, implies that $\overline{\mathrm{conv}}\, \Phi([0,1] \times C)$ is compact.

PROOF. If $C \subset \Omega_0$ is countable with (11.66), then we have for any β that (11.72) fails, and so $\overline{C}$ is compact. Since upper semicontinuous maps with compact values send compact sets into compact sets, we obtain that $\Phi([0,1] \times \overline{C})$ is compact. By Mazur's Theorem (1.33) in Chapter I.1, we conclude that also $\overline{\mathrm{conv}}\, \Phi([0,1] \times C)$ is compact.   □

It is important for certain applications that (11.72) is required only for countable sets C. Indeed, for differential and integral operators in Banach spaces, one can obtain good estimates for measures of noncompactness usually only for countable sets; see e.g. [AVV1], [Hei], [Moe1], [MoeHa], [Vae-M]. This is the reason why we passed to a countable set in Theorem (11.64).

(11.73) THEOREM (Continuation Principle). *Assume, that in addition to the hypotheses of Theorem (11.64), there is some nonempty open in K set $\Omega_1 \subset \Omega$ with $\Omega_1 \supset \mathrm{Fix}(\Phi(0, \cdot), \overline{\Omega})$ such that $\overline{\mathrm{conv}}\, \Phi(\{0\} \times \overline{\Omega}_1)$ is compact and contained in $\overline{\Omega}_1$. Then $\mathrm{Ind}_F(h(1, \cdot), \{1\} \times \Omega, \{1\} \times K) = 1$, and $\mathrm{Fix}(\Phi(1, \cdot), \Omega) \neq \emptyset$.*

PROOF. It suffices to prove, for $\varphi := h(0, \cdot)$, that

$$(11.74) \qquad\qquad \mathrm{Ind}_F(\varphi, \{0\} \times \Omega, \{0\} \times K) = 1.$$

Indeed, Theorem (11.64) then implies $\mathrm{Ind}_F(h(1, \cdot), \{1\} \times \Omega, \{1\} \times K) = 1$, and in view of the coincident point property and (11.63), we obtain $\mathrm{Fix}(\Phi(1, \cdot), \Omega) \neq \emptyset$.

The restriction property implies

$$\mathrm{Ind}_F(\varphi, \{0\} \times \Omega, \{0\} \times K) = \mathrm{Ind}_F(\varphi, \{0\} \times \Omega_1, \{0\} \times K).$$

Putting $K_0 := \overline{\mathrm{conv}}\Phi(\{0\} \times \overline{\Omega}_1)$, we have, in view of Proposition (11.62), that $\varphi(F^{-1}(\{0\} \times \Omega_1)) = \{0\} \times K_0$. Proposition (11.62) thus implies that

$$\mathrm{Ind}_F(\varphi, \{0\} \times \Omega_1, \{0\} \times K) = \mathrm{ind}_F(\varphi, \{0\} \times (\Omega_1 \cap K_0), \{0\} \times K_0).$$

By our assumptions, we have $K_0 \subset \overline{\Omega}_1$ and thus $\Phi(\{0\} \times K_0) \subset \Phi(\{0\} \times \overline{\Omega}_1) \subset K_0$. Proposition (11.62) thus implies in particular $\varphi(F^{-1}(\{0\} \times K_0)) \subset \{0\} \times K_0$. Since

$K_0 \in \mathcal{A}$, we conclude that $(\varphi, \{0\} \times K_0, \{0\} \times K_0) \in \mathcal{T}$. In view of our hypothesis, we obtain now by the excision property and Proposition (11.62) that

$$\mathrm{ind}_F(\varphi, \{0\} \times K_0, \{0\} \times K_0) = \mathrm{ind}_F(\varphi, \{0\} \times (\Omega_1 \cap K_0), \{0\} \times K_0).$$

Since $\Omega_1 \neq \emptyset$, we find some $z_0 \in \{0\} \times K_0$. Consider the convex homotopy $H(\lambda, \cdot) := \lambda\varphi + (1-\lambda)z_0$. The homotopy invariance and normalization property of ind_F now imply

$$\mathrm{ind}_F(\varphi, \{0\} \times K_0, \{0\} \times K_0) = \mathrm{ind}_F(y_0, \{0\} \times K_0, \{0\} \times K_0) = 1.$$

Combining the above equalities, (11.74) follows. □

(11.75) COROLLARY (Leray–Schauder Boundary Condition). *Let Φ be as described at the beginning of this subsection (i.e. after Example (11.61)), but such that $\Phi(y) = \Phi(t, y)$ is independent of t. Assume that Y is a closed convex subset of a locally convex metrizable space. If we have case 2(b)(ii) for Φ_n, we assume that Φ_n takes only compact values. Suppose that there is some open in K set $\Omega_0 \subset \Omega$ satisfying the pushing assumption*

$$\overline{\mathrm{conv}}(\Phi(\Omega_0)) \subset K$$

such that, for some $y_0 \in \Omega_0$, the Leray–Schauder boundary condition holds:

$$(11.76) \qquad \lambda(y - y_0) \notin \Phi(y) - y_0 \quad (\lambda \geq 1,\ y \in \overline{\Omega} \setminus \Omega_0).$$

Finally, assume that $\Phi_n(0, y)$ is separable, for each t and each $y \in (\Phi_{n-1,0} \circ \cdots \circ \Phi_{1,0})(\Omega_0)$, and that, for any countable $C \subset \Omega_0$, the relation

$$(11.77) \qquad \overline{C} = \overline{\Omega_0 \cap \mathrm{conv}(\Phi(C) \cup \{y_0\})}$$

implies that $\overline{\mathrm{conv}}\Phi(C)$ is compact. Then $\mathrm{Ind}_F(h(1, \cdot), \{1\} \times \Omega, \{1\} \times K) = 1$ (the index is well-defined), and $\mathrm{Fix}(\Phi, \Omega) \neq \emptyset$.

PROOF. Without any loss of generality, we can assume that each $\Phi_k = \Phi_{k,t}$ is independent of t. Moreover, we can assume that we do not have the case 2(b)(ii) for Φ_n, because otherwise, we consider $\Phi = \mathrm{id} \circ \Phi_n \circ \cdots \circ \Phi_1$, instead of Φ. Now, we replace Φ_n by the map $\Psi_n(t, \cdot) := t\Phi_n + (1-t)y_0$. Then we have for the corresponding composition $\Psi(t, \cdot) = t\Phi + (1-t)y_0$. The assumptions $y_0 \in \Omega_0$ and (11.76) are then equivalent to the inclusion $\mathrm{Fix}(\Psi, \overline{\Omega}) \subset \Omega_0$, and (11.77) is equivalent to $\overline{C} = \overline{\Omega_0 \cap \mathrm{conv}(\Psi([0,1] \times C))}$. Finally, note that if $\overline{\mathrm{conv}}\Phi(C)$ is compact, then also $\overline{\mathrm{conv}}\,\Psi([0,1] \times C) \subset g([0,1] \times \overline{\mathrm{conv}}\,\Phi(C))$ is compact, where $g(t, y) = ty + (1-t)y_0$. Hence, the claim follows by the continuation principle. □

Corollary (11.75) contains the two (single-valued) fixed point theorems of Mönch [Moe1] (see also [De3-M, Theorems 18.1 and 18.2] and [Vae2]) and their generalization to multivalued maps with convex values [ORP], [Vae4] as special cases.

We point out once more that the compactness condition of Corollary (11.75) is satisfied in Fréchet spaces if, for each countable $C \subset \Omega_0$ with $\alpha(C) \neq 0$ (where e.g. α denotes the Kuratowski measure of noncompactness) the relation

$$(11.78) \qquad\qquad \alpha(\Phi(C)) < \alpha(C)$$

holds. Moreover, in case $\Omega_0 = \Omega = K$ (i.e. if Φ is a self-map of K), the boundary condition (11.76) is satisfied automatically, and we can replace (11.76) by

$$\alpha(\Phi(C)) \neq \alpha(C).$$

Note that the latter holds in particular also if Φ "enlarges" the Kuratowski measure of noncompactness (instead of shrinking it as in the case of condensing maps).

II.12. Remarks and comments

II.1.

The question of a topological structure of fixed point sets was stimulated by the study of the solution sets of differential equations in the absence of uniqueness. N. Aronszajn's paper [Aro], where he practically proved the R_δ-structure, appeared in 1942. Our approach in the related papers [AGG3], [Go15] is mainly based on the technique of F. E. Browder and C. P. Gupta [BrGu]. For more details and many related references, see especially the survey paper [Go15] and [Cza1], [Cza2], [DMNZ-M], [DbMo1], [DbMo2], [LaR2], [Pet3], [Sz2], [Vid1]–[Vid3]. For multi-valued variants of the Browder–Gupta theorem, see also [Bog1], [CKZ], [COZ], [DMNZ-M], [Ga3], [Ge1], [Ge5], [Ge6], [Had3], [HP1-M], [KP], [Kry1-M], [LaR1].

II.2.

The existence of a fixed point of multivalued contractions (studied in (3.68)–(3.73) in Chapter I.3) with closed values in a complete metric space is guaranteed by the theorem of H. Covitz and S. B. Nadler [CN] from 1970. The topological structure of fixed point sets of multivalued contractions was treated e.g. in [AG1], [ADG], [BCF], [DMNZ-M], [Ge7], [Go5-M], [Go15], [GM], [GMS], [MS], [SR1]–[SR3], [Ri1]. In particular, in [GM], [GMS], the authors have proved that the fixed point set of a multivalued contraction having local extension property from a complete absolute retract into itself is an absolute retract, too. The topology of e.g. a Fréchet space may bring however some troubles in checking the contractivity of the operators. Even for an operator which is a contraction in every seminorm (with the same constant of contractivity), it seems to be impossible to prove its contractivity w.r.t. a metric in a Fréchet space.

This was a stimulation for applying the inverse limit method. This technique was initiated in our paper [AGG2] in 1999 and then it has been further developed in [AGG3], [Ga3].

II.3.

The first paper dealing with a lower bound of the topological dimension of a fixed point set of multivalued maps is due to J. Saint-Raymond [SR1]. Let us note that his result was conjectured by B. Ricceri (cf. [R1]). For further related references, see e.g. [ADG], [AG1], [BGKO], [DzGe], [DouOe], [FiMaPe], [Ge1], [Ge3], [Ge5]–[Ge7], [Mlj1], [Mlj2], [Ri1], [Ri6], [Ri7], [Smth], [SR1], [SR2].

II.4.

The notion of a topological essentiality for single-valued maps was introduced by A. Granas in [Gr4-M] and elaborated by M. Furi, M. Martelli and A. Vignoli in [FMV1]–[FMV3]. They call the related concept as zero-epi mappings (cf. also [HIR-M], where many further related references can be found. For some more recent references, see e.g. [Fr-M], [GGL-M], [RN-M], [AG1], [AVV2], [Fri2], [Fri3], [FriGr], [FGG], [FGK], [Gr7], [GS], [ORP], [Ski2], [Vae1], [Vae4].

II.5.

C. Bowszyc [Bow1] introduced in 1968 a relative Lefschetz number for a compact map on pairs of invariant ANR-spaces whose nontriviality implies the existence of a fixed point in the closure of the complement. For this topological invariant, defined in terms of the relative singular homology on the pair over the rationals $\mathbb{Q}$, the fixed points on the boundaries are allowed. The result of C. Bowszyc was generalized in various ways in [AGJ3], [GGr5], [Sr3], [Sr4]. Let us also note that the similar idea of admissible index pairs occurs in the Conley index theory, based on the celebrated Ważewski topological principle (cf. [Mr1], [Mr2], [Ryb-M], [Sr1], [Sr2]).

In 1986, H. Schirmer [Sch5] introduced the relative Nielsen number for a lower bound of the number of fixed points on the total space X, for compact maps on pairs of spaces $A \subset X$, $F : (X, A) \to (X, A)$, which can make better lower estimate than the standard Nielsen number. Leter on, other relative Nielsen numbers were defined, which provide lower bounds for either the number of fixed points on $X \setminus A$ or on $\overline{X \setminus A}$, in order to study the location of fixed points. The essential idea underlying the definitions, allowing us to make the estimates on the total space or on the complement, uses the concept of (weakly) common essential fixed point classes. For more details and the computation of the topological invariants, see [BrwS2], [Car], [CarWo], [Col], [GuoHe], [HeZh], [HeSY], [Je3], [NOW], [Sch5]–[Sch7], [Wng1], [Wng3], [Zha1]–[Zha4]. Our presentation is based on [AGJ3].

II.6.

Periodic point problems were systematically studied since the beginning of 50's (see e.g. [Bow2], [BrwJS], [BJ-M], [Ji1], [Ji2], [Col], [Fel-M], [Ful], [Haj1], [Haj2], [Hal2], [Hal3], [Hea], [HePY], [HeSY], [HeY], [HeZh], [JeMa], [Kam], [Kep], [Mat4], [Mat5], [McC-M], [McCoMiMr], [MrzPr], [MatShi], [Nu-M], [Pei], [Wng2], [You1], [You2]). Various methods were employed for obtaining the existence and multiplicity results, when using the homotopic invariants like the Lefschetz number [Bow2], [Haj1], the fixed point index or the Conley index (see e.g. [MaCoMiMr], [Nu-M]), and especially the Nielsen number (see e.g. [Ji1], [Ji2], [JiL], [BrwJS], [Col], [Hea], [Hal2], [HeKe], [HePY], [HeSY], [HeY], [HeZh], [JeMa], [Kep], [Kol], [Mat4], [Mat5], [MatShi], [Wng2], [You1], [You2]). In our paper [AGJ4], which is the basis for our presentation here, we deal with the multivalued case. However, also in the single-valued case some of our results are new.

II.7.

For further references on this topic, see [AKPRS-M], [AndBa], [AndVae], [Bad-M], [BrGMO3-M], [Cal], [Go5-M], [KOZ-M], [Pr1], [Sad2], [Vae2]–[Vae5]. Let us also note that, in metric spaces, we come back to the notion of a fixed point index for condensing maps in Chapters II.10 and II.11.

II.8.

The approximation methods, as treated in Chapter II.8, were initiated in 1935 by J. von Neumann [Neu]. The further development of his idea is related to the names of A. D. Myshkis, A. Cellina, A. Mas-Collel, F. D. Ancel, A. Granas, the second author, W. Kryszewski, M. Lassonde, H. Ben-El-Mechaiekh, and many others; for more historical remarks and the related references, see the survey paper [Kry8].

II.9.

From our point of view, the related references are especially [GGa-M], [GGK2], [GGK3], [Go2-M], [Go5-M], [Kry1-M], [Kry2-M].

II.10.

The fixed point index for compact or condensing maps in sufficiently general spaces, leading to continutation principles, was defined e.g. in [AGG1], [AndBa], [Bad-M], [BKr1], [BrGMO3-M], [Dnc], [Dz-M], [FPr1], [FPr2], [FV2], [FV3], [Gr6], [Ma-M], [Nag], [Nu1]–[Nu6], [Sad2], [Swr-M], [SeS]. Our presentation is based on the papers [AGG1] and [AndBa].

II.11.

The coincidence index, leading to continuation principles, was developed e.g. in [AndVae], [GaKr2], [GaM-M], [GK], [Ku1], [Ku2], [Maw2], [Maw3], [Vae3], [Vae5]. Our results come from [AndVae].

For more details concerning various aspects of the fixed point theory, see [AgMeOR-M], [BJ-M], [Bor-M], [Co-M], [Cu-M], [Cro-M], [Da-M], [DG-M], [EiFe-M], [Fel-M], [Fr-M], [FrGr-M], [FZ-M], [GD-M], [GGa-M], [GaM-M], [GG-M], [Goe-M], [GoeKi-M], [Go1-M], [Go4-M], [Go5-M], [Gr1-M]–[Gr4-M], [KOZ-M], [KiSi-M], [Kis-M], [Ks1-M], [Ks2-M], [KsZa-M], [KrwWu-M], [Kry1-M], [LR-M], [LeCal-M], [Ler-M], [Ll-M], [Ma-M], [Maw-M], [McC-M], [McGMe-M], [Ni-M], [Nu-M], [Pet-M], [PetPetRu-M], [Ptr-M], [Pr-M], [Rot-M], [RN-M], [Ryb-M], [Smr-M], [Swr-M], [Wa-M], [We-M], [Wie-M], [And23], [AGG1], [AG1], [AG2], [AGJ1]–[AGJ4], [AndVae], [BGK], [Bar], [BD], [BDG1], [BDG2], [BK1], [BK2], [CAl], [BoGl], [Bou1]–[Bou3], [Bow1], [BBrS1], [BBrS2], [Bro1]–[Bro5], [Brw1], [Brw2], [Brw6], [BrwGr], [BrwGS], [BrwJS], [BrwS1], [BrwS2], [BG1], [Cal], [Col], [CF], [CP], [CN], [DGP1]–[DGP3], [DeG], [Do], [EM], [Fad], [Fan1], [Fan2], [For], [Fo1]–[Fo4], [FG1]–[FG3], [FV1], [Fry5], [Ful], [FMV1]–[FMV3], [Ga1]–[Ga3], [Ge1]–[Ge3], [GO], [GG1], [GG2], [Go2]–[Go10], [Go12], [Go13], [Go17], [Go18], [GGr1]–[GGr5], [GGK1]–[GGK3], [GKr2], [GK], [GM], [GMS], [GPe], [GR], [Gr1]–[Gr7], [GrL1], [GrL2], [GF], [GuoHe], [HaS], [Haj1], [Haj2], [Hal2], [Hal3], [Hea], [HeKe], [HePY], [HeSY], [HeY], [HeZh], [Hi1], [IZ1], [IZ2], [JP], [Je1]–[Je3], [JeMa], [Ji1], [Ji2], [JiL], [Kuk], [Kep], [Kry3], [KM], [Ku1], [Ku2], [Lus], [LegWi], [Ler], [LS], [Lim], [MrzPr], [McL], [MTY], [NOW], [Ne1]–[Ne3], [NR], [ORP], [Pr1], [Pr2], [Pat], [Pei], [Po2], [Rho], [SR1]–[SR3], [Sc], [Sch1]–[Sch7], [Se4], [Se5], [Sh2], [Sko1]–[Sko4], [Vae1]–[Vae5], [Vid3], [You1], [You2], [Zha1]–[Zha4],...

CHAPTER III

APPLICATION TO DIFFERENTIAL
EQUATIONS AND INCLUSIONS

III.1. Topological approach to differential equations and inclusions

In this part, continuation principles in Chapter II.10 will be applied, jointly with some statements from the Nielsen Theory in Chapter I.11, to differential equations and inclusions. The results are based on our papers [AGG1], [AGJ2], [AndBa].

At first, we are interested in the existence problems for ordinary differential inclusions in Euclidean spaces on not necessarily compact intervals. Let us start with some definitions.

Let J be an interval in $\mathbb{R}$. We say that a map $x : J \to \mathbb{R}^n$ is *locally absolutely continuous* if x is absolutely continuous on every compact subset of J. The set of all locally absolutely continuous maps from J to $\mathbb{R}^n$ will be denoted by $AC_{loc}(J, \mathbb{R}^n)$.

Consider the inclusion

$$(1.1) \qquad \dot{x}(t) \in F(t, x(t)),$$

where F is a set-valued *u-Carathéodory map*, i.e. it has the following properties:

(C1) the set of values of $F(t, x)$ is nonempty, compact and convex, for all $(t, x) \in J \times \mathbb{R}^n$,
(C2) the map $F(t, \cdot)$ is u.s.c., for almost all $t \in J$,
(C3) the map $F(\cdot, x)$ is measurable, for all $x \in \mathbb{R}^n$.

By a *solution* of the inclusion (1.1), we mean a locally absolutely continuous function x such that (1.1) holds, for almost all $t \in J$.

We recall two known results which are needed in the sequel.

(1.2) THEOREM (cf. [AuC-M, Theorem 0.3.4]). *Assume that the sequence of absolutely continuous functions $x_k : K \to \mathbb{R}^n$ (K is a compact interval) satisfies the following conditions:*

(1.2.1) *the set $\{x_k(t) \mid k \in \mathbb{N}\}$ is bounded, for every $t \in K$,*

(1.2.2) *there is an integrable function (in the sense of Lebesgue)* $\alpha : K \to \mathbb{R}$ *such that*

$$|\dot{x}_k(t)| \leq \alpha(t), \quad \text{for a.a. } t \in K \text{ and for all } k \in \mathbb{N}.$$

Then there exists a subsequence (denoted just the same) $\{x_k\}$ *convergent to an absolutely continuous function* $x : K \to \mathbb{R}^n$ *in the following sense:*

(1.2.3) $\{x_k\}$ *uniformly converges to* x,

(1.2.3) $\{\dot{x}_k\}$ *weakly converges in* $L^1(K, \mathbb{R}^n)$ *to* $\dot{x}$.

The second one is the Mazur Theorem (1.33) in Chapter I.1.

The following result is crucial.

(1.4) PROPOSITION. *Let* $G : J \times \mathbb{R}^n \times \mathbb{R}^m \multimap \mathbb{R}^n$ *be a u-Carathéodory map and let* S *be a nonempty subset of* $\mathrm{AC}_{\mathrm{loc}}(J, \mathbb{R}^n)$. *Assume that*

(1.4.1) *there exists a subset* Q *of* $C(J, \mathbb{R}^n)$ *such that, for any* $q \in Q$, *the set* $T(q)$ *of all solutions of the boundary value problem*

$$\begin{cases} \dot{x} \in G(t, x(t), q(t)), & \text{for a.a. } t \in J, \\ x \in S \end{cases}$$

is nonempty,

(1.4.2) $T(Q)$ *is bounded in* $C(J, \mathbb{R}^n)$,

(1.4.3) *there exists a locally integrable function* $\alpha : J \to \mathbb{R}$ *such that*

$$|G(t, x(t), q(t))| = \sup\{|y| \mid y \in G(t, x(t), q(t))\} \leq \alpha(t), \quad \text{a.e. in } J,$$

for any pair $(q, x) \in \Gamma_T$, *where* Γ_T *denotes the graph of* T.

Then $T(Q)$ *is a relatively compact subset of* $C(J, \mathbb{R}^n)$. *Moreover, under the assumptions* (1.4.1)–(1.4.3), *the multivalued operator* $T : Q \multimap S$ *is u.s.c. with compact values if and only if the following condition is satisfied:*

(1.4.4) *given a sequence* $\{(q_k, x_k)\} \subset \Gamma_T$, *if* $\{(q_k, x_k)\}$ *converges to* (q, x) *with* $q \in Q$, *then* $x \in S$.

PROOF. For the relative compactness of $T(Q)$, it is sufficient to show that all elements of $T(Q)$ are equicontinuous.

By (1.4.3), for every $x \in T(Q)$, we have $|\dot{x}(t)| \leq \alpha(t)$, for a.a. $t \in J$, and

$$|x(t_1) - x(t_2)| \leq \left| \int_{t_1}^{t_2} \alpha(s) \, ds \right|.$$

This implies an equicontinuity of all $x \in T(Q)$.

We show that the set Γ_T is closed.

Let $\Gamma_T \supset \{(q_k, x_k)\} \to (q, x)$. Let K be an arbitrary compact interval such that α is integrable on K. By conditions (1.4.2) and (1.4.3), the sequence $\{x_k\}$ satisfies the assumptions of Theorem (1.2). Thus, there exists a subsequence (denoted just the same) $\{x_k\}$, uniformly convergent to x on K (because the limit is unique) and such that $\{\dot{x}_k\}$ weakly converges to $\dot{x}$ in L^1. Therefore, x belongs to the weak closure of the set $\text{conv}\{\dot{x}_m \mid m \geq k\}$, for every $k \geq 1$. By Theorem (1.33) in Chapter I.1, $\dot{x}$ also belongs to the strong closure of this set. Hence, for every $k \geq 1$, there is $z_k \in \text{conv}\{\dot{x}_m \mid m \geq k\}$ such that $\|z_k - \dot{x}\|_{L^1} \leq 1/k$. This implies that there exists a subsequence $z_{k_l} \to \dot{x}$ a.e. in K.

Let $s \in K$ be such that

(1.4.5) $G(s, \cdot, \cdot)$ is u.s.c.,

(1.4.6) $\lim_{l \to \infty} z_{k_l}(s) = \dot{x}(s)$,

(1.4.7) $\dot{x}_k(s) \in G(s, x_k(s), q_k(s))$.

Let $\varepsilon > 0$. There is $\delta > 0$ such that $G(s, z, p) \subset N_\varepsilon(G(s, x(s), q(s)))$, whenever $|x(s) - z| < \delta$ and $|q(s) - p| < \delta$. But we know that there exists $N \geq 1$ such that $|x(s) - x_m(s)| < \delta$ and $|q(s) - q_m(s)| < \delta$, for every $m \geq N$. Hence,

$$\dot{x}_k(s) \in G(s, x_k(s), q_k(s)) \subset N_\varepsilon G(s, x(s), q(s))).$$

By the convexity of $G(s, x(s), q(s))$, for $k_l \geq N$, we have

$$z_{k_l}(s) \in N_\varepsilon(G(s, x(s), q(s))).$$

Thus, $\dot{x}(s) \in \overline{N_\varepsilon(G(s, x(s), q(s)))}$, for every $\varepsilon > 0$. This implies

$$\dot{x}(s) \in G(s, x(s), q(s)).$$

Since K was arbitrary, $\dot{x}(t) \in G(t, x(t), q(t))$, a.e. in J. $\qquad\square$

We can now state one of the main results of this subsection.

(1.5) THEOREM. *Consider the boundary value problem*

(1.6) $\qquad \begin{cases} \dot{x} \in F(t, x(t)), & \text{for a.a. } t \in J, \\ x \in S, \end{cases}$

where J is a given real interval, $F : J \times \mathbb{R}^n \multimap \mathbb{R}^n$ is a u-Carathéodory map and S is a subset of $\text{AC}_{\text{loc}}(J, \mathbb{R}^n)$.

Let $G : J \times \mathbb{R}^n \times \mathbb{R}^n \times [0, 1] \multimap \mathbb{R}^n$ be a u-Carathéodory map such that

$$G(t, c, c, 1) \subset F(t, c), \quad \text{for all } (t, c) \in J \times \mathbb{R}^n.$$

Assume that

(1.5.1) *there exist a retract Q of $C(J, \mathbb{R}^n)$ and a closed bounded subset S_1 of S*
        such that the associated problem

(1.7)
$$\begin{cases} \dot{x} \in G(t, x(t), q(t), \lambda), & \text{for a.a. } t \in J, \\ x \in S_1 \end{cases}$$

is solvable with an R_δ-set of solutions, for each $(q, \lambda) \in Q \times [0, 1]$,

(1.5.2) *there exists a locally integrable function $\alpha : J \to \mathbb{R}$ such that*

$$|G(t, x(t), q(t), \lambda)| \leq \alpha(t), \quad a.e. \text{ in } J,$$

for any $(q, \lambda, x) \in \Gamma_T$, where T denotes the set-valued map which assigns
to any $(q, \lambda) \in Q \times [0, 1]$ the set of solutions of (1.7),

(1.5.3) $T(Q \times \{0\}) \subset Q$,

(1.5.4) *if $Q \ni q_j \to q \in Q$, $q \in T(q, \lambda)$, then there exists $j_0 \in \mathbb{N}$ such that, for*
        every $j \geq j_0$, $\theta \in [0, 1]$ and $x \in T(q_j, \theta)$, we have $x \in Q$.

Then the problem (1.6) has a solution.

PROOF. Consider the set

$$Q' = \{y \in C(J, \mathbb{R}^{n+1}) \mid y(f) = (q(t), \lambda), \ q \in Q, \ \lambda \in [0, 1]\}.$$

By Proposition (1.4), we obtain that the set-valued map $T : Q \times [0, 1] \multimap S_1$ is
u.s.c., and so it belongs to the class $J(Q \times [0, 1], C(J, \mathbb{R}^n))$. Moreover, it has a
relatively compact image. Assumption (1.5.4) implies that T is a homotopy in
$J_A(Q, C(J, \mathbb{R}^n))$. Corollary (10.4) in Chapter II.10 now gives the existence of a
fixed point of $T(\cdot, 1)$. However, by the hypothesis, it is a solution of (1.6).   □

Note that the conditions (1.5.3) and (1.5.4) in the above theorem hold if $S_1 \subset Q$.
This remark permits us to obtain the generalization of Theorem 1.2 in [CFM2],
where the result for single-valued right hand side of the equation and for a convex
set of parameters has been proved.

(1.8) COROLLARY. *Consider the boundary value problem*

(1.9)
$$\begin{cases} \dot{x} \in F(t, x(t)), & \text{for a.a. } t \in J, \\ x \in S, \end{cases}$$

where J is a given real interval, $F : J \times \mathbb{R}^n \multimap \mathbb{R}^n$ is a u-Carathéodory map and
S is a subset of $\mathrm{AC}_{\mathrm{loc}}(J, \mathbb{R}^n)$.

Let $G : J \times \mathbb{R}^n \times \mathbb{R}^n \multimap \mathbb{R}^n$ *be a u-Carathéodory map such that*

$$G(t, c, c) \subset F(t, c), \quad \text{for all } (t, c) \in J \times \mathbb{R}^n.$$

Assume that

(1.8.1) *there exists a retract* Q *of* $C(J, \mathbb{R}^n)$ *such that the associated problem*

$$(1.10) \qquad \begin{cases} \dot{x} \in G(t, x(t), q(t)), & \text{for a.a. } t \in J, \\ x \in S \cap Q \end{cases}$$

has an R_δ-*set of solutions, for each* $q \in Q$,

(1.8.2) *there exists a locally integrable function* $\alpha : J \to \mathbb{R}$ *such that*

$$|G(t, x(t), q(t))| \leq \alpha(t), \quad \text{a.e. in } J,$$

for any $(q, x) \in \Gamma_T$,

(1.8.3) $T(Q)$ *is bounded in* $C(J, \mathbb{R}^n)$ *and* $\overline{T(Q)} \subset S$.

Then the problem (1.9) *has a solution.*

Making use of the Eilenberg–Montgomery fixed point theorem (see [EM]) and modifying the proof of Theorem (1.5), we can easily obtain the generalization of Theorem 1.1 in [ACZ5].

(1.11) COROLLARY. *Consider problem* (1.9) *and assume that all the assumptions of Corollary* (1.8) *hold with the convex closed set* Q *and nonempty acyclic sets of solutions* (1.10). *Then the problem* (1.9) *has a solution.*

Let us note that in applications solution sets arc, in fact, R_δ-sets.

If, in particular, $J = [a, b]$ (i.e. compact), then Theorem (1.5) can be easily reformulated as follows.

(1.12) COROLLARY. *Consider the boundary value problem* (1.6), *where* $J = [a, b]$ *is a compact interval,* $F : J \times \mathbb{R}^n \multimap \mathbb{R}^n$ *is a u-Carathéodory map and* $S \subset \mathrm{AC}_{\mathrm{loc}}(J, \mathbb{R}^n)$.

Let $G : J \times \mathbb{R}^n \times \mathbb{R}^n \times [0, 1] \multimap \mathbb{R}^n$ *be a u-Carathéodory map such that* $G(t, c, c, 1) \subset F(t, c)$, *for all* $(t, c) \in J \times \mathbb{R}^n$. *Assume that*

(1.12.1) *there exist a (bounded) retract* Q *of* $C(J, \mathbb{R}^n)$ *such that* $Q \setminus \partial Q$ *is nonempty (open) and a closed bounded subset* S_1 *of* S *such that the associated problem* (1.7) *is solvable with an* R_δ-*set of solutions, for each* $(q, \lambda) \in Q \times [0, 1]$, *and* (1.5.2), (1.5.3) *hold true,*

(1.12.2) *the solution map* T *(defined in* (1.5.2)*) has no fixed points on the boundary* ∂Q *of* Q, *for every* $(q, \lambda) \in Q \times [0, 1]$.

Then the problem (1.6) has a solution.

Since $C^{(n-1)}(J)$ can be considered as a subspace of $C(J, \mathbb{R}^n)$, we can also apply the previous results to nth-order scalar differential inclusions. To solve an existence problem, one should check suitable a priori bounds for all the derivatives up to the order $n - 1$. Our technique simplifies a work. Let us describe it below.

We need the following lemma ([CFM2, Lemma 2.1]) related to the Banach space $H^{n,1}(I)^9$:

(1.13) LEMMA. *Let I be a compact real interval and let $a_0, a_1, \ldots, a_{n-1} : I \times \mathbb{R} \to \mathbb{R}$ be u-Carathéodory functions. Given any $q \in C^{(n-1)}(I)$, consider the following linear nth-order differential operator $L_q : H^{n,1}(I) \to L^1(I)$:*

$$L_q(x)(t) = x^{(n)}(t) + \sum_{i=0}^{n-1} a_i(t, q(t), \ldots, q^{(n-1)}(t)) x^{(i)}(t).$$

Assume there exists a subset Q of $C^{(n-1)}(I)$ and an L^1-function $\beta : I \to \mathbb{R}$ such that, for any $q \in Q$ and any $i = 0, 1, \ldots n - 1$, we have

$$|a_i(t, q(t), \ldots, q^{(n-1)}(t))| \leq \beta(t), \quad a.e. \text{ in } I.$$

Then the following two norms are equivalent in $H^{n,1}(I)$:

$$\|x\| = \sum_{i=0}^{n-1} \sup_{t \in I} |x^{(i)}(t)| + \int_I |x^{(n)}(t)| \, dt, \|x\|_Q = \sup_{t \in I} |x(t)| + \sup_{q \in Q} \int_I |L_q(x)(t)| \, dt.$$

(1.14) COROLLARY. *Consider the scalar problem*

$$(1.15) \quad \begin{cases} x^{(n)}(t) + \sum_{i=0}^{n-1} a_i(t, x(t), \ldots, x^{(n-1)}(t)) x^{(i)}(t) \\ \quad \in F(t, x(t), \ldots, x^{(n-1)}(t)), \quad \text{for a.a. } t \in J, \\ x \in S, \end{cases}$$

where $J \subset \mathbb{R}$, $S \subset C(J)$ and a_i, F are u-Carathéodory maps on $J \times \mathbb{R}^n$.

Suppose that there exists a u-Carathéodory map $G : J \times \mathbb{R}^n \times \mathbb{R}^n \times [0,1] \multimap \mathbb{R}^n$ such that, for every $c \in \mathbb{R}^n$ and $\lambda \in [0,1]$, $G(t, c, c, 1) \subset F(t, c)$, a.e. in J.

Then problem (1.15) has a solution, provided the following conditions are satisfied:

(1.14.1) *there is a retract Q of the space $C^{(n-1)}(J)$ such that, for every $(q, \lambda) \in Q \times [0,1]$, the following problem*

$$(1.16) \quad \begin{cases} x^{(n)}(t) + \sum_{i=0}^{n-1} a_i(t, q(t), \ldots, q^{(n-1)}(t)) x^{(i)}(t) \\ \quad \in G(t, x(t), \ldots, x^{(n-1)}(t), q(t), \ldots, q^{(n-1)}(t), \lambda), \quad \text{for a.a. } t \in J, \\ x \in S \cap Q, \end{cases}$$

---

[9] By $H^{n,1}(I)$, we denote the Banach space of all $C^{(n-1)}$-functions $x : I \to \mathbb{R}$, where I is a compact interval, with absolutely continuous $(n - 1)$th derivatives (see Example (1.32) in Chapter I.1).

has an R_δ-set of solutions,

(1.14.2) *there is a locally integrable function $\alpha : J \to \mathbb{R}$ such that, for every $i = 0, \ldots, n-1$:*

$$|a_i(t, q(t), \ldots, q^{(n-1)}(t))| \leq \alpha(t), \quad a.e. \text{ in } J,$$

and

$$|G(t, x(t), \ldots, x^{(n-1)}(t), q(t), \ldots, q^{(n-1)}(t), \lambda)| \leq \alpha(t), \quad \text{for a.e. } t \in J,$$

for each $(q, \lambda, x) \in Q \times [0, 1] \times C^{(n-1)}(J)$ satisfying (1.16),

(1.14.3) *$T(Q \times \{0\}) \subset Q$, where T denotes the set-valued map which assigns to any $(q, \lambda) \in Q \times [0, 1]$ the set of solutions of (1.16),*

(1.14.4) *the set $T(Q \times [0, 1])$ is bounded in $C(J)$ and its $C^{(n-1)}$-closure is contained in S (in particular, this holds if $S \cap C^{(n-1)}(J)$ is closed in $C^{(n-1)}(J)$),*

(1.14.5) *if $\{q_j\} \subset Q$ converges to $q \in Q$, $q \in T(q, \lambda)$ in $C^{(n-1)}(J)$, then there exists $j_0 \in \mathbb{N}$ such that, for every $j \geq j_0$, $\theta \in [0, 1]$ and $x \in T(q_j, \theta)$, we have $x \in Q$.*

PROOF. We construct a new problem in the following way:
Define $\widetilde{F} : J \times \mathbb{R}^n \multimap \mathbb{R}^n$,

$$\widetilde{F}(t, x(t), \ldots, x^{(n-1)}(t)) = F(t, x(t), \ldots, x^{(n-1)}(t))$$
$$- \sum_{i=0}^{n-1} a_i(t, x(t), \ldots, x^{(n-1)}(t)) x^{(i)}(t).$$

Denote $\bar{x}(t) = (x(t), \ldots, x^{(n-1)}(t)) \in \mathbb{R}^n$ and define $F' : J \times \mathbb{R}^n \multimap \mathbb{R}^n$,

$$F'(t, \bar{x}(t)) = \{\dot{x}(t), \ldots, x^{(n-1)}(t), y) \mid y \in \widetilde{F}(t, x(t), \ldots, x^{(n-1)}(t))\}.$$

So, we have a problem

(1.17)
$$\begin{cases} \dot{\bar{x}}(t) \in F'(t, \bar{x}(t)), & \text{for a.a. } t \in J, \\ \bar{x} \in \overline{S}, \end{cases}$$

where $\overline{S}$ is an image of $S \cap C^{(n-1)}(J)$ via the inclusion $i : C^{(n-1)}(J) \to C(J, \mathbb{R}^n)$.

Analogously, we find the associated problem

(1.18)
$$\begin{cases} \dot{\bar{x}}(t) \in G'(t, \bar{x}(t), \bar{q}(t), \lambda), & \text{for a.a. } t \in J, \\ \bar{x} \in \overline{S} \cap \overline{Q}. \end{cases}$$

Notice that

(1) $G'(t, \overline{x}(t), \overline{q}(t), 1) \subset F'(t, \overline{x}(t))$,
(2) the set $\overline{Q} = i(Q)$ is a retract of $C(J, \mathbb{R}^n)$,
(3) $\overline{S} \subset \mathrm{AC}_{\mathrm{loc}}(J, \mathbb{R}^n)$,
(4) for every $(q, \lambda) \in Q \times [0, 1]$, the sets of solutions of the problems (1.16) and (1.18) are the same,
(5) $\overline{T}(\overline{Q} \times [0, 1]) \subset \overline{S}$, where $\overline{T}$ is a suitable map corresponding to T

and

$$|G'(t, \overline{x}(t), \overline{q}(t), \lambda)| \leq |G(t, x(t), \dots, x^{(n-1)}, q(t), \dots, q^{(n-1)}(t), \lambda)|$$

$$+ \sum_{i=0}^{n-1} |a_i(t, q(t), \dots, q^{(n-1)}(t))| \, |x^{(i)}(t)|$$

$$\leq \alpha(t) + \alpha(t) \sum_{i=0}^{n-1} |x^{(i)}(t)|.$$

Since $T(Q \times [0, 1])$ is bounded in $C(J)$, there exists a positive continuous function $m : J \to \mathbb{R}$ such that $|x(t)| \leq m(t)$, for all $t \in J$ and any $x \in T(Q \times [0, 1])$. We will show that $T(Q \times [0, 1])$ is also bounded in $C^{(n-1)}(J)$. It is sufficient to prove that, for any compact subinterval I in J, there is a constant $M > 0$ such that

$$p_I(x) = \sum_{i=0}^{n-1} \sup |x^{(i)}(t)| \leq M, \quad \text{for all } x \in T(Q \times [0, 1]).$$

Let $I \subset J$ be an arbitrary compact interval. Using the notation in Lemma (1.13), we see that $p_I(x) \leq \|x\|$ and, by the equivalence of norms,

$$\|x\| \leq c\|x\|_Q \leq c \left(\max_{t \in I} m(f) + \int_I \alpha(t) \, dt \right) \leq M.$$

We conclude that $T(Q \times [0, 1])$ is bounded in $C^{(n-1)}(J)$ which implies that $\overline{T}(\overline{Q} \times [0, 1])$ is bounded in $C(J, \mathbb{R}^n)$. Moreover, there exists a continuous function $\varphi : J \to \mathbb{R}$ such that

$$|G'(t, \overline{x}(t), \overline{q}(t), \lambda)| \leq \alpha(t)(1 + \varphi(t)).$$

Obviously, the right-hand side of the above inequality is a locally integrable function.

Finally, an easy computation shows that the condition (1.5.4) in Theorem (1.5) holds for $\overline{Q}$ and $\overline{T}$. By Theorem (1.5), there exists a solution of (1.17) as well as the one of (1.15). $\qquad \square$

The same argument as in Corollary (1.8) shows how to generalize the analogous result in [CFM2] for the following scalar problem, namely

(1.19)
$$\begin{cases} x^{(n)}(t) + \sum_{i=0}^{n-1} a_i(t, x(t), \dots, x^{(n-1)}(t)) x^{(i)}(t) \\ \qquad \in F(t, x(t), \dots, x^{(n-1)}(t)), \qquad\qquad \text{for a.a. } t \in J, \\ x \in S, \end{cases}$$

where $J \subset \mathbb{R}$, $S \subset C(J)$ and a_i, F are u-Carathéodory maps on $J \times \mathbb{R}^n$, by means of the following linearized problem

(1.20)
$$\begin{cases} x^{(n)}(t) + \sum_{i=0}^{n-1} a_i(t, q(t), \dots, q^{(n-1)}(t)) x^{(i)}(t) \\ \qquad \in G(t, x(t), \dots, x^{(n-1)}(t), q(t), \dots, q^{(n-1)}(t)), \quad \text{for a.a. } t \in J, \\ x \in S \cap Q, \end{cases}$$

where Q is a retract of the space $C^{(n-1)}(J)$.

Theorem (10.7) in Chapter II.10 gives similar consequences as those of Theorem (10.3) in Chapter II.10. Unfortunately, the weakness of the assumption on solutions causes that we have to assume the convexity of the set Q. In spite of it, the results given below are important because of the applications.

(1.21) THEOREM. *Consider the boundary value problem*

(1.22)
$$\begin{cases} \dot{x} \in F(t, x(t)), & \text{for a.a. } t \in J, \\ x \in S, \end{cases}$$

where J is a given real interval, $F : J \times \mathbb{R}^n \multimap \mathbb{R}^n$ is a u-Carathéodory map and S is a subset of $\mathrm{AC}_{\mathrm{loc}}(J, \mathbb{R}^n)$.

Let $G : J \times \mathbb{R}^n \times \mathbb{R}^n \times [0,1] \multimap \mathbb{R}^n$ be as in Theorem (1.5). Assume that the assumptions (1.5.1)–(1.5.3) of Theorem (1.5) hold, with the convexity of the set Q, and

(1.21.1) *if $\partial Q \times [0,1] \supset \{(q_j, \lambda_j)\}$ converges to $(q, \lambda) \in \partial Q \times [0,1]$, $q \in T(q, \lambda)$, then there exists $j_0 \in \mathbb{N}$ such that, for every $j \geq j_0$, and $x_j \in T(q_j, \lambda_j)$, we have $x_j \in Q$.*

Then the problem (1.22) has a solution.

The proof can be obtained immediately by using our continuation principle presented in Theorem (10.7) in Chapter II.10.

(1.23) REMARK. If the associated problem (1.7) for G is uniquely solvable, for every $(q, \lambda) \in Q \times [0,1]$, then, by continuity of T, we can reformulate the condition (1.21.1) as follows:

(1.21.1') *if $\{(x_j, \lambda_j)\}$ is a sequence in $S_1 \times [0,1]$, with $\lambda_j \to \lambda \in [0,1)$ and x_j is converging to a solution $x \in Q$ of (1.7) (corresponding to (x, λ)), then x_j belongs to Q, for j sufficiently large.*

Thus, we have a generalization of Theorem 2.1 in [FP1].

Now, we are interested in the existence of several solutions of problem (1.6). For this, the Nielsen theory developed in Chapter I.11 will be applied. It will be convenient to use the following definition.

(1.24) DEFINITION. We say that the mapping $T : Q \multimap U$ is *retractible onto* Q, where U is an open subset of $C(J, \mathbb{R}^n)$ containing Q, if there is a (continuous) retraction $r : U \to Q$ and $p \in U \setminus Q$ with $r(p) = q$ implies that $p \notin T(q)$.

Its advantage consists in the fact that, for a retractible mapping $T : Q \multimap U$ onto Q with a retraction r in the sense of Definition (1.24), its composition with r, $r|_{T(Q)} \circ T : Q \multimap Q$, has a fixed point $\widehat{q} \in Q$ if and only if $\widehat{q}$ is a fixed point of T.

The following principal statement characterizes the matter.

(1.25) THEOREM. *Let the assumptions of Proposition* (1.4) *be satisfied, where* Q *is a closed connected subset of* $C(J, \mathbb{R}^n)$ *with a finitely generated abelian fundamental group and, instead of* (1.4.4), *we can simply assume that* $\overline{T(Q)} \subset S$. *Assume, furthermore, that the operator* $T : Q \multimap U$, *related to* (1.4.1), *is retractible onto* Q *with a retraction* r *in the sense of Definition* (1.24) *and with* R_δ-*values. At last, let*

$$(1.25.1) \qquad\qquad G(t, c, c) \subset F(t, c)$$

take place a.e. in J, *for any* $c \in \mathbb{R}^n$. *Then the original problem* (1.6) *admits at least* $N(r|_{T(Q)} \circ T(\cdot))$ *solutions belonging to* Q, *where* N *stands for the Nielsen number defined in Chapter* I.11.

PROOF. By the hypothesis, Q is a connected (metric) ANR-space with a finitely generated abelian fundamental group and $T(q)$ is an R_δ-mapping. Since T is also, according to Proposition (1.4), u.s.c. and such that $\overline{T(Q)}$ is compact, $r \circ T$ is compact, admissible, and consequently a CAC-mapping. This follows from the commutativity of the following diagram:

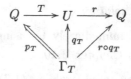

where (p_T, q_T) is a pair of natural projections of the graph Γ_T and p_T is Vietoris. Therefore, according to Therem (11.8) in Chapter I.11, $(p_T, r|_{T(Q)} \circ q_T)$ admits at least $N(r|_{T(Q)} \circ T(\cdot))$ coincidence points. Because of Definition (1.24), they represent the solutions of the problem in (1.4.1) and, in view of (1.25.1), they also satisfy the original problem (1.6). $\square$

Futhermore, we will consider boundary value problems on arbitrary (possibly infinite) intervals for differential inclusions in an arbitrary Banach space. We start with some definitions.

Let E be a Banach space with the norm $\| \cdot \|$. Denote by $C(J, E)$ the space of all continuous functions $x : J \to E$ with the locally convex topology generated by the one of a uniform convergence on compact subintervals of an arbitrary interval (possibly, the whole $\mathbb{R}$). As pointed out in Chapter I.1, this topology is completely metrizable, and thus $C(J, E)$ is a Fréchet space.

Recall that a mapping $x : J \to E$ is locally absolutely continuous if x is absolutely continuous on every compact subinterval of J. Unfortunately, in general, on each interval $[a, b] \subset J$, there need not exist $\dot{x}(t)$ (in the sense of Fréchet), for almost all (a.a.) $t \in [a, b]$ with $\dot{x} \in L^1([a, b], E)$ (the set of all Bochner integrable functions $[a, b] \to E$) and so need not be

$$x(t) = x_0 + \int_a^t \dot{x}(s)\, ds.$$

It is so if E satisfies the Radon–Nikodym property, in particular, if E is reflexive. Moreover, we have the following

(1.26) LEMMA. *Suppose $x : [a, b] \to E$ is absolutely continuous, $\dot{x}$ exists a.e., and*

$$\|\dot{x}(t)\| \leq y(t), \quad a.e., \text{ for some } y \in L^1([a, b], \mathbb{R}).$$

Then $\dot{x} \in L^1([a, b], E)$ and

$$(1.27) \qquad \int_\tau^t \dot{x}(s)\, ds = x(t) - x(\tau) \quad (t, \tau \in [a, b]).$$

PROOF.

$$\dot{x} = \lim_{n \to \infty} \frac{x\left(t + \frac{1}{n}\right) - x(t)}{\frac{1}{n}}$$

is a.e. the limit of a sequence of continuous (and so measurable) functions, and consequently measurable. Thus, $\dot{x} \in L^1([a, b], E)$.

If (1.27) fails, then, there is some $f \in E^*$ with

$$f\left(\int_\tau^t \dot{x}(s)\, ds \right) \neq f(x(t) - x(\tau)).$$

Let us note that the (uncountable) axiom of choice is not required for this conclusion, because by the measurability of x and $\dot{x}$, it is without any loss of generality to assume that E is separable.

Clearly, the scalar function $f \circ x$ is absolutely continuous, and so

$$f(x(t) - x(\tau)) = (f \circ x)(t) - (f \circ x)(\tau) = \int_\tau^t (f \circ x)(s)\, ds =$$

$$= \int_\tau^t f(\dot{x}(s))\, ds = f\left(\int_\tau^t \dot{x}(s)\, ds \right),$$

a contradiction. $\qquad\square$

The set of all locally absolutely continuous functions from J to E, satisfying all the above properties, will be denoted by $\mathrm{AC}_{\mathrm{loc}}(J, E)$.

Consider now the differential inclusion

$$(1.28) \qquad\qquad \dot{y}(t) \in F(t, y(t)).$$

By a *solution* of this differential inclusion we mean again a map $x \in \mathrm{AC}_{\mathrm{loc}}(J, E)$ satisfying (1.28), for a.a. $t \in J$. In (1.28), we assume that the set-valued map $F : J \times E \multimap E$ is a u-Carathéodory map, i.e.

(C1) $F(t, x)$ is nonempty, closed and convex, for every $(t, x) \in J \times E$,

(C2) $F(t, \cdot)$ is u.s.c., for a.a. $t \in J$,

(C3) $F(\cdot, x)$ is strongly measurable, on every compact interval $[a, b]$, for each $x \in E$.

To a u-Carathéodory map F, we associate the Nemytskiĭ (or superposition) operator $N_F : C(J, E) \multimap L^1_{\mathrm{loc}}(J, E)$ given by

$$N_F(x) := \{ f \in L^1_{\mathrm{loc}}(J, E) \mid f(t) \in F(t, x(t)), \quad \text{a.e. on J} \},$$

for each $x \in C(J, E)$.

In the sequel, we will need the following lemma (see e.g. [Vr-M, p. 88]).

(1.29) LEMMA. *Let $[a, b]$ be a compact interval. Let $F : [a, b] \times E \multimap E$ be a u-Carathéodory mapping and assume in addition that, for every nonempty bounded set $\Omega \subset E$, there exists $\nu = \nu(\Omega) \in L^1([a, b])$ such that*

$$\| F(t, x) \| := \sup\{ \| z \| \mid z \in F(t, x) \} \le \nu(t),$$

for a.e. $t \in [a, b]$ and every $x \in \Omega$. Then the Nemytskiĭ operator $N_F : C([a, b], E) \multimap L^1([a, b], E)$ has nonempty, convex values. Moreover, given sequences $\{x_n\} \subset C([a, b], E)$ and $\{f_n\} \subset L^1([a, b], E)$, $f_n \in N_F(x_n)$, $n \ge 1$, such that $x_n \to x$ in $C([a, b], E)$ and $f_n \to f$ weakly in $L^1([a, b], E)$, then $f \in N_F(x)$.

The following lemma extends Theorem 0.3.4 in [AuC-M] to infinite-dimensional spaces.

(1.30) LEMMA. *Assume that a sequence* $\{x_k \mid [a,b] \to E\}$ *of* AC-*maps satisfies the following conditions:*

(1.30.1) $\{x_k(t)\}$ *is relatively compact, for each* $t \in [a,b]$,

(1.30.2) *there exists* $\alpha \in L^1([a,b])$ *such that* $\|\dot{x}_k(t)\| \leq \alpha(t)$, *for a.a.* $t \in [a,b]$,

(1.30.3) $\{\dot{x}_k(t)\}$ *is weakly relatively compact, for a.a.* $t \in [a,b]$.

Then there exists a subsequence (again denoted by $\{x_k\}$*) that converges to an absolutely continuous map* $x : [a,b] \to E$ *in the following sense:*

(1.30.4) $x_k \to x$ *in* $C([a,b], E)$,

(1.30.5) $\dot{x}_k \to \dot{x}$ *weakly in* $L^1([a,b], E)$.

PROOF. Since

$$\|x_k(t) - x_k(s)\| \leq \int_s^t \|\dot{x}_k(\tau)\|\, d\tau \leq \int_s^t \alpha(\tau)\, d\tau,$$

we see that the sequence $\{x_k\}$ is a equicontinuous. Thus, the Arzelà–Ascoli lemma implies that $x_k \to x$ in $C([a,b], E)$. Using our assumptions (1.30.2) and (1.30.3), we infer that $\dot{x}_k \to f$, weakly in $L^1([a,b], E)$ (see Corollary 2.6 in [DRS]).

Let $t \in [a,b]$ be given. Then $\int_a^t : L^1([a,b], E) \to E$ is linear and continuous and so $x_k(t) = x_k(a) + \int_a^t \dot{x}_k(s)\, ds \to x(a) + \int_a^t f(s)\, ds$, weakly in E. But we already know that $x_k(t) \to x(t)$ in E. Thus, the uniqueness of the weak limit implies that $x(t) = x(a) + \int_a^t f(s)\, ds$. It follows that $\dot{x}(t) = f(t)$, for a.a. $t \in [a,b]$, and x is absolutely continuous. $\square$

(1.31) PROPOSITION. *Let* $G : J \times E \times E \multimap E$ *be a* u-*Carathéodory map and let* S *be a nonempty subset of* $\mathrm{AC}_{\mathrm{loc}}(J, E)$. *Assume that:*

(1.31.1) *there exists a closed* $Q \subset C(J, E)$ *such that, for any* $q \in Q$, *the boundary value problem*

$$\begin{cases} \dot{y}(t) \in G(t, y(t), q(t)), & \text{for a.a. } t \in J, \\ y \in S, \end{cases}$$

has a solution. Denote by $T : Q \multimap S$ *the solution mapping.*

(1.31.2) *There exist* $\alpha, \beta, \gamma \in L^1_{\mathrm{loc}}(J)$ *such that*

$$\|G(t, x, q)\| \leq \alpha(t) + \beta(t)\|x\|x + \gamma(t)\|x\|,$$

for a.a. $t \in J$ *and every* $x, q \in E$.

(1.31.3) *If* $\{(q_n, x_n)\}$ *is a sequence in the graph of* T *and* $(q_n, x_n) \to (q, x)$, *then* $x \in S$.

(1.31.4) $G(t, x, y)$ *is weakly relatively compact, for a.a.* $t \in J$ *and all* $(x,y) \in E^2$.

Then $T : Q \multimap S$ has a closed graph (S is endowed with the topology of $C(J, E)$).

PROOF. Let $\{(q_n, x_n)\}$ be an arbitrary sequence in the graph of T, i.e. $x_n \in T(q_n)$, for every $n \in \mathbb{N}$, and assume that $(q_n, x_n) \to (q_0, x_0)$. Thus, we see that

$$\dot{x}_n(t) \in G(t, x_n(t), q_n(t)), \quad \text{for a.a. } t \in J,$$

and $x_n \in S$. Then $q_0 \in Q$ and, by assumption (1.31.3), $x_0 \in S$.

Now, let $[a, b]$ be an interval in J. Using assumption (1.31.2) and (1.31.4), we see that the sequence $\{x_n\}$ satisfies the assumptions of Lemma (1.30). Thus, $\{x_n\}$ converges uniformly on $[a, b]$ to x_0 (because this limit is unique) and $\{\dot{x}_n\}$ converges to x_0, weakly in $L^1([a, b], E)$. Using Lemma (1.29), it follows that $\dot{x}_0(t) \in G(t, x_0(t), q_0(t))$, for a.a. $t \in [a, b]$. Since $[a, b]$ was arbitrary, we see that indeed $\dot{x}_0(t) \in G(t, x_0(t), q_0(t))$, for a.a. $t \in J$ and $x_0 \in T(q_0)$. $\qquad\square$

As another of the main results of this section, we formulate the following continuation principle.

(1.32) THEOREM. *Consider the boundary value problem*

$$(1.33) \qquad \begin{cases} \dot{y}(t) \in F(t, y(t)), & \text{for a.a. } t \in J, \\ y \in S, \end{cases}$$

where $F : J \times E \multimap E$ is a u-Carathéodory map and S is a subset of $\mathrm{AC}_{\mathrm{loc}}(J, E)$. Let $G : J \times E \times E \times [0, 1] \multimap E$ be a u-Carathéodory map with weakly relatively compact values, for a.a. $t \in J$ and all $(x, y, \lambda) \in E \times E \times [0, 1]$ such that

$$(1.34) \qquad G(t, c, c, 1) \subset F(t, c), \quad \text{for all } (t, c) \in J \times E.$$

Assume that:

(1.32.1) *there exists a closed, convex $Q \subset C(J, E)$ and a closed subset S_1 of S such that the problem*

$$\begin{cases} \dot{y}(t) \in G(t, y(t), q(t), \lambda), & \text{for a.a. } t \in J, \\ y \in S_1 \end{cases}$$

is solvable with an R_δ-set $T(q, \lambda)$, for each $(q, \lambda) \subset Q \times [0, 1]$.

(1.32.2) *There exist $\alpha, \beta, \gamma \in L^1_{\mathrm{loc}}(I)$ such that*

$$\|G(t, x, q, \lambda)\| \leq \alpha(t) + \beta(t)\|x\|x + \gamma(t)\|x\|,$$

for a.a. $t \in J$ and every $x, q \in E$ and every $\lambda \in [0, 1]$.

(1.32.3) *T is quasi-compact, i.e. T maps compact subsets onto compact subsets, and there exists a measure of noncompactness β in the sense of Definition (5.31) in Chapter I.5 such that, for each $\Omega \subset Q$, if*

$$\beta(T(\Omega \times [0,1])) \geq \beta(\Omega),$$

then Ω is relatively compact.

(1.32.4) $T(Q \times \{0\}) \subset Q$.

(1.32.5) *For each $\lambda_0 \in [0,1]$ and $q \in T(q_0, \lambda_0)$, if $q_n \to q_0$ in Q, then there is $n_0 \in \mathbb{N}$ such that, for each $n \geq n_0$, $\lambda \in [0,1]$ and $x \in T(q_n, \lambda)$, we have $x \in Q$.*

Then problem (1.33) has a solution.

PROOF. Using Proposition (1.31), we see that the map $T : Q \times [0,1] \multimap S_1$ has a closed graph. Since T is also quasi-compact (assumption (1.32.3)), we can easily derive that T is indeed an u.s.c. set-valued map (see e.g. [KOZ-M, Theorem 1.1.12]). From assumption (1.32.1), we get therefore that $T \in J(Q \times [0,1], C(\mathbb{R}, E))$ and assumption (1.32.3) implies that T is also β-condensing. By (1.32.5), we finally see that T is a homotopy in J_A, and thus Corollary (10.29) in Chapter II.10 implies the existence of a fixed point of $T(\,\cdot\,, 1)$. However, by the inclusion (1.34), it is a solution of (1.33). $\square$

(1.35) REMARK. As we can see, Theorem (1.32) extends Theorem (1.5) into the infinite-dimensional setting, when replacing $\mathbb{R}^n$ by a real Banach space. On the other hand, this is possible with some loss, namely Q is only convex and the solution operator T is assumed to be quasi-compact, additionally. Because of those restrictions, we are unfortunately unable to establish a full infinite dimensional analogy of Theorem (1.25).

Sometimes it is convenient to consider the asymptotic problem sequentially. For this purpose, it can be useful to employ

(1.36) PROPOSITION. *Let $J_1 \subset J_2 \subset \ldots$ be compact intervals such that $J = \bigcup_{m=1}^{\infty} J_m$ and $t_0 \in J_1$. Let $F : J \times E \multimap E$ be an (upper) Carathéodory mapping with nonempty, compact and convex values. Assume, furthermore, that*

(1.36.1) *There are $\alpha, \beta \in L^1_{loc}(J, \mathbb{R})$ with*

$$\|F(t,y)\| \leq \alpha(t) + \beta(t)\|y\|.$$

(1.36.2) *There is some u-Carathéodory mapping $g : J \times [0,\infty) \to [0,\infty)$ with $g(\,\cdot\,, 0)$ such that the only nonnegative measurable solution of*

$$x(t) \leq \left| \int_{t_0}^{t} g(s, x(s))\, ds \right|$$

is 0 (a.e.), and such that, for a.a. $t \in J$, $\gamma(F(\{t\}) \times C)) \leq g(t, \gamma(C))$, for countable, bounded subsets $C \subset E$, where γ denotes the Hausdorff measure of noncompactness.

(1.36.3) *E has the so called retraction property in the sense of [Vä-M], e.g. E is separable or reflexive.*

If $y_n \in AC(J_n, E)$ satisfies

$$\dot{y}_m(t) \in F(t, y_m(t)), \quad \text{for a.a. } t \in J_m, \ m \in \mathbb{N},$$

and $\{y_m(t_0) \mid m \in \mathbb{N}\}$ is a relatively compact set, then there is a solution $y \in AC_{loc}(J, E)$ of the inclusion

$$\dot{y}(t) \in F(t, y(t)), \quad \text{for a.a. } t \in J,$$

such that, for some subsequence,

$$y_{m_k} \to y, \quad \text{uniformly on each } J_m,$$

and

$$\dot{y}_m \to \dot{y}, \quad \text{weakly in } L^1(J, E).$$

If still

(1.36.4) $$\sup\{\|y_m(t)\| \mid m \in \mathbb{N}, \quad t \in J_m\} < \infty$$

and the values of y_m, $m \in \mathbb{N}$, are located in a given subdomain $\mathcal{D}$ of E, then there exists an entirely bounded solution y on J with $y(t) \in \mathcal{D}$, for all $t \in \mathbb{R}$.

PROOF. By (1.36.1) and the well-known Gronwall inequality (see e.g. [He-M]), we get the a priori estimates

$$\|y_m(t)\| \leq \widetilde{y}(t) \quad \text{and} \quad \|\dot{y}_m(t)\| \leq \widetilde{y}(t),$$

for some $\widetilde{y} \in L^1_{loc}(J, \mathbb{R})$.

We claim that $\{y_m(t) \mid m \geq m_t\}$ is a relatively compact set, for a.a. $t \in J$, where $m_t = \min\{m \mid t \in J_m\}$. To show it, put $h(t) := \gamma(\{y_m(t) \mid m \geq m_t\})$. Then h is measurable (for more details, see Proposition 11.12 in [Vae-M]). Moreover, by means of Proposition 11.12 in [Vae-M] and (1.36.2), we obtain

$$h(t) = \gamma(\{y_m(t) - y_m(t_0) \mid m \geq m_t\}) = \gamma\left(\left\{\int_{t_0}^t \dot{y}_m(s)\, ds \, \middle| \, m \geq m_t\right\}\right)$$

$$\overset{\text{[Vae-M]}}{\leq} \left|\int_{t_0}^t \gamma(\{\dot{x}_m(s) \mid m \geq m_t\})\, ds\right| \overset{(1.36.2)}{\leq} \left|\int_{t_0}^t g(s, h(s))\, ds\right|.$$

Applying (1.36.2) again, we arrive at $h(t) = 0$, for a.a. $t \in J$, as claimed.

Since $F(t, \cdot)$ maps compact sets into compact sets, $\{\dot{y}_m(t) \mid m \geq m_t\}$ becomes relatively compact as well, for a.a. $t \in J$. An application of the standard diagonalization argument implies, jointly with Lemma (1.30), the existence of a subsequence such that $y_{m_k} \to y$, uniformly on each J_m, and $\dot{y}_{m_k} \to \dot{y}$, weakly in $L^1(J, E)$, where $y \in \mathrm{AC}_{\mathrm{loc}}(J, E)$.

It follows from Lemma (1.29) that $\dot{y}(t) \in F(t, y(t))$. Since the remaining part of the assertion is implied by the foregoing one (just proved) and (1.36.4), the proof is completed. $\square$

If in particular, $E = \mathbb{R}^n$, then (1.36.2) and (1.33.3) hold automatically. Hence, Proposition (1.36) can be then simplified as follows.

(1.37) PROPOSITION. *Let* $F : \mathbb{R} \times \mathbb{R}^n \multimap \mathbb{R}^n$ *be an (upper) Carathéodory mapping with nonempty, compact and convex values, satisfying* (1.36.1), *for* $J = (-\infty, \infty)$. *Then, for every* $y_0 \in \mathbb{R}^n$, *there exists a solution* $y \in \mathrm{AC}_{\mathrm{loc}}(\mathbb{R}, \mathbb{R}^n)$ *of the Cauchy problem*

$$\begin{cases} \dot{y}(t) \in F(t, y(t)), & \text{for a.a. } t \in (-\infty, \infty), \\ y(t_0) = y_0. \end{cases}$$

Let $\{y_m(t)\}$ *be a sequence of absolutely continuous functions such that*

(1.37.1) *For every* $m \in \mathbb{N}$, $y_m \in \mathrm{AC}([-m, m], \mathbb{R}^n)$ *is a solution of*

$$\dot{y}(t) \in F(t, y(t)), \quad \text{for a.a. } t \in [-m, m],$$

(1.37.2) $\sup\{|y_m(t)| \mid m \in \mathbb{N}, \ t \in [-m, m]\} := M < \infty$ *and* $y_m(t) \in \mathcal{D} \subset \mathbb{R}^n$, *for every* $t \in [-m, m]$.

Then there exists an entirely bounded solution $y \in \mathrm{AC}_{\mathrm{loc}}(\mathbb{R}, \mathbb{R}^n)$ *of the inclusion*

$$\dot{y}(t) \in F(t, y(t)), \quad \text{for a.a. } t \in (-\infty, \infty),$$

such that

$$\sup_{t \in \mathbb{R}} |y(t)| \leq M \ (< \infty) \quad \text{and} \quad y(t) \in \mathcal{D}, \text{ for all } t \in \mathbb{R}.$$

III.2. Topological structure of solution sets: initial value problems

In this part, various methods for investigating the topological structure of solution sets will be presented. Only initial value problems will be considered. The results are mainly based on the papers [AGG1]–[AGG3], [COZ], [Ge7], [HP3].

The classical result, due to F. S. De Blasi and J. Myjak in [DBM2], deals with Cauchy problems for the (upper) Carathéodory differential inclusions in Euclidean spaces:

(2.1)
$$\begin{cases} \dot{x}(t) \in F(t, x(t)), \\ x(0) = x_0, \end{cases}$$

where $F : J \times \mathbb{R}^n \multimap \mathbb{R}^n$ is a multivalued mapping, satisfying conditions (C1)–(C3) in the foregoing Chapter III.1, and such that

(2.2) $$|F(t,x)| \leq \alpha + \beta|x|, \quad \text{for all } t \in J, \, x \in \mathbb{R}^n,$$

where α, β are nonnegative constants.

(2.3) THEOREM ([DBM2], cf. also [GP1]). *Problem* (2.1), *where J is a compact interval, has under the above assumptions an R_δ-set of solutions.*

We omit the proof of this theorem, because below we will prove its generalized version (cf. (2.12)).

We recall that a multivalued mapping $F : J \times \mathbb{R} \multimap \mathbb{R}^n$ is said to be *integrably bounded* (resp. *locally integrably bounded*) if there exists an integrable (resp. locally integrable) function $\mu : J \to [0, \infty)$ such that $|y| \leq \mu(t)$, for every $x \in \mathbb{R}^n$, $t \in J$ and $y \in F(t, x)$. We say that F has *at most a linear growth* (resp. *a local linear growth*) if there exist integrable (resp. locally integrable) functions $\mu, \nu : J \to [0, \infty)$ such that

$$|y| \leq \mu(t)|x| + \nu(t),$$

for every $x \in \mathbb{R}^n$, $t \in J$ and $y \in F(t, x)$.

It is obvious that F has at most a linear growth if there exists an integrable function $\mu : J \to [0, \infty)$ such that $|y| \leq \mu(t)(|x| + 1)$, for every $x \in \mathbb{R}^n$, $t \in J$ and $y \in F(t, x)$.

Let us also recall that a single-valued map $f : J \times \mathbb{R}^n \to \mathbb{R}^n$ is said to be *measurable-locally Lipschitz* if, for every $x \in \mathbb{R}^n$, there exists a neighbourhood V_x of x in $\mathbb{R}^n$ and an integrable function $L_x : J \to [0, \infty)$ such that

$$|f(t, x_1) - f(t, x_2)| \leq L_x(t)|x_1 - x_2|, \quad \text{for every } f \in J \text{ and } x_1, x_2 \in V_x,$$

where $f(\,\cdot\,, x)$ is measurable, for every $x \in \mathbb{R}^n$.

Now, for the considerations below, fix J as the halfline $[0, \infty)$ and assume that $F : J \times \mathbb{R}^n \multimap \mathbb{R}^n$ is again a multivalued u-Carathéodory map. Consider the Cauchy problem (2.1). By $S(F, 0, x_0)$, we denote the set of solutions of (2.1). For the characterization of the topological structure of $S(F, 0, x_0)$, it will be useful to recall the following well-known uniqueness criterium (see e.g. [Fi2-M, Theorem 1.1.2]).

(2.4) THEOREM. *If f is a single-valued, integrably bounded, measurable-locally Lipschitz map, then the set $S(f, 0, x_0)$ is a singleton, for every $x_0 \in \mathbb{R}^n$.*

The following result will be employed as well.

(2.5) THEOREM. *If F is locally integrably bounded, mLL-selectionable, then $S(F, 0, x_0)$ is contractible, for every $x_0 \in \mathbb{R}^n$.*

PROOF. Let $f \subset F$ be measurable – locally Lipschitz. By Theorem (2.4), the following Cauchy problem

$$(2.6) \qquad \begin{cases} \dot{x}(t) = f(t, x(t)), \\ x(t_0) = x_0, \end{cases}$$

has exactly one solution, for every to $t \in J$ and $x_0 \in \mathbb{R}^n$. For the proof, it is sufficient to define a homotopy $h : S(F, 0, x_0) \times [0, 1] \to S(F, 0, x_0)$ such that

$$h(x, s) = \begin{cases} x, & \text{for } s = 1 \text{ and } x \in S(F, 0, x_0), \\ \widetilde{x}, & \text{for } s = 0, \end{cases}$$

where $\widetilde{x} = S(f, 0, x_0)$ is exactly one solution of the problem (2.6).

Define $\gamma : [0, 1) \to [0, \infty)$, $\gamma(s) = \text{tg}(\pi s / 2)$ and put

$$h(x, s)(t) = \begin{cases} x(t), & \text{for } 0 \leq t \leq \gamma(s), \ s < 1, \\ S(f, \gamma(s), x(\gamma(s)))(t), & \text{for } \gamma(s) \leq t < \infty, \ s < 1, \\ x(t), & \text{for } 0 \leq t < \infty, \ s = 1. \end{cases}$$

Then h is a continuous homotopy, contracting $S(F, 0, x_0)$ to the point $S(f, 0, x_0)$. $\square$

Analogously, we can get the following result.

(2.7) THEOREM. *If F is locally integrably bounded, Ca-selectionable, or in particular c-selectionable, then $S(F, 0, x_0)$ is R_δ-contractible, for every $x_0 \in \mathbb{R}^n$.*

Observe that, if $F : J \times \mathbb{R}^n \multimap \mathbb{R}^n$ is an intersection of the decreasing sequence of $F_k : J \times \mathbb{R}^n \multimap \mathbb{R}^n$, $F(t, x) = \bigcap_{k=1}^{\infty} F_k(t, x)$ and $F_{k+1}(t, x) \subset F_k(t, x)$, for almost all $t \in J$ and for all $x \in \mathbb{R}^n$, then

$$(2.8) \qquad S(F, 0, x_0) = \bigcap_{k=1}^{\infty} S(F_k, 0, x_0).$$

From Theorems (2.5) and (2.7), we obtain

(2.9) THEOREM. *Let* $F : J \times \mathbb{R}^n \multimap \mathbb{R}^n$ *be a multivalued map with nonempty closed values.*

(2.9.1) *If* F *is* σ-*mLL-selectionable, then the set* $S(F, 0, x_0)$ *is an intersection of a decreasing sequence of contractible sets,*

(2.9.2) *if* F *is* σ-*Ca-selectionable, i.e., it is an intersection of a decreasing sequence of Ca-selectionable mappings, then the set* $S(F, 0, x_0)$ *is an intersection of a decreasing sequence of* R_δ-*contractible sets.*

Before formulating the following important theorem, recall that, for two metric spaces X, Y and the interval J, the multivalued map $F : J \times X \multimap Y$ is *almost upper semicontinuous* (a.u.s.c.), if for every $\varepsilon > 0$ there exists a measurable set $A_\varepsilon \subset J$ such that $m(J \setminus A_\varepsilon) < \varepsilon$ and the restriction $F|_{A_\varepsilon \times X}$ is u.s.c., where m stands for the Lebesgue measure.

It is clear that every a.u.s.c. map is u-Carathéodory. In general, the reverse is not true. The following Scorza–Dragoni type result describing possible regularizations of Carathéodory maps (see e.g. [JK1]) will be employed.

(2.10) PROPOSITION. *Let* X *be a separable metric space and* J *be an interval. Suppose that* $F : J \times X \multimap \mathbb{R}^n$ *is a nonempty, compact, convex valued u-Carathéodory map. Then there exists an a.u.s.c. map* $\psi : J \times X \multimap \mathbb{R}^n$ *with nonempty compact convex values and such that:*

(2.10.1) $\psi(t, x) \subset F(t, x)$, *for every* $(t, x) \in J \times X$,

(2.10.2) *if* $\Delta \subset J$ *is measurable,* $u : \Delta \to \mathbb{R}^n$ *and* $v : \Delta \to X$ *are measurable maps and* $u(t) \in \psi(f, u(t))$, *for almost all* $t \in \Delta$, *then* $u(t) \in \psi(t, v(t))$, *for almost all* $t \in \Delta$.

The proof of the following statement can be found in [Go5-M].

(2.11) PROPOSITION. *Let* E, E_1 *be two separable Banach spaces,* J *be an interval and* $F : J \times E \multimap E$ *be an a.u.s.c. map with compact convex values. Then* F *is* σ-*Ca-selectionable. The maps* $F_k : J \times E \multimap E_1$ *(see the Definition I.(3.32) of* σ-*selectionable maps) are a.u.s.c., and we have* $F_k(t, e) \subset \overline{\mathrm{conv}}(\bigcup_{x \in E} F(t, x))$, *for all* $(t, e) \in J \times E$. *Moreover, if* F *is integrably bounded, then* F *is* σ-*mLL-selectionable, i.e., it is an intersection of a decreasing sequence of mLL-selectionable mappings.*

Now, we are ready to give

(2.12) THEOREM. *If* $F : J \times \mathbb{R}^n \multimap \mathbb{R}^n$ *is a u-Carathéodory map with compact convex values having at most the linear growth, then* $S(F, 0, x_0)$ *is an* R_δ-*set, for every* $x_0 \in \mathbb{R}^n$.

PROOF. By the hypothesis, there exists an integrable function $\mu : J \to [0, \infty)$ such that $\sup\{|y| \mid y \in F(t, x)\} \leq \mu(t)(|x| + 1)$, for every $(t, x) \in J \times \mathbb{R}^n$. By means of the well-known Gronwall inequality (see [Har-M]), we obtain that $|x(t)| \leq$ $(|x_0| + \gamma) \exp(\gamma) = M$, where $x \in S(F, 0, x_0)$ and $\gamma = \int_0^\infty \mu(s)\,ds$.

Take $r > M$ and define $\widetilde{F} : J \times \mathbb{R}^n \multimap \mathbb{R}^n$ as follows

$$\widetilde{F}(t, x) = \begin{cases} F(t, x), & \text{if } |x| \leq r, \\ F\left(t, r\dfrac{x}{|x|}\right), & \text{if } |x| > r. \end{cases}$$

One can see that $\widetilde{F}$ is an integrably bounded u-Carathéodory map and $S(\widetilde{F}, 0, x_0) = S(F, 0, x_0)$. By Proposition (2.10), there exists an a.u.s.c. map $G : J \times \mathbb{R}^n \multimap \mathbb{R}^n$ with nonempty, convex, compact values such that $S(G, 0, x_0) = S(\widetilde{F}, 0, x_0)$. Applying Proposition (2.11) to the map G, we obtain the sequence of maps G_k. As in Theorem (2.9), we see that $S(G, 0, x_0)$ is an intersection of the decreasing sequence $S(G_k, 0, x_0)$ of contractible sets. By the well-known Arzelà–Ascoli lemma and Theorem (2.4), we obtain that, for every $k \in \mathbb{N}$, the set $S(G_k, 0, x_0)$ is compact and nonempty, which completes the proof. $\qquad\square$

Using the above results and the unified approach to the u.s.c. and l.s.c. case due to A. Bressan (cf. [Bre1], [Bre2]), we can obtain

(2.13) PROPOSITION. *Let* $G : J \times \mathbb{R}^n \multimap \mathbb{R}^n$ *be a l.s.c. bounded map with nonempty closed values. Then there exists an u.s.c. map* $F : J \times \mathbb{R}^n \multimap \mathbb{R}^n$ *with compact convex values such that, for any* $x_0 \in \mathbb{R}^n$, *the set* $S(G, 0, x_0)$ *contains an* R_δ-*set* $S(F, 0, x_0)$ *as a subset.*

Now, the same problem will be treated, when assuming that F is not necessarily convex-valued, i.e., when F satisfies the following conditions:

(A) $F : J \times \mathbb{R}^n \multimap \mathbb{R}^n$, F has non-empty, compact values, where J denotes the halfline $[0, \infty)$, and $F(\,\cdot\,, x)$ is measurable, for all $x \in \mathbb{R}^n$,

(B) there exists a locally integrable function $\nu : J \to J$ such that, for every $t \in J$ and all $x, y \in \mathbb{R}^n$, $d_H(F(t, x), F(t, y)) \leq \nu(t)|x - y|$,

(C) there exists a locally integrable function $\alpha : J \to J$ and a positive constant B such that, for every $x \in \mathbb{R}^n$ and for a.a. $t \in J$, $|F(t, x)| \leq \alpha(t)(B + |x|)$, where $|F(t, x)| = \sup\{|y| \mid y \in F(t, x)\}$.

(2.14) THEOREM. *Under the assumptions* (A)–(C), *the set of solutions to problem* (2.1) *is non-empty and can be obtained as a limit of an inverse system (see Chapter* II.2) *of AR-spaces, for every* $x_0 \in \mathbb{R}^n$.

PROOF. Fix $x_0 \in \mathbb{R}^n$ and denote

$$S = \{x \in \mathrm{AC}_{\mathrm{loc}} \mid \dot{x}(t) \in F(t, x(t)), \quad \text{for a.a. } t \in J \text{ and } x(0) = x_0\},$$

i.e. S is the set of all solutions to problem (2.1). A standard application of the well-known Gronwall inequality (see e.g. [Har-M]) allows us to find a map $G :$ $J \times \mathbb{R}^n \multimap \mathbb{R}^n$, which satisfies conditions (A), (B) and

(C') there exists a locally Lebesgue integrable function $\beta : J \to J$ such that, for every $x \in \mathbb{R}^n$ and for a.a. $t \in J$, $|G(t,x)| \leq \beta(t)$

and, moreover, the set of solutions to the Cauchy problem (2.1) with F replaced by G is equal to S. Thus,

$$S = \{x \in \mathrm{AC}_{\mathrm{loc}} \mid \dot{x}(t) \in G(t, x(t)), \quad \text{for a.a. } t \in J \text{ and } x(0) = x_0\}.$$

At first, we prove that S is non-empty. Define

$$\mathrm{AC}^0 = \{x \in \mathrm{AC}_{\mathrm{loc}} \mid x(0) = x_0\},$$
$$S = \mathrm{cl}_{C(J, \mathbb{R}^n)}\{x \in \mathrm{AC}^0 \mid |x(t)| \leq \beta(t)\}$$

and denote, for simplicity, $L_m^1 = L^1([0, m], \mathbb{R}^n)$. One can see that AC^0 is a closed, convex subset of the Fréchet space $\mathrm{AC}_{\mathrm{loc}}$ and, by the well-known Arzelà–Ascoli lemma, S (considered as a subset of $C(J, \mathbb{R}^n)$) is compact. Moreover, S is convex.

We shall define, by induction, multivalued lower semicontinuous maps $K_m :$ $S \multimap L_m^1$ with closed, decomposable values, for every $m \geq 1$. At first, by properties of the map G, for every $m \geq 1$ and $s \in S$, there exists a measurable selection of $G(\cdot, s(\cdot))$, restricted to $[0, m] \times \mathbb{R}^n$. Define $K_1 : S \multimap L_1^1$ by

$$K_1(s) = \{u \in L_1^1 \mid u(t) \in G(t, s(t)) \quad \text{for a.a. } t \in [0, 1]\}.$$

It is easy to see that K_1 has closed, decomposable values. Moreover, one can show (see, e.g., [Fry2]) that K_1 is lower semicontinuous. By the Fryszkowski selection theorem (see (3.81) in Chapter I.3), there exists a continuous selection k_1 of K_1, i.e. a map $k_1 : S \to L_1^1$ such that $k_1(s)(t) \in G(t, s(t))$, for all $s \in S$ and for a.a. $t \in [0, 1]$.

Now, suppose that there is defined $K_{m-1} : S \multimap L_{m-1}^1$ which is a lower semicontinuous map with closed, decomposable values, for some $m \geq 1$. Denote by k_{m-1} a continuous selection of K_{m-1} and define

$$K_m(s) = \{u \in L_m^1 \mid u(t) \in G(t, s(t)), \text{ for a.a. } t \in [0, m] \text{ and } u|_{[0, m-1]} = k_{m-1}(s)\}.$$

We see again that K_m is a lower semicontinuous map with closed, decomposable values and, therefore, there exists a continuous selection of K_m. We denote it by k_m.

Define a continuous map $k : S \to L_{\mathrm{loc}}^1(J, \mathbb{R}^n)$ by

$$k(s)(t) = k_m(s)(t) \quad (t \in [0, m]),$$

for every $s \in S$. It is obvious that $k(s)(t) \in G(t, s(t))$, for all $s \in S$ and for a.a. $t \in J$. The above consideration allows us to define a map $l : S \to S$ by

$$l(s)(t) = x_0 + \int_0^t k(s)(\tau) \, d\tau.$$

By continuity of k, we see that l is continuous as well. By the well-known Schauder–Tikhonov Fixed Point Theorem (1.38) from Chapter I.1, there exists a fixed point of l, i.e. a locally absolutely continuous function $x : J \to \mathbb{R}^n$ such that

$$x(t) = x_0 + \int_0^t k(x)(\tau) \, d\tau.$$

It means that $x(t)$ is a solution of problem (2.1).

In the second part of the proof, we study the structure of the solution set S. Let us make some further notations:

$$\mathrm{AC}_m = \{x \in \mathrm{AC}([0, m], \mathbb{R}^n) \mid x(0) = x_0\},$$

$$\varphi_m : \mathrm{AC}_m \multimap L_m^1, \quad \varphi(m) = \{u \in L_m^1 \mid u(t) \in G(t, x(t)), \text{ for a.a. } t \in [0, m]\},$$

$$T_m : L_m^1 \to \mathrm{AC}_m, \quad T_m(u)(t) = x_0 + \int_0^t u(s) \, ds,$$

$$\Phi_m : \mathrm{AC}_m \multimap \mathrm{AC}_m, \quad \Phi = T_m \circ \varphi_m.$$

Let $S_m \subset \mathrm{AC}_m$ denote the set of solutions to the Cauchy problem

$$(2.15) \qquad \begin{cases} \dot{x}(t) \in G(t, x(t)), & \text{for a.a. } t \in [0, m], \\ x(0) = x_0, \end{cases}$$

and

$$\mathcal{F}_m = \{\dot{x} \mid x \in S_m\}.$$

Observe that, for every $m \geq 1$, the set AC_m is a closed, convex subset of the Banach space $\mathrm{AC}([0, m], \mathbb{R}^n)$ and T_m is a homeomorphism. One can see that $S_m = \mathrm{Fix}\, \Phi_m$. Moreover, assumption (B) implies that φ_m is a Lipschitz map, by which φ is continuous. By [BCF, Theorem 2] the set $\mathcal{F}_m$ is an absolute retract. Since T_m is a homeomorphism, the set $S_m \in \mathrm{AR}$ is an absolute retract, too. Hence, it is acyclic.

We show that $\{\Phi_m\}$ is a map of the inverse system $\{\mathrm{AC}_m, \pi_m^p, \mathbb{N}\}$, where $\pi_m^p(x) = x|_{[0,m]}$ for every $x \in \mathrm{AC}_p$ and $p \geq m$. This easily follows from the equalities ($t \in [0, m]$):

$$\Phi_m \pi_m^p(x)(t) = \left\{ x_0 + \int_0^t u(s) \, ds \,\middle|\, u \in L_m^1 \text{ and } u(t) \in G(t, x(t)), \text{ for a.a. } t \in [0, m] \right\},$$

$$\pi_m^p \Phi_m(x)(t) = \left\{ x_0 + \int_0^t u(s) \, ds \,\middle|\, u \in L_p^1 \text{ and } u(t) \in G(t, x(t)), \text{ for a.a. } t \in [0, p] \right\}$$

and from the observation that

$$\{u \in L_m^1 \mid u(t) \in G(t, x(t)), \quad \text{for a.a. } t \in [0, m]\}$$
$$\{u|_{[0,m]} \mid u \in L_p^1 \quad \text{and} \quad u(t) \in G(t, x(t)), \text{ for a.a. } t \in [0, p]\}.$$

So the map $\{\Phi_m\}$ induces the limit one $\Phi : AC^0 \multimap AC^0$ such that $\Phi(x)|_{[0,m]} = \Phi_m(x|_{[0,m]})$. It means that the fixed point set of Φ is equal to S. By the proof of the first part of the present one, it follows that S is as required. $\qquad\square$

Replacing slightly the assumptions (A) and (B) in Theorem (2.14), the topological structure of the solution set to (2.1) can be expected to be slightly changed as well. Hence, let us assume that $F : J \times \mathbb{R}^n \multimap \mathbb{R}^n$ satisfies the conditions:

  (i) the values of F are nonempty, closed and connected,
  (ii) $F(t, \cdot)$ is u.s.c., for a.a. $t \in J$,
  (iii) $F(\cdot, x)$ is measurable, for every $x \in \mathbb{R}$,
  (iv) F has at most a linear growth, i.e. there is a locally integrable function $\mu : J \to J$ such that

$$|F(t, x)| \le \mu(x)(1 + |x|), \quad \text{for all } (t, x) \in J \times \mathbb{R}^n.$$

We can give

(2.16) THEOREM. *The set S of solutions to problem (2.1) is, under the above assumptions* (i)–(iv), *homeomorphic to a (possibly empty) intersection of absolute retracts. Moreover, if every Cauchy problem is solvable on every compact subinterval, then the set S is also nonempty.*

PROOF. We will proceed in several steps.

Step 1. By the recent result of C. Benassi and A. Gavioli (see [BeGa, Theorem 5.1)], there is a sequence of maps $F_k : J \times \mathbb{R}^n \multimap \mathbb{R}^n$ such that, for every $k \ge 1$,

  (a) F_k has nonempty, closed and connected values and the same growth as F,
  (b) $F_k(\cdot, x)$ is measurable, for every $x \in \mathbb{R}^n$,
  (c) $F_k(t, \cdot)$ is locally Lipschitzian, for a.a. $t \in J$,
  (d) $F(t, x) \subset F_{k+1}(t, x) \subset F_k(t, x)$, for all $(t, x) \in J \times \mathbb{R}^n$,
  (e) $\lim_{k \to \infty} d_H(F(t, x), F_k(t, x)) = 0$, for all $(t, x) \in J \times \mathbb{R}^n$.

Step 2. Consider, for every $k \ge 1$, the family of problems

$$(2.16.1) \qquad \begin{cases} \dot{x}(t) \in F_k(t, x(t)), & \text{for a.a. } t \in [0, m], \\ x(0) = x_0. \end{cases}$$

We show that the set S_k^m of solutions to the above problem is an absolute retract.

In fact, using the well-known Gronwall inequality (see e.g. [Har-M]), one can obtain by a standard manner that all solutions to (2.16.1) are bounded by the same constant D. Since the map $F_k(t, \cdot)$ is Lipschitzian on the ball $K(0, D) \subset \mathbb{R}^n$, we can assume, without any loss of generality, that F_k is globally Lipschitzian. Therefore, Theorem 2 in [BCF] applies, and the set S_k^m is an absolute retract.

Step 3. Define $\Phi^m, \Phi_k^m : AC([0, m], \mathbb{R}^n) \multimap AC([0, m], \mathbb{R}^n)$,

$$\Phi^m(x)(t) = x_0 + \int_0^t F(s, x(s))\, ds,$$

$$\Phi_k^m(x)(t) = x_0 + \int_0^t F_k(s, x(s))\, ds.$$

Note that $\text{Fix}(\Phi_k^m) = S_k^m \in \text{AR}$ and, furthermore, $\text{Fix}(\Phi^m) \subset \text{Fix}(\Phi_{k+1}^m) \subset \text{Fix}(\Phi_k^m)$, for every $k \geq 1$. One can also check that $S = \text{Fix}(\Phi^m)$ is the set of solutions to (2.1) on the interval $[0, m]$. We show that $S^m = \bigcap_{k=1}^\infty S_k^m$.

It is obvious that $S^m \subset \bigcap_{k=1}^\infty S_k^m$.

Let $x \in \bigcap_{k=1}^\infty S_k^m$. It means that there are $u_k \in L^1([0, m], \mathbb{R}^n)$, such that for any $k, k' \geq 1$ and $t \in [0, m]$,

$$\int_0^t (u_k(s) - u_{k'}(s))\, ds = 0,$$

and so the Lebesgue measure of the set

$$A = \{t \in [0, m] \mid u_k(t) \neq u_{k'}(t), \quad \text{for some } k, k' \geq 1\}$$

is equal to 0.

Let $u \in L^1([0, m], \mathbb{R}^n)$ be such that $u = u_1$, on $[0, m] \setminus A$. Then $u(s) \in F_k(s, x(s))$, for a.a. $s \in [0, m]$ and every $k \geq 1$. By conditions (d) and (e), and the fact that all values of F are closed, we have that $F(t, x) = \bigcap_{k=1}^\infty F_k(t, x)$, and so $u(s) \in F(s, x(s))$, for a.a. $s \in [0, m]$. Moreover, $x(t) = x_0 + \int_0^t u(s)\, ds$. Thus, $x \in S^m$.

Step 4. Observe that the maps Φ^m induce the limit map $\Phi : AC_{\text{loc}}(J, \mathbb{R}^n) \multimap AC_{\text{loc}}(J, \mathbb{R}^n)$, where $AC_{\text{loc}}(J, \mathbb{R}^n)$ is a limit of the inverse system

$$\{AC([0, m], \mathbb{R}), \pi_m^p, \mathbb{N}\},$$

as follows

$$\Phi(x)(t) = x_0 + \int_0^t F(s, x(s))\, ds.$$

It can be easily seen that $S = \text{Fix}(\Phi)$. Thus by Theorem (2.8) from Chapter II.2 $S = \lim_{\leftarrow}\{S\}$.

To see that $\mathcal{S}$ is homeomorphic to an intersection of absolute retracts, it is sufficient to consider the sets

$$Q_m = \left\{ (x_i) \in \prod_{i=1}^{\infty} \mathcal{S}^i \,\middle|\, x_i = x_m|_{[0,i]}, \quad \text{for } i \leq m \right\},$$

$$W_{mk} = \left\{ (x_i) \in \prod_{i=1}^{\infty} \mathcal{S}_k^i \,\middle|\, x_i = x_m|_{[0,i]}, \quad \text{for } i \leq m \right\} \approx \prod_{i=1}^{\infty} \mathcal{S}_k^i \in \mathrm{AR},$$

and realize that

$$\lim_{\leftarrow} \{\mathcal{S}^m\} = \bigcap_m Q_m = \bigcap_m \bigcap_k W_{mk}.$$

Since $W_{mk} \supset W_{m(k+1)}$ and $W_{mk} \supset W_{(m+1)k}$, for all $m, k \geq 1$, one obtains

$$\lim_{\leftarrow} \{\mathcal{S}^m\} = \bigcap_m W_{mm}.$$

which completes the main part of the proof.

For the nonemptiness of $\mathcal{S}$, it is sufficient to define $x(t) = x_m(t)$, for $t \in [m-1, m)$, where x_m is a solution to the Cauchy problem on $[m-1, m]$, starting from $x_{m-1}(m)$.   □

(2.17) REMARK. It seems that to prove "only" the connectedness of the solution set to (2.1) can be a difficult task, as can be seen from the following example.

(2.18) EXAMPLE. Namely, let $X_m = K_1 \cup K_2 \cup P_m$, where $K_i = \{(x,y) \in \mathbb{R}^2 \mid y = i\}$ and $P_m = \{(x,y) \in \mathbb{R}^2 \mid x \geq m \text{ and } y \in [1,2]\}$. One can easily check that each X_m is a closed absolute retract, but $X = \bigcap X_m = K_1 \cup K_2$ is not connected.

On the other hand, for (additionally) convex-valued right-hand sides, the compactness of the solution set to (2.1) can be proved by which the (compact) R_δ-structure can be deduced.

In this part, we are interested in the existence as well as in the structure of a solution set to an autonomous Cauchy problem with a discontinuous mutlivalued right-hand side, on the half-line $J = [0, \infty)$.

The classical Carathéodory problem can be regarded as a special case. Indeed, the problem

$$(2.19) \qquad \begin{cases} \dot{x}(t) = g(t, x(t)), & \text{for a.a. } t \in J, \\ x(0) = v, \end{cases}$$

where $g : J \times \mathbb{R}^n \to \mathbb{R}^n$ is a Carathéodory function, can be rewritten as follows. Define $f(x_0, x) = (1, g(x_0, x))$, where $x_0 = t$. Let $y = (x_0, x)$ be a new variable. Then we arrive at the autonomous problem

$$(2.20) \qquad \begin{cases} \dot{y}(t) = f(y(t)), & \text{for a.a. } t \in J, \\ y(0) = (0, v), \end{cases}$$

which is equivalent to (2.19).

The results below are generalizations of Theorems 1 and 2 in [BS] to the case of multivalued maps and noncompact intervals.

We will consider the following Cauchy problem:

$$(2.21) \qquad \begin{cases} \dot{x}(t) \in \varphi(x(t)), & \text{for a.a. } t \in J, \\ x(0) = v. \end{cases}$$

Here $\varphi : \mathbb{R}^n \multimap \mathbb{R}^n$ is a multivalued map of the form $\varphi(x) = \psi(\tau(x), x)$, where $\tau : \mathbb{R}^n \to \mathbb{R}$ is a single-valued map and $\psi : J \times \mathbb{R}^n \multimap \mathbb{R}^n$ is a multivalued one.

(2.22) THEOREM. *Assume that*

(2.22.1) $\tau : \mathbb{R}^n \to \mathbb{R}$ *is continuously differentiable and* $\psi : \mathbb{R} \times \mathbb{R}^n \multimap \mathbb{R}^n$ *is a u-Carathéodory map,*

(2.22.2) *for some compact, convex set* $K \subset \mathbb{R}^n$, *at every point* x, *one has*

$$(2.23) \qquad \varphi(x) \subset K, \quad \nabla\tau(x) \cdot z > 0, \quad \text{for every } z \in K.$$

Then the Cauchy problem (2.21) *has a solution. If, additionally,*

(2.22.3) *The gradient* $\nabla\tau$ *has a bounded directional variation*[10] *w.r.t. the cone*
$$\Gamma = \{\lambda z \mid \lambda \geq 0, \ z \in K\},$$

then problem (2.21) *has a nonempty, compact and acyclic set of solutions.*

(2.24) REMARK. We can understand the assumption (2.22.2) in Theorem (2.22) as a transversality assumption motivated by a classical Carathéodory single-valued problem. Indeed, the map f in (2.20) can jump across the hyperplanes of the form $x_0 = \text{const}$, which are transversal to f (the inner product of their normal vectors with f is equal to 1). In the above theorem, since $\dot{x} \in K$, for every solution x, transversality assumptions imply that all trajectories must cross transversally any hypersurface of the form $\tau(x) = \text{const}$.

(2.25) REMARK. The above theorem remains true with weaker assumptions on a regularity of the map T. Namely, it is sufficient to assume that τ is Lipschitzian and reformulate (2.23) e.g. in terms of the Clarke generalized gradients (see [Cl-M]).

For the proof of Theorem (2.22), we need some results describing possible regularizations of u-Carathéodory maps (see [JK1], [DBM2]).

---

[10]The map u has a bounded directional variation w.r.t. the cone Γ if

$$\sup\left\{ \sum_{i=1}^{N} |u(p_i) - u(p_{i-1})| \ \middle| \ N \geq 1, \ p_i - p_{i-1} \in \Gamma, \text{ for every } i \right\} < \infty.$$

(2.26) PROPOSITION. *Let J be an interval and $F : J \times \mathbb{R}^n \multimap \mathbb{R}^n$ be a nonempty, compact, convex valued u-Carathéodory map. Suppose that $\tau : \mathbb{R}^n \to \mathbb{R}$ satisfies assumptions (2.22.1) and (2.22.2) of Theorem (2.22). Then there exists an almost upper semicontinuous (see above) map $G : J \times \mathbb{R}^n \multimap \mathbb{R}^n$ with nonempty, compact, convex values and such that:*

(2.26.1) *$G(t,x) \subset F(t,x)$, for every $(t,x) \in J \times \mathbb{R}^n$,*

(2.26.2) *for every $T > 0$, if $\Delta \subset [0,T]$ is measurable, $u : \Delta \to \mathbb{R}^n$ is a measurable map, $v \in C_T$, where*

$$C_T = \left\{ v \in C([0,T], \mathbb{R}^n) \, \middle| \, v(0) = 0, \frac{v(t) - v(s)}{t - s} \in K \right\}$$

and $u(t) \in F(\tau(v(t)), v(t))$, for a.a. $t \in \Delta$, then $u(t) \in G(\tau(v(t)), v(t))$, for a.a. $t \in \Delta$.

PROOF. Due to [JK1], Theorem 1.5, and [DBM2], Lemma 1, there exists an a.u.s.c. map $G : J \times \mathbb{R}^n \multimap \mathbb{R}^n$ with nonempty, compact, convex values satisfying (2.26.1) and

if $\Delta \subset J$ is measurable, $u, v : \Delta \to \mathbb{R}^n$ are measurable maps and $u(t) \in F(t, v(t))$, for a.a. $t \in \Delta$, then $u(t) \in G(t, v(t))$, for a.a. $t \in \Delta$.

Following the proofs of Theorems 1.2, 1.3 and 1.5 in [JK1], one can check that (2.26.2) also holds.

Note that, in this consideration, we use the fact that, under the assumptions on τ,

(2.27) $\qquad m^*((\tau \circ v)^{-1}(B)) = m^*(B), \quad$ for every $B \subset \mathbb{R}$ with $m^*(B) = 0$,

where m^* stands for the outer Lebesgue measure on $\mathbb{R}$ and $v \in C_T$.

To proof (2.27), we define the set

$$X_T = \{x \in \mathbb{R}^n \mid |x| \leq T|K|\},$$

where $|K| = \sup_{x \in K} |x|$, and notice that, by the continuity of the gradient $\nabla \tau$ and the compactness of X and K, there exists $\delta > 0$ such that

(2.28) $\qquad \nabla \tau(t) \cdot z \geq \delta, \quad$ for every $x \in X$ and $z \in K$.

Take any $v \in C_T$. By the definition of the set C_T, we get that $v(t) \in X$, for every $t \in [0,T]$. Hence, condition (2.28) implies

$$\liminf_{t \to s^+} \frac{\tau(v(t)) - \tau(v(s))}{t - s} \geq \delta,$$

by which the map $\tau \circ v$ is strictly increasing and, moreover,

$$(2.29) \qquad \frac{1}{\delta}(\tau(v(b)) - \tau(v(a))) \geq b - a,$$

for all $a, b \in [0, T]$ with $b > a$.

Take a set $B \subset \mathbb{R}$ with $m^*(B) = 0$. We show that $m^*((\tau \circ v)^{-1}(B)) = 0$ or, equivalently, that for every $\varepsilon > 0$, there is a covering $\alpha = \{P_1, P_2, \dots\}$ of $(\tau \circ v)^{-1}(B)$ by intervals such that $\mathrm{vol}(\alpha) = \sum_{i=1}^{\infty} m(P_i) < \varepsilon$.

Let $\varepsilon > 0$ be arbitrary and $\beta = \{D_1, D_2, \dots\}$ be a covering of B by intervals such that $\mathrm{vol}(\beta) < \varepsilon \delta$. Since $\tau \circ v$ is continuous and increasing, all the sets $(\tau \circ v)^{-1}(D_i)$ are intervals and form a covering of $(\tau \circ v)^{-1}(B)$.

Putting $b_i - a_i = m(D_i)$, we get

$$\mathrm{vol}(\tau \circ v)^{-1}(B) = \sum_{i=1}^{\infty} \mathrm{vol}((\tau \circ v)^{-1}(D_i))$$

$$= \sum_{i=1}^{\infty}((\tau \circ v)^{-1}(b_i) - (\tau \circ v)^{-1}(a_i)) \leq \frac{1}{\delta}\sum_{i=1}^{\infty}(b_i - a_i) < \varepsilon,$$

which completes the proof of our proposition. $\qquad \square$

PROOF OF THEOREM (2.22). Without any loss of generality, we can assume that $v = 0$. In the first step, we describe a topological structure of the solution set of our problem considered on compact intervals.

Using Proposition (2.26), for every integer $m \geq 1$, we can find an a.u.s.c. map $G^m : J \times \mathbb{R}^n \multimap \mathbb{R}^n$ such that the set $S(G^m, r, v)$ of all solutions to the problem

$$(2.30) \qquad \begin{cases} \dot{x}(t) \in G^m(\tau(x(t)), x(t)), & \text{for a.a. } t \in [0, m], \\ x(0) = v \end{cases}$$

is equal to the set of solutions to

$$(2.31) \qquad \begin{cases} \dot{x}(t) \in \psi(\tau(x(t)), x(t)), & \text{for a.a. } t \in [0, m], \\ x(0) = v. \end{cases}$$

Proposition (2.11) implies the existence of a sequence of u-Carathéodory convex compact valued maps $G_k^m : J \times \mathbb{R}^n \multimap \mathbb{R}^n$, $k = 1, 2, \dots$, such that each G_k^m has a measurable – locally Lipschitzian selection and G^m is an intersection of G_k^m.

Consider (cf. Proposition (2.26)) the compact, convex set

$$C_m = \left\{ v \in C([0, m], \mathbb{R}^n) \,\middle|\, v(0) = 0, \; \frac{v(t) - v(s)}{t - s} \in K \right\}.$$

Each map from C_m has values in the compact set

$$X_m = \{x \in \mathbb{R}^n \mid |x| \geq m|K|\}.$$

Denote by $f_k^m : J \times \mathbb{R}^n \to \mathbb{R}^n$ a measurable – locally Lipschitzian selection of G_k^m and, by $g_k^m : J \times \mathbb{R}^n \to \mathbb{R}^n$, a map defined as follows:

$$g_k^m(t, x) = \begin{cases} f_k^m(t, x), & \text{for } x \in X_m, \\ f_k^m\left(t, m|K|\dfrac{x}{|x|}\right), & \text{for } x \notin X_m. \end{cases}$$

By the compactness of X_m, the map $g_k^m(t, \cdot)$ is globally Lipschitzian.

Theorem 1 in [BS] applied for g_k^m implies that the set $S(G_k^m, \tau, v)$ of all solutions to the problem on $[0, m]$ with G_k^m as a right-hand side is nonempty. By the well-known Arzelà–Ascoli lemma, this set is also compact.

Now, we show that, under assumption (2.22.3), it is contractible. We define a required homotopy $h : S(G_k^m, \tau, v) \times [0, 1] \to S(G_k^m, \tau, v)$ as follows:

$$h(u, \lambda)(t) = \begin{cases} u(t), & \text{for } 0 \leq t \leq \lambda m, \\ S(g_k^m, \tau, \lambda m, u(\lambda m)), & \text{for } \lambda m \leq t \leq m, \end{cases}$$

where $S(g_k^m, \tau, \lambda m, u(\lambda m))$ denotes a unique solution of the problem

$$(2.32) \qquad \begin{cases} \dot{x}(t) = g_k^m(\tau(x(t)), x(t)), & \text{for a.a. } t \in [0, m], \\ x(\lambda m) = u(\lambda m). \end{cases}$$

The existence and uniqueness result for problem (2.32) follows from Theorem 2 in [BS]. Thus, we obtain that $S(G_k^m, \tau, v)$ is compact and contractible. A standard computation shows that

$$S(G^m, \tau, v) = \bigcap_{k=1}^{\infty} S(G_k^m, \tau, v),$$

what implies that $S(G^m, \tau, v)$ is nonempty, compact and, under assumption (2.22.3), also acyclic (more precisely, R_δ).

Consider the following family of maps $\Psi_m : C([0, m], \mathbb{R}^n) \multimap C([0, m], \mathbb{R}^n)$, $m \geq l$,

$$\Psi_m(u)(t) = \left\{ \int_0^t v(s)\, ds \;\middle|\; v \in L^1([0, m], \mathbb{R}^n) \right.$$

$$\left. \text{and} \quad v(s) \in \psi(\tau(u(s)), u(s)), \text{ for a.a. } t \in [0, m] \right\},$$

It is easy to check (e.g. by means of Propositions (2.26), (2.11) and properties of measurable maps) that each Ψ_m has nonempty values. Moreover, by the convexity of K, Ψ_m maps C_m into C_m.

Denote $\Phi_m : C_m \multimap C_m$, $\Phi_m(u) = \Psi_m(u)$. It is easy to see that $S(G^m, \tau, v)$ is equal to the fixed point set of Φ_m.

Notice that the set

$$C = \left\{ v \in C(J, \mathbb{R}^n) \, \middle| \, v(0) = 0, \ \frac{v(t) - v(s)}{t - s} \in K \right\}$$

can be considered as a limit of the inverse system $\{C_m, \pi_m^p\}$, where $\pi_m^p : C_p \to C_m$ is a restricting map defined as follows $\pi_m^p(u) = u|_{[0,m]}$. Moreover, the map $\{\Phi_m\}$ of the above system induces the limit map $\Phi : C \multimap C$,

$$\Phi(u)(t) = \left\{ \int_0^t v(s)\, ds \, \middle| \, v \in L^1_{\mathrm{loc}}(J, \mathbb{R}^n) \right.$$
$$\left. \text{and} \quad v(s) \in \psi(\tau(u(s)), u(s)), \text{ for a.a. } t \in J \right\}.$$

Obviously, the fixed point set of Φ is equal to the solution set of problem (2.21). Hence, Proposition (2.1) and Theorem (2.8) in Chapter II.2 imply that the set of solutions to (2.21) is nonempty and compact under assumptions (2.22.1), (2.22.2), and, additionally, it is acyclic if we add assumption (2.22.3). The proof of Theorem (2.22) is so complete. $\square$

(2.33) REMARK. Note that Theorems 1 and 2 in [BS] can be reformulated and proved in the case of noncompact intervals (with a single-valued right-hand side), when using the same method as in [BS], under more restrictive assumptions on τ_i in Theorem 1 and τ in Theorem 2. Namely, one should assume that

$$\nabla \tau_i(x) \cdot z \geq \delta, \quad \text{for some } \delta > 0 \text{ and every } z \in K.$$

Now, we show how a topological essentiality in Chapter II.4 can be applied to differential inclusions. At first, let us give some preliminary assumptions.

Put $J_1 = [0, \infty)$ and let J_2 be some subinterval of $(-\infty, 0]$ such that $0 \in J_2$. Denote $J = J_2 \cup J_1$. Let $A_i : J_1 \to C(C(J, \mathbb{R}^n), C(J_2, \mathbb{R}^n))$ be continuous, for $i = 0, \ldots, k-1$ (e.g. $[A_i(t)(x)](s) = x(t+s)$). Assume still that

$$p_i : C^{(k-1)}(J, \mathbb{R}^n) \to C(J_2, \mathbb{R}^n), \quad i = 0, \ldots, k-2, \text{ are continuous,}$$
$$p_{k-1} : C^{(k-1)}(J, \mathbb{R}^n) \to C(J_2, \mathbb{R}^n), \quad p_{k-1}(x) = x^{(k-1)}|_{J_2},$$
$$\psi_i : C^{(k-1)}(J, \mathbb{R}^n) \multimap C(J_2, \mathbb{R}^n), \quad i = 0, \ldots, k-1 \text{ are admissible and compact.}$$

Denote $E_1 = C(J_2, \mathbb{R}^n) \times \ldots \times C(J_2, \mathbb{R}^n)$, $(k-1)$-times, $E_2 = C(J_2, \mathbb{R}^n) \times \ldots \times C(J_2, \mathbb{R}^n)$, k-times, and take some multivalued map $\varphi : J_1 \times E_2 \multimap \mathbb{R}^n$. Consider the family of problems

$$(2.34) \quad \begin{cases} x^{(k)}(t) \in \rho\varphi(t, A_0(t)x, A_1(t)\dot{x}, \ldots, A_{k-1}(t)x^{(k-1)}), & \text{for a.a. } t \in J_1, \\ p_0(x) \in \rho\psi_0(x), \\ \vdots \\ p_{k-1}(x) \in \rho\psi_{k-1}(x), \end{cases}$$

where $\rho \in [0,1]$. For every ρ, the set of solutions to (2.34) will be denoted by $S_\rho^A(\varphi)$.

In an ordinary case, when $J_2 = \{0\}$, we can put $[A_i(t)x](0) = x(t)$ and identify $C(J_2, \mathbb{R}^n) \equiv \mathbb{R}^n$. Hence we have $p_i : C^{(k-1)}(J_1, \mathbb{R}^n) \to \mathbb{R}^n$, $\psi_i : C^{(k-1)}(J_1, \mathbb{R}^n) \multimap \mathbb{R}^n$, for $i = 0, 1, \ldots, k-1$, and $p_{k-1}(x) = x^{(k-1)}(0)$. Thus, we get

$$(2.35) \quad \begin{cases} x^{(k)}(t) \in \rho\varphi(t, x(t), \dot{x}(t), \ldots, x^{(k-1)}(t)), & \text{for a.a. } t \in J_1, \\ p_0(x) \in \rho\psi_0(x), \\ \vdots \\ p_{k-1}(x) \in \rho\psi_{k-1}(x), & \rho \in [0,1]. \end{cases}$$

We denote by $S_\rho^k(\varphi)$ the set of solutions to (2.35).

The following special case of (2.35) will be considered below. We assume that $[A_i(t)x](s) = [A(t)x](s) = x(t+s)$, $p_i(x) = x^{(i)}|_{J_2}$, for every $i = 0, \ldots, k-1$. There is a map $b \in C^{(k-1)}(J_2, \mathbb{R}^n)$ such that $\psi_i(x) = \{b^{(i)}\}$, $i = 0, \ldots, k-1$. Under these assumptions, we get

$$(2.36) \quad \begin{cases} x^{(k)}(t) \in \rho\varphi(t, A(t)x, A(t)\dot{x}, \ldots, A(t)x^{(k-1)}), & \text{for a.a. } t \in J_1, \\ x|_{J_2} = \rho b, \\ \vdots \\ x^{(k-1)}|_{J_2} = \rho b^{(k-1)}, & \rho \in [0,1]. \end{cases}$$

The set of solutions to (2.36) will be denoted by $S_\rho(\varphi)$. One can see that, for $J_2 = \{0\}$, we obtain the family of Cauchy problems.

Now, we prove an existence result for (2.34) using a topological essentiality.

(2.37) THEOREM. *Assume that* $\varphi : J_1 \times E_2 \multimap \mathbb{R}^n$ *is a locally integrably bounded u.s.c. map with compact, convex values and there exists an open subset* $\Omega \subset C^{(k-1)}(J, \mathbb{R}^n)$ *such that*

(2.37.1) $S_\rho^A(\varphi) \subset \Omega$, *for every* $\rho \in [0,1]$,

(2.37.2) *the map* $g : C^{(k-1)}(J, \mathbb{R}^n) \to C(J, \mathbb{R}^n) \times E_1$,

$$g(x) = (x^{(k-1)}, p_0(x), \dots, p_{k-2}(x))$$

   is essential on $\overline{\Omega}$.

Then the set $S_1^A(\varphi)$ *is nonempty.*

PROOF. Denote $E = C^{(k-1)}(J, \mathbb{R}^n)$ and $F = C(J, \mathbb{R}^n)$. For every $x \in E$, $u \in \psi_{k-1}(x)$ and $z_x : J_1 \to \mathbb{R}^n$ such that

(2.38) $z_x(t) \in \varphi(t, A_0(t)x, A_1(t)\dot{x}, \dots, A_{k-1}(t)x^{(k-1)})$,   for a.a. $t \in J_1$

(such z_x exists, because φ is u.s.c.), we define the map $y_{x,u,z_x} : J \to \mathbb{R}^n$ as follows:

$$y_{x,u,z_x} = \begin{cases} u(0) + \displaystyle\int_0^t z_x(\tau)\, d\tau, & \text{for } t \in J_1, \\ u(t), & \text{for } t \in J_2, \end{cases}$$

Define $T : E \multimap F$,

$$T(x) = \{y_{x,u,z_x} \mid u \in \psi_{k-1}(x), \quad z_x \text{ satisfies (2.38)}\}.$$

This operator is a sum of ψ_{k-1} and an integral operator which is a compact map with compact, convex values (as a consequence of a local integrable boundedness of φ). Therefore, T is admissible and compact.

Consider the following map $\Phi : \overline{\Omega} \multimap F \times E_1$:

$$\Phi(x) = g(x) - T(x) \times \psi_0(x) \times \dots \times \psi_{k-2}.$$

One can see that $0 \notin \Phi(x)$, for every $x \in \partial\Omega$. Indeed, if $0 \in \Phi(x)$, then $x \in S_1^A(\varphi)$. Assumption (2.37.1) implies that $S_1^A(\varphi) \subset \Omega$; thus, $x \in \Omega$.

We want to prove that $0 \in \Phi(x)$, for some $x \in \Omega$. For this purpose, we show that Φ is essential. By assumption (2.37.2) we know that g is essential. Define the following homotopy $\kappa : \overline{\Omega} \times [0, 1] \multimap F \times E_1$,

$$\lambda(x, t) = g(x) - t(T(x) \times \psi_0(x) \times \dots \times \psi_{k-2}),$$

for every $x \in \overline{\Omega}$ and $t \in [0, 1]$. Applying assumption (2.37.1) again, we conclude that λ is a homotopy appropriate to use the coincidence Property (4.5) in Chapter II.4. This implies that $\Phi = \lambda(\cdot, 1)$ is essential. By the existence Property (4.3) in Chapter II.4, there is a point $x \in \Omega$ such that $0 \in \Phi(x)$, which completes the proof. $\square$

(2.39) COROLLARY. *Assume that $J = [0, \infty)$ and let $\varphi : J \times \mathbb{R}^{nk} \multimap \mathbb{R}^n$ be a locally integrably bounded u.s.c. map with compact, convex values. Suppose that there exists an open subset $\Omega \subset C^{(k-1)}(J, \mathbb{R}^n)$ such that*

(2.39.1) $S_\rho^k(\varphi) \subset \Omega$, *for every $\rho \in [0, 1]$,*
(2.39.2) *the map $g : C^{(k-1)}(J, \mathbb{R}^n) \to C(J, \mathbb{R}^n) \times E_1$,*

$$g(x) = (x^{(k-1)}, p_0(x), \dots, p_{k-2}(x))$$

is essential on $\overline{\Omega}$.

Then the set $S_1^k(\varphi)$ is nonempty.

(2.40) COROLLARY. *Assume that $\varphi : J_1 \times E_2 \multimap \mathbb{R}^n$ is a locally integrably bounded u.s.c. map with compact, convex values. Then the set S_1 is nonempty.*

PROOF. Define a closed subspace G of $C(J, \mathbb{R}^n) \times E_1$ (see the above notation) as follows:

$$G = \{(z, u_0, \dots, u_{k-2}) \subset C(J, \mathbb{R}^n) \times E_1 \mid \exists y \in C^{(k-1)}(J, \mathbb{R}^n) : y^{(k-1)} = z$$
$$\text{and } y^{(i)}|_{J_2} = u_i, \ i = 0, \dots, k-2\}.$$

The map $g : F \to G$ defined by the formula

$$g(x) = (x^{(k-1)}, x|_{J_2}, \dots, x^{(k-2)}|_{J_2})$$

is a linear isomorphism. Hence, by Example (2.18), g is essential on each open neighbourhood of the origin in F. By Theorem (2.37), we get a solution to (2.36). $\square$

(2.41) REMARK. Note that all the above results are true if φ is a measurable – continuous map. Moreover, in Corollary (2.40), we can assume that φ is only measurable – u.s.c.

Now, we are interested in a topological structure of the solution set of problem (2.36), Assume that $J_2 = [r, 0]$, for some $r < 0$.

(2.42) THEOREM. *Assume that $\varphi : J_1 \times E_2 \multimap \mathbb{R}^n$ is a locally integrably bounded map with compact, convex values satisfying the following conditions:*

(A) *for all $x_0, \dots, x_{k-1} \in C(J_2, \mathbb{R}^n)$, the map $\varphi(\cdot, x_0, \dots, x_{k-1})$ is measurable,*

(B) $\exists L \geq O \ \forall x_0, \dots, x_{k-1}, y_0, \dots, y_{k-1} \in C(J_2, \mathbb{R}^n) \ \forall t \in J_1$:

$$d_H(\varphi(t, x_0, \dots, x_{k-1}), \varphi(t, y_0, \dots, y_{k-1})) \leq L \sum_{i=0}^{k-1} \|x_i - y_i\|.$$

Then the set $S_1(\varphi)$ is a compact, acyclic subset of $C^{(k-1)}(J, \mathbb{R}^n)$.

PROOF. We can assume that $L \geq 1$. Consider a map $l : L^1_{\text{loc}}(J_1, \mathbb{R}^n) \to C^{(k-1)}(J, \mathbb{R}^n)$,

$$l(z)(t) = \begin{cases} \sum_{j=0}^{k-1} \dfrac{t^j}{j!} b^{(j)} + \displaystyle\int_0^t \int_0^{s_1} \cdots \int_0^{s_{k-1}} z(s)\, ds\, ds_{k-1} \ldots ds_1, & \text{for } t \in J_1, \\ b(t), & \text{for } t \in J_2, \end{cases}$$

and a sequence of maps $l_m : L^1([0, m], \mathbb{R}^n) \to C^{(k-1)}([r, m], \mathbb{R}^n)$,

$$l_m(z)(t) = \begin{cases} \sum_{j=0}^{k-1} \dfrac{t^j}{j!} b^{(j)} + \displaystyle\int_0^t \int_0^{s_1} \cdots \int_0^{s_{k-1}} z(s)\, ds\, ds_{k-1} \ldots ds_1, & \text{for } t \in [0, m], \\ b(t), & \text{for } t \in J_2, \end{cases}$$

Define the operator $\Phi : C^{(k-1)}(J, \mathbb{R}^n) \multimap C^{(k-1)}(J, \mathbb{R}^n)$ as follows:

$$\Phi(x) = \{y \in C^{(k-1)}(J, \mathbb{R}^n) \mid y(t) = l(z)(t)$$
$$\text{and } z(t) \in \varphi(t, A(t)x, A(t)\dot{x}, \ldots, A(t)x^{(k-1)}), \quad \text{for a.a. in } J_1\}.$$

In a similar way, we define a sequence of multivalued maps $\Phi_m : C^{(k-1)}([r, m], \mathbb{R}^n) \multimap C^{(k-1)}([r, m], \mathbb{R}^n)$ as follows:

$$\Phi_m(x) = \{y \in C^{(k-1)}([r, m], \mathbb{R}^n) \mid y(t) = l_m(z)(t)$$
$$\text{and } z(t) \in \varphi(t, A(t)x, A(t)\dot{x}, \ldots, A(t)x^{(k-1)}), \quad \text{a.a. in } [0, m]\}.$$

Observe that all values of $\Phi_m(x)$ are nonempty (since φ is measurable – continuous) and convex (since φ is convex valued). Moreover, because of the compactness and convexity of values and by the local integrable boundedness of φ, one can check that the values of Φ_m are compact, too.

It is easy to see that the fixed point set $\text{Fix}(\Phi)$ is equal to the one of all solutions to problem (2.36). Thus, we are interested in a topological structure of $\text{Fix}(\Phi)$. This will be studied, when applying the results from Chapter II.1.

Consider the equivalent norms in $C^{(k-1)}([r, m], \mathbb{R}^n)$,

$$q_m(x) = \sum_{i=0}^{k-1} \max_{t \in [r,m]} (|x^{(i)}(t)| e^{-Lkt}).$$

For every $x, y \in C^{(k-1)}([r, m], \mathbb{R}^n)$ and for every $t \in [r, m]$, we have

$$e^{-Lkt} \|A(t)x - A(t)y\| = \max\{e^{-Lkt} |x(t+s) - y(t+s)| \mid s \in [r, 0]\}$$
$$= \max\{e^{-Lk(t+s)} |x(t+s) - y(t+s)| \mid s \in [r, 0]\}$$
$$= \max\{e^{-Lkv} |x(v) - y(v)| \mid v \in [r, m]\} = q_m(x - y).$$

For any $x_1, x_2 \in C^{(k-1)}([r,m], \mathbb{R}^n)$ and $y_1 \in \Phi_m(x_1)$, $y_1(t) = l_m(z_1)(t)$, we can choose, according to the Kuratowski–Ryll–Nardzewski Selection Theorem (3.49) in Chapter I.3 (cf. [KRN]), $y_2 \in \Phi_m(x_2)$ such that, for every $t \in [r,m]$, we have $y_2(t) = l_m(z_2)(t)$ and

$$|z_1(t) - z_2(t)| = d_H(z_1(t), \varphi(t, A(t)x_2, A(t)\dot{x}_2, \ldots, A(t)x_2^{(k-1)})).$$

Now, for every $t \in [0,m]$ and $i = 0, \ldots, k-1$,

$$|y_1^{(i)}(t) - y_2^{(i)}(t)| = \left| \left(\int_0^t \int_0^{s_1} \cdots \int_0^{s_{k-1}} (z_1(s) - z_2(s)) \, ds \, ds_{k-1} \ldots ds_1 \right)^{(i)} \right|$$

$$\leq \int_0^t \int_0^{s_1} \cdots \int_0^{s_{k-1-i}} |z_1(s) - z_2(s)| \, ds \, ds_{k-1-i} \ldots ds_1$$

$$\leq L \int_0^t \int_0^{s_1} \cdots \int_0^{s_{k-1-i}} \sum_{j=0}^{k-1} \|A(s)x_1^{(j)} - A(s)x_2^{(j)}\| \, ds \, ds_{k-1-i} \ldots ds_1.$$

Multiplying it by e^{-Lkt}, we get

$$e^{-Lkt}|y_1^{(i)}(t) - y_2^{(i)}(t)|$$

$$\leq Le^{-Lkt} \int_0^t \int_0^{s_1} \cdots \int_0^{s_{k-1-i}} \sum_{j=0}^{k-1} e^{Lks} e^{-Lks} \|A(s)x_1^{(j)} - A(s)x_2^{(j)}\| \, ds \, ds_{k-1-i} \ldots ds_1$$

$$\leq q_m(x_1 - x_2) Le^{-Lkt} \int_0^t \int_0^{s_1} \cdots \int_0^{s_{k-1}} e^{Lks} \, ds \, ds_{k-1} \ldots ds_1$$

$$\leq q_m(x_1 - x_2) L \frac{1}{(Lk)^{k-i}} (e^{Lkt} - 1) e^{-Lkt} = q_m(x_1 - x_2) L \frac{1 - e^{-Lkt}}{(Lk)^{k-i-1} k}$$

$$\leq \frac{1 - e^{-Lkm}}{k} q_m(x_1 - x_2).$$

Thus, $q_m(y_1 - y_2) \leq L_m q_m(x_1 - x_2)$, where $L_m = 1 - e^{-Lkm} < 1$, what implies that Φ_m is a contraction.

Now, it is easy to check that, considering $C^{(k-1)}(J, \mathbb{R}^n)$ as a limit of an inverse system of Banach spaces endowed with the norms q_m, the map Φ is a limit map induced by a sequence $\{\Phi_m\}$. Using Corollary (2.10) in Chapter II.2, we get that the set $S_1(\varphi)$ (which is equal to $\mathrm{Fix}(\Phi)$) is nonempty, compact and acyclic.   □

Note that the case of unbounded interval J_2 brings some troubles and the related problem remains open.

Consider again problem (2.36), in an ordinary case, i.e.,

(2.43)
$$\begin{cases} x^{(k)}(t) \in \varphi(t, x(t), \dot{x}(t), \ldots, x^{(k-1)}(t)), & \text{for a.a. } t \in [0,a], \\ x(0) = x_0, \\ \vdots \\ x^{(k-1)}(0) = x_{k-1}, \end{cases}$$

but this time assume that the multivalued mapping $\varphi : [0, a] \times E^k \multimap E$ operates in a separable Banach space $E^k = E \times \ldots \times E$ (k-times). By a *solution* $x(t)$ we understand the Carathéodory one, i.e. $x(t)$ in $C^{(k-1)}([0, a], E)$ with $x^{(k-1)}(t) \in AC([0, a], E)$, satisfying (2.43), for almost all $t \in [0, a]$. Denote by $S(\varphi, x_0, \ldots, x_{k-1})$ the set of all solutions to (2.43). Observe that, for $k = 1$ and $E = \mathbb{R}^n$, problem (2.43) reduces to a classical Cauchy problem.

(2.44) THEOREM. *Let φ be a compact-valued mapping satisfying:*

(2.44.1) *φ is bounded, i.e. there is a number $M > 0$ such that $\|y\| \leq M$, for every $t \in [0, a]$, $X \in E^k$ and $y \in \varphi(t, X)$,*

(2.44.2) *$\varphi(\cdot, X)$ is measurable, for every $X \in E^k$,*

(2.44.3) *φ is Lipschitzian w.r.t. $X \in E^k$, i.e. there exists $L > 0$ such that, for almost all $t \in [0, a]$ and every $X_1, X_2 \in E^k$, we have*

$$d_H(\varphi(t, X_1), \varphi(t, X_2)) \leq L\|X_1 - X_2\|.$$

Then the set of solutions $S(\varphi, x_0, \ldots, x_{k-1})$ to problem (2.43) is an AR-space.

PROOF. Define (single-valued) mappings: $h_j : M([0, a]), E) \to AC^j$, $j = 0, \ldots, k - 1$, by putting

$$(h_j(z))(t) := x_0 + tx_1 + \ldots + (t^j/j!)x_j + \int_0^t \int_0^{s_1} \ldots \int_0^{s_j} z(s) \, ds \, ds_j \ldots ds_1,$$

where $M([0, a], E)$ denotes the Banach space of $C^{(j)}$-continuous mappings with essentially bounded $(j + 1)$th derivatives and

$$AC^j = \{u \in C^{(j)}([0, a], E) \mid u^{(j)} \text{ is absolutely continuous}\}.$$

For $u \in AC^j$, we put

$$\|u\| := \|u\|_{C^j} + \sup_{t \in [0, a]} \text{ess}\{\|u^{(j+1)}(t)\|\}.$$

Consider a multivalued mapping $\psi : M([0, a], E) \multimap M([0, a], E)$ defined as follows:

$$\psi(x) := \{z \in M([0, a], E) \mid z(t) \in \varphi(t, h_{k-1}(x)(t), \ldots, h_0(x)(t)),$$

$$\text{for a.a. } t \in [0, a]\}.$$

It follows from the well-known Kuratowski-Ryll-Nardzewski Selection Theorem (see (3.49) in Chapter I.3 or [KRN]) and (2.44.1) that ψ is well-defined (with closed decomposable values in $M([0, a], E)$), Moreover, it can be easily seen that

$h_{k-1}(\text{Fix}(\psi)) = S(\varphi, x_0, \dots, x_{k-1})$. Therefore, since h_{k-1} is a homeomorphism onto its image, according to Theorem (3.73) in Chapter I.3, it is sufficient to show that ψ is a contraction. We shall make it by applying the $M([0,a], E)$-version of Bielecki's method and the mentioned Kuratowski–Ryll–Nardzewski theorem.

In fact, it is enough to show that, for every $u, z \in M([0,a], E)$ and for every $y \in \psi(u)$, there is a $v \in \psi(z)$ such that

$$(2.45) \qquad \|y - v\|_1 \le \alpha \|u - z\|_1,$$

where $\alpha \in [0,1)$ and $\|w\|_1 = \sup \text{ess}_{t \in [0,a]} \{e^{-\text{Lakt}} \|w(t)\|\}$ is the Bielecki norm in $M([0,a], E)$.

Applying the Kuratowski–Ryll–Nardzewski Selection Theorem (3.49) in Chapter I.3 to ψ and z, we get a desired mapping $v \in \psi(z)$. Now, (2.45) follows directly from (2.44.3). This completes the proof. $\qquad \square$

Imposing more restrictions on φ, we can obtain a better information about the set $S(\varphi, x_0, \dots, x_{k-1})$, namely

(2.46) THEOREM. *Let $\varphi : [0,a] \times E^k \multimap E$ be a mapping with nonempty, convex, compact values such that (2.44.2) holds jointly with:*

(2.46.1) *$\varphi(t, \cdot)$ is completely continuous for a.a. $t \in [0,a]$, i.e. u.s.c. and maps bounded sets into compact sets,*

(2.46.2) *$\varphi(A)$ is compact, for every compact $A \subset [0,a] \times E^k$,*

(2.46.3) *$\|y\| \le \mu(t) + \nu(t)\|x\|$ holds, for a.a. $t \in [0,a]$ and all $x \in E^k$, $y \in \varphi(t,x)$, where $\mu, \nu : [0,a] \to [0,\infty)$ are suitable integrable functions.*

Then $S(\varphi, x_0, \dots, x_{k-1})$ is an R_δ-set.

Proof can be found in [Go5-M, pp. 357–358].

Now, we will consider again (cf. (2.34), (2.36)) the Cauchy problems for retarded functional differential inclusions. Hence, consider the system

$$(2.47) \qquad \dot{x} \in F(t, x_t), \quad x \in \mathbb{R}^n,$$

where $x_t(\cdot) = x(t+\cdot)$, for $t \in [0, \tau]$, denotes as usual a function from $[-\delta, 0]$, $\delta \ge 0$, into $\mathbb{R}^n$ and $F : [0, \tau] \times \mathcal{C} \multimap \mathbb{R}^n$, where $\mathcal{C} = \text{AC}([-\delta, 0], \mathbb{R}^n)$, is a u-Carathéodory multivalued function, i.e.

(C1$_f$) the set of values of $F(t, y)$ is nonempty, compact and convex, for all $(t, y) \in [0, \tau] \times \mathcal{C}$,

(C2$_f$) $F(t, \cdot)$ is u.s.c., for a.a. $t \in [0, \tau]$,

(C3$_f$) $F(\cdot, y)$ is measurable, for all $y \in \mathcal{C}$, i.e. for any closed $U \subset [0, \tau]$ and every $y \in \mathcal{C}$, the set $\{t \in [0, \tau] \mid F(\cdot, y) \cap U \ne \emptyset\}$ is measurable,

(C4$_f$) $|F(t, y)| \le \alpha + \beta \|y\|$, for every $y \in \mathcal{C}$ and a.a. $t \in [0, \tau]$, where α, β are suitable nonnegative constants.

For a nonempty, compact and convex set $K \subset \mathbb{R}^n$, the constraint, denote $\mathcal{K} = \{\xi \in \mathcal{C} \mid \xi(t) \in K, \text{ for } t \in [-\delta, 0]\}$ and assume that the Nagumo-type condition holds, namely

$$(2.48) \qquad F(t, y) \cap T_K(y(0)) \neq \emptyset, \quad \text{for all } (t, y) \in [0, \tau] \times \mathcal{K},$$

where $T_K(y(0)) = \{y \in \mathbb{R}^n \mid \liminf_{h \to 0+} d(y(0) + hy, K)/h = 0\}$ is the tangent cone (in the sense of Bouligand). Observe that $(C4_f)$ can be so reduced to

$(C4_f')$ sup ess$_{(t,y) \in [0,\tau] \times \mathcal{K}} |F(t, y)| < \infty.$

Then, for every $x_* \in \mathcal{K}$, there exists at least one Carathéodory solution $x(t, x_*)$ of (2.47) such that $x(t, x_*) = x_*$, for $t \in [-\delta, 0]$ and $x(t, x_*) \in K$, for $t \in [0, \tau]$. Let us recall that by a (Carathéodory) *solution* of (the initial problem to) (2.47), we mean as usual an absolutely continuous function $x(t) \in \text{AC}([-\delta, \tau], \mathbb{R})$ (with $x(t) = x_*, t \in [-\delta, 0]$), satisfying (2.47), for a.a. $t \in [-\delta, \tau]$.

The following result has been proved in [HP3].

(2.49) THEOREM ([HP3]). *Let the assumption* $(C1_f)$–$(C3_f)$, $(C4_f')$ *be satisfied jointly with* (2.48). *Then system* (2.47) *has an* R_δ-set of solutions $x(t, x_*)$ *such that* $x(t, x_*) = x_*$, *for* $t \in [-\delta, 0]$ *and* $x(t, x_*) \in K$, *for* $t \in [0, \tau]$.

(2.50) REMARK. For $\delta = 0$, Theorem (2.49) relates obviously to an ordinary system $\dot{x} \in F(t, x)$, $x \in \mathbb{R}^n$.

Consider finally the functional system

$$(2.51) \qquad \dot{x} + Ax \in F(t, x_t), \quad x \in \mathcal{B},$$

where $x_t(\cdot) = x(t + \cdot)$, for $t \in [0, \tau]$, denotes as above the mapping from $[-\delta, 0]$, $\delta \geq 0$, but this time into a real separable Banach space $\mathcal{B}$. Let, furthermore, the following assumptions be satisfied:

   (i) A is a closed, linear (not necessarily bounded) operator in $\mathcal{B}$, generating an analytic semigroup e^{At},
  (ii) the set of $F(t, y) : [0, \tau] \times \mathcal{C} \multimap \mathcal{B}$, where $\mathcal{C} = C([-\delta, 0], \mathcal{B})$ and $\delta \geq 0$, is nonempty, compact and convex, for all $(t, y) \in [0, \tau] \times \mathcal{C}$,
 (iii) $F(t, \cdot)$ is u.s.c., for a.a. $t \in [0, \tau]$,
  (iv) $F(\cdot, y)$ is measurable, for all $y \in \mathcal{C}$, i.e. for any open $U \subset \mathbb{R}^n$ and every $y \in \mathcal{C}$, the set $\{t \in [0, \tau] \mid F(\cdot, y) \cap U \neq \emptyset\}$ is measurable,
   (v) (cf. [COZ]) for every nonempty, bounded, equicontinuous set $D \subset \mathcal{C}$, we have

$$\gamma(F(t, D)) \leq g(t, \xi(D)), \quad \text{for a.a. } t \in [0, \tau],$$

where $\xi(D) \in C([-\delta, 0] \times [0, \infty))$, $\xi(D)(\emptyset) = \gamma(D(\theta))$, γ denotes the Hausdorff measure of noncompactness, and $g : [0, \tau] \times C([-\delta, 0] \times [0, \infty)) \to [0, \infty)$ is a Caratéodory-type function such that

(a) $g(t, \cdot)$ is nondecreasing, for a.a. $t \in [0, \tau]$, in the sense that if $\varphi, \psi \in C([-\delta, 0] \times [0, \infty))$ satisfy $\varphi(\theta) < \psi(\theta)$, for every $\theta \in [-\delta, 0]$, then $g(t, \varphi) < g(t, \psi)$,

(b) $|g(t, \varphi) - g(t, \psi)| < k(t) \|\varphi - \psi\|_1$, for a.a. $t \in [0, \tau]$ and for all $\varphi, \psi \in C([-\delta, 0] \times [0, \infty))$, where k is a Lebesgue measurable function and $\| \cdot \|_1$ denotes the norm in the space $C([-\delta, 0] \times [0, \infty))$,

(c) $g(t, 0) = 0$, for a.a. $t \in [0, \tau]$,

(vi) $\|F(t, y)\| \leq \alpha + \beta \|y\|_0$, for every $y \in \mathcal{C}$ and a.a. $t \in [0, \tau]$, where α, β are suitable positive constants and $\| \cdot \|_0$ denotes the norm in $\mathcal{C}$.

By a solution $x(t)$ of (the initial problem to) (2.51) we mean this time a *mild solution*, namely $x(t) \in C([-\delta, \tau], \mathcal{B})$ such that

$$x(t) = e^{At} x(0) + \int_0^t e^{A(t-s)} f(s) \, ds, \quad \text{for } t \in [0, \tau],$$

(with $x(t) = x_*$, for $t \in [-\delta, 0]$), where f is an (existing) measurable selection of $F(s, x_s(t))$, $t \in [-\delta, 0]$; such solutions exist on $[-\delta, \tau], \delta \geq 0$.

Hence, we can give the result presented is a slighty more general form in [COZ].

(2.52) THEOREM ([COZ]). *Under the assumptions* (i)–(vi), *the inclusion* (2.51) *has an R_δ-set of (mild) solutions* $x(t, x_*)$ *such that* $x(t, x_*) = x_* \in E$, *for* $t \in [-\delta, 0]$, *where E consists of equicontinuous functions.*

(2.53) REMARK. It can be seen in the proofs in [HP3] and [COZ] that system (2.47), satisfying $(C1_f)$–$(C4_f)$, has also an R_δ-set of solutions $x(t) \in AC([-\delta, \tau], \mathbb{R}^n)$ such that $x(t) = x_*$, $t \in [-\delta, 0]$.

The inverse system approach described above allows us to study the topological structure of solution sets of these differential problems on noncompact intervals. To illustrate it, consider the system (2.47).

Let $F : [0, \infty) \times \mathcal{K} \multimap \mathbb{R}^n$, where $\mathcal{K} = C([-\tau, 0], \mathbb{R}^n)$, $\tau > 0$, fulfil the following conditions:

(i) the values of F are nonempty, compact and convex, for all $(t, y) \in [0, \infty) \times \mathcal{K}$,

(ii) $F(t, \cdot)$ is u.s.c., for a.a. $t \in [0, \infty)$,

(iii) $F(\cdot, y)$ is measurable, for all $y \in \mathcal{K}$,

with

$$|F(t, y)| < \mu(t)(\|y\| + 1),$$

for every $y \in \mathcal{K}$ and a.a. $t \in [-0, \infty)$, where $\mu : [0, \infty) \to [0, \infty)$ is a suitable locally Lebesgue-integrable function and $\| \cdot \| = \max_{t \in [-\tau, 0]} | \cdot |$.

Consider the related Cauchy problem

$$(2.54) \qquad \begin{cases} \dot{x}(t) \in F(t, x_t), & \text{for a.a. } t \in [0, \infty), \\ x(t) = \varphi(t), & \text{for } t \in [-\tau, 0], \end{cases}$$

where $x_t(\cdot) \in C([-\tau, 0], \mathbb{R}^n)$ is defined by $x_t(\cdot) = x(t + \cdot)$.

Besides (2.54), consider still the family of Cauchy problems

$$(2.55) \qquad \begin{cases} \dot{x}(t) \in F(t, x_t), & \text{for a.a. } t \in [0, m], \\ x(t) = \varphi(t), & \text{for } t \in [-\tau, 0], \end{cases}$$

where $m \geq 1$. If follows from Remark (2.53) that, for every $m \geq 1$, the solution set $\mathcal{S}_m$ to (2.55) is a (nonempty, compact) R_δ-set in $C([-\tau, m], \mathbb{R}^n)$. Moreover, $\mathcal{S}_m$ is a fixed point set of the map

$$(2.56) \qquad \Psi_m : C([-\tau, m], \mathbb{R}^n) \multimap C([-\tau, m], \mathbb{R}^n),$$

where $\Psi_m(x)(t) = \varphi(t)$, for $t \in [-\tau, 0]$, and

$$\Psi_m(x)(t) = \left\{ \varphi(0) + \int_0^t u(s) \, ds \,\middle|\, u \in L^1([0, m], \mathbb{R}^n) \right.$$
$$\left. \text{and } u(s) \in F(s, x_s), \quad \text{for a.a. } s \in [0, m] \right\},$$

for $t \geq 0$.

One can readily check that $\{\Psi_m\}$ is a map of the inverse system

$$\{C([-\tau, m], \mathbb{R}^n), \pi_m^p, \mathbb{N}\},$$

where $\pi_m^p = x|_{[-\tau, m]}$, for very $x \in C([-\tau, p], \mathbb{R}^n)$. It induces the limit map on $C([-\tau, \infty), \mathbb{R}^n)$, namely $\Psi_m(x)(t) = \varphi(t)$, for $t \in [-\tau, 0]$, and

$$\Psi_m(x)(t) = \left\{ \psi(0) + \int_0^t u(s) \, ds \,\middle|\, u \in L^1_{\text{loc}}([0, \infty), \mathbb{R}^n) \right.$$
$$\left. \text{and } u(s) \in F(s, x_s), \quad \text{for a.a. } s \geq 0 \right\},$$

for $t \geq 0$, with the fixed point set $\mathcal{S}(y)$, the solution set to problem (2.54).

Hence a slightly modified version (see Proposition (2.1)(vi)) of Theorem (2.8) in Chapter II.2 applies, by which $\mathcal{S}(y)$ is (compact) R_δ, as required.

Thus, we can conclude by

(2.57) THEOREM. *Under the assumptions* (i)–(iii), *problem* (2.54) *has an* R_δ-*set of solutions.*

(2.58) REMARK. In the particular case (i.e. without a functional dependence in (2.54)), Theorem (2.57) reduces to Theorem (2.12), obtained by means of the Scorza–Dragoni type technique.

The last information concerns the covering (topological) dimension of solution sets to the Cauchy problem

$$(2.59) \qquad \begin{cases} \dot{x} \in F(t, x), \\ x(t_0) = x_0. \end{cases}$$

Let Ω be an open set in $\mathbb{R}^{n+1}$ such that $[t_0, t_0 + h] \times B(x_0, r) \subset \Omega$, where B denotes the closed ball centered at x_0 and with the radius r. Assume that $F : \Omega \multimap \mathbb{R}^n$ satisfies the following conditions:

(C1) the set of values of F is nonempty, compact and convex, for all $(t, x) \in \Omega$,

(C2) $F(t, \cdot) : B(x_0, r) \multimap \mathbb{R}^n$ is continuous, for a.a. $t \in [t_0, t_0 + h]$,

(C3) $F(\cdot, x) : [t_0, t_0 + h] \to \mathbb{R}^n$ is measurable, for all $x \in B(x_0, r)$,

(C4) there exist Lebesgue-integrable nonnegative functions $\alpha, \beta : [t_0, t_0 + h] \to [0, \infty)$ such that, for any $x \in B(x_0, r)$,

$$|F(t, x)| \le \alpha(t) + \beta(t)|x|, \quad \text{for a.a. } t \in [t_0, t_0 + h],$$

where $|F(t, x)| \le \sup\{|y| \mid y \in F(t, x)\}$.

Denote by $S([t_0, t_0 + d], x_0)$ the set of solutions $x \in \mathrm{AC}([t_0, t_0 + d], \mathbb{R}^n)$ of (2.59) on the interval $[t_0, t_0 + d]$, $0 < d \le h$.

The following two theorems are due to B. D. Gel'man [Ge7] (cf. Chapter II.3).

(2.60) THEOREM. *Let the assumption* (C1)–(C4) *be satisfied. Assume that the set*

$$A = \{t \in [t_0, t_0 + h] \mid \dim(F(t, x)) \ge 1, \quad \text{for any } x \in B(x_0, r)\}$$

is measurable and

$$\lim_{h \to 0} \frac{\mu(A \cap [t_0, t_0 + h])}{h} > 0,$$

where $\dim(\cdot)$ *denotes the covering dimension and* $\mu(\cdot)$ *stands for the Lebesgue measure. Then there exists a number* d_0 *such that, for any* $0 < d \le d_0$, *we have* $S = S([t_0, t_0 + d], x_0) \ne \{\emptyset\}$ *and* $\dim(S) = \infty$.

(2.61) THEOREM. *Let the assumptions of Theorem* (2.60) *be satisfied jointly with*

(C2') $F(t, \cdot) : B(x_0, r) \multimap \mathbb{R}^n$ *is Lipschitz-continuous.*

Then there exists a number d_0 *such that, for any* $0 < d \le d_0$, *any* $\varepsilon > 0$ *and any solution* $x \in S([t_0, t_0 + d], x_0)(\ne \{\emptyset\})$, *we have* $\dim(S_{x,\varepsilon}) = \infty$, *where* $S_{x,\varepsilon} = \{y \in S \mid \|x - y\| \le \varepsilon\}$.

III.3. Topological structure of solution sets: boundary value problems

In this part, the topological structure of solution sets to boundary value problems will be investigated. The results are mainly based on papers [And10], [AGG2], [AGG3], [ADG], [BP], [Mik], [Se3].

At first, we however study the reverse Cauchy problem when, instead of the origin, the value of solutions is prescribed at infinity, namely

$$(3.1) \quad \begin{cases} \dot{x}(t) \in F(t, x(t)), & \text{for a.a. } t \in [0, \infty), \\ \lim_{t \to \infty} x(t) = x_\infty \in \mathbb{R}^n, \end{cases}$$

where $F : [0, \infty) \times \mathbb{R}^n \multimap \mathbb{R}^n$ is a u-Carathéodory map, i.e.

(i) values of F are non-empty, compact and convex, for all $(t, x) \in [0, \infty) \times \mathbb{R}^n$,

(ii) $F(t, \cdot)$ is upper semicontinuous, for a.a. $t \in [0, \infty)$,

(iii) $F(\cdot, x)$ is measurable, for all $x \in \mathbb{R}^n$.

We will prove acyclicity of the solution set of problem (3.1).

Recalling that any contractible set is acyclic, we start with

(3.2) THEOREM. *Consider the target problem* (3.1), *where* $F : [0, \infty) \times \mathbb{R}^n \multimap \mathbb{R}^n$ *is a u-Carathéodory map and* $x_\infty \in \mathbb{R}^n$ *is arbitrary. Assume that there exists a globally integrable function* $\nu : [0, \infty) \to [0, \infty)$, *where* $\int_0^\infty \nu(t) \, dt = E < 1$, *such that*

$$(3.3) \quad d_H(F(t, x), F(t, y)) \leq \nu(t)|x - y|, \quad \text{for all } t \in [0, \infty) \text{ and } x, y \in \mathbb{R}^n.$$

Moreover, assume that $d_H(F(\cdot, 0), 0)$ *can be absolutely estimated by some globally integrable function. If* E *is a sufficiently small constant, then the set of solutions to problem* (3.1) *is compact and acyclic, for every* $x_\infty \subset \mathbb{R}^n$.

PROOF. Observe that condition (3.3) implies the existence a globally integrable function $\alpha : [0, \infty) \to [0, \infty)$ and a positive constant B such that

$$(3.4) \quad |F(t, x)| \leq \alpha(t)(B + |x|), \quad \text{for every } x \in \mathbb{R}^n \text{ and a.a. } t \in [0, \infty),$$

where $|F(t, x)| = \sup\{|y| \mid y \in F(t, x)\}$. Thus, problem (3.1) can be equivalently replaced by the problem

$$(3.5) \quad \begin{cases} \dot{x}(t) \in G(t, x(t)), & \text{for a.a. } t \in [0, \infty), \\ \lim_{t \to \infty} x(t) = x_\infty \in \mathbb{R}^n, \end{cases}$$

where G is a suitable u-Carathéodory map which can be estimated by a sufficiently large positive constant M, i.e.

$$|G(t, x)| \leq M, \quad \text{for every } x \in \mathbb{R}^n \text{ and a.a. } t \in [0, \infty),$$

and which satisfies condition (3.3) as well. In other words, the solution set $\mathcal{S}$ for problem (3.1) is the same as for problem (3.5), where

$$\mathcal{S} = \{x \in C([0, \infty), \mathbb{R}^n) \mid \dot{x}(t) \in F(t, x(t)), \quad \text{for a.a. } t \in [0, \infty) \text{ and } x(\infty) = x_\infty\}.$$

For the structure of $\mathcal{S}$, we will modify an approach from the begining of the foregoing Chapter III.2. Observe that, under the above assumptions, F as well G are well-known (see, Proposition (3.51) in Chapter I.3 or [ADTZ-M]) to be product-measurable, and subsequently having a Carathéodory selection $g \subset G$ which is Lipschitzian with a not necessarily same, but again sufficiently small constant (see, e.g., [HP1-M, pp. 101–103]). By the sufficiency we mean that, besides others,

$$|g(t, x) - g(t, y)| \leq \gamma(t)|x - y|$$

holds, for all $x, y \in \mathbb{R}^n$ and a.a. $t \in [0, \infty)$, with a Lebesgue integrable function $\gamma : [0, \infty) \to [0, \infty)$ such that $\int_0^\infty \gamma(t)\, dt < 1$.

Considering the single-valued problem $(g \subset G)$

$$(3.6) \qquad \begin{cases} \dot{x}(t) \in g(t, x(t)), & \text{for a.a. } t \in [0, \infty), \\ \lim_{t \to \infty} x(t) = x_\infty, \end{cases}$$

we can easily prove the existence of a unique solution $\overline{x}(t)$ of problem (3.6). The uniqueness can be verified in a standard manner by the contradiction, when assuming the existence of another solution $\overline{y}(t)$ of that problem, because so we would arrive at the false inequality

$$\sup_{t \in [0, \infty)} |\overline{x}(t) - \overline{y}(t)| = \sup_{t \in [0, \infty)} \left| \int_\infty^t g(s, \overline{x}(s))\, ds - \int_\infty^t g(s, \overline{y}(s))\, ds \right|$$

$$\leq \int_0^\infty |g(s, \overline{x}(s)) - g(s, \overline{y}(s))|\, dt$$

$$\leq \int_0^\infty \gamma(t) \sup_{t \in [0, \infty)} |\overline{x}(t) - \overline{y}(t)|\, dt$$

$$\leq \sup_{t \in [0, \infty)} |\overline{x}(t) - \overline{y}(t)| \int_0^\infty \gamma(t)\, dt$$

$$< \sup_{t \in [0, \infty)} |\overline{x}(t) - \overline{y}(t)|.$$

Hence, according to the definition of contractibility in Chapter I.4, it is sufficient to show that the solution set $\mathcal{S}$ of problem (3.5) is homotopic to a unique solution $\overline{x}(t)$ of problem (3.6), which is at the same time a solution of problem (3.5) as well. The desired homotopy reads ($\lambda \in [0, 1]$)

$$h(x, \lambda)(t) = \begin{cases} x(t), & \text{for } t \geq 1/\lambda - \lambda,\ \lambda \neq 0, \\ \overline{z}(t), & \text{for } 0 < t \leq 1/\lambda - \lambda,\ \lambda \neq 0, \\ \overline{x}(t), & \text{for } \lambda = 0, \end{cases}$$

where $\bar{z}$ is a unique solution to the reverse *Cauchy problem*

$$\begin{cases} \dot{x}(t) = g(t, z(t)), & \text{for a.a. } t \in [0, 1/\lambda - \lambda], \\ z(1/\lambda - \lambda) = x(1/\lambda - \lambda), \end{cases}$$

for each $\lambda \in [0, 1]$. Then h is a continuous homotopy such that $h(x, 0) = \bar{x}$, $h(x, 1) = x$, as required, and subsequently, the set S is acyclic. Using the convexity assumption on values of F, we can prove by the standard manner (Mazur's Theorem (1.33) in Chapter I.1) that S is closed in $C([0, \infty), \mathbb{R}^n)$. By Arzelà–Ascoli's lemma, this set is compact, and the proof is complete. $\square$

Now, Consider the problem

(3.7) $$\begin{cases} \dot{x}(t) + A(t)x(\alpha(t)) \in F(t, x(\beta(t))), & \text{for a.a. } t \in [0, T], \\ Lx = r, \end{cases}$$

where

(3.7.1) $A : [0, T] \to \mathcal{L}(\mathbb{R}^n, \mathbb{R}^n)$ is a measurable linear operator such that $|A(t)| \leq \gamma(t)$, for all $t \in [0, T]$ and some integrable function $\gamma : [0, T] \to [0, \infty)$,

(3.7.2) the associated homogeneous problem

$$\begin{cases} \dot{x}(t) + A(t)x(\alpha(t)) = 0, & \text{for a.a. } t \in [0, T], \ \alpha \in C([0, T], [0, T]), \\ Lx = 0 \end{cases}$$

has only the trivial solution and there exists $t_0 \in [0, T]$ such that

$$|\alpha(t) - t_0| \leq |t - t_0|, \quad \text{for } t \in [0, T],$$

(3.7.3) $F : [0, T] \times \mathbb{R}^n \multimap \mathbb{R}^n$ has nonempty, compact, convex values and $\beta \in C([0, T], [0, T])$,

(3.7.4) $F(\cdot, x)$ is measurable, for every $x \in \mathbb{R}^n$,

(3.7.5) there is a constant $M \geq 0$ such that

$$d_H(F(t, x), F(t, y)) \leq M|x - y|, \quad \text{for all } x, y \in \mathbb{R}^n \text{ and a.a. } t \in [0, T],$$

where d_H stands for the Hausdorff metric,

(3.7.6) there are two nonnegative Lebesgue-integrable functions $\delta_1, \delta_2 : [0, T] \to [0, \infty)$ such that,

$$|F(t, x)| \leq \delta_1(t) + \delta_2(t)|x|, \quad \text{for a.a. } t \in [0, T] \text{ and all } x \in \mathbb{R}^n,$$

where $|F(t, x)| = \sup\{|y| \mid y \in F(t, x)\}$.

In [ADG], the authors have proved the following theorem.

(3.8) THEOREM ([ADG]). *Under the assumptions* (3.7.1)–(3.7.6), *a certain "critical" value* λ *exists such that if* $M < \lambda$, *then the set of solutions of* (3.7) *is a (nonempty) compact AR-space. Moreover, if the Lebesgue measure of the set* $\{t \mid \dim F(t,x) < 1,\ for\ some\ x \in \mathbb{R}\}$ *is still zero, then the set of solutions of* (3.7) *is an infinite dimensional compact AR-space, where* $\dim X$ *denotes the covering (topological) dimension of a space* X.

(3.9) REMARK. Observe that for $A \equiv 0$ and $Lx = x(0)$, the related Cauchy problem has, under the assumptions of Theorem (3.8), infinitely many linearly independent solutions on the whole interval $[0, \tau]$ which is in a good accordance with the results in [DzGe], [Ge7].

The first assertion of Theorem (3.8) can be still improved. For the sake of simplicity, consider only the ordinary case in (3.7), namely

$$(3.10) \qquad \begin{cases} \dot{x}(t) + A(t)x(t) \in \alpha F(t, x(t)), & \text{for a.a. } t \in [0, T], \\ Lx = \theta, \end{cases}$$

and assume that the following conditions are satisfied:

(A) $F : [0, T] \times \mathbb{R}^n \multimap \mathbb{R}^n$ has nonempty, compact and convex values,

(B) $F(\cdot, x)$ is measurable, for every $x \in \mathbb{R}^n$,

(C) there is a constant $M \geq 0$ such that

$$d_H(F(t, x), F(t, y)) \leq M|x - y|, \quad \text{for all } x, y \in \mathbb{R}^n \text{ and a.a. } t \in [0, T],$$

where d_H again stands for the Hausdorff metric,

(D) there is an integrable function $\beta : [0, T] \to [0, \infty)$ such that

$$|F(t, x)| \leq \beta(t), \quad \text{for every } (t, x) \in [0, T] \times \mathbb{R}^n,$$

(E) $A : [0, T] \to \mathcal{L}(\mathbb{R}^n, \mathbb{R}^n)$ is a measurable function such that $|A(t)| \leq \gamma(t)$, for all $t \in [0, T]$ and some integrable function $\gamma : [0, T] \to [0, \infty)$,

(F) the associated homogenous problem

$$(3.11) \qquad \begin{cases} \dot{x}(t) + A(t)x(t) = 0, & \text{for a.a. } t \in [0, T], \\ Lx = 0 \end{cases}$$

has only the trivial solution.

In [ADG], the authors have proved that all solutions to problem (3.10) can be obtained as fixed points of the multivalued operator $\phi_\alpha : E \multimap E$, for $E = C([0, T], \mathbb{R}^n)$ with the Bielecki's norm

$$\|x\| = \sup_{t \in [0, T]} \left(|x(t)| \exp\left(-K \int_0^t \gamma(s)\, ds \right) \right),$$

(3.12) $$\phi_\alpha(x) = \overline{\mathcal{K}}T_1\mathcal{P}_F(x),$$

where

$$\mathcal{P}_F(x) = \{y \in L^1([O,T], \mathbb{R}^n) \mid y(t) \in F(t, x(t)) \quad \text{for a.a. } t \in [0, T]\},$$

$$T_1(y)(t) = \int_0^t y(s)\, ds$$

and $\overline{\mathcal{K}}$ is an affine operator in the space E of the form $\overline{\mathcal{K}}(x) = \mathcal{K}_1(\theta) + \mathcal{K}(x)$, where $\mathcal{K}_1 : \mathbb{R}^n \to E$ and $\mathcal{K} : E \to E$ are linear continuous.

According to (3.8), there is a certain critical number $\lambda > 0$ (namely, $\lambda = (\|\mathcal{K}\|MT)^{-1}$) such that, for every $0 \le \alpha < \lambda$, the solution set to problem (3.10) is a compact absolute retract as the fixed point set of a convex compact-valued contraction. One can also check that the constant α does not depend on the form of the right-hand side F, but only on the Lipschitz constant M and on the norms of A and the linear operator L.

Now, we are interested in a topological structure of the solution set to problem (3.10) for this critical value λ, i.e. for $\alpha = \lambda$, because then the associated operators can be no more contractions as above, but only nonexpansive. In the single-valued case, λ can be often expressed explicitly.

(3.13) THEOREM. *Under the asumptions* (A)–(F), *a certain "critical" value λ exists such that if $\alpha \le \lambda$, then the set of solutions of* (3.10) *is nonempty, compact and acyclic.*

PROOF. At first, by Arzelà–Ascoli's lemma and assumption (D), it follows that the operator $T_1\mathcal{P}_F$ is compact with the image in some closed ball $\widetilde{B}_0 \subset E$. Take the ball $\widetilde{B} = \widetilde{B}_0(1/\lambda + 2)$. Denote its radius by r (r should be greater than $(1/\lambda+2) \int_0^T \beta(t)\, dt$). Since $\overline{\mathcal{K}}$ is affine, one obtains that the image of ϕ_λ is included in the ball B with the radius $R = \lambda(\|\mathcal{K}\|r + \|\mathcal{K}_1\|(|\theta| + 1))$. Furthermore, observe that, for each $\alpha < \lambda$, $\phi_\alpha(E)$ is also included in B. Therefore, all fixed points of ϕ_λ and ϕ_α, $\alpha < \lambda$ (which are solutions to (3.10)) are in the same bounded set B.

Define the compact operators Φ_k, $\Phi : B \multimap E$,

$$\Phi_k(x)(t) = x(0) + Lx - \theta - \int_0^t A(s)x(s)\, ds + \lambda_k \int_0^t F(s, x(s))\, ds$$

and

$$\Phi(x)(t) = x(0) + Lx - \theta - \int_0^t A(s)x(s)\, ds + \lambda \int_0^t F(s, x(s))\, ds,$$

where $\lambda_k = \lambda - (k\|\mathcal{K}\| \int_0^T \beta(s)\,ds)^{-1}$ (without any loss of generality, we may assume that $\|\mathcal{K}\| \geq 1$ and $\int_0^T \beta(t)\,dt \geq 1$) and the integrals are understood in the sense of R. J. Aumann. One can readily check that the fixed points of Φ are the solutions of our problem (3.10), for $\alpha = \lambda$. Associate with the above maps suitable compact fields $\varphi_k = j - \Phi_k$ and $\varphi = j - \Phi$ (here j denotes the inclusion map).

We show that φ_k and φ satisfy all the assumptions of a slightly modified version of Theorem (1.4) in Chapter II.1) (see also Remark (1.5) in Chapter II.1).

Let us note that φ_k and φ can be verified, as in the single-valued case, to be proper.

To check condition (i) of Theorem (1.4) in Chapter II.1, take an arbitrary point $z \in \varphi_k(x)$. Then

$$z(t) = x(t) - x(0) - Lx + \theta + \int_0^t A(s)x(s)\,ds - \lambda_k \int_0^t u(s)\,ds,$$

for some $u \in L^1([0,T], \mathbb{R}^n)$ such that $u(s) \in F(s, x(s))$, for a.a. $s \in [0,T]$.

Put

$$y(t) = x(t) - x(0) - Lx + \theta + \int_0^t A(s)x(s)\,ds - \lambda \int_0^t u(s)\,ds.$$

Notice that $y \in \varphi(x)$ and

$$|z(t) - y(t)| \leq (\lambda - \lambda_k) \int_0^t |u(s)|\,ds$$

$$\leq \left(k\|\mathcal{K}\| \int_0^t \beta(s)\,ds \right)^{-1} \int_0^T \beta(s)\,ds \leq \frac{1}{k}.$$

Thus, $\|z - y\| \leq 1/k$, which implies that $\varphi_k(x) \subset N_{1/k}(\varphi(x))$, and assumption (i) is satisfied.

To check condition (ii) of Theorem (1.4) in Chapter II.1 (see also Remark (1.5) in Chapter II.1), define two sets:

$$C = \{x \in E \mid x \text{ is absolutely continuous}\},$$

$$V_k = \left\{ x \in C \mid \|x\| \leq \frac{1}{k} \quad \text{and} \quad |\dot{x}(t)| \leq \frac{1}{k}\beta(t), \quad \text{for a.a. } t \in [0,T] \right\}.$$

It can be easily seen that V_k is closed in E and $V_k \subset \overline{U_k} \cap C$, where $U_k \subset E$ is the closed ball with the center in the origin and the radius $1/k$.

Assume that $0 \in \varphi(x)$, for some $x \in B$. We show that $\varphi_k(x) \cap V_k \neq \emptyset$.

Since $x \in \Phi(x)$, one can see that $x \in C$. More precisely, x is of the form

$$x(t) = x(0) + Lx - \theta - \int_0^t A(s)x(s)\,ds + \lambda \int_0^t u(s)\,ds,$$

where $u \in L^1([0,T], \mathbb{R}^n)$ is such that $u(s) \in F(s, x(s))$, for a.a. $s \in [0,T]$. Take $y \in C$,

$$y(t) = x(0) + Lx - \theta - \int_0^t A(s)x(s)\,ds + \lambda_k \int_0^t u(s)\,ds.$$

We can show (as above) that $\|x - y\| \le 1/k$ and

$$|\dot{x}(t) - \dot{y}(t)| \le \frac{1}{k}|u(t)| \le \frac{1}{k}\beta(t).$$

This implies that $x - y \in \varphi_k(x)$ and $x - y \in V_k$, as required.

Now, we show that assumption (iii) of Theorem (1.4) in Chapter II.1 (see also Remark (1.5) in Chapter II.1) is also satisfied. To this end, suppose that $v \in V_k$ and note that x is a solution of the inclusion $u \in \varphi_k(x)$ if and only if

$$u(t) \in x(t) - x(0) - Lx + \theta + \int_0^t A(s)x(s)\,ds - \lambda_k \in \int_0^t F(s, x(s))\,ds$$

or, equivalently,

$$\begin{cases} \dot{x}(t) + A(t)x(t) \in \dot{u}(t) + \lambda_k F(t, x(t)), & \text{for a.a. } t \in [0,T], \\ Lx = \theta - u(0). \end{cases}$$

Defining $G(t, x) = \dot{u}(t)/\lambda_k + F(t, x)$, we obtain that the above problem can be rewritten as follows:

(3.14)
$$\begin{cases} \dot{x}(t) + A(t)x(t) \in \lambda_k G(t, x(t)), & \text{for a.a. } t \in [0,T], \\ Lx = \theta - u(0). \end{cases}$$

Since G is measurable - Lipschitz with the same constant M, it follows, by means of Theorem (3.8), that the solution set S_k to problem (3.14) is a compact absolute retract, and so acyclic. Finally, we have to prove that $S_k \subset B$. To this aim, taking the integral operator used in the proof of Theorem (3.8) (see [ADG]) such that S_k is its fixed point set, we obtain that

$$\|x\| \le \lambda_k \|\mathcal{K}\| \cdot \|T_1 \mathcal{P}_G(x)\| + \lambda_k \|\mathcal{K}_1\| \cdot |\theta - u(0)|,$$

for each $x \in S_k$.

Observe that, if $z \in T_1 \mathcal{P}_G(x)$, then there is a map $v \in L^1([0,T], \mathbb{R}^n)$ such that $v(s) \in F(s, x(s))$, for a.a. $t \in [0,T]$, and $z(t) = \int_0^t (\dot{u}(s)/\lambda_k + v(s))\,ds$. Therefore,

$$|z(t)| = \left| \int_0^t \left(\frac{1}{\lambda_k}\dot{u}(s) + v(s) \right) ds \right| \le \frac{1}{\lambda_k} \int_0^T |\dot{u}(s)|\,ds + \int_0^T \beta(s)\,ds$$

$$\le \left(\frac{1}{\lambda_k} + 1 \right) \int_0^T \beta(s)\,ds \le \left(\frac{1}{\lambda} + 2 \right) \int_0^T \beta(s)\,ds,$$

for a sufficiently large k. This implies that $z \in \widetilde{B}$ (see the assumption imposed on the radius of $\widetilde{B}$). Hence,

$$
\begin{aligned}
\|x\| &\leq \lambda_k (\|\mathcal{K}_1\| \cdot |\theta - u(0)| + \|\mathcal{K}\| \cdot \|T_1 \mathcal{P}_Q(x)\|) \\
&\leq \lambda(\|\mathcal{K}_1\|(|\theta| + |u(0)|) + \|\mathcal{K}\| r) \\
&\leq \lambda(\|\mathcal{K}_1\|(|\theta| + 1) + \|\mathcal{K}\| r) = R,
\end{aligned}
$$

which implies that $x \in B$.

Thus, assumption (iii) of Theorem (1.4) in Chapter II.1 (see also Remark (1.5) in Chapter II.1) is satisfied as well. Finally, applying this theorem, we obtain that the solution set to problem (3.10), for $\alpha = \lambda$, which is equal to $\varphi_+^{-1}(0)$, is compact and acyclic, as claimed. $\qquad \square$

(3.15) REMARK. Condition (D) in Theorem (3.13) can be replaced by $|F(t,x)| \leq \psi(x)$, for a.a. $t \in [0, \tau]$ and every $x \in \mathbb{R}$, where $\psi : \mathbb{R}^n \to [0, \infty)$ is a continuous function such that $\lim_{|x| \to \infty} \frac{\psi(x)}{|x|} = 0$. In fact, for the first assertion of Theorem (3.8), instead of $M < \lambda$, only $M \leq \lambda$, can be assumed in order the set of solutions of (3.7) to be nonempty, compact and acyclic.

(3.16) REMARK. In the single-valued case, the solution set in Theorem (3.8) consists obviously of a unique solution. On the other hand, it is not so in Theorem (3.13).

Another interesting result was obtained in [DBP5] and then improved in [Mik] for the *Dirichlet problem*

$$
(3.17) \qquad \begin{cases} \ddot{x} \in F(t, x, \dot{x}), \\ x(0) = x(T) = 0, \end{cases}
$$

where $F : [0, T] \times \mathbb{R}^{2n} \to \mathbb{R}^n$ satisfies the following conditions:

(3.17.1) F has nonempty, compact and convex values,

(3.17.2) $F(\,\cdot\,, x, y)$ is measurable, for all $(x, y) \in \mathbb{R}^{2n}$,

(3.17.3) $d_H(F(t, 0, 0), \{0\}) \leq \Phi(t)$, for a.a. $t \in [0, T]$, where $\int_0^T \Phi^2(t)\, dt < \infty$,

(3.17.4) there exist nonnegative constants A, B such that

$$
d_H(F(t, x_1, y_1), F(t, x_2, y_2)) \leq A|x_1 - x_2| + B|y_1 - y_2|
$$

holds, for a.a. $t \in [0, T]$ and all $x_1, x_2, y_1, y_2 \in \mathbb{R}^n$.

(3.18) THEOREM ([Mik], [DBP5]). *Let the assumptions* (3.17.1)–(3.17.4) *be satisfied. If still*

$$
(3.18.1) \qquad 0 < T < \frac{2\sqrt{2}}{\sqrt{A+B}},
$$

$$
(3.18.2) \qquad T < \frac{2}{A+B},
$$

then the set of solutions of (3.17) *is a nonempty, compact* AR-*space. Moreover, if* $F(t, x, y)$ *is independent of* y, *then condition* (3.18.2) *is superfluous.*

(3.19) REMARK. In particular, conditions (3.18.1), (3.18.2) are satisfied, when $T = 1$ and $A + B < 1$, as in [DBP5]. In the single-valued case, the solution set in Theorem (3.18) consists only of a unique solution.

In view of Remarks (3.16) and (3.19), a nontrivial structure of solution sets to single-valued boundary problems can be seen as a delicate problem. The following results in this field are rather rare.

(3.20) THEOREM ([BP]). *Consider the Floquet problem*

$$(3.21) \qquad \begin{cases} \dot{x} = f(t, x), & \text{for a.a. } t \in [a, b], \\ x(a) + \lambda x(b) = \xi & (\lambda > 0, \, \xi \in \mathbb{R}^n), \end{cases}$$

where $f : [a, b] \times \mathbb{R}^n \to \mathbb{R}^n$ *is a bounded Carathéodory function. Assume, furthermore, that* f *satisfies*

$$(3.22) \qquad |f(t, x) - f(t, y)| \le p(t)|x - y|, \quad \text{for a.a. } t \in [a, b] \text{ and } x, y \in \mathbb{R}^n,$$

where $p : [a, b] \to [0, \infty)$ *is a Lebesgue-integrable function such that*

$$(3.23) \qquad \int_a^b p(t)\, dt \le \sqrt{\pi^2 + \ln^2 \lambda}.$$

Then the set of solution to (3.21) *is an* R_δ-*set.*

(3.24) THEOREM ([BP]). *Consider the Cauchy–Nicoletti problem*

$$(3.25) \qquad \begin{cases} \dot{x}_i = f_i(t, x_1, \dots, x_n), & \text{for a.a. } t \in [a, b], \, (i = 1, \dots, n), \\ x_i(t_i) = \xi_i & (\xi_i \in \mathbb{R}^n, \, t_i \in [a, b], \\ & \quad i = 1, \dots, n, \, \xi = (\xi_1, \dots, \xi_n)), \end{cases}$$

where $f = (f_1, \dots, f_n) : [a, b] \times \mathbb{R}^n \to \mathbb{R}^n$ *is a bounded Carathéodory function. Assume, furthermore, that* f *satisfies* (3.22), *where* $p : [a, b] \to [0, \infty)$ *is a Lebesgue-integrable function satisfying*

$$(3.26) \qquad \int_a^b p(t)\, dt \le \frac{\pi}{2}.$$

Then the set of solutions to (3.25) *is an* R_δ-*set.*

(3.27) REMARK. As pointed out in [BP], if the sharp inequalities take place in (3.23) or (3.26), then problem (3.21) or (3.25) has a unique solution, respectively.

On the other hand, for non-sharp inequalities (3.23) or (3.26), problem (3.21) or (3.25) can possess more solutions, respectively.

The following boundary problem was considered in [Se3]:

$$(3.28) \quad \begin{cases} x^{(n)} + \gamma_1(t)x^{(n-1)} + \ldots \\ \quad + \gamma_n(t)x + f(t, x, \ldots, x^{(m)}) = q(t), \quad \text{for a.a. } t \in [a, b], \\ l_i(x) = 0, \quad\quad\quad\quad\quad\quad\quad\quad\quad i = 1, \ldots, n, \end{cases}$$

where $0 \le m \le n - 1$, $\gamma_k, q \in C([a, b], \mathbb{R})$ $(1 \le k \le n)$, $f \in C([a, b] \times \mathbb{R}^{m+1}, \mathbb{R})$ and $l_i \in C^{(n-1)}([a, b], \mathbb{R})$ $(1 \le i \le n)$ are linearly independent linear continuous functionals.

Consider still the associated homogeneous problem

$$(3.29) \quad \begin{cases} x^{(n)} + \gamma_1(t)x^{(n-1)} + \ldots + \gamma_n(t)x = 0, \quad t \in [a, b], \\ l_i(x) = 0, \quad\quad\quad\quad\quad\quad\quad\quad\quad i = 1, \ldots, n, \end{cases}$$

We say that a number j is *admissible* to (3.29) if it is defined in the following way:

(1) $0 \le j \le n - 2$, whenever $\gamma_k(t) \equiv r_k(t)$ on $[a, b]$, for $k = 1, \ldots, n - j - 1$, and $\gamma_{n-j}(t) \not\equiv r_{n-j}(t)$,

(2) $j = 0$, whenever $\gamma_k(t) \equiv r_k(t)$ on $[a, b]$, for $k = 1, \ldots, n$,

(3) $j = n - 1$, whenever $\gamma_1(t) \not\equiv r_1(t)$ on $[a, b]$,

where $r_k \in C([a, b], \mathbb{R})$, $k = 1, \ldots, n$, are coefficients of the linear equation in the problem

$$\begin{cases} x^{(n)} + r_1(t)x^{(n-1)} + \ldots + r_n(t)x = 0, \quad t \in [a, b], \\ l_i(x) = 0, \quad\quad\quad\quad\quad\quad\quad\quad\quad i = 1, \ldots, n, \end{cases}$$

having only the trivial solution. Such coefficients always exist, as pointed out in [Se3] (see also the references therein).

(3.30) THEOREM ([Se3]). *Assume that the following conditions are satisfied:*

(3.30.1) $\lim_{k \to \infty} f_k(t, x_1, \ldots, x_{m+1}) = f(t, x_1, \ldots, x_{m+1})$, *uniformly on* $[a, b] \times M$, *for each bounded closed subset* $M \subset \mathbb{R}^{m+1}$,

(3.30.2) *the boundary value problems*

$$(3.31) \quad \begin{cases} x^{(n)} + \gamma_1(t)x^{(n-1)} + \ldots \\ \quad + \gamma_n(t)x + f_k(t, x, \ldots, x^{(m)}) = q(t), \quad t \in [a, b], \\ l_i(x) = 0, \quad\quad\quad\quad\quad\quad\quad\quad\quad i = 1, \ldots, n, \end{cases}$$

where $f_k \in C([a, b] \times \mathbb{R}^{m+1}, \mathbb{R})$ $(k = 1, \ldots, n)$, *have at most one solution, for every* $q \in C([a, b] \times \mathbb{R}^{m+1}, \mathbb{R})$,

(3.30.3) *for every bounded $S \subset C([a, b] \times \mathbb{R}^{m+1}, \mathbb{R})$, there is an $R > 0$ such that all possible solutions $x(t)$ of problems (3.31), $k = 1, \dots, n$, with $q \in S$ satisfy the inequality*

$$\max_{k=0,\dots,r} \{\|x^{(k)}(t)\|\} \leq R, \quad for\ r = \max(m, j),$$

where j is the minimal admissible number to (3.29).

Then, for each $q \in C([a, b] \times \mathbb{R}^{m+1}, \mathbb{R})$, the set of solutions of problem (3.28) is an R_δ-set.

Unlike in the above theorems, the following problems can be regarded as those with "limiting" boundary conditions. In [ChL], the following result has been proved (as Theorem (3.1)) for the boundary value problem

$$(3.32) \qquad \begin{cases} \dot{x} = f(t, x), \\ Lx = r, \end{cases}$$

where $f : I \times \mathbb{R}^n \to \mathbb{R}^n$ is a continuous function, $L : C^1(I, \mathbb{R}^n) \to \mathbb{R}^n$ is a linear operator and $I = [a, b]$ is a compact interval.

(3.33) PROPOSITION ([ChL]). *Let $f : I \times \mathbb{R}^n \to \mathbb{R}^n$ be a fixed continuous function such that, for every $t_0 \in I$ and $x_0 \in \mathbb{R}^n$, there exists a unique (smooth) solution $x(t)$ of the equation $\dot{x} = f(t, x)$, satysfying $x(t_0) = x_0$.*

Let $\mathcal{U}$ be an open (in the norm topology) subset of the Banach space of all continuous linear operators $L : C^0_1(I, \mathbb{R}^n) \to \mathbb{R}^n$, where C^0_1 denotes the set $C^1(\mathbb{R}^n) \subset C^0(I, \mathbb{R}^n)$, topologized by the induced topology of $C^0(I, \mathbb{R}^n)$ $(:= C(I, \mathbb{R}))$.

If, for every $L \in \mathcal{U}$ and $r \in \mathbb{R}^n$, the boundary value problem (3.32) has at most one solution, then, for every $L \in \mathcal{U}$ and $r \in \mathbb{R}^n$, problem (3.32) has exactly one solution.

Our aim is to prove, by means of a slightly modified version of Theorem (1.4) in Chapter II.1 (see also Remark (1.5) in Chapter II.1), the following theorem.

(3.34) THEOREM. *Let $f : I \times \mathbb{R}^n \to \mathbb{R}^n$ be a fixed continuous function and $p : I \to \mathbb{R}^n$ be a continuous function such that, for every $t_0 \in I$, $x_0 \in \mathbb{R}^n$ and $p \in C(I, \mathbb{R}^n)$ with $\|p\| \leq 1$, there exists a unique solution $x(t)$ of*

$$(3.35) \qquad \dot{x} = f(t, x) + p(t),$$

satisfying $x(t_0) = x_0$.

Let $\mathcal{U}$ be an open (in the norm topology) subset of the Banach space of all continuous linear operators $L : C^0_1(I, \mathbb{R}^n) \to \mathbb{R}^n$, where C^0_1 has the same meaning as in Proposition (3.33).

Assume that, for every $L \in \mathcal{U}$, $r \in \mathbb{R}^n$ and $p \in C^1(\mathbb{R}^n)$ with $\|p\| \leq 1$, the boundary value problem

$$(3.36) \qquad \begin{cases} \dot{x} = f(t, x) + p(t), \\ Lx = r, \end{cases}$$

has at most one solution and that, for every $L \in \overline{\mathcal{U}}$, all solutions of problem (3.32) are uniformly (i.e. independently of $L \in \overline{\mathcal{U}}$) a priori bounded, where $\overline{\mathcal{U}}$ denotes the closure of $\mathcal{U}$ in the C_0^1-topology.

Then, for every $L \in \partial \mathcal{U}$ and $r \in \mathbb{R}^n$, where $\partial \mathcal{U}$ denotes the boundary of $\mathcal{U}$ in the C_1^0-topology, problem (3.32) has an R_δ-set of solutions.

PROOF. Since all assumptions of Proposition (3.33) are satisfied, problem (3.36) is solvable, for every $L \in \mathcal{U}$, $r \in \mathbb{R}^n$ and $p \in C(I, \mathbb{R}^n)$ with $\|p\| \leq 1$.

Furthermore, since $\overline{\mathcal{U}}$ is a closed subset of the Banach space of all continuous linear operators, each element $\widetilde{L} \in \partial \mathcal{U}$ can be regarded as a uniform limit of a suitable sequence $\{L_k\}$ such that $\widetilde{L} = \lim_{k \to \infty} L_k$, where $L_k \in \mathcal{U}(= \operatorname{int} \mathcal{U})$, for every $k \in \mathbb{N}$.

Fix such an $\widetilde{L} \in \partial \mathcal{U}$ and consider the compact operators $\Phi_k, \Phi : \mathcal{B} \to C_1^0(I, \mathbb{R}^n)$:

$$\Phi_k(x)(t) = x(a) + L_k x - r + \int_a^t f(s, x(s)) \, ds$$

and

$$\Phi(x)(t) = x(0) + \widetilde{L}x - r + \int_a^t f(s, x(s)) \, ds,$$

where $\mathcal{B} \subset C^0(I, \mathbb{R}^n)$ is a suitable closed ball centered at the origin, which is implied by the assumption of a uniform a priori boundedness of solutions. The compactness of operators follows directly by means of the well-known Arzelà–Ascoli lemma.

One can readily check the one-to-one correspondence between the fixed points of Φ and the solutions of problem (3.32) as well as those of Φ_k and the solutions of the equation $\dot{x} = f(t, x(t))$, satisfying

$$L_k x = r.$$

Thus, one can associate to Φ_k and Φ the proper maps $\varphi_k = \operatorname{id} - \Phi_k$ and $\varphi = \operatorname{id} - \Phi$, respectively, where id denotes the identity, namely

$$\varphi_k(x)(t) = x(t) - x(a) - L_k x + r - \int_a^t f(s, x(s)) \, ds$$

and

$$\varphi(x)(t) = x(t) - x(a) - \tilde{L}x + r - \int_a^t f(s, x(s))\, ds.$$

So, the nonempty kernel $\varphi^{-1}(0)$ of φ corresponds to the fixed points of Φ, and subsequently to solutions of (3.32), i.e.

$$x \in \Phi(x) \Leftrightarrow 0 \in x - \Phi(x) = (\mathrm{id} - \Phi)(x) = \varphi(x).$$

We can assume without any loss of generality that, for a sufficiently large $k \in \mathbb{N}$, we have

$$(3.37) \qquad |\varphi_k(x)(t) - \varphi(x)(t)| = |L_k x - \tilde{L}x| = |(L_k - \tilde{L})x| \leq \frac{1}{k},$$

because, otherwise, we can obviously select a subsequence with this property.

Since $\|\varphi_k(x)(t) - \varphi(x)(t)\| \leq 1/k$ holds, for every $x \in \mathcal{B}$, condition (i) of Theorem (1.4) in Chapter II.1 is satisfied.

In order to prove (ii) in Theorem (1.4) in Chapter II.1 (see also Remark (1.5) in Chapter II.1), it is sufficient to verify the following inequalities

$$|\varphi_k(x)(t)| \leq \frac{1}{k} \quad \text{and} \quad |(\varphi_k(x))^{\cdot}(t)| \leq \frac{1}{k}, \quad k \in \mathbb{N},$$

for every x with $\varphi(x) = 0$.

However, since $(\varphi_k(x))^{\cdot}(t) = \dot{x}(t) - f(t, x(t)) = 0$, $k \in \mathbb{N}$, and the first inequality follows from (3.37), we are done.

In order to verify (iii) in Theorem (1.4) in Chapter II.1 (see Remark (1.5) in Chapter II.1), we should realize that, for any $u \in V_k = \{u \in C^1(I, \mathbb{R}^n) : \|u\|_{C^0} \leq 1/k \text{ and } \|\dot{u}\|_{C^0} \leq 1/k\}$, for some $k \in \mathbb{N}$, $x(t)$ is a solution of the equation $u(t) = \varphi_k(x)(t)$, i.e.

$$u(t) = x(t) - x(a) - L_k x + r - \int_a^t f(s, x(s))\, ds,$$

if and only if it satisfies

$$(3.38) \qquad \begin{cases} \dot{x} = \dot{u}(t) + f(t, x), \\ L_k x = r - u(a). \end{cases}$$

By the hypothesis, problem (3.38) has a unique solution, for every $L_k \in \mathcal{U}$, $r \in \mathbb{R}^n$ and $u \in C^1(I, \mathbb{R}^n)$ with $\|u\| \leq 1$, as required. Therefore, applying Theorem (1.4) in Chapter II.1 (see Remark (1.5) in Chapter II.1), the set $\{\varphi(0)\}$ is R_δ. In other words, the solution set of the original problem (3.32) is R_δ as well. $\quad\square$

(3.39) REMARK. One can observe that the sole existence can be easily proved by means of the well-known Schauder fixed point theorem.

(3.40) EXAMPLE. According to Example 2 in [La1], problem

(3.41)
$$\begin{cases} \dot{x}_i = f_i(t, x_1, x_2) + p_i(t), & i = 1, 2, \\ ax_1(0) + x_2(0) = r_1, \quad bx_1(1) + x_2(1) = r_2, \end{cases}$$

is uniquely solvable, for every $a^2 < 1$, $b^2 > 1$, $r_i \in \mathbb{R}$, $(i = 1, 2)$ and $p = (p_1, p_2) \in C([0, 1], \mathbb{R})$, provided $f_i \in C^1([0, 1], \mathbb{R})$, $i = 1, 2$, and

$$\frac{\partial f_1}{\partial x_1} u_1^2 + \frac{\partial f_1}{\partial x_2} u_1 u_2 - \frac{\partial f_2}{\partial x_1} u_1 u_2 - \frac{\partial f_2}{\partial x_2} u_2^2 \geq 0, \quad (i = 1, 2),$$

for each triple $(t, x_1, x_2) \in [0, 1] \times \mathbb{R}^2$ and each double $(u_1, u_2) \in \mathbb{R}^2$.

Therefore, according to Theorem (3.34) (more precisely, according to its modified version, where the set of all continuous linear operators can be restricted (see [La1]) to the set of all real $(n \times n)$-matrices), problem

(3.42)
$$\begin{cases} \dot{x}_i = f_i(t, x_1, x_2), & i = 1, 2, \\ ax_1(0) + x_2(0) = r_1, \quad bx_1(1) + x_2(1) = r_2, \end{cases}$$

has an R_δ-set of solutions, for certain $(a, b) \in \mathbb{R}^2$ in a closed subset of $\mathbb{R}^2$ with $a^2 = 1$, $b^2 \geq 1$ or $a^2 \leq 1$, $b^2 = 1$ (r_1, r_2 can be arbitrary), whenever all solutions of problem (3.42) are uniformly a priori bounded, for such $a^2 \leq 1$, $b^2 \geq 1$.

This can be achieved for $a \neq b$, i.e. particularly with the exception of $a = b = 1$ or $a = b = -1$, and $b^2 \leq b_*^2$, for some $b_* > 1$, when e.g.

(3.43)
$$|f(t, x)| \leq \alpha|x| + \beta, \quad \text{for all } (t, x) \in [0, 1] \times \mathbb{R}^2,$$

where α, β are suitable nonnegative constants (α must be sufficiently small as below) and $x = (x_1, x_2)$, $f = (f_i, f_2)$.

Indeed. Since the linear homogeneous problem

$$\begin{cases} \dot{x}_i = 0, & i = 1, 2, \\ ax_1(0) + x_2(0) = 0, \quad bx_1(1) + x_2(1) = 0, \end{cases}$$

has for $a \neq b$ obviously only a trivial solution, every solution $x(t) = (x_1(t), x_2(t))$ of (3.42) takes the form (see e.g. [Co1] or Lemma (5.136) below)

$$x_i(t) = \int_0^1 G_i(t, s, a, b) f_i(s, x_1(s), x_2(s)) \, ds + \tilde{x}_i, \quad i = 1, 2,$$

where $G = (G_1, G_2)$ is the related Green function of the linearized problem (3.42), namely

$$\begin{cases} \dot{x}_i = f_i(t, x_1, x_2), & i = 1, 2, \\ ax_1(0) + x_2(0) = 0, \quad bx_1(1) + x_2(1) = 0, \end{cases}$$

i.e.

$$G(t,s,a,b) = \begin{cases} \dfrac{1}{b-a}\begin{pmatrix} b & 1 \\ -ab & -a \end{pmatrix}, & \text{for } 0 \le t \le s \le 1, \\[2ex] \dfrac{1}{b-a}\begin{pmatrix} a & 1 \\ -ab & -b \end{pmatrix}, & \text{for } 0 \le s \le t \le 1, \end{cases}$$

and $\widetilde{x} = (\widetilde{x}_1, \widetilde{x}_2)$ is a unique solution of the problem

$$\begin{cases} \dot{x}_i = 0, & i = 1,2, \\ ax_1(0) + x_2(0) = r_1, & bx_1(1) + x_2(1) = r_2, \end{cases}$$

i.e. $\widetilde{x}_1 = (r_2 - r_1)/(b-a)$, $\widetilde{x}_2 = r_1 - a(r_2 - r_1)/(b-a)$.

Let us fix (a,b) at the boundary $\partial \mathcal{U} = \{(a,b) \in \mathbb{R}^2 \mid a^2 = 1,\ b^2 \ge 1$ or $a^2 \le 1,\ b^2 = 1\}$ with $a \ne b$, for which we intend to get the result and cut off appropriately the corners with $a = b$, jointly with those (a,b) with $b^2 > b_*^2$ for some $b_* > 1$, as in Figure 1. The bold curve in Figure 1 so indicates the part of the boundary of our interest.

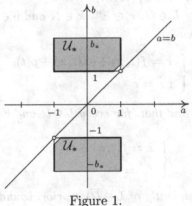

Figure 1.

Denoting $g = \max_{(a,b)\in\overline{\mathcal{U}}_*} g_{a,b}$, $g_{a,b} = \max_{t,s\in[0,1]} |G(t,s,a,b)|$, where $\overline{\mathcal{U}}_*$ is indicated in Figure 1 by the shaded region (observe that since $\overline{\mathcal{U}}_*$ is compact, g certainly exists), we obtain by means of (3.43) that

$$\|x(t)\| \le g(\alpha\|x(t)\| + \beta) + \max_{(a,b)\in\overline{\mathcal{U}}_*} |\widetilde{x}|,$$

i.e.

$$\|x(t)\| \le \frac{\beta g + \max_{(a,b)\in\overline{\mathcal{U}}_*} |\widetilde{x}|}{1 - \alpha g},$$

whenever $\alpha < g^{-1}$, as claimed. This completes the example.

(3.44) REMARK. One can easily check that, for fixed values of $(a,b) \in \partial \mathcal{U}$ with $a \ne b$, the condition $\alpha < g^{-1}$ can take e.g. the form

$$\alpha < \frac{|b-a|}{|b| + \max(1, |ab|)}.$$

A continuous function $f : I \times \mathbb{R}^n \to \mathbb{R}^n$ can be approximated with an arbitrary accuracy by locally Lipschitzian (in the second variable) functions (see Theorem (3.37) in Chapter I.3 or [Har-M]), say $(f + \varepsilon_k)$, $k \in \mathbb{N}$, such that $\lim_{k\to\infty} \|\varepsilon_k\| = 0$. Therefore, applying at first, for fixed $k \in \mathbb{N}$, Theorem (3.34) to the system

$$(3.45) \qquad\qquad \dot{x} = f(t, x) + \varepsilon_k(t, x),$$

and then proceeding quite analogously as in Theorem (3.13), we can avoid the uniqueness assumption in Theorem (3.34) as follows.

(3.46) THEOREM. *Let* $f : I \times \mathbb{R}^n \to \mathbb{R}^n$ *be a fixed continuous function and* $\varepsilon_k : I \times \mathbb{R}^n \to \mathbb{R}^n$, $k \in \mathbb{N}$, *be continuous functions with* $\|\varepsilon_k\| \leq \varepsilon$ (ε - *a sufficiently small constant) such that* $(f + \varepsilon_k)(t, \cdot) : \mathbb{R}^n \to \mathbb{R}^n$, $k \in \mathbb{N}$, *are locally Lipschitzian, for every* $t \in I$, *and* $p : I \to \mathbb{R}^n$ *be a continuous function with* $\|p\| \leq 1$.

Let $\mathcal{U}$ *be an open (in the norm topology) subset of the Banach space of all continuous linear operators* $L : C_1^0(I, \mathbb{R}^n) \to \mathbb{R}^n$.

Assume that, for every $L \in \mathcal{U}$, $r \in \mathbb{R}^n$, $k \in \mathbb{N}$ *and* $p \in C(I, \mathbb{R}^n)$ *with* $\|p\| \leq 1$, *the boundary value problem*

$$\begin{cases} \dot{x} = f(t, x) + \varepsilon_k(t, x) + p(t), \\ Lx = r \end{cases}$$

has at most one solution and that, for every $L \in \overline{\mathcal{U}}$ *and* $k \in \mathbb{N}$, *all solutions of the problem*

$$\begin{cases} \dot{x} = f(t, x) + \varepsilon_k(t, x), \\ Lx = r \end{cases}$$

are uniformly (i.e. independently of $L \in \overline{\mathcal{U}}$) *a priori bounded, where* $\overline{\mathcal{U}}$ *denotes the closure of* $\mathcal{U}$ *in the* C_1^0-*topology.*

Then, for every $L \in \partial\mathcal{U}$ *and* $r \in \mathbb{R}^n$, *where* $\partial\mathcal{U}$ *denotes the boundary of* $\mathcal{U}$ *in the* C_1^0-*topology, the problem* (3.32), *i.e.*

$$\begin{cases} \dot{x} = f(t, x), \\ Lx = r \end{cases}$$

has a (nonempty) compact acyclic set of solutions.

III.4. Poincaré operators

By the Poincaré operators, we mean the *translation operator along the trajectories* of the associated differential system and the *first return* (or *section*) *map* defined on the cross section of the torus by means of the flow generated by the vector field. The translation operator is sometimes also called as *Poincaré–Andronov* or *Levinson* or, simply, T-*operator*.

In the classical theory, both these operators are defined to be single-valued, when assuming among other things, the uniqueness of the initial value problems. At the absence of uniqueness, one usually approximates the right-hand sides of the given systems by the locally Lipschitzian ones (implying alredy uniquess), and then applies the *standard limiting argument* (for more details, see e.g. [Har-M], [Ks2-M]).

On the other hand, set-valued analysis allows us to handle directly with multivalued Poincaré operators which become, under suitable natural restrictions imposed on the right-hand sides of given differential systems, admissible in the sense of Definition (4.9) in Chapter I.4.

Below, the following cases will be considered separately:

  A. Translation multioperator for ordinary systems,
  B. Translation multioperator for functional systems,
  C. Translation multioperator for systems with constraints,
  D. Translation multioperator for systems in Banach spaces,
  E. Translation multioperator for random systems,
  F. Translation multioperator for directionally u.s.c. systems,
  G. First-return multimap for autonomous systems.

A. Translation multioperator for ordinary systems.
Consider the u-Carathéodory system

$$(4.1) \qquad \dot{x} \in F(t,x), \quad x \in \mathbb{R}^n,$$

where

  (i) the set of values of F is nonempty, compact and convex, for all $(t,x) \in [0,\tau] \times \mathbb{R}^n$,
  (ii) $F(\cdot,x)$ is u.s.c., for a.a. $t \in [0,\tau]$,
  (iii) $F(\cdot,x)$ is measurable, for every $x \in \mathbb{R}^n$, i.e. for any closed $U \subset [0,\tau]$ and every $x \in \mathbb{R}^n$, the set $\{t \in [0,\tau] \mid F(\cdot,x) \cap U \neq \emptyset\}$ is measurable,
  (iv) $|F(t,x)| \leq \alpha + \beta|x|$, for every $x \in \mathbb{R}^n$ and a.a. $t \in [0,\tau]$, where α, β are suitable positive constants.

By a *solution* $x(t)$ of (4.1), we mean an absolutely continuous function $x(t) \in AC([0,\tau],\mathbb{R}^n)$ satisfying (4.1), for a.a. $t \in [0,\tau]$, i.e. the one in the sense of Carathéodory, such solutions of (4.1) exist on $[0,\tau]$.

Hence, if $x(t,x_0) := x(t,0,x_0)$ is a solution of (4.1) with $x(0,x_0) = x_0 \in \mathbb{R}^n$, then the translation multioperator $T_\tau : \mathbb{R}^n \multimap \mathbb{R}^n$ at the time $\tau > 0$ along the trajectories of (4.1) is defined as follows:

$$(4.2) \quad T_\tau(x_0) := \{x(\tau,x_0) \mid x(\cdot,x_0) \text{ is a solution of (4.1) with } x(0,x_0) = x_0\}.$$

More precisely, T_τ can be considered as the composition of two maps, namely $T_\tau = \psi \circ \varphi$,

$$\mathbb{R}^n \xrightarrow{\varphi} AC([0,\tau], \mathbb{R}^n) \xrightarrow{\psi} \mathbb{R}^n,$$

where $\varphi(x_0) : x_0 \multimap \{x(t, x_0) \mid x(t, x_0)$ is a solution of (4.1) with $x(0, x_0) = x_0\}$ is well-known to be an R_δ-mapping (see Theorem (2.3)) and $\psi(y) : y(t) \to y(\tau)$ is obviously a continuous (single-valued) evaluation mapping.

In other words, we have the following commutative diagram:

$$
\begin{array}{ccc}
\mathbb{R}^n & \xrightarrow{\varphi} & AC([0,\tau], \mathbb{R}^n) \\
 & & \Big\downarrow{\psi} \\
 & \xrightarrow{T_\tau} & \mathbb{R}^n
\end{array}
$$

The following characterization of T_τ has been proved on various levels of abstraction in [LaR3], [De2], [DyG], [GP1], [Lew], etc.

(4.3) THEOREM. *T_τ defined by (4.2) is admissible and admissibly homotopic to identity. More precisely, T_τ is a composition of an R_δ-mapping and a continuous (single-valued) evaluation mapping.*

PROOF. According to Theorem (2.3), the mapping φ has an R_δ-set of values. We will show that it is u.s.c. by prooving the closedness of the graph Γ_φ of φ.

Let $(x_n, y_n) \in \Gamma_\varphi$, i.e. $y_n \in \varphi(x_n)$, and $(x_n, y_n) \to (x, y)$ as $n \to \infty$. Since the functions y_n are absolutely continuous on $[0, \tau]$, the application of the well-known Gronwall inequality (see e.g. [Har-M]) leads to the estimates (cf. (iv))

$$\|y_n\| \le M := \sup_{n \in \mathbb{N}}(|x_n| + \gamma\tau)\exp(\gamma\tau) \quad \text{and} \quad \|\dot{y}_n\| \le \gamma(1 + M),$$

where $\gamma = \max\{\alpha, \beta\}$. It follows that $\{y_n\}$ are equi-bounded.

Covergence Theorem (1.2) guarantues the existence of a sequence $\{y_n\}$ such that $y_n \to y$, uniformly, and $\dot{y}_n \to \dot{y}$, weakly in $L^1([0, \tau], \mathbb{R}^n)$. According to Mazur's Theorem (1.33) in Chapter I.1, $\dot{y}$ belongs to the strong closure $\dot{y} \in \overline{\text{conv}}\{\dot{y}_n \mid n \ge l\}$, for all $l \ge 1$. Thus, there also exists a subsequence $\{z_l\}$ such that $z_l \to \dot{y}$, in the L^1-topology, where $z_l \in \overline{\text{conv}}\{\dot{y}_n \mid n \ge l\}$. Moreover, there exists a subsequence (for the simplicity, denoted again by $\{z_l\}$) satisfying $z_l \to \dot{y}$, a.e. on $[0, \tau]$.

Let $I \subset [0, \tau]$ be a set of a full measure on $[0, \tau]$, i.e. $\mu(I) = \tau$, where μ denotes the Lebsesque measure, such that $z_l \to \dot{y}$ as $l \to \infty$, for all $t \in I$. It follows from the definition of z_l that $z_l(t) \in \sum_i \lambda_i F(t, y_{n_i}(t))$, where $\sum_i \lambda_i = 1$.

Since $F(t, \cdot)$ is u.s.c., for a.a. $t \in [0, \tau]$, and $y_{n_i}(t)$ is sufficiently close to $x(t)$ as well as $z_l(t)$ to $\dot{x}(t)$, we obtain $\dot{x}(t) \in \sum_i \lambda_i F(t, x(t)) + \varepsilon B$ for an arbitrary

$\varepsilon > 0$, where B is an open unit ball. This already means that $\dot{x}(t) \in F(t, x(t))$, and subsequently the graph Γ_φ of φ is closed. Since the arbitrary closed set $\{(x, y), (x_1, y_1), \ldots, (x_n, y_n), \ldots\}$ is, according to the well-known Arzelà–Ascoli lemma, compact, φ is u.s.c.

For the remaining part of the proof, it is sufficient to consider the admissible homotopy $T_{\lambda\tau}$, $\lambda \in [0, 1]$. $\qquad\qquad\qquad\qquad\qquad$ □

(4.4) REMARK. Since a composition of admissible maps is admissible as well (cf.(4.12) in Chapter I.4), T_τ can be still composed with further admissible maps ϕ such that $\phi \circ T_\tau$ becomes an (admissible) self-map on a compact ENR-space (i.e. homeomorphic to ANR in $\mathbb{R}^n$), for computation of the well-defined (cf. I.6) generalized generalized Lefschetz number:

$$\Lambda(\phi \circ T_\tau) = \Lambda(\phi).$$

T_τ considered on ENRs can be even composed e.g. with suitable homomorphisms $\mathcal{H}$ (again considered on ENRs), namely $\mathcal{H} \circ T_\tau$, for computation of the well-defined (cf. I.8), fixed point index:

$$\text{ind}(\mathcal{H} \circ T_\tau) = \text{ind}\,\mathcal{H},$$

provided the fixed point set of $\mathcal{H} \circ T_{\lambda\tau}$ is compact, for $\lambda \in [0, 1]$.

B. Translation multioperator for functional systems.
Consider the functional system

$$(4.5) \qquad\qquad \dot{x} \in F(t, x_t), \quad x \in \mathbb{R}^n,$$

where $x_t(\,\cdot\,) = x(t + \cdot\,)$, for $t \in [0, \tau]$, denotes as usual a function from $[-\delta, 0]$, $\delta \geq 0$, into $\mathbb{R}^n$ and $F : [0, \tau] \times \mathcal{C} \multimap \mathbb{R}^n$, where $\mathcal{C} = \text{AC}([-\delta, 0], \mathbb{R}^n)$, is a u-Carathéodory multivalued function, i.e.

(i) the set of values of $F(t, y)$ is nonempty, compact and convex, for all $(t, y) \in [0, \tau] \times \mathcal{C}$,

(ii) $F(t, \cdot)$ is u.s.c., for a.a. $t \in [0, \tau]$,

(iii) $F(\cdot, y)$ is measurable, for all $y \in \mathcal{C}$, i.e. for any closed $U \subset [0, \tau]$ and every $y \in \mathcal{C}$, the set $\{t \in [0, \tau] \mid F(\cdot, y) \cap U \neq \emptyset\}$ is measurable,

(iv) $|F(t, y)| \leq \alpha + \beta\|y\|$, for every $y \in \mathcal{C}$ and a.a. $t \in [0, \tau]$, where α, β are suitable positive constants.

By a *solution* $x(t)$ of (the initial problem to) (4.5), we mean again an absolutely continuous function $x(t) \in \text{AC}([-\delta, \tau], \mathbb{R}^n)$ (with $x(t) = x_*$, $t \in [-\delta, 0]$), satisfying (4.5), for a.a. $t \in [-\delta, \tau]$; such solutions exist on $[-\delta, \tau]$, $\delta > 0$.

Hence, if $x(t, x_*) := x(t, [-\delta, 0], x_*)$ is a solution of (4.5) with $x(0, x_*) = x_* \in E$ for $t \in [-\delta, 0]$, where E consists of equicontinuous functions, then the translation multioperator $T_\tau : \mathrm{AC}([-\delta, 0], \mathbb{R}^n) \multimap \mathrm{AC}([-\delta, 0], \mathbb{R}^n)$ at the time $\tau > 0$ along the trajectories of (4.5) is defined as follows:

(4.6) $T_\tau(x_*) := \{x(\tau, x_*) \mid x(\,\cdot\,, x_*)$ is a solution of (4.5) with

$$x(t, x_*) = x_*, \quad \text{for } t \in [-\delta, 0]\}.$$

More precisely, T_τ can be considered as the composition of two maps, namely $T_\tau = \psi \circ \varphi$,

$$\mathrm{AC}([-\delta, 0], \mathbb{R}^n) \overset{\varphi}{\multimap} \mathrm{AC}([-\delta, \tau], \mathbb{R}^n) \overset{\psi}{\longrightarrow} \mathrm{AC}([-\delta, 0], \mathbb{R}^n),$$

where $\varphi(x_*) : x_* \multimap \{x(t, x_*) \mid x(t, x_*)$ is a solution of (4.5) with $x(t, x_*) = x_*$, for $t \in [-\delta, 0]\}$ is known (cf. Remark (2.53) and [OZ]) to be an R_δ-mapping and $\psi(y) : y(t) \to y(\tau)$ is a continuous (single-valued) evaluation mapping.

In other words, we have the following commutative diagram:

$$\mathrm{AC}([-\delta, 0], \mathbb{R}^n) \xrightarrow{\ \varphi\ } \circ\, \mathrm{AC}([-\delta, \tau], \mathbb{R}^n)$$

$$\left\downarrow \psi \right.$$

$$\xrightarrow{\ T_\tau\ } \circ\, \mathrm{AC}([-\delta, 0], \mathbb{R}^n)$$

The following characterization of T_τ has been proved on various levels of abstraction in [DyJ], [HL], [HP3], [ObZe], etc.

(4.7) THEOREM. T_τ *defined by* (4.6) *is admissible and admissibly homotopic to identity. More precisely,* T_τ *is a composition of an* R_δ-*mapping and a continuous (single-valued) evaluation mapping.*

(4.8) REMARK. Theorem (4.7) reduces to Theorem (4.3), for $\delta = 0$, and Remark (4.4) can be appropriately modified here as well.

C. Translation multioperator for systems with constraints.

In view of Remark (4.8), consider again system (4.5), where $F : [0, \tau] \times \mathcal{C} \multimap \mathbb{R}^n$ is the same as above. For a nonempty, compact and convex set $K \subset \mathbb{R}^n$, the constraint, denote $\mathcal{K} = \{\xi \in \mathcal{C} \mid \xi(t) \in K, \text{ for } t \in [-\delta, 0]\}$ and assume that the Nagumo-type condition holds,

(4.9) $F(t, y) \cap T_K(y(0)) \neq \emptyset, \quad \text{for all } (t, y) \in [0, \tau] \times \mathcal{K},$

where

$$T_K(y(0)) = \left\{ z \in \mathbb{R}^n \,\middle|\, \liminf_{h \to 0^+} \frac{d(y(0) + h_z, K)}{h} = 0 \right\}$$

is the tangent cone (in the sense of Bouligand). Observe, that (iv) can be reduced to

(iv) $\sup \operatorname{ess}_{(t,y)\in[0,\tau]\times\mathcal{K}} |F(t,y)| < \infty.$

Then, for every $x_* \in \mathcal{K}$, there exists at least one Carathéodory solution $x(t, x_*)$ of (4.5) (see e.g. [HP3]) such that $x(t, x_*) = x_* \in E$, for $t \in [-\delta, 0]$, and $x(t, x_*) \in K$, for $t \in [0, \tau]$. Hence, we can define, under (4.9), the associated translation multioperator $T_\tau : \mathcal{K} \multimap \mathcal{K}$ at the time $\tau > 0$ along the trajectories of (4.5), which makes the set $\mathcal{K}$ invariant, as follows:

(4.10) $T_\tau(x_*) := \{x(\tau, x_*) \mid x(\,\cdot\,, x_*)$ is a solution of (4.5) with

$$x(t, x_*) = x_*, \text{ for } t \in [-\delta, 0], \text{ and } x(t, x_*) \in K, \text{ for } t \in [0, \tau]\}.$$

More precisely, T_τ can be considered as the composition of two maps, namely $T_\tau = \psi \circ \varphi$,

$$\mathcal{K} \xrightarrow{\varphi} \{z \in \mathrm{AC}([-\delta, \tau], \mathbb{R}^n) \mid z(t) \in K, \text{ for } t \in [-\delta, \tau]\} \xrightarrow{\psi} \mathcal{K},$$

where

$\varphi(x_*) = x_* \multimap \{x(\tau, x_*) \mid x(\,\cdot\,, x_*)$ is a solution of (4.5) with

$$x(t, x_*) = x_*, \text{ for } t \in [-\delta, 0], \text{ and } x(t, x_*) \in K, \text{ for } t \in [0, \tau]\}$$

is known (see Theorem (2.49) and [HP3]) to be an R_δ-mapping and $\psi(z) : z(t) \to z(\tau)$ is a continuous (single-valued) evaluation mapping.

In other words, we have the following commutative diagram:

$$\mathcal{K} \xrightarrow{\quad\varphi\quad} \{z \in \mathrm{AC}([-\delta, \tau], \mathbb{R}^n) \mid z(t) \in K, \text{ for } t \in [-\delta, \tau]\}$$
$$\Big\downarrow{\psi}$$
$$\xrightarrow{\quad T_\tau \quad} \mathcal{K}$$

The folowing characterization of T_τ has been proved on various levels of abstraction in [HP3], [Bad2], [HL], [Pl2], etc.

(4.11) THEOREM. T_τ *defined by (4.10) is, under (4.9), admissible and admissibly homotopic to identity. More precisely, T_τ is a composition of an R_δ-mapping and a continuous (single-valued) evaluation mapping.*

(4.12) REMARK. Theorem (4.11) coincides with Theorem (4.7), for $K = \mathbb{R}^n$, in spite of the fact that K is assumed to be bounded and closed. Remark (4.4) can be appropriately modified as well.

D. Translation multioperator for systems in Banach spaces.

Consider the functional system

$$\dot{x} + Ax \in F(t, x_t), \quad x \in \mathcal{B}, \tag{4.13}$$

where $x_t(\cdot) = x(t + \cdot)$, for $t \in [0, \tau]$, denotes as above the mapping from $[-\delta, 0]$, $\delta \geq 0$, into a real separable Banach space $\mathcal{B}$. Let, furthermore, the following assumptions be satisfied:

(i) A is a closed, linear (not necessarily bounded) operator in $\mathcal{B}$, generating an analytic semigroup e^{At},

(ii) the set of $F(t, y) : [0, \tau] \times \mathcal{C} \multimap \mathcal{B}$, where $\mathcal{C} = C([-\delta, 0], \mathcal{B})$ and $\delta \geq 0$, is nonempty, compact and convex, for all $(t, y) \in [0, \tau] \times \mathcal{C}$,

(iii) $F(t, \cdot)$ is u.s.c., for a.a. $t \in [0, \tau]$,

(iv) $F(\cdot, y)$ is measurable for all $y \in \mathcal{C}$, i.e. for any closed $U \subset [0, \tau]$ and every $y \in \mathcal{C}$, the set $\{t \in [0, \tau] \mid F(\cdot, y) \cap U \neq \emptyset\}$ is measurable,

(v) (cf. [ObZe]) for every nonempty, bounded, equicontinuous set $D \subset \mathcal{C}$, we have

$$\gamma(F(t, D)) \leq g(t, \xi(D)), \quad \text{for a.a. } t \in [0, \tau],$$

where γ denotes the Hausdorff measure of noncompactness,

$$\xi(D) \in C([-\delta, 0] \times [0, \infty)), \quad \xi(D)(\theta) = \gamma(D(\theta))$$

and

$$g : [0, \tau] \times C([-\delta, 0] \times [0, \infty)) \to [0, \infty)$$

is a Carathéodory-type function such that

(a) $g(t, \cdot)$ is nondecreasing, for a.a. $t \in [0, \tau]$, in the sense that if $\varphi, \psi \in C([-\delta, 0] \times [0, \infty))$ satisfy $\varphi(\theta) < \psi(0)$, for every $\theta \in [-\delta, 0]$, then $g(t, \varphi) < g(t, \psi)$,

(b) $|g(t, \varphi) - g(t, \psi)| < k(t)\|\varphi - \psi\|_1$, for a.a. $t \in [0, \tau]$ and for all $\varphi, \psi \in C([-\delta, 0] \times [0, \infty))$, where k is a Lebesgue measurable function and $\| \cdot \|_1$ denotes the norm in the space $C([-\delta, 0] \times [0, \infty))$,

(c) $g(t, 0) = 0$, for a.a. $t \in [0, \tau]$,

(vi) there exists a continuous bounded function $h : [0, \infty) \to [0, \infty)$ such that

$$\gamma(e^{At} S) \leq h(t), \quad \text{for } t \in [0, \infty),$$

where S denotes the unit sphere in $\mathcal{B}$, and

$$\sup_{t \in [0, \infty)} \int_0^t h(t - s) k(s) \, ds < 1,$$

(vii) the solutions of the problem

(4.14)
$$\begin{cases} w(t) = v(t), & t \in [-\delta, 0], \\ w(t) = \dfrac{1}{h(0)} h(t) w(0) + \displaystyle\int_0^t h(t-s) g(s, w_s)\, ds, & t \in [0, \tau], \end{cases}$$

are uniformly asymptotically bounded in the sense that there exists a function $\sigma : [0, \infty) \to [0, \infty)$ such that

$$\limsup_{t \to \infty} \sigma(t) < \frac{1}{h(0)}$$

and, for every solution $w(t, v)$ of (4.14), we have

$$\|w_t\|_1 \le \sigma(t) \|v\|_1 \quad \text{for } t \in [0, \infty),$$

(viii) $\|F(t, y)\| \le \alpha + \beta \|y\|_0$, for every $y \in C$ and a.a. $t \in [0, \tau]$, where α, β are suitable positive constants and $\| \cdot \|_0$ denotes the norm in C.

By a *solution* $x(t)$ of (the initial problem to) (4.13), we mean this time a *mild solution*, namely $x(t) \in C([-\delta, \tau], \mathcal{B})$ such that

$$x(t) = e^{At} x(0) + \int_0^t e^{A(t-s)} f(s)\, ds, \quad \text{for } t \in [0, \tau],$$

(with $x(t) = x_*$, for $t \in [-\delta, 0]$), where f is an (existing) measurable selection of $F(s, x_s(t))$, $t \in [-\delta, 0]$; such solutions exist on $[-\delta, \tau]$, $\delta \ge 0$.

Hence, if $x(t, x_*) = x_* \in E$, for $t \in [-\delta, 0]$, where E consists of equicontinuous functions, then the translation multioperator $T_\tau : C \to C$ at the time $\tau > 0$ along the trajectories of (4.13) is defined as follows:

(4.15) $\quad T_\tau(x_*) := \{x(\tau, x_*) \mid x(\,\cdot\,, x_*)$ is a solution of (4.13) with

$$x(t, x_*) = x_*, \text{ for } t \in [-\delta, 0]\}.$$

More precisely, T_τ can be considered as the composition of two maps, namely $T_\tau = \psi \circ \varphi$,

$$C([-\delta, 0], \mathcal{B}) \overset{\varphi}{\multimap} C([-\delta, \tau], \mathcal{B}) \overset{\psi}{\multimap} C([-\delta, 0], \mathcal{B}),$$

where $\varphi(x_*) : x_* \multimap \{x(t, x_*) \mid x(t, x_*)$ is a solution of (4.13) with $x(t, x_*) = x_*$, for $t \in [-\delta, 0]\}$ is known (see Theorem (2.52) and [ObZe]) to be an R_δ-mapping and $\psi(z) : z(t) \to z(\tau)$ is a continuous (single-valued) evaluation mapping.

In other words, we have the following commutative diagram:

$$
\begin{array}{ccc}
C([-\delta, 0], \mathcal{B}) & \overset{\varphi}{\longrightarrow\!\!\!\!\!\circ} & C([-\delta, \tau], \mathcal{B}) \\
& & \Big\downarrow \psi \\
& \overset{T_\tau}{\longrightarrow\!\!\!\!\!\circ} & C([-\delta, 0], \mathcal{B})
\end{array}
$$

The following characterization of T_τ has been proved on various levels of abstraction in [ObZe], [Bad2], [De2], [KOZ2], etc.

(4.16) THEOREM. T_τ defined by (4.15) is admissible and admissibly homotopic to identity. More precisely, T_τ is a composition of an R_δ-mapping and a continuous (single-valued) evaluation mapping, provided (i)–(v) hold. Under (i)–(viii), T_τ is γ-condensing on equicontinuous sets, i.e. with respect to the Hausdorff measure of noncompactness in $\mathcal{C}$, provided $\tau > \inf\{t' \mid \sigma(t) < 1/h(0), \text{ for all } t \geq t'\}$.

(4.17) REMARK. In the ordinary case ($\delta = 0$), the Banach space need not be necessarily separable (see e.g. [Bad2]), condition (v) can be weaken and condition (vii) can be avoided (see e.g. [Bad2], [KOZ2]).

(4.18) REMARK. It is a question whether Theorem (4.16) can be reformulated in an appropriate way for functional Carathéodory systems in Banach spaces with constraints, i.e. similarly as Theorem (4.11), but for $\mathbb{R}^n$ replaced by $\mathcal{B}$. So far, only particular cases were considered with this respect (see e.g. [Bad2] and the references therein).

E. Translation multioperator for random systems.

Consider the random system

$$(4.19) \qquad \dot{x}(\kappa, t) \in F(\kappa, t, x(\kappa, t)), \quad \kappa \in \Omega, \ x \in \mathbb{R}^n,$$

where Ω is a complete probability space and

- (i) the set of values of $F(\kappa, t, x)$ is nonempty, compact and convex, for all $(\kappa, t, x) \in \Omega \times [0, \tau] \times \mathbb{R}^n$,
- (ii) $F(\kappa, t, \cdot)$ is u.s.c., for a.a. $(\kappa, t) \in \Omega \times [0, \tau]$,
- (iii) $F(\cdot, \cdot, x)$ is measurable, for every $x \in \mathbb{R}^n$, i.e. for any closed $U \subset \mathbb{R}^n$ and every $x \in \mathbb{R}^n$, the set $\{(\kappa, t) \in \Omega \times [0, \tau] \mid F(\cdot, \cdot, x) \cap U \neq \emptyset\}$ is measurable,
- (iv) $F(\kappa, t, x) \leq \mu(\kappa, t)(1 + |x|)$, for a.a. $(\kappa, t) \in \Omega \times [0, \tau]$ and all $x \in \mathbb{R}^n$, where $\mu : \Omega \times [0, \tau] \to [0, \infty)$ is a map such that $\mu(\cdot, t)$ is measurable and $\mu(\kappa, \cdot)$ is Lebesgue integrable.

The operator F satisfying conditions (i)–(iv) is called a *random u-Carathéodory operator*. Similarly, for metric spaces X and Y we say that a multivalued mapping with nonempty closed values $\varphi : \Omega \times X \multimap Y$ is a *random operator* if φ is product-measurable and $\varphi(\kappa, \cdot)$ is u.s.c., for every $\kappa \in \Omega$. By a *random homotopy* $\chi : \Omega \times X \times [0, 1] \multimap Y$, we understand a product-measurable mapping with nonempty closed values which is u.s.c. w.r.t. the last variable and that, for every $\lambda \in [0, 1]$, $\chi(\cdot, \cdot, \lambda)$ is a random operator.

Furthermore, we say that a measurable map (a random variable) $\widehat{x} : \Omega \to X \cap Y$ is a *random fixed point* of a random operator $\varphi : \Omega \times X \multimap Y$ if $\widehat{x}(\kappa) \in \varphi(\kappa, \widehat{x}(\kappa))$, for a.a. $\kappa \in \Omega$.

The following proposition, proved in [Go5-M] (see Proposition 31.3), is crucial for further investigations.

(4.20) PROPOSITION. *Let* $\varphi : \Omega \times A \multimap X$, *where* A *is a closed subset of a metric space* X, *be a random operator with compact values such that, for every* $\kappa \in \Omega$, *the set of fixed points of* $\varphi(\kappa, \cdot)$ *is nonempty. Then* φ *has a random fixed point.*

Because of Proposition (4.20), we can define the random translation multioperator T_τ in a "deterministic" way. We can namely employ, for every $\kappa \in \Omega$ and $x_0 \in \mathbb{R}^n$, Carathéodory solutions $x(t, x_0)$ of the deterministic Cauchy problems

$$(4.21) \qquad \begin{cases} \dot{x} \in F_\kappa(t, x) = F(\kappa, t, x), \\ x(0, x_0) = x_0. \end{cases}$$

On the other hand, by a *solution* $x(\kappa, t)$ of (4.19), we mean a function such that $x(\cdot, t)$ is measurable, $x(\kappa, \cdot)$ is absolutely continuous and $x(\kappa, t)$ satisfies (4.19), for a.a. $(\kappa, t) \in \Omega \times [0, \tau]$; the derivative $\dot{x}(\kappa, t)$ is considered w.r.t. t.

Hence, the associated random translation multioperator $T_\tau : \Omega \times \mathbb{R}^n \multimap \mathbb{R}^n$ at the time $\tau > 0$ along the trajectories of the system $\dot{x} \in F_\kappa(t, x)$ is defined as follows:

$$(4.22) \qquad T_\tau(\kappa, x_0) := \{ x(\tau, x_0) \mid x(\cdot, x_0) \text{ is a solution of } (4.21) \}.$$

More precisely, T_τ can be considered as the composition of two maps, namely $T_\tau = \psi \circ \varphi$,

$$\Omega \times \mathbb{R}^n \overset{\varphi}{\multimap} AC([0, \tau], \mathbb{R}^n) \overset{\psi}{\longrightarrow} \mathbb{R}^n,$$

where $\varphi(\kappa, x_0) : (\kappa, x_0) \multimap \{ x(t, x_0) \mid x(t, x_0) \text{ is a solution of } (4.21) \}$ is, according to Theorem (4.3), an R_δ-mapping, for every $\kappa \in \Omega$, and the (single-valued) evaluation mapping $\psi(y) : y(t) \to y(\tau)$ is obviously continuous.

In other words, we have the following commutative diagram:

$$\Omega \times \mathbb{R}^n \overset{\varphi}{\longrightarrow} AC([0, \tau], \mathbb{R}^n)$$

$$\overset{T_\tau}{\multimap} \mathbb{R}^n \qquad \Big\downarrow \psi$$

Applying Proposition (4.20), one can show the following characterization of T_τ (for more details, see [DBGP3]).

(4.23) THEOREM. T_τ *defined by* (4.22) *is a random operator with compact values composed by a random operator with* R_δ-*values and a continuous (single-valued) evaluation mapping. Moreover, it is randomly homotopic to identity.*

Thus, $\hat{x}$ is a random fixed point of T_ω if and only if the original system (4.19) has a solution $x(\kappa, t)$ such that $x(\kappa, 0) = x(\kappa, \omega) = \hat{x}(\kappa)$, for a.a. $\kappa \in \Omega$.

(4.24) REMARK. In [Go5-M, pp. 156–157], the random degree theory is sketched, having quite anologous properties as in the deterministic case, and so it is available for proving the random fixed points of the random translation operator T_ω.

(4.25) REMARK. Theorem (4.23) reduces to Theorem (4.3) in the deterministic case, i.e. in the absence of Ω. Remark (4.4) can be appropriately modified here as well.

F. Translation multioperator for directionally u.s.c. systems.

Let $M \in \mathbb{R}$ and $\Gamma^M = \{(t,x) \in \mathbb{R} \times \mathbb{R}^n \mid |x| \leq Mt\} \subset \mathbb{R} \times \mathbb{R}^n$ be a closed, convex cone. Following [AJ1] (cf. also [Bre3]), we say that a multivalued mapping with nonempty closed values $F : \mathbb{R} \times \mathbb{R}^n \multimap \mathbb{R}^n$ is Γ^M-directionally u.s.c. if, at each point $(t_0, x_0) \in \mathbb{R} \times \mathbb{R}^n$, and for every $\varepsilon > 0$, there exists $\delta > 0$ such that, for all $(t,x) \in B((t_0, x_0), \delta)$ (i.e. (t,x) belonging to an open ball with the radius δ and centered at (t_0, x_0)) satisfying $|x - x_0| \leq M(t - t_0)$ holds $F(t,x) \subset F(t_0, x_0) + \varepsilon B$.

Consider the Γ^M-directionally u.s.c. system

$$(4.26) \qquad \dot{x} \in F(t,x), \quad x \in \mathbb{R}^n.$$

We will show that the solution set of (4.26) can be characterized by means of the Filippov-like regularization of (4.26). Subsequently, the related translation multioperator to (4.26) can be associated to the regularized system.

The proof of the following well-known lemma can be found e.g. in [Go5-M, Lemma 71.8].

(4.27) LEMMA. *Let $x(t)$ be a Carathéodory solution of differential inclusion (4.1) on the interval $[0, \tau]$. Consider the following conditions*

(i) $\dot{x}(t) \in F(t, x(t))$,

(ii) *there exists a decreasing sequence $t_n \to t$, for $n \to \infty$, such that $t_{n+1} < t_n$, $t_n \in [0, \tau]$, for all $n \in \mathbb{N}$, $\dot{x}(t_n) \to \dot{x}(t)$ and $\dot{x}(t_n) \in F(t_n, x(t_n))$, for all $n \in \mathbb{N}$.*

Define $J := \{t \in [0, \tau] \mid t \text{ satisfies (i) and (ii)}\} \subset \mathbb{R}$. Then $J \subset [0, \tau]$ is the set of a full measure, i.e. $\mu(J) = \tau$, where μ stands for the Lebesgue measure.

(4.28) DEFINITION ([Jar], cf. [Fi2-M]). Let $F(t,x) : [a,b] \times \mathbb{R}^n \multimap \mathbb{R}^n$ be (closed) convex-valued, locally bounded and measurable. Then the mapping

$$\phi(t,x) = \bigcap_{\delta > 0} \bigcap_{\substack{N \subset \mathbb{R}^{n+1} \\ \mu(N) = 0}} \overline{\mathrm{conv}}\, F(B((t,X), \delta) \setminus N),$$

called the *Filippov-like regularization*, satisfies the following properties

(i) $\phi(t,x)$ is u.s.c., for all $(t,x) \in [a,b] \times \mathbb{R}^n$,

(ii) $F(t, x) \subset \phi(t, x)$, for all $(t, x) \in [a, b] \times \mathbb{R}^n$,

(iii) ϕ is minimal in the following sense: if $\psi : [a, b] \times \mathbb{R}^n \multimap \mathbb{R}^n$ satisfies (i), (ii), then $\phi(t, x) \subset \psi(t, x)$, for all $(t, x) \subset [a, b] \times \mathbb{R}^n$,

where $\mu(N)$ stands for the Lebesgue measure of N and $\overline{\text{conv}}$ denotes the closed-convex hull of a set.

(4.29) PROPOSITION (cf. [Bre3]). *Let $\Omega \subset \mathbb{R} \times \mathbb{R}^n$ and $F : \Omega \multimap \mathbb{R}^n$ be a (closed) convex-valued, Γ^M-directionally u.s.c. and bounded in the following way $F(\Omega) \subset B(0, L)$, where $0 < L < M$. Let $\phi : \Omega \multimap \mathbb{R}^n$ be the regularization of F in the sense of Definition (4.28). Then every solution of the regularized inclusion*

$$(4.30) \qquad \dot{x}(t) \in \phi(t, x(t))$$

is the solution of the original inclusion (4.26), and vice versa.

PROOF. Let $x(t)$ be a solution of (4.30), on the interval $[0, \tau]$. Consider the following condition:

(4.31) there exists a decreasing sequence $t_n \to t$, for $n \to \infty$, such that $t_{n+1} < t_n$, $t_n \in [0, \tau]$, for all $n \in \mathbb{N}$, and $\dot{x}(t_n) \to \dot{x}(t)$ as well as $x(t_n) \in \phi(t, x(t_n))$, for all $n \in \mathbb{N}$.

Define $J := \{t \in [0, \tau] \mid t \text{ satisfies } (4.30) \text{ and } (4.31)\}$. According to Lemma (4.27), the set $J \subset [0, \tau]$ of points t, satisfying conditions (4.30) and (4.31), is the set of a full measure in $[0, \tau]$.

We will prove that

$$\dot{x}(t) \in F(t, x(t)), \quad \text{for all } t \in J.$$

Let, on the contrary, there be a $t \in J$ such that

$$(4.32) \qquad \varepsilon := d_H(\dot{x}(t), F(t, x(t))) > 0.$$

Since $F : \Omega \multimap \mathbb{R}^n$ is Γ^M-u.s.c. at the point $(t, x(t)) \in \Omega$, for a given $\varepsilon/2$, we can find a positive number $\delta > 0$ such that

$$(4.33) \qquad F(t', x') \subset F(t, x(t)) + \frac{\varepsilon}{2}B,$$

where $t \leq t' < t + \delta$, $|x' - x(t)| \leq M(t' - t)$, $(t', x') \in \Omega$.

Let t_n be the sequence satisfying (4.31). Then there exist $\xi > 0$ and $N \in \mathbb{N}$ such that

$$(4.34) \qquad \begin{aligned} & 0 < \xi < t_N - t < \frac{\delta}{2}, \\ & |\dot{x}(t_N) - \dot{x}(t)| < \frac{\varepsilon}{2}. \end{aligned}$$

The boundedness condition

$$F(t, x) \subset B(0, L), \quad \text{for all } (t, x) \in \Omega,$$

implies

$$F(t, x) = \bigcap_{\delta > 0} \overline{\text{conv}}\{\varphi(t', x') \mid (t', x') \in \Omega, \quad |(t, x) - (t' - x')| < \delta\} \subset B(0, L).$$

So, if $x(t)$ is a solution of (4.30), we have $|\dot{x}(t)| \leq L$, for almost all $t \in J$.

Applying the Mean Value Theorem, we have that $x(t)$ is Lipschitz-continuous with the Lipschitz constant L,

$$(4.35) \qquad |x(t') - x(t'')| \leq L|t' - t''|, \quad \text{for all } t', t'' \in [0, \tau].$$

Put

$$\psi := \min\left\{\frac{\xi}{2}, \frac{M - L}{L}\frac{\xi}{4}, (m - L)\frac{\xi}{4}\right\}.$$

If $|t' - t_N| < \nu$, then $|t' - t| \leq |t' - t_N| + |t_N - t| < \eta + \delta/2 < \delta$, and so

$$(4.36) \qquad t' < t + \delta.$$

On the other hand, $t' - t = t' - t_N + t_N - t > -\eta + \xi \geq -\xi/2 + \xi > 0$.

Similarly, if $|x' - x(t_N)| < \eta$, then

$$
\begin{aligned}
|x' - x(t')| &\leq |x' - x(t_N)| + |x(t_N) - x(t')| < \eta + L|t_N - t'| \\
(4.37) \qquad &< (M - L)\frac{\xi}{4} + L\eta \leq (M - L)\frac{\xi}{4} + L\frac{M - L}{L}\frac{\xi}{4} \\
&= (M - L)\frac{\xi}{2} < (M - L)(t' - t).
\end{aligned}
$$

Using (4.35) and (4.37), it follows that

$$
\begin{aligned}
(4.38) \qquad |x' - x(t')| &\leq |x'(t') - x(t)| \\
&< (M - L)(t' - t) + L(t' - t) = M(t' - t).
\end{aligned}
$$

In view of (4.36), inequality $t < t'$ and (4.38), we have proved that

$$(4.39) \qquad t \leq t' < t + \delta, \quad |x' - x(t)| \leq M(t' - t).$$

Thus, taking (4.33) into account, relation (4.39) leads to

$$F(t', x') \subset B\left(F(t, x(t)), \frac{\varepsilon}{2}\right),$$

and consequently

$$\phi(t_N, x(t_N)) \subset \overline{\text{conv}}\{F(t', x') \mid (t', x') \in \Omega, \ |t_N - t'| < \eta, \ |x(t_N) - x'| < \eta\}$$

$$\subset B\left(F(t, x(t)), \frac{\varepsilon}{2}\right).$$

This already means that

$$d_H(\dot{x}(t_N), F(t, x(t))) \leq \frac{\varepsilon}{2},$$

which jointly with (4.34) contradicts to (4.32), and consequently the function $x(t)$ is a Carathéodory solution of (4.26).

Since the reverse implication is obvious, the proof is complete. □

Hence, if $x(t, x_0) := x(t, 0, x_0)$ is a solution of (4.26) with $x(0, x_0) = x_0 \in \mathbb{R}^n$, then the translation multioperator $T_\tau : \mathbb{R}^n \multimap \mathbb{R}^n$, at the time $\tau > 0$ along the trajectories of (4.26) can be defined as follows:

(4.40) $T_\tau(x_0) := \{x(\tau, x_0) \mid x(\cdot, x_0) \text{ is a solution of (4.26) with } x(0, x_0) = x_0\}.$

In fact, according to Proposition (4.29), we also have

(4.41) $T_\tau(x_0) := \{x(\tau, x_0) \mid x(\cdot, x_0) \text{ is a solution of (4.30) with } x(0, x_0) = x_0\}.$

T_τ in (4.41) can be considered as the composition of two maps, namely $T_\tau = \psi \circ \varphi$,

$$\mathbb{R}^n \xrightarrow{\varphi} AC([0, \tau], \mathbb{R}^n) \xrightarrow{\psi} \mathbb{R}^n,$$

where $\varphi(x_0) : x_0 \multimap \{x(t, x_0) \mid x(t, x_0) \text{ is a solution of (4.30) with } x(0, x_0) = x_0\}$ is, according to Theorem (2.3) (cf. also Theorem (4.3)) an R_δ-mapping and $\psi(y) : y(t) \to y(\tau)$ is obviously a continuous (single-valued) evaluation mapping. The same is true, according to Proposition (4.29) for (4.40).

In other words, we have the following commutation diagram

$$\mathbb{R}^n \xrightarrow{\varphi} AC([0, \tau], \mathbb{R}^n)$$
$$\begin{array}{ccc} & & \downarrow \psi \\ & \xrightarrow{T_\tau} & \mathbb{R}^n \end{array}$$

Hence, we can summarize our investigations as follows.

(4.42) THEOREM. T_τ defined by (4.40) is admissible and admissibly homotopic to identify. More precisely, T_τ is a composition of an R_δ-mapping and a continuous (single-valued) evalution mapping.

(4.43) REMARK. An appropriately modified version of Remark (4.4) is true here as well.

G. First-return multimap for autonomous systems.

Consider the Marchaud system

$$(4.44) \qquad \dot{x} \in F(x), \quad x = (x_1, \ldots, x_n) \in \mathbb{T}^n = \mathbb{R}^n / \mathbb{Z}^n,$$

on the torus $\mathbb{T}^n = \mathbb{R}^n / \mathbb{Z}^n$, where $F = (f_1, \ldots, f_n)^T : \mathbb{T}^n \multimap \mathbb{R}^n$ is an u.s.c. mapping with nonempty, compact and convex values such that

$$(4.45) \qquad F(\ldots, x_{j+1}, \ldots) \equiv F(\ldots, x_j, \ldots), \quad j = 1, \ldots, n.$$

Consider still the $(n-1)$-dimensional subtorus $\Sigma \subset \mathbb{T}^n$ given by

$$\sum_{j=1}^n x_j = 0 \quad (\text{mod } 1)$$

and assume, additionally, that

$$(4.46) \qquad \inf_{x \in \mathbb{T}^n} \sum_{j=1}^n f_j(x) > 0 \quad \text{or} \quad \sup_{x \in \mathbb{T}^n} \sum_{j=1}^n f_j(x) < 0.$$

Then we can define on the cross section Σ the first-return multimap along the trajectories of (4.44) as follows:

$$(4.47) \qquad T_{\{\tau(x_0)\}}(x_0) : \Sigma \multimap \Sigma, \quad T_{\{\tau(x_0)\}}(x_0) := \{x(\tau(x_0), x_0)\},$$

where $T_0(x_0) = x_0 \in \Sigma$, $x(t, x_0) \in AC([0, \infty), \mathbb{T}^n)$ denotes the Carathéodory solution of (4.44) such that $x(0, x_0) = x_0$ and $\{\tau(x_0)\}$ denotes the least time for x_0 to return back to Σ, when taking into account each solution of $\{x(t, x_0)\}$. Indeed, (4.46) implies that

$$\sum_{j=1}^n \dot{x}_j(t, x_0) \neq 0, \quad \text{for every solution } x(t, x_0) \text{ of } (4.44)$$

and a.a. $t \geq 0$, by which the map $\tau(x_0) : \Sigma \multimap [1/E, 1/\varepsilon]$ is well-defined, where ε, E are positive constants such that

$$0 < \varepsilon \leq \inf_{(t,x) \in [0,\infty) \times \mathbb{T}^n} \left| \sum_{j=1}^n f_j(x) \right| \leq \sup_{(t,x) \in [0,\infty) \times \mathbb{T}^n} \left| \sum_{j=1}^n f_j(x) \right| \leq E.$$

Moreover, (4.46) means geometrically that the trajectories of (4.44), associated to (4.47), intersect Σ in a transversal way, which will be essential into the future.

We will show that $T_{\{\tau(x_0)\}}$ can be again considered as the composition of two maps, namely $T_{\{\tau(x_0)\}} = \psi \circ \varphi$. More precisely, we have

$$\Sigma \xrightarrow{\varphi} AC^* \left(\left[0, \frac{1}{\varepsilon}\right], \mathbb{T}^n\right) \xrightarrow{\psi} \Sigma,$$

where AC^* means the space of all absolutely continuous functions with the properties (cf. (4.46))

$$(4.48) \qquad E \geq \left| \sum_{j=1}^{n} \dot{y}_i(t, x_0) \right| \geq \varepsilon > 0, \quad \text{for a.a. } t \in \left[0, \frac{1}{\varepsilon}\right],$$

$$(4.49) \qquad \varepsilon t \leq \left| \sum_{j=1}^{n} y_i(t, x_0) \right| \leq Et, \quad \text{for } t \in \left[0, \frac{1}{\varepsilon}\right].$$

Here, $\varphi : x_0 \multimap \{x(t, x_0) \mid x(t, x_0) \text{ is solution of (4.44) with } x(0, x_0) = x_0\}$ is, according to Theorem (2.3) (see also the proof of Theorem (4.3)), an R_δ-mapping and $y(t, x_0) \to y(\tau(y), x_0) \in \Sigma$, which is obviously continuous, as far as $\tau(y)$ is so.

Observe that, because of the "asterisque" properties (4.48) and (4.49), $\tau(y)$ is again well-defined and, moreover,

$$(4.50) \qquad \sum_{j-1}^{n} y_j(\tau(y)) - y_i(0) = \pm 1.$$

Hence, applying to (4.50) a suitable implicit function theorem for maps without continuous differentialelity (see e.g. [AuE-M, Theorem 7.5.8]), the map $y \to \tau(y)$ can be easily verified to be, under (4.48), even Lipschitz-continuous, as required.

So, we have again the following commutative diagram

$$\Sigma \xrightarrow{\varphi} AC^* \left(\left[0, \frac{1}{\varepsilon}\right], \mathbb{R}^n\right)$$
$$\xrightarrow[T_{\{\tau\}}]{} \psi \downarrow$$
$$\Sigma$$

and we can summarize our investigations in the following.

(4.51) THEOREM. $T_{\{\tau\}}$ defined by (4.47) is, under (4.45), (4.46), admissible and admissibly homotopic to identity. More precisely, $T_{\{\tau\}}$ is a composition of an R_δ-mapping and a continuous (single-valued) evaluation mapping.

(4.52) REMARK. The admissible homotopy can be exhibited as follows

$$\chi_{\{\lambda\tau(x_0)\}}(x_0) = \Big\{ (x_1(\lambda\tau(x_0), x_0), \dots, x_{n-1}(\lambda\tau(x_0), x_0),$$

$$-\sum_{j=1}^{n-1} x_j(\lambda\tau(x_0), x_0)) \Big\},$$

where $\lambda \in [0, 1]$, because $\chi_0(x_0) = x_0$ and $\chi_{\{\tau(x_0)\}} = T_{\{\tau(x_0)\}}(x_0)$.

(4.53) REMARK. Since $\Lambda(T_{\{\tau(x_0)\}}) = \Lambda(T_0) = \Lambda(\mathrm{id}|_\Sigma) = \chi(\Sigma) = 0$, where Λ is the generalized Lefschetz number and $\chi(\Sigma)$ denotes the Euler–Poincaré characteristic of the subtorus Σ, it has no meaning to look for the fixed points of $T_{\{\tau\}}$. On the other hand, $\Lambda(\mathcal{H} \circ T_{\{\tau(x_0)\}}) = \Lambda(\mathcal{H})$, for any homeomorphism $\mathcal{H} : \Sigma \to \Sigma$.

III.5. Existence results

The results of this chapter are mainly based on the papers [And5], [And14], [AndBa], [AGG1], [AGG2], [AK1], [Kra1], [Kra2].

We start with the application of Theorem (1.32). Hence, let $(E, \|\cdot\|)$ be a Banach space and let $L(E)$ be the space of all linear continuous transformations in E. The Hausdorff measure of noncompactness (MNC) will be denoted by γ.

We are interested in the existence of a bounded solution to the semilinear differential inclusion

$$(5.1) \qquad \dot{x}(t) \in A(t)x(t) + F(t, x(t)), \quad \text{for a.a. } t \in \mathbb{R}$$

with $A(t) \in L(E)$ and a set-valued transformation F.

Our assumptions concerning the problem (5.1) will be the following:

(A1) $A : \mathbb{R} \to L(E)$ is strongly measurable and Bochner integrable, on every compact interval $[a, b]$.

(A2) Assume that

$$(5.2) \qquad \dot{x}(t) = A(t)x(t)$$

admits a regular exponential dichotomy. Denote by G the principal Green's function for (5.2).

(F1) Let $F : \mathbb{R} \times E \multimap E$ be a u-Carathéodory set-valued map such that

$$\|F(t, x)\| \leq m(t), \quad \text{for a.a. } t \in \mathbb{R}, \; x \in E.$$

Here $m \in L^1_{\mathrm{loc}}(\mathbb{R})$ is such that, for a constant M,

$$\sup \left\{ \int_t^{t+1} m(s)\, ds \,\Big|\, t \in \mathbb{R} \right\} < M.$$

(F2) Assume that

$$\gamma(F(t,\Omega)) \le g(t)h(\gamma(\Omega)), \quad \text{for a.a. } t \in \mathbb{R}$$

and each bounded $\Omega \subset E$, where g, h, are positive functions, g is measurable, h is nondecreasing such that

$$L := \sup\left\{ \int_{\mathbb{R}} \|G(t,s)\|_{L(E)} g(s)\, ds \,\Big|\, t \in \mathbb{R} \right\} < \infty$$

and $qh(t)L < t$, for each $t > 0$, with a constant $q = 1$, if E is separable, and $q = 2$, in the general case.

(5.3) THEOREM. *Under the assumptions* (A1), (A2), (F1), (F2), *the semilinear differential inclusion* (5.1) *admits a bounded solution on* $\mathbb{R}$.

The main obstruction in the application of Theorem (1.32) will be the estimation of a suitably chosen MNC. For this purpose, we recall the following rule of taking the MNC under the sign of the integral (see [KOZ-M, Corollary 4.2.5]).

(5.4) LEMMA. *Let* $\{f_n\} \subset L^1([a,b], E)$ *be a sequence of functions such that*

(i) $\|f_n(t)\| \le \nu(t)$, *for all* $n \in \mathbb{N}$ *and a.a.* $t \in [a,b]$, *where* $\nu \in L^1([a,b])$,

(ii) $\gamma(\{f_n(t)\}) \le c(t)$, *for a.a.* $t \in [a,b]$, *where* $c \in L^1([a,b])$.

Then we have the estimate

$$\gamma\left(\left\{ \int_a^t f_n(s)\, ds \right\}\right) \le q \int_a^t c(s)\, ds,$$

for each $t \in [a,b]$, *with* $q = 1$, *if* E *is separable, and* $q = 2$, *in general case.*

PROOF OF THEOREM (5.3). We carry out the proof in several steps.

(i) Let

$$Q := \left\{ x \in C(\mathbb{R}, E) \,\Big|\, \|x(t)\| \le K, \quad \text{for each } t \in \mathbb{R}, \ \|x(t_1) - x(t_2)\| \le \right.$$
$$\left. K \int_{t_1}^{t_2} \|A(s)\|_{L(E)}\, ds + \int_{t_1}^{t_2} m(s)\, ds, \quad \text{for all } t_1, t_2 \in \mathbb{R}, \ t_1 \le t_2 \right\}$$

with a constant K to be specified below. Clearly Q is a closed convex subset of $C(\mathbb{R}, E)$.

For a given $q \in Q$, we are interested in bounded solutions to the differenial inclusion

(5.5) $$\dot{y}(t) \in A(t)y(t) + F(t, q(t)), \quad \text{for a.a. } t \in \mathbb{R}.$$

Take $f \in N_F(q)$ (recall that such f exists, in view of Lemma (1.29), where N_F denotes the Nemytskiĭ operator. Since A admits an exponential dichotomy, we know that the problem

$$\dot{y}(t) = A(t)y(t) + f(t), \quad \text{for a.a. } t \in \mathbb{R},$$

has a unique, entirely bounded solution given by

$$x(f) = \int_{\mathbb{R}} G(t,s)f(s)\,ds$$

(see [DK-M], [MS-M]). Thus, problem (5.5) has a nonempty set of solutions $T(q)$. Using Lemma (1.29), it is also clear that this set is closed convex, and since its compactness will become clear in the subsequent steps of the proof, it is in fact an R_δ-set.

(ii) We will show that, for each $q \in Q$, we actually have $T(q) \subset Q$. Let $x \in T(q)$. Then, for suitable $f \in N_F(q)$, we have

$$
\begin{aligned}
(5.6) \quad \|x(t)\| &\le \int_{\mathbb{R}} \|G(t,s)\|_{L(E)}\|f(s)\|\,ds \\
&\le k\int_{-\infty}^{t} e^{-\mu(t-s)}m(s)\,ds + k\int_{t}^{\infty} e^{-\mu(s-t)}m(s)\,ds \\
&= k\sum_{j=0}^{\infty}\int_{j}^{j+1} e^{-\mu\sigma}m(t-\sigma)\,d\sigma + k\sum_{j=0}^{\infty}\int_{j}^{j+1} e^{-\mu\sigma}m(t+\sigma)\,d\sigma \\
&\le k\sum_{j=0}^{\infty}e^{-\mu j}\int_{j}^{j+1} m(t-\sigma)\,d\sigma + k\sum_{j=0}^{\infty}e^{-\mu j}\int_{j}^{j+1} m(t+\sigma)\,d\sigma \\
&\le k\sum_{j=0}^{\infty}e^{-\mu j}\int_{t-j-1}^{t-j} m(s)\,ds + k\sum_{j=0}^{\infty}e^{-\mu j}\int_{t+j}^{t+j+1} m(s)\,ds \\
&\le 2kM\sum_{j=0}^{\infty}e^{-\mu j} = 2kM(1-e^{-\mu})^{-1} =: K.
\end{aligned}
$$

In this estimation, we have used the fact that, by assumption (A2), there exist positive constants k, μ such that

$$\|G(t,s)\|_{L(e)} \le ke^{-\mu|t-s|}.$$

Now, let $t_1, t_2 \in \mathbb{R}$. Then

$$
\|x(t_1) - x(t_2)\| \le \int_{t_1}^{t_2} \|\dot{x}(s)\|\,ds \le \int_{t_1}^{t_2} \|A(s)\|_{L(E)}\|x(s)\|\,ds + \int_{t_1}^{t_2} \|f(s)\|\,ds
$$

$$
\le K\int_{t_1}^{t_2} \|A(s)\|_{L(E)}\,ds + \int_{t_1}^{t_2} m(s)\,ds.
$$

Consequently, $T(q) \subset Q$.

(iii) Let $\mathcal{M}$ be the power set of Q and define, for each $\Omega \in \mathcal{M}$, the real-valued MNC ψ by

$$\psi(\Omega) := \max_{D \in \mathcal{D}}(\Omega)(\sup_{t \in \mathbb{R}} \gamma(D(t))),$$

where $\mathcal{D}(Q)$ denotes the collection of all denumerable subsets of Ω and $D(t) = \{d(t) \mid d \in D\} \subset E$. Then ψ is well-defined and from the corresponding properties of γ it is clear that ψ has monotone and nonsingular properties of measure of nocompactness (see Definition (5.31) in Chapter I.5). Finally, observe that ψ is regular in view of the Arzelà–Ascoli lemma.

We wish to show that the mapping T given in step (i) is condensing w.r.t. the MNC ψ.

Take $\Omega \in \mathcal{M}$. Considering $T(\Omega)$, we see that by the definition of ψ there exists a sequence $\{x_n\} \subset T(\Omega)$ such that

$$\psi(T(\Omega)) = \sup_{t \in \mathbb{R}} \gamma(\{x_n(t)\}).$$

Thus, for each $n \in \mathbb{N}$, there is $z_n \in \Omega$ and $f_n \in N_F(z_n)$ such that

$$(5.7) \qquad x_n(t) = \int_{\mathbb{R}} G(t,s) f_n(s) \, ds,$$

Let $\varepsilon > 0$ be fixed. Choose a number $a > 0$ such that $Ke^{-\mu a} < \varepsilon$. Analogously to the estimation (5.6), one shows that

$$\left\| \int_{-\infty}^{t-a} G(t,s) f_n(s) \, ds + \int_{t+a}^{\infty} G(t,s) f_n(s) \, ds \right\| \leq \varepsilon,$$

for every $n \in \mathbb{N}$. Using (5.7), we thus infer for an arbitrary $t \in \mathbb{R}$ that

$$\gamma(\{x_n(t)\}) \leq \varepsilon + \gamma\left(\left\{ \int_{t-a}^{t+a} G(t,s) f_n(s) \, ds \right\}\right).$$

Now, from assumption (F1), we get that

$$\|G(t,s) f_n(s)\| \leq \|G(t,s)\|_{L(E)} m(s),$$

for a.a. $s \in \mathbb{R}$ and each $n \in \mathbb{N}$. Furthermore, using assumption (F2) and properties of γ (see Definition (5.31) in Chapter I.5), we see that the following estimate holds, namely

$$\gamma(\{G(t,s) f_n(s)\}) \leq \|G(t,s)\|_{L(E)} \gamma(\{f_n(s)\}) \leq \|G(t,s)\|_{L(E)} g(s) h(\gamma(\{z_n(s)\}))$$
$$\leq \|G(t,s)\|_{L(E)} g(s) h(\psi(\Omega)),$$

for a.a. $s \in \mathbb{R}$. Hence, an application of Lemma (5.4) gives us

$$\gamma(\{x_n(t)\}) \leq \varepsilon + qh(\psi(\Omega)) \int_{t-a}^{t+a} \|G(t,s)\|_{L(E)} g(s) \, ds.$$

It follows that

$$\psi(T(\Omega)) \leq \varepsilon + qh(\psi(\Omega))L,$$

and subsequently, since $\varepsilon > 0$ was arbitrary,

(5.8) $$\psi(T(\Omega)) \leq qh(\psi(\Omega))L.$$

Let us now assume that Ω is not relatively compact. Then $\psi(\Omega) > 0$ and so, by assumption (F2) and (5.8), we obtain

$$\psi(T(\Omega)) \leq \psi(\Omega).$$

Finally, observe that the estimate (5.8) also implies the quasi-compactness of the mapping T which subsequently justifies the compactness of the solution set to (5.5), as claimed.   □

Hence, we have verified all the assumptions of Theorem (1.32) and we can establish the existence of a bounded solution to problem (5.1).

(5.9) REMARK. One can easily check that the assumption (F1) can be replaced by a weaker one, namely

(5.9.1) $$\|F(t,x)\| \leq m(t) + K\|x\|, \quad \text{for a.a. } t \in \mathbb{R}, \ x \in E,$$

where $K \geq 0$ is a sufficiently small constant and m is the same as above.

(5.10) REMARK. If $A : E \to E$ is a linear, bounded operator whose spectrum does not intersect the imaginary axis, then the constant K in (5.9.1) can be easily taken as $K < 1/C(A)$, where

(5.10.1) $$\sup_{t \in \mathbb{R}} \left| \int_{-\infty}^{\infty} \|G(t,s)\| \, ds \right| \leq C(A), \quad G(t,s) = \begin{cases} e^{A(t-s)} P_-, & \text{for } t > s, \\ e^{A(t-s)} P_+, & \text{for } t < s \end{cases}$$

and P_-, P_+ stand for the coresponding spectral projections to the invariant subspaces of A.

(5.11) REMARK. For $E = \mathbb{R}^n$ ($\Rightarrow$ (F2) holds automatically), condition (A2) is satisfied, provided there exists a projection matrix P ($P = P^2$) and constants $k > 0$, $\lambda > 0$ such that

(5.11.1) $$\begin{cases} |X(t)PX^{-1}(s)| \leq k\exp(-\lambda(t-s)), & \text{for } s \leq t, \\ |X(t)(I-P)X^{-1}(s)| \leq k\exp(-\lambda(s-t)), & \text{for } t \leq s, \end{cases}$$

where $X(t)$ is the fundamental matrix of (5.2), satisfying $X(0) = I$, i.e., the unit matrix (cf. Lemma (5.136) below).

If A in (A1) is a piece-wise continuous and periodic, then it is well-known that (5.11.1) takes place, whenever all the associated Floquet multiplies lie off the unit cycle. If A in (A1) is (continuous and) almost-periodic, then it is enough (see [Pal3, p. 70] that (5.11.1) holds only on a half-line $[t_0, \infty)$ or even on a sufficiently long finite interval.

(5.12) REMARK. if $E = \mathbb{R}^n$ ($\Rightarrow$ (F2)) and $A = (a_{ij})$ satisfies the Gershgorin-type inequalities, i.e., either

$$\left(a_{ii} - \sum_{j=1,j\neq i}^{n} |a_{ij}| > 0 \quad \text{or} \quad a_{ii} + \sum_{j=1,j\neq i}^{n} |a_{ij}| < 0, \quad \text{for } i = 1,\dots,n \right)$$

or

$$\left(a_{jj} - \sum_{i=1,i\neq j}^{n} |a_{ij}| > 0 \quad \text{or} \quad a_{jj} + \sum_{i=1,i\neq j}^{n} |a_{ij}| < 0, \quad \text{for } j = 1,\dots,n \right),$$

by which the real parts of the associated eigenvalues of A are either all negative or positive ($\Rightarrow$ (A2)), then the constant $C(A)$ in (5.10.1) can be expressed explicitly in terms of the entries a_{ij}, namely $C(A) = \sum_{k=0}^{n-1} \frac{2^k \|A\|^k}{|\lambda|^{k+1}}$, where $|\lambda|$ denotes the minimum of the absolute values of the eigenvalues of A.

Now, the information concerning the topological structure of solution sets in III.2 and III.3 will be employed for obtaining existence criteria, on the basis of general methods established in III.1.

Applying Theorem (2.12) to Corollary (1.8), we can give

(5.13) THEOREM. *Consider the Cauchy problem*

$$(5.14) \qquad \begin{cases} \dot{x}(t) \in F(t, x(t)), & \text{for a.a. } t \in [0,\infty), \\ x(0) = x_0, \end{cases}$$

where $F : [0,\infty) \times \mathbb{R}^n \multimap \mathbb{R}^n$ is u-Carathéodory product-measurable map, satisfying:

(5.13.1) *There exists a globally integrable function $\alpha : [0,\infty) \to [0,\infty)$ and a positive constant B such that, for every $x \in \mathbb{R}^n$ and for a.a. $t \in [0,\infty)$, we have $|F(t,x)| \leq \alpha(t)(B + |x|)$, where $|F(t,x)| = \sup\{|y| \mid y \in F(t,x)\}$, and $\int_0^\infty \alpha(t)\, dt < \infty$.*

Then problem (5.14) admits a solution, for every $x_0 \in \mathbb{R}^n$,

PROF. Using the well-known Gronwall inequality (see e.g. [Har-M]), problem (5.14) can be equivalently replaced by

$$(5.15) \qquad \begin{cases} \dot{x}(t) \in G(t, x(t)), & \text{for a.a. } t \in [0,\infty), \\ x(0) = x_0, \end{cases}$$

where

$$g(t, x) = \begin{cases} F(t, x), & \text{for } |x| \leq D \text{ and } t \in [0, \infty), \\ F\left(t, D\dfrac{x}{|x|}\right), & \text{for } |x| \geq D \text{ and } t \in [0, \infty), \end{cases}$$

$$D \geq (|x_0| + AB) \exp A, \quad a = \int_0^\infty \alpha(t) \, dt < \infty.$$

Moreover, there certainly exists a positive constant γ such that

(5.16) $|x_0| + |G(t, x)| \leq |x_0| + A(B + D) \leq \gamma$,   for all $x \in \mathbb{R}^n$ and a.a. $t \in [0, \infty)$.

Hence, besides problem (5.15), consider still a one-parameter family of linear problems (notice that a product measurability of F implies the measurability of $G(t, q(t))$)

(5.17) $$\begin{cases} \dot{x}(t) \in G(t, q(t)), & \text{for a.a. } t \in [0, \infty), \, q \in Q, \\ x(t) \in Q \cap S, \end{cases}$$

where

$$S = \{x \in C([0, \infty), \mathbb{R}^n) \mid x(0) = x_0\},$$
$$Q = \mathrm{cl}_{C([0, \infty), \mathbb{R}^n)} \{x(t) \in AC_{\mathrm{loc}}([0, \infty), \mathbb{R}^n) \mid \sup_{t \in [0, \infty)} \mathrm{ess} \, |\dot{x}(t)| \leq \gamma\}.$$

It is well-known that problem (5.17) is equivalent to

$$x(t) \in x_0 + \int_0^t G(s, q(s)) \, ds := T(q),$$

where the integral is understood in the sense of R. J. Aumann (see [Aum], [HP1-M]). It follows from Theorem (2.12) that the problem (5.17) has, for each $q \in Q$, an R_δ-set of solutions and so conditions of Corollary (1.8) are satisfied, whenever $T(Q) \subset Q$. This, however, follows immediately from (5.16):

$$\sup_{t \in [0, \infty)} \left| x_0 + \int_0^t G(s, q(s)) \, ds \right| \leq |x_0| + \int_0^t |G(t, q(t))| \, dt$$

$$\leq |x_0| + (B + D) \int_0^\infty \alpha(t) \, dt$$

$$= |x_0| + A(B + D) < \infty.$$

This also implies the boundedness of solutions.   □

(5.18) REMARK. By the substitution $t := -\tau$, the conclusion of Theorem (5.13) is also true for the problem

$$\begin{cases} \dot{x}(t) \in F(t, x(t)), & \text{for a.a. } t \in (-\infty, 0], \\ x(0) = x_0 \in \mathbb{R}^n, \end{cases}$$

but provided $F : (-\infty, 0] \times \mathbb{R}^n \multimap \mathbb{R}^n$ is a u-Carathéodory product-measurable map satisfying the analogy of condition (5.13.1) on $(-\infty, 0]$ with $\int_{-\infty}^0 \alpha(t) \, dt < \infty$.

(5.19) EXAMPLE. Consider the system

$$
(5.20) \quad
\begin{cases}
\dot{x}_1 \in F_1(t, x_1, x_2)x_1 + F_2(t, x_1, x_2)x_2 + E_1(t, x_1, x_2), \\
\dot{x}_2 \in -F_2(t, x_1, x_2)x_1 + F_2(t, x_1, x_2)x_2 + E_2(t, x_1, x_2),
\end{cases}
$$

where $E_1, E_2, F_1, F_2 : [0, \infty) \times \mathbb{R}^2 \multimap \mathbb{R}^2$ are u-Carathéodory maps.

Assume, furthermore, the existence of positive constants $\widetilde{E}_1, \widetilde{E}_2, \widetilde{F}_1, \widetilde{F}_2, \lambda$ such that

$$
(5.19.1) \quad \sup_{t \in [0,\infty)} \mathrm{ess}[\sup_{|x_i| \le D, i=1,2} F_1(t, x_1, x_2)] \le -\lambda,
$$

$$
(5.19.2) \quad \sup_{t \in [0,\infty)} \mathrm{ess}[\sup_{|x_i| \le D, i=1,2} |F_1(t, x_1, x_2)|] \le \widetilde{F}_1,
$$

$$
(5.19.3) \quad \sup_{t \in [0,\infty)} \mathrm{ess}[\sup_{|x_i| \le D, i=1,2} |F_2(t, x_1, x_2)|] \le \widetilde{F}_2,
$$

$$
(5.19.4) \quad \sup_{t \in [0,\infty)} \mathrm{ess}[\sup_{|x_i| \le D, i=1,2} |E_1(t, x_1, x_2)|] \le \widetilde{E}_1,
$$

$$
(5.19.5) \quad \sup_{t \in [0,\infty)} \mathrm{ess}[\sup_{|x_i| \le D, i=1,2} |E_2(t, x_1, x_2)|] \le \widetilde{E}_2,
$$

where $D = 1/\lambda(\widetilde{E}_1, \widetilde{E}_2)$. Observe that, under the assumptions (5.19.2)–(5.19.5), we have

$$
(5.21) \quad \sup_{t \in [0,\infty)} \mathrm{ess}\, |\dot{x}_i(t)| \le D', \ i = 1, 2,
$$

where $D' = (\widetilde{F}_1 + \widetilde{F}_2)D + \max(\widetilde{E}_1, \widetilde{E}_2)$, so long as the solution $(x_1(t), x_2(t))$ of (5.20) satisfies

$$
(5.22) \quad \sup_{t \in [0,\infty)} |x_i(t)| \le D, \ i = 1, 2.
$$

Our aim is to prove, under the assumptions (5.19.1)–(5.19.5), the existence of a solution $x(t) = (x_1(t), x_2(t))$ satisfying

$$
(5.23) \quad x(0) = 0 \quad \text{and} \quad \sup_{t \in [0,\infty)} |x_i(t)| \le D, \ i = 1, 2.
$$

In order to apply Corollary (1.8) for this goal, define two sets

$$
Q := \{ r(t) = (r_1(t), r_2(t)) \in C([0, \infty) \times [0, \infty), \mathbb{R}^2) \mid \sup_{t \in [0,\infty)} |r_i(t)| \le D, \ i = 1, 2 \},
$$

$$
S := \{ s(t) = (s_1(t), s_2(t)) \in C([0, \infty) \times [0, \infty), \mathbb{R}^2) \cap Q \mid |s_i(t)| \le D't, \ i = 1, 2 \}
$$

(observe that $s(0) = 0$), where Q is a closed convex subset of $C([0, \infty) \times [0, \infty), \mathbb{R}^2)$ and S is a bounded closed subset of Q.

For $q(t) = (q_1(t), q_2(t)) \in Q$, consider still the family of systems

(5.24)
$$\begin{cases} \dot{x}_1 = p_1(t)x_1 + p_2(t)x_2 + r_1(t), \\ \dot{x}_2 = -p_2(t)x_1 + p_1(t)x_2 + r_2(t), \end{cases}$$

where $p_1(t) \subset F_1(t, q(t))$, $p_2(t) \subset F_2(t, q(t))$, $r_1(t) \subset E_1(t, q(t))$, $r_2(t) \subset E_2(t, q(t))$ are measurable selections.

To show the solvability of (5.20) and (5.23) by means of Corollary (1.8), we need to verify that, for each $q \in Q$, the linearized system

(5.25)
$$\begin{cases} \dot{x}_1 \in F_1(t, q(t))x_1 + F_2(t, q(t))x_2 + E_1(t, q(t)), \\ \dot{x}_2 \in -F_2(t, q(t))x_1 + F_1(t, q(t))x_2 + E_2(t, q(t)) \end{cases}$$

has an R_δ-set of solutions in S.

It is well-known that the general solution $x(t, 0, \xi)$ of (5.24), where $\xi = (\xi_1, \xi_2) \in \mathbb{R}^2$, reads as follows:

$$x_1(t, 0, \xi) = \left[\xi_1 \cos\left(\int_0^t p_2(s)\, ds \right) + \xi_2 \sin\left(\int_0^t p_2(s)\, ds \right) \right] \exp \int_0^t p_1(s)\, ds$$
$$+ \int_0^t \left[r_1(s) \exp \int_s^t p_1(w)\, dw \cos\left(\int_s^t p_2(w)\, dw \right) \right] ds$$
$$+ \int_0^t \left[r_2(s) \exp \int_s^t p_1(w)\, dw \sin\left(\int_s^t p_2(w)\, dw \right) \right] ds,$$

$$x_2(t, 0, \xi) = \left[-\xi_1 \sin\left(\int_0^t p_2(s)\, ds \right) + \xi_2 \cos\left(\int_0^t p_2(s)\, ds \right) \right] \exp \int_0^t p_1(s)\, ds$$
$$- \int_0^t \left[r_1(s) \exp \int_s^t p_1(w)\, dw \sin\left(\int_s^t p_2(w)\, dw \right) \right] ds$$
$$+ \int_0^t \left[r_2(s) \exp \int_s^t p_1(w)\, dw \cos\left(\int_s^t p_2(w)\, dw \right) \right] ds.$$

Because of (5.19.1), (5.19.4) and (5.19.5), we get

$$\sup_{t \in [0,\infty)} \left| \int_0^t \left[r_i(s)\, ds \exp \int_s^t p_1(w)\, dw \cos\left(\int_s^t p_2(w)\, dw \right) \right] ds \right|$$
$$\leq \tilde{E}_i \sup_{t \in [0,\infty)} \int_0^t \exp\left[-\int_s^t |p_1(w)|\, dw \right] ds \leq \frac{\tilde{E}_i}{\lambda},$$

$$\sup_{t \in [0,\infty)} \left| \int_0^t \left[r_i(s)\, ds \exp \int_s^t p_1(w)\, dw \sin\left(\int_s^t p_2(w)\, dw \right) \right] ds \right|$$
$$\leq \tilde{E}_i \sup_{t \in [0,\infty)} \int_0^t \exp\left[-\int_s^t |p_1(w)|\, dw \right] ds \leq \frac{\tilde{E}_i}{\lambda},$$

for $i = 1, 2$, and subsequently we arrive at

$$(5.26) \qquad \sup_{t \in [0,\infty)} |x_i(t, 0, \xi)| \le |\xi_1| + |\xi_2| + D, \quad i = 1, 2,$$

and $x(0, 0, \xi) = \xi$.

According to Theorem (2.12) (see also (5.26)), problem (5.25) $\cap$ (5.24) has the R_δ-set of solutions $x(t, 0, 0)$. Moreover, in view of the indicated implication ((5.22) $\Rightarrow$ (5.21)), these solutions $x(t, 0, 0)$ belong obviously to S, for every $q \in Q$, as required.

Thus, it follows from Corollary (1.8) that problem (5.20) $\cap$ (5.23) has, under the assumptions (5.19.1)–(5.19.5), at least one solution.

If the inequality (5.19.1) or both inequalities (5.19.4) and (5.19.5) are sharp, then the same conclusion is true for $x(0) = 0$ in (5.23) replaced by $x(0) = \xi$, where $|\xi|$ is sufficiently small. For bigger values of $|\xi|$, the above assumptions can be appropriately modifed as well.

(5.27) THEOREM. *Consider the target problem*

$$(5.28) \qquad \begin{cases} \dot{x}(t) \in F(t, x(t)), & \text{for a.a. } t \in [0, \infty), \\ \lim_{t \to \infty} x(t) = x_\infty \in \mathbb{R}^n \end{cases}$$

and assume that $F : [0, \infty) \times \mathbb{R}^n \multimap \mathbb{R}^n$ is a u-Carathéodory product-measurable map satisfying condition (5.13.1). Then problem (5.28) admits a (bounded) solution, for every $x_\infty \in \mathbb{R}^n$.

PROOF. Again, it is convenient to consider, instead of problem (5.28), the equivalent problem

$$(5.29) \qquad \begin{cases} \dot{x}(t) \in G(t, x(t)), & \text{for a.a. } t \in [0, \infty), \\ \lim_{t \to \infty} x(t) = x_\infty \in \mathbb{R}^n, \end{cases}$$

where

$$G(t, x) = \begin{cases} F(t, x), & \text{for } |x| \le D \text{ and } t \in [0, \infty), \\ F\left(t, D\dfrac{x}{|x|}\right), & \text{for } |x| \ge D \text{ and } t \in [0, \infty), \end{cases}$$

$$D \ge (|x_\infty| + AB)\exp A, \quad A = \int_0^\infty \alpha(t)\, dt < \infty \quad \text{and (5.16) holds.}$$

Besides problem (5.29), consider still a one-parameter family of linear problems (5.17), where this time

$$S = \{x \in C([0, \infty), \mathbb{R}^n) \mid \lim_{t \to \infty} x(t) = x_\infty\},$$

$$Q = \{q \in C([0, \infty), \mathbb{R}^n) \mid |q(t)| \le |x_\infty| + A(B + D), \quad \text{for } t \ge 0\}.$$

Consider the set

$$S_1 = \left\{ x \in Q \,\middle|\, |x(t) - x_\infty| \le (B + D) \int_t^\infty \alpha(s)\, ds, \quad \text{for } t \ge 0 \right\} \subset S.$$

It is evident that S_1 is a closed subset of S and all solutions to problem (5.17) belong to S_1.

At first, we assume that $G = g$ is single-valued. Then we have a single-valued continuous operator

$$T(q) = x_\infty + \int_\infty^t g(s, q(s))\, ds, \quad \text{for every } q \in Q.$$

Thus, to apply Corollary (1.11), only the condition $T(Q) \subset Q$ should be verified. But this follows immediately from (5.16), because

$$\sup_{t \in [0,\infty)} \left| x_\infty + \int_\infty^t G(s, q(s))\, ds \right| \le |x_\infty| + \int_0^\infty |G(t, q(t))|\, dt$$

$$\le |x_\infty| + (B + D) \int_0^\infty \alpha(t)\, dt = |x_\infty| + A(B + D) < \infty.$$

By Corollary (1.11), we obtain a solution to the problem with g as a right-hand side. This existence result can be used jointly with Theorem (3.3) which is needed to prove our statement in a general case. In fact, in view of the (just proved) existence result and Theorem (3.2), the map T which assigns to every $q \in Q$ the set of solutions to the linear problem (5.17), has nonempty, acyclic sets of values. Once more, we use Corollary (1.11), obtaining a solution to problem (5.28), and the proof is complete. □

(5.30) REMARK. By the substitution $t := -\tau$, the conclusion of Theorem (5.27) is also true for the problem

$$\begin{cases} \dot{x}(t) \in F(t, x(t)), & \text{for a.a. } t \in (-\infty, 0], \\ x(-\infty) = x_{-\infty} \in \mathbb{R}^n, \end{cases}$$

but provided $F : (-\infty, 0] \times \mathbb{R}^n \multimap \mathbb{R}^n$ is a u-Carathéodory product-measurable map satisfying the analogy of condition (5.13.1) on $(-\infty, 0]$ with $\int_{-\infty}^0 \alpha(t)\, dt < \infty$.

Summarizing Theorems (5.13) and (5.27), we can conclude, in view of Remarks (5.18) and (5.30), by the following

(5.31) COROLLARY. *Let* $F : \mathbb{R}^n \multimap \mathbb{R}^n$ *be a u-Carathéodory product-measurable map satisfying the analogy of condition* (5.13.1) *on* $(-\infty, \infty)$, *with* $\int_{-\infty}^\infty \alpha(t)\, dt < \infty$. *Then the inclusion*

$$\dot{x}(t) \in F(t, x(t)), \quad \text{for a.a. } t \in (-\infty, \infty),$$

admits an entirely bounded solution $x_-(t)$ *on* $(-\infty, \infty)$ *with* $x_-(-\infty) = x_{-\infty} \in \mathbb{R}^n$
and an entirely bounded solution $x_+(t)$ *on* $(-\infty, \infty)$ *with* $x_+(\infty) = x_\infty \in \mathbb{R}^n$. *If,*
additionally, $F(t, x) \equiv -F(-t, x)$ *holds, then at least one solution* $x(t)$ *exists with*
$x(-\infty) = x(\infty) = x_\infty$, *for each* $x_\infty \in \mathbb{R}^n$.

Now, we shall deal with boundary value problems of the type (3.11), i.e.

$$(5.32) \quad \begin{cases} \dot{x}(t) + A(t)x(t) \in F(t, x(t)), & \text{for a.a. } t \in [0, \tau], \\ Lx = \Theta, \end{cases}$$

where

(5.32.1) $A : [0, \tau] \to \mathcal{L}(\mathbb{R}^n, \mathbb{R}^n)$ is a measurable linear operator such that $|A(t)| \leq$
$\gamma(t)$, for all $t \in [0, \tau]$ and some integrable function $\gamma : [0, \tau] \to [0, \infty)$,

(5.32.2) the associated homogeneous problem

$$\begin{cases} \dot{x}(t) + A(t)x(t) = 0, & \text{for a.a. } t \in [0, \tau], \\ Lx = 0 \end{cases}$$

has only the trivial solution,

(5.32.3) $F : [0, \tau] \times \mathbb{R}^n \multimap \mathbb{R}^n$ is an (upper) Carathéodory mapping with nonempty,
compact and convex values,

(5.32.4) there are two nonnegative Lebesgue-integrable functions $\delta_1, \delta_2 : [0, \tau] \to$
$[0, \infty)$ such that

$$|F(t, x)| \leq \delta_1(t) + \delta_2(t)|x|, \quad \text{for a.a. } t \in [0, \tau] \text{ and all } x \in \mathbb{R}^n,$$

where $|F(t, x)| = \sup\{|y| \mid y \in F(t, x)\}$.

Applying Theorem (3.8) (cf. also (3.13)) to replace condition (1.12.1) in Corollary (1.12) for (5.32), we can immediately give

(5.33) PROPOSITION. *Consider problem* (5.32) *with* (5.32.1)–(5.32.4) *and let*
$G : [0, \tau] \times \mathbb{R}^n \times \mathbb{R}^n \times [0, 1] \to \mathbb{R}^n$ *be a product-measurable u-Carathéodory map*
such that

$$G(t, c, c, 1) \subset F(t, c), \quad \text{for all } (t, c) \times [0, \tau] \times \mathbb{R}^n.$$

Assume, furthermore, that

(5.33.1) *there exists a (bounded) retract* Q *of* $C([0, \tau], \mathbb{R}^n)$ *such that* $Q \setminus \partial Q$ *is*
nonempty (open) and such that $G(t, x, q(t), \lambda)$ *is Lipschitzian in* x *with a*
sufficiently small Lipschitz constant (see (3.8) *and* (3.13)), *for a.a.* $t \in [0, \tau]$
and each $(q, \lambda) \in Q \times [0, 1]$,

(5.33.2) *there exists a Lebesgue integrable function* $\alpha : [0, \tau] \to [0, \infty)$ *such that*

$$|G(t, x(t), q(t), \lambda)| \leq \alpha(t), \quad \text{a.e. in } [0, \tau]$$

for any $(x, q, \lambda) \in \Gamma_T$ (i.e. from the graph of T), where T denotes the set-valued map which assigns, to any $(q, \lambda) \in Q \times [0,1]$, the set of solutions of

$$\begin{cases} \dot{x}(t) + A(t)x(t) + G(t, x(t), q(t), \lambda), & \text{for a.a. } t \in [0,1], \\ Lx = \Theta, \end{cases}$$

(5.33.3) $T(Q \times \{0\}) \subset Q$ *holds and* ∂Q *is fixed point free w.r.t.* T, *for every* $(q, \lambda) \in Q \times [0,1]$.

Then problem (5.32) has a solution.

(5.34) REMARK. Rescaling t in (5.32), interval $[0, \tau]$ can be obviously replaced in Proposition (5.33) by any compact interval J, e.g. $J = [-m, m]$, $m \in \mathbb{N}$. Therefore, the second part of Proposition (1.37) can be still applied for obtaining an entirely bounded solution.

For periodic and anti-periodic problems, Proposition (5.33) can be easily simplified as follows (cf. (5.11)).

(5.35) COROLLARY. *Consider problem*

$$\begin{cases} \dot{x}(t) + A(t)x(t) \in F(t, x(t)), & \text{for a.a. } t \in [0, \tau], \\ x(0) = x(\tau), \end{cases}$$

where $F(t, x) \equiv F(t + \tau, x)$ *satisfies conditions (5.32.3) and (5.32.4). Let* $G : [0, \tau] \times \mathbb{R}^n \times \mathbb{R}^n \times [0,1] \multimap \mathbb{R}^n$ *be a product-measurable u-Carathéodory map such that*

$$G(t, c, c, 1) \subset F(t, c), \quad \text{for all } (t, c) \in [0, \tau] \times \mathbb{R}^n.$$

Assume that

(5.35.1) A *is a piece-wise continuous (single-valued) bounded τ-periodic $(n \times n)$-matrix whose Floquet multipliers lie off the unit cycle, jointly with (5.33.1) –(5.33.3), where $Lx = x(0) - x(\tau)$ and $\Theta = 0$.*

Then the inclusion $\dot{x} + A(t)x \in F(t, x)$ *admits a τ-periodic solution.*

(5.36) COROLLARY. *Consider problem*

$$\begin{cases} \dot{x}(t) \in F(t, x(t)), & \text{for a.a. } t \in [0, \tau], \\ x(0) = -x(\tau), \end{cases}$$

where $F(t, x) \equiv -F(t + \tau, -x)$ *satisfies conditions (5.32.3) and (5.32.4). Let* $G : [0, \tau] \times \mathbb{R}^n \times \mathbb{R}^n \times [0,1] \multimap \mathbb{R}^n$ *be a product-measurable u-Carathéodory map such that*

$$G(t, c, c, 1) \subset F(t, c), \quad \text{for all } (t, c) \in [0, \tau] \times \mathbb{R}^n.$$

Assume that (5.33.1)–(5.33.3) hold, where $Lx = x(0) + x(\tau)$ *and* $\Theta = 0$.

Then the inclusion $\dot{x} \in F(t, x)$ *admits a 2τ-periodic solution.*

(5.37) EXAMPLE. Consider problem (5.32). Assume that (5.32.1)–(5.32.3) are satisfied and that $F : [0, \tau] \times \mathbb{R}^n \multimap \mathbb{R}^n$. Taking

$$G(t, q(t)) = F(t, q(t)), \quad \text{for } q \in Q,$$

where $Q = \{\mu \in C([0, \tau], \mathbb{R}^n) \mid \max_{t \in [0,\tau]} |\mu(t)| \leq D\}$ and $D > 0$ is a sufficiently big constant which will be specified below, we can see that (5.33.1) holds trivially. Furtheromore, according to (5.32.4), we get

$$(5.37.1) \qquad |G(t, q(t))| \leq \delta_1(t) + \delta_2(t)D, \quad \text{for a.a. } t \in [0, \tau],$$

i.e. (5.33.2) holds as well with $\alpha(t) = \delta_1(t) + \delta_2(2)D$. At last, the associated linear problem

$$\begin{cases} \dot{x}(t) \in A(t)x(t) + F(t, x(t)), & \text{for a.a. } t \in [0, \tau], \\ Lx = \Theta \end{cases}$$

has, according to Theorem (3.8), for every $q \in Q$, an R_δ-set of solutions of the form (see e.g. [Co1] or Lemma (5.136) below)

$$T(q) = \int_0^\tau H(t, s) f(s, q(s)) \, ds,$$

where H is the related (to problem in (4.32.2)) Green function and $f \subset F$ is a measurable selection.

Therefore, in order to apply Proposition (5.33) for the solvability of (5.32), we only need to show (cf. (5.33.3)) that $T(Q) \subset Q$ (and that ∂Q is fixed point free w.r.t. T, for every $q \in Q$, which is, however, not necessary here). Hence, in view of (5.37.1), we have that

$$\max_{t \in [0,\tau]} |T(q)| = \max_{t \in [0,\tau]} \left| \int_0^\tau H(t, s) f(s, q(s)) \, ds \right|$$

$$\leq \max_{t \in [0,\tau]} \int_0^\tau |H(t, s)|(\delta_1(s) + \delta_2(s)D) \, ds$$

$$= \max_{t,s \in [0,\tau]} |H(t, s)| \left[\int_0^\tau \delta_1(t) \, dt + D \int_0^\tau \delta_2(t) \, dt \right],$$

and subsequently the above requirement holds for

$$D \geq \frac{\max_{t,s \in [0,\tau]} |H(t, s)| \int_0^\tau \delta_1(t) \, dt}{1 - \max_{t,s \in [0,\tau]} |H(t, s)| \int_0^\tau \delta_2(t) \, dt},$$

provided

$$\int_0^\tau \delta_2(t) \, dt < \frac{1}{\max_{t,s \in [0,\tau]} |H(t, s)|}.$$

(Observe that for D strictly bigger than the above quantity, ∂Q becomes fixed point free).

In the case of differential equations, Corollary (5.36) can be still improved, in view of Theorem (3.20), where $\lambda = 1$ and $\xi = 0$, as follows.

(5.38) COROLLARY. *Consider problem*

$$\begin{cases} \dot{x}(t) = f(t, x(t)), & \text{for a.a. } t \in [0, \tau], \\ x(0) = -x(\tau), \end{cases}$$

where $f(t, x) \equiv -f(t + \tau, -x)$ is a Carathéodory function. Let $g : [0, \tau] \times \mathbb{R}^n \times \mathbb{R}^n \times [0, 1] \to \mathbb{R}^n$ be a Carathéodory function such that

$$g(t, c, c, 1) = f(t, c), \quad \text{for all } (t, c) \in [0, \tau] \times \mathbb{R}^n.$$

Assume that

(5.38.1) *there exists a bounded retract Q of $C([0, \tau], \mathbb{R}^n)$ such that $Q \setminus \partial Q$ is non-empty (open) and such that $g(t, x, q(t), \lambda)$ satisfies*

$$|g(t, x, q(t), \lambda) - g(t, y, q(t), \lambda)| \le p(t)|x - y|, \quad x, y \in \mathbb{R}^n$$

for a.a. $t \in [0, \tau]$ and each $(q, \lambda) \in Q \times [0, 1]$, where $p : [0, \tau] \to [0, \infty)$ is a Lebesgue integrable function with (see (3.23))

$$\int_0^\tau p(t)\, dt \le \pi,$$

(5.38.2) *there exists a Lebesgue integrable function $\alpha : [0, \tau] \to [0, \infty)$ such that*

$$|g(t, x(t), q(t), \lambda)| \le \alpha(t), \quad \text{a.e. in } [0, \tau],$$

for any $(x, q, \lambda) \in \Gamma_T$, where T denotes the set-valued map which assigns, to any $(q, \lambda) \in Q \times [0, 1]$, the set of solutions of

$$\begin{cases} \dot{x}(t) = g(t, x(t), q(t), \lambda), & \text{for a.a. } t \in [0, \tau], \\ x(0) = -x(\tau), \end{cases}$$

(5.38.3) $T(Q \times \{0\}) \subset Q$ *holds and ∂Q is fixed point free w.r.t. T, for every $(q, \lambda) \in Q \times [0, 1]$.*

Then the equation $\dot{x} = f(t, x)$ admits a 2τ-periodic solution.

(5.39) REMARK. Since in Corollaries (5.35) and (5.36) the associated homogeneous problems (cf. (5.32.2)) have obviously only the trivial solution, the requirement $T(Q \times \{0\}) \subset Q$ reduces to $0 \in Q$, provided $G(t, x, q, \lambda) = \lambda G(t, x, \lambda)$, $\lambda \in [0, 1]$.

Now, consider the inclusion with constant coefficients $a_j > 0$, $j = 1, \dots, n$,

$$(5.40) \qquad x^{(n)} + \sum_{j=1}^n a_j x^{(n-j)} \in F(t, x, \dots, x^{(n-1)}),$$

where $F : \mathbb{R}^{n+1} \multimap \mathbb{R}^n$ is a u-Carathéodory mapping with nonempty, compact and convex values and $a_j \in \mathbb{R}$ are at least such that $\operatorname{Re}\lambda_j \neq 0$, $j = 1, \dots, n$, where λ_j are roots of the associated characteristic polynomial

$$(5.41) \qquad \lambda^n + \sum_{j=1}^{n} a_j \lambda^{n-j}.$$

Consider still the family of linearized inclusions

$$(5.42) \qquad x^{(n)} + \sum_{j=1}^{n} a_j x^{(n-j)} \in F(t, q(t), \dots, q^{(n-1)}(t)),$$

where the linear part is the same as above and

$$q \in Q := \{u(t) \in C^{(n-1)}(\mathbb{R}^n, \mathbb{R}^n) \mid \sup_{t \in (-\infty, \infty)} |u^{(k)}(t)| \leq D_k, \quad k = 0, \dots, n-1\}.$$

In order to apply Corollary (1.14) for obtaining the existence of an entirely bounded solution of (5.40), estimates of entirely bounded solutions (and their derivatives) of the inclusion (5.42) will be crucial. Consider therefore at first, the hyperbolic equation (i.e. the real parts of roots of the associated polynomial (5.41) are nonzero)

$$(5.43) \qquad x^{(n)} + \sum_{j=1}^{n} a_j x^{(n-j)} = p(t),$$

where $p : \mathbb{R} \multimap \mathbb{R}$ is a measurable function such that

$$(5.44) \qquad \sup_{t \in \mathbb{R}} \operatorname{ess} |p(t)| =: P < +\infty.$$

Writing roots of the characteristic polynomial (5.41) in the form $\lambda_j = \alpha_j + i\beta_j$, $\alpha_j \in \mathbb{R}$, $\alpha_j \neq 0$, $\beta_j \in \mathbb{R}$, for $j = 1, \dots, n$, we can denote

$$\Lambda_j := (\infty)\alpha_j, \quad j = 1, \dots, n.$$

In the sequel, we proceed by the similar way as in [AT1] or [Kra1].

(5.45) LEMMA. *The function $\varphi : \mathbb{R} \to \mathbb{R}$ given by the formula*

$$(5.46) \quad \varphi(t) = e^{\lambda_1 t} \int_{\Lambda_1}^{t} e^{(\lambda_2 - \lambda_1)t_1} \int_{\Lambda_2}^{t_1} e^{(\lambda_3 - \lambda_2)t_2}$$

$$\cdots \int_{\Lambda_{n-1}}^{t_{n-2}} e^{(\lambda_n - \lambda_{n-1})t_{n-1}} \int_{\Lambda_n}^{t_{n-1}} e^{-\lambda_n t_n} p(t_n) \, dt_n \cdots dt_1$$

represents a (Carathéodory) solution of the equation (5.43).

PROOF. Obviously, φ has a locally absolutely continuous $(n-1)$st derivative. By the straightforward computation, one can verify that, for $n=1$, the function

$$\varphi(t) = e^{\lambda_1 t} \int_{\Lambda_1}^t e^{-\lambda_1 t_1} p(t_1)\, dt_1$$

satisfies the differential equation $\dot{x} + a_1 x = p(t)$ (with the coefficient $a_1 = -\lambda_1$), a.e. in $\mathbb{R}$.

Now, suppose that the assertion of our lemma holds, for $n = k - 1 \geq 1$, and show its validity for $n = k$.

It follows from the hypothesis that the function φ given by (5.46), for $n = k$, and taking the form

$$\varphi(t) = e^{\lambda_1 t} \int_{\Lambda_1}^t \cdots \int_{\Lambda_{k-1}}^{t_{k-2}} e^{-\lambda_{k-1} t_{k-1}} \left(e^{\lambda_k t_{k-1}} \int_{\Lambda_k}^{t_{k-1}} e^{-\lambda_k t_k} p(t_k)\, dt_k \right) dt_{k-1} \ldots dt_1,$$

obeys the differential equation of the $(k-1)$st order

$$(5.47) \qquad x^{(k-1)} + \sum_{j=1}^{k-1} b_j x^{(k-1-j)} = e^{\lambda_k t} \int_{\Lambda_k}^t e^{-\lambda_k t_k} p(t_k)\, dt_k,$$

with the coefficients uniquely determined by the relation

$$(5.48) \qquad \lambda^{k-1} + \sum_{j=1}^{k-1} b_j \lambda^{k-1-j} = \prod_{j=1}^{k-1} (\lambda - \lambda_j).$$

The function $\varphi^{(k-1)}$ is locally absolutely continuous. Therefore, φ satisfies the equation

$$(5.49) \qquad x^{(k)} + \sum_{j=1}^{k-1} b_j x^{(k-j)} = \lambda_k e^{\lambda_k t} \int_{\Lambda_k}^t e^{-\lambda_k t_k} p(t_k)\, dt_k + p(t),$$

a.e. in $\mathbb{R}$. Subtracting the λ_k multiple of (5.47) from (5.49), we obtain that φ is a solution of the differential equation

$$(5.50) \qquad x^{(k)} + (b_1 - \lambda_k) x^{(k-1)} + \sum_{j=2}^{k-1} (b_j - \lambda_k b_{j-1}) x^{(k-j)} - \lambda_k b_{k-1} = p(t).$$

A trivial verification shows that

$$\lambda^k + (b_1 - \lambda_k) \lambda^{k-1} + \sum_{j=2}^{k-1} (b_j - \lambda_k b_{j-1}) \lambda^{k-j} - \lambda_k b_{k-1} = \prod_{j=1}^k (\lambda - \lambda_j)$$

holds for the characteristic polynomial of (5.50), which completes the proof. □

(5.51) LEMMA. *The equation (5.43) has exactly one entirely bounded solution. This solution is given by the formula (5.46) and satisfies the inequality*

$$(5.52) \qquad \sup_{t \in \mathbb{R}} |\varphi(t)| \leq \frac{P}{|\alpha_1 \ldots \alpha_n|}.$$

PROOF. Uniqueness follows immediately from the hyperbolicity of (5.43). For $n = 1$, we have

$$(5.53) \qquad |\varphi(t)| \leq |e^{\lambda_1 t}| \left| \int_{\lambda_1}^t |e^{-\lambda_1 t_1}| |p(t_1)| \, dt_1 \right|$$

$$\leq \frac{P}{|\alpha_1|} e^{\alpha_1 t} \lim_{s \to \Lambda_1} |a^{-\alpha_1 t} - e^{-\alpha_1 s}| = \frac{P}{|\alpha_1|}, \quad t \in \mathbb{R}.$$

Now, let (5.52) hold, for $n = k - 1 \geq 1$. The function

$$\varphi(t) = e^{\lambda_1 t} \int_{\Lambda_1}^t \cdots \int_{\Lambda_{k-1}}^{t_{k-2}} e^{-\lambda_{k-1} t_{k-1}} \left(e^{\lambda_k t_{k-1}} \int_{\Lambda_k}^{t_{k-1}} e^{-\lambda_k t_k} p(t_k) \, dt_k \right) dt_{k-1} \ldots dt_1$$

is then an entirely bounded solution of the equation (5.47). Applying the inductive assumption and (5.53), we can write

$$\sup_{t \in \mathbb{R}} |\varphi(t)| \leq \frac{\sup_{t \in \mathbb{R}} \left| e^{\lambda_k t} \int_{\lambda_k}^t e^{-\lambda_k t_k} p(t_k) \, dt_k \right|}{|\alpha_1 \ldots \alpha_{k-1}|} \leq \frac{1}{|\alpha_1 \ldots \alpha_{k-1}|} \frac{P}{|\alpha_k|},$$

which is the desired estimate. □

(5.54) CONSEQUENCE. *If the characteristic polynomial (5.41) has only real nonzero roots, then the estimate*

$$(5.55) \qquad \sup_{t \in \mathbb{R}} |\varphi(t)| \leq \frac{P}{|a_n|}.$$

holds for φ in (5.46).

PROOF. This result follows immediately from the Vieta formula

$$a_n = (-1)^n \prod_{j=1}^n \lambda_j.$$

□

(5.56) CONSEQUENCE. *Assume that the forcing term p in the hyperbolic equation (5.43) is a T-periodic function. Then (5.43) admits exactly one T-periodic solution. This solution is given by the formula (5.46). If p is a T-anti-periodic function, then the function φ in (5.43) represents exactly one T-anti-periodic solution of the equation (5.43).*

PROOF. Consider the unique entirely bounded solution φ of (5.43). If p is a T-periodic, then the function ψ, $\psi(t) := \varphi(t + T)$, represents also the entirely bounded weak solution of (5.43), and we conclude that $\psi = \varphi$, i.e. T-periodicity of φ. Writing $\chi(t) := -\varphi(t + T)$, we derive by the same manner the second assertion.   □

As we could see, there exists exactly one entirely bounded solution of the equation (5.43). Consequently, φ given by the formula (5.46) takes the form

$$(5.57) \quad \varphi(t) = e^{\lambda_{i_1} t} \int_{\Lambda_{i_1}}^{t} e^{(\lambda_{i_2} - \lambda_{i_1})t_1} \int_{\Lambda_{i_2}}^{t_1} e^{(\lambda_{i_3} - \lambda_{i_2})t_2}$$

$$\dots \int_{\Lambda_{i_{n-1}}}^{t_{n-2}} e^{(\lambda_{i_n} - \lambda_{i_{n-1}})t_{n-1}} \int_{\Lambda_{i_n}}^{t_{n-1}} e^{-\lambda_{i_n} t_n} p(t_n) \, dt_n \dots dt_1,$$

where $(i_1, \dots, i_n)$ is any permutation of $(1, \dots, n)$. Denoting briefly the right-hand side of (5.57) by $[i_1, \dots, i_n]$, we can write (5.57) in the form

$$(5.58) \qquad\qquad \sup_{t \in \mathbb{R}} |[i_1, \dots, i_n]| \le \frac{P}{|\alpha_{i_1}, \dots, \alpha_{i_n}|}.$$

(5.59) LEMMA.

$$(5.60) \qquad\qquad \frac{d([i_1, \dots, i_n])}{dt} = \lambda_{i_1}[i_1, \dots, i_n] + [i_2, \dots, i_n].$$

PROOF. The formula (5.60) follows immediately from the rule for derivating the product.   □

(5.61) LEMMA.

$$(5.62) \quad \frac{d^m([1, \dots, n])}{dt^m} = [m+1, \dots, n] + \sum_{c_1 = 1}^{m} \lambda_{c_1}[c_1, m+1, \dots, n]$$

$$\sum_{\substack{c_1, c_2 = 1 \\ c_1 < c_2}}^{m} \lambda_{c_1} \lambda_{c_2}[c_1, c_2, m+1, \dots, n] + \dots$$

$$+ \sum_{\substack{c_1, \dots, c_p = 1 \\ c_1 < \dots < c_p}}^{m} \left(\prod_{j=1}^{p} \lambda_{c_j} \right) [c_1, \dots, c_p, m+1, \dots, n] + \dots$$

$$+ \left(\prod_{j=1}^{m} \lambda_j \right) [1, \dots, n], \quad m = 1, \dots, n-1.$$

PROOF. We will proceed by the mathematical induction method. Since, for $m = 1$, the assertion follows from Lemma (5.59) (see (5.60)), we want to show its validity for $m = k$, provided it is true for $m = k - 1$ $((n - 2) \geq (k - 1) \geq 1)$.

Hence,

$$
\frac{d^k([1,\ldots,n])}{dt^k} = \frac{d}{dt}\left(\frac{d^{k-1}([1,\ldots,n])}{dt^{k-1}}\right) = ([k+1,\ldots,n])
$$

$$
+ \left(\lambda_k[k,\ldots,n] + \sum_{c_1=1}^{k-1}\lambda_{c_1}[c_1,k+1,\ldots,n]\right)
$$

$$
+ \sum_{c_1=1}^{k-1}\lambda_{c_1}\lambda_k[c_1,k,\ldots,n] + \ldots
$$

$$
+ \sum_{\substack{c_1,\ldots,c_{p-1}=1 \\ c_1<\ldots<c_{p-1}}}^{k-1}\left(\prod_{j=1}^{p-1}\lambda_{c_j}\right)[c_1,\ldots,c_{p-1},k+1,\ldots,n]
$$

$$
+ \left(\sum_{\substack{c_1,\ldots,c_{p-1}=1 \\ c_1<\ldots<c_{p-1}}}^{k-1}\left(\prod_{j=1}^{p-1}\lambda_{c_j}\right)\lambda_k[c_1,\ldots,c_{p-1},k,\ldots,n]\right.
$$

$$
+ \sum_{\substack{c_1,\ldots,c_p=1 \\ c_1<\ldots<c_p}}^{k-1}\left(\prod_{j=1}^{p}\lambda_{c_j}\right)[c_1,\ldots,c_p,k+1,\ldots,n]\Bigg)
$$

$$
+ \sum_{\substack{c_1,\ldots,c_p=1 \\ c_1<\ldots<c_p}}^{k-1}\left(\prod_{j=1}^{p}\lambda_{c_j}\right)\lambda_k[c_1,\ldots,c_p,k,\ldots,n] + \ldots
$$

$$
+ \left(\left(\prod_{j=1}^{k-1}\lambda_{c_j}\right)\lambda_k[1,\ldots,n]\right) = [k+1,\ldots,n]
$$

$$
+ \sum_{c_1=1}^{k}\lambda_{c_1}[c_1,k+1,\ldots,n] + \ldots
$$

$$
+ \sum_{\substack{c_1,\ldots,c_p=1 \\ c_1<\ldots<c_p}}^{k}\left(\prod_{j=1}^{p}\lambda_{c_j}\right)[c_1,\ldots,c_p,k+1,\ldots,n] + \ldots
$$

$$
+ \left(\prod_{j=1}^{k}\lambda_j\right)[1,\ldots,n].
$$

$\square$

(5.63) LEMMA. *If the characteristic polynomial (5.41) has only real nonzero* *roots, then the derivatives of φ in (5.46) satisfy the estimates*

$$(5.64) \qquad \sup_{t \in \mathbb{R}} \left| \frac{d^m \varphi(t)}{dt^m} \right| \leq \frac{2^m P}{|a_n|} \prod_{j=1}^{m} |\alpha_j|, \quad m = 1, \ldots, n-1.$$

PROOF. At first, we can estimate each term in (5.62) (see (5.58))

$$\left| \left(\prod_{j=1}^{p} \alpha_{c_j} \right) [c_1, \ldots, c_p, m+1, \ldots, n] \right|$$

$$\leq \left| \left(\prod_{j=1}^{p} \alpha_{c_j} \right) \right| \frac{P}{|(\prod_{j=1}^{p} \alpha_{c_j})(\prod_{j=m+1}^{n} \alpha_j)|} = \frac{P}{|\prod_{j=m+1}^{n} \alpha_j|} = \frac{P}{|a_n|} \left| \prod_{j=1}^{m} \alpha_j \right|.$$

Now, we can summarize these 2^m estimates to establish the desired formula.   □

(5.65) REMARK. Independence of the estimate (5.64), under the permutation of the roots, is evident and so we have

$$(5.66) \qquad \sup_{t \in \mathbb{R}} \left| \frac{d^m \varphi(t)}{dt^m} \right| \leq \frac{2^m P}{|a_n|} \prod_{j=1}^{m} |\alpha_{i_j}|, \quad m = 1, \ldots, n-1.$$

(5.67) LEMMA. *Let all roots of the characteristic polynomial (5.41) be negative and denote $a_0 := 1$. Then*

$$(5.68) \qquad \sup_{t \in \mathbb{R}} \left| \frac{d^m \varphi(t)}{dt^m} \right| \leq \frac{2^m a_m P}{\binom{n}{m} a_n}, \quad m = 0, \ldots, n-1.$$

PROOF. Consequence (5.56) ensures the validity of (5.68), for $m = 0$. Fix $m \in \{1, \ldots, n-1\}$. Then the inequality in (5.66) can be written in the form

$$(5.69) \qquad \sup_{t \in \mathbb{R}} \left| \frac{d^m \varphi(t)}{dt^m} \right| \leq \frac{2^m P}{a_n} (-1)^m \prod_{j=1}^{m} \alpha_{i_j}.$$

We have $\binom{n}{m}$ choices of the roots $\alpha_{i_1}, \ldots, \alpha_{i_m}$ for n roots of (5.41). Summarize the $\binom{n}{m}$ inequalities of the type (5.69) and divide by $\binom{n}{m}$. Now, to verify the lemma, it is sufficient to apply the Vieta formula

$$\sum_{\substack{c_1, \ldots, c_m = 1 \\ c_1 < \ldots < c_m}}^{n} (-1)^m \prod_{j=1}^{m} \alpha_{c_j} = a_m. \qquad\qquad □$$

(5.70) LEMMA. *Consider the sequence of the "shifted polynomials"*

$$\lambda^{n-p} + \sum_{j=1}^{n-p} a_j \lambda^{n-p-j}, \quad p = 0, \dots, n-1,$$

and assume that each of them has only real nonzero roots. Then the estimate

$$(5.71) \qquad \sup_{t \in \mathbb{R}} \left| \frac{d^m \varphi(t)}{dt^m} \right| \leq \frac{2^m P}{|a_{n-m}|}, \quad m = 0, \dots, n-1,$$

holds for φ in (5.46).

PROOF. As we could see, the derivatives of the solution φ of (5.43), up to the $(n-1)$th order, are entirely bounded. Substituting $y = \dot{x}$, we get the equation

$$(5.72) \qquad y^{(n-1)} + \sum_{j=1}^{n-1} a_j y^{(n-1-j)} = p(t) - a_n \varphi(t)$$

with exactly one entirely bounded solution $\dot{\varphi}$. Applying (5.55), we obtain

$$\sup_{t \in \mathbb{R}} \left| \frac{d\varphi(t)}{dt} \right| \leq \frac{\sup_{t \in \mathbb{R}} |p(t) - a_n \varphi(t)|}{|a_{n-1}|} \leq \frac{P + |a_n| \frac{P}{|a_m|}}{|a_{n-1}|} = \frac{2P}{|a_{n-1}|}.$$

Writing $z = \dot{y}$, we have the equation

$$z^{(n-2)} + \sum_{j=1}^{n-2} a_j z^{(n-2-j)} = p(t) - a_n \varphi(t) - a_{n-1} \dot{\varphi}(t)$$

with exactly one entirely bounded solution $\ddot{\varphi}$. Applying (5.55) again, we get

$$\sup_{t \in \mathbb{R}} |p(t) - a_n \varphi(t) - a_{n-1} \dot{\varphi}(t)| \leq P + \frac{|a_n| P}{|a_n|} + \frac{2|a_{n-1}| P}{|a_{n-1}|} = 4P,$$

and consequently

$$\sup_{t \in \mathbb{R}} \left| \frac{d^2 \varphi(t)}{dt^2} \right| \leq \frac{4P}{|a_{n-2}|},$$

by the same reason as above. Proceeding by the same way, till the substitution $w = \varphi^{(n-1)}$, leading to

$$\sup_{t \in \mathbb{R}} \left| \frac{d^{n-1} \varphi(t)}{dt^{n-1}} \right| \leq \frac{2^{n-1} P}{|a_1|},$$

we get successively all the estimates in (5.71), which proves the lemma. $\square$

(5.73) REMARK. It can be shown, for $m \neq 0$, that the estimate in (5.71) is better than the one in (5.68) (see [AT1]).

Applying Corollary (1.14), Lemma (5.67) and Consequence (5.56), we are ready to establish

(5.74) THEOREM. *Let all roots of the characteristic polynomial* (5.41) *be negative. Put* $a_0 := 1$ *and assume the existence of real nonnegative numbers* $C_0, \ldots, C_{n-1}$ *such that the inequalities*

$$(5.75) \qquad \underset{\substack{t \in \mathbb{R}, \; |x_j| \leq C_j \\ j=0,\ldots,n-1}}{\mathrm{sup\,ess}} |F(t, x_0, \ldots, x_{n-1})| \leq \binom{n}{m} \frac{a_n}{2^m a_m} C_m, \quad m = 0, \ldots, n-1,$$

hold for a u-*Carathéodory mapping* F *with nonempty, compact and convex values. Then the inclusion* (5.40) *admits an entirely bounded solution.*

If, additionally, F *is* T-*periodic in* t, *then there exists at least one* T-*periodic solution of* (5.40).

At last, under the assumption (5.75), (5.40) *admits a* T-*anti-periodic solution, provided*

$$F(t + T, -x_0, \ldots, -x_{n-1}) \equiv -F(t, x_0, \ldots, x_{n-1})$$

holds, for every $(t, x_0, \ldots, x_{n-1}) \in \mathbb{R}^{n+1}$.

PROOF. Define (the evidently nonempty closed, convex and bounded subset of the Fréchet space $C^{(n-1)}(\mathbb{R}, \mathbb{R})$):

$$Q = S := \{q \in C^{(n-1)}(\mathbb{R}, \mathbb{R}) \mid |q^m(t)| \leq C_m, \; t \in \mathbb{R}, \; m = 0, \ldots, n-1\}.$$

Choose any $q \in Q$ and put

$$p(t) := f(t, q(t), \ldots, q^{(n-1)}(t)) \subset F(t, q(t), \ldots, q^{n-1}(t)), \quad t \in \mathbb{R},$$

where f is a measurable selection. It follows from the proof of Theorem (5.3) that the associated linearized inclusion (5.42) has, for each $q \in Q$, an R_δ-set of entirely bounded solutions. Moreover, because of Lemmas (5.51) and (5.67), the problem $(5.42) \cap (x \in S)$ must have an R_δ-set of solutions as well. Since (5.75) immediately ensures also the validity of the condition (1.14.2), Corollary (1.14) implies the existence of an entirely bounded solution $x(t)$ of the inclusion (5.40) with $x \in S$.

Defining

$$S_1 := S \cap \{q \in C(\mathbb{R}, \mathbb{R}) \mid q(t) = q(t + T), \; t \in \mathbb{R}\}$$

and using, additionally, Consequence (5.56), we get by the same way the existence of a T-periodic solution $x(t)$ of (5.40) with $x \in S$.

To prove the third assertion, we put

$$S_2 := S \cap \{q \in C(\mathbb{R}, \mathbb{R}) \mid q(t) = -q(t + T), \; t \in \mathbb{R}\}$$

and repeat the previous procedure. $\qquad \square$

(5.76) THEOREM. *Consider the family of the "shifted polynomials"*

$$\lambda^{n-p} + \sum_{j=1}^{n-p} a_j \lambda^{n-p-j}, \quad p = 0, \ldots, n-1,$$

and assume that each of them has only real nonzero roots. Moreover, let there exist real nonnegative numbers $D_0, \ldots, D_{n-1}$ such the u-Carathéodory mapping F with nonempty, compact and convex values satisfies the following inequalities

$$(5.77) \qquad \sup_{\substack{t \in \mathbb{R}, \, |x_j| \leq D_j \\ j=0,\ldots,n-1}} \mathrm{ess} \, |F(t, x_0, \ldots, x_{n-1})| \leq \frac{|a_{n-m}|}{2^m} D_m, \quad m = 0, \ldots, n-1.$$

Then the same assertions as in Theorem (5.74) hold for the inclusion (5.40).

PROOF. Using the estimates (5.71), instead of (5.68), we can repeat the method of the previous proof. □

(5.78) THEOREM. *Let all the roots λ_j of the characteristic polynomial (5.41) have nonzero real parts, i.e. $\mathrm{Re}\lambda_j \neq 0$, $j = 1, \ldots, n$. Put $a_0 := 1$ and assume the existence of real nonnegative numbers $C_0, \ldots, C_{n-1}$ such that the inequalities*

$$\sup_{\substack{t \in \mathbb{R}, \, |x_j| \leq C_j \\ j=0,\ldots,n-1}} \mathrm{ess} \, |F(t, x_0, \ldots, x_{n-1})| \leq \frac{|\prod_{j=m+1}^{n} \mathrm{Re}\lambda_j|}{2^m} C_m, \quad m = 0, \ldots, n-1,$$

hold for a u-Carathéodory mapping F with nonempty, compact and convex values. Then the same assertion as in Theorem (5.74) is true for the inclusion (5.40).

PROOF. Using, instead of (5.68), the estimates (5.52) and

$$\sup_{t \in \mathbb{R}} \left| \frac{d^m \varphi(t)}{dt^m} \right| \leq \frac{2^m P}{|\prod_{j=m+1}^{n} \alpha_j|}, \quad m = 1, \ldots, n-1,$$

obtained quite analogously as (5.64) (see the proof of (5.63)), we can completely repeat the manner in the proof of Theorem (5.74). □

(5.79) REMARK. Clearly, another statement analogous to Theorem (5.74) can be given (see Consequence (5.56) and Lemma (5.63)), provided the characteristic polynomial (5.41) has only real nonzero roots. In this way, (5.75) can be replaced by

$$\sup_{\substack{t \in \mathbb{R}, \, |x_j| \leq C_j \\ j=0,\ldots,n-1}} \mathrm{ess} \, |F(t, x_0, \ldots, x_{n-1})| \leq \frac{|a_n|}{2^m \prod_{j=1}^{m} |\lambda_j|} C_m, \quad m = 0, \ldots, n-1,$$

where $\prod_{j=1}^{0} |\lambda_j| := 1$.

The equation

$$(5.80) \qquad x^{(n)} + \sum_{j=1}^{n-1} a_j x^{(n-j)} = p(t)$$

comes, after the substitution $y = \dot{x}$, to the one

$$(5.81) \qquad y^{(n-1)} + \sum_{j=1}^{n-1} a_j y^{(n-j-1)} = p(t)$$

with the associated characteristic polynomial

$$(5.82) \qquad \lambda^{n-1} + \sum_{j=1}^{n-1} a_j \lambda^{n-j-1}.$$

If all the roots λ_j of (5.82) have negative real parts, i.e. $\mathrm{Re}\lambda_j < 0$, for $j = 1, \dots, n-1$, then the general solution $y_0(t)$ of the homogeneous equation

$$y^{(n-1)} + \sum_{j=1}^{n-1} a_j y^{(n-j-1)} = 0$$

obviously satisfies

$$\lim_{t \to \infty} y_0^{(k)}(t) = 0, \quad k = 0, \dots, n-2, (n-1).$$

Since the particular solution of (5.81) can be taken as $y_p(t) = \dot{\varphi}(t)$, where φ is defined in (5.46), we can easily modify, under the assumption (5.44), Lemmas (5.51), (5.63), (5.67) and (5.70) as follows.

(5.83) LEMMA. *Let all the roots λ_j, $j = 1, \dots, n-1$, of the polynomal (5.82) have negative real parts and let (5.44) hold. Then every solution $x(t)$ of (5.80) satisfies*

$$(5.84) \qquad \limsup_{t \to \infty} |x^{(k)}(t)| \leq \frac{2^{k-1}P}{|\prod_{j=1}^{n-k} \mathrm{Re}\lambda_j|}, \quad k = 1, \dots, n-1.$$

If all the roots of (5.82) are still real, then (5.84) can be replaced by ($a_0 := 1$)

$$(5.85) \qquad \limsup_{t \to \infty} |x^{(k)}(t)| \leq \frac{2^{k-1} a_{k-1} P}{\binom{n-1}{k-1} a_{n-1}}, \quad k = 1, \dots, n-1.$$

Let all the roots of the "shifted polynomials"

$$(5.86) \qquad \lambda^{n-p} + \sum_{j=1}^{n-p} a_j \lambda^{n-p-j}, \quad p = 1, \ldots n - 1,$$

be negative and let (5.44) hold. Then every solution $x(t)$ of (5.80) satisfies

$$(5.87) \qquad \limsup_{t \to \infty} |x^{(k)}(t)| \leq \frac{2^{k-1} P}{a_{n-k}}, \quad k = 1, \ldots, n - 1.$$

Hence, every solution $x(t)$ of (5.80) satisfies, under the assumptions of Lemma (5.83), the inequalities

$$(5.88) \qquad \limsup_{t \to \infty} |x^{(k)}(t)| \leq \widetilde{D}_k P = D_k, \quad k = 1, \ldots, n - 1,$$

where the constants $\widetilde{D}_k$ are defined in (5.84) or (5.85) or (5.87), respectively.

Consider again the inclusion (5.40), but where this time the coefficients $a_j > 0$, $j = 1, \ldots n - 1$, satisfy the the respective assuptions of Lemma (5.83) and $a_n > 0$ is sufficiently small.

Let there exist nonnegative numbers $C_1, \ldots, C_{n-1}$ such that

$$(5.89) \qquad \underset{t \in \mathbb{R}}{\sup \operatorname{ess}}[\underset{\substack{x_0 \in \mathbb{R}, \, |x_j| \leq C_j \\ j=1,\ldots,n-1}}{\sup} |F(t, x_0, \ldots, x_{n-1})|] < \frac{C_m}{\widetilde{D}_m}, \quad m = 1, \ldots, n - 1,$$

where the constants $\widetilde{D}_m$ are defined in (5.84), (5.85), (5.87) and (5.88), respectively. Then every solution $x(t)$ of the associated linearized inclusion (5.42) such that

$$\sup_{t \in \mathbb{R}} |x(t)| \leq C_0$$

obviously satisfies, for every

$$q \in Q_x = \{u(t) \in C^{(n-1)}(\mathbb{R}, \mathbb{R}) \mid \sup_{t \in \mathbb{R}} |u^{(j)}(t)| \leq C_j, \, j = 1, \ldots, n - 1\},$$

the inequality

$$\limsup_{t \to \infty} |x^{(k)}(t)| < C_k + a_n C_0 \widetilde{D}_k, \quad k = 1, \ldots, n - 1.$$

In particular, every ω-periodic solution $x(t)$ of (5.42) with $\max_{t \in [0,\omega]} |x(t)| < C_0$ satisfies, under (5.89),

$$(5.90) \qquad \sup_{t \to \infty} |x^{(k)}(t)| < C_k + a_n C_0 \widetilde{D}_k, \quad k = 1, \ldots, n - 1.$$

If $a_n > 0$ is sufficiently small, then the real parts of all the roots of the polynomial (5.41) must be negative as well, because the related well-known Routh–Hurwitz conditions concern polynomials w.r.t. a_n, having near to the initial value (for $a_n \to 0^+$) positive values. Therefore, according to Theorem (3.9), the inclusion (5.42) has, for every ω-periodic $q \in Q_x$, an R_δ-set of ω-periodic solutions, provided

$$(5.91) \qquad F(t, x_0, \dots, x_{n-1}) \equiv F(t + \omega, x_0, \dots, x_{n-1}).$$

Moreover, if $\max_{t \in [0,\omega]} |x(t)| \leq C_0$, then the estimates (5.90) hold, under (5.89), for this set.

Hence, in order to apply Corollary (1.14), for obtaining the existence of an ω-periodic solution of the original inclusion (5.40), we only need to show that the ω-periodic solutions $x(t)$ of (5.42) with $q \in Q = \{u(t) \in C^{(n-1)}(\mathbb{R}, \mathbb{R}) \mid \sup_{t \in \mathbb{R}} |u^{(j)}(t)| \leq C_j, j = 0, \dots, n-1$ and $u(t) \equiv u(t + \omega)\}$ satisfy $x \in Q$ and, in particular,

$$(5.92) \qquad \sup_{t \in \mathbb{R}} |x(t)| \leq C_0.$$

Substituting such $x(t)$ into (5.42) and integrating the obtained inclusion from 0 to ω, we get

$$(5.93) \qquad \int_0^\omega [a_n x(t) - f_q(t)] dt = 0,$$

where $f_q(t) \subset F(t, q(t), \dots, q^{(n-1)}(t))$ is an ω-periodic measurable solection. Assuming still that a number $R > 0$ exists such that

$$(5.94) \qquad F(t, x_0, \dots, x_{n-1}) \operatorname{sgn} x_0 \geq 0, \quad \text{for } |x_0| \geq R, \ |x_j| \leq C_j,$$

one can readily check that (5.93) implies $t_0 \in [0, \omega]$ with

$$|x(t_0)| = \liminf_{t \to \infty} |x(t)| = \min_{t \in [0,\omega]} |x(t)| \leq R,$$

because otherwise a contradiction occurs.

Thus, we arrive at

$$\max_{t \in [0,\omega]} |x(t)| \leq |x(t_0)| + \int_{t_0}^{t_0+\omega} |\dot{x}(t)| dt \leq R + \omega(C_1 + a_n C_0 \tilde{D}_1).$$

Taking, for $a_n < 1/\tilde{D}_1$, $C_0 := (R + \omega C_1)/(1 - \omega a_n \tilde{D}_1)$, (5.92) easily follows, jointly with (cf. (5.90))

$$\sup_{x \in \mathbb{R}} |x^{(k)}(t)| \leq C_k + a_n \tilde{D}_k \frac{R + \omega C_1}{1 - \omega a_n \tilde{D}_1}.$$

It could be seen from here that a_n can be always taken small enough in order

$$\sup_{x \in \mathbb{R}} |x^{(k)}(t)| \leq C_k, \quad k = 1, \dots, n-1,$$

i.e., together with (5.92), $x \in Q$, as required. Corollary (1.14), whose all assumptions are satisfied, yields so the existence of an ω-periodic solution of (5.40).

We can formulate

(5.95) THEOREM. *Let the assumptions of Lemma (5.83), concerning the co-efficients $a_j > 0$, $j = 1, \ldots, n-1$, be satisfied. Assume that there exist real nonnegative numbers $C_1, \ldots, C_{n-1}$ such that (cf. (5.89))*

$$\sup_{\substack{(t,x_0)\in\mathbb{R}^2,\ |x_j|\leq C_j \\ j=1,\ldots,n-1}} \text{ess} \quad |F(t,x_0,\ldots,x_{n-1})| + a_n \frac{R + \omega C_1}{1 - \omega a_n \widetilde{D}_1} \leq \frac{C_m}{\widetilde{D}_m}, \quad m = 1,\ldots,n-1,$$

holds, for a product-measurable u-Carathéodory mapping $F : \mathbb{R}^{n+1} \multimap \mathbb{R}^n$ with nonempty, compact and convex values, satisfying (5.91) and (5.94), where the constants $\widetilde{D}_m$ are defined in (5.84), (5.85), (5.87) and (5.88). Then the inclusion (5.40) admits an ω-periodic solution, provided $a_n > 0$ is sufficiently small.

In the single-valued case, we can also consider a slightly more general equation than (5.40), namely

$$(5.96) \qquad x^{(n)} + \sum_{j=1}^{n-1} a_j x^{(n-j)} + h(x) = f(t, x, \ldots, x^{(n-1)}),$$

where the coefficients $a_j > 0$, $j = 1, \ldots, n-1$, satisfy the assumptions of Lemma (5.83), $h : \mathbb{R} \to \mathbb{R}$ is a continuous bounded function and $f : \mathbb{R}^{n+1} \to \mathbb{R}^n$ is a Carathéodory mapping.

Let, furthermore, nonnegative numbers $C_1, \ldots, C_{n-1}$ exist such that

$$(5.97) \qquad \sup_{x_0\in\mathbb{R}} |h(x_0)| + \sup_{t\in\mathbb{R}} \text{ess}[\sup_{\substack{x_0\in\mathbb{R},\ |x_j|\leq C_j \\ j=1,\ldots,n-1}} |f(t,x_0,\ldots,x_{n-1})|] \leq \frac{C_m}{\widetilde{D}_m},$$

$m = 1, \ldots, n-1$, where the constants $\widetilde{D}_m$ are again defined in (5.84), (5.85), (5.87) and (5.88). Assuming that

$$(5.98) \qquad f(t, x_0, \ldots, x_{n-1}) \equiv f(t + \omega, x_0, \ldots, x_{n-1}),$$

one can verify quite analogously as in the proof of Theorem (5.95) that every ω-periodic solution $x(t)$ of the associated linearized equation

$$(5.99) \qquad x^{(n)} + \sum_{j=1}^{n-1} a_j x^{(n-j)} + h(x) = f(t, q(t), \ldots, q^{(n-1)}(t))$$

satisfies, for each $q \in Q = \{u(t) \in C^{(n-1)}(\mathbb{R}, \mathbb{R}) \mid \max_{t\in\mathbb{R}} |u^{(j)}(t)| \leq C_j$, $j = 0, \ldots, n-1$ and $u(t) \equiv u(t+\omega)\}$, the inequalities (cf. (5.90))

$$(5.100) \qquad \sup_{t\in\mathbb{R}} |x^{(k)}(t)| \leq C_k, \quad k = 1, \ldots, n-1,$$

provided

$$(5.101) \qquad \sup_{t\in\mathbb{R}} |x(t) - \overline{x}| \leq C_0, \quad \text{for some } \overline{x} \in \mathbb{R}.$$

In order to verify (5.101), substitute $x(t)$ into (5.99) and integrate the obtained identity from 0 to ω. The resulting equality

$$\int_0^\omega [h(x(t)) - f(t, q(t), \dots, q^{(n-1)}(t))]dt = 0$$

leads, under

$$(5.102) \quad h(x)\mathrm{sgn}(x-\overline{x}) < - \sup_{\substack{(t,x)\in\mathbb{R}^2, |x_j|\leq C_j \\ j=1,\dots,n-1}} \mathrm{ess} \quad |f(t, x_0, \dots, x_{n-1})|, \quad \text{for } |x-\overline{x}| > R,$$

where $R > 0$ is a suitable number, to the existence of $t_0 \in [0, \omega]$ such that

$$|x(t_0) - \overline{x}| := \liminf_{t\to\infty} |x(t) - \overline{x}| = \min_{t\in[0,\omega]} |x(t) - \overline{x}| \leq R.$$

Moreover,

$$\max_{t\in[0,\omega]} |x(t) - \overline{x}| \leq |x(t_0) - \overline{x}| + \omega \max_{t\in[0,\omega]} |\dot{x}(t)| \leq R + \omega C_1.$$

Thus, taking $C_0 := R + \omega C_1$, (5.101) really holds.

If, additionally, $n/2$ is an odd integer and

$$(5.103) \qquad \sum_{k<n/4} a_{n-4k}\left(\frac{\omega}{2\pi}\right)^{n-4k} < 1, \quad \text{whenever } n > 3,$$

together with

$$(5.104) \quad [h(x) - h(y)](x - y) < 0, \quad \text{for all } x \neq y$$
$$\text{with } |x - \overline{x}| \leq R + \omega C_1 \text{ and } |y - \overline{x}| \leq R + \omega C_1,$$

then any ω-periodic solution $x(t)$ of (5.99) is unique on the domain $|x - \overline{x}| \leq C_0 := R + \omega C_1$.

Indeed. Let $x(t)$ and $[x(t) - y(t)]$ be two such solutions. Substituting them into (5.99) and subtracting the obtained identities, we have

$$y^{(n)}(t) + \sum_{j=0}^{n-1} a_j y^{(n-j)}(t) + h(x(t) + y(t)) - h(x(t)) = 0.$$

Multiplying this equation by $y(t)$ and intergating from 0 to ω, we obtain that

$$(-1)^{n/2} \int_0^\omega y^{(n/2)^2}(t) + \sum_{j=1}^{n-1} (-1)^{(n-j)/2} a_j \int_0^\omega y_+^{((n-j)/2)^2}(t)\, dt$$

$$+ \int_0^\omega ((h(x(t) + y(t)) - h(x(t)y(t)))\, dt = 0,$$

where

$$y_+^{(k/2)^2}(t) := \begin{cases} y^{(k/2)^2}(t), & \text{for } k\text{-even,} \\ 0, & \text{for } k\text{-odd.} \end{cases}$$

Using the well-known Wirtinger inequality (see e.g. [MW-M, p. 9]

$$\int_0^\omega \dot{y}^2(t)\, dt \le \left(\frac{\omega}{2\pi}\right)^2 \int_0^\omega \ddot{y}^2(t)\, dt,$$

one can easily check, under (5.103) and (5.104), that $y(t) \equiv 0$, as claimed.

Applying Corollary (1.14), whose all assumptions are satisfied, the equation (5.96) admits an ω-periodic solution, provided (5.97), (5.98), (5.102)–(5.104), and we can state

(5.105) THEOREM. *Let the coefficients $a_j > 0$ $j = 1, \ldots, n-1$, satisfy the assumptions of Lemma (5.83), jointly with (5.103), where $n/2$ is assumed to be an odd integer. Let, furthermore, nonnegative numbers $C_1, \ldots, C_{n-1}$ exist such that (5.97) holds for a continuous function $h : \mathbb{R} \to \mathbb{R}$ and a Carathéodory mapping $f : \mathbb{R}^{n+1} \to \mathbb{R}^n$, satisfying (5.98), where the constants $\tilde{D}_m$, $m = 1, \ldots, n-1$, are defined in (5.84), (5.85), (5.87) and (5.88). At last, let a number $R > 0$ exist such that (5.102) and (5.104) hold. Then the equation (5.96), where $n/2$ is an odd integer, admits an ω-periodic solution.*

In particular, for $f(t, x_0, \ldots, x_{n-1}) \equiv f(t)$, the equation (5.96) admits a unique ω-periodic solution.

The anti-periodic problem can be treated for the differential inclusion with non-complete linear part, namely ($p \in \mathbb{N}$)

$$(5.106) \qquad x^{(n)} + \sum_{j=1}^{n-p} a_j x^{(n-j)} \in F(t, x, \ldots, x^{(n-1)}), \quad 1 \le p \le n-1,$$

where $F : \mathbb{R}^{n+1} \multimap \mathbb{R}^n$ is a product-measurable u-Carathéodory mapping with nonempty, compact and convex values, satisfying

$$(5.107) \qquad F(t, x_0, \ldots, x_{n-1}) \equiv -F(t + \omega, -x_0, \ldots, -x_{n-1}).$$

Consider still the associated linearized system

$$(5.108) \qquad x^{(n)} + \sum_{j=1}^{n-p} a_j x^{(n-j)} \in F(t, q(t), \dots, q^{(n-1)}(t)), \quad 1 \le p \le n-1,$$

for $q \in Q = \{u(t) \in C^{(n-1)}(\mathbb{R}, \mathbb{R}) \mid \max|u^{(j)}(t)| \le C_j, \ j = 0, \dots, n-1$ and $u(t) \equiv -u(t+\omega)\}$.

Assuming that

$$(5.109) \qquad \operatorname*{sup\,ess}_{\substack{t \in \mathbb{R},\ |x_j| \le C_j \\ j=0,\dots,n-1}} |F(t, x_0, \dots, x_{n-1})| := P < \infty,$$

we can easily reformulate Lemma (5.83) for (5.108), as follows.

(5.110) LEMMA. *Let all the roots* λ_j, $j = 1, \dots, n-p$, $(1 \le p \le n-1)$ *of the polynomial*

$$(5.111) \qquad \lambda^{n-p} + \sum_{j=1}^{n-p} a_j \lambda^{n-j-p}, \quad 1 \le p \le n-1,$$

have negative real parts and let (5.109) *hold. Then every solution* $x(t)$ *of* (5.108) *satisfies*

$$(5.112) \qquad \limsup_{t \to \infty} |x^{(k)}(t)| \le \frac{2^{k-p} P}{|\prod_{j=1}^{n-k} \operatorname{Re} \lambda_j|}, \quad k = p, \dots, n-1.$$

If all the roots of (5.111) *are still real, then* (5.112) *can be replaced by* ($a_0 := 1$)

$$(5.113) \qquad \limsup_{t \to \infty} |x^{(k)}(t)| \le \frac{2^{k-p} a_{k-p} P}{\binom{n-p}{k-p} a_{n-p}}, \quad k = p, \dots, n-1.$$

Let all the roots of the "shifted polynomials"

$$(5.114) \quad \lambda^{n-p-r} + \sum_{j=1}^{n-p-r} a_j \lambda^{n-p-r-j}, \quad p = 1, \dots, n-1; \ r = 1, \dots, n-p-1,$$

be negative and let (5.109) *hold. Then every solution* $x(t)$ *of* (5.108) *satisfies*

$$(5.115) \qquad \limsup_{t \to \infty} |x^{(k)}(t)| \le \frac{2^{k-p} P}{a_{n-k}}, \quad k = p, \dots, n-1.$$

Hence, every ω-anti-periodic solution $x(t)$ of (5.108) satisfies, under the assumptions of Lemma (5.110), (5.109) and (5.107), the inequalities

$$(5.116) \qquad \max_{t \in [0,\omega]} |x^{(k)}(t)| \le \widetilde{D}_k P = D_k, \quad k = p, \dots, n-1,$$

where the constants $\widetilde{D}_k$ are defined in (5.112) or (5.113) or (5.115), respectively.

Because of anti-periodicity of $x(t)$, the estimates (5.116) also imply that

$$\max_{t\in[0,\omega]} |x^{(k)}(t)| \le \omega^{p-k} \max_{t\in[0,\omega]} |x^{(p)}(t)| := \omega^{p-k} D_p, \quad k=0,\dots,p-1,$$

i.e.

$$(5.117) \qquad \max_{t\in[0,\omega]} |x^{(k)}(t)| \le \omega^{p-k} \widetilde{D}_p P, \quad k=0,\dots,p-1, \ (1 \le p \le n-1).$$

Therefore, if nonnegative constants $C_0,\dots,C_{n-1}$ exist such that

$$(5.118) \qquad \operatorname*{sup\,ess}_{\substack{t\in\mathbb{R},\ |x_j|\le C_j \\ j=0,\dots,n-1}} |F(t,x_0,\dots,x_{n-1})| \le \frac{C_m}{\omega^{p-m}\widetilde{D}_p},$$

where $m=0,\dots,p-1,\ (1 \le p \le n-1)$, jointly with

$$(5.119) \qquad \operatorname*{sup\,ess}_{\substack{t\in\mathbb{R},\ |x_j|\le C_j \\ j=0,\dots,n-1}} |F(t,x_0,\dots,x_{n-1})| \le \frac{C_m}{\widetilde{D}_m}, \quad m=p,\dots,n-1,$$

then every ω-anti-periodic solution $x(t)$ of (5.108) belongs, under (5.107), to Q, i.e. $x \in Q$.

Moreover, since the associated homogeneous equation

$$x^{(n)} + \sum_{j=1}^{n-p} a_j x^{n-j} = 0, \quad 1 \le p \le n-1,$$

has, under the assumptions of Lemma (5.110), only trivial anti-periodic solutions, the linearized inclusion (5.109) has, according to Theorem (3.8), an R_δ-set of ω-anti-periodic solutions, $x(t) \in Q$, for every $q \in Q$, provided (5.107), (5.118) and (5.119).

Applying Corollary (1.14), whose all assumptions are satisfied, the inclusion (5.106) admits an ω-anti-periodic solution, provided (5.107), (5.118), (5.119), and we can state

(5.120) THEOREM. *Let the coefficients $a_j > 0$, $j=1,\dots,n-p$, $(1 \le p \le n-1)$ satisfy the assumptions of Lemma (5.110). Let, furthermore, nonnegative numbers $C_0,\dots,C_{n-1}$ exist such that (5.118) and (5.119) hold, for a product-measurable u-Carathéodory mapping $F : \mathbb{R}^{n+1} \multimap \mathbb{R}^n$ with nonempty, compact and convex values, satisfying (5.107), where the constans $\widetilde{D}_m$, $m=p,\dots,n-1$, are defined in (5.112), (5.113), (5.115)–(5.117). Then the inclusion (5.106) admits an ω-anti-periodic solution (i.e. a 2ω-periodic solution).*

(5.121) EXAMPLE. Consider

$$(5.122) \qquad x^{(5)} + a_1 x^{(4)} + a_2 x^{(3)} + a_3 x^{(2)} + a_4 x^{(1)} + a_5 x \in F(t, x, \dots, x^{(4)}),$$

where $F : \mathbb{R}^5 \multimap \mathbb{R}^4$ is a product-measurable u-Carathéodory mapping with non-empty, compact and convex values and $a_j > 0$, $j = 1, \dots, 5$.

If nonnegative constants $C_0, \dots, C_4$ exist such that ($a_0 := 1$)

$$(5.123) \qquad \sup_{\substack{t \in \mathbb{R},\, |x_j| \le C_j \\ j=0,\dots,4}} \mathrm{ess} \; |F(t, x_0, \dots, x_4)| \le \binom{5}{m} \frac{a_5}{2^m a_m} C_m, \quad m = 0, \dots, 4$$

and if all the roots of the polynomial

$$(5.124) \qquad \lambda^5 + a_1 \lambda^4 + a_2 \lambda^3 + a_3 \lambda^2 + a_4 \lambda^1 + a_5$$

are negative, then the inclusion (5.122) admits, according to Theorem (5.74), an entirely bounded solution or an ω-periodic solution, provided $F(t, x_0, \dots, x_4) \equiv F(t + \omega, x_0, \dots, x_4)$ or an ω-anti-periodic solution, provided

$$F(t, x_0, \dots, x_4) \equiv -F(t + \omega, -x_0, \dots, -x_4).$$

Let us note that the sufficient and necessary conditions for the negativity of all the roots of (5.124) read as follows (cf. [AT2]):

$$(5.125) \qquad \begin{cases} a_1 a_4 - 25 a_5 \ge 0, \; 4a_1^2 - 10 a_2 \ge 0, \\ A_0 \ge 0, \; A_2 \ge 0, \; B_0 \ge 0, \; B_1 \ge 0, \; C_0 \ge 0, \end{cases}$$

where

$$A_0 = 3a_1^2a_2^2 - 12a_2^3 - 8a_1^3a_3 + 38a_1a_2a_3 - 45a_3^2 - 16a_1^2a_4 + 40a_2a_4,$$

$$A_1 = 2a_1^2a_2a_3 - 8a_2^{2a_3} + 6a_1a_3^2 - 12a_1^3a_4 + 42a_1a_2a_4 - 60a_3a_4 - 20a_1^2a_5 + 50a_2a_5,$$

$$A_2 = a_1^2a_2a_4 - 4a_2^2a_4 + 3a_1a_3a_4 - 16a_1^{3a_5} + 55a_1a_2a_5 - 75a_3a_5,$$

$$\begin{aligned}
B_0 = 2(-2a_1^2 + 5a_2)^2 (&a_1^2a_2^2a_3^2 - 4a_2^3a_3^2 - -4a_1^3a_3^3 + 18a_1a_2a_3^3 \\
&- 27a_3^4 - 3a_1^2a_2^3a_4 + 12a_2^4a_4 + 14a_1^3a_2a_3a_4 - 62a_1a_2^2a_3a_4 - 6a_1^2a_3^2a_4 \\
&+ 117a_2a_3^2a_4 - 18a_1^4a_4^2 + 97a_1^2a_2a_4^2 - 88a_2^2a_4^2 - 132a_1a_3a_4^2 \\
&+ 160a_4^3 - 66a_1^2a_2a_3a_5 - 40a_2^2a_3a_5 + 120a_1a_3^2a_5 - 28a_1^3a_4a_5 + \\
&+ 130a_1a_2a_4a_5 - 300a_3a_4a_5 - 50a_1^2a_5^2 + 125a_2a_5^2),
\end{aligned}$$

$$\begin{aligned}
B_1 = (-2a_1^2 + 5a_2)^2 (&a_1^2a_2^2a_3a_4 - 4a_2^3a_3a_4 - 4a_1^3a_3^2a_4 + 18a_1a_2a_3^2a_4 \\
&- 27a_3^3a_4 + 3a_1^2a_2a_4^2 - 12a_1a_2^2a_4^2 - 7a_1^2a_3a_4^2 + 48a_2^2a_3a_4^2 - 16a_1a_4^3 - 9a_1^2a_2^2a_5 \\
&+ 36a_2^4a_5 + 32a_1^3a_2a_3a_5 - 146a_1^2a_2^2a_3a_5 + 4a_1^2a_3^2a_5 + 195a_2a_3^2a_5 \\
&- 48a_1^4a_4a_5 + 266a_1^2a_2a_4a_5 - 260a_2^2a_4a_5 - 290a_1a_3a_4a_5 + 400a_4^2a_5 \\
&- 80a_1^3a_5^2 + 275a_1a_2a_5^2 - 375a_3a_5^2),
\end{aligned}$$

$$\begin{aligned}
C_0 = (-2a_1^2 + 5a_2)^4 (&3a_1^2a_2^2 - 12a_2^3 - 8a_1^3a_3 + 38a_1a_2a_3 - 45a_3^2 - 16a_1^2a_4 + 40a_2a_4)^2 \\
\cdot (&a_1^2a_2^2a_3^2a_4^2 - 4a_2^3a_3^2a_4^2 - 4a_1^3a_3^3a_4^2 + 18a_1a_2a_3^3a_4^2 - 27a_3^4a_4^2 - 4a_1^2a_2^3a_4^3 \\
&+ 16a_2^4a_4^3 + 18a_1^3a_2a_3a_4^3 - 80a_1a_2^2a_3a_4^3 - 6a_1^2a_3^2a_4^3 + 144a_2a_3^2a_4^3 - 27a_1^4a_4^4 \\
&+ 144a_1^2a_2a_4^4 - 128a_2^2a_4^4 - 192a_1a_3a_4^4 + 256a_4^5 - 4a_1^2a_2^2a_3^3a_5 + 16a_2^3a_3^3a_5 \\
&+ 16a_1^3a_3^4a_5 - 72a_1a_2a_3^4a_5 + 108a_3^5a_5 + 18a_1^2a_2^3a_3a_4a_5 - 72a_2^4a_3a_4a_5 \\
&- 80a_1^3a_2a_3^2a_4a_5 + 356a_1a_2^2a_3^2a_4a_5 + 24a_1^2a_3^3a_4a_5 - 630a_2a_3^3a_4a_5 \\
&- 6a_1^3a_2^2a_4^2a_5 + 24a_1a_2^3a_4^2a_5 + 144a_1^4a_3a_4^2a_5 - 746a_1^2a_2a_3a_4^2a_5 \\
&+ 560a_2^2a_3a_4^2a_5 + 1020a_1a_3^2a_4^2a_5 - 36a_1^3a_3a_4^3a_5 + 160a_1a_2a_3a_4^3a_5 - 1600a_3a_4^3a_5 \\
&- 27a_1^2a_2^4a_5^2 + 180a_2^5a_5^2 + 144a_1^3a_2^2a_3a_5^2 - 630a_1a_2^3a_3a_5^2 \\
&- 128a_1^4a_3^2a_5^2 + 560a_1^2a_2a_3^2a_5^2 + 825a_2^2a_3^2a_5^2 - 900a_1a_3^3a_5^2 - 192a_1^4a_2a_4a_5^2 \\
&+ 1020a_1^2a_2^2a_4a_5^2 - 900a_2^3a_4a_5^2 + 160a_1^3a_3a_4a_5^2 - 2050a_1a_2a_3a_4a_5^2 \\
&+ 2250a_3^2a_4a_5^2 - 50a_1^2a_4^2a_5^2 + 2000a_2a_4^2a_5^2 + 256a_1^5a_5^3 - 1600a_1^3a_2a_5^3 \\
&+ 2250a_1a_2^2a_5^3 + 2000a_1^2a_3a_5^3 - 3750a_2a_3a_5^3 - 2500a_1a_4a_5^3 + 3125a_5^4).
\end{aligned}$$

If we would require that only negative real parts of all the roots of (5.124) are negative, then the above huge inequalities can be simplified to the sufficient and necessary (Routh–Hurwitz) conditions as follows (cf. [AT2]):

$$a_3a_4 - a_2a_5 > 0, \quad a_4(a_2a_3 + a_5 - a_1a_4) - a_2^2a_5 > 0,$$
$$a_4(a_1a_2a_3 + a_1a_5 - a_1^2a_4 - a_3^2) - a_5(a_1a_2^2 - a_2a_3 + a_5 - a_1a_4) > 0.$$

On the other hand, in order to get the same assertion, by means of Theorem (5.78), the inequalities (5.123) should be replaced by

$$\sup_{\substack{t \in \mathbb{R}, \, |x_j| \leq C_j \\ j=0,\dots,4}} \text{ess} \quad |F(t, x_0, \dots, x_4)| \leq \frac{|\prod_{j=m+1}^{4} \text{Re} \, \lambda_j|}{2^m} C_m, \quad m = 0, \dots, 4,$$

which is, however, much less explicit.

(5.126) EXAMPLE. Consider

$$(5.127) \quad x^{(6)} + a_1 x^{(5)} + a_2 x^{(4)} + a_3 x^{(3)} + a_4 x^{(2)} + a_5 x^{(1)} + a_6 x \in F(t, x, \dots, x^{(5)}),$$

where $F : \mathbb{R}^6 \multimap \mathbb{R}^5$ is a product-measurable u-Carathéodory mapping with nonempty, compact and convex values and $a_j > 0$, $j = 1, \dots, 6$.

We could see in the foregoing Example (5.121) that the application of Theorem (5.74) would be rather cumbersome and of Theorem (5.78) rather implicit. On the other hand, according to Theorem (5.95), the inclusion (5.127) admits, under (5.125), an ω-periodic solution, provided $F(t, x_0, \dots, x_5) \equiv F(t+\omega, x_0, \dots, x_5)$ and that nonnegative constants $C_1, \dots, C_5$ exist such that ($a_0 := 1$)

$$\sup_{\substack{(t,x_0) \in \mathbb{R}^2, \, |x_j| \leq C_j \\ j=0,\dots,5}} \text{ess} \quad |F(t, x_0, \dots, x_5)| + a_6 \frac{R + \omega C_1}{1 - \omega a_6 / a_5}$$

$$\leq \frac{\binom{5}{m-1} a_5 C_m}{2^{m-1} a_{m-1}}, \quad m = 1, \dots, 5,$$

where $R > 0$ is a number such that

$$F(t, x_0, \dots, x_5) \text{sgn} \, x_0 \geq 0, \quad \text{for } |x_0| \geq R, \, |x_j| \leq C_j, \, j = 1, \dots, 5.$$

However, $a_6 > 0$ must be still sufficiently small.

Moreover, according to Theorem (5.120), the "truncated" inclusion ($p = 4$)

$$(5.128) \qquad x^{(6)} + a_1 x^{(5)} + a_2 x^{(4)} \in F(t, x, \dots, x^{(5)}),$$

where $F : \mathbb{R}^6 \multimap \mathbb{R}^5$ has the same regularity as above and a_1, a_2 are positive numbers with $a_1^2 \geq 4a_2$, admits an ω-anti-periodic solution, provided $F(t, x_0, \dots, x_5) \equiv -F(t+\omega, -x_0, \dots, -x_5)$ and that nonnegative numbers $C_0, \dots, C_5$ exist such that (cf. (5.115))

$$(5.129) \qquad \sup_{\substack{t \in \mathbb{R}, \, |x_j| \leq C_j \\ j=0,\dots,5}} \text{ess} \quad |F(t, x_0, \dots, x_5)| \leq \frac{C_m a_2}{\omega^{4-m}}, \quad m = 0, \dots, 3,$$

$$(5.130) \qquad \sup_{\substack{t \in \mathbb{R}, \, |x_j| \leq C_j \\ j=0,\dots,5}} \text{ess} \quad |F(t, x_0, \dots, x_5)| \leq \frac{C_m a_{6-m}}{2^{m-4}}, \quad m = 4, 5.$$

Let us note that if $a_1^2 \geq 4a_2$ does not hold, then (5.129) and (5.130) can be replaced, for the same conclusion, by (cf. (5.112))

$$\sup_{\substack{t \in \mathbb{R}, \ |x_j| \leq C_j \\ j=0,\dots,5}} \text{ess} \ |F(t, x_0, \dots, x_5)| \leq \frac{C_m a_1^2}{\omega^{4-m}}, \quad m = 0, \dots, 3,$$

and

$$\sup_{\substack{t \in \mathbb{R}, \ |x_j| \leq C_j \\ j=0,\dots,5}} \text{ess} \ |F(t, x_0, \dots, x_5)| \leq \min\left(C_4 a_1^2, \frac{C_5 a_1}{2}\right),$$

respectively.

Proceeding sequentially, we would like to prove finally the existence of an entirely bounded solution of a Carathéodory system, whose values are located in a given domain. For the sake of simplicity, consider only the Carathéodory differential equation

$$(5.131) \qquad \qquad \dot{x} = f(t, x),$$

where $x = (x_1, \dots, x_n)$, $f = (f_1, \dots, f_n)^T \in \text{CAR}$, i.e.

(5.131.1) $f(\cdot, x) : \mathbb{R} \to \mathbb{R}^n$ is Lebesgue measurable, for all $x \in \mathbb{R}^n$,

(5.131.2) $f(t, \cdot) : \mathbb{R}^n \to \mathbb{R}$ is continuous, for a.a. $t \in \mathbb{R}$,

(5.131.3) for every compact interval $K \subset \mathbb{R}$ and every compact $V \subset \mathbb{R}^n$, there exists
$\rho : K \to [0, \infty)$ such that $\int_K \rho(t)\, dt < \infty$ and $|f(t, x)| \leq \rho(t)$, for a.a. $t \in K$ and all $x \in V$.

Our aim is to prove the existence of an (entirely bounded) solution $x(t)$ of (5.131) with $x \in Q$, where ($\mathcal{D}$ denotes a certain given domain).

$$Q = \{u \in C(\mathbb{R}, \mathbb{R}^n) \mid \sup_{t \in \mathbb{R}^n} |u(t)| \leq M \quad \text{and} \quad u(t) \in \mathcal{D} \subset \mathbb{R}^n, \text{ for all } t \in \mathbb{R}\}.$$

We want Q to be not necessarily invariant under the solution operator $\mathcal{T}$ of the associated linearized system.

Hence, consider still the one-parameter family of problems

$$(5.132) \qquad \qquad \begin{cases} \dot{x} = f(t, x), \\ x \in Q_{[k]}, \end{cases}$$

where $Q_{[k]} = \{q \in C([-k, k], \mathbb{R}^n) \mid q = u|_{[-k,k]}, \text{ for any } u \in Q\}$, $k \in \mathbb{N}$.

To ensure the solvability of problems (5.131), we apply a very special form of the continuation principle (10.7) developed in Chapter II.10 (cf. also [FPr1]).

(5.133) PROPOSITION. *Let $Q^*_{[k]} \subset C([-k,k], \mathbb{R}^n)$ be a nonempty, convex and closed set, $\mathcal{T} : Q^*_{[k]} \times [0,1] \mapsto C([-k,k], \mathbb{R}^n)$. Assume that*

(p₁) *$\mathcal{T}$ is a continuous operator with a relatively compact image $\mathcal{T}(Q^*_{[k]} \times [0,1])$,*

(p₂) *there exists $q_0 \in Q^*_{[k]}$ such that $\mathcal{T}(q, 0) = q_0$, for every $q \in Q^*_{[k]}$,*

(p₃) *the mapping $\mathcal{T}(\,\cdot\,, \lambda)$ has no fixed points on the boundary of $Q^*_{[k]}$ with respect to $\mathcal{T}(Q^*_{[k]} \times [0,1])$, i.e. if $(q, \lambda) \in Q^*_{[k]} \times [0,1]$ and $\mathcal{T}(q, \lambda) = q$, then*

$$q \notin \overline{\mathcal{T}(Q^*_{[k]} \times [0,1]) \setminus Q^*_{[k]}} \cap Q^*_{[k]}.$$

Then the equation $x = \mathcal{T}(x, 1)$ admits at least one solution.

The following two lemmas give a priori estimates to solutions of scalar equations.

(5.134) LEMMA. *Let x_f be a solution of (5.131) in $[-k, k]$, where $k > 0$ and CAR $\ni f : \mathbb{R}^2 \mapsto \mathbb{R}$. Assume that there exist $\widetilde{u} \in AC([-k, k], \mathbb{R})$, $\widetilde{v} \in AC([-k, k], \mathbb{R})$, satisfying*

(l₀) *$f(t, \cdot)$ is non-decreasing in $[\inf_{s \in [-k,k]} \widetilde{u}(s), \sup_{s \in [-k,k]} \widetilde{v}(s)]$, for a.a. $t \in [-k, k]$,*

(l₁) *$\widetilde{u}(t) < \widetilde{v}(t)$, for every $t \in [-k, k]$,*

(l₂) *$\widetilde{u}(-k) < x_f(-k) < \widetilde{v}(-k)$,*

(l₃) *$f(t, \widetilde{u}(t)) > \dot{\widetilde{u}}(t)$, for a.a. $t \in [-k, k]$,*

(l₄) *$f(t, \widetilde{v}(t)) < \dot{\widetilde{v}}(t)$, for a.a. $t \in [-k, k]$.*

Then $\widetilde{u}(t) < x_f(t) < \widetilde{v}(t)$, for every $t \in [-k, k]$.

PROOF. At first, we show that $x_f(t) > \widetilde{u}(t)$, for all $t \in [-k, k]$. Assume, contradictionally, the existence of a $t \in (-k, k]$ such that $x_f \leq \widetilde{u}(t)$. Denoting $t^* = \inf\{t \in (-k, k] \mid x_f(t) \leq \widetilde{u}(t)\}$, we have obviously $x_f(t^*) = \widetilde{u}(t^*)$, $x_f(t) > \widetilde{u}(t)$, for all $t \in [-k, t^*)$. Choose $\varepsilon > 0$ such that $x_f(t) < \widetilde{v}(t)$ holds, for every $t \in [t^* - \varepsilon, t^*]$, and put $\overline{t} = t^* - \varepsilon$. Since x_f is the solution of (5.131) in $[-k, k]$, it satisfies an integral equality

$$x_f(t) = x_f(\overline{t}) + \int_{\overline{t}}^{t} f(s, x_f(s)) \, ds, \quad \text{for every } t \in [-k, k].$$

Furthermore, (l₃) implies that

$$\widetilde{u}(t) < \widetilde{u}(\overline{t}) + \int_{\overline{t}}^{t} f(s, \widetilde{u}(s)) \, ds, \quad \text{for every } t \in (\overline{t}, k].$$

Taking into account that $f(t, \cdot)$ is non-decreasing, we obtain $\int_{\overline{t}}^{t^*} (f(s, x_f(s)) - f(s, \widetilde{u}(s))) \, ds \geq 0$. Consequently,

$$x_f(t^*) - \widetilde{u}(t^*) > x_f(\overline{t}) - \widetilde{u}(\overline{t}) + \int_{\overline{t}}^{t^*} (f(s, x_f(s)) - f(s, \widetilde{u}(s))) \, ds > 0.$$

Hence, $x_f(t^*) > \tilde{u}(t^*)$, which disagrees with the definition of t^*.

To prove the inequality $x_f(t) < \tilde{v}(t)$ in $[-k, k]$, we can involve the transformation $y = -x$ into (5.131) and apply the first part of our lemma to the equation

$$\dot{y} = -f(t, -y). \qquad \square$$

By the same manner, we can derive

(5.135) LEMMA. *Replacing* (l_0), (l_2), (l_3), (l_4) *by*

$(\tilde{l}_0)$ $f(t, \cdot)$ *is non-increasing in* $[\inf_{s \in [-k,k]} \tilde{u}(s), \sup_{s \in [-k,k]} \tilde{v}(s)]$, *for a.a.* $t \in [-k, k]$,

$(\tilde{l}_2)$ $\tilde{u}(k) < x_f(k) < \tilde{v}(k)$,

$(\tilde{l}_3)$ $f(t, \tilde{u}(t)) < \dot{\tilde{u}}(t)$, *for a.a.* $t \in [-k, k]$,

$(\tilde{l}_4)$ $f(t, \tilde{v}(t)) > \dot{\tilde{v}}(t)$, *for a.a.* $t \in [-k, k]$,

the same assertion as in the previous Lemma (5.135) *is true.*

Now, let us recall some properties of linear differential systems (cf. [AK1], [Co1]). A real constant matrix A of the type $m \times m$ is called *hyperbolic* if each of its eigenvalues has a nonzero real part.

(5.136) LEMMA. *Consider the system*

(5.137) $$\dot{x} = Ax + f(t),$$

where the constant real $(n \times n)$*-matrix* A *is hyperbolic and* $f : \mathbb{R} \mapsto \mathbb{R}^n$ *is a measurable function such that*

(5.138) $$\sup_{t \in \mathbb{R}} \operatorname{ess} |f(t)| = F < \infty.$$

Then there exists a matrix function $G : \mathbb{R} \mapsto \mathbb{R}^{n \cdot n}$ *with the following properties:*

(5.139) $G|_{(-\infty,0)} \in C((-\infty, 0), \mathbb{R}^{n \cdot n})$, $G|_{\mathbb{R}^+} \in C([0, \infty), \mathbb{R}^{n \cdot n})$,

$\lim_{t \to 0+} G(t) - \lim_{t \to 0-} G(t) = E$, *where* E *is the unit* $(n \times n)$*-matrix,*

(5.140) $(\exists c \in [0, \infty))$ $(\exists \alpha \in [0, \infty))$ $(\forall t \in \mathbb{R})$ $|G(t)| \leq ce^{-\alpha|t|}$,

(5.141) $(\forall t \in \mathbb{R} \setminus \{0\})$ $\dot{G}(t) = AG(t)$,

(5.142) *the function* $x_f \in C(\mathbb{R}, \mathbb{R})$, $(\forall t \in \mathbb{R})$ $x_f(t) = \int_{\infty}^{\infty} G(t-s)f(s)\,ds$, *represents the unique entirely bounded solution of the system* (5.137).

Furthermore, there exists a positive number $\alpha(A)$ *depending only on the matrix* A *and such that*

(5.143) $$\sup_{t \in \mathbb{R}} |x_f(t)| \leq \alpha(A) \sup_{t \in \mathbb{R}} \operatorname{ess} |f(t)|$$

holds, for every measurable f with $\sup \mathrm{ess}_{t \in \mathbb{R}} |f(t)| < \infty$.

PROOF. A can be written in the form $T(M_+, M_-)T^{-1}$, where T is a regular matrix and (M_+, M_-) is a block diagonal matrix, having the Jordan canonical form such that the eigenvalues of A with positive real parts make the diagonal of M_+ and those with negative real parts make the diagonal of M_-. Define $G : \mathbb{R} \mapsto \mathbb{R}^{n \cdot n}$ by the formula

$$(5.144) \qquad G(t) = \begin{cases} -T(e^{M_+ t}, 0)T^{-1}, & \text{for } t \leq 0, \\ T(0, e^{M_- t})T^{-1}, & \text{for } t > 0. \end{cases}$$

Evidently, (5.139) holds for G. Furthermore, denote by α_1 the smallest real part of eigenvalues of M_+, by α_2 the largest real part of eigenvalues of M_- and choose $\alpha = \min\{\alpha_1, \alpha_2\}$. Since there exist $c_1 \in [0, \infty)$, $c_2 \in [0, \infty)$ such that

$$(5.145) \qquad (\forall t \in (-\infty, 0)) \quad |e^{M_+ t}| \leq c_1 e^{\alpha_1 t},$$

$$(5.146) \qquad (\forall t \in (0, \infty)) \quad |e^{M_- t}| \leq c_2 e^{\alpha_2 t},$$

we can verify the validity of (5.140) for an appropriate positive constant c. Equality (5.141) can be derived by easy straightforward computations. Thus, it remains to prove (5.142) and (5.143). At first, we show that x_f is a solution of (5.137); since (5.139), (5.140) allow us to apply Fubini's theorem, we obtain for all $t \in [0, \infty)$:

$$\int_0^t (Ax_f(s) + f(s))\, ds = \int_0^t A\left(\int_{-\infty}^z G(z - s)f(s)\, ds \right) dz +$$

$$+ \int_0^t A\left(\int_0^\infty G(z - s)f(s)\, ds \right) dz + \int_0^t f(s)\, ds$$

$$= \int_{-\infty}^t \left(\int_0^t AG(z - s)\, dz \right) f(s)\, ds + \int_0^t \left(\int_s^t AG(z - s)\, dz \right) f(s)\, ds$$

$$+ \int_0^t \left(\int_0^s AG(z - s)\, dz \right) f(s)\, ds + \int_0^\infty \left(\int_0^t AG(z - s)\, dz \right) f(s)\, ds + \int_0^t f(s)\, ds$$

$$= \int_{-\infty}^0 (G(t - s) - G(-s))f(s)\, ds + \int_0^t \left(G(t - s) - \lim_{z \to 0^+} G(z) \right) f(s)\, ds$$

$$+ \int_0^t \left(\lim_{z \to 0^-} G(z) - G(-s) \right) f(s)\, ds + \int_t^\infty (G(t - s) - G(-s))f(s)\, ds + \int_0^t f(s)\, ds$$

$$= \int_{-\infty}^\infty G(t - s)f(s)\, ds - \int_{-\infty}^\infty G(-s)f(s)\, ds$$

$$= x_f(t) - x_f(0).$$

Analogously, for $t \in (-\infty, 0)$, the same equality can be obtained. Hence, x_f represents a solution of (5.137) which can be estimated in the following way:

$$(5.147) \qquad \sup_{t \in \mathbb{R}} |x_f(t)| \leq F \sup_{t \in \mathbb{R}} \int_{-\infty}^\infty |G(t - s)|\, ds \leq F\alpha(A),$$

where $\alpha(A) \in [0, \infty)$ depends only on A (apply (5.140)). Finally, the uniqueness follows immediately from the hyberbolicity of A. $\qquad\square$

We write $\mathcal{B}(A)$ for the set of all constants $\alpha(A)$ in the above lemma.

Let $q = (q_1, \ldots, q_n)$, $u = (u_1, \ldots, u_n)$ be functions in $C(I, \mathbb{R}^n)$. Then, for $i \in \{1, \ldots, n\}$, we define $q^{[u_i]} \in \mathcal{C}(J, \mathbb{R}^n)$ as

$$q^{[u_i]} = (q_1, \ldots, q_{i_1}, u_i, q_{i+1}, \ldots, q_n).$$

Consider the system of ODE-s in the form

(5.148)
$$\begin{cases} \dot{y} = Ay + g(t, y, z), \\ \dot{z} = h(t, y, z), \end{cases}$$

where $y = (y_i, \ldots, y_m)$, $z = (z_1, \ldots, z_s)$, A is a hyperbolic $(m \times m)$-matrix, $\mathrm{CAR} \ni g : \mathbb{R} \times \mathbb{R}^{m+s} \mapsto \mathbb{R}^m$ and $\mathrm{CAR} \ni h : \mathbb{R} \times \mathbb{R}^{m+s} \mapsto \mathbb{R}^s$. Let $\alpha \in [0, \infty)$, $\alpha(A) \in \mathcal{B}(A)$, $\xi = (\xi_1, \ldots, \xi_s) \in \mathbb{R}^s$, $\tilde{s} \in \{0, \ldots, s\}$ and $u_{m+1}, \ldots, u_{m+s}, v_{m+1}, \ldots, v_{m+s}$ be functions in $\mathrm{AC}_{\mathrm{loc}}(\mathbb{R}, \mathbb{R})$.

Assume that, for all $t \in \mathbb{R}$, $i \in \{1, \ldots, s\}$,

(c₁)
$$-\infty < \beta_{m+i} = \inf_{l \in \mathbb{R}} u_{m+i}(l) \le u_{m+i}(t) < \xi_i < v_{m+i}(t)$$
$$\le \sup_{l \in \mathbb{R}} v_{m+i}(l) = \gamma_{m+i} < \infty,$$

and set

$$Y = \{q \in C(\mathbb{R}, \mathbb{R}^{m+s}) \mid |(q_1(t), \ldots, q_m(t))| < \alpha, \quad \text{for all } t \in \mathbb{R}\},$$
$$Z = \{q \in C(\mathbb{R}, \mathbb{R}^{m+s}) \mid u_{m+i}(t) < q_{m+i}(t) < v_{m+i}(t),$$
$$\text{for } i = 1, \ldots, s \text{ and all } t \in \mathbb{R}\},$$
$$Q = \overline{Y} \cap \overline{Z}.$$

(5.149) THEOREM. *Suppose that*

(c₂)
$$\sup_{t \in \mathbb{R}} \mathrm{ess} \, |g(t, q(t))| \le \frac{\alpha}{\alpha(A)}, \quad \text{for every } q \in Q,$$

for every $q \in Q$ and a.a. $t \in \mathbb{R}$:

(c₃₁) $h_i(t, q^{[u_{m+i}]}(t)) > \dot{u}_{m+i}(t),$

(c₃₂) $h_i(t, q^{[v_{m+i}]}(t)) < \dot{v}_{m+i}(t),$

(c₃₃) *the function* $f_i : \mathbb{R} \mapsto \mathbb{R}$, $f_i(x) = h_i(t, q^{[x]}(t))$
is non-decreasing in $[\beta_{m+i}, \gamma_{m+i}]$, *provided* $i \in \{1, \ldots, \tilde{s}\}$, *and*

(c₃₄) $h_i(t, q^{[u_{m+i}]}(t)) < \dot{u}_{m+i}(t),$

(c₃₅) $h_i(t, q^{[v_{m+i}]}(t)) > \dot{v}_{m+i}(t),$

(c₃₆) *the function* $f_i : \mathbb{R} \mapsto \mathbb{R}$, $f_i(x) = h_i(t, q^{[x]}(t))$
is non-increasing in $[\beta_{m+i}, \gamma_{m+i}]$, *provided* $i \in \{\tilde{s}+1, \ldots, s\}$

Then (5.148) *admits at least one entirely bounded solution* $\varphi \in \overline{Y} \cap Z$.

PROOF. Evidently, Q is a nonempty, convex, closed and (globally) bounded set and the same is true for every $Q_{[k]}$. According to Lemmas (5.134) and (5.135), to prove the existence of an entirely bounded solution $\varphi \in Q$ of (5.148), it is sufficient to show the solvability of problems (5.148), $(y, z) \in Q_{[k]}$, where $k \in \mathbb{N}$. Fix $k \in \mathbb{N}$ and denote

$$Z^* = \{q \in C(\mathbb{R}, \mathbb{R}^{m+s}) \mid \beta_{m+i} - 1 < q_{m+i}(t) < \gamma_{m+i} + 1,$$
$$\text{for } i = 1, \ldots, s \quad \text{and all } t \in \mathbb{R}\},$$
$$Q^* = \overline{Y} \cap \overline{Z^*}.$$

Define $(i = 1, \ldots, s)$

$$\nu_{m+i}(t, x) = \begin{cases} u_{m+i}(t), & \text{for } x \in (-\infty, u_{m+i}(t)), \\ x, & \text{for } x \in [u_{m+i}(t), v_{m+i}(t)], \\ v_{m+i}(t), & \text{for } x \in (v_{m+i}(t), \infty), \end{cases}$$

$$g^*(t, y, z_1, \ldots, z_s) = g(t, y, \nu_{m+1}(t, z_1), \ldots, \nu_{m+s}(t, z_s)),$$
$$h^* = (h_1^*, \ldots, h_s^*),$$

$$h_i^*(t, y, z_1, \ldots, z_s) = \begin{cases} \dfrac{h_i(t, y, \nu_{m+1}(t, z_1), \ldots, \nu_{m+s}(t, z_s)) - 1}{u_{m+i}(t) - \beta_{m+i} + 1} \\ \cdot (z_i - \beta_{m+i} + 1) + 1, \quad \text{for } z_i \in (-\infty, u_{m+i}(t)), \\ h_i(t, y, \nu_{m+1}(t, z_1), \ldots, \nu_{m+s}(t, z_s)), \\ \quad \text{for } z_i \in [u_{m+i}(t), v_{m+i}(t)], \\ \dfrac{h_i(t, y, \nu_{m+1}(t, z_1), \ldots, \nu_{m+s}(t, z_s)) + 1}{v_{m+i}(t) - \gamma_{m+i} - 1} \\ \cdot (z_i - \gamma_{m+i} - 1) - 1, \quad \text{for } z_i \in (v_{m+i}(t), \infty), \end{cases}$$

provided $i \in \{1, \ldots, \widetilde{s}\}$,

$$h_i^*(t, y, z_1, \ldots, z_s) = \begin{cases} \dfrac{h_i(t, y, \nu_{m+1}(t, z_1), \ldots, \nu_{m+s}(t, z_s)) + 1}{u_{m+i}(t) - \beta_{m+i} + 1} \\ \cdot (z_i - \beta_{m+i} + 1) - 1, \text{ for } z_i \in (-\infty, u_{m+i}(t)), \\ h_i(t, y, \nu_{m+1}(t, z_1), \ldots, \nu_{m+s}(t, z_s)), \\ \quad \text{for } z_i \in [u_{m+i}(t), v_{m+i}(t)], \\ \dfrac{h_i(t, y, \nu_{m+1}(t, z_1), \ldots, \nu_{m+s}(t, z_s)) - 1}{v_{m+i}(t) - \gamma_{m+i} - 1} \\ \cdot (z_i - \gamma_{m+i} - 1) + 1, \quad \text{for } z_i \in (v_{m+i}(t), \infty), \end{cases}$$

provided $i \in \{\tilde{s}+1, \ldots, s\}$. Clearly, $\text{CAR} \ni g^* : \mathbb{R}^{1+m+s} \mapsto \mathbb{R}^m$, $\text{CAR} \ni h^* :$ $\mathbb{R}^{1+m+s} \mapsto \mathbb{R}^s$. Furthermore, if $q \in Q$, then

$$g^*(t, q(t)) = g(t, q(t)),$$
$$h^*(t, q(t)) = h(t, q(t)),$$

hold, for every $t \in \mathbb{R}$. Moreover,

(c_2^*) $\qquad \sup_{t \in \mathbb{R}} \text{ess} \, |g^*(t, q(t))| \leq \dfrac{\alpha}{\alpha(A)}, \quad$ for every $q \in Q^*$,

$(c_{3_1}^*)$ $h_i^*(t, q^{[\beta_{m+i}-1]}(t)) = 1 \quad (> 0)$,
$(c_{3_2}^*)$ $h_i^*(t, q^{[\gamma_{m+i}+1]}(t)) = -1 \quad (< 0)$,
$(c_{3_3}^*)$ the function $f_i : \mathbb{R} \mapsto \mathbb{R}$, $f_i(x) = h_i^*(t, q^{[x]}(t))$ is non-decreasing in $[\beta_{m+i} - 1, \gamma_{m+i} + 1]$, provided $i \in \{1, \ldots, \tilde{s}\}$, $q \in Q^*$, $t \in \mathbb{R}$,
$(c_{3_4}^*)$ $h_i^*(t, q^{[\beta_{m+i}-1]}(t)) = -1 \quad (< 0)$,
$(c_{3_5}^*)$ $h_i^*(t, q^{[\gamma_{m+i}+1]}(t)) = 1 \quad (> 0)$,
$(c_{3_6}^*)$ the function $f_i : \mathbb{R} \mapsto \mathbb{R}$, $f_i(x) = h_i^*(t, q^{[x]}(t))$ is non-increasing in $[\beta_{m+i} - 1, \gamma_{m+i} + 1]$, provided $i \in \{\tilde{s}+1, \ldots, s\}$, $q \in Q^*$, $t \in \mathbb{R}$

At first, we prove that the system

(5.150) $\qquad \begin{cases} \dot{y} = Ay + g^*(t, y, z), \\ \dot{z} = h^*(t, y, z) \end{cases}$

admits a solution $\varphi^* \in Q_{[k]}^*$.

Let $\mathcal{R} : Q_{[k]}^* \times [0, 1] \to Q^* \times [0, 1]$ be an operator defined by

$$\mathcal{R}(q, \lambda)(t) = \begin{cases} (q(-k), \lambda), & \text{for } t \in (-\infty, -k), \\ (q(t), \lambda), & \text{for } t \in [-k, k], \\ (q(k), \lambda), & \text{for } t \in (k, \infty). \end{cases}$$

Evidently, $\mathcal{R}$ is continuous and letting $\widetilde{Q} \times [0, 1] = \mathcal{R}(Q_{[k]}^* \times [0, 1])$, we can observe that $\widetilde{Q}$ is a nonempty and bounded subset of the space $C(\mathbb{R}, \mathbb{R}^{m+s})$. Denote by $\mathcal{S}$ the mapping $\mathcal{S} : \widetilde{Q} \times [0, 1] \to C(\mathbb{R}, \mathbb{R}^{m+s})$ which assigns, to any pair $(q, \lambda) \in \widetilde{Q} \times [0, 1]$, the (unique) entirely bounded solution $y_{q,\lambda}$ of the linearized system

$$\dot{y} = Ay + \lambda g^*(t, q(t)).$$

From the inequality (5.147) and condition (c_2^*), it follows that

(5.151) $\qquad \sup_{t \in \mathbb{R}} |y_{q,\lambda}(t)| \leq \alpha.$

Thus, (see Proposition (1.4), where $T = \mathcal{S}$), is the continuous operator with a relatively compact image $\mathcal{S}(\widetilde{Q} \times [0,1]) \subset C(\mathbb{R}, \mathbb{R}^m)$.

Introducing $\mathcal{P} : C(\mathbb{R}, \mathbb{R}^m) \to C([-k,k], \mathbb{R})$, $\mathcal{P}(q) = q|_{[-k,k]}$, we are ready to define the operator $\mathcal{G} : Q^*_{[k]} \times [0,1] \to C([-k,k], \mathbb{R}^m)$ as a composition $\mathcal{G} = \mathcal{P} \circ \mathcal{S} \circ \mathcal{R}$.

Now, we will deal with the second part of our system (5.150). Clearly, for every pair $(q, \lambda) \in Q^*_{[k]} \times [0,1]$, there exists exactly one solution $z_{q,\lambda} = \mathcal{H}(q, \lambda)$ of the two-point problem

$$\dot{z} = \lambda h^*(t, q(t)), \quad t \in [-k,k],$$
$$z_i(-k) = \xi_i, \qquad \text{for } i = 1, \dots, \widetilde{s},$$
$$z_j(k) = \xi_j, \qquad \text{for } j = \widetilde{s} + 1, \dots, s.$$

Applying Proposition (1.4) (where $T = \mathcal{H}$ and $Q = Q^*_{[k]}$), we have also the continuity of the operator $\mathcal{H} : Q^*_{k]} \times [0,1] \to C([-k,k], \mathbb{R}^s)$ and $\mathcal{H}(Q^*_{[k]} \times [0,1])$ is a relatively compact set.

Finally, we define the mapping $\mathcal{T} : Q^*_{[k]} \times [0,1] \to C([-k,k], \mathbb{R}^{m+s})$ as $\mathcal{T} = (\mathcal{G}, \mathcal{H})$ and use Proposition (5.133).

Observe that, by a construction of $\mathcal{T}$, the condition (r_i) is satisfied. Furthermore, letting $q_0 = (0, \dots, 0, \xi_1, \dots, \xi_s) \in \mathbb{R}^{m+s}$, we obtain that $q_0 \in \text{int} Q^*_{[k]}$, and subsequently (r_2) and the part of condition (r_3) (the one, related to $\lambda = 0$) hold.

Let us verify that (r_3) holds also for $\lambda \in (0,1]$. Assume the existence of $(q, \lambda) \in Q^*_{[k]} \times (0,1]$ such that

(5.152) $$\mathcal{T}(q, \lambda) = q,$$
(5.153) $$q \in \overline{\mathcal{T}(Q^*_{[k]} \times [0,1]) \setminus Q^*_{[k]}} \cap Q^*_{[k]}.$$

From (5.153) and (5.151), it follows that there exist $i \in \{1, \dots, s\}$ and $t^* \in [-k,k]$ satisfying

(5.154) $$q_{m+i}(t^*) = \beta_{m+i} - 1,$$

or

(5.155) $$q_{m+i}(t^*) = \gamma_{m+i} + 1.$$

However, (5.152) implies at the same time that q_{m+i} is a solution of (5.132) in $[-k,k]$, provided the right-hand side of the equation in (5.132) takes the form

$$f(t, x) = \lambda h_i^*(t, q_i(t), \dots, q_{m+i-1}(t), x, q_{m+i+1}(t), \dots, q_{m+s}(t)).$$

Hence, denoting $\tilde{u}_{m+i}(t) = \beta_{m+i} - 1$, $\tilde{v}_{m+i}(t) = \gamma_{m+1} + 1$ and using (c_{31}^*)–(c_{36}^*), for $i \in \{1, \ldots, \tilde{s}\}$ or $i \in \{\tilde{s}+1, \ldots, s\}$, we can apply Lemma (5.134) or Lemma (5.135), respectively, to obtain the inequality

$$\beta_{m+i} - 1 < q_{m+i}(t) < \gamma_{m+i} + 1, \quad \text{for every } t \in [-k, k],$$

which contradicts (5.154) and (5.155).

In view of Proposition (5.133), we have proved that the equation $\mathcal{T}(x, 1) = x$ has a solution $\varphi^* \in Q_{[k]}^*$. It follows from (5.151) that

$$\max_{t \in [-k,k]} |(\varphi_1^*(t), \ldots, \varphi_m^*(t))| \leq \alpha.$$

From the definition of $\mathcal{H}$ and condition (c_3), we obtain

$$u_{m+i}(t) < \varphi_{m+i}^*(t) < v_{m+i}(t), \quad \text{for } i = 1, \ldots, s \text{ and all } t \in [-k, k],$$

(see Lemmas (5.134), (5.135), where $\tilde{u}_{m+i} = u_{m+i}$, $v_{m+i} = v_{m+1}$). We conclude that $\varphi^* \in Q_{[k]}^*$, and so Proposition (1.37) implies the existence of an entirely bounded solution $\varphi \in Q = \overline{Y} \cap \overline{Z}$ of system (5.148).

Now, it is easy to check that, by means of Lemmas (5.134), (5.135) (applying again (c_{31})–(c_{36})) $\varphi \in Z$, which completes the proof. □

(5.156) REMARK. For a continuous map $h : \mathbb{R} \times \mathbb{R}^{m+s} \to \mathbb{R}^s$, conditions (c_{33}) and (c_{26}) in Theorem (5.149) can be omitted, because, for $f \in C(\mathbb{R}^2, \mathbb{R})$, conditions (l_0) in (5.134) and $(\tilde{l}_0)$ in (5.135) are superfluous (see [AK1]).

For constant seperating functions (u_{m+i}, v_{m+i}), Theorem (5.149) takes the following particular form.

(5.157) COROLLARY. *System (5.148) admits at least k entirely bounded solutions, provided the same as in Theorem (5.149) is true, but if, instead of a pair of separating functions (u_{m+i}, v_{m+i}) satisfying (c_1)–(c_3), we assume the existence of separating constants $\{(u_{m+i,j}, v_{m+i,j})\}_{j=1}^k$ with the following properties:*

(i) $u_{m+i,j} < v_{m+i,j}$, *for $i = 1, \ldots, s$ and $j = 2, \ldots, k$,*

(ii) $\text{int}\,(Q_p \cap Q_r) = \emptyset$, *for $p \neq r$, $p = 1, \ldots, k$ and $r = 1, \ldots, k$, where* $Q_j = \overline{Y} \cap \overline{Y}^{(j)}$, $\overline{Y} = \{q \in \mathbb{R}^{m+s} \mid |(q_1, \ldots, q_m)| \leq \alpha\}$, $Z^{(j)} = \{q \in \mathbb{R}^{m+s} \mid u_{m+i,j} < q_{m+i,j} < v_{m+i,j} \text{ for } i = 1, \ldots, s\}$,

(iii) $h_i(t, q^{[u_{m+i,j}]}) > 0$, $h_i(t, q^{[v_{m+i,j}]}) < 0$ *and the function $f_i : \mathbb{R} \to \mathbb{R}$, $f_i(x) = h_i(t, q^{[x]})$ is non-decreasing in $[u_{m+i,j}, v_{m+i,j}]$, or $h_i(t, q^{[u_{m+i,j}]}) < 0$, $h_i(t, q^{[v_{m+i,j}]}) > 0$, and f_i is non-increasing in $[u_{m+i,j}, v_{m+i,j}]$, for a.a. $t \in \mathbb{R}$, $q \in Q_j$, $i = 1, \ldots, s$ and $j = 1, \ldots, k$,*

(iv) $\sup \text{ess}_{t \in \mathbb{R}, \, q \in Q_j} |g(t, q(t))| \leq \frac{\alpha}{\alpha(A)}$ *for $j = 1, \ldots, k$.*

(5.158) EXAMPLE. One can easily check that system (5.148), where $m = 2$, $s = 2$, admits infinitely many entirely bounded solutions, when e.g.

$$A = \begin{pmatrix} 1 & 2 \\ 2 & 1 \end{pmatrix} \quad \text{and} \quad (n \in \mathbb{N})$$

$$g_1(t, y, z) = \sin y_2 + \sin z_1 + \sin z_2 + \arctan t,$$
$$g_2(t, y, z) = \sin y_1 + \sin z_1 + \sin z_2 + \arctan t,$$
$$h_1(t, y, z) = -(|y_1| + |y_2| + |z_1| + |z_2|)^n \sin z_1 + \arctan t,$$
$$h_2(t, y, z) = (|y_1| + |y_2| + |z_1| + |z_2|)^n \sin z_2 + \arctan t.$$

As a family of separating constants, we can take e.g.

$$(u_{2+i,j}, v_{2+i,j}) = \left(\left(j - \frac{1}{2} \right) \pi, \left(j + \frac{1}{2} \right) \pi \right), \quad \text{for } i = 1, 2, \text{ where } j \in \mathbb{Z}.$$

III.6. Multiplicity results

We could see in Corollary (5.157) and in Example (5.158) that multiplicity results can be simply obtained by "repeating" several times the existence results in disjoint domains. This will, however, not be the case in this section. More concretely, the existence of several (at least two) solutions of given problems will be studied either in the same domain (by means of the Nielsen number) or in disjoint domains, but when another solution can be deduced similarly as by means of the additivity properties of an index (in the frame of relative theories).

The results of this section rely mainly on the papers [And6], [And16], [AGJ1] –[AGJ4].

We start with the application of Theorem (1.25). Hence, consider the problem

(6.1)
$$\begin{cases} \dot{x} + A(t)x \in F(t, x), \\ Lx = \Theta. \end{cases}$$

Since the composed multivalued function $F(t, q(t))$, where $F : J \times \mathbb{R}^n \multimap \mathbb{R}^n$ is a product-measurable (upper) Carathéodory mapping with nonempty, compact and convex values and $q \in C(J, \mathbb{R}^n)$, is well-known (see [ADTZ-M, p. 34]) to be measurable, we can also employ Theorem (3.8) to the associated linearized system

(6.2)
$$\begin{cases} \dot{x} + A(t)x \in F(t, q(t)), \\ Lx = \Theta, \end{cases}$$

provided

(6.3)
$$|F(t, x)| \le \mu(t)(|x| + 1),$$

where $\mu : J \to [0, \infty)$ is a suitable (locally) Lebesgue integrable bounded function.

We can immediately give

(6.4) THEOREM. *Consider boundary value problem (6.1) on a compact inter-*
val J. Assume that $A : J \to \mathbb{R}^{n^2}$ is a single-valued continuous $(n \times n)$-matrix and
$F : J \times \mathbb{R} \multimap \mathbb{R}^n$ is a u-Carathéodory product-measurable mapping with nonempty,
compact and convex values satisfying (6.3). Furthermore, let $L : C(J, \mathbb{R}^n) \to \mathbb{R}^n$
be a linear operator such that the homogeneous problem

$$\begin{cases} \dot{x} + A(t)x = 0, \\ Lx = 0 \end{cases}$$

has only the trivial solution on J. Then the original problem (6.1) has $N(r|_{T(Q)} \circ$
$T(\cdot))$ solutions (for the definition of the Nielsen number N, see (10.19) in Chap-
ter I.10), provided there exists a closed connected subset Q of $C(J, \mathbb{R}^n)$ with a
finitely generated abelian fundamental group such that

(i) *$T(Q)$ is bounded,*
(ii) *$T(q)$ is retractible onto Q with a retraction r in the sense of Definition (1.24),*
(iii) *$\overline{T(Q)} \subset \{x \in AC(J, \mathbb{R}^n) \mid Lx = \Theta\}$,*

where $T(q)$ denotes the set of (existing) solutions to (6.2).

(6.5) REMARK. In the single-valued case, we can obviously assume the unique
solvability of the associated linearized problem. Moreover, Q need not then have
a finitely generated abelian fundamental group (see [And25]). In the multivalued
case, the latter is true, provided Q is compact and $\overline{T(Q)} \subset Q$ (see [And6]).

Before presenting a nontrivial example, it will be convenient to have the following reduction property.

(6.6) LEMMA (reduction). *Let X and its closed subset Y be ANR-spaces. As-*
sume that $f : X \to X$ is compact map, i.e. $\overline{f(X)}$ is a compact, such that $f(X) \subset Y$.
Denoting by $f' : Y \to Y$ the restriction of f, we have

(i) *$\text{Fix}(f') = \text{Fix}(f)$,*
(ii) *the Nielsen relations coincide,*
(iii) *$\text{ind}(C, f') = \text{ind}(C, f)$, for any Nielsen class $C \subset \text{Fix}(f)$.*
Thus, $N(f') = N(f)$.

PROOF. It is evident that $\text{Fix}(f') = \text{Fix}(f)$ and if x, y are Nielsen-related in
$\text{Fix}(f')$, then so they are in $\text{Fix}(f)$. To prove the converse, let us assume that
$x, y \in \text{Fix}(f)$ are Nielsen-related. This means that there is a path $u : [0, 1] \to X$
satisfying

$$H(t, 0) = u(t), \quad H(t, 1) = f \circ u(t), \quad H(0, s) = x, \quad H(1, s) = y$$

with a suitable continuous mapping H. However, then $u'(t) = f \circ u(t)$ is a path
in Y satisfying

$$H'(t, 0) = u'(t), \quad H'(t, 1) = f \circ u'(t), \quad H'(0, s) = f(x), \quad H'(1, s) = f(y),$$

where $H'(t, s) = f \circ H(t, s)$, which already means that x, y are Nielsen-related as the fixed points of $f : Y \to Y$.

Since the third property (iii) for the fixed point indices is well-known (see e.g. [BJ-M]), the proof is complete.   □

Consider the u-Carathéodory system (the functions e, f, g, h have the same regularity as in (6.4))

$$
(6.7) \qquad \begin{cases} \dot{x} + ax \in e(t, x, y)y^{(1/m)} + g(t, x, y), \\ \dot{y} + by \in f(t, x, y)y^{(1/n)} + h(t, x, y), \end{cases}
$$

where a, b are positive numbers and m, n are odd integers with $\min(m, n) \geq 3$. Let suitable positive constants E_0, F_0, G, H exist such that

$$
\begin{aligned}
|e(t, x, y)| \leq E_0, \quad |f(t, x, y)| \leq F_0, \\
|g(t, x, y)| \leq G, \quad |h(t, x, y)| \leq H,
\end{aligned}
$$

hold, for a.a. $t \in (-\infty, \infty)$ and all $(x, y) \in \mathbb{R}^2$.

Futhermore, assume the existence of positive constants e_0, f_0, δ_1, δ_2 such that

$$
(6.8) \qquad\qquad 0 < e_0 \leq e(t, x, y),
$$

for a.a. t, all x and $|y| \geq \delta_2$, jointly with

$$
(6.9) \qquad\qquad 0 < f_0 \leq f(t, x, y),
$$

for a.a. t, $|x| \geq \delta_1$ and all y.

As a constraint S, consider at first the periodic boundary condition

$$
(6.10) \qquad\qquad (x(0), y(0)) = (x(\omega), y(\omega)).
$$

More precisely, we take $S = Q = Q_1 \cap Q_2 \cap Q_3$, where

$$
Q_1 = \{q(t) \in C([0, \omega], \mathbb{R}^2) \mid \|q(t)\| := \max\{ \max_{t \in [0, \omega]} |q_1(t)|, \max_{t \in [0, \omega]} |q_2(t)|\} \leq D\},
$$

$$
Q_2 = \{q(t) \in C([0, \omega], \mathbb{R}^2) \mid \min_{t \in [0, \omega]} |q_1(t)| \geq \delta_1 > 0 \text{ or } \min_{t \in [0, \omega]} |q_2(t)| \geq \delta_2 > 0\},
$$

$$
Q_3 = \{q(t) \in C([0, \omega], \mathbb{R}^2) \mid q(0) = q(\omega)\},
$$

the constants δ_1, δ_2, D will be specified below. For $(Q_1 \cap Q_2) \cap \mathbb{R}^2$, the situation is schematically sketched in Figure 1.

Important properties of the set Q can be expressed as follows.

(6.11) LEMMA. *The set Q defined above satisfies:*

(i) *Q is a closed connected subset of $C([0, \omega], \mathbb{R}^2)$,*

(ii) *$Q \in ANR$,*

(iii) *$\pi_1(Q) = \mathbb{Z}$.*

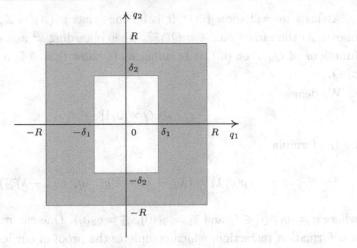

FIGURE 1

PROOF. Since Q is an intersection of closed sets Q_1, Q_2, Q_3, we conclude that Q is a closed subset of $C([0, \omega], \mathbb{R}^2)$ as well. The connectedness follows from the proof of (iii) below.

For (ii), it is enough to show that Q is a neighbourhood retract of Q_3. Hence, let $\varepsilon > 0$ be such that $\delta_1 - \varepsilon > 0$ and $\delta_2 - \varepsilon > 0$.

Defining

$$U = \{q \in Q_3 \mid \max\{\max_{t \in [0, \omega]} |q_1(t)|, \max_{t \in [0, \omega]} |q_2(t)|\} < D + \varepsilon$$
$$\text{and } [\min_{t \in [0, \omega]} |q_1(t)| > \delta_1 - \varepsilon \text{ or } \min_{t \in [0, \omega]} |q_2(t)| > \delta_2 - \varepsilon]\},$$

U is obviously an open neighbourhood of Q in Q_3.

Now, we will define the retraction $r : U \to Q$. Let us take

$$A = \{(x, y) \in \mathbb{R}^2 \mid \max(|x|, |y|) \leq D \quad \text{and} \quad [|x| \geq \delta_1 \text{ or } |y| \geq \delta_2]\}$$

and

$$V = \{(x, y) \in \mathbb{R}^2 \mid \max(|x|, |y|) \leq D + \varepsilon \quad \text{and} \quad [|x| > \delta_1 - \varepsilon \text{ or } |y| > \delta_2 - \varepsilon]\}.$$

There exists a retraction $r_0 : V \to A$.

Now, notice that, for every $q \in U$ and every $t \in [0, \omega]$, we have $q(t) \in Q$. Define $r : U \to Q$,

$$r(q)(t) = r_0(q(t)).$$

It is easy to see that r is a desired retraction and the proof of (ii) is complete.

At last, we will show (iii). It is obvious that $\pi_1(A) = \mathbb{Z}$, where A is defined above. At the same time, $A = Q \cap \mathbb{R}^2$, when regarding $\mathbb{R}^2$ as a subspace of constant functions of Q_3. For (iii), it is sufficient to show that A is a deformation retract of Q.

We define

$$\rho : Q \times [0,1] \to A$$

by the formula

$$\rho(q, \lambda) = (\lambda q_1 + (1 - \lambda)\overline{q_1}, \lambda q_2 + (1 - \lambda)\overline{q_2}),$$

where $q = (q_1, q_2) \in Q$ and $\overline{q_1} = q_1(0)$, $\overline{q_2} = q_2(0)$. One can readily check that ρ is a deformation retraction, which completes the proof of our lemma.   □

Besides (6.7) consider still its embedding into

$$(6.12) \qquad \begin{cases} \dot{x} + ax \in [(1 - \mu)e_0 + \mu e(t, x, y)]y^{1/m} + \mu g(t, x, y), \\ \dot{y} + by \in [(1 - \mu)f_0 + \mu f(t, x, y)]x^{1/n} + \mu h(t, x, y), \end{cases}$$

where $\mu \in [0, 1]$ and observe that (6.12) reduces to (6.7), for $\mu = 1$.

The associated linearized system to (6.12) takes, for $\mu \in [0, 1]$, the form

$$(6.13) \qquad \begin{cases} \dot{x} + ax \in [(1 - \mu)e_0 + \mu e(t, q_1(t), q_2(t))]q_2(t)^{1/m} + \mu g(t, q_1(t), q_2(t)), \\ \dot{y} + by \in [(1 - \mu)f_0 + \mu f(t, q_1(t), q_2(t))]q_1(t)^{1/n} + \mu h(t, q_1(t), q_2(t)), \end{cases}$$

or, equivalently,

$$(6.14) \qquad \begin{cases} \dot{x} + ax = [(1 - \mu)e_0 + \mu e_t]q_2(t)^{1/m} + \mu g_t, \\ \dot{y} + by = [(1 - \mu)f_0 + \mu f_t]q_1(t)^{1/n} + \mu h_t, \end{cases}$$

where $e_t \subset e(t, q_1(t), q_2(t))$, $f_t \subset f(t, q_1(t), q_2(t))$, $g_t \subset g(t, q_1(t), q_2(t))$, $h_t \subset h(t, q_1(t), q_2(t))$ are measurable selections. These exist, because the u-Carathéodory functions e, f, g, h are weakly selectionally measurable (see e.g. [ADTZ-M]).

It is well-known that problem $(6.14) \cap (6.10)$ has, for each $q(t) \in Q$ and every fixed quadruple of selections e_t, f_t, g_t, h_t, a unique solution $X(t) = (x(t), y(t))$, namely

$$X(t) = \begin{cases} x(t) = \displaystyle\int_0^\omega G_1(t, s)[((1 - \mu)e_0 + \mu e_s)q_2(s)^{1/m} + \mu g_s]\, ds, \\ y(t) = \displaystyle\int_0^\omega G_2(t, s)[((1 - \mu)f_0 + \mu f_s)q_1(s)^{1/n} + \mu h_s]\, ds, \end{cases}$$

where

$$G_1(t,s) = \begin{cases} \dfrac{e^{-a(t-s+\omega)}}{1-e^{-a\omega}}, & \text{for } 0 \leq t \leq s \leq \omega, \\[2mm] \dfrac{e^{-a(t-s)}}{1-e^{-a\omega}}, & \text{for } 0 \leq s \leq t \leq \omega, \end{cases}$$

$$G_2(t,s) = \begin{cases} \dfrac{e^{-b(t-s+\omega)}}{1-e^{-b\omega}}, & \text{for } 0 \leq t \leq s \leq \omega, \\[2mm] \dfrac{e^{-b(t-s)}}{1-e^{-b\omega}}, & \text{for } 0 \leq s \leq t \leq \omega. \end{cases}$$

In order to verify that $\overline{T_\mu(Q)} \subset S = Q$, where $T_\mu(\cdot)$ is the solution operator to $(6.13) \cap (6.10)$, it is sufficient just to prove that $T_\mu(Q) \subset Q$, $\mu \in [0,1]$, because $S = Q$ is closed. Hence, the Nielsen number $N(T_\mu)$ is well-defined, for every $\mu \in [0,1]$, provided only product-measurability of e, f, g, h and $T_\mu(Q) \subset Q$.

Since $X(0) = X(\omega)$, i.e. $T_\mu(Q) \subset Q_3$, it remains to prove that $T_\mu(Q) \subset Q_1$ as well as $T_\mu(Q) \subset Q_2$. Let us consider the first inclusion. In view of

$$\min_{t,s \in [0,\omega]} G_1(t,s) \geq \frac{e^{-a\omega}}{1-e^{-a\omega}} > 0 \quad \text{and} \quad \min_{t,s \in [0,\omega]} G_2(t,s) \geq \frac{e^{-b\omega}}{1-e^{-b\omega}} > 0,$$

we obtain, for the above solution $X(t)$, that

$$\max_{t \in [0,\omega]} |x(t)| \leq \max_{t \in [0,\omega]} \int_0^\omega |G_1(t,s)|[[(1-\mu)e_0 + \mu e_s]q_2(s)^{1/m} + \mu g_s]\, ds$$

$$\leq [(e_0 + E_0)D^{1/m} + G] \int_0^\omega G_1(t,s)\, ds = \frac{1}{a}[(e_0 + E_0)D^{1/m} + G]$$

and

$$\max_{t \in [0,\omega]} |y(t)| \leq \max_{t \in [0,\omega]} \int_0^\omega |G_2(t,s)|[[(1-\mu)f_0 + \mu f_s]q_1(s)^{1/n} + \mu h_s]\, ds$$

$$\leq [(f_0 + F_0)D^{1/n} + H] \int_0^\omega G_2(t,s)\, ds = \frac{1}{b}[(f_0 + F_0)D^{1/n} + H].$$

Because of

$$\|X(t)\| = \max\{ \max_{t \in [0,\omega]} |x(t)|, \max_{t \in [0,\omega]} |y(t)|\}$$

$$\leq \max \left\{ \frac{1}{a}[(e_0 + E_0)D^{1/m} + G], \frac{1}{b}[(f_0 + F_0)D^{1/n} + H] \right\},$$

a sufficiently large constant D certainly exists such that $\|X(t)\| \leq R$, i.e. $T_\mu(Q) \subset Q_1$, independently of $\mu \in [0,1]$ and e_t, f_t, g_t, h_t.

For the inclusion $T_\mu(Q) \subset Q_2$, we proceed quite analogously.
Assuming that $q(t) \in Q_2$, we have

$$\text{either} \quad \min_{t \in [0,\omega]} |q_1(t)| \geq \delta_1 > 0 \quad \text{or} \quad \min_{t \in [0,\omega]} |q_2(t)| \geq \delta_2 > 0.$$

Therefore, we obtain for the above solution $X(t)$ that (see (6.8))

$$\min_{t \in [0,\omega]} |x(t)| = \min_{t \in [0,\omega]} \int_0^\omega |G_1(t,s)| [[(1-\mu)e_0 + \mu e_s]q_2(s)^{1/m} + \mu g_s] \, ds$$

$$\geq |e_0 \delta_2^{1/m} - G| \int_0^\omega G_1(t,s) \, ds = \frac{1}{a} |e_0 \delta_2^{1/m} - G| > 0,$$

provided $G < e_0 \delta_2^{1/m}$, for $|q_2| \geq \delta_2$, or (see (6.9))

$$\min_{t \in [0,\omega]} |y(t)| = \min_{t \in [0,\omega]} \int_0^\omega |G_2(t,s)| [[(1-\mu)f_0 + \mu f_s]q_1(s)^{1/n} + \mu h_s] \, ds$$

$$\geq |f_0 \delta_1^{1/n} - H| \int_0^\omega G_2(t,s) \, ds = \frac{1}{b} |f_0 \delta_1^{1/n} - H| > 0,$$

provided $H < f_0 \delta_1^{1/n}$, for $|q_1| \geq \delta_1$.

So, in order to prove that $X(t) \in Q_2$, we need to fulfil simultaneously the following inequalities

$$(6.15) \qquad \begin{cases} (1/a)|e_0 \delta_2^{1/m} - G| \geq \delta_1 > (H/f_0)^n \\ (1/b)|f_0 \delta_1^{1/m} - H| \geq \delta_2 > (G/e_0)^m. \end{cases}$$

Let us observe that the "amplitudes" of the multivalued functions g, h must be sufficiently small. On the other hand, if e_0 and f_0 are sufficiently large (for fixed quantities a, b, G, H), then we can easily find δ_1, δ_2 satisfying (6.15).

After all, if there exist constants δ_1, δ_2 obeying (6.15), then we arrive at $X(t) \in Q_2$, i.e., $T_\mu(Q) \subset Q_2$, independently of $\mu \in [0,1]$ and e_t, f_t, g_t, h_t. This already means that $T_\mu(Q) \subset Q$, $\mu \in [0,1]$, as required.

Now, since all the assumptions of Theorem (6.4) are satisfied, problem $(6.12) \cap (6.10)$ possesses at least $N(T_\mu(\cdot))$ solutions belonging to Q, for every $\mu \in [0,1]$. In particular, problem $(6.7) \cap (6.10)$ has $N(T_1(\cdot))$ solutions, but according to the invariantness under homotopy, $N(T_1(\cdot)) = N(T_0(\cdot))$. So, it remains to compute the Nielsen number $N(T_0(\cdot))$ for the operator $T_0 : Q \to Q$, where

$$(6.16) \qquad T_0(q) = \left(e_0 \int_0^\omega G_1(t,s) q_2(s)^{1/m} \, ds, \, f_0 \int_0^\omega G_2(t,s) q_1(s)^{1/n} \, ds \right).$$

Hence, besides (6.16), consider still its embedding into the one-parameter family of operators

$$T^\nu(q) = \nu T_0(q) + (1-\nu) r \circ T_0(q), \quad \nu \in [0,1],$$

where $r(q) := (r(q_1), r(q_2))$ and

$$r(q_i) = q_i(0), \quad \text{for } i = 1, 2.$$

One can readily check that $r : Q \to Q \cap \mathbb{R}^2$ is a retraction and $T_0(\overline{q}) : Q \cap \mathbb{R}^2 \to Q$ is retractible onto $Q \cap \mathbb{R}^2$ with the retraction r in the sense of Definition (1.24). Thus, $r \circ T_0(\overline{q}) : Q \cap \mathbb{R}^2 \to Q \cap \mathbb{R}^2$ has a fixed point $\widehat{q} \in Q \cap \mathbb{R}^2$ if and only if $\widehat{q} = T_0(\widehat{q})$. Moreover, $r \circ T_0(q) : Q \to Q \cap \mathbb{R}^2$ has evidently a fixed point $\widehat{q} = Q \cap \mathbb{R}^2$ if and only if $\widehat{q} = T_0(\widehat{q})$. So, the investigation of fixed points for $T^0(q) = r \circ T_0(q)$ turns out to be equivalent with the one for $T^0(\overline{q}) : Q \cap \mathbb{R}^2 \to Q \cap \mathbb{R}^2$.

Since, in view of invariantness under homotopy, we have

$$N(T_1(\,\cdot\,)) = N(T_0(\,\cdot\,)) = N(T^1(\,\cdot\,)) = N(T^0(\,\cdot\,)),$$

where

$$T^0(q) = \left(\frac{e_0 e^{-a\omega}}{1 - e^{-a\omega}} \int_0^\omega e^{as} q_2(s)^{1/m}\, ds, \frac{f_0 e^{-b\omega}}{1 - e^{-b\omega}} \int_0^\omega e^{bs} q_1(s)^{1/n}\, ds \right)$$

and

$$T^0(\overline{q}) = \left(\frac{e_0}{a} \overline{q_2}^{(1/m)}, \frac{f_0}{b} \overline{q_1}^{(1/n)} \right), \quad \text{for } \overline{q}(\overline{q_1}, \overline{q_2}) = (q_1(0), q_2(0)) \in Q \cap \mathbb{R}^2,$$

it remains to estimate $N(T^0(\,\cdot\,))$. It will be useful to do it by passing to a simpler finite-dimensional analogy, namely by the direct computation of fixed points of the operator

$$T^0(\overline{q}) : Q \cap \mathbb{R}^2 \to Q \cap \mathbb{R}^2,$$

belonging to different Nielsen classes.

There are two fixed points $\widehat{q}_+ = (\widehat{q}_1, \widehat{q}_2)$ and $\widehat{q}_- = (-\widehat{q}_1, -\widehat{q}_2)$ in $Q \cap \mathbb{R}^2$, where

$$\widehat{q}_1 = \left(\frac{e_0}{a} \right)^{(mn/mn-1)} \left(\frac{f_0}{b} \right)^{(1/mn-1)},$$

$$\widehat{q}_2 = \left(\frac{e_0}{a} \right)^{(m/mn-1)} \left(\frac{f_0}{b} \right)^{(mn/mn-1)}.$$

These fixed points belong to different Nielsen classes, because any path u connecting them in $Q \cap \mathbb{R}^2$ and its image $T^0(u)$ are not homotopic in the space $Q \cap \mathbb{R}^2$, as it is schematically sketched in Fig. 2. Thus, according to the equivalent definition of the Nielsen number due to F. Wecken (see e.g. [BJ-M]), $N(T^0(\overline{q})) = 2$. By means of the reduction property which is true here (see Lemma (6.6)), we have, moreover,

$$N(T_1(\,\cdot\,)) = N(T^0(\,\cdot\,)) = N(T^0(\overline{q})) = 2$$

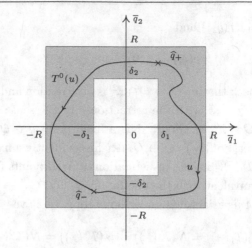

FIGURE 2

and so, according to Theorem (6.4), system (6.7) admits at least two solutions
belonging to Q, provided suitable positive constants δ_1, δ_2 exist satisfying (6.15)
and e, f, g, h are product-measurable.

In fact, system (6.9) possesses at least three solutions satisfying (6.10), when
the sharp inequalities appear in (6.15), by which the lower boundary of Q becomes
fixed point free. Indeed. Since

$$\Lambda(T_1(\,\cdot\,), Q) = \Lambda(T^0(\,\cdot\,), Q) = \lambda(T^0(\bar{q}), Q \cap \mathbb{R}^2)$$

holds for the generalized and ordinary Lefschetz numbers (see Chapter I.6) and
one can easily check that

$$|\lambda(T^0(\bar{q}), Q \cap \mathbb{R}^2)| = 2,$$

we obtain

$$|\Lambda(T_1(\,\cdot\,), Q)| = 2.$$

Futhermore, since for the self-map $T_1(\,\cdot\,)$ on the convex set $Q_1 \cap Q_3$ such that
$\overline{T_1(Q_1 \cap Q_3)}$ is compact we have

$$\Lambda(T_1(\,\cdot\,), Q_1 \cap Q_3) = 1$$

(see Chapter I.6), it follows from the additivity, excision and existence properties
of the fixed point index (see Chapter I.8) that the mapping $T_1(\,\cdot\,)$ has the third
coincidence point in $\overline{Q_1 \cap Q_3 \setminus Q}$ representing a solution of problem $(6.7) \cap (6.10)$
and belonging to $Q_1 \setminus Q$.

As we could see, problem $(6.7) \cap (6.10)$ admits at least two solutions in $Q_1 \cap Q_2$, for an arbitrary $\omega > 0$. Futhermore, because of rescaling (6.7), when replacing t by $t + (\omega/2)$, there are also two solutions of (6.7) satisfying $X(-\omega/2) = X(\omega/2)$, for an arbitrary $\omega > 0$, and belonging to $Q_1 \cap Q_2$.

Therefore, according to Proposition (1.37) and by obvious geometrical reasons, related to the appropriate subdomains of $Q_1 \cap Q_2$, system (6.7) possesses at least two entirely bounded solutions in $Q_1 \cap Q_2$.

Of course, because of replacing t by $(-t)$, the same result holds for (6.7) with negative constants a, b as well.

Finally, let us consider again system (6.7), where a, b, m, n are the same, but e, f, g, h are this time l.s.c. in (x, y), for a.a. $t \in (-\infty, \infty)$, multivalued functions with nonempty, convex, compact values and with the same estimates as above. Since each such mapping e, f, g, h has, under our regularity assumptions including the product-measurablity, a Carathéodory selection (see e.g. [Rbn1], [HP1-M]), the same assertion must be also true in this new situation.

So, after summing up the above conclusions, we can give finally

(6.17) THEOREM. *Let suitable positive constants δ_1, δ_2 exist such that the inequalities*

(6.18)
$$\begin{cases} \dfrac{1}{|a|} |e_0 \delta_2^{1/m} - G| \geq \delta_1 > \left(\dfrac{H}{f_0}\right)^n, \\[2mm] \dfrac{1}{|b|} |f_0 \delta_2^{1/n} - H| \geq \delta_2 > \left(\dfrac{G}{e_0}\right)^m \end{cases}$$

are satisfied for constants e_0, f_0, G, H estimating the product-measurable u-Carathéodory or l-Carathéodory multivalued functions (with nonempty, convex and compact values) e, f, g, h as above, for constants a, b with $ab > 0$ and for odd integers m, n with $\min(m, n) \geq 3$. Then system (6.7) admits at least two entirely bounded solutions. In particular, if multivalued functions e, f, g, h are still ω-periodic in t, then system (6.7) admits at least tree ω-periodic solutions, provided the sharp inequalities appear in (6.18).

(6.19) EXAMPLE. As a concrete choice of quantities satisfying (6.15) or (6.18), we can take $\delta_1 = 10^{-4}$, $\delta_2 = 10^{-2}$, $m = 3$, $n = 5$, $a = b = \omega = 1$, $e_0 = f_0 = 10$, $G = H = 1$. Thus, e.g. the system

$$\dot{x} + x = (15 + 3\sin x + 2\sin y)y^{1/3} + k\sin 2\pi t + (1 - k)\frac{2}{\pi}\arctan t,$$

$$\dot{y} + y = (15 + 2\sin x + 3\sin y)x^{1/5} + k\cos 2\pi t + (1 - k)\frac{2}{\pi}\arctan t$$

possesses, for $k \in [0, 1)$, at least two entirely bounded solutions and, for $k = 1$, at least there 1-periodic solutions, as it can be seen at the related phase-portrait in Fig. 3.

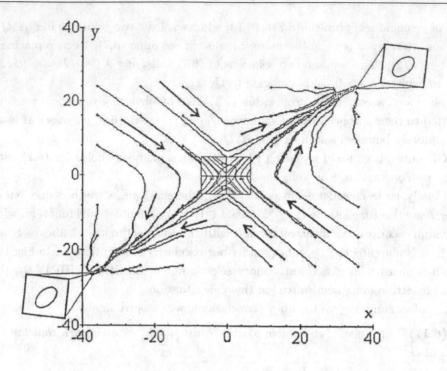

FIGURE 3

(6.20) REMARK. Unfortunately, because of the invariantness of the subdomains

$$\left\{ q(t) \in C\left(\left[-\frac{\omega}{2}, \frac{\omega}{2}, \mathbb{R}^2\right)\right) \,\middle|\, 0 < \delta_1 \leq q_1(t) \leq R \wedge 0 < \delta_2 \leq q_2(t) \leq R \right\}$$

and

$$\left\{ q(t) \in C\left(\left[-\frac{\omega}{2}, \frac{\omega}{2}, \mathbb{R}^2\right)\right) \,\middle|\, -R \leq q_1(t) \leq -\delta_1 < 0 \wedge -R \leq q_2(t) \leq -\delta_2 < 0 \right\},$$

for each $\omega \in (-\infty, \infty)$, the same result can also be obtained, for example, by means of the fixed point index.

Theorem (6.17) can be formally extended as follows. Consider the system of $2k$ inclusions.

$$(6.21) \quad \begin{cases} \dot{x}_i + a_i x_i \in e_i(t, X_i, Y_i) y_i^{(1/m_i)} + g_i(t, X_i, Y_i), \\ \dot{y}_i + b_i y_i \in f_i(t, X_i, Y_i) x_i^{(1/n_i)} + h_i(t, X_i, Y_i), \quad i = 1, \ldots, k, \end{cases}$$

where $X_i = (x_1, \ldots, x_i)$, $Y_i = (y_1, \ldots, y_i)$, a_i, b_i are nonzero numbers and m_i, n_i are odd integers with $\min(m_i, n_i) \geq 3$. Let suitable positive constants $E_{0,i}$, $F_{0,i}$,

G_i, H_i exist such that

$$|e_i(t, X_i, Y_i)| \leq E_{0,i}, \quad |f_i(t, X_i, Y_i)| \leq F_{0,i},$$
$$|g_i(t, X_i, Y_i)| \leq G_i, \quad |h_i(t, X_i, Y_i)| \leq H_i$$

hold, for a.a. $t \in (\infty, \infty)$ and all $(X_i, Y_i) \in \mathbb{R}^{2i}$, $i = 1, \ldots, k$.

Furthermore, assume the existence of positive constants $e_{0,i}$, $f_{0,i}$, $\delta_{1,i}$, $\delta_{2,i}$ ($i = 1, \ldots, k$) such that

$$0 < e_{0,i} \leq e_i(t, X_i, Y_i),$$

for all x_i, $|y_i| \geq \delta_{2,i}$ and a.a. t, jointly with

$$0 < f_{0,i} \leq f_i(t, X_i, Y_i),$$

for $|x_i| \geq \delta_{1,i}$, all y_i and a.a. t.

(6.22) THEOREM. *Let suitable positive constants $\delta_{1,i}$, $\delta_{2,i}$ exist such that the inequalities ($i = 1, \ldots, k$)*

(6.23)
$$\begin{cases} \dfrac{1}{|a_i|}|e_{0,i}\delta_{2,i}^{1/m_i} - G_i| \geq \delta_{1,i} > \left(\dfrac{H_i}{f_{0,i}}\right)^{n_i}, \\[2mm] \dfrac{1}{|b_i|}|f_{0,i}\delta_{2,i}^{1/n_i} - H_i| \geq \delta_{2,i} > \left(\dfrac{G_i}{e_{0,i}}\right)^{m_i} \end{cases}$$

are satisfied for constants $e_{0,i}$, $f_{0,i}$, G_i, H_i estimating the product-measurable u.s.c. or l.s.c. in $(X_i, Y_i) \in \mathbb{R}^{2i}$, for a.a. $t \in (-\infty, \infty)$, multivalued functions (with nonempty, convex and compact values) e_i, f_i, g_i, h_i as above, for constants a_i, b_i with $a_i b_i > 0$ and for odd integers m_i, n_i with $\min(m_i, n_i) \geq 3$. Then system (6.21) admits at least 2^k entirely bounded solutions. In particular, if multivalued functions e_i, f_i, g_i, h_i are still ω-periodic in t, then system (6.21) admits at least 3^k ω-periodic solutions, provided the sharp inequalities appear in (6.23).

(6.24) REMARK. Instead of Theorem (3.8), some other results concerning the topological structure of solution sets in Chapter III.3 can be obviously combined with Theorem (1.25) for obtaining the analogies of Theorem (6.4) as well.

Another approach is via Poincaré's first return maps treated in Chapter III.4. Consider the system

(6.25)
$$\dot{\Theta} \in F(\Theta),$$

where

$$\Theta = (\Theta_1, \ldots, \Theta_n) \in \mathbb{R}^n, \quad \dot{\Theta} = (\dot{\Theta}_1, \ldots, \dot{\Theta}_n)^T \text{ and } F(\Theta) = (F_1(\Theta), \ldots, F_n(\Theta))^T$$

is an u.s.c. (Marchaud) mapping with nonempty, compact and convex values.

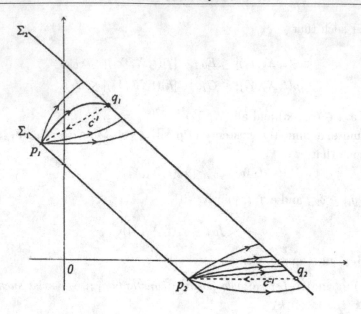

FIGURE 4

TARGET PROBLEM. *Given two codimension one manifolds $\Sigma_1 \subset \mathbb{R}^n$, $\Sigma_2 \subset \mathbb{R}^n$ and a constraint diffeomorphism $C : \Sigma_1 \to \Sigma_2$. Does there exist a solution $\Theta(t)$ of (6.25), where $\Theta(t_1) \in \Sigma_1$, such that $C(\Theta(t_1)) = \Theta(t_2) \subset \Sigma_2$, for some $t_2 > t_1$?*

For our convenience, a schematical sketch of the planar ($n = 2$) problem can be seen in Fig. 4, where,

- the set of admissible (under a constraint C) related initial points is denoted by $\{p_1, p_2, \dots\}$,
- the target set is $\{q_1, q_2, \dots\}$,
- the admissible trajectory reaching the target (a solution of the problem) is the bold curve.

As we will see in the sequel, we can give at least a partial answer to the above problem on the torus. Thus, for example, we will be only able to consider the situation in Fig. 4 mod $\sqrt{2}d$-like, where $|d|$ denotes the distance between Σ_1 and Σ_2.

Hence, consider (6.25) and assume that the set-valued map $F : \mathbb{R}^n \multimap \mathbb{R}^n$ has at most a linear growth in Θ. Then, according to Proposition (1.37), solutions of (6.25) entirely exist in the Carathéodory sense (i.e. are locally absolutely continuous and satisfy (6.25) a.e.).

Let, furthermore,

$$(6.26) \qquad \sum_{i=1}^{n} F_i(\Theta) \geq \varepsilon > 0 \quad \left(\text{or } \sum_{i=1}^{n} F_i(\Theta) \leq -\varepsilon < 0 \right),$$

by which

$$\lim_{t \to \infty} \sum_{i=1}^{n} \Theta_i(t) = \infty \quad \left(\text{or } \lim_{t \to \infty} \sum_{i=1}^{n} \Theta_i(t) = -\infty, \text{ respectively} \right).$$

Consider still the $(n-1)$-dimensional torus $\Sigma \subset \mathbb{T}^n = \mathbb{R}^n / \mathbb{Z}_b^n$ given by

$$(6.27) \qquad \sum_{i=1}^{n} \Theta_i(t) = 0 \bmod b,$$

where $\mathbb{Z}_b$ denotes the set of all integer multiples of b and $b \neq 0$ is an arbitrary constant. For better understanding of this – see Fig. 5, where the definition of two-dimensional torus $\Sigma \subset \mathbb{T}^3$ is illustrated.

Since (for our convenience) (6.25) will be considered on the torus $\mathbb{T}^n$, the natural restriction imposed on F is still

$$(6.28) \qquad F(\dots, \Theta_j + b, \dots) \equiv F(\dots, \Theta_j, \dots), \quad \text{for } j = 1, \dots, n.$$

(6.29) REMARK. The particular form of F (see (6.26) and (6.28)) will play an important role in the definition of the Poincaré set-valued map below. Under (6.26) and (6.28), this map has been proved in Theorem (4.51) to be admissible and admissibly homotopic to indentity.

We are in position to give

(6.30) THEOREM. *Let the above assumptions be satisfied jointly with* (6.26) *and* (6.28). *Assume* $C : \Sigma \to \Sigma$ *is a diffeomorphism having finitely many, but at least one, simple fixed points* $\gamma_1, \dots, \gamma_r$ *on the torus* $\Sigma \subset \mathbb{T}^n$ *(see* (6.27)*) and*

$$(6.31) \qquad \sum_{k=1}^{r} \operatorname{sgn} \det(I - dC_{\gamma_k}^{-1}) \neq 0,$$

where $dC_{\gamma_k}^{-1}$ *denotes the Jacobi matrix of* C^{-1} *at* $\gamma_k \in \Sigma$.

Then, for a given constant $b > 0$ *or* $b < 0$, *respectively (see* (6.26)*), there exist at least* $m \geq 1$ *solutions* $\Theta(t)$ *of* (6.25) *with* $\Theta(0) \in \Sigma$ *such that*

$$C(\Theta(0)) = \Theta(t^*) \in \Sigma, \quad \text{for some } t^* > 0,$$

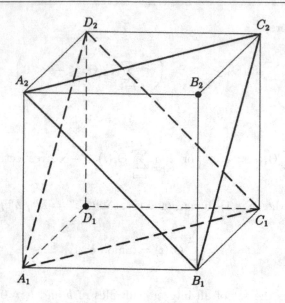

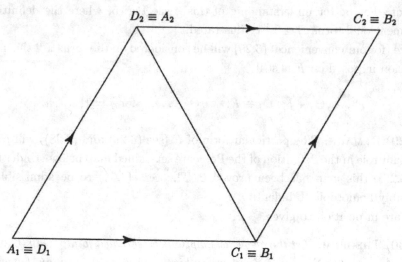

FIGURE 5

where $m = |\sum_{k=1}^{r} \text{sgn} \det(I - dC_{\gamma_k}^{-1})|$.

PROOF. Take into account only the first inequality in (6.26), the second one can be used quite analogously.

Because of (6.26) and (6.28), there is a well-defined (cf. Remark (6.29)) set-valued admissible Poincaré map Φ on Σ, namely

$$(6.32) \qquad \Phi_{\{\tau(p)\}}(p) : \Sigma \multimap \Sigma, \quad \Phi_{\{\tau(p)\}}(p) := \{\Theta(\tau(p))\},$$

where $\tau(p)$ denotes the least time for p to return back to Σ, when taking into account each branch of solution $\Theta(t)$ of (6.25) with $\Phi_0(p) = \Theta(0) = p \in \Sigma$.

Our problem is obviously solvable if we show $p \in \Sigma$ such that $C(p) \in \Phi(p)$ on Σ or, because of a diffeomorphism C, when we find a fixed point of $C^{-1}(\Phi(p))$ on Σ.

For this purpose, we apply the "multivalued analogy" of the Lefschetz trace formula. As pointed out in Remark (4.52), the map $\Phi : \Sigma \multimap \Sigma$ is homotopic, in the sense of multivalued admissible maps, to identity I.

The crucial step in applicatiom of the generalized Lefschetz fixed point theorem for the admissible (multivalued) map $C^{-1}(\Phi(p))$ on a compact $(n-1)$-dimensional manifold Σ consists in verifying the inequality $\Lambda(C^{-1}(\Phi(p))) \neq \{\emptyset\}$, where $\Lambda(\cdot)$ is the generalized generalized Lefschetz number. Because of the invariance under homotopy, it is however sufficient to show that $\Lambda(C^{-1}) \neq 0$.

Since C^{-1} is, by the hypothesis, smooth with finitely many simple fixed points, $\gamma_1, \ldots, \gamma_r$, on a compact manifold Σ, the Lefschetz number $\Lambda(C^{-1})$ can be simply calculated as the sum of the local indices, namely (see (6.58) below and cf. e.g. [BJ-M] or [Br1-M])

$$\Lambda(C^{-1}) = \sum_{k=1}^{r} \operatorname{sgn} \det(I - dC_{\gamma_k}^{-1}),$$

where $dC_{\gamma_k}^{-1}$ denotes the Jacobi matrix of C^{-1} on the tangent space $T_{\gamma_k}\Sigma$, which coincides here with Σ.

So because of (6.31), there exists at least one solution of our problem. The existence of $m \geq 1$ solutions follows from the formula (see Chapter I.10)

$$N(C^{-1}) = N(C^{-1} \circ \Phi) = |\Lambda(C^{-1} \circ \Phi)| = |\Lambda(C^{-1})|, \quad \text{on } \Sigma,$$

where N denotes the Nielsen number. $\qquad\square$

(6.33) REMARK. The time t^* to reach the target can be obviously estimated from above as $t^* \leq |b|/\varepsilon$. A lower estimate, $t^* \geq |b|/E$, holds provided, additionally,

$$\sum_{i=1}^{n} F_i(\Theta) \leq E \quad \left[\text{or } \sum_{i=1}^{n} F_i(\Theta) \geq -E\right].$$

In particular, for $E = \varepsilon \equiv |\sum_{i=1}^{n} F_i(\Theta)|$, we have the exact time, $t^* = |b|/\varepsilon$.

(6.34) REMARK. It can be easily checked that the conclusion of Theorem (6.30) is also true for a u-Carathéodory mapping $F(t, x)$, on the cylinder $[0, \infty) \times \mathbb{T}^n$, provided still $|\sum_{i=1}^{m} F_i(t, \Theta)| \equiv \varepsilon(t) \ldots$ single-valued, where $\int_0^{\omega} \varepsilon(t)\,dt = b > 0$. Namely, then at least $m \geq 1$ solutions $\Theta(t)$ with $\Theta(0) \in \Sigma$ exist such that $C(\Theta(0)) = \Theta(\omega) \in \Sigma$.

(6.35) EXAMPLE. Suppose, for simplicity, that $n = 3$, and C is linear, $C(\theta) := L\theta + c$, where

$$L = \begin{pmatrix} u_1 & u_2 & u_3 \\ v_1 & v_2 & v_3 \\ w_1 & w_2 & w_3 \end{pmatrix}$$

is a regular matrix such that $\mathcal{L} := u_1 + v_1 + w_1 = u_2 + v_2 + w_2 = u_3 + v_3 + w_3$, $\mathcal{L} \in \mathbb{Z}$, and $c = (c_1, c_2, c_3)^T$ is a nonzero vector such that $c_1 + c_2 + c_3 = 0 \bmod b$. Observe that the particular forms of L and c allow us to operate on Σ, where $\theta_1 + \theta_2 + \theta_3 = 0 \bmod b$.

Because of the regularity, there exists an inverse operator, namely $C^{-1}(\theta) = L^{-1}(\theta - c)$.

$C^{-1}(\theta)$ has exactly one fixed point γ_1 on Σ as far as $C(\theta)$ has, which is true if and only if

$$(6.36) \qquad 0 \neq -u_1 v_2 w_3 + u_1 v_3 w_2 + u_2 v_1 w_3 + u_2 v_3 w_1$$
$$+ u_3 v_1 w_2 + u_3 v_2 w_1 + u_1 v_2 + u_1 w_3 - u_2 v_1$$
$$- u_3 w_1 + v_2 w_3 - v_3 w_2 - u_1 - v_2 - w_3 + 1.$$

Then, under all the above assumptions,

$$|\Lambda(C^{-1}(\theta))| = |\operatorname{sgn} \det(I - dC_{\gamma_1}^{-1}(\theta))| = |\operatorname{sgn} \det(I - L^{-1})| = 1,$$

and we arrived at (6.31). So, Theorem (6.30) can be applied, provided still (6.26) and (6.28) for a suitable Marchaud map $F(\theta) = (F_1(\theta), F_2(\theta), F_3(\theta))^T$.

In Fig. 6, a trial of shooting to a target is demonstrated for $b = 6$ and $L = 2I$, $c = (1, 2, 3)^T$, which evidently satisfies (6.36) ($\Rightarrow$ (6.31)). Exactly one fixed point $\gamma_1 = -c$ of C^{-1} belongs to Σ.

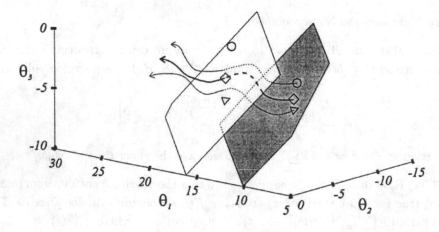

FIGURE 6

The targets related to the admissible initial states are indicated by the same marks. The trajectories starting at the initial states are generated here by a system of, for simplicity, differential equations (see Remark (6.34))

$$\dot{\theta}_1 = 10 + \sin\frac{\pi}{3}\theta_2 + \sin\frac{\pi}{3}\theta_3 + \cos 2\pi t,$$

$$\dot{\theta}_2 = -3 + \sin\frac{\pi}{3}\theta_1 - \sin\frac{\pi}{3}\theta_3 + 2\cos 2\pi t,$$

$$\dot{\theta}_3 = -1 - \sin\frac{\pi}{3}\theta_1 - \sin\frac{\pi}{3}\theta_2 - 3\cos 2\pi t.$$

Since $\varepsilon = E = 6$, the time t interval is chosen as $[0, 2]$. A special attention, however, should be paid to the value $t* = 1$, when the first hits to the target are expected. The bold trajectory, starting at the point $[9.5, -5.8, -3.7]$, hits the related target $[20, -9.6, -4.4]$ approximately for one correct decimal digit.

Now, we are in position to deal with the special sort of nonlinear oscillations of the equivariant system (6.25), namely with so called *splay-phase orbits*, i.e. those where the n coordinates are given by the same function, say $\varphi(t)$, but equally separated in phase:

$$\left(\varphi(t), \varphi\left(t + \frac{1}{n}T\right), \dots, \varphi\left(t + \frac{n+1}{n}T\right)\right),$$

and φ has period exactly T. Such solutions are sometimes called by the other authors as "*wagon-wheel*" or "*ponies on a merry-go-round*" solutions.

Consider the system of differential inclusions (6.25) on the torus $\mathbb{T}^n$, represented by the Cartesian product $[-\omega/2, \omega/2]^n$:

$$(6.37) \qquad \dot{\theta}_i \in F_i(\theta_1, \dots, \theta_n), \quad i = 1, \dots, n,$$

where $F = (F_1, \dots, F_n)^T$ is bounded and upper semicontinuous with nonempty, compact, convex values, by which all solutions of (6.37) entirely exist in the sense of Carathéodory (i.e. are absolutely continuous and satisfy (6.37) a.e.). Furthermore, let F_i be ω-periodic in each variable θ_j $(i, j = 1, \dots, n)$.

Define the shifts σ_+ and σ_- (shortly $(\sigma_{+,-} : \mathbb{T}^n \to \mathbb{T}^n)$ by the rule

$$\sigma_{+,-}(\theta_1, \dots, \theta_n) = \pm(\theta_2, \dots, \theta_n, \theta_1).$$

Let F_t be the time t flow for (6.37). We say that (6.37) is $\sigma_{+,-}$-*equivariant* if for all $t : F_t \circ \sigma_{+,-} = \sigma_{+,-} \circ F_t$. This means $(\theta_1(t), \dots, \theta_n(t))$ is an orbit for (6.37) if and only if $(\pm)^{i+1}(\theta_i(t), \dots, \theta_n(t), \theta_1(t), \dots, \theta_{i-1}(t))$ is, respectively. In terms of the components of the flow, the $\sigma_{+,-}$-equivariance means

$$(6.38\pm) \quad F_i(\theta_1, \dots, \theta_n) =$$

$$(\pm 1)^{i+1} F_i((\pm 1)^{i+1}\theta_i, \dots, (\pm 1)^{i+1}\theta_n, (\pm 1)^{i+1}\theta_1, \dots, (\pm 1)^{i+1}\theta_{i-1}),$$

respectively. We would like to deal also with the orbits of the form

$$\left(\varphi(t), -\varphi\left(t + \frac{1}{n}T\right), \ldots, \varphi\left(t + \frac{n-2}{n}T\right), -\varphi\left(t + \frac{n-1}{n}T\right)\right)$$

as the solutions of (6.37) with (6.38−). In the latter case, the cyclic permutation of coordinates brings the change of sign, by which the related periodic solutions might be analogously called "*anti-splay-phase orbits*". There is, however, one obvious restriction in the "anti"-case, namely n must be even and $n \neq 2$, when σ_- becomes an identity on Σ, where Σ has the same meaning as above. On the other hand, observe that the existence of one type of orbits can be simply transformed to the other in general, when $n \neq 2$ is just even. Indeed. For $n \neq 2$ even, it is clear that $(\theta_1(t), \ldots, \theta_n(t))$ is a splay-phase solution of σ_+-equivariant system (6.37) if and only if $(\theta_1(t), -\theta_2(t), \ldots, \theta_{n-1}(t), -\theta_n(t))$ is an anti-splay-phase one of σ_--equivariant system

$$\dot{\theta}_i \in (-1)^{i+1} F_i(\theta_1, -\theta_2, \ldots, \theta_{n-1}, -\theta_n), \quad i = 1, \ldots, n.$$

Thus, $(-1)^{i+1}((-1)^{i+1}\theta_i(t), \ldots, -\theta_n(t), \theta_1(t), \ldots, (-1)^i\theta_{i-1}(t))$ must also satisfy, for every i, the same σ_--equivariant system, and consequently

$$F_i(\theta_1, -\theta_2, \ldots, \theta_{n-1}, -\theta_n) =$$
$$(-1)^{i+1}F_i(\theta_i(t), \ldots, (-1)^i\theta_n(t), (-1)^{i+1}\theta_i(t), \ldots, -\theta_{i-1}(t)),$$

for all $i = 1, \ldots, n$. However, this is nothing else but (6.38−). In this light, we are in position to state (for the single-valued "+"-case, cf. [Mir]).

(6.39) THEOREM. *Let* (6.37) *be a* $\sigma_{+,-}$-*equivariant flow on* $\mathbb{T}^n$ *(see* (6.38±)). *Let, furthermore,* F_i *be* ω-*periodic in each variable* θ_j $(i, j = 1, \ldots, n)$ *and*

(6.40±) $$\sum_{i=1}^{n}(\pm 1)^i F_i(\theta_1, \ldots, \theta_n) > 0 \quad or \quad \sum_{i=1}^{n}(\pm 1)^i F_i(\theta_1, \ldots, \theta_n) < 0.$$

Then system (6.37) *has at least* n *nontrivial splay-phase or anti-splay-phase solutions* $\theta(t)$ *with* $\theta(0) \in \Sigma = \{\theta \in \mathbb{T}^n \mid \sum_{i=1}^{n}(\pm 1)^i\theta_i \equiv 0(\mathrm{mod}\,\omega)\}$, *respectively, provided* $n \neq 2$ *is even in the latter case.*

PROOF. Let us take only the first inequality in (6.40±), because the second one can be used quite analogously.

Consider the $(n-1)$-torus $\Sigma \subset \mathbb{T}^n$ given by $\sum_{i=1}^{n}(\pm 1)^i\theta \equiv 0 \bmod \omega$. Because of (6.40±), the (set-valued) Poincaré first-return map is well-defined on Σ which we denote by $\Phi : \Sigma \multimap \Sigma$.

It is obviously enough to verify the following two conditions:

(i) Suppose $p \in \Sigma$ satisfies $\Phi(p) \ni \sigma_{+,-}(p)$. Then the orbit of p includes a splay-phase or an anti-splay-phase solution, respectively.

(ii) The inclusion $\Phi(p) \ni \sigma_{+,-}(p)$ has solutions $p \in \Sigma$.

Ad (i) Assume $\Phi(p) \ni \sigma_{+,-}(p)$, for some $p \in \Sigma$. Let $\{T/n\} > 0$ be the time required for $p's$ first return to Σ. Recall that we denote the time t flow map $F_t : \mathbb{T}^n \multimap \mathbb{T}^n$. Now,

$$F_{\{T/n\}}(p) \supset \Phi(p) \ni \sigma_{+,-}(p),$$

and since $F_t \circ \sigma_{+,-} = \sigma_{+,-} \circ F_t$, for all t, we arrive at ($n \neq 2$ is even, in the second case)

$$F_{\{T\}}(p) = (F_{\{T/n\}})^n(p) \ni \sigma_{+,-}^n(p) = p.$$

Thus, the orbit of p contains a periodic branch, say $F_t^*(p)$, with period T/m, for some integer $m > 0$. But as we will show, T is the smallest period, i.e. $m = 1$.

Let

$$(\theta_1(t), \dots, \theta_n(t)) = F_t^*(p) \quad \text{and} \quad \varphi(t) = \theta_n(t).$$

Then, since for all t:

$$F_{t+T/n}^*(p) = F_t^*(\sigma_{+,-}(p)) = \sigma_{+,-}(F_t^*(p)),$$

we see that

$$(\pm 1)^i \theta_i(t) = \varphi\left(t + \frac{T}{n} i\right), \quad i = 1, \dots, n.$$

It remains to prove that $m = 1$. Because of

$$\sum_{i=1}^{n} (\pm 1)^i \dot\theta_i(t) = \sum_{i=1}^{n} \dot\varphi\left(t + \frac{T}{n} i\right),$$

we have

$$\int_0^{T/n} \sum_{i=1}^{n} (\pm 1)^i \dot\theta_i(t) \, dt = \int_0^{T/n} \sum_{i=1}^{n} \dot\varphi\left(t + \frac{T}{n} i\right) dt.$$

Thus,

$$\sum_{i=1}^{n} \int_0^{T/n} (\pm 1)^i \dot\theta_i(t) \, dt = \sum_{i=1}^{n} (\pm 1)^i \theta_i\left(\frac{T}{n}\right) - \sum_{i=1}^{n} (\pm 1)^i \theta_i(0) = \omega,$$

because T/n is the first-return time for p back to Σ, and therefore also

$$\varphi(T) - \varphi(0) = \omega.$$

But if φ has period T/n, then $[\varphi(t) - \varphi(0)]$ must be, in view of ω-periodicity of F_i ($i = 1, \dots n$), an integer multiple of $m\omega$, and consequently $m = 1$, indeed.

Ad (ii) It remains to prove that the map $\sigma_{+,-}^{-1} \circ \Phi$ has a fixed point on Σ. For this, we apply the Lefschetz fixed point theorem (6.15) in Chapter I.6. As pointed out in Remark (4.52), the map $\Phi : \Sigma \multimap \Sigma$ is homotopic, in the sense admissible maps, to the identity map.

Recalling the Lefschetz fixed point theorem (6.15) in Chapter I.6, the crucial step in its application consists in verifying the inequality $\Lambda(\sigma_{+,-}^{-1} \circ \Phi) \neq \{0\}$, where $\Lambda(\sigma_{+,-}^{-1} \circ \Phi)$ is the generalized Lefschetz number of an admissible compact map $\sigma_{+,-}^{-1} \circ \Phi$; for the appropriate definitions, see Chapter I.4. The fact that $\sigma_{+,-}^{-1} \circ \Phi$ is so follows from our assumptions above.

In our situation, since Φ is homotopic to the identity map, $\sigma_{+,-}^{-1} \circ \Phi$ is homotopic to $\sigma_{+,-}^{-1}$, so it suffices to show that $\Lambda(\sigma_{+,-}^{-1}) \neq 0$, because all the remaining assumptions of the Lefschetz fixed point theorem (6.15) in Chapter in I.6 are trivially satsified. Because of $|\Lambda(\sigma_{+,-})| = |\Lambda(\sigma_{+,-}^{-1})|$, we might as well work on $\sigma_{+,-}$. Since $\sigma_{+,-}$ is already a smooth map with finitely many fixed points p_k on Σ, namely ($n \neq 2$ is even; for n odd, analogously)

$$p_k = \left(\pm\frac{\omega l}{n}, \frac{\omega l}{n}, \dots, \pm\frac{\omega l}{n}, \frac{\omega l}{n} \right), \quad \text{for } l = -\left(\frac{n}{2}-1\right), \dots, -1, 0, 1, \dots, \left(\frac{n}{2}-1\right), \frac{n}{2},$$

where $k = l + n/2$, then also

$$\Lambda(\sigma_{+,-}) = \sum_{k=1}^{n} \operatorname{sgn} \det(I - d(\sigma_{+,-})_{p_k}),$$

where $d(\sigma_{+,-})_{p_k}$ is the Jacobi matrix of $\sigma_{+,-}$ on the tangent space $T_{p_k}\Sigma$.

All the fixed points p_k ($k = 0, 1, \dots, n-1$) have the same local structure, so it suffices to calculate the local index $\operatorname{sgn} \det(I - d(\sigma_{\sigma_{+,-}})_{p_k})$ at $p_0 = (0, \dots, 0)$. In local coordinates,

$$(\alpha_1, \dots, \alpha_{n-1}) \to \pm\left(\alpha_2, \dots, \alpha_{n-1}, -\sum_{i=1}^{n-1}(\pm1)^i \alpha_i \right)$$

near $(0, \dots, 0)$ on Σ, i.e. $d\sigma_{+,-}$ is given by the matrix

$$A_{+,-} = \begin{pmatrix} 0 & \pm1 & 0 & \dots & 0 \\ 0 & 0 & \pm1 & \dots & 0 \\ & & & \vdots & \\ 0 & 0 & 0 & \dots \pm1 \\ \pm1 & 1 & \pm1 & \dots & 1 \end{pmatrix},$$

namely

$$A_{+,-}(\alpha_1, \dots, \alpha_{n-1})^T = \pm\left(\alpha_2, \dots, \alpha_{n-1}, -\sum_{i=1}^{n-1}(\pm1)^i \alpha_i \right)^T.$$

The characteristic polynomial of this map is

$$\det(\mu I - A_+) = 1 + \mu + \mu^2 + \ldots + \mu^{n-3} + \mu^{n-2} + \mu^{n-1},$$
$$\det(\mu I - A_-) = 1 + \mu + \mu^2 + \ldots + \mu^{n-4} + \mu^{n-2},$$

i.e. for $\mu = 1 : \det(I - A_+) = n$ or $\det(I - A_-) = n - 2$, respectively. Therefore, $\Lambda(\sigma_+) = n$, while $\Lambda(\sigma_-) = n$, only for $n \geq 4$. Since

$$N(\sigma_{+,-}) = N(\sigma_{+,-}^{-1} \circ \varphi) = |\Lambda(\sigma_{+,-}^{-1} \circ \varphi)| = |\Lambda(\sigma_{+,-})| = n \quad \text{on } \Sigma,$$

whenever $n \neq 2$, where N denotes the Nielsen number (see Theorem (10.25) in Chapter I.10), the proof is complete. $\square$

(6.41) REMARK. One can readily check that Theorem (6.39) can be proved by means of Theorem (6.30), when taking

$$C := \sigma_{+,-} : \mathbb{T}^n \to \mathbb{T}^n.$$

(6.42) REMARK. The multiplicity of solutions can be easily deduced only by the definition of (anti-)splay-phase orbits, i.e. without an explicit usage of the Nielsen number.

(6.43) REMARK. Although Theorem (6.39) does not hold, in the "anti"-case, for $n = 2$, in general, the planar systems may have anti-splay-phase orbits.

For example, the system

$$\dot{\Theta}_1 = c_1 + c_2 \sin \Theta_1 - c_3 \sin \Theta_1 + c_3 \sin \Theta_2,$$
$$\dot{\Theta}_2 = -c_1 + c_2 \sin \Theta_2 + c_3 \sin \Theta_1 - c_3 \sin \Theta_2$$

can be reduced, for $\Theta_1 = -\Theta_2 := \Theta$, to

$$\dot{\Theta} = c_1 + (c_2 - 2c_3) \sin \Theta,$$
$$(\dot{\Theta} = c_1 + (c_2 - 2c_3) \sin \Theta,),$$

i.e. the single one.

For $|c_1| > |c_2 - 2c_3|$, we get that $\dot{\Theta} > 0$. Since $\Theta_1 = -\Theta_2$ on Σ, the inequality $\dot{\Theta} > 0$ implies, e.g. for the solution $\Theta(t, 0)$ with $\Theta(0, 0) = 0$, the time $t^* > 0$ such that $\Theta(t^*, 0) = 2\pi$, i.e. $\Theta(t^*, 0) = 0 \pmod{2\pi}$. Because of the evident uniqueness, we have $(\Theta_1(t), \Theta_2(t)) = (\Theta(t, 0), -\Theta(t, 0))$, and so the above system admits a

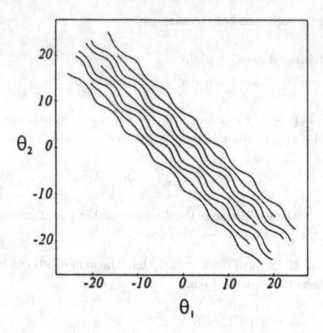

FIGURE 7

desired t^*-derivo-periodic solution (anti-splay-phase orbit) as well, as it can be seen in Fig. 7.

(6.44) EXAMPLE. As an example of the system satisfying (6.38$\pm$), consider for simplicity

$$(6.45\pm) \quad \dot{\theta}_i = (\pm 1)^{i+1} c_1 + c_2 \sin \theta_i + c_3 \sum_{j=1}^{n} (\pm 1)^{j+i+1} \sin \theta_j, \quad \text{for } i = 1, \ldots, n,$$

where c_1, c_2, c_3 are suitable constants. Condition (6.40$\pm$) reads here as

$$c_1 + \left(c_3 \pm \frac{c_2}{n} \right) \sum_{j=1}^{n} (\pm 1)^j \sin \theta_j \neq 0,$$

which can be easily satisfied for $|c_1| > |nc_3 \pm c_2|$.

(6.46) REMARK. By the numerical simulation, it seems that the existence results in Theorem (6.39) are stable under small (T/n)-periodic perturbations of the particular form. Namely, if they are either the same or only differing by the sign in each inclusion of the equivariant or anti-equivariant system, respectively. This can be seen in Fig. 8, where the phase portrait of (6.45+), for $n = 2$, $c_1 = 4$, $c_2 = c_3 = 1$, under the perturbation by $\cos t$ in both equations, is practically indifferent to the unperturbed one. It is, however, a problem to estimate with a sufficient accuracy the length of period T. Knowing it, we can apply, in the single-valued case, the perturbation technique, provided 1 is a simple characteristic multiplier of the variational system to (6.37) along the splay-phase or anti-splay-phase solutions. For $n = 2$, there is some chance to make it by means of the well-known trace formula for one multiplier λ (the second must be 1), namely $\lambda = \exp \int_0^T \operatorname{Tr} A(t)\, dt$, where $A(t)$ is the matrix of the linear variational system.

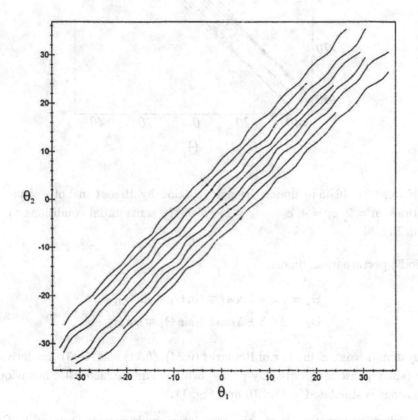

FIGURE 8. (6.45+) under the perturbation by $\cos t$ in both equations: $n = 2$, $c_1 = 4$, $c_2 = c_3 = 1$) (almost indifferent to the unperturbed one)

(6.47) REMARK. One may observe in the computer simulations for the behaviour of dynamical system (6.45$\pm$), perturbed in each equation by $(\pm 1)^i K \cos t$ $(i = 1, \ldots, n)$, where $K > 0$ is a sufficiently big constant, the combination of an irregular motion and (anti)symmetry. A typical situation is demonstrated in Fig. 9, where $n = 2$, $c_1 = 4$, $c_2 = c_3 = 1$ and $K = 10$.

(6.48) EXAMPLE. Consider the particular form of system (6.45$_-^+$), but under

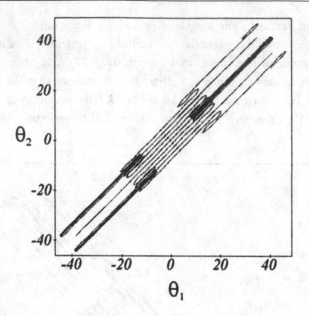

FIGURE 9. (6.45+) under the perturbation by $10\cos t$ in both equations: $n = 2$, $c_1 = 4$, $c_2 = c_3 = 1$ (with the same initial conditions as in Fig. 8)

a periodic perturbation, namely

$$\dot{\Theta}_1 = 0.5 + 5\cos t \mp \sin\Theta_1 + \sin\Theta_2,$$
$$\dot{\Theta}_2 = \pm 0.5 \pm 5\cos t + \sin\Theta_1 \mp \sin\Theta_2.$$

This system has again, in view of Remarks (6.34), (6.41) and (6.43), 2π-derivo-periodic (splay-phase and anti-splay-phase) orbits. The characteristic behaviour of its solutions is simulated in Fig. 10 and Fig. 11.

The following approach is via the translation multioperator treated in Chapter III.4. Hence, consider the system of differential inclusions

$$(6.49) \qquad\qquad \dot{x} \in F(t, x),$$

where $F : \mathbb{R}^{n+1} \multimap \mathbb{R}^n$ is a u-Carathéodory mapping with nonempty, compact and convex values, satisfying (6.3). Then all solutions of (6.49) exist in the sense of Carathéodory, namely they are locally absolutely continuous and satisfy (6.49) a.e.

If $x(t, x_0) := x(t, 0, x_0)$ is a solution of (6.49) with $x(0, x_0) = x_0$, then we can define the *Poincaré–Andronov map* (translation operator at the time $T > 0$) along

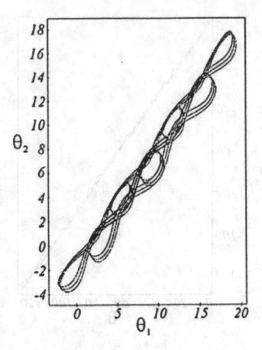

FIGURE 10

the trajectories of (6.49) as follows:

$$(6.50) \qquad \Phi_T : \mathbb{R}^n \multimap \mathbb{R}^n, \quad \Phi_T(x_0) := \{x(T, x_0) \mid x(t, x_0) \in (6.49)\}.$$

The goal is to represent the map Φ_T in terms of the pair (p, q). We let $\varphi : \mathbb{R}^n \multimap C([0, T], \mathbb{R}^n)$, where

$$\varphi(X) := \{x \in C([0, T], \mathbb{R}^n) \mid x(0) = X \quad \text{and} \quad x \in (6.49)\}$$

and $C([0, T], \mathbb{R}^n)$ is the Banach space of continuous maps. According to Theorem (4.3), φ is an R_δ-mapping.

Now, we let

$$e_T : C([0, T), \mathbb{R}^n) \to \mathbb{R}^n, \quad e_T(x) = x(T),$$

where $e_T(x)$ is evidently continuous.

One can readily check that $\Phi_T = e_T \circ \varphi$. Moreover, $\mathcal{H} \circ \Phi_T = \mathcal{H} \circ e_T \circ \varphi$ is admissible for any homeomorphism $\mathcal{H} : \mathbb{R}^n \to \mathbb{R}^n$.

In fact, we have the diagram

$$\mathbb{R}^n \xleftarrow{p_\varphi} \Gamma_\varphi \xrightarrow{q_\varphi} C([0, a], \mathbb{R}^n) \xrightarrow{e_T} \mathbb{R}^n \xrightarrow{\mathcal{H}} \mathbb{R}^n.$$

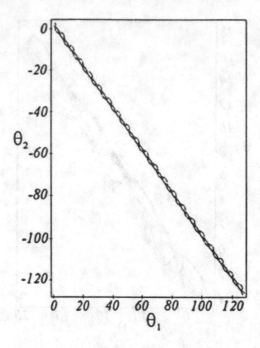

FIGURE 11

Hence, in what follows, we identify the Poincaré–Andronov map Φ_T or its composition with $\mathcal{H}$, that is $\mathcal{H} \circ \Phi_T$, with the pair $(p_\varphi, e_T \circ q_\varphi)$ or $(p_\varphi, \mathcal{H} \circ e_T \circ q_\varphi)$, i.e. we let

$$(6.51) \qquad \Phi_T = (p_\varphi, e_T \circ q_\varphi) \quad \text{or} \quad \mathcal{H} \circ \Phi_T = (p_\varphi, \mathcal{H} \circ e_T \circ q_\varphi),$$

respectively.

Let $C(\mathcal{H} \circ \Phi_T)$ denote the set of coincidence points of the pair $(p_\varphi, \mathcal{H} \circ e_T \circ q_\varphi)$, while $\mathrm{Fix}(\mathcal{H} \circ \Phi_T) = \{X \in \mathbb{R}^n \mid X \in \mathcal{H}(e_T(q_\varphi(p_\varphi^{-1}(X))))\}$.

A pair in (6.51) can be easily shown to be homotopic to the identity map, so that the pair $(p_\varphi, \mathcal{H} \circ e_T \circ q_\varphi)$ is homotopic to $\mathcal{H}$ (see Theorem (4.3)). Observe that a coincidence for the pair $(p_\varphi, \mathcal{H} \circ e_T \circ q_\varphi)$ is a point X and a solution $x(t)$ of (6.49) such that $x(0) = X$ and $\mathcal{H}(x(T)) = X$. Thus, we have a one-to-one correspondence between coincidence points and solutions. Since a coincidence point of $(p_\varphi, \mathcal{H} \circ e_T \circ q_\varphi)$ gives us in this way a solution $x(t)$ of (6.49) such that $x(0) = \mathcal{H}(x(T))$, the following proposition is self-evident.

(6.52) PROPOSITION. *If* $\#C(p_\varphi, \mathcal{H} \circ e_T \circ q_\varphi) = k$, *then system* (6.49) *has at least k solutions* $x_1(t), \dots, x_k(t)$ *such that* $x_i(0) = \mathcal{H}(x_i(T))$, $i = 1, \dots, k$.

The following lemma immediately follows from Theorem (4.3).

(6.53) LEMMA. *Assume that Y is a compact connected ANR-space such that* $\Phi_s(Y) \subset Y$, *for every $s \in [0,T]$ and $\mathcal{H}(Y) \subset Y$. Then the pair $(p_\varphi, \mathcal{H} \circ e_T \circ q_\varphi)$ restricted to Y is admissibly homotopic to $\mathcal{H}|_Y$.*

As a consequence of Theorems (10.21), (10.25) in Chapter I.10, Proposition (6.52) and Lemma (6.53), we have the following

(6.54) PROPOSITION. *Assume that Φ_s, for every $s \in [0,T]$ in (6.50) is a self-map on $\mathbb{T}^n$. Then system (6.49) has at least $N(\mathcal{H})$ solutions $x(t)$ such that $x(0) = \mathcal{H}(x(T))$ on $\mathbb{T}^n$, where $N(\mathcal{H})$ denotes the Nielsen number of a homeomorphism $\mathcal{H} : \mathbb{T}^n \to \mathbb{T}^n$.*

PROOF. The proof follows directly from Theorems (10.21), (10.25) in Chapter I.10, Proposition (6.52), Lemma (6.53) and the properties of the Nielsen number and coincidence points mentioned in Chapter I.10. □

If in particular $\mathcal{H} = \mathrm{id}$, then (see (10.25) in Chapter I.10 and cf. [BBPT])

$$N(\mathrm{id}) = |\lambda(\mathrm{id})| = |\chi(\,\cdot\,)|$$

holds on tori, and consequently the problems considered in Proposition (6.54) should have at least $|\chi(\,\cdot\,)|$ solutions, where λ is the Lefschetz number and χ denotes the Euler–Poincaré characteristic of $\mathbb{T}^n$. Since, unfortunately, $\chi(\,\cdot\,) = 0$ for tori, this is not a suitable case for applications.

On the other hand, we can say about the tori considered above the following

(6.55) LEMMA (cf. e.g. [BJ-M]). *Every single-valued self-map $\mathcal{H}$ on the torus $\mathbb{T}^m = \mathbb{R}^m/\mathbb{Z}^m$ is homotopic to a map induced by a homomorphism $\mathbb{R}^m \to \mathbb{R}^m$. Let A be the matrix representing this homomorphism. Then*

$$(6.56) \qquad N(\mathcal{H}) = |\lambda(\mathcal{H})| = |\det(I - A)|.$$

In, particular, for $m = 1$, any self-map $\mathcal{H}$ on the circle $\mathbb{R}/\mathbb{Z}$ is homotopic to $[t] \to [kt]$, for some $k \in \mathbb{Z}$, and so $N(\mathcal{H}) = |k - 1|$.

(6.57) LEMMA (cf. e.g. [BJ-M], [Br1-M]). *If a single-valued self-map $\mathcal{H}$ on the torus $\mathbb{T}^m$ has still finitely many, but at least one, simple fixed points $\gamma_1, \ldots, \gamma_r$ on $\mathbb{T}^m$, then formula (6.56) can be replaced by*

$$(6.58) \qquad N(\mathcal{H}) = |\lambda(\mathcal{H})| = \left| \sum_{k=1}^{r} \det(I - d\mathcal{H}_{\gamma_k}) \right|,$$

where $d\mathcal{H}_{\gamma_k}$ denotes the Jacobi matrix of $\mathcal{H}$ on the tangent space $T_{\gamma_k}\mathbb{T}^m$.

As the simplest application of Proposition (6.53), we can give immediately

(6.59) THEOREM. *Assume that*

(6.60) $$F(t, \dots, x_j + 1, \dots) \equiv F(t, \dots, x_j, \dots), \quad for \ j = 1, \dots, n,$$

where $x = (x_1, \dots, x_n)$ and consider system (6.49) on the set $[0, \infty) \times \mathbb{T}^n$, where $\mathbb{T}^n = \mathbb{R}^n / \mathbb{Z}^n$. Then system (6.49) admits, for every positive constant T, at least $N(\mathcal{H})$ solutions $x(t)$ such that

$$x(0) = \mathcal{H}(x(T)) \ (mod \ 1),$$

where $\mathcal{H}$ is a continuous self-map on $\mathbb{T}^n$ and $N(\mathcal{H})$ denotes the associated Nielsen number for which formulas (6.56) or (6.58) can be employed (the latter, under the appropriate additional assumptions).

As a consequence, we obtain for $\mathcal{H} = -\mathrm{id}$ easily

(6.61) COROLLARY. *If, in addition to the assumptions of Theorem (6.59),*

$$F(t + T, -x) \equiv -F(t, x),$$

then system (6.49) admits at least 2^n anti-T-periodic (or $2T$-periodic) solutions $x(t)$ on $\mathbb{T}^n$, namely $x(t + T) \equiv -x(t) \ (mod \ 1)$.

Let us conclude by a simple example.

(6.62) EXAMPLE. The system

$$\dot{x} + a \sin \pi x + b \sin \pi y \in P(t),$$
$$\dot{y} + c \sin \pi x + d \sin \pi y \in Q(t),$$

where a, b, c, d are arbitrary constants, $P(t)$, $Q(t)$ are essentially bounded measurable maps with nonempty, compact, convex values and $P(t + T) \equiv -P(t)$, $Q(t + T) \equiv = -Q(t)$, has according to Corollary (6.61) at least four $2T$-periodic solutions on $\mathbb{T}^2$. Moreover, one can check that (cf. (6.56))

$$|\det(I - (-I))| = \left| \det \begin{pmatrix} 2 & 0 \\ 0 & 2 \end{pmatrix} \right| = 4$$

or that (cf. (6.58))

$$\left| \sum_{i=1}^{4} \mathrm{sgn} \det(I - d(-\mathrm{id})) \right| = 4,$$

because $d\mathcal{H}_{[0,0]} = d\mathcal{H}_{[0,1/2]} = d\mathcal{H}_{[1/2,0]} = d\mathcal{H}_{[1/2,1/2]} = -d\mathrm{id} = -I$, where $[0, 0]$, $[0, 1/2]$, $[1/2, 0]$, $[1/2, 1/2]$ are the geometrically distinct fixed points of $-\mathrm{id}$ on $\mathbb{T}^2$.

Before applying the relative (Lefschetz and Nielsen) theories, it will be convenient to introduce suitable definitions of (periodic) [semi-] processes and proper pairs.

Assume that X is a metrizable space and $\Phi : D \multimap X$ is a $\mathbb{C}$AC-mapping (see Chapter I.5), where $D \subset \mathbb{R} \times X \times \mathbb{R}$ $[D \subset \mathbb{R} \times X \times [0, \infty)]$ is an open set. Denoting the mapping $\Phi(\sigma, \cdot, t)$ by $\Phi_{(\sigma, t)}$, we can give the definition of admissible (multivalued) [semi-] processes as follows.

(6.63) DEFINITION. Φ is called a *generalized local [semi-] process* (on the space X) if the following conditions are satisfied:

(i) $\forall \sigma \in \mathbb{R}, \ x \in X : \{t \in \mathbb{R}[t \geq 0] \mid (\sigma, t, x) \in D\}$ is an interval,

(ii) $\forall \sigma \in \mathbb{R} : \Phi_{(\sigma, 0)} = \mathrm{id}$,

(iii) $\forall \sigma, t, s \in \mathbb{R} \ [\forall \sigma \in \mathbb{R}, \ t \geq 0, \ s \geq 0] : \Phi_{(\sigma, s+t)} \subset [=] \Phi_{(\sigma+s, t)} \circ \Phi_{(\sigma, s)}$.

In the case $D = \mathbb{R} \times X \times \mathbb{R}$ $[\mathbb{R} \times X \times [0, \infty)]$, we call Φ a *generalized (global) [semi-] process*.

For $(\sigma, x) \in \mathbb{R} \times X$, under the assumption that

$$\Phi_{(\sigma, t)}(x) = \{y_{(\sigma, t)}(x) \in X \mid y_{(\sigma, \cdot)}(x) \text{ is a continuous function with}$$
$$y_{(\sigma, 0)}(x) = x \text{ and } (\sigma, x, t) \in D\},$$

the set

$$\{(\sigma + t, y_{(\sigma, t)}(x)) \in \mathbb{R} \times X \mid y_{(\sigma, t)}(x) \subset \Phi_{(\sigma, t)} \text{ is a single-valued}$$
$$\text{continuous selection and } (\sigma, x, t) \in D\},$$

is called the *set of trajectories in* (σ, x) *of* Φ.

If $T > 0$ is an integer and Φ still satisfies

(iv) $\forall \sigma, t \in \mathbb{R} \ [\forall \sigma \in \mathbb{R}, \ t \geq 0] : \Phi_{(\sigma, t)} = \Phi_{(\sigma+T, t)}$,

we call Φ a *T-periodic local* (or *global*) *generalized [semi-] process*.

For $\sigma = 0$, a generalized local (or global) [semi-] process $\Phi_{(0, t)}$ is called a *generalized local* (or *global*) *generalized [semi-] dynamical system*.

A local generalized process Φ on X determines a *local generalized [semi-] flow* Φ^* on $\mathbb{R} \times X$ by

$$\Phi_t^*(\sigma, x) = (\sigma + t, \Phi_{(\sigma, t)}(x)).$$

Denoting

$$Z(t) = \{x \in X \mid (t, x) \in Z\},$$

for every $Z \subset \mathbb{R} \times X$ and $\sigma \in \mathbb{R}$, we call the *set* Z to be *T-periodic* if, for every $t \in \mathbb{R}$,

$$Z(\sigma) \equiv Z(\sigma + T).$$

In this case, we put

$$Z = \left\{ \left(\exp\left(\frac{2\pi i \sigma}{T}\right), x \right) \in S^1 \times Z(\sigma) \right\},$$

i.e., there is a circumference by identifying σ and $\sigma + T$.

Assuming that (A, B) is a pair of subsets of $\mathbb{R} \times X$, where $A \subset B$, we can give

(6.64) DEFINITION. (A, B) is called a *proper pair* if

(i) $A(\sigma)$ and $B(\sigma)$ are ANR-spaces, for each $\sigma \in \mathbb{R}$,

(ii) there is a generalized [semi-] process Φ on X such that A and B are [positively-] invariant under the flow Φ^*, defined as above.

If still

(iii) A and B are T-periodic, then we speak about a *T-periodic proper pair*.

Now, consider the system

$$(6.65) \qquad\qquad \dot{y} \in F(t, y),$$

where $F : \mathbb{R} \times \Omega \multimap \mathbb{R}^n$ is an (upper) Carathéodory multivalued function, satisfying (6.3) and $\Omega \subset \mathbb{R}$ is an open subset (possibly, the whole space $\mathbb{R}^n$). Denoting by $y(t) = y(\sigma, x, t)$, $[t \geq \sigma]$, its (Carathéodory-like) solution, satisfying $y(\sigma) = x$, the *generalized [semi-] process φ generated by* (6.65) takes the form $[t \geq 0]$

$$\varphi_{(\sigma, t)}(x) = \varphi(\sigma, x, t) = \{y(\sigma, x, t + \sigma)\}.$$

In particular,

$$\varphi_{(0, t)}(x) = \{y(0, x, t)\}$$

or

$$\varphi_{(\sigma, t)}(x) = \{y(\sigma, x, t + \sigma)\} = \varphi_{(\sigma + T, t)}(x) = \{y(\sigma + T, x, t + \sigma + T)\},$$

provided $F(t, y) \equiv F(t + T, y)$, represents a *generalized [semi-] dynamical system* or a *generalized T-periodic [semi-] process generated* by (6.65), respectively.

It follows from the investigations in Chapter II.5 that if (A, B), where $B \subset A \subset \mathbb{R} \times X$, is a proper pair for a generalized [semi-] process $\varphi : (A(\sigma), B(\sigma)) \multimap (A(\sigma), B(\sigma))$, then the generalized Lefschetz number $\Lambda(\varphi)$ is well-defined, for every $\sigma \in \mathbb{R}$, satisfying

$$\Lambda(\varphi) = \Lambda(\varphi_{A(\sigma)}) - \Lambda(\varphi_{B(\sigma)}),$$

where $\varphi_{A(\sigma)} : A(\sigma) \multimap A(\sigma)$ and $\varphi_{B(\sigma)} : B(\sigma) \multimap B(\sigma)$ are particular [semi-] flows on $A(\sigma)$ and $B(\sigma)$, $\sigma \in \mathbb{R}$.

Moreover, for every $\sigma \in \mathbb{R}$, $t \in \mathbb{R}$ $[t \in [0, \infty)]$,

$$\varphi_{(\sigma,t)} \sim \varphi_{(\sigma,0)} = \mathrm{id};$$

a homotopy is given by $\varphi_{(\sigma,kt)}$, $k \in [0, 1]$.

In particular, for a T-periodic proper pair (A, B), $T > 0$

$$\varphi_0 \sim \varphi_T : (A(0), B(0)) \multimap (A(T), B(T)),$$

by which (in view of the invariance under homotopy)

$$\Lambda(\varphi_0^*) = \Lambda(\varphi_T) = \Lambda(\varphi_0) = \chi(A(0)) - \chi(B(0)),$$

where χ stands for the Euler–Poincaré characteristic.

Thus, if

$$\chi(A(0)) \neq \chi(B(0)),$$

then there exists a fixed point $z^* \in \{0\} \times \overline{A(0) \setminus B(0)}$ of the semi-flow φ_T^* : $\{0\} \times A(0) \multimap \{T\} \times A(T)$, i.e., $z^* \in \varphi_T^*(z^*)$, and subsequently also a T-periodic trajectory in $(0, z^*)$ of φ_A.

Hence, we can give:

(6.66) THEOREM. *If (A, B), where $B \subset A \subset \mathbb{R} \times X$, is a T-periodic proper pair for the T-periodic semi-flow $\varphi^* : (A, B) \multimap (A, B)$, determined by some T-periodic semi-process φ on $X \times X$, then there exists a T-periodic trajectory in some point $(0, z^*) \in \{0\} \times \overline{A(0) \setminus B(0)}$ of $\varphi_{A(0)} : A(0) \multimap A(T)$, provided*

$$\varphi_{A(0)}(x) = \{y_{(0,t)}(x) \in A(t) \mid y_{(0,\cdot)}(x) \text{ is a continuous function}$$
$$\text{with } y_{(0,0)}(x) = x \text{ and } (x, t) \in A(t) \times [0, T]\}.$$

In particular, if $z^* \notin fr(B(0))$ and $\chi(B(0)) \neq 0$, then there are at least two T-periodic trajectories: one in some point at the interior of $\{0\} \times A(0) \setminus B(0)$ and another one in some point at $\{0\} \times B(0)$. The same is true if an initial point of a T-periodic trajectory implied by $\chi(B(0)) \neq 0$ is at the interior of $B(0)$.

For the differential system (6.65), where $F(t, y) \equiv F(t + T, y)$, a T-periodic semi-process can be generated by means of the associated *Poincaré translation operator* along the trajectories of (6.65) at the time kT, $k \in [0, 1]$, defined as follows:

(6.67) $\quad \Phi_{kT} = \{y(x, kT) \mid y(x, \cdot) \text{ is solution of (6.65) with } y(x, 0) = x\}.$

According to Theorem (4.3) Φ is admissible. More precisely, it is a composition $\Phi = \psi_2 \circ \psi_1$ of an R_δ-mapping ψ_1 and a continuous (single-valued) evaluation mapping ψ_2, which is homotopic (in the same class of maps) to identity.

Therefore, if A_1, B_1, where $B_1 \subset A_1 \subset \Omega \subset \mathbb{R}^n$, are compact ENR-spaces such that

$$(6.68) \qquad \Phi_{kT}(A_1) \subset A_1 \quad \text{and} \quad \Phi_{kT}(B_1) \subset B_1 \quad \forall k \in [0,1],$$

then Φ_{kT} becomes a compact admissible mapping. We can even put

$$\Phi_{kT}^*(0, x) = (kT, \Phi_{(0,kT)}(x)) = (kT, \Phi_{kT}(x)),$$

in order to demonstrate the correspondence to a T-periodic generalized semi-flow in the above sense.

After all, assuming (6.68) and

$$(6.69) \qquad \chi(A_1) \neq \chi(B_1),$$

system (6.65) admits a T-periodic solution $y(t)$ with $y(0) \in \overline{A_1 \setminus B_1}$. In particular, if $y(0) \notin \mathrm{fr}(B_1)$ and $\chi(B_1) \neq 0$, then there are again at least two T-periodic solutions $y_1(t)$, $y_2(t)$ of (6.65) such that $y_1(0) := y(0) \in \mathrm{int}(A_1 \setminus B_1)$ and $y_2(0) \in B_1$. The same is true if an initial point of a T-periodic solution implied by $\chi(B(0)) \neq 0$ is in $\mathrm{int}B_1$.

Similarly, for the functional system

$$(6.70) \qquad \dot{x} \in F(t, x_t), \quad x \in \mathbb{R}^n,$$

where $x_t(\cdot) = x(t + \cdot)$, for $t \in [0, \omega]$, denotes as usual a function from $[-\delta, 0]$, $\delta \geq 0$, into $\mathbb{R}^n$ and $F : [0, \omega] \times \mathcal{C} \multimap \mathbb{R}^n$, where $\mathcal{C} = \mathrm{AC}([-\delta, 0], \mathbb{R}^n)$, is an (upper) Carathéodory mapping with nonempty, compact and convex values (see Chapter III.4, Part B), we can get

(6.71) THEOREM. *Let A and B, where $B \subset A \subset \mathcal{C}$, be compact ANR-spaces which are (strongly) semiflow invariant, under Poincaré's operator defined by (4.6), and such that $\chi(A) \neq \chi(B)$, where $\chi(\cdot)$ stands for the Euler–Poincaré characteristic. Then system (6.70), where $F(t, y) \equiv F(t + \omega, y)$, admits an ω-periodic solution $x(t, x_*)$ with $x_* \in \overline{A \setminus B}$. In particular, if $x(0, x_*) \notin \mathrm{fr}(B)$ and $\chi(B) \neq 0$, then there are at least two ω-periodic solutions $x(t, x_1)$, $x(t, x_2)$ of (6.70) such that $x(0, x_1) := x(0, x_*) \in \mathrm{int}(A \setminus B)$ and $x(0, x_2) \in B$. The same is true if an initial point of an ω-periodic solution implied by $\chi(B) \neq 0$ is in $\mathrm{int}B$.*

Now, we would like to make a nontrivial application of the relative Nielsen number. If $\mathcal{N}$ is, for example, a compact nilmanifold[11], then the well-known result

----

[11]$\mathcal{N}$ is a *nilmanifold* if $\mathcal{N} = S/\pi$, where S is a connected, simply connected nilpotent Lie group and π is a discrete subgroup, for more details, see e.g. [McC-M] and the references therein.

of D. Anosov (cf. e.g. [McC-M, p. 255]) asserts that, for a single-valued continuous self-map $f : \mathcal{N} \to \mathcal{N}$, we have $N(f) = |\Lambda(f)|$, where N and Λ stand for the (well-defined) Nielsen and Lefschetz numbers, respectively. If $\mathcal{N}$ is an AR-space (not necessarily a nilmanifold), then $N(f) = \Lambda(f) = 1$.

Furthermore, if $\mathcal{N}_1$, $\mathcal{N}_2$ are compact ANR-spaces such that $\mathcal{N}_2 \subset \mathcal{N}_1$ and $f : (\mathcal{N}_1, \mathcal{N}_2) \to (\mathcal{N}_1, \mathcal{N}_2)$, then (cf. e.g. Definitions 2.2 and 2.3 in [Sch7])

$$N(f; \mathcal{N}_1, \mathcal{N}_2) = N(f_{\mathcal{N}_1}) + N(f_{\mathcal{N}_2}) - N(f_{\mathcal{N}_1}, f_{\mathcal{N}_2}),$$

where $N(f_{\mathcal{N}_1}, f_{\mathcal{N}_2})$ is the relative Nielsen number of the map $f : (\mathcal{N}_1, \mathcal{N}_2) \to (\mathcal{N}_1, \mathcal{N}_2)$, $N(f_{\mathcal{N}_1})$ is the Nielsen number of $f|_{\mathcal{N}_1} : \mathcal{N}_1 \to \mathcal{N}_1$, $N(f_{\mathcal{N}_2})$ is the one of $f|_{\mathcal{N}_2} : \mathcal{N}_2 \to \mathcal{N}_2$ and $N(f_{\mathcal{N}_1}, f_{\mathcal{N}_2})$ is the number of essential common fixed point classes of $f|_{\mathcal{N}_1}$ and $f|_{\mathcal{N}_2}$. Let us note that, under suitable assumptions, P. Wong generalized in [Wng3] D. Anosov's result to $N(f; \mathcal{N}_1, \mathcal{N}_2)$, for compact nilmanifolds.

Therefore, if $(M_1, M_2) = (\mathbb{R} \times \mathcal{N}_1, \mathbb{R} \times \mathcal{N}_2)$, where $\mathcal{N}_2 \subset \mathcal{N}_1 \subset X$ are compact connected nilmanifolds or AR-spaces, is a proper pair for the semi-flow $\varphi^* : (M_1, M_2) \multimap (M_1, M_2)$, determined by a semi-process $\varphi = \psi_2 \circ \psi_1$ on $X \times X$, where ψ_1 is an R_δ-mapping and ψ_2 is a continuous (single-valued) mapping ($\Rightarrow$ A-property; see Chapter I.11 or II.5), then (in view of the obvious invariance under homotopy) we arrive at (see Chapter II.5)

$$N_H(p, q) + (\#S_H(p, q; \mathcal{N}_2)) = N_H(f \circ \varphi_{(0,T)}; \mathcal{N}_1, \mathcal{N}_2) = N(f \circ \mathrm{id}; \mathcal{N}_1, \mathcal{N}_2)$$
$$N(f; \mathcal{N}_1, \mathcal{N}_2) = N(f_{\mathcal{N}_1}) + N(f_{\mathcal{N}_2}) - N(f_{\mathcal{N}_1}, f_{\mathcal{N}_2})$$
$$= |\Lambda(f_{\mathcal{N}_1})| + |\Lambda(f_{\mathcal{N}_2})| - N(f_{\mathcal{N}_1}, f_{\mathcal{N}_2}),$$

where

$$\mathcal{N}_1 \xleftarrow{p} \Gamma_{f \circ \varphi_{(0,T)}} \xrightarrow{q} \mathcal{N}_1 \quad \text{and} \quad \mathcal{N}_2 \xleftarrow{p} \Gamma_{f \circ \varphi_{(0,T)}} \xrightarrow{q} \mathcal{N}_2.$$

In many situations,

$$N(f; \mathcal{N}_1, \mathcal{N}_2) = |\Lambda(f_{\mathcal{N}_1})|, \quad \text{when } \Lambda(f_{\mathcal{N}_1}) \neq 0,$$

according to P. Wong's result in [Wng3], by which "only"

$$N(f; \mathcal{N}_1, \mathcal{N}_2) = N(f_{\mathcal{N}_1}).$$

Thus, a nontrivial situation can be mostly expected, when $\Lambda(f_{\mathcal{N}_1}) = 0$. In our (more general, but not so explicit) situation, there exist at least

$$|\Lambda(f_{\mathcal{N}_1})| + |\Lambda(f_{\mathcal{N}_2})| - N(f_{\mathcal{N}_1}, f_{\mathcal{N}_2}) \quad (\geq |\Lambda(f_{\mathcal{N}_2})|)$$

coincidences of the pair (p, q) associated to the mapping $f \circ \varphi_{(0,T)} : \mathcal{N}_1 \multimap \mathcal{N}_1$.

Moreover, there is a one-to-one correspondence between the coincidences and the trajectories $(t, y_{(0,T)}(x^*)) \in [0, T] \times \mathcal{N}_1$ of $\varphi_{(0,t)}$, satisfying

$$y_{(0,0)}(x^*) = f(y_{(0,T)}(x^*)),$$

where x^* is a fixed point of $f \circ \varphi_{(0,T)}$, provided

$$\varphi_{(0,t)}(x) = \{y_{(0,t)}(x) \in \mathcal{N}_1 \mid y_{(0,\cdot)}(x) \subset \psi_1 \text{ is a continuous selection,}$$
$$\psi_1 = \cup y_{(0,\cdot)}(x) \text{ and } \varphi = \psi_2 \circ x\psi_1\}.$$

If the homotopy endomorphisms

$$(f_{\mathcal{N}_i})_* : H_*(\mathcal{N}_i, \mathbb{Q}) \to H_*(\mathcal{N}_i, \mathbb{Q}), \quad i = 1, 2,$$

are known for nilmanifolds $\mathcal{N}_i$ (for AR-spaces $\mathcal{N}_i$, we have $\Lambda(f_{\mathcal{N}_i}) = 1$), then still (cf. [McC-M])

$$(6.72) \qquad |\Lambda(f_{\mathcal{N}_i})| = \left| \sum_n (-1)^n \mathrm{tr}(f_{\mathcal{N}_i})_{n*} \right|, \quad i = 1, 2.$$

We can summarize our investigation about the relative Nielsen number (on the total space) as follows.

(6.73) THEOREM. *Assume that* $(M_1, M_2) = (\mathbb{R} \times \mathcal{N}_1, \mathbb{R} \times \mathcal{N}_2)$ *is a proper pair for the semi-flow* $\varphi^* : (M_1, M_2) \multimap (M_1, M_2)$, *determined by some generalized semi-process* $\varphi = \psi_2 \circ \psi_1$ *on* $X \times X$, *where* $\mathcal{N}_2 \subset \mathcal{N}_1 \subset X$ *are compact connected nilmanifolds or AR-spaces,* ψ_1 *is an* R_δ-*mapping and* ψ_2 *is a continuous (single-valued) mapping such that* $\varphi_{(0,t)}(x) = \{y_{(0,t)}(x) \in \mathcal{N}_1 \mid y_{(0,\cdot)}(x) \subset \psi_1$ *is a continuous selection,* $\psi_1 = \cup y_{(0,\cdot)}(x)$ *and* $\varphi = \psi_2 \circ \psi_1\}$.

Furthermore, let $f : (\mathcal{N}_1, \mathcal{N}_2) \to (\mathcal{N}_1, \mathcal{N}_2)$ *be a continuous (single-valued) mapping. Then there exist at least* $(|\Lambda(f_{\mathcal{N}_1})| + |\Lambda(f_{\mathcal{N}_2})| - N(f_{\mathcal{N}_1}, f_{\mathcal{N}_2}))$ *trajectories* $(t, y_{(0,t)}(x^*)) \in [0, T] \times \mathcal{N}_1$ *of* $\varphi_{(0,t)}$, *satisfying* $y_{(0,0)}(x^*) = f(y_{(0,T)}(x^*))$, *for some (one or more)* $x^* \in \mathcal{N}_1$. *For nilmanifolds* $\mathcal{N}_i$, *the generalized Lefschetz number* $|\Lambda(f_{\mathcal{N}_i})|$, $i = 1, 2$, *can be computed by means of (6.72) and* $N(f_{\mathcal{N}_1}, f_{\mathcal{N}_2})$ *denotes the number of essential common fixed point classes of* $f_{\mathcal{N}_1} = f|_{\mathcal{N}_1}$ *and* $f_{\mathcal{N}_2} = f|_{\mathcal{N}_2}$.

(6.74) REMARK. We have already pointed out that the translation operator in (6.67), associated to (6.65), generates a generalized semi-process $\Phi_{(\sigma, kT)}(x)$ as well as the generalized semi-flow $\Phi^*(\sigma, t) = (\sigma + kT, \Phi_{(\sigma, kT)}(x))$, which are exactly of the type as above. Moreover, there is a one-to-one correspondence between the coincidences and the solutions of (6.65).

(6.75) REMARK. In fact, in Theorem (6.73), the same amount of trajectories $(t + \sigma, y)$ exists, for every $\sigma \in \mathbb{R}$, satisfying

$$y_{(\sigma,0)}(x_\sigma^*) = f(y_{(\sigma,T)}(x_\sigma^*)).$$

Therefore, only the notion of local semi-dynamical systems was appropriate for conclusions in Theorem (6.73). Moreover, if e.g. $f = \mathrm{id}$, then we can get at least $(|\chi(\mathcal{N}_1)| + |\chi(\mathcal{N}_2)| - N(f_{\mathcal{N}_1}, f_{\mathcal{N}_2}))$ T-periodic trajectories, for every $\sigma \in \mathbb{R}$, under suitable T-periodicity assumptions, as in Theorem (6.66). For $f = -\mathrm{id}$, much more interesting multiplicity results can be obtained (cf. Corollary (6.61)) under suitable restrictions, for anti-periodic trajectories.

As we could see, the relative Lefschetz number was concerned to the existence of a fixed point on the closure of the complement, while the relative Nielsen number to the lower estimate of the number of coincidences on the total space. We conclude, therefore, this part by the application of a special relative Nielsen number (introduced for the first time in [Zha1]), estimating from below the number of coincidences just on the complement (see Chapter II.5 and, in the single-valued case, cf. the surplus number in [Zha2]).

If $\mathcal{N}_1$, $\mathcal{N}_2$ are ANR-spaces such that $\mathcal{N}_2 \subset \mathcal{N}_1$, then

$$(SN(f; \mathcal{N}_2) \geq) N(f; \mathcal{N}_1 \setminus \mathcal{N}_2) = N(f_{\mathcal{N}_1}) - E(f_{\mathcal{N}_1}, f_{\mathcal{N}_2})$$

is the so called *Nielsen number of essential weakly common fixed point classes* of $f_{\mathcal{N}_1}$ and $f_{\mathcal{N}_2}$, i.e., if there exists a path α from a point x_0 of an essential fixed point class to a point in $\mathcal{N}_2$ so that α is homotopic to $f \circ \alpha$, under a homotopy of the form $(I, 0, 1) \rightarrow (\mathcal{N}_1, x_0, \mathcal{N}_2)$, $I = [0, 1]$, for more details (including the computation of $N(f; \mathcal{N}_1 \setminus \mathcal{N}_2)$), cf. e.g. [Sch7], [Wng3], [Zha1], [Zha2].

Replacing the relative Nielsen number (on the total space) by the above Nielsen number on the complement; Theorem (6.73) can be reformulated as follows.

(6.76) COROLLARY. *Under the assumption of Theorem (6.73), there exist at least* $|\Lambda(f_{\mathcal{N}_1})| - E(f_{\mathcal{N}_1}, f_{\mathcal{N}_2})$ *trajectories* $(t, y_{(0,T)}(x^*)) \in [0, T] \times \mathcal{N}_1 \setminus \mathcal{N}_2$ *of* $\varphi_{(0,t)}$, *satisfying* $y_{(0,0)}(x^*) = f(y_{(0,T)}(x^*))$, *for some (one or more)* $x^* \in \mathcal{N}_1 \setminus \mathcal{N}_2$. *The number of essential weakly common fixed point classes* $E(f_{\mathcal{N}_1}, f_{\mathcal{N}_2})$ *of* $f_{\mathcal{N}_1}$ *and* $f_{\mathcal{N}_2}$ *is defined as above and, for a nilmanifold* $\mathcal{N}_1$, $\Lambda(f_{\mathcal{N}_1})$ *can be computed by means of* (6.72).

Let us still add two remarks. In the single-valued case, $\mathcal{N}_2$ need not be connected (cf. [Sch7]). Moreover, if $\mathcal{N}_1$ or $\mathcal{N}_2$ is particularly the 2-dimensional disk containing a finite number of fixed points on the boundary $\partial\mathcal{N}_1$ or $\partial\mathcal{N}_2$, then, under certain special additional restrictions, some further fixed points can be implied in the interior $\mathrm{int}\,\mathcal{N}_1$ or $\mathrm{int}\,\mathcal{N}_2$ (see e.g. [BBS], [BrwGr]).

If we would have used for our applications (in the single-valued case) continuous flows $\Phi_{(\sigma,t)}$, where t can take also negative values, then, for every $\sigma, t \in \mathbb{R}$,

$$\Phi \sim \widehat{\Phi} : (A(\sigma), B(\sigma)) \to (A(\sigma + t), B(\sigma + t));$$

a homotopy is given by

$$\widehat{\Phi}_{(\sigma+(1-s)t, st)} \circ \Phi_{(\sigma+t, -st)} \circ \Phi_{(\sigma,t)}, \quad \text{for } s \in [0,1].$$

Nevertheless, the semi-flows seem to be more appropriate for our applications, in the single-valued case, as well.

Finally, periodic point theorems will be applied for obtaining multiplicity results. It will be convenient to reformulate the periodic point theorems obtained in Chapter II.6 in terms of discrete multivalued semi-dynamical systems.

Let X be a metric ANR-space and denote by $\mathbb{Z}_0^+$ the set of nonnegative integers. By an *interval* in $\mathbb{Z}_0^+$, we mean the intersection of a closed real interval with $\mathbb{Z}_0^+$.

(6.77) DEFINITION. A multivalued map $\varphi : X \times \mathbb{Z}_0^+ \multimap X$ is called a *discrete multivalued semi-dynamical system* (dmss) if the following conditions are satisfied:

(i) $\varphi(x, 0) = \{x\}$, for all $x \in X$;
(ii) $\varphi(\varphi(x, n), m) = \varphi(x, n + m)$, for all $m, n \in \mathbb{Z}_0^+$ and all $x \in X$.

Using the notation $\varphi^n(x) := \varphi(x, n)$, one can see that

$$\varphi^n(x) = \underbrace{\varphi^1(x) \circ \cdots \circ \varphi^1(x)}_{n-\text{times}}, \quad n \in \mathbb{N}.$$

Observe also that in order to show, for φ satisfying (i), (ii), that $\varphi(.,n)$ is a $\mathbb{CAC}$-mapping (see Chapter I.5), it is sufficient to assume that $\varphi^1(x) \in \mathbb{CAC}(X)$. Therefore, such a φ^1 is a *generator* of a dmss.

(6.78) DEFINITION. Let I be an interval in $\mathbb{Z}_0^+$ containing the origin 0. A single-valued mapping $\sigma : I \to X$ is a *solution for φ through x* if $\sigma(n+1) \in \varphi(\sigma(n))$, for all $n, n+1 \in I$, and $\sigma(0) = x$, i.e. $\sigma(n) \in \varphi^n(\sigma(0))$, for all $n \in I$. A *solution* $\sigma : I \to X$ of φ through x is called *k-periodic*, $k \in \mathbb{N}$, on an interval I if $\sigma(n) = \sigma(n + k)$, for all $n, n + k \in I$, i.e. $\sigma(n) \in \varphi^k(\sigma(n))$ or, in particular, $x \in \varphi^k(x)$, and the related orbit $\{x, \sigma(1), \ldots, \sigma(k - 1)\}$ is then called a *k-orbit* of φ through x on an interval I.

Now, we can immediately reformulate the periodic point theorems in Chapter II.6 in terms of dmss as follows.

(6.79) THEOREM (cf. Theorem (6.7) in Chapter II.6). *Let $\varphi : X \times \mathbb{Z}_0^+ \multimap X$ be a dmss. If its generator $\varphi^1(x)$ satisfies $\chi(\varphi^1) \neq 0$ or $P(\varphi^1) \neq 0$, where $\chi(\varphi)$ stands for the Euler characteristic of φ and $P(\varphi)$ is an algebraic invariant associated with φ (both notions are defined in Chapter II.6), then there exists an n-periodic solution of φ, on an interval I, for some $m + 1 \leq n \leq P(\varphi^1)$ and an arbitrary natural $m \geq 1$, provided there is some $k \in I$ with $(k + P(\varphi^1)) \in I$.*

(6.80) THEOREM (cf. Theorem (6.7) in Chapter II.6). *Let $\varphi : (X, A) \times \mathbb{Z}_0^+ \multimap (X, A)$ be a dmss on pairs, where $A \subset X$. If its generator $\varphi^1(x, a) = (\varphi^1(x), \varphi^1(a))$ satisfies $\chi(\varphi^1) \neq 0$ or $P(\varphi^1) \neq 0$, where $\chi(\varphi)$ stands for the Euler characteristic of φ and $P(\varphi)$ is an algebraic invariant associated with φ (both notions are defined in Chapter II.6), then there exists an n-periodic solution σ of φ with $\sigma(0)$ in $\overline{X \setminus A}$, on an interval I, for some $m + 1 \leq n \leq P(\varphi^1)$ and an arbitrary natural $m \geq 1$, provided there is some $k \in I$ with $(k + P(\varphi^1)) \in I$.*

(6.81) THEOREM (cf. Theorem (6.24) in Chapter II.6). *Let $\varphi : X \times \mathbb{Z}_0^+ \multimap X$ be a dmss, where X is compact. If its generator $\varphi^1(x)$ satisfies condition (A) in Chapter I.11, then there exist at least $S_k(\widetilde{\varphi})$ k-periodic solutions of φ, on an interval I, forming irreducible cyclically different k-orbits on I, provided there is some $m \in I$ with $(m + k) \in I$. The Nielsen-type number S_k is defined in Chapter II.6.*

For single-valued maps $f : X \to X$, the following notation is standard:

$$P^n(f) := \text{Fix}(f^n) = \{x \in X \mid x = f^n(x)\},$$
$$P_n(f) := P^n(f) \setminus \bigcup_{k < n} P^k(f).$$

Defining the *Nielsen number of n-periodic points* $NP_n(f)$ by

$$NP_n(f) = n \cdot \mathcal{O}_n(f),$$

where $\mathcal{O}_n(f)$ is the number of irreducible essential n-orbits of f, it can be easily checked that

$$NP_n(f) \leq \sharp P_n(g), \quad \text{for every } g \sim f,$$

where $\sharp S$ denotes the cardinality of the set S.

Moreover, if f is a self-mapping on a torus (or, more generally, on a compact nilmanifold), then (see e.g. [HeKe], [Wng2])

$$N(f^n) = |\Lambda(f^n)| = \sum_{k \mid n} NP_k(f), \quad \text{provided } \Lambda(f^n) \neq 0,$$

and

$$NP_k(f) = \sum_{m|k} \mu(k/m)|\Lambda(f^m)|, \text{ provided } \Lambda(f^k) \neq 0,$$

where $\mu(d), d \in \mathbb{N}$, is the Möbius function, i.e.

$$\mu(d) = \begin{cases} 1 & \text{if } d = 1, \\ (-1)^k & \text{if } d \text{ is a product of } k \text{ distinct primes,} \\ 0 & \text{if } d \text{ is not square-free,} \end{cases}$$

and $m|n$ means that m divides n, or

$$NP_k(f) = \sum_{\tau \subset P(k)} (-1)^{\#\tau} N(f^{k:\tau}) = \sum_{\tau \subset P(k)} (-1)^{\#\tau}|\Lambda(f^{k:\tau})|,$$

provided $\Lambda(f^k) \neq 0$, where, $P(k)$ denotes the set of prime divisors of k and $k : \tau = k/\Pi_{p\in\tau}p$.

It has been proved in Theorem (10.24) in Chapter I.10 that *any admissible self-map (p,q) on a torus is admissibly homotopic to a pair representing a single-valued map*, say f. Moreover, it follows from the invariance under homotopy that

$$S_k(\widetilde{p}, \widetilde{q}) = \mathcal{O}_k(f) \geq \left[\frac{1}{k}NP_k(f)\right]^+ = \left[\frac{1}{k}\sum_{m|k} \mu(k/m)|\Lambda(f^m)|\right]^+$$

$$= \left[\frac{1}{k}\sum_{m|k} \mu(k/m)|\Lambda((p,q)^m)|\right]^+$$

$$= \left[\frac{1}{k}\sum_{\tau \subset P(k)} (-1)^{\#\tau}|\Lambda((p,q)^{k:\tau})|\right]^+,$$

provided $\Lambda((p,q)^k) \neq 0$, where $S_k(\widetilde{p}, \widetilde{q})$ is the number of irreducible and essential (cyclically different) k-orbits defined in Chapter II.6, $[r]^+ = [r] + \text{sgn}(r - [r])$ and $[r]$ denotes the integer part of r.

Hence, we can give

(6.82) COROLLARY. *The admissible pair (p,q) on a torus satisfies*

$$S_k(\widetilde{p}, \widetilde{q}) \geq \left[\frac{1}{k}\sum_{m|k} \mu(k/m)|\Lambda((p,q)^m)|\right]^+ = \left[\frac{1}{k}\sum_{\tau \subset P(k)} (-1)^{\#\tau}|\Lambda((p,q)^{k:\tau})|\right]^+,$$

provided $\Lambda((p,q)^k) \neq 0$, $k \in \mathbb{N}$, where μ is the Möbius function, $P(k)$ denotes the set of prime divisors of k and the used symbols have the same meaning as above.

(6.83) REMARK. Since on an n-torus, $n \in \mathbb{N}$, we have (see e.g. [JaMa])

$$\Lambda(f^m) = \det(I - A^m), \quad m \in \mathbb{N},$$

where A is the associated $(n \times n)$-matrix with integer coefficients, representing the induced endomorphism of the fundamental group, which corresponds to f and which is called the linearization of f, we obtain in fact that

$$S_k(\widetilde{p}, \widetilde{q}) \geq \left[\frac{1}{k} \sum_{m|k} \mu(k/m)|\det(I - A^m)| \right]^+ = \left[\frac{1}{k} \sum_{\tau \subset P(k)} (-1)^{\#\tau}|\det(I - A)^{k:\tau)}| \right]^+,$$

provided $\det(I - A^k) \neq 0$, $k \in \mathbb{N}$.

Now the obtained periodic theorems will be applied to differential inclusions on tori. On tori, the following theorem is true.

(6.84) THEOREM. *The multivalued self-map (p, q) on an n-torus, $n \in \mathbb{N}$, has at least*

$$S_k \geq \frac{1}{k} \left[\left| \sum_{m|k} \mu(k/m)\det(I - A^m) \right| \right]^+$$

$$= \left[\frac{1}{k} \sum_{\tau \subset P(k)} (-1)^{\#\tau}|\det(I - A)^{k:\tau)}| \right]^+$$

irreducible cyclically different k-orbits, provided $\det(I - A^k) \neq 0$, $k \in \mathbb{N}$, where μ is the Möbius functions, $P(k)$ denotes the set of prime divisors of k, $k : \tau = k/\prod_{p \in \tau}^k p$, $[r]^+ = [r] + \text{sgn}(r - [r])$, $[r]$ denotes the integer part of r, and A is the associated $(n \times n)$-matrix with integer coefficients, representing the induced endomorphism of the fundamental group.

If (p, q) satisfies $P(p, q) \neq 0$, where $P(p, q)$ is an algebraic invariant defined in Chapter II.6, then, it has at least one k-periodic point, for some $m + 1 \leq k \leq m + P(p, q)$ and an arbitrary natural number $m \geq 1$.

Working on the torus $\mathbb{T}^n = \mathbb{R}^n / \mathbb{Z}^n$, we shall still assume that F is bounded and

$$(6.85) \qquad F(t, \ldots, x_j, \ldots) \equiv F(t, \ldots, x_j + 1, \ldots), \qquad j = 1, \ldots, n,$$

where $x = (x_1, \ldots, x_n)$.

Then the Carathéodory solutions $x(t) \in \text{AC}_{\text{loc}}(\mathbb{R}, \mathbb{R}^n)$ exist on $\mathbb{R}$, satisfying (6.49), a.e.

If $x(t, x_0) := x(t, 0, x_0)$ is a solution of (6.49) with $x(0, x_0) = x_0$, then we can define the *Poincaré translation operator* (at the time $\omega > 0$) along the trajectories of (6.49) as follows (cf. (6.50)):

$$\Phi_\omega : \mathbb{R}^n \multimap \mathbb{R}^n, \quad \Phi_\omega(x_0) = \{x(\omega, x_0) \mid x(t, x_0) \text{ satisfies (6.49), a.e.}\}$$

According to Theorem (4.3), $\Phi_\omega = e_\omega \circ \varphi$, where φ is an R_δ-mapping (i.e. u.s.c. with R_δ-values) and e_ω is a continuous (single-valued) evaluation mapping. Thus, $\Phi_{\lambda\omega}$

considered on $\mathbb{T}^n$ satisfies condition (A) in Chapter II.6 as well as $\Phi_{\lambda\omega} \in \mathbb{CAC}(\mathbb{T}^n)$, (for the definition of $\mathbb{CAC}$-maps, see Chapter I.5), for every $\lambda \in [0, 1]$. Moreover, $\Phi_0(x_0) = x_0$, i.e., Φ_ω is homotopic to identity, and subsequently $\mathcal{H} \circ \Phi_\omega : \mathbb{T}^n \multimap \mathbb{T}^n$ is a $\mathbb{CAC}$-map, satisfying condition (A), which is homotopic to a (single-valued) homeomorphism $\mathcal{H} : \mathbb{T}^n \to \mathbb{T}^n$.

In what follows, we can identify Φ_ω or its composition with $\mathcal{H}$, $\mathcal{H} \circ \Phi_\omega$, with the pair $(p_\varphi, e_\omega \circ q_\varphi)$ or $(p_\varphi, \mathcal{H} \circ e_\omega \circ q_\varphi)$, i.e., we let

$$\Phi_\omega = (p_\varphi, e_\omega \circ q_\varphi), \text{ or } \mathcal{H} \circ \Phi_\omega = (p_\varphi, \mathcal{H} \circ e_\omega \circ q_\varphi), \text{ respectively.}$$

One can easily check that the k-periodic points x_0 to $(p_\varphi, \mathcal{H} \circ e_\omega \circ q_\varphi)$, i.e. fixed points x_0 of $(p_\varphi, \mathcal{H} \circ e_\omega \circ q_\varphi)^k \in \mathbb{CAC}(\mathbb{T}^n)$, are those such that

$$(6.86) \qquad \underbrace{\mathcal{H} \circ x(\omega; \mathcal{H} \circ x(\omega; \ldots \mathcal{H} \circ x(\omega; x(0, x_0) \ldots)))}_{k-\text{times}} = x(0, x_0) \pmod{1},$$

where $x(t, r)$ are solutions of (6.49) with $x(0, r) = r$.

Moreover, there is a one-to-one correspondence between the number of irreducible cyclically different k-orbits to $(p_\varphi, \mathcal{H} \circ e_\omega \circ q_\varphi)$ and the number of geometrically distinct k-tuples of solutions of (6.49) appearing in (6.86), where k is minimal.

In view of the above arguments, we can also immediately obtain the following multiplicity result, when applying Theorem (6.84) to system (6.49).

(6.87) THEOREM. *Assume that* $\det(I - A^k) \neq 0$, *for some* $k \in \mathbb{N}$, *where* A *is the associated* $(n \times n)$-*matrix with integer coefficients, representing the induced homomorphism of the fundamental group, which corresponds to* $\mathcal{H}$ *and which is called the linearization of* $\mathcal{H}$. *Then the number of geometrically distinct* k-*tuples of solutions of* (6.49) *with* (6.85), *satisfying* (6.86) *with the minimal* k, *is at least*

$$S_k \geq \left[\frac{1}{k} \sum_{m|k} \mu(k/m) |\det(I - A^m)| \right]^+ = \left[\frac{1}{k} \sum_{\tau \subset P(k)} (-1)^{\#\tau} |\det(I - A^{k:\tau})| \right]^+$$

where μ *is the Möbius function,* $P(k)$ *denotes the set of prime divisors of* k, $k : \tau = k/\Pi_{p \in \tau} p$, $[r]^+ = [r] + \text{sgn}(r - [r])$ *and* $[r]$ *denotes the integer part of* r, *and* $\omega > 0$ *in* (6.86) *is a given real number.*

(6.88) REMARK. *If* $P(\mathcal{H}) \neq 0$, *for a homeomorphism* $\mathcal{H} : \mathbb{T}^n \to \mathbb{T}^n$, *where* $P(\mathcal{H})$ *is an algebraic invariant defined in Chapter II.6, then the u-Carathéodory system* (6.49), *satisfying* (6.85), *admits, for some* $k \in \mathbb{N}$ *with* $m+1 \leq k \leq m+P(\mathcal{H})$ *and an arbitrary natural number* $m \geq 1$, *at least one* k-*tuple of solutions* $x(t)$ *of* (6.49) *such that* (6.86) *holds, where* $\omega > 0$ *is given real number.*

We conclude by the following simple examples.

(6.89) EXAMPLE. If

$$\mathcal{H} = A = \begin{pmatrix} 0 & 1 \\ 1 & 1 \end{pmatrix} \Rightarrow A^5 = \begin{pmatrix} 3 & 5 \\ 5 & 8 \end{pmatrix},$$

then $\det(I - A) = \det \begin{pmatrix} 1 & -1 \\ -1 & 0 \end{pmatrix} = -1$ and $\det(I - A^5) = \det \begin{pmatrix} -2 & -5 \\ -5 & -7 \end{pmatrix} = -11$.

Since $\mu(1) = 1$ and $\mu(5) = -1$, we obtain that

$$\frac{1}{5} \sum_{m|5} \mu(5/m)|\det(I - A^m)| = \frac{1}{5}(-|-1| - |-1| + |-11|) = 2.$$

Therefore, there exist at least two geometrically distinct 5-tuples of solutions of the planar system (6.49) with (6.85) such that

$$A \circ x(\omega; A \circ x(\omega; A \circ x(\omega; A \circ x(\omega; A \circ x(\omega; x(0, x_0)))))) = x(0, x_0) \ (\text{mod } 1),$$

where $\omega > 0$ is a given real number, and such that $A \circ x(\omega; x_0) \neq x_0$ (i.e. without the minimal period 5).

(6.90) EXAMPLE. Since, for $\mathcal{H} = -\text{id} : \mathbb{T}^n \to \mathbb{T}^n$, we have

$$\sum_{m|k} \mu(k/m)|\det(I - (-I)^m)| = 2^n \sum_{\substack{m|k \\ (m\text{-odd})}} \mu(k/m),$$

system (6.49) with (6.85) admits at least $[2^n/100]^+$ solutions $x(t)$ such that

$$\underbrace{-x(\omega; -x(\omega; \dots; -x(\omega; -x(\omega; x(0, x_0) \underbrace{\dots}_{100\times})) = x(0, x_0) \quad (\text{mod } 1),}_{100\times}$$

because, for $k = 100$, its odd divisors are 1, 5, 25, and so we get

$$\mu(100) + \mu(20) + \mu(4) = (-1)^2 + (-1)^2 + (-1) = 1.$$

So, if e.g. $n = 10$, then there are at least eleven such solutions, because

$$\left[\frac{2^{10}}{100}\right]^+ = \left[\frac{1024}{100}\right]^+ = [10.24]^+ = 10 + \text{sgn} \, 0.24 = 11.$$

For $k = 1$, we have obviously at least 2^n solutions $x(t)$ such that

$$-x(\omega) = x(0) \ (\text{mod } 1),$$

which corresponds to Corollary (6.61).

III.7. Ważewski-type results

The results of this chapter are based on the papers [And27], [AGZ], [AGL], [AK3].

Let us start with recalling the classical result of T. Ważewski in [Waz1] (cf. also [Har-M]). Hence, consider the system of differential equations

(7.1) $$\dot{x} = f(t, x),$$

where x, f are n-dimensional vector functions and f is continuous on a set Ω in $\mathbb{R}^{n+1}$.

For such a system, Tadeusz Ważewski proved in [Waz1] that if *solutions of initial value problem for* (7.1) *are unique, if* Ω_0 *is an open subset of* Ω *such that* $\Omega_0^e = \Omega_0^{se}$, *where* Ω_0^e *and* Ω_0^{se} *are the boundary points of* Ω_0 *in* Ω *which are respectively egress points and strict egress points of solutions trajectories of* (7.1), *and if* S *is a nonnull subset of* $\Omega_0 \cup \Omega_0^e$ *such that* $S \cap \Omega_0^e$ *is a retract of* Ω_0^e, *but not of* S, *then there is at least one* $(t_0, x_0) \in S$ *such that the solution of* (7.1) *with* $x(t_0) = x_0$ *remains in* Ω_0 *on its right maximal interval of existence.*

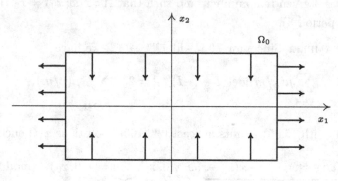

FIGURE 1. $(x = (x_1, x_2))$

A typical "planar" example of this so called topological principle of T. Ważewski is sketched in Fig. 1 and (strict) egress as well as ingress points are defined below in Definition (7.3).

Consider again the differential system (7.1), but where this time $f : [0,1] \times \mathbb{R}^n \to \mathbb{R}^n$ satisfies:

(i) $f(t, \cdot)$ is continuous, for a.a. $t \in [0,1]$,

(ii) $f(\cdot, x)$ is Lebesgue measurable, for every $x \in \mathbb{R}^n$,

(iii) for every $r \in [0, \infty)$, there exists $\rho_r \in L^1([0,1], [0,\infty))$ such that, for a.a. $t \in [0,1]$, the following implication holds, namely

$$|\xi| \leq r \Rightarrow |f(t, \xi)| \leq \rho_r(t).$$

We are interested in (Carathéodory) solutions $x \in AC([0,1], \mathbb{R}^n)$ of (7.1), satisfying

$$(7.2) \qquad\qquad x(0) = x(1),$$

whose values are located in a given set Θ.

(7.3) DEFINITION. Given a bounded open set $\emptyset \neq \Theta \subset \mathbb{R}^n$ with the boundary $\text{fr}(\Theta)$

  (i) we call a point $\xi \in \text{fr}(\Theta)$ an *ingress point* (into Θ) if, for every $t^* \in (0,1]$ and every solution γ of (7.1) in $[0,1]$, the implication

$$\gamma(t^*) = \xi \Rightarrow \exists \varepsilon \in \mathbb{R}^+ \; \forall t \in (t^* - \varepsilon, t^*) : \gamma(t) \in \mathbb{R}^n \setminus \text{cl}(\Theta)$$

  takes place.

  (ii) We call a point $\xi \in \text{fr}(\Theta)$ an *egress point* (from Θ) if, for every $t^* \in [0,1)$ and every solution γ of (7.1) in $[0,1]$, the implication

$$\gamma(t^*) = \xi \Rightarrow \exists \varepsilon \in \mathbb{R}^+ \; \forall t \in (t^*, t^* + \varepsilon) : \gamma(t) \in \mathbb{R}^n \setminus \text{cl}(\Theta)$$

  takes place.

The set of ingress or egress points of Θ will be denoted by Γ^+ or Γ^-, respectively.

(7.4) DEFINITION. Given a bounded open set $\emptyset \neq \Theta \subset \mathbb{R}^n$ with the boundary $\text{fr}(\Theta)$. We say that the quadruple (Θ, h, Ω, g) is *admissible* w.r.t. (7.1) if the following conditions are satisfied:

(7.4.1) every boundary point (i.e. of $\text{fr}(\Theta)$) is either an ingress or an egress point of Θ (w.r.t. solutions of (7.1)),

(7.4.2) h is a homeomorphism of $\mathbb{R}^n$ into itself and $h(\Omega) = \Theta$,

(7.4.3) $g : \mathbb{R}^n \to \mathbb{R}^n$ is a continuous mapping such that
  (a) if $h(\xi) \in \Theta \cup \Gamma^+$ and $\lambda \in [0,1)$, then $h(\xi - (1-\lambda)g(\xi)) \notin \Gamma^+$,
  (b) if $h(\xi) \in \Gamma^-$ and $\lambda \in [0,1)$, then $h(\xi - (1-\lambda)g(\xi)) \in \mathbb{R}^n \setminus \text{cl}(\Theta)$,
  (c) $\deg(g, \Omega, 0) \neq 0$, where deg stands for the Brouwer degree.

(7.5) REMARK. Denote $[-1,1]^0 \times [-1,1] = [-1,1] \times [-1,1]^0 = [-1,1]$ and assume that

$$h^{-1}(\Theta) = \Omega = (-1)^n, \quad s \in \{0, \ldots, n\},$$

$$h^{-1}(\Gamma^+) = \bigcup_{i=1}^{n-s} ([-1,1]^{s+i-1} \times \{-1,1\} \times [-1,1]^{n-s-i}),$$

$$h^{-1}(\Gamma^-) = \bigcup_{j=0}^{s-1} ([-1,1]^j \times \{-1,1\} \times [-1,1]^{n-j-1}).$$

Defining $g : \mathbb{R}^n \to \mathbb{R}^n$,

$$g(\xi_1, \dots, \xi) = \frac{1}{2}(-\xi_1, \dots, -\xi_s, \xi_{s+1}, \dots, \xi_n),$$

one can easily check that g satisfies conditions (7.4.3)(a), (b) and, since g is odd, also (7.4.3)(c).

Another possibility for satisfying (7.4.3)(c) is that Ω is an AR-space (in particular, a retract of a convex set) and g is a self-mapping (i.e. into).

(7.6) THEOREM. *Let* (Θ, h, Ω, g) *be an admissible quadruple w.r.t.* (7.1). *Then problem* (7.1) $\cap$ (7.2) *admits a solution with values located in* Θ.

PROOF. We proceed in eight steps.

(7.6.1) At first, define $r : C([0,1], \mathrm{cl}(\Omega)) \times \mathrm{cl}(\Omega) \times [0,1] \to C([0,1], \mathbb{R}^n)$ by

$$r(q, \xi, \lambda)(t) = \begin{cases} h^{-1}(\gamma(t)), & \text{for } t \in [0, \lambda], \\ h^{-1}(\gamma(\lambda)), & \text{for } t \in (\lambda, 1], \end{cases}$$

where $\gamma \in \mathrm{AC}([0,1], \mathbb{R}^n)$,

$$\gamma(t) = h(\xi) + \int_0^t f(s, h(q(s))) \, ds.$$

(7.6.2) Furthermore, define $s : C([0,1], \mathrm{cl}(\Omega)) \times \mathrm{cl}(\Omega) \times [0,1] \to \mathbb{R}^n$ by

$$s(q, \xi, \lambda) = g(r(g, \xi, \lambda)(\lambda)).$$

(7.6.3) At last, define $B = (B_1, B_2) : C([0,1], \mathrm{cl}(\Omega)) \times \mathrm{cl}(\Omega) \times [0,1] \to C([0,1], \mathbb{R}^n) \times \mathbb{R}^n$, by

$$B_1(q, \xi, \lambda) = q - r(q, \xi, \lambda),$$
$$B_2(q, \xi, \lambda) = \xi - [r(q, \xi, \lambda)(\lambda) - (1 - \lambda)s(q, \xi, \lambda)].$$

(7.6.4) Let us fix $\lambda \in [0,1]$, $q \in C([0,1], \mathrm{cl}(\Omega))$, $\xi \in \mathrm{cl}(\Omega)$ and denote $\gamma = h \circ q$. One can then observe that

$$B(q, \xi, \lambda) = 0 \Rightarrow$$

(7.7) $\dot{\gamma}(t) = f(t, \gamma(t)),$ for a.a. $t \in [0, \lambda]$,

(7.8) $\gamma(t) = \gamma(\lambda),$ for every $t \in (\lambda, 1]$,

(7.9) $q(0) = \xi = q(\lambda) - (1 - \lambda)g(q(\lambda)).$

In particular, for $\lambda = 1$, we have

$$\dot{\gamma}(t) = f(t, \gamma(t)), \quad \text{for a.a. } t \in [0, \lambda], \quad \gamma(0) = \gamma(1),$$

i.e. the solution of (7.1) ∩ (7.2). In order to verify the solvability of $B(q, \xi, 1) = 0$, we show that

$$\mathrm{Deg}(B(\,\cdot\,,\,\cdot\,, 1), C([0, 1], \Omega) \times \Omega, 0) \neq 0,$$

where Deg stands for the Leray–Schauder degree.

(7.6.5) Now, we show that, for every $\lambda \in [0, 1]$ and every pair $(q, \xi) \in \mathrm{fr}(C([0, 1], \Omega) \times \Omega)$, we have $B(q, \xi, \lambda) \neq 0$.

(a) At first, we prove, by contradiction, the implication

$$\lambda \in [0, 1], \quad q \in \mathrm{fr}(C([0, 1], \Omega)), \quad \xi \in \Omega \;\Rightarrow\; B(q, \xi, \lambda) \neq 0.$$

Let there exist $\lambda \in [0, 1]$, $q \in \mathrm{fr}(C([0, 1], \Omega))$ and $\xi \in \Omega$ such that $B(q, \xi, \lambda) = 0$. Then $q(\tau) \in \mathrm{fr}(\Omega)$, for some $\tau \in [0, 1]$ and, according to the assumption (i) in Definition (7.4), $h(q(\tau))$ is an ingress or an egress point of Θ.

- For $\tau \in (0, \lambda)$, (7.7) implies $\exists \varepsilon > 0$ such that either $h(q(t)) \in \mathbb{R}^n \setminus \mathrm{cl}(\Theta)$, for all $t \in (\tau - \varepsilon, \tau)$, or $h(q(t)) \in \mathbb{R}^n \setminus \mathrm{cl}(\Theta)$, for all $t \in (\tau, \tau + \varepsilon)$, takes place, respectively. This is, however, the contradiction with the choice of $q \in \mathrm{fr}(C([0, 1], \Omega))$.

- Let $\tau = 0$, $\lambda \in [0, 1)$. If $\lambda = 0$, then (7.9) implies $g(q(0)) = 0$, which is a contradiction with the condition (7.4.3)(c) in Definition (7.4) (actually, it is rather a contradiction with the conditions (7.4.3)(a) and (b) which guarantees that the degree in (7.4.3)(c) is defined). For $\lambda \in (0, 1)$ in (7.9), we know that $q(0) = q(\lambda) - (1 - \lambda)g(q(\lambda))$, where $q(\lambda) \in \mathrm{cl}(\Omega)$. If $h(q(0)) \in \Gamma^-$, then $\exists \varepsilon > 0$: $h(q(t)) \in \mathbb{R}^n \setminus \mathrm{cl}(\Omega)$, for every $t \in (0, \varepsilon)$, which is the contradiction with the choice of $q \in \mathrm{fr}(C([0, 1], \Omega))$. Therefore, we can assume that $h(q(0)) \in \Gamma^+$ which, however, contradicts conditions (7.4.3)(a), (b) in Defnition (7.4).

- Let $\tau = 0$, $\lambda = 1$, Since (7.9) implies $q(0) = q(1)$, we can proceed quite analogously as in the foregoing case. So, $\exists \varepsilon > 0$ such that either $h(q(t)) \in \mathbb{R}^n \setminus \mathrm{cl}(\Theta)$, for all $t \in (1 - \varepsilon, 1)$, or $h(q(t)) \in \mathbb{R}^n \setminus \mathrm{cl}(\Theta)$, for all $t \in (0, \varepsilon)$, takes place, respectively. This is, however, again the contradiction with the choice of $q \in \mathrm{fr}(C([0, 1], \Omega))$.

- Let $(0, 1] \ni \tau \geq \lambda \in [0, 1)$. If $\lambda = 0$, then τ can be taken, according to (7.8), equal 0 as well. Put $\lambda \in (0, 1)$. Then (7.8) implies that $q(\tau) = q(\lambda)$ and (7.7) implies that $h(q(\tau))$ is not an ingress point into Θ. Otherwise, $\exists \varepsilon > 0$ such that $h(q(t)) \in \mathbb{R}^n \setminus \mathrm{cl}(\Theta)$, for all $t \in (\lambda - \varepsilon, \lambda)$. If $h(q(\tau)) =$

$h(q(\lambda))$ is an egress point from Θ, then in view of (7.9) and condition (7.4.3)(b) in Definition (7.4), the initial value $h(q(0))$ does not belong to $\mathrm{cl}(\Theta)$, which is the contradiction with the choice of $q \in \mathrm{fr}(C([0,1],\Omega))$.

- Let $\tau = \lambda = 1$. Then (7.9) implies $q(0) = q(1)$ and (7.7) implies $\exists \varepsilon > 0$ such that either $h(q(\tau)) \in \mathbb{R}^n \setminus \mathrm{cl}(\Theta)$, for all $t \in (1 - \varepsilon, 1)$, or $h(q(t)) \in \mathbb{R}^n \setminus \mathrm{cl}(\Theta)$, for all $t \in (0, \varepsilon)$, takes place, respectively. This is, however, the contradiction with the choice of $q \in \mathrm{fr}(C([0,1],\Omega))$.

(b) Now, we prove again by contradiction the implication

$$\lambda \in [0,1], \quad q \in C([0,1], \mathrm{cl}(\Omega)), \quad \xi \in \mathrm{fr}(\Omega) \quad \Rightarrow \quad B(q,\xi,\lambda) \neq 0.$$

Hence, let there exist $\lambda \in [0,1]$, $q \in C([0,1], cl(\Omega))$, $\xi \in \mathrm{fr}(\Omega)$ such that $B(q,\xi,\lambda) = 0$. Let us employ that $\xi = q(0)$.

- Let $\lambda \in [0,1)$. If $\lambda = 0$, then (7.9) implies that $g(q(0)) = 0$, which contradicts to (7.4.2)(c) in Definition (7.4). For $\lambda \in (0,1)$, we know according to (7.9) that $\xi = q(0) = q(\lambda) - (1 - \lambda)g(q(\lambda))$, where $q(\lambda) \in \mathrm{cl}(\Omega)$. If $h(q(0)) \in \Gamma^-$, then $\exists \varepsilon > 0 : h(q(t)) \in \mathbb{R}^n \setminus \mathrm{cl}(\Theta)$, for all $t \in (0, \varepsilon)$, which is the contradiction with the choice of $q \in C([0,1], \mathrm{cl}(\Omega))$. Therefore, we can assume that $h(q(0)) \in \Gamma^+$, which is, however, a contradiction to conditions (7.4.3)(a), (b) in Definition (7.4).

- Let $\lambda = 1$. Then (7.9) implies $q(0) = q(1)$. Thus, $\exists \varepsilon > 0$ such that either $h(q(t)) \in \mathbb{R}^n \setminus \mathrm{cl}(\Theta)$, for all $t \in (1 - \varepsilon, 1)$, or $h(q(t)) \in \mathbb{R}^n \setminus \mathrm{cl}(\Theta)$, for all $t \in (0, \varepsilon)$, takes place, respectively, which is the contradiction with the choice of $q \in C([0,1], \mathrm{cl}(\Omega))$.

(7.6.6) Furthermore, we show that the image of the operator

$$U : C([0,1], \mathrm{cl}(\Omega)) \times \mathrm{cl}(\Omega) \times [0,1] \to C([0,1], \mathbb{R}^n) \times \mathbb{R}^n,$$

(7.9.1) $U(q,\xi,\lambda) = (q,\xi) - B(q,\xi,\lambda)$

is relatively compact.

Let us consider the sequence $\{(z_n, \sigma_n)\}_{n=1}^{\infty}$ of points from the image of U. Let $\{(q_n, \xi_n, \lambda_n)\}_{n=1}^{\infty}$ be a sequence such that $(q_n, \xi_n) - B(q_n, \xi_n, \lambda_n) = (z_n, \sigma_n)$, for all positive integers n.

(a) At first, let us consider the sequence $\{\gamma_n\}_{n=1}^{\infty}$, of absolutely continuous solutions of the related problems, i.e. the function sequence defined by

$$\gamma_n(t) = h(\xi_n) + \int_0^t f(s, h(q_n(s)))\, ds, \quad t \in [0,1].$$

We prove, that $\{\gamma_n \mid n \in \mathbb{N}\}$ is relatively compact. According to the well-known Arzelà–Ascoli lemma, it is sufficient to show that all functions from our set are uniformly bounded and equicontinuous. For a suitable $r \in [0, \infty)$, it obviously holds that

$$|\gamma_n(t)| = \left| h(\xi_n) + \int_0^t f(s, h(q_n(s))) \, ds \right| \leq |h(\xi_n)| + \int_0^1 |f(s, h(q_n(s)))| \, ds$$

$$\leq K_\Theta + \int_0^1 \rho_r(s) \, ds,$$

for all $t \in [0, 1]$, where $K_\Theta = \max\{|\theta| \mid \theta \in \mathrm{cl}(\Theta)\} < \infty$ and the Lebesgue integral on the right-hand side is, due to the Carathéodory conditions, finite. The equicontinuity follows easily as well, because of a finiteness of the integral of ρ_r, for every $\varepsilon > 0$, there is $\delta > 0$, such that, for all $n \in \mathbb{N}$ and t, p from $[0, 1]$ with $|t - p| < \delta$, we arrive at

$$|\gamma_n(t) - \gamma_n(p)| \leq \left| \int_p^t |f(s, h(q_n(s)))| \, ds \right| \leq \left| \int_p^t \rho_r(s) \, ds \right| < \varepsilon.$$

(b) Let $\{n_k\}_{k=1}^\infty$ be a sequence of positive integers such that

$$\lim_{k \to \infty} \gamma_{n_k} = \gamma, \quad \lim_{k \to \infty} \lambda_{n_k} = \lambda.$$

Then

$$z_{n_k}(t) = \begin{cases} h^{-1}(\gamma_{n_k}(t)), & \text{for } t \in [0, \lambda_{n_k}], \\ h^{-1}(\gamma_{n_k}(\lambda_{n_k})), & \text{for } t \in (\lambda_{n_k}, 1]. \end{cases}$$

Putting

$$z(t) = \begin{cases} h^{-1}(\gamma(t)), & \text{for } t \in [0, \lambda], \\ h^{-1}(\gamma(\lambda)), & \text{for } t \in (\lambda, 1] \end{cases}$$

we show that

$$\lim_{k \to \infty} z_{n_k} = z.$$

Choose $\varepsilon > 0$. The function $h^{-1} \circ \gamma$ is uniformly continuous on $[0,1]$, by which there exists $\delta > 0$ such that $|h^{-1}(\gamma(t)) - h^{-1}(\gamma(p))| < \varepsilon/2$, whenever $|t - p| < \delta$, $t \in [0, 1]$, $p \in [0, 1]$. Let $k_0 \in \mathbb{N}$ be such that

$$|\lambda_{n_k} - \lambda| < \delta, \quad \sup_{t \in [0,1]} |h^{-1}(\gamma_{n-k}(t)) - h^{-1}(\gamma(t))| < \frac{\varepsilon}{2},$$

for all $\mathbb{N} \ni k > k_0$. Observe that the latter inequality is correct, because the set $\{\gamma_{n_k} \mid k \in \mathbb{N}\}$ is bounded.

If $t \in [0, \min\{\lambda_{n_k}, \lambda\}]$, then

$$|z_{n_k}(t) - z(t)| = |h^{-1}(\gamma_{n_k}(t)) - h^{-1}(\gamma(t))| < \frac{\varepsilon}{2} < \varepsilon, \quad \text{for } \mathbb{N} \ni k > k_0.$$

If $t \in (\max\{\lambda_{n_k}, \lambda\}, 1]$, then

$$|z_{n_k}(t) - z(t)| = |h^{-1}(\gamma_{n_k}(\lambda_{n_k})) - h^{-1}(\gamma(\lambda))| \le |h^{-1}(\gamma_{n_k}(\lambda_{n_k})) - h^{-1}(\gamma(\lambda_{n_k}))|$$
$$+ |h^{-1}(\gamma(\lambda_{n_k})) - h^{-1}(\gamma(\lambda))| < \frac{\varepsilon}{2} + \frac{\varepsilon}{2} = \varepsilon, \quad \text{for } \mathbb{N} \ni k > k_0.$$

If $t \in [\lambda_{n_k}, \lambda]$, then $|\lambda_{n_k} - t| < \delta$, and subsequently

$$|z_{n_k}(t) - z(t)| = |h^{-1}(\gamma_{n_k}(\lambda_{n_k})) - h^{-1}(\gamma(t))| \le |h^{-1}(\gamma_{n_k}(\lambda_{n_k})) - h^{-1}(\gamma(\lambda_{n_k}))|$$
$$+ |h^{-1}(\gamma(\lambda_{n_k})) - h^{-1}(\gamma(t))| < \frac{\varepsilon}{2} + \frac{\varepsilon}{2} = \varepsilon, \quad \text{for } \mathbb{N} \ni k > k_0.$$

If $t \in [\lambda, \lambda_{n_k}]$, then $|\lambda - t| < \delta$, and subsequently

$$|z_{n_k}(t) - z(t)| = |h^{-1}(\gamma_{n_k}(t)) - h^{-1}(\gamma(\lambda))| \le |h^{-1}(\gamma_{n_k}(t)) - h^{-1}(\gamma(t))|$$
$$+ |h^{-1}(\gamma(t)) - h^{-1}(\gamma(\lambda))| < \frac{\varepsilon}{2} + \frac{\varepsilon}{2} = \varepsilon, \quad \text{for } \mathbb{N} \ni k > k_0.$$

Thus, we demonstrated that the set $\{z_n \mid n \in \mathbb{N}\}$ is relatively compact. It just remains to realize that ($\{|\gamma_n(\lambda_n)| \mid n \in \mathbb{N}\}$ is bounded) the set

$$\{\sigma_n \mid n \in \mathbb{N}\} = \{h^{-1}(\gamma_n(\lambda_n)) - (1 - \lambda_n)g(h^{-1}(\gamma_n(\lambda_n))) \mid n \in \mathbb{N}\}$$

is relatively compact as well.

(7.6.7) Here, we show the continuity of the operator U defined by means of (7.9.1). Let $\{(q_n, \xi_n, \lambda_n)\}_{n=1}^{\infty}$ be a sequence, in the domain of B, such that

$$\lim_{n\to\infty} (q_n, \xi_n, \lambda_n) = (q, \xi, \lambda).$$

We should prove that

$$\lim_{n\to\infty} ((q_n, \xi_n) - B(q_n, \xi_n, \lambda_n)) = (q, \xi) - B(q, \xi, \lambda).$$

(a) At first, let us consider the sequence $\{\gamma_n\}_{n=1}^{\infty}$ of absolutely continuous solutions of the related problems, i.e. the function sequence defined by

$$\gamma_n(t) = h(\xi_n) + \int_0^t f(s, h(q_n(s))) \, ds, \quad t \in [0, 1].$$

Observe that the functions $\hat{f}_n$, $\hat{f}_n(t) = f(t, h(q_n(t)))$, $t \in [0, 1]$, are measurable and have the common Lebesgue integrable majorant ρ_r. Moreover,

$$\lim_{n\to\infty} \hat{f}_n(t) = f(t, h(q(t))),$$

for almost all $t \in [0,1]$. Thus, it follows, according to the Lebesgue dominant convergence theorem, that

$$\lim_{n \to \infty} \gamma_n(t) = \gamma(t), \quad \text{for every } t \in [0,1],$$

whenever $\gamma(t) = h(\xi) + \int_0^t f(s, h(q(s))) \, ds$, $t \in [0,1]$. Since $\{\gamma_n \mid n \in \mathbb{N}\}$ is (see part $(7.6.6)(a)$) a bounded set of equicontinuous functions, our sequence tends to γ uniformly, on a compact interval $[0,1]$.

(b) The remaining part of the proof can be performed quite analogously as in the part $(7.6.6)(b)$.

$(7.6.8)$ At last, we will show the non-triviality of the degree for the operator $B(\cdot, \cdot, 0)$.

(a) At first, observe that $U(\cdot, \cdot, 0) : C([0,1], \mathrm{cl}(\Omega)) \times \mathrm{cl}(\Omega) \to \mathbb{R}^n \times \mathbb{R}^n$ and denote

$$\Delta = (C([0,1], \Omega) \times \Omega) \cap (\mathbb{R}^n \times \mathbb{R}^n),$$
$$D = B(\cdot, \cdot, 0)|_{\mathrm{cl}(\Delta)}.$$

By the contraction property of the degree (see Chapter I.8 or Chapter II.9), we have

$$\mathrm{Deg}(B(\cdot, \cdot, 0), \Omega, 0) = \deg(D, \Delta, 0).$$

(b) Following step by step the proof of the uniqueness of the degree in a finite dimensional space (see again Chapter I.8 or Chapter II.9 or [De3-M, pp. 5–12]), we see that we can assume, without any loss of generality, that $D \in C^{(1)}(\Delta, \mathbb{R}^{2n}) \cap C(\mathrm{cl}(\Delta), \mathbb{R}^{2n})$. By the same reasons, we can assume that 0 is a regular point of D, i.e. $\det(D'(x)) \neq 0$, for every solution x of the equation $D(x) = 0$. Recall that

$$D(\tau, \xi) = (\tau - \xi, g(\xi)), \quad \text{for } (\tau, \xi) \in \mathrm{cl}(\Delta).$$

So, if (τ, ξ) is a solution of our equation, then

$$\det(D'(\tau, \xi)) = \det \begin{pmatrix} I & M \\ (0) & g'(\xi) \end{pmatrix} = \det(g'(\xi)),$$

where I is a unit matrix of the type (n, n), (0) is a zero matrix of the same type as well as M. This, jointly with the assumption $(7.4.3)(c)$ in Definition (7.4), already implies that

$$\deg(D, \Delta, 0) = \deg(g, \Omega, 0) \neq 0,$$

as claimed. $\qquad \square$

Now, let us try to describe an appropriate given set for differential system (7.1), more explicitly. For the sake of simplicity, we restrict our demonstration to the systems with continuous right-hand sides, only. Thus, the following approach can be based on a simple lemma below which we state without the proof.

(7.10) LEMMA. *Consider system* (7.1), *where* f *is, additionally, continuous.
Let* φ *be a solution of* (7.1) *in an interval* $J \subset [0,1]$. *Assume that* $y : \mathbb{R}^n \to \mathbb{R}$ *is
a continuously differentiable function. Let* $y(\varphi(t^*)) = 0$, *for fixed* $t^* \in J$. *Then
there exists* $\varepsilon > 0$ *such that*

(7.10.1)
$$t^* > \inf J, \quad (\operatorname{grad} y)(\varphi(t^*)) \cdot f(t^*, \varphi(t^*)) < 0$$
$$\Rightarrow \forall t \in (t^* - \varepsilon, t^*) : y(\varphi(t)) > 0,$$

(7.10.2)
$$t^* < \sup J, \quad (\operatorname{grad} y)(\varphi(t^*)) \cdot f(t^*, \varphi(t^*)) > 0$$
$$\Rightarrow \forall t \in (t^*, t^* + \varepsilon) : y(\varphi(t)) > 0.$$

Now, we are ready to give a non-trivial application of Theorem (7.6).

(7.11) EXAMPLE. Consider the system

$$\dot{x}_{2i-1} = x_{2i},$$
$$\dot{x}_{2i} = f_{2i}(t, x_1, \ldots, x_{2n}),$$

where $i = 1, \ldots, n$ and $f = (x_2, f_2, x_4, f_4, \ldots, x_{2n}, f_{2n}) : [0,1] \times \mathbb{R}^{2n} \to \mathbb{R}^{2n}$ is a
continuous mapping. Denote $\Theta = \{(\alpha, \beta) \in \mathbb{R}^2 \mid |\alpha| + |\beta| < 1\}^n$. Assume that, for
all $t \in [0,1]$, $\xi \in \operatorname{cl}(\Theta)$, $i = 1, \ldots, n$, the following inequalities take place:

(7.12) $f_{2i}(t, \xi_1, \ldots; \xi_{2i-1}, 1 - \xi_{2i-1}, \xi_{2i+1}, \ldots, \xi_{2n}) > \xi_{2i-1} - 1,$
$$\text{for } \xi_{2i-1} \in [0,1],$$

(7.13) $f_{2i}(t, \xi_1, \ldots, \xi_{2i-1}, \xi_{2i-1} - 1, \xi_{2i+1}, \ldots, \xi_{2n}) > \xi_{2i-1} - 1,$
$$\text{for } \xi_{2i-1} \in [0,1],$$

(7.14) $f_{2i}(t, \xi_1, \ldots, \xi_{2i-1}, -1 - \xi_{2i-1}, \xi_{2i+1}, \ldots, \xi_{2n}) < \xi_{2i-1} + 1,$
$$\text{for } \xi_{2i-1} \in [-1,0],$$

(7.15) $f_{2i}(t, \xi_1, \ldots, \xi_{2i-1}, 1 + \xi_{2i-1}, \xi_{2i+1}, \ldots, \xi_{2n}) < \xi_{2i-1} + 1,$
$$\text{for } \xi_{2i-1} \in [-1,0],$$

Then the above system admits a solution satisfying (7.2), whose values are located
in Θ. Indeed. At first, observe that Θ can be obtained as the intersection

$$\Theta = \bigcap_{i=1}^n \{\xi \in \mathbb{R}^{2n} \mid y_{ji}(\xi) < 0, \quad \text{for } j = 1, \ldots, 4\},$$

where, for $i = 1, \ldots, n$,

$$y_{1i} : \mathbb{R}^{2n} \to \mathbb{R}, \quad y_{1i}(\xi) = \xi_{2i-1} + \xi_{2i} - 1,$$
$$y_{2i} : \mathbb{R}^{2n} \to \mathbb{R}, \quad y_{2i}(\xi) = \xi_{2i-1} - \xi_{2i} - 1,$$
$$y_{3i} : \mathbb{R}^{2n} \to \mathbb{R}, \quad y_{3i}(\xi) = -\xi_{2i-1} - \xi_{2i} - 1,$$
$$y_{4i} : \mathbb{R}^{2n} \to \mathbb{R}, \quad y_{4i}(\xi) = -\xi_{2i-1} + \xi_{2i} - 1.$$

Hence, $\xi \in \text{fr}(\Theta)$ if and only if there exist $j^* \in \{1, \ldots, 4\}$ and $i^* \in \{1, \ldots, n\}$ such that $y_{j^* i^*}(\xi) = 0$ and $y_{ji}(\xi) \leq 0$, for all

$$(j, i) \in (\{1, 2, 3, 4\} \times \{1, \ldots, n\}) \setminus (j^*, i^*).$$

Now, observe that if $\xi \in \text{fr}(\Theta)$ and $y_{1i}(\xi) = 0$, for some $i \in \{1, \ldots, n\}$, then $\xi_{2i} = 1 - \xi_{2i-1}$, $\xi_{2i-1} \in [0, 1]$ and, under the inequality (7.12),

$$(\text{grad } y_{1i})(\xi) \cdot f(t, \xi) = \xi_{2i} + f_{2i}(t, \xi) = 1 - \xi_{2i-1} + f_{2i}(t, \xi) > 0, \quad \text{for all } t \in [0, 1].$$

Analogously, applying inequalities (7.13)–(7.15), we obtain, in the sequel, for all

$$\xi \in \text{fr}(\Theta), \ y_{2i}(\xi) = 0 \Rightarrow \xi_{2i-1} \in [0, 1], \quad (\text{grad } y_{2i})(\xi) \cdot f(t, \xi) < 0,$$
$$\xi \in \text{fr}(\Theta), \ y_{3i}(\xi) = 0 \Rightarrow \xi_{2i-1} \in [-1, 0], \quad (\text{grad } y_{3i})(\xi) \cdot f(t, \xi) > 0,$$
$$\xi \in \text{fr}(\Theta), \ y_{4i}(\xi) = 0 \Rightarrow \xi_{2i-1} \in [-1, 0], \quad (\text{grad } y_{4i})(\xi) \cdot f(t, \xi) < 0.$$

Thus, under the foregoing lemma, the sets Γ^+, Γ^- can be described as follows:

$$\xi \in \Gamma^+ \Leftrightarrow \xi \in \text{cl}(\Theta) \text{ and there exists } i \in \{1, \ldots, n\} \text{ such that}$$
$$y_{2i}(\xi) = 0, \ \xi_{2i-1} \in [0, 1] \text{ or } y_{4i}(\xi) = 0, \ \xi_{2i-1} \in [-1, 0],$$
$$\xi \in \Gamma^- \Leftrightarrow \xi \in \text{cl}(\Theta) \text{ and there exists } i \in \{1, \ldots, n\} \text{ such that}$$
$$y_{1i}(\xi) = 0, \ \xi_{2i-1} \in [0, 1] \text{ or } y_{3i}(\xi) = 0, \ \xi_{2i-1} \in [-1, 0].$$

Observe that $\text{fr}(\Theta) = \Gamma^+ \cup \Gamma^-$.

Furthermore, denote $\Omega = (-1, 1)^{2n}$. It is easy to verify that, under the homeomorphism

$$h : \mathbb{R}^{2n} \to \mathbb{R}^{2n},$$
$$h(\xi) = \frac{1}{2}(\xi_1 + \xi_2, -\xi_1 + \xi_2, \ldots, \xi_{2i-1} + \xi_{2i}, -\xi_{2i-1} + \xi_{2i}, \ldots, -\xi_{2n-1} + \xi_{2n}),$$

$h(\Omega) = \Theta$ holds for Ω. Hence, we obtain the following equivalences for $\xi \in \text{cl}(\Omega)$:

$$\xi \in h^{-1}(\Gamma^+) \Leftrightarrow \text{there exists } i \in \{1, \ldots, n\} \text{ such that } \xi_{2i-1} \in \{-1, 1\},$$
$$\xi \in h^{-1}(\Gamma^-) \Leftrightarrow \text{there exists } i \in \{1, \ldots, n\} \text{ such that } \xi_{2i} \in \{-1, 1\}.$$

Finally, define

$$g : \mathbb{R}^{2n} \to \mathbb{R}^{2n}, \quad g(\xi) = \frac{1}{2}(\xi_1, -\xi_2, \ldots, \xi_{2i-1}, -\xi_{2i}, \ldots, \xi_{2n-1}, -\xi_{2n})$$

and observe that all assumptions in Definition (7.4) are satisfied for our quadruple (Θ, h, Ω, g).

(7.16) REMARK. One can easily formulate similar examples for higher-order systems.

(7.17) REMARK. Since conditions (7.4.3)(a) and (7.4.3)(b) roughly mean that the vector $-g(\xi)$ is directed outward Ω in the set $h^{-1}(\Gamma^-)$ and inward in the set $h^{-1}(\Gamma^+)$, we can apply the following formula due to R. Srzednicki (see [Sr2, p. 728]) for calculating the degree:

$$\deg(g, \Omega, 0) = (-1)^n \deg(-g, \Omega, 0) = \chi(\mathrm{cl}(\Theta)) - \chi(\Gamma^-),$$

where χ stands for the Euler characteristic. Thus, condition (7.4.3)(c) means that $\chi(\mathrm{cl}(\Theta)) - \chi(\Gamma^-) \neq 0$. This inequality appears in [Sr1, Theorems 1, 2] as well as in [Sr2, Corollary 7.4], instead of ours (7.4.3)(a)–(c), but in [Sr1], [Sr2] the right-hand in (7.1) is assumed to be continuous.

(7.18) REMARK. In this light, since in Example (7.11) we have that

$$\chi(\mathrm{cl}(\Theta)) - \chi(\Gamma^-) = (-1)^n,$$

the conclusion follows from the mentioned results in [Sr1], [Sr2] as well.

Now, we would like to avoid the transversality requirments imposed in Theorem (7.6) on the boundary $\mathrm{fr}(\Theta)$ of Θ. For this, let us recall some important facts.

The following approximation statement from Chapter II.8 is of a particular importance for defining a related fixed point index (see Remark (9.11) in Chapter II.9).

(7.19) PROPOSITION. *If $\varphi : X \multimap Y$ is a J-mapping or, in particular, an R_δ-mapping (i.e. u.s.c. with R_δ values) and X, Y are compact ANR-spaces, then, for every $\varepsilon > 0$, there exists a continuous map $f_\varepsilon : X \to Y$, which is an ε-approximation of φ. Moreover, there is an $\varepsilon_0 > 0$ such that every two ε_0-approximations of φ are homotopic.*

Assume we have a map $\varphi : X \multimap X$ with a decomposition:

$$D_\varphi : X = X_0 \overset{\varphi_1}{\multimap} X_1 \overset{\varphi_2}{\multimap} \ldots \overset{\varphi_n}{\multimap} X_n = X, \quad \varphi = \varphi_n \circ \ldots \circ \varphi_1,$$

where each φ_i is an R_δ-mapping on a compact ANR-space X_{i-1}, $i = 1, \ldots, n$. In such a case, we say that φ is *decomposable*. Let A be an open subset of X with no fixed points of φ on its boundary and let f_i, $i = 1, \ldots, n$, be ε-approximations of φ_i, for an $\varepsilon > 0$. Let us call the map $f = f_n \circ \ldots \circ f_1$ the ε-*decomposable approximation* of φ. Using Proposition (7.19), one can easily show that there exists an $\varepsilon_0 > 0$ such that every two ε_0-decomposable approximations of φ are homotopic with the homotopy $\chi : X \times [0,1] \multimap X$ such that:

$$\forall t \in [0,1] \; \forall x \in \partial A \quad x \neq \chi(x,t).$$

Now, following the arguments in Chapters II.9 and I.8, we can define an index of φ over X with respect to A:

(7.20) $$\operatorname{Ind}_X(D_\varphi, A) = \operatorname{ind}_X(f, A),$$

where ind indicates the ordinary fixed point index of maps on compact ANR's and f is an arbitrary ε-decomposable approximation of φ.

Below we recall some properties of Ind:

(7.21) PROPOSITION. *Let $\varphi, \psi : X \multimap X$ be decomposable maps such that $\operatorname{Ind}_X(D_\varphi, A)$ exists.*

(i) *(Existence) If $\operatorname{Ind}(D_\varphi, A) \neq 0$, then φ has a fixed point in A.*

(ii) *(Additivity) If A_j, $j = 1, \ldots, n$ are open, disjoint subsets of A and all fixed points of $\varphi|_A$ lie in $\bigcup_{j=1}^n A_j$, then $\operatorname{Ind}(D_\varphi, A_j)$, $j = 1, \ldots, n$, are well-defined and:*

$$\operatorname{Ind}_X(D_\varphi, A) = \sum_{j=1}^n \operatorname{Ind}_X(D_\varphi, A_j).$$

(iii) *(Homotopy invariance) Suppose that the decompositions D_φ and D_ψ are homotopic:*

$$D_\varphi : X = X_0 \xrightarrow{\varphi_1} X_1 \xrightarrow{\varphi_2} \ldots \xrightarrow{\varphi_n} X_n = X, \quad \varphi = \varphi_n \circ \ldots \circ \varphi_1,$$

$$D_\psi : X = X_0 \xrightarrow{\psi_1} X_1 \xrightarrow{\psi_2} \ldots \xrightarrow{\psi_n} X_n = X, \quad \psi = \psi_n \circ \ldots \circ \psi_1,$$

and there is a decomposable homotopy: $\chi : X \times [0,1] \multimap X$:

$$D_\chi : X \times [0,1] = X_0 \times [0,1] \xrightarrow{\overline{\chi}_1} X_1 \times [0,1] \xrightarrow{\overline{\chi}_2} \ldots \xrightarrow{\chi_n} X_n \times [0,1] = X,$$

where $\chi = \chi_n \circ \overline{\chi}_{n-1} \circ \ldots \circ \overline{\chi}_1$, $\overline{\chi}_i(x, \lambda) = \chi_i(x, \lambda) \times \{\lambda\}$, for $x \in X_{i-1}$, $\lambda \in [0,1]$, $i = 1, \ldots, n-1$, χ_i are u.s.c. with R_δ-values, $\chi_i(\cdot, 0) = \varphi_i$, $\chi_i(\cdot, 1) = \psi_i$, $i = 1, \ldots, n$, and $x \notin \chi(x, \lambda)$, for $x \in \partial A$ and $\lambda \in [0,1]$. Then $\operatorname{Ind}_X(D_\psi, A)$ is well-defined and $\operatorname{Ind}_X(D_\psi, A) = \operatorname{Ind}_X(D_\varphi, A)$.

(iv) *(Multiplicativity) If $\eta : Y \multimap Y$ is decomposable and $\operatorname{Ind}_Y(D_\eta, B)$ exists, then:*

$$\operatorname{Ind}_{X \times Y}(D_\varphi \times D_\eta, A \times B) = \operatorname{Ind}_X(D_\varphi, A) \circ \operatorname{Ind}_Y(D_\eta, B),$$

where:

$$D_\varphi : X = X_0 \overset{\varphi_1}{\multimap} X_1 \overset{\varphi_2}{\multimap} \ldots \overset{\varphi_n}{\multimap} X_n = X, \quad \varphi = \varphi_n \circ \ldots \circ \varphi_1,$$

$$D_\eta : Y = Y_0 \overset{\eta_1}{\multimap} Y_1 \overset{\eta_2}{\multimap} \ldots \overset{\eta_n}{\multimap} Y_n = Y, \quad \eta = \eta_n \circ \ldots \circ \eta_1,$$

and

$$D_\varphi \times D_\eta : X \times Y = X_0 \times Y_0 \overset{\varphi_1 \times \eta_1}{\multimap} X_1 \times Y_1 \overset{\varphi_2 \times \eta_2}{\multimap} \ldots \overset{\varphi_n \times \eta_n}{\multimap} X_n \times Y_n = X \times Y.$$

(v) (*Units*) *Suppose that φ is an R_δ-self-mapping on a compact ANR-space X. If φ is constant, i.e. for every $x \in X$, $\varphi(x) = B \subset X$, then:*

$$\mathrm{Ind}_X(D_\varphi, A) = \begin{cases} 1 & \text{if } A \cap B \neq \emptyset, \\ 0 & \text{if } A \cap B = \emptyset. \end{cases}$$

One can see that the Definition (7.20) depends on the decomposition D_φ. However, when it is clear which decomposition we mean, we usually write:

$$\mathrm{Ind}_X(\varphi, A),$$

instead of $\mathrm{Ind}_X(D_\varphi, A)$.

The following fixed point theorem will be applied to the Poincaré translation operator defined below.

(7.22) THEOREM. *Let E_1 and E_2 be two normed spaces, where E_1 is finite dimensional. Assume that*

$$\varphi : [0, \omega] \times (E_1 \times E_2) \multimap E_1,$$

$$\psi : [0, \omega] \times (E_1 \times E_2) \multimap E_2$$

are compositions of R_δ-maps with continuous (single-valued) maps such that the following conditions hold:

(i) *the maps $\varphi_s = \varphi(s, \cdot)$ and $\psi_s = \psi(s, \cdot)$ are projections onto the spaces E_1 and E_2, respectively, for $s \in [0, \omega]$, i.e. they make restrictions to the appropriate parts of components.*

Let $A \subset E_1$, $B \subset E_2$, A, B be open, bounded and let there exist a sufficiently large compact ball K_2 in E_2, centered at the origin and containing B, i.e. $B \subset K_2 \subset E_2$.

Assume, furthermore, that

(ii) *$A \cdot [0, 1] = A$, $B \cdot [0, 1] = B$, (that is: A, B are star-shaped with respect to the origins),*

(iii) *$\varphi_\omega(\partial A \times \overline{B}) \cap \overline{A} = \emptyset$, $\psi_\omega(\overline{A} \times \partial B) \subset B$,*

(iv) *$0 \notin \varphi([0, \omega] \times (\partial A \times \{0\}))$.*

Then the map $(\varphi_\omega, \psi_\omega) : E_1 \times E_2 \multimap E_1 \times E_2$, $(\varphi_\omega, \psi_\omega)(x) = \varphi_\omega(x) \times \psi_\omega(x)$ *has at least one fixed point in the set* $\mathcal{R} = A \times B$.

PROOF. Take K_1 and K_2 being the compact balls in E_1 and E_2 centered at the origins and large enough to contain the sets A and B, respectively. (We also demand that $\partial A \cap \partial K_1 = \partial B \cap \partial K_2 = \emptyset$). Such balls exist by the hypothesis.

Set $r_1 : E_1 \to K_1$, $r_2 : E_2 \to K_2$ to be the radial retractions onto K_1 and K_2, respectively. Consider the homotopy: $H : (K_1 \times K_2) \times [0,1] \multimap K_1 \times K_2$,

$$H((u,v), \lambda) = (r_1 \circ [(1-\lambda)u + \varphi_\omega(u, \lambda v)], \ r_2 \circ [\lambda \psi_\omega(\lambda u, v)]),$$

$$u \in K_1, \ v \in K_2, \lambda \in [0,1],$$

where $\varphi_\omega = \varphi(\omega, \cdot)$ and $\psi_\omega = \psi(\omega, \cdot)$. This homotopy is decomposable in the sense of (7.21)(iii). It is not difficult to see that the map

$$S_1 : (K_1 \times K_2) \times [0,1] \multimap E_1,$$

$$S_1((u,v), \lambda) = (1-\lambda)u + \varphi_\omega(u, \lambda v)$$

is a composition of an R_δ-map with a continuous (single-valued) map, so the convex hull of its image is a compact ANR contained in E_1. The same is true for the mapping

$$S_2 : (K_1 \times K_2) \times [0,1] \multimap E_2,$$

$$S_2((u,v), \lambda) = \lambda \psi_\omega(\lambda u, v).$$

Hence,

$$D_H : (K_1 \times K_2) \times [0,1] \overset{S_1 \times S_2}{\multimap} \overline{\text{conv}}((S_1 \times S_2)((K_1 \times K_2) \times [0,1])) \overset{r_1 \times r_2}{\multimap} K_1 \times K_2$$

is a decomposition of H. The fact that H has no fixed point on the boundary of $\mathcal{R}$, follows from (iii) and (iv). By the homotopy invariance of fixed point index (cf. (7.21)(iii)), we obtain the following equality:

$$\text{Ind}_{K_1 \times K_2}((r_1 \circ \varphi_\omega, r_2 \circ \psi_\omega), \mathcal{R}) = \text{Ind}_{K_1 \times K_2}((r_1 \circ [\cdot + \varphi_\omega(\cdot, 0)], 0), \mathcal{R}),$$

which by the multiplicativity and units properties (cf. (7.21)(iV) and (7.21)(v)) is equal to:

$$\text{Ind}_{K_1}(r_1 \circ [\cdot + \varphi_\omega(\cdot, 0)], A).$$

Define the homotopy: $\overline{H} : K_1 \times [0, \omega] \to K_1$ as

$$\overline{H}(u, t) = r_1 \circ [u + \varphi(t, (u, 0))], \quad u \in K_1, \ t \in [0, \omega].$$

Recalling (iv) and using the homotopy invariance property (7.21)(iii) once again, this time for the index on K_1, we have:

$$\text{Ind}_{K_1}(r_1 \circ [\,\cdot\, + \varphi_\omega(\,\cdot\,, 0)], A) = \text{Ind}_{K_1}(r_1 \circ [\,\cdot\, + \varphi_0(\,\cdot\,, 0)], A),$$

which, by (i) is equal to:

$$\text{Ind}_{K_1}(r_1 \circ [2 \cdot \text{id}_{K_1}], A) = \deg(-\text{id}_{E_1}, K, 0) = (-1)^j,$$

where deg denotes the Brouwer topological degree, K is a sufficiently small ball in E_1 and j indicates the dimension of E_1 (finite, under the hypothesis).

We are now in a position to write the following equality:

$$(7.23) \qquad \text{Ind}_{K_1 \times K_2}((r_1 \circ \varphi_\omega, r_2 \circ \psi_\omega), \mathcal{R}) = (-1)^j.$$

By the existence property (7.21)(i) of the fixed point index, the result follows. $\square$

(7.24) REMARK. Theorem (7.22) is also valid if we replace the assumption about the existence of a compact ball K_2 containing B by the total continuity of the map ψ. In both cases, E_1 must be finite dimensional.

Now, we shall apply Theorem (7.22) to the Poincaré translation operators associated to the functional system $\dot{x} \in S(t, x_t)$. The symbol x_t means, as usual, the function from $[-\tau, 0]$ into $\mathbb{R}^n$ defined by $x_t(\,\cdot\,) = x(t, \,\cdot\,)$, for $t \in [0, \omega]$, $\omega > 0$. Thus, we can study the existence of periodic solutions to the problem:

$$(7.25) \qquad \begin{cases} \dot{x} \in S(t, x_t), \\ x(0) = x(\omega), \end{cases}$$

where $S : [0, \omega] \times \mathcal{C} \multimap \mathbb{R}^n$ is a u-Carathéodory multivalued mapping (see III.4.B) and $\mathcal{C} = \text{AC}([-\tau, 0], \mathbb{R}^n)$, $\tau \geq 0$.

By a solution of (7.25), we mean the one in the sense of Carathéodory, i.e. the absolutely continuous function $x : [-\tau, \omega] \to \mathbb{R}^n$, satisfying $\dot{x} \in S(t, x_t)$, for a.a. $t \in [0, \omega]$, and such that $x(0) = x(\omega)$.

Consider the map $P : \mathcal{C} \to \text{AC}([-\tau, \omega], \mathbb{R}^n)$:

$$P(x_0) = \{x \in \text{AC}([-\tau, \omega], \mathbb{R}^n) \mid x \text{ is a solution of } \dot{x} \in S(t, x_t)$$
$$\text{with } x(t) = x_0, \text{ for } t \in [-\tau, 0]\},$$

for $x_0 \in \mathcal{C}$.

We define the evaluation map: $e_t : \text{AC}([-\tau, \omega], \mathbb{R}^n) \to \mathcal{C}$, $t \in [0, \omega]$,

$$e_t(x) = x(t), \quad x \in \text{AC}([-\tau, \omega], \mathbb{R}^n),$$

and have the following diagram:

$$\mathcal{C} \xrightarrow{P} \mathrm{AC}([-\tau,\omega],\mathbb{R}^n) \xrightarrow{e_t} \mathcal{C}.$$

The composition $Q_t = e_t \circ P$ is called the *Poincaré translation operator* whose properties are described in Theorem (4.7). Taking a compact, convex set $K \subset \mathcal{C}$ and a retraction $r : \mathcal{C} \to K$ onto K such that:

$$(7.26) \qquad\qquad r(\mathcal{C} \setminus K) \subset \partial K,$$

we obtain the decomposable map $r \circ Q_\omega : K \multimap K$, to which we can apply the fixed point index theory, described above. Namely, by the existence property, we have:

(7.27) PROPOSITION. *If $A \subset K$ is such that $\partial A \cap \partial K = \emptyset$ (the boundaries with respect to $\mathcal{C}$) and $\mathrm{Ind}_K(r \circ Q_\omega, A)$ is defined and different from 0, then (7.25) has a solution x with $x(t) \in A$, for $t \in [-\tau, 0]$.*

Consider still the one-parameter family of systems of differential inclusions, given by the u-Carathéodory map $S : [0, \omega] \times \mathcal{C} \times [0, 1] \to \mathbb{R}^n$,

$$(7.28) \qquad\qquad \begin{cases} \dot{x} \in S(t, x_t, \lambda), \\ x(0) = x(\omega), \end{cases}$$

with $\lambda \in [0, 1]$. For every $\lambda \in [0, 1]$, we can set the map P_λ:

$$P_\lambda(x_0) = \{x \in \mathrm{AC}([-\tau,\omega],\mathbb{R}^n) \mid x \text{ is a solution of } \dot{x} \in S(t, x_t, \lambda)$$
$$\text{with } x(t) = x_0, \text{ for } t \in [-\tau, 0]\},$$

for $x_0 \in \mathcal{C}$, and the map $Q_t^\lambda = e_t \circ P_\lambda$, which is the Poincaré operator for (7.28) and $\lambda \in [0, 1]$, whose properties are again described in Theorem (4.7).

Assume that S is a u-Carathéodory mapping (see conditions (i)–(iv) in III.4.B) and that P_1 $(\lambda = 1)$ splits in the following way:

$$P_1(x_0) = P_{11}(x_0) \times P_{12}(x_0),$$
$$P_{11} : \mathcal{C} \multimap \mathrm{AC}([-\tau,\omega],\mathbb{R}^j), \quad P_{12} : \mathcal{C} \multimap \mathrm{AC}([-\tau,\omega],\mathbb{R}^k), \quad j + k = n,$$

where P_{11} and P_{12} are R_δ-maps.

Define $\varphi(t, x_0) = e_t \circ P_{11}(x_0)$ and $\psi(t, x_0) = e_t \circ P_{12}(x_0)$, $x_0 \in \mathcal{C}$, $t \in [0, \omega]$. Then, under the assumptions (ii)–(iv) of Theorem (7.22), one can obtain (see the proof of Theorem (7.22)):

$$(7.29) \qquad\qquad \mathrm{Ind}_{K_1 \times K_2}((r_1 \circ \varphi_\omega, r_2 \circ \psi_\omega), \mathcal{R}) = (-1)^j,$$

$(K_1, K_2, r_1, r_2, \mathcal{R}$ are as in the proof of Theorem (7.22)).

We can reformulate (7.29) as follows:

$$(-1)^j = \mathrm{Ind}_{K_1 \times K_2}(r \circ (\varphi_\omega, \psi_\omega), \mathcal{R}) = \mathrm{Ind}_K(r \circ Q_\omega^1, \mathcal{R}),$$

where $r : \mathcal{C} \to K = K_1 \times K_2$ is a retraction for which (7.26) is valid. Letting $H : K \times [0,1] \multimap K$,

$$H(x, \lambda) = r \circ Q_\omega^\lambda(x),$$

H is a decomposable homotopy (cf. (7.21)(iii)), linking $r \circ Q_\omega^0$ with $r \circ Q_\omega^1$. If for each $\lambda \in [0,1)$ and every $x \in \partial\mathcal{R}$, $x \notin Q_\omega^\lambda(x)$, then by the homotopy invariance we obtain:

$$\mathrm{Ind}_K(r \circ Q_\omega^\lambda, \mathcal{R}) = (-1)^j,$$

which already implies, in view of Proposition (7.27), that the inclusion $\dot{x} \in S(t, x_t, 0)$ has at least one ω-periodic solution with the initial value in $\mathcal{R}$.

It will be convenient to summarize the above investigations for the functional (upper) Carathéodory system

(7.30) $$\dot{X} \in S(t, \widehat{X}_j, X_t),$$

where $X = (\widehat{X}_j, X_j)$, $\widehat{X}_j \in \mathbb{R}^j$, $X_j \in \mathbb{R}^{n-j}$ $(0 \leq j \leq n)$ and $X_t(\cdot) = X_j(t + \cdot)$.

Consider, in view of Proposition (7.27), an open set

$$\mathcal{R} = A \times B \subset \mathcal{U} = K_1 \times K_2 \subset \mathbb{R}^j \times C([-\tau, 0], \mathbb{R}^{n-j}), \quad \tau \geq 0,$$

where A, B are star-shaped w.r.t. the origins, open, bounded subset of K_1, K_2 such that $\partial A \cap \partial K_1 = \emptyset$, $\partial B \cap \partial K_2 = \emptyset$, K_1 is a bounded, closed subset of $\mathbb{R}^j$ and K_2 is a compact subset of uniformly bounded and equicontinuous functions from $C([-\tau, 0], \mathbb{R}^{n-j})$ such that $K_1 \times K_2$ is convex.

We can immediately state (see Fig. 2).

(7.31) THEOREM. *Let* $S : [0, \infty) \times \mathbb{R}^j \times C([-\tau, 0], \mathbb{R}^{n-j}) \multimap \mathbb{R}^j$ $(0 \leq j \leq n)$, $\tau \geq 0$, *be an (upper) Carathéodory mapping, satisfying conditions* (i)–(iv) *in* III.4.B. *Assume that:*

(7.31.1) *each trajectory of (7.30) with the initial values* $|\widehat{X}_j(0)| = a$, $|X_j(t)| \doteq 0$, *for* $t \in [-\tau, 0]$, *does not reach the origin, for any* $t \geq 0$,

(7.31.2) *each trajectory of (7.30), starting on the "vertical" sides of* $\mathcal{R}$, *i.e. those with* $|\widehat{X}_j(0)| = a$, $|X_j(t)| \geq b(t)$, *for* $t \in [-\tau, 0]$, *stays outside the domain, determined by the constants* $-a, a$ *for all future times bigger or equal than some* $t_0 > 0$,

(7.31.3) *each trajectory of (7.30), starting on the "horizontal" sides of* $\mathcal{R}$, *i.e. those with* $|\widehat{X}_j(0)| \leq a$, $|X_j(t)| = b(t)$, *for* $t \in [-\tau, 0]$, *stays in the domain, determined by the functions* $-b, b$, *for all future times bigger or equal than some* $t_0 > 0$,

(7.31.4) *each trajectory of* (7.30) *with initial values in* $\overline{\mathcal{R}}$ *and such that*

$$(\widehat{X}_j(0), X_j(t)) = (\widehat{X}_j(\widehat{t}), X_j(t + \widehat{t})), \quad \text{for any } \widehat{t} \geq t_0,$$

where $t \in [-\tau, 0]$, *stays a priori in* $\mathcal{U}$ *for all* $t \in [0, \widehat{t}]$.

Then there exists a number $k_o \in \mathbb{N}$ *such that, for any positive integer* $k \geq k_0$, *the system* (7.30) *admits at least one solution* $X(t)$ *with the values in* $\mathcal{U}$, *satisfying* $X(0) = X(k\omega)$ *and* $X(0) \in \mathcal{R}$, *where* $\omega > 0$ *is a given number.*

Applying Proposition (1.37), we have in the ordinary case ($\tau = 0$) still

(7.32) COROLLARY. *Under the assumptions of Theorem* (7.31), *the ordinary* ($\tau = 0$) *system* (7.30) *admits at least one bounded solution on* $[0, \infty)$, *whose values are in* $\mathcal{U}$.

(7.33) REMARK. Observe that, in Theorem (7.31) as well as in Corollary (7.32), no transversality in required and (in the single-valued case) no uniqueness is assumed. On the other hand, we can only obtain subharmonic $k\omega$-periodic solutions ($k \geq 1$), provided $S(t, Y) \equiv S(t + \omega, Y)$.

(7.34) REMARK. Theorem (7.31) holds obviously also when, in particular, $j = 0$ or $j = n$, by which the so called transformation theory of N. Levinson (cf. [Yos1-M], [Yos2-M]) is generalized, where the dissipativity (the uniform ultimate boundedness of solutions) implies the existence of (sub)harmonics.

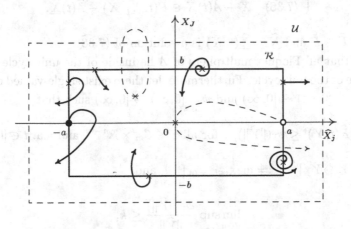

FIGURE 2. (the dashed trajectories are "forbidden")

Now, let let us try to guarantee the behaviour of trajectories as in Theorem (7.31), for the system (7.30). Taking this time $E_1 := \mathbb{R}^{n-m}$ ($m \leq n$),

$E_2 := \mathcal{C}_m = C([-\tau, 0], \mathbb{R}^m)$, i.e. $j = n - m$, $k = m$, the Poincaré operator $\phi_\omega = (\varphi_\omega^{(1)}, \varphi_\omega^{(2)})\,(= (\psi_\omega, \varphi_\omega))$ associated to (7.30) has, under the assumptions of Theorem (7.31), a fixed point corresponding to a solution $X(t)$ of (7.30) with $X(0) = X(k\omega)$, for some $k \in \mathbb{N}$.

For this aim, consider the particular form (observe that $\check{X}(t) = \check{X}_t(0)$) of inclusion (7.30), namely

$$(7.35) \qquad \dot{X} = A(t)X \in F(t, \widehat{X}_t, \check{X}),$$

where $X = \widehat{X} \oplus \check{X}$, $\widehat{X} \in \mathbb{R}^m$ $(m \le n)$, $A(t) \equiv A(t + \omega)$, $F(t, Y) \equiv F(t + \omega, Y)$, and the matrix A consists of the following blocks

$$A = \begin{pmatrix} \widehat{A} & 0 \\ \widetilde{A} & \check{A} \end{pmatrix}, \qquad \widehat{A} = \begin{pmatrix} a_{1,1} & \cdots & a_{1,m} \\ \vdots & & \vdots \\ a_{m,1} & \cdots & a_{m,m} \end{pmatrix},$$

$$\widetilde{A} = \begin{pmatrix} a_{m+1,1} & \cdots & a_{m+1,m} \\ \vdots & & \vdots \\ a_{n,1} & \cdots & a_{n,m} \end{pmatrix}, \quad \check{A} = \begin{pmatrix} a_{m+1,m+1} & \cdots & a_{m+1,n} \\ \vdots & & \vdots \\ a_{n,m+1} & \cdots & a_{n,n} \end{pmatrix}.$$

In the sequel, it will be also very useful to consider the associated one-parameter family of systems

$$(7.36) \qquad \begin{cases} (7.\widehat{36}) \quad \dot{\widehat{X}} + \widehat{A}(t)\widehat{X} \in \widehat{F}(t, \widehat{X}_t, \check{X}), \\ (7.\check{36}) \quad \dot{\check{X}} + \check{A}(t)\check{X} \in \check{F}(t, \underline{\widehat{X}}_t, \check{X}) - \widetilde{A}(t)\underline{\widehat{X}}, \end{cases}$$

where $F = \widehat{F} \oplus \check{F}$ and $\underline{\widehat{X}} = \underline{\widehat{X}}(t, \check{X})$ is a solution of $(7.\widehat{36})$.

Assume that all Floquet multipliers of $\widehat{A}$ lie inside of the unit cycle and those of $\check{A}$ outside of the unit cycle. Furthermore, let there exist single-valued continuous functions $\psi_1 : [0, \infty) \to [0, \infty)$ and $\psi_2 : [0, \infty) \to [0, \infty)$ such that

$$(7.37) \qquad |\widehat{F}(t, Y)| \le \psi_1(\|\widehat{Y}\|), \quad \text{for all } Y \in \mathcal{C}_m \times \mathbb{R}^{n-m} \text{ and a.a. } t \in [0, \infty),$$

where $Y = \widehat{Y} \oplus \check{Y}$, $\| \cdot \| := \max_{t \in [-\tau, 0]} | \cdot |$, and

$$(7.38) \qquad \limsup_{\|\widehat{Y}\| \to \infty} \frac{\psi_1(\|\widehat{Y}\|)}{\|\widehat{Y}\|} \le k,$$

holds with a sufficiently small constant k, and

$$(7.39) \qquad |\check{F}(t, Y)| \le \psi_2(\|\widehat{Y}\|), \quad \text{for all } Y \in \mathcal{C}_m \times \mathbb{R}^{n-m} \text{ and a.a. } t \in [0, \infty).$$

Then one can show the existence of positive constants a_1, a_2, λ_1, λ_2, R_1, R_2, R_3, S_0 such that the following estimates take place (cf. e.g. [Pl-M, Chapter I] and [ObZe])

$$(7.40) \qquad \|\widehat{\underline{X}}_t\| \leq a_1 \max\{R_1, e^{-\lambda_1 t} \sup_{t \in [-\tau, 0]} |\widehat{\underline{X}}_0(t)|\}, \quad t \geq 0,$$

for all solutions $\widehat{\underline{X}}$ of $(7.\widehat{36})$ with $\min_{t \in [-\tau, 0]} |\widehat{\underline{X}}_0(t)| \geq R_1 \max(1, a_1)$,

$$(7.41) \qquad |\check{\underline{X}}(t)| \geq a_2 e^{\lambda_2 t}[|\check{\underline{X}}(0)| - R_2] - R_3, \quad t \geq 0,$$

for all solutions $\check{\underline{X}}(t)$ of $(7.\check{36})$ with $|\check{\underline{X}}(0)| \geq R_2 + \frac{R_3}{a_2}$, provided $\max_{t \in [-\tau, 0]} |\widehat{\underline{X}}_0(t)| \leq S_0$. Therefore, applying Theorem (7.31), whose all assumptions are (in view of (7.40) and (7.41)) satisfied, we can give

(7.42) THEOREM. *Assume that:*

(i) $A : [0, \omega] \to \mathbb{R}^{n^2}$ *is a (single-valued), ω-periodic, continuous $(n \times n)$-matrix function having the same structure as above,*

(ii) *all Floquet multipliers of $\widehat{A}$ lie inside and those of $\check{A}$ outside of the unit cycle,*

(iii) $F : [0, \infty) \times C_m \times \mathbb{R}^{n-m} \multimap \mathbb{R}^n$ *is an (upper) Carathéodory multivalued function satisfying (7.37)–(7.39) and $F(t, Y) = F(t + \omega, Y)$.*

Then inclusion (7.35) admits a (subharmonic) periodic solution.

Applying Proposition (1.37) in the ordinary case $(\tau = 0)$ (cf. also Corollary (7.32)), we have still

(7.43) COROLLARY. *Let the assumptions of Theorem (7.42) be satisfied, but with an ordinary $(\tau = 0)$ multivalued function*

$$(7.44) \qquad F(t, X) \not\equiv F(t + \omega, X).$$

Then inclusion (7.35), where $\tau = 0$, admits a bounded solutions on $[0, \infty)$.

In order to study harmonics, consider the system of ordinary differential equations

$$(7.45) \qquad \dot{X} = F(t, X),$$

where $F \in \mathrm{CAR}(\mathbb{R}^{n+1}, \mathbb{R})$ and $F(t, X) \equiv F(t + \omega, X)$.

Consider still the Carathéodory one-parameter family of periodically perturbed autonomous systems

$$(7.45_\lambda) \qquad \dot{X} = F(t, X; \lambda), \quad \lambda \in [0, 1],$$

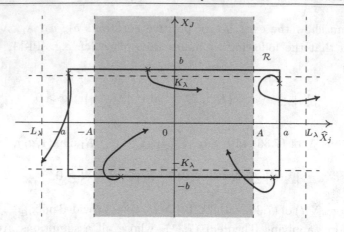

FIGURE 3. (the dark strip is "forbidden")

where $F(t, X; 0) \equiv G(X)$ and $F(t, X; 1) \equiv F(t, X)$, and observe that (7.45_1) becomes (7.45), while (7.45_0) becomes autonomous.

We can give (see Fig. 3)

(7.46) THEOREM. *System (7.45), where $X = (\widehat{X}_j, X_j)$, $\widehat{X} \in \mathbb{R}^j$, $X_j \in \mathbb{R}^{n-j}$ ($0 \leq j \leq n$) admits an ω-periodic solution starting in $\mathcal{R}$, provided:*

(7.46.1) *each trajectory of (7.45_λ) starting on the vertical sides of the (open) rectangle $\mathcal{R}$ does not reach the "dark strip" (determined by $-A, A$), for any $t \geq 0$,*

(7.46.2) *the same trajectory stays, for $t \geq t_0$, outside the domain, determined by constants $-L_\lambda, L_\lambda$ ($0 \leq \lambda < 1$), where $t_0 \geq \omega$ is a suitable number,*

(7.46.3) *each trajectory of (7.45_λ), starting on the horizontal sides of the rectanle $\mathcal{R}$, stays, for $t \geq t_0$, in the domain, determined by $-K_\lambda, K$ ($0 \leq \lambda < 1$),*

(7.46.4) *solutions of initial value problems, for (7.45_λ), ($0 \leq \lambda \leq 1$), are unique and globally extendable.*

Before we prove Theorem (7.46), we shall make some auxiliary computations.

Let $G : \mathbb{R}^n \to \mathbb{R}^n$ be a continuous function and suppose that uniqueness and global existence in the future for the solutions of the Cauchy problems associated to $\dot{X} = G(X)$, is ensured. We denote by $X(\,\cdot\,; z)$ the (unique) solution of the Cauchy problem ($\lambda = 0$)

$$(7.47) \qquad \dot{X} = G(X),$$

$$(7.48) \qquad X(0) = z,$$

for any $z \in \mathbb{R}^n$.

According to the assumptions, $X(t;z)$ is defined, for all $t \geq 0$. Moreover, the map $(t,z) \mapsto X(t;z)$ is well-known (see e.g. [Ks2-M] and cf. Chapter III.4.A) to be continuous on $[0,\infty) \times \mathbb{R}^n$. In particular, for every $t \geq 0$, the t-Poincaré operator

$$\mathcal{Q}_t : \mathbb{R}^n \to \mathbb{R}^n, \quad \mathcal{Q}_t : z \mapsto X(t;z)$$

is continuous, with

$$\mathcal{Q}_0 = \mathrm{id}_{\mathbb{R}^n} \quad (= \text{the identity in } \mathbb{R}^n).$$

We consider now the decomposition of $\mathbb{R}^n$ as

$$\mathbb{R}^n = \mathbb{R}^j \times \mathbb{R}^{n-j}$$

and, correspondingly, we set

$$z := (u,v), \quad \text{with } u \in \mathbb{R}^j, \ v \in \mathbb{R}^{n-j};$$
$$X(t;z) := (\varphi(t;z), \psi(t;z)), \quad \text{with } \varphi \in \mathbb{R}^j, \ \psi \in \mathbb{R}^{n-j}.$$

For simplicity, we restrict ourselves to the case:

$$1 \leq j \leq n-1.$$

If $j = 0$ or $j = n$, we can easily modify the arguments which are developed below; consequently, such situations are not examined in detail.

The following conditions are considered on the components φ and ψ of the solution $X(\cdot)$. There are constants

$$t_0 > 0, \quad 0 < A < a < L, \quad 0 < K < b,$$

such that, for each $z = (u,v) \in \mathbb{R}^n$, we have:

(i₁) $|u| = a, \quad |v| \leq b \Rightarrow |\varphi(t;z)| \geq A, \quad \forall t \geq 0,$

(i₂) $|u| = a, \quad |v| \leq b \Rightarrow |\varphi(t;z)| \geq L, \quad \forall t \geq t_0,$

(i₃) $|u| \leq a, \quad |v| = b \Rightarrow |\psi(t;z)| \leq K, \quad \forall t \geq t_0.$

Let $\mathcal{R} \subset \mathbb{R}^n = \mathbb{R}^j \times \mathbb{R}^{n-j}$, be the open rectangle:

$$\mathcal{R} := B(0,a) \times B(0,b),$$

where, as usual, $B(0,r)$ is the open ball of center at the origin and radius $r > 0$.

Assumptions (i₂) and (i₃) can be regarded as a kind of repulsivity of the origin along the u-component and attractivity with respect to the v-coordinate. Hypothesis (i₁) means that the components of the flow which are repelled, leave the ball $B(0,a)$ without passing through the origin before (see Example (7.58)).

Then we have:

(7.49) LEMMA. *Assume* (i_1), (i_2), (i_3). *Then*

$$\deg(G, \mathcal{R}, 0) = (-1)^{n-j}.$$

PROOF. First of all, we observe that by (i_1) and (i_2), the t-Poincaré map associated to (7.47) is fixed point free on the boundary $\partial \mathcal{R}$ of $\mathcal{R}$, for $t \geq t_0$, that is,

$$z \neq \mathcal{Q}_t(z), \quad t \geq t_0, \text{ for all } z \in \partial \mathcal{R}.$$

Hence, we also have $G(z) \neq 0$, for all $z \in \partial \mathcal{R}$.

Now, we can use the result in [CaMaZa, Corollary 2] according to which, we can conclude that

$$\deg(G, \mathcal{R}, 0) = (-1)^n \deg(\mathrm{id}_{\mathbb{R}^n} - \mathcal{Q}_t, \mathcal{R}, 0), \quad \text{for all } t \geq t_0.$$

Observe that, by our definitions,

$$\mathcal{Q}_t(z) = \mathcal{Q}_t(u, v) = (\varphi(t; (u, v)), \psi(t; (u, v))), \quad \text{for all } t \geq t_0.$$

Now, we introduce the (continuous) homotopy (for all $t \geq t_0$)

$$H : \mathbb{R}^n \times [0, 1] \to \mathbb{R}^n,$$
$$H : ((u, v), \lambda) \mapsto (\lambda u - \varphi(t; (u, \lambda v)), v - \lambda \psi(t; (\lambda u, v)))$$

and observe that

$$H(z, 1) = z - \mathcal{Q}_t(z), \quad H(z, 0) = (-\widetilde{\varphi}(t; u), v),$$

where in order to simplify the notations, we have set, for all $u \in \mathbb{R}^j$ and $t \geq 0$,

$$\widetilde{\varphi}(t; u) := \varphi(t; (u, 0)).$$

We claim that

(7.50) $$H(z, \lambda) \neq 0, \quad \text{for all } z \in \partial \mathcal{R}, \; \lambda \in [0, 1].$$

Indeed, assume by contradiction that there are $\overline{z} = (\overline{u}, \overline{v}) \in \partial \mathcal{R}$ and $\overline{\lambda} \in [0, 1]$, such that $H(\overline{z}, \overline{z}) = 0$. From the definition of $H(\cdot, \cdot)$, we immediately obtain that (for all $t \geq t_0$)

(7.51) $$\overline{\lambda} \overline{u} = \varphi(t; (\overline{u}, \overline{\lambda} \overline{v})), \quad \overline{v} = \overline{\lambda} \psi(t; (\overline{\lambda} \overline{u}, \overline{v})),$$

with

$$|\overline{u}| = a, \quad |\overline{\lambda v}| \le b \quad \text{or} \quad |\overline{\lambda u}| \le a, \quad |\overline{v}| = b.$$

In any case, passing to the norm in (7.51) and using, respectively (i_2) or (i_3), we obtain a contradiction. Thus, (7.50) is proved as well.

By the homotopic invariance of the degree, we get (for all $t \ge t_0$)

$$\begin{aligned}
\deg(\mathrm{id}_{\mathbb{R}^n} - \mathcal{Q}_t, \mathcal{R}, 0) &= \deg(H(\,\cdot\,, 0), \mathcal{R}, 0) \\
&= \deg(-\widetilde{\varphi}(t; \,\cdot\,), B(0, a), 0) \times \deg(\mathrm{id}_{\mathbb{R}^{n-1}}, B(0, b), 0) \\
&= (-1)^j \deg(\widetilde{\varphi}(t; \,\cdot\,), B(0, a), 0).
\end{aligned}$$

A comparison of the former identity for the degree with the latter one, yields (for all $t \ge t_0$)

$$(7.52) \qquad \deg(G, \mathcal{R}, 0) = (-1)^j \deg(\widetilde{\varphi}(t; \,\cdot\,), B(0, a), 0).$$

Our final step is the evaluation of $\deg(\widetilde{\varphi}(t; \,\cdot\,), B(0, a), 0)$, for all $t \ge t_0$. To this end, we consider the (continuous) retraction

$$r : \mathbb{R}^j \to B[0, a] := \overline{B(0, a)} \subset \mathbb{R}^j,$$

$$r(u) := \begin{cases} u, & \text{for } |u| \le a, \\ a\dfrac{u}{|u|}, & \text{for } |u| > a. \end{cases}$$

Observe that

$$r(u) = 0 \quad \text{if and only if } u = 0.$$

Next, we define the map (for all $t \ge t_0$)

$$r \circ \widetilde{\varphi} : \mathbb{R}^j \times [0, t] \to \overline{B(0, a)}.$$

We claim that

$$(7.53) \qquad r \circ \widetilde{\varphi}(t; u) \ne 0, \quad \text{for all } u \in \partial B(0, a) \text{ and } t \ge 0.$$

Indeed, it is sufficient to observe that

$$r \circ \widetilde{\varphi}(t; u) = 0 \Leftrightarrow \widetilde{\varphi}(t; u) = \varphi(t; (u, 0)) = 0$$

and then to use property (i_1) which ensures that

$$|\widetilde{\varphi}(t; u)| \ge A > 0, \quad \text{for } |u| = a \text{ and for all } t \ge 0.$$

Thus, (7.53) easily follows. Therefore, by the homotopic invariance of the degree and also recalling that

$$r \circ \widetilde{\varphi}(0; u) = r(\varphi(0; (u, 0))) = r(u) = u, \quad \text{for all } u \in B[0, a],$$

we get (for all $t \geq t_0$)

$$(7.54) \qquad \begin{aligned} \deg(r \circ \widetilde{\varphi}(t; \cdot), B(0, a), 0) &= \deg(\mathrm{id}_{B(0,a)}, B(0, a), 0) \\ &= \deg(\mathrm{id}_{\mathbb{R}^j}, B(0, a), 0) = 1. \end{aligned}$$

By the well-known Leray product formula (see e.g. [Ll-M], [Rot-M], and cf. the multiplicativity properties of degrees considered in Chapter I.8 or Chapter II.9), we can write[12]:

$$(7.55) \qquad \deg(r \circ \widetilde{\varphi}, B(0, a), 0) = \sum_k \deg(r, \Delta_k, 0) \times \deg(\widetilde{\varphi}, B(0, a), p_k),$$

where, the Δ_k's are the connected components of $\mathcal{M} \setminus \widetilde{\varphi}(\partial B(0, a))$, $p_k \in \Delta_k$, for each k, and $\mathcal{M}$ is an arbitrary open bounded set containing $\widetilde{\varphi}(B[0, a])$.

Let Δ_0 be the connected component of $\mathcal{M} \setminus \widetilde{\varphi}(\partial B(0, a))$ containing the origin of $\mathbb{R}^j$.

By assumption (i$_2$), it follows that

$$|\widetilde{\varphi}(u)| \geq L > a, \quad \text{for all } u \in \partial B(0, a),$$

so that

$$\widetilde{\varphi}(\partial B(0, a)) \cap B(0, a) = \emptyset.$$

Now, taking $\mathcal{M} = B(0, R)$, with $R > a$ and sufficiently large, we obtain that

$$B(0, a) \subset \mathcal{M} \setminus \widetilde{\varphi}(\partial B(0, a))$$

and, therefore, we have

$$B(0, a) \subset \Delta_0.$$

On the other hand, as $\Delta_k \cap \Delta_0 = \emptyset$, for each $k \neq 0$ (if any of such k exists), we have

$$B(0, a) \cap \Delta_k = \emptyset, \quad \text{for all } k \neq 0.$$

Since

$$\deg(r, \Delta_0, 0) = \deg(r, B(0, a), 0) = \deg(\mathrm{id}_{\mathbb{R}^j}, B(0, a), 0) = 1,$$

---

[12]In order to simplify further the notation, we set $\widetilde{\varphi}(\cdot) = \widetilde{\varphi}(t; \cdot)$.

while
$$\deg(r, \Delta_k, 0) = 0, \quad \text{for } k \neq 0,$$

(in fact, $r(u) \neq 0$, for all $u \in \Delta_k$, with $k \neq 0$), from (7.55) we finally obtain

$$(7.56) \qquad \deg(r \circ \widetilde{\varphi}, B(0, a), 0) = \deg(\widetilde{\varphi}, B(0, a), p_0) = \deg(\widetilde{\varphi}, B(0, a), 0),$$

(note that the choice $p_0 = 0$ is allowed as $0 \in \Delta_0$). Thus, from (7.52), (7.53) and (7.56), we get
$$\deg(G, \mathcal{R}, 0) = (-1)^{n-j}.$$

The proof of Lemma (7.49) is complete.   □

(7.57) REMARK. The assumption (i_1) is crucial; it is independent with respect to (i_2), (i_3) and cannot be completely omitted (cf. also condition (7.31.1)). An elementary example of this fact is the following:

(7.58) EXAMPLE. Let $\mathbb{R}^2 = \mathbb{R} \times \mathbb{R}$ and consider the equation $(X = (x_1, x_2))$

$$\dot{x}_1 = 1, \quad \dot{x}_2 = -x_2,$$

inducing the flow
$$X(t; z) = (u + t, ve^{-t}); \quad z = (u, v).$$

In this case, we have:

$$\varphi(t; z) = u + t, \quad \psi(t; z) = ve^{-t}.$$

In particular,

$$\lim_{t \to \infty} |\varphi(t; z)| = \infty,$$

uniformly in v and for u bounded, and

$$\lim_{t \to \infty} |\psi(t; z)| = 0,$$

uniformly in u and for v bounded. It is easy to check that, given arbitrary

$$0 < a < L, \quad 0 < K < b,$$

it is possible to find $t_0 > 0$ such that (i_2) and (i_3) are fulfilled. However, for any such t_0, (i_1) never holds. Note that in this case, for any open rectangle $\mathcal{R}$ containing the origin, we have

$$\deg(G, \mathcal{R}, 0) = 0,$$

for $G(x_1, x_2) = (1, -x_2)$, as $G(x) \neq 0$, for all $x \in \mathbb{R}^2$ (indeed, the degree is zero with respect to any open bounded subset of the plane).

The same example also shows that no condition of the form

$$\lim_{t \to \infty} |\varphi(t; z)| = \infty$$

(even if assumed to hold uniformly with respect to the v-coordinate) is itself sufficient to guarantee the existence of periodic solutions for systems like $\dot{X} = F(t, X)$ or $\dot{X} = G(x)$, without adding further conditions implying the validity of some property of the form like (i_1).

(7.60) REMARK. From the proof of Lemma (7.49) it is possible to check that (i_1) can be replaced by some weaker condition, like, for instance (cf. (7.31.1)):

(i_1^*) $z = (u, 0), \quad |u| = a \Rightarrow |\varphi(t; z)| \geq A, \quad$ for all $t \geq 0$.

Now, let $F = F(t, X, \lambda) : \mathbb{R} \times \mathbb{R}^n \times [0, 1] \to \mathbb{R}^n$ be a Carathéodory function which is periodic of a fixed period $\omega > 0$ in the t-variable and suppose that uniqueness and global existence in the future, for the solutions of the Cauchy problems associated to $\dot{X} = F(t, X, \lambda)$, is ensured.

We denote by $X(\cdot; z, \lambda)$ the (unique) solution of (7.45_λ), satisfying

$$X(0) = z \in \mathbb{R}^n,$$

which is defined on $[0, \infty)$.

We also set (cf. (7.45))

$$F(t, X) := F(t, X; 1).$$

As in above, we consider now the decomposition of the space $\mathbb{R}^n$ by

$$\mathbb{R}^n = \mathbb{R}^j \times \mathbb{R}^{n-j}$$

and, correspondingly, we set

$$z := (u, v), \quad \text{with } u \in \mathbb{R}^j, \ v \in \mathbb{R}^{n-j};$$
$$X(t; z, \lambda) := (\varphi(t; z, \lambda, \psi(t; z, \lambda)), \quad \text{with } \varphi \in \mathbb{R}^j, \ \psi \in \mathbb{R}^{n-j}.$$

The following conditions are now imposed on the components φ and ψ of the solution $X(\cdot)$.

For each $\lambda : 0 \leq \lambda < 1$, there are constants

$$t_0 \geq \omega > 0, \quad 0 < A < a < L_\lambda, \quad 0 < K_\lambda < b,$$

such that, for each $z = (u, v) \in \mathbb{R}^n$, we have:

(j_1) $|u| = a$, $|v| \leq b \Rightarrow |\varphi(t; z, 0)| \geq A$,   for all $t \geq 0$,

(actually, a weaker condition, in line with (i_1^*) could be assumed here)

(j_2) $|u| = a$, $|v| \leq b \Rightarrow |\varphi(t; z, \lambda)| \geq L_\lambda$, for all $0 \leq \lambda < 1$, and for all $t \geq t_0$,

(j_3) $|u| \leq a$, $|v| = b \Rightarrow |\psi(t; z, \lambda)| \leq K_\lambda$, for all $0 \leq \lambda < 1$, and for all $t \geq t_0$.

Observe that (j_1)–(j_3) correspond to conditions (7.46.1)–(7.46.3). Of course, nothing prevents the possibility that all the constants are independent of the parameter λ. However, assumptions (j_2), (j_3) allow us to deal with a slightly more general situation.

Let $\mathcal{R} \subset \mathbb{R}^n = \mathbb{R}^j \times \mathbb{R}^{n-j}$ be the open rectangle:

$$\mathcal{R} := B(0, a) \times B(0, b).$$

Now, we can prove Theorem (7.46).

PROOF OF THEOREM (7.46). We use a continuation theorem developed in [CaMaZa, Theorem 4], according to which for the validity of our statement it is sufficient to check that (cf. (7.45$_\lambda$))

$$\deg(G, \mathcal{R}, 0) \neq 0, \quad \text{where } G(X) = F(t, X; 0),$$

and

$$z \neq X(\omega; z, \lambda), \quad \text{for all } 0 \leq \lambda < 1 \text{ and for all } z \in \partial\mathcal{R}.$$

Now, the former property is satisfied, for $t_0 \geq \omega$, by Lemma (7.49), in virtue of the assumptions (j_1)–(j_3) (for $\lambda = 0$), i.e. by (7.46.1)–(7.46.3). With respect to the latter property, we argue as follows:

Take any $z = (u, v) \in \partial\mathcal{R}$ and consider $X(k\omega; z, \lambda) = (\varphi(k\omega; z, \lambda), \psi(k\omega; z, \lambda))$, for any $\lambda \in [0, 1)$. Then, if $u \in \partial B(0, a)$, by (j_2), we have that $|\varphi(k\omega; z, \lambda)| > a$, for a sufficiently big $k \in \mathbb{N}$, and therefore (because of uniqueness) $\varphi(\omega; z, \lambda) \neq u$. On the other hand, if $v \in \partial B(0, b)$, by (j_3) we have that $|\psi(l\omega; z, \lambda)| < b$, for a sufficiently big $l \in \mathbb{N}$, and therefore (again because of uniqueness) $\psi(\omega; z, \lambda) \neq v$.

In any case, we can conclude that $z \neq X(\omega; z, \lambda)$, as required. This completes the proof of Theorem (7.46). □

Theorem (7.46) can be rewritten in a more precise form as follows.

(7.61) THEOREM. *Let $F_0 : \mathbb{R}^n \to \mathbb{R}^n$ be a Carathéodory function such that*

$$F(t, X; 0) = F_0(X), \quad \text{for all } t \geq 0 \text{ and for all } X \in \mathbb{R}^n.$$

Assume that conditions (j_1)–(j_3) *hold, for some* $t_0 \geq \omega > 0$. *Then the equation* (*see* (7.45))

$$\dot{X} = F(t, X), \quad \text{where } F(t, X) \equiv F(t + \omega, X),$$

has at least one ω-periodic solution $X(t)$ such that $X(0) \in \mathcal{R}$.

Clearly, all the assumptions of our theorem are meaningful in the case:

$$1 \leq j \leq n - 1.$$

If $j = 0$ (or, respectively, $j = n$), we consider (j_1) and (j_2) (or, respectively (j_3)) to be vacuously satisfied.

(7.62) REMARK. One can readily check that no transversality on $\partial \mathcal{R}$ is again required. Moreover, uniqueness conditions in (7.46.4) can be avoided by means of the standard limiting argument (see e.g. [Har-M], [Ks2-M]).

(7.63) REMARK. Theorem (7.46) or (7.61) is another example (see Remark (7.34)) of a generalization of N. Levinson's transformation theory. Random versions of these theorems can be obtained by means of Theorem (4.23). In the ordinary case (i.e. for $\tau = 0$), the same is true for Theorems (7.31) and (7.42). For dissipative (i.e. for $m = n$) systems (7.35), an analogy in Banach spaces of Theorem (7.42) can be obtained (cf. [ObZe]), by means of Theorem (4.16), as well.

(7.64) EXAMPLE. Consider the quasi-linear (ordinary) system

$$(7.65) \qquad\qquad \dot{X} + A(t)X = F(t, X)$$

and assume that all assumptions of Theorem (7.42) are satisfied for $A \in C(\mathbb{R}, \mathbb{R}^{n^2})$ as well as for $F \in \mathrm{CAR}(\mathbb{R}^{n+1}, \mathbb{R}^n)$, where $\tau = 0$. Then one can easily check in the proof of Theorem (7.42) that Theorem (7.46) or (7.61) can be applied (see also Remark (7.62)), when considering

$$\dot{X} + A(t)X = \lambda F(t, X), \quad \lambda \in [0, 1].$$

Therefore, the system (7.65), obtained for $\lambda = 1$, admits an ω-periodic solution.

If $F(t, X) \not\equiv F(t + \omega, X)$, then system (7.65) admits (see Corollary (7.43)) a bounded solution on $[0, \infty)$.

Further effective existence criteria like those in Theorem (7.42), Corollary (7.43) and Example (7.64) can be obtained by means of Liapunov-like functions in the next chapter.

III.8. Bounding and guiding functions approach

In this section, Liapunov-like functions will be employed in order solutions of our problems to behave in a desired manner. The material is mainly taken from our papers [And1], [And12], [And14], [AGG1], [AGJ3], [AJ1], [AMT1], [AMT2], [DBGP1], [DBGP2].

Below, the following cases will be treated separately:

- (a) Bounding functions for globally u.s.c. r.h.s.;
- (b) Bounding functions for Carathéodory r.h.s.;
- (c) Guiding functions for globally u.s.c. r.h.s.;
- (d) Guiding functions for r.h.s. without convex values;
- (e) Guiding functions for discontinuous r.h.s.;
- (f) The usage of two guiding functions.

At first, we introduce the notion of a bound set, allowing us to express condition (5.33.3) in Proposition (5.33) in a more explicit way. Proposition (5.33) requires namely a suitable subset $Q \subset C([a, b], \mathbb{R}^n)$ of candidate solutions as well as the verification of the transversality condition (5.33.3), for each associated problem in (5.33.2).

In our opinion, a quite natural way how to construct Q is

$$Q = \{q \in C([a, b], \mathbb{R}^n) \mid q(t) \in \overline{K}(t), \quad \text{for all } t \in [a, b]\},$$

where $\{K(t)\}_{t \in J}$, $J = [a, b]$, denotes a one-parameter family of nonempty and open subsets of $\mathbb{R}^n$ and $\overline{K}(t)$ their closures. We shall always assume that $\{K(t)\}_{t \in J}$ is uniformly bounded, i.e. $|x| < R$, for each $t \in J$ and $x \in K(t)$, where R is a positive constant.

Hence, consider the problem

$$(8.1) \qquad \begin{cases} \dot{x}(t) \in F(t, x(t)), & \text{for a.a. } t \in J = [a, b], \\ x \in S, \end{cases}$$

where $F : J \times \mathbb{R}^n \multimap \mathbb{R}^n$ is a u-Carathéodory mapping with nonempty, convex and compact values and S is a subset of $AC(J, \mathbb{R}^n)$.

(8.2) DEFINITION. We say that $\{K(t)\}_{t \in J}$ is a *bound set* for problem (8.1) if there is no solution $x(t)$ of (8.1) such that $x(t) \in \overline{K}(t)$, for all $t \in J$, and $x(t) \in \partial K(\tau)$, for some $\tau \in J$.

We intend to develop at first a detailed theory for the *Floquet boundary value problem*

$$(8.3) \qquad \begin{cases} \dot{x}(t) \in F(t, x(t)), & \text{for a.a. } t \in [a, b], \\ x(b) = Mx(a), \end{cases}$$

where M denotes an $(n \times n)$ nonsingular matrix.

We shall distinguish separately the cases, where F is a globally u.s.c. or a u-Carathéodory right-hand side (r.h.s.).

(a) Bounding functions for globally u.s.c. r.h.s.

Hence, let $F : [a, b] \times \mathbb{R}^n \multimap \mathbb{R}^n$ be an u.s.c. mapping with nonempty, compact and convex values.

If X is an arbitrary metric space and $A \subset X$ its subset, we shall respectively denote by $\operatorname{int} A$, $\overline{A}$ and ∂A the interior, the closure and the boundary of A.

Moreover, the following notation will be used for a bound set $\{K(t)\}_{t \in J}$:

$$\Gamma_{\partial K} = \{(t, x) \mid t \in J \quad \text{and} \quad x \in \partial K(t)\},$$

and

$$\mathcal{K} = \{(t, x) \mid t \in J \quad \text{and} \quad x \in \overline{K}(t)\}$$

and

$$\partial \mathcal{K} = \{(t, x) \mid t \in J \quad \text{and} \quad x \in \partial K(t)\}.$$

For a real compact interval $J = [a, b]$, $C(J, \mathbb{R}^n)$ will be, as usual, the Banach space of continuous functions $x : J \to \mathbb{R}^n$ endowed with the usual sup norm.

For a point $x \in \mathbb{R}^n$ and a positive constant r, B_x^r denotes the closed ball centered in x and having the radius r; B denotes simply the unit closed ball.

Given a point $(\tau, \xi) \in \Gamma_{\partial K}$, we shall consider a continuous function $V_{(\tau, \xi)} : J \times \mathbb{R}^n \to \mathbb{R}$ such that

$$(8.4) \qquad\qquad V_{(\tau, \xi)}(\tau, \xi) = 0,$$

and

$$(8.5) \qquad\qquad V_{(\tau, \xi)|_{\mathcal{K}}} \le 0, \quad \text{in a neighbourhood of } (\tau, \xi).$$

(8.6) PROPOSITION. *Let $\{K(t)\}_{t \in J}$, $J = [a, b]$, be a nonempty, open and uniformly bounded family of subsets of $\mathbb{R}^n$. Assume that, for every $(\tau, \xi) \in \Gamma_{\partial K}$, there is a continuous function $V_{(\tau, \xi)}$ satisfying (8.4) and (8.5).*

Suppose moreover that, for all $\tau \in \operatorname{int} J$, $\xi \in \partial K(\tau)$ and $w_1, w_2 \in F(\tau, \xi)$ it holds

$$(8.7) \qquad 0 \notin \left[\liminf_{\substack{v \to w_1 \\ h \to 0^+}} \frac{V_{(\tau, \xi)}(\tau + h, \xi + hv)}{h}, \limsup_{\substack{v \to w_2 \\ h \to 0^-}} \frac{V_{(\tau, \xi)}(\tau + h, \xi + hv)}{h} \right].$$

Then all possible solutions $x(t)$ of (8.3) satisfying $x(t) \in \overline{K}(t)$, for every $t \in J$, are such that $x(t) \in K(t)$, for every $t \in$ int J.

PROOF. Assume by contradiction the existence of a solution $x(t)$ of (8.3) and the existence of a point $\tau \in$ int J satisfying $x(t) \in \overline{K}(t)$, for all $t \in J$ and $x(\tau) \in \partial K(\tau)$.

Let I be a compact interval such that $\tau \in I \subset J$. Since $x(I)$ is compact and F is upper semicontinuous, then $F(t, x(t))$ is bounded on I, and consequently x is a Lipschitz function on I. Let us denote by Ω the set of limit points of

$$\frac{x(\tau + h) - x(\tau)}{h}, \quad \text{when } h \to 0^+.$$

According to the lipschitzianity of x in a neighbourhood of the point τ and since we are in the Euclidean space $\mathbb{R}^n$, we get $\Omega \neq \emptyset$.

Taking $w_1 \in \Omega$, there exists a sequence $\{h_n\}_n$ of positive numbers such that $h_n \to 0^+$ and

$$(8.8) \qquad \frac{x(\tau + h_n) - x(\tau)}{h_n} \to w_1, \quad \text{when } n \to \infty.$$

As a consequence of the regularity assumption on F in the point $(\tau, x(\tau))$, we prove now that $w_1 \in F(\tau, x(\tau))$.

In fact, given $\varepsilon > 0$, it is possible to find $\sigma > 0$ such that, whenever $t \in J$ with $|t - \tau| < \sigma$ and $|x - x(\tau)| < \sigma$, then $F(t, x) \subset F(\tau, x(\tau)) + \varepsilon B = \bigcup_{w \in F(\tau, x(\tau))} B_w^\varepsilon$.

By the continuity of x, one can then find $0 < \eta \leq \sigma$ such that $|t - \tau| < \eta$ and $t \in J$ imply $F(t, x(t)) \subset F(\tau, x(\tau)) + \varepsilon B$.

Recalling that F is convex-valued, $F(\tau, x(\tau)) + \varepsilon B$ is a convex subset of $\mathbb{R}^n$. Therefore, for each m sufficiently big, we have

$$\frac{x(\tau + h_m) - x(\tau)}{h_m} = \frac{1}{h_m} \int_\tau^{\tau + h_m} \dot{x}(s)\, ds \in F(\tau, x(\tau)) + \varepsilon B.$$

Finally, since the set $F(\tau, x(\tau))$ is compact, it follows $w_1 \in F(\tau, x(\tau))$.

As a consequence of property (8.8), there exists a sequence $\{\Delta_m\}_m$ such that $\Delta_m \to 0$, as $m \to \infty$, and

$$x(\tau + h_m) = x(\tau) + h_m[w_1 + \Delta_m], \quad \text{for all } m \in \mathbb{N}.$$

Since, for all m, $x(\tau + h_m) \in \overline{K}(\tau + h_m)$, according to (8.5) we obtain, for m sufficiently big, that

$$0 \geq \frac{V_{(\tau, x(\tau))}(\tau + h_m, x(\tau + h_m))}{h_m} = \frac{V_{(\tau, x(\tau))}(\tau + h_m, x(\tau) + h_m[w_1 + \Delta_m])}{h_m}.$$

Therefore, recalling that $\Delta_m \to 0$, when $m \to \infty$, we have

$$(8.9) \qquad 0 \geq \liminf_{m \to \infty} \frac{V_{(\tau, x(\tau))}(\tau + h_m, x(\tau) + h_m[w_1 + \Delta_m])}{h_m}$$

$$\geq \liminf_{\substack{v \to w_1 \\ h \to 0^+}} \frac{V_{(\tau, x(\tau))}(\tau + h, x(\tau) + hv)}{h}.$$

By a similar reasoning, one can show the existence of a vector $w_2 \in F(\tau, x(\tau))$ and of a sequence $\{k_m\}_m$ such that $k_m \to 0^-$ and

$$(8.10) \qquad \frac{x(\tau + k_m) - x(\tau)}{k_m} \to w_2, \quad \text{when } m \to \infty.$$

This yields

$$(8.11) \qquad \liminf_{\substack{v \to w_2 \\ h \to 0^-}} \frac{V_{(\tau, x(\tau))}(\tau + h, x(\tau) + hv)}{h} \geq 0.$$

Notice that inequalities (8.9) and (8.11) are in contradiction with (8.7). Hence, the result is proven. $\qquad\qquad\qquad\qquad\qquad\qquad\qquad\qquad\qquad\qquad\qquad\qquad\square$

(8.12) REMARK. The technique we employed in order to prove the existence of vectors $w_i \in F(\tau, x(\tau))$, with $i = 1, 2$, respectively satisfying (8.8) and (8.10), was previously used by G. Haddad in his seminar paper [Had1] on the viability theory for differential inclusions.

More precisely, let $T_{\mathcal{K}}(\tau, x(\tau))$ be the Bouligand contingent cone of $\mathcal{K}$ (see Chapters III.2, III.4 and cf. [Had1]) in its point $(\tau, x(\tau))$. With our previous reasoning, we actually proved that

$$w_i \in F(\tau, x(\tau)) \cap T_{\mathcal{K}}(\tau, x(\tau)), \quad i = 1, 2,$$

and this is the necessary condition for a viability problem in a finite dimensional space.

(8.13) REMARK. Assume that $V_{(\tau, \xi)}(t, x)$ is locally Lipschitzian in x uniformly with respect to t in the point (τ, ξ), i.e. that there exists a positive constant $L_{(\tau, \xi)}$ such that

$$|V_{(\tau, \xi)}(t, x) - V_{(\tau, \xi)}(t, y)| \leq L_{(\tau, \xi)}|x - y|,$$

for all (t, x), (t, y) in a neighbourhood of (τ, ξ).

For such a function, we can respectively define in a standard manner the *upper right* and the *lower right Dini derivatives* at (τ, ξ) calculated in $(1, w)$ by

$$D^+ V_{(\tau, \xi)}(\tau, \xi)(1, w) = \limsup_{h \to 0^+} \frac{[V_{(\tau, \xi)}(\tau + h, \xi + hw) - V_{(\tau, \xi)}(\tau, \xi)]}{h},$$

$$D_+ V_{(\tau, \xi)}(\tau, \xi)(1, w) = \liminf_{h \to 0^+} \frac{[V_{(\tau, \xi)}(\tau + h, \xi + hw) - V_{(\tau, \xi)}(\tau, \xi)]}{h},$$

as well as the upper left and the lower left Dini derivatives $D^-V_{(\tau,\xi)}(\tau,\xi)(1,w)$ and $D_-V_{(\tau,\xi)}(\tau,\xi)(1,w)$, simply replacing $h \to 0^+$ by $h \to 0^-$ in previous definitions.

According to the lipschitzianity assumption on V, all these four quantities are real numbers. Moreover, for h small enough, it holds

$$V_{(\tau,\xi)}(\tau+h, \xi+hv) \le V_{(\tau,\xi)}(\tau+h, \xi+hw) + L_{(\tau,\xi)}|h| \cdot |v - w|.$$

Therefore, assuming that $x(t)$ is a solution of (8.3) such that $x(\tau) = \xi$, we can reformulate inequalities (8.9) as follows

$$
\begin{aligned}
0 \ge \liminf_{m\to\infty} & \frac{V_{(\tau,x(\tau))}(\tau+h_m, x(\tau)+h_m[w_1+\Delta_m])}{h_m} \\
\ge \liminf_{m\to\infty} & \left[\frac{V_{(\tau,x(\tau))}(\tau+h_m, x(\tau)+h_m w_1)}{h_m} - L_{(\tau,x(\tau))}|\Delta_m| \right] \\
\ge \liminf_{h\to 0^+} & \frac{V_{(\tau,x(\tau))}(\tau+h, x(\tau)+h w_1)}{h} = D_+V_{(\tau,x(\tau))}(\tau, x(\tau))(1, w_1),
\end{aligned}
$$

because of (8.4).

By a similar reasoning, we can replace (8.11) by

$$D_-V_{(\tau,x(\tau))}(\tau, x(\tau))(1, w_2) \le 0.$$

Therefore, when $V_{(\tau,\xi)}$ is locally Lipschitzian in x, a contradiction immediately follows by

(8.14) $0 \notin [D_+V_{(\tau,\xi)}(\tau,\xi)(1,w_1), D^-V_{(\tau,\xi)}(\tau,\xi)(1,w_2)]$, for all $w_1, w_2 \in F(\tau,\xi)$.

On the other hand, one has

$$
\begin{aligned}
\liminf_{\substack{v\to w \\ h\to 0^+}} \frac{V_{(\tau,\xi)}(\tau+h, \xi+hv)}{h} &\le \liminf_{h\to 0^+} \left[\frac{V_{(\tau,\xi)}(\tau+h, \xi+hw)}{h} + L_{V_{(\tau,\xi)}}|v-w| \right] \\
&= D_+V_{(\tau,\xi)}(\tau,\xi)(1,w).
\end{aligned}
$$

In an analogous way, one can prove that

$$\limsup_{\substack{v\to w \\ h\to 0^-}} \frac{V_{(\tau,\xi)}(\tau+h, \xi+hw)}{h} \ge D^-V_{(\tau,\xi)}(\tau,\xi)(1,w).$$

Hence, for every $w_1, w_2 \in F(\tau,\xi)$,

$$[D_+V_{(\tau,\xi)}(\tau,\xi)(1,w_1), D^-V_{(\tau,\xi)}(\tau,\xi)(1,w_2)]$$
$$\subset \left[\liminf_{\substack{v\to w_1 \\ h\to 0^+}} \frac{V_{(\tau,\xi)}(\tau+h, \xi+hv)}{h}, \limsup_{\substack{v\to w_2 \\ h\to 0^-}} \frac{V_{(\tau,\xi)}(\tau+h, \xi+hv)}{h} \right].$$

Consequently, when $V_{(\tau,\xi)}(t,x)$ is locally Lipschitzian in x, (8.14) is a more proper assumption than (8.7), because the regularity allows us to get a contradiction by means of a weaker condition than the one required in the general case.

Moreover, in [Tad] it was shown that, when F is single-valued, (8.14) can be replaced by the even more general one

$$0 \notin [D^+ V_{(\tau,\xi)}(\tau,\xi)(1, F(\tau,\xi)), D_- V_{(\tau,\xi)}(\tau,\xi)(1, F(\tau,\xi))].$$

Finally, when the function $V_{(\tau,\xi)}$ is of class C^1,

$$\lim_{h \to 0} \frac{V_{(\tau,\xi)}(\tau + h, \xi + hw)}{h} = (\nabla V_{(\tau,\xi)}(\tau,\xi)(1, w)),$$

where ∇ denotes the gradient, so condition (8.7) becomes

$$0 \notin [(\nabla V_{(\tau,\xi)}(\tau,\xi)(1, w_1)), (\nabla V_{(\tau,\xi)}(\tau,\xi)(1, w_2))], \quad \text{for all } w_1, w_2 \in F(\tau,\xi).$$

Since the set $F(\tau,\xi)$ is convex, this is equivalent to require

$$(\nabla V_{(\tau,\xi)}(\tau,\xi)(1, w)) \neq 0, \quad \text{for all } w \in F(\tau,\xi).$$

We are now able to give an existence theorem of a bound set for the Floquet problem (8.3).

In order to study the extremal points a and b, we need the following invariance condition

(8.15) $$M \partial K(a) = \{M\xi \mid \xi \in \partial K(a)\} = \partial K(b).$$

We point out that when the bound set is autonomous, i.e. when $K(t) \equiv K$, this is equivalent to the invariance of its boundary with respect to the subgroup of $GL^n(\mathbb{R})$ generated by M, which is a usual assumption in this setting.

(8.16) PROPOSITION. *Let* $\{K(t)\}_{t \in [a,b]}$, *be a nonempty, open and uniformly bounded family of subsets of* $\mathbb{R}^n$. *Assume that, for every* $(\tau,\xi) \in \Gamma_{\partial K}$, *there exists a continuous function* $V_{(\tau,\xi)}$ *satisfying* (8.4), (8.5) *and also* (8.7), *when* $\tau \in (a,b)$.

Suppose, furthermore, the invariance condition (8.15).

Finally, assume that, for any $\xi \in \partial K(a)$, $w_a \in F(a,\xi)$ *and* $w_b \in F(b, M\xi)$,

(8.17) $$0 \notin \left[\liminf_{\substack{v \to w_a \\ h \to 0^+}} \frac{V_{(a,\xi)}(a + h, \xi + hv)}{h}, \limsup_{\substack{v \to w_b \\ h \to 0^-}} \frac{V_{(b,\xi)}(b + h, M\xi + hv)}{h} \right].$$

Then $\{K(t)\}_{t \in [a,b]}$ *is a bound set for* (8.3).

PROOF. According to Proposition (8.6), we only need to show that, whenever (8.3) has a solution $x(t)$ such that $x(t) \in \overline{K}(t)$, for every $t \in [a,b]$, then $x(a) \in K(a)$ and $x(b) \in K(b)$.

Suppose, by contradiction, that this is false. Therefore, according to the invariance condition (8.15), both $x(a) \in \partial K(a)$ and $x(b) \in \partial K(b)$.

Following the same reasoning as in the proof of Proposition (8.6), there exist $w_a \in F(a, x(a))$ such that

$$\liminf_{\substack{v \to w_a \\ h \to 0^+}} \frac{V_{(a,x(a))}(a+h, x(a) + hv)}{h} \leq 0$$

and $w_b \in F(b, x(b))$ such that

$$\limsup_{\substack{v \to w_b \\ h \to 0^-}} \frac{V_{(b,x(b))}(b+h, x(b) + hv)}{h} \geq 0.$$

But $x(b) = Mx(a)$, by the boundary condition, and the contradiction follows from (8.17).   □

(8.18) REMARK. Take $V_{(\tau,\xi)}(t, x)$ locally Lipschitzian in x, uniformly with respect to t in the point (τ, ξ). Reasoning as in Remark (8.13), it is possible to show that (8.17) can be replaced by

$$0 \notin [D_+ V_{(a,\xi)}(a, \xi)(1, w_a), D^- V_{(b, M\xi)}(b, M\xi)(1, w_b)],$$

for all $w_a \in F(a, \xi)$ and $w_b \in F(b, M\xi)$.

Finally, in the C^1-case, it becomes

$$0 \notin [(\nabla V_{(a,\xi)}(a, \xi)(1, w_a)), (\nabla V_{(b, M\xi)}(b, M\xi)(1, w_b))].$$

(8.19) REMARK. In the theory of bound sets, a collection of continuous functions $V_{(\tau,\xi)}$, satisfying assumptions (8.4), (8.5), (8.7) and (8.17), is usually said to be a family of *bounding functions* for problem (8.3).

Existence results for Floquet problems, obtained by means of Proposition (5.33) in Chapter III.5, require in particular the knowledge of a subset $Q \subset C([a, b], \mathbb{R}^n)$ having no solution of the associated quasi-linearized problems on its boundary.

A quite natural way to construct such a Q is the following one:

(8.20) $Q = \{q \in C([a, b], \mathbb{R}^n) \mid q(t) \in \overline{K}(t), \quad \text{for all } t \in [a, b]\},$

where $\{K(t)\}_{t \in [a,b]}$ denotes a family of nonempty, open uniformly bounded subsets of $\mathbb{R}^n$. The following proposition gives sufficient conditions on such a Q in order that no solution of (8.3) belongs to its boundary. Since it makes use of a family of bounding functions, it is an application of previous investigations.

(8.21) PROPOSITION. *Let* $\{K(t)\}_{t \in [a,b]}$ *be a family of nonempty, open and uniformly bounded subsets of* $\mathbb{R}^n$ *and define* $Q \subset C([a,b], \mathbb{R}^n)$ *as in* (8.20). *Let* $\Gamma_{\partial K}$ *be closed in* $\mathbb{R}^{n+1}$ *and for each* $(\tau, \xi) \in \Gamma_{\partial K}$, *assume the existence of a continuous function* $V_{(\tau, \xi)} : [a,b] \times \mathbb{R}^n \to \mathbb{R}$ *satisfying* (8.4), (8.5), (8.7) *and* (8.17).

Then, Floquet problem (8.3) *has no solution on* ∂Q.

PROOF. First of all, we show that every function x in ∂Q admits at least a point $\tau = \tau_x \in [a,b]$ such that $x(\tau) \in \partial K(\tau)$. For this, suppose that $x(t) \in K(t)$, for every $t \in [a,b]$, and consider the function

$$d : [a,b] \to \mathbb{R}^+, \quad t \mapsto \text{dist}\{x(t), \partial K(t)\}.$$

Given $t_0 \in [a,b]$, take a sequence $\{t_m\}_m$ converging to t_0 and such that $d(t_m)$ converges to a real number l. Let us prove that $d(t_0) \leq l$.

By the boundedness of $\overline{K}(t)$, it follows the compactness of $\partial K(t)$, for each t. Hence, for every m, there exists $y_m \in \partial K(t_m)$ such that $d(t_m) = |x(t_m) - y_m|$. Moreover, since $\overline{K}(t)$ is uniformly bounded, then $\{y_m\}_m$ has a subsequence, again denoted for the sake of simplicity by $\{y_m\}_m$, which converges to a point $y_0 \in \mathbb{R}^n$. Notice that $(t_m, y_m) \in \Gamma_{\partial K}$, for each m; the closure of $\Gamma_{\partial K}$ then implies $y_0 \in \partial K(t_0)$.

Therefore, by the definition of the function d, we have

$$l = |x(t_0) - y_0| \geq d(t_0).$$

We have so obtained that d is a lower semicontinuous function on $[a,b]$. Hence, $d([a,b])$ has a minimum d_0, and since we assumed $x(t) \in K(t)$, for all t, d_0 must be positive. Therefore, $B_{x(t)}^{d_0} \subset K(t)$, for each $t \in [a,b]$, implying $y \in Q$, for any function $y \in C([a,b], \mathbb{R}^n)$ having $|x - y| < d_0$. Thus, $x \in \text{int } Q$.

Let us consider now a function x in ∂Q. According to its definition, Q is closed in the Banach space $C([a,b], \mathbb{R}^n)$. Hence, $x \in Q$ and so $x(t) \in \overline{K}(t)$, for each $t \in [a,b]$. As a consequence of the previous reasoning, there exists $\tau = \tau_x \in [a,b]$ such that $x(\tau) \in \partial K(\tau)$. By Proposition (8.16), the family $\{K(t)\}_{t \in [a,b]}$ is a bound set for (8.3). Thus, $x(t)$ cannot be a solution of (8.3), and the result is proved. $\square$

The present example deals with sufficient conditions for the existence of bound sets for the autonomous Floquet boundary value problem

$$(8.22) \qquad \begin{cases} \dot{x} \in F(x), & \text{for a.a. } t \in [a,b], \\ x(b) = Mx(a), \end{cases}$$

where $F : \mathbb{R}^n \multimap \mathbb{R}^n$ is an upper semicontinuous multivalued function with nonempty, compact and convex values.

Since the problem is autonomous, in our opinion, it is natural to look for a bound set K constant in time. Therefore, the family of bounding functions will be also taken independent of time. We shall study separately the case of a convex set K, when a family of C^1-bounding functions arises naturally, and the opposite one, where we have to look for a less regular bounding function.

More precisely, for each $\xi \in \partial K$, we shall consider a continuous function $V_\xi :$ $\mathbb{R}^n \to \mathbb{R}$ such that

$$(8.23) \qquad\qquad\qquad V_\xi(\xi) = 0$$

and

$$(8.24) \qquad\qquad V_\xi|_{\overline{K}} \leq 0, \quad \text{in a neighbourhood of } \xi.$$

Hence, Proposition (8.16) can be reformulated as follows.

(8.25) PROPOSITION. *Let K be a nonempty, open and bounded subset of $\mathbb{R}^n$ having an invariant boundary with respect to the subgroup of $GL^n(\mathbb{R})$ generated by M. Assume that, for every $\xi \in \partial K$, there exists a continuous function V_ξ satisfying* (8.23) *and* (8.24).

Suppose, moreover, that, for all $\xi \in \partial K$ and $w_1, w_2 \in F(\xi)$, it holds

$$(8.26) \qquad 0 \notin \left[\liminf_{\substack{v \to w_1 \\ h \to 0^+}} \frac{V_\xi(\xi + hv)}{h}, \limsup_{\substack{v \to w_2 \\ h \to 0^-}} \frac{V_\xi(\xi + hv)}{h} \right].$$

Finally, assume that, for any $\xi \in \partial K$, $w_a \in F(\xi)$ and $w_b \in F(M\xi)$,

$$(8.27) \qquad 0 \notin \left[\liminf_{\substack{v \to w_a \\ h \to 0^+}} \frac{V_\xi(\xi + hv)}{h}, \limsup_{\substack{v \to w_b \\ h \to 0^-}} \frac{V_{M\xi}(M\xi + hv)}{h} \right].$$

Then K is a bound set for (8.22).

We first consider the case in which K is convex. Geometrically, this means that, for every $\xi \in \partial K$, there exist a vector n_ξ, not necessarily unique, and a neighbourhood U_ξ of ξ such that

$$(8.28) \qquad\qquad (n_\xi, (x - \xi)) \leq 0, \quad \text{for all } x \in U_\xi \cap \overline{K}.$$

Let V_ξ be the C^1-function defined by

$$V_\xi : \mathbb{R}^n \to \mathbb{R}, \quad x \mapsto (n_\xi, (x - \xi)).$$

It follows immediately from (8.28) that V_ξ satisfies (8.23) and (8.24).

Moreover, $\nabla V_\xi(\xi) = n_\xi$. Hence, recalling the discussion in Remark (8.13) and Remark (8.18) for a C^1-bounding function, (8.26) and (8.27) are respectively equivalent to

$$(8.29) \qquad (n_\xi, w) \neq 0, \quad \text{for all } w \in F(\xi),$$

and

$$(8.30) \qquad 0 \notin [(n_\xi, w_a), (n_{M\xi}, w_b)], \quad \text{for all } w_a \in F(\xi), \ w_b \in F(M\xi).$$

Consider now the case when K is not locally convex, in some ξ of its boundary. Then, as shown in [Tad, Example 4.2], for differential equations in $\mathbb{R}^2$, in general it is not possible to satisfy for a C^1-function V_ξ at the same time (8.23), (8.24) and (8.26).

On the other hand, take the Lipschitzian function

$$V_\xi(x) = \text{dist}(x, \overline{K}),$$

which trivially verifies both (8.23) and (8.24). Again, by Remark (8.13), (8.26) is equivalent to

$$(8.31) \qquad 0 \notin \left[\liminf_{h \to 0^+} \frac{\text{dist}(\xi + hw_1, \overline{K})}{h}, \limsup_{h \to 0^-} \frac{\text{dist}(\xi + hw_2, \overline{K})}{h} \right],$$

for all $w_1, w_2 \in F(\xi)$. The nonnegativity of the distance function implies that the previous condition is satisfied if and only if at least one between the left and the right extremes of the interval is different from zero, since they are always respectively nonnegative and nonpositive.

It is easy to prove that

$$\limsup_{h \to 0^-} \frac{\text{dist}(\xi + hw_2, \overline{K})}{h} \neq 0$$

if and only if

$$\liminf_{h \to 0^+} \frac{\text{dist}(\xi - hw_2, \overline{K})}{h} \neq 0.$$

Therefore, recalling the definition of the Bouligand contingent cone $T_{\overline{K}}(\xi)$ of $\overline{K}$ in ξ (see (2.48), (4.9), cf. also [Had1]), (8.31) is equivalent to

$$w_1 \notin T_{\overline{K}}(\xi) \quad \text{or} \quad (-w_2) \notin T_{\overline{K}}(\xi), \quad \text{for all } w_1, w_2 \in F(\xi).$$

If $w \notin T_{\overline{K}}(\xi)$, for all $w \in F(\xi)$, then (8.31) holds.

On the contrary, if there exists $w_1 \in F(\xi) \cap T_{\overline{K}}(\xi)$, (8.31) is verified if and only if $(-w) \notin T_{\overline{K}}(\xi)$, for all $w \in F(\xi)$.

Hence, we get that (8.26) is equivalent to

$$(8.32) \qquad F(\xi) \cap T_{\overline{K}}(\xi) = \emptyset \quad \text{or} \quad (-F(\xi)) \cap T_{\overline{K}}(\xi) = \emptyset.$$

Finally, if K is not locally convex also in $M\xi$, we can define $V_{M\xi}$, identically equal to V_ξ and, by a reasoning similar to the previous one, taking into account Remark (8.18), one can show that (8.27) is equivalent to

$$(8.33) \qquad F(\xi) \cap T_{\overline{K}}(\xi) = \emptyset \quad \text{or} \quad (-F(M\xi)) \cap T_{\overline{K}}(M\xi) = \emptyset.$$

As already pointed out, we are interested in studying retracts of the space $C([a,b],\mathbb{R}^n)$. We consider, in particular, subsets Q of $C([a,b],\mathbb{R}^n)$ of the type given by (8.20) and we give sufficient conditions on the sets $K(t)$ in order that Q is such a retract.

(8.34) LEMMA. *Let $\{\overline{K}(t)\}_{t \in [a,b]}$ be a family of nonempty subsets of $\mathbb{R}^n$.*

Assume that the graph $\mathcal{K}$ of the map $t \multimap \overline{K}(t)$ is a retract of $[a,b] \times \mathbb{R}^n$ and admits a retraction $\varphi : [a,b] \times \mathbb{R}^n \to \mathcal{K}$ which is the identity on its first component, i.e. that

$$\varphi(t,x) = (t, \widetilde{\varphi}(t,x)), \quad \text{for all } (t,x) \in [a,b] \times \mathbb{R}^n,$$

where $\widetilde{\varphi} : [a,b] \times \mathbb{R}^n \to \mathbb{R}^n$ is a continuous function. Then $Q = \{q \in C([a,b],\mathbb{R}^n) \mid q(t) \in \overline{K}(t), \text{ for all } t \in [a,b]\}$ is a retract of the space $C([a,b],\mathbb{R}^n)$.

PROOF. Let us consider the function

$$\varphi' : C([a,b],\mathbb{R}^n) \to Q,$$
$$q \to \varphi'(q) : [a,b] \to \mathbb{R}^n,$$
$$t \to \widetilde{\varphi}(t,q(t)).$$

From the definition of Q and the properties of φ, it immediately follows that φ' is well-defined and that $\varphi'(q) = q$, for every $q \in Q$.

Take now $\{q_n\}_n$ converging to $q \in C([a,b],\mathbb{R}^n)$. Then there exists a compact subset $T \subset \mathbb{R}^n$ such that $q(t)$ and $q_m(t)$ belong to T, for every t and every m. The continuity of $\widetilde{\varphi}$ implies its uniform continuity in $[a,b] \times T$. Hence, by the convergence of q_m to q, we get the convergence of $\varphi'(q_m)$ to $\varphi'(q)$ and also the continuity of φ' is proved. $\qquad \square$

(8.35) REMARK. If $\overline{K}(t)$ is convex (which trivially implies the same property for the set Q), then the associated bounding functions can be taken smooth in x, as pointed out both in previous section and in [GaM-M, p. 43] (cf. also the main converse theorem in [ClkLS]).

We consider now the following Floquet boundary value problem

$$(8.36) \qquad \begin{cases} \dot{x} + A(t)x \in F(t,x), & \text{for a.a. } t \in [a,b], \\ x(b) = Mx(a), \end{cases}$$

where $A : [a,b] \to \mathbb{R}^n \times \mathbb{R}^n$ is a continuous $(n \times n)$ matrix, $F : [a,b] \times \mathbb{R}^n \to \mathbb{R}^n$ is a globally upper semicontinuous multivalued function with nonempty, compact and convex values and M is a regular $(n \times n)$ matrix. Under appropriate conditions on A and F and with the fixed point technique described in Proposition (5.33), we prove its solvability. Notice, in particular, that we define the set Q of candidate solutions (see condition (d) below) as in (8.20) and use the bounding functions approach developed above in order to show the transversality condition (5.33.3) of Proposition (5.33).

(8.37) THEOREM. *Let us consider the Floquet boundary value problem* (8.26). *Assume that*

(a) *the associated homogeneous problem*

$$\begin{cases} \dot{x} + A(t)x = 0, & \text{for a.a. } t \in [a,b], \\ x(b) = Mx(a) \end{cases}$$

has only the trivial solution,

(b) *there exists a globally upper semicontinuous multivalued function* $G : [a,b] \times \mathbb{R}^n \times \mathbb{R}^n \times [0,1] \multimap \mathbb{R}^n$ *with nonempty, compact and convex values such that*

$$G(t,c,c,1) \subset F(t,c), \quad \text{for all } (t,c) \in [a,b] \times \mathbb{R}^n,$$

(c) *there exists a family* $\{K(t)\}_{t \in [a,b]}$ *of nonempty, open and uniformly bounded subsets of* $\mathbb{R}^n$ *such that the graph* $\mathcal{K}$ *of the map* $t \multimap \overline{K}(t)$ *is a retract of* $[a,b] \times \mathbb{R}^n$ *with a retraction* $\varphi : [a,b] \times \mathbb{R}^n \to \mathcal{K}$ *which is the identity on its first component, i.e.* $\varphi(t,x) = (t, \widetilde{\varphi}(t,x))$; *assume, moreover, that the graph of its boundary* $\Gamma_{\partial K} = \{(t,x) \mid t \in [a,b], \ x \in \partial K(t)\}$ *is closed,*

(d) $G(t, \cdot, q, \lambda)$ *is Lipschitzian with a sufficiently small Lipschitz constant, for each* $t \in [a,b]$ *and* $(q,\lambda) \in Q \times [0,1]$, *where* $Q = \{q \in C([a,b], \mathbb{R}^n) \mid q(t) \in \overline{K}(t), \text{ for all } t \in [a,b]\}$,

(e) *there exists a Lebesgue integrable function* $\alpha : [a,b] \to \mathbb{R}$ *such that*

$$|G(t,x(t),q(t),\lambda)| \le \alpha(t), \quad \text{for a.a. } t \in [a,b],$$

for any $(x, q, \lambda) \in \Gamma_T$, where T is the multivalued function which assigns to any $(q, \lambda) \in Q \times [0, 1]$ the set of solutions of

$$\begin{cases} \dot{x} + A(t)x \in G(t, x, q(t), \lambda), & \text{for a.a. } t \in [a, b], \\ x(b) = Mx(a) \end{cases}$$

and Γ_T its graph,

(f) $T(Q \times \{0\}) \subset Q$ *and ∂Q is fixed point free, i.e.*

$$\{q \in Q \mid q \in T(q, 0)\} \cap \partial Q = \emptyset,$$

(g) *for every $(\tau, \xi) \in \Gamma_{\partial K}$, there exists a continuous function $V_{(\tau, \xi)} : [a, b] \times \mathbb{R}^n \to \mathbb{R}$ satisfying (8.4) and (8.5),*

(h) *for every $\tau \in (a, b)$, $\xi \in \partial K(\tau)$, $(q, \lambda) \in Q \times (0, 1]$ and $w_1, w_2 \in G(\tau, \xi, q(\tau), \lambda) - A(\tau)\xi$, one has*

$$0 \notin \left[\liminf_{\substack{v \to w_1 \\ h \to 0^+}} \frac{V_{(\tau, \xi)}(\tau + h, \xi + hv)}{h}, \limsup_{\substack{v \to w_2 \\ h \to 0^-}} \frac{V_{(\tau, \xi)}(\tau + h, \xi + hv)}{h} \right],$$

(i) $M\partial K(a) = \{M\xi \mid \xi \in \partial K(a)\} = \partial K(b)$,

(j) *for any $\xi \in \partial K(a)$, $(q, \lambda) \in Q \times (0, 1]$, $w_a \in G(a, \xi, q(a), \lambda) - A(a)\xi$ and $w_b \in G(b, M\xi, q(b), \lambda) - A(b)M\xi$, it holds*

$$0 \notin \left[\liminf_{\substack{v \to w_a \\ h \to 0^+}} \frac{V_{(a, \xi)}(a + h, \xi + hv)}{h}, \limsup_{\substack{v \to w_b \\ h \to 0^-}} \frac{V_{(b, M\xi)}(b + h, M\xi + hv)}{h} \right].$$

Then (8.26) admits a solution.

PROOF. Let us prove that all the assumptions of Proposition (5.33) are satisfied (see also Remark (5.34)). Define the set Q as in (d). The continuity of A and the global upper semicontinuity of G are sufficient to get the global upper semicontinuity of $G(t, x, q(t), \lambda) - A(t)x$, for every fixed $(q, \lambda) \in Q \times [0, 1]$. Therefore, also thanks to conditions (a) and (d) and the continuity of A, we are able to apply Theorem (3.8) to each quasi-linearized associated problem

$$\begin{cases} \dot{x} \in G(t, x, q(t), \lambda) - A(t)x, & \text{for a.a. } t \in [a, b], \\ x(b) = Mx(a) \end{cases}$$

and to assure its solvability with an R_δ-set of solutions, i.e. that the set of solutions is, in particular, nonempty, compact and connected (hence, lying in some B_0^R, where R is sufficiently big). Moreover, it follows from the proof of Theorem (3.8) (see [ADG]) that the ball B_0^R can be taken the same, for all $q \in Q$.

Taking $S_1 = B_0^R \cap \{x \in AC([a,b], \mathbb{R}^n) \mid x(b) = Mx(a)\}$, the boundedness of B_0^R implies the same property for S_1. Moreover, according to (c) and Lemma (8.34), Q is a retract of $C([a,b], \mathbb{R}^n)$. Since $Q \setminus \partial Q$ is nonempty, condition (5.33.1) of Proposition (5.33) holds.

Assumptions from (g) to (j) guarantee that $\{K(t)\}_{t \in [a,b]}$ is a bound set for each problem

$$\begin{cases} \dot{x} \in G(t, x, q(t), \lambda) - A(t)x, & \text{for a.a. } t \in [a,b], \\ x(b) = Mx(a). \end{cases}$$

Therefore, since by (c) $\Gamma_{\partial K}$ is closed, according to Proposition (8.21), for $\lambda \in (0,1]$, we have $T(Q \times [0,1]) \cap \partial Q) = \emptyset$. This implies condition (5.33.3) of Proposition (5.33) (for $\lambda = 0$, it follows from (f)) and the proof is complete. $\square$

(8.38) REMARK. Because of the method used to solve problem (8.36), we obtained in particular solutions belonging to the set Q. Consequently, previous theorem gives an existence result for the Floquet viability problem

$$\begin{cases} \dot{x} + A(t) \in F(t,x), & \text{for a.a. } t \in [a,b], \\ x(a) = Mx(b), \\ x(t) \in K(t), & \text{for all } t \in [a,b]. \end{cases}$$

(8.39) REMARK. Theorem (8.37) can be reformulated in the sense that, instead of (c) and (f), we can assume respectively (c') an (f') or (c") an (f") as follows:

(c') $\{K(t)\}_{t \in [a,b]}$ is a suitable (i.e. w.r.t. conditions (d)–(j)) family of nonempty, open and uniformly bounded subsets of $\mathbb{R}^n$ having $\Gamma_{\partial K}$ closed in $\mathbb{R}^{n+1}$,

(f') $T(Q \times \{0\}) = \{q_0\} \subset Q \setminus \partial Q$,

or

(c") $\{K(t)\}_{t \in [a,b]}$ is a suitable family of nonempty, open and uniformly bounded and convex subsets of $\mathbb{R}^n$ with $\Gamma_{\partial K}$ closed in $\mathbb{R}^{n+1}$,

(f") $T(Q \times \{0\}) \subset Q$ and ∂Q is fixed point free.

(8.40) EXAMPLE. Consider the anti-periodic problem

(8.41)
$$\begin{cases} \dot{x} \in F_1(t,x) + F_2(t,x), \\ x(a) = -x(b), \end{cases}$$

where $x = (x_1, \ldots, x_n)$, $F = F_1 + F_2 = (f_{11}, \ldots, f_{1n}) + (f_{21}, \ldots, f_{2n})$, $F_1, F_2 : [a,b] \times \mathbb{R}^n \multimap \mathbb{R}^n$ are globally upper semicontinuous multivalued functions which are bounded in $t \in [a,b]$, for every $x \in \mathbb{R}^n$, and linearly bounded in $x \in \mathbb{R}^n$, for every $t \in [a,b]$.

Assume, furthermore, that there exist positive constants R_i, $i = 1, \ldots, n$ such that

$$(8.42) \qquad |f_{1i}(t, x(\pm R_i))| > \max_{\substack{t \in [a,b] \\ x \in \overline{K}}} |f_{2i}(t, x)|, \quad i = 1, \ldots n, \ t \in (a, b),$$

where $x(\pm R_i) = (x_1, \ldots, x_{i-1}, \pm R_i, x_{i+1}, \ldots, x_n)$, $|x_j| \leq R_j$ and $K = \{x \in \mathbb{R}^n \mid |x_i| < R_i, \ i = 1, \ldots, n\}$,

$$(8.43) \quad [f_{1i}(a, x(\pm R_i)) + f_{2i}(a, y)] \cdot [f_{1i}(b, -x(\pm R_i)) + f_{2i}(b, z)] < 0, \quad i = 1, \ldots, n,$$

where $x, y, z \in \overline{K}$,

$$(8.44) \qquad F_1(t, \cdot) \text{ is Lipschitzian with a sufficiently small constant } L,$$

for every $t \in [a, b]$; in the single-valued case, when $F \in C([a, b] \times \mathbb{R}^n, \mathbb{R}^n)$, it is enough to take $L \leq \pi/(b - a)$ (cf. [And14]).

In order to apply Theorem (8.37) for the solvability of (8.41), let us still consider the enlarged family of problems

$$(8.41_q) \qquad \begin{cases} \dot{x} \in \lambda F_1(t, x) + \lambda F_2(t, q(t)), \quad \lambda \in [0, 1], \\ x(a) = -x(b), \end{cases}$$

where $q \in Q = \{\tilde{q} \in C([a, b], \mathbb{R}^n) \mid \tilde{q}(t) \in \overline{K}, \text{ for all } t \in [a, b]\}$.

Observe that if $\xi \in \partial K$, then

$$\xi = \xi(\pm R_i) = (\xi_1, \ldots, \xi_{i-1}, \pm R_i, \xi_{i+1}, \ldots, \xi_n),$$

for some i and $|\xi_j| \leq R_j$ for all $j \neq i$. Therefore, let us define for (8.41_q) the bounding functions as (cf. [GaM-M, p. 78]) $V_\xi(x) = \pm x_i - R_i$, $i = 1, \ldots, n$ where $\xi = \xi(\pm R_i) \in \partial K$.

One can easily check that K is a bound set for (8.41_q), provided (8.42)–(8.44) hold. Indeed, making also use of the discussions both in Remark (8.13) and in Remark (8.18), we have

ad (g) $V_\xi(\xi) = 0$ and $V_\xi(\xi) \leq 0$, for $x \in \overline{K}$,

ad (h) $(\nabla V_\xi(\xi)(\lambda F_1(t, \xi) + \lambda F_2(t, q(t)))) = \pm \lambda[f_{1i}(t, \xi) + f_{2i}(t, q(t))] \neq 0$, for all $t \in (a, b)$, where $\lambda \in (0, 1]$, according to (8.44),

ad (j) $(\nabla V_\xi(\xi)(\lambda F_1(a, \xi) + \lambda F_2(a, q(a)))) \cdot (\nabla V_\xi(\xi)(\lambda F_1(b, -\xi) + \lambda F_2(b, q(b)))) = -\lambda^2[f_{1i}(a, \xi) \cdot f_{2i}(a, q(a))] \cdot [f_{1i}(b, -\xi) + f_{2i}(b, q(b))] > 0$, where $\lambda \in (0, 1]$, according to (8.43).

The other conditions in Theorem (8.37) are satisfied as follows:

ad (a), (f) the associated homogeneous problem ($\lambda = 0$)

$$(8.41_0) \qquad \begin{cases} \dot{x} = 0, \\ x(a) = -x(b) \end{cases}$$

has only the trivial solution $x(t) = T(q, \{0\}) \equiv 0$, by which $T(Q \times \{0\}) \equiv 0 \in Q$,

ad (b) $G(t, x, q, \lambda) = \lambda F_1(t, x) + \lambda F_2(t, q)$; hence, $F_1(t, c) + F_2(t, c) = F(t, c)$,

ad (c) since K is convex, it is an absolute retract, and so a retract of $\mathbb{R}^n$,

ad (d) it follows immediately from (8.44),

ad (e) it is satisfied by the hypothesis concerning the growth restrictions imposed on F_1 and F_2,

ad (i) $\partial K(a) = -\partial K(b)$, where $\partial K = \{\xi \in \mathbb{R}^n \mid (\xi_1, \dots, \xi_{i-1}, \pm R_i, \xi_{i+1}, \dots, \xi_n), |\xi_j| \leq R_j\}$.

Hence, applying Theorem (8.37), anti-periodic problem (8.41) admits a solution, provided (8.42)–(8.44) take place jointly with the above growth restrictions on F_1 and F_2.

(b) Bounding functions for Carathéodory r.h.s..

In this part, we are interested in finding verifiable conditions assuring that $\{K(t)\}_{t \in [a,b]}$ is a bound set for the boundary value problem (8.2), where $F : [a,b] \times \mathbb{R}^n \multimap \mathbb{R}^n$ is this time an (upper) Carathéodory mapping with nonempty, compact and convex values. For this aim, we shall introduce again a family of so called bounding functions $\{V_{(\tau, \xi)}\}_{(\tau, \xi) \in \partial \mathcal{K}}$, satisfying suitable transversality conditions. Let us recall that, given $\tau \in [a, b]$ and $\xi \in \partial K(\tau)$, as usual, by a *bounding function* we mean a continuous function $V_{(\tau, \xi)} : [a, b] \times \mathbb{R}^n \to \mathbb{R}$ satisfying

$$(8.45) \qquad \begin{aligned} & V_{(\tau, \xi)}(\tau, \xi) = 0, \\ & V_{(\tau, \xi)}(t, x) \leq 0, \quad \text{for all } (t, x) \in \overline{\mathcal{K}}, \text{ in a neighbourhood of } (\tau, \xi). \end{aligned}$$

We reason by a contradiction, i.e. we assume the existence of a solution of (8.2), having the graph entirely contained in $\overline{\mathcal{K}}$ and touching $\partial \mathcal{K}$ at some point $(\tau, x(\tau))$. Denoting by f the composed function $V_{(\tau, x(\tau))}(\cdot, x(\cdot))$, we have, according to (8.45), that f is non-positive in a suitable neighbourhood of the point τ and $f(\tau) = 0$. Assuming various transversality conditions on the bounding functions, we then get various monotonicity properties on f and this yields, in all cases, the required contradiction with (8.45).

In particular, when the multivalued function F is globally upper semicontinuous, then a local monotonicity in τ is sufficient, as one could see in the previous part. On the contrary, when F is u-Carathéodory, we need to assume the decreasing monotonicity of f in a left neighbourhood of τ or the increasing monotonicity in a right one. In conclusion, the theory of bound sets is based on the investigation of local, i.e. in one point, or global monotonicity properties of real continuous functions.

Consider, therefore, a u-Carathéodory r.h.s. F. Again, when the bounding funtions $V_{(\tau,\xi)}(t,x)$ are more regular, i.e. at least locally Lipschitzian in all variables, a sign condition on their Dini directional derivatives, i.e. (8.45), guarantees that $\{K(t)\}_{t\in[a,b]}$ is a bound set. We investigate this situation in Proposition (8.50). On the contrary, in the case of general continuous bounding functions, we need to impose certain growth restrictions on F. We also need a monotonicity result on continuous real functions. With these two ingredients, we are then able to show suitable general transversality conditions on $V_{(\tau,\xi)}$, again assuring that $\{K(t)\}_{t\in[a,b]}$ is a bound set for the Floquet boundary value problem. When the growth restriction on F does not depend on t, then the criterion can be improved, and the existence of a bound set simply follows by sign conditions on suitable contingent derivatives of $V_{(\tau,\xi)}$.

As usual in this setting, we introduce the following invariance condition

$$(8.49) \qquad M\partial K(a) = \{M\xi \mid \xi \in \partial K(a)\} = \partial K(b).$$

(8.50) PROPOSITION. *Let $\{K(t)\}_{t\in[a,b]}$ be a family of non-empty, open and uniformly bounded subsets of $\mathbb{R}^n$, satisfying the invariance condition (8.49). Assume that, for every $(\tau,\xi) \in \partial\mathcal{K}$, there is a function $V_{(\tau,\xi)}$, locally Lipschitzian in all variables, satisfying (8.45).*

Suppose, moreover, that for all $\tau \in (a,b], t \leq \tau, x \in \overline{K(t)}$, with (t,x) in a neighbourhood of (τ,ξ),

$$(8.51) \qquad \liminf_{h\to 0^-} \frac{V_{(\tau,\xi)}(t+h, x+hw) - V_{(\tau,\xi)}(t,x)}{h} < 0, \quad \text{for all } w \in F(t,x).$$

Then $\{K(t)\}_{t\in[a,b]}$ is a bound set for (8.3).

PROOF. Let $x : [a,b] \to \mathbb{R}^n$ be a solution of (8.3) satisfying $x(t) \in \overline{K(t)}$, for every $t \in [a,b]$, and assume by a contradiction the existence of $\tau \in [a,b]$ such that $x(\tau) \in \partial K(\tau)$.

According to the invariance condition (8.49), $x(a) \in \partial K(a)$ if and only if $x(b) \in \partial K(b)$. Therefore, there is no loss of generality to assume that $\tau \in (a,b]$.

Let us consider $V_{(\tau,x(\tau))}$ and denote it, throughout this proof, simply by V. By (8.45) and the absolute continuity both of V and of x, it follows the existence of $\overline{h} < 0$ such that

$$(8.52) \qquad 0 \leq -V(\tau+\overline{h}, x(\tau+\overline{h})) = \int_{\tau+\overline{h}}^{\tau} \frac{d}{dt} V(s, x(s)) \, ds.$$

Moreover, since x is a solution of (8.3), for almost every $t \in [\tau + \overline{h}, \tau]$, there exist both a function $\Delta(h)$, which in fact depends on t, such that $\Delta(h) \to 0$, when $h \to 0$, and a vector $w \in F(t, x(t))$ satisfying, for h small enough,

$$(8.53) \qquad x(t + h) = x(t) + h[w + \Delta(h)].$$

Hence, the local lipschitzianity of V combined with (8.51) implies that, for almost every $t \in [\tau + \overline{h}, \tau]$,

$$\frac{d}{dt} V(t, x(t)) = \lim_{h \to 0^-} \frac{V(t + h, x(t) + h[w + \Delta(h)]) - V(t, x(t))}{h} =$$

$$= \liminf_{h \to 0^-} \frac{V(t + h, x(t) + h[w + \Delta(h)]) - V(t, x(t))}{h} \leq$$

$$\leq \liminf_{h \to 0^-} \left[\frac{V(t + h, x(t) + hw) - V(t, x(t))}{h} + L_{(\tau, x(\tau))} |\Delta(h)| \right] < 0,$$

and this leads to a contradiction with (8.52).   □

(8.54) REMARK. For $\tau < b$, the conclusion of Proposition (8.50) also holds, when replacing the assumption (8.51) by

$$(8.51') \qquad \limsup_{h \to 0^+} \frac{V_{(\tau, \xi)}(t + h, x + hw) - V_{(\tau, \xi)}(t, x)}{h} > 0, \quad \text{for all } w \in F(t, x),$$

for all $t \geq \tau$ and $x \in \overline{K(t)}$, with (t, x) in a neighbourhood of (τ, ξ). It is in fact possible to prove, similarly as in the previous theorem, that (8.51') implies the increasing monotonicity of $V_{(\tau, x(\tau))}(\cdot, x(\cdot))$ in a right neighbourhood of τ.

(8.55) REMARK. Notice that, when $V_{(\tau, \xi)}$ is a function from the class C^1, then both (8.51) and (8.51') reduce to

$$(8.51'') \qquad (\nabla V_{(\tau, \xi)}(t, x), (1, w)) \neq 0, \quad \text{for all } w \in F(t, x).$$

In [MW2, Theorem 2.1], an existence result for the periodic boundary value problem associated to a Carathéodory differential equation is given by means of a C^1-guiding-like function. They consider an autonomous bound set defined as the counter image of the negative real line through a unique bounding function $V : \mathbb{R}^n \to \mathbb{R}$ from the class C^1 and they ask V to satisfy the following condition

$$(8.56) \qquad (\nabla V(x), f(t, x)) < 0, \quad \text{for all } (t, x) \in [a, b] \times V^{-1}([-\varepsilon, 0]),$$

where ε is a positive constant.

Since we can reformulate (8.51'') as $(\nabla V(x), w) \neq 0$, for all $t \in (a, b]$, $x \in \overline{K}$ in a neighbourhood of ∂K and $w \in F(t, x)$, (8.51'') becomes trivially a generalization of (8.56) to the multivalued case and to non-autonomous bound sets defined by means of a family of bounding functions, instead of a guiding function.

(8.57) REMARK. In the previous part the existence of a bound set is proved, for the Floquet problem (8.3), when F is globally upper semicontinuous. In Remark (8.13), a bound set is obtained with assumption (8.51) replaced by

$$0 \notin \left[\liminf_{h \to 0^+} \frac{V_{(\tau,\xi)}(\tau + h, \xi + hw_1)}{h}, \limsup_{h \to 0^-} \frac{V_{(\tau,\xi)}(\tau + h, \xi + hw_2)}{h} \right],$$

for all $\tau \in (a,b)$, $\xi \in \partial K(\tau)$ and $w_1, w_2 \in F(\tau, \xi)$, and

$$0 \notin \left[\liminf_{h \to 0^+} \frac{V_{(a,\xi)}(a + h, \xi + hw_a)}{h}, \limsup_{h \to 0^-} \frac{V_{(b,M\xi)}(b + h, M\xi + hw_b)}{h} \right],$$

for all $\xi \in \partial K(a)$, $w_a \in F(a, \xi)$ and $w_b \in F(b, M\xi)$.

In order to make a comparison of the above conditions there with those in Proposition (8.50), we should point out that, unlike in Proposition (8.50), the above conditions are only assumed to hold in one point. On the other hand, if conditions (8.51) and (8.51') are localized into one point, then they are implied by their analogues above.

Locally Lipschitzian Liapunov functions were used in [Del-M, Proposition 14.1], to obtain necessary and sufficient conditions for the strong stability of the zero solution of a differential inclusion. For this aim, condition (8.51) is globally assumed to guarantee the decreasing monotonicity of the Liapunov functions along all the solutions of the inclusion.

The same problem was recently analyzed in [BCM] for autonomous continuous inclusions, possibly Lipschitz-continuous. The authors provide a characterization of the monotonicity of a Liapunov function along all the trajectories, both in terms of proximal subdifferential and of contingent (epi)derivatives.

(8.58) EXAMPLE. Consider the anti-periodic problem

$$(8.59) \qquad \begin{cases} \dot{x} \in F(t,x), \\ x(b) = -x(a), \end{cases}$$

where $F : [a,b] \times \mathbb{R}^n \multimap \mathbb{R}^n$ is a u-Carathéodory multivalued function.

Assume that there exist positive constants ε and $R_j, j = 1, \ldots, n$, such that, denoted by

$$K = \Pi_{j=1}^n (-R_j, R_j)$$

and by

$$\partial K_j = \{\xi \in \partial K \mid \xi_j = \pm R_j\},$$

one has, for all $j = 1, \ldots, n$, $\xi \in \partial K_j$, $x \in \overline{K} \cap B_\xi^\varepsilon$, $t > a$ and $w \in F(t, x)$,

$$(\text{sign}\, \xi_j \cdot w_j) \neq 0.$$

Let us consider, for every $j = 1, \ldots, n$ and $\xi \in \partial K_j$, the C^1-function defined by

$$V_\xi(x) = \text{sign}\, \xi_j \cdot x_j - R_j.$$

We have

$$V_\xi(\xi) = 0$$

and

$$V_\xi(x) \leq 0,$$

for all $x \in \overline{K}$. Moreover,

$$\nabla V_\xi(x) \equiv \text{sign}\, \xi_j \cdot e_j.$$

Hence, for every $j = 1, \ldots, n$, $\xi \in \partial K_j$, $x \in \overline{K} \cap B_\xi^\varepsilon$, $t > a$ and $w \in F(t, x)$,

$$(\nabla V_\xi(x), w) = \text{sign}\, \xi_j \cdot w_j \neq 0.$$

Therefore, recalling also Remark (8.55), K satisfies all the hypotheses to be a bound set for (8.59), because (8.49) is equivalent to the symmetry of ∂K with respect to the origin.

We propose now a general theory on bound sets valid for arbitrary continuous bounding functions. The following theorem, involving Dini derivatives, is an appropriate version of a known result (see [KK-M, Theorem 5.2.3], cf. also [Th-M, Theorem 55.10]) on monotonicity properties for real continuous functions.

For this purpose, given a continuous function $f : [a, b] \to \mathbb{R}$ and a point $t_0 \in (a, b)$, we respectively denote by

$$D_- f(t_0) = \liminf_{h \to 0^-} \frac{f(t_0 + h) - f(t_0)}{h}$$

and

$$D^- f(t_0) = \limsup_{h \to 0^-} \frac{f(t_0 + h) - f(t_0)}{h}$$

the lower and upper left Dini derivatives in t_0. Similarly, letting $h \to 0^+$, we obtain the right Dini derivatives $D_+ f(t_0)$ and $D^+ f(t_0)$ of f in t_0.

(8.60) LEMMA. *Let* $f : [a, b] \to \mathbb{R}$ *be a continuous function such that*

$$(8.61) \qquad\qquad D_-f(t) < 0, \quad \text{for a.a. } t,$$

and

$$(8.62) \qquad D_-f(t) < \infty, \quad \text{for all } t \text{ except at most a countable set.}$$

Then f *is monotone decreasing in* $[a, b]$.

PROOF. Let us consider a positive real number ε and take $m \in \mathbb{N}$. Then (8.61) implies the existence of an open subset Z_n^ε of $[a, b]$ such that

$$\{t \in [a, b] \mid D_-f(t) \geq 0\} \subset Z_m^\varepsilon \quad \text{and} \quad \lambda(Z_m^\varepsilon) < \frac{\varepsilon}{2^m},$$

where λ denotes the Lebesgue measure on $[a, b]$.

The properties of λ yield, for every $t \in [a, b]$,

$$\sum_{m=1}^{\infty} \lambda([t, b] \cap Z_m^\varepsilon) \leq \sum_{m=1}^{\infty} \lambda(Z_m^\varepsilon) < \sum_{m=1}^{\infty} \frac{\varepsilon}{2^m} = \varepsilon.$$

Therefore, the function

$$Z : [a, b] \to \mathbb{R}, \quad t \to f(t) + \sum_{m=1}^{\infty} \lambda([t, b] \cap Z_m^\varepsilon)$$

is well-defined.

Given $\delta > 0$, take $\overline{m}$ such that

$$\sum_{m=\overline{m}+1}^{\infty} \lambda(Z_m^\varepsilon) \leq \frac{\delta}{2}.$$

For every $t < t'$, with $t' - t \leq \frac{\delta}{2\overline{m}}$, by the properties of the Lebesgue measure, it follows that

$$0 \leq \sum_{m=1}^{\infty} \lambda([t, b] \cap Z_m^\varepsilon) - \sum_{m=1}^{\infty} \lambda([t', b] \cap Z_m^\varepsilon) = \sum_{m=1}^{\infty} \lambda([t, t'] \cap Z_m^\varepsilon)$$

$$= \sum_{m=1}^{\overline{m}} \lambda([t, t'] \cap Z_m^\varepsilon) + \sum_{m=\overline{m}+1}^{\infty} \lambda([t, t'] \cap Z_m^\varepsilon) \leq \overline{m}(t' - t) + \frac{\delta}{2} \leq \delta.$$

We have so proved that

$$\left| \sum_{m=1}^{\infty} \lambda([t, b] \cap Z_m^\varepsilon) - \sum_{m=1}^{\infty} \lambda([t', b] \cap Z_m^\varepsilon) \right| \leq \delta, \quad \text{whenever } |t - t'| \leq \frac{\delta}{2\overline{m}}.$$

Consequently, the function $t \to \sum_{m=1}^{\infty} \lambda([t, b] \cap Z_m^{\varepsilon})$ is uniformly continuous on $[a, b]$ and, because of the continuity of f, this implies the continuity of Z.

Since, by the definition, $t \in Z_m^{\varepsilon}$, for every t such that $D_- f(t) \geq 0$, and Z_m^{ε} is open, it follows that, for every $m \in \mathbb{N}$, there exists $h_m < 0$ such that $[t + h_m, t] \subset Z_m^{\varepsilon}$ and $h_m \to 0^-$, when $m \to \infty$.

For every $m \in \mathbb{N}$, define $\rho_m = \max_{i=1,\ldots,m} h_i$ and take $h \in [\rho_m, 0)$. It follows that

$$\frac{\sum_{i=1}^{\infty} \lambda([t + h, b] \cap Z_i^{\varepsilon}) - \sum_{i=1}^{\infty} \lambda([t, b] \cap Z_i^{\varepsilon})}{h}$$

$$= \sum_{i=1}^{\infty} \frac{\lambda([t + h, t] \cap Z_i^{\varepsilon})}{h} \leq \sum_{i=1}^{m} \frac{\lambda([t + h, t])}{h} = -m.$$

Therefore, since m is arbitrary, we have that

$$\frac{d}{dt} \sum_{i=1}^{\infty} \lambda([t, b] \cap Z_m^{\varepsilon})(t) = -\infty \quad \text{for all } t \in [a, b] : D_- f(t) \geq 0.$$

Hence, according to condition (8.62),

$$D_- Z(t) = -\infty, \quad \text{for all } t \in [a, b] : 0 \leq D_- f(t) < \infty.$$

Moreover, since the function $t \to \sum_{m=1}^{\infty} \lambda([t, b] \cap Z_m^{\varepsilon})$ is monotone non-increasing, recalling (8.62), it follows that $D_- Z(t) < 0$, for all $t \in [a, b]$ with the possible exception of a countable set, implying that $\{Z(t) \mid t \in [a, b], D_- Z(t) \geq 0\}$ is at most countable.

Let us now suppose that there exist $t_1 < t_2$ satisfying $Z(t_1) < Z(t_2)$. Since Z is continuous, the interval $[Z(t_1), Z(t_2)]$ contains a continuum of points. Hence, it is then possible to find $r \in [Z(t_1), Z(t_2)]$ and $\bar{t} \in [t_1, t_2]$ such that

$$\bar{t} = \min\{t \in [t_1, t_2] \mid Z(t) = r\}$$

and

(8.63) $$D_- Z(\bar{t}) < 0.$$

Again, by the continuity of Z, it follows that

$$Z(t) < r \quad \text{for all } t \in [t_1, \bar{t}],$$

i.e. that

$$D_- Z(\bar{t}) \geq 0,$$

which is a contradiction with (8.63).

Thus, Z is monotone non-increasing, which yields, for every $t_1 < t_2$ and every $\varepsilon > 0$,

(8.64)

$$f(t_1) - f(t_2) \geq \sum_{m=1}^{\infty} \lambda([t_2, b] \cap Z_m^\varepsilon) - \sum_{m=1}^{\infty} \lambda([t_1, b] \cap Z_m^\varepsilon)$$

$$= -\sum_{m=1}^{\infty} \lambda([t_1, t_2] \cap Z_m^\varepsilon) \geq -\sum_{m=1}^{\infty} \lambda(Z_m^\varepsilon) = -\varepsilon,$$

implying that f is monotone non-increasing.

Finally, if there exist $t_1 < t_2$ such that $f(t_1) = f(t_2)$, then f is constant in $[t_1, t_2]$, which contradicts (8.52), and so the proof is complete. $\qquad\square$

We are now able to state the result concerning the existence of bound sets defined by means of continuous bounding functions.

For this aim, we need to assume certain growth conditions on the multivalued function F, namely the existence of $c \in L^1([a, b])$ such that

(8.65) $$|F(t, x)| \leq c(t), \quad \text{for all } (t, x) \in \overline{\mathcal{K}}.$$

(8.66) PROPOSITION. *Let $\{K(t)\}_{t\in[a,b]}$ be a family of non-empty and open subsets of $\mathbb{R}^n$, uniformly bounded by B_0^R and satisfying the invariance condition (8.49). Assume that, for every $(\tau, \xi) \in \partial\mathcal{K}$, there is a continuous function $V_{(\tau,\xi)}$ satisfying (8.45).*

Suppose, moreover, that for all $\tau > a$, $t \leq \tau$, $x \in \overline{K(t)}$, with (t, x) in a neighbourhood of (τ, ξ),

(8.67) $$\limsup_{\substack{h\to 0^- \\ v\to w}} \frac{V_{(\tau,\xi)}(t + h, x + hv) - V_{(\tau,\xi)}(t, x)}{h} < 0, \quad \text{for all } w \in F(t, x),$$

and

(8.68) $$\limsup_{\substack{h\to 0^- \\ y\to x \\ |y-x|\leq \int_{t+h}^{t} c(s)\, ds}} \frac{V_{(\tau,\xi)}(t + h, y) - V_{(\tau,\xi)}(t, x)}{h} < \infty.$$

Then $\{K(t)\}_{t\in[a,b]}$ is a bound set for (8.3), provided (8.65) holds.

PROOF. Reasoning as in Proposition (8.50), if we suppose that there exists a solution x of (8.3) whose graph is entirely contained in $\overline{\mathcal{K}}$ and touches its boundary in τ, we can assume without any loss of generality that $\tau \in (a, b]$. Take $V_{(\tau, x(\tau))}$ and, for the sake of simplicity, let us denote it by V.

From (8.67) and (8.54), we get

$$\limsup_{h \to 0^-} \frac{V(t+h, x(t+h)) - V(t, x(t))}{h}$$

$$= \limsup_{h \to 0^-} \frac{V(t+h, x(t) + h(w + \Delta(h))) - V(t, x(t))}{h}$$

$$\leq \limsup_{\substack{h \to 0^- \\ v \to w}} \frac{V(t+h, x(t) + hv) - V(t, x(t))}{h} < 0,$$

for almost all $t \in [\tau + \overline{h}, \tau]$ with $\overline{h} < 0$ sufficiently small, implying

$$D^- V(t, x(t)) < 0, \quad \text{for a.e. } t \in [\tau + \overline{h}, \tau].$$

Moreover, since x is a solution of (8.3) and $x(t) \in K(t)$, for all $t \in [\tau + \overline{h}, \tau]$ and h small enough, it follows that

$$|x(t+h) - x(t)| \leq \int_{t+h}^{t} |\dot{x}(s)| \, ds \leq \int_{t+h}^{t} c(s) \, ds.$$

Thus,

$$\limsup_{h \to 0^-} \frac{V(t+h, x(t+h)) - V(t, x(t))}{h} \leq$$

$$\limsup_{\substack{h \to 0^- \\ y \to x \\ |y-x| \leq \int_{t+h}^{t} c(s) \, ds}} \frac{V(t+h, y) - V(t, x)}{h} < \infty,$$

because of (8.68).

We have so proved that

$$D^- V(t, x(t)) < \infty, \quad \text{for all } t \in [\tau + \overline{h}, \tau].$$

Hence, Lemma (8.60) implies the decreasing monotonicity of $t \to V(t, x(t))$ in a left neighbourhood of τ which leads to a contradiction with (8.45). Therefore, $x(t) \in K(t)$, for all $t \in [a, b]$. $\qquad \square$

(8.69) REMARK. Taken $\tau < b$, the conclusion of Proposition (8.66) holds true also if we replace the assumptions (8.67) and (8.68) by

$$(8.67') \quad \liminf_{\substack{h \to 0^+ \\ v \to w}} \frac{V_{(\tau, \xi)}(t+h, x+hv) - V_{(\tau, \xi)}(t, x)}{h} > 0, \quad \text{for all } w \in F(t, x),$$

and

$$(8.68') \qquad . \quad \liminf_{\substack{h \to 0^+ \\ y \to x \\ |y-x| \leq \int_{t+h}^{t} c(s)\, ds}} \frac{V_{(\tau,\xi)}(t+h,y) - V_{(\tau,\xi)}(t,x)}{h} > -\infty,$$

for all $t \geq \tau$ and $x \in \overline{K(t)}$, with (t,x) in a neighbourhood of (τ,ξ). It is in fact sufficient to make use of the increasing monotonicity of $t \to V_{(\tau,x(\tau))}(t,x(t))$ in a right neighbourhood of τ.

(8.70) REMARK. Observe that if $V_{(\tau,\xi)}$ is locally Lipschitzian in all variables, then it is possible to prove, as in the previous part, that (8.51) and (8.51') are respectively weaker than (8.67) and (8.67'). Moreover, (8.68) and (8.68') are not necessary to get a bound set, like Proposition (8.50) shows.

In the previous part, a bound set is obtained without requiring hyphothesis (8.68) and with assumption (8.67) replaced by the conditions

$$0 \notin \left[\liminf_{\substack{v \to w_1 \\ h \to 0^+}} \frac{V_{(\tau,\xi)}(\tau+h, \xi+hv)}{h}, \limsup_{\substack{v \to w_2 \\ h \to 0^-}} \frac{V_{(\tau,\xi)}(\tau+h, \xi+hv)}{h} \right],$$

for all $\tau \in (a,b)$, $\xi \in \partial K(\tau)$ and $w_1, w_2 \in F(\tau,\xi)$, and

$$0 \notin \left[\liminf_{\substack{v \to w_a \\ h \to 0^+}} \frac{V_{(a,\xi)}(a+h, \xi+hv)}{h}, \limsup_{\substack{v \to w_b \\ h \to 0^-}} \frac{V_{(b,M\xi)}(b+h, M\xi+hv)}{h} \right],$$

for all $\xi \in \partial K(a)$, $w_a \in F(a,\xi)$ and $w_b \in F(b, M\xi)$.

Notice that both (8.67) or (8.67') imply the above assumptions. Hence, in the globally upper semicontinuous case, weaker conditions than those required in Proposition (8.66) allow us to get the existence of a bound set.

If the growth conditions imposed on F do not depend on t, precisely if there exists a real constant c such that

$$(8.71) \qquad |F(t,x)| \leq c \quad \forall (t,x) \in [a,b] \times \overline{K},$$

then (8.68) can be weakened as shown in the following result.

(8.72) PROPOSITION. *Let $\{K(t)\}_{t \in [a,b]}$ be a family of non-empty and open subsets of $\mathbb{R}^n$, uniformly bounded by B_0^R and satisfying the invariance condition* (8.49). *Assume that, for every $(\tau,\xi) \in \partial \mathcal{K}$, there is a continuous function $V_{(\tau,\xi)}$ satisfying* (8.45).

Suppose, moreover, that for all $\tau > a$, $t \le \tau$, $x \in \overline{K(t)}$, with (t, x) in a neighbourhood of (τ, ξ), (8.67) holds jointly with

$$(8.73) \qquad \limsup_{\substack{h \to 0^- \\ v \to w}} \frac{V_{(\tau, \xi)}(t + h, x + hv) - V_{(\tau, \xi)}(t, x)}{h} < \infty, \quad \text{for all } w \in B_0^c.$$

Then $\{K(t)\}_{t \in [a,b]}$ is a bound set for (8.3).

PROOF. Reasoning as in the proof of Proposition (8.66), if x is a solution of (8.3) whose graph entirely lays in $\overline{\mathcal{K}}$ and touches its boundary in $\tau \in (a, b]$, we get a sufficiently small negative $\overline{h}$ such that

$$D^- V_{(\tau, x(\tau))}(t, x(t)) < 0, \quad \text{for a.a. } t \in [\tau + \overline{h}, \tau].$$

Moreover, according to the growth condition (8.71), it holds

$$|\dot{x}(t)| \le c,$$

implying that x is a Lipschitzian function with the Lipschitz constant c.

Take $t \in [\tau + \overline{h}, \tau]$, $\{h_m\}_{m \in \mathbb{N}} \subset [\tau - t + \overline{h}, 0)$ and consider the sequence

$$\left\{ \frac{x(t + h_m) - x(t)}{h_m} \right\}_{m \in \mathbb{N}}.$$

Since every bounded sequence has a convergent subsequence, again denoted as the sequence, there exist $\{\Delta_m\}_{m \in \mathbb{N}}$, with $\Delta_m \to 0$ as $m \to \infty$, and $w \in B_0^c$ such that

$$x(t + h_m) = x(t) + h_m[w + \Delta_m], \quad \text{for all } m \in \mathbb{N}.$$

Therefore, (8.73) yields

$$\limsup_{m \to \infty} \frac{V_{(\tau, x(\tau))}(t + h_m, x(t + h_m)) - V_{(\tau, x(\tau))}(t, x(t))}{h_m}$$

$$\le \limsup_{\substack{h \to 0^- \\ v \to w}} \frac{V_{(\tau, x(\tau))}(t + h, x(t) + hv) - V_{(\tau, x(\tau))}(t, x(t))}{h} < \infty,$$

i.e.

$$D^- V_{(\tau, x(\tau))}(t, x(t)) < \infty, \quad \text{for all } t \in [\tau + \overline{h}, \tau].$$

Hence, Lemma (8.60) implies the decreasing monotonicity of $t \to V_{(\tau, x(\tau))}(t, x(t))$ in $[\tau + \overline{h}, \tau]$ which leads to a contradiction with (8.45). Therefore, $x(t) \in K(t)$, for all $t \in [a, b]$. $\qquad \square$

(8.74) REMARK. Observe that condition (8.68) is, under hypothesis (8.71), equivalent to

$$\limsup_{\substack{(h,v)\to(0^-,x)\\|y-x|\le hc}} \frac{V_{(\tau,\xi)}(t+h,y)-V_{(\tau,\xi)}(t,x)}{h} < \infty.$$

Therefore, since for every $w \in B_0^c$:

$$\limsup_{\substack{h\to0^-\\v\to w}} \frac{V_{(\tau,\xi)}(t+h,x+hv)-V_{(\tau,\xi)}(t,x)}{h} \le \limsup_{\substack{h\to0^-,\\y\to x\\|y-x|\le hc}} \frac{V_{(\tau,\xi)}(t+h,y)-V_{(\tau,\xi)}(t,x)}{h},$$

the previous theorem shows that if we assume a stronger condition on the growth behaviour of the multivalued function, it is possible to get the existence of a bound set by means of weaker assumptions on the bounding functions.

(8.75) REMARK. Notice that the conclusion of Proposition (8.72) also holds if we replace (8.67) and (8.73) by (8.67') and

$$(8.73') \qquad \liminf_{\substack{h\to0^+\\v\to w}} \frac{V_{(\tau,\xi)}(t+h,x+hv)-V_{(\tau,\xi)}(t,x)}{h} > -\infty, \quad \text{for all } w \in B_0^c.$$

Moreover, if $V_{(\tau,\xi)}$ is locally Lipschitzian in all variables, with the Lipschitz constant L, for $t \in [a,b]$, $w \in B_0^c$ and $v \in B_w^1$, then one has that

$$\left|\frac{V_{(\tau,\xi)}(t+h,x+hv)-V_{(\tau,\xi)}(t,x)}{h}\right| \le L|(1,v)| \le L(1+|v|) \le L(2+c).$$

Hence, all the contingent derivatives are finite and, in particular, condition (8.73) trivially holds.

Nevertheless, in the locally Lipschitzian case, the conditions of Proposition (8.50) are weaker than those required in the previous theorem (see Remark (8.70)).

If $V_{(\tau,\xi)}$ is only locally Lipschitzian in x, uniformly w.r.t., then (8.73) or (8.73') reduces to

$$\limsup_{h\to0^-} \frac{V_{(\tau,\xi)}(t+h,x)-V_{(\tau,\xi)}(t,x)}{h} < \infty$$

or

$$\liminf_{h\to0^+} \frac{V_{(\tau,\xi)}(t+h,x)-V_{(\tau,\xi)}(t,x)}{h} > -\infty.$$

Thus, (8.73) or (8.73') is implied by (8.67) or (8.67'), respectively.

Now, we consider the Floquet boundary value problem

$$(8.76) \qquad \begin{cases} \dot{x}+A(t)x \in F(t,x), & \text{for a.a. } t \in [a,b],\\ x(b)=Mx(a), \end{cases}$$

where $F : [a, b] \times \mathbb{R}^n \multimap \mathbb{R}^n$ is a u-Carathéodory multivalued function with non-empty, compact and convex values, $A : [a, b] \to \mathbb{R}^n \times \mathbb{R}^n$ is a continuous $n \times n$ matrix and M is a regular $n \times n$ matrix.

We will combine the bound sets approach, developed above, with Proposition (5.33) (see also Remark (5.34)) in order to solve (8.76).

As usual in this setting, we shall proceed by the following way to construct the set Q,

$$Q = \{q \in C([a, b], \mathbb{R}^n) \mid q(t) \in \overline{K(t)}, \quad \text{for all } t \in [a, b]\}.$$

We recall Lemma (8.34), for sufficient conditions in order that Q is a retract of $C([a, b], \mathbb{R}^n)$.

We are now able to state a result about the solvability of (8.76).

(8.77) THEOREM. *Let us consider the Floquet boundary value problem* (8.76). *Assume that*

(8.77.1) *the associated homogeneous problem*

$$\begin{cases} \dot{x} + A(t)x = 0, & \text{for a.a. } t \in [a, b], \\ x(b) = Mx(a) \end{cases}$$

has only the trivial solution,

(8.77.2) *there exists a u-Carathéodory multivalued function* $G : [a, b] \times \mathbb{R}^n \times \mathbb{R}^n \times [0, 1] \multimap \mathbb{R}^n$ *with non-empty, compact and convex values such that*

$$G(t, p, p, 1) \subset F(t, p), \text{ for all } (t, p) \in [a, b] \times \mathbb{R}^n,$$

(8.77.3) *there exists a family* $\{K(t)\}_{t \in [a,b]}$ *of non-empty and open subsets of* $\mathbb{R}^n$, *uniformly bounded by* B_0^R, *such that* $\overline{K}$ *is a retract of* $[a, b] \times \mathbb{R}^n$ *with a retraction* $\phi : [a, b] \times \mathbb{R}^n \to \overline{K}$ *which is the identity on its first component, i.e.* $\phi(t, x) = (t, \widetilde{\phi}(t, x))$ *and* ∂K *is closed,*

(8.77.4) $G(t, \cdot, q, \lambda)$ *is Lipschitzian with a sufficiently small Lipschitz constant, for each* $(t, \lambda) \in [a, b] \times [0, 1]$ *and* $q \in \overline{K(t)}$,

(8.77.5) *there exists an integrable function* $c : [a, b] \to \mathbb{R}$ *such that*

$$|G(t, x(t), q(t), \lambda)| \leq c(t), \quad \text{a.e. in } [a, b],$$

for any $(q, \lambda, x) \in \Gamma_T$, *where* $Q = \{q \in C([a, b], \mathbb{R}^n) \mid q(t) \in \overline{K(t)}, \text{ for all } t \in [a, b]\}$, T *denotes the map which assigns to any* $(q, \lambda) \in Q \times [0, 1]$ *the set of solutions of*

$$(8.78) \quad \begin{cases} \dot{x} + A(t)x \in G(t, x, q(t), \lambda), & \text{for a.a. } t \in [a, b], \\ x(b) = Mx(a) \end{cases}$$

and Γ_T *its graph,*

(8.77.6) $T(Q \times \{0\}) \subset Q$ *and* ∂Q *is fixed point free, i.e.* $\{q \in Q : q \in T(q,0)\} \cap \partial Q = \emptyset$,

(8.77.7) *for every* $(\tau, \xi) \in \partial K$, *there exists a function* $V_{(\tau,\xi)} : [a,b] \times \mathbb{R}^n \to \mathbb{R}$, *locally Lipschitzian in all variables and satisfying* (8.45),

(8.77.8) *for every* $\tau \in (a,b]$, $\xi \in \partial K(\tau)$, $t \leq \tau$, $x \in \overline{K(t)}$, *with* (t,x) *in a neighbourhood of* (τ, ξ), $(q, \lambda) \in Q \times (0,1]$, *one has*

$$(8.79) \quad \liminf_{h \to 0^-} \frac{V_{(\tau,\xi)}(t+h, x+hw)}{h} < 0, \quad \text{for all } w \in G(t,x,q(t),\lambda) - A(t)x,$$

(8.77.9) $M \partial K(a) = \{M\xi \mid \xi \in \partial K(a)\} = \partial K(b)$.

Then (8.76) *admits a solution.*

PROOF. Define Q as in (8.77.5). Hence, Q is a closed, bounded subset of $C([a,b], \mathbb{R}^n)$. From assumption (8.77.3) and Lemma (8.34), we get that Q is a bounded retract of $C([a,b], \mathbb{R}^n)$. In addition, since ∂K is closed, we can reason as in the proof of Proposition (8.21) and obtain that $q \in \partial Q$ if and only if $q(\bar{t}) \in \partial K(\bar{t})$, for some $\bar{t} \in [a,b]$. Thus, $Q \backslash \partial Q$ is non-empty.

According to conditions (8.77.1), (8.77.2) and (8.77.4), we are able to apply Theorem (3.8) to each problem (8.78) with $q \in Q$ and $\lambda \in [0,1]$ in order to assure its solvability with an R_δ-set of solutions. In particular, each solution set lies in some ball B_0^ρ. Moreover, it follows from the proof of Theorem (3.8) (see [ADG]) that B_0^ρ can be taken the same for all $q \in Q$. Therefore, defining

$$S_1 = B_0^\rho \cap \{x \in \text{AC}([a,b], \mathbb{R}^n) \mid x(b) = Mx(a)\},$$

we obtain that S_1 is bounded.

Thanks to Proposition (8.50), assumptions (8.77.7)–(8.77.9) guarantee that $\{K(t)\}_{t \in [a,b]}$ is a bound set for each problem (8.78) with $q \in Q$ and $\lambda \in (0,1]$. Hence, any solution $x \in Q$ of (8.78) satisfies $x(t) \in K(t)$, for all $t \in [a,b]$, so $x \notin \partial Q$. Therefore, T is fixed point free on ∂Q, for all $\lambda \neq 0$. By assumption (8.77.6), T is also fixed point free on ∂Q, for $\lambda = 0$.

In conclusion, all the requirements of Proposition (5.33) are satisfied and (8.76) has a solution x such that $x(t) \in \overline{K(t)}$, for each $t \in [a,b]$. $\square$

(8.80) REMARK. Since all the sets $K(t)$ are uniformly bounded, in the previous theorem, we can replace assumption (8.77.5) by the following one

$$|G(t, x(t), q(t), \lambda)| \leq d(t)(1 + |x(t)|),$$

for all $t \in [a,b]$, $(q, \lambda, x) \in \Gamma_T$ and with $d : [a,b] \to \mathbb{R}$ integrable.

Similar versions of Theorem (8.77) hold, when we consider continuous bounding functions. In this case, condition (8.51) should be replaced by (8.67) and (8.68) or by (8.67) and (8.73) and the existence of a bound set derives, respectively, from Proposition (8.66) and Proposition (8.72). Notice, in particular, that when using (8.67) and (8.73), we need to replace assumption (8.77.5) by

$$|G(t, x(t), q(t), \lambda)| \leq c,$$

for all $t \in [a, b]$ and $(q, \lambda, x) \in \Gamma_T$, where c is a positive constant. Alternatively, we can take condition (8.79), but with $d(t) \equiv d$, that is constant.

Now, we will combine Theorem (8.77) with a classical sequential approach (see Proposition (1.37)) in order to obtain the existence of a solution for the inclusion

$$(8.81) \qquad \dot{x} \in F(t, x),$$

which is bounded on all the real line. Here, we shall assume the multivalued function F to satisfy the growth assumption

$$(8.82) \qquad |F(t, x)| \leq c(t), \quad \text{for all } (t, x) \in \mathbb{R} \times B_0^R, \text{ where } c \in L_{\text{loc}}^1(\mathbb{R}).$$

For the sake of completeness, we can reformulate Proposition (1.37) as follows.

(8.83) LEMMA. *Let $F : \mathbb{R} \times \mathbb{R}^n \multimap \mathbb{R}^n$ be a u-Carathéodory set-valued map with non-empty, compact and convex values satisfying (8.82) and let $\{x_n\}_n$ be a sequence of absolutely continuous functions such that*

(8.83.1) *for every $m \in \mathbb{N}$, x_m is a solution of (8.81) defined in $[-m, m]$,*
(8.83.2) $\sup\{|x_m(t)| \mid m \in \mathbb{N}, \, t \in [-m, m]\} = R < \infty$.

Then inclusion (8.81) has a bounded solution on $\mathbb{R}$, whose values are contained in B_0^R.

Now, we are ready to give sufficient conditions for the existence of a bounded solution of (8.81).

(8.84) THEOREM. *Let us consider inclusion (8.81). Assume that*

(8.84.1) *there exists a bounded, non-empty and open set $K \subset \mathbb{R}^n$ such that K is symmetric with respect to the origin and $\overline{K}$ is a retract of $\mathbb{R}^n$, for every $\xi \in \partial K$, there exists a locally Lipschitzian function $V_\xi : \mathbb{R}^n \to \mathbb{R}$ satisfying*

$$(8.85) \qquad \begin{aligned} &V_\xi(\xi) = 0, \\ &V_\xi(x) \leq 0, \quad \text{for all } x \in \overline{K} \text{ in a neighbourhood of } \xi, \end{aligned}$$

(8.84.2) *there exists a u-Carathéodory multivalued function $G : \mathbb{R} \times \mathbb{R}^n \times \mathbb{R}^n \times$*
$[0,1] \multimap \mathbb{R}^n$ with non-empty, compact and convex values such that

$$G(t, p, p, 1) \subset F(t, p), \quad for\ all\ (t, p) \in \mathbb{R} \times \mathbb{R}^n,$$

(8.84.3) *for every $t \in \mathbb{R}$, $q \in \overline{K}$ and $\lambda \in [0, 1]$, $G(t, \cdot, q, \lambda)$ is Lipschitzian with a*
sufficiently small Lipschitz constant,

(8.84.4) *there exists a locally integrable function $c : \mathbb{R} \to \mathbb{R}$ satisfying*

$$|G(t, x, y, \lambda)| \le c(t)(1 + |x|), \quad for\ all\ (t, x, y, \lambda) \in \mathbb{R} \times \mathbb{R}^n \times \overline{K} \times [0, 1],$$

where $c \in L_{loc}^1(\mathbb{R})$,

(8.84.5) *for every $m \in \mathbb{N}$, $T_m(Q \times \{0\}) \subset Q$, where $Q = \{q \in C(\mathbb{R}, \mathbb{R}^n) \mid q(t) \in \overline{K}$,*
for all $t\}$, T_m denotes the map which assigns to all $(q, \lambda) \in Q \times [0, 1]$ the
set of solutions of

$$\begin{cases} \dot{x} \in G(t, x, q(t), \lambda), & for\ a.a.\ t \in [-m, m], \\ x(-m) = -x(m), \end{cases}$$

and ∂Q is fixed point free, i.e. $\{q \in Q \mid q \in T_m(q, 0)\} \cap \partial Q = \emptyset$,

(8.84.6) *for every $t \in \mathbb{R}$, $\xi \in \partial K$, $x \in \overline{K}$ in a neighbourhood of ξ, $(q, \lambda) \in Q \times (0, 1]$,*
one has

$$(8.86) \qquad \liminf_{h \to 0^-} \frac{V_\xi(x + hw) - V_\xi(x)}{h} < 0, \quad for\ all\ w \in G(t, x, q(t), \lambda).$$

Then inclusion (8.81) admits a bounded solution on $\mathbb{R}$.

PROOF. Given $m \in \mathbb{N}$, let us consider the following anti-periodic boundary
value problem on $[-m, m]$:

$$(8.87_m) \qquad \begin{cases} \dot{x} \in F(t, x), & for\ a.a.\ t \in [-m, m], \\ x(-m) = -x(m), \end{cases}$$

i.e. problem (8.76) with $M = -I$ and $A \equiv 0$, whose associated homogeneous
problem has only the trivial solution.

Since $\overline{\mathcal{K}} = [-m, m] \times \overline{K}$ is a retract of $[-m, m] \times \mathbb{R}^n$, by condition (8.84.1), and
$\partial \mathcal{K} = [-m, m] \times \partial K$ is closed, then K is an autonomous bound set for (8.87_m),
for each $m \in \mathbb{N}$, because (8.49) is equivalent to the symmetry of ∂K with respect
to the origin.

Therefore, Theorem (8.77), whose condition (8.77.5) follows immediately from
the present assumption (8.77.4), by means of the arguments in the proof of Theorem (3.8) (see [ADG]), implies the existence of a solution of (8.87_m), i.e., in particular, of an absolutely continuous function $x_m : [-m, m] \to \mathbb{R}$, satisfying (8.81).

Moreover, $x_m(t) \in \overline{K}$, for each $m \in \mathbb{N}$ and $t \in [-m, m]$, and the conclusion follows by Lemma (8.83), because K is bounded. $\square$

(8.88) REMARK. In line of Remark (8.80), we leave to the reader the formulation of an existence result for bounded solutions for (8.81) using continuous bounding functions.

(8.89) REMARK. Similar results can be obtained sequentially, when replacing the anti-periodic boundary value problems by a one-parameter family of different Floquet problems.

(8.90) REMARK. In the globally upper semicontinuous case, condition (8.84.6) can be replaced by those recalled in Remark (8.57) which are only localized in one point.

(8.91) EXAMPLE. Consider the differential inclusion

$$(8.92) \qquad \dot{x} \in F_1(t, x) + F_2(t, x),$$

where $F_1, F_2 : \mathbb{R} \times \mathbb{R}^n \multimap \mathbb{R}^n$ are u-Carathéodory multivalued functions with non-empty convex and compact values such that there exist c_1 and $c_2 \in L^1_{\mathrm{loc}}(\mathbb{R})$, satisfying

$$|F_1(t, 0)| \leq c_1(t), \quad \text{for all } t \in \mathbb{R},$$
$$|F_2(t, x)| \leq c_2(t), \quad \text{for all } (t, x) \in \mathbb{R} \times \overline{K}$$

(K is defined below) and $F_1(t, \cdot)$ is Lipschitzian, with a sufficiently small Lipschitz constant L, for all $t \in \mathbb{R}$ ($\Rightarrow |F_1(t, x)| \leq L|x| + |F_1(t, 0)| \leq L|x| + c_1(t) \leq (L + c_1(t))(1 + |x|)$, for all $(t, x) \in \mathbb{R} \times \mathbb{R}$).

Assume, furthermore, the existence of positive constants ε and R_j, $j = 1, \ldots, n$, such that

$$K = \Pi_{j=1}^n (-R_j, R_j), \quad \partial K_j = \{\xi \in \partial K \mid \xi_j = \pm R_j\}$$

and

$$Q = \{q \in C(\mathbb{R}) \mid q(t) \in \overline{K}, \quad \text{for all } t \in \mathbb{R}\},$$

and take, for all $j = 1, \ldots, n$, $\xi \in \partial K_j$, $x \in \overline{K} \cap B^\varepsilon_\xi$, $t \in \mathbb{R}$, $q \in Q$ and $w \in F_1(t, x) + F_2(t, q(t))$, satisfying

$$(\operatorname{sign} \xi_j \cdot w_j) \neq 0.$$

Let us consider the family of multivalued functions defined as

$$G(t, x, q, \lambda) = \lambda(F_1(t, x) + F_2(t, q)),$$

which, recalling the growth conditions imposed on F_1 and F_2 and the boundedness of K, satisfy the assumptions (8.84.2)–(8.84.4) of Theorem (8.84), with

$$c(t) = L + c_1(t) + c_2(t).$$

Moreover, assumption (8.84.5) is trivially satisfied, because the only solution of

$$\begin{cases} \dot{x} = 0, \\ x(-m) = -x(m) \end{cases}$$

is $x \equiv 0 \in \operatorname{int} Q$.

Finally, recalling Example (8.58), K is a bound set for each problem

$$\begin{cases} \dot{x} \in G(t, x, q(t), \lambda), & \text{for a.a. } t \in [-m, m], \\ x(-m) = -x(m), \end{cases}$$

with $\lambda \in (0, 1]$, and we get the existence of a bounded solution for (8.73), by Theorem (8.84).

Now, we would like to reformulate in terms of bounding functions condition (6.68), playing the fundamental role in the application of the relative Lefschetz number in Chapter III.6.

Consider the inclusion

(8.93) $\dot{y} \in F(t, y),$

where $F : \mathbb{R} \times \Omega \multimap \mathbb{R}^n$ is an (upper) Carathéodory mapping which is essentially bounded in t and linearly bounded in y, and $\Omega \subset \mathbb{R}^n$ is an open subset, i.e.

$$|F(t, y)| \le \alpha + \beta|y|, \quad \text{for a.a. } t \in \mathbb{R} \text{ and for all } y \in \Omega,$$

where α, β are nonnegative constants.

The following lemma can be simply obtained by a slight modification of the proof of Proposition (8.72) (see also Remark (8.75)).

(8.94) LEMMA. *Let* $V_u(t, y) \equiv V_u(t + T, y) \in C([0, T] \times \Omega, \mathbb{R})$ *be a family of (bounding) functions and* $c \in \mathbb{R}$. *Set* $M = [V_u \le c] = \{y \in 0 \mid V_u(t, y) \le c\}$; *the set* $[V_u > c]$ *is defined analogously. Assume that* M *is bounded, for each* $u \in \partial M$ *and* $t \in [0, T]$, *there exists* $\varepsilon > 0$ *such that* V_u *is locally Lipschitzian in* y *on* $[V_u > c] \cap B(u, \varepsilon)$, *uniformly w.r.t.* $t \in [0, T]$,

$$\limsup_{h \to 0^+} \frac{1}{h}[V_u(t + h, y + hf) - V_u(t, y)] \le 0,$$

for every $y \in [V_u > c] \cap B(u, \varepsilon)$ *and* $f \in F(t, y)$.

Then M *is positively flow-invariant for* (8.93), *i.e.* $\Phi_{kT}(M) \subset M$, $k \in [0, 1]$, *where* Φ *is the associated translation operator in* (6.67)

We are in position to reformulate Theorem (6.66) in terms of bounding functions as follows.

(8.95) THEOREM. *Let A_1 and B_1, where $B_1 \subset A_1 \subset \Omega \subset \mathbb{R}^n$, be compact ENR-spaces such that $\chi(A_1) \neq \chi(B_1)$ holds, where χ stands for the Euler–Poincaré characteristic. Assume the existence of a family of bounding functions. $V_u(t,u)$, $W_u(t,y)$ and constants c_1, c_2 satisfying the conditions of Lemma (8.94), for $A_1 = [V_u \leq c]$ and $B_1 = [W_u \leq c_2]$. Then the system (8.93), where $F(t,y) \equiv F(t+T,y)$, admits a T-periodic solution $y(t)$ with $y(0) \in \overline{A_1 \setminus B_1}$.*

(8.96) REMARK. Instead of Lemma (8.94), we could employ other criteria ensuring the strong positive flow-invariance of A_1, B_1, when the boundaries ∂A_1 and ∂B_1 are not reached by a solution from the interior of A_1 and B_1. Then, the additivity, excision and existence properties of the fixed point index could be also applied for the same aim. However, since fixed points of the translation operator in (6.67) are allowed in Theorem (8.95), for any $k \in [0,1]$, on the boundaries ∂A_1 and ∂B_1, Theorem (8.95) can be regarded as a nontrivial example of an application of the relative Lefschetz number. The same is all the better true for Theorem (6.71).

Similarly, as a nontrivial application of the relative Nielsen number defined in Chapter II.5, Theorem (6.73) can be immediately reformulated in terms of bounding functions as follows.

(8.97) THEOREM. *Let A_1 and B_1, where $B_1 \subset A_1 \subset \Omega \subset \mathbb{R}^n$, be compact connected either nilmanifolds and ENR-spaces or AR-spaces. Furthermore, let $f : (A_1, B_1) \to (A_1, B_1)$ be a continuous (single-valued) mapping. At last, assume the existence of a family of bounding functions $V_u(t,u)$, $W_u(t,u)$ and constants c_1, c_2, satisfying the conditions of Lemma (8.94), for $A_1 = [V_u \leq c_1]$ and $B_1 = [W_u \leq c_1]$. Then system (8.93) admits at least $(|\Lambda(f_{A_1})| + |\Lambda(f_{A_2})| - N(f_{A_1}, f_{A_2}))$ solutions $y(t)$, with $y(0) = f(y(T))$, $y(0) \in A_1$. For nilmanifolds A_i, the generalized Lefschetz number $\Lambda(f_{A_i})$, $i = 1, 2$, can be computed by means of (6.72) and $N(f_{A_1}, f_{A_2})$ denotes the number of essential common fixed point classes of $f_{A_1} = f|_{A_1}$ and $f_{A_2} = f|_{A_2}$ (see Chapter II.5).*

(8.98) REMARK. If A_1 is an AR-space, then $|\Lambda(f_{A_1})| \geq N(f_{A_1}, f_{A_2})$ by which the lower estimate of solutions $y(t)$ of (8.93) with $y(0) = f(y(T))$ reduces either to 1 or to $|\Lambda(f_{A_2})|$.

(c) Guiding functions for globally u.s.c. r.h.s.

The guiding function (or guiding potential) method has been mainly developed to study periodic problems for differential equations and inclusions. The guiding potential V was supposed there mostly to be C^1. Here, we shall assume V again only locally Lipschitzian.

For A, B, nonempty compact subsets of $\mathbb{R}^n$, we define the lower inner product

$\langle A, B \rangle^-$ of A and B by

$$\langle A, B \rangle^- = \inf\{\langle a, b \rangle \mid a \in A, \ b \in B\}.$$

We also use the standard notation

$$|A|^- = \inf\{|a| \mid a \in A\}, \quad |A|^+ = \sup\{|a| \mid a \in A\},$$

where $A \subset \mathbb{R}^n$ is nonempty and bounded.

(8.99) LEMMA. *Let A, B, C, A_1, B_1 be nonempty compact subsets of $\mathbb{R}^n$. We have:*

(i) $\langle A, B \rangle^- = \langle B, A \rangle^-$,

(ii) $\langle \lambda A, B \rangle^- = \lambda \langle B, A \rangle^{-1}$, $\lambda \geq 0$,

(iii) $\langle A, B \rangle^- > 0$ *implies* $0 \notin A$ *and* $0 \notin B$,

(iv) $\langle A + B, C \rangle^- \geq \langle A, C \rangle^- + \langle B, C \rangle^-$,

(v) $\langle A, B \rangle^- \geq \langle A_1, B_1 \rangle^-$ *if* $A \subset A_1$, $B \subset B_1$,

(vi) $|\langle A, B \rangle^-| \leq |A|^+ |B|^+$,

(vii) $\langle A + C, B + C \rangle^- \geq \langle A, B \rangle^- - |C|^+(|A|^+ + |B|^+ + |C|^+)$,

(viii) $\langle A, B \rangle^- = \langle \overline{\operatorname{conv}} A, \overline{\operatorname{conv}} B \rangle^-$,

(ix) $\langle \int_a^b F(s)\, ds, C \rangle^- \geq \int_a^b \langle F(s), C \rangle^-\, ds$, *where $F : [a, b] \multimap \mathbb{R}^n$ is an integrable multivalued map with nonempty compact values.*

PROOF. We consider only (viii) and (ix) (the remaining ones are obvious).

To prove (viii), let $\overline{a} \in \overline{\operatorname{conv}} A$ and $\overline{b} \in \overline{\operatorname{conv}} B$ be such that

$$\langle \overline{a}, \overline{b} \rangle = \langle \overline{\operatorname{conv}} A, \overline{\operatorname{conv}} B \rangle.$$

By the well-known Carathéodory theorem, there exist $n + 1$ points $a_i \in A$, $b_j \in B$ and corresponding $n + 1$ numbers $\lambda_i \geq 0$, $\mu_j \geq 0$, with $\lambda_1 + \ldots + \lambda_{n+1} = 1$ and $\mu_1 + \ldots + \mu_{n+1} = 1$, such that $\overline{a} = \lambda_1 a_1 + \ldots + \lambda_{n+1} a_{n+1}$ and $\overline{b} = \mu_1 b_1 + \ldots + \mu_{n+1} b_{n+1}$. It follows that

$$\langle \overline{a}, \overline{b} \rangle = \sum_{i,j=1}^{n+1} \lambda_i \mu_j \langle a_i, b_j \rangle \geq \sum_{i,j=1}^{n+1} \lambda_i \mu_j \langle A, B \rangle^- = \langle A, B \rangle^-,$$

and so $\langle \overline{\operatorname{conv}} A, \overline{\operatorname{conv}} B \rangle^- \geq \langle A, B \rangle^-$. The reverse inequality is evident, and so (viii) is verified.

To see (ix), observe that the function $s \to \langle F(s), C \rangle^-$ is measurable and integrable if F is so. Then (ix) follows at once in view of the convexity inequality

$$\left\langle \sum_{i=1}^k \lambda_i A_i, B \right\rangle^- \geq \sum_{i=1}^k \lambda_i \langle A_i, B \rangle^-,$$

where $A_1, \ldots, A_k, B$ are nonempty compact sets and $\lambda_1, \ldots, \lambda_k$ are nonnegative numbers whose sum is one, and $k \in \mathbb{N}$ is arbitrary. Thus, also (ix) is verified, completing the proof. $\qquad\square$

We recall some notions of the Clarke generalized gradient calculus (see [Cl-M]), we shall need later.

Let $V : \mathbb{R}^n \to \mathbb{R}$ be a locally Lipschitzian function. For $x_0 \in \mathbb{R}^n$ and $v \in \mathbb{R}^n$, let $V^0(x_0, v)$ be the *generalized directional derivative* of V at x_0 in the direction v, that is

$$V^0(x_0, v) = \limsup_{\substack{x \to x_0 \\ t \to 0^+}} \frac{V(x + tv) - V(x)}{t}.$$

Then the *generalized gradient* $\partial V(x_0)$ of V at x_0 is defined by

$$\partial V(x_0) = \{y \in \mathbb{R}^n \mid \langle y, v \rangle \leq V^0(x_0, v), \text{ for every } v \in \mathbb{R}^n\}.$$

The map $\partial V : \mathbb{R}^n \multimap \mathbb{R}^n$ defined by the above equality is u.s.c. with nonempty compact convex values (see [Cl-M, p. 27–29]). If, in addition, V is also convex, then $\partial V(x_0)$ coincides with the subdifferential of V at x_0 in the sense of convex analysis (see [Cl-M, p. 36]), i.e.

$$\partial V(x_0) = \{y \in \mathbb{R}^n \mid \langle y, x - x_0 \rangle \leq V(x) - V(x_0), \text{ for every } x \in \mathbb{R}^n\}.$$

If V is C^1, then $\partial V(x_0)$ reduces to the singleton set $\{\operatorname{grad} V(x_0)\}$ (see [Cl-M, p. 33]).

(8.100) DEFINITION. Let $V : \mathbb{R}^n \to \mathbb{R}^n$ be locally Lipschitzian. If, for some $r_0 > 0$, V satisfies

$$(8.101) \qquad \langle \partial V(x), \partial V(x) \rangle^- \to 0, \quad \text{for every } |x| \geq r_0,$$

then V is called a *direct potential*. If F is also convex, and instead of (8.101), satisfies

$$(8.102) \qquad 0 \notin \partial V(x), \quad \text{for every } |x| \geq r_0,$$

then V is called a *direct convex potential*.

Observe that (8.101) implies (8.102), the converse is not true in general. Moreover, if V is C^1, then either (8.101) or (8.102) is equivalent to saying that

$$\operatorname{grad} V(x) \neq 0, \quad \text{for every } |x| \geq r_0.$$

In view of that, the above definition can be interpreted as a generalization of the definition of a direct potential V in the sense of M. A. Krasnosel'skiĭ [Ks2-M] (see also [KsZa-M]), where V is assumed to be a C^1-function.

Let $V : \mathbb{R}^n \to \mathbb{R}^n$ be a direct potential. Observe that the topological degree (see Chapters I.8 and II.9) $\deg(\partial V)$ is well-defined on B_0^r. Hence, it makes sense to define the index $\mathrm{ind}(F)$ of the direct potential V, by putting

$$\mathrm{ind}(V) = \deg(\partial V),$$

where ∂V stands for the restriction of ∂V to B_0^r, $r \geq r_0$. Of course, the definition of $\mathrm{ind}(V)$ is analogous if V is a direct convex potential.

The following proposition is proved in [BoKl-M, Theorem 14.2, p. 126].

(8.103) PROPOSITION. *If $V : \mathbb{R}^n \to \mathbb{R}^n$ is a direct potential (or a direct convex potential) satisfying $\lim_{|x| \to \infty} V(x) = \infty$, then $\mathrm{ind}(V) = 1$.*

Some other cases of direct potentials with nonzero index can be found in [KsZa-M].

Let $V : \mathbb{R}^n \to \mathbb{R}^n$ be a direct potential. For $k \in \mathbb{N}$, set $M_k = \sup\{|\partial V(x)|^+ \mid x \in B_0^k\}$, and observe that M_k is finite, because ∂V is u.s.c. Now, define $\eta : \mathbb{R}^n \to \mathbb{R}$ by

$$\eta(x) = 1 + (|x| - k)M_{k+2} + (k + 1 - |x|)M_{k+1}, \quad \text{for } k \leq |x| \leq k+1, \ k = 0, 1, \ldots$$

Clearly η is continuous, and satisfies

$$\eta(x) \geq \max\{1, |\partial V(x)|^+\}, \quad \text{for every } x \in \mathbb{R}^n.$$

Thus, we have

(8.104) PROPOSITION. *Let $V : \mathbb{R}^n \to \mathbb{R}$ be a direct potential (or a direct convex potential) and let $W : \mathbb{R}^n \to \mathbb{R}$ be given by*

$$(8.105) \qquad\qquad W(x) = \frac{\partial V(x)}{\eta(x)}.$$

Then W is u.s.c., with nonempty compact convex values, and satisfies $|W(x)|^+ \leq 1$, for every $x \in \mathbb{R}^n$.

A connection between the notion of a direct potential and differential inclusions is given by the following:

(8.106) DEFINITION. Let $F : [0, a] \times \mathbb{R}^n \multimap \mathbb{R}^n$ be a map with nonempty compact values. A direct potential $V : \mathbb{R}^n \to \mathbb{R}^n$ is called a *guiding potential* for F, if there is $r_0 > 0$ such that:

(8.106.1) for every $(t, x) \in [0, a] \times \mathbb{R}^n$, with $|x| \geq r_0$, we have $\langle F(t, x), \partial V(x) \rangle^- \geq 0$, while V is called a *weakly guiding potential* for F, if there is $r_0 > 0$ such that:

(8.106.2) for every $(t, x) \in [0, a] \times \mathbb{R}^n$, with $|x| \geq r_0$, there is $y \in F(t, x)$ such that $\langle y, \partial V(x) \rangle^- \geq 0$.

By replacing "potential" with "convex potential" we have analogous definitions. Any guiding potential is a weakly guiding potential. The converse is not true, in general. The following result was obtained in [GPl].

(8.107) THEOREM. *Let $F : [0, a] \times \mathbb{R}^n \multimap \mathbb{R}^n$ be an (upper) Carathéodory map with nonempty, compact and convex values, and let $V : \mathbb{R}^n \to \mathbb{R}$ be a C^1-weakly guiding potential for F, with $\mathrm{ind}(V) \neq 0$. Then the periodic problem*

(8.108)
$$\begin{cases} \dot{x} \in F(t, x), \\ x(0) = x(a) \end{cases}$$

has a solution.

In the sequel, we discuss some versions of Theorem (8.107) in which the guiding potential is not necessarily C^1. For F, we shall use the following assumptions:

(8.109) $F : [0, a] \times \mathbb{R}^n \multimap \mathbb{R}^n$ is u.s.c., with nonempty, compact, convex values,

(8.110) $|F(t, x)|^+ \leq \mu(x)(1 + |x|)$, for every $(t, x) \in [0, a] \times \mathbb{R}^n$, where $\mu : [0, a] \to [0, \infty)$ is an integrable function.

(8.111) LEMMA. *Let $F : [0, a] \times \mathbb{R}^n \multimap \mathbb{R}^n$ satisfy (8.109)–(8.110). Let $x : [0, a] \to \mathbb{R}^n$ be any solution of the Cauchy problem*

(8.112)
$$\begin{cases} \dot{x} \in F(t, x), \\ x(0) = x_0. \end{cases}$$

Then, for every $t \in [0, a]$, we have

(8.113)
$$\partial x(t) \subset F(t, x(t)).$$

PROOF. Let $0 < t < a$ (if $t = 0$ or $t = a$, the argument is similar). Let $\varepsilon > 0$. Since F is u.s.c., there is an open interval $(t - \delta, t + \delta)$ contained in $[0, a]$ such that

$$F(s, x(s)) \subset F(t, x(t)) + B_0^\varepsilon, \quad \text{for every } s \in (t - \delta, t + \delta).$$

F is locally bounded, therefore, x is locally Lipschitzian. By virtue of [Cl-M, Theorem 2.5.1, p. 63], we have

$$\partial x(t) = \overline{\mathrm{conv}} \{ \lim_{k \to \infty} \dot{x}(t_k) \mid t_k \to t,\ t_k \in [0,a] \setminus \Omega \},$$

where Ω is a set of the Lebesgue measure zero containing the set of points at which x is not differentiable. Hence, $\partial x(t) \subset F(t,x(t)) + B_0^\varepsilon$. Since $\varepsilon > 0$ is arbitrary, (8.113) follows, completing the proof. $\square$

Now, consider the Cauchy problem

$$(8.114) \qquad \begin{cases} \dot{x} \in W(x), \\ x(0) = x_0. \end{cases}$$

(8.115) LEMMA. *Let* $V : \mathbb{R}^n \to \mathbb{R}$ *be a direct potential and let* $W(x) = \partial V(r)/\eta(x)$, $x \in \mathbb{R}^n$. *Then there is* $\tilde{r} > 0$ *for which the following property is satisfied: for each* $r > \tilde{r}$, *there exists* $t_r \in (0,a]$ *such that, for every* $(x_0, \lambda) \in S^{n-1}(r) \times [0,1]$, *where* $S^{n-1}(r)$ *denotes the* $(n-1)$-*dimensional sphere with the radius* r, *centered at the origin, and any solution* $x : [0,a] \to \mathbb{R}^n$ *of (8.114), we have*

$$0 \notin \lambda(x(t) - x_0) + (1-\lambda)\partial V(x_0), \quad \text{for every } t \in (0, t_r].$$

PROOF. Take $\tilde{r} = r_0$, where r_0 is given by (8.101). Let $r > \tilde{r}$ be arbitrary.

On the sphere $S^{n-1}(r)$, the function $x_0 \mapsto \langle W(x_0), W(x_0) \rangle^-$ is strictly positive and lower semicontinuous (∂V is u.s.c.). Thus, there is $0 < \delta < 1$ such that

$$(8.116) \qquad \langle W(x_0), W(x_0) \rangle^- \geq \delta, \quad \text{for every } x_0 \in S^{n-1}(r).$$

By Lemma (8.99)(vii), for each $x \in S^{n-1}(r)$, we have

$$(8.117)\ \langle W(x_0) + B_0^{\delta/5}, W(x_0) + B_0^{\delta/5} \rangle^- \geq \langle W(x_0), W(x_0) \rangle^- - \frac{\delta}{5} - \frac{\delta}{5} - \frac{\delta^2}{25} > \frac{\delta}{2},$$

for $|W(x_0)|^+ \leq 1$.

W is u.s.c., thus, for each $x_0 \in S^{n-1}(r)$, there is $\delta(x_0) > 0$ such that

$$(8.118) \qquad W(x) \subset W(x_0) + B_0^{\delta/5}, \quad \text{for every } x \in B_{x_0}^{\delta(x_0)} \text{ (the ball is open)}.$$

The family $\{B_{x_0}^{\delta(x_0)}\}_{x_0 \in S^{n-1}(r)}$ is an open covering of $S^{n-1}(r)$.

Let $\{B_{x_0^i}^{\delta(x_0^i)}\}_{i=1}^k$ be a finite subcovering of $S^{n-1}(r)$, and let $\delta^* > 0$ be a Lebesgue number associated to it.

Now, let $x_0 \in S^{n-1}(r)$ be arbitrary. As $B_{x_0}^{\delta^*} \subset B_{x_0^i}^{\delta(x_0^i)}$, for some $1 \le i \le k$, by virtue of (8.118) (with x_0^i in the place of x_0), we have

$$(8.119) \qquad W(x) \subset W(x_0^i) + B_0^{\delta/5}, \quad \text{for every } x \in B_{x_0}^{\delta(x^*)}$$

By virtue of (8.119) and (8.117), it follows

$$(8.120) \qquad \langle W(x), W(x_0) \rangle^- \ge \langle W(x_0^i) + B_0^{\delta/5}, W(x_0) + B_0^{\delta/5} \rangle^- > \frac{\delta}{2},$$

for every $x \in B_{x_0}^{\delta^*}$. Since W is bounded, there exists $0 < t_r \le a$ such that, for each $x_0 \in S^{n-1}(r)$ and any solution $x : [0, a] \to \mathbb{R}^n$ of (8.114), we have

$$(8.121) \qquad x(t) \in B_0^{\delta^*}, \quad \text{for every } t \in [0, t_r].$$

With this choice of t_r the statement of the lemma is satisfied. To see this, let $(x_0, \lambda) \in S^{n-1}(r) \times [0, 1]$ be arbitrary, and let $x : [0, a] \to \mathbb{R}^n$ be any solution of (8.114). In view of Lemma (8.99), for every $t \in (0, t_r]$, we have

$$\langle \lambda(x(t) - x_0) + (1 - \lambda)\partial V(x_0), \partial V(x_0) \rangle^-$$
$$\ge \lambda \langle x(t) - x_0, \partial V(x_0) \rangle^- + (1 - \lambda)\langle \partial V(x_0), \partial V(x_0) \rangle^-$$
$$= \lambda \left\langle \int_0^t \dot{x}(s)\, ds, \eta(x_0)W(x_0) \right\rangle^- + (1 - \lambda)\langle \eta(x_0)W(x_0), \eta(x_0)W(x_0) \rangle^-$$
$$\ge \lambda\eta(x_0) \int_0^t \langle \dot{x}(s), W(x_0) \rangle^-\, ds + (1 - \lambda)\eta^2(x_0)\langle W(x_0), W(x_0) \rangle^-$$
$$\ge \lambda\eta(x_0) \int_0^t \langle W(x(s)), W(x_0) \rangle^-\, ds + (1 - \lambda)\eta^2(x_0)\delta \quad \text{by (8.116)}$$
$$\ge \lambda\eta(x_0)\frac{\delta}{2}t + (1 - \lambda)\eta^2(x_0)\delta \quad \text{by (8.121) and (8.120)}$$
$$\ge \lambda\frac{\delta}{2}t + (1 - \lambda)\delta \quad \text{(for } \eta(x_0) \ge 1)$$

The last quantity is strictly positive, for $t \in (0, t_r]$. Hence, by Lemma (8.99), we have $0 \notin \lambda(x(t) - x_0) + (1 - \lambda)\partial V(x_0)$, for every $t \in (0, t_r]$, completing the proof.$\Box$

For any $\lambda \in [0, 1]$, let us consider the Cauchy problem

$$(8.122) \qquad \begin{cases} \dot{x} \in (1 - \lambda)W(x) + \lambda F(t, x), \\ x(0) = x_0. \end{cases}$$

(8.123) LEMMA. *Let* $F : [0,a] \times \mathbb{R}^n \multimap \mathbb{R}^n$ *satisfy* (8.109)–(8.110). *Let* $V :$ $\mathbb{R}^n \to \mathbb{R}$ *be a guiding potential for* F *and let* $W(x) = \partial V(x)/\eta(x)$, $x \in \mathbb{R}^n$. *There is* $\tilde{r} > 0$, *for which the following property is satisfied: for each* $r > \tilde{r}$, *for each* $(x_0, \lambda) \in S^{n-1}(r) \times [0,1)$ *and any solution* $x : [0,a] \to \mathbb{R}^n$ *of* (8.122), *we have*

$$x(t) - x_0 \neq 0, \quad \text{for every } t \in (0,a].$$

PROOF. Let $r_0 > 0$ be so that (8.101) and (8.106.1) are satisfied. One can check (see Proposition 2.2 in [DBGP2] and cf. [GP1]) that there is $\tilde{r} > r_0$ such that, for each $(x_0, \lambda) \in \mathbb{R}^n \times [0,1]$ with $|x_0| > \tilde{r}$, and each solution $x : [0,a] \to \mathbb{R}^n$ of (8.122), we have

(8.124) $$|x(t)| > r_0, \quad \text{for every } t \in [0,a].$$

With this choice of $\tilde{r}$, the statement of the lemma is satisfied. Indeed, in the opposite case, there exists $r > \tilde{r}$, $(x_0, \lambda) \in S^{n-1}(r) \times [0,1)$ and a solution $x :$ $[0,a] \to \mathbb{R}^n$ of (8.122) such that $x(\hat{t}) = x_0$, for some $\hat{t} \in (0,a]$. Thus,

$$(V \circ x)(\hat{t}) = (V \circ x)(0).$$

The function $V \circ x$ is continuous in $[0,\hat{t}]$. Hence, it has a global extremum at some point $\tau \in (0,\hat{t})$. Since $V \circ x$ is locally Lipschitzian, by [Cl-M, Proposition 2.3.2, p. 38], we have

(8.125) $$0 \in \partial(V \circ x)(\tau).$$

By the chain rule [Cl-M, Theorem 2.3.9, p. 12],

$$\partial V \circ x(\tau) \subset \text{conv}\{\langle \alpha, \beta \rangle \mid \alpha \in \partial V(x(\tau)), \ \beta \in \partial x(\tau)\}.$$

On the other hand, the map $(1 - \lambda)W + \lambda F$ is u.s.c. and so, by Lemma (8.111),

(8.126) $$\partial x(\tau) \in (1 - \lambda)W(x(\tau)) + \lambda F(\tau, x(\tau)).$$

Now, let $\alpha \in \partial V(x(\tau))$ and $\beta \in \partial x(\tau)$ be arbitrary. In view of Lemma (8.99), we have

$$\langle \alpha, \beta \rangle \geq \langle \partial V(x(\tau)), \partial x(\tau) \rangle^-$$
$$\geq \langle \partial V(x(\tau)), (1 - \lambda)W(x(\tau)) + \lambda F(\tau, x(\tau)) \rangle^- \quad \text{(by (8.126))}$$
$$\geq (1 - \lambda)\langle \partial V(x(\tau)), W(x(\tau)) \rangle^- + \lambda \langle \partial V(x(\tau)), F(\tau, x(\tau)) \rangle^-$$
$$\geq \frac{(1 - \lambda)\Theta}{\eta(x(\tau))} \quad \text{(by (8.124))},$$

where $\Theta = \langle \partial V(x(\tau))\partial V(x(\tau)) \rangle^- > 0$. Hence,

$$\partial(V \circ x)(\tau) \subset \left[\frac{(1 - \lambda)\Theta}{\eta(x(\tau))}, \infty \right)$$

From this and (8.125) a contradiction follows, completing the proof. $\square$

(8.127) THEOREM. *Let $F : [0,a] \times \mathbb{R}^n \multimap \mathbb{R}^n$ satisfy* (8.109)–(8.110). *Let $V : \mathbb{R}^n \to \mathbb{R}$ be a guiding potential for F with* $\mathrm{ind}\,(V) \neq 0$. *Then the periodic problem* (8.108) *has a solution.*

PROOF. Let $r_0 > 0$ be such that (8.101) and (8.106.1) are fulfilled. Take $\tilde{r} > r_0$ so that the statements of Lemma (8.115) and (8.123) are both satisfied.

Let $r > \tilde{r}$ be arbitrary. Let $t_r \in (0,a]$ be given by Lemma (8.115). Define the map $\Psi : B_0^r \times [0,1] \multimap \mathbb{R}^n$ by

$$\Psi(x_0, \lambda) = \lambda P_{t_r}^W(x_0) + (1 - \lambda)\partial V(x_0),$$

where $P_{t_r}^W$ is the Poincaré map (see Chapter III.4) associated to the Cauchy problem (8.124). Observe that, by Lemma (8.123), $0 \notin \Psi(x_0, \lambda)$, for every $(x_0, \lambda) \in S^{n-1} \times [0,1]$. Hence, by the invariance under homotopy of the degree (see Chapters I.8 and II.9), $\deg(P_{t_r}^W) = \deg(\partial V)$. Thus,

$$(8.128) \qquad\qquad \deg(P_{t_r}^W) = \mathrm{ind}\,(V).$$

Let P_a^F be the Poincaré map (see Chapter III.4) associated to (8.108). We claim that $P_{t_r}^W$ and $P_{t_a}^F$ are admissibly homotopic. To show this, set

$$k(\lambda) = \begin{cases} 1 & \text{if } \lambda \in [0, 1/2], \\ 2 - 2\lambda & \text{if } \lambda \in [1/2, 1], \end{cases} \qquad h(\lambda) = \begin{cases} t_r + 2(a - t_r)\lambda & \text{if } \lambda \in [0, 1/2], \\ a & \text{if } \lambda \in [1/2, 1], \end{cases}$$

and define the map $G : [0,a] \times \mathbb{R}^n \times [0,1] \multimap \mathbb{R}^n$ by

$$G(t, x, \lambda) = k(\lambda)W(x) + (1 - k(\lambda))F(t, x).$$

Clearly, the map $\chi : B_0^r \times [0,1] \multimap C([0,a], \mathbb{R}^n) \times [0,1]$ defined by

$$\chi(x_0, \lambda) = S^{G(\cdot\,,\,\cdot\,,\lambda)}(x_0) \times \{h(\lambda)\},$$

where S^G denotes the solution set of $\dot{x} \in G(t, x, \lambda)$, $x(0) = x_0$, is u.s.c. with R_δ-values (see Theorem (4.3) in III.4), while the map $e : C([0,a], \mathbb{R}^n) \to \mathbb{R}^n$ given by

$$e(x, \lambda) = x(h(\lambda)) - x(0)$$

is continuous. Now, define the map $H : B_0^r \times [0,1] \to \mathbb{R}^n$ by

$$H(x_0, \lambda) = e(\chi(x_0, \lambda), \lambda).$$

We have $H(x_0, 0) = P_{t_r}^W(x_0)$ and $H(x_0, 1) = P_a^F(x_0)$, for every $x_0 \in B_0^r$. Furthermore, suppose that $0 \notin H(x_0, 1)$, for every $x_0 \in S^{n-1}(r)$ (otherwise, (8.108) has a solution and there is nothing to prove). We have:

(i) $\chi(x_0, 0) = S^W(x_0) \times \{t_r\}$, $\chi(x_0, 1) = S^F(x_0) \times \{a\}$, for every $x_0 \in B_0^r$, where S^W and S^F denote the respective solution sets of the Cauchy problems,

(ii) $e(x, 0) = x(t_r) - x(0)$, $e(x, 1) = x(a) - x(0)$, for every $x \in C([0,a], \mathbb{R}^n)$,

(iii) for every $(x_0, \lambda) \in S^{n-1}(r) \times [0,1]$ and $x \in \chi(x_0, \lambda)$, we have $e(x, \lambda) \neq 0$.

Properties (i) and (ii) are trivially verified. To show (iii), let $(x_0, \lambda) \in S^{n-1} \times [0, 1]$ and $x \in \chi(x_0, \lambda)$ be arbitrary. If $0 \leq \lambda < 1$, we have $e(x, \lambda) = x(h(\lambda)) - x_0 \neq 0$, by Lemma (8.123), while if $\lambda = 1$, $e(x, 1) \neq 0$, from the hypothesis, and so also (iii) is satisfied. Hence $P_{t_r}^F$ and P_a^F are homotopic in the class of admissible maps and, by the homotopy property of the degree (see Chapters I.8 nad II.9), we have

$$\deg(P_{t_r}^W) = \deg(P_a^F).$$

Combining the last equality and (8.128) gives $\deg(P_a^F) \neq 0$. By the existence property of the degree (see again Chapters I.8 and II.9), there is $x_0 \in B_0^r$ such that $0 \in P_a^F(x_0)$ and, subsequently, the periodic problem (8.108) has a solution. This completes the proof. $\qquad\square$

(8.129) REMARK. In Theorem (8.127), the existence of solutions to the periodic problem (8.108) has been established by showing that the degree $\deg(P_a^F)$ of the Poincaré map P_a^F associated to (8.112) is non zero on some ball B_0^r in $\mathbb{R}^n$. The calculation of $\deg(P_a^F)$ carried out in the proof of Theorem (8.127) relies on the guiding potential (the guiding function) method.

Now, we discuss the periodic problem (8.108) under the hypothesis that F admits a guiding convex potential V. Thus, V satisfies, instead of inequality (8.101), the condition (8.102).

(8.130) LEMMA. *The statement of Lemma (8.115) remains valid with a direct convex potential $V : \mathbb{R}^n \to \mathbb{R}$, instead of a direct potential.*

PROOF. Take $\tilde{r} = r_0$, where r_0 is given by (8.101). Let $r > \tilde{r}$ be arbitrary. On the sphere $S^{n-1}(r)$, the function $x_0 \mapsto |W(x_0)|^-$, where $W(x_0) = \partial V(x_0)/\eta(x_0)$, is strictly positive and l.s.c. (W is u.s.c.). Hence, there is $\sigma > 0$ such that

$$|W(x_0)|^- > \sigma, \quad \text{for every } x_0 \in S^{n-1}(r).$$

Since W is u.s.c., for every $x_0 \in S^{n-1}(r)$, there is $\delta(x_0) > 0$ such that

$$W(x) \subset W(x_0) + B_0^{\sigma/2}, \quad \text{for every } x \in B_{x_0}^{\delta(x_0)} \text{(the ball is open)}.$$

By a compactness argument, a finite family, say $\{B_{x_0^i}^{\delta(x_0^i)}\}_{i=1}^k$, is a cover of $S^{n-1}(r)$. Therefore, there is $\delta^* > 0$ such that, for each $x_0 \in S^{n-1}(r)$ and some $1 \leq i \leq k$, we have $B_{x_0}^{\delta^*} \subset B_{x_0^i}^{\delta(x_0^i)}$, and so

(8.131) $$W(x) \subset W(x_0^i) + B_0^{\sigma/2}, \quad \text{for every } x \in B_{x_0}^{\delta^*}.$$

Furthermore, as W is bounded, there is $0 < t_r < \min\{a, 1\}$ such that, for each $x_0 \in S^{n-1}(r)$ and any solution $x : [0, a] \to \mathbb{R}^n$ of (8.114), we have

(8.132) $$x(t) \in B_{x_0}^{\delta^*}, \quad \text{for every } t \in [0, t_r].$$

With this choice of t_r, the statement of the lemma is satisfied.

To see this, let $(x_0, \lambda) \in S^{n-1}(r) \times [0, 1]$ be arbitrary. Let $x : [0, a] \to \mathbb{R}^n$ be any solution of (8.114). In view of (8.132) and (8.131), for any $t \in [0, t_r]$, we have

$$(8.133) \qquad x(t) - x_0 = \int_0^t \dot{x}(s)\, ds \in \int_0^t W(x(s))\, ds \subset t[W(x_0^i) + B_0^{\sigma/2}].$$

Let $t \in (0, t_r]$ be arbitrary. Then,

$$\lambda(x(t) - x_0) + (1 - \lambda)\partial V(x_0)$$
$$\subset \lambda t[W(x_0^i) + B_0^{\sigma/2}] + (1 - \lambda)\eta(x_0)W(x_0) \quad \text{(by (8.133))}$$
$$\subset \lambda t[W(x_0^i) + B_0^{\sigma/2}] + (1 - \lambda)\eta(x_0)[W(x_0^i) + B_0^{\sigma/2}] \quad \text{(by (8.131))}$$
$$= (\lambda t + (1 - \lambda)\eta(x_0))[W(x_0^i) + B_0^{\sigma/2}].$$

Since $\lambda t + (1 - \lambda)\eta(x_0) \geq \lambda t + 1 - \lambda \geq t$ and any $u \in W(x_0^i) + B_0^{\sigma/2}$ satisfies $|u| > \sigma/2$, we have $(\lambda t + (1 - \lambda)\eta(x_0))|u| > t\sigma/2$. Hence,

- $\quad 0 \notin \lambda(x(t) - x_0) + (1 - \lambda)\partial V(x_0), \quad$ for every $t \in (0, t_r]$.

As $(x_0, \lambda) \in S^{n-1}(r) \times [0, 1]$ is arbitrary, the proof is complete. $\qquad \square$

(8.134) LEMMA. *The statement of Lemma (8.123) remains valid with a direct convex potential* $V : \mathbb{R}^n \to \mathbb{R}$, *instead of a direct potential.*

PROOF. Let $r_0 > 0$ be so that (8.105) and (8.106.4) are satisfied. One can check (see Proposition 2.2 in [DBGP2] and cf. [GPl]), that there is $\tilde{r} > r_0$ such that, for each $\lambda \in [0, 1]$ and each solution $x : [0, a] \to \mathbb{R}^n$ of (8.122) with $|x_0| > \tilde{r}$, we have

$$(8.135) \qquad |x(t)| > r_0, \quad \text{for every } t \in [0, a].$$

Let $r > \tilde{r}$ be arbitrary. Let $(x_0, \lambda) \in S^{n-1}(r) \times [0, 1)$ be arbitrary. Let $x : [0, a] \to \mathbb{R}^n$ be a solution of (8.122). We claim that

$$(8.136) \qquad x(t) - x_0 \neq 0, \quad \text{for every } t \in (0, a].$$

Indeed, as $\lambda \in [0, 1)$ and $|x_0| > r_0$, for some $\sigma > 0$, we have

$$(8.137) \qquad |(1 - \lambda)W(x_0)|^- > \sigma.$$

The maps W and F are u.s.c. Thus, there is $0 < \delta < a$ such that $(t, x) \in [0, \delta) \times B_{x_0}^{\delta}$ implies

$$(8.138) \qquad (1 - \lambda)W(x) + \lambda F(t, x) \subset (1 - \lambda)W(x_0) + \lambda F(0, x_0) + B_0^{\sigma/2}.$$

Now, fix $\tau \in (0, \delta)$ in order to have

(8.139) $\qquad\qquad (t, x(t)) \in [0, \delta) \times B_{x_0}^\delta, \quad$ for every $t \in [0, \tau)$.

Let $t \in (0, \tau]$ be arbitrary. We have

$$x(t) - x_0 = \int_0^t \dot{x}(s)\, ds \in \int_0^t [(1-\lambda)W(x(s)) + \lambda F(s, x(s))]\, ds$$

$$\subset t[(1-\lambda)W(x_0) + \lambda F(0, x_0) + B_0^{\sigma/2}] \quad \text{(by (8.139) and (8.138))}.$$

Hence, for some $y_t \in \partial V(x_0)$, $z_t \in F(0, x_0)$ and $v_t \in B_0^{\sigma/2}$, we have

(8.140) $\qquad\qquad x(t) - x_0 = t\left[(1-\lambda)\dfrac{y_t}{\eta(x_0)} + \lambda z_t + v_t\right].$

Since V is convex, $\partial V(x_0)$ coincides with the subdifferential of V at x_0, in the sense of convex analysis. Thus, we have

$$V(x(t)) - V(x_0) \geq \langle y_t, x(t) - x_0 \rangle \quad \text{(as } y_t \in \partial V(x_0))$$

$$= t\left[\frac{1-\lambda}{\eta(x_0)}|y_t|^2 + \lambda\langle y_t, z_t\rangle + \langle y_t, v_t\rangle\right] \quad \text{(by (8.140))}$$

$$\geq t\left[\frac{1-\lambda}{\eta(x_0)}|y_t|^2 + \lambda\langle\partial V(x_0), F(0, x_0)\rangle^- - |y_t|\frac{\sigma}{2}\right]$$

$$\geq t|y_t|\left[(1-\lambda)\frac{|y_t|}{\eta(x_0)} - \frac{\sigma}{2}\right] \quad \text{(by (8.106.1))}$$

$$> t|y_t|\frac{\sigma}{2} \quad \text{(by (8.137))}.$$

Hence, for every $t \in (0, \tau]$,

$$V(x(t)) - V(x_0) > t|y_t|\frac{\sigma}{2}.$$

Now, set $T = \{\tau \in (0, a] \mid V(x(t)) > V(x_0), \text{ for every } t \in (0, \tau]\}$, and let $\tilde{t} = \sup T$. Clearly, $V(x(\tilde{t})) \geq V(x_0)$. If the equality holds, the function $t \mapsto V(x(t))$ has a global maximum at some point $t^* \in (0, \tilde{t})$ and, by (8.135), $|x^*| > r_0$, where $x^* = x(t^*)$. By the previous argument, with t^*, x^* in place of 0, x_0, respectively, one obtains a contradiction. It follows that $V(x(\tilde{t})) > V(x_0)$, and thus $\tilde{t} \in T$. Supposing $\tilde{t} < a$ gives easily a contradiction, by the previous argument, and so $\tilde{t} = a$. Consequently,

$$V(x(t)) > V(x_0), \quad \text{for every } t \in (0, a],$$

which implies (8.136), completing the proof. $\qquad\qquad\qquad\qquad\qquad\qquad\square$

By the argument of Theorem (8.127), using Lemma (8.130) and Lemma (8.134), one can prove:

(8.141) THEOREM. *Let* $F : [0, a] \times \mathbb{R}^n \multimap \mathbb{R}^n$ *satisfy* (8.109)–(8.110). *Let* $V : \mathbb{R}^n \to \mathbb{R}$ *be a guiding convex potential for* F *with* $\mathrm{ind}\,(V) \neq 0$. *Then the periodic problem* (8.108) *has a solution.*

(d) Guiding function for r.h.s. without convex values.

Now, we will consider the periodic problem

(8.142)
$$\begin{cases} \dot{x} \in R(t, x), \\ x(0) = x(a) \end{cases}$$

with R being an u.s.c. map with nonempty, compact, but not necessarily convex values.

To study the problem (8.142), we introduce a suitable class of nonconvex valued maps R, that we shall call regular. In order to simplify the presentation, we consider only the case, when R is bounded.

(8.143) DEFINITION. A bounded multivalued map $R : [0, a] \times \mathbb{R}^n \multimap \mathbb{R}^n$ with nonempty values is said to be *regular* if there is a map $F : [0, a] \times \mathbb{R}^n \multimap \mathbb{R}^n$, called a *regular quasi selection* of R, satisfying the properties:

(i) F is (upper) Carathéodory with nonempty, compact, convex values,

(ii) $F(t, x) \cap R(t, x) \neq \emptyset$, for every $(t, x) \in [0, a] \times \mathbb{R}^n$,

(iii) each solution $x : [0, a] \to \mathbb{R}^n$ of the differential inclusion $\dot{x}(t) \in F(t, x(t))$ is also a solution of $\dot{x}(t) \in R(t, x(t))$.

For any regular map $R : [0, a] \times \mathbb{R}^n \multimap \mathbb{R}^n$, we set

(8.144) $\mathcal{U}(R) = \{F : [0, a] \times \mathbb{R}^n \multimap \mathbb{R}^n \mid F \text{ is a regular quasi-selection of } R\}.$

Clearly, any map $F : [0, a] \times \mathbb{R}^n \multimap \mathbb{R}^n$ with nonempty, compact, convex values, which is bounded and (upper) Carathéodory, is regular. However, there are some important classes of nonconvex valued maps R which are regular. Before describing these classes, we recall some additional notions (see Chapter I.3 or [ADTZ-M]).

Let $R : [0, a] \times \mathbb{R}^n \multimap \mathbb{R}^n$ be a multivalued map with nonempty compact values. R is called *Scorza–Dragoni upper* (resp. *lower*) *semicontinuous*, for short SD-u.s.c., (resp. SD-l.s.c.) if, for every $\varepsilon > 0$, there is a closed set $J \subset [0, a]$ with $m([0, a] - J) < \varepsilon$, m the Lebesgue measure, such that the restriction of R to $J \times \mathbb{R}^n$ is u.s.c. (resp. l.s.c.). A map R which is both SD-u.s.c. and SD-l.s.c. is called *Scorza–Dragoni*, SD, for short.

(8.145) REMARK. Let $R : [0, a] \times \mathbb{R}^n \multimap \mathbb{R}^n$ be bounded, with nonempty compact values. As is well-known (see Chapter I.3 or [ADTZ-M]), we have:

(i) R is SD if and only if R is Carathéodory,

(ii) if R is SD-u.s.c. (SD-l.s.c.), then R is (upper) Carathéodory ((lower) Carathéodory).

The following result, due to A. Bressan [Bre3], shows the significance of the class of regular maps.

(8.146) THEOREM. *Let $R : [0,a] \times \mathbb{R}^n \multimap \mathbb{R}^n$ be a bounded multivalued map with nonempty compact values. If, in addition, R is SD-l.s.c., then R is regular. In particular, any bounded Carathéodory map $R : [0,a] \times \mathbb{R}^n \multimap \mathbb{R}^n$ with nonempty compact values is regular.*

Now, let us consider the periodic problem (8.142) with R not necessarily convex valued. Even in the simplest cases, say when R is continuous with nonempty compact values, the solution set map S^R is no longer upper (or lower) semicontinuous and, furthermore, S^R takes values which are not necessarily closed. Thus, the construction of a degree for the Poincaré map $P_a^F = e_a \circ S^F$ associated to (8.112) cannot be carried out as in the convex case, but some modifications are needed. This will be discussed in the following section.

We state below some immediate properties of regular maps, we shall need later.

(8.147) PROPOSITION. *Let $R : [0,a] \times \mathbb{R}^n \multimap \mathbb{R}^n$ be regular. Then, for each $F \in \mathcal{U}(R)$, the solution set map $S^F : \mathbb{R}^n \multimap C([0,a], \mathbb{R}^n)$ is u.s.c. with R_δ-values, and satisfies*

$$S^F(x_0) \subset S^R(x_0), \quad \text{for every } x_0 \in \mathbb{R}^n.$$

(8.148) PROPOSITION. *Let $R : [0,a] \times \mathbb{R}^n \multimap \mathbb{R}^n$ be regular, and let $F_1, F_2 \in \mathcal{U}(R)$. Then, for every $t_0 \in [0,a]$, the map $H_{t_0} : [0,a] \times \mathbb{R}^n \multimap \mathbb{R}^n$ given by*

$$H_{t_0}(t,x) = \begin{cases} F_1(t,x), & 0 \le t < t_0, \\ \overline{\text{conv}}\{F_1(t_0,x) \cup F_2(t_0,x)\}, & t = t_0, \\ F_2(t,x), & t_0 < t \le a \end{cases}$$

is also in $\mathcal{U}(R)$.

(8.149) PROPOSITION. *Let $R : [0,a] \times \mathbb{R}^n \multimap \mathbb{R}^n$ be regular, and let $F_1, F_2 \in \mathcal{U}(R)$. Then the map $H : [0,a] \times \mathbb{R}^n \times [0,1] \multimap \mathbb{R}^n$ given by*

$$H(t,x,\lambda) = \begin{cases} F_1(t,x), & 0 \le t < \lambda a, \\ \overline{\text{conv}}\{F_1(\lambda a, x) \cup F_2(\lambda a, x)\}, & t = \lambda a, \\ F_2(t,x), & \lambda a < t \le a \end{cases}$$

satisfies the properties:

  (i) $H(\cdot, \cdot, \lambda) \in \mathcal{U}(R)$, *for every $\lambda \in [0,1]$,*
  (ii) $H(t,x,0) = F_2(t,x)$, *for every $(t,x) \in (0,a] \times \mathbb{R}^n$,*
  (iii) $H(t,x,1) = F_1(t,x)$, *for every $(t,x) \in [0,a) \times \mathbb{R}^n$.*

(8.150) PROPOSITION. *Let* $R : [0, a] \times \mathbb{R}^n \multimap \mathbb{R}^n$ *be regular. The periodic problem* (8.142) *has a solution if and only if, for some* $x_0 \in \mathbb{R}^n$, *we have* $0 \in P_a^R(x_0)$, *where* P_a^R *is the Poincaré map associated to*

(8.151)
$$\begin{cases} \dot{x} \in R(t, x), \\ x(0) = x_0. \end{cases}$$

We need to define a topological degree for a class of multivalued maps which covers the case of the Poincaré map P_a^R occurring in periodic problems (8.142), with R regular. The construction is a modification of that considered in Chapter II.9 (cf. also Chapter I.8).

(8.152) DEFINITION. Let $S : B_0^r \multimap X$ be an arbitrary multivalued map, with $X \in \mathrm{ANR}$. We say that $S \in K(B_0^r, X)$ if the following two properties are fulfilled:

(i) There exists a nonempty set $\mathcal{V}(S) \subset J(B_0^r, X) := \{F : B_0^r \multimap X \mid F$ is u.s.c. with R_δ-values$\}$ such that each $s \in \mathcal{V}(S)$ satisfies

$$s(u) \subset S(u), \quad \text{for every } u \in B_0^r.$$

(ii) If $s_1, s_2 \in \mathcal{V}(S)$, there exists an u.s.c. homotopy $\chi : B_0^r \times [0, 1] \multimap X$ with R_δ-values such that

(8.153) $\chi(u, 0) = s_1(u)$ $\chi(u, 1) = s_2(u)$,   for every $u \in B_0^r$,

(8.154) $\chi(u, \lambda) \subset S(u)$,   for every $(u, \lambda) \in B_0^r \times [0, 1]$.

For any continuous map $f : X \to \mathbb{R}^n$, $X \in \mathrm{ANR}$, we set

$$K_f(B_0^r, \mathbb{R}^n) = \{\varphi : B_0^r \multimap \mathbb{R}^n \mid \varphi = f \circ S, \text{ for some}$$
$$S \in K(B_0^r, \mathbb{R}^n), \text{ and } \varphi(S^{n-1}(r)) \subset \mathbb{R}^n \setminus \{0\}\},$$

$$CK(B_0^r, \mathbb{R}^n) = \bigcup \{K_f(B_0^r, \mathbb{R}^n) \mid f : X \to \mathbb{R}^n \text{ is continuous, where } X \in \mathrm{ANR}\}.$$

Observe that $J(B_0^r, X) \subset K(B_0^r, \mathbb{R}^n)$ if we take $\mathcal{V}(S) = \{S\}$, for any $S \in J(B_0^r, X)$. Similarly, $J_f(B_0^r, \mathbb{R}^n) := \{\varphi : B_0^r \to \mathbb{R}^n \mid \varphi = f \circ F, \text{ for some } F \in J(B_0^r, X), \text{ and } \varphi(S^{n-1}(r)) \subset \mathbb{R}^n \setminus \{0\}\} \subset K_f(B_0^r, \mathbb{R}^n)$, for any continuous map $f : X \to \mathbb{R}$, and so

$$CJ(B_0^r, \mathbb{R}^n) := \bigcup \{J_f(B_0^r, \mathbb{R}^n) \mid f : X \to \mathbb{R}^n \text{ is continuous, where } X \in \mathrm{ANR}\}$$
$$\subset CK(B_0^r, \mathbb{R}^n).$$

Now, our purpose is to extend the definition of the topological degree, introduced in Chapter II.9, for maps $\varphi \in CJ(B_0^r, \mathbb{R}^n)$, to the larger class of maps $\varphi \in CK(B_0^r, \mathbb{R}^n)$. To this end, we need an appropriate definition of homotopy in $CK(B_0^r, \mathbb{R}^n)$.

(8.155) DEFINITION. Let $\varphi_1, \varphi_2 \in CK(B_0^r, \mathbb{R}^n)$ be two maps of the form:

$$\varphi_1 = f_1 \circ S_1, \quad B_0^r \xrightarrow{S_1} X \xrightarrow{f_1} \mathbb{R}^n,$$

$$\varphi_2 = f_2 \circ S_2, \quad B_0^r \xrightarrow{S_2} X \xrightarrow{f_2} \mathbb{R}^n,$$

where $S_1, S_2 \in K(B_0^r, X)$, and $f_1, f_2 : X \to \mathbb{R}^n$ are continuous. We say that φ_1 and φ_2 are *homotopic* in $CK(B_0^r, \mathbb{R}^n)$ if there exist $s_1 \in \mathcal{V}(S_1)$ and $s_2 \in \mathcal{V}(S_2)$ such that the maps

$$\psi_1 = f_1 \circ s_1 \quad \text{and} \quad \psi_2 = f_2 \circ s_2$$

are homotopic in $CJ(B_0^r, \mathbb{R}^n)$.

(8.156) DEFINITION. Let $\varphi \in CK(B_0^r, \mathbb{R}^n)$ be a map of the form

$$\varphi = f \circ S, \quad B_0^r \xrightarrow{S} X \xrightarrow{f} \mathbb{R}^n,$$

where $S \in K(B_0^r, X)$ and $f : X \to \mathbb{R}^n$ is continuous. Let $\mathcal{V}(S)$ correspond to S according to Definition (8.152). We define the *topological degree* $\deg(\varphi, \mathcal{V}(S))$ of φ with respect to $\mathcal{V}(S)$ by

$$\deg(\varphi, \mathcal{V}(S)) = \deg(\psi),$$

where $\psi \in CJ(B_0^r, \mathbb{R}^n)$ is any map of the form $\psi = f \circ s$, for some $s \in \mathcal{V}(S)$, and $\deg(\psi)$ is the topological degree defined in Chapter II.9.

Definition (8.156) is meaningful in view of the following:

(8.157) PROPOSITION. *Let* $\varphi \in CK(B_0^r, \mathbb{R}^n)$ *be as in Definition* (8.156). *Let* $\psi_1, \psi_2 \in CJ(B_0^r, \mathbb{R}^n)$ *be two maps of the form* $\psi_1 = f \circ s_1, \psi_2 = f \circ s_2$, *for some* $s_1, s_2 \in \mathcal{V}(S)$. *Then we have* $\deg(\psi_1) = \deg(\psi_2)$.

PROOF. Since $S \in K(B_0^r, X)$ and $s_1, s_2 \in \mathcal{V}(S)$, there is an u.s.c. map $\chi : B_0^r \times [0, 1] \multimap X$ with R_δ-values satisfying (8.153) and (8.154). It follows that ψ_1 and ψ_2 are homotopic in $CJ(B_0^r, \mathbb{R}^n)$, (for more details, see [DBGP2]), and subsequently $\deg(\psi_1) = \deg(\psi_2)$. This completes the proof. $\square$

(8.158) REMARK. The topological degree given by Definition (8.156) may depend upon the choice of the family $\mathcal{V}(S)$. To see that, let $S : B_0^r \multimap \mathbb{R}^2$ be given by

$$S(x) = S^1(1),$$

and let $f = \mathrm{id}_{\mathbb{R}^2}$. Define $\varphi : B_0^r \multimap \mathbb{R}^2$ by

$$\varphi(x) = (f \circ S)(x).$$

For $k \in \mathbb{N}$, let $\mathcal{V}_k(S) = \{f_k\}$, where $f_k(z) = z^k$ (we identify $\mathbb{R}^2$ with the field of complex numbers). Then, for each $k \in \mathbb{N}$, we have

$$\deg(\varphi, \mathcal{V}(S)) = k.$$

Some properties of the topological degree we have just defined are listed in the following.

(8.159) THEOREM. *We have*

(i) *If* $\varphi = f \circ S \in CJ(B_0^r, \mathbb{R}^n)$, *then*

$$\deg(\varphi, \{S\}) = \deg(\varphi),$$

where $\deg(\varphi)$ *is defined in Chapter II.9.*

(ii) *If* $\varphi = f \circ S \in CK(B_0^r, \mathbb{R}^n)$ *satisfies* $\deg(\varphi, \mathcal{V}(S)) \neq 0$, *then there is* $u \in B_0^r$ *such that* $0 \in \varphi(u)$.

(iii) *If* $\varphi_1 = f_1 \circ S_1$ *and* $\varphi_2 = f_2 \circ S_2$ *are homotopic in* $CK(B_0^r, \mathbb{R}^n)$, *then*

$$\deg(\varphi_1, \mathcal{V}(S_1)) = \deg(\varphi_2, \mathcal{V}(S_2)).$$

(iv) *If* $\varphi = f \circ S \in CK(B_0^r, \mathbb{R}^n)$ *and* $\{u \in B_0^r \mid 0 \in \varphi(u)\} \subset B_0^{\widetilde{r}}$, *where* $0 < \widetilde{r} < r$, *then*

$$\deg(\widetilde{\varphi}, \mathcal{V}(\widetilde{S})) = \deg(\varphi, \mathcal{V}(S)),$$

where $\widetilde{\varphi}$, $\widetilde{S}$ *stands, respectively, for the restriction of* φ, S *to* $B_0^{\widetilde{r}}$.

(v) *Let* $\varphi_1, \varphi_2 \in CK(B_0^r, \mathbb{R}^n)$ *be two maps of the form*

$$\varphi_1 = f_1 \circ S_1, \quad B_0^r \overset{S_1}{\longrightarrow} X \overset{f_1}{\longrightarrow} \mathbb{R}^n,$$
$$\varphi_2 = f_2 \circ S_2, \quad B_0^r \overset{S_2}{\longrightarrow} Y \overset{f_2}{\longrightarrow} \mathbb{R}^n,$$

where $X, Y \in \mathrm{ANR}$. *If moreover,*

$$0 \notin \lambda \varphi_1(u) + (1 - \lambda)\varphi_2(u), \quad \text{for every } (u, \lambda) \in S^{n-1}(r) \times [0, 1],$$

then

$$\deg(\varphi_1, \mathcal{V}(S_1)) = \deg(\varphi_2, \mathcal{V}(S_2)).$$

PROOF. (i) is obvious. To show (ii), let $\psi \in CJ(B_0^r, \mathbb{R}^n)$ be any map of the form $\psi = f \circ s$, for some $s \in \mathcal{V}(S)$. Since $\deg(\psi) \neq 0$, there exists $u \in B_0^r$ such that $0 \in \psi(u)$. As $\psi(u) \subset \varphi(u)$, we have $0 \in \varphi(u)$, proving (ii).

Consider (iii), and let φ_1, φ_2 be homotopic in $CK(B_0^r, \mathbb{R}^n)$. By Definition (8.155), there exist $s_1 \in \mathcal{V}(S_1)$ and $s_2 \in \mathcal{V}(S_2)$ such that the maps $\psi_1 = f_1 \circ s_1$ and $\psi_2 = f_1 \circ s_2$ are homotopic in $CJ(B_0^r, \mathbb{R}^n)$. Because of $\deg(\psi_1) = \deg(\psi_2)$, we get $\deg(\varphi_1, \mathcal{V}(S_1)) = \deg(\varphi_2, \mathcal{V}(S_2))$, proving (iii).

Now, suppose $\varphi_1 \in CK(B_0^r, \mathbb{R}^n)$ satisfies the assumption in (iv). Let $\psi = f \circ s \in CJ(B_0^r, \mathbb{R}^n)$, for some $s \in \mathcal{V}(S)$. Denote by $\widetilde{s}$ (resp. $\widetilde{\psi}$) the restriction of s (resp. ψ) to $B_0^{\widetilde{r}}$. Clearly, $\widetilde{\psi} = f \circ \widetilde{s}$ and $\widetilde{s} \in \mathcal{V}(\widetilde{S})$. Because of $\deg(\widetilde{\psi}) = \deg(\psi)$, we get $\deg(\widetilde{\psi}, \mathcal{V}(\widetilde{S})) = \deg(\psi, \mathcal{V}(S))$, proving (iv).

To show (v), let $\varphi_1, \varphi_2 \in CK(B_0^r, \mathbb{R}^n)$ satisfy the assumption in (v). Let $\psi_1, \psi_2 \in CJ(B_0^r, \mathbb{R}^n)$ be given by $\psi_1 = f_1 \circ s_1$, $\psi_2 = f_2 \circ s_2$, for some $s_1 \in \mathcal{V}(S_1)$, $s_2 \in \mathcal{V}(S_2)$. From the hypothesis,

$$0 \notin \lambda\psi_1(u) + (1 - \lambda)\psi_2(u), \quad \text{for every } (u, \lambda) \in S^{n-1}(r) \times [0, 1].$$

Because of $\deg(\psi_1) = \deg(\psi_2)$, we get

$$\deg(\varphi_1, \mathcal{V}(S_1)) = \deg(\varphi_2, \mathcal{V}(S_2)),$$

proving (v). This completes the proof. $\qquad\square$

Now, the generalized topological degree defined above is ready to be used in the study of periodic problems (8.142), with R not necessarily convex valued.

Consider the periodic problem (8.142), where $R : [0, a] \times \mathbb{R}^n \multimap \mathbb{R}^n$ is a regular map. We recall that, for $t \in [0, a]$, the Poincaré map $P_t^R : \mathbb{R}^n \multimap \mathbb{R}^n$ associated to (8.151) is given by

$$P_t^R = e_t \circ S^R, \quad \mathbb{R}^n \overset{S^R}{\multimap} C([0, a], \mathbb{R}^n) \overset{e_t}{\multimap} \mathbb{R}^n,$$

where S^R is the solution set map of the corresponding Cauchy problem (8.151), and $e_t : C([0, a], \mathbb{R}^n) \multimap \mathbb{R}^n$ is given by $e_t(x) = x(0) - x(t)$.

(8.160) PROPOSITION. *Let $R : [0, a] \times \mathbb{R}^n \multimap \mathbb{R}^n$ be a regular map. We have*

(i) $P_a^F(u) \subset P_a^R(u)$, *for each* $F \in \mathcal{U}(R)$ *and* $u \in \mathbb{R}^n$,
(ii) $S^R \in K(B_0^r, C([0, a], \mathbb{R}^n))$, *with* $\mathcal{V}(S^R) = \{S^F \mid F \in \mathcal{U}(R)\}$.

If, in addition, $P_a^R(S^{n-1}(r)) \subset \mathbb{R}^n \setminus \{0\}$, for some $r > 0$, we have

(iii) $P_a^F \in CJ(B_0^r, \mathbb{R}^n)$, *for every* $F \in \mathcal{U}(R)$,
(iv) $P_a^R \in CK(B_0^r, \mathbb{R}^n)$.

PROOF. (i) Since R is regular, for each $F \in \mathcal{U}(R)$ and $u \in \mathbb{R}^n$, we have $S^F(u) \subset S^R(u)$, which yields $P_a^F(u) \subset P_a^R(u)$, proving (i).

(ii) Let R be regular, and let $\mathcal{U}(R)$ correspond, according to (8.144). Clearly, any $S^F \in \mathcal{V}(S^R)$, where $F \in \mathcal{U}(R)$, is also in $J(B_0^r, C([0,a], \mathbb{R}^n))$ and satisfies $S^F(u) \subset S^R(u)$, for every $u \in B_0^r$. Furthermore, let $S^{F_1}, S^{F_2} \in \mathcal{V}(S^R)$, for some $F_1, F_2 \in \mathcal{U}(R)$. Let $H : [0,a] \times \mathbb{R}^n \times [0,1] \multimap \mathbb{R}^n$ be as in Proposition (8.149). Then the map $\chi : B_0^r \times [0,1] \multimap C([0,a], \mathbb{R}^n)$ given by

$$\chi(u,\lambda) = S^{H(\,\cdot\,,\,\cdot\,,\lambda)}(u)$$

is u.s.c. with R_δ-values, and satisfies:

$$\chi(u,0) = S^{F_2}(u), \quad \chi(u,1) = S^{F_1}(u), \quad \text{for every } u \in B_0^r,$$
$$\chi(u,\lambda) \subset S^R(u), \qquad\qquad\qquad \text{for every } (u,\lambda) \in B_0^r \times [0,1].$$

Thus, $S^R \in K(B_0^r, C([0,a], \mathbb{R}^n))$ and (ii) is proved.

(iii) It follows from Theorem 4.3 in Chapter III.4.

(iv) The map $P_a^R = e_a \circ S^R$ is in $CK(B_0^r, \mathbb{R}^n)$, because $S^R \in K(B_0^r, C([0,a], \mathbb{R}^n))$ and $P_a^R(S^{n-1}(r)) \subset \mathbb{R}^n \setminus \{0\}$. This completes the proof. $\qquad\square$

(8.161) THEOREM. *Let $R : [0,a] \times \mathbb{R}^n \multimap \mathbb{R}^n$ be a regular map. Let $P_a^R(r) : \mathbb{R}^n \multimap \mathbb{R}^n$, the Poincaré map associated with (8.151), satisfy $P_a^R(r) \subset \mathbb{R}^n \setminus \{0\}$, for some $r > 0$. If, in addition, $\deg(P_a^R, \mathcal{V}(S^R)) \neq 0$, where $\mathcal{V}(S^R) = \{S^F \mid F \in \mathcal{V}(R)\}$, then the periodic problem (8.142) has a solution.*

PROOF. By Proposition (8.160)(iv), $P_a^R \subset CK(B_0^r, \mathbb{R}^n)$. Thus, by Theorem (8.159)(ii), there is $x_0 \in B_0^r$ such that $0 \in P_a^R(x_0)$. By Proposition (8.150), the periodic problem (8.142) has a solution, completing the proof. $\qquad\square$

To apply Theorem (8.161), one should know the topological degree $\deg(P_a^R, \mathcal{V}(S^R))$ of the Poincaré map P_a^R, with respect to $\mathcal{V}(S^R)$. In some cases, the following result can be useful.

(8.162) THEOREM. *Let $R : [0,a] \times \mathbb{R}^n \multimap \mathbb{R}^n$ be a regular map. Assume that $V : \mathbb{R}^n \to \mathbb{R}$ is a C^1-guiding potential for R with $\mathrm{ind}(V) \neq 0$. Then the periodic problem (8.142) has a solution.*

PROOF. Let $F \in \mathcal{U}(R)$. From the definition of $\mathcal{U}(R)$, we have:

(j) F is (upper) Carathéodory, with nonempty, compact, convex values,

(jj) $F(t,x) \cap R(t,x) \neq \emptyset$, for every $(t,x) \in [0,a] \times \mathbb{R}^n$,

(jjj) $S^F(x_0) \subset S^R(x_0)$, for every $x_0 \in \mathbb{R}^n$.

In view of (jjj), it suffices to show that the periodic problem (8.142) has a solution.

Observe that, in general, V is not a guiding potential for F, because F is not necessarily a selection of R. To overcome this difficulty, we introduce an auxiliary map.

Hence, fix $r_0 > 0$ so that (8.101) and (8.106.1) (with $\operatorname{grad} V$ in place of ∂V) are both satisfied. Define $\widetilde{F} : [0, a] \times \mathbb{R}^n \multimap \mathbb{R}^n$ by

$$\widetilde{F}(t, x) = \begin{cases} F(t, x) & \text{if } (t, x) \in [0, a] \times B_0^{r_0}, \\ F(t, x) \cap H(x) & \text{if } (t, x) \in [0, a] \times (\mathbb{R}^n \setminus B_0^{r_0}), \end{cases}$$

where $H(x) = \{ y \in \mathbb{R}^n \mid \langle y, \operatorname{grad} V(x) \rangle \geq 0 \}$. It is a routine to check that $\widetilde{F}$ is (upper) Carathéodory, with nonempty, compact, convex values. Moreover, let $(t, x) \in [0, a] \times \mathbb{R}^n$, $|x| > r_0$ be arbitrary. Take $y \in F(t, x) \cap R(t, x)$, a nonempty set by (jj). Since X is a C^1-guiding potential for R, we have

$$\langle y, \operatorname{grad} V(x) \rangle \geq \langle R(t, x), \operatorname{grad} V(x) \rangle^- \geq 0;$$

thus, $y \in \widetilde{F}(t, x)$. Hence, V is C^1-weakly guiding potential for $\widetilde{F}$. By Theorem (8.107), the periodic problem

$$\begin{cases} \dot{x} \in \widetilde{F}(t, x), \\ x(0) = x(a) \end{cases}$$

has a solution which, clearly, is a solution of (8.108) and also of (8.142). This completes the proof. $\qquad\square$

In the following theorem, the guiding potential V associated to F is supposed to be only locally Lipschitzian, instead of C^1.

Recall that, if $f : [0, a] \times \mathbb{R}^n \to \mathbb{R}^n$ is a bounded map, the *Filippov regularization* of f is the map $F : [0, a] \times \mathbb{R}^n \multimap \mathbb{R}^n$ given by

$$F(t, x) = \bigcap_{\delta > 0} \overline{\operatorname{conv}} f(B((t, x), \delta)),$$

where $B((t, x), \delta) = \{ (s, y) \in [0, a] \times \mathbb{R}^n \mid |s - t| < \delta \text{ and } |y - x| < \delta \}$. Observe that F is a bounded u.s.c. map with nonempty, compact, convex values, and that f is a selection of F.

(8.163) THEOREM. *Let $R : [0, a] \times \mathbb{R}^n \multimap \mathbb{R}^n$ be a bounded continuous map with nonempty compact values. Let $V : \mathbb{R}^n \multimap \mathbb{R}$ be a locally Lipschitzian guiding potential for R with $\operatorname{ind}(V) \neq 0$. Then the periodic problem (8.142) has a solution.*

PROOF. According to [Bre3, 'Corollary 2.2 and Lemma 3.2], there exists a single-valued map $f : [0, a] \times \mathbb{R}^n \to \mathbb{R}^n$ such that

(j) $f(t, x) \in R(t, x)$, for every $(t, x) \in [0, a] \times \mathbb{R}^n$,
(jj) $S^f(x_0) = S^F(x_0)$, for every $x_0 \in \mathbb{R}^n$,

where F denotes the Filippov regularization of f. We have

(8.164) $F(t,x) \subset \overline{\text{conv}}\, R(t,x)$,   for every $(t,x) \in [0,a] \times \mathbb{R}^n$.

Let $(t,x) \in [0,a] \times \mathbb{R}^n$ and let $\varepsilon > 0$ be arbitrary. Since R is continuous, there is $\delta > 0$ such that $(s,y) \in B((t,r),\delta)$ implies $R(s,y) \subset R(t,x) + B_0^\varepsilon$. Hence, in view of (j), we have

$$F(t,x) \subset \overline{\text{conv}}\, f(B((t,x),\delta)) \subset \overline{\text{conv}}\, R(t,x) + B_0^\varepsilon$$

and, since $\varepsilon > 0$ is arbitrary, (8.164) is proved.

By the hypothesis, V is a guiding potential for R. Let $r_0 > 0$ correspond, according to Definition (8.106). For each $(t,x) \in [0,a] \times \mathbb{R}^n$, with $|x| > r_0$, we have:

$$\langle F(t,x), \partial V(x) \rangle^- \geq \langle \overline{\text{conv}}\, R(t,x), \partial V(x) \rangle^- \quad \text{(by (8.164))}$$
$$\geq \langle R(t,x), \partial V(x) \rangle^- \qquad \text{(by Lemma (8.99)(viii)).}$$

Thus, V is a guiding potential for F. Furthermore, F is a bounded u.s.c. map with nonempty, compact, convex values. Thus, (8.108) has a solution, by Theorem (8.127), and subsequently also the periodic problem (8.142) has a solution, in view of (j) and (jj). This completes the proof.   □

(8.165) REMARK. The full analogy of Theorem (8.127) was established for the retarted functional differential inclusion

(8.166) $\dot{x}(t) \in F(t, x(t-\tau_1), \ldots, x(t-\tau_m))$,   for a.a. $t \in [0,a]$,

in [GaPi]. For a C^1-guiding potential, inclusion (8.166) can be considered there even (upper-) Carathéodory (cf. Theorem (8.106)). Although the analogy of Theorem (8.162) seems to be available for the nonconvex-valued r.h.s. in inclusion (8.166), it is a question whether or not the same is true for Theorem (8.163).

(8.167) REMARK. For a single-valued integrably bounded Carathéodory r.h.s. $f : [0,a] \times \mathbb{R}^n \to \mathbb{R}^n$, if a locally Lipschitzian guiding potential $V(x)$ is coercive, i.e. $\lim_{|x| \to \infty} V(x) = \infty$, then $\text{ind}(V) = 1$ (see [Lew]), and subsequently periodic problem

$$\begin{cases} \dot{x} = f(t,x), \\ x(0) = x(a) \end{cases}$$

admits a solution.

(e) Guiding functions for discontinuous r.h.s.

The inclusion with Γ^M-directionally u.s.c. r.h.s. $\varphi : [a, b] \times \mathbb{R}^n \multimap \mathbb{R}^n$,

$$\dot{x} \in \varphi(t, x), \tag{8.168}$$

was already treated in Chapters III.2 and III.4. Now, we would like to elaborate the obtained results by means of (for the sake of simplicity) C^1-guiding functions.

Before, it will be convenient to recall some auxiliary results (see [GPl] and cf. (c)).

(8.169) LEMMA. *Let V be a C^1-direct potential with a constant $r_0 > 0$. Then, for every $r > r_0 + a$, there exists a $t_r \in (0, a]$ such that any solution $x(t) : [0, a] \to \mathbb{R}^n$, $|x_0| = r$, of the differential equation*

$$\dot{x}(t) = W_V(x(t)), \tag{8.170}$$

where $(\nabla V = \mathrm{grad}\, V)$

$$W_V(x) = \begin{cases} \nabla V(x) & \text{if } |\nabla V(x)| \leq 1, \\ \dfrac{\nabla V(x)}{|\nabla V(x)|} & \text{if } |\nabla V(x)| > 1, \end{cases} \tag{8.171}$$

satisfies the following conditions

(i) $\langle x(t) - x(0), \nabla V(x, (0)) \rangle > 0$, *for all $t \in (0, t_t]$,*

(ii) $x(t) - x(0) \neq 0$, *for all $t \in (0, a]$.*

(8.172) LEMMA. *Let $x(t)$ be a solution of the inclusion (8.168) with a linearly bounded right-hand side φ, i.e. $|y| \leq \eta(t)(1 + |x|)$; for all $y \in \varphi(t, x)$ and let $r_0 > 0$ be given. If $x(t)$ satisfies*

$$|x(0)| > \left(r_0 + \int_0^a \eta(\tau)\, d\tau \right) \exp \left(\int_0^a \eta(\tau)\, d\tau \right) = r_\eta,$$

then $|x(t)| > r_0$, for every $t \in [0, a]$.

PROOF. Suppose that there exists a solution $x(t)$ and $t_0 \in [0, a]$ such that $|x_0| > r_\eta$ and $|x(t_0)| \leq r_0$. Define $y(t) := x(t_0 - t)$, $\xi(t) := \mu(t_0 - t)$, $\psi(t, x) := -\varphi(t_0 - t, x)$, for every $t \in [0, t_0]$. Obviously, $\dot{y}(t) \in \psi(t, y(t))$. Since, for every $z \in \psi(t, y)$, $|z| \leq \xi(t)(1 + |y|)$, we obtain, by using the Gronwall inequality (see [Har-M]),

$$|y(t)| \leq \left(|y(0)| + \int_0^t \xi(\tau)\, d\tau \right) \exp \left(\int_0^t \xi(\tau)\, d\tau \right) \leq r_\eta,$$

for every $t \in [0, t_0]$. Thus, $|x(0)| = |y(t_0)| \leq r_\eta$, and we get a contradiction. $\qquad \square$

(8.173) THEOREM. *Let $\varphi : [0, a] \times \mathbb{R}^n \multimap \mathbb{R}^n$ be a Γ^M-directionally upper-semicontinuous map with bounded convex values, $\varphi(t, x) \subset B_0^L$, for all $(t, x) \in [0, a] \times \mathbb{R}^n$, $L < M$. Let there exist a C^1-guiding function $V : \mathbb{R}^n \to \mathbb{R}$, for the map φ, such that $\operatorname{ind} V \neq 0$. Then the periodic problem*

(8.174)
$$\begin{cases} \dot{x}(t) \in \varphi(t, x(t)), \\ x(0) = x(a) \end{cases}$$

has a solution.

PROOF. Let us choose $r_0 > 0$, according to (8.101) and (8.106.1). Define the function $\alpha : \mathbb{R}^n \to \mathbb{R}^n$, $\alpha(x) := 0$, for $|x| \leq r_0$, and $\alpha(x) := 1$, for $|x| > r_0$. Define also the map $A : \mathbb{R}^n \multimap \mathbb{R}^n$ by

$$A(x) := \{y \in \mathbb{R}^n \mid \langle y, \alpha(x)\nabla V(x)\rangle \geq 0\}.$$

According to Definition (8.106), it follows immediately that the graph Γ_A of A is closed. Furthermore, the map $\varphi_V(t, x) = \varphi(t, x) \cap A(x)$ is defined correctly (i.e. it has nonempty values). Moreover, φ_V is bounded and Γ^M-directionally u.s.c.

To fulfil the assumptions of Lemma (8.172), we can take $\eta(t) = L$, for all $t \in [0, a]$. So there exists $r \geq r_0 + a$ such that, for every $\kappa \in [0, 1]$, $|x| \geq r$, and for every solution $x : [0, a] \to \mathbb{R}^n$ of the Cauchy problem

$$\begin{cases} \dot{x}(t) \in \kappa W_V(x(t)) + (1 - \kappa)\varphi_V(t, x(t)), \\ x(0) = x_0, \end{cases}$$

the following inequality holds

$$|x(t)| \geq r_0, \quad \text{for every } t \in [0, a].$$

According to Lemma (8.169), let us choose $t_r > 0$ and define a decomposable homotopy $H_1 : \overline{B_0^r} \times [0, 1] \multimap \mathbb{R}^n$,

$$H_1(x_0, \lambda) = (1 - \lambda)\nabla V(x_0) + \lambda e_{t_r}(\mathcal{S}_{W_v}(x_0)),$$

where $\mathcal{S}_{W_v}$ denotes the solution set map to (8.170).

From Lemma (8.169) and from the properties of the Brouwer degree, we obtain

$$\deg(e_{t_r} \circ \mathcal{S}_{W_V}, B_0^r) = \operatorname{ind} V.$$

Let us define

$$k(\lambda) := \begin{cases} 1, & \text{for } \lambda \in [0, 1/2), \\ 2 - 2\lambda, & \text{for } \lambda \in [1/2, 1), \end{cases}$$

$$h(\lambda) := \begin{cases} 2(a - t_r)\lambda + t_r, & \text{for } \lambda \in [0, 1/2), \\ a, & \text{for } \lambda \in [1/2, 1). \end{cases}$$

The map $G : [0, a] \times \mathbb{R}^n \times [0, 1] \multimap \mathbb{R}^n$,

$$(8.175) \qquad G(t, x, \lambda) = k(\lambda)W_V(x) + (1 - k(\lambda))\varphi_V(t, x),$$

satisfies the assumptions of Theorem (4.42) in Chapter III.4. Therefore, the map $\chi : \overline{B_0^r} \times [0, 1] \multimap C([0, a], \mathbb{R}^n) \times [0, a]$,

$$(8.176) \qquad \chi(x_0, \lambda) = S_{G(\cdot, \cdot, \lambda)}(x_0) \times \{h(\lambda)\},$$

is u.s.c. and $\chi(x_0, \lambda)$ is an R_δ-set, for every $(x_0, \lambda) \in \mathbb{R}^n \times [0, 1]$. Therefore, the homotopy $H : \overline{B_0^r} \times [0, 1] \multimap \mathbb{R}^n$ given by $H = e \circ \chi$ is decomposable.

Now, we show that $0 \notin H(x_0, \lambda)$, for $|x_0| = r$ and $\lambda \in [0, 1]$.

If $\lambda \in [0, 1/2]$ and $z \in H(x_0, \lambda)$, then $k(\lambda) = 1$, and there exists a solution $x(t) : [0, a] \to \mathbb{R}^n$ of the Cauchy problem

$$\begin{cases} \dot{x}(t) = W_V(x(t)), \\ x(0) = x_0 \end{cases}$$

such that $z = x(h(\lambda)) - x_0$ (see the definition of the evaluation mapping $e(\cdot, \cdot)$ in Chapter III.4.F). By Lemma (8.169)(ii), we have that $z \neq 0$.

If $\lambda \in [1/2, 1)$ and $z \in H(x_0, \lambda)$, then there exists a solution $x(t) : [0, a] \to \mathbb{R}^n$ of the Cauchy problem

$$\begin{cases} \dot{x}(t) \in G(t, x(t), \lambda), \\ x(0) = x_0 \end{cases}$$

such that $z = x(a) - x_0$. From (8.169) and (8.175), we have

$$\langle k(\lambda)W_V(x(t)) + (1 - k(\lambda))y, \nabla V(x(t)) \rangle > 0,$$

for every $y \in \varphi_v(t, x(t))$, and consequently

$$\langle \dot{x}(t), \nabla V(x(t)) \rangle > 0, \quad \text{for almost all } t \in [0, a].$$

Therefore,

$$V(x(a)) - V(x(0)) = \int_0^a \langle \dot{x}(t), \nabla V(x(t)) \rangle \, dt > 0,$$

and so $0 \notin H(x, \lambda)$, for every $|x| = r$ and $\lambda \in [0, 1)$.

If there is an $x \in \partial B_0^r$ such that $0 \in H(x, 1)$, then the conclusion of the theorem holds true. If there is no such x, then H is a homotopy and, from the invariance of the degree under homotopy (see Chapter II.9), we can deduce $0 \neq \text{ind } V = \deg(H(x, 1))$. Hence, we infer that $0 \in H(x, 1)$, for some $x \in \overline{B_0^r}$, which completes the proof. $\square$

Let us still consider the differential equation

$$(8.177) \qquad \dot{x}(t) = F(g_1(\tau_1(x(t)), x(t)), \dots, g_m(\tau_m(x(t)), x(t))) + h(t),$$

shortly $\dot{x}(t) = f(x(t)) + h(t)$, where

(i) each map $\tau_i : \mathbb{R}^n \to \mathbb{R}$, $i = l, \dots, m$ is a C^1-function, each map $g_i : \mathbb{R} \times \mathbb{R}^n \to \mathbb{R}$, $i = 1, \dots, m$, is a Carathéodory function and $F : \mathbb{R}^n \to \mathbb{R}^n$ is a continuous function,

(ii) for some compact set $K \subset \mathbb{R}^n$, at every point x, one has

$$f(x) \in K_{-a}, \ a \geq 0, \ \langle \nabla \tau_i(x), z \rangle > b \geq 0, \quad \text{for every } z \in K, \ a + b > 0,$$

where $K_{-a} := \{x \in \mathbb{R}^n \mid \text{dist}(x, \partial K) \geq a\}$,

(iii) $h(t) : [0, a] \to \mathbb{R}^n$ is an essentially bounded function satisfying $\|h\|_{L^\infty} \leq c$, $c > 0$.

(8.178) PROPOSITION (for $h \equiv 0$, cf. [BS, Theorem 1]). *Suppose that the positive numbers a, b, c are related through*

$$c < a + \sup\{\alpha > 0 \mid \langle \nabla \tau_i, z \rangle > 0, \quad \text{for all } z \in K_\alpha\},$$

where $K_\alpha := \{x \in \mathbb{R}^n \mid \text{dist}(x, K) \leq \alpha\}$. Then the Cauchy problem (8.177), $x(0) = x_0$, has at least one (Carathéodory) solution, provided (i)–(iii) hold.

PROOF. Assume, for simplicity, that $x_0 = 0$ and K is convex (the general situation can be treated analogously only under a slight technical modification). Following the ideas in [BS], to prove the existence of a solution of the Cauchy problem, we should construct a certain continuous operator $\mathcal{F}$ mapping some compact set Ω of functions into itself.

For a given compact set $K \subset \mathbb{R}^n$, there exists a real number $\delta > 0$ such that $\langle \nabla \tau_i, z \rangle \geq b + \delta > 0$, for all $z \in K$. From the continuity of the scalar product, there exists a positive constant $d := \sup\{\alpha > 0 \mid \langle \nabla \tau_i, z \rangle \geq \delta > 0, \text{ for all } z \in K_\alpha\} > 0$.

Let Ω be the compact set of continuous functions defined as follows:

$$\Omega := \left\{ u \in C([0, a], \mathbb{R}^n) \mid u(0) = 0, \ \frac{u(t) - u(s)}{t - s} \in K_d, \text{ for all } t > s \right\}.$$

Every function $u \in \Omega$ takes the values in the closed ball $\overline{B_0^{T|K_d|}}$, where $|K_d| := \max_{z \in K_d} |z|$. Since the derivatives of the maps $\tau_i \circ u : t \to \tau_i(u(t))$ are positive, for every $u \in \Omega$, $i = 1, \dots, m$,

$$(8.179) \qquad \frac{d\tau_i(u(t))}{dt} = \langle \nabla \tau_i(u(t)), \dot{u}(t) \rangle \geq \delta > 0, \quad \text{for all } t \in [0, a],$$

it follows that the maps $\tau_i \circ u$ are strictly increasing on $[0, a]$.

Define the *Picard operator* $\mathcal{F} : 0 \to C([0, a], \mathbb{R}^n)$:

$$(\mathcal{F}u)(t) := \int_0^t F(g_1(\tau_i(u(s)), u(s)), \dots, g_m(\tau_m(u(s))) + h(s)] \, ds.$$

For an arbitrary $\varepsilon > 0$, and for every map g_i, $i = 1, \dots m$, there exists a closed set J_i, $\mu(\mathbb{R} \setminus J_i) \leq \varepsilon$, such that g_i is continuous, when restricted to $J_i \times \mathbb{R}^n$. Moreover, the composed map f is continuous, when restricted to A, where $A := \{x \in \mathbb{R}^n \mid \tau_i(x) \in J_i, \text{ for every } i = 1, \dots, m\}$.

For an arbitrary $u \in \Omega$, the measure of the set $I_u := \{t \in [0, a] \mid \tau_i(u(t)) \notin J_i, \text{ for some } i = 1, \dots, m\}$ satisfies, in view of (8.179), the inequality

$$(8.180) \qquad \mu(I_u) \leq \frac{m\varepsilon}{\delta}.$$

By the extension theorem, there exists a continuous map $\tilde{f} : \mathbb{R}^n \to K_{-a}$ such that $f = \tilde{f}$ on A. Since the Picard operator $\tilde{\mathcal{F}}$, corresponding to the function $\tilde{f}$,

$$(\tilde{\mathcal{F}}u)(t) := \int_0^t \tilde{f}(u(s)) + h(s) \, ds,$$

is continuous, for every fixed $u \in \Omega$, there exists $\eta > 0$ such that

$$(8.181) \qquad |\tilde{\mathcal{F}}v - \tilde{\mathcal{F}}u| \leq \varepsilon, \quad \text{for every } v \in \Omega, \ |v - u| \leq \eta.$$

From (8.180) it follows that

$$(8.182) \qquad |\tilde{\mathcal{F}}v - \mathcal{F}v| \leq \mu(I_v) \sup_{x \in X} |f(x) - \tilde{f}(x)| \leq 2|K_{-a}| \frac{m\varepsilon}{\delta}.$$

Inequalities (8.181) and (8.182) yield

$$(8.183) \qquad |\mathcal{F}v - \mathcal{F}u| \leq |\mathcal{F}v - \tilde{\mathcal{F}}v| + |\tilde{\mathcal{F}}v - \tilde{\mathcal{F}}u| + |\tilde{\mathcal{F}}u - \mathcal{F}u| \leq \varepsilon + 4|K| \frac{m\varepsilon}{\delta},$$

for every $v \in \Omega$, $|v - u| \leq \eta$. Since ε in (8.183) can be taken arbitrarily small, the Picard operator $\mathcal{F} : u \to \mathcal{F}u$ is continuous and maps the compact sets Ω into itself:

$$\frac{(\mathcal{F}u)(t) - (\mathcal{F}u)(s)}{t - s} = \frac{1}{t - s} \int_s^t f(u(\tau)) \, d\tau + \frac{1}{t - s} \int_s^t h(s) \, ds$$

$$\frac{1}{t - s} \int_s^t f(u(\tau)) \, d\tau + \overline{B_0^c} \subset K_{-a} + \overline{B_0^c} \subset K_d.$$

The Schauder fixed point theorem implies the existence of a solution of the Cauchy problem (8.177), $x(0) = x_0$. $\qquad \square$

Let us still consider the differential equation

(8.184) $$\dot{x}(t) = g(\tau(x(t)), x(t)) + h(t),$$

shortly $\dot{x}(t) = f(x(t)) + h(t)$, where

(i) the function $\tau : \mathbb{R}^n \to \mathbb{R}$ is continuously differentiable, the map $g(t,x) :$ $\mathbb{R} \times \mathbb{R}^n \to \mathbb{R}^n$ is measurable in t and uniformly Lipschitz-continuous in x,

(ii) there exists a compact set $K \subset \mathbb{R}^n$, at every point x, such that (K_{-a} is defined as above)

$$f(x) \in K_{-a}, \ a \ge 0, \langle \nabla\tau(x), z \rangle > b \ge 0, \quad \text{for every } z \in K, \ a + b > 0,$$

(iii) $h(t) : [0, a] \to \mathbb{R}^n$ is an essentially bounded function satisfying $\|h\|_{L^\infty} \le c$,

(iv) the gradient $\nabla\tau$ has a bounded directional variation w.r.t. the cone $\Gamma = \{\lambda x \mid \lambda \ge 0, \ x \in K\}$.

(8.185) PROPOSITION (for $h \equiv 0$, cf. [BS, Theorem 2]). *Suppose that the positive numbers a, b, c are related through (K_α is defined as in Proposition (8.178))*

$$c < a + \sup\{\alpha > 0 \mid \langle \nabla\tau, z \rangle > 0, \quad \text{for all } z \in K_\alpha\}.$$

Then the Cauchy problem (8.184), $x(0) = x_0$ has a unique solution, depending locally Lipschitz-continuously on the initial value, provided (i)–(iv) hold.

PROOF. In [BS], it was proved that if $\tilde{x}(t)$, $\tilde{y}(t)$ are two (distinct) solutions of the autonomous discontinuous differential equation (8.184), where $h \equiv 0$, such that $|\tilde{x}(0)| \le R$, $|\tilde{y}(0)| \le R$, then the inequality

$$|\tilde{x}(t) - \tilde{y}(t)| \le C_R|\tilde{x}(0) - \tilde{y}(0)|, \quad t \in [0, a].$$

holds true, where the constant C_R depends only on f and R. The uniqueness of a solution of the Cauchy problem is a direct consequence of this inequality.

Let $x(t)$, $y(t)$ be two solutions of the nonautonomously perturbed discontinuous differential equation (8.184) such that $|x(0)| \le R$, $|y(0)| \le R$. Since every solution is understood in the Carathéodory sense, i.e. as an absolutely continuous function satisfying (8.184) almost everywhere, we have

(8.186) $$|x(t) - y(t)| = |x(0) + \int_0^t (f(x(\tau)) + h(\tau)) \, d\tau - y(0)$$
$$- \int_0^t (f(y(\tau)) + h(\tau)) \, d\tau|$$
$$= |\tilde{x}(t) - \tilde{y}(t)| \le C_R|\tilde{x}(0) - \tilde{y}(0)|$$
$$= C_R|x(0) - y(0)|, \quad t \in [0, a].$$

The functions $\tilde{x}(t)$, $\tilde{y}(t)$ are solutions of the autonomous discontinuous differential equation (8.184), where $h \equiv 0$, (the conditions imposed on f are stronger then those required in [BS, Theorem 2]). The claim follows immediately from (8.186). $\square$

Now, we can already give

(8.187) THEOREM. *Let $f : \mathbb{R}^n \to \mathbb{R}^n$ and $h : [0, a] \to \mathbb{R}^n$ be the functions satisfying the assumptions of Proposition (8.185), i.e. $f(x)$ is discontinuous in the above mentioned sense and $h(t)$ is essentially bounded. Let $V : \mathbb{R}^n \to \mathbb{R}$ be a guiding function for the map f such that $\operatorname{ind} V \neq 0$ and $\nabla V(x)$ be uniformly Lipschitz-countinuous. Then the periodic problem (8.184), $x(a) = x(0)$, has a solution.*

PROOF. To prove this statement, we can follow the steps in the proof of Theorem (8.173) The only difference is that, instead of a multivalued right-hand side, we have a single-valued right-hand side and the notion of upper semicontinuity of the associated Poincaré operator is just replaced, in view of Proposition (8.185), by the continuity. Lemma (8.172) holds true for bounded differential equations as well. This is our case, because the function $f(x)$ takes its values in the compact set and the perturbation h is essentially bounded, $|f(x) + h(t)|_{L^1} < |k| + a\|h\|_{L^\infty}$ on $[0, a]$.

When dealing with the equation similar to (8.178),

$$(8.188) \qquad G(t, x, \lambda) = k(\lambda)W_V(x) + (1 - k(\lambda))f_V(t, x),$$

function G satisfies the assumptions of Proposition (8.185), because so does f_V and $|W_V| \leq 1$. Therefore, the map $\chi : \overline{B_0^r} \times [0, 1] \to C([0, a], \mathbb{R}^n) \times [0, a]$, given by (8.176), has again the properties suitable for the usage of the Brouwer degree and the same argument gives us the existence of a periodic solution. $\qquad \square$

(f) The usage of two guiding functions.

In this part, we would like to reformulate, at first, Corollary (7.32) and an ordinary case ($\tau = 0$) of Theorem (7.31) in Chapter III.7, for the inclusions

$$(8.189) \qquad \dot{X} \in F(t, X),$$

in terms of two guiding functions.

Hence, let us recall Corollary (7.32):

(8.190) COROLLARY. *Let F be an (upper-)Carathéodory map which is essentially bounded with respect to $t \in [0, \infty)$ and bounded w.r.t. the variables from $\mathbb{R}^j$ ($1 \leq j \leq n - 1$) and which has nonempty, compact, convex values. Assume some part of components associated with all solutions $X(t, 0)$ of (8.189) is uniformly and ultimately bounded (i.e. is uniformly partially dissipative in the sense of Levinson, see Chapter III.7) and another part (related just to $\mathbb{R}^j$), starting outside of some neighbourhood of the origin, which behaves as a repeller, is uniformly tending (in the appropriate norm) to infinity. Then inclusion (8.189) admits a bounded solution on the positive ray.*

If, additionally,

$$F(t, X) \equiv F(t + \omega, X), \quad \omega > 0,$$

then (8.189) has, for some $k \in \mathbb{N}$, a $k\omega$-periodic solution.

In order to reformulate Corollary (8.190) and an ordinary case ($\tau = 0$) of Theorem (7.31) in terms of two guiding functions, which is very convenient for applications, let $\pi_j X$ and $\pi_{n-j} X$ denote the natural projections of the vector $X \in \mathbb{R}^n$ onto the spaces $\mathbb{R}^j$ and $\mathbb{R}^{n-j}$, respectively.

As a direct consequence of the above investigations, it will be also useful to state

(8.191) LEMMA. *If $V : [0, \infty) \times \mathbb{R}^n \to \mathbb{R}$ is a locally Lipschitzian in all variables function satisfying, for a.a. $t \in I \subset \mathbb{R}$:*

$$(8.192) \liminf_{h \to 0^+} \frac{1}{h}[V(t + h, X + hw) - V(t, X)] > 0, \quad \text{for all } w \in F(t, X),$$

$$(\limsup_{h \to 0^+} \frac{1}{h}[V(t + h, X + hw) - V(t, X)] < 0, \quad \text{for all } w \in F(t, X)),$$

then $V(\cdot, X(\cdot))$ is strictly monotone increasing (decreasing) on the interval I, where (8.192) holds.

(8.193) THEOREM. *Assume F is an (upper-)Carathéodory map which is essentially bounded with respect to $t \in [0, \infty)$ and bounded w.r.t. the variables from $\mathbb{R}^j$ ($1 \leq j \leq n - 1$) and which has nonempty, compact, convex values.*

Let two locally Lipschitzian in (t, x) guiding functions $V(t, X)$ and $W(t, X)$ exist such that

$$(8.194) \quad \begin{cases} a(|\pi_{n-j}X|) \leq V(t, X) \leq b(|\pi_{n-j}X|), \\ \limsup_{h \to 0^+} \dfrac{[V(t + h, X + hY) - V(t, X)]}{h} \leq -C(|\pi_{n-j}X|), \\ \text{for a.a. } t \in [0, \infty), \text{ for all } Y \in F(t, X), \\ \text{and for } |\pi_{n-j}X| \geq R_2, \end{cases}$$

and

$$(8.195) \quad \begin{cases} A(|\pi_j X|) \leq W(t, X) \leq B(|\pi_j X|), \quad \text{for } |\pi_{n-j}| \leq R_3, \\ \liminf_{h \to 0^+} \dfrac{[W(t + h, X + hY) - W(t, X)]}{h} \geq D(|\pi_j X|), \\ \text{for a.a. } t \in [0, \infty), \text{ for all } Y \in F(t, X), \\ \text{and for } |\pi_j X| \geq R_1, \ |\pi_{n-j}X| \leq R_3, \end{cases}$$

where $R_1, R_2 \leq R_3$ are suitable positive constants which may be large enough, the wedges $a(r), b(r), A(r), B(r)$ are continuous increasing functions such that both $a(r) \to \infty$ and $A(r) \to \infty$ as $r \to \infty$ and $C(r), D(r)$ are positive continuous functions not vanishing at infinity. Then inclusion (8.189) admits a bounded solution $X(t)$ such that

$$(8.196) \qquad \sup_{t \in [0,\infty)} |\pi_{n-j}X(t)| \leq R_3, \qquad \sup_{t \in [0,\infty)} |\pi_j X(t)| \leq R_4,$$

where $R_4 (\geq R_1)$ is a sufficiently big constant.

If, additionally,

$$F(t, X) \equiv F(t + \omega, X), \quad \omega > 0,$$

then there exists a $k\omega$-periodic solution of (8.189), satisfying (8.196), for some positive integer k.

PROOF. Condition (8.192) implies, according to Lemma (8.191), that all solutions $X(t)$ of (8.189) are partially uniformly bounded and partially uniformly ultimately bounded, i.e.

$$\limsup_{t \to \infty} |\pi_{n-j}X(t)| \leq R_3 \quad \text{(cf. (8.196))}.$$

For more details, see [Yos1-M], [Yos2-M], where however only single-valued F is considered.

On the other hand, condition (8.195) implies the existence of a constant $R_4 > R_1$ such that

$$(8.197) \qquad \inf_{|\pi_j X| \geq R_4} W(t, X) \geq A(R_4) > B(R_1) \geq \sup_{|\pi_j X| = R_1} W(t, X),$$

uniformly for $t \geq 0$, because of $|\pi_{n-j}X| \leq R_3$.

Let $X(t)$ be a solution of (8.189) satisfying

$$|\pi_j X(t_0)| > R_4 \quad \text{and} \quad |\pi_{n-j}X(t_0)| \leq R_3$$

$X(t)$ exists, by the hypothesis, on the whole ray $[t_0, \infty)$ and, moreover,

$$|\pi_{n-j}X(t)| \leq R_3, \quad \text{for } t \geq t_0 \text{ (cf. [Yos1-M], [Yos2-M])}.$$

Therefore, the existence of suitable positive constants ε, δ is, furthermore, implied by (8.195) such that

$$(8.198) \qquad \liminf_{h \to 0^+} \frac{1}{h}[W(t + h, X + hw) - W(t, X)] > \varepsilon,$$

$$\text{for all } w \in F(t, X), \text{ for a.a. } t \in [t_0, t_0 + \delta),$$

because of $|\pi_j X(t)| > R_1$, on the same interwal.

Hence, if $\delta > 0$ is such that $|\pi_j X(t_0 + \delta)| = R_1$, then it should be

$$W(t_0 + \delta, X(t_0, \delta)) \leq W(t_0, X(t_0)),$$

according to (8.197), a contradiction, because (8.198) implies by Lemma (8.191) the opposite. Thus, (8.193) holds on the whole half-line $[t_0, \infty)$, and consequently

$$\lim_{t \to \infty} W(t, X(t)) = 0.$$

This is, however, possible only if $|\pi_j X(t)| \to \infty$, whenever $t \to \infty$, according to (8.195).

Application of Corollary (8.190), whose all assumptions are satisfied, completes the proof. $\square$

(8.199) REMARK. For C^1-functions $V(t, X)$ and $W(t, X)$, conditions (8.192) and (8.195) can be rewritten into the form

$$\lim V(t, X)_{|\pi_{n-j}X| \to \infty} = \infty, \quad \langle \operatorname{grad} V(t, X), (1, Y) \rangle \leq -\varepsilon_1 < 0,$$
$$\text{for a.a. } t \in [0, \infty), \text{ for all } Y \in F(t, X)$$
$$\text{and } |\pi_{n-j}X| \geq R_2,$$

and

$$\lim W(t, X)_{|\pi_j X| \to \infty} = \infty, \quad \langle \operatorname{grad} W(t, X), (1, Y) \rangle \geq \varepsilon_2 > 0,$$
$$\text{for a.a. } t \in [0, \infty), \text{ for all } Y \in F(t, X)$$
$$\text{and } |\pi_j X| \geq R_1, |\pi_{n-j}X| \leq R_3,$$

where $\varepsilon_1, \varepsilon_2$ are suitable positive numbers.

Now, Theorem (8.193) will be applied to the quasi-linear inclusion

(8.200) $$\dot{X} \in AX + P(t, X),$$

where A is the real $n \times n$ matrix of the form

$$A = \begin{pmatrix} A_j & 0 \\ \cdot & 0 \end{pmatrix}, \quad A_j = \begin{pmatrix} a_{11} & \cdots & a_{1j} \\ a_{j1} & \cdots & a_{jj} \end{pmatrix}, \quad 1 \leq j \leq n,$$

and

$$X = (x_1, \ldots, x_n) := (X_j, \widehat{X}_j), \quad \text{for } j < n,$$

(we do not distinguish, for the sake of simplicity, between X and its transpose X^T), where $X_j = (x_1, \ldots, x_j)$, $\widehat{X}_j = (x_{j+1}, \ldots, x_n)$.

Similarly, denote $(1 \leq j \leq n)$

$$P(t,X) = (p_1(t,X),\dots,p_n(t,X)) := (P_j(t,X),\widehat{P}_j(t,X)),$$
$$P_j(t,X) = (p_1(t,X),\dots,p_n(t,X)), \quad \widehat{P}_j(t,X) = (p_{j+1}(t,X),\dots,p_n(t,X)).$$

The only presumption we require is that $P : [0,\infty) \times \mathbb{R}^n \to \mathbb{R}^n$ is (upper-) Carathéodory with nonempty, compact and convex values, for $t \geq 0$, $X \in \mathbb{R}^n$, and such that the global existence of all solutions to (8.200) of the associated Cauchy problems is guaranteed.

Consider still the linear system

(8.201) $$\dot{Y} \in A_j Y + Q(t),$$

where $Y = (y,\dots,y_j)$, $Q(t) = (q_1(t),\dots,q_j(t))$, $1 \leq j \leq n$, $Q : [0,\infty) \multimap \mathbb{R}^n$ is measurable map with nonempty, compact and convex values.

(8.202) LEMMA. *If A_j is a stable matrix (all its eigenvalues have negative real parts) and a positive constant $\overline{Q}$ exists such that*

(8.203) $$(\infty >) \quad \overline{Q} \geq \sup_{t \geq 0} \operatorname{ess} |r(t)| \quad \forall r \subset Q$$

(r is a single-valued measurable selection), all solutions $Y(t)$ of (8.201) are uniformly bounded and

(8.204) $$\limsup_{t \to \infty} |Y(t)| \leq \overline{Q} \sum_{k=0}^{j-1} \frac{2^k |A_j|^k}{(-\lambda)^{k+1}} := D_j,$$

where λ is the maximum of the real parts of the eigenvalues of A (for the upper estimates of λ, see, e.g., [Ju-M], [Kos]) and $|A|$ is the matrix norm corresponding to the vector norm employed.

PROOF. It is well known (see, e.g. [BVGN-M]) that $Y(t)$ can be expressed as

$$Y(t) = e^{A_j(t-t_0)} Y(t_0) + \int_{t_0}^{t} e^{A_j(t-s)} r(s)\,ds, \quad \text{for } t_0 \geq 0,$$

where $r \subset Q$ is a (single-valued) measurable selection. Applying the refined version (see [BVGN-M]) of the Gel'fand–Shilov inequality, namely

$$e^{A_j t} \leq e^{\lambda t} \sum_{k=0}^{j-1} \frac{2t |A_j|^k}{k!}, \quad t \geq 0,$$

and (8.203), we can write immediately

$$|Y(t)| \leq |e^{A_j(t-t_0)}||Y(t_0)| + \int_{t_0}^{t} |e^{A_j(t-s)}| \sup_{s \geq t_0} |r(s)| \, ds$$

$$\leq |Y(t_0)||e^{\lambda(t-t_0)}| \sum_{k=0}^{j-1} \frac{(2(t-t_0)|A_j|)^k}{k!}$$

$$+ \overline{Q} \int_{t_0}^{t} e^{\lambda(t-s)} \sum_{k=0}^{j-1} \frac{(2(t-s)|A_j|)^k}{k!} \, ds,$$

and consequently,

$$\limsup_{t \to \infty} |Y(t)| \leq \overline{Q} \limsup_{t \to \infty} \int_{t_0}^{t} e^{\lambda(t-s)} \sum_{k=0}^{j-1} \frac{(2(t-s)|A_j|)^k}{k!} \, ds.$$

Since we have

$$\int_{t_0}^{t} e^{\lambda(t-s)} \sum_{k=0}^{n-1} \frac{(2(t-s)|A_j|)^k}{k!} \, ds$$

$$= \sum_{k=0}^{n-1} \frac{2^k}{k!} |A_j| \int_{t_0-t}^{0} e^{-\lambda u} (-u)^k \, du$$

$$= \sum_{k=0}^{n-1} 2^k |A_j|^k \left(\frac{1 - e^{\lambda(t-t_0)}}{(-\lambda)^{k+1}} - \sum_{m=0}^{j-1} \frac{(t-t_0)^{k-m} e^{\lambda(t-t_0)}}{(k-m)!(-\lambda)^{m+1}} \right),$$

when substituting $u := s - t$ and integrating by parts, we can conclude immediately that (8.204) holds. This completes the proof. □

(8.205) CONSEQUENCE. *If A_j is a stable matrix, then also all solutions $X(t)$ of (8.200) satisfy (cf. (8.204))*

(8.206) $$\limsup_{t \to \infty} |X_j(t)| \leq D_j,$$

where $X_j(t)$ are uniformly bounded, provided

(8.207) $$\overline{Q} \geq \sup_{t \geq 0} \text{ess}[\sup_{X \in \mathbb{R}^n} |P_j(t, X)|].$$

PROOF. Since every solution $X(t)$ of (8.200) satisfies the (linearized) system

$$\dot{X}_j \in A_j X_j + P_j(t, X(t)), \quad 1 \leq j \leq n,$$

it is sufficient to take $Y := X_j$, $Q(t) := P_j(t, X(t))$, and the assertion follows immediately. □

(8.208) CONSEQUENCE. *Under the assumptions of Consequence* (8.205), *a locally Lipschitzian Liapunov function* $V(X)$ *exists, for* $t \geq 0$, $|x| \geq R > 0$ (R *being a constant), such that*

(i) $a(|X_j|) \leq V(X) \leq b(|X_j|)$, *where the wedges* $a(r)$, $b(r)$ *are continuous increasing functions and* $a(r) \to \infty$ *as* $r \to \infty$,

(ii) $\limsup_{h \to 0+} \frac{V(X+hw)-V(t,X)}{h} \leq -c(|X_j|)$, *for a.a.* $t \in [0,\infty)$, *for all* $w \in F(t,X)$, *for* $|X_j| \geq R_j > 0$ (R_j *being a suitable constant which may be large), where* $c(r)$ *is a positive continuous function.*

PROOF. Because of the right-hand side in (8.202) and restriction (8.207), it can be shown by means of the well-known Liapunov theorem (for more details, see, e.g. [BT-M, p. 287]) that $V(X)$ exists taking only the quadratic form (and so it is autonomous), for system

$$\dot{X}_j \in A_j X_j + P_j(t, X_j, \widehat{X}_j(t)).$$

For (8.200), the same must be true in view of (8.207).   □

(8.209) LEMMA. *Let the assumptions of Consequence* (8.205) *be satisfied. If still*

(8.210) $$\lim_{|x_k| \to \infty} (t, X) \operatorname{sgn} x_k = \infty, \quad where \ k = j+1, \ldots, n,$$

for $|X_j| \leq \overline{D}_j$ ($\overline{D}_j$ *is sufficiently large, see* R_3 *in* (8.195)), *uniformly with respect to a.a.* $t \geq 0$ *and the variables* x_p, *where* $j+1 \leq p \neq k$, *then a guiding function* $W(t,X)$ *exists, satisfying condition* (8.195) *in Theorem* (8.193), *and subsequently the family of solutions* $X(t)$ *of* (8.200) *exists such that*

$$\lim_{t \to \infty} |\widehat{X}_j(t)| = \infty.$$

PROOF. It follows from the proof of Theorem (8.193) that it is sufficient to construct the Liapunov function $W(t,X)$ satisfying (8.195).

Hence, define

$$W(t,X) \equiv W(\widehat{X}_j) := \widehat{X}_j \widehat{X}_j = \sum_{k=j+1}^{n} x_k^2.$$

Since the first inequality in (8.195) is evidently fulfilled, let us go to verifying the second one in (8.195). Taking into account (8.207) and (8.210), we arrive at

$$\liminf_{h \to 0+} \frac{W(\widehat{X}_j + h\widehat{P}_j(t,X)) - W(\widehat{X}_j)}{h} = 2\widehat{X}_j \dot{\widehat{X}}_j = 2\widehat{X}_j \widehat{P}_j(t,X) + 2 \sum_{k=j+1}^{n} \sum_{p=1}^{j} a_{kp} x_k x_p$$

$$\geq 2 \sum_{k=j+1}^{n} |x_k| \left(|p_k(t,X)| - \sum_{p=1}^{j} |a_{kp} x_p| \right) \geq \varepsilon > 0,$$

for a.a. $t \in [0, \infty)$, $|X_j| \leq \overline{D}_j$ and $|\widehat{X}_j| \geq S$, a suitable positive constant implied by (8.210). This completes the proof.   □

In view of Theorem (8.193), Consequence (8.208) and Lemma (8.209), we can give immediately

(8.211) THEOREM. *Let A_j, $1 \leq j \leq n$, be a (stable) matrix whose all eight-values have negative real parts. Assume that $P : [0, \infty) \times \mathbb{R}^n \multimap \mathbb{R}^n$ is an (upper) Carathéodory mapping with empty, compact and convex values, satisfying conditions (8.207) and (8.210). Then inclusion (8.200) admits a bounded solution on $[0, \infty)$.*

If, in particular, $F(t, X) \equiv F(t + \omega, X)$, $\omega > 0$, then there exists a $k\omega$-periodic solution of (8.200), for some $k \in \mathbb{N}$.

In the single-valued case, Theorem (7.61) (see also Theorem (7.46)) can be, quite analogusly to Theorem (8.193) in Part (f), expressed in terms of two guiding functions as follows:

(8.212) THEOREM. *Let the assumptions of Theorem (8.193) hold for one-parameter family of single-valued Carathéodory functions $F(t, X) = G(X) + \lambda P(t, X)$, $\lambda \in [0, 1]$, where $P(t, X) \equiv P(t + \omega, X)$, $\omega > 0$. Then the system of equations*

$$(8.213) \qquad \dot{X} = G(X) + P(t, X)$$

admits an ω-periodic solution.

Taking, in particular, $G(X) := AX$ in (8.213), equation (8.213) becomes (8.200). Therefore, Theorem (8.211) can be still specified, according to Theorem (8.212), into

(8.214) COROLLARY. *Let the assumptions of Theorem (8.211) be satisfied for a one-parameter family of single-valued Carathéodory functions $F(t, X) = \lambda P(t, X)$, $\lambda \in [0, 1]$, where $P(t, x) \equiv P(t + \omega, X)$, $\omega > 0$. Then the system of equations*

$$\dot{X} = AX + P(t, X)$$

admits an ω-periodic solution.

Finally, consider the Carathéodory equation

$$(8.215) \qquad x^{(n)} + \sum_{j=1}^{n-1} a_j x^{(n-j)} + h(x) = p(t) \quad (n > 1),$$

where $h \in C(\mathbb{R}, \mathbb{R})$, $p \in L_{\text{loc}}^1([0, \infty), \mathbb{R})$ and $a_j > 0$, $j = 1, \dots, n - 1$ are constants.

Consider still the equation

$$(8.216) \qquad x^{(n)} + \sum_{j=1}^{n-1} a_j x^{(n-j)} = g(t, x), \quad (n > 1),$$

where a_j, $j = 1, \ldots, n-1$, are positive constants and $g(t, x)$ is a Carathéodory bounded function for all $(t, x) \in [0, \infty) \times \mathbb{R}$, i.e.

$$\infty > G := \sup_{t \in \mathbb{R}} \operatorname{ess}[\sup_{x \in \mathbb{R}} |g(t, x)|].$$

(8.217) LEMMA. *Let the roots of the polynomials*

$$\lambda^{n-p} + \sum_{j=1}^{n-p} a_j \lambda^{n-j-p}$$

be negative for all $p = 1, \ldots, n-1$. Then the derivatives $x^{(k)}(t)$, $k = 1, \ldots, n-1$, of all solutions $x(t)$ of (8.216) are uniformly (ultimately) bounded and

$$(8.218) \qquad \limsup_{t \to \infty} |x^{(k)}(t)| \leq D_k := \frac{2^{k-1}}{a_{n-k}} G, \quad (k = 1, \ldots, n-1).$$

PROOF. Since every solution $x(t)$ of (8.216) satisfies the equation

$$x^{(n)} + \sum_{j=1}^{n-1} a_j x^{(n-j)} = g(t, x(t)), \quad (n > 1),$$

i.e.

$$y^{(n-1)} + \sum_{j=1}^{n-1} a_j y^{(n-j-1)} = g(t, x(t)), \quad \text{for } y := \dot{x},$$

we can employ the formula (5.46) in Chapter III.5 for its particular solution $(\dot{x}_p(t) =) y_p(t)$, namely

$$(\dot{x}_p(t) =) \, y_p(t) = e^{\lambda_1 t} \int_{-\infty}^{t} e^{(\lambda_2 - \lambda_1)t_1} \int_{-\infty}^{t_1} e^{(\lambda_3 - \lambda_2)t_2} \ldots$$
$$\int_{-\infty}^{t_{n-2}} e^{-\lambda_{n-1}t_{n-1}} g(t, x(t_{n-1})) \, dt_{n-1} \ldots dt_1,$$

where λ_j, $j = 1, \ldots, n-1$, are the negative roots of the characteristic polynomial

$$\lambda^{n-1} + \sum_{j=1}^{n-1} a_j \lambda^{n-j-1}.$$

Thus, the related general solution $(\dot{x}(t) =)y(t)$ reads $y(t) = y_0(t) + y_p(t)$, where $y_0(t)$ is the "homogeneous" part such that $\lim_{t\to\infty} y_0(t) = 0$, and consequently

$$\limsup_{t\to\infty} |y(t)| \leq G \left| e^{\lambda_1 t} \int_{-\infty}^{t} e^{(\lambda_2-\lambda_1)t_1} \int_{-\infty}^{t_1} e^{(\lambda_3-\lambda_2)t_2} \cdots \right.$$
$$\left. \int_{-\infty}^{t_{n-2}} e^{-\lambda_{n-1}t_{n-1}} \, dt_{n-1} \cdots dt_1 \right| \leq \frac{G}{|\lambda_{n-1}\lambda_{n-2}\ldots\lambda_1|} = \frac{G}{a_{n-1}}.$$

Similarly, since every solution $x(t)$ of (8.216) satisfies the equation

$$x^{(n)} + \sum_{j=1}^{n-2} a_j x^{(n-j)} = g(t, x(t)) - a_{n-1}\dot{x}(t),$$

i.e.

$$z^{(n-2)} + \sum_{j=1}^{n-2} a_j z^{(n-j-2)} = g(t, x(t)) - a_{n-1}y(t), \quad \text{for } z := \ddot{x},$$

we can employ the same formula for its particular solution $(\ddot{x}_p(t) =)z_p(t)$, namely

$$(\ddot{x}_p(t) =) \; z_p(t) = e^{\lambda_1 t} \int_{-\infty}^{t} e^{(\lambda_2-\lambda_1)t_1} \int_{-\infty}^{t_1} e^{(\lambda_3-\lambda_2)t_2} \cdots$$
$$\int_{-\infty}^{t_{n-3}} e^{-\lambda_{n-2}t_{n-2}} [g(t, x(t_{n-2})) - a_{n-1}y(t)] \, dt_{n-2} \cdots dt_1,$$

where λ_j, $j = 1, \ldots, n-2$, are the roots of the characteristic polynomial

$$\lambda^{n-2} + \sum_{j=1}^{n-2} a_j \lambda^{n-j-2}.$$

Thus, the related general solution $(\ddot{x}(t) =)z(t)$ reads $z(t) = z_0(t) + z_p(t)$, where $z_0(t)$ is the "homogeneous" part such that $\lim_{t\to\infty} z_0(t) = 0$, and consequently

$$\limsup_{t\to\infty} |z(t)| \leq 2G \left| e^{\lambda_1 t} \int_{-\infty}^{t} e^{(\lambda_2-\lambda_1)t_1} \int_{-\infty}^{t_1} e^{(\lambda_3-\lambda_2)t_2} \cdots \right.$$
$$\left. \int_{-\infty}^{t_{n-3}} e^{-\lambda_{n-2}t_{n-2}} \, dt_{n-2} \cdots dt_1 \right| \leq \frac{2G}{|\lambda_{n-2}\lambda_{n-3}\ldots\lambda_1|} = \frac{2G}{a_{n-2}},$$

when using the above asymptotic formula for $y(t)$.

Proceeding by the same way, we get successively the sequence of estimates with the final one

$$\limsup_{t\to\infty} |x^{(n-1)}(t)| \leq \frac{2^{n-2}}{a_1} G,$$

which completes the proof. $\square$

(8.219) REMARK. Formula (8.218) follows also from Lemma (5.67) in Chapter III.5. Moreover, according T. Yoshizawa's converse theorem [Yos1-M, p. 107], a guiding function V always exists satisfying condition like (8.194). Under weaker restrictions imposed on the coefficients a_j, $j = 1, \ldots, n-1$, the obtained estimates in Chapter III.5 can be used alternatively, instead of (8.218).

For the sake of convenience, consider still $(n > 1)$,

$$(8.220) \qquad x^{(n)} + \sum_{j=1}^{n-1} a_j x^{(n-j)} + h^*(x) = p(t), \quad p(t) \equiv p(t+T),$$

where (R will be specified below)

$$h^*(x) := \begin{cases} h(x + \bar{x}), & \text{for } |x| \leq R, \\ h(R \operatorname{sgn} x + \bar{x}), & \text{for } |x| \geq R, \end{cases} \quad \int_0^T p(t)\, dt = 0,$$

and make the following notation $(\sum_{j=1}^k \cdot := 0, \text{ for } k \leq 0)$:

$$\Delta(R) := \max \left\{ \frac{1}{a_{n-1}} (D_{n-1} + \sum_{j=1}^{n-3} a_j D_{n-j-1} + P_0), \right.$$
$$\left. \frac{1}{a_{n-1}^2} [M + ((M - a_{n-1}^2 R)_+^2 + 2 a_{n-1}^2 N)^{\frac{1}{2}}] \right\},$$

where $(\cdot)_+$ denotes the "positive part",

$$M := a_{n-2} H + 2 a_{n-1} (D_{n-1} + \sum_{j=1}^{n-2} a_j D_{n-j-1} + P_0),$$

$$N := a_{n-1} R (D_{n-1} + \sum_{j=1}^{n-2} a_j D_{n-j-1} + P_0) + \frac{3}{2} D_{n-1}^2 + 2 D_{n-1} P_0 + \frac{3}{2} P_0^2$$
$$+ 2(D_{n-1} + P_0) \sum_{j=1}^{n-2} a_j D_{n-j-1} + \frac{3}{2} (\sum_{j=1}^{n-2} a_j D_{n-j-1})^2,$$

$$D_k := \frac{2^{k-1}}{a_{n-k}} (H + P), \quad \text{for } k = 1, \ldots, n-1,$$

$$H := \max_{|x - \bar{x}| \leq R} |h(x)|, \quad P := \operatorname*{sup\,ess}_{t \in [0,T]} |p(t)|,$$

$$P_0 := \min_{t_1 \in [0,T]} \left[\max_{t_2 \in [0,T]} \left| \int_{t_1}^{t_2} p(t)\, dt \right| \right] = \max_{t \in [0,T]} \left| \int_{t_0}^t p(s)\, ds \right|.$$

(8.221) THEOREM. *Let the assumptions of Lemma (8.217) be satisfied. If there exist a positive constant R and a point $\bar{x}$ such that*

(8.222) $$h(\bar{x} + R) < 0, \quad h(\bar{x} - R) > 0,$$

(8.223) $$R \geq \Delta(R),$$

then equation (8.215) admits a T-periodic solution, provided $p(t) \equiv p(t + T)$ and $\int_0^T p(t)\, dt = 0$.

PROOF. Equation (8.220) is well-known (see e.g. [Wa1]) to possess, under our assumptions (even without (8.223)), a T-periodic solution, say $x(t)$.

Therefore, our purpose is to prove the same for (8.215) by showing that $|x(t)| \leq R$. More precisely, we will show that each $x(t)$ with the property $|x(t^*)| > R$, for any $t^* \geq t_0 \geq 0$, yields contradictionally

(8.224) $$\lim_{t \to \infty} |x(t)| = \infty.$$

This will be done by means of the following guiding function

$$W(t, x, \dot{x}, \dots, x^{(n-1)})$$
$$:= a_{n-2} \int_0^x h^*(s)\, ds + \frac{1}{2} \left[x^{(n-1)} + \sum_{j=1}^{n-1} a_j x^{(n-j-1)} - \int_{t_0}^t p(s)\, ds \right]^2.$$

Since we have for the time-derivative of W along (8.220) that $(\sum_{j=1}^{n-3} \cdot := 0$, for $n \leq 3)$,

$$\dot{W}_{(8.220)}(t, x, \dot{x}, \dots, x^{(n-1)}) := \left(\operatorname{grad} W, \left(1, p(t) - h^*(x) - \sum_{j=1}^{n-1} a_j x^{n-j} \right) \right)$$
$$= -h^*(x) \left\{ a_{n-1} x + \left[x^{(n-1)} + \sum_{j=1}^{n-3} a_j x^{(n-j-1)} - \int_{t_0}^t p(s)\, ds \right] \right\}$$
$$\geq |h^*(x)| \left(a_{n-1} |x| - D_{n-1} - \sum_{j=1}^{n-3} a_j D_{n-j-1} - P_0 \right), \quad \text{for } |x| \geq R + \varepsilon,$$

because of (8.223) (see the notation above) and (8.218) with $G = H + P$, the composed function $W(t, x(t), \dot{x}(t), \dots, x^{(n-1)}(t))$ is not only strictly increasing for $|x(t)| \geq R + \varepsilon$, $\varepsilon > 0$ being infinitesimally small, but also

(8.225) $$\lim_{t \to \infty} W(t, x(t), \dot{x}(t), \dots, x^{(n-1)}(t)) = \infty, \quad \text{for } |x(t)| \geq R + \varepsilon.$$

Since there is still

$$W(t, x, \dot{x}, \ldots, x^{(n-1)}) = W(t^* + \Delta_t, x_0 + \Delta_x, \dot{x}_0 + \Delta_{\dot{x}}, \ldots, x_0^{(n-1)} + \Delta_{x^{(n-1)}})$$
$$= W(t^*, x_0, \dot{x}_0, \ldots, x_0^{(n-1)}) + \Delta_W,$$

where

$$\Delta_W = a_{n-2} \int_{x_0}^{x_0 + \Delta_x} h^*(x)\, dx + x_0^{(n-1)} \Delta_{x^{(n-1)}} + \frac{1}{2} \Delta_{x^{(n-1)}}^2 - x_0^{(n-1)} \int_{t^*}^{t^* + \Delta_t} p(t)\, dt$$
$$- \Delta_{x^{(n-1)}} \int_{t_0}^{t^* + \Delta_t} p(t)\, dt + \int_{t_0}^{t^*} p(t)\, dt \int_{t^*}^{t^* + \Delta_t} p(t)\, dt + \frac{1}{2} \left[\int_{t^*}^{t^* + \Delta_t} p(t)\, dt \right]^2$$
$$+ x_0^{(n-1)} \sum_{j=1}^{n-1} a_j \Delta_{x^{(n-j-1)}} + \Delta_{x^{(n-1)}} \sum_{j=1}^{n-1} a_j (x_0^{(n-j-1)} + \Delta_{x^{(n-j-1)}})$$
$$- \int_{t_0}^{t^* + \Delta_t} p(t)\, dt \sum_{j=1}^{n-1} a_j \Delta_{x^{(n-j-1)}} - \int_{t^*}^{t^* + \Delta_t} p(t)\, dt \sum_{j=1}^{n-1} a_j x_0^{(n-j-1)}$$
$$+ \left[\sum_{j=1}^{n-1} a_j x_0^{(n-j-1)} \right] \cdot \left[\sum_{j=1}^{n-1} a_j \Delta_{x^{(n-j-1)}} \right] + \frac{1}{2} \left[\sum_{j=1}^{n-1} a_j \Delta_{x^{(n-j-1)}} \right]^2,$$

we have (take $\Delta_x = -|\Delta_x| \operatorname{sgn} x_0$ and put $\sum_{j=1}^{n-2} \cdot = 0$, for $n \le 2$),

$$\Delta_W \le -a_{n-1}^2 |x_0 \Delta_x| + \frac{1}{2} a_{n-1}^2 \Delta_x^2 + a_{n-2} H |\Delta_x|$$
$$+ a_{n-1} \left(D_{n-1} + \sum_{j=1}^{n-2} a_j D_{n-j-1} + P_0 \right) \cdot (|\Delta_x| + |x_0|)$$
$$+ \frac{3}{2} D_{n-1}^2 + 2 D_{n-1} P_0 + \frac{3}{2} P_0^2 + 2(D_{n-1} + P_0) \sum_{j=1}^{n-2} a_j D_{n-j-1}$$
$$+ \frac{3}{2} \left(\sum_{j=1}^{n-2} a_j D_{n-j-1} \right)^2.$$

Therefore (this time, take $x_0 = R + \varepsilon + |\Delta_x|$ and put $\sum_{j=1}^{n-2} \cdot = 0$, for $n \leq 2$),

$$
\Delta_W \leq -\frac{1}{2} a_{n-1}^2 \Delta_x^2
$$
$$
+ |\Delta_x| \left[-a_{n-1}^2 (R+\varepsilon) + a_{n-2} H + 2a_{n-1} \left(D_{n-1} + \sum_{j=1}^{n-2} a_j D_{n-j-1} + P_0 \right) \right]
$$
$$
+ a_{n-1}(R+\varepsilon) \left(D_{n-1} + \sum_{j=1}^{n-2} a_j D_{n-j-1} + P_0 \right)
$$
$$
+ \frac{3}{2} D_{n-1}^2 + 2D_{n-1} P_0 + \frac{3}{2} P_0^2 + 2(D_{n-1} + P_0) \sum_{j=1}^{n-2} a_j D_{n-j-1}
$$
$$
+ \frac{3}{2} \left(\sum_{j=1}^{n-2} a_j D_{n-j-1} \right)^2,
$$

and consequently $\Delta_W \leq 0$ as far as (see the notation above)

$$
|\Delta_x| \geq \frac{1}{a_{n-1}^2} \left\{ M + \left[(M - a_{n-1}^2 (R+\varepsilon))_+^2 \right. \right.
$$
$$
\left. \left. + 2a_{n-1}^2 \left(N + a_{n-1}\varepsilon \left(D_{n-1} + \sum_{j=1}^{n-2} a_j D_{n-j-1} + P_0 \right) \right) \right]^{1/2} \right\}
$$
$$
- (R+\varepsilon) := \Delta_\varepsilon(R) - (R+\varepsilon).
$$

So, if there is $|x(t^*)| = |x_0| \geq R + \varepsilon + |\Delta_x| \geq \Delta_\varepsilon(R)$, for some $t^* \geq t_0 \geq 0$, then also $|x(t)| \geq R+\varepsilon$, for all $t \geq t^*$, because otherwise there must appear a point $t_1(> t^*)$ such that the following two inequalities would hold simultaneously, namely

$$
W(t_1, x(t_1), \dots, x^{(n-1)}(t_1)) > W(t^*, x(t^*), \dots, x^{(n-1)}(t^*))
$$

as well as

$$
W(t_1, x(t_1), \dots, x^{(n-1)}(t_1)) \leq W(t^*, x(t^*), \dots, x^{(n-1)}(t^*)),
$$

which is a contradiction.

Moreover, (8.225) takes place which immediately implies the inequality (8.224), because $W(t, x, \dot{x}, \dots, x^{(n-1)})$ is bounded in the t-variable and, according to Lemma (8.217), so are the derivatives of $x(t)$ up to the $(n-1)$th order.

Hence, we can only get that $|x(t)| < \Delta_\varepsilon(R)$, and so, because of (8.223) and the definition of $\Delta_\varepsilon(R)$, we arrive at the desired inequality $|x(t)| \leq R$.

This completes the proof. $\square$

(8.226) REMARK. It can be easily deduced from our result in [AndS] that the assertion of Theorem (8.221) remains valid if (8.222) is replaced by (this time, however, $P_0 := \max_{t_1, t_2 \in [0,T]} | \int_{t_1}^{t_2} p(t)\, dt|$)

$$(8.222') \qquad\qquad h(\bar{x} + R) > 0, \quad h(\bar{x} - R) < 0.$$

Furthermore, because of the transformation $t := -\tau$, all of these is also true for the equation

$$(8.227) \qquad (-1)^n x^{(n)} + \sum_{j=1}^{n-1} (-1)^{(n-j)} a_j x^{(n-j)} + h(x) = p(t), \quad (n > 1).$$

(8.228) EXAMPLE. For $\pi > \Delta(\pi)$, $(R = \pi, H = b)$, the pendulum-type equation

$$(8.229) \qquad x^{(n)} + \sum_{j=1}^{n-1} a_j x^{(n-j)} + b \sin x = p(t), \quad (n > 1)$$

has at least two geometrically distinct (for $\bar{x} = \pi$ and $\bar{x} = 0$) T-periodic solutions, provided the coefficients a_j, $j = 1, \ldots, n-1$, have the same Hurwitzean structure as in Lemma (8.217) and $p(t)$ is as in Remark (8.226).

Let us still study the existence of bounded solutions on $[0, \infty)$. Instead of (8.215), consider at first (for the sake of convenience, again)

$$(8.230) \qquad x^{(n)} + \sum_{j=1}^{n-1} a_j x^{(n-j)} + h_*(x) = p(t), \quad (n > 1),$$

where ($R \leq R_0$ will be specified below)

$$h_*(x) := \begin{cases} h(x + \bar{x}), & \text{for } |x| \leq R, \\ \text{any nonincreasing continuous extension} \\ \text{of } h(x + \bar{x}) \text{ such that} \\ h_*(R_0 + \bar{x}) < -P, \ h_*(-R_0 + \bar{x}) > P, & \text{for } R_0 \geq |x| \geq R, \\ h_*(R_0 \operatorname{sgn} x + \bar{x}), & \text{for } |x| \geq R_0, \end{cases}$$

$$\infty > P := \sup_{t \in [t_0, \infty)} \operatorname{ess} |p(t)|, \quad \infty > P_0 := \sup_{t \in [t_0, \infty)} \left| \int_{t_0}^{t} p(s)\, ds \right|.$$

The notation above will be still (i.e. besides P, P_0) slightly adjusted for H defined as

$$H := \left[P, \max_{|x - \bar{x}| \leq R} |h(x)| \right].$$

(8.231) THEOREM. *Let the assumptions of Lemma (8.217) be satisfied. If there exist a positive constant R and a point $\bar{x}$ such that (8.222) and*

$$(8.223') \qquad\qquad R > \Delta(R)$$

hold, then equation (8.215) admits a bounded solution on a positive ray, provided $p(t)$ and $\int_{t_0}^{t} p(s)\,ds$ are bounded on the interval $[t_0, \infty)$, where t_0 may be very large.

PROOF. Equation (8.230) possesses (see [AV]), under our assumptions (even without (8.223')), a bounded solution on a positive ray, say $x(t)$.

Similarly as in the proof of Theorem (8.221), we will show that, under our assumptions, $|x(t)| \leq R$. Using the same approach as for Lemma (8.217) (only the meaning of some symbols is appropriately adjusted; see above), we arrive at the relation $|x(t)| < \Delta_\varepsilon(R) + \hat{\varepsilon}$, where $\hat{\varepsilon} > 0$ is a suitable arbitrarily small number due to the replacement of the constants D_k estimating ultimately $|x^{(k)}(t)|$, $k = 1, \ldots, n-1$, by $(D_k + \varepsilon_k) \geq |x^{(k)}(t)|$, for $t \geq t_1 \geq t_0$, t_1 being sufficiently big, according to Lemma (8.217) and the definition of H.

Using (8.223'), we get obviously the desired inequality $|x(t)| \leq R$, for $t \geq t_1$. This completes the proof. □

(8.232) REMARK. The assertion of Theorem (8.231) remains valid for P, P_0 defined in the following way:

$$(\infty >)P := \lim_{t\to\infty} \left[\operatorname*{sup\,ess}_{s \in [t,\infty)} |p(t)| \right], \quad (\infty >)P_0 := \limsup_{t\to\infty} \left| \int_{t_0}^{t} p(s)\,ds \right|,$$

where t_0 may be very big.

(8.233) REMARK. It can be easily deduced from our result in [AndS] that the assertion of Theorem (8.231) remains valid if (8.222) is replaced by (8.222') and P, P_0 are defined as in Remark (8.232).

(8.234) EXAMPLE. For $\pi > \Delta(\pi)$, $(R = \pi$, $H = \max(b, P))$, the pendulum-type equation (8.229) has at least two geometrically distinct (for $\bar{x} = \pi$ and $\bar{x} = 0$) bounded solutions on a positive ray, provided the coefficients a_j, $j = 1, \ldots, n-1$, have the same Hurwitzean structure as in Lemma (8.217) and $p(t)$ is as in Remark (8.233); in fact, equation (8.229) is stable in the sense of Lagrange (see [AndS]).

III.9. Infinitely many subharmonics

In Theorems (8.193) and (8.211) of the foregoing paragraph, the existence of $k\omega$-periodic, where $k \in \mathbb{N}$, i.e. subharmionic periodic solutions was established. Here, we would like to show how infinitely many subharmonics can be implied by finitely many. This goal is strictly related to the study of Sharkovskiĭ-type theorems. Our results are based on the papers [And18], [AFJ], [AJP], [Ji1], [LiYo], [Mat1], [Mat2], [Mat3], [Sh1].

In 1964, A. N. Sharkovskiĭ introduced a new ordering of all positive integers,

$$3 \triangleright 5 \triangleright 7 \triangleright 9 \triangleright \cdots \triangleright 2 \cdot 3 \triangleright 2 \cdot 5 \triangleright \cdots \triangleright 2^2 \cdot 3 \triangleright 2^2 \cdot 5 \triangleright \cdots$$
$$\cdots 2^n \cdot 3 \triangleright 2^n \cdot 5 \triangleright \cdots \triangleright 2^{n+1} \cdot 3 \triangleright 2^{n+1} \cdot 5 \triangleright \cdots \triangleright 2^{n+1} \triangleright 2^n \triangleright \cdots \triangleright 2^2 \triangleright 2 \triangleright 1,$$

to state his celebrated theorem in [Sh1]

(9.1) THEOREM (A. N. Sharkovskiĭ). *Let $f : \mathbb{R} \to \mathbb{R}$ be a continuous function. If f has a point of period n with $n \triangleright k$ (in the above Sharkovskiĭ ordering), then it has also a point of period k.*

By a period, we mean the least period, i.e. a point $a \in \mathbb{R}$ is a *periodic point of period n* (shortly, n-periodic point) of f if $f^n(a) = a$ and $f^j(a) \neq a$, for $0 < j < n$.

Independently, in 1975, T. Y. Li and J. Yorke rediscovered the following weaker form of Theorem (9.1) in [LiYo].

(9.2) THEOREM (T. Y. Li–J. Yorke). *Assume $f : \mathbb{R} \to \mathbb{R}$ is continuous, and there is a point a such that either*

(i) $f^3(a) \leq a < f(a) < f^2(a)$ *or* (ii) $f^3(a) \geq a > f(a) > f^2(a)$.

Then, f has points of all periods.

Influence of both these theorems to the development of the modern theory of dynamical systems is enormous and well-documented (see e.g. [Rob-M]).

Consider the scalar equation

(9.3) $$\dot{x} = F(t,x), \quad F(t,x) \equiv F(t + \omega, x), \quad \omega > 0,$$

where $F \in C(\mathbb{R}^2, \mathbb{R})$ ensures the global existence of all solutions on a positive half-line and satisfies one of conditions guaranteeing the unique solvability of Cauchy (initial value) problems.

Uniqueness conditions are well-known (e.g. those of Lipschitz, Nagumo, Osgood, Kamke, Kibenko, Borůvka — see eg.[Har-M], [Ks2-M] and the references therein).

Using the well-known one-to-one correspondence between (subharmonic) periodic solutions of differential equations, with a uniqueness assumption, and periodic points of the associated Poincaré (translation) operators (see Chapter III.4), Theorem (9.1) might seem to be simply reformulated in terms of (9.3).

More precisely, if we could assume that equation (9.3) admits a periodic solution with minimal period $n\omega$, then (9.3) would have, for each k such that $n \triangleright k$ in the Sharkovskiĭ ordering, a $k\omega$-periodic solution $x(t)$, i.e. $x(t) \equiv x(t + k\omega)$ and

$x(t) \not\equiv x(t + j\omega)$, for $0 < j < k$. In particular, for $n = 3$, equation (9.3) would possess subharmonics of all minimal periods $k\omega$, $k \in \mathbb{N}$.

Unfortunately, such a statement would be empty, because every bounded solution is, under the uniqueness assumption, either ω-periodic or asymptotically ω-periodic (see e.g. [Pl-M, p. 120–122]).

(9.4) REMARK. According to the well-known theorem of J. L. Massera (see e.g. [Yos2-M, p. 164]), if there exists a bounded solution on $[0, \infty)$, then there is a solution $x(t) \equiv x(t + \omega)$. However, this does not mean that ω is necessarily the minimal period. For example, in the linear case, $\dot{x} = a(t)x$, where $a(t) \equiv a(t + \omega)$ and $a < 0$, the only periodic solution is $x(t) \equiv 0$.

On the other hand, e.g. the linear inclusion

$$\dot{x} + cx \in P(t),$$

where $P(t) = [0, |\sin(\pi t)|]$, $t \in (-\infty, \infty)$, admits, for $c \neq 0$, k-periodic solutions, for every $k \in \mathbb{N}$, because P possesses the k-periodic selections $p_k \subset P$, $k \in \mathbb{N}$, where

$$p_k(t) = \begin{cases} 0, & \text{for } t \in [0, k-1], \\ |\sin(\pi t)|, & \text{for } t \in [k-1, k]. \end{cases}$$

The associated Poincaré translation operators become, however, multivalued.

Thus, a natural question arises, whether or not a suitable (multivalued) analogy of Theorem (9.1) or Theorem (9.2) exists, which could be expressed (via the associated Poincaré operators) not vacuously in terms of (9.3), in the lack of uniqueness, or more generally, in terms of scalar diferential inclusions.

As we shall see, this is unfortunately impossible in general. Nevertheless, we are able to give particular multivalued versions of Theorem (9.1) and Theorem (9.2), which might be suitable for desired applications (in this spirit) to equation (9.3) as well as to the (upper) Carathéodory inclusions

(9.5) $\dot{x} \in F(t, x).$

Recall that $F : \mathbb{R}^2 \multimap \mathbb{R}$, is called an (*upper*) *Carathéodory mapping* if the following conditions are satisfied:

(i) the set of values of F is nonempty, compact and convex for all $(t, x) \in \mathbb{R}^2$,
(ii) $F(t, \cdot)$ is upper semicontinuous (u.s.c.), for a.a. $t \in \mathbb{R}$,
(iii) $F(\cdot, x)$ is measurable, for all $x \in \mathbb{R}$.

Solutions $x(t)$ of (9.5) are also understood *in the sense of Carathéodory*, i.e. $x(t) \in \mathrm{AC}_{\mathrm{loc}}(\mathbb{R}, \mathbb{R})$ satisfy (9.5) almost everywhere (a.e.).

Furthermore, we know (see Theorem (4.3)) that the associated *Poincaré translation operator* Φ to (9.5), where F is essentially bounded in t and linearly bounded in x, namely

$$\Phi_{m\omega}(x_0) := \{x(m\omega, x_0) \mid x(\,\cdot\,, x_0) \text{ is a solution of (9.5) with } x(0, x_0) = x_0\},$$

where $m \in \mathbb{N}$, $\omega > 0$, is *admissible* in the sense of Definition (4.9) in Chapter I.4. More precisely, it is an u.s.c. composition of an R_δ-map and a single-valued continuous function. The set of values of admissible maps is moreover nonempty, compact and connected. Thus, in $\mathbb{R}$, *Poincaré operators are u.s.c. maps, whose values are either single points or closed intervals, and subsequently the related graphs are compact and connected (i.e. continua).*

So, although the shape of values is extremely simple, the shape of graph can be rather complicated. Moreover, since a composition of admissible maps is admissible as well (see (4.12) in Chapter I.4), the same must be all the better true for iterates. Therefore, *u.s.c. maps whose values are either single points or closed intervals preserve their character in $\mathbb{R}$ under iteration.*

After all, we have got a good candidate for a multivalued generalization of Theorem (9.1) and Theorem (9.2) with respect to indicated applications.

In view of the above conclusion, we will assume in the entire section that all maps under consideration in $\mathbb{R}$ are u.s.c. and the set of their values are either single points or closed intervals. For brevity, these maps will be denoted by M-*maps*.

Defining an n-*periodic point* $a \in \mathbb{R}$ of an M-*map* f as the one with $a \in f^n(a)$, but $a \notin f^j(a)$, for $0 < j < n$, we show the first counter-example, which is due to D. Miklaszewski.

(9.6) COUNTER-EXAMPLE. Consider the tent map

$$f : [0,1] \to [0,1], \text{ where } f(x) := \begin{cases} 2x, & \text{for } x \in [0, \frac{1}{2}], \\ 2 - 2x, & \text{for } x \in [\frac{1}{2}, 1]. \end{cases}$$

One can easily observe that f has two 1-periodic (fixed) points (see Fig. 1), two 2-periodic points (i.e. 2 fixed points of f^2) (see Fig. 2), six 3-periodic points (i.e. 6 fixed points of f^3) (see Fig. 3), etc.

Taking a sufficiently small $\varepsilon > 0$ (in Fig. 4, $\varepsilon = \frac{1}{40} = 0.025$), we can define an M-map (see Fig. 4) $\varphi : [0,1] \multimap [0,1]$, where

$$\varphi(x) := \begin{cases} [2(1-\varepsilon)x, 2(1-\varepsilon)x + \varepsilon], & \text{for } x \in [0, \frac{1}{2}], \\ [-2(1-\varepsilon)x + 2(1-\varepsilon), -2(1-\varepsilon)x + 2 - \varepsilon], & \text{for } x \in [\frac{1}{2}, 1]. \end{cases}$$

such that $\varphi \supset f$ has again at least six 3-periodic points.

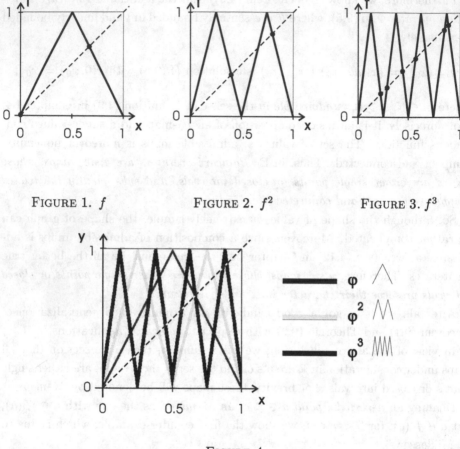

FIGURE 1. f FIGURE 2. f^2 FIGURE 3. f^3

FIGURE 4

Hence, if a desired multivalued version of Theorem (9.1) is true, then φ should admit k-periodic points, for every $k \in \mathbb{N}$.

However, one can find a sufficiently large $n = n(\varepsilon)$ such that $\varphi^m(x) = [0,1]$, for all $x \in [0,1]$ and $m \geq n$, by which there are no m-periodic points, for $m > n$, in the above sense.

Indeed, for each $x \in [0,1]$, $\{\varphi(x)\}$ is an interval of the length ε. Obviously, $\{\varphi^2(\varphi(x))\} = \{\varphi^3(x)\}$ is an interval of the length at least $4\varepsilon/2 = 2\varepsilon$ (see Fig. 4), $\{\varphi^2(\varphi^3(x))\} = \{\varphi^5(x)\}$ has the length at least $4(2\varepsilon/2) = 4\varepsilon$ (see Fig. 5–Fig. 7), etc.

Thus, for $m \geq n \geq 1/\varepsilon + 1$, we arrive at the indicated mapping having, for $m \geq 1/\varepsilon + 2$, no m-periodic points, as pointed out. In our case, for $\varepsilon = 1/40$, it was in fact, when already $m > 7$.

On the other hand, although the assumptions of a desired multivalued version

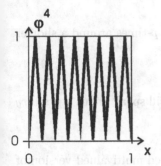

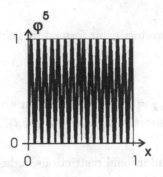

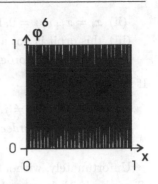

FIGURE 5. φ^4 FIGURE 6. φ^5 FIGURE 7. φ^6

of Theorem (9.1) are not satisfied for an M-map

$$\varphi(x) := [0, 1], \quad \text{for all } x \in [0, 1],$$

φ obviously possesses e.g. a continuous logistic selection

$$f(x) = 3.84x(1 - x),$$

which is well-known to have an approximative orbit $\mathcal{O} = \overline{\{0.488, 0.959, 0.149\}}$, consisting of 3-periodic points (see Fig. 8).

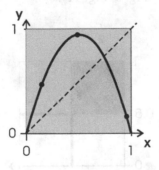

FIGURE 8

Subsequently, Theorem (9.1) applies, and so f admits periodic points of all periods.

This leads us to the conclusion that, rather than periodic points, a suitable notion of periodic orbits is more appropriate to M-maps.

By an *orbit of the kth-order* (shortly, a *k-orbit*) to an M-map φ, we mean a sequence $\{x_i\}_{i=0}^{\infty}$ such that

(i) $x_{i+1} \in \varphi(x_i)$, $i = 0, 1, \dots$,

(ii) $x_i = x_{i+k}$, $i = 0, 1, \ldots,$

(iii) this orbit is not a product orbit formed by going p-times around a shorter orbit of mth-order, where $mp = k$.

If still

(iv) $x_i \neq x_j$, for $i \neq j$, $i, j = 0, 1, \ldots, k-1$, then we will speak about a *primary orbit of kth-order* (shortly, a *primary k-orbit*).

Unfortunately, without additional restrictions, a desired multivalued version of Theorem (9.1) does not hold for $n = 3$, as can be seen from the following

(9.7) COUNTER-EXAMPLE. Consider the M-mapping (see Fig. 9)

$$\varphi : [0,1] \multimap [0,1], \quad \text{where } \varphi(x) := \begin{cases} \dfrac{1}{2}, & \text{for } x \in [0, 1/2), \\[2mm] \left[\dfrac{1}{2}, 1\right], & \text{for } x = 1/2, \\[2mm] -x + 1, & \text{for } x \in (1/2, 1]. \end{cases}$$

One can observe that, although φ has infinitely many (primary) 3-orbits (even 3-periodic points), there is (unlike in Counter-Example (9.6)) no 2-orbit, but the sole fixed point $a = 1/2$ (see Fig. 9–Fig. 11).

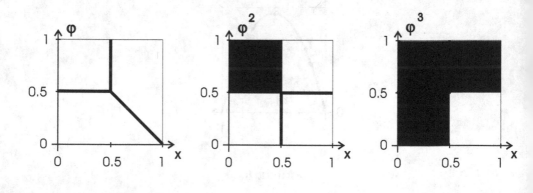

FIGURE 9. φ FIGURE 10. φ^2 FIGURE 11. φ^3

Therefore, there is really no straightforward multivalued version of Theorem (9.1), for $n = 3$. The following counter-example, due to T. Matsuoka, demonstrates similar obstructions for $n > 3$, but the absence of a 4-orbit is exceptional.

(9.8) COUNTER-EXAMPLE. Consider the M-mapping (see Fig. 12)

$$\varphi : [0,4] \multimap [0,4], \quad \text{where } \varphi(x) := \begin{cases} x+2, & \text{for } x \in [0,1), \\ [3,4], & \text{for } x = 1, \\ 3, & \text{for } x \in (1,2], \\ -2x+7, & \text{for } x \in (2,3], \\ -x+4, & \text{for } x \in (3,4]. \end{cases}$$

One can observe that, although φ has a primary 5-orbit $\overline{\{0,2,3,1,4\}}$ (see Fig. 16) and no 3-orbit (see Fig. 14), there is also no 4-orbit (see Fig. 15), because $7/3, 1$ and 3 are the only fixed points of φ^4. Moreover, $\frac{7}{3}$ is a fixed point of φ. Hence, the desired 4-orbit does not exist, because it is formed by going 2-times around the 2-orbit $\overline{\{1,3\}}$ (see Fig. 13 and Fig. 15), as explained below. φ^6 in Fig. 17 indicates a further development of iterates.

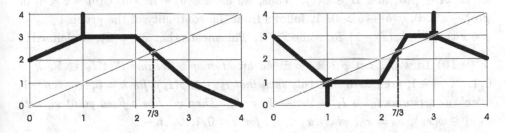

FIGURE 12. φ FIGURE 13. φ^2

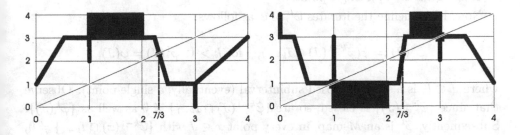

FIGURE 14. φ^3 FIGURE 15. φ^4

In order to state a multivalued analogy of Theorem (9.2), we proceed by a standard manner (cf. e.g. [Rob-M] and the references therein).

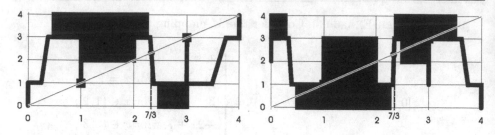

FIGURE 16. φ^5 FIGURE 17. φ^6

(9.9) LEMMA. *Let $\varphi : I \multimap \mathbb{R}$ be an M-map, where $I = [a, b] \subset \mathbb{R}$ is a closed interval such that $I \subset \varphi(I)$. If there are points $A \in \varphi(a)$ and $B \in \varphi(b)$ such that $a < A$, $B < b$ or $a > A$, $B > b$, then there exists a fixed point $\widehat{x} \in I$ of φ.*

PROOF. If a or b are not fixed points, then $a < A$, $B < b$ or $a > A$, $B > b$, for all $A \in \varphi(a)$ and $B \in \varphi(b)$. Thus, we have $\varphi(a) - a > 0$, $\varphi(b) - b < 0$ or $\varphi(a) - a < 0$, $\varphi(b) - b > 0$. It follows from the continuity of the graph $\Gamma_{\varphi-x}$ of $\varphi - x$ that $0 \in [\varphi(\widehat{x}) - \widehat{x}]$, for some $\widehat{x} \in I$. This means that $\widehat{x} \in \varphi(\widehat{x})$, as claimed.□

(9.10) LEMMA. *Let $\varphi : \mathbb{R} \multimap \mathbb{R}$ be an M-map. Assume that $I_k \subset \mathbb{R}$, $k = 0, 1, \ldots, n-1$, are closed intervals such that $I_{k+1} \subset \varphi(I_k)$, for $k = 0, 1, \ldots, n-1$, which we write as $I_0 \to I_1 \to \ldots \to I_n = I_0$. Then φ^n has a fixed point x_0 with $x_{k+1} \in \varphi(x_k)$, $x_n = x_0$, where $x_k \in I_k$, for $k = 0, 1, \ldots, n-1$.*

PROOF. Letting $I_0 \to I_1$, i.e. $I_1 \subset \varphi(I_0)$, there certainly exists a subinterval $K_1 \subset I_0$ such that $I_1 \subset \varphi(K_1)$ and K_1 is minimal in the sense that, for each $L_1 \subset K_1$, $L_1 \neq K_1$, there exists an $x \in I_1$ with $x \notin \varphi(L_1)$. Moreover, K_1 is closed or can degenerate to a single point.

Now, let us define the iterates $\widetilde{\varphi}^k$ of φ as follows:

$$\widetilde{\varphi}^k(J) := \varphi(\widetilde{\varphi}^{k-1}(J) \cap I_{k-1}), \quad \text{for } k > 1, \ \widetilde{\varphi}^1(J) = \varphi(J),$$

where $J \subset I_0$ is a suitable closed subinterval (eventually, a single point). Observe that since $\{\widetilde{\varphi}^{k-1}(J)\}$ is closed, so are $\{\widetilde{\varphi}^{k-1}(J) \cap I_{k-1}\} \neq \emptyset$ as well as $\{\widetilde{\varphi}^k(J)\}$. Subsequently, $\widetilde{\varphi}^k$ is an M-map, in every point $x \in J$ with $\{\widetilde{\varphi}^{k-1}(x) \cap I_{k-1}\} \neq \emptyset$, because the graph $\Gamma_{\widetilde{\varphi}^k}$ of $\widetilde{\varphi}^k$ is there compact.

If $I_1 \to I_2$, i.e. $\widetilde{\varphi}^2(K_1) = \varphi(\varphi(K_1) \cap I_1) = \varphi(I_1) \supset I_2$, then there is similarly a minimal subinterval $K_2 \subset K_1$ such that $I_2 \subset \widetilde{\varphi}^2(K_2)$, etc. Finally, for an m-th iterate, we arrive at

$$\widetilde{\varphi}^m(K_{m-1}) = \varphi(\widetilde{\varphi}^{m-1}(K_{m-1}) \cap I_{m-1}) = \varphi(I_{m-1}) \supset I_m,$$

which implies the existence of a minimal subinterval $K_m \subset K_{m-1}$ with $I_m \subset \widetilde{\varphi}^m(K_m)$, for $m = 1, \dots, n$.

In particular, for $m = n$, we obtain a minimal interval K_n such that $\widetilde{\varphi}^n(K_n) \supset I_0 \supset K_n$ which yields, according to Lemma (9.9), an $x_0 \in K_n$ with $x_0 \in \widetilde{\varphi}^n(x_0)$.

Furthermore, by the minimality of $K_n \subset K_{n-1} \subset \dots \subset K_1$, we have that $\widetilde{\varphi}^k(x) \cap I_k \neq \emptyset$, for every $x \in K_n$; otherwise, K_n would not be minimal. Hence, $x_0 \in \widetilde{\varphi}^n(x_0) \cap I_n$.

If $x_0 \notin \varphi(y) \cap I_n$, for all $y \in \widetilde{\varphi}^{n-1}(x_0) \cap I_{n-1}$, we get that $x_0 \notin \varphi(\widetilde{\varphi}^{n-1}(x_0) \cap I_{n-1}) \cap I_n = \widetilde{\varphi}^n(x_0) \cap I_n$, which is a contradiction. Thus, there exists an $x_{n-1} \in \widetilde{\varphi}^{n-1}(x_0) \cap I_{n-1}$ such that $x_0 \in \widetilde{\varphi}(x_{n-1}) \cap I_n$.

Similarly, if $x_{n-1} \notin \varphi(y) \cap I_{n-1}$, for all $y \in \widetilde{\varphi}^{n-2}(x_0) \cap I_{n-2}$, we get that $x_{n-1} \notin \varphi(\widetilde{\varphi}^{n-2}(x_0) \cap I_{n-2}) \cap I_{n-1} = \widetilde{\varphi}^{n-1}(x_0) \cap I_{n-1}$, which is again a contradiction. Thus, there exists an $x_{n-2} \in \widetilde{\varphi}^{n-2}(x_0) \cap I_{n-2}$ such that $x_{n-1} \in \varphi(x_{n-2}) \cap I_{n-1}$, etc. Finally, we arrive at the existence of an $x_1 \in \varphi(x_0) \cap I_1$ such that $x_2 \in \varphi(x_1) \cap I_2$.

At last, it follows from the definition of $\widetilde{\varphi}^k$ that $\widetilde{\varphi}^k(J) \subset \varphi^k(J)$. Subsequently, a fixed point of $\widetilde{\varphi}^n$ is a fixed point of φ^n as well. At the same time, points $x_0, x_1, \dots, x_n = x_0$ of an n-orbit $\mathcal{O} = \overline{\{x_0, x_1, \dots, x_{n-1}\}}$ of φ satisfy $x_k \in I_k$, $i = 0, 1, \dots, n-1$. This completes the proof. $\square$

Now, we can already give the first result.

(9.11) PROPOSITION. *Let $\varphi : \mathbb{R} \multimap \mathbb{R}$ be an M-map. Assume the existence of a (primary) 3-orbit $\overline{\{a, b, c\}}$ of φ, where $b \in \varphi(a)$, $c \in \varphi(b)$, $a \in \varphi(c)$ such that there is a $d \in \varphi(c)$ with $d \leq a < \min_{x \in \varphi(a)} x \leq \max_{x \in \varphi(a)} x < \min_{y \in \varphi(b)} y$ or $d \geq a > \max_{x \in \varphi(a)} x \geq \min_{x \in \varphi(a)} x > \max_{y \in \varphi(b)} y$. Then, for each prime number $k \in \mathbb{N}$, φ has a k-orbit.*

PROOF. Let us take into account only the first inequalities (the other case can be obtained by a reflection on the line). Put $I_1 = [a, b]$ and $I_2 = [b, c]$. By the hypothesis, $I_2 \subset \varphi(I_1)$ and $I_1 \cup I_2 \subset \varphi(I_2)$.

Consider $I_1 \to \underbrace{I_2 \to \dots \to I_2}_{(q-1)\text{-times}} \to I_1$, where $q \in \mathbb{N}$ is a prime number. In view of Lemma (9.10), we have a q-orbit $\mathcal{O} = \overline{\{x_1, \dots, x_q\}}$ such that $x_1 \in I_1$ and $x_i \in I_2$, for $i = 2, \dots, q$.

Assume that $\mathcal{O}$ is formed by going r-times around a p-orbit (i.e. if $q = pr$); otherwise, we are done. Since q is prime, it turns out that $p = 1$, by which $q = r$. Therefore, $x_1 = \dots = x_q = b$. Subsequently, $b \in \varphi(b)$ at the same time as $b \in \varphi(a)$, which is a contradiction with the assumptions. So, the points $x_1, \dots, x_q$ must form a q-orbit, as claimed. $\square$

(9.12) REMARK. It can be seen in Counter-Example (9.7) that the inequalities in the assumptions of Proposition (9.11) cannot be omitted, in general. On the other hand, if a 3-orbit is not primary, we can easily obtain, without these restrictions, orbits of all integer orders.

Before extending Propostion (9.11), it will be useful to recall the following simple property of M-maps.

(9.13) LEMMA. *Let $\varphi : \mathbb{R} \multimap \mathbb{R}$ be an M-map and let there exist $a, b \in \mathbb{R}$ such that $a < b$, $A \in \varphi(a)$, $B \in \varphi(b)$. If $C \in [A, B]$, then there exists a point $c \in [a, b]$ such that $C \in \varphi(c)$.*

PROOF. Assume that $A > B$ and denote $s = \sup\{t \in [a, b] \mid \exists T \in \varphi(t) : T \geq C\}$. Supposing the existence of $S_1 \in \varphi(s)$, $S_1 > C$, and the absence of $S_2 \in \varphi(s)$, $S_2 \leq C$, we obtain a contradiction with the upper semicontinuity of φ, because $\varphi(s)$ is compact and $\varphi(t) < C$, for every $t \in (s, b]$. Assuming the existence of $S_2 \in \varphi(s)$, $S_2 < C$, and the absence of $S_1 \in \varphi(s)$, $S_1 \geq C$, and using the compactness of $\varphi(s)$ jointly with the definition of s, we obtain again a contradiction with the upper-semicontinuity of φ.

Thus, there exist $S_1 \in \varphi(s)$, $S_1 \geq C$, and $S_2 \in \varphi(s)$, $S_2 \leq C$. Since $\varphi(s)$ is a convex set, it contains C.

If $A < B$ (the case $A = B$ is trivial), then we can define s as $\sup\{t \in [a, b] \mid \exists T \in \varphi(t) : T < C\}$ and then proceed analogously. $\square$

Now, Proposition (9.11) can be extended into the following multivalued version of the Li–Yorke Theorem (9.2).

(9.14) THEOREM. *Let $\varphi : \mathbb{R} \multimap \mathbb{R}$ be an M-map. Assume the existence of a (primary) 3-orbit $\overline{\{a, b, c\}}$ of φ, where $b \in \varphi(a)$, $c \in \varphi(b)$, $a \in \varphi(c)$ such that*

$$a < \min_{x \in \varphi(a)} x \leq \max_{x \in \varphi(a)} x < \min_{y \in \varphi(b)} y$$

or

$$a > \max_{x \in \varphi(a)} x \geq \min_{x \in \varphi(a)} x > \max_{y \in \varphi(b)} y.$$

Then φ has a k-orbit for every $k \in \mathbb{N} \setminus \{4, 6\}$.

PROOF. Let us consider only the first inequalities (the second case can be obtained by a reflection on the line). We note that these inequalities imply $a < b < c$.

Using Lemma (9.9), we obtain a fixed point x_2 of φ such that $x_2 \in (b, c]$, i.e. $\{x_2\}$ is a 1-orbit. Lemma (9.13) shows that there exists $x_3 \in (b, c]$, $x_2 \leq x_3$, satisfying $b \in \varphi(x_3)$.

Considering

$$[a, b] \to [b, c] \to [a, b]$$

and applying Lemma (9.10), we obtain, with respect to $b \notin \varphi(b)$, the existence of a 2-orbit.

Besides $k = 1, 2$, we prove still the existence of further desired k-orbits. The following possibilities can occur:

Case 1. $x_2 = x_3 = c$.

Fix an arbitrary $x_4 \in (b, c)$ and $k \in \mathbb{N}$, $k \geq 3$. Let us consider

$$[a, b] \to \underbrace{[x_4, c] \to \cdots \to [x_4, c]}_{(k-1)\text{-times}} \to [a, b].$$

By Lemma (9.10), we have the existence of a k-orbit $\overline{\{y_1, \ldots, y_k\}}$. Indeed, since we start and finish in the interval $[a, b]$ and $[a, b] \cap [x_4, c] = \emptyset$, a sequence $\{y_1, \ldots, y_k\}$ cannot be formed by going p-times around a shorter m-orbit, where $k = mp$.

Case 2. $c > x_2 \geq \min_{y \in \varphi(b)} y$.

Because of the upper semicontinuity of φ and the inequality $\max_{x \in \varphi(a)} x < \min_{y \in \varphi(b)} y$, there exist $x_1 \in (a, b)$ and $x_{10} \in (b, \min_{y \in \varphi(b)} y)$ such that $[b, x_{10}] \subset \varphi[a, x_1]$.

(a) If $k \geq 3$ is odd, then we consider

$$[a, x_1] \to \underbrace{[b, x_{10}] \to [x_2, c] \to \cdots \to [b, x_{10}] \to [x_2, c]}_{((k-1)/2)\text{-times}} \to [a, x_1].$$

By Lemma (9.10) and by the same reasoning as in Case 1, we obtain a k-orbit.

(b) If $k \geq 8$ is even, then we consider

$$[a, x_1] \to \underbrace{[b, x_{10}] \to [x_2, c] \to \cdots \to [b, x_{10}] \to [x_2, c]}_{(k/2-2)\text{-times}} \to [a, x_1]$$

$$\to [b, x_{10}] \to [x_2, c] \to [a, x_1].$$

Using Lemma (9.10) again, there exists a k-orbit $\overline{\{y_1, \ldots, y_k\}}$. We note that a sequence $\{y_1, \ldots, y_k\}$ cannot be formed by going p-times around a shorter m-orbit, where $k = mp$, because we begin in the interval $[a, x_1]$, which does not appear again until the second half of the process, and

$$[a, x_1] \cap [b, x_{10}] = \emptyset, \quad [a, x_1] \cap [x_2, c] = \emptyset.$$

Case 3. $x_2 < \min_{y \in \varphi(b)} y$, $x_2 = x_3$.

Thanks to Lemma (9.13) and the inequality $\max_{x \in \varphi(a)} x < \min_{y \in \varphi(b)} y$, there exists $x_1 \in (a, b)$ with the property $[b, x_2] \subset \varphi[a, x_1]$. For $k \geq 3$, we consider

$$[a, x_1] \to \underbrace{[b, x_2] \to \cdots \to [b, x_2]}_{(k-1)\text{-times}} \to [a, x_1].$$

Using Lemma (9.10) and reasoning as in Case 1, we obtain a k-orbit.

Case 4. $x_2 < \min_{y \in \varphi(b)} y$, $x_2 \neq x_3$.

The same arguments as in Case 3 yield the existence of $x_1 \in (a, b)$ satisfying $[b, x_2] \subset \varphi[a, x_1]$.

(a) If $k \geq 3$ is odd, then we consider

$$[a, x_1] \to \underbrace{[b, x_2] \to [x_2, x_3] \to \ldots \to [b, x_2] \to [x_2, x_3]}_{((k-3)/2)\text{-times}} \to [b, x_2] \to [x_2, c] \to [a, x_1].$$

In an analogous way as in Case 1, we have a k-orbit.

(b) If $k \geq 8$ is even, then we consider

$$[a, x_1] \to \underbrace{[b, x_2] \to [x_2, x_3] \to \ldots \to [b, x_2] \to [x_2, x_3]}_{(k/2-3)\text{-times}} \to [b, x_2]$$

$$\to [x_2, c] \to [a, x_1] \to [b, x_2] \to [x_2, c] \to [a, x_1].$$

In an analogous way as in Case 2(b), we obtain a k-orbit.   □

Theorem (9.14) cannot be improved, as follows from the following

(9.15) COUNTER-EXAMPLE. Consider the M-mapping (see Fig. 18) $\varphi : [0, 4] \multimap [0, 4]$, where

$$\varphi(x) := \begin{cases} x + 2, & \text{for } x \in [0, 1), \\ 3, & \text{for } x \in (1, 2] \cup (2, 4), \\ [3, 4], & \text{for } x = 2, \\ [0, 3], & \text{for } x = 4. \end{cases}$$

One can observe that φ has a primary 3-orbit $\overline{\{0, 2, 4\}}$ which satisfies the first inequalities in Theorem (9.14) (see Fig. 20). On the other hand, φ has no 4-orbit (see Fig. 21). Indeed, 2, 3 and 4 are the only fixed points of φ^4 and, moreover, 3 is a fixed point of φ. Hence, we can easily show that both possible candidates for a 4-orbit are formed by going 2-times around the 2-orbit (see Fig. 19).

φ has also no 6-orbit (see Fig. 22). Now, all possible candidates (for example, $\overline{\{0, 2, 4, 0, 2, 4\}}$ or $\overline{\{4, 2, 4, 2, 4, 2\}}$) are formed by going 2-times around the 3-orbit (see Fig. 18) or 3-times around the 2-orbit (see Fig. 19).

Notice that in Theorem (9.1) there are no assumptions on the inequalities for the given 3-orbit as in the Li–Yorke Theorem (9.2), i.e. "three implies everything" there even at the absence of these inequalities.

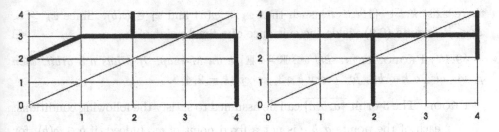

FIGURE 18. φ FIGURE 19. φ^2

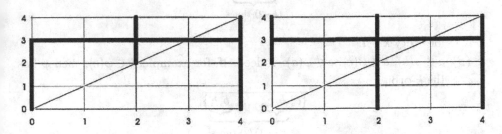

FIGURE 20. φ^3 FIGURE 21. φ^4

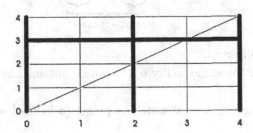

FIGURE 22. $\varphi^5 = \varphi^6$

Thus, a natural question arises what happens when we omit the assumption of the inequalities for the given 3-orbit in Theorem (9.14).

The following statement will help us to answer this question.

(9.16) PROPOSITION. *If an M-map has an n-orbit, for some $n \in \mathbb{N}$, then it has also a 1-orbit (a fixed point).*

PROOF. Let us denote $a = \min_{i=1,\ldots,n} x_i$ and $b = \max_{i=1,\ldots,n} x_i$, where an n-orbit of the given M-map is $\overline{\{x_1,\ldots,x_n\}}$. Then, for $I = [a,b]$, $I \subset \varphi(I)$, and there exist $k, l \in \{1, \ldots, n\}$ such that $x_k \in \varphi(a)$ and $x_l \in \varphi(b)$. Since $x_k \geq a$, $x_k \leq b$, Lemma (9.9) yields the existence of a fixed point of φ. □

(9.17) PROPOSITION. *Let* $\varphi : \mathbb{R} \multimap \mathbb{R}$ *be an* M-*map. If* φ *has a 3-orbit, then* φ *has also a* k-*orbit, for every* $k \lhd 3$, *except* $k = 2, 4, 6$.

PROOF. The 3-orbit $\overline{\{a, b, c\}}$ can be assumed to satisfy the following conditions:

(α) each of the points a, b, c is not a fixed point of φ. Indeed, if $b \in \varphi(b)$, for example, then we have the k-orbit

$$\{a, \underbrace{b, \ldots, b}_{(k-2)\text{-times}}, c\},$$

for every $k \lhd n$.

(β) $a \notin \varphi(b)$, $b \notin \varphi(c)$, $c \notin \varphi(a)$. Indeed. If, for example, $b \in \varphi(c)$, then φ has the k-orbit

$$\{a, \underbrace{b, c, \ldots, b, c}_{((k-1)/2)\text{-times}}\},$$

where $k \geq 5$ is odd and the k-orbit

$$\{a, b, c, a, \underbrace{b, c, \ldots, b, c}_{(k/2-2)\text{-times}}\},$$

where $k \geq 8$ is even.

(γ) $a < b$. The reverse inequality $b < a$ can be obtained by a reflection in the point.

(δ) $a < c$. Otherwise, we redenote the points (a instead of c, b instead of a, c instead of b).

Now, there are two possibilities: (1) $b < c$ and (2) $b > c$. In both cases, φ admits a 1-orbit, according to Proposition (9.16).

Case 1. $b < c$.

(a) $\varphi(a) \cap \varphi(b) = \emptyset$.

In view of the assumptions (α)–(δ), one has

$$a < \min_{x \in \varphi(a)} x \leq \max_{x \in \varphi(a)} x < \min_{y \in \varphi(b)} y,$$

but it is just Theorem (9.14). Therefore, φ has a k-orbit, for every $k \lhd n$, except $k = 4, 6$.

(b) $\varphi(a) \cap \varphi(b) \neq \emptyset$.

Since $b \notin \varphi(b)$ and $c \notin \varphi(a)$ (see the assumptions (α) and (β), respectively), there exists $b_2 \in (b, c)$ satisfying $b_2 \in \varphi(a)$ and $b_2 \in \varphi(b)$.

Using $c \in \varphi(b)$, $a \in \varphi(c)$ and $b \notin \varphi(b)$, $b \notin \varphi(c)$ (see the assumptions (α), (β), respectively), Lemma (9.13) yields the existence of $x_3 \in (b, c)$ with the property $b \in \varphi(x_3)$. Case 1(b) now splits into the following subcases:

(A) $x_3 < b_2$.

Fix an arbitrary $x_1 \in (a, b)$ and $k \in \mathbb{N}$, $k \geq 4$. We consider

$$[a, x_1] \to \underbrace{[b, x_3] \to \ldots \to [b, x_3]}_{(k-2)\text{-times}} \to [x_3, c] \to [a, x_1].$$

By Lemma (9.10), we have the existence of a k-orbit $\overline{\{y_1, \ldots, y_k\}}$. Indeed, since we start and finish in the interval $[a, x_1]$ and $[a, x_1] \cap [b, x_3] = \emptyset$, a sequence $\{y_1, \ldots, y_k\}$ cannot be formed by going p-times around a shorter m-orbit, where $k = mp$.

(B) $x_3 \geq b_2$.

Thanks to the convexity of $\varphi(b)$, one has $x_3 \in \varphi(b)$. If $k \geq 5$ is odd, then we have the k-orbit

$$\overline{\{a, b, x_3, \ldots, \underbrace{b, x_3}, b, c\}},$$
$$\underbrace{\qquad\qquad\qquad}_{((k-3)/2)\text{-times}}$$

Finally, if $k \geq 8$ is even, then there exists the k-orbit

$$\overline{\{a, b, c, a, b, x_3, \ldots, \underbrace{b, x_3}, b, c\}}.$$
$$\underbrace{\qquad\qquad\qquad}_{(k/2-3)\text{-times}}$$

Case 2. $c < b$.

Using Lemma (9,9) and the assumption (α), there exists the fixed point $x_2 \in (a, c)$. The following possibilities can then occur.

(a) $x_2 \in \varphi(c)$.

With respect to $c \notin \varphi(a)$ (see the assumption (β)), Lemma (9.13) yields the existence of $x_1 \in (a, x_2]$, satisfying $c \in \varphi(x_1)$. Since $a \in \varphi(c)$, $x_2 \in \varphi(c)$ and $\varphi(c)$ is a convex set, it holds also $x_1 \in \varphi(c)$. If $k \geq 5$ is odd, then we have the k-orbit

$$\overline{\{a, b, c, x_1, c, \ldots, \underbrace{x_1, c}\}}.$$
$$\underbrace{\qquad\qquad\qquad}_{((k-3)/2)\text{-times}}$$

If $k \geq 8$ is even, then we have the k-orbit

$$\overline{\{a, b, c, a, b, c, \underbrace{x_1, c, \ldots, x_1, c}_{(k/2-3)\text{-times}}\}}.$$

(b) $x_2 \notin \varphi(c)$.

Using Lemma (9.13) again, we obtain $x_3 \in (c, b)$, satisfying $[x_2, c] \subset \varphi[x_3, b]$. Moreover, $[x_3, b] \subset \varphi[a, x_2]$. If $k \geq 5$ is odd, then we consider

$$[x_3, b] \to \underbrace{[x_2, c] \to [a, x_2] \to \ldots \to [x_2, c] \to [a, x_2]}_{((k-1)/2)\text{-times}} \to [x_3, b].$$

If $k \geq 8$ is even, then we consider

$$[x_3, b] \to \underbrace{[x_2, c] \to [a, x_2] \to \ldots \to [x_2, c] \to [a, x_2]}_{(k/2-2)\text{-times}} \to [x_3, b]$$

$$\to [x_2, c] \to [a, x_2] \to [x_3, b].$$

By Lemma (9.10) and by the same reasoning as in the proof of Theorem (9.14), we obtain the existence of the desired k-orbit. $\square$

For obtaining a particular multivalued version of the Sharkovskiĭ Theorem (9.1), we also apply an approximative technique in [BGK]. For M-maps, the related statement (Lemma 4.5 and Remark 4.6 in [BGK]) takes the following special form.

(9.18) LEMMA. *Let $\varphi : I \multimap I$ be a composition of M-maps $\varphi_i : I_{i-1} \multimap I_i$, $i = 1, \ldots, n$, i.e. $\varphi = \varphi_n \circ \ldots \circ \varphi_1$, where $I_0 = I_n = I \subset \mathbb{R}$ and $I, I_1, \ldots, I_{n-1}$ are closed intervals. Assume that a_k are fixed points of φ, $a_k \in \varphi(a_k)$, $k = 1, \ldots, m$.*

Then, for every $\varepsilon > 0$, there exists a continuous ε-approximation $f = f_n \circ \ldots \circ f_1$ of φ (on the graph of φ), namely $\Gamma_f \subset N_\varepsilon(\Gamma_\varphi)$, where $N_\varepsilon(\Gamma)$ denotes an open neighbourhood of Γ in $\mathbb{R}^2$ such that f_i are continuous $\delta(\varepsilon)$-approximations of φ_i, for every $i = 1, \ldots, n$, with $\lim_{\varepsilon \to 0} \delta(\varepsilon) = 0$ and $a_k = f(a_k)$, $k = 1, \ldots, m$.

Moreover, there exist a_{ik} with $a_{ik} \in \varphi_i(a_{i-1,k})$, $a_{0k} = a_k \in \varphi_n(a_{n-1,k})$ such that $a_{ik} = f_i(a_{i-1,k})$, $a_{0k} = a_k = f_n(a_{n-1,k})$, for every $i = 1, \ldots, n-1$ and $k = 1, \ldots, m$.

Hence, we can give

(9.19) PROPOSITION. *Assume that an M-map $\varphi : \mathbb{R} \multimap \mathbb{R}$ has a primary n-orbit $\mathcal{O} = \overline{\{x_1, \ldots, x_n\}}$, where $n > 3$ is an odd integer, but φ has no l-orbit with $l \triangleright n$. Then φ admits a k-orbit, for each k such that $n \triangleright k$, except $k = 4$.*

PROOF. Assume for a moment that φ is bounded. Then there is a closed interval I containing $\mathcal{O}$ such that $\varphi^i : I \multimap I$, for every $i = 1, \ldots, n$.

Applying Lemma (9.18), there is an ε-approximation f_ε^n of φ^n, for every sufficiently small $\varepsilon > 0$, such that $x_i \in \varphi^n(x_i)$, $x_i = f_\varepsilon^n(x_i)$, for $i = 1, \ldots, n$. Furthermore, there is a $\delta(\varepsilon)$-approximation f_δ of φ such that $x_{i+1} \in \varphi(x_i)$, $x_{i+1} = f_\delta(x_i)$, for every $i = 1, \ldots, n$, where $\lim_{\varepsilon \to 0} \delta(\varepsilon) = 0$.

Since $\mathcal{O}$ becomes, for every $\delta = \delta(\varepsilon)$, an n-orbit of f_δ as well, it follows from Theorem (9.1) that f_δ has also k-orbits, for every k such that $n \triangleright k$. Requiring, however, k-orbits with a special pattern on the line, we follow the well-known idea of P. Štefan (cf. also Remark 3.1.4 in [BGK]).

Since φ has, by the hypothesis, no l-orbit with $l \triangleright n$, the only fixed points of φ^l are just those of φ. Thus, there is a δ-approximation f_δ of φ having no l-orbit with $l \triangleright n$, but an n-orbit $\mathcal{O}$, for any sufficiently small $\delta(\varepsilon)$. Indeed. According to Lemma (9.18), φ can be approximated with an arbitrary accuracy keeping fixed points. Therefore, for a given ε ($\Rightarrow \varepsilon_l(\varepsilon)$), there certainly exists a Δ-approximation f_Δ^l of φ^l with no more fixed points outside $\dfrac{\varepsilon_l}{2}$-neighbourhoods of fixed points of φ^l, where $\Delta \ll \dfrac{\varepsilon_l}{2}$. Since, in $\dfrac{\delta}{2}$-neighbourhoods of fixed points of φ, f_Δ can be always continuously replaced in a desired manner (for more details, see [BGK]), we get an ε_l-approximation $f_{\varepsilon_l}^l$ of φ^l with no more fixed points, as pointed out.

Denote by $J = [\min_{i=1,\ldots,n} x_i, \max_{i=1,\ldots,n} x_i]$ and let Λ be the partition of J by the elements x_i of $\mathcal{O}$, namely

$$x_n < x_{n-2} < \ldots < x_3 < x_1 < x_2 < x_4 < \ldots < x_{n-1},$$

or reversely. Applying Lemma 3.1.6 in [Rob-M], whose assumptions are so satisfied, the associated transition graph of f_δ for Λ contains a transition subgraph (so called Štefan's subgraph) of the following form:

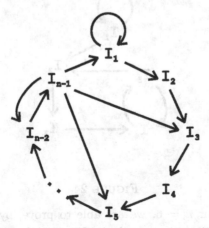

FIGURE 23. Štefan's subgraph

where $I_1 = [x_1, x_2]$, $I_2 = [x_3, x_1]$, $I_3 = [x_2, x_4]$, $I_{2j} = [x_{2j+1}, x_{2j-1}]$ and $I_{2j-1} = [x_{2j-2}, x_{2j}]$, for $j = 2, \ldots, (n-1)/2$.

Now, in order to find a desired k-orbit $\{y_1^\delta, \ldots, y_k^\delta\}$ of f_δ, we can always employ, according to Lemma (9.10), a suitable sequence of intervals, e.g.

$$I_{n-1} \to I_{n-k} \to I_{n-k+1} \to \cdots \to I_{n-1}, \quad \text{for } k \text{ even } (k < n),$$

or

$$I_1 \to I_2 \to \cdots \to I_{n-1} \to I_1 \to I_1 \to \cdots \to I_1, \quad \text{for } k \text{ odd } (k > n),$$

or some others, for k even $(k > n)$.

One can observe that, because of Lemma 3.1.6 in [Rob-M], for all sufficiently small ε (and so δ), the associated transition subgraphs of f_δ are same. Therefore, because of the boundedness of sets $\{y_1^\delta\}_\delta, \ldots, \{y_k^\delta\}_\delta$, convergent sequences can be chosen, for $\varepsilon \to 0$ $(\Rightarrow \delta \to 0)$, such that the limiting orbit $\overline{\{y_1, \ldots, y_k\}}$ cannot be formed by going p-times around a shorter orbit of mth-order, where $k = mp$ (see Fig. 23), except $k = 4$. The case, when $k = 4$, is exceptional, because it corresponds to the sequence $I_{n-1} \to I_{n-4} \to I_{n-3} \to I_{n-2} \to I_{n-1}$, and 4-orbits of approximating maps may converge to the two points x_{n-3} and x_{n-2} (see Counter-Example (9.8). For the other cases, it is impossible. So, we have got a k-orbit of φ, $k \neq 4$.

Since, for a fixed k such that $n \rhd k, k \neq 4$, any M-map $\varphi : \mathbb{R} \multimap \mathbb{R}$ satisfying our assumptions can be replaced by a suitable bounded M-map, which is identical to the original one on a sufficiently large bounded closed interval, involving both an n-orbit and a k-orbit, the proof is complete. □

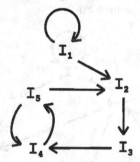

FIGURE 24

(9.20) REMARK. For $n = 6$, we are able to prove, by means of the related transition subgraph in Fig. 24, an analogy of Proposition (9.19), provided there is a primary 6-orbit, but no l-orbit with $l \rhd 6$. Unfortunately, for n such that $6 \rhd n$, the related cases consist in a rather cumbersome verification of $(n-1)!/2$ possibilities.

The following counter-example demonstrates that there is an M-map, satisfying the assumptions of Theorem (9.14), which admits a primary 6-orbit, but which has no 4-orbit.

(9.21) COUNTER-EXAMPLE. Consider the M-mapping (see Fig. 25) $\varphi : [0,4] \multimap [0,4]$ where

$$\varphi(x) := \begin{cases} [1,2], & \text{for } x = 0, \\ x+2, & \text{for } x \in (0,1), \\ [3,4], & \text{for } x = 1, \\ 3, & \text{for } x \in (1,2], \\ -2x+7, & \text{for } x \in (2,3], \\ -x+4, & \text{for } x \in (3,4]. \end{cases}$$

One can observe that φ has a primary 3-orbit $\overline{\{0,1,4\}}$ satisfies the first inequalities in Theorem (9.14) (see Fig. 25 and Fig. 27) and that φ has infinitely many primary 6-orbits (see Fig. 30), as for example $\overline{\{1,7/2,1/2,5/2,2,3\}}$.

On the other hand, there is no 4-orbit of φ (see Fig. 28), because 1, 7/3 and 3 are the only fixed points of φ^4 and moreover, 7/3 is a fixed point of φ. Thus, the desired 4-orbit does not exist, because both possible candidates are formed by going 2-times around the 2-orbit (see Fig. 26).

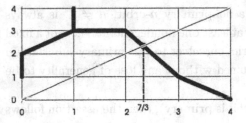
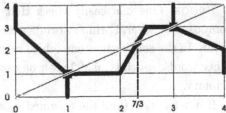

FIGURE 25. φ FIGURE 26 φ^2

We saw in Counter-Example (9.7) that a 3-orbit need not imply a 2-orbit. On the other hand, according to Proposition (9.19), primary odd-orbits imply, at the absence of 3-orbits, everything smaller (in the Sharkovskiĭ ordering), but 4-orbits. Thus, another natural question arises whether or not "a 2-orbit is implied by everything, but fixed points, at the absence of 3-orbits".

(9.22) PROPOSITION. *Let an M-map φ have no primary 3-orbits. If φ has an n-orbit, where $n \neq 1$, then φ admits a 2-orbit.*

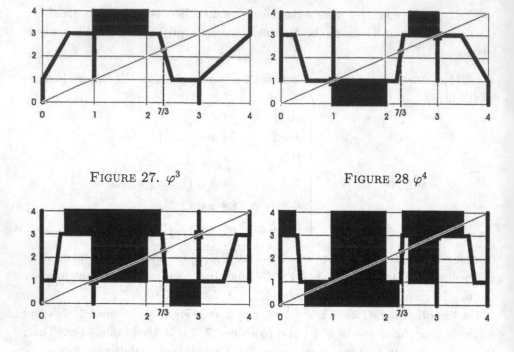

FIGURE 27. φ^3 FIGURE 28 φ^4

FIGURE 29. φ^5 FIGURE 30 φ^6

PROOF. One can easily check that a nonprimary n-orbit, $n \neq 1$, is always composed by several primary orbits with at least one nondegenerate, i.e. not a fixed point. In the sequel, it is enough to restrict ourselves to this primary suborbit. In other words, n-orbits, $n \neq 1$, can be assumed without any loss of generality to be primary.

If $n > 3$ is odd (and the assumed n-orbit is primary), then the assertion follows directly from Proposition (9.19).

Otherwise, let n be the smallest even integer in the usual ordering such that φ has a (primary) n-orbit $\mathcal{O} = \overline{\{x_1, x_2, \dots, x_n\}}$. Denote by $\varphi_{\mathcal{O}}(x_i)$ the successor of x_i on $\mathcal{O}$, i.e. $\varphi_{\mathcal{O}}(x_i) = x_{i+i}$, for $i = 1, \dots, n-1$, and $\varphi_{\mathcal{O}}(x_n) = x_1$. Let $a = \max_{x_i \in \mathcal{O}}\{x_i \mid x_{i+1} = \varphi_{\mathcal{O}}(x_i) > x_i\}$ and $b = \min_{x_i \in \mathcal{O}}\{x_i \mid x_i > a\}$.

Put $I_1 = [a, b]$ and define $I = [\min_{x_i \in \mathcal{O}} x_i, \max_{x_i \in \mathcal{O}} x_i] = [x, y]$.

If there is no interval K satisfying $K \neq I_1$ and $\varphi(K) \supset I_1$, then all points x_j such that $x_j \leq a$ are mapped to b or to the right-hand side of b, while all points x_j such that $x_j \geq b$ either to a or to the left. Since there is a point in $\mathcal{O}$ mapped to b as well as another one mapped to y, we have that $b, y \in \varphi([x, a])$, and so $\varphi([x, a]) \supset [b, y]$. Similarly, $\varphi([b, y]) \supset [x, a]$, and subsequently, we have a cycle of two disjoint intervals. Applying Lemma (9.10), φ admits a 2-orbit.

If there exists an interval K, $K \neq I_1$ and $\varphi(K) \supset I_1$, then denote, again by K, the interval $L \neq I_1$, $L = [r, s]$, $r, s \in \mathcal{O}$ and $\mathcal{O} \cap (r, s) = \emptyset$, satisfying $\varphi(L) \supset I_1$, provided there is no other interval with these properties between L and I_1.

If K and I_1 are not disjoint, e.g. if $K \cap I_1 = \{b\}$, then $K = [b, c]$, where $c \in \mathcal{O}$, $c > b$ and c is the next after b, in the ordering of the real line. Moreover, $\varphi_{\mathcal{O}}(c) = b$, $\varphi_{\mathcal{O}}(b) \leq a$ and $\varphi_{\mathcal{O}}(a) \geq c$, where at least one of the inequalities is strict, because $n \neq 3$.

The following possibilities can then occur:

(i) If $d = \varphi_{\mathcal{O}}(a) > c$, then, by the definition of I_1, $\varphi_{\mathcal{O}}(d) < d$.

(α) If $\varphi_{\mathcal{O}}(d) \leq a$, then $[a, b]$ and $[c, d]$ form the couple of disjoint intervals such that $[a, b] \to [c, d] \to [a, b]$, and so Lemma (9.10) implies a 2-orbit.

(β) If $\varphi_{\mathcal{O}}(d) \geq c$, then $[a, b]$, $[b, c]$ and $[c, d]$ form the triplet of overlapped intervals $[b, c] \to [a, b] \to [c, d] \to [b, c]$ and, according to Lemma (9.10), there exists a 3-orbit. By the hypothesis, it is nonprimary, and hence it is composed by a desired 2-orbit and a fixed point.

(ii) Otherwise, $\varphi_{\mathcal{O}}(a) = c$ and $\varphi_{\mathcal{O}}(b) < a$.

(α) If $a = \varphi_{\mathcal{O}}(e)$, $f = \varphi_{\mathcal{O}}(b)$, $e < a$ and $f \leq e$, then we have the 3-cycle $[e, a] \to [b, c] \to [a, b] \to [e, a]$ and, by the same reasoning as in (i)(β), we obtain a 2-orbit.

(β) If $e < f$ or $e > c$, then the existence of a 2-cycle or a 3-cycle need not be straightforward, in general. Nevertheless, we can construct, for some $m \leq n - 2$, the $(m + 1)$-cycle of intervals $[e, a] \to [b, c] \to I_2 \to I_3 \to \ldots \to I_m \to [e, a]$, when $e < f$, or the $(m + 1)$-cycle $[c, e] \to I_1 \to I_2 \to \ldots I_m \to [c, e]$, when $c < e$. Lemma (9.10), therefore, implies an $(m + 1)$-suborbit, which is "at most" an $(n - 1)$-suborbit (being not a fixed point). Taking into account what we have already proved, we can recurrently apply the assertion of our proposition to this suborbit, to get a 2-orbit. The recurrence is namely always finite, and any 4-orbit implies always a 2-orbit (the complete discussion of this fact consists of checking six different cycles, as in the single-valued case, cf. e.g. [Rob-M]).

The cases $K \cap A = \{a\}$ and $K \cap I_1 = \emptyset$ can be treated analogously, including the recurrence argument. $\square$

Using Proposition (9.17), we are able to improve partially Proposition (9.19).

(9.23) PROPOSITION. *Let $\varphi : \mathbb{R} \multimap \mathbb{R}$ be an M-map. If φ has an n-orbit, where n is odd, then φ has also a k-orbit for every $k \vartriangleleft n$, except $k = 2, 4, 6$.*

PROOF. Because of Proposition (9.17) and Proposition (9.19), without much effort, our proposition becomes clear in the cases when $n = 1, 3$ or when the given n-orbit is primary, but φ has no l-orbit with $l \vartriangleright n$.

Otherwise, we can assume that φ has an n'-orbit $\overline{\{x_i, x_2, \ldots, x_{n'}\}}$, where $n' \neq 3, 1$, $n' = n$ or $n' \rhd n$, which is maximal in the Sharkovskiĭ ordering, i.e. φ has no l-orbit with $l \rhd n$.

If the considered n'-orbit is primary, then it is just the case of Proposition (9.19). Otherwise, there is neither $i \in \{2, 3, \ldots, n' - 2\}$ such that $x_j = x_{j+i}$, for some $j \in \{1, 2, \ldots, n' - 2\}$, nor more than one $i \in \{1, 2, \ldots, n' - 1\}$ satisfying $x_{i+1} = x_i$, because supposing the converse, we obtain a shorter odd orbit. This is, however, a contradiction with the assumption of the maximality of n'.

So, it suffices to examine an n'-orbit ordered, without any loss of generality, as $\overline{\{x_1, x_2, \ldots, x_{n'-1}, x_1\}}$, where $x_i \neq x_j$, for $i \neq j$, $i, j \in \{1, 2, \ldots, n' - 1\}$. Setting $p = n' - 1$, one has that φ admits a fixed point $\{x_1\}$ and an even primary p-orbit. Although one can see the existence of a k-orbit, for every $k \geq p$, it seems to be nontrivial to show the existence of a k-orbit, when k is even, $k \leq p$, $k \neq 2, 4, 6$.

We can assume that $x_1 = 0$ and $x_1 < x_2$ (the other cases can be obtained by a translation and a reflection in the point, respectively). Furthermore, $x_i \notin [x_1, x_2]$, for every $i \in \{3, 4, \ldots, p - 1\}$, because otherwise, we have either an odd orbit $\overline{\{x_1, x_i, x_{i+1}, \ldots, x_p\}}$ or an odd orbit $\overline{\{x_1, x_1, x_i, x_{i+1}, \ldots, x_p\}}$, in the dependence on i. This is, however, a contradiction with the assumption of the maximality of n'.

It also holds that $\mathrm{sgn}(x_i) = \mathrm{sgn}(x_{i+1})$ implies $\mathrm{sgn}(x_{i+2}) = \mathrm{sgn}(x_i)$, for every $i \in \{2, 3, \ldots, p - 2\}$. Indeed, suppose conversely that there exists an $i \in \{2, 3, \ldots, p - 2\}$ with the property $\mathrm{sgn}(x_i) = \mathrm{sgn}(x_{i+1})$, and $\mathrm{sgn}(x_{i+2}) \neq \mathrm{sgn}(x_i)$. Then we consider

$$[x_1, x_2] \to [x_2, x_3] \to \ldots \to [x_i, x_{i+1}] \to [x_1, x_2],$$

if i is odd, and

$$[x_1, x_2] \to [x_1, x_2] \to [x_2, x_3] \to \ldots \to [x_i, x_{i+1}] \to [x_1, x_2],$$

if j is even.

In both cases, applying Lemma (9.10) and respecting $[x_i, x_{i+1}] \cap [x_1, x_2] = \emptyset$, for $i \in \{3, \ldots, p - 2\}$ (for $i = 2$, it should be treated separately, but in an analogous way), we obtain a shorter odd orbit, contradictionally.

Now, the proof splits into the following cases. We show that only Case 3 is actual, because Case 1 and Case 2 imply the existence of odd shorter orbit which is again a contradiction.

Case 1. $x_i > x_2$, for every $i \in \{3, 4, \ldots, p\}$. There are two possibilities:

(i) $x_p > x_{p-1}$.

Using $x_p \in \varphi(x_{p-1})$, $x_1 \in \varphi(x_p)$, $x_p > x_{p-1}$, and Lemma (9.9), there exists an $a \in [x_{p-1}, x_p]$ such that a is a fixed point of φ. If $a = x_p$, then we have a 3-orbit $\overline{\{x_p, x_p, x_{p-1}\}}$ (we note only that the convexity of $\varphi(x_p)$ and $x_1 \in \varphi(x_p)$, $x_p \in \varphi(x_p)$ yield $x_{p-1} \in \varphi(x_p)$). If $a \in (x_{p-1}, x_p)$, then we can consider

$$[x_1, x_{p-1}] \to [x_{p-1}, a] \to [a, x_p] \to [x_1, x_{p-1}]$$

and apply Lemma (9.10) to complete the proof. Finally, we discuss the case $a = x_{p-1}$. Thanks to Lemma (9.13), there exists a $b \in [x_{p-1}, x_p]$ with the property $x_2 \in \varphi(b)$. Subsequently, we have an odd $(p-1)$-orbit $\overline{\{x_{p-1}, x_2, x_3, \dots, x_{p-2}\}}$.

(ii) $x_p < x_{p-1}$.

It is clear that there exists $j \in \{2, 3, \dots, p-2\}$ with the property $x_j < x_p < x_{j+i}$. Considering

$$[x_1, x_j] \to [x_1, x_j] \to [x_j, x_p] \to [x_1, x_j],$$

we obtain either a 3-orbit of φ or $x_j \in \varphi(x_j)$. In the second case, the convexity of $\varphi(x_j)$ and $x_{j+1} \in \varphi(x_j)$ imply $x_p \in \varphi(x_j)$, and we have either a $(j+1)$-orbit $\overline{\{x_1, x_2, \dots, x_j, x_p\}}$, if j is even, or a $(j+2)$-orbit $\overline{\{x_1, x_1, x_2, \dots, x_j, x_p\}}$, if j is odd.

Case 2. $x_i < 0$, for every $i \in \{3, 4, \dots, p\}$.

If $x_3 < x_p$, then there exists $\varepsilon \in (x_p, x_1)$ satisfying $[\varepsilon, x_1] \subset \varphi[x_3, x_p]$. Considering

$$[x_3, x_p] \to [\varepsilon, x_1] \to [x_1, x_2] \to [x_3, x_p]$$

and applying Lemma (9.10), we obtain a 3-orbit. Otherwise, there are two possibilities again:

(i) $x_p < x_{p-1}$.

We can proceed like in Case 1(i). So, using $x_p \in \varphi(x_{p-1})$, $x_1 \in \varphi(x_p)$, $x_p < x_{p-1}$ and Lemma (9.9), there exists an $a \in [x_p, x_{p-1}]$ such that a is a fixed point of φ. If $a = x_p$, then we have the 3-orbit $\overline{\{x_p, x_p, x_{p-1}\}}$. If $a \in (x_p, x_{p-1})$, then we consider

$$[x_{p-1}, x_1] \to [a, x_{p-1}] \to [x_p, a] \to [x_{p-1}, x_1]$$

and apply Lemma (9.10), to complete the proof. Finally, we discuss the case $a = x_{p-1}$. Thanks to Lemma (9.13), there exists a $b \in [x_p, x_{p-1}]$ with the property $x_3 \in \varphi(b)$. Subsequently, we have an odd $(p-1)$-orbit $\overline{\{x_{p-1}, x_{p-1}, b, x_3, x_4, \dots, x_{p-2}\}}$.

(ii) $x_p > x_{p-1}$.

It is obvious that there exists a $j \in \{3, \dots, p-2\}$ with the property $x_j > x_p > x_{j+1}$. Considering

$$[x_j, x_1] \to [x_j, x_1] \to [x_p, x_j] \to [x_j, x_1],$$

we obtain either a 3-orbit of φ or $x_j \in \varphi(x_j)$. In the second case, we have either a $(j+1)$-orbit $\overline{\{x_1, x_2, \ldots, x_j, x_p\}}$, it j is even, or a $(j+2)$-orbit $\overline{\{x_1, x_1, x_2, \ldots, x_j, x_p\}}$, if j is odd.

Case 3. $\mathrm{sgn}(x_i) = -\mathrm{sgn}(x_{i+1})$, for every $i \in \{2, 3, \ldots, p-1\}$.

We redenote the set $\overline{\{x_1, x_2, \ldots, x_p\}}$ into $\overline{\{a_1, a_2, \ldots, a_p\}}$ in order the set $\overline{\{a_1, a_2, \ldots, a_p\}}$ can be ordered as $a_{p-1} < \ldots < a_3 < a_1 < a_2 < \ldots < a_p$, and we consider a map $a : \overline{\{x_1, x_2, \ldots, x_p\}} \to \overline{\{a_1, a_2, \ldots, a_p\}}$ such that $a(x_i) = a_j$ if and only if $a_j = x_i$, whenever $i, j \in \{1, 2, \ldots, p\}$. Thanks to the previous assumptions, we have $a_1 = x_1$, $a_2 = x_2$ and $a_i = a(x_p) > x_2$. Furthermore, we consider an $a_m \in [x_2, a_l]$ with the property $s(a^{-1}a_m) = \min\{s(a^{-1}(a_i)) \mid i = 2, 4, 6, \ldots, l-2\}$, where $s(x_i) = x_{i+1}$, for every $i = 2, 4, \ldots, l-2$.

It can be readily checked that $[a_1, a_{i+1}] \subset \varphi[a_1, a_i]$, for every $i \in \{2, 3, \ldots, p-1\}$. Subsequently, it holds $[a_m, a_l] \subset \varphi[a_{l-1}, a_1]$, $[a_{l-1}, a_1] \subset \varphi[a_1, a_m]$, and $[a_{l-1}, a_1] \subset \varphi[a_m, a_l]$.

Now, if $k = 2$, then we consider

$$[a_m, a_l] \to [a_{l-1}, a_1] \to [a_m, a_l].$$

If $k \geq 6$ is even, $k < p$, $k < m+4$, then we consider

$$[a_m, a_l] \to [a_{l-1}, a_1] \to [a_1, a_s] \to [a_1, a_{s+1}] \to [a_1, a_{s+2}] \to \ldots$$
$$\to [a_1, a_m] \to [a_1, a_{l-1}] \to [a_m, a_l],$$

where $s = m + 4 - k$.

At last, if $k > m + 4$ is even, $k < p$, then we take an arbitrary $b \in [a_1, a_2)$. Let us consider

$$[a_m, a_l] \to [a_{l-1}, a_1] \to \underbrace{[a_1, b] \to \ldots \to [a_1, b]}_{(k-m-2)\text{-times}} \to [a_1, a_2]$$
$$\to [a_1, a_3] \to \ldots \to [a_1, a_m] \to [a_1, a_{l-1}] \to [a_m, a_l].$$

In all the previous cases, by Lemma (9.10) and by the analogous considerations as in Theorem (9.14), we obtain the existence of a k-orbit.

So, respecting the previous conclusions, φ has a k-orbit, for every $k \triangleleft n$, except $k = 4$. $\square$

A further extension of Proposition (9.19) requires to study the following problem. Let us consider an M-map $\varphi : \mathbb{R} \multimap \mathbb{R}$ and an M-map $g : \mathbb{R} \multimap \mathbb{R}$ such that $g = \varphi^l$, where $l \in \mathbb{N}$, and suppose that $q \in \mathbb{N}$. It seems that the existence of a ql-orbit of φ is not guaranteed by the existence of a q-orbit of g; see the sequence

$$\overline{\{x_1, a_1, x_4, a_2, x_3, a_3, x_2, a_4 \mid x_1, a_1, x_4, a_2, x_3, a_3, x_2, a_4 \mid x_1, a_1, x_4, a_2, x_3, a_3, x_2, a_4\}}.$$

Nevertheless, we can state the following lemma.

(9.24) LEMMA. *Let $\varphi : \mathbb{R} \multimap \mathbb{R}$ and $g : \mathbb{R} \multimap \mathbb{R}$ be M-maps satisfying $g = \varphi^l$, where $l = 2^s$, for some $s \in \mathbb{N}$.*

(9.24.1) *If g has a q-orbit, where q is odd, then φ has also a q-orbit or an lq-orbit.*

(9.24.2) *If g has a q-orbit, where $q = 2^r$, for some $r \in \mathbb{N}$, then φ has a 2^{r+s}-orbit.*

PROOF. By the hypothesis, there exists a sequence
$$\mathcal{A} = \overline{\{x_1, x_2, \ldots, x_l, x_{l+1}, \ldots, x_{(q-1)l}, x_{(q-1)l+1}, \ldots, x_{ql}\}}$$
such that $x_{i+1} \in \varphi(x_i)$, for every $i = 1, 2, \ldots, ql$, where $x_{ql+1} = x_1$, and such that a sequence
$$\mathcal{B} = \overline{\{x_1, x_{l+1}, \ldots, x_{(q-1)l+1}\}}$$
is a q-orbit of g.

$\mathcal{A}$ cannot be formed by going lq/m-times around an m-orbit, where l is divided by m, because then $x_1 = x_{l+1} = \cdots = x_{(q-1)l+1}$, which means that $\mathcal{B}$ is formed by going q-times around x_1, a contradiction.

Furthermore, let us suppose that $\mathcal{A}$ is formed by going lq/m-times around an m-orbit, where q is divided by m, but $q \neq m$. Subsequently, $x_1 = x_{lm+1} = x_{2lm+1} = \cdots, x_{l+1} = x_{l(m+1)+1} = x_{l(2m+1)+1} = \cdots, x_{l(m-1)+1} = x_{l(2m-1)+1} = \cdots = x_{l(q-1)+1}$. $\mathcal{B}$ is then formed by going q/m-times around an m-orbit, a contradiction. Hence, φ has either a q-orbit or an lq-orbit.

Finally, if g has a 2^r-orbit, then supposing that $\mathcal{A}$ is formed by going around a 2^p-orbit, $0 \leq p < r + s$, we obtain that $\mathcal{B}$ is formed by going around a shorter orbit, a contradiction. Therefore, φ has an 2^{r+s}-orbit. $\square$

Let us solve the converse of the problem described before Lemma (9.24). It seems that the existence of a q-orbit of g is not guaranteed by the existence of an lq-orbit of φ; see the sequence
$$\overline{\{x_1, a_1, x_2, a_2, x_1, a_3, x_2, a_1, x_1, a_2, x_2, a_3\}},$$
where we consider $l = 2$, $q = 6$.

Nevertheless, the following lemma holds.

(9.25) LEMMA. *Let $\varphi : \mathbb{R} \multimap \mathbb{R}$ and $g : \mathbb{R} \multimap \mathbb{R}$ be M-maps satisfying $g = \varphi^l$, where $l \in \mathbb{N}$, and let $q \in \mathbb{N}$. If φ has an lq-orbit, then g has an m-orbit, where q is divided by m and $m \neq 1$.*

PROOF. Let us suppose that φ has an lq-orbit
$$\mathcal{O} = \overline{\{x_1, x_2, \ldots, x_l, \ldots, x_{2l}, \ldots, x_{ql}\}}$$

and consider the sequence $\mathcal{O}_1 = \overline{\{x_1, x_{l+1}, x_{2l+1}, \dots, x_{(q-1)l+1}\}}$. If $\mathcal{O}_1$ is formed by going p-times around an m-orbit, where $q = mp$ and $m \neq 1$, then the proof of our lemma is complete. Otherwise, $\mathcal{O}_1$ is formed by going q-times around $\{x_1\}$ and we proceed by considering the sequences $\mathcal{O}_k = \overline{\{x_k, x_{l+k}, x_{2l+k}, \dots, x_{(q-1)l+k}\}}$, for $k = 2, 3, \dots, l$, respectively. Repeating the previous considerations for $\mathcal{O}_k$, instead of $\mathcal{O}_1$, for $k = 2, 3, \dots, l$, respectively, we either finish the proof or we get that $\mathcal{O}$ is formed by going q-times around the l-orbit $\overline{\{x_1, x_2, \dots, x_l\}}$, a contradiction. $\square$

Proposition (9.23) and Lemma (9.25) imply immediately

(9.26) LEMMA. *Let* $\varphi : \mathbb{R} \multimap \mathbb{R}$ *and* $g : \mathbb{R} \multimap \mathbb{R}$ *be* M-*maps satisfying* $g = \varphi^l$, *where* $l \in \mathbb{N}$, *and let* q *be odd. If* φ *has an* lq-*orbit, then* g *has a* q-*orbit.*

Now, we can give the main result of this section, the extension of Proposition (9.19), representing a multivalued version of the Sharkovskiĭ Theorem (9.1).

(9.27) THEOREM. *Let an* M-*map* $\varphi : \mathbb{R} \multimap \mathbb{R}$ *have an* n-*orbit,* $n \in \mathbb{N}$. *Then* φ *has also a* k-*orbit, for every* $k \lhd n$, *with the exception of at most three orbits.*

(9.28) REMARK. The absent three orbits in Theorem (9.27) can be detected as follows from the proof, see Corollary (9.29).

PROOF OF THEOREM (9.27). Let $n = 2^m q$, where $m \in \mathbb{N}_0 = \mathbb{N} \cup \{0\}$ and $q \in \mathbb{N}$ is odd. The proof of our theorem now splits into the following cases:
Case 1. $q > 1$.
This case can be proved by the induction on $\mathbb{N}_0$. By Proposition (9.23), our theorem is true, when $m = 0$. Assume that $n = 2^m q$, $m \geq 1$, and that the existence of a $2^s p$-orbit, where $s < m$, $s \in \mathbb{N}_0$ and $p \in \mathbb{N}$ is odd, implies the existence of a k-orbit, for every $k \lhd 2^s p$, with the exception of at most three orbits.

Respecting the induction assumption, φ can be assumed to have no k-orbit, for any $k = 2^s p$, where $s \in \mathbb{N}_0$, $s < m$ and $p \in \mathbb{N}$ is odd.

Let us consider an M-map $g : \mathbb{R} \multimap \mathbb{R}$ such that $g = \varphi^{2^m}$. By Lemma (9.26), g has a q-orbit. Thanks to Proposition (9.23), g has also a k-orbit, for every $k \lhd q$, except of $k = 2, 4, 6$. Subsequently, using Lemma (9.25) and the fact that φ has no odd orbit, φ has a $2^m p$-orbit, for every odd $p \lhd q$, and a 2^r-orbit, for every $r \in \mathbb{N}$, $r > m + 2$. Furthermore, considering M-maps $g_k = g^{2^k}$, for $k = 1, 2, \dots$, respectively, we obtain by Lemma (9.24) that $g_1 = \varphi^{2^{m+1}}$ has a 5-orbit, and that $g_k = \varphi^{2^{m+k}}$ has a 3-orbit, for $k = 2, 3, \dots$, respectively. Applying again Lemma (9.24) and the fact that φ has no odd orbit, φ has a $2^{m+1} p$-orbit for $p = 5$ and for every $p \lhd 5$, p is odd, and a $2^{m+k} p$-orbit for $p = 3$, and for every $p \lhd 3$, p is odd, whenever $k = 2, 3, \dots$, respectively.

Now, since the existence of a fixed point follows from Proposition (9.17), it suffices to show that φ has a 2^s-orbit, for every $s \in \mathbb{N}$, $s \leq m$. We fix such an arbitrary s and consider an M-map $g : \mathbb{R} \multimap \mathbb{R}$, $g = \varphi^{2^{s-1}}$. Using the fact that φ has neither 3-orbit nor $2^{s-1}3$-orbit, g has no 3-orbit by Lemma (9.24). On the other hand, φ has a $2^m q$-orbit. Hence, because of

$$\varphi^{2^m q} = \varphi^{2^{s-1} \cdot 2^{m-s+1} \cdot q},$$

and using Lemma (9.25), g has some r-orbit, where $2^{m-s+1} q$ is divided by r and $r \neq 1$. Applying Proposition (9.22), g has a 2-orbit. Thus, by Lemma (9.24), φ has a 2^s-orbit.

Case 2. $q = 1$.

With respect to Case 1, we can assume that φ has no $2^s p$-orbit, where $s \in \mathbb{N}$ and p is odd. We fix an arbitrary $s < m$ and consider an M-map $g : \mathbb{R} \multimap \mathbb{R}$, $g = \varphi^{2^{s-1}}$. Analogously as in the second part of Case 1, we obtain that g has no 3-orbit, but it admits an even orbit. Therefore, by Proposition (9.22), g has a 2-orbit and Lemma (9.24) implies the existence of a 2^s-orbit. Since $s < m$ was arbitrary and φ has a fixed point by Proposition (9.17), φ has a k-orbit, for every $k \lhd 2^m$. $\qquad\square$

(9.29) COROLLARY. *Let φ have an n-orbit, where $n = 2^m q$, $m \in \mathbb{N}_0$ and q is odd, and let n be maximal in the Sharkovskiĭ ordering.*

(9.29.1) *If $q > 3$, then φ has a k-orbit, for every $k \lhd n$, except $k = 2^{m+2}$.*
(9.29.2) *If $q = 3$, then φ has a k-orbit, for every $k \lhd n$, except $k = 2^{m+1}3, 2^{m+2}, 2^{m+1}$.*
(9.29.3) *If $q = 1$, then φ has a k-orbit, for every $k \lhd n$.*

For the first positive integers $1, 2, 3, 4, 5$, the Sharkovskiĭ-like relationship can be schematically sketched as in Fig. 31.

Now, we are in position to apply Theorem (9.27), Proposition (9.23) (and Proposition (9.16)) to differential equations (without the uniqueness property)

$$(9.30) \qquad \dot{x} = f(t, x), \quad f(t, x) \equiv f(t + \omega, x), \quad \omega > 0,$$

or, more generally, to the (upper) Carathéodory differential inclusions

$$(9.31) \qquad \dot{x} \in F(t, x), \quad F(t, x) \equiv F(t + \omega, x), \quad \omega > 0,$$

where $F : \mathbb{R}^2 \multimap \mathbb{R}$ has nonempty, compact and convex values, for all $(t, x) \in \mathbb{R}^2$, $F(\cdot, x)$ is measurable, for every $x \in \mathbb{R}$ (i.e., for any open $U \in \mathbb{R}$ and every $x \in \mathbb{R}$, the set $\{t \in (-\infty, \infty) \mid F(\cdot, x) \subset U\}$ is measurable), $F(t, \cdot)$ is u.s.c. (i.e., for any open $V \subset \mathbb{R}$ and a.a. $t \in \mathbb{R}$, the set $\{x \in \mathbb{R} \mid F(f, \cdot) \subset V\}$ is open) and $|F(t, x)| \leq \alpha + \beta|x|$ holds, for all $(t, x) \in \mathbb{R}^2$, with suitable constants α, β.

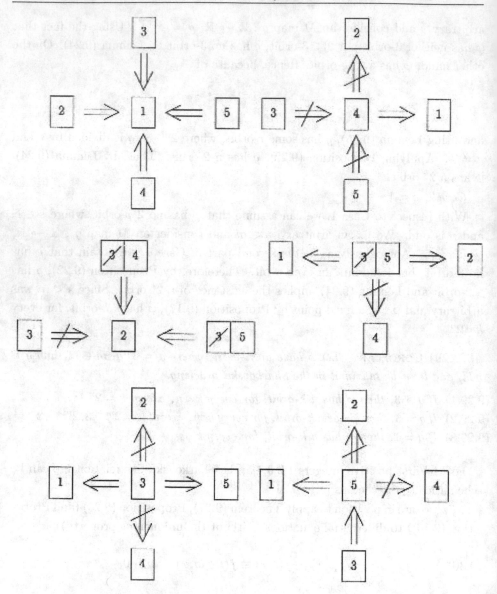

FIGURE 31

The Poincaré translation operator related to (9.31) takes the form (cf. Chapter III.4)

(9.32) $\Phi_{k\omega}(x_0) := \{x(k\omega, x_0) \mid x(\,\cdot\,, x_0)$ is a solution

of (9.31) with $x(0, x_0) = x_0\}$.

As pointed out at the begining of this chapter $\Phi_{k\omega}$ is an M-map satisfying $\Phi_\omega^k = \Phi_{k\omega}$, and, obviously, every k-periodic point of $\Phi_{k\omega}$ determines a $k\omega$-periodic solution of (9.31). Thus, Theorem (9.27), Proposition (9.23) (or Proposition (9.16)) applied to $\varphi := \Phi_\omega$ will lead to the existence of a subharmonic periodic solution $x(t)$ of (9.31), satisfying $x(t) \equiv x(t+k\omega)$ (or to the existence of harmonics of (9.31), satisfying $x(t) \equiv x(t+\omega)$).

Let us note that the non-uniqueness property is rather exceptional for continuous (as well as Carathéodory) equations (9.30). More precisely, W. Orlicz proved in [Orl] that the set of continuous functions f in (9.30), for which the equation (9.30) does not have uniqueness of all solutions satisfying the initial conditions $x(0) = x_0$, is meager, i.e. a set of the first Baire category.

Therefore, it is more reasonable to assume that $f : \mathbb{R} \multimap \mathbb{R}$ in (9.30) is only measurable (not necessarily continuous in x) and satisfying $|f(t,x)| \leq \alpha + \beta|x|$, for all $(t,x) \in \mathbb{R}^2$, because then we can study periodic solutions in the sense of A. F. Filippov (see [Fi2-M]). More precisely, by a *Filippov solution* of (9.30), we mean a Carathéodory solution $x(t) \in \mathrm{AC}_{\mathrm{loc}}(\mathbb{R}, \mathbb{R})$ of the inclusion

$$(9.33) \qquad \dot{x} \in F(t,x) := \bigcap_{\substack{\delta>0}} \bigcap_{\substack{r \subset \mathbb{R} \\ \mu(r)=0}} \overline{\mathrm{conv}} f(t, B_x^\delta) \setminus r),$$

where $\mu(r)$ is the Lebesgue measure of the set r.

Since the right-hand side in (9.33) is an (upper) Carathéodory multivalued function with nonempty, compact and convex values (see [Fi2-M], [Jar]), the related Poincaré translation operator is again an M-map taking the form (9.32). Subsequently, there is the same correspondence between $k\omega$-periodic (Filippov) solutions of (9.30) and k-periodic points of (9.32), as for the inclusion (9.31).

We can immediately reformulate the most important part of Theorem (9.27) jointly with Proposition (9.23), in terms of differential inclusions (9.31), as follows.

(9.34) THEOREM. *Let the above regularity assumptions and linear growth restrictions be satisfied for the right-hand side of* (9.31). *If the inclusion* (9.31) *has an $n\omega$-periodic solution, $n \in \mathbb{N}$, then it also admits $k\omega$-periodic solutions, for every $k \triangleleft n$, with the exception of at most three periodic solutions. If n is additionally odd, then* (9.31) *has $k\omega$-periodic solutions, for every $k \triangleleft n$, except $k \neq 2, 4, 6$. In particular, if $n \neq 2^m$, for any $m \in \mathbb{N}$, then* (9.31) *admits infinitely many (subharmonic) periodic solutions.*

In view of the above discussion, we have still

(9.35) COROLLARY. *Let the above regularity assumptions and linear growth restrictions be satisfied for the right-hand side of (9.30). If the equation (9.30) has an $n\omega$-periodic solution in the sense of Filippov, $n \in \mathbb{N}$, then it also admits $k\omega$-periodic Filippov solutions, for every $k \lhd n$, with the exception of at most three periodic Filippov solutions. If n is additionally odd, then (9.30) has $k\omega$-periodic Filippov solutions, for every $k \lhd n$, except $k \neq 2, 4, 6$. In particular, if $n \neq 2^m$, for any $m \in \mathbb{N}$, then (9.30) admits infinitely many (subharmonic) periodic Filippov solutions.*

(9.36) EXAMPLE. Consider the inclusion

$$(9.37) \qquad\qquad \dot{x} \in F(x) + p(t),$$

where

$$F(x) := \begin{cases} 0, & \text{for } x \in \mathbb{R} \setminus \{0\}, \\ [-1, 1], & \text{for } x = 0, \end{cases}$$

$p(t) := \cos t$, for $t \in [0, \pi)$ and $p(t) \equiv p(t + \pi)$.

One can easily check that (9.37) admits, for example, a (subharmonic) 3π-periodic solution $x_3(t)$, where

$$x_3(t) := \begin{cases} 0, & \text{for } t \in [0, 2\pi], \\ \sin t, & \text{for } t \in (2\pi, 3\pi), \end{cases}$$

and subsequently, according to Proposition (9.23), 3π-periodic solutions $x_k(t) \equiv x_k(t + k\pi)$, for every $k \in \mathbb{N} \setminus \{2, 4, 6\}$.

On the other hand, one can also check (by substitution) that inclusion (9.37) has, for every $k \in \mathbb{N}$, the $k\pi$-periodic solutions $x_k(t)$, where

$$x_k(t) := \begin{cases} 0, & \text{for } t \in [0, (k-1)\pi], \\ \sin t, & \text{for } t \in ((k-1)\pi, k\pi). \end{cases}$$

Similarly, Proposition (9.16) can be reformulated as follows.

(9.38) COROLLARY. *Let the above regularity assumptions and linear growth restrictions be satisfied for the right-hand sides of (9.31) (or (9.30)). If the inclusion (9.31) has a $k\omega$-periodic Carathéodory solution (or the equation (9.30) has a $k\omega$-periodic Filippov solution), for some $k \in \mathbb{N}$, then (9.31) has also an ω-periodic Carathéodory solution (or (9.30) has also an ω-periodic Filippov solution), respectively.*

If all solutions of (9.31) (or (9.30) are uniformly bounded and uniformly ultimately bounded, which means that there is a sufficiently large compact subinterval of $\mathbb{R}$, invariant under the related Poincaré translation operator $\Phi_{k\omega}$ in (9.32), for sufficiently big values $k \geq k_0$, $k \in \mathbb{N}$, then inclusion (9.31) (or equation (9.30)) admits a harmonic periodic Carathéodory (or Filippov) solution, respectively. It is namely enough to apply a suitable multivalued version of the Brouwer fixed point theorem (see e.g. (9.7) in Chapter II.9 or (10.6) in Chapter II.10), implying the existence of a $k\omega$-periodic solution, for some $k \geq k_0$, and then Corollary (9.35), for the existence of a (harmonic) ω-periodic solution.

Thus, Corollary (9.35) has still the following consequence.

(9.39) COROLLARY. *Let all Carathéodory solutions of* (9.31) (*or Filippov solutions of* (9.30)) *be, under the above regularity assumptions, uniformly bounded* (*i.e., for each* $B_1 > 0$, *there exists* $B_2 > 0$ *such that* $[t_0 \in \mathbb{R}, |x_0| < B_1, t \geq t_0]$ *imply that* $|x(t, t_0, x_0)| < B_2$) *and uniformly ultimately bounded* (*i.e. there exists some* $B > 0$ *such that if, for each* $B_3 > 0$, *there exists* $K \geq 0$ *with* $[t_0 \in \mathbb{R}, |x_0| < B_3, t \geq t_0 + K]$, *then* $|x(t, t_0, x_0)| < B$). *Then the inclusion* (9.31) *admits an* ω-*periodic Carathéodory solution* (*or the equation* (9.30) *admits an* ω-*periodic Filippov solution*), *respectively.*

(9.40) EXAMPLE. Assume the existence of a constant $B > 0$ such that

$$(9.41) \qquad F(t, x)x < 0 \quad (\text{or } f(t, x)x < 0), \quad \text{for } |x| \geq B.$$

Since every Carathéodory solution $x(t)$ of (9.31) (or Filippov solution $x(t)$ of (9.30)) satisfies $x^2(t) > 0$, jointly with $\frac{d}{dt}x^2(t) = 2x(t)f(t) < 0$, whenever $|x(t)| \geq B$, where $f(t) \subset F(t, x(t))$ is a measurable selection, one can easily verify that all Carathéodory solutions of (9.31) (or Filippov solutions of (9.30)) must be uniformly bounded as well as uniformly ultimately bounded. Otherwise, we would obviously get a contradiction. Therefore, inclusion (9.31) (or equation (9.30)) has, under (9.41), an ω-periodic Carathéodory (or Filippov) solution, respectively. On the other hand, a suitable multivalued version of the Brouwer fixed point theorem (see e.g. (9.7) in Chapter II.9 or (10.6) in Chapter II.10) can be applied here directly for obtaining ω-periodic solutions.

(9.42) REMARK. Since the associated Poincaré translation operators become M-maps also for further various classes of scalar differential equations and inclusions like directionally u.s.c. ones (see Theorem (4.42) in Chapter III.4) Theorem (9.27) can be also reformulated in terms of them.

All the above Sharkovskiĭ-type theorems and their applications are, in principle, only one-dimensional. In order to proceed to $\mathbb{R}^2$, we need another approach like the Artin braid group theory (see e.g. [Bi-M]). The obtained results are, however, only "generic" in the sense of this theory. Moreover, this theory is strictly single-valued, dealing with isotopic homeomorphisms. Thus, the single-valued approximations with required properties of admissible multivalued maps are rather delicate. Nevertheless, we would like at least to indicate the possibilities in this field.

It follows from Boju Jiang's interpretation [Ji1] of T. Matsuoka's results [Mat1]–[Mat3] that planar Carathéodory periodic (in time) systems of ordinary differential equations, satisfying the uniqueness condition and having at least three harmonics, possess "generically" infinitely many subharmonics. The "genericity" is understood in terms of the Artin braid group theory, i.e. with the exception of certain simplest braids, representing three given harmonics.

In the lack of uniqueness, the associated Poincaré (translation) operators become multivalued, by which the employed one-to-one correspondence between periodic points and subharmonics is lost. A natural question therefore arises, whether a similar "generic" result as above still holds true. As we shall see, there are many related obstructions in the multivalued case.

In the sequel, all spaces are metric and all mappings have (nonempty) compact values.

Periodic solutions will be determined by fixed points of the iterates of given multivalued maps. For multivalued maps, it is however, as we could see above, more appropriate to deal with k-orbits rather than k-periodic points, namely $\overline{\{x_1, \ldots, x_k\}}$, where $x_1 \in \Phi^k(x_1)$ and $x_j \in \Phi^{j-1}(x_1)$, $2 \le j \le k$, $k \in \mathbb{N}$.

Following [Bi-M], [Ji1], we recall at first some basic notions from the Artin theory of braids. Let X be a connected surface and let $F : X \to X$ be an isotopic homeomorphism, i.e. there exists

$$H = \{h_t\}_{t \in I} : \mathrm{id} \sim F, \ I = [0, 1],$$

such that $h_0 = \mathrm{id}$ and $h_1 = F$, or vice versa.

Assuming that $P \subset \mathrm{int} X$ with $F(P) = P$ is a finite subset, the set

$$\{(h_t(x), t) \in X \times I \mid x \in P\}$$

forms a *geometric braid* in $X \times I$, representing a *braid* σ_P in the *r-string braid group* $\pi_1 B_{0,r} \mathbb{R}^2$ of X, where π_1 stands for the fundamental group (for more details, see [Bi-M]).

If, in particular, $X = \mathbb{R}^2$, $P = \{x_1, x_2, x_3\}$ and $H : \mathrm{id} \sim F : X \to X$, then $\pi_1 B_{0,3} \mathbb{R}^2$ becomes Artin's braid group

$$B_3 = \langle \sigma_1, \sigma_2 \mid \sigma_1 \sigma_2 \sigma_1 = \sigma_2 \sigma_1 \sigma_2 \rangle$$

with the generators σ_1, σ_2:

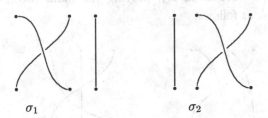

$$\sigma_1 \qquad\qquad \sigma_2$$

Its center is $Z = \langle (\sigma_1\sigma_2)^3 \rangle$, the infinite cyclic group generated by $(\sigma_1\sigma_2)^3 = (\sigma_1\sigma_2\sigma_1)^2$. It is a full twist, corresponding to a 2π-rotation of the plane and having no effect on F. Thus, we can employ the factor group B_3/Z.

For a compact connected surface $X \subset \mathbb{R}^2$, the isotopy H determines an element of B_3 as the image under the homomorphism from the braid group of X to the braid group of $\mathbb{R}^2$, induced by the natural inclusion of X into $\mathbb{R}^2$. However, since the strings corresponding to the curves $h_t(x_i)$, $i = 1, 2, 3$, determine a usual braid in $\mathbb{R}^2$, we can employ B_3/Z for $X \subset \mathbb{R}^2$ as well.

Before recalling Boju Jiang's interpretation [Ji1] of T. Matsuoka's results in [Mat1]–[Mat3], let us note that

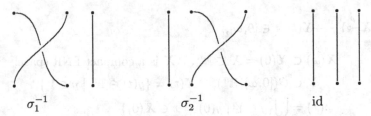

$$\sigma_1^{-1} \qquad\qquad \sigma_2^{-1} \qquad\qquad \text{id}$$

are the inverse elements $(\sigma_i^{-1}\sigma_i = \sigma_i\sigma_i^{-1} = \text{id}, i = 1, 2)$ of σ_1, σ_2 and id is the trivial braid.

(9.43) THEOREM. *Let $\{h_t\}_{t\in I} : \text{id} \sim F : X \to X$ be an isotopic homeomorphism, $X \subset \mathbb{R}^2$ be a compact connected surface and $P = \{x_1, x_2, x_3 \in \text{int} X \mid x_i = F(x_i), i = 1, 2, 3\}$. If the generated braid σ_P is not conjugated to σ_1^m, $(\sigma_1^2\sigma_2^2)^m$, $(\sigma_1\sigma_2)^m$ or $(\sigma_1\sigma_2\sigma_1)^m$ in B_3/Z, for any integer $m \in \mathbb{N}$, then the number of n-orbits of F grows exponentially in n. In particular, there are infinitely many periodic orbits.*

The "simplest" braids in B_3, playing the main (exceptional) role in Theorem (9.43), look as follows:

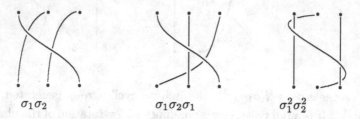

$$\sigma_1\sigma_2 \qquad\qquad \sigma_1\sigma_2\sigma_1 \qquad\qquad \sigma_1^2\sigma_2^2$$

(9.44) REMARK. In fact, it is sufficient that σ_P is not conjugated only to σ_1^m in B_3/Z, for any $m \in \mathbb{N}$, because $\sigma_1^2\sigma_2^2$ is conjugated to σ_1^2 up to multiplication by a full twist braid (see [Mat3, p. 201]) and, instead of $(\sigma_1\sigma_2)^m$, $(\sigma_1\sigma_2\sigma_1)^m$, it is enough to take only $m = 1, 2$ and $m = 1$, respectively. However, these braids are not pure, and so they do not actually appear.

Now, let us consider admissible (multivalued) iterates $(k \in \mathbb{N})$

$$\Phi^k = (\psi_\omega \circ \varphi)^k = (\psi_\omega \circ \varphi) \underbrace{\circ \ldots \circ}_{k-\text{times}} (\psi_\omega \circ \varphi), \quad \Phi^k : X \multimap X,$$

where $X \xrightarrow{\varphi} Y \xrightarrow{\psi_t} X(t)$, $t \in [0, \omega]$,

$$X(\omega) \subset X(0) = X \subset \mathbb{R}^n, \quad X \text{ is a compact ENR-space,}$$
$$Y \subset C([0, \omega], \mathbb{R}^n), \quad X(t) = \{y(t) \in \mathbb{R}^n \mid y \in Y\},$$
$$\varphi(x) = \bigcup \{y \in Y \mid y(0) = x \in X(0)\}$$

is assumed to be an R_δ-mapping (i.e. u.s.c. with R_δ-values) and $\psi_t(y) = y(t)$, $y \in Y$, is a continuous (single-valued) evaluation map, for each $t \in [0, \omega]$.

The following lemma can be proved by the same manner as Lemma 4.5 in [BGK] (cf. also Lemma (9.18)).

(9.45) LEMMA. *Let Φ^k be a composition as above and assume that $x_p \in X$ are fixed points of Φ^k, $x_p \in \Phi^k(x_p)$, $p = 1, \ldots, r$. Then, for every $\varepsilon > 0$, there exists a continuous (single-valued) ε-approximation $F^k = (\psi_\omega \circ f)^k : X \to X$ (on the graph of Φ^k), namely $\Gamma_{F^k} \subset N_\varepsilon(\Gamma_{\Phi^k})$, where $N_\varepsilon(\Gamma)$ denotes an (open) neighbourhood of Γ in X^2, such that f is a continuous (single-valued) $\delta(\varepsilon)$-approximation of φ, with $\delta(\varepsilon) \to 0$ whenever $\varepsilon \to 0$, and $x_p = F^k(x_p)$, $p = 1, \ldots, r$. Furthermore, there exist $x_{ip} \in (\psi_\omega \circ \varphi)^i(x_{i-1,p})$, $x_{0p} = x_p \in (\psi_\omega \circ \varphi)^k(x_{k-1,p})$, such that $x_{ip} = (\psi_\omega \circ f)^i(x_{i-1,p})$, $x_{0p} = x_p = (\psi_\omega \circ f)^k(x_{k-1,p})$, for every $i = 1, \ldots k - 1$ and $p = 1, \ldots, r$. At last, $X \xrightarrow{f} Y \xrightarrow{\psi_{\lambda\omega}} X(\lambda\omega)$, $\lambda \in [0, 1]$: id $\sim \psi_\omega \circ f$, is a continuous (single-valued) isotopy.*

In order to apply the braid invariants (Theorem (9.43), we need still F to be injective and $X \xrightarrow{f} Y \xrightarrow{\psi_{\lambda\omega}} X(\lambda\omega)$, $\lambda \in [0,1]$, to be an isotopic homeomorphism, which seems to be a difficult task, in general.

Therefore, we state this condition only as a hypothesis:

(H1) $\begin{cases} \text{an approximation } F = (\psi_\omega \circ f) \text{ of } \Phi \text{ is a homeomorphism} \\ \text{and } X \xrightarrow{f} Y \xrightarrow{\psi_{\lambda\omega}} X(\lambda\omega), \ \lambda \in [0,1], \\ \text{is an isotopic homeomorphism, in Lemma (9.45)} \end{cases}$

If, in particular, $X \subset \mathbb{R}^2$ is a compact connected surface and x_1, x_2, x_3 are (internal) fixed points of Φ, then the set $\{((\psi_t \circ f)(x), t) \in X \times [0, \omega] \mid x \in P = \{x_1, x_2, x_3\}\}$ forms, under hypothesis (H1), a geometric braid in $X \times [0, \omega]$, representing an Artin's braid σ_P in B_3.

Hence, Theorem (9.43) applies and, under its assumptions, there exists a one-parameter ($\varepsilon > 0$) family of k-periodic points of F, i.e. fixed points $x_k(\varepsilon)$ of $(\psi_\omega \circ f)^k$, for infinitely many $k \in \mathbb{N}$. Since this is true for every sufficiently small $\varepsilon > 0$, a subsequence $x_k(\varepsilon_m)$, with $\varepsilon_m \to 0$ whenever $m \to \infty$, can be selected such that

$$\lim_{m \to \infty} x_k(\varepsilon_m) = x_k(0),$$

where $x_k(0)$ is a fixed point of Φ^k, $k \in \mathbb{N}$. On the other hand, for certain prime (and all the better other) integers $k \in \mathbb{N}$, $x_k(0)$ can be equal to some fixed point of Φ.

Since this cannot be unfortunately avoided, in general, we state the second hypothesis:

(H2) $\begin{cases} \text{fixed points } x_k(0) \text{ of } \Phi^k \text{ form a } k\text{-orbit which does not consist only} \\ \text{it of the same fixed points of } \Phi, \text{ for prime integers } k > 1. \end{cases}$

After all, if $X \subset \mathbb{R}^2$ is a compact connected surface and x_1, x_2, x_3 are fixed points (in int X) of Φ, then, under the assumptions of (H1), (H2) and Theorem (9.43), there exist k-orbits of Φ, for infinitely many prime $k \in \mathbb{N}$.

(9.46) DEFINITION. We say that fixed points $x_1, x_2, x_3 \in \text{int} X \subset \mathbb{R}^2$ of $\Phi : X \multimap X$ imply *"generically"* the existence of k-orbits of Φ, for infinitely many prime $k \in \mathbb{N}$, if it is so under the condition that these orbits exist in the sense of (H2) and that the associated braid $\sigma_{\{x_1, x_2, x_3\}}$ is not conjugated to σ_1^m in B_3/Z, for any integer $m \in \mathbb{N}$, provided (H1) holds.

Using Definition (9.46), we can reformulate our conclusion as follows.

(9.47) THEOREM. *Let* $\Phi : X \multimap X$, *where* $X \subset \mathbb{R}^2$ *is a compact connected surface, be an admissible mapping as above, having at least three fixed points in* int$X$. *Then* Φ *admits "generically" k-orbits, for infinitely many prime $k \in \mathbb{N}$, provided* (H1) *holds.*

In the single-valued case, where (H1) and (H2) are trivially fulfilled, Theorem (9.47) reduces obviously to a special form of Theorem (9.43). Although for some classes of multimaps, e.g. maximal monotone multimaps (cf. [AuF-M]), condition (H1) might be verified, it remains an open problem how to guarantee (H1) in more general situations, e.g. for multivalued Poincaré operators treated below.

Now, consider the differential inclusion

$$(9.48) \qquad \dot{y} \in g(t, y), \quad g(t, y) \equiv g(t + \omega, y), \quad \omega > 0,$$

where $g : [0, \omega] \times \mathbb{R}^n \multimap \mathbb{R}^n$ is an (upper) Carathéodory mapping.

A *solution* $y(t)$ of (9.48) will be understood in the *Carathéodory sense*, namely a locally absolutely continuous function $y(t) \in \mathrm{AC}_{\mathrm{loc}}(\mathbb{R}, \mathbb{R}^n)$ satisfying (9.48), for a.a. $t \in \mathbb{R}$.

Assuming that $g(\cdot, y)$ is essentially bounded, for all $y \in \mathbb{R}^2$, and $g(t, \cdot)$ is linearly bounded, for a.a. $t \in [0, \omega]$, the global existence (i.e. on the whole line $\mathbb{R}$) of solutions is ensured.

Thus, we can define the *Poincaré translation operator* along the trajectories of (9.48) at the time $\lambda \omega$, $\lambda \in [0, 1]$, as follows:

$$(9.49) \qquad \begin{cases} \Phi_{\lambda\omega}(x) = \{y(\lambda\omega, x) \in \mathbb{R}^n \mid y(t, x) \text{ is a solution of } (9.48) \\ \quad \text{with } y(0, x) = x \in \mathbb{R}^n\}. \end{cases}$$

According to Theorem (4.3) in Chapter III.4, (9.49) is admissible, more precisely, it is a composition

$$\Phi_{\lambda\omega} = \psi_{\lambda\omega} \circ \varphi : \mathbb{R}^n \xrightarrow{\varphi} \mathrm{AC}([0, \omega], \mathbb{R}^n) \xrightarrow{\psi_{\lambda\omega}} \mathbb{R}^n, \quad \lambda \in [0, 1],$$

where

$$\varphi(x) : \mathbb{R}^n \ni x \multimap \{y(t, x) \in AC([0, \omega], \mathbb{R}^n) \mid y(t, x) \text{ is a solution of } (9.48)$$
$$\text{with } y(0, x) = x\}$$

is an R_δ-mapping (i.e. u.s.c. with R_δ-values) and

$$\psi_{\lambda\omega}(y) : y(t) \in \mathrm{AC}([0, \omega], \ \mathbb{R}^n) \to y(\lambda\omega)$$

is obviously a continuous (single-valued) evaluation mapping, for every $\lambda \in [0, 1]$.

Let us note that, although it is not usual for compositions in general, the iterates $\Phi_\omega^k = \Phi_\omega \underbrace{\circ \ldots \circ}_{k-\text{times}} \Phi_\omega : \mathbb{R}^n \multimap \mathbb{R}^n$, $k \in \mathbb{N}$, are again compositions of R_δ-maps and continuous (single-valued) maps, because $\Phi_\omega^k = \Phi_{k\omega}$, for every $k \in \mathbb{N}$.

In fact, $\Phi_{\lambda k \omega} : \text{id} \sim \Phi_{k\omega}$, $\lambda \in [0,1]$, $k \in \mathbb{N}$, is an isotopy which is compact on any bounded subset of $\mathbb{R}^n$, generating a multivalued semi-dynamical system.

In the single-valued case, i.e. for the Carathéodory differential equations

$$(9.50) \qquad \dot{y} = g(t,y), \quad g(t,y) \equiv g(t+\omega, y), \quad \omega > 0,$$

having the same growth restrictions, but a unique solvability of all initial-value problems (implying the continuous dependence of solutions of initial values), the associated Poincaré translation operators $F_{k\omega} = F_\omega^k$, $k \in \mathbb{N}$, are homeomorphisms (see Chapter III.4 and cf. [Ks2-M]). Moreover, $F_{\lambda k \omega} : \text{id} \sim F_{k\lambda}$, $\lambda \in [0,1]$, $k \in \mathbb{N}$, becomes an isotopic homeomorphism.

Hence, assuming that

$$\Phi_{k\omega} : X \multimap X \quad \text{or} \quad F_{k\omega} : X \to X, \quad k \geq k_0, \ k \in \mathbb{N},$$

where $X \subset \mathbb{R}^2$ is a compact connected surface, we can apply (for $n = 2$) Theorems (9.43) and (9.47) in the following way (in the single-valued case, cf. [Mat1], [Mat2]).

(9.51) THEOREM. *Assume that the planar ($n = 2$) u-Carathéodory inclusion (9.48), where g has the growth restrictions as above, admits three geometrically distinct isolated ω-periodic solutions with values in* int$X$, *where $X \subset \mathbb{R}^2$ is a compact connected surface. Then (9.48) possesses "generically" $k\omega$-subharmonics (i.e. $k\omega$-periodic solutions with a minimal period $k\omega$), for infinitely many prime $k \geq k_0$, provided (H1) holds w.r.t. the associated translation operator $\Phi_{k\omega} : X \multimap X$, $k \geq k_0$.*

In particular, for the planar ($n = 2$) Carathéodory equation (9.50), satisfying (locally) a uniqueness condition for all initial-value problems in X ($\Rightarrow$ (H1), (H2) hold automatically) three harmonics (i.e. ω-periodic solutions) with values in int$X$ *imply "generically" the existence of $k\omega$-subharmonics, for infinitely many integer $k \geq k_0$.*

(9.52) REMARK. In fact, there exist at least $(k-1)$ geometrically distinct $k\omega$-periodic orbits of equation (9.50). This follows from Theorem 1 in [Mat1], when applying the pseudo-Anosov theory (see e.g. [CB-M]).

In order to find effective criteria for the existence of three harmonics of (9.48) or (9.50) such that the associated Artin braid in B_3 would be nontrivial is also a rather difficult problem. It seems to us that the most appropriate approach of it can be Theorem (6.4) in Chapter III.6, based on the Nielsen fixed point theory.

Because of a rather unpleasant restriction (H1), it seems reasonable to look for an alternative approach. It is known [Srv] that the right-hand side of (9.48) can be approximated with an arbitrary accuracy (in the Hausdorff metric) by a Carathéodory single-valued function which is locally Lipschitzian in a y-variable, for a.a. $t \in [0, \omega]$. Of course, because of an ω-periodicity of g in t, a Carathéodory selection can be found to be ω-periodic in t as well.

Hence, assuming that all planar $(n = 2)$ equations with ε-approximative right-hand sides as above have three ω-periodic solutions, they possess "generically," according to the second part of Theorem (9.51), $k\omega$-subharmonics $x_{k,\varepsilon}(t)$, for infinitely many $k \in \mathbb{N}$ and every $\varepsilon > 0$, because we can take here $X = \mathbb{R}^2 \cup \{\infty\}$, as e.g. in [Ji1]. Passing to the limit $(\varepsilon \to 0)$, a converging subsequence $\{x_{k,1/m}(t)\}$ of a (by the hypothesis) uniformly bounded set $x_{k,\varepsilon}(t)$ can be found, when applying the standard limiting argument (similarly as in the single-valued case, see e.g. [Har-M], [Ks2-M]), such that $x_{k,0}(t) = \lim_{m \to \infty} x_{k,1/m}(t)$ are $k\omega$-periodic solutions of (9.48). Unfortunately, $k\omega$ need not be the least period of these solutions, in general. Nevertheless, regarding the situation when $k\omega$ is not the least period as exceptional, we can also speak about the *"generic" results*, provided still that the Artin braids related to given three harmonics of all approximative locally Lipschitzian equations are not the "simplest" ones, as indicated above.

Thus, we can conclude by

(9.53) THEOREM. *Let all the (existing) approximative to (9.48) planar $(n = 2)$ equations (with ω-periodic in t, for all $y \in \mathbb{R}^2$, and locally Lipschitzian in y, for a.a. $t \in [0, \omega]$, right-hand sides) have at least three ω-periodic solutions. Then the planar $(n = 2)$ u-Carathéodory inclusion (9.48) admits "generically" $k\omega$-periodic solutions, for infinitely many $k \in \mathbb{N}$, provided the set $x_{k,\varepsilon}(t)$ of $k\omega$-periodic solutions of equations with ε-approximative right-hand sides is uniformly bounded, for $k \in \mathbb{N}$.*

III.10. Almost-periodic problems

In this chapter, bounded solutions of differential equations and inclusions, whose existence was proved in Chapter III.5, will be shown, under suitable natural assumptions, almost-periodic (a.p.). Since the notion of almost-periodicity can be understood in many different ways, we added Apendix 1 to clarify the subject. The results are based on the papers [And5], [And19], [And20], [AndBe], [AndBeL].

We would like to employ various approaches. For differential inclusions in a Banach space (i.e. for evolution inclusions), two fixed point principles are available.

One of them is the Banach-like fixed point theorem for multivalued contractions which is due to H. Covitz and S. B. Nadler [CN], namely (see Theorem (3.68) in Chapter I.3).

(10.1) PROPOSITION ([CN]). *If X is a complete metric space and $F: X \multimap X$ is a (multivalued) contraction with nonempty closed values, namely*

$$d_H(F(x), F(y)) \leq L d(x, y), \quad \text{for all } x, y \in X,$$

where $L \in [0, 1)$ and d_H stands for the Hausdorff metric, then F has a fixed point, i.e. there exists $\widehat{x} \in X$ such that $\widehat{x} \in F(\widehat{x})$.

Since a closed subset of a complete metric space is complete, Proposition (10.1) can be immediately reformulated as follows.

(10.1') PROPOSITION. *If X is a closed subset of a complete metric space and $F: X \multimap X$ is a contraction with nonempty closed values, then F has a fixed point.*

The following Schauder–Tikhonov-like fixed point theorem for condensing multivalued mappings in Fréchet spaces represents a particular case of Theorem (7.9) in Chapter II.7 or Lemma (10.24) in Chapter II.10 (see also [AndBa] and cf. [Pr1], [Pr2]).

(10.2) PROPOSITION. *Let X be a closed convex subset of a Fréchet space E and let $F: X \multimap X$ be an R_δ-mapping (i.e. upper semicontinuous mapping with R_δ-values) which is β-condensing in the sense of Definition (5.31) in Chapter I.5. Then F has a fixed point, i.e. there exists $\widehat{x} \in X$ such that $\widehat{x} \in F(\widehat{x})$.*

Assuming that $F: X \multimap X$ is additionally a contraction (w.r.t. all seminorms of E which, however, does not automatically mean the contraction in a metric of E, see [Ga3]) on a closed bounded subset X of a Fréchet space E, it follows that F is γ-condensing w.r.t. the Hausdorff measure of noncompactness γ. So, we can still give

(10.2') PROPOSITION. *Let X be a closed, bounded and convex subset of a Fréchet space. Let $F: X \multimap X$ be a contraction with R_δ-values (i.e., in particular, with nonempty, compact and connected values). Then F has a fixed point.*

Comparing Proposition (10.2') to Proposition (10.1'), the assumptions of Proposition (10.2') might seem to be (in spite of the fact that, in Proposition (10.1'), a contraction is w.r.t. the metric of X) rather restrictive and partially superfluous. Moreover, in order to apply Proposition (10.2) without further restrictions (like contractivity), verifying that F is mapped into a suitable closed subset of a complete space of almost-periodic functions is rather difficult. On the other hand, there are situations, when Proposition (10.2') applies, but not Proposition (10.1').

Before applying Propositions (10.1) and (10.2) to differential inclusions, some further auxiliary results, definitions, notations, etc., must be presented; see also Appendix 1.

The notion of almost-periodicity will be understood now in the sense of V. V. Stepanov. We say that a locally Bochner integrable function $f \in L_{loc}(\mathbb{R}, B)$ (= $L^1_{loc}(\mathbb{R}, B)$, where B is a real separable Banach space, is *Stepanov almost-periodic* (S_{ap}) if the following is true:

- for every $\varepsilon > 0$, there exists $k > 0$ and, for every $a \in \mathbb{R}$, there exists $\tau \in [a, a + k]$ such that

$$D_S(f(t + \tau), f(t)) < \varepsilon,$$

where $D_S(f, g) := \sup_{a \in \mathbb{R}} \int_a^{a+1} |f(t) - g(t)| \, dt$ stands for the Stepanov metric and $| \cdot |$ is the norm in B.

It is well-known (cf. Appendix 1) that the space S of S_{ap}-functions is Banach and that, for a uniformly continuous $f \in C(\mathbb{R}, B)$, S_{ap} means *uniform almost-periodicity* (u.a.p.), namely

- for every $\varepsilon > 0$, there exists $k > 0$ and, for every $a \in \mathbb{R}$, there exists $\tau \in [a, a + k]$ such that

$$\|f(t + \tau) - f(t)\| < \varepsilon,$$

where $\| \cdot \| := \sup_{t \in \mathbb{R}} | \cdot |$.

Denoting $\|f\|_S = D_S(f, 0)$, let us still consider the real Banach space $BS = \{f \in L_{loc}(\mathbb{R}, B) | \|f\|_S < \infty\}$. Obviously, $S \subset BS$.

The *Bochner transform* (cf. Appendix 1)

$$f^b(t) := f(t + \eta), \quad \eta \in [0, 1], \quad t \in \mathbb{R},$$

associates to each $t \in \mathbb{R}$ a function defined on $[0, 1]$ and

$$f^b \in L_{loc}(\mathbb{R}, L([0, 1])), \quad \text{whenever } f \in L_{loc}(\mathbb{R}, B).$$

Thus, $BS = \{f \in L_{loc}(\mathbb{R}, B) | f^b \in L^\infty(\mathbb{R}, L([0, 1]))\}$, because

$$\|f\|_S = \|f^b\|_{L^\infty} = \sup_{t \in \mathbb{R}} \text{ess} \, \|f^b(t)\|_{L([0,1])} = \sup_{t \in \mathbb{R}} \text{ess} \int_0^1 |f(t + \eta)| \, d\eta.$$

Since still (cf. again Appendix 1)

$$f^b \in C(\mathbb{R}, L([0, 1])), \quad \text{where } f \in L_{loc}(\mathbb{R}, B),$$

we arrive at

$$BS = \{f \in L_{\text{loc}}(\mathbb{R}, B)|f^b \in BC(\mathbb{R}, L([0,1]))\},$$

where BC denotes the space of bounded and continuous functions.

S. Bochner has shown for the space S of S_{ap}-functions that (cf. Appendix 1)

$$S = \{f \in L_{\text{loc}}(\mathbb{R}, B)|f^b \in C_{\text{ap}}(\mathbb{R}, L([0,1]))\},$$

where C_{ap} means the space of uniformly almost-periodic functions. This important property, jointly with obvious relations

$$\|f\|_S = \|f^b\|_{BC(\mathbb{R}, L([0,1]))}$$

and $f_n \xrightarrow{S} f$ if and only if $f_n^b \xrightarrow{BC} f^b$, will play an important role in the sequel.

Defining, for a given S_{ap}-function $f \in S$, the sets

$$\Omega_f := \{(\varepsilon, k, a, \tau) \in \mathbb{R}^4 \mid \tau \in [a, a+k] \text{ and } \|f(t+\tau) - f(t)\|_S < \varepsilon\},$$

$$M_f := \{g \in C(\mathbb{R}, B) \mid (\varepsilon, k, a, \tau) \in \Omega_f \Rightarrow \text{ for all } t \in \mathbb{R} \; |g(t+\tau) - g(t)| < \varepsilon\},$$

we can state the following lemma (observe that if $g \in M_f$, then $g \in C_{\text{ap}}$), which is essential in applying Proposition (10.2) (or Proposition (10.2')).

(10.3) LEMMA. *M_f is a closed subset (in the topology of the uniform convergence on compact subintervals of $\mathbb{R}$) of $C(\mathbb{R}, B)$.*

PROOF. Let $M_f \ni g_l \xrightarrow{\text{loc}} g$ hold on $\mathbb{R}$, by which $g \in C(\mathbb{R}, B)$. Assume that $(\varepsilon, k, a, \tau) \in \Omega_f$, for all $t \in \mathbb{R}$. Then, for each $\delta > 0$, there exists l_0 such that, for all $l > l_0$, we have

$$|g(t+\tau) - g_l(t+\tau)| < \delta/2 \text{ and } |g(t) - g_l(t)| < \delta/2.$$

It follows from the inequality

$$|g(t+\tau) - g(t)| \leq |g(t+\tau) - g_l(t+\tau)| + |g_l(t+\tau) - g_l(t)| + |g_l(t) - g(t)|$$

that

$$l > l_0 : |g(t+\tau) - g(t)| < \delta + \varepsilon.$$

Since δ can be chosen arbitrarily small, we arrive at $|g(t+\tau) - g(t)| < \varepsilon$. □

(10.4) REMARK. Using the Bochner transform, one can prove quite analogously that M_f' is a closed subset of $L_{\text{loc}}(\mathbb{R}, B)$, where

$$\Omega_f' := \{(\varepsilon, k, a, \tau) \in \mathbb{R}^4 \mid \text{ for all } t \in \mathbb{R}, \; \tau \in [a, a+k]$$
$$\text{and } |f^b(t+\tau) - f^b(t)| < \varepsilon\},$$
$$M_f' := \{g \in L_{\text{loc}}(\mathbb{R}, B) \mid (\varepsilon, k, a, \tau) \in \Omega_f'$$
$$\Rightarrow \text{ for all } t \in \mathbb{R}, \; |g^b(t+\tau) - g^b(t)| < \varepsilon\}.$$

Moreover, M_f as well as M'_f can be proved to be convex (cf. [AndBeL]).

Observe that if $\alpha \in [-1, 1]$, $\beta \in \mathbb{R}$, then $\alpha M_f + \beta \subset M_f$, and subsequently $M_f \cap Q_C$ is a convex, closed subset of $C(\mathbb{R}, B)$, where $Q_C := \{g \in C(\mathbb{R}, B) | \sup_{t \in \mathbb{R}} |g(t)| \leq C\}$.

The following Bohr–Neugebauer-type statement is true.

(10.5) LEMMA. *Consider the linear evolution equation in the Banach space B:*

$$(10.6) \qquad \dot{X} + AX = P(t),$$

where $A \colon B \to B$ is a linear, bounded operator whose spectrum does not intersect the imaginary axis and $P \in S$ is an essentially bounded S_{ap}-function. Then (10.6) possesses a unique uniformly a.p. solution $X(t) \in \mathrm{AC}_{\mathrm{loc}}(\mathbb{R}, B)$ of the form

$$(10.7) \qquad X(t) = \int_{-\infty}^{\infty} G(t - s) P(s) \, ds,$$

where $G(t - s)$ is the principal Green function for (10.6), which takes the form

$$G(t - s) = \begin{cases} e^{A(t-s)} P_-, & \text{for } t > s, \\ -e^{A(t-s)} P_+, & \text{for } t < s, \end{cases}$$

and P_-, P_+ stand for the corresponding spectral projections to the invariant subspaces of A (for more details, see Chapter III.5 and cf. [DK-M, p. 79–81]).

PROOF. It is well-known (see e.g. the proof of Theorem (5.3) or Lemma (5.136) and cf. [DK-M], [MS-M]) that, under the above assumptions, equation (10.6) has exactly one solution $X(t)$ of the form (10.7).

In order to prove that $X(t)$ is uniformly a.p., we will equivalently show (see Appendix 1) that the set of functions $X_\tau(t) := X(t + \tau)$, $\tau \in \mathbb{R}$, is precompact in the topology $\|X\|_S = \|X^b\|_{BC(\mathbb{R}, L([0,1]))}$.

Since $P(t)$ is S_{ap}, we can choose from the sequence $\{P_{-\tau_k}(t)\}$ a Cauchy subsequence $\{P_{-\tau_{k_j}}(t)\}$. Having apparently

$$X_{\tau_{k_j}}(t) = X(t + \tau_{k_j}) = \int_{-\infty}^{\infty} G(t - s) P(s - \tau_{k_j}) ds = \int_{-\infty}^{\infty} G(t - s) P_{-\tau_{k_j}}(s) ds,$$

it follows that $X_{\tau_{k_j}}(t)$ is a Cauchy sequence (in the BC-topology) as well. In fact,

$$\|X^b_{\tau_{k_j}}(t) - X^b_{\tau_{k_i}}(t)\|_{BC}$$

$$= \left\| \int_{-\infty}^{\infty} G^b(t - s)[P_{-\tau_{k_j}}(s) - P_{-\tau_{k_i}}(s)] \, ds \right\|_{BC}$$

$$\leq \sup_{t \in \mathbb{R}} \left| \int_{-\infty}^{\infty} |G(t - s)| \|P^b_{-\tau_{k_j}}(s) - P^b_{-\tau_{k_i}}(s)\|_{BC} \, ds \right|$$

$$\leq \|P^b_{-\tau_{k_j}}(t) - P^b_{-\tau_{k_i}}(t)\|_{BC} \sup_{t \in \mathbb{R}} \left| \int_{-\infty}^{\infty} |G(t - s)| \, ds \right|$$

$$\leq C(A) \|P^b_{-\tau_{k_j}}(t) - P^b_{-\tau_{k_i}}(t)\|_{BC},$$

where $C(A)$ is a finite constant depending only on A.

This already means that the set of functions $X_\tau(t)$ is precompact, which completes the proof. $\square$

(10.8) REMARK. Another Bohr–Neugebauer-type theorem has been proved in [Co4]. Although this theorem even deals with (10.6), where $A = A(t, X)$ can be time-dependent and nonlinear, it only applies to our situation in particular cases.

(10.9) REMARK. For $B = \mathbb{R}^n$, the $(n \times n)$-matrix A can be arbitrary in order every entirely bounded solution of (10.6) to be (uniformly) a.p. (see e.g. [Fin-M, p. 86]). In $\mathbb{R}^2$, A can be, more generally, a maximal monotone operator for the same goal (see [Har]).

Now, Proposition (10.1) will be applied to the differential inclusion in a real separable Banach space with the norm $|\cdot|$, namely

$$(10.10) \qquad \dot{X} + AX \in F(X) + \Sigma(t),$$

where $A: B \to B$ is again a (single-valued) bounded, linear operator whose spectrum does not intersect the imaginary axis, $F: B \multimap B$ is a Lipschitz-continuous multivalued function with bounded, closed, convex values and $\Sigma: \mathbb{R} \to B$ is an essentially bounded S_{ap}-multivalued function with closed, convex values. By a *solution* $X(t)$ of (10.10) we mean everywhere the function belonging to the class $AC_{loc}(\mathbb{R}, B)$ and satisfying (10.10) almost everywhere.

Let us recall that by the *Lipschitz-continuity* of F we mean:

$$\exists L \in [0, \infty) : d_H(F(X), F(Y)) \leq L|X - Y|, \quad \text{for all } X, Y \in B,$$

where $d_H(\cdot, \cdot)$ stands for the Hausdorff metric, and by an S_{ap}-*multivalued function* Σ the measurable one (i.e. $\{t \in \mathbb{R} \mid \Sigma(t) \subset U\}$ is a measurable set, for each open $U \in B$) satisfying that, for every $\varepsilon > 0$, there exists a positive number $k = k(\varepsilon)$ such that, in each interval of the length k, there is at least one number τ with

$$\sup_{a \in \mathbb{R}} \int_a^{a+1} d_H(\Sigma(t), \Sigma(t + \tau)) \, dt < \varepsilon.$$

Let us note that F admits, under our assumptions, a Lipschitz-continuous selection $(F \supset) f: B \to B$ if and only if B is finite dimensional (see [HP1-M, p. 101]). However, even for $B = \mathbb{R}^n$, the Lipschitz constant need not be the same (for the related estimates and more details, see [HP1-M, p. 101–103]). On the other hand, although a uniformly a.p. multivalued function need not admit a uniformly a.p. selection (see [BVLL]), $S(t)$ possesses (see [Dan2], [DoSh]) an S_{ap}-selection $\sigma \subset \Sigma$.

Hence, consider still a one-parameter family of linear inclusions

$$(10.11) \qquad \dot{X} + AX \in F(q(t)) + \sigma(t), \quad q \in Q,$$

where $\sigma \subset \Sigma$ is an (existing) S_{ap}-selection and Q is the Banach space of uniformly a.p. functions $q \in C(\mathbb{R}, B)$.

Since one can easily check the composition $F(q)$ to be S_{ap} (see e.g. [AG1], [Dan2]), it can be Castaing-like represented in the form (see [Dan2], [DoSh])

$$F(q(t)) = \overline{\bigcup_{n \in \mathbb{N}} f_n(q(t))},$$

where $f_n(q)$, $n \in \mathbb{N}$, are related S_{ap}-selections. Therefore, denoting (cf. (10.7) and (10.11))

$$T(q) := \int_{-\infty}^{\infty} G(t - s) \left[\bigcup_{n \in \mathbb{N}} f_n(q(s)) + \sigma(s) \right] ds, \quad q \in Q,$$

where the integral is understood in the sense of R. J. Aumann (see [Aum] and cf. [HP1-M]), one can already discuss the possibility of applying Proposition (10.1). It is required that

(i) Q is complete,

(ii) $T: Q \multimap Q$ is a Lipschitz-continuous multivalued function with nonempty, closed values, having a Lipschitz constant $L_0 \in [0, 1)$.

Since Q is (by the hypothesis) Banach, only (ii) remains to be verified.

Taking into account the well-known elementary properties of S_{ap} functions (the S-limit of a sequence of S_{ap} functions is an S_{ap} function and the sum of two S_{ap} functions is an S_{ap} function as well) and applying Lemma (10.5) to (10.11) (when taking separately the indicated S_{ap} selections on the right-hand side of (10.11)), we get that $T(Q) \subset Q$. Moreover, the set of values of T can be verified quite analogously as in Theorem (5.3) to be nonempty, closed and convex, for every $q \in Q$. Thus, we only need to show that T is a contraction.

If F is Lipschitzian with a sufficiently small Lipschitz constant $L \in [0, 1)$, then we obtain (cf. [HP1-M, p. 199])

$$\sup_{t \in \mathbb{R}} d_H(T(q_1), T(q_2))$$

$$= \sup_{t \in \mathbb{R}} d_H \left(\int_{-\infty}^{\infty} G(t - s) F(q_1(s)) \, ds, \int_{-\infty}^{\infty} G(t - s) F(q_2(s)) \, ds \right)$$

$$\leq \sup_{t \in \mathbb{R}} \left| \int_{-\infty}^{\infty} |G(t - s)| d_H(F(q_1(s)), F(q_2(s))) \, ds \right|$$

$$\leq L \sup_{t \in \mathbb{R}} \left| \int_{-\infty}^{\infty} |G(t - s)| \sup_{t \in \mathbb{R}} |q_1(t) - q_2(t)| \, ds \right|$$

$$\leq LC(A) \sup_{t \in \mathbb{R}} |q_1(t) - q_2(t)| = LC(A) d(q_1, q_2),$$

where $C(A)$ is a constant depending only on A (cf. [DK-M]).

So, the desired contraction takes place, when $L_0 := LC(A) < 1$.

We are in position to give the first theorem.

(10.12) THEOREM. *Let the above assumptions be satisfied. Then inclusion* (10.10) *admits a uniformly a.p. solution, provided the Lipschitz constant L satisfies the inequality $L < 1/C(A)$, where $C(A)$ is a constant depending only on A such that*

$$\sup_{t \in \mathbb{R}} \left| \int_{-\infty}^{\infty} |G(t-s)| ds \right| \leq C(A),$$

(*G denotes the principal Green function for* (10.6)).

(10.13) REMARK. For $B = \mathbb{R}^n$, the explicit estimate of $C(A)$ can be found, under some additional restrictions in Remark (5.12) (cf. also [AK2]).

Consider again inclusion (10.10), but this time assume, for a moment, that F is no longer Lipschitz-continuous, but upper semicontinuous (i.e. for any open subset $U \subset B$, the set $\{X \in B | F(X) \subset U\}$ is open) and such that

(10.14) $|F(X)| \leq L|X| + M,$

where $0 \leq L < 1/C(A)$ and $M \geq 0$ is an arbitrary constant. Let all the other assumptions be satisfied.

Applying Proposition (10.2) to (10.10), one can establish quite analogously as in Theorem (5.3) in Chapter III.5 the following statement.

(10.15) PROPOSITION. *Let all the above assumptions be satisfied (jointly with* (10.14), *where $L < \delta/C(A)$ and $\delta \leq 1$ is a given constant related to the fact that the moduls of frequencies of S_{ap}-multivalued functions involve those of their S_{ap}-selections (see Appendix 1 and cf. [Dan2], [DoSh]). Assume, furthermore, that*

(10.16) $\gamma(F(\Omega)) < \dfrac{1}{C(A)} \gamma(\Omega),$ *for every bounded $\Omega \subset B$,*

where γ stands for the Hausdorff measure of noncompactness (and $C(A)$ has the same meaning as above). Then inclusion (10.10) *admits a uniformly a.p. solution, provided:*

(10.15.1) $F(q) \in \left\{ G : \mathbb{R} \to B \text{ is measurable} \ \middle| \ (\varepsilon, k, a, \tau) \in \widetilde{\Omega}_0' \right.$

$\left. \Rightarrow \sup_{b \in \mathbb{R}} \int_b^{b+1} d_H(G(t), G(t+\tau)) \, dt < \delta L_1 \varepsilon \right\},$

for every $q \in \widetilde{M}_\sigma := \{g \in C(\mathbb{R}, B) | (\varepsilon, k, a, \tau) \in \widetilde{\Omega}_\sigma' \Rightarrow \|g(t+\tau) - g(t)\|_S < \varepsilon\}$, where $\widetilde{\Omega}_\sigma' := \{(\varepsilon, k, a, \tau) \in \mathbb{R}^4 \mid \tau \in [a, a+k] \text{ and } \|\sigma(t+\tau) - \sigma(t)\|_S < \varepsilon/\Delta\}$, $L_1 < 1/C(A)$ and $\Delta \gg 1$ is sufficiently big,

(10.15.2) $\delta = \delta(\varepsilon)$ *in* (10.15.1) *is independent of $\varepsilon > 0$,*

(10.15.3) $T(Q) \subset \widetilde{M}_\sigma \ (\Rightarrow T(Q) \subset Q),$

where

$$T(q) := \int_{-\infty}^{\infty} G(t-s) \left[\bigcup_{n \in \mathbb{N}} \widetilde{f}_n(q(s)) + \sigma(s) \right] ds,$$

$$\widetilde{f}_n(q) \in \widetilde{M}'_\sigma := \{ g \in L_{\mathrm{loc}}(\mathbb{R}, B) \mid (\varepsilon, k, a, \tau) \in \widetilde{\Omega}'_\sigma \Rightarrow \|g(t+\tau) - g(t)\|_S < L\varepsilon/\delta \},$$

and $\widetilde{f}_n(q) \subset F(q)$, *for every* $n \in \mathbb{N}$,

$$Q := \widetilde{M}_\sigma \cap Q_C$$

(observe that Q is again a closed, convex subset of $C(\mathbb{R}, B)$),

$$Q_C := \{ g \in C(\mathbb{R}, B) \mid \sup_{t \in \mathbb{R}} |g(t)| \leq C \},$$

and $C > 0$ is a constant such that

$$C \geq \frac{C(A)}{1 - C(A)L_1} (M + \sup_{t \in \mathbb{R}} \mathrm{ess} |\Sigma(t)|).$$

(10.17) REMARK. Conditions (10.15.1), (10.15.2) imply (see [Dan2], [DoSh]) that $F(q)$ can be Castaing-like represented in the form

$$F(q(t)) = \overline{\bigcup_{n \in \mathbb{N}} \widetilde{f}_n(q(t))}, \quad \text{for every } n \in \mathbb{N},$$

where $\widetilde{f}_n(q) \in \widetilde{M}'_\sigma$ and $\widetilde{f}_n(q) \subset F(q)$, for every $n \in \mathbb{N}$.

Since satisfying conditions (10.15.1)–(10.15.3) without a Lipschitz-continuity of F seems to be, even in particular single-valued cases, a difficult task, we still give

(10.18) PROPOSITION. *Assume* (10.15.2) *and let the assumptions of Theorem* (10.12) *hold with $L < \delta/C(A)$, where $\delta \leq 1$ is a given constant. Then all conditions in Proposition* (10.15) *are satisfied.*

PROOF. It is well-known (see e.g. [KOZ-M, p. 85] and cf. Chapter I.5) that, under the above assumptions, the Lipschitz-continuity of F with the constant $L < \delta/C(A)$, $\delta \leq 1$, implies (10.16). As (10.14) follows immediately, we restrict ourselves to checking only (10.15.1) and (10.15.3).

Since the Lipschitz-continuity of F implies

$$\sup_{b \in \mathbb{R}} \int_b^{b+1} d_H \left(F(q(t)), F(q(t+\tau)) \right) dt \leq L \sup_{b \in \mathbb{R}} \int_b^{b+1} |q(t+\tau) - q(t)| \, dt < \delta L_1 \varepsilon,$$

for every $q \in \widetilde{M}_\sigma$, hypothesis (10.15.1) is satisfied and so, in view of Remark (10.17), $\widetilde{f}_n(q) \in \widetilde{M}'_\sigma$, for every $n \in \mathbb{N}$.

As concerns (10.15.3), consider a uniformly a.p. solution $X(t)$ of the equation

$$\dot{X} + AX = \widetilde{f}(q(t)) + \sigma(t), \quad q \in Q \ (= \widetilde{M}_\sigma \cap Q_C),$$

where $\widetilde{f}(q) \in \widetilde{M}'_\sigma$ and $\widetilde{f}(q) \subset F(q)$. We have that

$$\|X^b(t+\tau) - X^b(t)\|_{BC}$$
$$= \left\| \int_{-\infty}^\infty G^b(t-s)[\widetilde{f}(g(s+\tau)) - \widetilde{f}(g(s)) + \sigma(s+t) - \sigma(s)]\,ds \right\|_{BC}$$
$$\leq \sup_{t\in\mathbb{R}} \left| \int_{-\infty}^\infty |G(t-s)| \|\widetilde{f}^b(q(s+\tau)) - \widetilde{f}^b(q(s)) + \sigma^b(s+\tau) - \sigma^b(s)\|_{BC}\,ds \right|$$
$$\leq \|\widetilde{f}^b(q(t+\tau)) - \widetilde{f}^b(q(t)) + \sigma^b(t+\tau) - \sigma^b(t)\|_{BC} \sup_{t\in\mathbb{R}} \left| \int_{-\infty}^\infty |G(t-s)|\,ds \right|$$
$$\leq C(A)\|\widetilde{f}^b(q(t+\tau)) - \widetilde{f}^b(q(t)) + \sigma^b(t+\tau) - \sigma^b(t)\|_{BC}$$
$$< C(A)\left(\frac{L\varepsilon}{\delta} + \frac{\varepsilon}{\Delta}\right) = \varepsilon C(A)\left(\frac{L}{\delta} + \frac{1}{\Delta}\right) < \varepsilon,$$

by the hypothesis $L < \delta/C(A)$ and $\Delta \gg 1$. Thus, $X(t) = T(q) \in \widetilde{M}_\sigma$, for every $q \in Q$, which completes the proof. $\square$

(10.19) REMARK. If F is single-valued, then one can obviously take $\delta = 1$. If $B = \mathbb{R}^n$, then a lower estimate for δ can be obtained explicitly for a Lipschitz-continuous F, namely $\delta \leq 1/n(12\sqrt{3}/5 + 1)$ (see [HP1-M, p. 101–103]). In the both cases, (10.15.2) holds automatically.

Hence, we can give the second theorem.

(10.20) THEOREM. *Let the assumptions of Theorem (10.12) be satisfied, where F is single-valued or $B = \mathbb{R}^n$. Then inclusion (10.10) admits (on the basis of Proposition (10.2)) a uniformly a.p. solution belonging to the set Q, provided $L < 1/C(A)$ or $L < 1/C(A)n(12\sqrt{3}/5 + 1)$, respectively.*

We could see that if F is a contraction with $L < \delta/C(A)$, $\delta \leq 1$, then we arrived, under (10.15.2), at the same results, when applying Theorem (10.12) or Theorem (10.20). In fact, due to (10.15.3), Theorem (10.20) gives us a bit more, namely that an a.p. solution belongs to Q. Thus, the lipschitzianity of F can be restricted to the set $Q \cap B$. On the other hand, it is a question, whether or not conditions (10.15.1)–(10.15.3) can be satisfied in particular (especially, single-valued) cases without the Lipschitz-continuity of F. If so, then (in spite of the fact that the application of Proposition (10.1) is apparently more straightforward) Proposition (10.2) might be more efficient in this field.

In the single-valued case, the problem reduces to verifying only (10.15.1), because (10.15.2) holds trivially (see Remark (10.19) and (10.15.3) is then satisfied, whenever $C(A) < 1$. As a simplest example for $C(A) < 1$, we can take $B = \mathbb{R}^2$ and $A = \mathrm{diag}(a_{11}, a_{22})$, where $a_{11} > 0 > a_{22}$ and $(1/a_{11} - 1/a_{22}) < 1$ (see Remark (10.39) below, where more examples can be found). Moreover, according to Corollary to Lemma 3 in [Dan2], $F(q) = \tilde{f}(q)$ is an S_{ap}-function whose modul of frequencies is involved in the one of any $q \in \widetilde{M}_\sigma$. This, however, does not yet mean (10.15.1).

Nevertheless, for differential inclusions in Banach spaces, Theorem (10.12) seems to be a new result. On the other hand, the conclusion of Theorem (10.20) can be also obtained by different techniques (see e.g. [And5], [Pn-M]).

In order to deal with more general sorts of a.p. solutions, let us give some new definitions. For the sake of simplicity, only differential inclusions in Euclidean spaces will considered. Furthermore, denote by D_G any of the following metrics:

(Stepanov) $$D_S(f, g) := \sup_{x \in \mathbb{R}} \frac{1}{L} \int_x^{x+L} |f(t) - g(t)|\, dt,$$

(Weyl) $$D_W(f, g) := \lim_{L \to \infty} \sup_{x \in \mathbb{R}} \frac{1}{L} \int_x^{x+L} |f(t) - g(t)|\, dt$$
$$= \lim_{L \to \infty} D_S(f, g),$$

(Besicovitch) $$D_B(f, g) := \limsup_{T \to \infty} \frac{1}{2T} \int_{-T}^{T} |f(t) - g(t)|\, dt,$$

where $f, g \in L^1_{\mathrm{loc}}(\mathbb{R}, \mathbb{R}^n)$ and $|\cdot|$ denotes the norm in $\mathbb{R}^n$.

The metric space (G, D_G) denotes similarly any of the spaces endowed with the metric D_S or D_W or D_B.

(10.21) DEFINITION. A function $f \in L^1_{\mathrm{loc}}(\mathbb{R}, \mathbb{R}^n)$ is said to be G-*almost-periodic in the sense of Bohr* (G_{ap}) if, for every $\varepsilon > 0$, there exists a relatively dense set $\{\tau\}_\varepsilon$ of almost-periods (i.e. there exists an $l > 0$ such that every interval of the length l contains at least one point of $\{\tau\}_\varepsilon$) such that

$$D_G(f(t + \tau), f(t)) < \varepsilon, \quad \text{for every } \tau \in \{\tau\}_\varepsilon.$$

(10.22) DEFINITION. A function $f \in L^1_{\mathrm{loc}}(\mathbb{R}, \mathbb{R}^n)$ is said to be G-*almost-periodic in the sense of Bochner* (G-*normal*) if the family of functions $\{f(t + h) \mid h \in \mathbb{R}\}$ is G-precompact, i.e. if, for each sequence $\{f(t + h_n)\}$, we can select a G-fundamental subsequence.

(10.23) DEFINITION. A function $f \in L^1_{\mathrm{loc}}(\mathbb{R}, \mathbb{R}^n)$ is said to be *G-almost-periodic w.r.t. approximations by trigonometric polynomials* (G) if, for every $\varepsilon > 0$, there exists a vector of trigonometric polynomials

$$P_\varepsilon(t) = \left(\sum_{k_1=1}^{m_1} a_{k_1} e^{i\lambda_{k_1} t}, \ldots, \sum_{k_n=1}^{m_n} a_{k_n} e^{i\lambda_{k_n} t} \right),$$

where $m_j \in \mathbb{N}$, $a_{k_j} \in \mathbb{R}$, $\lambda_{k_j} \in \mathbb{R}$ $(j = 1, \ldots, n)$; such that

$$D_G(f(t), P_\varepsilon(t)) < \varepsilon.$$

Hence, we can define various sorts of a.p. functions by means of the following

	a. periods	normal	approx.
Stepanov	S_{ap}	S-normal	S
Weyl	W_{ap}	W-normal	W
Besicovitch	B_{ap}	B-normal	B

(10.24) Table

Let us note that the classes of a.p. functions in Table (10.24) are not equivalent. Their hierarchy, jointly with the relationship to other classes of a.p. functions, can be found in Table 1 in (A1.5) in Appendix 1. In view of the indicated implications in Table 1 in (A1.5) and because of practical reasons, it is convenient for applications to differential equations and inclusions to be concentrated especially to the G_{ap}-classes of solutions.

Furthermore, in view of a possible Castaing-like representation (for more details, see Remark (10.17) and [And5], [And19], [AndBeL]), the definitions of G_{ap}-functions can be direcly extended to measurable, essentially bounded G_{ap}-multivalued functions with nonempty, closed values, when just replacing the Euclidean norm by the Hausdorff distance. On the other hand, although, for convex and compact valued S_{ap}-multivalued functions there exist S_{ap}-(single-valued) selections (see [Dan2], [DoSh]), we do not know whether the same is true for W_{ap} or B_{ap}-multivalued functions (for uniformly a.p. multivalued functions, it is not so [BVLL]).

Therefore, it seems to be reasonable to define only multivalued S_{ap}-perturbations of (single-valued) W_{ap} and B_{ap}-functions as follows.

(10.25) DEFINITION. A multivalued function $\varphi : \mathbb{R} \multimap \mathbb{R}^n$ is called a *multivalued S_{ap}-perturbation of a W_{ap} or B_{ap}-function* if it takes the form $\varphi = \varphi_S + f_W$ or

$\varphi = \varphi_S + f_B$, where $\varphi_S : \mathbb{R} \multimap \mathbb{R}^n$ is an essentially bounded multivalued mapping with nonempty, convex and compact values, satisfying Definition (10.21) with

$$D_G := \sup_{x \in \mathbb{R}} \frac{1}{L} \int_x^{x+L} d_H(\varphi(t+\tau), \varphi(t))\, dt < \varepsilon,$$

($d_H(\cdot, \cdot)$ denotes the Hausdorff distance), and $f_W \in W_{ap}$ or $f_B \in B_{ap}$, respectively.

Since the spaces W_{ap} or B_{ap} do not seem to be linear, some uniform-W or B-continuity restrictions should be imposed in order the sum of two W_{ap}-functions to be W_{ap} or B_{ap}-functions to be B_{ap}, respectively. Fortunately, one can prove in a slightly modified way as e.g. in [Bes-M, Chapter 1, pp. 1–5], or in [Lev-M, Theorem 5.2.4, pp. 203–204], that the sum $f_1 + f_2$ or a uniformly G-continuous function $f_1 \in G_{ap}$ and a function $f_2 \in G_{ap}$ belongs also to the class G_{ap}, where G_{ap} means here W_{ap} or G_{ap}, respectively.

Furthermore, since the W_{ap}-space is incomplete and since we do not know whether or not the B_{ap}-space is complete, we cannot proceed as above in order to prove the existence of W_{ap}-solutions or B_{ap}-solutions to differential inclusions (in Euclidean spaces) of the form (10.10).

The following useful statement is a direct consequence of Theorem (5.3) or Lemma (5.136) in Chapter III.5 (cf. also Remarks (10.13) and (10.19)).

(10.26) LEMMA. *Under the assumptions of Theorem* (10.12), *inclusion* (10.10), *considered in* $\mathbb{R}^n$ *(i.e.* $B = \mathbb{R}^n$), *admits an entirely bounded solution* $X(t)$ *such that*

$$(10.27) \qquad X(t) = \int_{-\infty}^{\infty} G(t,s)[\sigma(s) + f(X(s))]\, ds,$$

where G *is Green function related to the equation* (10.6), *where* $P(t) := \sigma(t) + f(X(t))$, *satisfying*

$$(10.28) \qquad |G(t,s)| \le k e^{-\lambda|t-s|} \quad \left(\Rightarrow \sup_{t \in \mathbb{R}} \left| \int_{-\infty}^{\infty} |G(t,s)|\, ds \right| \le \frac{2k}{\lambda} \right),$$

with suitable positive constants k, λ; $\sigma \subset \Sigma$ *is a (single-valued) measurable selection and* $f \subset F$ *is a (single-valued) Lipschitz-continuous selection with the constant* $L_1 < \lambda n (12\sqrt{3}/5 + 1)/(2k)$.

We are ready to give

(10.29) THEOREM. *Let* A *be a constant* $(n \times n)$-*matrix whose all eigenvalues have nonzero real parts. Assume, furthermore, that* $\Sigma : \mathbb{R} \multimap \mathbb{R}^n$ *is a multivalued*

S_{ap}-perturbation of a W_{ap}-function. Let, at last, $F : \mathbb{R}^n \multimap \mathbb{R}^n$ be a Lipschitz-continuous multivalued mapping with nonempty, compact and convex values, having
the constant $L < \lambda/2kn(12\sqrt{3}/5 + 1)$, where $\lambda > 0$, $k > 0$ are suitable constants
depending only on A such that (see (10.28))

$$(10.30) \qquad \sup_{t \in \mathbb{R}} \left| \int_{-\infty}^{\infty} |G(t, s)| \, ds \right| \leq \frac{2k}{\lambda},$$

and G denotes the Green function related to the equation (10.6).

Then the inclusion (10.10) $(B = \mathbb{R}^n)$ admits a W_{ap}-solution.

PROOF. It follows from the foregoing investigations that Σ possesses a (single-valued) W_{ap}-selection $\sigma \subset \Sigma$ as well as F has (see Remark (10.19)) a (single-valued) Lipschitz-continuous selection $f \subset F$ with the constant L_0 such that
$Ln(12\sqrt{3}/5 + 1) = L_0 < \lambda/(2k)$. Thus, instead of (10.10), we can only consider
the equation

$$(10.31) \qquad \dot{X} + AX = f(X) + \sigma(t).$$

According to Lemma (10.26) or Lemma (5.136), equation (10.31) admits an entirely bounded solution $X(t)$ of the form (10.27). Since for $X(t)$ in (10.27) we also
have

$$(10.32) \qquad X(t + \tau) = \int_{-\infty}^{\infty} G(t + \tau, s)[\sigma(s) + f(X(s))] \, ds$$
$$= \int_{-\infty}^{\infty} G(t, s)[\sigma(s + \tau) + f(X(s + \tau))] \, ds,$$

one can deal with $\int_a^{a+l} |X(t + \tau) - X(t)| \, dt$, where $l > 0$ is an arbitrary constant,
to prove the W_{ap}-periodicity of $X(t)$, i.e., in the Weyl metric used in Definition (10.21). Hence, applying (after several steps) the well-known Fubini theorem
and using the norm $\| \cdot \| = \sup \operatorname{ess}_{t \in \mathbb{R}} | \cdot |$ and compatible vector and metrix norms,

we obtain successively (see (10.28))

$$\int_a^{a+l} |X(t+\tau) - X(t)|\, dt$$

$$= \int_a^{a+l} \left| \int_{-\infty}^{\infty} G(t,s)[f(X(s+\tau)) - f(X(s)) + \sigma(s+\tau) - \sigma(s)]\, ds \right| dt$$

$$\leq \int_a^{a+l} dt \int_{-\infty}^{\infty} |G(t,s)| |f(X(s+\tau)) - f(X(s)) + \sigma(s+\tau) - \sigma(s)|\, ds$$

$$\leq \int_a^{a+l} dt \left[\int_{-\infty}^{t} ke^{-\lambda(t-s)} |f(X(s+\tau)) - f(X(s)) + \sigma(s+\tau) - \sigma(s)|\, ds \right.$$

$$\left. + \int_{t}^{\infty} ke^{\lambda(t-s)} |f(X(s+\tau)) - f(X(s)) + \sigma(s+\tau) - \sigma(s)|\, ds \right]$$

$$= k \int_{-\infty}^{a} |f(X(s+\tau)) - f(X(s)) + \sigma(s+\tau) - \sigma(s)|\, ds \int_a^{a+l} e^{-\lambda(t-s)}\, dt$$

$$+ k \int_a^{a+l} |f(X(s+\tau)) - f(X(s)) + \sigma(s+\tau) - \sigma(s)|\, ds \int_s^{a+l} e^{-\lambda(t-s)}\, dt$$

$$+ k \int_a^{a+l} |f(X(s+\tau)) - f(X(s)) + \sigma(s+\tau) - \sigma(s)|\, ds \int_a^{s} e^{\lambda(t-s)}\, dt$$

$$+ k \int_{a+l}^{\infty} |f(X(s+\tau)) - f(X(s)) + \sigma(s+\tau) - \sigma(s)|\, ds \int_a^{a+l} e^{\lambda(t-s)}\, dt$$

$$= \frac{k}{\lambda} e^{-\lambda a}(1 - e^{-\lambda l}) \int_{-\infty}^{a} e^{\lambda s} |f(X(s+\tau)) - f(X(s)) + \sigma(s+\tau) - \sigma(s)|\, ds$$

$$+ \frac{k}{\lambda} \int_a^{a+l} (1 - e^{-\lambda(a+l-s)}) |f(X(s+\tau)) - f(X(s)) + \sigma(s+\tau) - \sigma(s)|\, ds$$

$$+ \frac{k}{\lambda} \int_a^{a+l} (1 - e^{\lambda(a-s)}) |f(X(s+\tau)) - f(X(s)) + \sigma(s+\tau) - \sigma(s)|\, ds$$

$$+ \frac{k}{\lambda} e^{\lambda a}(e^{\lambda l} - 1) \int_{a+l}^{\infty} e^{-\lambda s} |f(X(s+\tau)) - f(X(s)) + \sigma(s+\tau) - \sigma(s)|\, ds$$

$$\leq 2\frac{k}{\lambda} \int_a^{a+l} |f(X(s+\tau)) - f(X(s)) + \sigma(s+\tau) - \sigma(s)|\, ds$$

$$+ 2\frac{k}{\lambda^2} \|f(X(t+\tau)) - f(X(t)) + \sigma(s+\tau) - \sigma(s)\|.$$

Since $X(t)$ is bounded, and subsequently $\|f(X(t))\| < \infty$, the last term vanishes in the Weyl metric used in Definition (10.21). Thus, since by the hypothesis

$$\lim_{l\to\infty} \left[\sup_{a\in\mathbb{R}} \frac{1}{l} \left\{ \int_a^{a+l} |\sigma(t+\tau) - \sigma(t)|\, dt \right\} \right] < \varepsilon,$$

we get furthermore, by means of the Lipschitz property of f, that

$$\lim_{l \to \infty} \left[\sup_{a \in \mathbb{R}} \frac{1}{l} \left\{ \int_a^{a+l} |X(t+\tau) - X(t)| \, dt \right\} \right]$$

$$\leq 2\frac{k}{\lambda} \lim_{l \to \infty} \left[\sup_{a \in \mathbb{R}} \frac{1}{l} \left\{ \int_a^{a+l} |f(X(t+\tau)) - f(X(t)) + \sigma(t+\tau) - \sigma(t)| \, dt \right\} \right]$$

$$< 2\varepsilon \frac{k}{\lambda} + 2L_0 \frac{k}{\lambda} \lim_{l \to \infty} \left[\sup_{a \in \mathbb{R}} \frac{1}{l} \left\{ \int_a^{a+l} |X(t+\tau) - X(t)| \, dt \right\} \right].$$

After all, we arrive at

$$\lim_{l \to \infty} \left[\sup_{a \in \mathbb{R}} \frac{1}{l} \left\{ \int_a^{a+l} |X(t+\tau) - X(t)| \, dt \right\} \right] < \frac{2\varepsilon k}{\lambda - 2L_0 k},$$

as far as $L \cdot n(12\sqrt{3}/5 + 1) = L_0 < \lambda/(2k)$, which already verifies, according to Definition (10.21), the desired W_{ap}-periodicity of $X(t)$. So, the proof is complete. $\square$

(10.33) REMARK. For the existence of a W_{ap}-solution of inclusion (10.10) $(B = \mathbb{R}^n)$, any of the following classes is available, under the assumptions of Theorem (10.29), for the single-valued part of the forcing term $\Sigma(t)$ (see Appendix 1): u.a.p., u-normal, C_{ap}, S^1, S^1-normal, S^1_{ap}, e-W^1, e-W^1-normal, e-W^1_{ap}, W^1, W^1-normal, W^1_{ap}. In fact, W_{ap}-solutions in Theorem (10.29) become W-normal (see Appendix 1).

Similarly, we can state

(10.34) THEOREM. *Let A be a constant $(n \times n)$-matrix whose all eigenvalues have nonzero real parts. Assume, furthermore, that $\Sigma : \mathbb{R} \multimap \mathbb{R}^n$ is a multivalued S_{ap}-perturbation of a B_{ap}-function. Let, at last, $F : \mathbb{R}^n \multimap \mathbb{R}^n$ be a Lipschitz-continuous multivalued mapping with nonempty, compact and convex values, having the constant $L < \lambda/2kn(12\sqrt{3}/5 + 1)$, where $\lambda > 0$, $k > 0$ are suitable constants depending only on A such that (10.30) takes place.*

Then the inclusion (10.10) $(B = \mathbb{R}^n)$ admits a B_{ap}-solution.

PROOF. It follows from the foregoing investigation that Σ possesses a (single-valued) B_{ap}-selection $\sigma \subset \Sigma$ as well as F has a (single-valued) Lipschitz-continuous selection $f \subset F$ with the constant $L_0 < \lambda/(2k)$. Thus, instead of (10.10), we can again consider only the equation (10.31) having, according to Lemma (10.26) or Lemma (5.136), an entirely bounded solution $X(t)$ satisfying (10.27) and (10.32). Thus, one can deal with $\int_{-T}^T |X(t+\tau) - X(t)| \, dt$, to prove the B_{ap}-periodicity of $X(t)$ in the Besicovitch metric used in Definition (10.21). Hence, applying (after several steps) the well-known Fubini theorem and using the norm $\| \cdot \| =$

sup ess$_{t\in\mathbb{R}}|\cdot|$ and compatible vector and metrix norms, we obtain successively

$$\int_{-T}^{T}|X(t+\tau)-X(t)|\,dt$$

$$=\int_{-T}^{T}\left|\int_{-\infty}^{\infty}G(t,s)[f(X(s+\tau))-f(X(s))+\sigma(s+\tau)-\sigma(s)]\,ds\right|dt$$

$$\leq\int_{-T}^{T}dt\int_{-\infty}^{\infty}|G(t,s)||f(X(s+\tau))-f(X(s))+\sigma(s+\tau)-\sigma(s)|\,ds$$

$$\leq\int_{-T}^{T}dt\left[\int_{-\infty}^{t}ke^{-\lambda(t-s)}|f(X(s+\tau))-f(X(s))+\sigma(s+\tau)-\sigma(s)|\,ds\right.$$
$$\left.+\int_{t}^{\infty}ke^{\lambda(t-s)}|f(X(s+\tau))-f(X(s))+\sigma(s+\tau)-\sigma(s)|\,ds\right]$$

$$=k\int_{-\infty}^{a}|f(X(s+\tau))-f(X(s))+\sigma(s+\tau)-\sigma(s)|\,ds\int_{-T}^{T}e^{-\lambda(t-s)}\,dt$$

$$+k\int_{-T}^{T}|f(X(s+\tau))-f(X(s))+\sigma(s+\tau)-\sigma(s)|\,ds\int_{s}^{T}e^{-\lambda(t-s)}\,dt$$

$$+k\int_{-T}^{T}|f(X(s+\tau))-f(X(s))+\sigma(s+\tau)-\sigma(s)|\,ds\int_{-T}^{s}e^{\lambda(t-s)}\,dt$$

$$+k\int_{T}^{\infty}|f(X(s+\tau))-f(X(s))+\sigma(s+\tau)-\sigma(s)|\,ds\int_{-T}^{T}e^{\lambda(t-s)}\,dt$$

$$=\frac{k}{\lambda}(e^{\lambda T}-e^{-\lambda t})\int_{-\infty}^{-T}e^{\lambda s}|f(X(s+\tau))-f(X(s))+\sigma(s+\tau)-\sigma(s)|\,ds$$

$$+\frac{k}{\lambda}\int_{-T}^{T}(1-e^{-\lambda(T-s)})|f(X(s+\tau))-f(X(s))+\sigma(s+\tau)-\sigma(s)|\,ds$$

$$+\frac{k}{\lambda}\int_{-T}^{T}(1-e^{-\lambda(T+s)})|f(X(s+\tau))-f(X(s))+\sigma(s+\tau)-\sigma(s)|\,ds$$

$$+\frac{k}{\lambda}(e^{\lambda T}-e^{-\lambda T})\int_{T}^{\infty}e^{-\lambda s}|f(X(s+\tau))-f(X(s))+\sigma(s+\tau)-\sigma(s)|\,ds$$

$$\leq2\frac{k}{\lambda}\int_{-T}^{T}|f(X(s+\tau))-f(X(s))+\sigma(s+\tau)-\sigma(s)|\,ds$$

$$+2\frac{k}{\lambda^{2}}\|f(X(t+\tau))-f(X(t))+\sigma(s+\tau)-\sigma(s)\|.$$

Since $X(t)$ is bounded, and subsequently $\|f(X(t))\|<\infty$, the last term vanishes in the Besicovitch metric used in Definition (10.21). Thus, since by the hypothesis

$$\lim_{T\to\infty}\frac{1}{2T}\left[\int_{-T}^{T}|\sigma(t+\tau)-\sigma(t)|\,dt\right]<\varepsilon,$$

we get furthermore, by means the Lipschitz property of f, that

$$\lim_{T \to \infty} \frac{1}{2T} \left[\int_{-T}^{T} |X(t + \tau) - X(t)| \, dt \right]$$

$$\leq 2\frac{k}{\lambda} \lim_{T \to \infty} \frac{1}{2T} \left[\int_{-T}^{T} |f(X(t + \tau)) - f(X(t)) + \sigma(t + \tau) - \sigma(t)| \, dt \right]$$

$$< 2\varepsilon\frac{k}{\lambda} + 2L\frac{k}{\lambda} \lim_{T \to \infty} \frac{1}{2T} \left[\int_{-T}^{T} |X(t + \tau) - X(t)| \, dt \right].$$

After all, we arrive at

$$\lim_{T \to \infty} \frac{1}{2T} \left[\int_{-T}^{T} |X(t + \tau) - X(t)| \, dt \right] < \frac{2\varepsilon k}{\lambda - 2L_0 k},$$

as far as $L \cdot n(12\sqrt{3}/5 + 1) = L_0 < \lambda/(2k)$, which already verifies, according to Definition (10.21), the desired B_{ap}-periodicity of $X(t)$. So, the proof is complete.□

(10.35) REMARK. For the existence of a B_{ap}-solution of inclusion (10.10) $(B = \mathbb{R}^n)$, any of the following classes is available, under the assumption of Theorem (10.34), for the single-valued part of the forcing term $\Sigma(t)$ (see Appendix 1): u.a.p., u-normal, C_{ap}, S^1, S^1-normal, S^1_{ap}, e-W^1, e-W^1-normal, e-W^1_{ap}, W^1, W^1-normal, W^1_{ap}, B^1, B^1-normal, B^1_{ap}. In fact, B_{ap}-solutions in Theorem (10.34) become B-normal (see Appendix 1).

The last three main theorems can be summarized as follows.

(10.36) COROLLARY. *Let A be a constant $(n \times n)$-matrix whose all eigenvalues have nonzero real parts. Assume, furthermore, that $\Sigma : \mathbb{R} \multimap \mathbb{R}^n$ is a multivalued S_{ap}-perturbation of a G_{ap}-function. Let, at last, $F : \mathbb{R}^n \multimap \mathbb{R}^n$ be a Lipschitz-continuous multivalued mapping with nonempty, compact and convex values, having the constant $L < \lambda/2kn(12\sqrt{3}/5 + 1)$, where $\lambda > 0$, $k > 0$ are suitable constants depending only on A such that (10.30) takes place.*

Then the inclusion (10.10) $(B = \mathbb{R}^n)$ admits a G_{ap}-solution, (equivalently, a G-normal solution) where G_{ap} means S_{ap} or W_{ap} or B_{ap}, respectively. In the single-valued case, the (bounded $\Rightarrow$) G_{ap}-solution is unique, provided only $L < \lambda/(2k)$, where G_{ap} means again S_{ap} or W_{ap} or B_{ap}, respectively.

PROOF. The existence of a (bounded) G_{ap}-solution is guaranted directly by Theorems (10.20), (10.29) and (10.34) (see also Remarks (10.33) and (10.35)).

So, it is enough to show the uniqueness of an entirely bounded solution $X(t)$ of (10.10). Assume, on the contrary, that there exists another bounded (not necessarily G_{ap}) solution of (10.10), namely (cf. (10.27))

$$Y(t) = \int_{-\infty}^{\infty} G(t, s)[\sigma(s) + f(Y(s))] \, ds.$$

Because of the lipschitzianity of f, i.e.

$$|f(X) - f(Y)| \leq L|X - Y|, \quad \text{for all } X, Y \in \mathbb{R}^n,$$

we get

$$
\begin{aligned}
\|X(t) - Y(t)\| &= \left\| \int_{-\infty}^{\infty} G(t,s)[f(X(s)) - f(Y(s))]\, ds \right\| \\
&\leq \left\| \int_{-\infty}^{\infty} |G(t,s)| |f(X(s)) - f(Y(s))|\, ds \right\| \\
&\leq L \left\| \int_{-\infty}^{\infty} |G(t,s)| |X(s) - Y(s)|\, ds \right\| \\
&\leq L\|X(s) - Y(s)\| \int_{-\infty}^{\infty} |G(t,s)|\, ds \leq L\frac{2k}{\lambda}\|X(s) - Y(s)\|,
\end{aligned}
$$

where $\lambda > 0$, $k > 0$ are constants in (10.30).

This already leads to

$$\|X(s) - Y(s)\| \left(1 - L\frac{2k}{\lambda} \right) \leq 0, \quad \text{whenever } L < \frac{\lambda}{2k},$$

by which $X(t) \equiv Y(t)$ as claimed. □

(10.37) REMARK. The lipschitzianity of F can be again restricted to the set $Q \cap \mathbb{R}^n$, where Q has the same meaning as in Theorem (10.20). The "uniqueness" part of Corollary (10.36) represents the Bohr–Nongebauer type theorem, where "boundedness implies almost-periodicity"; cf. Lemma (10.5).

(10.38) REMARK (cf. [AK2]). If A is stable (Hurwitzean), then we can take in Corollary (10.36) more explicitly

$$L < 1/n\left(\frac{12}{5}\sqrt{3} + 1 \right) C(A),$$

where $C(a) = \sum_{k=0}^{n-1} \frac{2^k |A|^k}{|\mathrm{Re}\lambda|^{k+1}}$, $|A|$ denotes a (compatible) matrix norm and $|\mathrm{Re}\lambda|$ means the minimum of absolute values of real parts of the eigenvalues associated to A.

For a single-valued mapping F, because of the absence of the "barycentric subcondivision" factor $n(12\sqrt{3}/5 + 1)$, L can be only taken as $L < 1/C(A)$.

If A is a (2×2)-matrix with the eigenvalues λ_1, λ_2, then the above $C(A)$ can be replaced by $C_1(A)$, where

(i) $\lambda_1 = \lambda_2 \neq 0 \Rightarrow C_1(A) = \dfrac{|E|}{|\lambda_1|} + \dfrac{|A - \lambda_1 E|}{\lambda_1^2}$,

(ii)
$$\mathrm{sgn}(\mathrm{Re}\lambda_1) = \mathrm{sgn}(\mathrm{Re}\lambda_2) \neq 0,\ \lambda_1 \neq \lambda_2 \Rightarrow$$
$$C_1(A) = \frac{1}{|\lambda_1 - \lambda_2|} \left(\frac{|A - \lambda_2 E|}{|\mathrm{Re}\lambda_1|} + \frac{|A - \lambda_1 E|}{|\mathrm{Re}\lambda_2|} \right),$$

(iii)
$$\lambda_1 > 0 > \lambda_2,\ a_{21} \neq 0 \Rightarrow$$
$$C_1(A) = \frac{1}{|a_{21}|(\lambda_1 - \lambda_2)}$$
$$\cdot \left(\frac{\left| \begin{matrix} a_{21}(a_{11}-\lambda_2)\ (\lambda_2 - a_{11}(a_{11}-\lambda_1) \\ a_{21}^2 \qquad a_{21}(\lambda_1 - a_{11}) \end{matrix} \right|}{\lambda_1} - \frac{\left| \begin{matrix} a_{21}(\lambda_1 - a_{11})\ (a_{11}-\lambda_2)(a_{11}-\lambda_1) \\ -a_{21}^2 \qquad a_{21}(a_{11}-\lambda_2) \end{matrix} \right|}{\lambda_2} \right),$$

$$\lambda_1 > 0 > \lambda_2,\ a_{12} \neq 0 \Rightarrow$$
$$C_1(A) = \frac{1}{|a_{12}|(\lambda_1 - \lambda_2)}$$
$$\cdot \left(\frac{\left| \begin{matrix} a_{12}(a_{22}-\lambda_1) \qquad -a_{12}^2 \\ (a_{22}-\lambda_2)(a_{22}-\lambda_1)\ a_{12}(a_2 - a_{22}) \end{matrix} \right|}{\lambda_1} - \frac{\left| \begin{matrix} a_{12}(\lambda_2 - a_{22}) \qquad a_{12}^2 \\ (a_{22}-\lambda_2)(\lambda_1 - a_{22})\ a_{12}(a_{22}-\lambda_1) \end{matrix} \right|}{\lambda_2} \right),$$

$$\lambda_1 > 0 > \lambda_2,\ a_{12} = a_{21} = 0 \Rightarrow C_1(A) = \frac{1}{\lambda_1} - \frac{1}{\lambda_2}.$$

(One can use row, column, spectral or Schmidt norms above).

Better inequalities for L can be however obtained for differential equations. Therefore, consider still the equation

(10.39) $$x^{(n)} + \sum_{j=1}^{n} a_j x^{(n-j)} = f(t, x),$$

where a_j, $j = 1, \ldots, n$, are real constants and $f(t, x) \in C(\mathbb{R}^2, \mathbb{R})$. Assume that
(10.40) all roots λ_j, $j = 1, \ldots, n$, of the characteristic polynomial

$$\lambda^n + \sum_{j=1}^{n} a_j \lambda^{n-j}$$

be nonzero reals,
(10.41) a positive constant D_0 exists such that

$$\sup_{t \in \mathbb{R},\ |x| \leq D_0} |f(t, x)| \leq |a_n| D_0.$$

It follows from Theorem (5.78) and (5.79) in Chapter III.5 that, under (10.40) and (10.41), the equation (10.39) admits an antirely bounded solution $x(t)$ such that

$$(10.42) \qquad \sup_{t \in \mathbb{R}} |x^{(k)}(t)| \leq 2^k D_0 \Lambda^k, \quad \text{for } k = 0, \ldots, n-1,$$

where Λ is the spectral radius to the polynomial in (10.40), satisfying (see [AT2])

$$\Lambda \leq \min \left[\max(|a_1| + 1, \ldots, |a_{n-1}| + 1, |a_n|), \right.$$

$$\left. \max(1, |a_1| + \ldots + |a_n|), \max \left(|a_1|, \left| \frac{a_2}{a_1} \right|, \ldots, \left| \frac{a_n}{a_{n-1}} \right| \right) \right].$$

In particular, for $k = 0$, (10.42) simplifies into

$$(10.43) \qquad \sup_{t \in \mathbb{R}} |x(t)| \leq D_0.$$

Assume still that

(10.44) $f(t, x)$ is Lipschitzian in x, with $|x| \leq D_0$, uniformly w.r.t. $t \in \mathbb{R}$, with a constant $L \leq |a_n|$, i.e.

$$|f(t, x) - f(t, y)| \leq L|x - y|, \quad \text{for } t \in \mathbb{R}, \ |x| \leq D_0, \ |y| \leq D_0,$$

and that

(10.45) $f(t, x)$ is uniformly a.p. (u.a.p.) in t, uniformly w.r.t. $|x| < D$, where $D > D_0$, i.e. for any $\varepsilon > 0$ and any compact set $K \subset \{x \in \mathbb{R} \mid |x| < D\}$, there exists a positive number $l(\varepsilon, K)$ such that any interval of the length $l(\varepsilon, K)$ contains τ, for which

$$|f(t + \tau, x) - f(t, x)| \leq \varepsilon,$$

for all $t \in \mathbb{R}$ and $x \in K$.

Then the above bounded solution becomes unique and u.a.p., as relates the following

(10.46) THEOREM. *Let the conditions* (10.40), (10.41), (10.44) *and* (10.45) *be satisfied. Then equation* (10.39) *admits a u.a.p. solution* $x(t)$ *with u.a.p. derivatives, up to the* $(n-1)th$ *order, such that* (10.42) *takes place. Moreover, this solution is unique on the domain* $\{x \in \mathbb{R} \mid |x| \leq D_0\}$.

PROOF. As pointed out, it follows from the proof of Theorem (5.78) and (5.79) in Chapter III.5 (cf. also Lemma (5.51)) that, under (10.40) and (10.41), the equation (10.39) admits a solution $x(t)$ of the form (see (5.46))

$$x(t) = e^{\lambda_1 t} \int_{\Lambda_1}^{t} e^{(\lambda_2 - \lambda_1)t_1} \int_{\Lambda_2}^{t_1} \cdots$$
$$\int_{\Lambda_{n-1}}^{t_{n-2}} e^{(\lambda_n - \lambda_{n-1})t_{n-1}} \int_{\Lambda_n}^{t_{n-1}} e^{-\lambda_n t_n} f(t_n, x(t_n)) \, dt_n \ldots dt_1,$$

satisfying (10.42), where $\Lambda_j := (\infty)\alpha_j$, $\lambda_j = \alpha_j + i\beta_j$, $j = 1, \ldots, n$. Thus,

$$x(t+\tau) = e^{\lambda_1 t} \int_{\Lambda_1}^{t} e^{(\lambda_2 - \lambda_1)t_1} \int_{\Lambda_2}^{t_1} \cdots$$
$$\int_{\Lambda_{n-1}}^{t_{n-2}} e^{(\lambda_n - \lambda_{n-1})t_{n-1}} \int_{\Lambda_n}^{t_{n-1}} e^{-\lambda_n t_n} f(t_n + \tau, x(t_n + \tau)) \, dt_n \ldots dt_1,$$

where $t = \tau(\varepsilon, D)$ is an almost-period of $f(t, x)$ in t, for ε, D.

Applying still (10.44) and (10.45), we get successively

$$|x(t+\tau) - x(t)| = \left| e^{\lambda_1 t} \int_{\Lambda_1}^{t} e^{(\lambda_2 - \lambda_1)t_1} \int_{\Lambda_2}^{t_1} \cdots \right.$$
$$\left. \int_{\Lambda_n}^{t_{n-1}} e^{-\lambda_n t_n} [f(t_n + \tau, x(t_n + \tau)) - f(t_n, x(t_n))] \, dt_n \ldots dt_1 \right|$$
$$\leq \frac{1}{|a_n|} \sup_{t \in (-\infty, \infty)} |f(t+\tau, x(t+\tau)) - f(t, x(t))|$$
$$\leq \frac{1}{|a_n|} \sup_{t \in (-\infty, \infty)} [|f(t+\tau, x(t+\tau)) - f(t, x(t+\tau))|$$
$$+ |f(t, x(t+\tau)) - f(t, x(t))|]$$
$$\leq \frac{1}{|a_n|} [\sup_{\substack{t \in (-\infty, \infty) \\ |x| \leq D_0}} |f(t+\tau, x) - f(t, x)| + L \sup_{t \in (-\infty, \infty)} |x(t+\tau) - x(t)|]$$
$$\leq \frac{1}{|a_n|} [\varepsilon + L \sup_{t \in (-\infty, \infty)} |x(t+\tau) - x(t)|],$$

and consequently

$$\sup_{t \in (-\infty, \infty)} |x(t+\tau) - x(t)| \leq \frac{\varepsilon}{|a_n| - L}, \quad \text{as far as } L < |a_n|.$$

This already means the uniform almost-periodicity of $x(t)$.

Because of (10.42), the derivatives $\dot{x}(t), \ldots, x^{(n-1)}(t)$ are bounded as well, which implies (after the substitution into (10.39) the same for $x^{(n)}(t)$. Thus, they are uniformly continuous on $\mathbb{R}$ which is necessary and sufficient for their uniform almost periodicity (see e.g. [KsBuKo-M, p. 12]). This completes the existence part of the proof.

For the uniqueness, assume that we have two u.a.p. solutions of (10.39), namely

$$x(t) = e^{\lambda_1 t} \int_{\Lambda_1}^{t} e^{(\lambda_2 - \lambda_1)t_1} \int_{\Lambda_2}^{t_1} \cdots \int_{\Lambda_n}^{t_{n-1}} e^{-\lambda_n t_n} f(t_n, x(t_n)) \, dt_n \ldots dt_1$$

and

$$y(t) = e^{\lambda_1 t} \int_{\Lambda_1}^{t} e^{(\lambda_2 - \lambda_1)t_1} \int_{\Lambda_2}^{t_1} \cdots \int_{\Lambda_n}^{t_{n-1}} e^{-\lambda_n t_n} f(t_n, y(t_n)) \, dt_n \ldots dt_1$$

where

$$\sup_{t \in (-\infty, \infty)} |x(t)| \leq D_0 \quad \text{and} \quad \sup_{t \in (-\infty, \infty)} |y(t)| \leq D_0.$$

We will show that, under our assumptions, they are identical on the domain $|x| \leq D_0$. Hence, by the same reasons as above, we have (see (10.44))

$$|x(t) - y(t)| \leq \sup_{t \in (-\infty, \infty)} |f(t, x(t)) - f(t, y(t))|$$

$$\left| e^{\lambda_1 t} \int_{\Lambda_1}^{t} e^{(\lambda_2 - \lambda_1)t_1} \int_{\Lambda_2}^{t_1} \cdots \int_{\Lambda_n}^{t_{n-1}} e^{-\lambda_n t_n} \, dt_n \ldots dt_1 \right|$$

$$\leq \frac{L}{|a_n|} \sup_{t \in (-\infty, \infty)} |x(t) - y(t)|,$$

i.e.

$$\left(1 - \frac{L}{|a_n|}\right) \sup_{t \in (-\infty, \infty)} |x(t) - y(t)| \leq 0.$$

Since, by the hypothesis, $L < |a_n|$, we arrive at the desired identity $x(t) \equiv y(t)$. This completes the whole proof.   □

(10.47) REMARK. In particular, for f periodic in t, we can prove quite analogously the existence of a harmonic. On the other hand, the sole boundedness result is obviously true without the almost-periodicity restrictions on f.

Now, because of an accuracy of the criterium for L, we can finally consider the pendulum-type equation, namely

(10.48) $\ddot{x} + a\dot{x} + b \sin x = p(t),$

where a, b are positive constants such that $a^2 \geq 4b$ and $p(t)$ is a u.a.p. function.

Rewriting (10.48) into the form

(10.49) $\ddot{x} + a\dot{x} + bx = b(x - \sin x) + p(t),$

and considering still the equation

(10.50) $\ddot{x} + a\dot{x} - bx = -b[x + \sin(x - \pi)] + p(t),$

we can use for both (10.49) and (10.50) the result obtained in Theorem (10.46).

(10.51) COROLLARY. *Equation* (10.49) *admits, under the above assumptions, at least two u.a.p. solutions* $x_1(t)$ *and* $x_2(t)$ *with u.a.p. derivatives such that*

$$(10.52) \qquad \sup_{t\in(-\infty,\infty)} |x_1(t)| < \frac{\pi}{2}, \qquad \sup_{t\in(-\infty,\infty)} |x_2(t) - \pi| < \frac{\pi}{2},$$

provided only

$$(10.53) \qquad P_0 := \sup_{t\in(-\infty,\infty)} |p(t)| < b.$$

PROOF. Putting $D_0 = \pi/2 - \Delta$ into (10.41), (10.43) and (10.44), where $\Delta > 0$ is a sufficiently small constant, both equations (10.49) and (10.50) possess, according to Theorem (10.46), the desired u.a.p. solutions, provided

$$P_0 + b \max_{|x|\leq\pi/2-\Delta} |x - \sin x| \leq b\left(\frac{\pi}{2} - \Delta\right)$$

and

$$L := b \max_{|x|\leq\pi/2-\Delta} |1 - \cos x| \leq b.$$

Since, however,

$$\max_{|x|\leq\pi/2} |x - \sin x| \leq \frac{\pi}{2} - 1 \quad \text{and} \quad \max_{|x|\leq\pi/2-\Delta} |1 - \cos x| < 1,$$

both conditions can be really satisfied for a sufficiently small $\Delta > 0$, when assuming (10.53).

Since equation (10.48) admits so, under (10.53), at least two u.a.p. solutions satisfying (10.52), the proof is complete. □

(10.54) REMARK. Because of the substitution $t := -\hat{t}$, the same is obviously true for nonzero coefficients a, b in (10.48) such that $a^2 \geq 4|b|$. Then, condition (10.53) takes the appropriate form $P_0 < |b|$. Without the condition $a^2 \geq 4b$, equation (10.48) admits evidently at least one u.a.p. solution.

(10.55) REMARK. Theorem (10.46) as well as Corollary (10.51) can be also proved by means of the Banach contraction principle. Moreover, since the "separation condition" of L. Amerio is obviously satisfied, the results already follow from the sole uniquenes (for more details, see e.g. [AP-M], [Fin-M], [KsBuKo-M]).

III.11. Some further applications

In this chapter, problems related to implicit differential equations, those to derivo-periodic solutions and optimal control problems will be investigated. We also show how fractals can be implicitly generated by means of differential equations. The material is mainly taken from the papers [And22], [AGG1], [AFi1], [BiG1], [BiG2], [GN1], [GN3].

At first, we wolud like to show that many problems for differential equations whose right-hand sides involve the highest-order derivative can be easily transformed to those for certain differential inclusions with the right-hand sides no longer depending on these highest-order derivatives. We will apply this technique to the ordinary differential equations of first or higher order, but some other applications are possible, e.g. for partial differential equations (see e.g. [BiG2]).

Below, by X we mean the closed ball in $\mathbb{R}^n$ or the whole space $\mathbb{R}^n$. Furthermore, for a compact subset A of X, by the dimension, $\dim A$, we understand the topological covering dimension.

Following [AGG1], [BiG2], we recall:

(11.1) PROPOSITION. *Let A be a compact subset of X such that $\dim A = 0$. Then, for every $x \in A$ and for every open neighbourhood U of x in X, there exists an open neighbourhood $V \subset U$ of x in X such that $\partial V \cap A = \emptyset$.*

In the Euclidean space $\mathbb{R}^n$, we can identify the notion of the Brouwer degree with the fixed point index (cf. Chapters I.8 and II.9).

Namely, let U be an open bounded subset of $\mathbb{R}^n$ and let $g : \overline{U} \to \mathbb{R}^n$ be a continuous single-valued map such that $\mathrm{Fix}(g) \cap \partial U = \emptyset$. We let $\widetilde{g} : U \to \mathbb{R}^n$,

$$\widetilde{g}(x) = x - g(x), \quad x \in \overline{U},$$

and

(11.2) $$\mathrm{ind}(g, U) = \deg(\widetilde{g}, U),$$

where $\deg(\widetilde{g}, U)$ denotes the Brouwer degree of $\widetilde{g}$ with respect to U; then $\mathrm{ind}(g, U)$ is called the *fixed point index* of g with respect to U.

Now, all the properties of the Brouwer degree can be reformulated in terms of the fixed point index.

The proof of the following fact can be found in [Go5-M].

(11.3) PROPOSITION. *Let $g : X \to X$ be a compact map. Assume, furthermore, that the following two conditions are satisfied:*

(11.3.1) $\dim \mathrm{Fix}(g) = 0$,

(11.3.2) *there exists an open subset $U \subset X$ such that $\partial U \cap \mathrm{Fix}(g) = \emptyset$ and $\mathrm{ind}(g, U) \neq 0$.*

Then there exists a point $z \in \mathrm{Fix}(g)$, for which we have:

(11.3.3) *for every open neighbourhood U_z of z in X, there exists an open neighbourhood V_z of z in X such that: $V_z \subset U_z$, $\partial V_z \cap \mathrm{Fix}(g) = \emptyset$ and $\mathrm{ind}(g, V_z) \neq 0$.*

Now, let Y be a locally arcwise-connected space and let $f : Y \times X \to X$ be a compact map. Define, for every $y \in Y$, a map $f_y : X \to X$ by putting $f_y(x) = f(y, x)$, for every $x \in X$. Since X is an absolute retract, $\text{Fix}(f_y) \neq \emptyset$, for every $y \in Y$. It is easy to see that the following condition automatically holds:

(11.4) $\qquad \forall y \in Y \; \exists U_y$ open in X and $\text{ind}(f_y, U_y) \neq 0.$

Thus, we can associate with a map $f : Y \times X \to X$ the following multivalued map:

$$\varphi_f : Y \multimap X, \quad \varphi_f(y) = \text{Fix}(f_y).$$

We immediately obtain (cf. [Go5-M]):

(11.5) PROPOSITION. *Under all the above assumptions, the map* $\varphi_y : Y \multimap X$ *is u.s.c.*

Let us remark that, in general, φ_f is not a l.s.c. map. Below, we would like to formulate a sufficient condition which guarantees that φ_f has a l.s.c. selection. To this end, we assume that f satisfies the following condition:

(11.6) $\qquad \forall y \in Y : \dim \text{Fix}(f_y) = 0.$

Note that the condition (11.6) is satisfied for several classes of maps. Namely, for some classes of maps, the fixed point set $\text{Fix}(f_y)$ is a singleton, for every $y \in Y$, e.g., when f_y is a k-set contraction with $0 < k < 1$, or when the following assumption is satisfied (see [BiG1]):

$$\langle f(y, x_1) - f(y, x_2), x_1 - x_2 \rangle \leq k|x_1 - x_2|, \quad 0 < k < 1, \; y \in Y, x_1, x_2 \in X.$$

Now, in view of (11.4) and (11.6), we are able to define the map $\psi_f : Y \multimap X$ by putting $\psi_f(y) = \text{cl}\,\{z \in \text{Fix}(f_y) \mid \text{for } z \text{ condition (11.3.3) from Proposition (11.3)}$ is satisfied$\}$, for every $y \in Y$.

(11.7) THEOREM (see [AGG1], [BiG2]). *Under all the above assumptions, we have:*

(11.7.1) ψ_f *is a selection of* φ_f,

(11.7.2) ψ_f *is a l.s.c. map.*

Observe that condition (11.6) is rather restrictive. Therefore, it is interesting to characterize the topological structure of all mappings satisfying (11.6). We shall do it in the case when $Y = A$ is a closed subset of $\mathbb{R}^m$ and $X = \mathbb{R}^n$.

By $C_c(A \times \mathbb{R}^n, \mathbb{R}^n)$ we denote the Banach space of all compact (single-valued) maps from $A \times \mathbb{R}^n$ into $\mathbb{R}^n$ with the usual supremum norm. Let

$$Q = \{f \in C_c(A \times \mathbb{R}^n, \mathbb{R}^n) \mid f \text{ satisfies (11.6)}\}.$$

We have (cf. [AGG1], [BiG2]):

(11.8) THEOREM. *The set Q is dense in $C_c(A \times \mathbb{R}^n, \mathbb{R}^n)$.*

Let us note that all the above results remain true for X being the arbitrary ANR-space (see [BiG2]).

Now, we shall show how to apply the above results.

We start with ordinary differential equations of the first order. According to the above consideration, we let $Y = J \times \mathbb{R}^n$, where J is a closed half-line (possibly a closed interval), $X = \mathbb{R}^n$ and let $f : Y \times X \to X$ be a compact map. Then f satisfies condition (11.4) automatically, and so we can assume only (11.6). Let us consider the following equation:

$$(11.9) \qquad \dot{x}(t) = f(t, x(t), \dot{x}(t)),$$

where the solutions are understood in the sense of Carathéodory, i.e.

$$x(t) \in \mathrm{AC}_{\mathrm{loc}}(J, \mathbb{R}^n)$$

satisfy (11.9), for a.a. $t \in J$.

We associate with (11.9) the following two differential inclusions:

$$(11.10) \qquad \dot{x}(t) \in \varphi_f(t, x(t))$$

and

$$(11.11) \qquad \dot{x}(t) \in \psi_f(t, x(t)),$$

where φ_f and ψ_f are defined as before and by a solution of (11.10) or (11.11), we mean again a locally absolutely continuous function which satisfies (11.10) (resp. (11.11)) a.e. in J.

Denote by $S(f), S(\varphi_f)$ and $S(\psi_f)$ the sets of all solutions of (11.9), (11.10) and (11.11), respectively. Then we get:

$$S(\psi_f) \subset S(f) = S(\varphi_f).$$

However, the map ψ_f is a bounded, l.s.c. one with closed values, so by Proposition (2.13) in Chapter III.2, we obtain, that $S(\psi_f)$ contains an R_δ-set as a subset.

In particular, we have proved:

$$\emptyset \neq S(\psi_f) \subset S(\varphi_f) = S(f).$$

Observe that in (11.10) and (11.11) the right-hand sides do not depend on the derivative.

In an analogous way, we can consider ordinary differential equations of higher order. Let $Y = J \times \mathbb{R}^{kn}$, $X = \mathbb{R}^n$ and let $f : Y \times X \to X$ be a compact map. To study the existence problem for the following equation:

$$x^{(k)}(t) = f(t, x(t), \dot{x}(t), \ldots, x^{(k)}(t)),$$

we consider the following two differential inclusions:

$$x^{(k)}(t) \in \varphi_f(f(t, x(t), \dot{x}(t), \ldots, x^{(k-1)}(t)))$$

and

$$x^{(k)}(t) \in \psi_f(f(t, x(t), \dot{x}(t), \ldots, x^{(k-1)}(t))).$$

Thus, we can get the analogous conclusions.

Furthermore, we will study derivo-periodic solutions of the differential inclusion in $\mathbb{R}^n$:

$$(11.12) \qquad\qquad \dot{X} \in F(X) + P(t),$$

where $X = (x_1, \ldots, x_n)$, $F(X) = (F_1(X), \ldots, F_n(X))$ and $P(t) = (P_1(t), \ldots, P_n(t))$.

(11.13) DEFINITION. We say that $X(t) \in AC_{\text{loc}}(\mathbb{R}, \mathbb{R}^n)$ is an ω-*derivo-periodic* *solution* of (11.12) if $\dot{X}(t) = \dot{X}(t + \omega)$, for a.a. $t \in \mathbb{R}$.

Assume, we have a suitable notion of a multivalued derivative (see Appendix 2), say D^*, or multivalued partial derivatives $D^*_{x_j}$ (w.r.t. x_j), $j = 1, \ldots, n$, such that $F : \mathbb{R}^n \multimap \mathbb{R}^n$ is $\alpha\omega$-D^*-*periodic*, $\alpha \in \mathbb{R}^n$, $\omega > 0$, i.e.

$$(11.14) \qquad\qquad D^*F(X) \equiv D^*F(X + \alpha\omega),$$

where D^*F means the Jacobi matrix, namely $D^*F = (D^*(F_i)_{x_j})_{i,j=1}^n$.

(11.15) HYPOTHESIS. *The $\alpha\omega$-D^*-periodicity of F, i.e. (11.14), implies that F can be written as*

$$(11.16) \qquad\qquad F(X) = F_0(X) - AX,$$

where F_0 is $\alpha\omega$-periodic, i.e.

$$(11.17) \qquad\qquad F_0(X) \equiv F_0(X + \alpha\omega),$$

and A is a suitable $(n \times n)$-matrix.

If so, then (11.12) with an $\alpha\omega$-D^*-periodic F would take the following quasi-linear form:

$$(11.18) \qquad\qquad \dot{X} + AX \in F_0(X) + P(t),$$

where F_0 satisfies (11.17).

Moreover, $X_0(t)$ is obviously an ω-periodic solution of

(11.19) $$\dot{X} + AX \in F_0(X + \alpha t) + P(t),$$

where F_0 satisfies (11.17) and P is ω-periodic, i.e.

(11.20) $$P(t) \equiv P(t + \omega),$$

iff $X(t) = X_0(t) + \alpha t$ is an ω-derivo-periodic solution of

(11.21) $$\dot{X} + AX \in F_0(X) + [P(t) + (tA + E)\alpha],$$

where E denotes the unit matrix.

(11.22) HYPOTHESIS. $[P(t) + (tA + E)\alpha]$ *is ω-D^*-periodic, i.e.*

(11.23) $$D^*\{P(t + \omega) + [(t + \omega)A + E]\alpha\} \equiv D^*[P(t) + (tA + E)\alpha]$$
$$= D^*P(t) + A\alpha \equiv D^*P(t + \omega) + A\alpha$$

holds, provided (11.20) *takes place.*

More important is, however, the "reverse" formulation: $X(t) = X_0(t) + \alpha t$ is an ω-derivo-periodic solution of (11.18) iff $X_0(t)$ is an ω-periodic solution of

(11.24) $$\dot{X} + AX \in F_0(X + \alpha t) + [P(t) - (tA + E)\alpha],$$

where F_0 satisfies (11.17).

(11.25) HYPOTHESIS. *An ω-D^*-periodic P implies that*

(11.26) $$P(t) = P_0(t) + (tA + E)\alpha$$

holds with an ω-periodic P_0, i.e.

(11.27) $$P_0(t) \equiv P_0(t + \omega).$$

If so, then (11.24) would take the form

(11.28) $$\dot{X} + AX \in F_0(X + \alpha t) + P_0(t),$$

where, in view of (11.17),

$$F_0(X + \alpha(t + \omega)) \equiv F_0(X + \alpha t) =: F_1(t, X) \equiv F_1(t + \omega, X).$$

For (11.28), we can easily find sufficient conditions for the existence of at least one ω-periodic solution, e.g. (see Corollary (5.35) in Chapter III.5) if only all the eigenvalues of A have nonzero real parts, provided an u.s.c. F_0 with nonempty, convex and compact values is $\alpha\omega$-periodic, i.e. (11.17), and a measurable, essentially bounded P_0 with nonempty, convex and compact values is ω-periodic, i.e. (11.27).

Therefore, we can give immediately

(11.29) PROPOSITION. *If Hypothesis (11.25) is satisfied for a D^*-differentiable,* ω-D^*-periodic P, *i.e. if (11.26) is implied, then (11.28) with the same α can be written in the form of (11.24), and subsequently (11.18) admits an ω-derivo-periodic solution, provided all the eigenvalues of A have nonzero real parts and (11.17) takes place for an u.s.c. multivalued function F_0 with nonempty, convex and compact values.*

(11.30) PROPOSITION. *Let the assumptions of Proposition (11.29) be satisfied. If Hypothesis (11.15) is still satisfied, for an $\alpha\omega$-D^*-periodic, D^*-differentiable multivalued function F with nonempty, convex and compact values and with the same A (see (11.6)) as in (11.18), then even inclusion (11.12) admits an ω-derivo-periodic solution.*

Although we have to our disposal several different notions of multivalued derivatives (see Appendix 2), to satisfy Hypothesis (11.15) or (11.25), which are assumed in Propositions (11.29) and (11.30), only the usage of a multivalued derivative due to F. S. De Blasi is available, namely

(11.31) DEFINITION. A multivalued function $G : M_1 \multimap M_2$, with nonempty, bounded, closed values, where M_1, M_2 are Banach spaces, is said to be *De Blasi-like differentiable* at $x \in M_1$ if there exist an u.s.c. mapping $D_x : M_1 \multimap M_2$ with nonempty, bounded, closed, convex values which is positively homogeneous and a number $\delta > 0$ such that

$$d_H(G(x + h), G(x) + D_x(h)) = o(h), \quad \text{whenever } \|h\| \leq \delta,$$

where $o(h)$ denotes a nonnegative function such that

$$\lim_{h \to 0} \frac{o(h)}{\|h\|} = 0,$$

$d_H(\cdot, \cdot)$ stands for the Hausdorff metric. D_x is called the (multivalued) *differential of G at x*. Of course, G is said to be *De Blasi-like differentiable on M_1* (or simply, *De Blasi-like differentiable*) if it is so at every point $x \in M_1$.

Moreover, F in Hypothesis (11.15) must have a special form to satisfy (11.16), e.g. such that

$$(11.32) \qquad D^*(F_i(X))_{x_j} = D^*(F_i(x_j))_{x_j}, \quad \text{for all } i, j = 1, \ldots, n.$$

According to Theorem (A2.18) in Appendix 2, we can define equivalently a De Blasi-like differentiable function as a sum of single-valued continuous function having right-hand side and left-hand side derivatives plus a multivalued constant.

Therefore, it is natural to call $F(X)$ with (11.32) or $P(t)$ *De Blasi-like continuously differentiable* on $\mathbb{R}^n$ or $\mathbb{R}$ if there is a single-valued continuously differentiable function $f \in C^1(\mathbb{R}^n, \mathbb{R}^n)$ or $p \in C^1(\mathbb{R}, \mathbb{R}^n)$ such that $F(X) = f(X) + I_1$ with

$$(11.33) \qquad \frac{\partial f_i(X)}{\partial x_j} = \frac{\partial f_i(x_j)}{\partial x_j}, \quad \text{for all } i, j = 1, \dots, n,$$

or $P(t) = p(t) + I_2$, where $I_1 \subset \mathbb{R}^n$ or $I_2 \subset \mathbb{R}^n$ are vectors of bounded intervals (vectors of multivalued constants), i.e.

$$D^* F(X) = f'(X) = \left(\frac{\partial f_i}{\partial x_j} \right)_{i,j=1}^n \quad \text{or} \quad D^* P(t) = \dot{p}(t), \text{ respectively.}$$

According to Theorem (A2.20) in Appendix 2, a De Blasi-like continuously differentiable multivalued function F with (11.32) or P is ω-D^*-periodic iff $f \in C^1(\mathbb{R}^n, \mathbb{R}^n)$ with (11.33) or $p \in C^1(\mathbb{R}, \mathbb{R}^n)$ is so, respectively. Thus, according to Theorems (A2.3) and (A2.18) in Appendix 2, a De Blasi-like continuously differentiable multivalued function F with (11.32) takes the form (11.16) with F_0 satisfying (11.17) iff (11.14) takes place jointly with

$$(11.34) \qquad F_i(\alpha_j \omega) = F_i(0) - a_{ij}\alpha_j, \quad \text{for all } i, j = 1, \dots, n,$$

where $A = (a_{ij})_{i,j=1}^n$ and $\alpha = (\alpha_1, \dots, \alpha_n)$.

Similarly, a De Blasi-like continuously differentiable multivalued function P takes the form (11.26) with P_0 satisfying (11.27) if P is ω-D^*-periodic, i.e.

$$(11.35) \qquad\qquad D^* P(t) \equiv D^* P(t + \omega),$$

jointly with

$$(11.36) \qquad\qquad P(\omega) = P(0) + (\omega A + E)\alpha.$$

It is not difficult to show that a multivalued function $\varphi : \mathbb{R} \multimap \mathbb{R}$ is De Blasi-like continuously differentiable iff the (single-valued!) function $x \to D_x \varphi$ (see (11.31)) is additionally continuous.

Hence, we are ready to give the following two theorems, when applying Propositions (11.29) and (11.30).

(11.37) THEOREM. *Let all the eigenvalues of a real $(n \times n)$-matrix A have nonzero real parts. Assume, furthermore, that $F_0 : \mathbb{R}^n \multimap \mathbb{R}^n$ is an u.s.c. multivalued function with nonempty, convex and compact values which is $\alpha\omega$-periodic, i.e. (11.17), where $\alpha \in \mathbb{R}^n$, $\omega > 0$. Let, at last, $P : \mathbb{R} \multimap \mathbb{R}^n$ be a De Blasi-like continuously differentiable multivalued function with nonempty, convex and compact values which is ω-derivo-periodic, i.e. (11.35), satisfying (11.36), where E is a unit matrix. Then the inclusion (11.18) admits an ω-derivo-periodic solution in the sense of Definition (11.13).*

(11.38) THEOREM. *Let $F : \mathbb{R}^n \multimap \mathbb{R}^n$ be a De Blasi-like continuously differentiable multivalued function with nonempty, convex and compact values which is $\alpha\omega$-derivo-periodic, i.e. (11.14), satisfying (11.32) and (11.34), where A is a real $(n \times n)$-matrix whose all eigenvalues have nonzero real parts, E is a unit matrix and $\alpha \in \mathbb{R}^n$, $\omega > 0$. Assume, furthermore, that $P : \mathbb{R} \multimap \mathbb{R}^n$ is a De Blasi-like continuously differentiable multivalued function with nonempty, convex and compact values which is ω-derivo-periodic, i.e. (11.35), satisfying (11.36). Then the inclusion (11.12) admits an ω-derivo-periodic solution in the sense of Definition (11.13).*

One can readily check that an appropriate type (on the graph) of Hypothesis (11.22) can be also satisfied for another notion of differentiability considered in Appendix 2, namely the contingent derivative (see (A2.30)). Thus, we can also give

(11.39) THEOREM. *Let $F : \mathbb{R}^n \multimap \mathbb{R}^n$ be a De Blasi-like continuously differentiable multivalued function with nonempty, convex and compact values which is $\alpha\omega$-derivo-periodic, i.e. (11.14), satisfying (11.32) and (11.34), where A is a real $(n \times n)$-matrix whose all eigenvalues have nonzero parts, E is a unit matrix and $\alpha \in \mathbb{R}^n$, $\omega > 0$. Then the inclusion*

$$(11.40) \qquad \dot{X} \in F(X) + [P(t) + (tA + E)\alpha]$$

admits an ω-derivo-periodic solution in the sense of Definition (11.13), provided an essentially bounded measurable multivalued function $P : \mathbb{R} \multimap \mathbb{R}^n$ with nonempty, convex and compact values is ω-periodic, i.e. (11.20).

Assume, additionally, that P is a differentiable multivalued function in the sense of Definition (A2.8) or (A2.13) (see Appendix 2). Then the multivalued function $[P(t) + (tA + E)\alpha]$ in (11.40) is ω-derivo-periodic, i.e. (11.23), in the respective sense (see Appendix 2).

(11.41) REMARK. The existence of derivo-periodic solutions of given differential equations or inclusions can be proved directly, on the basis of Theorem (A2.3) (see Appendix 2), by means of the appropriately modified Wirtinger-type inequalities (see [And22]). Special sorts of derivo-periodic solutions were already treated in Theorem (6.39) in Chapter III.6.

Now, optimal (feedback) control problems will be treated. At first, let us consider a nonlinear control system with deterministic uncertain dynamics modelled by the differential inclusion of the form:

$$(11.42) \qquad \dot{x} \in F(t, x, u), \quad t \in [0, T], \ u \in U \subset \mathbb{R}^m, \ x \in \mathbb{R}^n,$$

where U is a connected compact set.

In the entire text, a system dynamics f will be a continuous function $f : D = [0,T] \times \mathbb{R}^n \times U \to \mathbb{R}^n$ such that $f(t,x,u) \in F(t,x,u)$, for any $(t,x,u) \in D$.

We assume the following conditions on the dynamics $F : D \multimap \mathbb{R}^n$.

(11.43) $F(t,x,u)$ is a nonempty, closed, convex set, for any $(t,x,u) \in D$, F is (t,x,u)-continuous in the Hausdorff metric.

Observe that, by (11.43) and the Michael selection theorem I.(3.30) (see Chapter I.3), the set of the system dynamics f is nonempty.

FORMULATION OF THE PROBLEM. *Given a nonempty closed set $K \subset \mathbb{R}^n$. To find a state feedback control $\overline{u}(t,x)$ defined in $[0,T] \times \mathbb{R}^n$ such that, for every dynamics $f(t,x,\overline{u}(t,x)) \in F(t,x,\overline{u}(t,x))$ and every $x_0 \in K$, any solution $x(t)$, $t \in [0,T]$, of the Cauchy problem*

$$(11.44) \qquad \begin{cases} \dot{x} = f(t,x,\overline{u}(t,x)), \\ x(0) = x_0 \end{cases}$$

satisfies $x(t) \in K$, for each $t \in [0,T]$, i.e. it is viable. In other words, $\overline{u}(t,x)$ makes K invariant under every dynamics $f(t,x,\overline{u}(t,x))$.

Since, in general, the control law $(t,x) \to \overline{u}(t,x)$ will not be continuous in the variable x, but only measurable in the pair (t,x), we cannot expect that (11.44) possesses Carathéodory solutions, i.e. absolutely continuous functions $x(t)$ satisfying (11.44), for a.a. $t \in [0,T]$. This is the reason why we will consider two regularizations of the function $g(t,x) := f(t,x,\overline{u}(t,x))$, called respectively the Krasovskiĭ regularization and the Filippov regularization of g, (see e.g. [AuC-M]). They are defined as follows:

$$K(g)(t,x) := \bigcap_{\delta > 0} \overline{\text{conv}}\, g(B((t,x),\delta)),$$

$$F(g)(t,x) := \bigcap_{\delta > 0} \bigcap_{\mu(N)=0} \overline{\text{conv}}\, g(B((t,x),\delta) \setminus N),$$

where μ denotes the Lebesgue measure in $\mathbb{R}^{n+1}$, $\overline{\text{conv}}$ the closed, convex hull and $B((t,x),\delta)$ the open ball centered at the point (t,x) with radius δ.

Any absolutely continuous solution of the multivalued Cauchy problems

$$(11.45) \qquad\qquad \dot{x} \in K(g)(t,x), \quad x(0) = x_0,$$

$$(11.46) \qquad\qquad \dot{x} \in F(g)(t,x), \quad x(0) = x_0,$$

on the time interval $[0,T]$ is called a *Krasovskiĭ (Filippov) solution* to (11.44).

Obviously, any Filippov solution is a Krasovskiĭ solution. Thus, if we solve the proposed problem for (11.45), we have also solved the problem for system (11.46).

In what follows, we will impose conditions on the dynamics f and on the set K which, together (11.43), will ensure that any solution of (11.45) is viable, whenever $x_0 \in K$. For this, we give the following:

(11.47) DEFINITION ([Au-M], [AuC-M], [AuF-M]). Let K be a nonempty closed subset of $\mathbb{R}^n$. The set $T_K(x) \subset \mathbb{R}^n$, $x \in K$, defined by

$$T_K(x) = \left\{ y \in \mathbb{R}^n \ \middle|\ \liminf_{\tau \to 0^+} \frac{1}{\tau} \operatorname{dist}(x + \tau y, K) = 0 \right\}$$

where $\operatorname{dist}(u, K) = \inf\{ |u - x| \mid x \in K \}$, is called the *Bouligand (contingent) cone* to K at $x \in K$.

Our problem requires that $K \subset \mathbb{R}^n$ is invariant under any dynamics $f(t, x, \overline{u}(t, x))$ $\in F(t, x, \overline{u}(t, x))$, $(t, x) \in [0, T] \times \mathbb{R}^n$. The invariance property involves the behaviour of F outside of K. Therefore, we need to extend the notion of contingent cone to K to a set $A \supset K$. For this purpose, following [Au-M], we introduce the concept of external contingent cone to K.

(11.48) DEFINITION. Let $K \subset A$ be a nonempty closed set of $\mathbb{R}^n$. Let $x \in A$, the set $\widehat{T}_K(x) \subset \mathbb{R}^n$ defined by

$$\widehat{T}_K(x) = \left\{ y \in \mathbb{R}^n \ \middle|\ \liminf_{\tau \to 0^+} \frac{1}{\tau} [\operatorname{dist}(x + \tau y, K) - \operatorname{dist}(x, K)] \leq 0 \right\}$$

is called the *external contingent cone* to K at the point $x \in A$.

Obviously, $\widehat{T}_K(x) = T_K(x)$, for any $x \in K$. Let $x \in A$, denote by $\Pi_K(x)$ the set of projections of x onto K, i.e. the set of points $z \in K$ such that $|x - z| = \operatorname{dist}(x, K)$. We have the following result:

(11.49) PROPOSITION ([Au-M, Lemma 5.1.2]). *Let K be a nonempty closed subset of $\mathbb{R}^n$. Then*

$$T_K(\Pi_K(x)) \subset \widehat{T}_K(x).$$

We introduce now a particular class of nonempty closed sets K by means of the following definition.

(11.50) DEFINITION ([Au-M], [AuF-M]). A nonempty closed subset $K \subset \mathbb{R}^n$ is said to be *sleek* if the multivalued map $T_K : K \multimap \mathbb{R}^n$ is lower semicontinuous.

It is well-known that any nonempty closed convex set K is sleek ([AuF-M, Theorem 4.22]). Furthermore, if K is sleek, then the multivalued map $T_K : K \multimap \mathbb{R}^n$ is convex-valued ([AuF-M, Theorem 4.1.8]).

A larger class of sleek sets is represented by the so-called proximate retracts (see [BGP], [Pl1]).

(11.51) DEFINITION. A nonempty closed subset K of $\mathbb{R}^n$ is said to be a *proximate retract* if there exists an open neighbourhood I of K in $\mathbb{R}^n$ and a continuous map $r : I \to K$ (called a metric retraction) such that the following two conditions are satisfied:

$$r(x) = x, \qquad \text{for each } x \in K,$$
$$|r(y) - y| = \text{dist}(y, K), \quad \text{for each } y \in I.$$

It is known that any C^2-manifold of $\mathbb{R}^n$ is a proximate retract (see [Pl1]).

We can now formulate the following assumptions:

(11.52) There exists $\gamma > 0$ such that, for any $(t, x) \in [0, T] \times \mathbb{R}^n$, there exists $u \in U$, for which

$$F(t, x, u) + \gamma B_1 \subset T_K(\Pi_K(x)),$$

where B_1 is the unitary open ball centered at the origin.

(11.53) For any $(t, x) \in [0, T] \times \mathbb{R}^n$, $u, v \in U$ and $\theta \in [0, 1]$, there exists $\widetilde{u} \in U$ such that

$$\theta F(t, x, u) + (1 - \theta) F(t, x, v) = F(t, x, \widetilde{u}).$$

Furthermore, there exist constants $a, b > 0$ such that

$$a|u - v| \le d_H(F(t, x, u), F(t, x, v)) \le b|u - v|,$$

for any $(t, x) \in [0, T] \times \mathbb{R}^n$ and any pair $u, v \in U$.

Here $d_H(A, B)$ denotes the Hausdorff distance of the two sets A, B. We recall that d is a metric in the family of all nonempty, bounded and closed subsets of a Banach space. We give the following

(11.54) DEFINITION. The multivalued map $R : [0, T] \times \mathbb{R}^n \multimap U$ defined as follows

$$R(t, x) = \{u \in U \mid F(t, x, u) \subset T_K(\Pi_K(x))\}$$

is called the *regulation map*.

In the sequel, we will denote $T_K(\Pi_K(x))$ simply by $T(x)$.

Finally, we assume the following condition.

(11.55) $\Pi_K : \mathbb{R}^n \to K$ is a single-valued, continuous function which will be denoted by π_K.

We are now in the position to prove the following.

(11.56) THEOREM. *Let $K \subset \mathbb{R}^n$ be a sleek subset of $\mathbb{R}^n$. Assume* (11.43), (11.52), (11.53) *and* (11.55). *Then the regulation map R has nonempty closed values and it is lower semicontinuous.*

PROOF. The proof follows the lines of that in ([AuC-M, Theorem 3]), where a single-valued autonomous dynamics, affine in the control, was considered.

In virtue of (11.52), the regulation map $R(t,x)$ has nonempty values, for any $(t,x) \in [0,T] \times \mathbb{R}^n$. Furthermore, $R(t,x)$ is closed by (11.43). We have now to prove that, for fixed $(t,x) \in [0,T] \times \mathbb{R}^n$, $u \in R(t,x)$ and $\varepsilon > 0$, there exists a neighbourhood $N(t,x)$ of (t,x) such that, for any $(t',x') \in N(t,x)$, there exists $\tilde{u} \in R(t',x') \cap (u + \varepsilon B_1)$.

Let $\delta > \operatorname{diam} U$ such that $\delta > (2\gamma + \varepsilon'a)/2a$, with $\varepsilon' = 2a\varepsilon/(a+b)$. Let $\alpha = \gamma\varepsilon'/(2\delta - \varepsilon') > 0$. Since F is continuous in the Hausdorff metric, U is compact and T is lower semicontinuous, there exists a neighbourhood $N(t,x)$ of (t,x) and $\eta < \varepsilon'$ such that, for any $(t',x') \in N(t,x)$, we have:

(1) for any u_1, u_2 with $|u_1 - u_2| < \eta$, we have $d(F(t',x',u_1), F(t,x,u_2)) \le \alpha/2$,

(2) $F(t,x,u) \subset T(x') + (\alpha/2)B_1$.

Therefore, for any $(t',x') \in N(t,x)$, there exists $v' \in U$ with $|u - v'| < \eta$ such that

$$(11.57) \qquad F(t',x',v') \subset T(x') + \alpha B_1.$$

Let $0 < \theta = \gamma/(\alpha + \gamma) < 1$. Since $\theta\alpha = (1-\theta)\gamma$, by multiplying (11.57) by θ and taking into account that the sets $T(x')$ and αB_1 are convex, we obtain

$$(11.58) \qquad \theta F(t',x',v') \subset \theta T(x') + (1-\theta)\gamma B_1.$$

By (11.52), we have that, for any $(t',x') \in N(t,x)$, there exists $u' \in U$ such that

$$(11.59) \qquad (1-\theta)F(t',x',u') + (1-\theta)\gamma B_1 \subset (1-\theta)T(x').$$

Summing (11.58) and (11.59), we get

$$\theta F(t',x',v') + (1-\theta)F'(t',x',u') + (1-\theta)\gamma B_1$$
$$\subset \theta T(x') + (1-\theta)T(x') + (1-\theta)\gamma B_1,$$

and by the convexity of $T(x')$, we have

$$\theta F(t',x',v') + (1-\theta)F(t',x',u') + (1-\theta)\gamma B_1 \subset T(x') + (1-\theta)\gamma B_1.$$

Thus,

$$\theta F(t',x',v') + (1-\theta)F(t',x',u') \subset T(x').$$

By (11.53), there exists $\tilde{u} \in U$ such that

$$\theta F(t', x', v') + (1 - \theta)F(t', x', u') = F(t', x', \tilde{u}),$$

that is $\tilde{u} \in R(t', x')$. Now, to finish the proof, we must show that $|u - \tilde{u}| < \varepsilon$.
    For this, we consider

$$d(F(t, x, u), F(t', x', \tilde{u})) = d(F(t, x, u), \theta F(t', x', v') + (1 - \theta)F(t', x', u')).$$

From the properties of the Hausdorff metric (see e.g. [Ki-M]), we obtain

$$d(F(t, x, u), F(t', x', \tilde{u})) \leq \theta d(F(t, x, u), F(t', x', v'))$$
$$+ (1 - \theta)d(F(t, x, u), F(t', x', u')).$$

On the other hand,

$$d(F(t, x, u), F(t', x', u')) \leq d(F(t, x, u), F(t, x, u')) + d(F(t, x, u'), F(t', x', u')).$$

Using again (11.53), we obtain

$$(11.60) \qquad d(F(t, x, u), F(t', x', u')) \leq \frac{\alpha}{2} + (1 - \theta)b|u - u'| \leq \frac{\alpha}{2} + b\frac{\varepsilon'}{2}.$$

Assume $(1 - \theta)|u - u'| \leq \varepsilon'/2$. However,

(11.61)
$$d(F(t, x, u), F(t', x', \tilde{u})) \geq d(F(t, x, u), F(t, x, \tilde{u})) - d(F(t, x, \tilde{u}), F(t', x', \tilde{u}))$$
$$\geq a|u - \tilde{u}| - \frac{\alpha}{2}.$$

In conclusion, combining (11.60) and (11.61), we get

$$|u - \tilde{u}| \leq \frac{\alpha}{a} + \frac{b\varepsilon'}{2a} \leq \frac{\varepsilon'}{2} + \frac{b\varepsilon'}{2a} < \frac{(a + b)}{2a}\varepsilon' < \varepsilon. \qquad \square$$

(11.62) REMARK. We have assumed, for simplicity, the continuity of F with
respect to the time t and that the control set U is independent of (t, x). Indeed,
for the validity of Theorem (11.56), we can only assume the measurability of F
with respect to t and $U = U(t, x)$. The multivalued map $U : [0, T] \times \mathbb{R}^n \multimap W$,
where W is a compact subset of $\mathbb{R}^m$, is assumed to satisfy the following conditions:

(U$_1$) $U(t, x)$ is a nonempty, convex, closed set, for any $(t, x) \in [0, T] \times \mathbb{R}^n$.
(U$_2$) U is measurable.
(U$_3$) $U(t, \cdot)$ is lower semicontinuous, for a.a. $t \in [0, T]$.

In fact, under these assumptions, following the arguments employed in ([NOZ1, Theorem 3.4]) and adapting the proof of Theorem (11.56), we can show that the regulation map R has nonempty closed values, it is measurable with respect to (t, x) and lower semicontinuous in x.

Some comments on the assumptions are useful. First of all, observe that (11.43) and the Michael selection theorem (3.30) in Chapter I.3 imply that F has the continuous selection property. That is, for any $p_0 \in F(t_0, x_0, u_0)$ with $(t_0, x_0, u_0) \in D$, there exists a continuous selection f of F such that $p_0 = f(t_0, x_0, u_0)$. In fact, for any $(t_0, x_0, u_0) \in D$ and $p_0 \in F(t_0, x_0, u_0)$, it is sufficient to consider the multivalued map $G : D \multimap \mathbb{R}^n$ defined as follows

$$G(t, x, u) = \begin{cases} F(t, x, u), & \text{for } (t, x, u) \neq (t_0, x_0, u_0), \\ p_0, & \text{for } (t, x, u) = (t_0, x_0, u_0), \end{cases}$$

which satisfies the assumptions of the Michael selection theorem (3.30) in Chapter I.3.

Furthermore, if we assume a more restrictive condition on the continuity of F on D, then any continuous selection f of F can be expressed as a convex combination of extreme points of the set $F(t, x, u)$; see [SZ].

Concerning (11.52), observe that even if F were a single-valued continuous map, condition (11.52) could not be weakened to the following (tangentiality) condition

- for any $(t, x) \in [0, T] \times \mathbb{R}^n$, there exists $u \in U$ such that $F(t, x, u) \subset T(x)$.

In fact, in this case, the map $R(t, x)$ could be not longer lower semicontinuous as the following simple example shows (see [GNZ]).

(11.63) EXAMPLE. Let $f : [0, 2] \times [0, 1] \times [0, 1] \to \mathbb{R}$ be the map defined as follows

$$f(t, x, u) = \begin{cases} \max\{1 - u, t\} - 1, & \text{for } t \leq 1, \\ \max\{u, 2 - t\} - 1, & \text{for } t \geq 1. \end{cases}$$

Then f satisfies our conditions, $K = [0, 1]$ and the feedback map $R : [0, 2] \times [0, 1] \multimap \mathbb{R}$ is given by

$$R(t, x) = \begin{cases} \{0\}, & \text{for } x = 0 \text{ and } t < 1, \\ \{1\}, & \text{for } x = 0 \text{ and } t > 1, \\ [0, 1], & \text{for } x = 0 \text{ and } t = 1 \text{ or } x > 0, \end{cases}$$

which is not lower semicontinuous.

We exhibit now a class of uncertain control dynamics F for which assumption (11.53) is satisfied.

(11.64) EXAMPLE. Consider $F : [0, T] \times \mathbb{R}^n \times U \multimap \mathbb{R}^n$ of the form

$$F(t, x, u) = \varphi(t, x) + f(t, x)h(u),$$

where

(i) $\varphi : [0, T] \times \mathbb{R}^n \multimap \mathbb{R}^n$ has nonempty, closed, convex values and it is bounded and continuous in the Hausdorff metric.

(ii) $f : [0, T] \times \mathbb{R}^n \to \mathbb{R}^n$ is a continuous function for which there exist constants $m_1, m_2 > 0$ such that

$$m_1 \leq |f(t, x)| \leq m_2, \quad \text{for any } (t, x) \in [0, T] \times \mathbb{R}^n.$$

(iii) $h : \mathbb{R}^m \to \mathbb{R}$ is a continuous map. Moreover, there exist constants $\lambda, \Lambda > 0$ such that

$$\lambda |u - v| \leq |h(u) - h(v)| \leq \Lambda |u - v|,$$

for any pair $u, v \in U$, with U connected compact set of $\mathbb{R}^n$.

It is easy to verify that F satisfies assumption (11.53).

(11.65) REMARK. By Remark (11.62) and considering the restriction of F to $[0, T] \times K$, it follows that Theorem (11.56) is a generalization of ([AuC-M, Theorem 3]) and ([Au-M, Theorem 6.3.1]) even in the single-valued case. In fact, it reduces to this result if the dynamics is affine in u and the process is autonomous. Moreover, observe that in the multivalued case even if F is affine in the control, we cannot use the argument employed in the proof of ([AuC-M, Theorem 3]) in order to prove our result. For instance, if $F(t, x, u) = \psi(t, x)u$, where $u \in \mathbb{R}$ and $\psi(t, x) \subset \mathbb{R}^n$ is a closed convex set, the set $\theta F(t, x, u) + (1 - \theta)F(t, x, v)$ does not necessarily coincide with the set $F(t, x, \theta u + (1 - \theta)v)$ unless $u \in U = [a, b]$, $a > 0$. Finally, in (11.53), we have assumed the existence of constants $a, b > 0$ such that

$$a|u - v| \leq d(F(t, x, u), F(t, x, v)) \leq b|u - v|,$$

for any $(t, x) \in [0, T] \times \mathbb{R}^n$ and any pair $u, v \in U$. This condition appears reasonable from the point of view of the effectiveness of the control variable u on the dynamical system. In fact, the dynamics is sensitive with respect to the variations of the control via the parameter a. On the other hand, the parameter b provides a necessary upper bound on this sensitivity.

(11.66) REMARK. Condition (11.55) is satisfied, for instance, if K is a nonempty, closed, convex set or if K is a proximate retract with $I = \mathbb{R}^n$.

Now, we state and prove the result which solves the proposed problem. For this, we also assume the following condition.

(11.67) There exists a positive constant M such that

$$|F(t, x, u)| \leq M, \quad \text{for any } (t, x, u) \in D.$$

Here $|F(t, x, u)| = \sup \{|y| \mid y \in F(t, x, u)\}$.

We give in the sequel a definition and a result which is a direct consequence of ([Bre3, Theorem 1]).

(11.68) DEFINITION. Consider the cone

$$\Gamma^M := \{(t,x) \in \mathbb{R} \times \mathbb{R}^n \mid |x| \leq Mt\}.$$

We say that a map $f : \mathbb{R} \times \mathbb{R}^n \to \mathbb{R}^n$ is *directionally Γ^M-continuous at a point* $(\bar{t}, \bar{x})$ if $f(t_n, x_n) \to f(\bar{t}, \bar{x})$, for every sequence $(t_n, x_n) \to (\bar{t}, \bar{x})$ with $(t_n - \bar{t}, x_n - \bar{x}) \in \Gamma^M$, for any $n \geq 1$. Moreover, we say that it is Γ^M-continuous on a set $Q \subset \mathbb{R}^{n+1}$ if it is Γ^M-continuous at every point $(\bar{t}, \bar{x}) \in Q$.

(11.69) PROPOSITION. *Under assumptions* (11.43), (11.52)–(11.56), (11.67), *the regulation map R admits a Γ^M-continuous selection $\bar{u}$ defined on $[0,T] \times \mathbb{R}^n$.*

Proposition (11.69) guarantees that any dynamics $f(t, x, \bar{u}(t, x))$ of $F(t, x, \bar{u}(t, x))$ is Γ^M-continuous with respect to $(t, x) \in [0, T] \times \mathbb{R}^n$. Observe that the Krasovskiĭ regularization $K(g)(t, x)$ of $g(t, x) := f(t, x, \bar{u}(t, x))$ is not necessarily contained in $F(t, x, \bar{\mu}(t, x))$.

We prove now the main related result.

(11.70) THEOREM. *Under assumptions* (11.43), (11.52), (11.53), (11.55), (11.67), *any Krasovskiĭ solution of the Cauchy problem*

$$\begin{cases} \dot{x} = f(t, x, \bar{u}(t, x)), \\ x(0) = x_0 \in K \end{cases}$$

satisfies $x(t) \in K$, for any $t \in [0, T]$.

PROOF. At first, we show that if $x(t)$, $t \in [0, T]$, is an absolutely continuous function such that

$$\begin{cases} \dot{x}(t) \in K(g)(t, x(t)), & \text{for a.a. } t \in [0, T], \\ x(0) = x_0, \end{cases}$$

where $x_0 \in \mathbb{R}^n$, then $x(t)$, $t \in [0, T]$, satisfies the Cauchy problem

$$\begin{cases} \dot{x}(t) = g(t, x(t)), & \text{for a.a. } t \in [0, T], \\ x(0) = x_0. \end{cases}$$

For this, we use the arguments of [Bre3] and [HP1] which combine the Γ^M-continuity with standard techniques.

By means of Lusin's Theorem there exists $Q_n \subset [0, T]$, $n \in \mathbb{N}$, measurable sets such that $\dot{x}$ restricted to Q_n is continuous, $\dot{x}(t) \in K(g)(t, x(t))$, for any $t \in Q_n$ and

$\mu([0,T] \setminus \cup_{n \in \mathbb{N}} Q_n) = 0$. Moreover, the Lebesgue density theorem (see [Ox-M]) ensures, for any $n \in \mathbb{N}$, the existence of a measurable set $N_n \subset Q_n$, with $\mu(N_n) = 0$ such that any point $t \in Q_n \setminus N_n$ is a density point for Q_n. Let $t \in Q_n \setminus N_n$, we can find a sequence $\{t_m\} \subset Q_n \setminus N_n$ with $t_m > t$ and $t_m \to t$ decreasing as $m \to \infty$.

Hence $\dot{x}(t_m) \to \dot{x}(t)$. On the other hand, $|x(t) - x(s)| \leq M|t - s|$, for all $t, s \in [0,T]$, because $|K(g)(t,x)| \leq M$ by (11.67), the fact that g is a selection of F and the definition of $K(g)$. Let $\varepsilon > 0$. We have $\dot{x}(t_m) \in K(g)(t_m, x(t_m)) \subset g(t_m, x(t_m)) + \frac{\varepsilon}{2} B_1$, for any $m \in \mathbb{N}$. However, g is Γ^M-continuous in $[0,T] \times \mathbb{R}^n$, and so there exists $m_0(\varepsilon) \in \mathbb{N}$ such, that for any $m \geq m_0$, we have $|g(t_m, x(t_m)) - g(t, x(t))| < \varepsilon/2$, that is

$$g(t_m, x(t_m)) \in g(t, x(t)) + \frac{\varepsilon}{2} B_1.$$

Therefore, for $m \geq m_0$, we have $\dot{x}(t_m) \in g(t, x(t)) + \varepsilon B_1$, and so $\dot{x}(t) \in g(t, x(t)) + \varepsilon B_1$.

For $\varepsilon \to 0$, we get $\dot{x}(t) = g(t, x(t))$, $x(0) = x_0$, for all $t \in \widehat{Q} = \bigcup_{n \in \mathbb{N}} (Q_n \setminus N_n)$ with $\mu([0,T] \setminus \widehat{Q}) = 0$.

In conclusion, $x(t)$ is an absolutely continuous function such that

$$\dot{x}(t) \in T(x(t)), \quad \text{for a.a. } t \in [0,T].$$

Now, if we take $x_0 \in K$, then $x(t) \in K$, for a.a. $t \in [0,T]$. To show this, let $t \in [0,T]$ such that there exists $\dot{x}(t)$ and $\dot{d}(t)$, where $d(t) := \text{dist}(x(t), K)$ is absolutely continuous. Consider $x(t + \tau)$, with $\tau > 0$, there exists $\varepsilon(\tau)$ with $\varepsilon(\tau) \to 0$, when $\tau \to 0^+$, such that

$$x(t + \tau) = x(t) + \tau \dot{x}(t) + \tau \varepsilon(\tau).$$

Therefore,

$$\dot{d}(t) = \lim_{\tau \to 0^+} \frac{\text{dist}(x(t) + \tau \dot{x}(t) + \tau \varepsilon(\tau), K) - \text{dist}(x(t), K)}{\tau}.$$

On the other hand, by Proposition (11.49), we have $\dot{x}(t) \in T(x(t)) \subset \widehat{T}_K(x(t))$. Thus, $\dot{d}(t) \leq 0$, for a.a. $t \in [0,T]$. From this, if $t \in [0,T]$ is such that $d(t) > 0$, we have

$$0 < d(t) = \int_0^t \dot{d}(s)\, ds \leq 0,$$

which is a contradiction. This concludes the proof. $\square$

(11.71) REMARK. From the proof of Theorem (11.70), it turns out that any Krasovskiĭ solution $x(t)$ of (11.44), whenever $f(t, x, \overline{u}(t,x)) \in F(t, x, \overline{u}(t,x))$, is such that the function $\text{dist}(x(t), K)$ is nonincreasing in $[0, T]$. Therefore, if the initial condition $x(0) = x_0$ is sufficiently close to K, say $\text{dist}(x_0, K) < \varepsilon$, then $\text{dist}(x(t), K) < \varepsilon$, for any $t \in [0, T]$. Furthermore, if K is bounded and the initial conditions x_0 are taken in a bounded neighbourhood $V \subset B(0, r)$ of K, then condition (11.67) can be replaced by the weaker condition:

(11.67') There exist constants $L, M > 0$, $L < M$, such that

$$|F(t, x, u)| \le M,$$

for any $t \in [0, T]$, any $|x| < r + LT$ and any $u \in U$.

In Remark (11.62), we pointed out that if F is Carathéodory and $U = U(t, x)$ satisfies (U_1), (U_2) and (U_3), then the regulation map R is measurable in t and lower semicontinuous in x with nonempty closed values. Thus, if we introduce, for $\overline{u}(t, x)$, the notion of Scorza–Dragoni Γ^M-continuity (see e.g. [HP1]), we can prove Theorem (11.70) under these assumptions. Furthermore, under some regularity assumption, we could also allow the set K to depend on t, and the existence of periodic solutions of (11.44) in K could be also investigated.

An important control problem for system (11.43) that we can tackle by means of the approach presented here is illustrated in the sequel.

Consider two control systems

(11.72) $$\dot{x} \in F(t, x, u), \quad t \in [0, T], \ u \in U \subset \mathbb{R}^m, \ x \in \mathbb{R}^n$$

and

(11.73) $$\dot{y} = \psi(t, y, v), \quad t \in [0, T], \ v \in V \subset \mathbb{R}^m, \ y \in \mathbb{R}^n,$$

where U, V are connected compact sets of $\mathbb{R}^m$ and ψ is continuous.

The initial state $x(0) = x_0$ is also uncertain, but bounded, i.e., $x_0 \in B(0, r)$, $r > 0$. The initial condition $y(0) = y_0$ is taken in a given ball $B(0, \rho)$, $\rho > 0$.

Introduce now a multivalued map $Q : \mathbb{R}^n \multimap \mathbb{R}^n$, where the domain represents the states y and the range represents the states x. Assume the following conditions:

(11.74) $Q(y)$ is a nonempty, convex, compact set of $\mathbb{R}^n$, for any $y \in \mathbb{R}^n$, and Q is upper semicontinuous.

(11.75) Γ_Q is a convex set of $\mathbb{R}^n \times \mathbb{R}^n$ containing the set $B(0, \rho) \times B(0, r)$.

The map Q^{-1} is often called the observation map.

We consider the following problem.

(11.76) Given a state-control pair $(y(t), v(t))$ of (11.73), the problem that we want to solve is that of determining a feedback control $\overline{u}(t, x)$, $(t, x) \in [0, T] \times \mathbb{R}^n$ such that, for any dynamics $f(t, x, \overline{u}(t, x)) \in F(t, x, \overline{u}(t, x))$, and for any

$x_0 \in B(0, r)$, we have $x(t) \in Q(y(t))$, for any $t \in [0, T]$, where $x(t)$ is any Krasovskiǐ's solution of the Cauchy problem

$$\begin{cases} \dot{x} = f(t, x, \overline{u}(t, x)), \\ x(0) = x_0. \end{cases}$$

In other words, given a nonlinear control model (11.73), we want to construct a feedback control $\overline{u}(t, x)$ under which any solution, in the Krasovskiǐ sense, of (11.72) is such that the pair $(y(\cdot), x(\cdot))$ is viable in the graph of Q.

Let $K := \Gamma_Q$. Since K is a closed convex subset of $\mathbb{R}^n \times \mathbb{R}^n$, there exists a continuous (non-expansive) projector $\pi_K : \mathbb{R}^n \times \mathbb{R}^n \to K$. Define now the regulation map $R_Q : [0, T] \times \mathbb{R}^n \times \mathbb{R}^n \times V \multimap U$ as follows:

$$R_Q(t, x, y, v) = \{u \in U \mid (\psi(t, y, v), F(t, x, u)) \subset T_K(\pi_K(y, x))\},$$

if $v \in V$ and $R_Q(t, x, y, v) = \emptyset$ if $v \notin V$.

We have the following result.

(11.77) PROPOSITION. *Under assumptions* (11.43), (11.52), (11.53), (11.55), (11.67) *and* (11.74), (11.75), *problem* (11.76) *is solvable.*

PROOF. For any fixed state-control pair $(y(t), v(t))$ of the reference model (11.73), by Theorem (11.56), we get that the map

$$(t, x) \to \widetilde{R}_Q(t, x) := R_Q(t, x, y(t), v(t))$$

is lower semicontinuous with nonempty closed values. Then we can apply the first part of the proof of Theorem (11.70) to conclude that

$$(\dot{y}(t), \dot{x}(t)) \in T_K(\pi_K(y(t), x(t))).$$

On the other hand, $(y_0, x_0) \in K$, and so using the second part of the proof of Theorem (11.70), we obtain that

$$(y(t), x(t)) \in K, \quad \text{for any } t \in [0, T]. \qquad \square$$

Observe that the condition

$$(\psi(t, y, v), F(t, x, u)) \subset T_K(\pi_K(y, x))$$

can be rewritten in terms of the contingent derivative $DQ(y, x)$ of the map Q as follows

$$F(t, x, u) \subset DQ(y, x)(\psi(t, y, v)).$$

In fact,

$$\Gamma_{DQ(y,x)} := T_{\Gamma_{Q(y,x)}}.$$

Now, it is time to consider the second feedback control problem. More precisely, we will consider an autonomous nonlinear control system described by the differential equations

$$(11.78) \qquad\qquad \dot{x} = f(x,u),$$

where f is a continuous function. The state variable x belongs to $\mathbb{R}^n$ and the control variable u belongs to a set $U(x) \subset \mathbb{R}^m$, depending on x. The assumptions on the multivalued map $x \multimap U(x)$ will be precised later.

The control problem for system (11.78) that we consider is in the class of control problems involving sliding manifolds. Specifically, the problem that we deal with here can be formulated as follows.

FORMULATION OF THE PROBLEM. *Given a "sufficiently" smooth, nonempty, closed set $K \subset \mathbb{R}^n$, a suitable neighborhood I of K and an initial state $x_0 \in I \setminus K$. We want, by means of a control u which takes value in $U(x)$, $x \in I \setminus K$, to steer in finite time and then hold the state x of system (11.78) in a prescribed ε-neighborhood of K.*

The set K represents the required behaviour of the controlled dynamics. In the case when K can be defined as the zeros of a continuously differentiable map $s : \mathbb{R}^n \to \mathbb{R}^m$, $m \le n$, $s = (s_j)_{j=1}^m$ and one uses feedback control laws $u = u(x)$ which are discontinuous along the surfaces $s_j(x) = 0$, usually $u_j(x) = -\operatorname{sgn} s_j(x)$, $j = 1, \ldots, m$, then this is the classical nonlinear variable structure control problem for which there is a very broad literature, see e.g. the monographs [Ut1-M] and [Ut2-M].

We start with some definitions.

Recall that if the cone-valued map $x \multimap T_K(x)$ in Definition (11.47) is lower semicontinuous at any $x \in K$, then the set K is said to be *sleek* (see Definition (11.50)). Furthermore, the *polar cone* of $T_K(x)$ is the normal cone to K, denoted by $N_K(x)$.

(11.79) DEFINITION. Let $K \subset \mathbb{R}^n$ be a nonempty, closed set. Denote by $\pi_K(\cdot)$ the metric projection on K defined by

$$\pi_K(x) := \{y \in K \mid |x - y| = d_K(x)\}.$$

We assume the following conditions on K:

(11.80) there exists an open neighborhood I of K in $\mathbb{R}^n$ such that, for any $x \in I \setminus K$ the metric projection $\pi_K(x)$ is single-valued.

(11.81) REMARK. In ([Fed, Theorem 4.8]), for a nonempty, closed set $K \subset \mathbb{R}^n$ satisfying (11.80), it is shown that $x \to \pi_K(x)$ is continuous in I and locally Lipschitz in $I \backslash K$ together with grad $d_K(x)$. Furthermore, in ([Fed, Theorem 4.18]) it is proved, in an implicit way, that condition (11.80) is equivalent to the following property:

(11.82) there exists a continuous function $\varphi : K \times K \to [0, \infty)$ such that, for all $x, y \in K$, $v \in N_K(x)$, we have that

$$\langle v, y - x \rangle \leq \varphi(x,y)|v|\,|y - x|^2.$$

Here $\langle \cdot, \cdot \rangle$ denotes the inner product in $\mathbb{R}^n$. A nonempty, closed set satisfying this property is called φ-convex set (see [Can]). In [CoGo1], it is shown that a φ-convex set of an infinite dimensional Hilbert space satifies (11.80) with $x \to \pi_K(x)$ continuous in I. Furthermore, a nonempty, closed set K of a Banach space which satisfies (11.80) with $\pi_K(x) : I \to K$ continuous (metric retraction) is called *proximate retract* (see Definition (11.51)). Therefore, in a finite dimensional space, the class of φ-convex sets coincides with that of proximate retracts. It is still an open question if a proximate retract in an infinite dimensional Hilbert space is φ-convex.

For any $x \in I$, consider the set $K(x)$ defined as follows:

$$K(x) := K + d_K(x)B_1,$$

where $B_1 \subset \mathbb{R}^n$ denotes the unit closed ball centered at the origin. We pose the following problem.

(11.83) To show that the cone-valued map defined in I by

$$x \multimap T_{K(x)}(x)$$

is lower semicontinuous, at any $x \in I$, i.e. the set $K(x)$ is sleek, at any $x \in I$.

To solve problem (11.83) is one of our aims in this part.

(11.84) REMARK. Note that by ([AuF-M, Theorem 4.1]), we have that if $x \multimap T_{K(x)}(x)$ is lower semicontinuous at x, then the cone $T_{K(x)}(x)$ is convex.

The first step to solve (11.83) is to prove the following result which is known in the case when K is convex ([AuF-M, p. 141]).

(11.85) LEMMA. *Assume that K is a nonempty, closed set satisfying condition* (11.80). *Then*

$$T_{K(x)}(x) = T_K(\pi_K(x)) + T_{d_K(x)B_1}(x - \pi_K(x)), \quad x \in I.$$

PROOF. Obviously, if $x \in K$, then there is nothing to prove. Therefore let $x \in I \setminus K$ and assume that $v \in T_{K(x)}(x)$, this means that there exist sequences $\tau_n \to 0+$, $v_n \to v$ such that $x_n := x + \tau_n v_n \in K(x)$, i.e. $d_{K(x)}(x_n) = 0$. By property (11.80) and Remark (11.81), there exists $l = l(x) > 0$ such that $|y - y_n| \le l|x_n - x|$, for $n \in \mathbb{N}$ sufficiently large with $y = \pi_K(x)$ and $y_n = \pi_K(x_n)$. On the other hand, for $n \in \mathbb{N}$, there exists $b_n \in B_1$ such that $x_n = y_n + |x - y|b_n$, because $|x_n - y_n| = d_K(x_n) \le d_K(x)$. In fact, from $x_n \in K(x)$ it follows that there exist $\widehat{y}_n \in K$ and $\widehat{b}_n \in B_1$ such that $x_n = \widehat{y}_n + d_K(x)\widehat{b}_n$. Thus, $d_K(x_n) = d_K(\widehat{y}_n + d_K(x)\widehat{b}_n) \le d_K(\widehat{y}_n) + d_K(x)|\widehat{b}_n| \le d_K(x)$. Furthermore, by passing to a subsequence if necessary, we have that $b_n \to b$, where $b = \dfrac{(x - y)}{|x - y|}$. Rewrite $x_n = x + \tau_n v_n$ as follows

$$(11.86) \qquad x + \tau_n v_n = y + |y_n - y|\frac{y_n - y}{|y_n - y|} + (x - y) + |x - y||b_n - b|\frac{b_n - b}{|b_n - b|},$$

where $b_n := \dfrac{x_n - y_n}{|x - y|}$. Observe that

$$y_n = y + |y_n - y|\frac{y_n - y}{|y_n - y|} \in K,$$

with $|y_n - y| \to 0+$ and

$$|x - y|b_n = (x - y) + |x - y||b_n - b|\frac{b_n - b}{|b_n - b|} \in d_K(x)B_1$$

with $|x - y||b_n - b| \to 0+$. Then

$$w_0 := \lim_{n \to \infty} \frac{y_n - y}{|y_n - y|} \in T_K(y) \quad \text{and} \quad z_0 := \lim_{n \to \infty} \frac{b_n - b}{|b_n - b|} \in T_{d_K(x)B_1}(x - y).$$

On the other hand, from (11.86), we have

$$(11.87) \qquad v_n = \frac{|y_n - y|}{\tau_n}\frac{y_n - y}{|y_n - y|} + \frac{|x - y||b_n - b|}{\tau_n}\frac{b_n - b}{|b_n - b|},$$

with $\dfrac{|y_n - y|}{\tau_n} \le l|v_n| \le M$, for n sufficiently large. Thus by passing to a subsequence if necessary, we have

$$\frac{|y_n - y|}{\tau_n} \to \alpha_0 \ge 0 \quad \text{and} \quad \frac{|x - y||b_n - b|}{\tau_n} \to \beta_0 \ge 0,$$

obtaining from (11.87)

$$v = \alpha_0 w_0 + \beta_0 z_0 \in T_K(y) + T_{d_K(x)B_1}(x - y).$$

Vice versa, let us now prove that from $w \in T_K(y)$ and $z \in T_{d_K(x)B_1}(x - y)$, where $y = \pi_K(x)$ and $x \in I \setminus K$, it follows that $w + z \in T_{K(x)}(x)$. Observe that, without loss of generality, we can assume that $|z| = |x - y|$. Assuming $w \in T_K(y)$, there exist sequences $\widehat{\tau}_n \to 0+$ and $w_n \to w$ such that $y + \widehat{\tau}_n w_n \in K$. We show now the existence of a sequence $z_n \to z$ such that $|z_n| = |x - y|$ and

$$(x - y) + \widehat{\tau}_n z_n \in d_K(x)B_1.$$

In fact, for any $z \in T_{d_K(x)B_1}(x - y)$, with $|z| = |x - y|$, there exists a sequence $z_n \to z$ such that $|z_n| = |x - y|$, for any $n \in \mathbb{N}$. Therefore, for any sequence $\tau_n \to 0^+$, we have

$$(x - y) + \tau_n z_n \in d_K(x)B_1,$$

for n sufficiently large. In particular,

$$(x - y) + \widehat{\tau}_n z_n \in d_K(x)B_1.$$

In conclusion,

$$y + \widehat{\tau}_n w_n + (x - y) + \widehat{\tau}_n z_n = x + \widehat{\tau}_n(w_n + z_n) \in K(x)$$

with $\widehat{\tau}_n \to 0+$. Thus, $w + z \in T_{K(x)}(x)$. This completes the proof. $\square$

(11.88) REMARK. As a consequence of Lemma (11.85) and the closedness of $T_{K(x)}(x)$ we have that, in our case, the set

$$T_K(\pi_K(x)) + T_{d_K(x)B_1}(x - \pi_K(x)), \quad x \in I,$$

is closed.

(11.89) DEFINITION. Let $\gamma > 0$ and K be as in Lemma (11.85). For any $x \in I \setminus K$, we define the *closed affine cone* $\widetilde{T}_{K,\gamma}(x)$ as follows

$$\widetilde{T}_{K,\gamma}(x) := \left\{ v \in \mathbb{R}^n \mid \liminf_{\tau \to 0+} \frac{d_K(x + \tau v) - d_K(x)}{\tau} \leq -\gamma \right\}.$$

We assume that, for γ sufficiently small, $\widetilde{T}_{K,\gamma}(x)$ is nonempty. For simplicity, we denote by $\widetilde{T}_K(x)$ the external contingent Bouligand cone $\widetilde{T}_{K,0}(x)$. Observe that if $x \in K$, then $\widetilde{T}_{K,\gamma}(x)$ reduces to $T_K(x)$.

We can now prove the following result.

(11.90) LEMMA. $\tilde{T}_K(x) = T_{K(x)}(x)$, for any $x \in I$.

PROOF. Let $v \in T_{K(x)}(x)$, i.e. $\liminf_{\tau \to 0+} \dfrac{d_{K(x)}(x + \tau v)}{\tau} = 0$. On the other

hand, $d_{K(x)}(x + \tau v) = \inf_{z \in K(x)} d(x + \tau v, z)$, where $z = \zeta + d_K(x)b$, $\zeta \in K, b \in B_1$. Thus,

$$d_{K(x)}(x + \tau v) \geq d_K(x + \tau v) - d_K(x),$$

because $d(x + \tau v, \zeta + d_K(x)b) \geq d(x + \tau v, \zeta) - d_K(x)$, for any $\zeta \in K$ and any $b \in B_1$. In conclusion,

$$0 \geq \liminf_{\tau \to 0+} \frac{d_K(x + \tau v) - d_K(x)}{\tau},$$

which implies $v \in \tilde{T}_K(x)$.

Vice versa, assume that $v \in \tilde{T}_K(x)$. Thus, there exist sequences $\tau_n \to 0+$, $v_n \to v$ such that

$$(11.91) \qquad d_K(x + \tau_n v_n) \leq d_K(x).$$

Let $y_n = \pi_K(x + \tau_n v_n)$. Thus, $|x - y_n + \tau_n v_n| = d_K(x + \tau_n v_n)$ and by (11.91), we get $|x - y_n + \tau_n v_n| \leq |x - y|$, for any $n \in \mathbb{N}$. This implies that $x + \tau_n v_n \in K(x)$, because $x + \tau_n v_n \in y_n + d_K(x)B_1$. Consequently, we obtain $d_{K(x)}(x + \tau_n v_n) = 0$, for any $n \in \mathbb{N}$. Hence, $v \in T_{K(x)}(x)$. This completes the proof. $\square$

We have the following result.

(11.92) PROPOSITION. Assume that $K \subset \mathbb{R}^n$ is a nonempty, closed set which satisfies condition (11.80). Then the cone-valued map $x \multimap \tilde{T}_K(x)$ is lower semi-continuous, at any $x \in I$, with nonempty, closed, convex values.

PROOF. The proof easily follows from Lemmas (11.85) and (11.90), taking into account Remarks (11.81) and (11.84). $\square$

(11.93) DEFINITION. Let $\psi : \text{Dom}(\psi) \to \mathbb{R}$ be a function and $x \in \text{Dom}(\psi)$. We define the *contingent epiderivative* (sometimes also called the *upper contingent derivative*) $D_\uparrow \psi(x)(v)$ of ψ at x in the direction v as follows:

$$D_\uparrow \psi(x)(v) := \liminf_{\substack{\tau \to 0+ \\ v' \to v}} \frac{\psi(x + \tau v') - \psi(x)}{\tau}.$$

As pointed out in Remark (11.81), if K is a nonempty, closed set satisfying condition (11.80), then $x \to d_K(x)$ is continuously differentiable in $I \setminus K$. Therefore, in this case, we have that

$$D_\uparrow d_K(x)(v) = \langle \text{grad } d_K(x), v \rangle,$$

for any $x \in I \setminus K$ and $v \in \mathbb{R}^n$. Furthermore, if K is any nonempty, closed set of $\mathbb{R}^n$ and $x_0 \in \mathbb{R}^n \setminus K$ is such that $x \to d_K(x)$ is differentiable at x_0, then $\pi_K(x_0)$ is a singleton (see [Fe]). Finally, since the distance function $x \to d_K(x)$, $x \in \mathbb{R}^n$, is Lipschitz of constant 1, from ([AuF-M, Proposition 6.1.7]) it follows that, for any $x \in \mathbb{R}^n$,

(i) $D_\uparrow d_K(x)(v) = \liminf_{\tau \to 0+} \dfrac{d_K(x + \tau v) - d_K(x)}{\tau}$, for any $v \in \mathbb{R}^n$.

(ii) $|D_\uparrow d_K(x)(v)| \leq |v|$, for any $v \in \mathbb{R}^n$.

In particular, from (ii) it follows that d_K is contingently epidifferentiable, i.e. $D_\uparrow d_K(x)(0) = 0$, at any $x \in \mathbb{R}^n$, or equivalently, $D_\uparrow d_K(x)(v) > -\infty$, at any $x \in \mathbb{R}^n$, whenever $v \in \mathbb{R}^n$. Finally, observe that if the epigraph of $d_K : E_p(d_K)$ is sleek with $\mathrm{Dom}\, d_K = D$, i.e. $x \multimap T_{E_p(d_K)}(x, d_K(x))$ is lower semicontinuous, at any $x \in D$, then the cone-valued map $x \multimap E_p(D_\uparrow d_K(x))$ is also lower semicontinuous with closed, convex values, at any $x \in D$; in fact, $T_{E_p(d_K)}(x, d_K(x)) = E_p(D_\uparrow d_K(x))$, see ([AuF-M, Proposition 6.1.4]).

The following two results are in the spirit of those of ([Au-M, Section 9.4.4]).

(11.94) PROPOSITION. *Assume that $K \subset \mathbb{R}^n$ is a nonempty, closed set which satisfies condition (11.80). Moreover, assume that, for γ sufficiently small and any $x \in I \setminus K$, there exists $\overline{v} \in \mathbb{R}^n$ such that $D_\uparrow d_K(x)(\overline{v}) < -\gamma$. Then the cone-valued map $x \multimap \widetilde{T}_{K,\gamma}(x)$, $x \in I \setminus K$, is lower semicontinuous with nonempty, closed, convex values.*

PROOF. For $x \in I \setminus K$, let us define the map
$$\widehat{T}_{K,\gamma}(x) := \{v \in \mathbb{R}^n \mid D_\uparrow d_K(x)(v) < -\gamma\}.$$
By the hypothesis, this set is nonempty. We show that the graph of $\widehat{T}_{K,\gamma}$: $\Gamma_{\widehat{T}_{K,\gamma}}$ is open, where $\mathrm{Dom}\,\widehat{T}_{K,\gamma} = I \setminus K$. For this, letting $(x, v) \in \Gamma_{\widehat{T}_{K,\gamma}}$, for any sequence $(x_n, v_n) \to (x, v)$, we have that
$$\lim_{n \to \infty} D_\uparrow d_K(x_n)(v_n) = D_\uparrow d_K(x)(v) < -\gamma.$$
Furthermore, the set $\widehat{T}_{K,\gamma}(x)$ is convex, because $D_\uparrow d_K(x)(v) = \langle \mathrm{grad}\, d_K(x), v \rangle$. Finally, consider the map
$$x \multimap \widetilde{T}_K(x) \cap \widehat{T}_{K,\gamma}(x) = \widehat{T}_{K,\gamma}(x), \quad x \in I \setminus K.$$
It is lower semicontinuous, because it is the intersection of the lower semicontinuous map $x \multimap \widetilde{T}_K(x)$ (Proposition (11.92)) with a map having an open graph. Furthermore, it is convex valued and so $x \multimap \widetilde{T}_{K,\gamma}(x) = \overline{\widehat{T}_{K,\gamma}(x)}$ is also lower semicontinuous. As a consequence, $x \multimap \widetilde{T}_{K,\gamma}(x)$ is lower semicontinuous, at any $x \in I \setminus K$. $\qquad \square$

Finally, we can also prove the following result, which provides the same conclusion of Proposition (11.94), under different assumptions.

(11.95) PROPOSITION. *Let $K \subset \mathbb{R}^n$ be a nonempty, closed set. Let I be a neighborhood of K. Assume that $E_p(d_K)$ is sleek with $\mathrm{Dom}\, d_K = I$, assume that, for $\gamma > 0$ sufficiently small and any $x \in I \setminus K$, there exists $\bar{v} \in \mathbb{R}^n$ such that $D_\uparrow d_K(x)(\bar{v}) < -\gamma$. Then the cone-valued map $x \multimap \widetilde{T}_{K,\gamma}(x)$, $x \in I \setminus K$, is lower semicontinuous with nonempty, closed, convex values.*

PROOF. Let $x_n \to x$ and $v \in \widetilde{T}_{K,\gamma}(x)$. Hence, $(v, -\gamma) \in E_p(D_\uparrow d_K(x))$. By the lower semicontinuity of the map $x \multimap E_p(D_\uparrow d_K(x))$, there exist sequences $v_n \to v$ and $\varepsilon_n \geq 0$, $\varepsilon_n \to 0$ such that

$$(v_n, -\gamma - \varepsilon_n) \in E_p(D_\uparrow d_K(x_n)).$$

Let $\bar{v} \in \widetilde{T}_{K,\gamma}(x)$. Hence, $\bar{u} = -\gamma - D_\uparrow d_K(x)(\bar{v}) > 0$, and so $(\bar{v}, -\gamma - \bar{u}) \in E_p(D_\uparrow d_K(x))$. Therefore, there exist sequences $\bar{v}_n \to \bar{v}$ and $\varepsilon'_n > 0$, $\varepsilon'_n \to 0$ such that

$$(\bar{v}_n, -\gamma - \bar{u} + \varepsilon'_n) \in E_p(D_\uparrow d_K(x_n)).$$

Define now $w_n = (1 - \theta_n)v_n + \theta_n \bar{v}_n$, where $\theta_n = \dfrac{\varepsilon_n}{2(\varepsilon_n + \varepsilon'_n)} \in [0,1]$. We have that

$$(w_n, -\gamma - \varepsilon_n/2) = (1 - \theta_n)(v_n, -\gamma - \varepsilon_n) + \theta_n(\bar{v}_n, -\gamma - \bar{u} + \varepsilon_n).$$

Since $E_p(D_\uparrow d_K(x_n))$ is convex, it follows that $(w_n, -\gamma - \varepsilon_n/2) \in E_p(D_\uparrow d_K(x_n))$, that is $D_\uparrow d_K(x_n)(v_n) \leq -\gamma - \varepsilon_n/2$, and so $w_n \in \widetilde{T}_{K,\gamma}(x_n)$ with $w_n \to v$. Moreover, $\widetilde{T}_{K,\gamma}(x)$ is convex. Indeed, the fact that $E_p(d_K)$ is sleek implies the upper semicontinuity of the map $(x, v) \multimap D_\uparrow d_K(x)(v)$, $x \in I \setminus K$ and $v \in \mathbb{R}^n$. Now, for $v_1, v_2 \in \widetilde{T}_{K,\gamma}(x)$, define $E = \{\theta \in [0,1] : D_\uparrow d_K(x)(\theta v_1 + (1 - \theta)v_2) < -\gamma\}$. It is nonempty by the hypothesis. Moreover, it is easy to verify that E is both closed and open relatively to $[0,1]$, and so $E = [0,1]$. $\qquad \square$

(11.96) REMARK. To our best knowledge (see also [GN3]), the characterization of the nonempty, closed sets K for which $E_p(d_K)$ is sleek is an open question.

Now, we consider the autonomous nonlinear control system (11.78). We assume the following conditions:

(11.97) $f : \bar{I} \times \mathbb{R}^m \to \mathbb{R}^n$ is a continuous map and I is an open set of $\mathbb{R}^n$. The control parameter u belongs to the set $U(x)$, depending on $x \in I$. The set-valued map $x \multimap U(x) \subset \mathbb{R}^m$ is upper semicontinuous with nonempty, compact, convex values.

We aim at solving the following control problem for system (11.78).

(11.98) FORMULATION OF THE PROBLEM. *Given a nonempty, closed set $K \subset \mathbb{R}^n$ satisfying condition (11.80). Given $x_0 \in I \setminus K$, with I a neighborhood of K assigned by condition (11.80). We want to show the existence of a trajectory*

$x = x(t)$, $t \geq 0$, $x(0) = x_0 \in I \setminus K$, *of the control system* (11.78) *corresponding to a control* $u(t) \in U(x(t))$, $t \geq 0$, *which reaches a prescribed neighborhood of* K *in finite time and remains in it for all future times.*

Observe that here problem (11.98) is formulated having in mind Propositions (11.92) and (11.94) as tools to solve it. Clearly, if one wants to use Proposition (11.95), then K and I are the sets satisfying the conditions of Proposition (11.95). Moreover, in this case, we also assume that K is sleek.

Consider the set of velocities

$$F(x) := f(x, U(x)), \quad x \in I.$$

It results that the map $x \multimap F(x)$, $x \in I$, is upper semicontinuous with nonempty, compact values. Furthermore, we assume the following condition.

(11.99) $F(x)$ is a convex set, for any $x \in I$.

(11.100) DEFINITION. For $x \in I \setminus K$, we define the *regulation map* associated to (11.78) as follows

$$R_{K,\gamma}(x) := F(x) \cap \widetilde{T}_{K,\gamma}(x).$$

We assume that

(11.101) $R_{K,\gamma}(x)$ is nonempty, for any $x \in I \setminus K$ and for $\gamma > 0$ sufficiently small.

(11.102) PROPOSITION. *Under assumptions* (11.97), (11.99), (11.101) *and the assumptions of Proposition* (11.94), *for any* $\delta > 0$, *there exists a continuous map* $g : I \to \mathbb{R}^n$ *such that*

(i) $g(x) \in \widetilde{T}_{K,\gamma}(x)$, *for any* $x \in I \setminus K$;

(ii) g *is a* δ-*approximation in the graph of the map* F. *That is* $\Gamma_{(g+\delta B_1)} \subset \Gamma_F + \delta B_1$.

PROOF. Under our assumptions, the regulation map $x \multimap R_{K,\gamma}(x)$, $x \in I \setminus K$, has nonempty, compact, convex values. Thus, the proof is a direct consequence of Lemma 5.1 and Remark 5.2 in [BK2]. □

The following result provides the relevant behaviour of the trajectories of the dynamical system $\dot{x} = g(x)$.

(11.103) THEOREM. *Under the assumptions of Proposition* (11.102), *any solution of the initial value problem*

(11.104) $$\begin{cases} \dot{x} = g(x), \\ x(0) = x_0, \quad x_0 \in I \setminus K \end{cases}$$

reaches the set K in a finite time $t_0 > 0$ and remains in it, for all $t \geq t_0$. Moreover,
$t_0 \leq d_K(x_0)/\gamma.$

PROOF. Let $x = x(t)$ be any solution of the initial value problem (11.104). Let $t \geq 0$ and $d(t) := d_K(x(t))$. Observe that $t \to d_K(x(t))$ is an absolutely continuous function. For $\tau > 0$, we can write $x(t+\tau) = x(t) + \tau \dot{x}(t) + \tau \alpha(\tau)$, where $\alpha(\tau) \to 0$ as $\tau \to 0$. We have that

$$\dot{d}(t) = \lim_{\tau \to 0+} \frac{d_K(x(t+\tau)) - d_K(x(t))}{\tau}$$

$$= \lim_{\tau \to 0+} \frac{d_K(x(t) + \tau \dot{x}(t) + \tau \alpha(\tau)) - d_K(x(t))}{\tau} \leq -\gamma.$$

Since $\dot{x}(t) = g(x(t)) \in \widetilde{T}_{K,\gamma}(x(t))$. Thus, for $t \geq 0$, such that $x(t) \in I \setminus K$, we have proved, that $\dot{d}(t) \leq -\gamma$, namely

$$d(t) \leq d(0) - \gamma t = d_K(x_0) - \gamma t.$$

Let $t_0 > 0$ such that $d(t_0) = 0$ and $d(t) > 0$, for all $0 \leq t < t_0$, then $t_0 \leq \dfrac{d_K(x_0)}{\gamma}$. $\square$

As a straightforward consequence of ([Fi2-M, Theorem 1]) and ([AuF-M, Theorem 8.2.10]), we can derive the following

(11.105) THEOREM. *For any $\varepsilon > 0$, there exists $\delta > 0$ such that if $g : I \setminus K \to \mathbb{R}^n$ is a continuous δ-approximation in the graph of the multivalued map F, then, for any solution $y = y(t)$, $t \geq 0$, of the initial-value problem (11.104), there exists a solution $x = x(t)$, $t \geq 0$, of the initial-value problem*

$$\begin{cases} \dot{x} \in F(x), \\ x(0) = x_0 \end{cases}$$

such that, for any time interval $[0, a]$, we have

$$\max_{t \in [0,a]} |x(t) - y(t)| \leq \varepsilon.$$

Moreover, there exists a measurable control $u(t) \in U(x(t))$, such that $\dot{x}(t) = f(x(t), u(t))$, for a.a. $t \geq 0$.

It is now evident that, as a consequence of Theorems (11.103) and (11.105), the trajectory $x = x(t)$, $t \geq 0$, reaches in a finite time the ε-neighborhood of K and remains in it for all the future times. Hence, problem (11.98) is solved.

(11.106) REMARK. If $x \multimap F(x)$ is lower semicontinuous, then the Michael theorem (3.30) in Chapter I.3 ensures the existence of a continuous selection $g(x)$

of the regulation map $x \multimap R_{K,\gamma}(x)$, $x \in I \setminus K$, and so we can take $x(t) = y(t)$, $t \geq 0$. Note that, under our assumptions, $x \multimap F(x)$ is lower semicontinuous if so is $x \multimap U(x)$.

(11.107) REMARK. Observe that if, in Theorem (11.103), we assume the conditions of Proposition (11.92), then by means of the above arguments we can show the existence of a trajectory $x = x(t)$, $t \geq 0$, of the control system which does not leave a prescribed ε-neighborhood of I. Finally, observe that the condition $F(x) \subset \tilde{T}_{K,\gamma}(x)$, or $F(x) \subset \tilde{T}_K(x)$, allows us to prove the same behaviour for all the trajectories of the control system.

(11.108) REMARK. Both feedback control problems considered above can be reformulated in a more general setting on proximate retracts in Hilbert spaces (see [GN4]).

Finally, we will show how fractals can be generated by means of the Poincaré translation operators of differential equations, on the basis of the existence theorems developed in Appendix 3.

Consider the not necessarily coupled system of ordinary differential equations

$$(11.109) \qquad \dot{X} = F_i(t, X), \quad i = 1, \ldots, n,$$

where $F_i : [t_0, \infty) \times \mathbb{R} \to \mathbb{R}$ are Carathéodory maps, i.e.

  (i) $F_i(\cdot, X) : [t_0, \infty) \to \mathbb{R}^n$ are measurable, for every $X \in \mathbb{R}^n$,
  (ii) $F_i(t, \cdot) : \mathbb{R}^n \to \mathbb{R}^n$ are continuous, for a.a. $t \in [t_0, \infty)$,
  (iii) $|F_i(t, X)| \leq C_i$, for a.a. $t \geq t_0$ and $|X| \leq D_i$, where $C_i = C_i(D_i)$ and D_i are nonnegative constants.

Assuming the global existence and uniqueness of Carathéodory solutions $X(t) \in AC_{\mathrm{loc}}([t_0, \infty), \mathbb{R}^n)$ of (11.109), it is well-known (see e.g. [Ks2-M] and Theorem (4.3) in Chapter III.4) that the (well-defined) associated Poincaré translation operators

$$(11.110) \quad {}_iT_\omega(X_0) := \{X_i(\omega, t_0, X_0) \mid X_i(t, t_0, X_0) \text{ is a solution of (11.109)}$$
$$\text{with } X_i(t_0, t_0, X_0) = X_0\}$$

are continuous.

Furthermore, if all solutions $X_i(t)$ of the systems in (11.109) are, for every $i = 1, \ldots, n$, uniformly ultimately bounded, i.e.
(11.111)
$$\forall B_1 > 0 \; \exists K > 0 : [t_0 \in \mathbb{R}, \; |X_0| < B_1, \; t \geq t_0 + K] \Rightarrow |X_i(t, t_0, X_0)| < B_2,$$

where $B_2 > 0$ is a common constant, for all $B_1 > 0$, $i = 1, \ldots, n$, then ${}_iT_\omega$ in (11.110) become self-maps, because a ball $B \subset \mathbb{R}^n$ and a number $\omega^* > 0$

certainly exist such that $_iT_\omega(B) \subset B$, for $\omega \geq \omega^*$. Moreover, $\overline{_iT_\omega(B)}$ is compact (bounded and closed in $\mathbb{R}^n$).

Let us note that, according to the result of N. Pavel [Pav], $(11.111) \Rightarrow (11.112)$, provided $F_i(t, X) \equiv F_i(t + \overline{\omega}, X)$, for some $\overline{\omega} > 0$, $i = 1, \ldots, n$, where

$$(11.112) \quad \forall B_3 > 0 \; \exists B_4 > 0 : [t_0 \in \mathbb{R}, \; |X_0| < B_3, \; t \geq t_0] \Rightarrow |X_i(t, t_0, X_0)| < B_4.$$

In fact, for periodic systems, the sole ultimate boundedness implies (11.112), and subsequently (11.111) (see [Pav]).

Conditions (11.111), (11.112) can be satisfied e.g. by means of locally Lipschitzian (in all variables) Liapunov functions $_iV(t, X)$, defined for $t \geq t_0$, $|X| \geq R$, where $R > 0$ can be large, such that (see the arguments in Chapter III.8, especially the proof of Theorem (8.193))

$$(11.113) \qquad {_iW_1}(|X|) \leq {_iV}(t, X) \leq {_iW_2}(|X|),$$

$$(11.114) \; {_i\dot{V}_{(11.109)}}(t, X) := \limsup_{h \to 0+} \frac{1}{h}[{_iV}(t + h, X + hF_i(t, X)) - {_iV}(t, X)]$$
$$\leq -{_iW_3}(|X|),$$

where the wedges $_iW_1(r)$, $_iW_2(r)$ are continuous increasing functions such that $_iW_1(r) \to \infty$, whenever $r \to \infty$, and $_iW_3(r)$ are positive continuous functions, $i = 1, \ldots, n$.

Defining the related Hutchinson–Barnsley map (see (A3.15))

$$(11.115) \qquad F(X_0) := \bigcup_{i=1}^{n} {_iT_\omega}(X_0), \quad X_0 \in B \subset \mathbb{R}^n, \; \omega \geq \omega^*,$$

and the Hutchinson–Barnsley operator (see (A3.16))

$$(11.116) \qquad F^*(A) := \bigcup_{X_0 \in A} F(X_0), \quad A \in C(B),$$

where $C(B)$ denotes the hyperspace of compact subsets of B, endowed with the Hausdorff metric d_H, we can apply Theorem (A3.52) (see also Remark (A3.53)) in Appendix 3, for obtaining

(11.117) THEOREM. *Assume the existence of locally Lipschitzian (in all variables) Liapunov functions $_iV(t, X)$, $i = 1, \ldots, n$, defined for $t \geq t_0$, $|X| \geq R$, where R can be large, such that conditions (11.113), (11.114) are satisfied for (11.109). Then the Hutchinson–Barnsley operator F^* in (11.116) admits a fixed point $A^* \in (C(X), d_H)$, representing a compact invariant subset of $\mathcal{B} = \{X \in \mathbb{R}^n \mid |X| \leq B = B(\omega^*)$, a sufficiently large number$\}$, under the Hutchinson–Barnsley* *map F in (11.115), called a fractal. Moreover, the (maximal) fractal A^* can be obtained as $\bigcap_{n \in \mathbb{N}} \overline{\bigcup_{m \geq n} \bigcup_{x \in \mathcal{B}} F^m(x)}$.*

(11.118) REMARK. Analogical statements can be formulated for not necessarily coupled systems of functional or random differential equations, because compact subinvariant subsets of the related phase-spaces w.r.t. them exist, provided the right-hand sides are linearly bounded in the space variables, such that the associated Poincaré translation operators are again compact (continuous) self-maps on these subsets (see Chapter III.4).

(11.119) EXAMPLE. Consider the systems

(11.120)
$$\begin{cases} \dot{x} = -2x - 2y + a & (a \in \mathbb{R}), \\ \dot{y} = 2x - 2y + b & (b \in \mathbb{R}), \end{cases}$$

and

(11.121)
$$\begin{cases} \dot{x} = -x - y - xy^2 - x^3, \\ \dot{y} = x - y - x^2y - y^3. \end{cases}$$

The general solution of the linear system (11.120) takes the form $X(t) = (x(t), y(t))$, where

(11.122)
$$\begin{cases} x(t) = e^{-2t}(C_1 \cos 2t - C_2 \sin 2t) + (a - b)/4, \\ y(t) = e^{-2t}(C_1 \sin 2t - C_2 \cos 2t) + (a + b)/4. \end{cases}$$

One can readily check that all solutions of (11.120) tend (uniformly) to the equilibrium point $((a - b)/4, (a + b)/4)$ which is so globally asymptotically stable. Moreover, according to the well-known converse theorem due to A. M. Liapunov (cf. [Yos1-M, Theorem 18.2]), a positive definite quadratic form $V(X) = V(x, y)$ exists such that conditions (11.113), (11.114) hold w.r.t. (11.120), for all $a, b \in \mathbb{R}$.

Furthermore, defining $V(X) = V(x, y) := x^2 + y^2$, we obtain

$$\begin{aligned} \dot{V}_{(11.121)}(x, y) &= 2x(-x - y - xy^2 - x^3) + 2y(x - y - x^2y - y^3) \\ &= 2(-x^2 - xy - x^2y^2 - x^4 + xy - y^2 - x^2y^2 - y^4) \\ &= -2(x^2 + y^2 + (x^4 + 2x^2y^2 + y^4)) = -2(x^2 + y^2 + (x^2 + y^2)^2), \end{aligned}$$

and subsequently

$$\dot{V}_{(11.121)}(x, y) < 0, \quad \text{for every } x, y \neq (0, 0), \text{ as well as } \dot{V}_{(11.121)}(0, 0) = 0.$$

Observe also that, on any circle, centered at the origin and having the radius $r > 0$, i.e. $x^2 + y^2 = r^2$, the time derivative

(11.122) $\dot{V}_{(11.121)}(x, y) = -2(x^2 + y^2 + (x^2 + y^2)^2) = -2(r^2 + r^4)$

is constant, by which the ball $\{(x, y) \in \mathbb{R}^2 \mid x^2 + y^2 \le r^2\}$ is subinvariant, under the associated Poincaré translation operator (11.110).

In any case, conditions (11.113), (11.114) are satisfied w.r.t. (11.121), too.

Hence, we can define the following Poincaré translation operators (cf. (11.110)):

$$(11.123) \quad \begin{cases} {}_1T_\omega(x_0, y_0) := \{e^{-2\omega}[(x_0 - 1)\cos 2\omega - y_0 \sin 2\omega] + 1, \\ \qquad e^{-2\omega}[(x_0 - 1)\sin 2\omega + y_0 \cos 2\omega]\}, \\ \qquad \text{for } t_0 = 0 \text{ and } a = +2, b = -2 \text{ in (11.120)}, \\ {}_2T_\omega(x_0, y_0) := \{e^{-2\omega}[(x_0 + 1)\cos 2\omega - y_0 \sin 2\omega] - 1, \\ \qquad e^{-2\omega}[(x_0 + 1)\sin 2\omega + y_0 \cos 2\omega]\}, \\ \qquad \text{for } t_0 = 0 \text{ and } a = -2, b = +2 \text{ in (11.120)}, \\ {}_3T_\omega(x_0, y_0) := \{e^{-2\omega}[x_0 \cos 2\omega - (y_0 - 1)\sin 2\omega], \\ \qquad e^{-2\omega}[x_0 \sin 2\omega + (y_0 - 1)\cos 2\omega] + 1\}, \\ \qquad \text{for } t_0 = 0 \text{ and } a = +2, b = +2 \text{ in (11.120)}, \\ {}_4T_\omega(x_0, y_0) := \{e^{-2\omega}[x_0 \cos 2\omega - (y_0 + 1)\sin 2\omega], \\ \qquad e^{-2\omega}[x_0 \sin 2\omega + (y_0 + 1)\cos 2\omega] - 1\}, \\ \qquad \text{for } t_0 = 0 \text{ and } a = -2, b = -2 \text{ in (11.120)}, \end{cases}$$

$$(11.124) \qquad {}_5T_\omega(x_0, y_0) := \{x(\omega, 0, (x_0, y_0)), y(\omega, 0, (x_0, y_0))\},$$

where $[x(t, 0, (x_0, y_0)), y(t, 0, (x_0, y_0))]$ is the solution of (11.121) with $[x(0, 0, (x_0, y_0)), y(0, 0, (x_0, y_0))] = (x_0, y_0)$.

For ${}_iT_\omega$, $i = 1, 2, 3, 4$, we have:

$$d({}_iT_\omega(x_1, y_1), {}_iT_\omega(x_2, y_2))$$
$$= \{[e^{-2\omega}(x_2 \cos 2\omega - y_2 \sin 2\omega) - e^{-2\omega}(x_1 \cos 2\omega - y_1 \sin 2\omega)]^2$$
$$\quad + [e^{-2\omega}(x_2 \sin 2\omega + y_2 \cos 2\omega) - e^{-2\omega}(x_1 \sin 2\omega + y_1 \cos 2\omega)]^2\}^{1/2}$$
$$= \{[e^{-2\omega}(x_2 \cos 2\omega - y_2 \sin 2\omega - x_1 \cos 2\omega + y_1 \sin 2\omega)]^2$$
$$\quad + [e^{-2\omega}(x_2 \sin 2\omega + y_2 \cos 2\omega - x_1 \sin 2\omega - y_1 \cos 2\omega)]^2\}^{1/2}$$
$$= \{[e^{-2\omega}((x_2 - x_1)\cos 2\omega + (y_1 - y_2)\sin 2\omega)]^2$$
$$\quad + [e^{-2\omega}((x_2 - x_1)\sin 2\omega - (y_1 - y_2)\cos 2\omega)]^2\}^{1/2}$$
$$= e^{-2\omega}[(x_2 - x_1)^2 \cos^2 2\omega + 2(x_2 - x_1)(y_1 - y_2)\cos 2\omega \sin 2\omega + (y_1 - y_2)^2 \sin^2 2\omega$$
$$\quad + (x_2 - x_1)^2 \sin^2 2\omega - 2(x_2 - x_1)(y_1 - y_2)\sin 2\omega \cos 2\omega + (y_1 - y_2)^2 \cos^2 2\omega]^{1/2}$$
$$= e^{-2\omega}[(x_2 - x_1)^2(\cos^2 2\omega + \sin^2 2\omega) + (y_1 - y_2)^2(\sin^2 2\omega + \cos^2 2\omega)]^{1/2}$$
$$= e^{-2\omega}[(x_2 - x_1)^2 + (y_1 - y_2)^2]^{1/2} = e^{-2\omega}d((x_1, y_1), (x_2, y_2)).$$

Thus, $_iT_\omega$, where $i = 1, 2, 3, 4$, are contractions, because $e^{-2\omega} \in (0, 1)$, whenever $\omega > 0$.

Therefore, we can take (cf. also (11.122) any $\omega > 0$, e.g. $\omega = \frac{1}{2}$, in $_iT_\omega$. Let us also note that the values of the operator $_5T_{\frac{1}{2}}(x_0, y_0)$ in (11.124) can be simulated numerically.

The numerical approximation of the fractal obtained, on the basis of Theorem (11.117), as the fixed point of F^* in (11.116), is plotted in Figure 1.

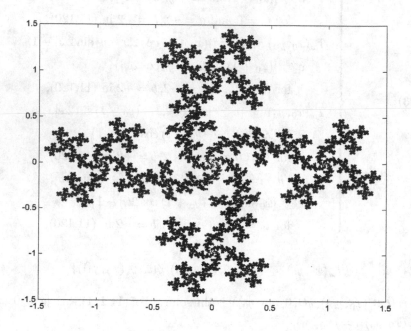

FIGURE 1. Invariant subset of F in (11.115) for (11.123), (11.124).

(11.125) REMARK. Since the translation operators $_iT_\omega$, in (11.123) $i = 1, 2, 3, 4$, of the sole system (11.120) were shown to be contractions, for any $\omega > 0$, one can directly apply Corollary (A3.28) in Appendix 3, instead of Theorem (11.117).

The resulting object, i.e. the fixed point of F^* in (11.117), is plotted in Figure 2.

III.12. Remarks and comments

III.1.

As pointed out, our continuation principles for differential inclusions are taken from [AGG1] (in Euclidean spaces) and from [AndBa] (in Banach spaces). Theorem (1.25), allowing us to make a lower estimate of the number of solutions of boundary value problems, is taken from [AGJ2] (cf. also [And6], [And19]).

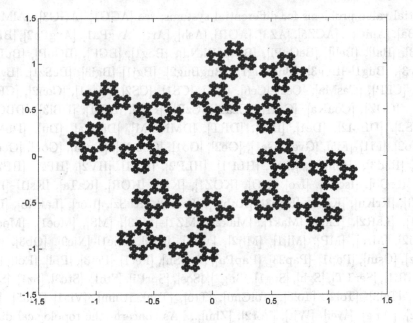

FIGURE 2. Invariant subset of F in (11.115) for (11.123).

Propositions (1.36) and (1.37), dealing with entirely bounded solutions, obtained sequentially, are new, but intuitively clear (cf. Lemma 2 in [AMT2]).

For alternative continuation principles for differential equations, see e.g. the books by M. A. Krasnosel'skiĭ et al. [Ks1-M], [Ks2-M], [KsZa-M], J. Mawhin et al. [GaM-M], [Maw-M], [MW-M] and some others [AGN-M], [Fr-M], [FrGr-M], [FZ-M], [Fuc-M], [Gr1-M], [Gr2-M], [GG-M], [GGL-M], [KOZ-M], [KrwWu-M], [Ptr-M], [ReSaCo-M].

III.2.

W. Orlicz [Orl] showed in 1932 that the generic continuous (single-valued) Cauchy problem is solvable (according to the Peano theorem [Pea1]) in a unique way. On the other hand, H. Kneser [Kne2] proved already in 1923 that the sets of solutions to Cauchy problems for continuous systems are, at every time, continua (i.e. compact and connected). His result was improved later by M. Hukuhara [Hu1] who proved that the solution set itself is a continuum in $C(I, \mathbb{R}^n)$, where I is a compact interval. N. Aroszajn [Aro] specified in 1942 that these continua are R_δ-sets. The same is true for u-Carathéodory systems of ordinary differential inclusions by the result of F. S. De Blasi and J. Myjak [DBM2] even on an arbitrary (e.g. infinite) interval [AGG1].

Now, there is a vaste literature about the topological structure of solution sets to initial value problems for differential systems; see [AGG1]–[AGG3], [AlMaPe], [AngBa], [AniCo], [ACZ5], [AZ3], [AOR], [Ash], [Art], [AvgPa1], [AvgPa2], [BKr1]–[BKr3], [Bal], [Beb], [BeGoPl], [BP], [BhNd], [Bog1], [BCF], [BGoP], [BuLy1]–[BuLy3], [Bug1]–[Bug3], [BugBug1], [BugBug2], [Bgj1]–[Bgj3], [BgSz1], [BgSz2], [Car], [CFP], [CasMa], [Ce2]–[Ce6], [CK], [CS1], [CS2], [Con1], [Con2], [CFRS], [CKZ], [COZ], [CouKa], [Cza1], [Cza2], [CzP], [Dav], [DBl1], [DBl3], [DBGP2], [DBPS2], [DBM2], [De1], [De5], [DRT], [DMNZ-M], [DiaWa], [Du], [DbMo1], [DbMo2], [Fi1]–[Fi5], [Gve], [Ge1], [Ge2], [Go1], [Go2-M], [Go15], [Gon], [GoeRz], [GPl], [Had3], [HlmHe], [Heu], [HLP1], [HLP2], [HV1], [HV2], [HP2], [HP2-M], [Hy], [IbrGo], [IscYu], [Izo], [KO], [KOZ1], [Kar], [KOR], [KaTa], [Ksl1]–[Ksl5], [KlbFi], [KrKu], [Kub1]–[Kub3], [Kubi1], [Kubi2], [KbSz], [Kor], [LakNo], [LaY], [LaR1], [LaR2], [Lim], [Mak1], [Mak2], [MZ1]–[MZ3], [MS], [Moe1], [MoeHa], [MaVi2], [Mrk], [MP], [Mlj1], [Mlj2], [Muh], [Nas], [Nie1]–[Nie9], [Ob3], [Ol5], [OrlSz], [Osm], [Pap1]–[Pap23], [PapPa], [PapSh], [Pet1]–[Pet3], [Ppl], [Pol], [Pia], [Pl2], [Ri1], [SerTT], [Sgh], [Sng1], [Sng2], [Sos], [SprBi], [Stc1]–[Stc3], [Sz1]–[Sz19], [Tal1], [Tal2], [Tol2], [Tol5], [TolChu], [Tto], [Tua], [Umn], [Via1]–[Via5], [Vel], [Waz2], [Wng], [Wei], [Wjt], [Yor2], [Zhu],... As concerns the topological dimension of the solution sets, see [BGKO], [DouOe], [DzGe], [FiMaPe], [Ge5]–[Ge7], [Mlj1], [Mlj2], [Ri6], [Ri7], [Smth].

III.3.

Unlike for initial value problems, there are only several results concerning the topological structure of solutiuon sets to boundary value problems; see [Akh], [And10], [And14], [AGG2], [AGG3], [ADG], [BeM], [BaF2], [BeGoPr], [BP], [DBP5], [FPr3], [GPr], [Kmn], [Ke3], [MZ1]–[MZ2], [Mik], [Nie4], [Nie6], [Per], [Rss], [Se3], [Zvy]. Hence, our presentation is the first collection in this field. In [ADG], there is also the information about the topological dimension of the solution sets to BVPs.

III.4.

Poincaré's idea of the translation operator along the trajectories of differential systems comes back to the end of the nineteenth century [Poi]. Since it was effectively applied for investigating periodic orbits by A. A. Andronov [Adv] in the late 20's and by N. Levinson [Lev] in 1944, its name is sometimes also related to them. This topic became popular due to the monographs [Ks2-M], [Pl-M] of M. A. Krasnosel'skiĭ and V. A. Pliss, dealing with ODEs, and [Hal2-M] of J. K. Hale, dealing with functional differential equations.

It is difficult to recognize when the idea of a multivalued translation operator appeared for the first time. Perhaps in Remark 12 of [LaR3] saying that "*La multi-application $\{x_0 \to x(t) \mid x \in \mathcal{T}, x(0) = x_0\}$ est un example naturel de multi-application pseudo-acyclique*", where $\mathcal{T}$ denotes the set of solutions of a Carathéodory system of inclusions. In Chapter III of [De2], entitled *Existence* *without uniqueness*, where [LaR3] is quoted, the multivalued translation operator was proved to be pseudoacyclic (= admissible). The systematic study of admissible maps in [Go1-M] allowed us to obtain deeper results in this field. Since 1976, i.e. the year of publication of both [LaR3] and [Go1-M], Poincaré's operator has been treated and applied many times on various levels of abstraction; see e.g. [And3], [And8], [And18], [And24], [AGZ], [AGJ1], [AGJ3], [AGJ4], [AGL], [AJ1], [Bad2], [BKr2], [BGoP1], [DBGP2], [DyG], [DyJ], [Go16], [Go5-M], [GP1], [HL], [HP3], [KOZ2], [Lew-M], [ObZe], [Pl2], [Zan2].

III.5.

Although there is an enormous amount of existence results for (even multivalued) boundary value problems, those obtained by means of the solution sets (i.e. when using the Schauder (quasi) linearization or, more precisely, parametrization) are rather rich; see [And5], [And14], [And19], [And27], [AndBa], [AndBe], [AGG1], [AGG2], [AMT1], [AMT2], [ACZ5], [BeM], [DMNZ-M], [MZ1], [MZ2], and the references therein.

The history of boundary value problems on infinite intervals, which we especially focussed on, starts at the end of the nineteenth century with the pioneering work of H. Kneser [Kne1] about monotone solutions and their derivatives on $[0, \infty)$ for second-order ordinary differential equations. The Kneser-type results were then followed by A. Mambriani [Mam] in 1929 and others from the beginning of the 50's untill now (see e.g. [Gr], [HW], [Wo], [Sa], [Be], [Se3], [KR], [KS-M], [R1]–[R3]).

At the beginning of the 50's the study of bounded solutions via BVPs was initiated by C. Corduneanu [Co2], [Co3], who considered second-order BVPs on the positive ray as well as on the whole real line. Since the 60's similar problems have been studied using mostly the lower and upper solutions technique (see e.g. [Avr], [Be1], [BeJ], [FJ], [MalMa], [Scr2], [Scr3]).

Since the beginning of the 70's BVPs on infinite interwals have been studied systematically; see [Ab], [Ag-M], [Ah1], [And1], [And2], [And5]–[And9], [And11]–[And13], [And19], [AndBa], [AGL], [AGG1]–[AGG3], [AK1], [AK2], [AndS], [AulVM4], [AMT2], [AMP], [AT1], [AV], [ACZ5], [AS], [AZ3], [Avr], [Bax], [BeJ], [Bel], [Bess], [Bo], [CFM1], [CFM2], [CFMP], [CMZ1]–[CMZ4], [Ci], [Co1]–[Co5], [CP], [DR], [Dnc], [DK-M], [FonZa], [FJ], [Fri1], [FrP1], [FrP2], [GGL0], [Gr], [Gud], [HW], [Hy], [I], [Ka1]–[Ka6], [Krp], [KMP], [KMKP], [Kra1], [Kra2], [Kub1]–[Kub3], [Ke1]–[Ke3], [Ki], [Kry7], [KR], [KS-M], [Ks2-M], [Lia], [Lo], [MN], [Mar2], [MW1], [MW2], [Mal], [MalMa], [Maw6]–[Maw8], [Ort], [ORe1]–[ORe3], [OrtTi], [PG], [Pn-M], [P1]–[P3], [R1]–[R3], [Rz], [Sac], [Scr1], [Scr2], [Se1]–[Se3], [Sr2], [Scu1], [Scu2], [Spe], [St1], [St2], [Sv1], [Sv2], [Sz3], [U], [UV], [Wat], [Wa1]–[Wa8], [Wo], [ZZ],...

III.6.

For multiplicity results, the additivity properties of the topological degree or the fixed point index have been usually employed. The application of the relative Lefschetz number here is in the same spirit. Sometimes also various versions of the Poincaré–Birkhoff fixed point theorem (the last Poincaré's geometric theorem) (cf. e.g. [LeCal-M], [McGMe-M]), possessing at least two given solutions, or the Leggett–Williams fixed point theorem [LegWi], possessing at least three given solutions were applied. There are, however, not so many multiple fixed point theorems like those, apart from the Nielsen theory.

The multiplicity results for differential equations (inclusions), obtained by means of the Nielsen number, are related to the problem of Jean Leray [Le] posed at the first International Congress of Mathematics hold after World War II in Cambridge, Mass., in 1950. Namely, he suggested the problem of adapting the Nielsen theory to the needs of nonlinear analysis and, in particular, of *its application to differential systems for obtaining multiplicity results*. Since that time only several papers have been devoted to this problem; see [And4], [And6], [And18], [And19], [And25], [And26], [AG2], [AGJ1]–[AGJ4], [Ji1], [BoKuMa1], [BoKuMa2], [Brw3]–[Brw5], [BrwZe], [Fec1]–[Fec5], [McC-M], [Mat1]–[Mat3], [Wjc].

III.7.

T. Ważewski formulated his classical principle [Waz1] in 1947. This principle was generalized several times in the sense that either the uniqueness (see e.g. [BeKe], [BeSchu], [Crd1], [Crd2], [JacKla]) or the transversality restrictions on the boundary of given domains (see e.g. [And13], [AGL], [AGZ]) were relaxed. Let us note that the notion of the Conley index (see e.g. [Co-M], [Ryb-M], [Sm-M]) is strictly related and allows us to avoid the transversality restrictions as well. For some further applications of the Ważewski-type methods, see e.g. [And27], [KLY], [Ryb], [Sr1], [Sr5], [Sr6].

Let us also note that a part of our results, based on [AGL], [AGZ], can be regarded as a generalization of the transformation theory of N. Levinson [Lev], because (partial) dissipativity implies there the existence of periodic solutions. For further related results from this point of view, see [And2], [CapGa], [HlLaSl], [HlLo], [Hof], [Sel2], [Yos2-M], [Zan1].

III.8.

The second (direct) method of A. M. Liapunov, based on the construction of Liapunov functions, appeared as a part of his thesis, under the leadership of P. L. Chebyshev, in 1892. The systemmatic application of Liapunov functions, sometimes called according to the contexts as guiding functions, started from the 60's (cf. [Ks2-M], [KsZa-M], [Yos1-M], [Yos2-M]). The related bound sets theory, based on constructing the bounding functions, was initiated a bit later by R. E. Gaines and J. Mawhin [GaM-M]. The further development in both directions was mainly related either to the usage of smooth Liapunov functions or to smooth solutions; for the exceptions, see e.g. [And3], [And14], [AGG1], [AMT1], [AMT2], [Au-M], [AuC-M], [BCM], [ClkLeS], [ClkStWo], [DBGP2], [Del-M], [KMP], [KKM], [KMKP], [Lew], [MrsMi], [Maw5], [MW2], [Rox], [Tad], [Yos1-M], [Yos2-M], [Zan3].

III.9.

Although A. N. Sharkovskiĭ published his cycle coexisting theorem [Sh1] already in 1964, it became curiously well-known after appearing the paper [LiYo] of T.-Y. Li and J. A. Yorke, dealing with its particular case (implications of period 3), more than ten years after. Since that time the Sharkovskiĭ theorem was several times reproved in a modern way and generalized in various directions (cf. [ABLM-M], [Rob-M], [Bog], [Dem], [Ye], and the references therein). The real line was replaced by linear continua in [ABLM-M], [AndPa], [Bae], [BGM4], [Sch8]. Multidemensional versions were studied in [And31], [AndPa], [ABLM-M], [FrnMi], [Gua1], [Gua2], [GST], [Han], [Kam], [KSS], [Klo], [Kol], [LlMcK], [Mat6], [Zgl1], [Zgl2].

Multivalued versions, allowing applications to differential equations and inclusions, were developed in [AFJ], [AJ2], [AJP], [AndPa]. Let us note that the only reference of the other authors related to applications to differential equations is the paper [ZhWu], written unfortunately in Chinese. The straightforward implications of the Sharkovskiĭ-type theorems in terms of the deterministic chaos theory can be done by means of the standard aguments (cf. [ASY-M]).

The Artin braid group theory (cf. [Atn], [Bi-M], [BiLi-M], [Han-M], [Mor-M], [MurKu]) was applied for obtaining infinitely many subharmonics in [And18], [Ji1], [HuaJa], [Mat1]–[Mat3]. For further applications of the theory of braids to differential equations, see e.g. [BW1], [BW2], [Hlm], [HlmW].

III.10.

The first paper dealing with a special sort of quasi-periodic solutions of differential equations in Euclidean spaces is due to P. Bohr [Bhr], whose method however works also for almost periodic solutions. P. Bohl published his paper in 1906. Another paper dealing with quasi-periodic solutions is [Esc] published in 1915 by E. Esclangon. In 1926, H. Bohr and O. Neugebauer [BohNe] considered already (uniformly) almost periodically forced linear differential equations with constant coefficients and showed that all bounded solutions are also (uniformly) almost periodic. Their result was later extended by C. Corduneanu [Cor-M] to linear systems with constant coefficients and (uniformly) almost periodic forcing. The Bohr–Neugebauer-type results (boundedness implies almost-periodicity) can be extended to Banach spaces (cf. [BstTs]) and holds also for almost periodic functions in other metrics (cf. [Rad]).

For nonhomogeneous linear systems with almost-periodic coefficients and forcing, the Bohr–Neugebauer-like aguments fail (cf. e.g. [Yos2-M] and the references therein) The adequate method is related to the name of J. Favard (cf. [Fav-M], [ZhLe]), which originates from the paper [Fav] published in 1921. Although the almost-periodic analogy of the Floquet theory for homogeneous linear systems is impossible in the sense that the fundamental matrix should be expressed in the form of an almost-periodic matrix times $\exp(Rt)$, where R is a constant, regular, square matrix, there are some variants of the Floquet theory like "almost-reducibility" results, those related to the Floquet exponents, various spectra, etc. For some related results, see e.g. [GchJo], [Joh1], [Joh6], [Joh7], [Mil], [SckSe1]–[SckSe4], and the references therein.

For nonlinear systems, the theory is still much more delicate. Although the appopriate Levinson transformation theory (ultimate boundedness implies almost periodicity) fails here, according to [FinFr], the Massera-type result (existence of a bounded solution implies the existence of an almost periodic solution) was recently obtained in [OrTa]. Obviously, the almost-periodicity was understood there in a rather general way and, moreover, solutions are allowed to be discontinuous. The results concerning non-uniformly almost-periodic solutions are quite rare, see [And5], [And19], [AndBe], [AndBeL], [BelFoSD1]–[BelFoSD3], [BelFoHa], [LZ-M], [Pn-M], [ZL], [Zha]. Almost-periodic solutions of differential inclusions are treated in [And5], [And19], [AndBe], [AndBeL]. A very powerful method is due to L. Amerio (cf. [Ame2], [AP-M]).

For some further results about almost-periodic oscillations, see [Ah1], [AhTi], [And20], [AK2], [AriBa], [AulVM1]–[AulVM4], [Ava], [Bas4], [Bir], [BrgCh1], [BrgCh2], [BloMaCi], [Blo1], [Blo2], [Cam], [Cop], [Cor4]–[Cor7], [Co-M], [De-HeSh], [Fin-M], [Fin1], [Fin2], [Fis2], [FabJoPa], [FSW], [Har], [Joh2]–[Joh5], [JohMo], [KsBuKo-M], [Kha-M], [Lil], [NG-M], [Pal1]–[Pal3], [Sei3], [Sei4], [Sel1], [SckSe-M], [SchtWa], [SheYi-M], [Wa8], [Zai-M].

III.11.

Besides studying implicit differential equations by means of explicit differential inclusions as we consider here (cf. also [AGG1], [BiG1], [BiG2], [FrKaKr], [FGK], [FK], [Ric3]), there are also employed for the same goal the properties of k-set contractions, condensing maps, A-proper maps, or monotone maps; see e.g. [Fec5], [Kry7], [Li], [Mar4], [Pet4], [PeYu1], [PeYu2], [St3], and the references therein.

The study of *derivo-periodic* orbits or *second kind* periodic orbits (in the Minorsky terminology [Min-M]) or *rotations*, as they are sometimes also called, was initiated in H. Poincaré's celestial mechanics investigations; see [Poi-M, p. 80].

They all mean that the functions should have periodic derivatives, but need not be periodic themselves. Physically, such a solution can correspond, for example, to a periodic velocity [Bar], [Poi-M], a subsynchronous level of performance of the motor [EdFo], [Pus], a "slalom orbit" of an electron beam [CLY] or a motion of particles in a sinusoidal potential related to a free-electron laser [Mey], characterized by M. Meystre as "marbles rolling on a corrugated rooftop". So, one can find many applications in astronomy, engineering, laser physics, etc. In spirit of the mentioned practical importance, there are not so many contributions concerning this problem; see [AlAu], [Ame2], [And4], [And16], [And22], [And28]–[And30], [ArGoKr], [ArGoMP], [Bur], [CLY], [ChiZe], [EdFo], [ElO], [Far1]–[Far3], [Far-M], [Fel], [GLY-M], [Gia], [GolShSt-M], [GolSt], [Leo], [Lvi1]–[Lvi4], [Luo], [Mmn], [MmrSe], [Maw1], [Mir], [Mos], [Naz], [Sed], [Sei1], [Sei2], [Tri], [Vej1], [Vej2], [Wen], [Youj], [ZhnCha]. Let us note that a special sort of derivo-periodic orbits was already studied in Chapter III.6.

Besides optimal control problems considered above, many others can be investigated by means of our methods; see e.g. [GN4], [GNZ], and the references therein. For an alternative approach, see e.g. [AGN-M], [BoKl-M], [KiSi-M], [Kur1-M], [Kur2-M], [Ut1-M], [Ut2-M], [Ani2], [AF1], [AF3], [Au3], [AZ1], [AZ2], [Brm], [BCL], [BT1], [BT2], [BGN1], [BGN2], [Bh1], [Bh2], [BO1], [BO2], [CDMN], [CoLe], [FNPZ], [GaP1], [GN1]–[GN4], [Her], [KNOZ], [KaPa], [La], [LaO2], [Nis], [NOZ1], [NOZ2],...

Although fractal structures often appear w.r.t. chaotic dynamics of differential equations (cf. e.g. [ASY-M]), our idea to generate fractals (implicitly) by means of families of differential equations is new.

For further results and techniques concerning differential inclusions, see [Au-M], [AuC-M], [De1-M], [Fi1-M], [Fi2-M], [HP2-M], [KiSi-M], [Kun-M], [MM-M], [Pet-M], [Smr-M], [To-M], [Ut1-M], [AGG2], [AGG3], [AGJ1]–[AGJ4], [AJ1], [AMT1], [AMT2], [AZ1]–[AZ3], [Au1]–[Au3], [AF1], [BCM], [Bad1], [Bad2], [BaP], [BaF1], [BaF2], [BGoP1], [Bog1], [Bog2], [Bor], [Bot], [BC2], [Crd1], [Crd2], [Cas], [C2]–[C6], [CC1], [CC2], [CM], [ClOr], [CK], [Clk], [Con1], [Con2], [CKZ], [CMN], [COZ], [Cor], [CouKa], [De1]–[De5], [DR], [DiaWa], [DBl1], [DBl2], [DGP1]–[DGP3], [DBP1]–[DBP7], [DBPS1], [DBPS2], [Fi1]–[Fi5], [Fri2], [Fri3], [FriGr], [FGK], [Fry3], [FryGo], [GaKr1], [GaPi], [GaP2], [Ge5]–[Ge7], [Gon], [Go1], [Go15], [GNO], [GP1], [GS], [Had1]–[Had3], [HL], [HLP1], [HLP2], [HP1]–[HP3], [IbrGo], [JK1], [KaPa], [KaTa], [Ksl1]–[Ksl5], [MNZ1]–[MNZ3], [Mar1], [Mar3], [Mar4], [MS], [MZ1]–[MZ3], [Muh], [Ob1]–[Ob3], [ObZe], [Ol5], [Pap1]–[Pap24], [PapPa], [Pp1], [Pl1], [Pl2], [Pol], [PG], [Sz1]–[Sz19], [Tol1]–[Tol5], [TolChu], [Umn], [Zar], [Zhu], [ZZ],...

In particular, differential equations and inclusions in abstract spaces were considered in: [Ba-M], [DK-M], [De2-M], [HP2-M], [KOZ-M], [Pn-M], [To-M], [Vr-M], [AndBa], [AngBa], [Ani1], [Ani2], [AVV1], [AulVM1], [Bug2], [BugBgj2], [Bgj1], [BgjSz1], [Crn], [CS1], [CS2], [CK], [COZ], [Co4], [DBl1], [DBl2], [DR], [De3], [De4], [DbMo1], [DbMo2], [God1], [God2], [GoeRz], [HL1], [Kmn], [KO], [KOZ1], [KOZ2], [Kar], [Ksl3], [Kubi1], [Kubi2], [KbSz], [LaY], [L], [Li], [MZ1]–[MZ3], [Moe1], [MoeHa], [Muh], [Ob1]–[Ob3], [ObZe], [Ol5], [Pia], [Sgh], [Shi], [Sng1], [Sng2], [Sz1]–[Sz19], [SzSz], [Tol1]–[Tol5], [TolChu], [Umn], [Yor3], [Zhu],...

APPENDIX 1

ALMOST-PERIODIC SINGLE-VALUED
AND MULTIVALUED FUNCTIONS

In this supplement, the hierarchy of almost-periodic function spaces is mainly clarified. The various types of definitions of almost-periodic functions are examined and compared. Apart from the standard definitions, we also introduce new classes and comment some, less traditional, definitions, to make a picture as much as possible complete. All is related to possible applications to nonlinear almost-periodic oscillations. The material for this exposition is based on the papers [AndBeGr], [AndBeL].

In the theory of almost-periodic (a.p.) functions, there are used many various definitions, mostly related to the names of H. Bohr, S. Bochner, V. V. Stepanov, H. Weyl and A. S. Besicovitch ([AP-M], [BH-M], [BJM-M], [Bes-M], [BeBo], [Bo-M], [BF], [BPS-M], [Cor-M], [Fav-M], [GKL-M], [Lev-M], [LZ-M], [Muc], [NG-M], [Pn-M]).

On the other hand, it is sometimes difficult to recognize whether these definitions are equivalent or if one follows from another. It is well-known that, for example, the definitions of uniformly a.p. (u.a.p. or Bohr-type a.p.) functions, done in terms of a relative density of the set of almost-periods (the Bohr-type criterion), a compactness of the set of translates (the Bochner-type criterion, sometimes called normality), the closure of the set of trigonometric polynomials in the sup-norm metric, are equivalent (see, for example, [Bes-M], [Cor-M]).

The same is true for the Stepanov class of a.p. functions ([Bes-M], [BF], [GKL-M], [Pn-M]), but if we would like to make some analogy for, e.g., the Besicovitch class of a.p. functions, the equivalence is no longer true.

For the Weyl class, the situation seems to be even more complicated, because in the standard (Bohr-type) definition, the Stepanov-type metric is used, curiously, instead of the Weyl one.

Moreover, the space of the Weyl a.p. functions is well-known [BF], unlike the other classes, to be incomplete in the Weyl metric.

In [Fol], E. Fœlner already pointed out these considerations, without arriving at a clarification of the hierarchies.

Besides these definitions, there exists a lot of further characterizations, done, e.g., by J.-P. Bertrandias [Ber3], R. Doss [Dos1], [Dos2], [Dos3], A. S. Kovanko [Kov10], [Kov7], [Kov1], [Kov12], [Kov13], [Kov6], [Kov2], B. M. Levitan [Lev-M], A. A. Pankov [Pn-M] and the references therein, A. C. Zaanen [Zaa-M], which are, sometimes, difficult to compare with more standard ones. Moreover, because of possible applications, the relations collected in Table 1 in Part 5 are crucial.

This appendix is divided into the following parts:

(A1.1) Uniformly almost-periodicity definitions and horizontal hierarchies,
(A1.2) Stepanov almost-periodicity definitions and horizontal hierarchies,
(A1.3) Weyl and equi-Weyl almost-periodicity definitions and horizontal hierarchies,
(A1.4) Besicovitch almost-periodicity definitions and horizontal hierarchies,
(A1.5) Vertical hierarchies,
(A1.6) Some further generalizations.

(A1.1) Uniformly almost-periodicity definitions and horizontal hierarchies.

The theory of a.p. functions was created by H. Bohr in the twenties, but it was restricted to the class of uniformly continuous functions.

Let us consider the space $C(\mathbb{R}, \mathbb{R})$ of all continuous functions, defined on $\mathbb{R}$ and with the values in $\mathbb{R}$, endowed with the usual sup-norm.

In this part, the definitions of almost-periodicity will be based on this norm.

(A1.1.1) DEFINITION. A set $X \subset \mathbb{R}$ is said to be *relatively dense* (r.d.) if there exists a number $l > 0$ (called the *inclusion interval*), s.t. every interval $[a, a + l]$ contains at least one point of X.

(A1.1.2) DEFINITION (see, for example, [AP-M, p. 3], [Bes-M, p. 2], [GKL-M, p. 170])[*Bohr-type definition*]. A function $f \in C(\mathbb{R}, \mathbb{R})$ is said to be *uniformly almost-periodic* (u.a.p.) if, for every $\varepsilon > 0$, there corresponds a r.d. set $\{\tau\}_\varepsilon$ s.t.

$$(A1.1.3) \quad \|f(x + \tau) - f(x)\|_{C^0} = \sup_{x \in \mathbb{R}} |f(x + \tau) - f(x)| < \varepsilon, \quad \text{for all } \tau \in \{\tau\}_\varepsilon.$$

Each number $\tau \in \{\tau\}_\varepsilon$ is called an *ε-uniformly almost-period* (or a *uniformly ε-translation number*) of $f(x)$.

(A1.1.4) PROPOSITION ([AP-M, p. 5], [Bes-M, p. 2], [Kat-M, p. 155]). *Every u.a.p. function is uniformly continuous.*

(A1.1.5) PROPOSITION ([AP-M, p. 5], [Bes-M, p. 2], [Kat-M, p. 155]). *Every u.a.p. function is uniformly bounded.*

(A1.1.6) PROPOSITION ([AP-M, p. 6], [Bes-M, p. 3]). *If a sequence of u.a.p. functions $f_n(x)$ converges uniformly in $\mathbb{R}$ to a function $f(x)$, then $f(x)$ is u.a.p., too.*

In other words, the set of u.a.p. functions is closed w.r.t. the uniform convergence. Since it is a closed subset of a Banach space, it is Banach, too.

Actually, it is easy to show that the space is a commutative Banach algebra, w.r.t. the usual product of functions (see, for example, [Sh-M, pp. 186–188]).

(A1.1.7) DEFINITION ([Bes-M, p. 10], [Cor-M, p. 14], [LZ-M, p. 4]) [*normality or Bochner-type definition*]. A function $f \in C(\mathbb{R}, \mathbb{R})$ is called *uniformly normal* if, for every sequence $\{h_i\}$ of real numbers, there corresponds a subsequence $\{h_{n_i}\}$ s.t. the sequence of functions $\{f(x + h_{n_i})\}$ is uniformly convergent.

The numbers h_i are called *translation numbers* and the functions $f(x + h_i)$ are called *translates*.

In other words, f is uniformly normal if the set of translates is precompact in $C_b := C \cap L^\infty$ (see [AP-M], [Pn-M]).

Let us recall that a metric space X is *compact* (*precompact*) if every sequence $\{x_n\}_{n \in \mathbb{N}}$ of elements belonging to X contains a convergent (fundamental) subsequence. Obviously, in a complete metric space X, the notions of precompactness, relative compactness (i.e. the closure is compact) and compactness coincide. The necessary (and, in a complete metric space, also sufficient) condition for the compactness, or an equivalent condition for the precompactness, of a set X can be characterized by means of (see, for example, [KolF-M], [LS-M]):

the total boundedness (Hausdorff Theorem): for every $\varepsilon > 0$, there exists a finite number of points $\{x_k\}_{k=1,\dots,n}$ s.t.

$$X \subset \bigcup_{k=1}^{n} (x_k, \varepsilon),$$

where (x_k, ε) denotes a spherical neighbourhood of x_k with radius ε; the set $\{x_k\}_{k=1,\dots,n}$ is called an ε-*net* for X.

(A1.1.8) REMARK. Every trigonometric polynomial

$$P(x) = \sum_{k=1}^{n} a_k e^{i\lambda_k x}, \quad (a_k \in \mathbb{R}, \lambda_k \in \mathbb{R})$$

is u.a.p. Then, according to Proposition (A1.1.6), every function $f(x)$, obtained as the limit of a uniformly convergent sequence of trigonometric polynomials, is u.a.p.

It is then natural to introduce the third definition:

(A1.1.9) DEFINITION ([BeBo, p. 224], [BF, p. 36], [Cor-M, p. 9]) [*approxima-tion*]. We call $C_{ap}(\mathbb{R}, \mathbb{R})$ the (Banach) space obtained as the closure of the space $\mathcal{P}(\mathbb{R}, \mathbb{R})$ of all trigonometric polynomials w.r.t. the sup-norm.

(A1.1.10) REMARK. Definition (A1.1.9) may be expressed in other words: A function $f(t)$ belongs to $C_{ap}(\mathbb{R}, \mathbb{R})$ if, for any $\varepsilon > 0$, there exists a trigonometric polynomial $T_\varepsilon(t)$, s.t.

$$\sup_{x \in \mathbb{R}} |f(x) - T_\varepsilon(x)| < \varepsilon.$$

It is easy to show that C_{ap}, like C, is invariant under translations, that is C_{ap} contains, together with $f(x)$, the functions $f^t(x) := f(x + t)$, for all $t \in \mathbb{R}$ (see, for example, [LZ-M, p. 4]).

The three main definitions, (A1.1.2), (A1.1.7) and (A1.1.9), are shown to be equivalent:

(A1.1.11) THEOREM ([AP-M, p. 8], [Bes-M, pp. 11–12], [Lev-M, pp. 23–27], [LZ-M, p. 4], [Pn-M, pp. 7–8]) [*Bochner criterion*]. *A continuous function $f(x)$ is u.a.p. iff it is uniformly normal.*

(A1.1.12) THEOREM ([BeBo, p. 226], [Pn-M, p. 9]). *A continuous function $f(x)$ is u.a.p. iff it belongs to $C_{ap}(\mathbb{R}, \mathbb{R})$.*

(A1.1.13) REMARK. To show the equivalence among Definitions (A1.1.2), (A1.1.7) and (A1.1.9), in his book [Cor-M, pp. 15–23], C. Corduneanu follows another way that will be very useful in the following Sections: he shows that

$$(A1.1.9) \Longrightarrow (A1.1.7) \Longrightarrow (A1.1.2) \Longrightarrow (A1.1.9).$$

In order to satisfy the S. Bochner criterion, L. A. Lusternik has proved an Ascoli–Arzelà-type lemma, introducing the notion of equi-almost-periodicity.

(A1.1.14) THEOREM ([Cor-M, p. 143], [LZ-M, p. 7], [LS-M, pp. 72–74]) [*Lus-ternik*]. *The necessary and sufficient condition for a family $\mathcal{F}$ of u.a.p. functions to be precompact is that*

(1) $\mathcal{F}$ *is equicontinuous, i.e. for any $\varepsilon > 0$, there exists $\delta(\varepsilon) > 0$ s.t.*

$$|f(x_1) - f(x_2)| < \varepsilon \quad \text{if } |x_1 - x_2| < \delta(\varepsilon), \text{ for all } f \in \mathcal{F},$$

(2) $\mathcal{F}$ *is equi-almost-periodic, i.e. for any $\varepsilon > 0$ there exists $l(\varepsilon) > 0$ s.t., for any interval whose length is $l(\varepsilon)$, there exists $\xi < l(\varepsilon)$ s.t.*

$$|f(x + \xi) - f(x)| < \varepsilon, \quad \text{for all } f \in \mathcal{F}, \quad x \in \mathbb{R},$$

(3) *for any $x \in \mathbb{R}$, the set of values $f(x)$ of all the functions in $\mathcal{F}$ is precompact.*

(A1.1.15) REMARK. As already seen, for numerical almost-periodic functions, condition (3) in the Lusternik theorem coincides with the following:

(3') for any $x \in \mathbb{R}$, the set of values $f(x)$ of all the functions in $\mathcal{F}$ is uniformly bounded.

The u.a.p. functions, like the periodic ones, can be represented by their Fourier series.

(A1.1.16) DEFINITION. For every function $f(x)$, we will call as the *mean value* of $f(x)$ the number

(A1.1.17)
$$M[f] = \lim_{T \to \infty} \frac{1}{2T} \int_{-T}^{T} f(x)\,dx =: \fint f(x)\,dx.$$

(A1.1.18) THEOREM ([Bes-M, pp. 12–15], [BF, p. 45], [LZ-M, pp. 22–23]) [*Mean value theorem*]. *The mean value of every u.a.p. function $f(x)$ exists and*

A1.1.19) (a) $M[f] = \lim\limits_{T \to \infty} \dfrac{1}{T} \int_{0}^{T} f(x)\,dx = \lim\limits_{T \to \infty} \dfrac{1}{T} \int_{-T}^{0} f(x)\,dx,$

A1.1.19) (b) $M[f] = \lim\limits_{T \to \infty} \dfrac{1}{2T} \int_{a-T}^{a+T} f(x)\,dx,$ *uniformly w.r.t. $a \in \mathbb{R}$.*

(A1.1.20) REMARK. Every even function satisfies (A1.1.19), while necessary condition for an odd function to be u.a.p. is that $M[f] = 0$. Furthermore, since, for every u.a.p. function $f(x)$ and for every real number λ, the function $f(x)e^{-i\lambda x}$ is still a u.a.p. function, the number

(A1.1.21)
$$a(\lambda, f) := M[f(x)e^{-i\lambda x}]$$

always exists.

(A1.1.22) THEOREM ([Bes-M, p. 18], [LZ-M, pp. 23–24]). *For every u.a.p. function $f(x)$, there always exists at most a countable infinite set of values λ (called the Bohr–Fourier exponents or frequencies) for which $a(\lambda) \neq 0$.*

The numbers $a(\lambda, f)$ are called the *Bohr–Fourier coefficients* and the set

$$\sigma(f) := \{\lambda_n | a(\lambda_n, f) \neq 0\}$$

is called the *spectrum* of f.

The formal series $\sum_n a(\lambda_n, f)e^{-i\lambda x}$ is called the *Bohr–Fourier series* of $f(x)$ and we write

(A1.1.23)
$$f(x) \sim \sum_n a(\lambda_n, f)e^{-i\lambda x}.$$

Let us now consider the connection between the Bohr–Fourier exponents and the almost-periods. To this aim, we recall the so-called *Kronecker Theorem* about the diophantine approximation (see, for example, [AP-M, pp. 30–38], [Bes-M, p. 35], [Cor-M, pp. 146–150]).

(A1.1.24) LEMMA ([Cor-M, pp. 146–147]). *Let*

$$f(x) \sim \sum_{k=1}^{\infty} a(\lambda_k, f) e^{i\lambda_k x}$$

be a u.a.p. function. For every $\varepsilon > 0$, there correspond $n \in \mathbb{N}$ and $\delta \in \mathbb{R}$, $0 < \delta < \pi$, s.t. any real number τ which is a solution of the system of diophantine (or congruencial) inequalities

$$|\lambda_k \tau| < \delta \ (\text{mod } 2\pi), \quad k = 1, \dots, n,$$

is an ε-almost-period for $f(x)$.

(A1.1.25) THEOREM ([AP-M, pp. 31–33], [Bes-M, p. 35], [Cor-M, pp. 147–149]) [*Kronecker Theorem*]. *Let λ_k, θ_k ($k = 1, \dots, n$) be arbitrary real numbers. The system of diophantine inequalities*

$$|\lambda_k \tau - \theta_k| < \delta(\text{mod } 2\pi), \quad k = 1, \dots, n$$

has solutions $\tau_\delta \in \mathbb{R}$, for any $\delta > 0$, iff every relation

$$\sum_{k=1}^{n} m_k \lambda_k = 0, \quad m_k \in \mathbb{N},$$

implies

$$\sum_{k=1}^{n} m_k \theta_k \equiv 0 \ (\text{mod } 2\pi), \quad m_k \in \mathbb{N}.$$

(A1.1.26) LEMMA ([Cor-M, p. 149]). *Let*

$$f(x) \sim \sum_{k=1}^{\infty} a(\lambda_k, f) e^{i\lambda_k x}$$

be a u.a.p. function and λ a real number which is rationally linearly independent of the Bohr–Fourier exponents λ_k. For every $\varepsilon > 0$, there correspond a number $\delta \in \mathbb{R}$, $0 < \delta < \pi/2$ and $n \in \mathbb{N}$, s.t. there exists an ε-almost-period τ which satisfies the system of inequalities

$$|\lambda_k \tau| < \delta(\text{mod } 2\pi), \quad |\lambda \tau - \pi| < \delta(\text{mod } 2\pi), \quad k = 1, \dots, n.$$

(A1.1.27) PROPOSITION ([Bes-M, p. 18], [LZ-M, pp. 31–33]) [*Bohr Fundamental Theorem*]. *The Parseval equation*

$$\sum_{n} |a(\lambda_n, f)|^2 = M\{|f(x)|^2\}$$

is true for every u.a.p. function.

(A1.1.28) PROPOSITION ([Bes-M, p. 27], [LZ-M, p. 24]) [*Uniqueness Theorem*]. *If two u.a.p. functions have the same Fourier series, then they are identical.*

In other words, two different elements belonging to C_{ap} cannot have the same Bohr–Fourier series.

It is worthwhile to introduce a further definition of the u.a.p. functions, which may be useful, by a heuristic way, to understand more deeply the structure of the space C_{ap} (see [Pn-M, pp. 5–9]).

(A1.1.29) DEFINITION. The *Bohr compactification*, or the *compact hull*, of $\mathbb{R}$ is a pair $(\mathbb{R}_B, i_B)$, where $\mathbb{R}_B$ is a compact group and $i_B : \mathbb{R} \to \mathbb{R}_B$ is a group homomorphism, s.t. for any homomorphism $\Phi : \mathbb{R} \to \Gamma$ into a compact group Γ there exists a unique homomorphism $\Phi_B : \mathbb{R}_B \to \Gamma$ s.t. $\Phi = \Phi_B \circ i_B$.

The Bohr compactification of a given group is always uniquely determined up to isomorphisms. Since $\mathbb{R}$ is a locally compact abelian group, its Bohr compactification can be constructed by means of the group $\mathbb{R}'$ of the characters of $\mathbb{R}$ (that is, the group of all the homomorphisms χ from $\mathbb{R}$ into the circumference $T = \{z \in \mathbb{C} \mid |z| = 1\}$), that can be written as

$$\chi(x) = e^{i\xi x}, \quad x \in \mathbb{R}, \ \xi \in \mathbb{R}.$$

Since the map $\xi \to e^{i\xi x}$ defines an isomorphism between $\mathbb{R}$ and $\mathbb{R}'$, we can identify $\mathbb{R}'$ with $\mathbb{R}$. In other words, the Bohr compactification may be interpreted as an isomorphism between $\mathbb{R}$ and a subgroup of the cartesian product (with the power of continuum) of the circumference T: if $T_\lambda \equiv T$ for all $\lambda \in \mathbb{R}$,

$$T^C = \prod_{\lambda \in \mathbb{R}} T_\lambda$$

endowed with an appropriate topology (for further information, see, e.g., [AH], [AnKa], [Ava], [BH-M], [BJM-M], [Hew], [HR-M], [Hol], [Kah], [Shu], [Wei-M]).

(A1.1.30) THEOREM ([Pn-M, p. 7]). $f(x) \in C_{ap}$ *iff there exists a function* $\widetilde{f} \in C(\mathbb{R}_B, \mathbb{R})$ *s.t.*

$$f = \widetilde{f} \circ i_B =: i_B^* \widetilde{f}$$

(*i.e. f can be extended to a continuous function on $\mathbb{R}_B$*).

(A1.1.31) REMARK. The extension $\widetilde{f}$ is unique and it satisfies

$$\sup_{x \in \mathbb{R}} |f(x)| = \sup_{y \in \mathbb{R}_B} |\widetilde{f}(y)|.$$

Thus, we can establish an isometric isomorphism

$$i_B^* : C(\mathbb{R}_B, \mathbb{R}) \sim C_{ap}(\mathbb{R}, \mathbb{R})$$

and every u.a.p. function can be identified with a *continuous* function defined on $\mathbb{R}_B$. This isometry allows us to deduce many properties of C_{ap} by means of the properties of C (see [DS-M], [HR-M], [Pn-M]).

The importance of the Bohr compactification will be more clear, when we study the Besicovitch-like a.p. functions, in Part 4.

The possibility to generalize the notion of almost-periodicity in the framework of continuous functions was studied by B. M. Levitan, who introduced the notion of N-almost-periodicity (see [GKL-M], [Lev-M], [LZ-M]), in terms of a diophantine approximation.

(A1.1.32) DEFINITION. A number $\tau = \tau(\varepsilon, N)$ is said to be an (ε, N)-*almost-period* of a function $f(x) \in C(\mathbb{R}, \mathbb{R})$ if, for every x s.t. $|x| < N$,

(A1.1.33) $|f(x + \tau) - f(x)| < \varepsilon.$

(A1.1.34) DEFINITION. A function $f(x) \in C(\mathbb{R}, \mathbb{R})$ is said to be an N-*almost-periodic* (N-a.p.) if we can find a countable set of real numbers $\{\Lambda_n\}_{n \in \mathbb{N}}$, depending on f and possessing the property that, for every choice of ε and N, we can find two positive numbers $n = n(\varepsilon, N)$ and $\delta = \delta(\varepsilon, N)$ s.t. each real number τ, satisfying the system of inequalities

$$|\Lambda_k \tau| < \delta \pmod{2\pi}, \quad k = 1, \ldots, n,$$

is an (ε, N)-almost-period of the function $f(x)$, i.e. satisfies inequality (A1.1.33).

Although every u.a.p. function is N-a.p., the converse is not true.

(A1.1.35) EXAMPLE ([GKL-M, p. 185], [LZ-M, pp. 58–59]). Given the function

$$p(x) = 2 + \cos x + \cos(\sqrt{2}x),$$

we have $\inf_{x \in \mathbb{R}} p(x) = 0$, then the function $q(x) = 1/p(x)$ is unbounded, and consequently it is not u.a.p. On the other hand, the function $q(x)$ is N-a.p.

Although this class of functions preserves many properties of the u.a.p. functions, many other properties do not hold anymore. For example, the mean value (A1.1.17), in general, does not exist, even for bounded functions. Furthermore, we can associate, to every N-a.p. function, different Fourier series (see [Lev-M, pp. 150–153], [LZ-M, p. 62]). Nevertheless, this space is very useful to obtain generalizations of classical results in the theory of ordinary differential equations with almost-periodic coefficients (see [Lev-M]).

In [Boc1], [Vee] (for a more recent reference, see also [NG-M]), a further class of almost-periodic functions, called *almost-automorphic*, was introduced. It can be shown that this class is a subset of the space of N-almost-periodic functions. This class was further generalized in [Bol], [Rei], where it is shown that this more general space of almost-automorphic functions coincides with the class of N-almost-periodic functions.

The theory of u.a.p. functions can be generalized to spaces of functions defined in $\mathbb{R}^n$ or, more generally, on abelian groups (see, for example, [AH], [BH-M], [BJM-M], [Fol1], [Lo-M], [Shu], [Wei-M]), and with the values in $\mathbb{R}^n$, in $\mathbb{C}$ or, more generally, on a metric, on a Banach or on a Hilbert space (see, for example, [AP-M], [Boc1], [Cor-M], [LZ-M], [Pn-M]).

These generalizations can be very useful to introduce and to study the Stepanov-like a.p. functions, described in the next part, based on the necessity to generalize the notion of almost-periodicity to discontinuous functions which must be, in any case, locally integrable.

One of the most important goals of this first generalization to discontinuous functions, as much as of the other spaces studied in the next parts, is to find a Parseval-like relation for the coefficients of the Bohr–Fourier series related to the functions belonging to these spaces and, consequently, to find approximation theorems for these spaces, which generalize Theorem (A1.1.12).

(A1.2) Stepanov almost-periodicity definitions and horizontal hierarchies.

Since all the various extensions of the definition of a.p. functions will involve also discontinuous functions, by means of integrals on bounded intervals, it is natural to work with locally integrable functions, i.e. $f \in L^p_{\text{loc}}(\mathbb{R}, \mathbb{R})$.

First of all, let us introduce the following Stepanov norms and distances:

$$(A1.2.1) \qquad \|f\|_{S^p_L} = \sup_{x \in \mathbb{R}} \left[\frac{1}{L} \int_x^{x+L} |f(t)|^p \, dt \right]^{1/p},$$

$$(A1.2.2) \quad D_{S^p_L}[f,g] = \|f - g\|_{S^p_L} = \sup_{x \in \mathbb{R}} \left[\frac{1}{L} \int_x^{x+L} |f(t) - g(t)|^p \, dt \right]^{1/p}.$$

Since L is a fixed positive number, we might expect infinite Stepanov norms, but it can be trivially shown that, for every $L_1, L_2 \in \mathbb{R}^+$, there exist $k_1, k_2 \in \mathbb{R}^+$ s.t.

$$k_1 \|f\|_{S^p_{L_1}} \leq \|f\|_{S^p_{L_2}} \leq k_2 \|f\|_{S^p_{L_1}}$$

i.e. all the Stepanov norms are equivalent.

Due to this equivalence, we can replace in formula (A1.2.1) L by an arbitrary positive number. In particular, we can consider the norm, where $L = 1$.

(A1.2.3) DEFINITION ([AP-M, pp. 76–77], [Bes-M, p. 77], [Cor-M, p. 156], [GKL-M, p. 189], [Lev-M, p. 200], [LZ-M, p. 33]) [*Bohr-type definition*]. A function

$f \in L^p_{\text{loc}}(\mathbb{R}, \mathbb{R})$ is said to be *almost-periodic in the sense of Stepanov* (S^p_{ap}) if, for every $\varepsilon > 0$, there corresponds a r.d. set $\{\tau\}_\varepsilon$ s.t.

$$(\text{A}1.2.4) \qquad \sup_{x \in \mathbb{R}} \left[\int_x^{x+1} |f(t+\tau) - f(t)|^p \, dt \right]^{1/p} < \varepsilon, \quad \text{for all } \tau \in \{\tau\}_\varepsilon.$$

Each number $\tau \in \{\tau\}_\varepsilon$ is called an *ε-Stepanov almost-period* (or *Stepanov ε-translation number* of $f(x)$).

Originally, V. V. Stepanov [Stp1], [Stp2] called the spaces S^1_{ap} and S^2_{ap} respectively "the class of almost-periodic functions of the second and the third type". N. Wiener [Wie] called the space S^2_{ap} "the space of pseudoperiodic functions". P. Franklin [Fra] called the spaces S^1_{ap} and S^2_{ap} respectively *apS* (almost-periodic summable functions) and *apSsq* (almost-periodic functions with a summable square).

The space S^1_{ap} will be shortly indicated as S_{ap}.

(A1.2.5) THEOREM ([GKL-M, p. 189], [Lev-M, Theorem 5.2.3]). *Every S^p_{ap}-function is*

(a) *S^p-bounded and*

(b) *S^p-uniformly continuous, i.e.*

$$\forall \varepsilon > 0 \exists \delta = \delta(\varepsilon) \quad \text{s.t. if } |h| < \delta, \quad \text{then } D_{S^p}[f(x+h), f(x)] < \varepsilon.$$

(A1.2.6) DEFINITION ([GKL-M, p. 189], [Kov7], [Urs]) [*S^p-normality*]. A function $f \in L^p_{\text{loc}}(\mathbb{R}, \mathbb{R})$ is said *S^p-normal* if the family of functions $\{f(x+h)\}$ (h is an arbitrary real number) is *S^p-precompact*, i.e. if for each sequence $f(x+h_1), f(x+h_2), \dots$, we can choose an S^p-convergent sequence.

(A1.2.7) DEFINITION ([BeBo, p. 224], [BF, p. 36]) [*approximation*]. We will call $S^p(\mathbb{R}, \mathbb{R})$ the space obtained as the closure of the space $\mathcal{P}(\mathbb{R}, \mathbb{R})$ of all trigonometric polynomials w.r.t. the norm (A1.2.1).

Using the appropriate implications (see [BF, Theorem 1], [GKL-M, Theorems 7 and 4] or, analogously, [Bes-M, pp. 88–91], [Pn-M, pp. 26–27]), we can show the main

(A1.2.8) THEOREM. *The three spaces, defined by the definitions* (A1.2.1), (A1.2.4), (A1.2.7), *are equivalent.*

(A1.2.9) THEOREM ([BF, pp. 51–53], [GKL-M, Theorem 6]). *The spaces S^p_{ap} are complete w.r.t. the norm* (A1.2.1).

An important contribution to the study of the equivalence of the different definitions for the spaces of the Stepanov, the (equi-)Weyl and the Besicovitch type, came from A. S. Kovanko. Unfortunately, many of his papers (written in Russian or in Ukrainian) were published in rather obscure journals; furthermore, many of his results were written without any proof. It is however useful to quote these results, in order to clarify the several hierarchies. Since the notion of normality is related to precompactness, A. S. Kovanko ([Kov10], [Kov7]) studied the necessary and sufficient conditions to guarantee the precompactness of some subclasses of the spaces S_{ap}^p, by means of a Lusternik-type theorem, introducing the notion of S^p-equi-almost-periodicity.

(A1.2.10) DEFINITION. Let $E \in \mathbb{R}$ be a measurable set and, for every closed interval $[a, b]$, let $E(a, b) := E \cap [a, b]$. Given two measurable functions f, g, let us define, for every $a \geq 0$,

$$E_a := \{x \in \mathbb{R} \quad \text{s.t.} \quad |f(x) - g(x)| \geq a\},$$

(A1.2.11) $$\delta_S^L(E) := \sup_{x \in \mathbb{R}} \frac{\mu[E(x, x + L)]}{L},$$

(where $\mu(X)$ represents the usual Lebesgue measure of a set X);

(A1.2.12) $$D_{S_L^p}^E[f(x), g(x)] := \sup_{x \in \mathbb{R}} \left[\frac{1}{L} \int_{E(x, x+L)} |f(t) - g(t)|^p \, dt \right]^{1/p},$$

(A1.2.13) $$\mathcal{D}_S^L[f(x), g(x)] := \inf_{0 < a < \infty} [a, \delta_S^L(E_a)],$$

$$\overline{f_L(x)} := \frac{1}{L} \int_x^{x+L} f(t) \, dt.$$

(A1.2.14) THEOREM ([Kov10], [Kov7]). *The necessary and sufficient condition for a family $\mathcal{F}$ of S_{ap}^p-functions to be S^p-precompact for every value L is that, for every $\varepsilon > 0$, $L > 0$,*

(1) *there exists $\sigma = \sigma(\varepsilon, L) > 0$ s.t.*

$$D_{S_L^p}^E[f(x), 0] < \varepsilon \quad \text{if} \quad \delta_S^L(E) < \sigma, \quad \text{for all } f \in \mathcal{F},$$

(2) *there exists $\rho = \rho(\varepsilon, L)$ s.t.*

$$D_{S_L^p}[f(x), \overline{f_h}] < \varepsilon \quad \text{for all } 0 < h < \rho, \text{ for all } f \in \mathcal{F},$$

(3) *there exists a r.d. set of S^p-almost-periods $\{\tau(\varepsilon, L)\}$, common to all the elements of $\mathcal{F}$, i.e.*

$$D_{S_L^p}[f(x + \tau), f(x)] < \varepsilon, \quad \text{for all } f \in \mathcal{F}.$$

(A1.2.15) REMARK. In Theorem (A1.2.14), conditions (2) and (3) can be respectively replaced by the conditions:

(2') for every $\varepsilon > 0$, $L > 0$, there exists $\delta = \delta(\varepsilon, L)$, s.t.

$$\mathcal{D}_S^L[f(x+h), f(x)] < \varepsilon, \quad \text{for all} \quad 0 < h < \delta, \text{ for all } f \in \mathcal{F},$$

(3') there exists a r.d. set (w.r.t. the distance (A1.2.13)) of almost-periods $\{\tau(\varepsilon, L)\}$, common to every $f \in \mathcal{F}$, s.t.

$$\mathcal{D}_S^L[f(x+\tau), f(x)] < \varepsilon, \quad \text{for all } f \in \mathcal{F}.$$

(A1.2.16) REMARK. Since the spaces S_{ap}^p are subspaces of $L_{\text{loc}}^p(\mathbb{R}, \mathbb{R})$, they must be regarded as quotient spaces, where each element is an equivalence class w.r.t. the relation

$$f(x) \sim g(x) \iff D_{S^p}[f(x), g(x)] = 0.$$

Consequently, two different functions belong to the same class iff they differ from each other by a function with S^p-norm equal to 0. This fact occurs when the two functions differ only on a set of the zero Lebesgue measure.

The theory of the S_{ap}^p-spaces can be included in the theory of C_{ap}-spaces with the values in a Banach space (see [AndBe], [AP-M, pp. 7, 76–78], [Cor-M, p. 137], [LZ-M, pp. 33–34], [Pn-M, pp. 24–28]), by means of the so-called *Bochner transform*, that will be briefly recalled here.

Defining the Banach space

$$BS^p := \{f \in L_{\text{loc}}^p(\mathbb{R}, \mathbb{R}) \mid \|f\|_{S^p} < \infty\},$$

we have, by virtue of Theorem (A1.2.5), $S_{ap}^p \subset BS^p$.

The Bochner-transform

$$f^b(x) = f(x+\eta), \quad \eta \in [0,1], \ x \in \mathbb{R},$$

associates, to each $x \in \mathbb{R}$, a function defined on $[0,1]$.

Thus, if $f \in L_{\text{loc}}^p(\mathbb{R}, \mathbb{R})$, then $f^b \in L_{\text{loc}}^p(\mathbb{R}, L^p([0,1]))$.

Consequently,

$$BS^p = \{f \in L_{\text{loc}}^p(\mathbb{R}, \mathbb{R}) \mid f^b \in L^\infty(\mathbb{R}, L^p([0,1]))\},$$

because $\|f\|_{S^p}^p = \|f^b\|_{L^\infty}$:

$$\|f^b\|_{L^\infty} = \sup_{x \in \mathbb{R}} \text{ess} \|f^b\|_{L^p([0,1))} = \sup_{x \in \mathbb{R}} \text{ess} \left[\int_0^1 |f(x+\eta)|^p \, d\eta \right]^{1/p}.$$

Moreover, since for every $f \in L^p_{\text{loc}}(\mathbb{R}, \mathbb{R})$, $f^b \in C(\mathbb{R}, L^p([0,1]))$, then

$$BS^p = \{f \in L^p_{\text{loc}}(\mathbb{R}, \mathbb{R}) \mid f^b \in BC(\mathbb{R}, L^p([0,1]))\},$$

where BC denotes the space of bounded continuous functions.

S. Bochner has shown (see [AP-M, pp. 76–78]) that

$$S^p = \{f \in L^p_{\text{loc}}(\mathbb{R}, \mathbb{R}) \mid f^b \in C_{ap}(\mathbb{R}, L^p([0,1]))\}.$$

(A1.2.17) REMARK. Since

$$\|f\|^p_{S^p} = \|f^b\|_{BC(\mathbb{R}, L^p([0,1]))},$$

we have

$$f_n \to f \text{ in } S^p_{ap} \iff f^b_n \to f^b \text{ in } C_{ap}(\mathbb{R}, L^p([0,1])).$$

The possibility to relate the spaces S^p_{ap} to the space $C_{ap}(\mathbb{R}, L^p([0,1]))$ enables us to explain the similarity of the results obtained for S^p_{ap} and C_{ap}, in particular, for the equivalence of the three definitions of almost-periodicity.

In [Fra], [Stp1] and [Stp2], a very wide generalization of the spaces S^p_{ap} to measurable functions is shown.

(A1.2.18) DEFINITION. A measurable function f is said to be *measurable almost-periodic* (M_{ap}) if, for every $\varepsilon > 0$, there exists a r.d. set $\{\tau_\varepsilon\}$ s.t., for a fixed number d,

$$|f(x + \tau) - f(x)| < \varepsilon,$$

for every x except a set whose Lebesgue exterior measure in every interval of length d is less than $d\varepsilon$, or whose density on every interval of length d is less than ε.

Originally, V. V. Stepanov [Stp1], [Stp2] called this space "class of almost periodic of the first type".

As remarked by P. Franklin ([Fra]), the definition remains essentially unchanged if d is not fixed, but may be arbitrary; in this case, L depends on both d and ε.

(A1.2.19) THEOREM ([Fra], [Stp2]). *For every $\varepsilon > 0$, any function $f(x) \in M_{ap}$ is bounded except a set of density less than ε in every interval of length d.*

The space M_{ap} can be also defined by means of an approximation theorem.

(A1.2.20) THEOREM ([Fra]). *A measurable function $f(x)$ belongs to M_{ap} iff there exists a sequence of trigonometric polynomials $\{P_\varepsilon\}$ s.t., for every $\varepsilon > 0$,*

$|f(x) - P_\varepsilon(x)| < \varepsilon$, *for every* x *except a set of density less than* ε *in every interval of length* d.

It is important to underline that, while changing values of every u.a.p. function in every non-empty bounded set gives a function which cannot be u.a.p., for the functions belonging to S_{ap}^p or to M_{ap} an analogous property holds if we modify a function in a set with a nonzero Lebesgue measure.

On the other hand, V. V. Stepanov [Stp2] has shown that, since the following inclusions hold (see Formula (A1.5.16))

$$(A1.2.21) \qquad C_{ap} \subset S_{ap}^{p_1} \subset S_{ap}^{p_2} \subset S_{ap}^1 \subset M_{ap}, \quad \text{for all } p_1 > p_2 > 1,$$

if a function belonging to one of the last four spaces in the sequence (A1.2.21) is respectively uniformly continuous, p_1-integrable, p_2-integrable, uniformly integrable, then it belongs to the space of the corresponding earlier type.

Let us recall that (see, for example, [Kov15], [Stp2]) a measurable function $f(x)$ is said to be *uniformly integrable* if, for every $\varepsilon > 0$ and $d > 0$, there corresponds a number $\eta > 0$ s.t.

$$\int_E |f(x)|\, dx < \varepsilon,$$

for every set E s.t. $\mu(E) < \eta$ and $\operatorname{diam}(E) \leq d$.

The difficulty related to the space M_{ap} consists in the definition of frequencies. In fact, if a measurable function is not integrable, then the quantities (A1.1.21) need not exist, in general. The problem can be overcome by considering a sequence of *cut-off functions*

$$g_n(x) = \begin{cases} f(x), & \text{for } |f(x)| \leq n, \\ n\dfrac{f(x)}{|f(x)|}, & \text{for } |f(x)| > n. \end{cases}$$

In fact, the functions g_n are uniformly integrable, and consequently S_{ap}^1 (see [Stp2]). So, by virtue of Theorem (A1.5.7), we have a countable set of frequencies, given by the union of all the frequencies $a(\lambda, g_n)$ of the functions g_n. Rejecting all the frequencies s.t. $\lim_{n\to\infty} a(\lambda, g_n) = 0$, this set can be interpreted as the spectrum of the measurable function $f(x)$, even if the limit of some sequence of frequencies is not finite or does not exist at all. It can be shown ([Fra]) that the spectrum of $f(x)$ does not depend on the choice of the sequence of cut-off functions (instead of $a_n = \{n\}$, we could consider another increasing sequence $a_k = \{n_k\}$, s.t. $\lim_{k\to\infty} n_k = \infty$ and s.t. there exists $K > 0$, for which $n_{k+1} - n_k < K$).

Let us remark that, when we restrict ourselves to uniformly integrable functions, this definition of the spectrum coincides with the classical one for the S_{ap}^1-functions (see [Fra]).

(A1.3) Weyl and equi-Weyl almost-periodicity definitions and horizontal hierarchies.

Although the three definitions of the C_{ap} and S_{ap}^p-spaces are related to the same norms (respectively, the sup-norm and (A1.2.1)), the classical definitions of Weyl spaces are using two different norms: (A1.2.1) and the Weyl norm

$$(A1.3.1) \qquad \|f\|_{W^p} = \lim_{L \to \infty} \sup_{x \in \mathbb{R}} \left[\frac{1}{L} \int_x^{x+L} |f(t)|^p \, dt \right]^{1/p} = \lim_{L \to \infty} \|f\|_{S_L^p},$$

induced by the distance

$$(A1.3.2) \qquad D_{W^p}[f,g] = \lim_{L \to \infty} \sup_{x \in \mathbb{R}} \left[\frac{1}{L} \int_x^{x+L} |f(t) - g(t)|^p \, dt \right]^{1/p} = \lim_{L \to \infty} D_{S_L^p}[f,g].$$

It can be easily shown that these limits always exist (see [Bes-M, pp. 72–73], [Lev-M, pp. 221–222]).

In order to clarify the reason of the usage of two different norms, let us introduce in a "naive" way six definitions.

(A1.3.3) DEFINITION ([AndBeL], [Bes-M, p. 77], [BeBo, pp. 226–227], [GKL-M, p. 190], [Lev-M, p. 200]) [*Bohr-type definition*]. A function $f \in L_{\text{loc}}^p(\mathbb{R}, \mathbb{R})$ is said to be *equi-almost-periodic in the sense of Weyl* ($e - W_{ap}^p$) if, for every $\varepsilon > 0$, there correspond a r.d. set $\{\tau\}_\varepsilon$ and a number $L_0 = L_0(\varepsilon)$ s.t.

$$(A1.3.4) \qquad \sup_{x \in \mathbb{R}} \left[\frac{1}{L} \int_x^{x+L} |f(t+\tau) - f(t)|^p \, dt \right]^{1/p} < \varepsilon,$$

$$\text{for all } \tau \in \{\tau\}_\varepsilon, \text{ for all } L \geq L_0(\varepsilon).$$

Each number $\tau \in \{\tau\}_\varepsilon$ is called an *ε-equi-Weyl almost-period* (or *equi-Weyl ε-translation number of $f(x)$*).

(A1.3.5) DEFINITION [*equi-W^p-normality*]. A function $f \in L_{\text{loc}}^p(\mathbb{R}, \mathbb{R})$ is said to be *equi-W^p-normal* if the family of functions $\{f(x+h)\}$ (h is an arbitrary real number) is S_L^p-precompact for sufficiently large L, i.e. if, for each sequence $f(x+h_1), f(x+h_2), \ldots$, we can choose an S_L^p-fundamental subsequence, for a sufficiently large L.

(A1.3.6) DEFINITION [*approximation*]. We will denote by *equi-$W^p(\mathbb{R}, \mathbb{R})$* the space obtained as the closure of the space $\mathcal{P}(\mathbb{R}, \mathbb{R})$ of all trigonometric polynomials w.r.t. the norm (A1.2.1) for sufficiently large L, i.e. for every $f(x) \in e - W^p$ and for every $\varepsilon > 0$ there exist $L_0 = L_0(\varepsilon)$ and a trigonometric polynomial T_ε s.t.

$$D_{S_{ap}^p}[f(x), T_\varepsilon] < \varepsilon, \quad \text{for all } L \geq L_0(\varepsilon).$$

(A1.3.7) DEFINITION ([AndBeL], [Kov1]) [*Bohr-type definition*]. A function $f \in L^p_{\text{loc}}(\mathbb{R}, \mathbb{R})$ is said to be *almost-periodic in the sense of Weyl* (W^p_{ap}) if, for every $\varepsilon > 0$, there corresponds a r.d. set $\{\tau\}_\varepsilon$ s.t.

$$(A1.3.8) \qquad \lim_{L \to \infty} \sup_{x \in \mathbb{R}} \left[\frac{1}{L} \int_x^{x+L} |f(t+\tau) - f(t)|^p \, dt \right]^{1/p} < \varepsilon.$$

Each number $\tau \in \{\tau\}_\varepsilon$ is called an ε-*Weyl almost-period* (or a *Weyl ε-translation number* of $f(x)$).

(A1.3.9) DEFINITION [W^p-*normality*]. A function $f \in L^p_{\text{loc}}(\mathbb{R}, \mathbb{R})$ is said to be W^p-*normal* if the family of functions $\{f(x + h)\}$ (h is arbitrary real number) is W^p-precompact, i.e. if for each sequence $f(x + h_1), f(x + h_2), \ldots$, we can choose a W^p-fundamental subsequence.

(A1.3.10) DEFINITION ([Bes-M, pp. 74–75], [BeBo, p. 225], [BF, pp. 35–36]) [*approximation*]. We denote by $W^p(\mathbb{R}, \mathbb{R})$ the space obtained as the closure of the space $\mathcal{P}(\mathbb{R}, \mathbb{R})$ of all trigonometric polynomials w.r.t. the norm (A1.3.1).

The spaces $e - W^1_{ap}$ and W^1_{ap} will be shortly indicated as $e - W_{ap}$ and W_{ap}.

Definition (A1.3.7) has been used in [AndBeL], but, as already pointed out by the authors, it was introduced by A. S. Kovanko in the paper without proofs [Kov1].

Due to the equivalence of all the S^p_L-norms, to find a number L_1 s.t., by means of Definition (A1.3.6), a sequence of polynomials converges in the norm $S^p_{L_1}$ implies that the sequence converges in every S^p_L-norm; it follows that the spaces given by Definition (A1.3.6) coincide with the spaces S^p_{ap}.

On the other hand, the following theorem holds.

(A1.3.11) THEOREM ([Bes-M, pp. 82–83], [BF, Theorem 2]). *A function $f \in W^p$ satisfies Definition (A1.3.3).*

Consequently, we cannot expect the equivalence of the Definitions for each type of spaces. As shown in [AndBeL], the space defined by means of Definition (A1.3.7) is an intermediate space between S^p_{ap} and W^p_{ap} and the inclusion is strict (see [AndBeL, Example 1]).

Analogously to the Stepanov spaces, we can introduce the space

$$BW^p := \{ f \in L^p_{\text{loc}}(\mathbb{R}, \mathbb{R}) \quad \text{s.t.} \quad \|f\|_{W^p} < \infty \}.$$

(A1.3.12) THEOREM ([Bes-M, p. 83], [BeBo, pp. 232–233], [GKL-M, p. 190], [Lev-M, pp. 222–223]). *A function $f \in e - W^p_{ap}$ belongs to BW^p.*

(A1.3.13) REMARK. It can be easily shown (see [BF, p. 37]) that the sets BS^p and BW^p coincide, but the different norms imply a big difference between the two spaces. In fact, although BS^p is complete w.r.t. the Stepanov norm (see [BF, pp. 51–53]), the space BW^p is incomplete w.r.t. the Weyl norm (see [BF, pp. 58–61]).

On the other hand, since the set of S^p-bounded functions coincides with the set of W^p-bounded functions, every $e - W_{ap}^p$-function is also $e - W^p$-bounded and S^p-bounded.

(A1.3.14) THEOREM ([Bes-M, p. 84], [BeBo, pp. 233–234], [GKL-M, p. 190], [Lev-M, pp. 223–224]). *A function $f \in e - W_{ap}^p$ is equi-W^p-uniformly continuous, i.e. for any $\varepsilon > 0$ there exist two positive numbers $L_0 = L_0(\varepsilon)$ and $\delta = \delta(\varepsilon)$ s.t., if $|h| < \delta$, then*

$$D_{S_L^p}\{f(x+h) - f(x)\} < \varepsilon, \quad \text{for all } L \geq L_0(\varepsilon).$$

(A1.3.15) THEOREM ([GKL-M, p. 191]). *For every function $f \in e - W_{ap}^p$ and every $\varepsilon > 0$, we can find a trigonometric polynomial P_ε, satisfying the inequality*

$$D_{W^p}[f, P_\varepsilon] < \varepsilon.$$

The meaning of the last theorem is that Definition (A1.3.3) $\Longrightarrow$ Definition (A1.3.10).

Consequently, by Theorems (A1.3.11) and (A1.3.15), we have shown that Definition (A1.3.3) is equivalent to Definition (A1.3.10). The same result was obtained in [Bes-M, pp. 82–91], [BeBo, pp. 231–241], [Fra].

(A1.3.16) THEOREM. *The space of $e - W^p$-normal functions is equivalent to $e - W_{ap}^p$.*

PROOF. The proof is based partly on [AndBeL], [Lev-M] and [Sto1].

Sufficiency: fix $\varepsilon > 0$. Since $\{f^h \text{ s.t. } h \in \mathbb{R}\}$ is $e - W^p$-precompact, there exists $L_0 = L_0(\varepsilon)$ s.t.

(A1.3.17) $\quad \forall L \geq L_0(\varepsilon) \; \forall h \in \mathbb{R} \; \exists j = 1, \ldots, n$

$$\text{s.t. } D_{S_L^p}[f^{h-h_j}, \, f] = D_{S_L^p}[f^{h_j}, f^h] < \varepsilon.$$

Thus, the numbers $\tau = h - h_j$ are $S_L^p - \varepsilon$-almost periods. Take

(A1.3.18) $$k = \max_{j=1,\ldots,n} |h_j|$$

and let $a \in \mathbb{R}$ be arbitrary. If $h = a + k$ and h_j satisfy (A1.3.17), we obtain, due to (A1.3.18), that $h - h_j \in [a, a + 2k]$. Thus, each interval of length $2k$ contains an

$e - W^p - \varepsilon$-almost period of f and the number $2k$ is a constant of relative density to the set

$$\{\tau \text{ s.t. } \tau = h - h_j, h \in \mathbb{R}, \; j = 1, \dots, n, \; D_{S_L^p}[f^{h_j}, f^h] < \varepsilon\},$$

which is consequently r.d.

Necessity: assume that f is an $e - W_{ap}^p$ function and fix $\varepsilon > 0$. By virtue of Theorem (A1.3.14), the function is $e - W^p$-uniformly continuous, i.e.

(A1.3.19) $\exists L_0 = L_0(\varepsilon)$ s.t. for all $L \geq L_0$ exists $\delta > 0$

$$\text{s.t. for all } |w| < \delta \quad D_{S_L^p}[f, f^w] < \frac{\varepsilon}{2}.$$

Let k be a constant of relative density to the set $\{\tau \text{ s.t. } D_{S_{L_1}^p}[f, f^w] < \frac{\varepsilon}{2}\}$, (i.e., for every interval I of length k there exists $\tau \in I$ s.t. $D_{S_L^p}[f, f^\tau] < \frac{\varepsilon}{2}$, for every $L \geq L_1$. To these numbers k and δ we associate a positive integer n s.t.

(A1.3.20) $$n\delta \leq k < (n+1)\delta$$

and put $h_j = j \cdot \delta$, $j = 1, \dots, n$. For any $h \in \mathbb{R}$, in the interval $[-h, -h+k]$ of length k we find some $S_{L_1}^p - \frac{\varepsilon}{2}$-almost period τ, s.t.

(A1.3.21) $$D_{S_L^p}[f^\tau, f] < \frac{\varepsilon}{2}, \quad \text{for all } L \geq L_1.$$

Futhermore, we choose h and τ in such a way that

(A1.3.22) $$|h + \tau - h_j| < \delta$$

(this is possible because of (A1.3.20) and the fact that $\tau \in [-h, -h+k]$). Take $L_2 = \max\{L_0, L_1\}$. By means of (A1.3.19), (A1.3.21) and (A1.3.22), we write, for every $L \geq L_2$,

$$D_{S_L^p}[f^h, f^{h_j}] \leq D_{S_L^p}[f^h, f^{h+\tau}] + D_{S_L^p}[f^{h+\tau}, f^{h_j}]$$
$$= D_{S_L^p}[f^\tau, f] + D_{S_L^p}[f^{h+\tau-h_j}, f] < \frac{\varepsilon}{2} + \frac{\varepsilon}{2} = \varepsilon.$$

This shows that $\{f^{h_j} \mid j = 1, \dots, n\}$ is a finite ε-net to $\{f^h \mid h \in \mathbb{R}\}$, w.r.t. the equi-Weyl metric. $\square$

Due to the fact that the spaces C_{ap} and S_{ap}^p are complete, it is possible to state the Bochner criterion in terms of compactness instead of pre-compactness (see [BF, pp. 51–53], [Lev-M, pp. 23–27, 199–200, 216–220], [ZL, pp. 10–11, 38]). Surprisingly, the spaces BW^p and W^p are not complete w.r.t. the Weyl norm (see [AndBeL], [BF, pp. 58–61], [Kov1], [Lev-M, pp. 242–247]).

As for the Stepanov spaces, A. S. Kovanko ([Kov1], [Kov6], [Kov13]) studied the necessary and sufficient conditions to guarantee the compactness of some sub-classes of the spaces $e - W_{ap}^p$ and W_{ap}^p, by means of a Lusternik-type theorem.

In order to find necessary and sufficient conditions for an $e - W_{ap}^p$ function to be $e - W^p$-normal, let us introduce another definition.

(A1.3.23) DEFINITION ([Kov13], [Kov6]). A sequence of W^p-bounded functions $\{f_n(x)\}$ is called

(1) $e - W^p$-*uniformly fundamental* if, for every $\varepsilon > 0$, there exists $L_0(\varepsilon)$ s.t.

$$\limsup_{m,n \to \infty} D_{S_L^p}[f_m, f_n] < \varepsilon \quad \text{for all } L \geq L_0;$$

(2) $e - W^p$-*uniformly convergent* if there exists a function $f \in BW^p$ s.t., for every $\varepsilon > 0$, there exists $L_0(\varepsilon)$ s.t.

$$\limsup_{n \to \infty} D_{S_L^p}[f, f_n] < \varepsilon \quad \text{for all } L \geq L_0.$$

(A1.3.24) THEOREM ([Kov13]). *A sequence of functions belonging to BW^p is $e - W^p$-uniformly fundamental iff it is $e - W^p$-uniformly convergent. It means that the space BW^p, endowed with the norm of $e - W^p$-uniform convergence, is complete.*

(A1.3.25) THEOREM ([Kov6]). *A set $\mathcal{M}$ of $e - W_{ap}^p$ functions is compact, w.r.t. the $e - W^p$-uniform convergence, if for every $\varepsilon > 0$,*

(i) $\exists \sigma > 0, T_1 > 0$ s.t.

$$D_{S_T^p}^E[f, 0] < \varepsilon \quad \text{if} \quad \delta_S^T(E) < \sigma, \text{ for all } T \geq T_1, \text{ for all } f \in \mathcal{M},$$

(ii) *($e - W^p$-equicontinuity)* $\exists \eta > 0, T_2 > 0$ s.t.

$$D_{S_T^p}[f(x+h) - f(x)] < \varepsilon \quad \text{if} \quad |h| < \eta, \quad \text{for all } T \geq T_2, \text{ for all } f \in \mathcal{M},$$

(iii) *($e - W^p$-equi-almost-periodicity)* $\exists T_3 > 0$ and a r.d. set $\{\tau_\varepsilon\}$ of real numbers s.t.

$$D_{S_T^p}[f(x+\tau) - f(x)] < \varepsilon \quad \text{if} \quad \tau \in \{\tau_\varepsilon\}, \quad \text{for all } T \geq T_3, \text{ for all } f \in \mathcal{M},$$

where $D_{S_T^p}$, $D_{S_T^p}^E$ and $\delta_S^T(E)$ are respectively given by (A1.2.2), (A1.2.11), (A1.2.12).

(A1.3.26) THEOREM ([Kov6]). *The necessary and sufficient condition in order to have $f(x) \in e - W_{ap}^p$ is that the set of all the translates $\{f(x+\tau)\}$ be compact in the sense of $e - W^p$-uniform convergence.*

To show the second theorem about normality, we need some introductory definitions, too.

(A1.3.27) DEFINITION. Given a Lebesgue-measurable set $E \subset \mathbb{R}$, let $E(a, b) := E \cap (a, b)$, for every interval (a, b), and $|E(a, b)|$ its Lebesgue measure. Let us denote

$$\delta_W = \lim_{T \to \infty} \left[\sup_{a \in \mathbb{R}} \frac{|E(a - T, a + T)|}{2T} \right].$$

(A1.3.28) DEFINITION. For every $f, \phi \in L_{loc}^p(\mathbb{R}, \mathbb{R})$, let us introduce the distance

$$D_{W^p}^E(f, \varphi) := \lim_{T \to \infty} \left\{ \sup_{a \in \mathbb{R}} \left[\frac{1}{2T} \int_{E(a-T, a+T)} |f - \varphi|^p \, dx \right]^{1/p} \right\}.$$

This distance, when $E \equiv \mathbb{R}$, coincides with the Weyl distance $D_{W^p}(f, \varphi)$.

In order to avoid any confusion about the concept of compactness, A. S. Kovanko introduced the so-called *ideal limits* of every Cauchy sequence of W_{ap}^p-functions. The distance between two ideal limits f and g is defined in the following way:

$$D_{W^p}[f, g] := \lim_{m, n \to \infty} D_{W^p}[f_m, g_n],$$

where the sequences $\{f_m\}, \{g_n\}$ are two Cauchy sequences whose ideal limits are respectively f and g.

We are now ready to state the following Lusternik-type theorem:

(A1.3.29) THEOREM ([Kov1]). *The necessary and sufficient condition for the compactness in the Weyl norm of a class* $\mathcal{M}(f)$ *of functions* $f \in W_{ap}^p$ *is that, for every* $\varepsilon > 0$,

(i) *there exists a number* $\sigma > 0$ *s.t.* $D_{W^p}^E[f, 0] < \varepsilon$ *if* $\delta_W(E) < \sigma$, *for every function* $f(x) \in \mathcal{M}$,

(ii) (W^p-*equicontinuity) there exists a number* $\eta > 0$ *s.t.*

$$D_{W^p}[f(x + h), f(x)] < \varepsilon \quad \text{if } |h| < \eta,$$

for every function $f \in \mathcal{M}$,

(iii) (W^p-*equi-almost-periodicity) there exists a r.d. set of almost-periods* $\{\tau_\varepsilon\}$ *s.t.*

$$D_{W^p}[f(x + \tau), f(x)] < \varepsilon,$$

for every function $f \in \mathcal{M}$.

(A1.3.30) REMARK. In both Theorems (A1.3.25) and (A1.3.29), conditions (ii) and (iii) are the integral versions of the corresponding hypotheses in the Lusternik theorem for C_{ap}-functions; on the other hand, in the original Lusternik theorem the first condition is related to the Ascoli-Arzelà lemma; in Theorems (A1.3.25) and (A1.3.29) it is substituted by a condition that recalls the L^p-version of the Ascoli-Arzelà lemma, given by M. Riesz, M. Fréchet and A. N. Kolmogorov (see, for example, [Bre-M, Theorem IV.25 and Corollary IV.26]).

(A1.3.31) THEOREM ([Kov1]). *The spaces of* W^p-*normal functions in the sense of Kovanko and* W_{ap}^p *are equivalent.*

If we weaken the hypothesis on compactness and ask only the pre-compactness for the set of translates $\{f^h\}$, we need an auxiliary condition:

(A1.3.32) HYPOTHESIS ([AndBeL]). *Let* $f \in L^1_{\text{loc}}(\mathbb{R}, \mathbb{R})$, *with* $D_{W^p}(f) < \infty$, *be uniformly continuous in the mean, i.e.*

$$\forall \frac{\varepsilon}{3} > 0 \ \exists \delta > 0 \ \forall |h| < \delta : \frac{1}{l} \int_0^l |f^h(t) - f(t)| \, dt < \frac{\varepsilon}{3},$$

uniformly w.r.t. $l \in (0, \infty)$.

(A1.3.33) THEOREM ([AndBeL]). *If a* W^p_{ap} *function satisfies the Hypothesis, then it is* W^p-*uniformly continuous, i.e. for any* $\varepsilon > 0$ *there exists* $\delta = \delta(\varepsilon)$ *s.t., if* $|h| < \delta$, *then*

(A1.3.34) $$D_{W^p}\{f(x + h) - f(x)\} < \varepsilon.$$

(A1.3.35) THEOREM ([AndBeL]). *Let* $f \in L^1_{\text{loc}}(\mathbb{R}, \mathbb{R})$ *be a* W^p-*function satisfying the Hypothesis* (A1.3.32). *Then* $f \in W^p_{ap}$ *iff it is* W^p-*normal.*

(A1.3.36) COROLLARY. *Every* W^p-*normal function is* W^p_{ap}.

Following analogous proofs to u.a.p. functions (see [Cor-M, Theorem 10] or [Bes-M, pp. 11–12]), it is possible to show the following

(A1.3.37) THEOREM. *Every* W^p-*function is* W^p-*normal.*

PROOF. Let us consider an arbitrary W^p-function $f(x)$ and a sequence of trigonometric polynomials $\{T_n\}$, W^p-converging to $f(x)$. Let us take a sequence of real numbers $\{h_n\}$ and a subsequence $\{h_{1n}\}$ s.t. $\{T_1(x + h_{1n})\}$ is W^p-convergent. Then, we can extract from $\{h_{1n}\}$ a subsequence $\{h_{2n}\}$ s.t. $\{T_2(x + h_{2n})\}$ is W^p-convergent, too, and so on. In this way, we construct a subsequence $\{h_{rn}\}$, for every $r \in \mathbb{N}$ s.t. $\{T_q(x + h_{rn})\}$ is W^p-convergent, for every $q \leq r$. Let us take the subsequence $\{h_{rr}\}$, which is a subsequence of every sequence $\{h_{qn}\}$, with the exception of at most a finite number of terms. Consequently, the sequence $\{T_n(x + h_{rr})\}$ is W^p-convergent, for every $n \in \mathbb{N}$. Let $\varepsilon > 0$ be sufficiently large so that

(A1.3.38) $$D_{W^p}[f(x) - T_n(x)] < \frac{\varepsilon}{3}.$$

There exists $N(\varepsilon) > 0$ s.t.

(A1.3.39) $\quad D_{W^p}[f(x + h_{rr}) - f(x + h_{qq})] \leq D_{W^p}[f(x + h_{rr}) - T_n(x + h_{rr})]$

$\qquad + D_{W^p}[T_n(x + h_{rr}) - T_n(x + h_{qq})] + D_{W^p}[T_n(x + h_{qq}) - f(x + h_{qq})] < \varepsilon,$

for all $q, r \geq N(\varepsilon)$. Thus, the sequence $\{f(x + h_{rr})\}$ is W^p-convergent, and consequently the function $f(x)$ is W^p-normal. $\qquad\square$

(A1.3.40) REMARK. The analogy of Theorem (A1.3.37) for $e - W^p$ spaces is guaranteed by the fact that

    (i) the spaces $e - W^p$ coincide with the spaces S^p,

    (ii) the spaces S^p coincide with the spaces S^p-normal (see Theorem (A1.2.8)),

    (iii) the spaces S^p-normal are included in the spaces $e - W^p$-normal (see Formula (A1.5.16)).

The converse of Corollary (A1.3.36) or Theorem (A1.3.37) is, in general, not true.

(A1.3.41) EXAMPLE (cf. [St-M, pp. 20–21]) (*Example of an equi-W^1-normal function which is not an equi-W^1-function*). Let us consider the function, defined on $\mathbb{R}$,

$$f(x) = \begin{cases} 1, & \text{for } 0 < x < \tfrac{1}{2}, \\ 0, & \text{elsewhere.} \end{cases}$$

For every $L, \tau \in \mathbb{R}$, $L \geq 1$, we have

$$\int_x^{x+L} |f(t+\tau) - f(t)|\, dt \leq 1.$$

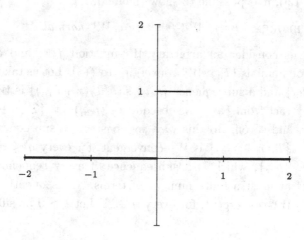

FIGURE 1

Thus,

$$(A1.3.42) \qquad D_{S_L}[f(x+\tau), f(x)] = \sup_{x \in \mathbb{R}} \left\{ \frac{1}{L} \int_x^{x+L} |f(t+\tau) - f(t)|\, dt \right\} \leq \frac{1}{L}.$$

For every $\varepsilon > 0$, there exists $L \geq 1$, s.t.

$$D_{S_L}[f(x+\tau), f(x)] \leq \varepsilon, \quad \text{for all } \tau \in \mathbb{R}.$$

Consequently, the function belongs to $e - W_{ap}$.

From Theorem (A1.3.16), we conclude that the function is $e - W$-normal.

On the other hand, there always exists $x \in \mathbb{R}$ such that, for every $\tau > \varepsilon$, where $0 < \varepsilon < \frac{1}{2}$, we have $(L = 1)$

$$\int_x^{x+1} |f(t+\tau) - f(t)|\, dt > \varepsilon.$$

Therefore, if $\varepsilon < \frac{1}{2}$, then $(L = 1)$

$$D_{S_1}[f(x+\tau), f(x)] > \varepsilon, \quad \text{for all } \tau > \varepsilon.$$

For $\tau \geq L - \frac{1}{2}$, we get even

$$D_{S_L}[f(x+\tau), f(x)] \geq \frac{1}{2L}.$$

So, the function is not S_{ap}. Since the sets S_{ap} and $e - W^1$ coincide, we have the claim.

(A1.3.43) EXAMPLE (*Example of a W_{ap}^1-function which is not a W^1-normal function*). The example is partly based on [St-M, pp. 42–47].

Let us consider the function

$$f(x) = \begin{cases} 0 & \text{if } x \in (-\infty, 0], \\ \sqrt{\frac{n}{2}} & \text{if } x \in (n-2, n-1], \, n = 2, 4, 6, \ldots, \\ -\sqrt{\frac{n}{2}} & \text{if } x \in (n-1, n], \, n = 2, 4, 6, \ldots, \end{cases}$$

Let us show that this function is a W_{ap}-function. To this aim, let us consider the set $\{x + 2k, k \in \mathbb{Z}\}$ and let us show that

$$D_W[f(x+2k), f(x)] = 0, \quad \text{for all } k \in \mathbb{Z}.$$

If $k = 0$, the proof is trivial. Furthermore, if $k < 0$, $k = -m$, then

$$D_W[f(x+2k), f(x)] = D_W[f(x), f(x-2k)] = D_W[f(x+2m), f(x)].$$

It will be then sufficient to study the case $k > 0$. Since we will consider the limit for $L \to \infty$, let us take $L > 2k$. There exists an integer i such that

(A1.3.44) $$2k \leq 2i \leq L < 2(i+1).$$

Let us compute

(A1.3.45) $$D_{S_L}[f(x+2k), f(x)] = \sup_{x \in \mathbb{R}} \left\{ \frac{1}{L} \int_x^{x+L} |f(t+2k) - f(t)|\, dt \right\}.$$

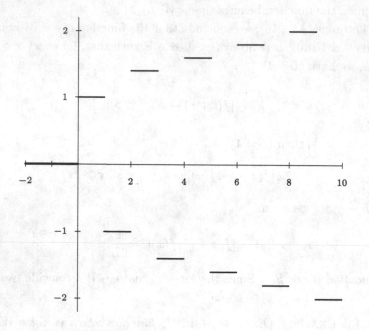

FIGURE 2

Since, in the interval $(-\infty, -2i)$, we have $|f(x + 2k) - f(x)| = 0$; in the interval $(-2i, 0)$, the function $|f(x + 2k) - f(x)|$ is increasing; in the interval $(0, \infty)$, the function $|f(x + 2k) - f(x)|$ is decreasing, the maximum value for the integral in (A1.3.42) is obtained in an interval including 0. Considering (A1.3.44), we can write

$$D_{S_L}[f(x + 2k), f(x)] = \sup_{x \in \mathbb{R}} \{\frac{1}{L} \int_x^{x+L} |f(t + 2k) - f(t)| \, dt\}$$

$$\leq \frac{1}{2i} \int_{-2(i+1)}^{2(i+1)} |f(t + 2k) - f(t)| \, dt = \frac{1}{2i} \int_{-2(i+1)}^{0} |f(t + 2k) - f(t)| \, dt$$

$$+ \frac{1}{2i} \int_0^{2(i+1)} |f(t + 2k) - f(t)| \, dt = \frac{1}{i} \sum_{j=1}^k \sqrt{j} + \frac{1}{i} \sum_{j=1}^{i+1} [\sqrt{j+k} - \sqrt{j}]$$

$$= \frac{1}{i} \sum_{j=1}^k \sqrt{j} + \frac{1}{i} \sum_{j=1}^{i+1} \left[\frac{k}{\sqrt{j+k} + \sqrt{j}}\right] \leq \frac{1}{i} \sum_{j=1}^k \sqrt{j} + \frac{1}{i} \sum_{j=1}^{i+1} \frac{k}{\sqrt{j}}.$$

By virtue of the Cauchy integral criterion for positive series, or by induction, it can be shown that

$$\sum_{j=1}^l \frac{1}{\sqrt{j}} \leq 2\sqrt{l}, \quad \text{for all } l \in \mathbb{N}.$$

Consequently,

$$D_{S_L}[f(x+2k), f(x)] \le \left[\frac{1}{i} \sum_{j=1}^{k} \sqrt{j}\right] + \frac{2k\sqrt{i+1}}{i}.$$

Passing to the limit for $L \to \infty$, we obtain

$$D_W[f(x+2k), f(x)] = \lim_{L \to \infty} D_{S_L}[f(x+2k), f(x)]$$

$$\le \lim_{L \to \infty} \left[\frac{1}{i} \sum_{j=1}^{k} \sqrt{j} + \frac{2k\sqrt{i+1}}{i}\right] = 0.$$

Then the set $\{2k \mid k \in \mathbb{Z}\}$ represents, for every $\varepsilon > 0$, a set of $W_{ap}^1 - \varepsilon$-almost-periods for the function $f(x)$, which is, consequently, W^1-almost-periodic. In Example (A1.4.45), it will be shown that this function is not B^1-normal. Furthermore, in Part 5 we will show that the space of W^1-normal functions is included in the space of B^1-normal functions. Consequently, this function is not W^1-normal.

Let us observe that this function does not satisfy both the conditions of the Hypothesis (A1.3.32). In fact,

$$\|f\|_{S_L} = \sup_{x \in \mathbb{R}} \frac{1}{L} \int_x^{x+L} |f(t)| \, dt \ge \frac{1}{L} \int_0^L |f(t)| \, dt.$$

For every L, there exists $k > 0$ s.t. $2k \le L < 2k+1$. Then

$$\|f\|_{S_L} \ge \frac{1}{2k+1} \int_0^{2k} |f(t)| \, dt = \frac{1}{2k+1} \left[2 \sum_{j=1}^{k} \sqrt{j}\right] \ge \frac{4}{3(2k+1)} k^{\frac{3}{2}},$$

where the last inequality is obtained by virtue of the Cauchy integral criterion of convergence. Consequently, letting $L \to \infty$, we obtain that the function is unbounded in the W^1-norm. Furthermore, the function is not uniformly continuous in the mean. In fact,

$$\frac{1}{L} \int_0^L |f(t+h) - f(t)| \, dt$$

$$\ge \frac{1}{2k+1} \left\{ \sum_{j=1}^{k} \left[\int_{2j-1-h}^{2j-1} 2\sqrt{j} \, dt + \int_{2j-h}^{2j} |\sqrt{j+1} + \sqrt{j}| \, dt \right] \right\}$$

$$= \frac{h}{2k} \left[3 + \sqrt{k+1} + \sum_{j=2}^{k} 4\sqrt{j}\right] \ge \frac{h}{2k} \left[3 + \sqrt{k+1} + \frac{8}{3}\left(k^{\frac{3}{2}} - 1\right)\right] \ge \frac{4}{3} h\sqrt{k},$$

where we have again used the Cauchy integral criterion.

The nonuniformity follows immediately. Furthermore, the function is not W^1-continuous, since

$$\|f(x+h) - f(x)\|_{S^1} \geq \sup_{x \in \mathbb{R}} \left[\frac{4}{3}h\sqrt{k}\right]$$

and

$$\|f(x+h) - f(x)\|_{W^1} \geq \lim_{k \to \infty} \sup_{x \in \mathbb{R}} \left[\frac{4}{3}h\sqrt{k}\right] = \infty.$$

The previous example shows that Theorems (A1.3.12) and (A1.3.14) cannot be extended to W_{ap}^p-functions because, in general, a W_{ap}^p-function is neither BW^p nor W^1-continuous.

(A1.3.46) EXAMPLE ([AndBeL], [St-M, p. 48]) (*Example of a W^1-normal function which is not a W^1-function*). In [AndBeL], the Heaviside step function

$$H(x) = \begin{cases} 0 & \text{if } x < 0, \\ 1 & \text{if } x \geq 0 \end{cases}$$

is shown to be W_{ap}, but not $e - W_{ap}$, that is, by virtue of Theorems (A1.3.11) and (A1.3.15), not W^1. On the other hand, J. Stryja [St-M, p. 48] has shown that the function is W-normal.

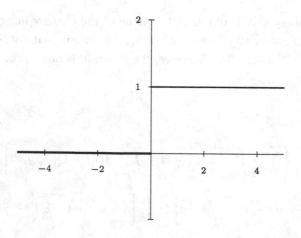

FIGURE 3

In fact, since, from the almost-periodicity of $f(x)$, for any $\tau \in \mathbb{R}$,

$$D_W[f(x+\tau), f(x)] = 0,$$

then we have that, for every $\varepsilon > 0$ and for every set of translates $\{f(x+a) \mid a \in \mathbb{R}\}$, there exists a finite ε-net w.r.t. the distance D_W, given by the only value $f(x)$. The W-normality follows immediately.

Let us finally observe that, as can be easily seen, the Heaviside function is BW^1 and uniformly continuous in the mean, which are the two sufficient conditions to guarantee the W-normality of a W_{ap}-function.

As already observed in [AndBeL], since the spaces BW^p and W^p are incomplete in the Weyl norm, the between lying spaces $W_{ap}^p \cap BW^p$ and W^p-normal $\cap BW^p$ are incomplete, too. Furthermore, the spaces $e - W^p$, equivalent to S_{ap}^p, are complete in the equi-Weyl norm. However, it is an open question, whether the spaces $e - W_{ap}^p \cap BS^p$ are complete or not.

A. S. Kovanko has generalized the definition of almost-periodic functions in the sense of Weyl, by means of a Bohr-like definition.

(A1.3.47) DEFINITION ([Kov15]). A measurable function is said to be *asymptotically almost-periodic* (a.a.p.) if, for every $\varepsilon > 0$, there correspond two positive numbers $l = l(\varepsilon), T_0 = T_0(\varepsilon)$ s.t., in every interval of length $l(\varepsilon)$, there exists an ε-asymptotic almost-period $\tau(\varepsilon)$ s.t.

$$|f(x + \tau) - f(x)| < \varepsilon,$$

for every $x \in \mathbb{R}$, except a set of density less than ε, on every interval of length greater than T_0.

(A1.3.48) THEOREM ([Kov15]). *Every a.a.p. function $f(x)$ s.t. $f^p(x)$ is uniformly integrable is W^p.*

For the a.a.p. function, it is possible to state an approximation theorem.

(A1.3.49) THEOREM ([Kov15]). *An integrable function $f(x)$ is a.a.p. iff, for every $\varepsilon > 0$, there exists a trigonometric polynomial P_ε and a positive number $T = T_\varepsilon$ s.t.*

$$|f(x) - P_\varepsilon| < \varepsilon,$$

for every $x \in \mathbb{R}$, except a set of density less than ε on every interval of length greater than T_0.

In [Urs], H. D. Ursell introduced four new definitions in terms of normality and almost-periods. He called the first two classes respectively W-normal and W_{ap}. Here, in order to avoid any confusion with Definitions (A1.3.3), (A1.3.5), (A1.3.7) and (A1.3.9), we will call these classes respectively $\overline{W}$-normal and $\overline{W}_{ap}$.

(A1.3.50) DEFINITION ([Urs]). A function $f(x)$ is said to be $\overline{W}$-*normal* if, for every sequence $\{f(x + h_n)\}$, there exists a subsequence $\{f(x + h_{n_k})\}$ s.t.

$$\lim_{j,k\to\infty}\left[\lim_{T\to\infty}\frac{1}{T}\int_0^T |f(x+h_{n_j}) - f(x+h_{n_k})|\, dx\right] = 0.$$

This definition is evidently more general than Definition (A1.3.9) of the W^p-normality, for $p = 1$ (put $x = 0$ in (A1.3.8)). The limit is namely not made uniformly w.r.t. every interval $[a, a+T]$, but only on the interval $[0, T]$. However, we do not know whether or not H. D. Ursell's $\overline{W}$-normal space is complete.

(A1.3.51) DEFINITION ([Urs]). A function $f(x)$ is said to be $\overline{W}_{ap}$ if, for every $\varepsilon > 0$, there exist a r.d. set of numbers τ and a number $T_0 = T_0(\varepsilon)$ s.t.

$$\frac{1}{T}\int_0^T |f(x+\tau) - f(x)|\, dx, \quad \text{for all } T > T_0.$$

H. D. Ursell shows that Definition (A1.3.50) implies Definition (A1.3.51). He also claims that the converse is true, but he does not prove the statement.

Furthermore, he introduces the space W_{*-ap}, which is equivalent to the $e - W_{ap}$ space, and the W_*-normal space.

(A1.3.52) DEFINITION ([Urs]). A function $f(x)$ is said to be W_*-*normal* if, for every sequence $\{f(x + h_n)\}$, there exists a subsequence $\{f(x + h_{n_k})\}$ s.t.

$$\lim_{j,k\to\infty}\left\{\lim_{T\to\infty}\sup_{a\in\mathbb{R}}\left[\frac{1}{T}\int_a^{a+T} |f(x+h_{n_j}) - f(x+h_{n_k})|\, dx\right]\right\} = 0.$$

This space is evidently the same as the W-normal one. H. D. Ursell shows that the $e - W_{ap}$ space is contained in the W_*-normal one and he concludes that the $\overline{W}_{ap}$, $\overline{W}$-normal, $e - W_{ap}$ and W_*-normal spaces are equivalent. This last statement is again not proved and is a bit surprising.

It seems to us that H. D. Ursell's statement is false, because it is easy to show that the Heaviside step function, which is not $e - W_{ap}$, is $\overline{W}$-normal.

As already observed, due to their continuity, the elements of C_{ap} are in fact real functions; furthermore, since every space S_L^p is a subspace of $L_{loc}^p(\mathbb{R}, \mathbb{R})$, it is obtained as a quotient space, w.r.t. the usual equivalence relation for (Bochner-) Lebesgue integrable functions:

$$f \sim g \in S_{ap}^p \iff \mu\{x \in \mathbb{R} \mid f(x) \neq g(x)\} = 0,$$

where μ is the usual Lebesgue measure.

On the other hand, the elements of the space W_{ap}^p are more general classes of equivalence. In fact, two different functions, belonging to the same class, may differ even on a set with Lebesgue measure greater than 0 (even infinite), provided

$$f - g \in L_{loc}^p(\mathbb{R}, \mathbb{R}).$$

Consequently, to handle elements in W_{ap}^p (and, *a fortiori*, as will be shown in the next part, in B_{ap}^p), is less convenient than to work with S_{ap}^p-functions.

(A1.4) Besicovitch almost-periodicity definitions and horizontal hierarchies.

As already pointed out, the structure of the Weyl spaces is more intricated than S_{ap}^p and C_{ap}, because every element of the space is a class of $L_{loc}^p(\mathbb{R}, \mathbb{R})$ functions, which may differ from each other even on a set of an infinite Lebesgue measure. We have not deepened the question in the previous section, but it is necessary to talk about this fact for the Besicovitch spaces.

Following [Pn-M], let us consider the *Marcinkiewicz spaces*

$$\mathcal{M}^p(\mathbb{R}) = \left\{ f : \mathbb{R} \to \mathbb{R}, f \in L_{loc}^p(\mathbb{R}, \mathbb{R}), \text{ s.t.} \right.$$

$$\left. \|f\|_p = \limsup_{T \to \infty} \left(\frac{1}{2T} \int_{-T}^{T} |f(x)|^p \, dx \right)^{1/p} < \infty \right\}, \text{ for all } p \geq 1.$$

For the case $p = \infty$, we have

$$\mathcal{M}^\infty(\mathbb{R}) = \{ f : \mathbb{R} \to \mathbb{R}, f \in L_{loc}^1, \text{ s.t. } \|f\|_\infty = \|f\|_{L^\infty} < \infty \}.$$

$\mathcal{M}^p$, endowed with the seminorm

$$(\text{A1.4.1}) \qquad \|f\|_p = \begin{cases} \limsup_{T \to \infty} \left(\frac{1}{2T} \int_{-T}^{T} |f(x)|^p \, dx \right)^{1/p} & \text{if } 1 \leq p < \infty, \\ \|f\|_{L^\infty} = \sup \text{ess}_{x \in \mathbb{R}} |f(x)| & \text{if } p = \infty \end{cases}$$

is a seminormed space.

Sometimes it is convenient to use the seminorm ([Bes2], [BF, p. 42])

$$\|f(x)\|_p^* = \max \left\{ \limsup_{T \to \infty} \left[\frac{1}{T} \int_0^T |f(x)|^p \, dx \right]^{1/p}, \limsup_{T \to \infty} \left[\frac{1}{T} \int_{-T}^0 |f(x)|^p \, dx \right]^{1/p} \right\},$$

which is equivalent to the seminorm (A1.4.1), because

$$\left(\frac{1}{2} \right)^{1/p} \|f(x)\|_p^* \leq \|f\|_p \leq \|f(x)\|_p^*.$$

(A1.4.2) THEOREM ([Mrc]) [*Marcinkiewicz*]. *The space $\mathcal{M}^p$ is a Fréchet space, i.e. a topological seminormed complete space.*

The proof is essentially based on the following

(A1.4.3) LEMMA. *For a seminormed space $(X, \|\cdot\|)$, the following conditions are equivalent:*

  (i) *X is complete,*
  (ii) *every absolutely convergent series is convergent (i.e. for all $\{x_n\}_{n \in \mathbb{N}} \subset X$ s.t. $\sum_{n=1}^{\infty} \|x_n\| < \infty$, there exists $x \in X$ s.t. $\lim_{N \to \infty} \|x - \sum_{n=1}^{N} x_n\| = 0$).*

Let us note that the limits in the Marcinkiewicz space are not uniquely determined. In fact, two different functions, differing from each other (even on an infinite set) by a function belonging to L^p, can be the limits of the same Cauchy sequence of elements in $\mathcal{M}^p$. Following the standard procedure, let us consider the kernel of the seminorm (A1.4.1)

$$K_p = \{f \in \mathcal{M}^p \quad \text{s.t.} \quad \|f\|_p = 0\}.$$

Let us consider the equivalence relation

(A1.4.4) $f \sim g \iff \|f - g\|_p = 0, \quad f, g \in \mathcal{M}^p$

and the quotient space

$$M^p(\mathbb{R}) = \mathcal{M}^p / K_p,$$

denoting by $\widehat{f}$ the element belonging to $\mathcal{M}^p$, corresponding to the function f.

Since $\mathcal{M}^p$ is a seminormed space and K_p is a subspace, then (A1.4.1) represents a norm on $\mathcal{M}^p$. Since $\mathcal{M}^p$ is complete, then M^p is a Banach space. This fact follows from the well-known

(A1.4.5) LEMMA. *Let $(X, \|\cdot\|)$ be a semi-normed space. Then*

  (i) *the kernel $K = \{x \in X \text{ s.t. } \|x\| = 0\}$ is a subspace of X,*
  (ii) *if $[x]$ is an equivalence class, then $\|[x]\| := \|x\|$ defines a norm on the quotient space X/K,*
  (iii) *if X is complete, then X/K is a Banach space.*

Let us now consider the class

$$W^p = \left\{ f \in L^p_{\text{loc}}(\mathbb{R}, \mathbb{R}) \text{ s.t. } \exists \lim_{T \to \infty} \left(\frac{1}{2T} \int_{-T}^{T} |f(x)|^p \, dx \right)^{1/p} \right\} \subset \mathcal{M}^p.$$

It is possible to show that this class is not a linear space, because it is not closed w.r.t. the summation.

(A1.4.6) EXAMPLE. Let us consider the functions

$$f_1(x) = \begin{cases} 0, & \text{for } x \leq 1, \\ x + \sqrt{2 + \sin \log x + \cos \log x}, & \text{for } x > 1 \end{cases}$$

and

$$f_2(x) = \begin{cases} 0, & \text{for } x \leq 1, \\ -x, & \text{for } x > 1. \end{cases}$$

We have

$$\frac{1}{2T} \int_{-T}^{T} f_1^2(x) \, dx$$

$$= \frac{1}{2T} \left[\frac{T^3 - 1}{3} + T(2 + \sin \log T) - 2 + \int_1^T 2x\sqrt{2 + \sin \log x + \cos \log x} \, dx \right].$$

Then

$$\int f_1^2(x) \, dx = \lim_{T \to \infty} \left[\frac{T^2}{6} - \frac{1}{6T} + \frac{2 + \sin \log T}{2} - \frac{1}{T} \right]$$

$$+ \lim_{T \to \infty} \frac{\int_1^T x\sqrt{2 + \sin \log x + \cos \log x} \, dx}{T}$$

(applying de L'Hôspital's rule)

$$= \lim_{T \to \infty} \frac{T^2}{6} + \lim_{T \to \infty} \frac{2 + \sin \log T}{2} + \lim_{T \to \infty} T\sqrt{2 + \sin \log T + \cos \log T}$$

$$\geq \lim_{T \to \infty} \left[\frac{T^2}{6} + \frac{1}{2} \right] = \infty.$$

On the other hand,

$$\int f_2^2(x) \, dx = \lim_{T \to \infty} \frac{1}{2T} \int_1^T x^2 \, dx = \lim_{T \to \infty} \frac{1}{2T} \left[\frac{x^3}{3} \right]_1^T = \infty.$$

However, the function

$$g(x) = f_1(x) + f_2(x) = \begin{cases} 0, & \text{for } x \leq 1, \\ \sqrt{2 + \sin \log x + \cos \log x}, & \text{for } x > 1 \end{cases}$$

is such that

$$\int_{-T}^{T} g^2(x) \, dx = \int_1^T [2 + \sin \log x + \cos \log x] \, dx = T(2 + \sin \log T) - 2,$$

and consequently

$$\int g^2(x) \, dx = \lim_{T \to \infty} \left[\frac{2 + \sin \log T}{2} - \frac{1}{T} \right] = \lim_{T \to \infty} \left[1 + \frac{\sin \log T}{2} \right]$$

does not exist.

(A1.4.7) DEFINITION ([BF, p. 36], [GKL-M, p. 192], [Zaa-M, pp. 103–108])
[*approximation*]. We will denote by $B^p(\mathbb{R})$ the Besicovitch space obtained as the
closure in $\mathcal{M}^p$ of the space $\mathcal{P}(\mathbb{R}, \mathbb{R})$ of all trigonometric polynomials.

In other words, an element in B^p can be represented by a function $f \in L^p_{\text{loc}}(\mathbb{R}, \mathbb{R})$
s.t., for every $\varepsilon > 0$, there exists $P_\varepsilon \in \mathcal{P}$ s.t.

$$\limsup_{T \to \infty} \left(\frac{1}{2T} \int_{-T}^{T} |f(x) - P_\varepsilon(x)|^p \, dx \right)^{1/p} < \varepsilon.$$

(A1.4.8) PROPOSITION ([BF, p. 45], [Lau]). *The space B^p is a closed subspace
of $\mathcal{W}^p$.*

Consequently, since B^p is a closed subset of the complete space M^p, it is complete, too.

It is possible to introduce another space as the completion of the space $\mathcal{P}$.

(A1.4.9) DEFINITION ([AndBeI]). We will denote by $\mathcal{B}^p$ the space obtained as
the *abstract* completion of the space $\mathcal{P}$ w.r.t. the norm

$$(A1.4.10) \qquad |||P|||_p = \lim_{T \to \infty} \left(\frac{1}{2T} \int_{-T}^{T} |P(x)|^p \, dx \right)^{1/p}, \quad P \in \mathcal{P}.$$

By the definition, $\mathcal{B}^p$ is a Banach space and its elements are classes of Cauchy
sequences of trigonometric polynomials w.r.t. the norm (A1.4.10). Thus, according
to this definition, it is rather difficult to understand the meaning of an element
of $\mathcal{B}^p$. The following theorem allows us to assign to every element of this space a
real function.

(A1.4.11) THEOREM. $B^p \equiv \mathcal{B}^p$.

PROOF. First of all, let us remark that, for every element of the space $\mathcal{P}$, the
norms (A1.4.1) and (A1.4.10) coincide.

Both B^p and $\mathcal{B}^p$ contain a subspace isomorphic to the space $\mathcal{P}$. Let us identify these subspaces. Let $\widehat{P} \in \mathcal{B}^p$ be an equivalence class of Cauchy sequences
of trigonometric polynomials w.r.t. the norm (A1.4.10). Then, every sequence
$\{P_n\}_{n \in \mathbb{N}} \in \widehat{P}$ is such that $\|P_n - P_m\|_p \to 0$, for $m, n \to \infty$. It follows that
$\{P_n\}_{n \in \mathbb{N}}$ is a Cauchy sequence in B^p, and consequently there exists $\widehat{f} \in B^p$ such
that $\|P_n - \widehat{f}\|_p \to 0$, for $n \to \infty$. In other words, the class $\widehat{P}$ uniquely determines
a class $\widehat{f}$ of functions belonging to $L^p_{\text{loc}}(\mathbb{R}, \mathbb{R})$ w.r.t. the equivalence relation

$$f, g \in \widehat{f} \Longleftrightarrow \|f - g\|_p = 0.$$

In fact, if the sequence $\{Q_n\}_{n\in\mathbb{N}} \in \widehat{P}$ were such that $\|Q_n - g\|_p \to 0$, then we should have

$$\|f - g\|_p \leq \|f - P_n\|_p + \|P_n - Q_n\|_p + \|Q_n - g\|_p \to 0, \quad \text{whenever } n \to \infty,$$

because, if $\{P_n\}_{n\in\mathbb{N}}, \{Q_n\}_{n\in\mathbb{N}} \in \widehat{P}$, then $\|P_n - Q_n\|_p \to 0$, whenever $n \to \infty$.

On the other hand, a class $\widehat{f} \in B^p$ uniquely determines a class of trigonometric polynomials $\widehat{P} \in \mathcal{B}^p$. In fact, let $\{P_n\}_{n\in\mathbb{N}}, \{Q_n\}_{n\in\mathbb{N}} \subset B^p(\mathbb{R})$ be such that $\|P_n - \widehat{f}\|_p \to 0$, $\|Q_n - \widehat{f}\|_p \to 0$, whenever $n \to \infty$ (i.e. there exist $f, g \in \widehat{f}$ such that $\|f - P_n\|_p \to 0$ and $\|Q_n - g\|_p \to 0$). Let us show that there exists $\widehat{P} \in \mathcal{B}^p$ such that $\{P_n\}_{n\in\mathbb{N}}, \{Q_n\}_{n\in\mathbb{N}} \in \widehat{P}$. In fact,

$$\|P_n - Q_n\|_p \leq \|f - P_n\|_p + \|Q_n - f\|_p \to 0, \quad \text{whenever } n \to \infty.$$

Then the two sequences are equivalent and belong to the same class of the space B^p_{ap}. It is easy to show that the equivalence between the two spaces is an isometry. $\square$

(A1.4.12) REMARK. Considering the closure of the space $\mathcal{P}$ w.r.t. the norm (A1.4.1) in the space $\mathcal{M}^p$, we obtain a space $\widetilde{B}^p$, which is still a seminormed, complete space, whose elements are still functions; its quotient space, w.r.t. the equivalence relation (A1.4.4), is B^p.

Definition (A1.4.7) is obtained as an approximation definition. It is possible to show that this definition is equivalent to a *Bohr-like* one, provided we introduce a new property of numerical sets.

(A1.4.13) DEFINITION ([Bes-M, pp. 77–78], [BeBo, p. 227], [Fol]). A set $X \subset \mathbb{R}$ is said to be *satisfactorily uniform* (s.u.) if there exists a positive number l such that the ratio r of the maximum number of elements of X included in an interval of length l to the minimum number is less than 2.

Every s.u. set is r.d. The converse is, in general, not true.

Although, for example, the set $\mathbb{Z}$ is r.d. and s.u. in $\mathbb{R}$, the set $X = \mathbb{Z} \cup \{\frac{1}{n}\}_{n\in\mathbb{N}}$ is r.d., but it is not s.u.: in fact, due to the presence of the accumulation point 0, $r = \infty$, for all $l > 0$. Thus, a r.d. set, in order to be s.u., cannot have any finite accumulation point.

(A1.4.14) DEFINITION ([Bes-M, p. 78], [BeBo, p. 227], [Fol, p. 6]). A function $f \in L^p_{loc}(\mathbb{R}, \mathbb{R})$ is said to be *almost-periodic in the sense of Besicovitch* ($\mathcal{B}^p_{ap}$) if, for every $\varepsilon > 0$, there corresponds a s.u. set $\{\tau_k\}_{k\in\mathbb{Z}}$ ($\tau_j < \tau_i$ if $j < i$) s.t., for each i,

$$(A1.4.15) \qquad \limsup_{T\to\infty} \left(\frac{1}{2T} \int_{-T}^{T} |f(x + \tau_i) - f(x)|^p \, dx \right)^{1/p} < \varepsilon,$$

and, for every $c > 0$,

$$(A1.4.16) \quad \limsup_{T \to \infty} \left(\frac{1}{2T} \int_{-T}^{T} \left[\limsup_{n \to \infty} \frac{1}{2n+1} \right. \right.$$
$$\left. \left. \sum_{i=-n}^{n} \frac{1}{c} \int_{x}^{x+c} |f(t+\tau_i) - f(t)|^p \, dt \right] dx \right)^{1/p} < \varepsilon.$$

The space $\mathcal{B}_{ap}^1$ will be shortly indicated by $\mathcal{B}_{ap}$.

(A1.4.17) THEOREM ([Bes-M, pp. 95–97, 100–101], [BeBo, pp. 247–257]). *The spaces $\mathcal{B}_{ap}^p$ and B^p are equivalent.*

It can be readily checked that Definition (A1.4.14) is rather cumbersome, even in its simplified form, obtained substituting conditions (A1.4.15) and (A1.4.16) with the simplest one [Bes2]

$$(A1.4.18) \quad \limsup_{T \to \infty} \left(\frac{1}{2T} \int_{-T}^{T} \left[\limsup_{n \to \infty} \frac{1}{2n+1} \sum_{i=-n}^{n} |f(x+\tau_i) - f(x)|^p \right] dx \right)^{1/p} < \varepsilon.$$

It can be be shown [Bes-M], [BeBo] that the spaces given by these two different definitions are equivalent.

A. S. Besicovitch introduced even a simpler definition, which permits us to introduce another space.

(A1.4.19) DEFINITION ([Bes-M, p. 112], [BeBo, p. 267]). A function $f(x) \in L_{\text{loc}}^1$ is said to be $\overline{B}_{ap}^1$ if

$$\liminf_{T \to \infty} \frac{1}{2T} \int_{-T}^{T} |f(x)| \, dx < \infty$$

and, for every $\varepsilon > 0$, there corresponds a s.u. set of numbers τ_i s.t., for each i, (A1.4.15) and (A1.4.16) are satisfied.

(A1.4.20) THEOREM ([Bes-M, pp. 113–123], [BeBo, pp. 268–269]). $\overline{B}_{ap}^1 \subset \mathcal{B}_{ap}$.

The inclusion is strict, as shown in [Bes-M, pp. 126–129], [BeBo, pp. 286–291].

It is worthwhile to observe that, although $\overline{B}_{ap}^1$ is strictly contained in $\mathcal{B}_{ap}$, to every function in $\overline{B}_{ap}^1$ there corresponds a function $\mathcal{B}_{ap}$ with the same Bohr–Fourier series. This property is related to the following

(A1.4.21) THEOREM ([Bes-M, p. 123], [BeBo, pp. 281–282]). *To every function $f(x) \in \mathcal{B}_{ap}$, there corresponds a $\overline{B}_{ap}^1$-function differing from $f(x)$ by a function the mean value of whose modulus is zero.*

Due to the difficulty of the original definition, several authors have studied alternative (and simpler) definitions of the Besicovitch spaces, each of them based on

Bohr-like or Bochner-like properties in the Besicovitch norm. It is then convenient to consider the norm (A1.4.1) rather than the norm given by Definition (A1.4.14). To this aim, we need some preliminary definitions in terms of (A1.4.1).

(A1.4.22) DEFINITION ([AndBeL], [Dos1], [Ber3, p. 69]) [*Bohr-type defini-tion*]. A function $f \in L^p_{\mathrm{loc}}(\mathbb{R}, \mathbb{R})$ is said to be *almost-periodic in the sense of Doss* (B^p_{ap}) if, for every $\varepsilon > 0$, there corresponds a r.d. set $\{\tau\}_\varepsilon$ s.t.

$$\limsup_{T \to \infty} \left(\frac{1}{2T} \int_{-T}^{T} |f(x + \tau) - f(x)|^p \, dx \right)^{1/p} < \varepsilon, \quad \text{for all } \tau \in \{\tau\}_\varepsilon.$$

Each number $\tau \in \{\tau\}_\varepsilon$ is called an ε-B^p *almost-period* (or a $B^p - \varepsilon$-*translation number*) of $f(x)$.

(A1.4.23) DEFINITION ([Dos2], [Dos4]) [*normality or Bochner-type definition*].

A function $f \in L^p_{\mathrm{loc}}(\mathbb{R}, \mathbb{R})$ is called B^p-*normal* if, for every sequence $\{h_i\}$ of real numbers, there corresponds a subsequence $\{h_{n_i}\}$ s.t. the sequence of functions $\{f(x + h_{n_i})\}$ is B^p-convergent, i.e.

$$\lim_{n \to \infty} \limsup_{T \to \infty} \frac{1}{2T} \int_{-T}^{T} |f(x + h_n)|^p \, dx = 0.$$

(A1.4.24) DEFINITION ([Ber3, p. 15], [Dos1], [Dos2]) [*continuity*]. A function $f \in L^p_{\mathrm{loc}}(\mathbb{R}, \mathbb{R})$ is called B^p-*continuous* if

$$\lim_{\tau \to 0} \limsup_{T \to \infty} \frac{1}{2T} \int_{-T}^{T} |f(x + \tau) - f(x)|^p \, dx = 0.$$

The space of all the B^p-continuous functions will be indicated with B^p_c. Clearly, it is a (complete) subspace of $\mathcal{M}^p$.

(A1.4.25) DEFINITION ([Ber3, p. 15]) [*regularity*]. A function $f \in L^p_{\mathrm{loc}}(\mathbb{R}, \mathbb{R})$ is called B^p-*regular* if, for every $l \in \mathbb{R}$,

(A1.4.26) $$\limsup_{T \to \pm\infty} \frac{1}{2T} \int_{T-l}^{T} |f(x)|^p \, dx = 0.$$

This condition implies that a B^p-regular function cannot assume too large values in finite intervals. The space of all the B^p-regular functions will be indicated by B^p_r. Clearly, it is a (complete) subspace of $\mathcal{M}^p$ (see [Ber3, p. 16]). Besides, since

$$\limsup_{T \to \pm\infty} \frac{1}{2T} \int_{T-l}^{T} |f(x)|^p \, dx \leq \|f\|_p^p,$$

it follows that every null function in the Besicovitch norm is regular.

(A1.4.27) THEOREM ([Dos2]). *A function $f \in L^p_{loc}(\mathbb{R}, \mathbb{R})$ belongs to the space B^p iff*

(i) $f(x)$ *is B^p-bounded, i.e. it belongs to $\mathcal{M}^p$,*

(ii) $f(x)$ *is B^p-continuous,*

(iii) $f(x)$ *is B^p-normal,*

(iv) *for every $\lambda \in \mathbb{R}$,*

$$\lim_{L \to \infty} \limsup_{T \to \infty} \int_{-T}^{T} \left| \frac{1}{L} \int_x^{x+L} f(t)e^{i\lambda t}\, dt - \frac{1}{L} \int_0^L f(t)e^{i\lambda t}\, dt \right| dx = 0.$$

(A1.4.28) REMARK. Condition (iv) is, actually, formed by infinite conditions, each one for each value of λ. Each of them is independent of the others. For example, it can be proved (see [Dos2]) that, for every λ_0, the functions $f(x) = e^{i\lambda_0 x} \operatorname{sign} x$ satisfy conditions (i)–(iv), for every value $\lambda \neq \lambda_0$.

Condition (iv) can be replaced by the following condition ([Dos2]):

(iv') to every $\lambda \in \mathbb{R}$, there corresponds a number $a(\lambda)$ s.t.

$$\lim_{L \to \infty} \limsup_{T \to \infty} \int_{-T}^{T} \left| \frac{1}{L} \int_x^{x+L} f(t)e^{i\lambda t}\, dt - a(\lambda) \right| dx = 0,$$

or by (see [Dos2])

(iv") for every $a \in \mathbb{R}$, there exists a function $f^{(a)}(x) \in L^p$, a-periodic and s.t.

$$(A1.4.29) \qquad \lim_{n \to \infty} \left\{ \limsup_{T \to \infty} \frac{1}{2T} \int_{-T}^{T} \left[\frac{1}{n} \sum_{k=0}^{n-1} |f(x+ka) - f^{(a)}(x)|^p\, dx \right] \right\} = 0.$$

Moreover, condition (iii) can be replaced by a Bohr-like condition ([Dos1]):

(iii') for every ε, the set of numbers τ for which

$$\limsup_{T \to \infty} \frac{1}{2T} \int_{-T}^{T} |f(x+\tau) - f(x)|\, dx < \varepsilon$$

is r.d.

It follows that, under conditions (i), (ii) and (iv), a function is B^p-normal iff it is almost-periodic in the sense of R. Doss.

Comparing Theorems (A1.4.17) and (A1.4.27), let us note that introducing a Bohr-like definition as in condition (iii') represents a weaker structural characterization than Definition (A1.4.14).

J.-P. Bertrandias has restricted his analysis to B^p_c-functions (see Definition (A1.4.24)), showing the equivalence of the different definitions.

(A1.4.30) DEFINITION ([Ber3, p. 69]). A function $f(x) \in B_c^p$ is called $\mathcal{M}^p$-almost-periodic ($\mathcal{M}_{ap}^p$) if, for every $\varepsilon > 0$, there exists an r.d. set $\{\tau_\varepsilon\}$ s.t.

$$\limsup_{T \to \infty} \left(\frac{1}{2T} \int_{-T}^{T} |f(x+\tau) - f(x)|^p \, dx \right)^{1/p} < \varepsilon, \quad \text{for all } \tau \in \{\tau\}_\varepsilon.$$

(A1.4.31) DEFINITION ([Ber3, p. 69]) [*normality*]. A function $f(x) \in B_c^p$ is called $\mathcal{M}^p$-normal if the set $\{f^\tau\}$ of its translates is B^p-precompact.

(A1.4.32) THEOREM ([Ber3, p. 69]). *Definitions* (A1.4.30) *and* (A1.4.31) *are equivalent.*

In order to show the equivalence with the third type of definition, we need a preliminary definition.

(A1.4.33) DEFINITION ([Ber3, p. 50]). Given a family $\{k_\lambda(x)\}_{\lambda \in \mathbb{R}}$ of B^p-constant functions (see Definition (A1.5.11)), we will call by a *generalized trigonometric polynomial*, or by a *trigonometric polynomial with B^p-constant coefficients*, the function

$$\sum_{\lambda \in \mathbb{R}} k_\lambda(x) e^{i\lambda x}.$$

The class of generalized trigonometric polynomials will be denoted by P^p. It is easy to show that this is a linear subspace of B_c^p. Obviously, the space $\mathcal{P}$ is a subspace of P^p.

(A1.4.34) THEOREM ([Ber3, p. 71]). *A function $f(x) \in B_c^p$ is $\mathcal{M}_{ap}^p$ iff it is the B^p-limit of a sequence of generalized trigonometric polynomials. In other words, $\mathcal{M}_{ap}^p$ is the closure, w.r.t. the Besicovitch norm, of the space P^p.*

Since $\mathcal{M}_{ap}^p$ is a closed subspace of the complete space B_c^p, it is complete, too. The space B^p is a complete subspace of $\mathcal{M}_{ap}^p$.

(A1.4.35) THEOREM ([Ber3, p. 72]). *The space $\mathcal{M}_{ap}^p$ is a complete subspace of B_r^p.*

(A1.4.36) THEOREM ([Ber3, p. 72]) [*Uniqueness Theorem*]. *If two functions belonging to $\mathcal{M}_{ap}^p$ have the same generalized Bohr–Fourier coefficients, they are equivalent in the Besicovitch norm.*

Comparing Theorem (A1.4.27), Remark (A1.4.28) and Definitions (A1.4.30), (A1.4.27), it is clear that Condition (iv) in Theorem (A1.4.27), or the equivalent (iv'), is the necessary and sufficient condition in order a $\mathcal{M}_{ap}^p$-function to be a B^p-function. For example, the functions $f_\lambda(x) = e^{i\lambda x} \operatorname{sign} x$, are $\mathcal{M}_{ap}^p$-functions (since sign$x$ is a B^p-constant), but they are not B^p-functions, as already shown in Remark (A1.4.28).

J.-P. Bertrandias has also proved a further characterization of the Besicovitch functions, in terms of *correlation functions*, whose discussion would bring us far from the goal of this chapter. For more information, see [Ber3, pp. 70–71].

On the other hand, A. S. Kovanko has introduced a new class of functions and has proved its equivalence with the space B^p_{ap}.

(A1.4.37) DEFINITION ([Kov2]). Given a function $f(x) \in L^p_{loc}(\mathbb{R}, \mathbb{R})$ and a set $E \subset \mathbb{R}$, let us define

$$\overline{M}^E\{|f(x)|^p\} = \{D^E_{B^p}[f(x), 0]\}^p := \limsup_{T \to \infty} \left[\frac{1}{2T} \int_{E \cap (-T,T)} |f(x)|^p \, dx \right]$$

and

$$\overline{\delta}E := \limsup_{T \to \infty} \frac{|E(-T,T)|}{2T},$$

where

$$|E(-T,T)| = \mu[E \cap (-T,T)].$$

Observe that that, if $E = \mathbb{R}$,

$$\overline{M}^E\{|f(x)|^p\} = \|f\|^p_p.$$

$f(x)$ is said to be B^p-*uniformly integrable* ($f(x) \in B^p_{u.i.}$) if, for every $\varepsilon > 0$, there exists $\eta(\varepsilon) > 0$ s.t.

$$\overline{M}^E\{|f(x)|^p\} < \varepsilon, \quad \text{whenever } \overline{\delta}E < \eta.$$

(A1.4.38) DEFINITION ([Kov2], [Kov5]). A function $f(x)$ is said to belong to the *class A_p* if

(i) $f(x) \in B^p_{u.i.}$,

(ii) for every $\varepsilon > 0$, there exists $\eta > 0$ and an r.d. set of ε-almost periods τ s.t.

$$|f(x+t) - f(x)| < \varepsilon, \quad \text{for } \tau - \eta < t < \tau + \eta,$$

for arbitrary $x \in \mathbb{R}$, possibly with an exception of a set E_t, s.t. $\overline{\delta}E_t < \eta$,

(iii) for every $a > 0$ there exists an a-periodic function $f^{(a)}(x)$ which is a.e. bounded and s.t. (A1.4.29) holds.

(A1.4.39) REMARK. In [Kov2], A. S. Kovanko shows that, if a function $f(x)$ belongs to A_p, then it belongs to L^p. Thus, condition (iv") in Remark (A1.4.28) is a consequence of condition (iii) in Definition (A1.4.38).

(A1.4.40) THEOREM ([Kov2]). $B^p \equiv A_p$.

G. Bruno and F. R. Grande proved a Lusternik-type theorem, very similar to the corresponding theorem for C_{ap}-functions.

(A1.4.41) THEOREM ([BuGr1]). *Let $\mathcal{F}$ be a family of elements belonging to B_{ap}^p, $1 \le p < \infty$, closed and bounded. Then the following statements are equivalent:*

  (i) *$\mathcal{F}$ is compact in the B^p-norm,*

 (ii) *$\mathcal{F}$ is B^p-equicontinuous, i.e. for any $\varepsilon > 0$ there exists $\delta = \delta(\varepsilon)$ s.t., if $|h| < \delta$, then*

$$D_{B^p}[f(x+h) - f(x)] < \varepsilon, \quad \text{for all } f \in \mathcal{F}$$

and B^p-equi-almost-periodic, i.e., for any $\varepsilon > 0$, there exists $l(\varepsilon) > 0$ s.t., for any interval whose lenghth is $l(\varepsilon)$, there exists $\xi < l(\varepsilon)$ s.t.

$$D_{B^p}[f(x+\xi), f(x)] < \varepsilon, \quad \text{for all } f \in \mathcal{F}.$$

(A1.4.42) THEOREM ([BuGr1]). *Every B^p-function is B^p-normal.*

(A1.4.43) REMARK. Theorem (A1.4.42) is also a corollary of Theorem (A1.4.27), by means of which we also prove that every B^p-function is B_{ap}^p, B^p-bounded and B^p-continuous.

(A1.4.44) THEOREM ([BuGr1]). *Every B^p-normal function is B_{ap}^p.*

For both Theorems (A1.4.42) and (A1.4.44), the converse is not true.

(A1.4.45) EXAMPLE (*Example of a function which is B_{ap}^p, but not B^p-normal*). The example is based partly on [St-M, pp. 42–47]. In Example (A1.3.43), it has been shown that the function

$$f(x) = \begin{cases} 0 & \text{if } x \in (-\infty, 0], \\ \sqrt{\frac{n}{2}} & \text{if } x \in (n-2, n-1], \ n = 2, 4, 6, \ldots, \\ -\sqrt{\frac{n}{2}} & \text{if } x \in (n-1, n], \ n = 2, 4, 6, \ldots, \end{cases}$$

is a W_{ap}^1-function. Furthermore, in Part 5, it will be shown that every W_{ap}^p-function is B_{ap}^p. Consequently, since the function is W_{ap}^1, it is B_{ap}^1. Now, we want to show that it is not B^1-normal. Let us take $c \in \mathbb{R}$, $c \ne 2k$, $k \in \mathbb{Z}$. Without any loss of generality, we can suppose $c > 0$. In fact, if $c < 0$, $(c = -d)$, then

$$D_B[f(x+d), f(x)] = D_B[f(x), f(x-d)] = D_B[f(x+c), f(x)].$$

Since we will have to consider the limit for $T \to \infty$, let us take $T > c$. Then there exist $i, l \in \mathbb{N}$ s.t.

$$2i \leq c < 2(i + 1), \quad 2l \leq T < 2(l + 1).$$

Put $\delta := c - 2i$. We distinguish two cases:

(a) $0 \leq \delta \leq 1$,

(b) $1 < \delta < 2$.

Since we have to evaluate

$$\frac{1}{2T} \int_{-T}^{T} |f(t + c) - f(t)| \, dt,$$

let us compute the difference $|f(t + c) - f(t)|$ in intervals whose union is strictly included in the interval $[-T, T]$. For the case (a), we will take the intervals $(j - 1 - \delta, j - 1)$, $j = 2, 4, 6, \ldots$, for the case (b), we will take the intervals $(j, j + \delta)$, $j = 2, 4, 6, \ldots$ We have respectively

$$|f(t + c) - f(t)| = \sqrt{\frac{j}{2}} + \sqrt{\frac{j + 2i}{2}},$$

$$|f(t + c) - f(t)| = \sqrt{\frac{j + 2}{2}} + \sqrt{\frac{j + 2 + 2i}{2}}.$$

Since the second equality can be obtained from the first one by means of a variable shifting, let us consider only the case (a). We have

$$\frac{1}{2T} \int_{-T}^{T} |f(t + c) - f(t)| \, dt \geq \frac{1}{2l} \sum_{n=1}^{l} \int_{2n-1-\delta}^{2n-1} |f(t + c) - f(t)| \, dt$$

$$= \frac{1}{2l} \sum_{n=1}^{l} \delta(\sqrt{n + i} + \sqrt{n})$$

$$\geq \frac{1}{l} \sum_{n=1}^{l} \delta \sqrt{n} \geq \frac{\delta}{l} \frac{2}{3} l^{3/2},$$

where the last inequality is obtained by virtue of the Cauchy integral criterion of convergence. Thus, we get

$$\frac{1}{2T} \int_{-T}^{T} |f(t + c) - f(t)| \, dt \geq \frac{2\delta}{3} \sqrt{l}.$$

Passing to the limit for $T \to \infty$, i.e. for $l \to \infty$, provided $c \neq 2k$ ($k \in \mathbb{Z}$), we obtain

$$D_B[f(x + c), f(x)] = \limsup_{T \to \infty} \frac{1}{2T} \int_{-T}^{T} |f(t + c) - f(t)| \, dt \geq \frac{\delta}{3} \lim_{l \to \infty} \sqrt{l} = \infty.$$

If we fix a number $a \in \mathbb{R}$ and a sequence $\{a_i\}_{i \in \mathbb{N}}$ s.t. $a_i - a \neq 2k$, $k \in \mathbb{Z}$, then

$$D_B[f(x+a), f(x+a_i)] = D_B[f(x+a-a_i), f(x)] = \infty.$$

We conclude that the sequence of translates $\{f(x+a_i)\}$ is not relatively compact and consequently the function is not B^1-normal.

Let us note that the function cannot satisfy all the conditions (i)–(iii) in Theorem (A1.4.27) (otherwise, according to Remark (A1.4.28), it would be B^1-normal). Let us show that $f(x)$ is not B^1-continuous. In fact,

$$|f(x+a) - f(x)| = \begin{cases} \sqrt{2n} & \text{if } x \in (n-1-\alpha, n-1], \\ \sqrt{\frac{n}{2}} + \sqrt{\frac{n-2}{2}} & \text{if } x \in (n-2-\alpha, n-2], \end{cases}$$

for every $n = 2, 4, 6, \ldots$

Let us take $T = 2l$. Then

$$\int_{-T}^{T} |f(x+a) - f(x)| \, dx = \sum_{k=1}^{l} [\sqrt{k} + \sqrt{k-1}]\alpha + \sum_{k=1}^{l} 2\sqrt{k}\alpha$$

$$= \left[3\sqrt{l} + \sum_{k=1}^{l-1} 4\sqrt{k} \right] \alpha.$$

Passing to the limit,

$$\lim_{T \to \infty} \frac{1}{2T} \int_{-T}^{T} |f(x+a) - f(x)| \, dx \geq \lim_{l \to \infty} \frac{\alpha}{4l} \left[3\sqrt{l} + \sum_{k=1}^{l-1} 4\sqrt{k} \right]$$

$$\geq \alpha \lim_{l \to \infty} \left[\frac{3}{4\sqrt{l}} + \frac{2}{3l}(l-1)^{\frac{3}{2}} \right] = \infty,$$

and the claim follows.

(A1.4.46) EXAMPLE (*Example of a function which is B^p-normal, but not B^p*). The example is partly based on [BF, p. 107] and [Fol, p. 5]. Let us consider the function

$$\text{sign}(x) = \begin{cases} -1 & \text{if } x < 0, \\ 0 & \text{if } x = 0, \\ 1 & \text{if } x > 0. \end{cases}$$

As shown in [Fol, p. 5], this function is not B^p: it is sufficient to recall that, by virtue of Theorem (A1.5.2), for every B^p-function, there exists the mean value (A1.1.17) and

$$M[f] = \lim_{T \to \infty} \frac{1}{T} \int_{0}^{T} f(x) \, dx = \lim_{T \to \infty} \frac{1}{T} \int_{-T}^{0} f(x) \, dx.$$

However,

$$\lim_{T\to\infty} \frac{1}{T} \int_0^T \operatorname{sign}x \, dx = 1,$$

$$\lim_{T\to\infty} \frac{1}{T} \int_{-T}^0 \operatorname{sign}x \, dx = -1.$$

Thus, Theorem (A1.5.2) is not fulfilled. On the other hand, $\operatorname{sign}x$ is a B_{ap}^p-function, because

$$|\operatorname{sign}(x+\tau) - \operatorname{sign}(x)| = \begin{cases} 2 & \text{if } x \in [-\tau, 0), \\ 0 & \text{elsewhere}, \end{cases}$$

and

$$\|\operatorname{sign}(x+\tau) - \operatorname{sign}(x)\|_p = 0, \quad \text{for all } \tau \in \mathbb{R}.$$

Let us show that it is B^p-normal, too.

In fact, for every choice of h_m, h_n (let us choose, without any loss of generality, $h_m > h_n > 0$), we have

$$\int_{-T}^T |\operatorname{sign}(x+h_m) - \operatorname{sign}(x+h_n)|^p \, dx = \int_{-h_m}^{-h_n} dx = h_m - h_n,$$

and consequently

$$\|\operatorname{sign}(x+h_m) - \operatorname{sign}(x+h_n)\|_p = 0$$

as well as the B^p-normality.

More generally, we can consider a function $f(x) = e^{-i\lambda x} \operatorname{sign}x$, which is not B^p, for every $\lambda \in \mathbb{R}$ (see Remark (A1.4.28)).

On the other hand, each of the functions $e^{-i\lambda x} \operatorname{sign}x$ is B^p-normal. In fact, for every choice of h_m, h_n ($h_m > h_n$), we have

$$\int_{-T}^T |f(x+h_m) - f(x+h_n)| \, dx = \int_{-h_m}^{-h_n} 1 \, dx = h_m - h_n,$$

and consequently

$$\oint |e^{i\lambda(x+h_m)} - e^{i\lambda(x+h_n)}| \, dx = 0, \quad \text{for all } h_m, h_n \in \mathbb{R}.$$

So, the claim follows.

However, for these functions, Formula (A1.1.19) holds. In fact

$$\lim_{T\to\infty} \frac{1}{T} \int_0^T e^{-i\lambda x} \operatorname{sign}x \, dx = \lim_{T\to\infty} \frac{i}{\lambda T}[e^{-i\lambda T} - 1] = 0,$$

$$\lim_{T\to\infty} \frac{1}{T} \int_{-T}^0 e^{-i\lambda x} \operatorname{sign}x \, dx = \lim_{T\to\infty} \frac{i}{\lambda T}[e^{i\lambda T} - 1] = 0.$$

This example shows that Theorem (A1.5.2) in Part 5, characterizing B^p-functions, is, in general, not satisfied by B^p_{ap} and B^p-normal functions.

The importance and the properties of the spaces B^p_{ap} can be better understood applying again the Bohr compactification. Let us introduce on $\mathbb{R}_B$ the normalized *Haar measure* μ (i.e. the positive regular Borel measure s.t. $\mu(U) = \mu(U + s)$, for every Borel subset $U \subset \mathbb{R}_B$ and for every $s \in \mathbb{R}_B$ (invariance property) and s.t. $\mu(\mathbb{R}_B) = 1$ (normality property)).

It is possible [Pn-M] to show that the space B^p_{ap} is isomorphic to the space $L^p(\mathbb{R}_B, \mathbb{R})$, where L^p is taken w.r.t. the Haar measure defined on $\mathbb{R}_B$. It follows that

$$\|f\|^p_{B^p} = \begin{cases} \|\widetilde{f}\|^p_{L^p(\mathbb{R}_B,\mu)} = \int_{\mathbb{R}_B} |\widetilde{f}(x)|^p \, d\mu(x) & \text{if } 1 \le p < \infty, \\ \sup\operatorname{ess}_{x \in \mathbb{R}_B} |\widetilde{f}(x)| & \text{if } p = \infty, \end{cases}$$

where $\widetilde{f}$ is the extension by continuity of f from $\mathbb{R}$ to $\mathbb{R}_B$.

From this isomorphism, many properties for the spaces B^p_{ap} can be obtained. For example, two functions differing from each other even on the whole real axis can belong to the same Besicovitch class, because two functions belonging to the same $L^p(\mathbb{R}_B, \mathbb{R})$-class may differ from each other on a set of the Haar measure zero and the real numbers are embedded in the Bohr compactification as a dense set of the Haar measure zero. Furthermore, recalling the inclusions among the spaces L^p on compact sets, we have

$$B^\infty_{ap} \subset B^{p_1}_{ap} \subset B^{p_2}_{ap} \subset B^1_{ap}, \quad \text{for all } p_1 > p_2 > 1,$$

where

$$B^\infty_{ap} = \bigcap_{p \in \mathbb{N}} B^p_{ap}.$$

Furthermore ([Fol5], [Ian1], [Ian2]), the spaces B^p_{ap}, $1 \le p < \infty$, are reflexive spaces and their duals are given by B^q_{ap}, where q is s.t. $\frac{1}{q} + \frac{1}{p} = 1$. The spaces B^p_{ap} are not separable (see, for example, [Zaa-M, p. 108]). In particular, the space B^2_{ap} is a non-separable Hilbert space, in which the exponents $e^{i\lambda x}$ ($\lambda \in \mathbb{R}$) form an orthonormal basis. Other properties can be found in [AndBeI], [Fol5], [Ian1], [Ian2], [Pn-M, pp. 11–12], [Pit].

(A1.5) Vertical hierarchies.

In the previous sections, we have shown that, although for the spaces C_{ap} and S^p_{ap}, the three definitions, in terms of relative density, normality and polynomial approximation, are equivalent, for the remaining spaces (e-W^p_{ap}, W^p_{ap}, B^p_{ap}) the equivalence does not hold anymore.

It is then important to check the relationships among every definition obtained w.r.t. one norm and the less restrictive definitions related to more general classes.

Before studying these relations and the vertical hierarchies among the spaces up to now studied, let us recall the most important properties that are common to all these spaces.

In fact, many of the properties of the u.a.p. functions can be satisfied by the functions belonging to the spaces of generalized a.p. functions. For the sake of simplicity, let us indicate with G^p the either (generic) space S_{ap}^p, e-W_{ap}^p, W_{ap}^p or B_{ap}^p. If not otherwise stated, the following theorems will be valid for any of the spaces studied.

First of all, let us underline the connection of almost-periodic functions with the trigonometric series.

(A1.5.1) THEOREM ([Bes-M, p. 104], [BeBo, p. 262], [BF, p. 45], [GKL-M, p. 191, 193]). *Every G^p-functions can be represented by its Fourier series, given by formula* (A1.1.23).

(A1.5.2) THEOREM ([Bes-M, p. 93], [BeBo, pp. 244–245], [BF, p. 45], [GKL-M, p. 191] [*Mean value theorem*]. *The mean value* (A1.1.17) *of every G^p-function $f(x)$ exists and*

$$(A1.5.3) \qquad \text{(a)} \quad M[f] = \lim_{T \to \infty} \frac{1}{T} \int_0^T f(x)\,dx = \lim_{T \to \infty} \frac{1}{T} \int_{-T}^0 f(x)\,dx,$$

$$(A1.5.4) \qquad \text{(b)} \quad M[f] = \lim_{T \to \infty} \frac{1}{2T} \int_{a-T}^{a+T} f(x)\,dx,$$

where the last limit exists uniformly w.r.t. $a \in \mathbb{R}$, for every function in S^p, in $e - W^p$ and in W^p.

Theorem (A1.5.2) is related to a property of the S^p-norm, stated by S. Koizumi.

(A1.5.5) THEOREM ([Koi]). *A function $f(x) \in L_{\text{loc}}^p(\mathbb{R}, \mathbb{R})$ belongs to BS^p iff there exists a positive constant K' s.t.*

$$\limsup_{T \to \infty} \frac{1}{2T} \int_{-T}^T |f(x+t)|^p\,dt \leq K', \quad \text{uniformly w.r.t. } x \in \mathbb{R}.$$

(A1.5.6) REMARK. Repeating the considerations done in Remark (A1.1.20), for every $f \in G^p$, the quantities $a(\lambda)$, given by (A1.1.21), are finite, for every $\lambda \in \mathbb{R}$.

This fact is no longer true, in general, for G_{ap}^p-functions, as shown in Example (A1.6.8).

(A1.5.7) THEOREM ([Bes-M, p. 104], [BeBo, p. 262], [BF, p. 45]). *For every G^p-function $f(x)$, there always exists at most a countable infinite set of the Bohr-Fourier exponents λ, for which $a(\lambda) \neq 0$, where $a(\lambda)$ are given by (A1.1.21).*

(A1.5.8) THEOREM ([Bes-M, p. 109], [BF, p. 47]) [*Bohr Fundamental Theorem*]. *The Parseval equation*

$$\sum_n |a(\lambda_n, f)|^2 = M\{|f(x)|^2\}$$

is true for every G^2-function.

(A1.5.9) THEOREM ([Bes-M, p. 109], [BeBo, p. 266], [BF, p. 45]) [*Uniqueness Theorem*]. *If two G^p-functions f, g have the same Fourier series, then they are identical, i.e.*

$$D_G[f(x), g(x)] = 0.$$

In other words, two different elements belonging to G^p cannot have the same Bohr–Fourier series.

The functions whose G^p-norm is equal to zero are called G^p-zero functions ([BF, p. 38]).

(A1.5.10) PROPOSITION. *Every G^p-zero function is a G^p-function and belongs to the class of the function $f(x) \equiv 0$.*

(A1.5.11) DEFINITION ([Ber3]). A G^p-bounded function is said G^p-*constant* if, for every real number τ,

$$\|f(x + \tau) - f(x)\|_{G^p} = 0.$$

(A1.5.12) PROPOSITION. *Every G^p-zero function is G^p-constant.*

PROOF. In fact, for every $\tau \in \mathbb{R}$, $\|f^\tau - f\|_{G^p} \le \|f^\tau\|_{G^p} + \|f\|_{G^p} = 0$. $\qquad \square$

The converse is, in general, not true. For example, the function $f(x) \equiv 1$ is, obviously, G^p-constant, but

$$\|f\|_p = \limsup_{T \to \infty} \frac{1}{2T} \int_{-T}^{T} 1 \, dx = 1 \ne 0.$$

(A1.5.13) PROPOSITION. *Every G^p-constant is G^p_{ap} and G^p-normal.*

The last property cannot be extended to the G^p-functions. For example, the function $f(x) = \text{sign} x$ is B^p-constant. It is B^p_{ap}, but it is not B^p (see Example (A1.4.46)).

(A1.5.14) THEOREM ([Bes-M, pp. 110–112], [BF, p. 47]) [*Riesz–Fischer Theorem*]. *To any series $\sum a_n e^{i\lambda_n x}$, for which $\sum |a_n|^2$ converges, corresponds a G^2-function having this series as its Bohr–Fourier series.*

In order to establish the desired vertical hierarchies, let us recall the most important relationships among the norms (A1.1.3), (A1.2.1), (A1.3.1), (A1.4.1) (see, for example, [Bes-M, pp. 72–76], [BeBo, pp. 220–224], [BF, pp. 36–37]).

For every $f \in L^p_{\text{loc}}(\mathbb{R}, \mathbb{R})$ and for every $p \geq 1$, the following inequalities hold:

$$\|f\|_{C^0} \geq \|f\|_{S^p_L} \geq \|f\|_{W^p} \geq \|f\|_p.$$

Consequently, we obtain, for every $p \geq 1$,

(A1.5.15)

$$C_{ap} \subset S^p \subset W^p \subset B^p,$$

$$C_{ap} \subset S^p - normal \subset e - W^p - normal \subset W^p - normal \subset B^p - normal,$$

$$C_{ap} \subset S^p_{ap} \subset e - W^p_{ap} \subset W^p_{ap} \subset B^p_{ap}.$$

Furthermore, denoting by G^p one of the spaces S^p, W^p or B^p, it is easy to show, by virtue of the Hölder inequality, that, for every $1 \leq p_1 < p_2$,

$$\|f\|_{G^{p_1}} \leq \|f\|_{G^{p_2}},$$

and, consequently,

(A1.5.16) $$C_{ap} \subset G^{p_2} \subset G^{p_1} \subset G^1 \subset L^1_{\text{loc}}.$$

From formula (A1.5.16) and from Definition (A1.1.9), the following theorem holds.

(A1.5.17) THEOREM ([Bes-M, pp. 75–76], [BF, p. 38]). *The spaces G^p coincide with the spaces obtained as the closures of the space C_{ap} w.r.t. the norms (A1.2.1), (A1.3.1), (A1.4.1).*

Furthermore, the following properties for bounded functions hold:

(A1.5.18) THEOREM ([BF, p. 37]). *Every G^p-function is G^p-bounded.*

(A1.5.19) REMARK. Theorem (A1.5.18) does not hold, in general, for the spaces W^p_{ap} and B^p_{ap}, as shown by Example (A1.3.43).

(A1.5.20) THEOREM ([BF, pp. 62–63]). *Every bounded function belonging to G^1 belongs to every space G^p, for all $p > 1$.*

Inclusions (A1.5.15) can be improved, recalling some theorems and examples.

(A1.5.21) THEOREM ([Dos1]). *If there exists a number M s.t. the exponents of a S_{ap}^p-function $f(x)$ are less in modulus than M, then $f(x)$ is equivalent to a u.a.p. function.*

(A1.5.22) THEOREM ([AP-M, Theorem VII], [Bes-M, pp. 81–82], [Cor-M, Theorem 6.16]) [*Bochner*]. *If $f(x) \in S_{ap}^p$ is uniformly continuous, then f is u.a.p.*

H. D. Ursell [Urs] has shown an example of a continuous S_{ap}^p-function, which is not uniformly continuous and which is not C_{ap}.

On the other hand, it is not difficult to show that the space C_{ap} is strictly contained in S_{ap}^p, for every p.

In his book [Lev-M], B. M. Levitan shows two interesting examples.

(A1.5.23) EXAMPLE ([Lev-M, pp. 209–210]). Let $f(x) \in C_{ap}$. Then the function

$$F(x) = \text{sign}(f(x)) = \begin{cases} 1 & \text{if } f(x) > 0, \\ 0 & \text{if } f(x) = 0, \\ -1 & \text{if } f(x) < 0 \end{cases}$$

is S_{ap}^1.

(A1.5.24) EXAMPLE ([Lev-M, pp. 212–213]). Given the quasi-periodic (and, a *fortiori*, almost-periodic) function $\varphi(x) = 2 + \cos x + \cos \sqrt{2}x$, the function

$$f(x) = \sin\left(\frac{1}{\varphi(x)}\right)$$

is S_{ap}^1.

However, in order to have S_{ap}^p-functions which are not in C_{ap}^0, it would be sufficient to consider, for example, the functions obtained modifying the values of an $f \in C_{ap}$ on all the relative integers, because the elements of $L_{\text{loc}}^p(\mathbb{R}, \mathbb{R})$ (and, consequently, of S_{ap}^p) are the classes of functions obtained by means of the equivalence relation $f \sim g$ if $f = g$, a.e. in $\mathbb{R}$.

The following example shows a function $f \in S_{ap}^p$ which is unbounded.

(A1.5.25) EXAMPLE. The function

$$f(x) = \begin{cases} \cos x & \text{if } x \neq k\pi, \\ k & \text{if } x = k\pi, \end{cases}$$

is not continuous and it is unbounded, but it belongs to the same class of $L_{\text{loc}}^p(\mathbb{R}, \mathbb{R})$ as the function $g(x) = \cos x$, which is, obviously, u.a.p. Thus, it is S_{ap}^p, for every p.

(A1.5.26) THEOREM ([Bes-M, p. 77], [GKL-M, p. 190], [Lev-M, p. 222]). *If, in the norm* (A1.3.1), $\limsup_{\varepsilon \to 0} L(\varepsilon)$ *is finite, then a function* $f(x) \in e\text{-}W_{ap}^p$ *is an* S_{ap}^p*-function.*

In other words, the spaces S_{ap}^p can be interpreted as uniform W^p-spaces. The spaces S_{ap}^p are strictly included in $e - W_{ap}^p$.

(A1.5.27) EXAMPLE (*Example of an* $e - W^p$*-normal function which does not belong to* S_{ap}^p). The example is partly based on [St-M, pp. 20–21]. In Example (A1.3.41), we have already proved that the function, defined on $\mathbb{R}$,

$$f(x) = \begin{cases} 1, & \text{for } 0 < x < 1/2, \\ 0, & \text{elsewhere,} \end{cases}$$

is $e - W$-normal, but not S_{ap}.

Let us note that, by Formula (A1.3.32), for every $L \geq 1$,

$$D_{S_L}[f(x+\tau) - f(x)] < \varepsilon \quad \text{if } L = L(\varepsilon) \geq \frac{1}{\varepsilon}.$$

Thus, $\limsup_{\varepsilon \to 0} L(\varepsilon) = \infty$, and the hypothesis in Theorem (A1.4.26) is not satisfied.

Another example can be found in [BF, Main Example 2 and Main Example II]. In Example (A1.3.46), we have shown that the Heaviside step function

$$H(x) = \begin{cases} 0 & \text{if } x < 0, \\ 1 & \text{if } x \geq 0 \end{cases}$$

is a W_{ap}^1-function, but not $e - W_{ap}^1$. Since the space $e - W_{ap}^1$ corresponds to the space W^1 (see Theorems (A1.3.11), (A1.3.15)) and the spaces W^p are included in the spaces W^p-normal, Example (A1.3.43) shows a second example of a W_{ap}^p-function which is not an $e - W_{ap}^p$-function, other than [AndBeL, Example 1].

As already shown for the W_{ap}^p and $e - W_{ap}^p$ spaces, the three definitions for a more general space can be inserted among those of a more restrictive space. This situation is very clear when we compare the W^p and B^p-definitions.

(A1.5.28) EXAMPLE (*Example of a* B^p*-function which is not a* W_{ap}^p*-function*). Let us take the function

$$f(x) = \begin{cases} n^{1/(2p)} & \text{if } n^2 \leq x < n^2 + \sqrt{n}, \\ 0 \cdot & \text{elsewhere,} \end{cases}$$

where $p \in \mathbb{R}$, $p > 1$ and $n \in \mathbb{N}$, $n \geq 1$.

As pointed out in [Ber3, p. 42], this function is unbounded, B^p-bounded and B^p-constant. Consequently, by virtue of Proposition (A1.5.13), it is B_{ap}^p.

Let us compute $\|f\|_q$, for every $q \geq 1$. It is sufficient to take $T = N^2 + \sqrt{N}$, $N \in \mathbb{N}$.

$$\|f\|_q^q = \lim_{T \to \infty} \sup \frac{1}{2T} \int_{-T}^{T} |f(x)|^q \, dx = \lim_{N \to \infty} \sup \frac{1}{2(N^2 + \sqrt{N})} \int_0^{N^2 + \sqrt{N}} |f(x)|^q$$

$$= \lim_{N \to \infty} \sup \frac{1}{2(N^2 + \sqrt{N})} \sum_{k=1}^{N} \sqrt{k} k^{q/(2p)}.$$

For $q = p$, we have

$$\|f\|_p^p = \lim_{N \to \infty} \frac{1}{2(N^2 + \sqrt{N})} \frac{N(N+1)}{2} = \frac{1}{4}.$$

For $q > p$, we have

$$\|f\|_q^q \geq \lim_{N \to \infty} \sup \frac{1}{2(N^2 + \sqrt{N})} \frac{2p}{3p + q} N^{1/2(3 + q/p)} = \infty,$$

because $1/2(3 + q/p) > 2$.

For $q < p$, we have

$$\|f\|_q^q \leq \lim_{N \to \infty} \sup \frac{1}{2(N^2 + \sqrt{N})} \frac{2p}{3p + q} (N + 1)^{1/2(3 + q/p)} = 0,$$

because $1/2(3 + q/p) < 2$.

In the last two cases, we have used the Cauchy integral criterion.

It follows that $f(x) \in B^q$, for all $q < p$, because it is a B^q-zero function.

Let us show that $f(x) \notin W_{ap}^q$.

Without any loss of generality, we can take $\tau > 0$ and $T > \tau$. There exists a real number M s.t. $\sqrt{M} > T > \tau$. Thus,

$$\sup_{x \in \mathbb{R}} \frac{1}{T} \int_x^{x+T} |f(x + \tau) - f(x)|^q \, dx \geq \frac{1}{T} \int_{N^2 + \sqrt{N} - \tau}^{N^2 + \sqrt{N}} N^{q/(2p)} \, dx = \frac{\tau}{T} N^{q/(2p)},$$

for every $N \geq M$. Consequently, taking the limit for $N \to \infty$, we arrive at

$$\|f(x + \tau) - f(x)\|_{W^q} = \infty \quad \text{for all } q \geq 1, \text{ for all } \tau > 0, \text{ for all } T > \tau.$$

Let us observe that, since $B^q \subset B^q$-normal and W^q-normal $\subset W_{ap}^q$, this example shows simultaneously a function which is B^q, but not W^q-normal, and a function which is B^q-normal, but not W_{ap}^q.

Let us finally remark that this function does not satisfy Condition (A1.5.4) in Theorem (A1.5.2), uniformly w.r.t. a. In fact, for every $T > 0$, there exists $N \in \mathbb{N}$ s.t. $\sqrt{N} > 2T$. Consequently,

$$\left| \sup_{a \in \mathbb{R}} \frac{1}{2T} \int_{a-T}^{a+T} f(x)\, dx - M[f] \right| \geq \frac{1}{2T} \int_{N^2+\sqrt{N}/2-T}^{N^2+\sqrt{N}/2+T} N^{1/(2p)}\, dx$$

$$= N^{1/(2p)} \to_{N \to \infty} \infty,$$

and the claim follows.

(A1.5.29) EXAMPLE (*Example of a W^1-normal function which is not a B^1-function*). The example is partly based on [St-M, p. 48]. As shown in Example (A1.3.46), the function $H(x)$ is W^1-normal. On the other hand, it does not satisfy (A1.5.3), because

$$\lim_{T \to \infty} \frac{1}{T} \int_0^T H(x)\, dx = 1 \neq \lim_{T \to \infty} \frac{1}{T} \int_{-T}^0 H(x)\, dx = 0.$$

Consequently, it is not B^1.

It means that $H(x)$ does not satisfy all the conditions of Theorem (A1.4.27). In fact, while, for every $\lambda \neq 0$ and for L sufficiently large,

$$\frac{1}{2TL} \left[\int_{-T}^T \left| \int_x^{x+L} H(t)e^{i\lambda t}\, dt - \int_0^L H(t)e^{i\lambda t}\, dt \right| dx \right]$$

$$= \frac{1}{2TL} \left[\int_{-T}^0 \left| \int_0^{x+L} e^{i\lambda t}\, dt - \int_0^L e^{i\lambda t}\, dt \right| dx \right.$$

$$\left. + \int_0^T \left| \int_x^{x+L} e^{i\lambda t}\, dt - \int_0^L e^{i\lambda t}\, dt \right| dx \right]$$

$$= \frac{1}{2TL|\lambda|} \left[\int_{-T}^0 \left| e^{i\lambda x} - 1 \right| dx + \int_0^T |e^{i\lambda x} - 1||e^{i\lambda L} - 1|\, dx \right] \leq \frac{3}{L|\lambda|},$$

and consequently,

$$\lim_{L \to \infty} \lim_{T \to \infty} \frac{1}{2TL} \left[\int_{-T}^T \left| \int_x^{x+L} H(t)e^{i\lambda t}\, dt - \int_0^L H(t)e^{i\lambda t}\, dt \right| dx \right]$$

$$= \lim_{L \to \infty} \frac{3}{L|\lambda|} = 0, \quad \text{for } \lambda = 0,$$

$$\lim_{L \to \infty} \lim_{T \to \infty} \frac{1}{2TL} \left[\int_{-T}^T \left| \int_x^{x+L} H(t)e^{i\lambda t}\, dt - \int_0^L H(t)e^{i\lambda t}\, dt \right| dx \right]$$

$$= \lim_{L \to \infty} \lim_{T \to \infty} \frac{1}{2TL} \left[\int_{-T}^0 |x|\, dx \right] = \lim_{L \to \infty} \lim_{T \to \infty} \frac{T}{4L} = \infty,$$

and the function does not satisfy Condition (iv), for every $\lambda \in \mathbb{R}$.

(A1.5.30) EXAMPLE ([St-M, pp. 42–47]) (*Example of a W_{ap}^1-function which is not a B^1-normal function*). In Example (A1.3.43), we have shown that the function

$$f(x) = \begin{cases} 0 & \text{if } x \in (-\infty, 0], \\ \sqrt{\frac{n}{2}} & \text{if } x \in (n-2, n-1], \, n = 2, 4, 6, \ldots, \\ -\sqrt{\frac{n}{2}}, & \text{if } x \in (n-1, n], \, n = 2, 4, 6, \ldots, \end{cases}$$

belongs to $e - W_{ap}^1$. On the other hand, in Example (A1.4.45), the function is shown not to be B^1-normal.

(A1.5.31) EXAMPLE ([BF, Example 3b], [St-M, pp. 34–38]) (*Example of a B^p-function which is not a W^p-function*). The function

$$f(x) = \begin{cases} 1 & \text{if } x \in [n-1/2, n+1/2), \, n \in \mathbb{Z}, \, n \bmod 2 = 0 \text{ but } n \bmod 2^2 \neq 0, \\ 2 & \text{if } x \in [n-1/2, n+1/2), \, n \in \mathbb{Z}, \, n \bmod 2^2 = 0 \text{ but } n \bmod 2^3 \neq 0, \\ 3 & \text{if } x \in [n-1/2, n+1/2), \, n \in \mathbb{Z}, \, n \bmod 2^3 = 0 \text{ but } n \bmod 2^4 \neq 0, \\ \vdots \\ 0 & \text{elsewhere,} \end{cases}$$

is a B^p-function which is not a W^p-function.

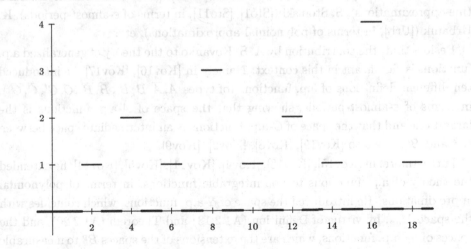

FIGURE 4

Taking into account the last four examples, we can conclude that there does not exist any inclusive relation between the spaces W_{ap}^p and B^p.

The previous theorems and examples allow us to write down Table 1.

	a. periods		normal		approx.
Bohr	C_{ap} ⇕⇑	⇔	u. normal ⇕⇑	⇔	u.a.p. ⇕⇑
Stepanov	S_{ap}^p ⇕⇑	⇔	S^p-normal ⇕⇑	⇔	S^p ⇕
equi-Weyl	$e-W_{ap}^p$ ⇕⇑	⇔ ↖	$e-W^p$-normal ......(⇕⇑)......	⇔ ↘	$e-W^p$ ⇕⇑
Weyl	W_{ap}^p ⇕⇑	⇐ ⇗⇖	W^p-normal ......(⇕⇑)......	⇐ ⇗⇖	W^p ⇕⇑
Besicovitch	B_{ap}^p	⇐ ⇏	B^p-normal	⇐ ⇏	B^p

Table 1

(A1.6) Some further generalizations.

As for the case of the C_{ap}-functions, it is possible to generalize the theory of the G^p-functions to spaces of functions defined on arbitrary groups (see, for example, [Fol6], [Fol8], [Fol7]).

Many authors have furthermore generalized in different directions the notion of almost-periodicity; for example, R. Doss ([Dos5], [Dos6], in terms of diophantine approximations), S. Stoinski ([Sto1]–[Sto11], in terms of ε-almost-periods), K. Urbanik ([Urb], in terms of polynomial approximations), etc.

Besides that, the contribution by A. S. Kovanko to the theory of generalized a.p. functions is significant in this context. Namely in [Kov16], [Kov17], he introduced ten different definitions of a.p. functions (of types $A, \overline{A}, B, B', \overline{B}, \overline{B}', C, C', \overline{C}, \overline{C}'$), in terms of ε-almost-periods, showing that the space of A-a.p. functions is the largest one and that the space of C-a.p. functions is an intermediate space between W^2 and B^2. See also [Kov19], [Kov8], [Kov2], [Kov9].

Furthermore, in [Kov15], [Kov18], [Kov4], [Kov11], [Kov5], [Kov14], he extended the theory of a.p. functions to non-integrable functions, in terms of polynomial approximations. He introduced the space of α-a.p. functions, which coincides with the space M_{ap}, by virtue of Definition (A1.2.18) and Theorem (A1.2.20), and the spaces of α_k-a.p. functions, which are the extensions of the spaces B^p to measurable functions. These spaces are included in the space of α-a.p. functions. Moreover, it is possible to prove a Bohr-like property for the spaces of α_k-a.p. functions and to show that, for every α_1-a.p. function, the mean value exists and the set of values $a(\lambda)$, defined by (A1.1.21), is at most countable.

In [Kov11], the author introduces the space of β-a.p. functions, in terms of a Bohr-like definition, where the set of almost-periods is satisfactorily uniform, like in Definition (A1.4.13), and shows that this space coincides with the space of α-a.p. functions. Thus, he shows the equivalence of two definitions (Bohr-like and approximation) for these spaces. Moreover, he proves that the space B^1 is included in this space. Finally, he states a necessary and sufficient condition in order a B^1-function to be α-a.p., in terms of the so-called asymptotic uniform integrability (see [Kov11]). E. Fœlner [Fol] has specialized the study of these spaces, considering B^p-bounded functions and bounded functions, proving interesting relationships with the spaces B^p.

Generalizing the theory of weakly a.p. functions (see [Eb1], [Eb2]), J.-P. Bertrandias [Ber3, pp. 64–68, 71] has introduced the spaces of B^p-weakly a.p. functions, showing that every B^p-function is B^p-weakly a.p. (see [Ber3, p. 71]).

In this part, we will only concentrate our attention to the generalizations given by C. Ryll-Nardzewski, S. Hartman, J. P. Kahane (see [Kah] and the references therein), which are related to the Bohr-Fourier coefficients, not considering the almost-periodicity of the functions, and C. Zhang [Zh] (cf. [ADA]). We will also mention possible multivalued extensions in [And19], [And5], [AndBeL], [BVLL], [Dan1], [Dan2], [DoSh].

(A1.6.1) DEFINITION ([Kah]). A function $f(x) \in L^p_{\text{loc}}(\mathbb{R}, \mathbb{R})$ is called *almost-periodic in the sense of Hartman* (shortly, H^1_{ap}) if, for every $\lambda \in \mathbb{R}$, the number

$$a_f(\lambda) = \lim_{T \to \infty} \frac{1}{2T} \int_{-T}^{T} f(x) e^{-i\lambda x} \, dx$$

is finite.

(A1.6.2) DEFINITION ([Kah]). A function $f(x) \in L^p_{\text{loc}}(\mathbb{R}, \mathbb{R})$ is called *almost-periodic in the sense of Ryll-Nardzewski* (shortly, R^1_{ap}) if, for every $\lambda \in \mathbb{R}$, the number

$$(A1.6.3) \qquad b_f(\lambda) = \lim_{T \to \infty} \frac{1}{T} \int_{X}^{X+T} f(x) e^{-i\lambda x} \, dx, \quad \text{uniformly w.r.t. } x \in \mathbb{R},$$

is finite.

Every R^1_{ap}-function is, obviously, H^1_{ap} (and, for every λ and for every $f(x) \in R^1_{ap}$, $a_f(\lambda) = b_f(\lambda)$). The converse, in general, is not true.

(A1.6.4) EXAMPLE (*Example of a H^1_{ap}-function, neither belonging to R^1_{ap} nor to B^1*). We have already shown in Example (A1.4.46) that the function $f(x) = \text{sign} x$ does not belong to B^1. Nevertheless, $f(x) \in H^1_{ap}$ and its spectrum is empty. In fact, for every $\lambda \neq 0$,

$$\int_{-T}^{T} f(x) e^{-i\lambda x} \, dx = -\int_{-T}^{0} e^{-i\lambda x} \, dx + \int_{0}^{T} e^{-i\lambda x} \, dx = \frac{2}{i\lambda} [1 - \cos(\lambda T)].$$

Thus,

$$a_f(\lambda) = \lim_{T \to \infty} \frac{1}{2T} \int_{-T}^{T} f(x) e^{-i\lambda x} \, dx = \lim_{T \to \infty} \frac{1}{i\lambda T}[1 - \cos(\lambda T)] = 0, \quad \text{for all } \lambda \neq 0.$$

If $\lambda = 0$,

$$a_f(0) = \lim_{T \to \infty} \frac{1}{2T} \int_{-T}^{T} f(x) \, dx = \lim_{T \to \infty} \frac{1}{2T} \left[-\int_{-T}^{0} dx + \int_{0}^{T} dx \right] = 0.$$

Finally, let us prove that $f(x) \notin R_{ap}^1$. It is sufficient to show that property (A1.6.3) does not hold for $\lambda = 0$. In fact,

$$\int_{X}^{X+T} \operatorname{sign} x \, dx = \begin{cases} -\int_{X}^{0} dx + \int_{0}^{X+T} dx, & \text{for } X < 0, \\ \int_{X}^{X+T} dx, & \text{for } X \geq 0, \end{cases}$$

$$= \begin{cases} 2X + T, & \text{for } X < 0, \\ T, & \text{for } X \geq 0. \end{cases}$$

Then

$$b_f(0) = \lim_{T \to \infty} \frac{1}{T} \int_{X}^{X+T} \operatorname{sign} x \, dx = 1.$$

Since

$$\left| \frac{1}{T} \int_{X}^{X+T} \operatorname{sign} x \, dx - 1 \right| = \begin{cases} \left| \dfrac{2X}{T} \right|, & \text{for } X < 0, \\ 0, & \text{for } X \geq 0, \end{cases}$$

we have

$$\left| \frac{1}{T} \int_{X}^{X+T} \operatorname{sign} x \, dx - b_f(0) \right| < \varepsilon,$$

whenever $|2X/T| < \varepsilon$, i.e. for all $T > 2|X|/\varepsilon$. Consequently, the limit $b_f(0)$ is not uniform w.r.t. X and $f(x) = \operatorname{sign} x \notin R_{ap}^1$.

It can be observed that, in this case, $a_f(0) = 0 \neq b_f(0) = 1$. On the other hand, it can be easily shown that $b_f(0) = 0$ for all $\lambda \neq 0$, uniformly w.r.t. $X \in \mathbb{R}$.

As pointed out in [Kah] and from Theorem (A1.5.2) and Remark (A1.5.4), every S_{ap}^p and W_{ap}^p-function is R_{ap}^1 and every B^p-function is H_{ap}^1, while there is no relation between the spaces B_{ap}^p and R_{ap}^1 and between the spaces B_{ap}^p and H_{ap}^1.

(A1.6.5) EXAMPLE (*Example of a function belonging to B_{ap}^1, but not to R_{ap}^1*). In Example (A1.5.28), we could see a function which is B_{ap}^1, but not W_{ap}^1, showing that (A1.5.4) does not hold uniformly w.r.t. $a \in \mathbb{R}$. Consequently, the function does not belong to R_{ap}^1.

Analogously to the spaces G^p, we can introduce the *spectrum* $\sigma(f)$ for a H_{ap}^1-function $f(x)$ as

$$\sigma(f) = \{\lambda \in \mathbb{R} \text{ s.t. } a_f(\lambda) \neq 0\}.$$

(A1.6.6) THEOREM ([Kah]). *The spectrum of every function* $f(x) \in H^1_{ap}$ *is at most countable.*

(A1.6.7) REMARK. Every nonnegative H^1_{ap}-function is B^1-bounded, because, in this case, $a_f(0) = \|f\|_1$. This property is no longer true for general H^1_{ap}-functions.

(A1.6.8) EXAMPLE (*Example of a function which is* B^p_{ap}, *for every* $p \geq 1$, *but not* H^1_{ap}). Let us take the function, with values in $\mathbb{C}$,

$$f(x) = e^{i \log |x|}.$$

Clearly,

$$\|f\|^p_p = \int |e^{i \log |x|}|^p \, dx = 1.$$

Besides that, J.-P. Bertrandias [Ber3, p. 42] has shown that $f(x)$ is B^p-constant and, consequently, by virtue of Proposition (A1.5.13), it is B^p_{ap}, for every $p \geq 1$.

Let us show that this function does not belong to H^1_{ap}, in particular, that it does not have a mean value. In fact,

$$M[f] = \int e^{i \log |x|} \, dx = \lim_{T \to \infty} \frac{1}{T} \int_0^T e^{i \log x} \, dx = \frac{(1-i)}{2} \lim_{T \to \infty} e^{i \log T}.$$

Since the limit in the last equality does not exist, the claim follows.

(A1.6.9) THEOREM ([Kah]). *Every function* $f(x) \in H^1_{ap}$, *belonging to some Marcinkiewicz space* $\mathcal{M}^p$, $p > 1$, *is the sum of a* B^p_{ap}-*function and of a* H^1_{ap}-*function whose spectrum is empty.*

J. Bass [Bas1]–[Bas3] and J.-P. Bertrandias [Ber1], [Ber2], [Ber3] introduced the spaces of pseudo-random functions, in terms of correlation functions, showing that these spaces are included in B^p_c and in the space of Hartman functions whose spectrum is empty (see [Ber2]). There is no relation between the spaces B^p and the spaces of pseudo-random functions, but some theorems concerning operations involving B^p and pseudo-random functions can be proved (see, for example, [Bas2, pp. 28–31]).

In the framework of the resolution of systems of ordinary differential equations, C. Zhang has introduced in [Zha-M] a new class of a.p. functions.

(A1.6.10) DEFINITION ([ADA], [Zha-M]). Set

$$PAP_0(\mathbb{R}) = \left\{ \varphi \in C(\mathbb{R}) \quad \text{s.t.} \quad \lim_{T \to \infty} \frac{1}{2T} \int_{-T}^T |\varphi(x)| \, dx = 0 \right\}.$$

A function $f \in C(\mathbb{R})$ is called a *pseudo almost-periodic function* if it is the sum of a function $g(x) \in C_{ap}$ and of a function $\varphi \in PAP_0(\mathbb{R})$.

$g(x)$ is called the *almost-periodic component* of $f(x)$ and $\varphi(x)$ is the *ergodic perturbation*.

E. Ait Dads and O. Arino [ADA] have furtherly generalized these spaces to measurable functions, introducing the spaces $\widetilde{P}AP$. As remarked in [ADA], the mean value, the Bohr-Fourier coefficients and the Bohr-Fourier exponents of every pseudo a.p. function are the same of its a.p. component.

As concerns almost-periodic multifunctions (considered in [And19], [And5], [AndBeL]; cf. also [BVLL], [Dan2], [DoSh]), let us introduce the following metrics:

(Bohr) $$D(\varphi, \psi) := \sup_{t \in \mathbb{R}} d_H(\varphi(t), \psi(t)),$$

(Stepanov) $$D_{S_L^p}(\varphi, \psi) := \sup_{t \in \mathbb{R}} \left[\frac{1}{L} \int_x^{x+L} d_H(\varphi(t), \psi(t))^p \, dt \right]^{1/p},$$

$$D_{W^p}(\varphi, \psi) := \lim_{L \to \infty} \sup_{t \in \mathbb{R}} \left[\frac{1}{L} \int_x^{x+L} d_H(\varphi(t), \psi(t))^p \, dt \right]^{1/p}$$

(Weyl) $$= \lim_{L \to \infty} D_{S_L^p}(\varphi, \psi),$$

(Besicovitch) $$D_{B^p}(\varphi, \psi) := \limsup_{L \to \infty} \left[\frac{1}{2T} \int_{-T}^T d_H(\varphi(t), \psi(t))^p \, dt \right]^{1/p},$$

where $\varphi, \psi : \mathbb{R} \multimap \mathbb{R}$ are measurable multifunctions with nonempty, bounded, closed values and $d_H(\cdot, \cdot)$ stands for the Hausdorff metric.

Since every multifunction, say $P : \mathbb{R} \multimap \mathbb{R}$, is well-known (see e.g. [Dan2], [DoSh]) to be measurable if and only if there exists a sequence $\{p_n(t)\}$ of measurable (single-valued) selections of P, i.e. $p_n \subset P$ for all $n \in \mathbb{N}$, such that P can be Castaing-like represented as follows

$$P(t) = \overline{\bigcup_{n \in \mathbb{N}} p_n(t)},$$

the standard (single-valued) measure-theoretic arguments make the distance $d_H(\varphi(t), \psi(t))$ to become a single-valued measurable function.

Therefore, replacing the metrics in the appropriate definitions in the single-valued case (cf. Table 1) by the related ones above, we have correct definitions of almost-periodic multifunctions.

(A1.6.11) DEFINITION. We say that a measurable multifunction $\varphi : \mathbb{R} \multimap \mathbb{R}$ with nonempty, bounded, closed values is *G-almost-periodic* if G means any of the respective classes in Table 1 with metrics replaced by the above ones, i.e. those involving the Hausdorff metric.

(A1.6.12) REMARK. Although S_{ap}^p-multifunctions with nonempty (convex) compact values possess (single-valued) S_{ap}^p-selections (see [Dan2], [DoSh]), the same is not true for C_{ap}-multifunctions (see [BVLL]). It is an open problem whether or not (equi-)W_{ap}^p or B_{ap}^p-multifunctions possess the respective (single-valued) selections.

APPENDIX 2

DERIVO-PERIODIC SINGLE-VALUED
AND MULTIVALUED FUNCTIONS

We would like to collect the most important properties of derivo-periodic functions and multifunctions. In the latter case, we must recall the notion of a suitable multivalued derivative. The exposition relies mainly on the results of the papers [And21] and [Jut].

(A2.1) DEFINITION. By a $D_\theta^{(r)}$-periodic function (simply, $D^{(r)}$-periodic function) $y(t) \in C^{(n)}([0, \theta])$, we mean the one with

(A2.2) $y^{(r)}(t + \theta) \equiv y^{(r)}(t)$ $(\Rightarrow y^{(r+m)}(t + \theta) \equiv y^{(r+m)}(t), \ 0 \leq r \leq r + m \leq n).$

Besides the obvious implication indicated in (A2.2), the following well-known result is crucial (see e.g. [Far-M, p. 235])

(A2.3) THEOREM. *The function* $y \in C^1(\mathbb{R})$ *is* D_θ^1-periodic, *i.e. the derivo-periodic with period* $\theta > 0$, *if and only if there exist a constant* $a \in \mathbb{R}$ *and a* θ-periodic *function* $x \in C^1(\mathbb{R})$ *such that* $y(t) \equiv at + x(t)$.

PROOF. The sufficiency is obvious. For the necessity, assume that y is D_θ^1-periodic, i.e. that $\dot{y}(t + \theta) \equiv \dot{y}(t)$ holds.

Integrating both sides from 0 to t, we get

$$y(t + \theta) \equiv y(t) + y(\theta) - y(0).$$

Denoting

$$a := \frac{1}{\theta}[y(\theta) - y(0)] = \frac{1}{\theta} \int_0^\theta \dot{y}(t) \, dt$$

and defining a function x by $x(t) := y(t) - at$, we get that $y \in C^1(\mathbb{R})$ and

$$x(t + \theta) = y(t + \theta) - at - a\theta = y(t) + y(\theta) - y(0) - at - y(\theta) + y(0) = x(t).$$

Thus, x is θ-periodic, as claimed.   □

By a mathematical induction argument, Theorem (A2.3) immediately implies

(A2.4) COROLLARY. *A function $y(t) \in C^{(r)}(\mathbb{R})$, $r \in \mathbb{N}$, is $D_\theta^{(r)}$-periodic, i.e. satisfying (A2.2) if and only if it takes the form*

$$y(t) = y_0(t) + P_r(t),$$

where $y_0(t) \in C^{(r)}(\mathbb{R})$ is a θ-periodic function and $P_r(t)$ is a real polynomial of the degree less or equal r.

Applying Theorem (A2.3) and Corollary (A2.4), the following relations can be verified.

(A2.5) LEMMA ([And21]). *$y(t) \in C^{(n)}([0, \theta])$ satisfies*

$$\begin{cases} y^{(k)}(\theta) - y^{(k)}(0) = \gamma_k, & \text{for } k = 0, \dots, p-1, \\ y^{(k)}(\theta) = y^{(k)}(0), & \text{for } k = p, \dots, n-1, \end{cases}$$

or $(p = n)$ $y^{(k)}(\theta) - y^{(k)}(0) = \gamma_k$, for all $k = 0, \dots, n-1$, if it takes the form

$$y(t) = y_0(t) + P_p^{(0)}(t),$$

where $y_0(t)$ fulfils

$$y_0^{(k)}(\theta) = y_0^{(k)}(0), \quad \text{for } k = 0, \dots, n-1,$$

and

$$\begin{cases} P_p^{(k)}(t) = \left[\dfrac{1 - \operatorname{sgn}(1-p)}{2} \right]^{\operatorname{sgn}(k-p)} \displaystyle\sum_{j=k}^{p} \dfrac{j!}{(j-k)!} t^{j-k} \alpha_{p-j} \quad (k, p = 1, \dots, n), \\ \alpha_0 = \gamma_{p-1}/\theta p!, \quad \alpha_p \dots \text{ an arbitrary constant}, \\ \alpha_k = \dfrac{1}{\theta(p-k)!} \left\{ \gamma_{p-k-1} + \displaystyle\sum_{s=2}^{k+1} \prod_{r=s}^{k+1} \left[-\binom{p}{r} \dfrac{\theta^{r-s}}{(p-r+2)(p-r+1)} \right] \gamma_{p-s-1} \right\} \\ (k = 0, \dots, p-1). \end{cases}$$

(A2.6) REMARK. In particular, for $p = 1$, $y(t) \in C^1([0, \theta])$ is D_θ^1-periodic if and only if it takes the form

$$y(t) = y_0(t) + \gamma_0 t/\theta,$$

where $\gamma_0(t) \in C^1([0, \theta])$ is θ-periodic.

Proceeding to multifunctions, let us recall the following facts (cf. [MaVi1]). Let φ be a multivalued map from a linear space X into a linear space Y. Assume that the following linearity condition holds

$$\varphi(\alpha x + \beta y) = \alpha\varphi(x) + \beta\varphi(y), \quad \text{for any } x, y \in X, \text{ and } \alpha, \beta \in \mathbb{R}.$$

Then φ is single-valued. In fact, for $\alpha = -\beta = 1$ and $x = y$, we have $\varphi(x) - \varphi(x) = \varphi(0)$. Since $\varphi(0) = 0$, it follows that $\varphi(x)$ is a singleton, for any $x \in X$.

Therefore, the definition of differentiability for single-valued maps cannot be generalized to the multivalued case in a word by a word fashion.

The majority of approaches to the problem of differentiability of multifunctions has some common features. The mostly frequent ones are:

(i) applying the usual differentiability of $\varphi : X \multimap Y$ in special space, especially in a real normed linear space, whose elements are subsets of Y, or

(ii) reducing the requirements imposed on the differential, namely (α) the multivalued differential is required to be positively homogeneous, or (β) using the tangency concept, following the fact that the tangent space to the graph of a function $f : \mathbb{R} \to \mathbb{R}$, at any point $x \in \mathbb{R}$, is the line with the slope $f'(x)$ passing through $(x, f(x))$.

Let B denote the unit ball $B(0, 1)$ in a Banach space X and let $\mathcal{B}(X)(\mathcal{B}_c(X))$ be the family of all non-empty bounded, closed (bounded, closed, convex) subsets of X. Let $\overline{A}$ denote the closure of $A \subset X$. Define

$$d_H(M, N) := \inf\{t > 0 \mid M \subset N + tB, N \subset M + tB\}$$

or, equivalently,

$$d_H(M, N) := \max\{\sup_{x \in M} \text{dist}(x, N), \sup_{x \in N} \text{dist}(x, M)\}.$$

d_H is obviously the Hausdorff metric in the space $\mathcal{B}(X)$.

(A2.7) DEFINITION. A multivalued map $\varphi : X \multimap Y$ is said to be *positively homogeneous* if $\varphi(tx) = t\varphi(x)$, $t \geq 0$, $x \in X$.

(A2.8) DEFINITION ([DBl3]). A multivalued map $\varphi : X \to \mathcal{B}(Y)$ is said to be *differentiable* at $x \in X$ if there exist a map $D_x : X \to \mathcal{B}_c(Y)$, which is u.s.c. and positively homogeneous, and a number $\delta > 0$ such that

(A2.9) $\qquad d_H(\varphi(x + h), \varphi(x) + D_x(h)) = o(h)$, whenever $\|h\| \leq \delta$.

D_x is called the (*multivalued*) *differential* of φ at x.

In the sequel we call this type of differential as the De Blasi-like differential.

(A2.10) REMARK. The motivation of defining such a type of multivalued differential is based on the incrementary property. The differential is required to be only u.s.c. and positively homogeneous (instead of continuous and linear). If this multivalued differential exists, it is unique ([DBl3]). On the other hand, the class of multivalued mappings having such a differential at a point $x_0 \in \text{Dom}\,\varphi$ (or even on some interval $[a, b]$) is rather narrow in comparison with other classes for other definitions of differentiability.

For this reason, the notion of such a differential can be weakened, for example, to the *upper differential* ψ of φ at x, where ψ is u.s.c. and positively homogeneous and

$$\varphi(x + h) \subset \varphi(x) + \psi(h), \quad \text{for } \|h\| < \delta, \ \delta > 0.$$

This differential is not unique. Denote by $\mathbb{F}$ the sets of all upper differentials of φ at x, and observe that $\mathbb{F}$ can be empty. If φ is Lipshitzian, i.e. $d_H(\varphi(x+h), \varphi(x)) \leq L\|h\|$, for sufficiently small h, then $\mathbb{F} \neq \emptyset$, and it leads to the Lasota–Strauss notion of a differential Δ_x, defined by $\Delta_x(h) := \bigcap_{\psi \in \mathbb{F}} \psi(h)$.

The Lasota–Strauss differential is not the only possibility, how to weaken requirements imposed on the differential introduced in (A2.9), cf. [MaVi1]. In [FV1], the presented multivalued differential (at a point) is not necessarily unique. The above mentioned definitions of multivalued differential extend the class of differentiable multifunctions. For our aims, the differential introduced in Definition (A2.8) is convenient for the sake of lucidity (see Chapter III.11).

Another definition of differentiability of multivalued maps can be found in [BaJa] as a usual derivative in a special space.

If X has reasonable properties, e.g. when X is reflexive, then there is a real normed linear space $\mathbb{B}(X)$ (or, simply $\mathbb{B}$) and an isometric mapping $\pi : \mathcal{B}_c \to \mathbb{B}$, where $\mathcal{B}_c$ is metrized by d_H, such $\pi(\mathcal{B}_c)$ is a convex cone in $\mathbb{B}$ and $\mathbb{B}$ is minimal.

(A2.11) DEFINITION. A multivalued map $\varphi : X \to \mathcal{B}_c(X)$ is said to be π-*differentiable* at $x_0 \in X$ if the induced mapping $\widehat{\varphi} : X \to \mathbb{B}(X)$ is differentiable at $x_0 \in X$, i.e. if there exists a continuous linear mapping $\widehat{\varphi}'(x_0) : X \to \mathbb{B}$ such that

$$\widehat{\varphi}(x) - \widehat{\varphi}(x_0) - \widehat{\varphi}'(x_0)(x - x_0) = o(\|x - x_0\|).$$

This definition is not a "real" multivalued definition of differentiability. Therefore, Definition (A2.11) is indicated here only for the sake of completeness, and we will not work with it.

The third possible approach is based on a property of tangency. In the single-valued case, the graph of the derivative at a point x_0 is also a tangent space to the graph at that point. So, we can define multivalued differential of a multivalued map at (x, y) to be a multivalued map whose graph is a tangent cone to the original graph at (x, y).

(A2.12) DEFINITION (cf. [AuF-M]). Let K be a nonempty subset of a Hilbert space X. We define the *Bouligand* (*contingent*) *cone* $T_K(x)$ to K at x as follows

$$v \in T_K(x) \quad \text{if} \quad \liminf_{h \to 0^+} \frac{d_K(x + hv)}{h} := \liminf_{h \to 0^+} \frac{d(x + hv, K)}{h} = 0.$$

(A2.13) DEFINITION. Let $\varphi : X \multimap X$ be a multivalued map. Denote by $D\varphi(x, y)(\cdot) : X \multimap X$ the multivalued map, whose graph is the contingent cone $T_{\Gamma_\varphi}((x, y))$ to the graph Γ_φ of φ at (x, y). We shall say that $D\varphi(x, y)$ is the *contingent derivative* of φ at $x \in X$ and $y \in \varphi(x)$.

(A2.14) LEMMA ([AuF-M]). *Let* $\varphi : X \multimap X$ *be a multivalued map. Then* $v \in D\varphi(x, y)(u)$ *holds if and only if there exist sequences of positive numbers* h_n *and of elements* $u_n \in X$, $v_n \in X$ *satisfying* $\lim_{n \to \infty} h_n = 0$, $\lim_{n \to \infty} u_n = u$, $\lim_{n \to \infty} v_n = v$ *and* $y + h_n v_n \in \varphi(x + h_n u_n)$.

(A2.15) REMARK. Contingent cones, and subsequently contingent derivatives, are not the only possibilities how to define multivalued derivatives. Another cones (for example, Dubovitskiĭ–Milijutin's, adjacent, Clarke's tangent cones, etc.) can be considered and corresponding derivatives can be treated (see e.g. [AuF-M]).

Now, let us proceed to possible definitions of a periodic multifunction. The following definition is the most convenient for our purposes (see Chapter III.11).

(A2.16) DEFINITION. A multifunction $\varphi : \mathbb{R} \multimap \mathbb{R}$ is said to be *periodic* with a period $T > 0$ if $\varphi(t) \equiv \varphi(t + T)$.

The following definition and theorem investigate the case of De Blasi-like differentiable multifunctions (cf. Definition (A2.8)).

Note that if there is a continuously differentiable function $a : I \to \mathbb{R}$ and a bounded interval $U \subset \mathbb{R}$ such that $\varphi(x) = a(x) + U$, for all $x \in I$, then it is easy to check that the differential D_x takes the form of $D_x(h) = a'(x)h$, for all $x \in I$.

At first, we recall the notion of the Dini derivatives.

(A2.17) DEFINITION. Let $a(x) : \mathbb{R} \to \mathbb{R}$ be a continuous function. The notation $D^+a(x), D^-a(x), D_+a(x), D_-a(x)$ will be used for the *right and left, upper and lower Dini derivatives* of $a(x)$

$$D^+a(x) = \limsup_{h \to 0^+} \frac{a(x + h) - a(x)}{h}, \quad D_+a(x) = \liminf_{h \to 0^+} \frac{a(x + h) - a(x)}{h},$$

$$D^-a(x) = \limsup_{h \to 0^-} \frac{a(x + h) - a(x)}{h}, \quad D_-a(x) = \liminf_{h \to 0^-} \frac{a(x + h) - a(x)}{h},$$

respectively.

(A2.18) THEOREM. *A multifunction $\varphi : \mathbb{R} \multimap \mathbb{R}$ is De Blasi-like differentiable on an interval I, $I \subset \mathbb{R}$, if and only if there exist a function $a : I \to \mathbb{R}$, $a \in \mathcal{A}$, where $\mathcal{A}$ denotes the set of all continuous functions*

$$\mathcal{A} := \{a(x) \mid a(x) \in C(I, \mathbb{R}), \ D^+ a(x) = D_+ a(x) = a'_+(x) \in \mathbb{R},$$
$$D^- a(x) = D_- a(x) = a'_-(x) \in \mathbb{R}, \ for \ all \ x \in I\},$$

and a bounded interval $U \subset \mathbb{R}$, such that $\varphi(x) = a(x) + U$, for all $x \in I$.

PROOF. It immediately follows that if $\varphi(x)$ cannot be written in the form of $\varphi(x) = a(x) + U$ on I, U is bounded, then $\varphi(x)$ is not De Blasi-like differentiable on I. Note that if diam $\varphi(x)$ is not constant on I, then $\varphi(x)$ is not De Blasi-like differentiable. Similarly, if $\varphi(x + \varepsilon) \not\subseteq \varphi(x)$, for $\varepsilon > 0$ small enough, and diam $\varphi(x)$ is constant, then $\varphi(x)$ is not De Blasi-like differentiable, too.

If $a(x) \notin \mathcal{A}$ is continuous (at $x \in I$) and if there exists D_x satisfying condition (A2.9), then D_x is not positively homogeneous and therefore $\varphi(x)$ is not De Blasi-like differentiable.

Let $a(x)$ be continuous and let D^+, D_+, D^-, D_- be the Dini derivatives of $a(\cdot)$ at x. Suppose that there exists an u.s.c. and positively homogeneous D_x satisfying (A2.9). Suppose $D^+ \neq D_+$, for example, then there exist $h_n \to 0+$ such that

$$D^+ = \lim_{n \to \infty} \frac{a(x + h_n) - a(x)}{h_n}$$

and $h'_n \to 0+$ such that

$$D_+ = \lim_{n \to \infty} \frac{a(x + h'_n) - a(x)}{h'_n}.$$

Since $o(h_n) = d_H(a(x + h_n) + U, a(x) + U + D_x(h_n)) = d_H(a(x) + h_n D^+ + o(h_n) + U, a(x) + U + D_x(h_n)) = d_H(a(x) + h_n D^+ + U, a(x) + U + D_x(h_n)) + o(h_n)$, we have $D_x(h_n) = D^+ h_n$, where $\{z_n\} \in o(h_n)$, whenever $\lim_{n \to \infty} z_n / h_n = 0$, for $n \to \infty$. Similarly, $D_x(h'_n) = D_+ h'_n$.

Since D_x is positively homogeneous and since $h_n = h'_n h_n / h'_n$, we have

$$D^+ h_n = D_x(h_n) = D_x(h'_n \frac{h_n}{h'_n}) = \frac{h_n}{h'_n} D_x(h'_n) = \frac{h_n}{h'_n} D_+ h'_n = D_+ h_n,$$

which implies $D^+ = D_+$.

The cases $D^+ = D_+ = \pm\infty$ and $D^- = D_- = \pm\infty$ are not possible, because then $D_x(h)$, satisfying the assumptions in Definition (A2.8), would not exist (it can be shown, for example, that $D_x(h)$ has an unbounded value, for $h = 0$).

If $D^+ = D_+ = D+ \in \mathbb{R}$ and $D^- = D_- = D- \in \mathbb{R}$, then it immediately follows that

$$D_x(h) = \begin{cases} (D+)h & h > 0, \\ (D-)h & h < 0, \\ 0 & h = 0. \end{cases}$$

$\square$

(A2.19) DEFINITION. Let $\varphi : \mathbb{R} \multimap \mathbb{R}$ be a De Blasi-like differentiable multifunction. We say that φ is *derivo-periodic* with a period T if the mapping $D(x) := D_x$ (see above) is T-periodic.

(A2.20) THEOREM. *Let $\varphi : \mathbb{R} \multimap \mathbb{R}$ be a De Blasi-like differentiable multifunction. Then φ is derivo-periodic with a period T if and only if $a(x) \in \mathcal{A}$ satisfies*

$$(A2.21) \qquad a'_+(x) = a'_+(x+T) \quad and \quad a'_-(x) = a'_-(x+T), \quad for\ all\ x \in \mathbb{R},$$

(i.e. $a(x)$ is derivo-periodic with period T, whenever $a(x)$ is (continuously) differentiable).

PROOF. Let φ be a T-derivo-periodic multifunction. Since φ is differentiable, we have $\varphi(x) = a(x) + U$, $a(x)$ satisfies (A2.21). Since $D_x(h) = a'_+(x)h$, $h \geq 0$ and $D_x(h) = a'_-(x)h$, $h < 0$ is T-periodic (in x), we obtain

$$a'_+(x)h = D_x(h) = D_{x+T}(h) = a'_+(x+T)h, \quad h \geq 0,$$

and similarly for $h < 0$.

Let $a(x)$, satisfying (A2.21), be a T-derivo-periodic function. Since

$$D_x(h) = a'_+(x)h = a'_+(x+T)h = D_{x+T}(h), \quad h \geq 0,$$

and similarly for $h < 0$, the multifunction $\varphi(x)$ is derivo-periodic with period T. $\square$

The presented differential is a common single-valued function and the above mentioned class of differentiable multifunctions is rather narrow (see the definition of $\mathcal{A}$).

The contingent cones lead to another convenient definitions of derivo-periodicity for multifunctions. Recall, we say that $D\varphi(x, y)$ is the contingent derivative of φ at Γ_φ if the graph Γ_φ of $D\varphi(x, y)(\cdot)$ is the contingent cone $T_{\Gamma_\varphi}(x, y)$.

(A2.22) DEFINITION. A multifunction $\varphi : \mathbb{R} \multimap \mathbb{R}$ is called *derivo-periodic* with a period T if the differential $D\varphi(x, y)$ is periodic (in x) with period T, i.e. $D\varphi(x, y) = D\varphi(x + T, z)$, for all $(x, y) \in \mathbb{R} \times \mathbb{R}$, where $z \in \varphi(x + T)$ is a value corresponding to y (they are both boundary points, for example).

(A2.23) REMARK. Note that the derivative of a multifunction is defined at any point of its graph and not as a point which belongs to the closure of the graph, but not to the graph. For this reason, different multivalued maps having the same closure of their graphs, can have different derivatives.

(A2.24) REMARK. The main difficulty arising when we handle with a derivative, is its additivity. It is not difficult to find, for example, an u.s.c. mapping and a l.s.c. mapping, for which the linearity of derivative does not hold true, so additional conditions must be satisfied.

The additivity of the derivative at $(x, y) \in \Gamma_{\varphi+\psi}$ can be defined in the following way.

(A2.25) DEFINITION. We say that the pair of multifunctions $\varphi, \psi : \mathbb{R} \multimap \mathbb{R}$ satisfies the *additivity condition of the contingent derivative* if

$$(A2.26) \qquad D(\varphi + \psi)(x, y)(u) = \bigcup_{z \in Z_{(x,y)}} (D\varphi(x, z)(u) + D\psi(x, y - z)(u)),$$

where $Z_{(x,y)}$ is the set containing all points $z \in \mathbb{R}$ such that $(x, z) \in \Gamma_\varphi$, $(x, y-z) \in \Gamma_\psi$.

Equation (A2.26) is not satisfied for arbitrary pairs of functions $\varphi, \psi : \mathbb{R} \multimap \mathbb{R}$, as we show in the following example.

(A2.27) COUNTER-EXAMPLE. Graphs of the functions φ, ψ are plotted in the following figure. Additivity of the contingent derivative is not satisfied at the point (x, y), $y \in (\varphi + \psi)(x)$, see the marked points ●,

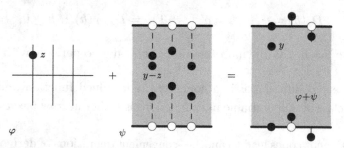

FIGURE 1

It is true that

$$D\varphi(x, z)(u) = \begin{cases} \emptyset & u < 0, \\ \mathbb{R} & u = 0, \\ \emptyset & u > 0, \end{cases}$$

and $D\psi(x, y - z)(u) = \mathbb{R}$, for all $u \in \mathbb{R}$, so

$$\bigcup_{z \in Z_{(x,y)}} (D\varphi(x,z)(u) + D\psi(x, y - z)(u)) = \begin{cases} \emptyset & u < 0, \\ \mathbb{R} & u = 0, \\ \emptyset & u > 0, \end{cases}$$

where the fact $\emptyset + A = \emptyset$ was used. This, however, differs from the mapping

$$D(\varphi + \psi)(x,y)(u) = \mathbb{R}, \quad \text{for all } u \in \mathbb{R}.$$

(A2.28) THEOREM (cf. [AuF-M]). *Let* $\varphi : \mathbb{R} \multimap \mathbb{R}$ *be an (arbitrary) multi-function and* $f : \mathbb{R} \to \mathbb{R}$ *be continuously differentiable at* $x \in \mathbb{R}$. *Then, for all* $y \in (f + \varphi)(x)$, *the equality*

$$D(f + \varphi)(x,y)(u) = f'(x)(u) + D\varphi(x, y - f(x))(u)$$

holds true.

PROOF. "$\subset$" If $v \in D(f + \varphi)(x,y)(u)$, then there exist $h_n \to 0+$, $u_n \to u$ and $v_n \to v$ such that $y + h_n v_n \in (f + \varphi)(x + h_n u_n)$. Since f is continuously differentiable

$$\begin{aligned} f(x + h_n u_n) &= f(x) + h_n(f'(x)(u_n) + \widetilde{\varepsilon}(h_n)) \\ &= f(x) + h_n(f'(x)(u) + f'(x)(u_n - u) + \widetilde{\varepsilon}(h_n)) \\ &= f(x) + h_n(f'(x)(u) + \varepsilon(h_n)), \end{aligned}$$

where $\varepsilon(h_n) \to 0$. Thus,

$$y - f(x) + h_n(v_n - f'(x)(u) - \varepsilon(h_n)) \in \varphi(x + h_n u_n).$$

This means $v - f'(x)(u) \in D\varphi(x, y - f(x))(u)$.
"$\supset$" If $v \in f'(x)(u) + D\varphi(x, y - f(x))(u)$, then

$$v - f'(x)(u) \in D\varphi(x, y - f(x))(u).$$

So, there exist $w_n \to v - f'(x)(u)$, $u_n \to u$ and $h_n \to 0+$ such that

$$y - f(x) + h_n w_n \in \varphi(x + h_n u_n).$$

Thus, there exists $\varepsilon(h_n) \to 0$ such that

$$w_n + f'(x)(u) + \varepsilon(h_n) \to v.$$

From the definition of the continuous differentiability,

$$f'(x)(u) = \frac{1}{h_n}(f(x + h_n u_n) - f(x)) - \varepsilon(h_n),$$

and from the condition imposed on w_n, we have

$$w_n + \frac{1}{h_n}(f(x + h_n u_n) - f(x)) \to v.$$

Defining $v_n = w_n + (f(x + h_n u_n) - f(x))/h_n$, then

$$y - f(x) - h_n \frac{1}{h_n}(f(x + h_n u_n) - f(x)) + h_n v_n$$
$$= y - f(x + h_n u_n) + h_n v_n \in \varphi(x + h_n u_n).$$

Consequently, $y + h_n v_n \in (f + \varphi)(x + h_n u_n)$ and $v \in D(f + \varphi)(x, y)(u)$, because $h_n \to 0+$, $u_n \to u$, $v_n \to v$.   □

(A2.29) THEOREM. *Let $\varphi : \mathbb{R} \multimap \mathbb{R}$ be a T-periodic multifunction. Then $D\varphi(x, y)$ is T-periodic (in x), for all $y \in \varphi(x)$.*

PROOF. If $v \in D\varphi(x, y)(u)$, then there exist $v_n \to v$, $u_n \to u$ and $h_n \to 0+$ such that $y + h_n v_n \in \varphi(x + h_n u_n)$. Since φ is T-periodic, we have $y + h_n v_n \in \varphi(x + T + h_n u_n)$, so $v \in D\varphi(x + T, y)$, and the conclusion holds true.   □

The following theorem is our first theorem treating derivo-periodicity of multi-valued functions, when using contingent derivatives.

(A2.30) THEOREM. *Let $\varphi(x) = \psi(x) + ax$, where $a \in \mathbb{R}$ and $\psi(x) : \mathbb{R} \multimap \mathbb{R}$ be T-periodic. Then $\varphi(x)$ is derivo-periodic with the period T and $D\varphi(x, y) = a + D\psi(x, y - ax)$.*

PROOF. From the additivity of a multivalued derivative, proved already for such a class of multifunctions, we have $D\varphi(x, y) = a + D\psi(x, y - ax)$, where $y \in \psi(x) + ax$, and $D\varphi(x + T, z) = a + D\psi(x + T, z - ax - aT)$, where $z \in \psi(x + T) + ax + aT$. Since ψ is T-periodic, we have $z - aT \in \psi(x) + ax$, and since there is a one-to-one correspondence between y and $z = y + aT$, we have that, $D\varphi(x + T, y + aT) = D\varphi(x + T, z) = a + D\psi(x + T, z - ax - aT) = a + D\psi(x, y - ax) = D\varphi(x, y)$, because $D\psi(x, y)$ is T-periodic.   □

The main disadvantage of this theorem is that the "linear" part is strictly single-valued. The following theorem allows us to use, instead of a linear function, some "zone" map. As usual, some additional requirements must be satisfied.

By a *zone map*, we understand a multivalued mapping $\psi : \mathbb{R} \multimap \mathbb{R}$ such that $\psi(x) = [a_1(x), a_2(x)]$, where $a_1, a_2 \in \mathcal{A}$, $a_1(x) \leq a_2(x)$, for all $x \in \mathbb{R}$.

(A2.31) LEMMA. *Let* $\psi : \mathbb{R} \multimap \mathbb{R}$ *be a zone multifunction. If* $y \in \text{int}\, \psi(x)$, *then* $D\psi(x,y)(u) = \mathbb{R}$, *for all* $u \in \mathbb{R}$. *If* $y = a_1(x) \neq a_2(x)$, *resp.* $y = a_1(x) = a_2(x)$, *then*

$$D\psi(x,y)(u) = \begin{cases} [a'_{1-}(x)u, \infty) & u < 0, \\ [0, \infty) & u = 0, \\ [a'_{1+}(x)u, \infty) & u > 0, \end{cases}$$

resp.

$$D\psi(x,y)(u) = \begin{cases} [a'_{1-}(x)u, a'_{2-}(x)u] & u < 0, \\ 0 & u = 0, \\ [a'_{1+}(x)u, a'_{2+}(x)u] & u > 0, \end{cases}$$

The proof is straightforward, and the result can be easily extended to continuous functions $a_1(x)$ and $a_2(x)$ having their one-sided derivatives.

(A2.32) THEOREM. *Let* $\varphi : \mathbb{R} \multimap \mathbb{R}$ *be an (arbitrary) multifunction. Let* $\psi : \mathbb{R} \multimap \mathbb{R}$ *be a zone multifunction. Then*

$$(A2.33) \qquad \cup_{z \in Z_{(x,y)}} (D\varphi(x,z)(u) + D\psi(x, y - z)(u)) \subset D(\varphi + \psi)(x,y)(u).$$

PROOF. Let $z \in Z_{(x,y)}$ be arbitrary and let $v \in D\varphi(x,z)(u)$. Then there exist $h_n \to 0+$, $u_n \to u$ and $v_n \to v$ such that $z + h_n v_n \in \varphi(x + h_n u_n)$.

Let $w \in D\psi(x, y - z)(u)$.

If $y - z \in \text{int}\, \psi(x)$, then, for all $h_n \to 0+$ for all $u_n \to u$ and for all $w_n \to w$, we have that $y - z + h_n w_n \in \psi(x + h_n u_n)$, for n large enough, and consequently $v + w \in D(\varphi + \psi)(x,y)(u)$.

If $y - z \in \partial\psi(x)$ ($y - z = a_1(x)$, for example), then there exist $h_n \to 0+$, $u'_n \to u$ and $w_n \to w$ such that $y - z + h_n w_n \in \psi(x + h_n u'_n)$.

Since the left $a'_{1-}(x)$ and right $a'_{1+}(x)$ derivatives exist, it is easy to see that a sequence $h_n \to 0+$ can be arbitrary, and similarly for $y - z = a_2(x)$. So we take the same sequence h_n as for $v \in D\varphi(x,z)(u)$.

Note that $\psi(x + h_n u'_n) = \psi(x + h_n u_n + h_n(u'_n - u_n))$ and $u'_n - u_n \to 0$, for $n \to \infty$. Since ψ is continuous, we have $\psi(x + h_n u_n) \subset \psi(x + h_n u'_n) + B(0, o(h_n))$. Therefore, there exists $\varepsilon_n \to 0$, for $n \to \infty$, such that $y - z + h_n(w_n + \varepsilon_n) \in \psi(x + h_n u_n)$, and the conclusion immediately follows. □

(A2.34) DEFINITION. We say that a continuous multivalued map $\varphi(x)$ has a *smooth boundary* if, for every open interval $C \subset \mathbb{R}$, there exist an open subset $U \subset \mathbb{R}$ and C^1-functions $b_i : U \to \mathbb{R}$, $i = 1, \ldots, n$, such that

$$\partial\Gamma_{\varphi|_C} = \bigcup_{i=1}^{n} \Gamma_{b_i}.$$

The function b_k, having at a point t a vertical tangent, for example, for $t_1 > t$, having two values $b_{k1}(t_1)$, $b_{k2}(t_1)$ and, for $t_2 < t$, an empty set of values, is allowed. To avoid such a situation, some rotated coordinate systems are appropriate. However, such a definition and the proof of the following theorem can become confusing.

So, note that the boundary of graph of such a function consists of smooth functions.

(A2.35) THEOREM. *Let $\varphi : \mathbb{R} \multimap \mathbb{R}$ be a continuous multivalued map with smooth boundary and closed graph. Let $\psi : \mathbb{R} \multimap \mathbb{R}$ be a zone map, $\psi(x) = [a_1(x), a_2(x)]$, where $a_1, a_2 \in \mathcal{A}$, $a_1(x) \leq a_2(x)$, for all $x \in \mathbb{R}$. Then*

$$D(\varphi + \psi)(x, y)(u) = \bigcup_{z \in Z_{(x,y)}} (D\varphi(x, z)(u) + D\psi(x, y - z)(u)),$$

where $Z_{(x,y)}$ is the set containing all points $z \in \mathbb{R}$ such that $(x, z) \in \Gamma_\varphi$, $(x, y - z) \in \Gamma_\psi$.

The proof will be done, for sake of simplicity, only for continuously differentiable functions $a_1(x)$, $a_2(x)$. For general functions $a_1(x)$, $a_2(x) \in \mathcal{A}$, the proof is quite analogous and we leave it to the reader.

PROOF. If $y \in \text{int}\,(\varphi + \psi)(x)$, then from the continuity assumptions, we have $(x, y) \in \text{int}\,\Gamma_{\varphi+\psi}$ and $D(\varphi + \psi)(x, y)(u) = \mathbb{R}$, for all $u \in \mathbb{R}$.

Now, three situations can appear. At first, there exists a point $z \in Z_{(x,y)}$, satisfying $z \in \text{int}\,\varphi(x)$ and $y - z \in \text{int}\,\psi(x)$ (if one point exists, infinitely many such points exist). Thus, $D\varphi(x, z)(u) = \mathbb{R} = D\psi(x, y - z)(u)$, for all $u \in \mathbb{R}$, and equality (A2.26) holds true.

For the second possibility, there exists a point $z \in Z_{(x,y)}$, satisfying $z \in \partial\varphi(x)$ and $y - z \in \text{int}\,\psi(x)$. Thus, $D\varphi(x, y)(u) \neq \emptyset$, $D\psi(x, y - z(u)) = \mathbb{R}$, and the equality (A2.26) holds true.

At third, there are two points $z_1, z_2 \in \mathbb{R}$ such that $z_1, z_2 \in \partial\varphi(x)$, and $a_1(x) = y - z_1$, $a_2(x) = y - z_2$. By the hypothesis, we have $(b_i + a_2)'(x) = 0$ and $(b_j + a_1)'(x) = 0$, where b_i, b_j are appropriate smooth functions according to Definition (A2.34). Suppose, for example, that z_1 is a "local maximum" of the values $\varphi(x)$, i.e. if $w \in \varphi(x)$, $w \in V$, where V is a sufficiently small neighbourhood of z_1, then $w \leq z_1$. Let us compute $D\varphi(x, z_1)$. The tangent to the boundary of φ at a point (x, z_1) has $b_i'(x) \in \mathbb{R}$ as its directive. Thus, $D\varphi(x, z_1)(u) = (-\infty, b_i'(x)u]$, analogously $D\psi(x, y - z_1)(u) = (-\infty, a_2'(x)u]$, because $y - z_1$ is a similar "maximum" of the values $\psi(x)$, $D\varphi(x, z_1)(u) + D\psi(x, y - z_1)(u) = (-\infty, 0]$. The other point z_2 can be treated analogously, $D\varphi(x, z_2)(u) + D\psi(x, y - z_2)(u) = [0, \infty)$, and finally equality (A2.26) holds.

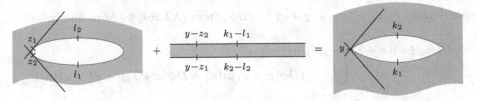

FIGURE 2. Illustration of the idea of the proof

Figure 2 gives us some illustration about this approach.

If the zone map is thicker (after collapsing of the points k_1, k_2 into one point k, which is just the point y, in this proof), points l_1, l_2 will correspond to points z_1, z_2, in the same part of the proof.

If $y \in \partial(\varphi + \psi)(x)$, then there exist two points $z_1, z_2 \in Z$ satisfying similar properties as mentioned above (see Figure 2, and points y, z_1, z_2, $y - z_1, y - z_2$, for illustration). The difference in the proof presented before is that $(b_i + a_2)'(x)$, $(b_j + a_1)'(x)$ can take arbitrary values. Also the values $\pm\infty$ are now admissible; the procedure is similar. If $(b_i + a_2)'(x) = \infty$, for example, then the corresponding derivative takes the values $D(\varphi + \psi)(x, y)(u) = \emptyset$, for example, for $u < 0$, and $D(\varphi + \psi)(x, y)(u) = \mathbb{R}$, for $u \geq 0$.

The last possibility is that there is exactly one point $z \in Z$, and the conclusion is immediate. In Figure 2, see points $k_1, l_1, k_1 - l_1$, for example. $\square$

(A2.36) THEOREM. *Let* $\varphi(x) : \mathbb{R} \multimap \mathbb{R}$ *be a* T-*periodic continuous set-valued map with a smooth boundary. Let* $\psi : \mathbb{R} \multimap \mathbb{R}$ *be a zone map,* $\psi(x) = a(x) + U$, *where* U *is a closed interval and* $a(x) \in A$ *satisfies* (A2.21). *Then* $(\varphi + \psi)(x)$ *is* T-*derivo-periodic.*

PROOF. Since φ is T-periodic, we have $D\varphi(x + T, z) = D\varphi(x, z)$, for every $z \in Z_{(x,y)}$. For the inner points $y \in (\varphi + \psi)(x)$, the conclusion follows easily. For boundary points, the proof is continued.

If U is not a singleton, then $D\psi(x, y + aT - z)(u) = (-\infty, a'_+(x + T)(u)]$ or $[a'_+(x + T)(u), \infty)$, for $u \geq 0$, and analogously, for $u < 0$. Otherwise, we obtain $D\psi(x, y + aT - z)(u) = \{a'_+(x + T)(u)\}$, for $u \geq 0$, and similarly, for $u < 0$. Condition (A2.21) (i.e. the T-derivo-periodicity of the continuously differentiable function $a(x)$) implies $a'_\pm(x + T)(u) = a'_\pm(x)(u)$. Observe that the point corresponding to (x, y) is $(x + T, y + cT)$. Thus, from (A2.26), we have finally

$$D(\varphi + \psi)(x + T, y + cT)(u)$$
$$= \bigcup_{z \in Z_{(x+T, y+cT)}} (D\varphi(x + T, z)(u) + D\psi(x + T, y + cT - z)(u))$$
$$= \bigcup_{z \in Z_{(x,y)}} (D\varphi(x, z)(u) + D\psi(x, y - z)(u))$$
$$= D(\varphi + \psi)(x, y)(u).$$

$\square$

One can also use another definition of periodicity. Nevertheless, Definition (A2.16) and the following Definition (A2.37) coincide, for single-valued maps.

(A2.37) DEFINITION. A multifunction $\varphi : \mathbb{R} \multimap \mathbb{R}$ is called *forward periodic* with a period T if $\varphi(t) \subset \varphi(t + T)$, for all $t \in \mathbb{R}$.

The disadvantage of this definition is that the periodicity condition holds only in a forward direction and not in the opposite, backward, direction.

For this definition, Theorem (A2.20) holds true in the same meaning, because the classes of differentiable multifunctions are identical. The analogy of Theorem (A2.29), and subsequently of Theorem (A2.30) and Theorem (A2.32) hold for a larger class of maps. The proofs are similar, only the equality symbols "=" are replaced by the inclusion ones "$\subset$".

(A2.38) REMARK. Sometimes, the closed convex processes, i.e. multivalued mappings having closed convex cones as their graphs, are considered as a multivalued analogy of linear maps. For example the Open Mapping Theorem and the Closed Graph Theorem hold for such maps (see [AuF-M]). The derivative of such a map need not be, however, a constant. A suitable candidate can be the above mentioned zone map (which is, however, not a process).

(A2.39) REMARK. The mapping $(x, y, u) \in \Gamma_\varphi \times \mathbb{R} \multimap D\varphi(x, y)(u)$ is, under certain natural assumptions, l.s.c. Note that if there exists some inner point $(x, y) \in$ int Γ_φ, then $D\varphi(x, y)(u) = \mathbb{R}$, for all $u \in \mathbb{R}$. So, a modification of Theorem (A2.35) for l.s.c. maps would be useful, for example, for computing the second contingent derivatives and studying second (or higher) order derivo-periodic multifunctions.

APPENDIX 3

FRACTALS AND MULTIVALUED FRACTALS

We present the existence results concerning fractals and multivalued fractals as (sub)invariant subsets of certain union map (called the Hutchinson–Bernsley map), generated by a multifunction system. These results are obtained by means of various fixed point theorems. Weakly contractive and compact multifunction systems are considered, but systems of more general multifunctions are discussed as well. The presentation relies mainly on the papers [AFi2], [And23], [AG1].

Iterated function systems (IFS) were introduced by J. E. Hutchinson [Hut] and popularized by M. Barnsley [Bar-M], as a natural generalization of the well-known Banach contraction principle. They represent one way of defining fractals as attractors of certain discrete dynamical systems. Moreover, they can be effectively applied to fractal image compressions. Therefore, no wonder that they attract more and more the attention of mathematicians and computer experts (see e.g. [BnDe], [BlKe], [DvEmHu], [Fis-M], [Kie-M], [Mch], [Wic-M]).

As observed recently in [And23], [AG1], [AFi2], the same approach can be naturally extended to *iterated multifunction systems* (IMS) with resulting objects called by ourselves *multivalued fractals*. Originally, we wanted to call them *multifractals*, but this name was already reserved for fractals with inhomogeneous fractal dimensions, usually in the context of self similar measures (cf. [EvMa]). One can also distinguish between deterministic and stochastic IFS (IMS) (see e.g. [Bar-M], [Fis-M]).

Since the contraction requirement for generating (multi)functions might be unpleasant because of many reasons (verification, too restrictive admissible classes, etc.), a natural question arises whether also non-contractive systems can be used for the same goal. The affirmative answer is related to the application of different fixed point theorems: the Lifschitz theorem in [And23], the Tarski–Kantorovitch theorem in [Hay], [JGP], the Knaster–Tarski theorem in [Les], or to considering fractals or semifractals as extremal sets (i.e. as Kuratowski's limits), provided IFS are asymptotically stable or semistable in [Fis-M], [Kie-M], [LaMy1]–[LaMy5].

In this supplement, the existence of multivalued fractals is proved by means of the Banach-like fixed point theorem for weakly contractive maps due to B. E. Rhoades [Rho], and the Schauder-like fixed point theorem for compact maps due to A. Granas [Gr1]. The related resulting objects are therefore called here respectively as *metric* and *topological multivalued fractals* to distinguish between the applied metric and topological fixed point theories (cf. [GoKi-M], [KiSi-M], [Go5-M]). Moreover, since no iteration processes are present, we are not speaking about IFS (IMS), but only about (*multivalued*) *function systems*. The application of the Knaster–Tarski theorem (cf. [DG]), and the Lifschitz theorem (cf. [GoKi-M]), are discussed as well. The auxiliary results for hyperspaces and related maps are presented in the following parts.

Auxiliary results for spaces and hyperspaces.

In the sequel, all spaces under consideration will be, for the sake of simplicity, at least metric. The notion of a hyperspace, associated to a metric space X, comes back to F. Hausdorff, who considered those of nonempty, compact subsets of X, endowed with the Hausdorff metric. Later on, different subsets where also treated by means of various metrics by K. Kuratowski, L. Vietoris, and many others. For the recent state of the research in this field, see e.g. the monographs [IlNa-M], [Nad-M], [Wic-M].

Let (X, d) be a complete metric space. For our convenience, we collect below some important properties of the following hyperspaces:

$$\mathrm{Cl}(X) := \{A \subset X \mid A \text{ is nonempty and closed}\},$$
$$B(X) := \{A \subset X \mid A \text{ is nonempty, bounded and closed}\},$$
$$C(X) := \{A \subset X \mid A \text{ is nonempty and compact}\},$$

provided they are endowed with the induced Hausdorff metric d_H. For the basic properties of the Hausdorff metric, see Chapter I.3 or [Ki-M]. Obviously, the following inclusion takes place:

$$C(X) \subset B(X) \subset \mathrm{Cl}(X).$$

First of all, it is well-known (see e.g. [HP1-M], [IlNa-M], [Nad-M]) that $(\mathrm{Cl}(X), d_H)$, $(B(X), d_H)$ and $(C(X), d_H)$ become complete. Moreover, if (X, d) is Polish (i.e. separable and complete), then $(C(X), d_H)$ is Polish as well (see the same references).

Letting $I = [0, 1]$, R. M. Schori and J. E. West [SchWe] proved that $(\mathrm{Cl}(I), d_H)$ is homeomorphic to the Hilbert cube I^∞, where

$$I^\infty = [0, 1]^\infty = \underbrace{[0, 1] \times [0, 1] \times \dots}_{\text{countably many}}$$

Every homeomorphic image of the Hilbert cube is an example of a (metric) *absolute retract* (AR) space (see e.g. [Go5-M]). For more details concerning AR-spaces, see Chapter I.2.

Moreover, compact AR-apaces are, up to homeomorphisms, the retract images of the Hilbert cubes and the Hilbert cubes are compact subsets of l^2-spaces (see e.g. [Go5-M]). On the other hand, the homeomorphic image can be rather far from the original spaces, as shown e.g. in [PeTo], where the homeomorphic image of a Hilbert space was found to be possibly not locally convex. There is even a long standing conjecture, whether or not *"every complete linear metric space is homeomorphic to a Hilbert space"*.

Furthermore, D. W. Curtis [Cur] has shown that (X, d) is a connected, locally connected and nowhere locally compact Polish space if and only if $(C(X), d_H)$ is homeomorphic to an l^2-space.

At last, it follows from the results of M. Wojdyslawski [Woj] and U. Tasmetov [Tas] that *if (X, d) is a connected and locally connected Polish space, then $(C(X), d_H)$ is a Polish AR-space.* In particular, if (X, d) is a Peano's continuum (i.e. a locally connected continuum), then $(\text{Cl}(X), d_H)$ is also a complete AR-space.

Auxiliary result for maps and multimaps.

Since the multivalued mappings acting in suitable spaces will become (more precisely, they induce those being) single-valued in the related hyperspaces, we can apply standard fixed point theorems for single-valued maps. Everybody knows the Banach contraction principle and the Schauder fixed point theorem. For our purposes, the application of their respective generalizations for weakly contractive maps due to B. E. Rhoades [Rho], and for compact maps on AR-spaces due to A. Granas [Gr1] (cf. (6.20) in Chapter I.6) will be essential.

(A3.1) DEFINITION. Assume that (X, d) is a complete metric space and let h be a function such that

   (i) $h : [0, \infty) \to [0, \infty)$ is continuous and nondecreasing,
   (ii) $h(t) = 0 \Leftrightarrow t = 0$ (i.e. $h(t) > 0$, for $t \in (0, \infty)$),
   (iii) $\lim_{t \to \infty} h(t) = \infty$.

A mapping $f : X \to X$ is said to be *weakly contractive* if, for any $x, y \in X$,

$$d(f(x), f(y)) \leq d(x, y) - h(d(x, y)).$$

(A3.2) PROPOSITION (Banach-like [Rho]). *Let (X, d) be a complete metric space and $f : X \to X$ be a weakly contractive map. Then f has exactly one fixed point in X.*

(A3.3) PROPOSITION (Schauder-like [Gr1]). *Let (X, d) be a (metric) AR-space and $f : X \to X$ be a compact (continuous) map (i.e. $\overline{f(X)}$ is compact). Then f has a fixed point in X.*

It is important that continuity can be equivalently expressed by means of the Hausdorff metric (see (3.64) in Chapter I.3):

A multivalued mapping $\varphi : X \to C(X)(Y)$ is continuous if and only if it is *Hausdorff-continuous*, i.e. if and only if it is continuous w.r.t. the metric d in (X, d) and the induced Hausdorff metric d_H in $C(X)$. However, for $\varphi : X \to B(X)(Y)$, the Hausdorff-continuity implies only l.s.c. (see (3.65), (3.66) in Chapter I.3).

On the other hand, if $\varphi : X \multimap Y$ is *Lipschitz-continuous* with the constant L, i.e. if, for any $x, y \in X$,

$$d_H(\varphi(x), \varphi(y)) \leq L d(x, y),$$

then φ is obviously continuous.

(A3.4) DEFINITION. Assume that (X, d) is a complete metric space and let h be the same function as in Definition (A3.1), i.e. satisfying conditions

(i) $h : [0, \infty) \to [0, \infty)$ is continuous and nondecreasing,
(ii) $h(t) = 0 \Leftrightarrow t = 0$ (i.e. $h(t) > 0$, for $t \in (0, \infty)$),
(iii) $\lim_{t \to \infty} h(t) = \infty$.

A multivalued mapping $\varphi : X \multimap X$ with nonempty closed values is said to be *weakly contractive* if, for any $x, y \in X$,

(A3.5) $$d_H(\varphi(x), \varphi(y)) \leq d(x, y) - h(d(x, y)).$$

Definition (A3.4) can be still equivalently reformulated as follows:

(A3.6) DEFINITION. Assume that (X, d) is a complete metric space and let $\overline{h}$ be a function such that

(a) $\overline{h} : [0, \infty) \to [0, \infty)$ is continuous and nondecreasing,
(b) $\overline{h}(0) = 0$, and $0 < \overline{h}(t) < t$, for $t > 0$,
(c) $\lim_{t \to \infty} t - \overline{h}(t) = \infty$.

A multivalued mapping $\varphi : X \multimap X$ with nonempty closed values is said to be *weakly contractive* if, for any $x, y \in X$,

(A3.7) $$d_H(\varphi(x), \varphi(y)) \leq \overline{h}(d(x, y)).$$

(A3.8) LEMMA. *Definitions* (A3.4) *and* (A3.6) *are equivalent.*

PROOF. Definition (A3.4) $\Rightarrow$ Definition (A3.6), i.e. that, for every h and φ satisfying the conditions (i)–(iii) and (A3.5) of Definition (A3.4), there exists $\overline{h}$ such that $\overline{h}$ and φ satisfy the conditions (a)–(c) and (A3.7) of Definition (A3.6).

To obtain an appropriate $\overline{h}$, we define at first the function $\widehat{h} : [0, \infty) \to [0, \infty)$ by the formula

$$(A3.9) \qquad \widehat{h}(t) := \sup_{0 \le s \le t} \{s - h(s)\}.$$

It can be easily seen that $\widehat{h}$ is nondecreasing, $\widehat{h}(0) = 0$ and $0 \le \widehat{h}(t) < t$, for all $t \in (0, \infty)$.

Now, we can construct the function $\overline{h}$ in the following way:

Case 1. If $\widehat{h}(t) \equiv 0$, for all $t \in [0, \infty)$, then for positive real numbers a, b, $a > b$, we define

$$\overline{h}(t) = \begin{cases} \dfrac{b}{a} t, & \text{for } t \le a, \\ b, & \text{for } t > a. \end{cases}$$

Case 2. If $\widehat{h}(t) = 0$, only for $t = 0$, then

$$\overline{h}(t) = \widehat{h}(t), \quad \text{for } t \in (0, \infty).$$

Case 3. If $\widehat{h}(t) = 0$, on some nondegenerated subinterval of $(0, \infty)$, then for a positive real number a, for which $\widehat{h}(a) > 0$, and still for

$$b := \sup\{t \mid \widehat{h}(t) = 0\},$$

we define

$$\overline{h}(t) := \begin{cases} \dfrac{\widehat{h}(a)}{b} t, & \text{for } t < b, \\ \widehat{h}(a), & \text{for } b \le t < a, \\ \widehat{h}(t), & \text{for } t \ge a. \end{cases}$$

We will show that $\overline{h}$ and φ satisfy the conditions of Definition (A3.6).

ad (a) It follows from the definition of $\overline{h}$ that $\overline{h}$ is nonnegative, continuous and nondecreasing.

ad (b) $\overline{h}(0) = 0$ and $0 < \overline{h}(t) < t$ (this follows from the inequality $0 < h(t) \le t$).

ad (c) *Case* 1.

$$\lim_{t \to \infty} (t - \overline{h}(t)) = \lim_{t \to \infty} (t - b) = \infty.$$

Cases 2 and 3.

$$\lim_{t \to \infty} (t - \overline{h}(t)) = \lim_{t \to \infty} (t - \widehat{h}(t)) = \lim_{t \to \infty} (t - \sup_{0 \le s \le t} \{s - h(s)\})$$
$$\ge \lim_{t \to \infty} (t - \sup_{0 \le s \le t} \{t - h(s)\}) = \lim_{t \to \infty} (t - t + \sup_{0 \le s \le t} \{h(s)\})$$
$$= \lim_{t \to \infty} (\sup_{0 \le s \le t} \{h(s)\}) = \lim_{t \to \infty} (h(t)) = \infty.$$

ad (A3.7) We show that $d_H(\varphi(x), \varphi(y)) \le d(x, y) - h(d(x, y))$ implies $d_H(\varphi(x), \varphi(y)) \le \overline{h}(d(x, y))$. Since

$$\overline{h}(t) \ge \widehat{h}(t) = \sup_{0 \le s \le t} \{s - h(s)\} \ge t - h(t),$$

we obtain

$$d_H(\varphi(x), \varphi(y)) \le d(x, y) - h(d(x, y)) \le \overline{h}(d(x, y)).$$

This completes the first part of the proof.

Definition (A3.6) $\Rightarrow$ Definition (A3.4), i.e. that, for every $\overline{h}$ and φ satisfying the conditions (a)–(c) and (A3.7) of Definition (A3.6), there exists h such that h and φ satisfy the conditions (i)–(iii) and (A3.5) of Definition (A3.4).

Now, we can define the function h by the following formula

(A3.10) $$h(t) = \sup_{0 \le s \le t} \{s - \overline{h}(s)\}.$$

The required properties of h can be verified as follows:

ad (i) From the definition of h, it is clear that h is nonnegative, continuous and nondecreasing,

ad (ii) $h(t) = 0$ only for $t = 0$, and from $\overline{h}(t) < t$, for $t > 0$, it follows that $h(t) > 0$, for $t > 0$.

ad (iii)

$$\lim_{t \to \infty} h(t) = \lim_{t \to \infty} \sup_{0 \le s \le t} \{s - \overline{h}(s)\} \ge \lim_{t \to \infty} t - \overline{h}(t) = \infty.$$

ad (A3.5) We show that $d_H(\varphi(x), \varphi(y)) \le \overline{h}(d(x, y))$ implies

$$d_H(\varphi(x), \varphi(y)) \le d(x, y) - h(d(x, y)).$$

Since

$$t - h(t) = t - \sup_{0 \le s \le t} \{s - \overline{h}(s)\} \ge t - (t - \overline{h}(t)) = \overline{h}(t),$$

we obtain

$$d_H(\varphi(x), \varphi(y)) \le \overline{h}(d(x, y)) \le d(x, y) - h(d(x, y)).$$

This completes the proof.   □

The following well-known properties of multivalued maps will be useful (see Chapter I.3).

(A3.11) LEMMA. *If $\varphi : X \multimap Y$ is u.s.c. with compact values and K is a compact subset of X, then $\varphi(K)$ is a compact subset of Y.*

If φ is, in particular, a compact u.s.c. mapping (i.e. $\overline{\varphi(X)}$ is compact, and so closed sets of values become compact), then the same conclusion is true.

(A3.12) LEMMA. *If $\varphi : X \multimap Y$ is a Lipschitz-continuous mapping with bounded values and B is a bounded subset of X, then $\varphi(B)$ is a bounded subset of Y.*

PROOF. One can show, similarly as in the proof of Proposition (A3.20) below, that the inequality
$$d_H(\varphi(a), \varphi(B)) \leq L d_H(a, B)$$
holds with some constant L, for any bounded $B \subset X$ and $a \in X$. This already implies the boundedness of φ, because, otherwise, we would get a contradiction.□

Metric multivalued fractals.

In the following text, $D(X)$ denotes $\mathrm{Cl}(X)$ or $B(X)$ or $C(X)$, and N is a positive integer. At first, we will define and consider weakly contractive multifunction systems.

(A3.13) DEFINITION. Let (X, d) be a complete metric space and $\varphi_i : (X, d) \to (D(X), d_H)$, $i = 1, \ldots, N$, be continuous (i.e. u.s.c. and l.s.c.) multifunctions. The system

(A3.14) $\{\varphi_i : (X, d) \to (D(X), d_H), \quad i = 1, \ldots, N\}$

will be called the *multifunction system* (MS), the associated map F given by the formula

(A3.15) $$F(x) := \bigcup_{i=1}^{N} \varphi_i(x), \quad x \in X,$$

will be called the *Hutchinson–Barnsley map*, and the associated operator F^* given by the formula

(A3.16) $$F^*(A) := \overline{\bigcup_{x \in A} F(x)}, \quad A \in D(X),$$

will be called the *Hutchinson–Barnsley operator*.

(A3.17) DEFINITION. Let (X, d) be a complete metric space and let $D(X)$ denote $Cl(X)$ or $B(X)$ or $C(X)$. Furthermore, let functions $\bar{h}_i$, $i = 1, \ldots, N$, satisfy conditions (a)–(c) of Definition (A3.6) and multifunctions $\varphi_i : (X, d) \to (D(X), d_H)$, $i = 1, \ldots, N$ satisfy inequality (A3.7), namely

(A3.18) $d_H(\varphi_i(x), \varphi_i(y)) \leq \bar{h}_i(d(x, y))$, $x, y \in X$, $i = 1, \ldots, N$.

The system

(A3.19) $\{\varphi_i : (X, d) \to (D(X), d_H), \quad i = 1, \ldots, N\}$,

will be called the *weakly contractive multifunction system* (WCMS).

The following Proposition will be given for the Hutchinson–Barnsley operator of WCMS (A3.19).

(A3.20) PROPOSITION. *The Hutchinson–Barnsley operator F^* of a weakly contractive multifunction system (A3.19) is weakly contractive.*

PROOF. We claim that there exists $\bar{h}^*$, satisfying the conditions of Definition (A3.6), such that inequality (A3.7) holds, for all $A, B \in D(X)$, namely

(A3.21) $d_H(F^*(A), F^*(B)) \leq \bar{h}^*(d_H(A, B))$, $A, B \in D(X)$.

The following estimate

(A3.22) $d_H(F^*(A), F^*(B)) = d_H\left(\overline{\bigcup_{i=1}^{N} \varphi_i(A)}, \overline{\bigcup_{i=1}^{N} \varphi_i(B)}\right)$

$$\leq \max_{i=1,\ldots,N}\{d_H(\varphi_i(A), \varphi_i(B))\},$$

follows from the well-known inequality for the Hausdorff distance (see e.g. [Bar-M]).

We recall that

(A3.23) $d_H(\varphi_i(A), \varphi_i(B)) = \max\{e(\varphi_i(A), \varphi_i(B)), e(\varphi_i(B), \varphi_i(A))\}$,

where

$$e(A, B) := \sup_{x \in A}\{d(x, B)\} := \sup_{x \in A}\{\inf_{y \in B} d(x, y)\}.$$

Estimating (A3.23), we obtain

$e(\varphi_i(A), \varphi_i(B)) = \sup_{x \in A}\{e(\varphi_i(x), \varphi_i(B))\}$

$= \sup_{x \in A}\{\inf_{y \in B}\{e(\varphi_i(x), \varphi_i(y))\}\} \leq \sup_{x \in A}\{\inf_{y \in B}\{d_H(\varphi_i(x), \varphi_i(y))\}\}$

$\leq \sup_{x \in A}\{\inf_{y \in B}\{\bar{h}_i(d(x, y))\}\} = \sup_{x \in A}\{\bar{h}_i(\inf_{y \in B}\{d(x, y)\})\}$

$= \sup_{x \in A}\{\bar{h}_i(d(x, B))\} = \bar{h}_i(\sup_{x \in A}\{d(x, B)\}) = \bar{h}_i(e(A, B))$, where we used the obvious inequality

$$e(\varphi_i(x), \varphi_i(y)) \le d_H(\varphi_i(x), \varphi_i(y)),$$

and the fact that $\bar{h}_i$ is nondecreasing.

Similarly, we can obtain

(A3.24) $$e(\varphi_i(B), \varphi_i(A)) \le \bar{h}_i(e(B, A)),$$

and so

$$
\begin{aligned}
d_H(\varphi_i(A), \varphi_i(B)) &= \max\{e(\varphi_i(A), \varphi_i(B)), e(\varphi_i(B), \varphi_i(A))\} \\
&\le \max\{\bar{h}_i(e(A, B)), \bar{h}_i(e(B, A))\} = \bar{h}_i(\max\{e(A, B), e(B, A)\}) \\
&= \bar{h}_i(d_H(A, B)).
\end{aligned}
$$

This estimate holds for all $i = 1, \ldots, N$.

Thus, we can return to (A3.22):

$$
\begin{aligned}
d_H(F^*(A), F^*(B)) &\le \max_{i=1,\ldots,N}\{d_H(\varphi_i(A), \varphi_i(B))\} \\
&\le \max_{i=1,\ldots,N}\{\bar{h}_i(d_H(A, B))\} = \bar{h}^*(d_H(A, B)),
\end{aligned}
$$

where the function

$$\bar{h}^*(t) := \max_{i=1,\ldots,N}\{\bar{h}_i(t)\}$$

evidently satisfies all conditions of Definition (A3.6).

Therefore, the claim (A3.21) holds, which completes the proof. $\square$

(A3.25) LEMMA. *The Hutchinson–Barnsley map F of MS (A3.14) operates from (X, d) into $(D(X), d_H)$.*

PROOF. The proof is obvious, because a finite union of closed sets is a closed set, a finite union of bounded closed sets is a bounded closed set, and a finite union of compact sets is a compact set. $\square$

(A3.26) LEMMA. *The Hutchinson–Barnsley operator F^* of WCMS (A3.19) is a self map on $(D(X), d_H)$.*

PROOF. For $D(X) = \text{Cl}(X)$, the proof is trivial.

Recalling that every weakly contractive map is Lipschitz-continuous, the result follows, for $D(X) = B(X)(X)$, immediatelly from Proposition (A3.20) and Lemma (A3.12).

For $D(X) = C(X)$, the result is a direct consequence of Lemmas (A3.11) and (A3.25). $\square$

We are ready to formulate

(A3.27) THEOREM. *The Hutchinson–Barnsley operator F^* in (A3.16) of a weakly contractive multivalued system (A3.19) has exactly one fixed point $A^* \in (D(X), d_H)$.*

This fixed point, representing a closed or a bounded, closed or a compact (sub)invariant subset of X, under the Hutchinson–Barnsley map F in (A3.15), is called a metric multivalued fractal.

PROOF. At first, we recall that $(D(X), d_H)$ is complete, because so is (X, d). Furthermore, according to Proposition (A3.20), F^* is weakly contractive. At last, according to Lemma (A3.26), F^* is a self-map on $(D(X), d_H)$.

Thus, also the assumptions of Proposition (A3.2) are satisfied (for F^*), by which exactly one fixed point of F^* exists in $(D(X), d_H)$.   □

The following corollary represents evidently a particular case of Theorem (A3.27).

(A3.28) COROLLARY. *There is exactly one closed or bounded, closed or compact (sub)invariant subset of a complete metric space X (a metric multivalued fractal), under the Hutchinson–Barnsley map $F(x) = \bigcup_{i=1}^{N} \varphi_i(x)$, $x \in X$, generated by multivalued contractions (i.e. by Lipschitz-continuous maps with constants $L_i \in [0, 1)$) $\varphi_i : (X, d) \to (D(X), d_H)$, $i = 1, \ldots, N$, where $D(X)$ denotes $\mathrm{Cl}(X)$ or $B(X)$ or $C(X)$, respectively.*

(A3.29) REMARK. One can easily check that the fractals in Theorem (A3.27) and Corollary (A3.28) are always compact, provided the sets of values of all maps φ_i, $i = 1, \ldots, N$, in MS (A3.19) are compact.

As an example of the application of Corollary (A3.28), we can give

(A3.30) EXAMPLE.

In Fig. 1, one can see Barnsley's fern (a single-valued fractal) generated by means of the transformations (see [Bar-M])

$$\begin{cases} f_1(x, y) = (0.849x + 0.037y + 0, \; -0.037x + 0.849y + 1.6), \\ f_2(x, y) = (0.197x - 0.257y + 0, \; 0.226x + 0.223y + 1.6), \\ f_3(x, y) = (-0.15x + 0.283y + 0, \; 0.26x + 0.238y + 0.44), \\ f_4(x, y) = (0x + 0y + 0, \; 0x + 0.16y + 0). \end{cases}$$

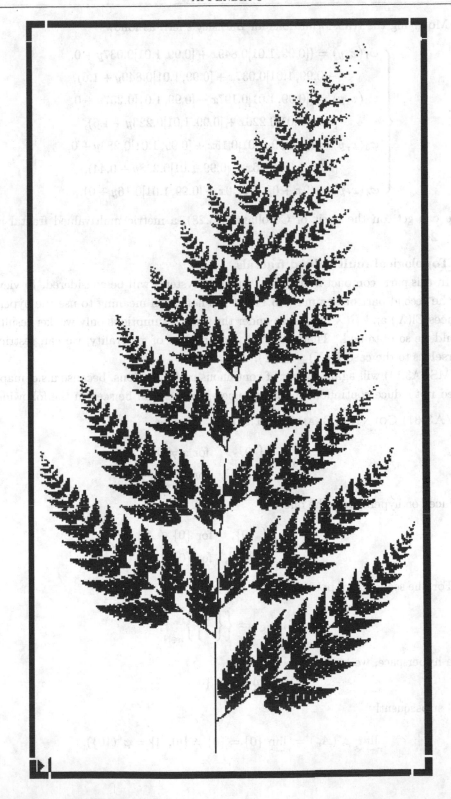

FIGURE 1

Modifying the transformations for Barnsley's fern as follows

$$
\begin{cases}
\varphi_1(x,y) = ([0.99, 1.01]0.849x + [0.99, 1.01]0.037y + 0, \\
\qquad -[0.99, 1.01]0.037x + [0.99, 1.01]0.849y + 1.6), \\
\varphi_2(x,y) = ([0.99, 1.01]0.197x - [0.99, 1.01]0.257y + 0, \\
\qquad [0.99, 1.01]0.226x + [0.99, 1.01]0.223y + 1.6), \\
\varphi_3(x,y) = (-[0.99, 1.01]0.15x + [0.99, 1.01]0.283y + 0, \\
\qquad [0.99, 1.01]0.26x + [0.99, 1.01]0.238y + 0.44), \\
\varphi_4(x,y) = (0x + 0y + 0, \ 0x + [0.99, 1.01]0.16y + 0),
\end{cases}
$$

one can get, on the basis of Corollary (A3.28), a metric multivalued fractal in Fig. 2.

Topological multivalued fractals.

In this part, compact multivalued function systems will be considered. In view of the second part of Lemma (A3.11), it has not much meaning to use the hyperspaces $\mathrm{Cl}(X)$ and $B(X)$, because under the same assumptions only weaker results could be so obtained. Therefore, without any loss of generality, we can restrict ourselves to the case $D(X) = C(X)$.

MS (A3.14) will again consist of continuous multifunctions, because u.s.c. maps need not induce continuous maps on hyperspaces, as can be seen in the following

(A3.31) COUNTER-EXAMPLE. The u.s.c. mapping

$$
\varphi(x) = \begin{cases} 0, & \text{for } x \neq 0, \\ [0, 1], & \text{for } x = 0, \end{cases}
$$

induces on hyperspaces the one

$$
\varphi^*(A) = \begin{cases} \{0\}, & \text{for } \{0\} \notin A, \\ [0, 1], & \text{for } \{0\} \in A. \end{cases}
$$

For the sequence of points

$$
\{A_n\}_{n \in \mathbb{N}} = \left\{ \left\{ \frac{1}{n} \right\} \right\}_{n \in \mathbb{N}},
$$

in a hyperspace, we have

$$
\lim_{n \to \infty} A_n = \{0\},
$$

and subsequently

$$
\lim_{n \to \infty} \varphi^*(A_n) = \lim_{n \to \infty} \{0\} = \{0\} \neq \{[0, 1]\} = \varphi^*(\{0\}).
$$

FIGURE 2

This inequality demonstrates the discontinuity of φ^*.

Although we have for our disposal more general versions of Proposition (A3.3) e.g. for CAC-maps (compact absorbing contractions), involving only a less amount of compactness (see Chapters I.6 and I.7), its application might be difficult, as follows from

(A3.32) COUNTER-EXAMPLE. Consider the functions

$$f_1(x) = \begin{cases} 0, & \text{for } x \leq 0, \\ -2x, & \text{for } x > 0, \end{cases}$$

and

$$f_2(x) = \begin{cases} -2x, & \text{for } x \leq 0, \\ 0, & \text{for } x > 0. \end{cases}$$

Since their second iterates f_1^2, f_2^2 are identically equal to zero, $f_1^2(x) \equiv 0$ and $f_2^2(x) \equiv 0$, they are eventually compact. Moreover, they are evidently locally compact. The class of eventually compact (i.e. some of their iterates are compact) and locally compact maps is, besides compact maps, the simplest subclass of CAC-maps (see Chapter I.5).

For the iterates of the Hutchinson–Barnsley map $F(x) = f_1(x) \cup f_2(x)$ of the MS $\{f_1 : \mathbb{R} \to \mathbb{R}, \ i = 1, 2\}$, we have

$$F : \{x\} \to \{0, -2x\} \to \{0, 4x\} \to \ldots \to \{0, (-2)^k x\} \to \ldots$$

whenever $x \neq 0$ and

$$F : \{0\} \to \{0\} \to \ldots \to \{0\} \to \ldots$$

Although zero is apparently the fixed point (the fractal) of F, the divergent component $(-2)^k x$, for $x \neq 0$, excludes the eventual compactness of F.

(A3.33) DEFINITION. Let (X, d) be a metric space such that $(C(X), d_H)$ is an AR-space. Let, furthermore, MS

(A3.34) $\{\varphi_i : (X, d) \to (\text{Cl}(X), d_H), \quad i = 1, \ldots, N\}$

consist of continuous and compact multifunctions φ_i, $i = 1, \ldots, N$.

Then MS ((A3.34)) will be called the *compact multifunction system* (CMS).

In order to apply Proposition (A3.3), we must prove that the Hutchinson–Barnsley operator F^* in (A3.16) of CMS (A3.34) is (I) a self-map which is (II) continuous and (III) compact.

As we will show, conditions (I) and (II) can be deduced from their analogies for F in (A3.15), i.e. when F is continuous and compact-valued. For this implication, the following two lemmas will be useful.

(A3.35) LEMMA. *Let (X, d) be a metric space and $F : X \to C(X)$ be continuous. Then the mapping $\widehat{F}$, defined as $\widehat{F}(A) := \{F(x) \mid x \in A\}$, $A \in C(X)$,*

   (i) *operates from $(C(X), d_H)$ into $(C(X)(C(X)), d_{d_H})$, where d_{d_H} is the Hausdorff metric in $C(X)(C(X))$, induced by the one d_H in $C(X)$, and*

   (ii) *it is continuous.*

PROOF. (i) Since F is continuous, $\widehat{F}(A) := \{F(x) \mid x \in A\}$ is by Lemma (A3.11) compact, for any compact A, i.e. $\widehat{F}(A) \in C(X)(C(X))$.

(ii) The continuity of F implies that, for any sequence of $\{x_n\}$, $x_n \in X$, $n \in \mathbb{N}$, and $x_0 \in X$, we have

(A3.36) $x_n \xrightarrow{d} x_0 \Rightarrow F(x_n) \xrightarrow{d_H} F(x_0)$.

Now, we proceed by a contradiction. Assuming the existence of $A_0 \in C(X)$ and $\{A_n\}$, $A_n \in C(X)$, such that

$$(A_n \xrightarrow{d_H} A_0) \quad \text{and} \quad (\widehat{F}(A_n) \xrightarrow{d_{d_H}} \widehat{F}(A_0))$$

means

$$(\exists \varepsilon_0 > 0)(\forall n \in \mathbb{N}) : \quad d_{d_H}(\widehat{F}(A_n), \widehat{F}(A_0)) > \varepsilon_0.$$

Recall that, for any $\mathcal{A}, \mathcal{B} \in C(X)(C(X))$:

$$d_{d_H}(\mathcal{A}, \mathcal{B}) = \max\{e_{d_H}(\mathcal{A}, \mathcal{B}), e_{d_H}(\mathcal{B}, \mathcal{A})\},$$

where

$$e_{d_H}(\mathcal{A}, \mathcal{B}) := \sup_{A \in \mathcal{A}}\{d_H(A, \mathcal{B})\} = \sup_{A \in \mathcal{A}}\{\inf_{B \in \mathcal{B}}\{d_H(A, B)\}\},$$
$$e_{d_H}(\mathcal{B}, \mathcal{A}) := \sup_{B \in \mathcal{B}}\{d_H(B, \mathcal{A})\} = \sup_{B \in \mathcal{B}}\{\inf_{A \in \mathcal{A}}\{d_H(B, A)\}\}.$$

Hence, let (the second case can be treated quite analogously)

$$d_{d_H}(\widehat{F}(A_n), \widehat{F}(A_0)) = \max\{e_{d_H}(\widehat{F}(A_n), \widehat{F}(A_0)), e_{d_H}(\widehat{F}(A_0), \widehat{F}(A_n))\}$$
$$= e_{d_H}(\widehat{F}(A_n), \widehat{F}(A_0)) > \varepsilon_0,$$

where, more precisely,

$$e_{d_H}(\widehat{F}(A_n), \widehat{F}(A_0)) = e_{d_H}(\{F(x) \mid x \in A_n\}, \{F(y) \mid y \in A_0\})$$
$$= \sup_{F(x) \in \widehat{F}(A_n)}\{\inf_{F(y) \in \widehat{F}(A_0)} d_H(F(x), F(y))\}$$
$$= \sup_{x \in A_n}\{\inf_{y \in A_0}\{d_H(F(x), F(y))\}\} > \varepsilon_0 > 0, \quad n \in \mathbb{N}.$$

From here, it follows, for every $n \in \mathbb{N}$, the existence of $x_n \in A_n$ such that

$$(A3.37) \qquad \inf_{y \in A_0} \{d_H(F(x_n), F(y))\} = d_H(\widehat{F}(x_n), \widehat{F}(A_0)) > \varepsilon_0 > 0.$$

Because of $x_n \in A_n$, $n \in \mathbb{N}$, and

$$A_n \xrightarrow{d_H} A_0, \quad \text{for } n \to \infty,$$

one gets

$$x_n \xrightarrow{d} A_0, \quad \text{for } n \to \infty,$$

i.e. there exists a sequence $\{y_n\}_{n \in \mathbb{N}}$ of points of A_0 such that

$$x_n \xrightarrow{d} y_n, \quad \text{for } n \to \infty.$$

Since $\{y_n\}_{n \in \mathbb{N}}$ belongs to a compact set, its converging subsequence $\{y_{n_k}\}_{k \in \mathbb{N}}$ exists such that

$$y_{n_k} \xrightarrow{d} y_0 \in A_0, \quad \text{for } k \to \infty.$$

The last two relations imply that also

$$x_{n_k} \xrightarrow{d} y_0 \in A_0, \quad \text{for } k \to \infty,$$

and subsequently, in view of (A3.36), that

$$F(x_{n_k}) \xrightarrow{d_H} F(y_0) \in \widehat{F}(A_0), \quad \text{for } k \to \infty.$$

This, however, means that

$$d_H(\widehat{F}(x_{n_k}), \widehat{F}(A_0)) \to 0, \quad \text{for } k \to \infty,$$

a contradiction with (A3.37), which completes the proof. $\qquad\square$

For the union mapping

$$(A3.38) \qquad u : \mathcal{B} \in 2^{2^X} \to \bigcup_{B \in \mathcal{B}} B,$$

the following properties are true.

(A3.39) LEMMA. u defined by (A3.38):

(i) operates from $C(X)(C(X))$ into $C(X)$,
(ii) it is continuous (in $C(X)(C(X))$).

PROOF. (i) For $\mathcal{B} \in C(X)(C(X))$ and $\{x_n\}_{n \in \mathbb{N}}$ in $u(\mathcal{B})$, we have

$$(\forall n \in \mathbb{N})(\exists B_n \in \mathcal{B}) : x_n \in B_n,$$

which defines the sequence $\{B_n\}_{n \in \mathbb{N}}$ of $\mathcal{B}$. Since $\mathcal{B}$ is compact, there exists a subsequence $\{B_{n_k}\}_{k \in \mathbb{N}}$ and $B_0 \in \mathcal{B}$ such that

$$B_{n_k} \xrightarrow{d_H} B_0, \quad \text{for } k \to \infty.$$

Because of $x_n \in B_n, n \in \mathbb{N}$, one still has

$$x_{n_k} \xrightarrow{d} B_0, \quad \text{for } k \to \infty,$$

which implies the existence of $\{y_{n_k}\}_{k \in \mathbb{N}}$ in B_0 with

$$x_{n_k} \xrightarrow{d} y_{n_k}, \quad \text{for } k \to \infty.$$

Since B_0 is compact, there exists a converging subsequence $\{y_{n_{k_l}}\}_{l \in \mathbb{N}}$ and its limit $y_0 \in B_0$, i.e.

$$y_{n_{k_l}} \xrightarrow{d} y_0, \quad \text{for } l \to \infty.$$

The last two relations imply that

$$x_{n_{k_l}} \xrightarrow{d} y_0 \in B_0 \subset u(\mathcal{B}), \quad \text{for } l \to \infty.$$

After all, for every sequence in $u(\mathcal{B})$, there exists a converging subsequence in $u(\mathcal{B})$, and so

$$u : C(X)(C(X)) \to C(X).$$

(ii) We will show that u is non-expansive (i.e. Lipschitzian with the constant 1). For any compact $\mathcal{A}, \mathcal{B} \in C(X)(C(X))$, we have

$$
\begin{aligned}
(A3.40) \quad d_{d_H}(\mathcal{A}, \mathcal{B}) &= \max\{e_{d_H}(\mathcal{A}, \mathcal{B}), e_{d_H}(\mathcal{B}, \mathcal{A})\} \\
&= \max\{\sup_{A \in \mathcal{A}}\{d_H(A, \mathcal{B})\}, \sup_{B \in \mathcal{B}}\{d_H(B, \mathcal{A})\}\} \\
&= \max\{\sup_{A \in \mathcal{A}}\{\inf_{B \in \mathcal{B}} d_H(A, B)\}, \sup_{B \in \mathcal{B}}\{\inf_{A \in \mathcal{A}} d_H(B, A)\}\}.
\end{aligned}
$$

Similarly, for their images in $C(X)$, we have

$$(A3.41) \quad d_H(u(\mathcal{A}), u(\mathcal{B})) = \max\{e_d(u(\mathcal{A}), u(\mathcal{B})), e_d(u(\mathcal{B}), u(\mathcal{A}))\}$$

$$= \max\left\{ e_d\left(\bigcup_{A\in\mathcal{A}} A, \bigcup_{B\in\mathcal{B}} B \right), e_d\left(\bigcup_{B\in\mathcal{B}} B, \bigcup_{A\in\mathcal{A}} A \right) \right\}$$

$$= \max\left\{ \sup_{x\in\bigcup_{A\in\mathcal{A}} A}\left\{ d\left(x, \bigcup_{B\in\mathcal{B}} B \right)\right\}, \sup_{y\in\bigcup_{B\in\mathcal{B}} B}\left\{ d\left(y, \bigcup_{A\in\mathcal{A}} A \right)\right\} \right\}$$

$$= \max\left\{ \sup_{A\in\mathcal{A}}\left\{ \sup_{x\in A}\left\{ d\left(x, \bigcup_{B\in\mathcal{B}} B \right)\right\}\right\},\right.$$

$$\left. \sup_{B\in\mathcal{B}}\left\{ \sup_{y\in B}\left\{ d\left(y, \bigcup_{A\in\mathcal{A}} A \right)\right\}\right\} \right\}$$

$$= \max\left\{ \sup_{A\in\mathcal{A}} e_d\left(A, \bigcup_{B\in\mathcal{B}} B \right), \sup_{B\in\mathcal{B}} e_d\left(B, \bigcup_{A\in\mathcal{A}} A \right) \right\}$$

$$= \max\left\{ \sup_{A\in\mathcal{A}}\left\{ \inf_{y\in\bigcup_{B\in\mathcal{B}} B}\{e_d(A, y)\}\right\},\right.$$

$$\left. \sup_{B\in\mathcal{B}}\left\{ \inf_{x\in\bigcup_{A\in\mathcal{A}} A}\{e_d(B, x)\}\right\} \right\}$$

$$= \max\{\sup_{A\in\mathcal{A}}\{ \inf_{B\in\mathcal{B}}\{ \inf_{y\in B}\{e_d(A, y)\}\}\}, \sup_{B\in\mathcal{B}}\{ \inf_{A\in\mathcal{A}}\{ \inf_{x\in A}\{e_d(B, x)\}\}\}\}$$

$$= \max\{\sup_{A\in\mathcal{A}}\{ \inf_{B\in\mathcal{B}}\{e_d(A, B)\}\}, \sup_{B\in\mathcal{B}}\{ \inf_{A\in\mathcal{A}}\{e_d(B, A)\}\}\}$$

$$(A3.42) \quad \leq \max\{\sup_{A\in\mathcal{A}}\{ \inf_{B\in\mathcal{B}}\{d_H(A, B)\}\}, \sup_{B\in\mathcal{B}}\{ \inf_{A\in\mathcal{A}}\{d_H(B, A)\}\}\},$$

because

$$d_H(A, B) = \max\{e_d(A, B), e_d(B, A)\},$$

and subsequently

$$e_d(A, B) \leq d_H(A, B) \quad \text{and} \quad e_d(B, A) \leq d_H(A, B).$$

From (A3.41) and (A3.42), the inequality

$$d_H(u(\mathcal{A}), u(\mathcal{B})) \leq \max\{\sup_{A\in\mathcal{A}}\{ \inf_{B\in\mathcal{B}}\{d_H(A, B)\}\}, \sup_{B\in\mathcal{B}}\{ \inf_{A\in\mathcal{A}}\{d_H(B, A)\}\}\},$$

whose right-hand side equals (A3.40), follows.

After all,

$$d_H(u(\mathcal{A}), u(\mathcal{B})) \leq d_{d_H}(\mathcal{A}, \mathcal{B}),$$

i.e. u is non-expansive and, in particular, continuous. $\square$

Summarizing Lemmas (A3.35) and (A3.39), we can give

(A3.43) PROPOSITION. *If $F : (X,d) \to (C(X), d_H)$ is continuous, then the induced mapping*

$$F^*(A) = \bigcup_{x \in A} F(x), \quad A \in C(X)$$

(i) *is a self-map on $(C(X), d_H)$ (i.e. compact-valued),*

(ii) *it is continuous.*

PROOF. F^* takes the form of composition:

$$F^* : u \circ \widehat{F},$$

where $\widehat{F}$ is defined in Lemma (A3.35) and u in (A3.38).

Applying Lemmas (A3.35) and (A3.37), we get directly that

(i) F^* is compact-valued, because

$$F^* : C(X) \xrightarrow{\widehat{F}} C(X)(C(X)) \xrightarrow{u} C(X),$$

and

(ii) F^* is continuous as a composition of two continuous maps.

$\square$

In order to satisfy (I) and (II), by means of Proposition (A3.43), we need

(A3.44) LEMMA. *The Hutchinson–Barnsley map F in (A3.15) of CMS (A3.34) is compact-valued and continuous.*

PROOF. At first, we will show that the set of values of F is relatively compact. Because of

$$\overline{\varphi_i(X)} = \overline{\bigcup_{x \in X} \varphi_i(x)} := K_i \in C(X), \quad i = 1, \dots, N,$$

we get

$$\overline{F(X)} = \overline{\bigcup_{x \in X} F(x)} = \overline{\bigcup_{x \in X} \bigcup_{i=1}^{N} \varphi_i(X)} = \bigcup_{i=1}^{N} \overline{\varphi_i(X)} = \bigcup_{i=1}^{N} \overline{\varphi_i(X)}$$

$$= \bigcup_{i=1}^{N} K_i := K \in C(X), \quad \text{i.e. } F \text{ is compact.}$$

In view of the second part of Lemma (A3.11), we arrive at

(A3.45) $F : X \to C(K) \quad (\subset C(X) \subset \mathrm{Cl}(X)),$

as required.

Now, we will check the continuity of F. Applying the well-known inequality (see e.g. [Bar-M])

$$d_H\left(\bigcup_{i=1}^{N} A_i, \bigcup_{i=1}^{N} B_i\right) \leq \max_{i=1,\ldots,N}\{d_H(A_i, B_i)\},$$

we obtain

$$d_H(F(x), F(y)) = d_H\left(\bigcup_{i=1}^{N} \varphi_i(x), \bigcup_{i=1}^{N} \varphi_i(y)\right)$$

(A3.46)

$$\leq \max_{i=1,\ldots,N}\{d_H(\varphi_i(x), \varphi_i(y))\},$$

for every $x, y \in X$.

The continuity of φ_i means that

$$(\forall \varepsilon_i > 0)(\exists \delta_i > 0): d(x, y) < \delta_i \Rightarrow d_H(\varphi_i(x), \varphi_i(y)) < \varepsilon_i, \quad i = 1, \ldots, N.$$

Taking

$$\delta = \min_{i=1,\ldots,N} \delta_i \quad \text{and} \quad \varepsilon = \max_{i=1,\ldots,N} \varepsilon_i,$$

we so have

$$(\forall \varepsilon > 0)(\exists \delta > 0): d(x, y) < \delta \Rightarrow d_H(\varphi_i(x), \varphi_i(y)) < \varepsilon, \quad i = 1, \ldots, N,$$

and subsequently

$$(\forall \varepsilon > 0)(\exists \delta > 0): d(x, y) < \delta \Rightarrow \max_{i=1,\ldots,N}\{d_H(\varphi_i(x), \varphi_i(y))\} < \varepsilon.$$

By means of (A3.46), we already get the desired continuity of F, namely

$$(\forall \varepsilon > 0)(\exists \delta > 0):$$

$$d(x, y) < \delta \Rightarrow d_H(F(x), F(y)) \leq \max_{i=1,\ldots,N}\{d_H(\varphi_i(x), \varphi_i(y))\} < \varepsilon.$$

$\square$

In order to satisfy (III), we need still

(A3.47) LEMMA. *The Hutchinson–Barnsley operator F^* in (A3.16) of CMS (A3.34) is compact, i.e.*

(A3.48) $$\overline{F^*(C(X))} \in C(X).$$

PROOF. Lemmas (A3.11) and (A3.44) imply, in view of (A3.45), in fact that

$$F^* : K_k(X) \to C(X)(K).$$

Recalling that, for any compact $K \subset X$, $C(X)(K)$ is a compact subset of $C(X)$ (see e.g. [IlNa-M], [Nad-M]), we therefore have

$$\overline{F^*(C(X))} = \overline{\{F^*(A), A \in C(X)\}} \subset \overline{C(X)(K)} = C(X)(K),$$

i.e. that $\overline{F^*(C(X))}$ is a closed subset of the compact set $C(X)(K)$. Therefore, (A3.48) is true as well.   □

Hence, we are ready to formulate

(A3.49) THEOREM. *The Hutchinson–Barnsley operator F^* in* (A3.16) *of compact multivalued system* (A3.34) *admits a fixed point $A^* \in (C(X), d_H)$.*

This fixed point, representing a compact invariant subset of X, under the Hutchinson–Barnsley map F in (A3.15), *is called a topological multivalued fractal.*

PROOF. The assertion follows directly, on the basis of Proposition (A3.3), from Proposition (A3.43) and Lemmas (A3.44) and (A3.47).   □

In view of the mentioned results of M. Wojdyslawski [Woj] and U. Tasmetov [Tas], Theorem (A3.49) has the following consequence.

(A3.50) COROLLARY. *Let (X, d) be a connected and locally connected Polish space (e.g. a Peano's continuum) and let $\varphi_i : (X, d) \to (\mathrm{Cl}(X), d_H)$, $i = 1, \ldots, N$, be continuous and compact multifunctions. Then there exists a compact invariant subset of X (a topological multivalued fractal), under the Hutchinson–Barnsley map $F(x) = \bigcup_{i=1}^{N} \varphi_i(x)$, $x \in X$.*

Some further possibilities.

Applying, instead of (Schauder-like) Proposition (A3.3), a particular form of the well-known Knaster–Tarski fixed point theorem (see e.g. [DG]), the assumptions of Theorem (A3.49) can be relaxed.

Let (X, d) be a metric space. Let us recall that for $\varphi : X \multimap X$, which maps compact sets into compact sets (i.e. $\varphi^* : C(X) \to C(X)$), we can define the following classes of elements of $C(X)$:

(1) $\mathcal{I}nvt(\varphi) := \{A \in C(X) \mid \varphi(A) = A\}$, i.e. the class of *invariant sets* of φ or of fixed points of φ^*;

(2) $\mathcal{M}in\mathcal{I}nvt(\varphi) = \{B \in C(X) \mid \forall A \in \mathcal{I}nvt(\varphi) : A \subset B \Rightarrow A = B\}$, i.e. the class of *minimal invariant sets* of φ.

Let us also recall that A is *invariant (subinvariant)* w.r.t. φ if $\varphi(A) = A$ ($\varphi(A) \subset A$).

(A3.51) PROPOSITION (Knaster–Tarski, cf. [DG-M]). *If, for a map $\varphi : X \multimap X$ such that $\varphi^* : C(X) \to C(X)$, there exists a subinvariant set $A_* \in C(X)$, i.e. $\varphi(A_*) \subset A_*$, then $Min\mathcal{I}nvt(\varphi) \neq \emptyset$ and, in particular, $\mathcal{I}nvt(\varphi) \neq \emptyset$.*

Hence, applying Proposition (A3.51), we can state

(A3.52) THEOREM. *Let (X, d) be a metric space and let*

$$\{\varphi_i : X \to \mathrm{Cl}(X), \quad i = 1, \ldots, N\}$$

be a system of continuous and compact multifunctions. Then the associated Hutchinson–Barnsley operator F^* in (A3.16) admits a fixed point $A^* \in (C(X), d_H)$, called a multivalued fractal.

PROOF. First of all, $F^* : C(X) \to C(X)$, according to Proposition (A3.43) and Lemma (A3.44) (observe that the restrictions, imposed on X in Definition (A3.33), can be omitted here).

The existence of a subinvariant set $A_* \in C(X)$ w.r.t. F can be detected from the proof of Lemma (A3.44), namely $A_* := K$.

Thus, according to Proposition (A3.51), whose all assumptions are satisfied, there exists a compact set A^* which is invariant under F, i.e. a fixed point of F^*.□

(A3.53) REMARK. In [Kie-M], the analogy of Theorem (A3.52) is presented, where the space X can be topological (Hausdorff), but the generating functions are single-valued. Of course, instead of the Hausdorff topology, the more general Vietoris one had to be employed there. The (maximal) fractal A^* can be obtained there as

$$\bigcap_{n \in \mathbb{N}} \overline{\bigcup_{m \geq n} \bigcup_{x \in X} F^m(x)}.$$

In [Les], the author discussed the usage of less regular (e.g. u.s.c.) and not necessarily compact (e.g. condensing) multivalued maps for the same aim, again on the basis of Proposition (A3.51).

(A3.54) REMARK. Although the application of Proposition (A3.51) looks more efficient than of Proposition (A3.3), the approach based on Proposition (A3.3) might seem to be (at least formally) extendable, when replacing Proposition (A3.3) by some related continuation principle, i.e. when using the homotopy arguments (for the candidates of such principles, see Chapter II).

(A3.55) EXAMPLE. In Fig. 3 and Fig. 4, one can see the multivalued Sierpiński-like triangles, representing the attractors of the following MS:

$$\begin{cases} \varphi_1(x,y) = ([0.5x, 0.5x+0.05], [0.5y, 0.5y+0.05]), \\ \varphi_2(x,y) = (0.5x+0.5, 0.5y), \\ \varphi_3(x,y) = (0.5x+0.25, 0.5y+0.5), \end{cases}$$

and

$$\begin{cases} \varphi_1(x,y) = ([0.5x, 0.55x], [0.5y, 0.55y]), \\ \varphi_2(x,y) = (0.5x+0.5, 0.5y), \\ \varphi_3(x,y) = (0.5x+0.25, 0.5y+0.5), \end{cases}$$

respectively.

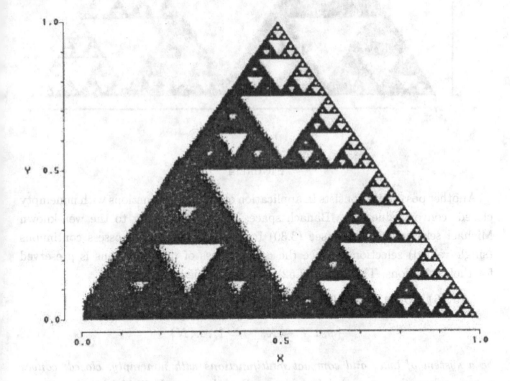

FIGURE 3

Although these multivalued fractals, which are (unlike the one in Fig. 2) also (*multivalued*) *multifractals*, were obtained on the basis of Corollary (A3.28), they can be also regarded (jointly with the one in Fig. 2) as applications of Theorem (A3.52) or Corollary (A3.50), because a sufficienlty large ball $B \subset \mathbb{R}^n$ certainly exists such that φ_i, $i = 1, 2, 3$, are compact, continuous self-maps on B.

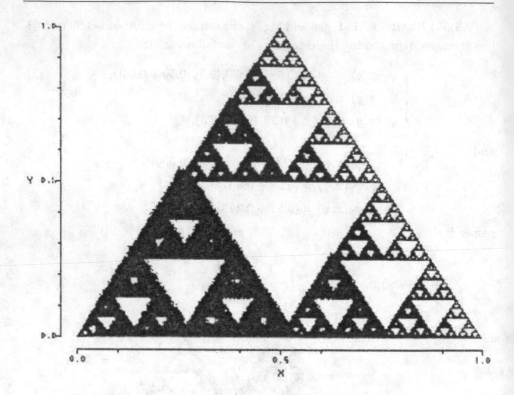

FIGURE 4

Another possibility consists in application of l.s.c. multifunctions with nonempty closed, convex values in a Banach space, because according to the well-known Michael selection theorem (see (3.30) in Chapter I.3), they possess continuous (single-valued) selections. Since the compactness of multifunctions is preserved for their selections, Theorem (A3.52) can be modified as follows.

(A3.56) COROLLARY. *Let (X, d) be a Banach space and let*

$$\{\varphi_i : X \multimap X, \quad i = 1, \ldots, N\}$$

be a system of l.s.c. and compact multifunctions with nonempty, closed, convex values. Then the associated Hutchinson–Barnsley map F in (A3.15) possesses a continuous (multivalued) selection $F_0 \subset F$ such that a compact invariant subset $A^* \in (C(X), d_H)$ exists w.r.t. F_0, i.e. $F_0(A^*) = A^*$, called a (single-valued) sub-fractal.

Now, we would like to indicate the possibility of constructing a continuation method for obtaining fractals (see Remark (A3.54)). This approach is completely new and we believe that it will deserve the future interest.

We say that a metric space (X, d) is a *hyper absolute neighbourhood retract* (a *hyper absolute retract*) if $(C(X), d_H)$ is an absolute neighbourhood retract (an absolute retract), written $X \in$ HANR ($X \in$ HAR).

Of course if $X \in$ HAR, then $X \in$ HANR. Nontrivial examples of HAR-spaces can be found above.

Let $\chi : X \times [0, 1] \multimap X$ be a Hausdorff continuous compact map. Then we have the induced operator

$$\chi^* : C(X) \times [0, 1] \to C(X).$$

One can prove as above that χ^* is also a compact operator.

We let χ_0, χ_1 by putting

$$\chi_0(x) = \chi(x, 0), \quad \chi_1(x) = \chi(x, 1).$$

Then the operators $\chi_0^*, \chi_1^* : C(X) \to C(X)$ are induced by χ_0 and χ_1, respectively.

We are in position to formulate the following continuation method for fractals.

(A3.57) THEOREM. *Let $X \in$ HANR and $\chi : X \times [0, 1] \multimap X$ be as above. Then there exists an essential fractal for χ_0 (i.e. $\Lambda(\chi_0^*) \neq 0$) if and only if the same is true for χ_1.*

PROOF. By the hypothesis, it follows that χ_0^* and χ_1^* are compact and homotopic mappings on an ANR-space $C(X)$ into itself.

In view of Theorem (6.20) in Chapter I.6, we obtain that the generalized Lefschetz numbers $\Lambda(\chi_0^*)$ and $\Lambda(\chi_1^*)$ are well-defined and, moreover, we have

$$\Lambda(\chi_0^*) = \Lambda(\chi_1^*).$$

Applying again Theorem (6.20), we obtain that $\mathrm{Fix}(\chi_0^*) \neq \emptyset$ if and only if $\mathrm{Fix}(\chi_1^*) \neq \emptyset$, which completes the proof. $\square$

(A3.58) COROLLARY. *If $X \in$ HAR, then any compact Hausdorff continuous map $\varphi : X \multimap X$ has a fractal.*

(A3.59) REMARK. Theorem (A3.57) can be applied in the following situation:

$$\text{let } \chi_1, \ldots, \chi_n : X \times [0, 1] \multimap X$$

be compact and Hausdorff continuous mappings. We define

$$\chi : X \times [0, 1] \multimap X$$

by letting

$$\chi(x, t) = \bigcup_{i=1}^{n} \chi_i(x, t).$$

Then one can prove (see: (A3.43), (A3.44), (A3.47)), that χ is compact and Hausdorff continuous, and subsequently, we are in the situation of Theorem (A3.57).

As concerns metric multivalued fractals, Theorem (A3.27) can be certainly modified, when applying another metric fixed point theorem (see e.g. [GoKi-M], [KiSi-M]), instead of Proposition (A3.2).

In [And23], Theorem (A3.27) (or, more precisely, Corollary (A3.28)) was reformulated, on the basis of the Lifschitz fixed point theorem (see [KsZa-M] or [GoKi-M]), for a system of the Lipschitz-continuous multivalued functions with the constants less or equal than the so called Lifschitz characteristic of the related hyperspaces $D(X)$. Although this characteristic can be greater than 1 (e.g., for the Hilbert space, it is greater or equal to $\sqrt{2}$), its computation for the hyperspace $D(X)$ is cumbersome. According to the mentioned result of D. W. Curtis [Cur], $(C(X), d_H)$ is homeomorphic to the l^2-space (i.e. the Hilbert space), provided (X, d) is a connected, locally connected and nowhere locally compact Polish space, but the homeomorphism is not explicitly known.

As a concrete example of such an X (see [Cur]), consider $X = \bigcup_{i=1}^{\infty} s_i$, where s_i are copies of $s = \prod_1^{\infty}(-1, 1)$, meeting at a single point θ. X is endowed at θ with a uniform topology (i.e. if $f : X \to s$ is a function whose restriction on every s_i is a homeomorphism onto s, then the basic neighbourhoods of θ in X take the form $f^{-1}(U)$, where U is a neighbourhood of $f(\theta)$ in s). Thus, $C(X) \approx s \approx l^2$. However, even this concrete example does not allow us to compute the desired Lifschitz characteristic of $(C(X), d_H)$, because $C(X)$ can be still rather far from the Hilbert space l^2.

On the other hand, if the composed map

$$h^{-1} \circ F^* \circ h : \mathcal{H} \xrightarrow{h} D(X) \xrightarrow{F^*} D(X) \xrightarrow{h^{-1}} \mathcal{H},$$

where $\mathcal{H}$ denotes a Hilbert space (e.g. l^2) and h is the mentioned homeomorphism, preserves the lipschitzianity of generating multifunctions $\varphi_i : X \to D(X)$, $i = 1, \ldots, N$, and subsequently of F^*, then we would define a *multivalued pseudofractal* as a fixed point, say $\alpha^* \in \mathcal{H}$, of $h^{-1} \circ F^* \circ h$, namely $\alpha^* = h^{-1} \circ F^* \circ h(\alpha^*)$. But the problem concerning the explicit knowledge of h occurs here as well.

We omitted here the structure of (multivalued) fractals. To have an idea, let us only note that the fractal dimension D of multivalued fractals in Fig. 3 and Fig. 4 can be estimated (for the definition and more details see e.g. [Hut], [Bar-M]) as $\log 3 / \log 2 \leq D \leq 2$, where $\log 3 / \log 2$ denotes the fractal dimension of the single-valued Sierpiński triangle (on the right hand-side) and 2 is the dimension in the left-hand corner.

References

I. Monographs, lecture notes and surveys

[Ag-M]   R. P. Agarwal, *Boundary Value Problems for Higher Order Differential Equations*, World Scientific, Singapore, 1986.

[AgMeOR-M]   R. P. Agarwal, M. Meehan and D. O'Regan, *Fixed Point Theory and Applications*, Cambridge University Press, Cambridge, 2001.

[AKPRS-M]   R. R. Akhmerov, M. I. Kamenskiĭ, A. S. Potapov, A. E. Rodkina and B. N. Sadovskiĭ, *Measures of Noncompactness and Condensing Operators*, Birkhäuser, Basel, 1992.

[AlPa-M]   P. Alexandrov and B. Pasynkov, *Introduction to the Theory of Dimension*, Nauka, Moscow, 1973. (in Russian)

[ASY-M]   K. T. Alligood, T. D. Sauer and J. A. Yorke, *Chaos (An Introduction to Dynamical Systems)*, Springer, Berlin, 1997.

[AP-M]   L. Amerio and G. Prouse, *Almost-Periodic Functions and Functional Equations*, Van Nostrand Reinhold Co., New York, 1971.

[AGN-M]   J. Andres, L. Górniewicz and P. Nistri (eds.), *Differential Inclusions and Optimal Control*, Lecture Notes in Nonlinear Analysis, Nicolaus Copernicus University, Toruń, 1997.

[ADTZ-M]   J. Appell, E. De Pascale, N. II. Thái and P. P. Zabreiko, *Multi-Valued Superpositions*, Diss. Math., vol. 345, PWN, Warsaw, 1995.

[ABLM-M]   L. Alseda, F. Balibrea, J. Llibre and M. Misiurewicz, *Thirty Years after Sharkovskii's Theorem: New Perspectives*, vol. 5, Internat. J. Bifur. Chaos Appl. Sci. Engrg., 1995.

[Au-M]   J. P. Aubin, *Viability Theory*, Birkhäuser, Basel, 1991.

[AuC-M]   J. P. Aubin and A. Cellina, *Differential Inclusions*, Springer-Verlag, Berlin, 1984.

[AuE-M]   J. P. Aubin and I. Ekeland, *Applieded Nonlinear Analysis*, XIV, 584, John Wiley and Sons, New York, 1984.

[AuF-M]   J. P. Aubin and H. Frankowska, *Set-Valued Analysis*, Birkhäuser, Basel, 1990.

[ABI-M]   A. Avantaggiati, G. Bruno and R. Iannacci, *Classical and New Results on Besicovitch Spaces of Almost-Periodic Functions and Their Duals*, Quaderni Dip. Me. Mo. Mat., "La Sapienza" – Roma, 1993, pp. 1–50.

[Bad-M]   R. Bader, *Fixpunktindextheorie mengenwertiger Abbildungen und einige Anwendungen*, Ph. D. Thesis, University of Munich, 1995. (in German)

[BaGo-M]   J. Banaś and K. Goebel, *Measures of Noncompactness in Banach Spaces*, M. Dekker, New York, 1980.

[BT-M]   E. A. Barbashin and V. A. Tabueva, *Dynamical Systems with Cylindrial Phase Space*, Nauka, Moscow, 1969. (in Russian)

[Ba-M]   V. Barbu, *Nonlinear Semigroups and Differential Equations in Banach Spaces*, Nordhoff, Leyden, 1976.

[Bar-M]   M. F. Barnsley, *Fractals Everywhere*, Academic Press, New York, 1988.

[Be-M]   C. Berge, *Espaces Topologiques, Fonctions Multivoques*, Dunod, Paris, 1959.

[BH-M]   J. F. Berglund and K. H. Hofmann, *Compact Semitopological Semigroups and Weakly Almost Periodic Functions*, Springer-Verlag, Berlin, 1967.

[BJM-M]   J. F. Berglund, H. D. Junghenn and P. Milnes, *Compact Right Topological Semigroups and Generalizations of Almost Periodicity*, Springer-Verlag, Berlin, 1978.

[Bes-M]   A. S. Besicovitch, *Almost Periodic Functions*, Cambridge Univ. Press, Cambridge, 1932; reprinted: Dover, New York, 1954.

[BPe-M]   C. Bessaga and A. Pełczyński, *Selected Topics in Infinite-Dimensional Topology*, Monograf. Mat, vol. 58, PWN, Warsaw, 1975.

[Bi-M]   J. S. Birman, *Braids, Links and Mapping Class Groups*, vol. 82, Ann. Math. Studies, Princeton Univ. Press, Princeton, 1974.

[BiLi-M]   J. S. Birman and A. Libgober (Eds.), *Braids. Proceedings of a Research Conference*, Contemp. Math., vol. 78, Amer. Math. Soc, Providence, R. I., 1988.

[BoKl-M]   N. A. Bobylev and V. S. Klimov, *Methods of Nonlinear Analysis in Nonsmooth Optimization*, Nauka, Moscow, 1992. (in Russian)

[Bo-M]   H. Bohr, *Almost Periodic Functions*, Chelsea, New York, 1956.

[BJ-M]   Boju Jiang, *Lectures on Nielsen Fixed Point Theory*, Contemp. Math., vol. 14, Amer. Math. Soc., Providence, R. I., 1983.

[Bor-M]   K. C. Border, *Fixed Point Theorems with Applications to Economics and Game Theory*, Cambridge University Press, Cambridge, 1985.

[BrGMO1-M]   Yu. G. Borisovitch, B. D. Gelman, A. D. Myshkis and V. V. Obuhovskii, *Introduction to the Theory of Multivalued Mappings*, Voronezh, Voronezh. Gos. Univ., 1986. (in Russian)

[BrGMO2-M]   _____, *Multivalued Analysis and Operator Inclusions*, Itogi Nauki-Seriya "Matematika" (Progress in Science-Mathematical Series) **29** (1986), 151–211 (in Russian); Engl. transl., J. Soviet Math. **39** (1987), 2772–2811.

[BrGMO3-M]   Yu. G. Borisovich, B. D. Gel'man, A. D. Myshkis and V. V. Obukhovskiĭ, *Topological Methods in the Fixed-Point Theory of Multi-Valued Maps*, Uspekhi Mat. Nauk **35** (1980), no. 1, 59–126 (in Russian); Russian Math. Surveys **35** (1980), no. 1, 65–143. (Engl. transl.)

[BrGMO4-M]   _____, *New Results in the Theory of Multivalued Mappings* I: *Topological Characteristics and Solvability of Operator Relations*, Mathematical Analysis, vol. 25, Itogi Nauki i Tekhniki, 1987, pp. 123–197. (in Russian)

[BrGMO5-M]   _____, *Multivalued Maps*, Itogi Nauki-Seriya "Matematika" (Progress in Science-Mathematical Series) **19** (1982), 127–230 (in Russian); J. Soviet Math. **24** (1984), 719–791. (Engl. transl.)

[Brs-M]   K. Borsuk, *Theory of Retracts*, vol. 44, Monografie Matematyczne, PWN, Warsaw, 1967.

[Bre-M]   H. Brezis, *Analyse fonctionelle – Théorie et applications*, Masson, Paris, 1983.

[Br1-M]   R. F. Brown, *The Lefschetz Fixed Point Theorem*, Scott, Foresman and Co., Glenview, Ill., 1971.

[Br2-M]   _____, *A Topological Introduction to Nonlinear Analysis*, Birkhäuser, Boston, 1993.

[BPS-M]   G. Bruno, A. Pankov and I. Spitkovsky, *Acta Appl. Math.*, vol. 65; *(Special Issue Dedicated to Antonio Avantaggiati on the Occasion of his 70th Birthday)*, Kluwer Acad. Publ., Dordrecht, 2001.

[BVGN-M]   B. F. Bylov, R. E. Vinograd, D. M. Grobman and V. V. Nemytskiĭ, *Theory of Liapunov Exponents*, Nauka, Moscow, 1966. (in Russian)

[CB-M]   A. J. Casson and S. A. Bleiler, *Automorphisms of Surfaces After Nielsen and Thurston*, vol. 9, London Math. Soc. Student Texts, Cambridge Univ. Press, Cambridge, 1988.

[CV-M]   C. Castaing and M. Valadier, *Convex Analysis and Measurable Multifunctions*, Lect. Notes in Math., vol. 580, Springer, Berlin, 1977.

[Cl-M]   F. H. Clarke, *Optimization and Nonsmooth Analysis*, Wiley, New York, 1983.

[Cu-M]   B. Cong Cuong, *Some Fixed Point Theorems for Multifuncions with Applications in Game Theory*, Diss. Math., vol. 245, PWN, Warsaw, 1985.

[Co-M]   C. Conley, *Isolated Invariant Sets and the Morse Index*, CBMS Regional Conference Ser. Math., vol. 38, Amer. Math. Soc., Providence, R. I., 1978.

[Cor-M]   C. Corduneanu, *Almost Periodic Functions*, Wiley, New York, 1968; Chelsea, New York, 1989.

[Cro-M]   J. Cronin, *Fixed Points and Topological Degree in Nonlinear Analysis*, vol. 11, Math. Surveys and Monogr., Amer. Math. Soc., Providence, R. I., 1964.

[DK-M]   J. L. Daleckiĭ and M. G. Krein, *Stability of Solutions of Differential Equations and Banach Space*, Transl. Math. Monogr., vol. 43, Amer. Math. Soc., Providence, R. I., 1974.

[Da-M]   A. Dawidowicz, *Methodes homologiques dans la theorie des applicationes et des champs de vecteurs spheriques dans les espaces de Banach*, Diss. Math., vol. 324, Warszawa, 1993.

[De1-M]   K. Deimling, *Multivalued Differential Equations*, W. De Gruyter, Berlin, 1992.

[De2-M]   _____, *Ordinary Differential Equations in Banach Spaces*, Lecture Notes in Math, vol. 596, Springer-Verlag, Berlin, New York, 1977.

[De3-M]   _____, *Nonlinear Functional Analysis*, Springer-Verlag, Berlin, 1985.

[Do-M]   A. Dold, *Lectures on Algebraic Topology*, Springer-Verlag, Berlin, 1972.

[DMNZ-M]   R. Dragoni, J. W. Macki, P. Nistri and P. Zecca, *Solution Sets of Differential Equations in Abstract Spaces*, Longman, Harlow, 1996.

[Du-M]   J. Dugundij, *Topology, 8th ed.*, Allyn and Bacon, Boston, 1973.

[DG-M]   J. Dugundij and A. Granas, *Fixed Point Theory*, I, Monograf. Mat., vol. 61, PWN, Warsaw, 1982.

[DS-M]   N. Dunford and J. T. Schwartz, *Linear Operators*, I, II, III, Wiley, New York, 1958, 1963, 1971.

[Dz-M]   Z. Dzedzej, *Fixed Point Index Theory for a Class of Nonacyclic Multivalued Maps*, Diss. Math., vol. 253, PWN, Warsaw, 1985.

[ES-M]   S. Eilenberg and N. Steenrod, *Foundations of Algebraic Topology*, Princeton Univ. Press, Princeton, N.J., 1952.

[EiFe-M]   G. Eisenack and C. Fenske, *Fixpunkttheorie*, BI Wissenschaftsverlag, Mannheim, Wien, Zürich, 1978.

[Eng1-M]   R. Engelking, *Outline of General Topology*, PWN, Warsaw, North-Holland, Amsterdam, 1968.

[Eng2-M]   _____, *Theory of Dimensions, Finite and Infinite*, Helderman, Leipzig, 1995.

[Far-M]   M. Farkas, *Periodic Motions*, Springer, Berlin, 1994.

[Fav-M]   J. Favard, *Leçons sur les fonctions presque-périodique*, Gauthier–Villars, Paris, 1933.

[Fel-M]   A. Fel'shtyn, *Dynamical Zeta Function*, Nielsen Theory and Reidemeister Torsion. MEMO, vol. 147, Amer. Math. Soc., Providence, R. I., 2000.

[Fi2-M]   A. F. Filippov, *Differential Equations with Discontinuous Right-Hand Sides*, vol. 224, Nauka, Moscow, 1985 (in Russian); Kluwer, Dordrecht, 1988.

[Fi1-M]   V. V. Filippov, *Solution Space of Ordinary Differential Equations*, Moscow University Publishers, Moscow, 1993. (in Russian)

[Fin-M]   M. Fink, *Almost Periodic Differential Equations*, vol. 377, LNM, Springer, Berlin, 1974.

[Fis-M]   J. Fišer, *Iterated Function and Multifunction Systems, Attractors and Their Basins of Attraction*, Ph.D. Thesis, Palacký University, Olomouc, 2002. (in Czech)

[FMMN-M]   P. M. Fitzpatrick, M. Martelli, J. Mawhin and R. Nussbaum, *Topological Methods for Ordinary Differential Equations* (M. Furi and P. Zecca, eds.), Springer, Berlin, 1993.

[Fr-M]   M. Frigon, *Application de la theorie de la transverssalite topologique a des problems non linearies pour des equations differentiales ordinaire*, Diss. Math., vol. 296, Warszawa, 1990.

[FrGr-M]   M. Frigon and A. Granas (eds.), *Topological Methods in Differential Equations and Inclusions*, Kluwer, Dordrecht, 1995.

[Fuc-M]   S. Fučík, *Solvability of Nonlinear Equations and Boundary Value Problems*, Reidel, Dordrecht, 1980.

[FZ-M]   M. Furi and P. Zecca (eds.), *Topological Methods in the Theory of Ordinary Differential Equations*, Lect. Notes Math., Springer, Berlin, 1993.

[GD-M]   D. Gabor, *Coincidence Points of Fredholm and Noncompact Multivalued Operators, PhD Thesis*, Nicolaus Copernicus University, Toruń, 2000. (in Polish)

[GGa-M]   G. Gabor, *Fixed Points of Multivalued Maps of Subsets of Locally Convex Spaces, PhD Thesis*, Nicolaus Copernicus University, Toruń, 1997. (in Polish)

[GaM-M]   R. G. Gaines and J. Mawhin, *Coincidence Degree and Nonlinear Differential Equations*, vol. 568, Lecture Notes in Math., Springer-Verlag, Berlin, 1977.

[GG-M]   K. Gęba and L. Górniewicz (eds.), *Topology in Nonlinear Analysis*, Banach Center Publications, vol. 35, Warsaw, 1996.

[GLY-M]   A. Kh. Gelig, G. A. Leonov and A. Yakubovich, *Stability of Nonlinear Systems with the Nonunique Equilibrium State*, Nauka, Moscow, 1978. (in Russian)

[Goe-M]   K. Goebel, *Coincise Course on Fixed Point Theorems*, Yokohama Publishers, Yokohama, 2002.

[GoeKi-M]   K. Goebel and W. A. Kirk, *Topics in Metric Fixed Point Theory*, vol. 28, Cambridge Studies in Advanced Mathematics, Cambridge University Press, Cambridge, 1990.

[GolShSt-M] M. Golubitsky, D. G. Shaeffer and I. N. Stewart, *Singularities and Bifurcation Theory*, II, Appl. Math. Sci. Scr., vol. 69, Springer, Berlin, 1988.

[Go1-M]   L. Górniewicz, *Homological Methods in Fixed Point Theory of Multivalued Maps*, Diss. Math., vol. 129, Warsaw, 1976.

[Go2-M]   _____ , *Topological Approach to Differential Inclusions*, Topological Methods in Differential Equations and Inclusions, NATO ASI Series C 472 (A. Granas, M. Frigon, eds.), Kluwer, Dordrecht, 1995, pp. 343–382.

[Go3-M]   _____ , *Recent Results on the Solution Sets of Differential Inclusions*, Sem. de Math. Super, vol. 110, Univ of Montreal, Montreal, 1990.

[Go4-M]   _____ , *Topological Degree of Morphisms and Its Applications to Differential Inclusions*, Raccolta di Sem. del. Dip. Mat., vol. 5, Dell'Univ. Studi della Calabria, Cosenza, 1985.

[Go5-M]   _____ , *Topological Fixed Point Theory of Multivalued Mappings*, Kluwer, Dordrecht, 1999.

[Gr1-M]   A. Granas, *Topics in the Fixed Point Theory*, Sem. J. Leray, Paris, 1969/70.

[Gr2-M]   _____ , *Points fixes pour les applications compactes en topologie et analyse fonctionnelle*, Sem. de Math. Superieures, Montreal, 1973.

[Gr3-M]   _____ , *Points fixes pour les applications compactes: Espaces de Lefschetz et la theorie del'indice*, SMS, vol. 68, Montreal, 1980.

[Gr4-M]   _____ , *The Theory of Compact Vector Fields and Some of Its Applications*, Diss. Math., vol. 30, Warsaw, 1962.

[GGL-M]   A. Granas, R. Guenther and J. Lee, *Nonlinear Boundary Value Problems for Ordinary Differential Equations*, Diss. Math., vol. 244, Warsaw, 1985.

[GKL-M]   R. S. Guter, L. D. Kudryavtsev and B. M. Levitan, *Elements of the Theory of Functions*, Pergamon Press, Oxford, 1966.

[Hal1-M]   J. K. Hale, *Asymptotic Behavior of Dissipative Systems*, MSM, vol. 25, AMS, Providence, R. I., 1988.

[Hal2-M]   _____ , *Theory of Functional Differential Equations*, Springer, Berlin, 1977.

[Han-M]   V. L. Hansen, *Braids and Coverings: Selected Topics*, London Math. Soc. Student Texts, vol. 18, Cambridge Univ. Press, Cambridge, 1989.

[Har-M]   P. Hartman, *Ordinary Differential Equations*, Willey, New York-London-Sydney, 1964.

[He-M]   D. Henry, *Geometric Theory of Semilinear Parabolic Equations*. LNM, vol. 840, Springer, Berlin, 1981.

[HR-M]   E. Hewitt and K. A. Ross, *Abstract Harmonic Analysis*, Springer-Verlag, Berlin, 1976.

[Hu1-M]   S.-T. Hu, *Homotopy Theory*, Academic Press, New York, 1959.

[Hu2-M]   _____ , *Theory of Retracts*, Wayne State Univer. Press, Detroit, 1965.

[HP1-M]   S. Hu and N. S. Papageorgiou, *Handbook of Multivalued Analysis, Theory*, vol. 1, Kluwer, Dordrecht, 1997.

[HP2-M]   S. Hu and N. S. Papageorgiou, *Handbook of Multivalued Analysis, Applications*, vol. 2, Kluwer, Dordrecht, 2000.

[HW-M]   W. Hurewicz and H. Wallman, *Dimension Theory*, Princeton Univ. Press, Princeton, 1941.

[HIR-M]   D. H. Hyers, G. Isak and T. M. Rassias, *Zero-Epi Mappings*, Topics in Nonlinear Analysis and Applications, World Scientific Publ., Singapore, 1997, pp. 241–324.

[IlNa-M]   A. Illanes and S. B. Nadler Jr., *Hyperspaces: Fundamentals and Recent Advances*, Pure and Applied Mathematics, vol. 216, Marcel Dekker, New York, 1999.

[Ju-M]   E. I. Jury, *Inners and Stability of Dynamic Systems*, Wiley, New York, 1974.

[KOZ-M]   M. I. Kamenskiĭ, V. V. Obukovskiĭ and P. Zecca, *Condensing Multiovalued Maps and Semilinear Differential Inclusions in Banach Spaces*, De Gruyter, Berlin, 2001.

[KK-M]   R. Kannan and C. King Krueger, *Advanced Analysis on the Real Line*, Springer, Berlin, 1996.

[Kat-M]   Y. Katznelson, *An Introduction to Harmonic Analysis*, Dover, New York, 1976.

[KTs-M]   Kiang Tsai-han, *The Theory of Fixed Point Classes*, Springer, Berlin, 1989.

[Kha-M]   V. Kh. Kharasakhal, *Almost-Periodic Solutions of Ordinary Differential Equations*, Nauka, Alma-Ata, 1970. (in Russian)

[Kie-M]   B. Kieninger, *Iterated Function Systems on Compact Hausdorff Spaces*, Ph.D. Thesis, Institut für Mathematik der naturwissenschaftlichen Fakultät der Universität Augsburg, Augsburg, 2002.

[KS-M]   I. T. Kiguradze and B. L. Shekhter, *Boundary Value Problems for Systems of Ordinary Differential Equations*, Singular Boundary Value Problems for the Second-Order Ordinary Differential Equations, Itogi Nauki i Tekh., Ser. Sovrem. Probl. Mat. 30, VINITI, Moscow, 1987. (in Russian)

[KiSi-M]   W. A. Kirk and B. Sims (eds.), *Handbook of Metric Fixed Point Theory*, Kluwer Academic Publishers, Dordrecht, 2001.

[Ki-M]   M. Kisielewicz, *Differential Inclusions and Optimal Control*, Polish Sc. Publishers and Kluwer Academic Publishers, Dordrecht, 1991.

[KolF-M]   A. N. Kolmogorov and S. V. Fomin, *Elements of the Theory of Functions and Functional Analysis*, Dover Publ., Mineola, 1999.

[Kö-M]   G. Köthe, *Topological Vector Spaces* I, Springer, Berlin, 1969.

[Kr-M]   B. Krajc, *Bounded Solutions of Differential Systems, PhD Thesis*, Palacký Univ, Olomouc, 1999.

[Ks1-M]   M. A. Krasnosel'skiĭ, *Topological Methods in the Theory of Nonlinear Integral Equations*, Gos. Izdat. Tehn.- Trov. Lit., Moscow, 1956 (in Russian); English translation in Pergamon Press, Oxford, 1963.

[Ks2-M]   ———, *Translation Operator Along the Trajectories of Differential Equations*, Nauka, Moscow, 1966 (in Russian); Amer. Math. Soc., Providence, R. I., 1968.

[KsBuKo-M] M. A. Krasnosel'skiĭ, V. Sh. Burd and Yn. S. Kolesov, *Nonlinear Almost Periodic Oscillations*, Nauka, Moscow, 1970. (in Russian)

[KsZa-M]   M. A. Krasnosel'skĭ and P. P. Zabreiko, *Geometrical Methods of Nonlinear Analysis*, Springer Verlag, Berlin, 1984.

[KrwWu-M]   W. Krawcewicz and J. Wu, *Theory of Degrees with Applications to Bifurcations and Differential Equations*, Wiley, New York, 1997.

[Kry1-M]   W. Kryszewski, *Topological and Approximation Methods in the Degree Theory of Set-Valued Maps*, Diss. Math., vol. 336, Warsaw, 1994, pp. 1–102.

[Kry2-M]   ———, *Homotopy Properties of Set-Valued Mappings*, UMK, Toruń, 1997.

[Kun-M]   M. Kunze, *Non-Smooth Dynamical Systems*, LNM, vol. 1744, Springer, Berlin, 2000.

[KR-M]   H. Kurland and J. Robin, *Infinite Codimension and Tranversality*, Lectures Notes in Math, vol. 468, Springer, Berlin, 1974.

[Kur1-M]   A. B. Kurzhanskiĭ, *Advances in Nonlinear Dynamics and Control: A Report from Russia*, Springer-Verlag, Berlin, 1993.

[Kur2-M]   ———, *Control and Observation Under Conditions of Uncertainty*, Nauka, Moscow, 1997. (in Russian)

[LR-M]   J. M. Lasry and R. Robert, *Analyse non lineaire multivoque*, Publication no 7611, Centre de Recherche de Mathematique de la Decision Ceremade, Univ. de Paris, Dauphine, 1977.

[LeCal-M]   P. LeCalvez, *Dynamical Properties of Diffeomorphisms of the Annulus and of the Torus*, vol. 4, SMF/AMS Texts and Monographs, Amer. Math. Soc., Providence, R. I., 2000.

[Ler-M]   J. Leray, *La théorie des pointes fixés et ses applications en analyse*, Proc. International Congress of Math., vol. 2, Amer. Math. Soc., Providence, R. I., 1952.

[Lev-M]   B. M. Levitan, *Almost-Periodic Functions*, G.I.T.-T.L., Moscow, 1959. (in Russian)

[LZ-M]   B. M. Levitan and V. V. Zhikov, *Almost Periodic Functions and Differential Equations*, Cambridge Univ. Press, Cambridge, 1982.

[Lew-M]   M. Lewicka, *Multivalued Poincaré Operator*, Magister Thesis, University of Gdańsk, Gdańsk, 1996. (in Polish)

[Ll-M]   N. G. Lloyd, *Degree Theory*, Cambridge Univ. Press, Cambridge, 1978.

[Lo-M]   L. H. Loomis, *An Introduction to Abstract Harmonic Analysis*, Van Nostrand, Co., New York, 1953.

[Lu-M]   J. Lukeš, *Notes in Functional Analysis*, Karolineum, Charles Univ., Prague, 1998. (in Czech)

[LS-M]   L. A. Lusternik and V. J. Sobolev, *Elements of Functional Analysis*, Gordon and Breach, New York, 1968.

[Ma-M]   T.-W. Ma, *Topological Degrees of Set-Valued Compact Fields in Locally Convex Spaces*, vol. 92, Diss. Math., Warsaw, 1972.

[Maa-M]   M. Maak, *Fastperiodische Funktionen*, Springer, Berlin, 1950.

[MM-M]   M. D. P. Monteiro Marques, *Differential Inclusins in Nonsmooth Mechanical Problems (Shocks and Dry Friction)*, Birkhäuser, Basel, 1993.

[MS-M]   J. L. Massera and J. J. Schaeffer, *Linear Differential Equations and Function Spaces*, Academic Press, New York, 1966.

[Maw-M]   J. Mawhin, *Topological Degree Methods in Nonlinear Boundary Value Problems. CBMS*, American Math. Soc., Providence, R. I., 1979.

[MW-M]   J. Mawhin and M. Willem, *Critical Point Theory and Hamiltonian Systems*, Springer, Berlin, 1989.

[McC-M]   C. K. McCord (ed.), *Nielsen Theory and Dynamical Systems.*, Contenp. Math., vol. 152, Amer. Math. Soc., Providence, R. I., 1993.

[McGMe-M]   R. McGehee and K. R. Meyer (eds.), *Twist Mappings and Their Applications*, The IMA Vol. in Math. Appl., vol. 44, Springer, Berlin, 1992.

[Min-M]   N. Minorsky, *Introduction to Non-Linear Mechanics*, I, J. W. Edwards, Ann Arbor, 1947.

[Mor-M]   S. Moran, *The Mathematical Theory of Knots and Braids: An Introduction*, vol. 82, North-Holland Math. Studies, Horth-Holland, Amsterdam, 1983.

[MurKu-M]   K. Murasugi and B. I. Kurpita, *A Study of Braids*, Kluwer, Dordrecht, 1999.

[Mu-M]   J. Musielak, *Introduction to Functional Analysis*, PWN, Warsaw, 1976. (in Polish)

[Nad-M]   S. B. Nadler, Jr., *Hyperspaces of Sets. A Text With Research Questions*, vol. 49, Monographs and Textbooks in Pure and Applied Mathematics, Marcel Dekker, New York - Basel, 1978.

[NG-M]   G. M. N'Guerékata, *Almost Automorphic and Almost Periodic Functions in Abstract Spaces*, Kluwer Acad. Publ., Dordrecht, 2001.

[Ni-M]   L. Nirenberg, *Topics in Nonlinear Functional Analysis*, New York Univ. Press, New York, 1975.

[Nu-M]   R. D. Nussbaum, *The Fixed Point Index and Some Applications*, Séminaire de Mathematiques Supérieures, vol. 94, Les Presses de l'Université Montreal, C. P. 6128, succ. "A", Montreal, Quebec, Canada H3C3J7, 1985.

[Ox-M]   J. Oxtoby, *Measure and Category*, Springer-Verlag, Berlin, 1971.

[Pn-M]   A. A. Pankov, *Bounded and Almost Periodic Solutions of Nonlinear Operator Differential Equations*, Kluwer, Dordrecht, 1990.

[Pet-M]   A. Petrusel, *Operational Inclusions*, House of the Book of Science, Cluj-Napoca, 2002.

[PetPetRu-M] A. Petrusel, G. Petrusel and I. A. Rus, *Fixed Point Theory* 1950–2000: *Romanian Contributions*, House of the Book of Science, Cluj-Napoca, 2002.

[Ptr-M]   W. V. Petryshyn, *Generalized Topological Degree and Semilinear Equations*, Cambridge Univ. Press, Cambridge, 1995.

[Pl-M]   V. A. Pliss, *Nonlocal Problems of the Oscillation Theory*, Nauka, Moscow, 1964. (in Russian)

[Poi-M]   H. Poincaré, *Lés méthodes nouvelles de la mécanique céleste*, Gauthiers–Villars, Paris, 1892.

[Pr-M]   T. Pruszko, *Some Applications of the Degree Theory to Multi-Valed Boundary Value Problems*, Diss. Math., vol. 229, Warsaw, 1984.

[ReSaCo-M] R. Reissig, G. Sansone and R. Conti, *Nichtlineare Differentialgleichungen höherer Ordnung*, Cremonese, Roma, 1969; English transl. in Noordhoff, Leyden, 1974.

[ReSe-M]   D. Repovš and P. Semenov, *Continuous Selections of Multivalued Mappings*, vol. 455, Math. Appl., Kluwer, Dordrecht, 1998.

[RoRo-M]   A. P. Robertson and W. Robertson, *Topological Vector Spaces*, Cambridge Univ. Press, Cambridge, 1964.

[Rob-M]   C. Robinson, *Dynamical Systems (Stability, Symbolic Dynamics, and Chaos)*, CRC Press, Boca Raton, Fl., 1995.

[Rot-M]   E. H. Rothe, *Introduction to Various Aspects of Degree Theory in Banach Spaces*, Mathematical Surveys and Monographs, vol. 23, AMS, Providence, R. I., 1986.

[RN-M]   D. Rozpłoch-Nowakowska, *Fixed Points of Morphisms of Admissible Spaces in the Sense of Klee, PhD Thesis*, Nicolaus Copernicus University, Toruń, 2000. (in Polish)

[Ru-M]   W. Rudin, *Functional Analysis*, Mc Graw-Hill Book Comp., New York, 1973.

[Ryb-M]   K. P. Rybakowski, *The Homotopy Index and Partial Differential Equations*, Springer, Berlin, 1987.

[SckSe-M]   R. J. Sacker and G. R. Sell, *Lifting Properties in Skew-Product Flows with Applications to Differential Equations*, Memoirs of the AMS, vol. 190, Amer. Math. Soc., Providence, R. I., 1977.

[Sch-M]   H. H. Schaeffer, *Topological Vector Spaces*, The Macmillan Company, New York, Collier-Macmillan Ltd., London, 1966.

[SheYi-M]   Wenxian Shen and Yingfei Yi, *Almost Automorphic and Almost Periodic Dynamics in Skew-Product Semiflows*, Memoirs of AMS, vol. 136, American Math. Soc., Providence, R. I., 1998, pp. 1–93.

[Sh-M]   M. A. Shubin, *Partial Differential Equations VII*, Encyclopaedia of Mathematical Sciences, vol. 64, Springer-Verlag, Berlin, 1994.

[Sm-M]   J. Smoller, *Shock Waves and Reaction Diffusion Equations*, Springer-Verlag, Berlin, 1980.

[Smr-M]   G. V. Smirnov, *Introduction to the Theory of Differential Inclusions*, Grad. Stud. Math., vol. 41, Amer. Math. Soc., Providence, R. I., 2002.

[Sp-M]   E. H. Spanier, *Algebraic Topology*, McGraw-Hill, New York, 1966.

[Sr-M]   M. Srebrny, *Measurable Selectors of PCA Multifunctions with Applications*, Mem. AMS, vol. 52, 311, Amer. Math. Soc., Providence, R. I., 1984.

[St-M]   J. Stryja, *Analysis of Almost-Periodic Functions*, Mgr. Thesis, Palacký University, Olomouc, 2001. (in Czech)

[Swr-M]   J. T. Schwartz, *Nonlinear Functional Analysis*, Gordon and Breach, New York, 1969.

[Th-M]   B. S. Thompson, *Real Functions*, Springer, Berlin, 1985.

[To-M]   A. A. Tolstonogov, *Differential Inclusions in Banach Spaces*, Sc. Acad. of Sciences, Siberian Branch, Novosibirsk, 1986. (in Russsian)

[Ut1-M]   V. I. Utkin, *Sliding Modes and Their Applications in Variable Structure Systems*, MIR, Moscow, 1978. (in Russsian)

[Ut2-M]   _____, *Sliding Modes in Control Optimization*, Springer-Verlag, Berlin, 1997.

[Vae-M]   M. Vaeth, *Volterra and Integral Equations of Vector Functions*, M. Dekker, Inc, New York, 2000.

[Vr-M]   I. I. Vrabie, *Compactness Methods for Nonlinear Evolutions*, 2nd ed., Longman House, Burn Mill, Harlow, 1990.

[Wa-M]   T. van der Walt, *Fixed and Almost Fixed Points*, North-Holland, Amsterdam, 1963.

[Wei-M]   A. Weil, *L'intégration dans les groupes topologiques*, Hermann, Paris, 1965.

[We-M]   R. Węgrzyk, *Fixed-Point Theorems for Multi-Valued Functions and Their Applications to Functional Equations*, Diss. Math., vol. 201, Warsaw, 1982.

[Wic-M]   K. R. Wicks, *Fractals and Hyperspaces*, vol. 1492, Lecture Notes in Mathematics, Springer-Verlag, Berlin, 1991.

[Wie-M]   A. Wieczorek, *Survey of Results on Kakutani Property of Spaces with generalized Convexity*, Fixed Point Theory and its Applications, vol. 252, Pitman Research Notes in Mathematics, 1990.

[Ysd-M]   K. Yoshida, *Functional Analysis*, Springer, Berlin, 1968.

[Yos1-M]   T. Yoshizawa, *Stability Theory by Liapunov's Second Method*, Math. Soc. Japan, Tokyo, 1966.

[Yos2-M]   ――――, *Stability Theory and Existence of Periodic Solutions and Almost Periodic Solutions*, Springer, Berlin, 1975.

[Zaa-M]   A. C. Zaanen, *Linear Analysis*, North-Holland Publ. Co., Amsterdam, 1964.

[Zai-M]   S. Zaidman, *Almost Periodic Functions in Abstract Spaces*, Chapman and Hall/-CRC, Boca Raton, Fl, 1999.

II. Articles

[Ab]   Kh. Abduvaitov, *Sufficient conditions for the existence of periodic and bounded solutions of the second-order nonlinear differential equations*, Diff. Urav. **21** (1985), no. 12, 2027–2036. (in Russian)

[Ad]   M. Adamczak, $C^{(n)}$-*almost periodic functions*, Commentat. Math. **37** (1997), 1–12.

[AOR]   R. P. Agarwal and D. O'Regan, *The solutions set of integral inclusions on the half line*, Analysis (2000), 1–7.

[Ah1]   S. Ahmad, *On almost periodic solutions of the competing species problems*, Proceed. of Amer. Math. Soc. **102** (1988), no. 4, 855–861.

[Ah2]   ――――, *A nonstandard resonance problem for ordinary differential equations*, Trans. Amer. Math. Soc. **323** (1991), no. 2, 857–875.

[AhTi]   S. Ahmad and A. Tineo, *Almost periodic solutions of second order systems*, Appl. Analysis **63** (1996), 385–395.

[ADA]   E. Ait Dads and O. Arino, *Exponential dichotomy and existence of pseudo almost-periodic solutions of some differential equations*, Nonlin. Anal. **27** (1996), 369–386.

[Akh]   R. R. Akhmerov, *The structure of the solution set of a boundary value problem for a one-dimensional stationary equation of variable type*, Chisl. Metody Mekh. Sploshn. Sredy **15** (1984), 20–30.

[Al]   J. C. Alexander, *A primer on connectivity*, Proc. on Fixed Point theory **886** (1980), Sherbrooke, 455–483.

[AlAu]   J. C. Alexander and J. F. G. Auchmuty, *Global bifurcations of phase-locked oscillators*, Arch. Rat. Mech. Anal. **93** (1986), 253–270.

[AlMaPe]   J. C. Alexander, I. Massabò and J. Pejsachowicz, *On the connectivity properties of the solution set of infinitely-parametrized families of vector fields*, Boll. Un. Mat. Ital. A **1** (1982), 309–312.

[AH]   E. M. Alfsen, P. Holm, *A note on compact representations and almost periodicity in topological spaces*, Math. Scand. **10** (1962), 127–136.

[CAl]   J. Carmona Alvárez, *Measure of noncompactness and fixed points of nonexpansive condensing mappings in locally convex spaces*, Rev. Real-Acad. Exact. Fis. Natur. Madrid **79** (1985), 53–66.

[Ame1]   L. Amerio, *Studio asintotico del molo di un punto su una linea chiusa, per azione di forze indipendenti dal tempo*, Ann. Sc. Norm. Sup. di Pisa (1950), 19–57.

[Ame2]   ――――, *Soluzioni quasi-periodiche, o limitate, di sistemi diffenziali non lineari-periodici, o limitati*, Ann. Mat. Pura Appl. **39** (1955), 97–119.

[EAH]   K. El Amri and C. Hess, *On the Pettis integral of closed valued multifunctions*, Set-Valued Anal. **8** (1999), no. 4, 153–165.

[And1]   J. Andres, *Asymptotic properties of solutions to quasi-linear differential systems*, J. Comput. Appl. Math. **41** (1991), 56–64.

[And2]   _____, *Transformation theory for nondissipative systems. Some remarks and simple application in examples*, Acta Univ. Palacki. Olomuc. **32** (1993), 125–132.

[And3]   _____, *On the multivalued Poincaré operators*, Topol. Methods Nonlinear Anal. **10** (1997), 171–182.

[And4]   _____, *Nonlinear rotations*, Nonlin. Anal. TMNA **30** (1997), no. 1, 495–503.

[And5]   _____, *Almost-periodic and bounded solutions of Carathéodory differential inclusions*, Differential Integral Equations **12** (1999), 887–912.

[And6]   _____, *Multiple bounded solutions of differential inclusions. The Nielsen theory approach*, J. Diff. Eqns **155** (1999), 285–310.

[And7]   _____, *Boundedness results of solutions to the equation $x''' + ax'' + g(x)x' + h(x) = p(t)$ without the hypothesis $h(x)\mathrm{sgn}\,x \geq 0$ for $|x| > R$*, Atti Accad. Naz. Lincei **80** (1987), 533–539.

[And8]   _____, *Four-point and asymptotic boundary value problems via a possible modification of Poincaré's mapping*, Math. Nachr. **149** (1990), 155–162.

[And9]   _____, *Note to the asymptotic behaviour of solutions of damped pendulum equations under forcing*, Nonlin. Anal. TMA **18** (1992), no. 8, 705–712.

[And10]   _____, *Ordinary differential equations in the lack of uniqueness*, Atti. Sem. Mat. Fis. Univ. Modena **49** (2001), 247–267.

[And11]   _____, *Existence of periodic and bounded solutions of the generalized Liénard equation under forcing*, Rep. Math. Phys. **39** (1997), no. 1, 91–98.

[And12]   _____, *Large-period forced oscillations to higher-order pendulum-type equations*, Diff. Eqns. and Dyn. Syst. **3** (1995), no. 4, 407–421.

[And13]   _____, *Ważewski-type results without transversality*, Proceed. of Equadiff 95 (Lisbon, July 24–29), World Scientific, Singapore, 1998, pp. 233–238.

[And14]   _____, *Using the integral manifolds to solvability of boundary value problems*, Set-Valued Mappings with Application in Nonlinear Analysis (R. P. Agarwal and D-O'Regan, ed.), vol. 4, Ser. Math. Anal. Appl., Taylor and Francis, Singapore, 2002, pp. 27–38.

[And15]   _____, *A target problem for differential inclusions with state-space constraints*, Demonstr. Math. **30** (1997), no. 4, 783–790.

[And16]   _____, *Splay-phase and anti-splay-phase orbits for equivariant set-valued flows on tori*, Diff. Equations Dynam. Syst. **4** (1996), no. 1, 89–98.

[And17]   _____, *Further remarks on problem of Moser and conjecture of Mawhin*, Topol. Meth. Nonlin. Anal. **6** (1996), no. 1, 163–174.

[And18]   _____, *Nielsen number, Artin braids, Poincaré operators and multiple nonlinear oscillations*, Nonlin. Anal. **47** (2001), 1017–1028.

[And19]   _____, *Bounded, almost-periodic and periodic solutions of quasi-linear differential inclusions*, Lecture Notes in Nonlin. Anal. **2** (1998), N. Copernicus Univ., Toruń, 35–50.

[And20]   _____, *Existence of two almost periodic solutions of pendulum-type equations*, Nonlin. Anal. **37** (1999), 797–804.

[And21]   _____, *Periodic derivatives and the Wirtinger-type inequalities*, Acta Univ. Palacki. Olomuc. **116** (1994), 177–192.

[And22]   _____, *Derivo-periodic boundary value problems for nonautonomous ordinary differential equations*, Riv. Mat. Pura Appl. **13** (1993), 63–90.

[And23]   _____, *Some standard fixed-point theorems revisited*, Atti Sem. Mat. Fis. Univ. Modena **49** (2001), 455–471.

[And24]   _____, *Poincaré's translation multioperator revisited*, LN in Nonlin. Anal. **3** (2002), N. Copernicus Univ., Toruń, 7–22.

[And25]   _____, *A nontrivial example of application of the Nielsen fixed-point theory to differential systems: problem of Jean Leray*, Proceed. Amer. Math. Soc. **128** (2000), 2921–2931.

[And26]   _____, *Nielsen number and multiplicity results for multivalued boundary value problems*, Nonlinear Analysis and Differential Equations (M. R. Grossinho, M. Ramos, C. Rebelo and L. Sanchez, eds.), Birkhäuser, Basel, 2001, pp. 175–187.

[And27]   _____, *Periodic solutions of quasi-linear functional differential inclusions*, Funct. Differential Equations **5**, no. 1998, 287–296.

[And28]   _____, *Higher kind periodic orbits*, Acta Univ. Palacki. Olomuc. 88, Phys. **26** (1987), 87–92.

[And29]   _____, *Solution with peridic second derivative of a certain third order differential equation*, Math. Slovaca **37** (1987), no. 3, 239–245.

[And30]   _____, *Periodic derivative of solutions to nonlinear differential equations*, Czech. Math. J. **40** (1990), no. 3, 353–360.

[And31]   _____, *Period three implications for expansive maps in* $\mathbb{R}^N$, J. Difference Eqns Appl. (to appear).

[AndBa]   J. Andres and R. Bader, *Asymptotic boundary value problems in Banach spaces*, J. Math. Anal. Appl. **247** (2002), no. 1, 437–457.

[AndBe]   J. Andres and A. M. Bersani, *Almost-periodicity problem as a fixed-point problem for evolution inclusions*, Topol. Meth. Nonlin. Anal **18** (2001), 337–350.

[AndBeGr]   J. Andres, A. M. Bersani and R. F. Grande, *Hierarchy of almost-periodic function spaces*, Preprint (2002).

[AndBeL]   J. Andres, A. M. Bersani and K. Leśniak, *On some almost-periodicity problems in various metrics*, Acta Appl. Math. **65** (2001), 35–57.

[AFi1]   J. Andres and J. Fišer, *Fractals generated by differential equations*, Dynam. Syst. Appl. **11** (2002), no. 4, 471–480.

[AFi2]   _____, *Metric and topological multivalued fractals*, Int. J. Bifurc. Chaos (to appear).

[AFJ]   J. Andres, J. Fišer and L. Jüttner, *On a multivalued version of the Sharkovskiĭ theorem and its application to differential inclusions*, Set-Valued Anal. **10** (2002), no. 1, 1–14.

[AGG1]   J. Andres, G. Gabor and L. Górniewicz, *Boundary value problems on infinite intervals*, Trans. Amer. Math. Soc. **351** (1999), 4861–4903.

[AGG2]   _____, *Topological structure of solution sets to multivalued asymptotic problems*, Zeit. Anal. Anwendungen **18** (1999), no. 4, 1–20.

[AGG3]   _____, *Acyclicity of solution sets to functional inclusions*, Nonlinear Analysis TMA **49** (2002), 671–688.

[AGZ]   J. Andres, M. Gaudenzi and F. Zanolin, *A transformation theorem for periodic solutions of nondissipative systems*, Rend. Sem. Mat. Univ. Politc. Torino **48(2)** (1990), 171–186.

[AG1]   J. Andres and L. Górniewicz, *On the Banach contraction principle for multivalued mappings*, In: Approximation, Optimization and Mathematical Economics (Marc Lassonde, ed.), Physica-Verlag, Springer, Berlin, 2001, pp. 1–22.

[AG2]   _____, *From the Schauder fixed-point theorem to the applied multivalued Nielsen theory*, Topol. Math. Nonlin. Anal. **14** (1999), no. 2, 229–238.

[AGJ1]   J. Andres, L. Górniewicz and J. Jezierski, *A generalized Nielsen number and multiplicity results for differential inclusions*, Topology Appl. **100** (2000), 193–209.

[AGJ2]   _____, *Noncompact version of the multivalued Nielsen theory and its application to differential inclusions*, In: Differential Inclusions and Optimal Control (J. Andres, L. Górniewicz and P. Nistri, eds.), vol. 2, Lecture Notes Nonlinear Anal., N. Copernicus Univ., Toruń, 1998, pp. 33–50.

[AGJ3]   _____, *Relative versions of the multivalued Lefschetz and Nielsen theorems and their application to admissible semi-flows*, Topol. Meth. Nonlin. Anal. **16** (2000), 73–92.

[AGJ4] ———, *Periodic points of multivalued mappings with applications to differential inclusions on tori*, Topology Appl. **127** (2003), no. 3, 447–472.

[AGL] J. Andres, L. Górniewicz and M. Lewicka, *Partially dissipative periodic processes*, Topology in Nonlinear Analysis, vol. 35, Banach Center Publ., 1996, pp. 109–118.

[AJ1] J. Andres and L. Jüttner, *Periodic solutions of discontinuous differential systems*, Nonlin. Anal. Forum **6** (2001), 391–407.

[AJ2] ———, *Period three plays a negative role in a multivalued version of Sharkovskiĭ's theorem*, Nonlin. Anal. **51** (2002), 1101–1104.

[AJP] J. Andres, L. Jüttner and K. Pastor, *On a multivalued version of the Sharkovskiĭ theorem and its application to differential inclusions*, II, Preprint (2002).

[AK1] J. Andres and B. Krajc, *Bounded solutions in a given set of Carathéodory differential systems*, J. Comput. Appl. Math. **113** (2000), 73–82.

[AK2] ———, *Unified approach to bounded, periodic and almost-periodic solutions of differential systems*, Ann. Math. Sil. **11** (1997), 39–53.

[AK3] ———, *Periodic solutions in a given set of differential systems*, J. Math. Anal. Appl. **264** (2001), 495–509.

[AMT1] J. Andres, L. Malaguti and V. Taddei, *Floquet boundary value problems for differential inclusions: a bound sets approach*, Z. Anal. Anwend. **20** (2001), 709–725.

[AMT2] ———, *Bounded solutions of Carathéodory differential inclusions: a bound sets approach*, Abstr. Appl. Anal. (to appear).

[AMP] J. Andres, J. Mikolajski and J. Palát, *Über die Trichotomie von Lösungen einer nichtlinearen Vektordifferentialgleichung zweiter Ordnung*, Acta Univ. Palacki. Olomuc. 91, Math. **27** (1988), 211–224.

[AndPa] J. Andres and K. Pastor, *On a multivalued version of the Sharkovskiĭ theorem and its application to differential inclusions*, III, Preprint (2002).

[AndS] J. Andres and S. Staněk, *Note to the Lagrange stability of excited pendulum-type equations*, Math. Slovaca **43** (1993), 617–630.

[AT1] J. Andres and T. Turský, *Asymptotic estimates of solutions and their derivatives of nth-order nonhomogeneous ordinary differential equations with constant coefficients*, Discuss. Math.-Diff. Inclusions **16** (1996), no. 1, 75–89.

[AT2] ———, *On the method of Esclangon*, Acta Univ. Palacki. Olomuc. Math. **35** (1996), 7–20.

[AndVae] J. Andres and M. Vaeth, *Coincidence index for noncompact mappings on nonconvex sets*, Nonlinear. Funct. Anal. Appl. **7** (2002), no. 4, 619–658.

[AV] J. Andres and V. Vlček, *Asymptotic behaviour of solutions to the nth-order nonlinear differential equation under forcing*, Rend. Ist. Matem. Univ. Trieste **21** (1989), no. 1, 128–143.

[Adv] A. A. Andronov, *Poincaré's limit cycles and the theory of oscillations*, 6th Meeting of Russian Physicists, Gosud. Izd., Moscow (1928), 23–24 (in Russian); *Les cycles limites de Poincaré et la théorie des oscillations autoen tretenues*, C. R. Acad. Sci. Ser. Paris Ser. I **189** (1929), 559–562.

[AngBa] A. Anguraj and K. Balachandran, *On the solution sets of differential inclusions in Banach spaces*, Tamkang J. Math. **23** (1992), 59–65.

[Ani1] G. Anichini, *Approximate selections for non convex set valued mapping*, Boll. Un. Mat. Ital. B (7) **4** (1990), no. 2, 313–326.

[Ani2] ———, *Boundary value problem for multivalued differential equations and controllability*, J. Math. Anal. Appl. **105** (1985), no. 2, 372–382.

[AniCo] G. Anichini and G. Conti, *How to make use of the solution set to solve boundary value problems*, Recent Trends in Nonlinear Analysis, Birkhäuser, Basel, 2000, pp. 15–25.

[ACZ1] G. Anichini, G. Conti and P. Zecca, *Un teorema di selezione per applicazioni multivoche a valori non convessi*, Boll. Un. Mat Ital. C (5), vol. 1, 1986, pp. 315–320.

[ACZ2] ———, *A further result on the approximation of nonconvex set valued mappings*, Boll. Un. Mat. Ital. C (4), vol. 1, 1985, pp. 155–171.

[ACZ3] ———, *Approximation and selection theorem for nonconvex multifunctions in infinite dimensional space*, Boll. Un. Mat. Ital. B (7), vol. 4, 1990, pp. 411–422.

[ACZ4] ———, *Approximation of nonconvex set valued mappings*, Boll. Un. Mat. Ital. C (6), vol. 4, 1985, pp. 145–154.

[ACZ5] ———, *Using solution sets for solving boundary value problems for ordinary differential equations*, Nonlinear Anal., vol. 17, 1991, pp. 465–472.

[AS] G. Anichini and J. D. Schuur, *Using a fixed point theorem to describe the asymptotic behaviour of solutions of nonlinear ordinary differential equations*, Atti de Convegno "EQUADIFF '78" (R. Conti, G. Sestini, G. Villari, eds.), Centro 2P, Firenze, 1978, pp. 245–256.

[AZ3] G. Anichini and P. Zecca, *Problemi ai limiti per equazioni differenziali multivoche su intervalli non compatti*, Riv. Mat. Univ. Parma 1 (1975), 199–212.

[AZ1] G. Anichini and P. Zecca, *Multivalued differential equations and control problems*, Houston J. Math. 10 (1984), no. 3, 307–313.

[AZ2] ———, *Multivalued differential equations in Banach space, An application to control theory*, J. Optim. Theory Appl. 21 (1977), 477–486.

[AnC] H. A. Antosiewicz and A. Cellina, *Continuous selections and differential relations*, J. Differential Equations 19 (1975), 386–398.

[AnKa] H. Anzai and S. Kakutani, *Bohr compactifications of a locally compact abelian group, I and II*, Proc. Imp. Ac. Tokyo 19 (1943), 476–480 and 533–539.

[Ap] J. Appell, *Multifunctions of two variables: Examples and counterexamples*, Topology in Nonlinear Analysis, Banach Center Publ. 35 (1996), Warsaw, 119–128.

[AVV1] J. Appell, M. Väth and A. Vignoli, *Compactness and existence results for ordinary differential equations in Banach spaces*, Z. Anal. Anwendungen 18 (1999), no. 3, 569–584.

[AVV2] ———, *$\mathcal{F}$-epi maps*, Topol. Meth. Nonlin. Anal. 18 (2001), no. 2, 373–393.

[ArEe] R. F. Arens and J. J. Eells, *On embedding uniform and topological spaces*, Pacific J. Math. 6 (1956), 397–403.

[AriBa] O. Arino and M. Bahaj, *Periodic and almost periodic solutions of differential equations in Banach spaces*, Nonlin. Anal, T. M. A. 26 (1996), no. 2, 335–341.

[ArGoKr] D. G. Aronson, M. Golubitsky and M. Krupa, *Coupled arrays of Josephson junctions and bifurcation of maps with S_N symmetry*, Nonlinearity 4 (1991), 861–902.

[ArGoMP] D. G. Aronson, M. Golubitsky and J. Mallet-Paret, *Ponies on a merry-go-round in large arrays of Josephson junctions*, Nonlinearity 4 (1991), 903–910.

[Aro] N. Aronszajn, *Le correspondant topologique de l'unicité dans la théorie des équations différentielles*, Ann. Math. 43 (1942), 730–738.

[Atn] E. Artin, *Theorie der Zopfe*, Hamburg Abh. 4 (1925), 47–72.

[Art] Z. Artstein, *Continuous dependence on parameters of solutions of operator equations*, Trans. Amer. Math. Soc. 231 (1977), 143–166.

[ArtBy] Z. Artstein and J. Byrns, *Integration of compact set-valued functions*, Pacific J. Math. 58 (1975), 297–307.

[Ash] M. T. Ashordiya, *The structure of the solution set of the Cauchy problem for a system of generalized ordinary differential equations*, Tbiliss. Gos. Univ. Inst. Prikl. Mat. Trudy 17 (1986), 5–16. (in Russian)

[Au1] J.-P. Aubin, *Contingent derivatives of set-valued maps and existence of solutions to nonlinear inclusions and differential inclusions*, J. Math. Anal. Appl. (1981), 159–229.

[Au2] ———, *Heavy viable trajectories of a decentralized allocation mechanism*, Lecture Notes in Control and Inform. Sci. 62 (1984), 487–501.

[Au3] ———, *Slow and heavy trajectories of controlled problems*, Lect. Notes in Math. 1091 (1984), 105–116.

[AF1] J. P. Aubin and H. Frankowska, *Trajectoires lourdes de systemes controles*, C. R. Acad. Sci. Sér. I Math. 20 (1984), no. 298, 521–524.

[AF2] _____, *On inverse function theorems for set-valued maps*, J. Math. Pures Appl. **65** (1987), no. 1, 71–89.

[AF3] _____, *Observability of systems under uncertainty*, SIAM J. Control and Optimization **27** (1989), no. 5, 949–975.

[ADG] A. Augustynowicz, Z. Dzedzej and B. D. Gel'man, *The solution set to BVP for some functional differential inclusions*, Set-Valued Analysis **6** (1998), 257–263.

[AulVM1] B. Aulbach and N. van Minh, *Semigroups and differential equations with almost periodic coefficients*, Nonlin. Anal., T. M. A. **32** (1998), no. 2, 287–297.

[AulVM2] _____, *Almost periodic mild solutions of a class of partial functional differential equations*, Abstr. Appl. Anal. **3** (1998), no. 3–4, 425–436.

[AulVM3] _____, *A sufficient condition for almost periodicity of solutions of nonautonomous nonlinear evolution equations*, Nonlin. Anal. (to appear).

[AulVM4] _____, *Bounded and almost periodic solutions and evolution semigroups associated with nonautonomous functional differential equations*, Abst. Appl. Anal. (to appear).

[Aum] R. J. Aumann, *Intergals of set-valued fuctions*, J. Math. Anal. Appl. **12** (1965), 1–12.

[Ava] A. Avantaggiati, *Soluzioni analitiche quasi periodiche delle equazioni alle derivate parziali a coefficienti costanti*, Ann. Univ. Ferrara, Sez. VII, Sc. Mat., Suppl. **XLV** (1999), 21–43.

[AvgPa1] E. P. Avgerinos and N. S. Papageorgiou, *On the solution set of maximal monotone differential inclusions in $\mathbb{R}^n$*, Math. Japon. **38** (1993), 91–110.

[AvgPa2] _____, *Topological properties of the solution set of integrodifferential inclusions*, Comment. Math. Univ. Carolin. **36** (1995), 429–442.

[Avr] C. Avramescu, *Sur l'existence des solutions convergentes des systèmes d'équations différentielles non linéaires*, Ann. Mat. Pura Appl. **4** (1969), no. 81, 147–168.

[BCM] A. Bacciotti, F. Ceragioli and L. Mazzi, *Differential inclusions and monotonicity conditions for nonsmooth Liapunov functions*, Set-Valued Anal. **8** (2000), 299–309.

[BDP] M. Bachir, A. Danilidis and J.-P. Penot, *Lower subdifferentiability and integration*, Set-Valued Anal. **10** (2002), no. 1, 89-108.

[Bad1] R. Bader, *A sufficient condition for the existence of multiple periodic solutions of differential inclusions*, Topology in Nonlinear Analysis, Banach Center Publ. **35** (1996), Warsaw, 129–138.

[Bad2] _____, *On the semilinear multi-valued flow under constraints and the periodic problem*, Differential Inclusions and Optimal Control, vol. 2, Lecture Notes in Nonlinear Anal., N. Copernicus Univ., Toruń, 1998, pp. 51–55.

[BGK] R. Bader, G. Gabor and W. Kryszewski, *On the extention of approximations for set-valued maps and the repulsive fixed points*, Boll. Un. Mat. Ital. **10** (1996), 399–416.

[BGKO] R. Bader, B. D. Gel'man, M. Kamenskiĭ and V. V. Obukhovskiĭ, *On the extention of approximations for set-valued maps and the repulsive fixed points*, Boll. Un. Mat. Ital. **10** (1996), 399–416.

[BKr1] R. Bader and W. Kryszewski, *Fixed-point index for compositions of set-valued maps with proximally ∞-connected values on arbitrary ANR's*, Set-Valued Anal. **2** (1994), 459–480.

[BKr2] _____, *On the solution sets of constrained differential inclusions with applications*, Set-Valued Anal. **9** (2001), 289–313.

[BKr3] _____, *On the solution sets of differential inclusions and the periodic problem in Banach spaces* (to appear).

[Bae] J. S. Bae, *An example of a partially ordered Sharkovskiĭ space*, Bull. Korean. Math. Soc. **27** (1990), no. 2, 127–131.

[Bal] M. E. Ballotti, *Aronszajn's theorem for a parabolic partial differential equation*, Nonlinear Anal. TMA **9** (1985), no. 11, 1183–1187.

[BaJa]   H. T. Banks and M. Q. Jacobs, *A differential calculus for multifunctions*, J. Math. Anal. Appl. **29** (1970), 246–272.

[Bar]   E. Barcz, *Some fixed points theorems for multi-valued mappings*, Demonstratio Math. **16(3)** (1983), 735–744.

[BaP]   C. Bardaro and P. Pucci, *Some contributions to the theory of multivalued differential equations*, Atti. Sem. Mat. Fis. Univ. Modena **32(1)** (1983), 175–202.

[Brm]   B. R. Barmish, *Stabilization of uncertain systems via linear control*, IEEE Trans. Autom. Contr. **AC-28** (1983), no. 8, 848–850.

[BCL]   B. R. Barmish, M. Corless and G. Leitmann, *A new class of stabilizing controllers for uncertain dynamical systems*, SIAM J. Control and Optimazation **21** (1983), no. 2, 246–255.

[BnDe]   M. F. Barnsley and S. Demko, *Iterated function systems and the global construction of fractals*, Proc. R. Soc. Lond. **399** (1985), 243–275.

[Brr]   R. B. Barrar, *Existence of periodic orbits of the second kind in the restricted problem of three bodies*, Astronom. J. **70** (1965), no. 1, 3–4.

[BT1]   G. Bartolini and T. Tolezzi, *Asymptotic linearization of uncertain systems by variable structure control*, System and Control Letters **10** (1988), 111–117.

[BT2]   ———, *Some new application of V. S. S. theory control of uncertain nonlinear systems*, Proc. 27-th IEEE Conference on Decision and Control (1988), Austin.

[BaF1]   G. Bartuzel and A. Fryszkowski, *On existence of solutions for inclusion $\Delta u \in F(x, \nabla u)$*, In: Proc. Fourth Conf. Numer. Treatment of Ordinary Diff. Eqns (R. März, ed.), Humboldt Univ., Berlin, 1984, pp. 1–7.

[BaF2]   ———, *A topological property of the solution set to the Sturm–Liouville differential inclusions*, Demonstratio Math. **28** (1995), 903–914.

[BstTs]   R. Boles Basit and L. Tsend, *The generalized Bohr–Neugebauer theorem*, Diff. Eqns **8** (1972), no. 8, 1031–1035.

[Bas1]   J. Bass, *Espaces vectoriels de fonctions pseudo-aléatoires*, C. R. Acad. Sci. Paris **255** (1962), 2353–2355.

[Bas2]   ———, *Les fonctions pseudo-aléatoires*, Mémor. Sci. Math., Fasc. **53** (1962), Gauthier–Villars, Paris.

[Bas3]   ———, *Espaces de Besicovitch, fonctions presque-périodiques, fonctions pseudo-aléatoires*, Bull. Soc. Math. France **91** (1963), 39–61.

[Bas4]   ———, *Solutions preque-périodiques l'un systéme différential nonlinéaire*, C. R. Acad. Sc. Paris **288** (1979), no. 5, 351–353.

[Bax]   J. V. Baxley, *Existence and uniqueness for nonlinear boundary value problems on infinite intervals*, J. Math. Anal. Appl. **147** (1990), no. 1, 122–133.

[Beb]   J. W. Bebernes, *Solution set properties for some nonlinear parabolic differential equations*, Equadiff IV, Proc. Czechoslovak Conf. Differential Equations and Their Applications, Springer, Berlin, 1979, pp. 25–30.

[BeJ]   J. W. Bebernes and L. K. Jackson, *Infinite interval boundary value problems for 2nd order differential equations*, Duke Math. J. **34** (1967), no. 1, 39–48.

[BeKe]   J. W. Bebernes and W. Kelley, *Some boundary value problems for generalized differential equations*, SIAM J. Appl. Math. **20** (1973), 16–23.

[BeM]   J. W. Bebernes and M. Martelli, *On the structure of the solution set for periodic boundary value problems*, Nonlinear Anal. TMA **4** (1980), no. 4, 821–830.

[BeSchm]   J. W. Bebernes and K. Schmitt, *Invariant sets and the Hukuhara–Kneser property for systems of parabolic partial differential equations*, Rocky Mount. J. Math. **7** (1967), no. 3, 557–567.

[BeSchu]   J. W. Bebernes and J. D. Schuur, *The Ważewski topological methods for contingent equations*, Ann. Mat. Pura Appl. **87** (1970), 271–280.

[Bee1]   G. Beer, *The approximation of upper semicontinuous multifunctions by step multifunctions*, Pacific. J. Math. **87(1)** (1980), 11–19.

[Bee2]   ———, *On functions, that approximate relations*, Proc. Amer. Math. Soc. **88(4)** (1983), 643–647.

[Bee3]   ———, *Dense selections*, J. Math. Anal. Appl. **95(2)** (1983), 416–427.

[Bee4]   _____, *Approximate selections for upper semicontinuous convex valued multi-functions*, J. Approx. Theory **39(2)** (1983), 172–184.

[Bee5]   _____, *On a theorem of Cellina for set valued functions*, Rocky Mountain J. Math. **18** (1988), 37–47.

[Beg]   E. G. Begle, *The Vietoris mapping theorem for bicompact spaces*, Ann. of Math. **51** (1950), 534–543.

[BelFoSD1]   J. M. Belley, G. Fournier and K. Saadi Drissi, *Almost periodic weak solutions to forced pendulum type equations without friction*, Aequationes Math. **44** (1992), 100–108.

[BelFoSD2]   _____, *Solutions faibles presque périodiques d'équation différentialle du type du pendule forcé*, Acad. Roy. Belg. Bull. Cl. Sci. **6** (1992), no. 3, 173–186.

[BelFoSD3]   _____, *Solutions presque périodiques du systéme différential du type du pendule forcé*, Acad. Roy. Belg. Bull. Cl. Sci. **6** (1992), no. 3, 265–278.

[BelFoHa]   J. M. Belley, G. Fournier and J. Hayes, *Existence of almost periodic weak type solutions for the conservative forced perdulum equation*, J. Diff. Eqns **124** (1996), 205–224.

[Be1]   M. M. Belova, *Bounded solutions of non-linear differential equations of second order*, Mat. Sbornik **56** (1962), 469–503. (in Russian)

[Be2]   H. Ben-El-Mechaiekh, *Continuous approximations of multifunctions, fixed points and coincidences*, Approximation and Optimization in the Carribean II (M. Florenzano et al., eds.), Verlag Peter Lang, Frankfurt, 1995, pp. 69–97.

[BD]   H. Ben-El-Mechaiekh and P. Deguire, *Approachability and fixed points for non-convex set-valued maps*, J. Math. Anal. Appl. **170** (1992), 477–500.

[BDG1]   H. Ben-El-Mechaiekh, P. Dequire and A. Granas, *Une alternative nonlineaire en analyse convexe et applications*, C. R. Acad. Sci. Paris Sér. I Math. **295(3)** (1982), 257–259.

[BDG2]   _____, *Points fixes et coincidences pour les functions multivoques*, III (Applications de type M^* et M), C. R. Acad. Sci. Paris Sér. I Math. **305(9)** (1987), 381–384.

[BK1]   H. Ben-El-Mechaiekh and W. Kryszewski, *Équilibres dans les ensembles non-convexes*, C. R. Acad. Sci. Paris Sér. I Math. **320** (1995), 573–576.

[BK2]   _____, *Equilibria of set-valued maps on nonconvex domains*, Trans. Amer. Math. Soc. **349** (1997), 4159–4179.

[BeGa]   C. Benassi and A. Gavioli, *Approximation from the exterior of Carathéodory multifunctions*, Acta Univ. Palacki. Olomuc. **39** (2000), 17–35.

[BGN1]   M. Benchohra, L. Górniewicz and S. Ntouyas, *Controllability on infinite time horizon for first and second order functional differential inclusions in Banach spaces*, Discuss. Math. **21** (2001), 261–282.

[BGN2]   _____, *Controllability neutral functional differential and integrodifferential inclusions in Banach spaces with nonlocal conditions*, Nonlinear Anal. Forum **7** (2002), 39–54.

[BrgCh1]   M. S. Berger and Y. Y. Chen, *Forced quasi periodic and almost periodic oscillations of nonlinear Duffing equations*, Nonlinear Anal. T. M. A. **19** (1992), no. 3, 249–257.

[BrgCh2]   _____, *Forced quasi periodic and almost periodic solution for nonlinear systems*, Nonlinear Anal. T. M. A. **21** (1993), no. 12, 949–965.

[Ber1]   J.-P. Bertrandias, *Sur l'analyse harmonique généralisée des fonctions pseudo-aléatoires*, C. R. Acad. Sci. Paris **253** (1961), 2829–2831.

[Ber2]   _____, *Fonctions pseudo-aléatoires et fonctions presque périodiques*, C. R. Acad. Sci. Paris **255** (1962), 2226–2228.

[Ber3]   _____, *Espaces de fonctions bornees et continues en moyenne asymptotique d'ordre p*, Bull. Soc. Math. France, Mémoire n. 5 (1966), 1–106.

[Bes1]   A. S. Besicovitch, *Sur quelques points de le théorie des fonctions presque-périodiques*, C. R. Acad. Sci. Paris **180** (1925), 394–397.

[Bes2]   ———, *Analysis of conditions of almost periodicity*, Acta Math. **58** (1932), 217–230.

[BeBo]   A. S. Besicovitch and H. Bohr, *Almost periodicity and general trigonometric series*, Acta Math. **57** (1931), 203–292.

[Bess]   U. Bessi, *A variational proof of a Silnikov-like theorem*, Nonlin. Anal. TMA **20** (1993), no. 11, 1303–1318.

[Bi1]   R. Bielawski, *Simplicial convexity and its applications*, J. Math. Anal. Appl. **127** (1987), 155–171.

[Bi2]   ———, *The fixed point index for acyclic maps on ENR-s*, Bull. Polish Acad. Sci. Math. **35(7-8)** (1987), 487–499.

[Bi3]   ———, *A selection theorem for open-graph multifunctions*, Fund. Math. **133** (1989), 97–100.

[BiG1]   R. Bielawski and L. Górniewicz, *A fixed point index approach to some differential equations*, Topological Fixed Point Theory and Applications (Boju Jiang, ed.), vol. 1411, Lecture notes in Math., Springer-Verlag, Berlin-Heidelberg-New York, 1989, pp. 9–14.

[BiG2]   ———, *Some applications of the Leray–Schauder alternative to differential equations*, In Nonlinear Fucntional Analysis and Its Applications (S. P. Singh, ed.), D. Reidel Publ. Comp., Dordrecht, 1986, pp. 187–194.

[BGoPl]   D. Bielawski, L. Górniewicz and S. Plaskacz, *Topological approach to differential inclusions on closed sets of* $\mathbb{R}^n$, Dynamics Reported **1** (1992), 225–250.

[BP]   D. Bielawski and T. Pruszko, *On the structure of the set of solutions of a functional equation with application to boundary value problems*, Ann. Polon. Math. **53** (1991), 201–209.

[Bir]   G. I. Birjuk, *On a theorem of the existence of almost periodic solutions of some systems of nonlinear differential equations with a small parameter*, Dokl. Akad. Nauk SSSR **96** (1954), no. 1, 5–7. (in Russian)

[BW1]   J. S. Birman and R. F. Williams, *Knotted periodic orbits in dynamical systems I: Lorenz's equations*, Topology **22** (1983), 47–82.

[BW2]   ———, *Knotted periodic orbits in dynamical systems II: Knot holders for fibred knots*, Contemp. Math. **20** (1983), 1–60.

[Bh1]   V. I. Blagodatskikh, *Local controllability of differential inclusions*, Differentsial'nye Uravneniya **9** (1973), 361–362. (in Russian)

[Bh2]   ———, *Sufficient conditions for optimality in problems with state constraints*, Appl. Math. Optim. **7** (1981), no. 2, 149–157.

[BhNd]   V. I. Blagodatskikh and P. Ndiĭ, *Convexity of the solution set of a differential inclusion*, Vestnik Moskov. Univ. Ser. XV Vychisl. Mat. Kibernet. (1998), 21–22. (in Russian)

[BlKe]   L. Block and J. Keesling, *Iterated function systems and the code space*, Topol. Appl. **122** (2002), 65–75.

[BlGuMiYo]   L. Block, J. Guckenheimer, M. Misiurewicz and L. S. Young, *Periodic points and topological entropy of one-dimensional maps*, Lect. Notes in Math., vol. 819, Springer, Berlin, 1983, pp. 18–24.

[Blo1]   J. Blot, *Une méthode hilbertienne pour les trajectories presque périodique*, C. R. Acad. Sci. Paris Sér. I Math. **313** (1991), 487–490.

[Blo2]   ———, *Almost periodically forced pendulum*, Funkcial. Ekvac **36** (1993), 235–250.

[BloMaCi]   J. Blot, J. Mawhin and P. Cieutat, *Almost periodic oscillations of monotone second order systems*, Adv. Diff. Eqns **2** (1997), 693–714.

[Boc1]   S. Bochner, *Abstrakte fastperiodische Funktionen*, Acta Math. **61** (1933), 149–184.

[Boc2]   ———, *A new approach to almost periodicity*, Proc. Nat. Ac. Sc. USA **48** (1962), 195–205.

[Bog]   S. A. Bogatyĭ, *Iteration indices of multivalued mapping*, C.R. Acad. Bulgare Sci. **41** (1988), no. 2, 13–16. (in Russian)

[Bog1] A. V. Bogatyrev, *Fixed points and properties of solutions of differential inclusions*, Izv. Akad. Nauk SSSR **47** (1983), 895–909. (in Russian)

[Bog2] ———, *Continuous branches of multivalued mappings with nonconvex right-hand side*, Mat. Sb. **120** (1983), no. 3, 344–353. (in Russian)

[BoK] H. F. Bohnenblust and S. Karlin, *On a theorem of Ville. Contributions to the Theorey of Games* I, Ann. of Math. Stud., Princeton Univ. Press, Princeton, 1950, pp. 155–160.

[Bhr] P. Bohr, *Über eine Differentialgleichung der Störungs theorie*, J. f. reine u. angew. Math. **131** (1906), 268–321.

[BF] H. Bohr and E. Fœlner, *On some types of functional spaces*, Acta Math. **76** (1945), 31–155.

[BohNe] H. Bohr and O. Neugebauer, *Über lineare Differentialgleichungen mit konstanten Koeffizienten und fastperiodischer rechter Seite*, Nachr. Ges. Wiss. Göttingen, Math-Phys. Klasse (1926), 8–22.

[Bon] G. Bonanno, *Two theorems on the Scorza–Dragoni property for multifunctions*, Atti Accad. Naz. Lincei Cl. Sci. Fis. Mat. Natur. Rend. **93** (1989), 51–56.

[Bol] R. Boles Basit, *Connection between the almost-periodic functions of Levitan and almost automorphic functions*, Vestnik Moskovskogo Univ., Ser. 1, Mat. Mek. **26** (1971), 11–15. (in Russian)

[Bor] Yu. G. Borisovich, *Global analysis and some problems for differential equations*, Differencial equations and applications I, II, Tech. Univ., Russe, 1982, pp. 98–118. (in Russian)

[BoGl] Yu. G. Borisovich and Yu. E. Gliklikh, *Fixed points of mappings of Banach manifolds and some applications*, Nonlin. Anal. T.M.A. **4** (1980), no. 1, 165–192.

[BoKuMa1] Yu. G. Borisovich, Z. Kucharski and W. Marzantowicz, *Nielsen numbers and lower estimates for the number of solutions to a certain system of nonlinear integral equations*, New Developments in Global Analysis Series (1994), Voronezh Univ. Press, Voronezh, 3–10.

[BoKuMa2] ———, *A multiplicity result for a system of real integral equations by use of the Nielsen number*, Banach Center Publ. **49** (1999), Inst. of Math, Polish Acad. Sci..

[BO1] Borisovich and V. V. Obuhovskii, *On an optimization problem for control systems of parabolic type*, Trudy Mat. Ins. Steklov **211** (1995), 95–101. (in Russian)

[BO2] ———, *An optimization problem for control systems described by differential equations of parabolic type*, Algebraic Problems in Analysis and Topology, Novoe Global. Anal., Voronezh, 1990, pp. 105–109. (in Russian)

[Bot] D. Bothe, *Upper semicontinuous perturbations of m-accrentive inclusions with dissipative right-hand side*, Topology in Nonlinear Analysis Banach Center Publ., vol. 35, Inst. of Math. Polish Acad. of Sci., Warsaw, 1996, pp. 139–148.

[BBS] K. Boucher, M. Brown and E. E. Slaminka, *A Nielsen-type theorems for area-preserving homeomorphisms of the two disc*, Continuum Theory and Dynamical Systems (T. West, ed.), vol. 149, LNPAM, M. Dekker, Inc, New York, pp. 43–50.

[Bo] M. A. Boudourides, *On bounded solutions of ordinary differential equations*, Comment. Math. Univ. Carolin **22** (1981), 15–26.

[Bou1] D. G. Bourgin, *A generalization of the mappings degree*, Canad. J. Math. **26** (1974), 1109–1117.

[Bou2] ———, *A degree for nonacyclic multi-valued transformations*, Bull. Amer. Math. Soc. **80** (1974), 59–61.

[Bou3] ———, *Cones and Vietoris-Begle type theorems*, Trans. Amer. Math. Soc. **174** (1972), 155–183.

[Bow1] C. Bowszyc, *Fixed points theorem for the pairs of spaces*, Bull. Polish Acad. Sci. Math. **16** (1968), 845–851.

[Bow2] ———, *On the Euler–Poincaré characteristic of a map and the existence of periodic points*, Bull. Acad. Polon. Sci. **17** (1969), 367–372.

[BrFH]   T. R. Brahana, M. K. Fort and W. G. Horstman, *Homotopy for cellular set-valued functions*, Proc. Amer. Math. Soc. **16** (1965), 455–459.

[Bre1]   A. Bressan, *Directionally continuous selections and differential inclusions*, Funkcial. Ekvac. **31** (1988), 459–470.

[Bre2]   ———, *On the qualitative theory of lower semicontinuous differential inclusions*, J. Differential Equations **77** (1989), 379–391.

[Bre3]   ———, *Upper and lower semicontinuous differential inclusions: a unified approach*, Nonlinear Controllability and Optimal Control (H. Sussmann, ed.), M. Dekker, New York, 1988, pp. 21–31.

[Bre4]   ———, *On differential relations with lower continuous right hand side. An existence theorem.*, J. Differential Equations **37** (1980), no. 1, 89–97.

[Bre5]   ———, *Solutions of lower semicontinuous differential inclusions on closed sets*, Rend. Sem. Mat. Univ. Padova **69** (1983).

[Bre6]   ———, *Sulla funzione tempo minimo nei sistemi non lineari*, Atti. Accad. Naz. Lincei Cl. Sci. Fis. Mat. Mem. **5** (1979), 383–388.

[BCF]   A. Bressan, A. Cellina and A. Fryszkowski, *A class of absolute retracts in spaces of integral functions*, Proc. Amer. Math. Soc. **112** (1991), 413–418.

[BC1]   A. Bressan and G. Colombo, *Extensions and selections of maps with decomposable values*, Studia Math. **40** (1988), 69–86.

[BC2]   ———, *Generalized Baire category and differential inclusions in Banach spaces*, J. Differential Equations **76(1)** (1988), 135–158.

[BS]   A. Bressan and W. Shen, *On discontinuous differential equations*, Differential Inclusions and Optimal Control (J. Andres, L. Górniewicz and P. Nistri, eds.), vol. 2, Lecture Notes Nonlinear. Anal., N. Copernicus Univ., Toruń, 1998, pp. 73–87.

[BBPT]   R. B. S. Brooks, R. F. Brown, J. Pak and D. H. Taylor, *Nielsen numbers of maps of tori*, Proc. Amer. Math. Soc. **52** (1975), 398–400.

[BBrS1]   R. B. S. Brooks, R. Brown and H. Schirmer, *The absolute degree and Nielsen root number of compositions and cartesian products of maps*, Topol. Appl. **116** (2001), 5–27.

[BBrS2]   ———, *The absolute degree and Nielsen root number of a fiber-preserving map*, Topol. Appl. **125** (2002), no. 1, 1–46.

[Bro1]   F. E. Browder, *Coincidence theorems, minimax theorems, and variational inequalites*, Contemp. Math. **26** (1984), 67–80.

[Bro2]   ———, *Degree of mapping for nonlinear mappings of monotone type: densely defined mapping*, Proc. Nat. Acad. Sci. U.S.A. **80** (1983), no. 8, 2405–2407.

[Bro3]   ———, *The fixed point theory of multi-valued mappings in topological vector spaces*, Math. Ann. **177** (1968), 283–301.

[Bro4]   ———, *Another generalization of the Schauder Fixed Point Theorem*, Duke Math. J. **32** (1965), 399–406.

[Bro5]   ———, *A futher generalization of the Schauder Fixed Point Theorem*, Duke Math. J. **32** (1965), 575–578.

[BrGu]   F. E. Browder and C. P. Gupta, *Topological degree and nonlinear mappings of analytic type in Banach spaces*, J. Math. Anal. Appl. **26** (1969), 390–402.

[Brw1]   R. F. Brown, *Nielsen fixed point theory of manifolds*, Preprint (1999).

[Brw2]   ———, *On the Nielsen fixed point theorem for noncompact maps*, Duke Math. J. **36** (1969), 699–708.

[Brw3]   ———, *Multiple solutions to parametrized nonlinear differential systems from Nielsen fixed point theory*, Nonlinear Analysis (World Scientific, Singapore) (1987), 89–98.

[Brw4]   ———, *Nielsen fixed point theory and parametrized differential equations*, Amer. Math. Soc. 72 (1988), Providence, R. I., 33–45.

[Brw5]   ———, *Topological identification of multiple solutions to parametrized nonlinear equations*, Pacific J. Math. **131** (1988), no. 1, 51–69.

[Brw6]   ———, *The absolute degree*, Preprint (2002).

[BrwGr]   R. F. Brown and R. E. Green, *An interior fixed point property of the disc*, Amer. Math. Monthly **101** (1994), 39–47.

[BrwGS]   R. F. Brown, R. E. Green and H. Schirmer, *Fixed points of map extensions*, Topological Fixed Point Theory and Applications, vol. 141, Lecture Notes in Math., Springer, Berlin, 1989, pp. 24–45.

[BrwJS]   R. F. Brown, Boju Jiang and H. Schirmer, *Roots of iterates of maps*, Topology Appl. **66** (1995), 129–157.

[BrwS1]   R. F. Brown and H. Schirmer, *Nielsen root theory and Hopf degree theory*, Pacific J. Math. **170** (1995), 405–420.

[BrwS2]   ———, *Nielsen theory of roots of maps of pairs*, Topology Appl. **92** (1999), 247–274.

[BrwZe]   R. F. Brown and P. Zecca, *Multiple local solutions to nonlinear control processes*, J. Optim. Theory Appl. **67** (1990), no. 3, 463–485.

[BuGr1]   G. Bruno and F. R. Grande, *A compactness criterion in B^q_{ap} spaces*, Rend. Accad. Naz. Sci. XL, Mem. Mat. Appl. **114** (1996), no. 20, 1, 95–121.

[BuGr2]   ———, *Compact embedding theorems for Sobolev–Besicovitch spaces of almost periodic functions*, Rend. Accad. Naz. Sci. XL, Mem. Mat. Appl. **114** (1996), no. 20, 1, 157–173.

[Bry]   J. Bryszewski, *On a class of multi-valued vector fields in Banach spaces*, Fund. Math. **97** (1977), 79–94.

[BG1]   J. Bryszewski and L. Górniewicz, *A Poincaré type coincidence theorem for multivalued maps*, Bull. Polish Acad. Sci. Math. **24** (1976), 593–598.

[BG2]   ———, *Multi-valued maps of subsets of Euclidean spaces*, Bull. Polish Acad. Sci. Math. **90** (1976), 233–251.

[BGoPr]   J. Bryszewski, L. Górniewicz and T. Pruszko, *An application of the topological degree theory to the study of the Darboux problem for hyperbolic equations*, J. Math. Anal. Appl. **76** (1980), 107–115.

[Bug1]   D. Bugajewska, *On the equation of nth order and the Denjoy integral*, Nonlinear Anal. **34** (1998), 1111–1115.

[Bug2]   ———, *A note on the global solutions of the Cauchy problem in Banach spaces*, Acta Math. Hung. **88** (2000), 341–346.

[Bug3]   ———, *On the structure of solution sets of differential equations in Banach spaces*, Math. Slovaca **50** (2000), 463–471.

[BugBgj1]   D. Bugajewska and D. Bugajewski, *On the equation $x^{(n)}_{ap} = f(t,x)$*, Czech. Math. Journal **46** (1996), 325–330.

[BugBgj2]   ———, *On nonlinear equations in Banach spaces and axiomatic measures of noncompactness*, Funct. Differ. Eqns **5** (1998), 57–68.

[Bgj1]   D. Bugajewski, *On the structure of the L^{p_1, p_2}-solution sets of Volterra integral equations in Banach spaces*, Comment. Math. **30** (1991), 253–260.

[Bgj2]   ———, *On differential and integral equations in locally convex spaces*, Demonstr. Math. **28** (1995), 961–966.

[Bgj3]   ———, *On the structure of solution sets of differential and integral equations and the Perron integral*, Proceedings of the Prague Mathematical Conference 1996, Icarus, Prague, 1997, pp. 47–51.

[BgjSz1]   D. Bugajewski and S. Szufla, *Kneser's theorem for weak solutions of the Darboux problem in Banach spaces*, Nonlinear Anal. **20** (1993), 169–173.

[BgjSz2]   ———, *On the Aronszajn property for differential equations and the Denjoy integral*, Comment. Math. **35** (1995), 61–69.

[BuLy1]   A. I. Bulgakov and L. N. Lyapin, *Some properties of the set of solutions of a Volterra–Hammerstein integral inclusion*, Diff. Uravn. **14** (1978), no. 8, 1043–1048. (in Russian)

[BuLy2]   ———, *Certain properties of the set of solutions of the Volterra–Hammerstein integral inclusion*, Diff. Uravn. **14** (1978), no. 8, 1465–1472. (in Russian)

[BuLy3]   _____, *On the connectedness of sets of solutions of functional inclusions*, Mat. Sbornik **119** (1982), no. 2, 295–300. (in Russian)

[Bur]   I. M. Burkin, *On the existence of limit cycles of the second kind to the high-dimensional phase locking system*, Diff. Urav. **12** (1991), no. 27, 2044–2049. (in Russian)

[BZ]   T. A. Burton and Shunian Zhang, *Unified boundedness, periodicity, and stability in ordinary and functional differential equations*, Annali di Matematica Pura ed Applicata (IV) **CXLV** (1986), 129–158.

[BVLL]   B. F. Bylov, R. E. Vinograd, V. Ya. Lin and O. O. Lokutsievskiĭ, *On the topological reasons for the anomalous behaviour of certain almost periodic systems*, Problems in the Asymptotic Theory of Nonlinear Oscillations, Naukova Dumka, Kiev, 1977, pp. 54–61. (in Russian)

[Byr]   C. Byrne, *Remarks on the set-valued integrals of Debreu and Aumann*, J. Math. Anal. Appl. **62** (1978), 243–246.

[Cal]   B. D. Calvert, *The local fixed point index for multivalued transformations in Banach space*, Math. Ann. **190** (1970), 119–128.

[Cam]   R. H. Cameron, *Almost periodic properties of bounded solutions of linear differential equations with almost periodic coefficients*, J. Math. Phys. **15** (1936), 73–81.

[Can]   A. Canino, *On φ-convex sets and geodesics*, J. Diff. Eqns **75** (1988), 118–157.

[CapGa]   A. Capietto and B. M. Garay, *Saturated invariant sets and boundary behaviour of differential systems*, J. Math. Anal. Appl. **176** (1993), no. 1, 166–181.

[CapMaZa]   A. Capietto, J. Mawhin and F. Zanolin, *Continuation theorems for periodic perturbations of autonomous systems*, Trans. Amer. Math. Soc. **329** (1992), 41–72.

[Crd1]   P. Cardaliaguet, *Conditions suffisantes de non-vacuité du noyan de viabilité*, C. R. Acad. Sci Paris Sér I **1314** (1992), no. 11, 797–800.

[Crd2]   _____ P. Cardaliaguet, *Sufficient conditions of nonemptiness of the viability kernel*, Preprint no. 9363, CEREMADE (1993), Univ. de Paris-Dauphine, Paris.

[Crn]   T. Cardinali, *On the structure of the solution set of evolution inclusions with Fréchet subdifferentials*, J. Appl. Math. Stochastic Anal. **13** (2000), 51–72.

[CrnFP]   T. Cardinali, A. Fiacca and N. S. Papageorgiou, *On the solution set of nonlinear integrodifferential inclusions in* $\mathbb{R}^N$, Math. Japon. **46** (1997), 117–127.

[Car]   F. S. P. Cardona, *Reidemeister theory for maps of pairs*, Far East J. Math. Sci. Special Volume, Part I, Geometry and Topology (1999), 109–136.

[CarWo]   F. S. P. Cardona and P. N.-S. Wong, *On the computation of the relative Nielsen number*, Topology Appl. (to appear).

[Cas]   C. Castaing, *Sur les equations differentielles multivoques*, C. R. Acad. Sci. Paris Sér. I Math. **263** (1966), 63–66.

[CasMa]   C. Castaing and M. Marques, *Topological properties of solution sets for sweeping processes with delay*, Portugal. Math. **54** (1997), 485–507.

[CDMN]   A. Cavallo, G. De Maria and P. Nistri, *Some control problems solved via a sliding manifold approach*, Diff. Eqns and Dyn. Sys. **1** (1993), 295–310.

[CFM1]   M. Cecchi, M. Furi and M. Marini, *On continuity and compactness of some nonlinear operators associated with differential equations in noncompact intervals*, Nonlin. Anal. TMA **9** (1985), no. 2, 171–180.

[CFM2]   _____, *About solvability of ordinary differential equations with asymptotic boundary conditions*, Boll. U.M.I. **VI** (1985), 329–345.

[CFMP]   M. Cecchi, M. Furi, M. Marini and M. P. Pera, *Fredholm linear operators associated with ordinary differential equations on noncompact intervals*, Electronic J. Diff. Eqns **44** (1999), 1–16.

[CMZ1]   M. Cecchi, M. Marini and P. L. Zezza, *Linear boundary value problems for systems of ordinary differential equations on non-compact intervals, Parts 1–2*, Ann. Mat. Pura Appl. **4** (1980), no. 123, 267–285; **4** (1980), no. 124, 367–379.

[CMZ2]   _____, *Asymptotic properties of the solutions of non-linear equations with dichotomies and applications*, Boll. U.M.I. **6** (1982), no. 1-C, 209–234.

[CMZ3]   _____ M. Cecchi, M. Marini and P. L. Zezza, *Boundary value problems on* $[a, b)$ *and singular perturbations*, Annal. Polon. Math. **64** (1984), 73–80.

[CMZ4]   _____ , *Existence of bounded solutions for multivalued differential systems*, Nonlin. Anal. TMA **9** (1985), no. 8, 775–786.

[Ce1]   A. Cellina, *A selection theorem*, Rend. Sem. Mat. Univ. Padova **55** (1976), 99–107.

[Ce2]   _____ , *On the differential inclusion* $x' \in [-1, 1 - A]$, Atti. Accad. Naz. Lincei Cl. Sci. Fis. Mat. Natur. Rend. Lincei **69** (1980, 1981), no. 1–2, 1–6.

[Ce3]   _____ , *On the set of solutions to Lipschitzean differential equations*, Differential Integral Equations **1** (1988), 495–500.

[Ce4]   _____ , *On the existence of solutions of ordinary differential equations in a Banach space*, Funkc. Ekvac. **14** (1971), 129–136.

[Ce5]   _____ , *On the local existence of solutions of ordinary differential equations*, Bull. Acad. Polon. Sci. **20** (1972), 293–296.

[Ce6]   _____ , *On the nonexistence of solutions of differential equations in nonreflexive spaces*, Bull. Amer. Math. Soc. **78** (1972), 1069–1072.

[CC2]   A. Cellina and G. Colombo, *An existence result for differential inclusions with non-convex right-hand side*, Funkcial. Ekvac. **32** (1989), 407–416.

[CC1]   A. Cellina and R. M. Colombo, *Some qualitative and quantitive results on a differential inclusions*, Set-Valued Analysis and Differential Inclusions (A. B. Kurzhanski and V. M. Veliov, eds.), vol. 16, PSCT, 1993, pp. 43–60.

[CCF]   A. Cellina, G. Colombo and A. Fonda, *Approximate selections and fixed points for upper semicontinuous maps with decomposable values*, Proc. Amer. Math. Soc. **98** (1986), no. 4, 663–666.

[CL1]   A. Cellina and A. Lasota, *A new approach to the definition of topological degree for multivalued mappings*, Atti. Accad. Naz. Lincei Cl. Sci. Fis. Mat. Natur. Rend. Lincei **8** (1969), 434–440.

[CM]   A. Cellina and M. V. Marchi, *Non-convex perturbations of maximal monote differential inclusions*, Israel J. Math. **46** (1983), no. 1–2, 1–11.

[ClOr]   A. Cellina and A. Ornelas, *Convexity and the closure of the solution set to differential inclusions*, Boll. Un. Mat. Ital. B **7** (1990), no. 4, 255–263.

[ChiZe]   L. Chiercia and E. Zehnder, *On asymptotic expansion of quasiperiodic solutions*, Ann. Sc. Norm. Sup. Pisa **4** (1989), no. 16, 245–258.

[ChL]   S. N. Chow and A. Lasota, *On boundary value problems for ordinary differential equations*, J. Diff. Equations **14** (1973), 326–337.

[CS1]   S. N. Chow and J. D. Schuur, *An existence theorem for ordinary differential equations in Banach spaces*, Bull. Amer. Math. Soc. **77** (1971), 1018–1020.

[CS2]   _____ , *Fundamental theory of contingent differential equations in Banach spaces*, Trans. Amer. Math. Soc. **179** (1973), 133–144.

[Ci]   M. Cichoń, *Trichotomy and bounded solutions of nonlinear differential equations*, Math. Bohemica **119** (1994), no. 3, 275–284.

[CK]   M. Cichoń and I. Kubiaczyk, *Some remarks on the structure of the solutions set for differential inclusions in Banach spaces*, J. Math. Anal. Appl. **233** (1999), 597–606.

[Clp]   M. H. Clapp, *On a generalization of absolute neighbourhood retracts*, Fund. Math. **70** (1971), 117–130.

[Clk]   F. H. Clarke, *Periodic solutions to Hamiltonian inclusions*, J. Diff. Eqns **40** (1981), no. 1, 1–6.

[ClkLS]   F. H. Clarke, Yu. S. Ledyaev and R. J. Stern, *Asymptotic stability and smooth Lyapunov functions*, J. Diff. Eqns **149** (1988), 69–114.

[ClkStWo]   F. H. Clarke, R. J. Stern and P. R. Wolenski, *Subgradient criteria for monotonicity, the Lipschitz condition and convexity*, Can. J. Eqns **45** (1993), 1167–1183.

[Col]   P. Collins, *Relative periodic point theory*, Topol. Appl. **115** (2002), 97–114.

[CFRS]   R. M. Colombo, A. Fryszkowski, T. Rzezuchowski and V. Staicu, *Continuous selections of solution sets of Lipschitzean differential inclusions*, Funkcial. Ekvac. **34** (1991), 321–330.

[CG1]   G. Colombo and V. V. Goncharov, *A class of non-convex sets with locally well-posed projection*, Report S.I.S.S.A. **78/99/M**, 1–6.

[CG2]   ———, *Variational properties of closed sets in Hilbert spaces*, Report S.I.S.S.A. **57/99/M**, 1–16.

[CF]   P. E. Conner and E. E. Floyd, *Fixed point free involutions and equivariant maps*, Bull. Amer. Math. Soc. **66** (1960), 416–441.

[Con1]   A. Constantin, *On the stability of solution sets for operational differential inclusions*, An. Univ. Timişoara Ser. Ştiinţ. Mat. **29** (1991), 115–124.

[Con2]   ———, *Stability of solution sets of differential equations with multivalued right hand side*, J. Diff. Eqns **114** (1994), 243–252.

[Co1]   R. Conti, *Recent trends in the theory of boundary value problems for ordinary differential equations*, Boll. U.M.I. **22** (1967), 135–178.

[CKZ]   G. Conti, W. Kryszewski and P. Zecca, *On the solvability of systems of noncompact inclusions*, Ann. Mat. Pura Appl. **160** (1991), 371–408.

[CMN]   G. Conti, I. Massabo and P. Nistri, *Set-valued perturbations of differential equations at resonance*, Nonlinear Anal. **4** (1980), no. 6, 1031–1041.

[COZ]   G. Conti, V. Obukhovskii and P. Zecca, *On the topological structure of the solution set for a semilinear functional-differential inclusion in a Banach space*, Topology in Nonlinear Analysis Banach Center Publ., vol. 35, Inst. of Math. Polish Acad. of Sci., Warsaw, 1996, pp. 159–169.

[CP]   G. Conti and J. Pejsachowicz, *Fixed point theorems for multivalued weighted maps.*, Ann. Mat. Pura Appl. **126** (1980).

[CLY]   J. S. Cook, W. H. Louisell and W. H. Yocom, *Stability of an electron beam on a slalom orbit*, J. Appl. Phys. **29** (1958), 583–587.

[Cop]   W. A. Coppel, *Almost periodic properties of ordinary differential equations*, Ann. Mat. Pura Appl. **76** (1967), 27–49.

[Co2]   C. Corduneanu, *Citive probleme globale referitoare la ecuatiile differentiale ne-lineare de ordinne al doilea*, Acad. Rep. Pop. Rom., Fil. Iasi, Stud. Cer. St., Mat. **7** (1956), 1–7.

[Co3]   ———, *Existenta solutiilar marginuite pentru unele ecuatii differentiale de ordinue al doilea*, Acad. Rep. Pop. Rom., Fil. Iasi, Stud. Cer. St., Mat. **7** (1957), 127–134.

[Co4]   ———, *Almost periodic solutions to differential equations in abstract spaces*, Rev. Roumaine Math. Pures Appl. **42** (1997), 9–10.

[Co5]   ———, *Bounded and almost periodic solutions of certain nonlinear parabolic equations*, Libertas Math. **2** (1982), 131–139.

[Co6]   ———, *Some almost periodicity criteria for ordinary differential equations*, Libertas Math. **3** (1983), 21–43.

[Co7]   ———, *Two qualitative inequalities*, J. Diff. Eqns **64** (1986), 16–25.

[CoLe]   M. J. Corless and G. Leitmann, *Continuous state feedback guaranteeing uniform ultimate boundedness for uncertain dynamic systems*, IEEE Trans. Autom. Contr. **AC-26** (1981), no. 25, 1139–1144.

[Cor]   B. Cornet, *Existence of slow solutions for a class of differential inclusions*, J. Math. Anal. Appl. **96** (1983), no. 1, 130–147.

[CouKa]   J.-F. Couchouron and M. Kamenskiĭ, *Perturbations d'inclusions paraboliques par des opérateurs condensants*, C. R. Acad. Sci. Paris **320** (1995), 1–6.

[CN]   H. Covitz, S. B. Nadler, Jr., *Multi-valued contraction mappings in generalized metric spaces*, Israel J. Math. **8** (1970), 5–11.

[Cur]   D. W. Curtis, *Hyperspaces homeomorphic to Hilbert space*, Proc. Am. Math. Soc. **75** (1979), 126–130.

[Cza1] K. Czarnowski, *Structure of the set of solutions of an initial-boundary value problem for a parabolic partial differential equations in an unbounded domain*, Nonlinear Anal. TMA **27** (1996), 723–729.

[Cza2] _____, *On the structure of fixed point sets of 'k-set contractions' in B_0 spaces*, Demonstratio Math. **30** (1997), 233–244.

[CzP] K. Czarnowski and T. Pruszko, *On the structure of fixed point sets of compact maps in B_0 spaces with applications to integral and differential equations in unbounded domain*, J. Math. Anal. Appl. **54** (1991), 151–163.

[Cze] S. Czerwik, *Continuous solutions of a system of functional inequalities*, Glas. Mat. **19** (1984), no. 1, 105–109.

[Dnc] E. N. Dancer, *Boundary-value problems for differential equations on infinite intervals*, Proc. London Math. Soc. **3** (1975), no. 30, 76–94.

[Dan1] L. I. Danilov, *Almost periodic selections of multivalued maps*, Izv. Otdela Mat. Inform. Udmurtsk. Gos. Univ. **1** (1993), Izhevsk, 16–78. (in Russian)

[Dan2] _____, *Measure-valued almost periodic functions and almost periodic selections of multivalued maps*, Mat. Sb. **188** (1997), 3–24 (in Russian); Sbornik: Mathematics **188** (1997), 1417–1438.

[Dar] S. Darbo, *Theoria de l'omologia in una categoria die mappe pluriralenti ponderati*, Rend. Sem. Math. Univ. Padova **28** (1958), 188–220.

[Dav] J. L. Davy, *Properties of the solution set of a generalized differential equation*, Bull. Austral. Math. Soc. **6** (1972), 379–398.

[Daw] A. Dawidowicz, *Spherical maps*, Fund. Math. **127**, **3** (1987), 187–196.

[DR] M. Dawidowski and B. Rzepecki, *On bounded solutions of nonlinear differential equations in Banach spaces*, Demonstr. Math. **18** (1985), no. 1, 91–102.

[Day] M. M. Day, *On the basis problem in normed spaces*, Proc. Amer. Math. Soc. **13** (1962), 655–658.

[DBl1] F. S. De Blasi, *Existence and stability of solutions for autonomous multivalued differential equations in Banach spaces*, Rend. Accad. Naz. Lincei, Serie VII **60** (1976), 767–774.

[DBl2] _____, *Characterizations of certain classes of semicontinuous multifunctions by continuous approximations*, J. Math. Anal. Appl. **196** (1985), no. 1, 1–18.

[DBl3] _____, *On the differentiability of multifunctions*, Pacific J. Math. **66** (1976), 67–81.

[DBl4] _____, *On a property of the unit sphere in a Banach space*, Bull. Soc. Math. R. S. Roumaine **21** (1977), 259–262.

[DBGP1] F. S. De Blasi, L. Górniewicz and G. Pianigiani, *Topological degree and periodic solutions of differential inclusions*, Centro Vito Volterra, vol 263, Universita Degli Studi Di Roma "Torvergata", 1996, pp. 1–26.

[DBGP2] _____, *Topological degree and periodic solutions of differential inclusions*, Nonlinear Anal. TMA **37** (1999), 217–245.

[DBGP3] _____, *Random topological degree and random differential inclusions*, Unpublished manuscript (1999).

[DBLa1] F. S. De Blasi and A. Lasota, *Daniell's method in the theory of the Aumann-Hukuhara integral of set-valued functions*, Atti Acad. Naz. Lincei. Rend. **45** (1968), 252–256.

[DBLa2] _____, *Characterization of the integral of set-valued functions*, Atti Acad. Naz. Lincei. Rend. **46** (1969), 154–157.

[DBM1] F. S. De Blasi and J. Myjak, *Sur l'existence de selections continues*, C. R. Accad. Sci. Paris Sér I Math. **296** (1983), no. 17, 737–739.

[DBM2] _____, *On the solution sets for differential inclusions*, Bull. Polish Acad. Sci. Math. **12** (1985), 17–23.

[DBM3] _____, *A remark on the definition of topological degree for set-valued mappings*, J. Math. Anal. Appl. **92** (1983), 445–451.

[DBM4] _____, *On continuous approximation for multifunctions*, Pacific J. Math. **123** (1986), 9–30.

[DBM5]   ———, *Continuous selections for weakly Hausdorff lower semicontinuous mul-tifunctions*, Proc. Amer. Math. Soc., vol. 93, 1985, pp. 369–372.

[DBM6]   ———, *On the structure of the set of solutions of the Darboux problem for hyperbolic equations*, Proc. Edinburgh Math. Soc., Ser. 2 **29** (1986), no. 1, 7–14.

[DBP1]   F. S. De Blasi and G. Pianigiani, *Topological properties of nonconvex differential inclusions*, Nonlinear Anal. **20** (1993), 871–894.

[DBP2]   ———, *A Baire category approach to the existence of solutions of multivalued differential equations in Banach spaces*, Funkcial. Ekvac. **25** (1982), 153–162.

[DBP3]   ———, *On the density of extremal solutions of differential inclusions*, Ana. Polon. Math. **56** (1992), 133–142.

[DBP4]   ———, *Differential inclusions in Banach spaces*, J. Differential Equations **1987** (66), 208–229.

[DBP5]   ———, *Solution sets of boundary value problems for nonconvex differential inclusions*, Topol. Methods Nonlinear Anal. **1** (1993), 303–314.

[DBP6]   ———, *Remarks on Hausdorff continuous multifunctions and selections*, Comment. Math. Univ. Carolin., vol. 24, 1983, pp. 553–561.

[DBP7]   ———, *The Baire category method in existence problem for a class of multivalued differential equations with nonconvex right hand side*, Funkcial. Ekvac. **28** (1985), no. 2, 139–156.

[DBP8]   ———, *On the solution sets of nonconvex differential inclusions*, J. Diff. Eqns **128** (1996), 541–555.

[DBPS1]   F. S. De Blasi, G. Pianigiani and V. Staicu, *Topological properties of nonconvex differential inclusions of evolution type*, Nonlinear Anal. TMA **24** (1995), 711–720.

[DBPS2]   ———, *On the solution sets of some nonconvex hyperbolic differential inclusions*, Czechoslovak Math. J. **45** (1995), 107–116.

[Deb]   G. Debreu, *Integration of correspondences*, Proceed. Fifth Berkeley Symp. Math. Stat. Probab., vol. II, Univ. California Press, Berkeley, 1967, pp. 351–372.

[DeG]   P. Deguire and A. Granas, *Sur une certaine alternative nonlineaire en analyse convexe*, Studia Math. **83** (1986), no. 2, 127–138.

[De1]   K. Deimling, *On solution sets of multivalued differential equations*, Appl. Anal. **30** (1988), 129–135.

[De2]   ———, *Cone-valued periodic solutions of ordinary differential equations*, Proceedings of International Conference on Applications Nonlinear Analysis (1978), Academic Press, New York, 127–142.

[De3]   ———, *Open problems for ordinary differential equations in a Banach space*, Equazioni Differenziali, Florence, 1978.

[De4]   ———, *Periodic solutions of differential equations in Banach spaces*, Man. Math. **24** (1978), 31–44.

[De5]   ———, *Bounds for solution sets of multivalued ODEs*, Recent Trends in Differential Equations, World Sci. Publishing, River Edge, NJ, 1992, pp. 127–134.

[DeHeSh]   K. Deimling, G. Hetzer and Wenxian Shen, *Almost periodicity enforced by Coulomb friction*, Advances Diff. Eqns **1** (1996), no. 2, 265–281.

[DeRa]   K. Deimling and M. R. M. Rao, *On solution sets of multivalued differential equations*, Appl. Anal. **28** (1988), 129–135.

[Dem]   A. I. Demin, *Coexisting of periodic, almost-periodic and recurrent points of transformations of n-od*, Vestnik Mosk. Univ **1** (1996), no. 3, 84–87. (in Russian)

[DiaWa]   P. Diamond and P. Watson, *Regularity of solution sets for differential inclusions quasi-concave in a parameter*, Appl. Math. Lett. **13** (2000), 31–35.

[DRS]   J. Diestel, W. M. Ruess and W. Schachermaer, *Weak compactness in $L^1(\mu, X)$*, Proceed. Amer. Math. Soc. **118** (1993), 447–453.

[DbrKu]   R. Dobrenko and Z. Kucharski, *On the generalization of the Nielsen number*, Fund. Math. **134** (1990), 1–14.

[DoSh]   A. M. Dolbilov and I. Ya. Shneiberg, *Almost periodic multifunctions and their selections*, Sibirsk. Math. Zh. **32** (1991), 172–175. (in Russian)

[Do]   A. Dold, *Fixed point index and fixed point theorem for Euclidean neighbourhood retracts*, Topology **4** (1965), 1–8.

[Dos1]   R. Doss, *On generalized almost periodic functions*, Annals of Math. **59** (1954), 477–489.

[Dos2]   _____, *On generalized almost periodic functions – II*, J. London Math. Soc. **37** (1962), 133–140.

[Dos3]   _____, *The integral of a generalized almost periodic function*, Duke Math. J. **30** (1963), 39–46.

[Dos4]   _____, *Groupes compacts et fonctions presque périodiques généralisées*, Bull. Sci. Math. **77** (1953), 186–194.

[Dos5]   _____, *Sur une nouvelle classe de fonctions presque-périodiques*, C. R. Acad. Sci. Paris **238** (1954), 317–318.

[Dos6]   _____, *On Riemann integrability and almost periodic functions*, Compositio Math. **12** (1956), 271–283.

[DouOe]   A. Douady and J. Oesterlé, *Dimension de Hausdorff des attractors*, C. R. Acad. Sci. Paris Sér I Math. **290** (1980), 1135–1138.

[Du]   Y. H. Du, *The structure of the solution set of a class of nonlinear eigenvalue problems*, J. Math. Anal. Appl. **170** (1992), 567–580.

[DbMo1]   J. Dubois and P. Morales, *On the Hukuhara–Kneser property for some Cauchy problems in locally convex topological vector spaces*, Lecture Notes in Math., vol. 964, Springer, Berlin, 1982, pp. 162–170.

[DbMo2]   _____, *Structure de l'ensemble des solutions du probléme de Cauchy sous le conditions de Carathéodory*, Ann. Sci. Math. Quebec **7** (1983), 5–27.

[Dug1]   J. Dugundji, *An extension of Tietze's theorem*, Pacific J. Math. **1** (1951), 353–367.

[Dug2]   _____, *Modified Vietoris theorems for homotopy*, Fund. Math. **66** (1970), 223–235.

[DvEmHu]   P. F. Duvall, Jr., J. W. Emert, and L. S. Husch, *Iterated function systems, compact semigroups, and topological contractions*, Continuum Theory and Dynamical Systems (T. West, ed.), Marcel Dekker, Inc., New York; Lect. Notes Pure Appl. Math. **149** (1993), 113–155.

[DyG]   G. Dylawerski and L. Górniewicz, *A remark on the Krasnosielskiĭ's translation operator*, Serdica **9** (1983), 102–107.

[DyJ]   G. Dylawerski and J. Jodel, *On Poincaré's translation operator for ordinary equations with retards*, Serdica **9** (1983), 396–399.

[DzGe]   Z. Dzedzej and B. Gel'man, *Dimension of the solution set for differential inclusions*, Demonstration Math. **26** (1993), 149–158.

[Eb1]   W. F. Eberlein, *Abstract ergodic theorems and weak almost periodic functions*, Trans. Amer. Math. Soc. **67** (1949), 217–240.

[Eb2]   _____, *The point spectrum of weakly almost periodic functions*, Mich. J. Math. **3** (1956), 137–139.

[EdFo]   H. E. Edgerton and P. Fourmarier, *The pulling- into-step of a salient – pole synchronous motor*, Trans. A. I. E. E. **50** (1931), 769–778.

[EM]   S. Eilenberg and D. Montgomery, *Fixed point theorems for multivalued transformations*, Amer. Math. J. **58** (1946), 214–222.

[ElO]   H. El-Owaidy, *On stability of derivo-periodic solutions of non holonomic systems*, J. Natur. Sci. Math. **1** (1981), no. 21, 115–125.

[ErKaKr]   L. H. Erbe, T. Kaczyński and W. Krawcewicz, *Solvability of two-point boundary value problems for systems of nonlinear differential equations of the form* $y'' = g(t, y, y', y'')$, Rocky Mount. J. Math. **20** (1990), 899–907.

[Esc]   E. Esclangon, *Sur les integrales quasi-périodique d'une equation différetialle linéaire*, C. R. Acad. Sci. Paris **160** (1915), 488–489, 652–653.

[EvMa]   C. J. G. Evertsz and B. B. Mandelbrot, *Multifractals measures*, Chaos and Fractals. New Frontiers of Science (H.-O. Peitgen, H. Jürgens, and D. Saupe, eds.), Springer-Verlag, Berlin, 1992, pp. 849–881.

[FabJoPa]   R. Fabbri, R. Johnson and R. Pavani, *On the nature of the spectrum of the quasi-periodic Schrödinger operator*, Preprint (1999).

[Fad]   E. Fadell, *Recent results in the fixed point theory of continuous maps*, Bull. Amer. Math. Soc. **1** (1970), 10–29.

[Fan1]   K. Fan, *Fixed point and minimax theorems in locally convex topological linear spaces*, Proc. Nat. Acad. Sci. U.S.A. **38** (1952), 271–275.

[Fan2]   ———, *Some properties of convex sets related to fixed point theorems*, Math. Ann. **266** (1984), no. 4, 519–537.

[Far1]   M. Farkas, *On stability and geodesies*, Ann. Univ. Sci. Budapest. Rol. Eotvos Nom. **11** (1968), 145–159.

[Far2]   ———, *Controllably periodic perturbations of autonomous systems*, Acta Math. Acad. Sci. Hungar. **22** (1971), no. 3–4, 337–348.

[Far3]   ———, *Determination of controllably periodic perturbed solutions by Poincaré's method*, Stud. Sci. Math. Hungar. **7** (1972), 257–266.

[Fav]   J. Favard, *Sur les équations différentialles á coefficients presque périodiques*, Acta Math. **51** (1927), 31–81.

[Fec1]   M. Fečkan, *Multiple perturbed solutions near nondegenerate manifolds of solutions*, Comment. Math. Univ. Carolin. **34** (1993), no. 4, 635–643.

[Fec2]   ———, *Nielsen fixed point theory and nonlinear equations*, J. Differ. Eqns **106** (1993), no. 2, 312–331.

[Fec3]   ———, *Multiple periodic solutions of small vector fields on differentiable manifolds*, J. Differ. Eqns. **113** (1994), no. 1, 189–200.

[Fec4]   ———, *Multiple solutions of nonlinear equations via Nielsen fixed-point theory: a survey*, Nonlinear Anal. in Geometry and Topology (Th. M. Rassias, eds.), Hadronic Press, Palm Harbor, 2000, pp. 77–97.

[Fec5]   ———, *Existence results for implicit differential equations*, Math. Slovaca **48** (1998), no. 1, 35–42.

[Fed]   H. Federer, *Curvature measures*, Trans. Amer. Math. Soc. **93** (1959), 418–491.

[Fel]   P. L. Felmer, *Rotation type solutions for spatially periodic Hamiltonian systems*, Nonlinear. Anal. T. M. A. **19** (1992), no. 5, 409–425.

[FP1]   C. C. Fenske and H. O. Peitgen, *Attractors and the fixed point index for a class of multivalued mappings* I, Bull. Polish Acad. Sci. Math. **25** (1977), 477–482.

[FP2]   ———, *Attractors and the fixed point index for a class of multivalued mappings* II, Bull. Polish Acad. Sci. Math. **25** (1977), 483–487.

[FerZa]   M. L. Fernandes and F. Zanolin, *Remarks on strongly flow-invariant sets*, J. Math. Anal. Appl. **128** (1987), 176–188.

[Fi1]   V. V. Filippov, *Basic topological structures of the theory of ordinary differential equations*, Topology in Nonlinear Analysis Banach Center Publ., vol. 35, Inst. Acad. of Sci., Warsaw, 1996, pp. 171–192.

[Fi2]   ———, *On Luzin's and Scorza-Dragoni's theorem*, Vestnik Moskov. Univ. Ser. I Mat. Mekh. **42** (1987), 66–68 (in Russian); Moscow Univ. Math. Bull. **42** (1987), 61–63.

[Fi3]   ———, *On Luzin's theorem and right-hand sides of differential inclusions*, Mat. Zametki **37** (1985), 93–98 (in Russian); Math. Notes **37** (1985), 53–56.

[Fi4]   ———, *Topological structure of solution spaces of ordinary differential equations*, Uspekhi Mat. Nauk **48** (1993), no. 1, 103–154. (in Russian)

[Fi5]   ———, *On the acyclicity of solution sets of ordinary differential equations*, Dokl. Akad. Nauk **352** (1997), 28–31. (in Russian)

[Fin1]   A. M. Fink, *Uniqueness theorems and almost periodic solutions to second order differential equations*, J. Diff. Eqns **4** (1968), 543–548.

[Fin2]   ———, *Almost automorphic and almost periodic solutions which minimize functionals*, Tohoku Math. J. **20** (1968), 323–332.

[FinFr]   A. M. Fink and P. Frederickson, *Ultimate boundedness does not imply almost periodicity*, J. Diff. Eqns **9** (1971), 280–284.

[Fis1]   A. Fischer, *Approximation of almost periodic functions by periodic ones*, Czechoslovak Math. J. **48** (1998), 193–205.

[Fis2]   ———, *Almost periodic solutions with a prescribed spectrum of systems of linear and quasilinear differential equations with almost periodic coefficients and constant time lag (Cauchy integral)*, Math. Bohemica **124** (1999), no. 4, 351–379.

[FiMaPe]   P. M. Fitzpatrick, I. Masabó and J. Pejsachowicz, *On the covering dimension of the set of solutions of some nonlinear equations*, Trans. Amer. Math. Soc. **296** (1986), 777–798.

[FiPe1]   P. M. Fitzpatrick and W. V. Petryshyn, *A degree theory, fixed point theorems and mappings for multivalued noncompact mappings*, Trans. Amer. Math. Soc. **194** (1974), 1–25.

[FiPe2]   ———, *Fixed point theorems and the fixed-point index for multi-valued mappings in cones*, J. London Math. Soc. **12** (1975), 75–85.

[Fol]   E. Fœlner, *On the structure of generalized almost periodic functions*, Danske Vid. Selsk. Math. Phys., Medd. **21** (1945), no. 11, 1–30.

[Fol1]   ———, *A proof of the main theorem for almost periodic functions in an abelian group*, Ann. Math. **50** (1949), 559–569.

[Fol2]   ———, *Almost periodic functions on Abelian groups*, Mat. Tidsskr. B (1946), 153–161.

[Fol3]   ———, *Note on the definition of almost periodic functions in groups*, Mat. Tidsskr. B. (1950), 58–62.

[Fol4]   ———, *Generalization of a theorem of Bogolioùboff to topological abelian groups*, Math. Scand. **2** (1954), 5–18.

[Fol5]   ———, *On the dual spaces of the Besicovitch almost periodic spaces*, Danske Vid. Selsk. Mat.-Fys. Medd. **29** (1954), 1–27.

[Fol6]   ———, *W-almost periodic functions in arbitrary groups*, C.R. Dixième Congrès Math. Scand. 1946, Jul. Gjellerups Forlag (1947), 356–362.

[Fol7]   ———, *Generalization of the general diophantine approximation theorem of Kronecker*, Mat. Scand. **68** (1991), 148–160.

[Fol8]   ———, *Besicovitch almost periodic functions in arbitrary groups*, Math. Scand. **5** (1957), 47–53.

[FonZa]   A. Fonda and F. Zanolin, *Bounded solutions of nonlinear second order ordinary differential equations*, Discr. Contin. Dynam. Syst. **4** (1998), no. 1, 91–98.

[For]   M. K. Fort, *Essential and nonessential fixed points*, Amer. J. Math. **72** (1950), 315–322.

[FJ]   L. Fountain and L. K. Jackson, *A generalized solution of the boundary value problem for $y'' = f(x, y, y')$*, Pacific J. Math. **12** (1962), 1251–1272.

[Fo1]   G. Fournier, *A simplicial approach to the fixed point index*, Lecture Notes in Math. **886** (1981), 73–102.

[Fo2]   ———, *Théoréme de Lefschetz I-Applications éventuellement compactes*, Bull. Polish Acad. Sci. Math. **6** (1975), 693–701.

[Fo3]   ———, *Théoréme de Lefschetz II-Applications d'attraction compacte*, Bull. Polish Acad. Sci. Math. **6** (1975), 701–706.

[Fo4]   ———, *Théoréme de Lefschetz III-Applicationes asymptotiquement compactes*, Bull. Polish Acad. Sci. Math. **6** (1975), 707–713.

[FG1]   G. Fournier and L. Górniewicz, *The Lefschetz fixed point theorem for multivalued maps of non-metrizable spaces*, Fund. Math. **42** (1976), 213–222.

[FG2]   ———, *The Lefschetz fixed point theorem for some noncompact multivalued maps*, Fund. Math. **94** (1977), 245–254.

[FG3]   ———, *Survey of some applications of the fixed point index*, Sem. Math. Superiore **96** (1985), Montreal, 95–136.

[FM]   G. Fournier and M. Martelli, *Set-valued transformations of the unit sphere*, Lett. Math. Phys. **10** (1985), no. 2-3, 125–134.

[FSW]   G. Fournier, A. Szulkin and M. Willem, *Semilinear elliptic equation in* $\mathbb{R}^N$ *with almost periodic or unbounded forcing term*, SIAM J. Math. Anal. **27** (1996), 1653–1660.

[FV1]   G. Fournier and D. Violette, *A fixed point theorem for a class of multi-valued continuously differentiable maps*, Ann. Polon Math. **47** (1987), no. 3, 381–402.

[FV2]   ――――, *A fixed point index for compositions of acyclic multivalued maps in Banach spaces*, The MSRI-Korea Publications **1** (1966), 139–158.

[FV3]   ――――, *Un indice de point fixe pour les composées de fonctions multivoques acycliques dans des espaces de Banach*, Ann. Sci. Math. Québec **22** (1998), no. 2, 225–244.

[Fra]   P. Franklin, *Approximation theorems for generalized almost-periodic functions*, Math. Zeitsch. **29** (1928), 70–87.

[FrnMi]   J. Franks and M. Misiurewicz, *Cycles for disk homomorphisms and thick trees*, Contemp. Math. **152** (1993), Amer. Math. Soc., Providence, R. I., 69–139.

[Fri1]   M. Frigon, *Theoreme d'existence de solutions faibles pour un probleme aux limites du second ordre dans l'intervale* $(0, \infty)$, Preprint. no. 85-18 (Juin 1985), Univ. de Montreal. Dépt. de. Math. et de Statistique.

[Fri2]   ――――, *On continuation methods for contractive and nonexpansive mappings*, Recent Advances on Metric Fixed Point Theory (T. D. Benavides, ed.), Univ. Sevilla, Sevilla, 1996, pp. 19–30.

[Fri3]   ――――, *Fixed point results for generalized contractions in gauge spaces and applications*, Proceed. Amer. Math. Soc. (to appear).

[FriGr]   M. Frigon and A. Granas, *Resultats de type Leray–Schauder pour des contractions sur des espaces de Fréchet*, Ann. Sci. Math. Québec **22** (1998), no. 2, 261–268.

[FGG]   M. Frigon, A. Granas and E. A. Guennoun, *Alternative non lineaire pour les applications contractantes*, Ann. Sci. Math. Québec **19** (1995), no. 1, 65–68.

[FGK]   M. Frigon, L. Górniewicz and T. Kaczyński, *Differential inclusions and implicit equations on closed subsets of* $\mathbb{R}^n$, Proceed of the First WCNA (V. Laksmikantham, ed.), W. de Gruyter, Berlin, 1996, pp. 1197–1806.

[FK]   M. Frigon and T. Kaczyński, *Boundary value problems for systems of implicit differential equations* **179** (1993), J. Math. Anal. Appl., 317–326.

[Fry1]   A. Fryszkowski, *Carathéodory type selectors of set-valued maps of two variables*, Bull. Polish Acad. Sci. Math. **25** (1977), 41–46.

[Fry2]   ――――, *Continuous selections for a class of non-convex multivalued maps*, Studia Math. **76** (1983), no. 2, 163–174.

[Fry3]   ――――, *Existence of solutions of functional-differential inclusion in nonconvex case*, Ann. Polon Math. **45** (1985), no. 2, 121–124.

[Fry4]   ――――, *Properties of the solutions of orientor equation*, J. Differ. Equations Appl. **181** (1982), no. 2, Russe, 750–763.

[Fry5]   ――――, *The generalization of Cellina's fixed point theorem*, Studia Math. **78** (1984), no. 2, 213–215.

[FryGo]   A. Fryszkowski and L. Górniewicz, *Mixed semicontinuous mappings and their applications to differential inclusions*, Set-Valued Anal. **8** (2000), 203–217.

[FryRz]   A. Fryszkowski and T. Rzeżuchowski, *Continuous version of Filippov-Ważewski theorem*, J. Differential Equations **94** (1991), 254–265.

[Ful]   F. G. Fuller, *The existence of periodic points*, Ann. Math. **57** (1953), 229–230.

[FMV1]   M. Furi, M. Martelli and A. Vignoli, *Stable-solvable operators in Banach spaces*, Atti. Accad. Naz. Lincei-Rend. Sci. **1** (1976), 21–26.

[FMV2]   ――――, *Contributions to the spectral theory for nonlinear operators in Banach spaces*, Ann. Mat. Pura Appl. **118** (1978), 229–294.

[FMV3]   ――――, *On the solvability of nonlinear operator equations in normed spaces*, Ann. Mat. Pura Appl. **124** (1980), 321–343.

[FNPZ]   M. Furi, P. Nistri, P. Pera and P. Zecca, *Topological methods for the global controllability of nonlinear systems*, J. Optim. Theory Appl. **45** (1985), 231–256.

[FPr1]   M. Furi and P. Pera, *A continuation method on locally convex spaces and applications to ordinary differential equations on noncompact intervals*, Ann. Polon. Math. **47** (1987), 331–346.

[FPr2]   _____, *On the fixed point index in locally convex spaces*, Proc. Roy. Soc. Edinb. **106** (1987), 161–168.

[FPr3]   _____, *On the existence of an unbounded connected set of solutions for nonlinear equations in Banach spaces*, Atti Accad. Nar. Lincei-Rend. Sci **67** (1979), no. 1–2, 31–38.

[Ga1]   G. Gabor, *Fixed points of set-valued maps with closed proximally ∞-connected values*, Discuss. Math. **15** (1995), 163–185.

[Ga2]   _____, *On the classification of fixed points*, Math. Japon. **40** (1994), 361–369.

[Ga3]   _____, *On the acyclicity of fixed point sets of multivalued maps*, Topol. Meth. Nonlin. Anal. **14** (1999), 327–343.

[GaKr1]   D. Gabor and W. Kryszewski, *On the solvability of systems of noncompact and nonconvex inclusions*, Differential Equations Dynam. Systems **1** (1998), 135–152.

[GaKr2]   _____, *A coincidence theory involving Fredholm operators of nonnegative index*, Topol. Meth. Nonlin. Anal. (to appear).

[GaPi]   G. Gabor and R. Pietkun, *Periodic solutions of differential inclusions with retards*, Topol. Meth. Nonlin. Anal. **16** (2000), 103–123.

[GaP1]   R. G. Gaines and J. K. Peterson, *Degree theoretic methods in optimal control*, J. Math. Anal. Appl. **94** (1983), no. 1, 44–77.

[GaP2]   _____, *Periodic solutions of differential inclusions*, Nonlinear Anal. **5** (1981), no. 10, 1109–1131.

[GST]   J.-M. Gambaudo, S. van Strien and C. Tresser, *Vers un ordre de Sharkovskiĭ pour les plongements du disque préservant l'orientation*, C. R. Acad. Sci. Paris, Sér. I Math. **310** (1990), 291–294.

[Ge1]   B. D. Gelman, *The topological characteristic of multivalued mappings and a fixed point theorem*, Soviet Math. Dokl. **16** (1975), 260–264.

[Ge2]   _____, *On Kakutani-type fixed-point theorems for multivalued mappings*, Global Analysis and Nonlinear Equations, Voronezh, 1988, pp. 117–119. (in Russian)

[Ge3]   _____, *Some problems in the theory of fixed points of multivalued mappings*, Topol. and Geom. Methods of Analysis, Global. Anal., Voronezh, 1989, pp. 90–105. (in Rusian)

[Ge4]   _____, *On continuous selections of multivalued mappings having no coincidence points*, Pr. Uniw. Gdański, Inst. Mat. **74** (1990), 1–18. (in Russian)

[Ge5]   _____, *On the structure of the set of solutions for inclusions with multivalued operators*, Lecture Notes in Math. **1334** (1987), Springer, Berlin, 60–78.

[Ge6]   _____, *Topological properties of fixed point sets of multivalued maps*, Mat. Sb. **188** (1997), 33–56. (in Russian)

[Ge7]   _____, *On topological dimension of a set of solutions of functional inclusions*, Differential Inclusions and Optimal Control (J. Andres. L. Górniewicz and P. Nistri, eds.), vol. 2, Lecture Notes Nonlinear Anal., 1998, pp. 163–178.

[GO]   B. D. Gelman and V. V. Obuhovskii, *Some fixed-point theorems for condensing-type multivalued mappings*, Algebraic Problems in Analysis and Topology, Novoe Global. Anal., Voronezh, 1990, pp. 110–115. (in Russian)

[GG1]   K. Gęba and L. Górniewicz, *On the Bourgin–Yang theorem for multivalued maps. I*, Bull. Polish Acad. Sci. Math. **34** (1986), 315–322.

[GG2]   _____, *On the Bourgin–Yang theorem for multivalued maps. II*, Bull. Polish Acad. Sci. Math. **34** (1986), no. 5-6, 323–329.

[Gia]   T. V. Giap, *Stability of D-periodic solutions of third-order non-linear differential equations*, Studia Sci. Math. Hungar. **9** (1974), 355–368.

[GchJo]   R. Giachetti nad R. A. Johnson, *The Floquet exponent for two-dimmensional linear systems with bounded coefficients*, J. Math. Pures Appl. **65** (1986), 93–117.

[Gvl]   A. Gavioli, *On the solution set of the nonconvex sweeping process*, Discuss. Math., Differential Incl. **19** (1999), 45–65.

[Gir]   J. Girolo, *Approximating compact sets in normed linear spaces*, Pacific J. Math. **98** (1982), 81–89.

[God1]   A. N. Godunov, *A counter example to Peano's Theorem in an infinite dimensional Hilbert space*, Vestnik Mosk. Gos. Univ., Ser. Mat. Mek. **5** (1972), 31–34. (in Russian)

[God2]   ———, *Peano's Theorem in an infinite dimensional Hilbert space is false even in a weakened form*, Math. Notes **15** (1974), 273–279.

[GoeRz]   K. Goebel and W. Rzymowski, *An existence theorem for the equation $x' = f(t, x)$ in Banach spaces*, Bull. Acad. Polon. Math. **18** (1970), 367–370.

[GolSt]   M. Golubitsky and I. N. Stewart, *Hopf bifurcation with dihedral group symmetry: coupled nonlinear oscillators*, Contemp. Math. **56** (1986), Providence, R. I., AMS, 131–173.

[Gon]   V. V. Goncharov, *Co-density and other properties of the solution set of differential inclusions with noncompact right-hand side*, Discuss. Math. Differential Incl. **16** (1996), 103–120.

[Go1]   L. Górniewicz, *On the solution set of differential inclusions*, J. Math. Anal. Appl. **113** (1986), 235–244.

[Go2]   ———, *Remarks on the Lefschetz fixed point theorem*, Bull. Polish Acad. Sci. Math. **11** (1973), 993–999.

[Go3]   ———, *Some consequences of the Lefschetz fixed point theorem for multi-valued maps*, Bull. Polish Acad. Sci. Math. **2** (1973), 165–170.

[Go4]   ———, *On the Lefschetz coincidence theorem*, Lect. Notes in Math. **886** (1981), Springer, Berlin, 116–139.

[Go5]   ———, *A Lefschetz-type fixed point theorem*, Fund. Math. **88** (1975), 103–115.

[Go6]   ———, *Asymptotic fixed point theorems for multi-valued maps*, Symp. of Topology, Tbilisi, 1972.

[Go7]   ———, *Fixed point theorem for multi-valued mappings of approximative ANR-s*, Fund. Math. **18** (1970), 431–436.

[Go8]   ———, *Fixed point theorems for multi-valued maps of special ANR-s*, University of Gdańsk **1** (1971), 39–49.

[Go9]   ———, *Fixed point theorems for multi-valued maps of subsets of Euclidean spaces*, Bull. Polish Acad. Sci. Math. **27** (1979), 111–115.

[Go10]   ———, *On a Coincidence Theorem*, Bull. Polish Acad. Sci. Math. **2** (1974), 277–283.

[Go11]   ———, *On non-acyclic multi-valued maps of subsets of Euclidean spaces*, Bull. Polish Acad. Sci. Math. **5** (1972), 379–385.

[Go12]   ———, *On the Birkhoff–Kellogg theorem*, Proc. Internat. Conf. on Geometric Topology, PWN, Warsaw, 1980, pp. 155–160.

[Go13]   ———, *Repulsive fixed points of compact maps of topologically complete ANR-s*, Zesz. Nauk. Univ. Gdańskiego **3** (1976), 59–66.

[Go14]   ———, *On homotopically simple multi-valued mappings*, WSP Słupsk **3** (1984), 11-22.

[Go15]   ———, *Topological structure of solution sets: Current results*, Arch. Math. **36** (2000), 343–382.

[Go16]   ———, *Periodic problems for ODE's via multivalued Poincaré operators*, Arch. Math. (Brno) **34** (1998), 93–104.

[Go17]   ———, *On the Lefschetz fixed point theorem*, Math. Slovaca **52** (2002), no. 2, 221–233.

[Go18]   ———, *Present state of the Brouwer fixed point theory for multivalued mappings*, Ann. Sci. Math. Quebec **22** (1998), 169–179.

[GGr1]   L. Górniewicz and A. Granas, *Fixed point theorems for multi-valued maps of ANR-s*, J. Math. Pures Appl. **49** (1970), 381–395.

[GGr2]   _____, *Some general theorems in coincidence theory*, J. Math. Pures Appl. **60** (1981), 361–373.

[GGr3]   _____, *The Lefschetz fixed point theorem for multi-valued mappings*, Coll. Inter. Theorie du Point Fixe et Appl. (1989), 69–72.

[GGr4]   _____, *Topology of morphisms and fixed point problems for setvalued operators*, Pitman Research Notes in Math. **252** (1990), 173–192.

[GGr5]   _____, *On a theorem of C. Bowszyc concerning the relative version of the Lefschetz fixed point theorem*, Bull. Inst. Math. Acad. Sinica **13** (1985), 137-142.

[GGK1]   L. Górniewicz, A. Granas and W. Kryszewski, *Sur la methode de l'homotopic dans la theorie des points fixes* Partie I, Transversalite topologique **307** (1988), 489–492.

[GGK2]   _____, *Sur la methode de l'homotopie dans la theorie des points fixes pour les applications multivoques* Partie II *L'indice dans les* ANR-*s compacts*, C. R. Acad. Sci. Paris Sér I Math. **308** (1989), no. 14, 449–452.

[GGK3]   _____, *On the homotopy method in the fixed point index theory of multi-valued mappings of compact absolute neighbourhood retracts*, J. Math. Anal. Appl. **161** (1991), 457–473.

[GKr1]   L. Górniewicz and W. Kryszewski, *Bifurcation invariants for acyclic mappings*, Rep. Math. Phys. **31** (1992), 217–229.

[GKr2]   _____, *Topological degree theory for acyclic mappings related to the bifurcation problem*, Boll. Un. Mat. Ital. **7** (1992), 579–595.

[GK]   L. Górniewicz and Z. Kucharski, *Coincidence of k-set contraction pairs*, J. Math. Anal. Appl. **107** (1985), no. 1, 1–15.

[GL]   L. Górniewicz and M. Lassonde, *Approximation and fixed points for compositions of R_δ-maps*, Topology Appl. **55** (1994), 239–250.

[GM]   L. Górniewicz and S. A. Marano, *On the fixed point set of multi-valued contractions*, Rend. Circ. Mat. Palermo **40** (1996), 139–145.

[GMS]   L. Górniewicz, S. A. Marano and M. Ślosarski, *Fixed points of contractive multi-valued maps*, Proc. Amer. Math. Soc. **124** (1996), 2675–2683.

[GN1]   L. Górniewicz and P. Nistri, *An invariance problem for control systems with deterministic uncertainty*, Topology in Nonlinear Analysis (K. Geba and L. Górniewicz, eds.), vol. 35, Banach Center Publications, Warsaw, 1996, pp. 193–205.

[GN2]   _____, *Topological essentiality and nonlinear boundary value control problems*, Topol. Methods Nonlinear Anal. **13** (1999), 53–72.

[GN3]   _____, *Feedback stability of closed sets for nonlinear control systems*, Progress in Nonlinear Differential Equations and Their Applications (J. Appel, ed.), vol. 40, Birkhäuser, Basel, 2000, pp. 165–175.

[GN4]   _____, *Two nonlinear feedback control problems on proximate retracts of Hilbert spaces*, Nonlinear Anal. **47** (2001), 1003–1015.

[GNO]   L. Górniewicz, P. Nistri and V. Obukhovskiĭ, *Differential inclusions on proximate retracts of Hilbert spaces*, Int. J. Nonlinear Diff. Eqns TMA **3** (1997), 13–26.

[GNZ]   L. Górniewicz, P. Nistri and P. Zecca, *Control problems in closed subset of $\mathbb{R}^n$ via feedback controls*, Topol. Methods Nonlinear Anal. **2** (1993), 163–178.

[GPe]   L. Górniewicz and H. O. Peitgen, *Degeneracy, non-ejective fixed points and the fixed point index*, J. Math. Pures Appl. **58** (1979), 217–228.

[GPl]   L. Górniewicz and S. Plaskacz, *Periodic solutions of differential inclusions in $\mathbb{R}^n$*, Boll. Un. Mat. Ital. A **7** (1993), 409–420.

[GPr]   L. Górniewicz and T. Pruszko, *On the set of solutions of the Darboux problem for some hyperbolic equations*, Bull. Acad. Polon. Math. **28** (1980), no. 5–6, 279–286.

[GR]   L. Górniewicz and D. Rozpłoch-Nowakowska, *On the Schauder fixed point theorem*, Topology in Nonlinear Analysis Banach Center Publ., vol. 35, Inst. of Math. Polish Acad. of Sci., Warsaw, 1996, pp. 207–219.

[GSc]   L. Górniewicz and M. Schmidt, *Bifurcations of compactifying maps*, Reports of Mathematical Physics **34** (1994), 241–248.

[GS]   L. Górniewicz and M. Ślosarski, *Topological essentiality and differential inclu-sions*, Bull. Austr. Math. Soc. **45** (1992), 177–193.

[Gra]   S. Graf, *Selected results on measurable selections*, Rend. Circ. Mat. Palermo **31** (1982), no. 2 suppl., 87–122.

[Gr1]   A. Granas, *Generalizing the Hopf–Lefschetz fixed point theorem for non-compact ANR-s*, Symp. Inf. Dim. Topol., Báton-Rouge, 1967; Ann. Math. Studies **69** (1972), 119–130.

[Gr2]   ———, *Theorem on antipodes and fixed points for a certain class of multivalued maps in Banach spaces*, Bull. Polish Acad. Sci. Math. **7** (1959), 271–275.

[Gr3]   ———, *Fixed point theorems for approximative ANR-s*, Bull. Polish Acad. Sci. Math. **16** (1968), 15–19.

[Gr4]   ———, *Some theorems in fixed point theory. The Leray–Schauder index and Lefschetz number*, Bull. Polish Acad. Sci. Math. **16** (1968), 131–137.

[Gr5]   ———, *Sur la notion du degree topologique pour une certaine classe de transfor-mations multivalentes dans les espaces de Banach*, Bull. Polish Acad. Sci. Math. **7** (1959), 181–194.

[Gr6]   ———, *The Leray–Schauder index and the fixed point theory for arbitrary ANR's*, Bull. Soc. Math. France **100** (1972), 209–228.

[Gr7]   ———, *Continuation method for contractive maps*, Topol. Meth. Nonlin. Anal. **3** (1994), 375–379.

[GGLO]   A. Granas, R. B. Guenther, J. W. Lee and D. O'Regan, *Boundary value problems on infinite intervals and semiconductor devices*, J. Math. Anal. Appl. **116** (1986), no. 2, 335–348.

[GrL1]   A. Granas and M. Lassonde, *Sur un principe geometrique an analyse convexe*, Studia Math. **101** (1991), 1–18.

[GrL2]   ———, *Some elementary general principles of convex analysis*, Topol. Methods Nonlinear Anal. **5** (1995), 23–38.

[GF]   A. Granas and Fon-Che-Liu, *Théorémes du minimax*, C. R. Acad. Sci. Paris Sér. I Math. **298** (1984), no. 1, 329–332.

[GJ1]   A. Granas and J. Jaworowski, *Some theorems of multivalued maps of subsets of the Euclidean space*, Bull. Polish Soc. Math. **6** (1965), 277–283.

[Gr]   O. A. Gross, *The boundary value problem on an infinite interval*, J. Math. Anal. Appl. **7** (1963), 100–109.

[Gua1]   V. V. Guaschi, *Pseudo-Anosov braid types of the disc or sphere of low cardinality imply all periods*, J. London Math. Soc. **2** (1994), no. 50, 594–608.

[Gua2]   ———, *Nielsen theory, braids and fixed points of surface homomorphisms*, Topol. Appl. **117** (2002), 199–230.

[Gud]   V. V. Gudkov, *On finite and infinite interval boundary value problems*, Diff. Urav. **12** (1976), no. 3, 555–557. (in Russian)

[GuoHe]   J. Guo and P. R. Heath, *Coincidence theory on the complement*, Topol. Appl. **95** (1999), 229–250.

[Had1]   G. Haddad, *Monotone trajectories of differential inclusions and functional dif-ferential inclusions with memory*, Israel J. Math. **39** (1981), no. 1-2, 83–100.

[Had2]   ———, *Monotone viable trajectories for functional differential inclusions*, J. Dif-ferential Equations **42** (1981), no. 1, 1–24.

[Had3]   ———, *Topological properties of the sets of solutions for functional differential equations*, Nonlinear Anal. **5** (1981), 1349–1366.

[HL]   G. Haddad and J. M. Lasry, *Periodic solutions of functional differential inclu-sions and fixed points of σ-selectionable correspondences*, J. Math. Anal. Appl. **96** (1983), no. 2, 295–312.

[HaS]   F. von Haeseler and G. Skordev, *Borsuk–Ulam theorem, fixed point index and chain approximations for maps with multiplicity*, Pac. J. Math. **153** (1992), 369–395.

[Haj1]   O. Hájek, *Homological fixed point theorems*, Comment. Math. Univ. Carolin. **5** (1964), 13–31; **6** (1965), 157–164.

[Haj2] _____, *Flows and periodic motions*, Comment. Math. Univ. Carolin. **6** (1965), 165–178.

[HlLaSl] J. K. Hale, J. P. LaSalle and M. Slemrod, *Theory of general class of dissipative processes*, J. Math. Anal. Appl. **39** (1972), no. 1, 117–191.

[HlLo] J. K. Hale and O. Lopes, *Fixed-point theorems and dissipative processes*, J. Diff. Eqns **13** (1973), 391–402.

[HlmHe] T. G. Hallam and J. W. Heidel, *Structure of the solution set of some first order differential equations of comparison type*, Trans. Amer. Math. Soc. **160** (1971), 501–512.

[Hal1] B. Halpern, *Algebraic topology and set valued maps*, Lecture Notes in Math. **171** (1970), 23–33.

[Hal2] _____, *Periodic points on tori*, Pacific J. Math. **83** (1979), 117–133.

[Hal3] _____, *Nielsen type numbers for periodic points*, Unpublished.

[Han] M. Handel, *The forcing partial order on three times punctured disk*, Ergod. Th. Dynam. Sys. **17** (1997), 593–610.

[Har] A. Haraux, *Asymptotic behavior for two-dimensional, quasi-autonomous, almost-periodic evolution equations*, J. Differential Equations **66** (1987), 62–70.

[HW] P. Hartman and A. Wintner, *On the non-increasing solutions of $y'' = f(x, y, y')$*, Amer. J. Math. **73** (1951), 390–404.

[Hay] S. Hayashi, *Self-similar sets as Tarski's fixed points*, Publ. Res. Inst. Math. Sci. **21** (1985), 1059–1066.

[Hea] P. R. Heath, *A survey of Nielsen periodic point theory (fixed n)*, Banach Center Publ. **49** (1999), Inst. of Math, Polish Acad. Sci., Warsaw, 159–188.

[HeKe] P. R. Heath and E. Keppelmann, *Fibre techniques in Nielsen periodic point theory on nil and solvmanifolds I, II*, Topology Appl. **76** (1997), 217–247; **106** (2000), no. 2, 149–167.

[HePY] P. R. Heath, R. Piccinini and C. Y. You, *Nielsen type number for periodic points I*, Topological Fixed Point Theory and Its Applications, vol. 1411, LNM, Springer, Berlin, 1989, pp. 68–106.

[HeSY] P. R. Heath, H. Schirmer and C. You, *Nielsen type numbers for periodic points of pairs of spaces*, Topology Appl. **63** (1995), 117–138.

[HeY] P. R. Heath and C. You, *Nielsen type numbers for periodic points II*, Topology Appl. **43** (1992), 219–236.

[HeZh] P. R. Heath and Xuezhi Zhao, *Periodic points on the complement*, Topology Appl. **102** (2000), 253–277.

[Hei] H.-P. Heinz, *On the behaviour of measures of noncompactness with respect to differentiation and integration of vector-valued functions*, Nonlinear Anal. **7** (1983), no. 12, 1351–1371.

[Her] M. Hermes, *Calculus of set-valued functions and control*, J. Math. Mech. **18** (1968), 47–59.

[HL1] G. Herzog and R. Lemmert, *Ordinary differential equations in Fréchet spaces*, Proceed. of the Third Internat. (D. Bainov and V. Covachev, eds.), Colloq. on Diff. Eqns, held in Plovdiv, Bulgaria, August 1992, VSP, Zeist, 1993.

[HL2] _____, *On the structure of the solution set of $u'' = f(t, u)$, $u(0) = u(1) = 0$*, Math. Nachr. **215** (2000), 103–105.

[Heu] A. J. Heunis, *Continuous dependence of the solutions of an ordinary differential equation*, J. Diff. Eqns **54** (1984), 121–138.

[Hew] E. Hewitt, *Linear functionals on almost periodic functions*, Trans. Amer. Math. Soc. **74** (1953), 303–322.

[HiaUm] F. Hiai and H. Umegaki, *Integrals, conditional expectations, and martingales of multivalued functions*, J. Multivariate Anal. **4** (1977), 149–182.

[Hi1] C. J. Himmelberg, *Fixed points of compact multi-valued maps*, J. Math. Anal. Appl. **38** (1972), 205–209.

[Hi2] _____, *Measurable relations*, Fund. Math. **87** (1975), 59–71.

[HPV]   C. J. Himmelberg, T. Parthasarathy and F. S. van Vleck, *On measurable relations*, Fund. Math. **111** (1981), no. 2, 161–167.

[HPVV]   C. J. Himmelberg, K. Prikry and F. S. van Vleck, *The Hausdorff metric and measurable selections*, Topology Appl. **2** (1985), 121–133.

[HV1]   C. J. Himmelberg and F. van Vleck, *A note on the solution sets for differential inclusions*, Rocky Mountain J. Math. **12** (1982), 621–625.

[HV2]   ――――, *An extension of Brunovsky's Scorza-Dragoni type theorem for unbounded set-valued function*, Math. Slovaca **26** (1976), 47–52.

[HV3]   ――――, *Existence of solutions for generalized differential equations with unbounded right-hand side*, J. Diff. Eqns **61** (1986), no. 3, 295–320.

[HV4]   ――――, *On the topological triviality of solution sets*, Rocky. Mountain J. Math. **10** (1980), 247–252.

[Hof]   J. Hofbauer, *An index theorem for dissipative semiflows*, Rocky Mountain J. Math. **20** (1993), no. 4, 1017–1031.

[Hol]   P. Holm, *On the Bohr compactification*, Math. Annalen **156** (1964), 34–36.

[Hlm]   P. J. Holmes, *Knotted periodic orbits in suspensions of annulus maps*, Proc. R. Soc. London A **411** (1987), 351–378.

[HlmW]   P. J. Holmes and R. F. Williams, *Knotted periodic orbits in suspensions of Smale's horseshoe: torus knots abd bifurcation sequences*, Arch. Rat. Mechanics and Analysis **90** (1985), no. 2, 115–194.

[LVH1]   Le van Hot, *On the differentiability of multivalued mappings* I, Comm. Math. Univ. Carol. **22** (1981), 267–280.

[LVH2]   ――――, *On the differentiability of multivalued mappings* II, Comm. Math. Univ. Carol. **22** (1981), 337–350.

[Hov]   Ch. Hovarth, *Some results on multivalued mappings and inequalities without convexity*, Nonlinear and Convex Anal, Proc. Hon. Ky Fan Santa Barbara, Calif., June, 23–26, 1985; Basel, New York, N. Y., 1987, pp. 99–106.

[HLP1]   S. C. Hu, V. Lakshmikantham and N. S. Papageorgiou, *On the solution set of nonlinear evolution inclusions*, Dynamic Systems Appl. **1** (1992), 71–82.

[HLP2]   ――――, *On the properties of the solution set of semilinear evolution inclusions*, Nonlinear Anal. **24** (1995), 1683–1712.

[HP1]   S. Hu and N. S. Papageorgiou, *On the existence of periodic solutions for nonconvex valued differential inclusions in* $\mathbb{R}^n$, Proceed. Amer. Math. Soc. **123** (1995), 3043–3050.

[HP2]   ――――, *On the topological regularity of the solution set of differential inclusions with constraints*, J. Diff. Equat. **107** (1994), 280–289.

[HP3]   ――――, *Delay differential inclusions with constraints*, Proceed. Amer. Math. Soc. **123** (1995), 2141–2150.

[HuaJa]   Hai-Hua Huang and Boju Jiang, *Braids and periodic solutions*, Topological Fixed Point Theory and Applications, vol. 1411, Lecture Notes, Springer, Berlin, 1989, pp. 107–123.

[Hu1]   M. Hukuhara, *Sur les systémes des équations differentielles ordinaires*, Jap. J. Math. **5** (1928), 345–350.

[Hu2]   ――――, *Sur l'application semi-continue dont la veleur est un compact convex*, Funkcial. Ekvac. **10** (1967), 43–66.

[Hu3]   ――――, *Intégration des applications measurables dont la valeur est un compact convexe*, Funkcial. Ekvac. **10** (1967), 205–229.

[Hut]   J. E. Hutchinson, *Fractals and self similarity*, Indiana Univ. Math. J. **30** (1981), 713–747.

[Hy]   D. M. Hyman, *On decreasing sequence of compact absolute retracts*, Fund. Math. **64** (1959), 91–97.

[Ian1]   R. Iannacci, *Besicovitch spaces of almost periodic vector-valued functions and reflexivity*, Confer. Sem. Mat. Univ. Bari **251** (1993), 1–12.

[Ian2]   ――――, *About reflexivity of the* B_{ap}^q *spaces of almost periodic functions*, Rend. Mat. Appl. **13** (1993), 543–559.

[IbrGo] A. G. Ibrahim and A. M. Gomaa, *Topological properties of the solution sets of some differential inclusions*, Pure Math. Appl. **10** (1999), 197–223.

[IscYu] G. Isac and G. X.-Z. Yuan, *Essential components and connectedness of solution set for complementarity problems*, Fixed Point Theory and Applications, Nova Sci. Publ, Huntington, NY, 2000, pp. 35–46.

[Iv] O. V. Ivanov, *Some remarks on the Bohr compactification on the number line*, Ukrainian Math. J. **38** (1986), 154–158.

[Izo] N. A. Izobov, *The measure of the solution set of a linear system with the largest lower exponent*, Differentsial'nye Uravneniya **24** (1988), 2168–2170, 2207. (in Russian)

[Iz1] M. Izydorek, *On the Bourgin-Yang theorem for multi-valued maps in the non-symmetric case* II, Serdica **13** (1987), no. 4, 420–422.

[Iz2] _____, *Remarks on Borsuk-Ulam theorem for multi-valued maps*, Bull. Polish Acad. Sci. Math. **35** (1987), no. 7-8, 501–504.

[I] D. V. Izyumova, *Positive bounded solutions of second-order nonlinear ordinary differential equations*, Tbilis. Gos. Univ. Inst. Prikl. Mat. Trudy **22** (1987), 100–105. (in Russian)

[JGP] J. Jachymski, L. Gajek, and P. Pokarowski, *The Tarski-Kantorovitch principle and the theory of iterated function systems*, Bull. Austral. Math. Soc. **61** (2000), 247–261.

[JacKla] L. K. Jackson and G. Klaasen, *A variation of the topological method of Ważewski*, SIAM J. Appl. Math. **20** (1971), no. 1, 124–130.

[Jar] J. Jarník, *Multivalued mappings and Filippov's operation*, Czechoslov. Math. J. **31** (1981), 275–288.

[JK1] J. Jarník and J. Kurzweil, *On conditions on right hand sides of differential relations*, Časopis Pest. Math. **102** (1977), 334–349.

[JK2] _____, *Integral of multivalued mappings and its connection with differential relations*, Časopis Pest. Math. **108** (1983), 8–28.

[Ja1] J. Jaworowski, *Set-valued fixed point theorems of approximative retracts*, Lecture Notes in Math. **171** (1970), 34–39.

[Ja2] _____, *Some consequences of the Vietoris mapping theorem*, Fund. Math. **45** (1958), 261–272.

[Ja3] _____, *Continuous homology properties of approximative retracts*, Bull. Polish Acad. Sci. Math. **18** (1970), 359–362.

[JP] J. W. Jaworowski and M. Powers, *Λ-spaces and fixed point theorems*, Fund. Math. **64** (1969), 157–162.

[Je1] J. Jezierski, *An example of finitely-valued fixed point free map*, Zesz. Nauk. **6** (1987), University of Gdańsk, 87–93.

[Je2] _____, *The Nielsen relation for multivalued maps*, Serdica **13** (1987), 147–181.

[Je3] _____, *A modification of the relative Nielsen number of H. Schirmer*, Topology Appl. **62** (1995), 45–63.

[JeMa] J. Jezierski and W. Marzantowicz, *Homotopy minimal periods for nilmanifold maps*, Preprint (1999).

[Ji1] Boju Jiang, *Nielsen theory for periodic orbits and applications to dynamical systems*, Contemp. Math., vol. 152, Amer. Math. Soc., Providence, R. I., 1993, pp. 183–202.

[Ji2] _____, *Estimation of the number of periodic orbits*, Pacific J. Math. **172** (1996), 151–185.

[JiL] Boju Jiang and J. Llibre, *Minimal sets of periods for torus maps*, Discrete Cont. Dynam. Syst. **4** (1998), no. 2, 301–320.

[Joh1] R. A. Johnson, *On a Floquet theory for almost-periodic, two-dimensional linear systems*, J. Diff. Eqns **37** (1980), no. 2, 184–205.

[Joh2] _____, *A linear, almost-periodic equation with an almost automorphic solution*, Proceed. Amer. Math. Soc. **82** (1981), no. 2, 199–205.

[Joh3]   ———, *A review of recent work on almost periodic differential and difference operators*, Acta. Appl. Math. **1** (1983), 241–248.

[Joh4]   ———, *Un'equazione lineare quasi-periodica con una proprietà inconsueta*, Boll. U. M. I. **6** (1983), 115–121.

[Joh5]   ———, *Lyapunov numbers for the almost periodic Schrödinger equation*, Illinois J. Math. **28** (1984), no. 3, 397–419.

[Joh6]   ———, *The Oseledec and Sacker-Sell spectra for almost periodic linear systems: an example*, Proceed. Amer. Math. Soc. **99** (1987), no. 2, 261–267.

[Joh7]   ———, *m-functions and Floquet exponents for linear differential systems*, Ann. Mat. Pura Appl. **6** (1987), no. 147, 211–248.

[JohMo]   R. A. Johnson and J. Moser, *The rotation number for almost periodic potentials*, Comm. Math. Phys. **84** (1982), 403–438.

[JHG]   I. Joong Ha and E. G. Gilbert, *Robust tracking in nonlinear systems*, IEEE Trans. Autom. Contr. **AC-32** (1987), no. 9, 763–771.

[Jut]   L. Jüttner, *On derivo-periodic multifunctions*, Discuss. Math.-Diff. Incl. **21** (2001), 81–95.

[Kac]   T. Kaczyński, *Unbounded multivalued Nemytskii operators in Sobolev spaces and their applications to discontinuous nonlinearity*, Rocky Mountain J. Math. **22** (1992), 635–643.

[KaM]   T. Kaczyński and M. Mrozek, *Conley index for discrete multivalued dynamical systems*, Topology Appl. **65** (1995), 83–96.

[Kah]   J. P. Kahane, *Sur les fonctions presque-périodiques généralisées dont le spectre est vide*, Stud. Math. **21** (1962), 231–236.

[Kak]   S. Kakutani, *A generalization of Brouwer's fixed point theorem*, Duke Math. J. **8** (1941), 457–459.

[Kmn]   M. I. Kamenskiĭ, *On the Peano theorem in infinite dimensional spaces*, Mat. Zametki **11** (1972), no. 5, 569–576.

[KNOZ]   M. I. Kamenskiĭ, P. Nistri, V. V. Obukhovskiĭ and P. Zecca, *Optimal feedback control for a semilinear evolution equation*, J. Optim. Theory Appl. **82** (1994), no. 3, 503–517.

[KO]   M. I. Kamenskiĭ and V. V. Obukhovskiĭ, *Condensing multioperators and periodic solutions of parabolic functional — differential inclusions in Banach spaces*, Nonlinear Anal. TMA **20** (1991), 781–792.

[KOZ1]   M. I. Kamenskiĭ, V. V. Obukhovskiĭ and P. Zecca, *Method of the solution sets for a quasilinear functional-differential inclusion in a Banach space*, Differential Equations Dynam. Systems **4** (1996), 339–350.

[KOZ2]   ———, *On the translation multioperators along the solutions of semilinear differential inclusions in Banach spaces*, Canad. Appl. Math. Quart. **6** (1998), 139–155.

[Kmn]   T. Kaminogo, *Kneser's property and boundary value problems for some related functional differential equations*, Tōhoku Math. J. **30** (1978), 471–486.

[Kam]   J. Kampen, *On fixed points of maps and itereated maps and applications*, Nonlin. Anal. **42** (2000), no. 3, 509–532.

[KaPa]   D. A. Kandilakis and N. S. Papageorgiou, *On the properties of the Aumann integral with applications to differential inclusions and control systems*, Czechoslovak Math. J. **39** (1989), no. 1, 1–15.

[KOR]   R. Kannan and D. O'Regan, *A note on the solution set of integral inclusions*, J. Integral Equations Appl. **12** (2000), 85–94.

[KSS]   V. Kannan, P. V. S. P. Saradhi and S. P. Seshasai, *A generalization of Sharkovskiĭ theorem to higher dimensions*, J. Nat. Acad. Math. India **11** (1995), 69–82.

[KaTa]   Z. Kánnai and P. Tallos, *Stability of solution sets of differential inclusions*, Acta Sci. Math. (Szeged) **61** (1995), 197–207.

[KLY]   J. Kaplan, A. Lasota and J. A. Yorke, *An application of the Ważewski retract method to boundary value problems*, Zesz. Nauk. Uniw. Jagiel. **356** (1974), no. 16, 7–14.

[Kar]   A. Kari, *On Peano's theorem in locally convex spaces*, Studia Math. **73** (1982), no. 3, 213–223.

[Krp]   W. Karpińska, *A note on bounded solutions of second order differential equations at resonance*, Topol. Meth. Nonlin. Anal. **14** (1999), no. 2, 371–384.

[Ka1]   A. G. Kartsatos, *The Leray–Schauder theorem and the existence of solutions to boundary value problems on infinite intervals*, Indiana Un. Math. J. **23** (1974), no. 11, 1021–1029.

[Ka2]   ———, *A stability property of the solutions to a boundary value problem on an infinite interval*, Math. Jap. **19** (1974), 187–194.

[Ka3]   ———, *A boundary value problem on an infinite interval*, Proc. Edinb. Math. Soc. **2** (1974/75), no. 19, 245–252.

[Ka4]   ———, *The Hildebrandt–Graves Theorem and the existence of solutions of boundary value problems on infinite intervals*, Math. Nachr. **67** (1975), 91–100.

[Ka5]   ———, *Locally invertible operators and existence problems in differential systems*, Tôhoku Math. J. **28** (1976), 167–176.

[Ka6]   ———, *The existence of bounded solutions on the real line of perturbed nonlinear evolution equations in general Banach spaces*, Nonlin. Anal. TMA **17** (1991), no. 11, 1085–1092.

[Ke1]   M. Kečkemétyová, *On the existence of a solution for nonlinear operator equations in Fréchet spaces*, Math. Slovaca **42** (1992), no. 1, 43–54.

[Ke2]   ———, *Continuous solutions of nonlinear boundary value problems for ODEs on unbounded intervals*, Math. Slovaca **42** (1992), no. 3, 279–297.

[Ke3]   ———, *On the structure of the set of solutions of nonlinear boundary value problems for ODE's on unbounded domains*, Preprint (2000).

[Kel]   W. G. Kelley, *A Kneser theorem for Volterra integral equations*, Proc. Amer. Math. Soc. **40** (1973), no. 1, 183–190.

[Kep]   E. Keppelmann, *Periodic points on nilmanifolds and solvmanifolds*, Pacific J. Math. **164** (1994), 105–128.

[Ki]   I. T. Kiguradze, *On bounded and periodic solutions of linear higher order differential equations*, Mat. Zametki **37** (1985), no. 1, 46–62. (in Russian)

[KR]   I. T. Kiguradze and I. Rachůnková, *On a certain non-linear problem for two-dimensional differential systems*, Arch. Math. Brno **16** (1980), 15–38.

[Ksl1]   M. Kisielewicz, *Continuous dependence of solution sets for generalized differential cquations of neutral type*, Atti Accad. Sci. Istit. Bologna Cl. Sci. Fis. Rend. **13** (1980/81), no. 8, 191–195.

[Ksl2]   ———, *Compactness and upper semicontinuity of solution set uf generalized differential equation in a separable Banach space*, Demonstratio Math. **15** (1982), 753–761.

[Ksl3]   ———, *Multivalued differential equations in separable Banach spaces*, J. Optim. Th. Appl. **37** (1982), no. 2, 231–249.

[Ksl4]   ———, *Properties of solution set of stochastic inclusions*, J. Appl. Math. Stochastic Anal. **6** (1993), 217–235.

[Ksl5]   ———, *Quasi-retractive representation of solution sets to stochastic inclusions*, J. Appl. Math. Stochastic Anal. **10** (1997), 227–238.

[KlbFi]   B. S. Klebanov and V. V. Filippov, *On the acyclicity of the solution set of the Cauchy problem for differential equations*, Mat. Zametki **62** (1997). (in Russian)

[Kle]   V. Klee, *Leray–Shauder theory without local convexity*, Math. Annalen **141** (1960), 286–296.

[Klo]   P. E. Kloeden, *On Sharkovskii's cycle coexisting ordering*, Bull. Austral. Math. Soc. **20** (1979), 171–177.

[Kne1]   H. Kneser, *Untersuchung und asymptotische Darstellung der Integrale gewisser Differentialgleichungen bei grossen Werthen des Arguments*, J. Reine Angew. Math **1** (1896), no. 116, 178–212.

[Kne2]   _____, *Über die Lösungen eine Systeme gewöhnlicher differential Gleichungen, das der lipschitzchen Bedingung nicht genügt*, S. B. Preuss. Akad. Wiss. Phys. Math. Kl. **4** (1923), 171–174.

[Koi]   S. Koizumi, *Hilbert transform in the Stepanoff space*, Proc. Japan Acad. **38** (1962), 735–740.

[Kol]   B. Kolev, *Periodic points of period 3 in the disc.*, Nonlinearity **7** (1994), 1067–1072.

[Kor]   P. Korman, *The global solution set for a class of semilinear problems*, J. Math. Anal. Appl. **226** (1998), 101–120.

[Kos]   T. V. Kostova, *Estimates of the real parts of eigenvalues of complex matrices*, Dokl. Bolg. Akad. Nauk. (C. R. Acad. Bulgare Sci.) **38** (1985), 15–18.

[Kov1]   A. S. Kovanko, *Sur la compacité des systèmes de fonctions presque périodiques généralisées de H. Weyl*, C. R. (Doklady) Ac. Sc. URSS **43** (1944), 275–276.

[Kov2]   _____, *On a certain property and a new definition of generalized almost periodic functions of A. S. Besicovitch*, Ukrainian Math. J. **8** (1956), 273–288. (in Russian)

[Kov3]   _____, *On compactness of systems of Levitan's almost-periodic functions*, Utchen. Zap. Univ. Lwow **29** (1954), 45–49. (in Russian)

[Kov4]   _____, *Sur les classes de fonctions presque-périodiques généralisées*, Ann. Mat. **9** (1931), 1–24.

[Kov5]   _____, *Sur une propriété périodique des fonctions $\widetilde{B}$ presque-périodiques*, Ann. Pol. Math. **VIII** (1960), 271–275.

[Kov6]   _____, *On compactness of systems of generalized almost-periodic functions of Weyl*, Ukrainian Math. J. **5** (1953), 185–195. (in Russian)

[Kov7]   _____, *Sur les systèmes compacts de fonctions présque périodiques généralisées de W. Stepanoff*, Rec. Math. (Mat. Sbornik) N. S. **9** (1941), 389–401.

[Kov8]   _____, *Sur la correspondence entre les classes diverses de fonctions presque-périodiques généralisées*, Bull. (Izvestiya) Math. Mech. Inst. Univ. Tomsk **3** (1946), 1–36. (in Russian)

[Kov9]   _____, *A certain relation of two classes of generalized almost periodic functions*, Visnik Lviv Derz. Univ. Ser. Meh.-Mat. **6** (1971), 3–4. (in Ukrainian)

[Kov10]   _____, *Sur les systèmes compacts de fonctions presque périodiques généralisées de W. Stepanoff*, C. R. (Doklady) Ac. Sc. URSS **26** (1940), 211–213.

[Kov11]   _____, *Sur la structure de fonctions presque périodiques généralisées*, C. R. Acad. Sc. Paris **198** (1934), 792–794.

[Kov12]   _____, *Sur la compacité des systèmes de fonctions presque périodiques généralisées de A. Besicovitch*, Rec. Math. **16** (1945), 365–382.

[Kov13]   _____, *On convergence of sequences of functions in the sense of Weyl's metric D_{W_ω}*, Ukrainian Math. J. **3** (1951), 465–476. (in Russian)

[Kov14]   _____, *A criterion for compactness of systems $\widetilde{B}$ and B_p almost periodic functions*, Izv. Akad. Nauk Azerbajdzan. SSR, Baku (1961), 143–145. (in Russian)

[Kov15]   _____, *Sur l'approximation des fonctions presque périodiques généralisées*, C. R. Acad. Sci. Paris **188** (1929), 142–143.

[Kov16]   _____, *Sur une généralisation des fonctions presque-périodiques*, C. R. Acad. Sci. Paris **186** (1928), 354–355.

[Kov17]   _____, *Sur quelques généralisations des fonctions presque-périodiques*, C. R. Acad. Sci. Paris **186** (1928), 729–730.

[Kov18]   _____, *Sur une classe de fonctions presque-périodiques qui engendre les classes de fonctions p.p. de W. Stepanoff, H. Weyl et A. Besicovitch*, C. R. Acad. Sci. Paris **189** (1929), 393–394.

[Kov19]   _____, *Sur quelques espaces de fonctions presque-périodiques généralisées*, Bull. (Izvestija) Acad. Sci. SSSR **1** (1935), 75–96.

[Kra1]   B. Krajc, *A note on existence of bounded solutions of an n-th order ODE on the real line*, Acta Univ. Palacki. Olomuc., Math. **37** (1998), 57–67.

[Kra2]   _____, *Bounded solutions in a given set of Carathéodory differential systems*, Trans. VSB-Techn. Univ. Ostrava **1** (2001), 97–106.

[KMP]   M. A. Krasnosel'skiĭ, J. Mawhin and A. V. Pokrovskiĭ, *New theorems on forced periodic oscillations and bounded solutions*, Doklady AN SSSR **321** (1991), no. 3, 491–495. (in Russian)

[KKM]   A. M. Krasnosel'skiĭ, M. A. Krasnosel'skiĭ and J. Mawhin, *On some conditions of forced periodic oscillations*, Differential Integral Equations **5(6)** (1992), 1267–1273.

[KMKP]   A. M. Krasnosel'skiĭ, J. Mawhin, M. A. Krasnosel'skiĭ and A. V. Pokrovskiĭ, *Generalized guiding functions in a problem of high frequency forced oscillations*, Nonlin. Anal. **22** (1994), 1357–1371.

[KP]   M. A. Krasnosel'skiĭ and A. I. Perov, *On existence of solutions of some nonlinear functional equations*, Dokl. Acad. Nauk SSSR **126** (1959), 15–18. (in Russian)

[KSZ]   M. A. Krasnosel'skiĭ, V. V. Strygin and P. P. Zabreiko, *Invariance of rotation principles*, Izv. Vyssh. Uchebn. Zaved. Mat. **120** (1972), no. 5, 51–57. (in Russian)

[KrKu]   P. Krbec and J. Kurzweil, *Kneser's theorem for multivalued, differential delay equations*, Časopis pro Pest. Mat. **104** (1979), no. 1, 1–8.

[Kry1]   W. Kryszewski, *Homotopy invariants for set-valued maps homotopy - approximation approach, Fixed point theory and applications*, Pitman Res. Notes Math. Ser. **252** (1991), 269–284.

[Kry2]   _____, *Some homotopy classification and extension theorems for the class of compositions of acyclic set-valued maps*, Bull. Soc. Math. France **119** (1995), 21–48.

[Kry3]   _____, *The fixed-point index for the class of compositions of acyclic set-valued maps of ANR-s*, Bull. Soc. Math. France **120** (1996), 129–151.

[Kry4]   _____, *Graph approximation of set-valued maps on noncompact spaces*, Topology Appl. **83** (1997), 1–21.

[Kry5]   _____, *Remarks to the Vietoris theorem*, Topol. Methods Nonlinear Anal. **8** (1997), 383–405.

[Kry6]   _____, *The Lefschetz type theorem for a class of noncompact maps*, Rend. Circ. Mat. Palermo, Suppl. **14** (1987), 365–384.

[Kry7]   _____, *An application of A-mapping theory to boundary value problems for ordinary differential equations*, Nonlin. Anal. TMA **15** (1990), no. 8, 697–717.

[Kry8]   _____, *Graph-approximation of set-valued maps. A survey.*, L. N. Nonlin. Anal. **2** (1998), N. Copernicus Univ., Toruń, 223–235.

[KM]   W. Kryszewski and D. Miklaszewski, *The Nielsen number of set-valued maps. An approximation approach*, Serdica **15** (1989), no. 4, 336–344.

[Kub1]   Z. Kubáček, *A generalization of N. Aronszajn's theorem on connectedness of the fixed point set of a compact mapping*, Czech. Math. J. **37** (1987), no. 112, 415–423.

[Kub2]   _____, *On the structure of fixed point sets of some compact maps in the Fréchet space*, Math. Bohemica **118** (1993), no. 4, 343–358.

[Kub3]   _____, *On the structure of the solution set of a functional differential system on an unbounded interval*, Arch. Math. (Brno) **35** (1999), 215–228.

[Kubi1]   I. Kubiaczyk, *Structure of the sets of weak solutions of an ordinary differential equation in a Banach space*, Ann. Polon. Math. **44** (1980), no. 1, 67–72.

[Kubi2]   _____ I. Kubiaczyk, *Kneser's theorem for differential equations in Banach spaces*, J. Diff. Eqns **45** (1982), no. 2, 139–147.

[KbSz]   I. Kubiaczyk and S. Szufla, *Kneser's theorem for weak solutions of ordinary differential equations in Banach spaces*, Publ. Inst. Math. (Beograd) (NS) **32** (1982), no. 46, 99–103.

[Ku1]   Z. Kucharski, *A coincidence index*, Bull. Polish Acad. Sci. Math. **24** (1976), 245–252.

[Ku2]   _____, *Two consequences of the coincidence index*, Bull. Polish Acad. Sci. Math. **24** (1976), 437–444.

[Kuc1]   A. Kucia, *Extending Carathéodory functions*, Bull. Polish Acad. Sci. Math. **36** (1988), 593–601.

[Kuc2]   _____, *On the existence of Carathéodory selectors*, Bull. Polish Acad. Sci. Math. **32** (1984), 233–241.

[Kuc3]   _____, *Scorza–Dragoni type theorems*, Fund. Math. **138** (1991), 197–203.

[KN]   A. Kucia and A. Nowak, *Carathéodory type selectors in a Hilbert space*, Ann. Math. Sil. **14** (1986), 47–52.

[KRN]   K. Kuratowski and C. Ryll–Nardzewski, *A general theorem on selectors*, Bull. Polish Acad. Sci. Math. **13** (1965), 397–403.

[LadLak]   G. S. Ladde and V. Lakshmikantham, *On flow-invariant sets*, J. Math. Anal. Appl. **128** (1987), 176–188.

[LakNo]   A. V. Lakeev and S. I. Noskov, *Description of the solution set of a linear equation with an interval-defined operator and right-hand side*, Dokl. Akad. Nauk **330** (1993), 430–433. (in Russian)

[La1]   A. Lasota, *On the existence and uniqueness of solutions of a multipoint boundary value problem*, Annal. Polon. Math. **38** (1980), 205–310.

[La2]   _____, *Applications of generalized functions to contingent equations and control theory*, Inst. Dynamics Appl. Math. Lecture, vol. 51, Univ. of Maryland, 1970-1971, pp. 41–52.

[LaMy1]   A. Lasota and J. Myjak, *Semifractals*, Bull. Pol. Ac. Math. **44** (1996), 5–21.

[LaMy2]   _____, *Markov operators and fractals*, Bull. Pol. Ac. Math. **45** (1997), 197–210.

[LaMy3]   _____, *Semifractals on Polish Spaces*, Bull. Pol. Ac. Math. **46** (1998), 179–196.

[LaMy4]   _____, *Fractals, semifractals and Markov operators*, Int. J. Bifurcation Chaos Appl. Sci. Eng. **9** (1999), 307–325.

[LaMy5]   _____, *Attractors of multifunctions*, Bull. Pol. Ac. Math. **48** (2000), 319–334.

[LaO1]   A. Lasota and Z. Opial, *An application of the Kakutani–Ky Fan theorem in the theory of ordinary differential equations*, Bull. Polish Acad. Sci. Math. **13** (1965), 781–786.

[LaO2]   _____, *Fixed point theorem for multivalued mappings and optimal control problems*, Bull. Polish Acad. Sci. Math. **16** (1968), 645–649.

[LaY]   A. Lasota and J. A. Yorke, *The generic property of existence of solutions of differential equations in Banach spaces*, J. Diff. Eqns **13** (1973), 1–12.

[LaR1]   J.-M. Lasry and R. Robert, *Degré pour les fonctions multivoques et applications*, C. R. Acad. Sci. Paris Sér. I Math. **280** (1975), 1435–1438.

[LaR2]   _____, *Acyclicité de l'ensemble des solutions de certaines équations fonctionnelles*, C. R. Acad. Sci. Paris, A-B **282** (1976), 1283–1286.

[LaR3]   _____, *Degré topologique pour certains couples de fonctions et applications aux équations differetielles multivoques*, C. R. Acad. Sci. Ser Paris Ser. I **289** (1976), 163–166.

[Las]   M. Lassonde, *On the use of KKM multifunctions in fixed point theory and related topics*, J. Math. Anal. Appl. **97** (1983), no. 1, 151–201.

[Lau]   K. S. Lau, *On the Banach spaces of functions with bounded upper means*, Pac. J. Math. **91** (1980), 153–172.

[LVC]   G. T. LaVarnway and R. Cooke, *A characterization of Fourier series of Stepanov almost-periodic functions*, J. Fourier Anal. Appl. **7** (2001), 127–142.

[LegWi]   R. W. Leggett and L. R. Williams, *Multiple positive fixed points on nonlinear operators on ordered Banach spaces*, Indiana Univ. Math. J. **28** (1979), 673–689.

[Leo]   G. A. Leonov, *On a problem of Barbashin*, Vestnik Leningrad Univ. Math. **13** (1981), 293–297.

[Ler]   J. Leray, *Théorie des pointes fixés: indice total et nombre de Lefschetz*, Bull. Soc. Math. France **87** (1959), 221–233.

[LS]   J. Leray and J. P. Schauder, *Topologie et équations functionnalles*, Ann. Sci. École Norm. Sup. **65** (1934), 45–78.

[Les]   K. Leśniak, *Extremal sets as fractals*, Nonlin. Anal. Forum (to appear).

[Lvi1]   M. Levi, *Beating modes in the Josephson junction*, Chaos in Nonlinear Dynamical Systems (J. Chandra, ed.), SIAM, Philadelphia, 1984, pp. 56–73.

[Lvi2]   _____, *Caterpillar solutions in coupled pendula*, Ergod. Th. and Dynam. Sys. **8** (1988), 153–174.

[Lvi3]   _____, *KAM theory for particles in periodic potentials*, Ergod. Th. and Dynam. Sys. **10** (1990), 777–785.

[Lvi4]   _____, *Quasiperiodic motions in superquadratic time-periodic potentials*, Comm. Math. Phys. **143** (1991), 43–83.

[Lev]   N. Levinson, *Transformation theory of nonlinear differential equations of the second order*, Ann. of Math. **45** (1944), 723–737.

[Lew]   M. Lewicka, *Locally lipschitzian guiding function method for ODE's*, Nonlin. Anal. **33** (1998), 747–758.

[Li]   Desheng Li, *Peano's theorems for implicit differential equations*, J. Math. Anal. Appl. **258** (2001), 591–616.

[LiYo]   T.-Y. Li and J. A. Yorke, *Period three implies chaos*, Amer. Math. Monthly **82** (1975), 985–992.

[Lia]   Z. C. Liang, *Limit boundary value problems for nonlinear differential equations of the second order*, Acta Math. Sinica N. S. **1** (1985), no. 2, 119–125.

[Lil]   J. C. Lillo, *On almost periodic solutions of differential equations*, Ann. Math. **69** (1959), no. 2, 467–485.

[Lim]   T. C. Lim, *On fixed point stability for set valued contractive mappings with applications to generalized differential equations*, J. Math. Anal. Appl. **110** (1985), 436–441.

[LlMcK]   J. Llibre and R. S. MacKay, *Pseudo-Anosov homeomorphisms on a sphere with four puncteres have all periods*, Math. Proc. Cambridge Phil. Soc. **112** (1992), 539–550.

[L]   S. G. Lobanov, *Peano's theorem is invalid for any infinite-dimensional Fréchet space*, Mat. Sbornik **184** (1993), no. 2, 83–86. (in Russian)

[Lo]   D. L. Lovelady, *Bounded solutions of whole-line differential equations*, Bull. AMS **79** (1972), 752–753.

[Luo]   D. Luo, *On periodic orbits of the second kind of toral differential equations*, J. Nanjing Univ. Math. Biquart **1** (1984), no. 2, 137–144.

[MNZ1]   J. W. Macki, P. Nistri and P. Zecca, *Corrigendum: A tracking problem for uncertain vector systems*, Nonlin. Anal. **20** (1993), 191–192.

[MNZ2]   _____, *The existence of periodic solutions to nonautonomus differential inclusions*, Proc. Amer. Math. Soc. **104** (1988), no. 3, 840–844.

[MNZ3]   _____, *A tracking problem for uncertain vector systems*, Nonlin. Anal. **14** (1990), 319–328.

[Mak1]   V. P. Maksimov, *On the parametrization of the solution set of a functional-differential equation*, Funct. Differ. Eqns (1988), Perm. Politekh. Inst., Perm, 14–21. (in Russian)

[Mak2]   _____, *On the parametrization of the solution set of a functional-differential equation*, Funct. Differ. Eqns **3** (1996), 371–378.

[Mal]   L. Malaguti, *Bounded solutions for a class of second order nonlinear differential equations*, Diff. Eqns Dynam. Syst. **3** (1995), 175–188.

[MalMa]   L. Malaguti and C. Marcelli, *Existence of bounded trajectories via upper and lower solutions*, Discr. Contin. Dynam Syst. **6** (2000), no. 3, 575–590.

[Mam]   A. Mambriani, *Su un teoreme relativo alle equazioni differenziali ordinarie del 2^0 ordine*, Atti Accad. Naz. Lincei, Cl. Sci. Fis. Mat. Nat. **9** (1929), 620–622.

[Mmn]   S. Mamonov, *Limit cycles of the second kind to a system of ordinary differential equations*, Diff. Urav. (Rjazan) (1986), 46–51. (in Russian)

[MmrSe]   J. Mamrilla and S. Sedziwy, *The existence of periodic solutions of a certain dynamical system in a cylindrical space*, Boll. U. M. I. **4** (1971), no. 4, 119–122.

[Mar1]   S. A. Marano, *Classical solutions of partial differential inclusions in Banach spaces*, Appl. Anal. **42** (1991), 127–143.

[Mar2]   _____, *Fixed point of multivalued contractions with nonclosed nonconvex values*, Rend. Accad. Naz. Lincei **9** (1994), 203–212.

[Mar3]   _____, *Generalized solutions of partial differential inclusions depending on a parameter*, Rend. Accad. Naz. Sci. XL **13** (1989), 281–295.

[Mar4]   _____, *On a boundary value problem for the differential equation* $f(t, x, x', x'') = 0$, J. Math. Anal. Appl. **182** (1994), 309–319.

[MS]   S. Marano and V. Staicu, *On the set of solutions to a class of nonconvex non-closed differential inclusions*, Acta Math. Hungarica **76** (1997), 287–301.

[Mch]   M. V. Marchi, *Self-similarity in spaces of homogeneous type*, Adv. Math. Sci. Appl. **9** (1999), 851–870.

[Mrc]   J. Marcinkiewicz, *Une remarque sur les espaces de M. Besikowitch*, C. R. Acad. Sc. Paris **208** (1939), 157–159.

[MrsMi]   M. Marcus and V. J. Mizel, *Absolute continuity on tracks and mappings of Sobolev spaces*, Arch. Rational Mech. Anal. **45** (1972), no. 21, 294–320.

[MN]   A. Margheri and P. Nistri, *An existence result for a class of asymptotic boundary value problem*, Diff. Integral Eqns. **6** (1993), no. 6, 1337–1347.

[MZ1]   A. Margheri and P. Zecca, *Solution sets and boundary value problems in Banach spaces*, Topol. Methods Nonlinear Anal. **2** (1993), 179–188.

[MZ2]   _____, *Solution sets of multivalued Sturm–Liouville problems in Banach spaces*, Atti Accad. Naz. Lincei Cl. Sci. Fis. Mat. Natur. Rend. Lincei, 9 Mat. Appl. **5** (1994), 161–166.

[MZ3]   _____, *A note on the topological structure of solution sets of Sturm–Liouville problems in Banach spaces*, Atti Accad. Naz. Lincei, Rend. Cl. Sci. Fis. Math. Nat. (to appear).

[Mrk]   J. T. Markin, *Stability of solution sets for generalized differential equations*, J. Math. Anal. Appl. **46** (1974), 289–291.

[Mrt]   M. Martelli, *Continuation principles and boundary value problems*, Topological Methods for Ordinary Differential Equations (M. Furi and P. Zecca, eds.), vol. 1537, LNM, Springer, Berlin, 1991, pp. 32–73.

[MaVi1]   M. Martelli and A. Vignoli, *On differentiability of multi-valued maps*, Boll. U.M.I. **4** (1974), 701–712.

[MaVi2]   _____, *On the structure of the solution set of nonlinear equations*, Nonlinear Anal. **7** (1983), 685–693.

[MarSa1]   A. Martelloti and A. R. Sambucini, *Multivalued integration: the weakly compact case*, Preprint of Univ. Perugia (2000).

[MarSa2]   _____, *On the comparison between Aumann and Bochner integrals*, J. Math. Anal. Appl. **260** (2001), no. 1, 6–17.

[MrzPr]   T. Marzantowicz and P. M. Przygodzki, *Finding periodic points of a map by use of a k-adic expansion*, Discrete Cont. Dynam. Syst. **5** (1999), no. 3, 495–514.

[MC]   A. Mas-Colell, *A note on a theorem of F. Browder*, Math. Programming **6** (1974), 229–233.

[MasNi]   I. Massabò and P. Nistri, *A topological degree for multivalued A-proper maps in Banach spaces*, Boll. Un. Mat. Ital. B **13** (1976), 672–685.

[MP]   I. Massabò and J. Pejsachowicz, *On the connectivity properties of the solution set of parametrized families of compact vector fields*, J. Funct. Anal. **59** (1984), 151–166.

[Mat1]   T. Matsuoka, *The number and linking of periodic solutions of periodic systems*, Invent. Math. **70** (1983), 319–340.

[Mat2]   _____, *Waveform in the dynamical study of ordinary differential equations*, Japan J. Appl. Math. **1** (1984), 417–434.

[Mat3]   _____, *The number and linking of periodic solutions of non-dissipative systems*, J. Diff. Eqns **76** (1988), 190–201.

[Mat4]   _____, *The number of periodic points of smooth maps*, Ergod. Th. Dynam. Sys. **8** (1989), 153–163.

[Mat5]   _____, *Periodic points of disk homomorphism having a pseudo-Anosov component*, Hokkaido Math. J. **27** (1998), no. 2, 423–455.

[Mat6]   _____, *Braids of periodic points and a 2-dimensional analogue of Sharkovskii's ordering*, Dynamical Systems and Nonlinear Oscillations (G. Ikegami, ed.), World Sci. Press, Singapore, 1986, pp. 58–72.

[MatShi]   T. Matsuoka and H. Shiraki, *Smooth maps with finitely many periodic points*, Mem. Fac. Sci. Kochi Univ., Ser. A Math. **11** (1990), 1–6.

[Maw1]   J. Mawhin, *From Tricomi's equation for synchronous motors to the periodically forced pendulum*, In Tricomi's Ideas and Contemporary Applied Mathematics, vol. 147, Atti Conv. Lincei, Accad. Naz. Lincei (Roma), 1998, pp. 251–269.

[Maw2]   _____, *Equivalence theorems for nonlinear operator equations and coincidence degree theory for some mappings in locally convex topological vector spaces*, J. Differential Equations **12** (1972), 610–636.

[Maw3]   _____, *Topological degree and boundary value problems for nonlinear differential equations*, Topological Methods for Ordinary Differential Equations (M. Furi and P. Zecca, eds.), vol. 1537, LNM, Springer, Berlin, 1991, pp. 74–142.

[Maw4]   _____, *Leray–Schauder degree: A half century of extensions and applications*, Topol. Methods Nonlinear Anal. **14** (1999), 195–228.

[Maw5]   _____, *Bound sets and Floquet boundary value problems for nonlinear differential equations*, Jagell. Univ. Acta Math. **36** (1998), 41–53.

[Maw6]   _____, *Bounded and almost periodic solutions of nonlinear differential equations: Variational vs. non-variational approach*, Calculus of Variations and Differential Equations (A. Ioffe, S. Reich and I. Shafrir, eds.), vol. 410, Chapman & Hall/CRC, Boca Raton, 1999.

[Maw7]   _____, *Remarques sur les solutions bornées on presque périodiques de l'equation du pendule forcé*, Ann. Sci. Math. Québec **22** (1998), no. 2, 213–224.

[Maw8]   _____, *Bounded solutions of nonlinear ordinary differential equations*, CISM Cources and Lectures **371** (1996), Springer, Vienna, 121–147.

[MW1]   J. Mawhin and J. R. Ward Jr., *Bounded solutions of some second order nonlinear differential equations*, J. London Math. Soc. **13** (1998), 733–747.

[MW2]   _____, *Guiding-like functions for periodic or bounded solutions of ordinary differential equations* (to appear).

[McCl1]   J. F. McClendon, *Minimax theorem for ANR's*, Proc. Amer. Math. Soc. **35** (1984), 244–250.

[McCl2]   _____, *Subopen multifunctions and selections*, Fund. Math. **121** (1984), 25–30.

[McCo1]   C. K. McCord, *Computing Nielsen numbers*, Nielsen Theory and Dynamical Systems, Contemp. Math. (C. K. McCord, ed.), vol. 152, Amer. Math. Soc., Providence, R. I., 1993, pp. 249–267.

[McCo2]   _____, *The three faces of Nielsen: Coincidences, intersections and preimages*, Topol. Appl. **103** (2000), 155–177.

[McCoMiMr]   C. K. McCord, K. Mischaikow and M. Mrozek, *Zeta functions, periodic trajectories and the Conley index*, J. Diff. Eqns **121** (1995), no. 2, 258–292.

[McL]   A. McLennan, *Fixed points of contractible valued correspondences*, Internat. J. Game Theory **18** (1989), 175–184.

[MTY]   G. Mehta, K.-K. Tan and X.-Z. Yuan, *Fixed points, maximal elements and equilibria of generalized games*, Nonlinear Anal. **28** (1997), 689–699.

[MW]   M. Memory and J. R. Ward, Jr., *Conley index and the method of averaging*, J. Math. Anal. Appl. **158** (1991), 509–518.

[Mey]   P. Meystre, *Free-electron lasers: An introduction*, Laser Physics (D. F. Walls and J. D. Harvey, eds.), Academic Press, New York, 1980.

[Mi1]   E. A. Michael, *Continuous selections* I, Ann. Math. **63** (1956), no. 2, 361–381.

[Mi2]   _____, *A generalization of a theorem on continuous selections*, Proc. Amer. Math. Soc. **105** (1989), no. 1, 236–243.

[Mi3]   _____, *Continuous selections: A guide for avoiding obstacles*, Gen. Topol. and Relat. Mod. Anal. and Algebra, vol. 6, Proc. the Prague Topol. Symp., Aug. 25 - 29, 1986; Berlin, 1988, pp. 344–349.

[Mik]   D. Miklaszewski, *The two-point problem for nonlinear ordinary differential equations and differential inclusions*, Univ. Iagell Acta Math. **36** (1998), 127–132.

[Mil]   V. Millionshchikov, *Proof of the existence of non-irreducible systems of linear differential equations with almost periodic coefficients*, J. Diff. Eqns **4** (1968), 203–205.

[Mlj1]   P. S. Milojević, *On the index and the covering dimension of the solution set of semilinear equations*, Nonlinear Functional Analysis and its Applications, Part 2, vol. 1986, Amer. Math. Soc., Providence, R. I., pp. 183–205.

[Mlj2]   _____, *On the dimension and the index of the solution set of nonlinear equations*, Trans. Amer. Math. Soc. **347** (1995), 835–856.

[Mir]   R. E. Mirollo, *Splay-phase orbits for equivariant flows on tori*, SIAM J. Math. Anal. **25** (1994), no. 4, 1176–1180.

[Mos]   J. K. Moser, *Quasi-periodic solutions of nonlinear elliptic partial differential equations*, Bol. Soc. Bras. Mat. **20** (1989), no. 1, 29–45.

[Moe1]   H. Mönch, *Boundary value problems for nonlinear ordinary differential equations in Banach spaces*, Nonlinear Anal. **4** (1980), 985–999.

[MoeHa]   H. Mönch and G.-F. von Harten, *On the Cauchy problem for ordinary differential equations in Banach spaces*, Arch. Math. (Basel) **39** ((1982)), 153–160.

[Mr1]   M. Mrozek, *A cohomological index of Conley type for multivalued admissible flows*, J. Diff. Eqns **84** (1990), 15–51.

[Mr2]   _____, *Leray functor and cohomological Conley index for discrete dynamical systems*, Trans. Amer. Math. Soc. **318** (1990), 149–178.

[Muc]   C. F. Muckenhoupt, *Almost-periodic functions and vibrating systems*, J. Math. Phys. **8** (1928/1929), 163–199.

[Muh]   A. M. Muhsinov, *On differential inclusions in a Banach space*, Soviet Math. Dokl. **15** (1974), 1122–1125.

[My]   J. Myjak, *A remark on Scorza–Dragoni theorem for differential inclusions*, Čas. Pěstování Mat. **114** (1989), 294–298.

[Nag]   M. Nagumo, *Degree of mappings in convex, linear topological spaces*, Amer. J. Math. **73** (1951), 497–511.

[Nas]   O. Naselli, *On the solution set of an equation of the type* $f(t, \Phi(u)(t)) = 0$, Set-Valued Anal. **4** (1996), 399–405.

[Naz]   E. A. Nazarov, *Conditions for the existence of periodic trajectories in dynamical systems with the cylindrical phase space*, Diff. Urav. **6** (1970), 377–380. (in Russian)

[Neu]   J. von Neumann, *A model of general economic equilibrium*, Collected Works, vol. VI, Pergamon Press, Oxford, 1963, pp. 29–37.

[Nie1]   J. J. Nieto, *On the structure of the solution set for first order differential equations*, Appl. Math. Comput. **16** (1985), 177–187.

[Nie2]   _____, *Periodic solutions of nonlinear parabolic equations*, J. Diff. Eqns **60** (1985), no. 1, 90–102.

[Nie3]   _____, *Nonuniqueness of solutions of semilinear elliptic equations at resonance*, Boll. Un. Mat. Ital. **6, 5-A** (1986), no. 2, 205–210.

[Nie4]   _____, *Hukuhara–Kneser property for a nonlinear Dirichlet problem*, J. Math. Anal. Appl. **128** (1987), 57–63.

[Nie5]   _____, *Structure of the solution set for semilinear elliptic equations*, Colloq. Math. Soc. Janos Bolyai **47** (1987), 799–807.

[Nie6]   _____, *Aronszajn's theorem for some nonlinear Dirichlet problem*, Proc. Edinburg Math. Soc. **31** (1988), 345–351.

[Nie7] _____, *Decreasing sequences of compact absolute retracts and nonlinear problems*, Boll. Un. Mat. Ital. **2-B** (1988), no. 7, 497–507.

[Nie8] _____, *Nonlinear second order periodic value problems with Carathéodory functions*, Appl. Anal. **34** (1989), 111–128.

[Nie9] _____, *Periodic Neumann boundary value problem for nonlinear parabolic equations and application to an elliptic equation*, Ann. Polon. Math. **54** (1991), no. 2, 111–116.

[Nis] P. Nistri, *Optimal control problems via a direct method*, Ann. Mat. Pura Appl. **69** (1991), 295–314.

[NOZ1] _____, *Viability for feedback control systems in Banach spaces via Carathéodory closed-loop controls*, Differential Equations Dynam. Systems **4** (1996), 367–378.

[NOZ2] _____, *On the solvability of systems inclusions involving noncompact operators*, Trans. Amer. Math. Soc. **342** (1994), no. 2, 543–562.

[NOW] B. Norton-Odenthal and P. Wong, *On the computation of the relative Nielsen number*, Topology Appl. **56** (1994), 141–157.

[Nu1] R. D. Nussbaum, *Generalizing the fixed point index*, Math. Ann. **228** (1979), 259–278.

[Nu2] _____, *A geometric approach to the fixed point index*, Pac. J. Mat. **39** (1971), 751–766.

[Nu3] _____, *The fixed-point index for local condensing maps*, Ann. Mat. Pura Appl. **89** (1971), 217–258.

[Nu4] _____, *Some fixed-point theorems*, Bull. Amer. Math. Soc. **77** (1971), 360–365.

[Nu5] _____, *Degree theory for local condensing maps*, J. Math. Anal. Appl. **37** (1972), 741–766.

[Nu6] _____, *The fixed-point index and fixed-point theorems*, Topological Methods for Ordinary Differential Equations (M. Furi and P. Zecca, eds.), vol. 1537, LNM, Springer, Berlin, pp. 149–205.

[Ob1] V. V. Obukhovskiĭ, *On evolution differential inclusions in a Banach space*, IV Conference on Differential Equations and Applications – Russe, 1989.

[Ob2] _____, *On periodic solutions of differential equations with multivalued right-hand side*, Trudy Mat. Fak. Voronezh. Gos. Univ., vol. 10, Voronezh, 1973, pp. 74–82. (in Russian)

[Ob3] _____, *Semilinear functional differential inclusions in a Banach space and controlled parabolic systems*, Soviet J. Automat. Inform. Sci. **24** (1991), no. 3, 71–79.

[ObZe] V. V. Obukhovskiĭ and P. Zecca, *On some properties of dissipative functional diffrential inclusions in a Banach space*, Topol. Methods Nonlinear Anal. **15** (2000), 369–384.

[Ol1] C. Olech, *Existence of solutions of non convex orientor field*, Boll. Un. Mat. Ital. **4** (1975), 189–197.

[Ol2] _____, *Boundary solutions of differential inclusions*, Lecture Notes in Math. **979** (1983), 236–239.

[Ol3] _____, *Decomposability as a substitute for convexity*, Lecture Notes in Math. **1091** (1984), 193–205.

[Ol4] _____, *Lectures on the integration of set-valued functions*, Preprint of ISAS no. 44/87/M (1987), Trieste.

[Ol5] _____, *On the existence and uniqueness of solutions of an ordinary differential equation in the case of a Banach space*, Bull. Acad. Polon. Math. **8** (1969), 667–673.

[Ne1] B. O'Neill, *A fixed point theorem for multivalued functions*, Duke Math. J. **14** (1947), 689–693.

[Ne2] _____, *Essential sets and fixed points*, Amer. J. Math. **75** (1953), 497–509.

[Ne3] _____, *Induced homology homomorphisms for set-valued maps*, Pacific J. Math. **7** (1957), 1179–1184.

[ORe1] D. O'Regan, *Positive solutions for a class of boundary value problems on infinite intervals*, Nonlin. Diff. Eqns Appl. **9** (1994), 203–228.

[ORe2] _____, *Boundary value problems on noncompact intervals*, Proceed. Royal Soc. Edinburgh **125A** (1995), 777–799.

[ORe3] _____, *Singular nonlinear differential equations on the half line*, Topol. Meth. Nonlin. Anal. **8** (1996), 137–159.

[ORP] D. O'Regan and R. Precup, *Fixed point theorems for set-valued maps and existence principles for integral inclusions*, J. Math. Anal. Appl. **245** (2000), 594–612.

[Orl] W. Orlicz, *Zur Theorie der Differentialgleichung $y' = f(x, y)$*, Bull. Akad. Polon. Sci, Sér A **00** (1932), 221–228.

[OrlSz] W. Orlicz and S. Szufla, *On the structure of L^φ-solution sets of integral equations in Banach spaces*, Studia Math. **77** (1984), 465–477.

[Ort] R. Ortega, *A boundedness result of Landesman–Lazer type*, Diff. Integral Eqns **8** (1995), no. 4, 729–734.

[OrtTa] R. Ortega and M. Tarallo, *Massera's theorem for quasi-periodic differential equations*, Topol. Meth. Nonlin. Anal. **19** (2002), no. 1, 39–61.

[OrtTi] R. Ortega and A. Tineo, *Resonance and non-resonance in a problem of boundedness*, Proceed. Amer. Math. Soc. **124** (1996), 2089–2096.

[Osm] V. G. Osmolovskiĭ, *The local structure of the solution set of a first-order nonlinear boundary value problem with constraints at points*, Sibirsk. Mat. Zh. **27** (1986), 140–154, 206. (in Russian)

[Pal1] K. J. Palmer, *On the reducibility of almost periodic systems of linear differential equations*, J. Diff. Eqns **36** (1980), no. 3, 374–390.

[Pal2] _____, *On bounded solutions of almost periodic linear differential systems*, J. Math. Anal. Appl. **103** (1984), no. 1, 16–25.

[Pal3] _____, *Exponential dichotomies for almost-periodic equations*, Proceed. Amer. Math. Soc. **101** (1987), 293–298.

[Pap1] N. S. Papageorgiou, *A property of the solution set of differential inclusions in Banach spaces with Carathéodory orientor field*, Appl. Anal. **27** (1988), 279–287.

[Pap2] _____, *Boundary value problems for evolution inclusions*, Comment. Math. Univ. Carolin. **29** (1988), no. 2, 355–363.

[Pap3] _____, *Fixed point theorems for multifunctions in metric and vector spaces*, Nonlinear Anal. **7** (1983), 763–770.

[Pap4] _____, *Nonsmooth analysis on partially ordered vector spaces*, Part 2: *Nonconvex case, Clarke's theory*, Pacif. J. Math. **109** (1983), no. 2, 463–495.

[Pap5] _____, *On multivalued evolution equations and differential inclusions in Banach spaces*, Comment. Math. Univ. St. Paul. **36** (1987), no. 1, 21–39.

[Pap6] _____, *On multivalued semilinear evolution equations*, Boll. Un. Mat. Ital. B **3** (1989), no. 1, 1–16.

[Pap7] _____, *On the solution evolution set of differential inclusions in Banach spaces*, Appl. Anal. **25** (1987), 319–329.

[Pap8] _____, *On Fatou's lemma and parametric integrals for set-valued functions*, J. Math. Anal. Appl. **187** (1994), no. 3, 809–825.

[Pap9] _____, *Kneser's theorem for differential equations in Banach spaces*, Bull. Austral. Math. Soc. **33** (1986), no. 3, 419–434.

[Pap10] _____, *On the solution set of differential inclusions in a Banach space*, Appl. Anal. **25** (1987), no. 4, 319–329.

[Pap11] _____, *On the solution set of differential inclusions with state constraints*, Appl. Anal. **31** (1989), 279–289.

[Pap12] _____, *Convexity of the orientor field and the solution set of a class of evolution inclusions*, Math. Slovaca **43** (1993), no. 5, 593–615.

[Pap13] _____, *On the properties of the solution set of nonconvex evolution inclusions of the subdifferential type*, Comment. Math. Univ. Carolin. **34** (1993), no. 4, 673–687.

[Pap14] _____, *A property of the solution set of nonlinear evolution inclusions with state constraints*, Math. Japon. **38** (1993), no. 3, 559–569.

[Pap15]   _____, *On the solution set of nonlinear evolution inclusions depending on a parameter*, Publ. Math. Debrecen **44** (1994), no. 1–2, 31–49.

[Pap16]   _____, *On the solution set of nonconvex subdifferential evolution inclusions*, Czechoslovak Math. J. **44** (1994), no. 3, 481–500.

[Pap17]   _____, *On the topological regularity of the solution set of differential inclusions with constraints*, J. Diff. Eqns **107** (1994), no. 2, 280–289.

[Pap18]   _____, *On the topological properties of the solution set of evolution inclusions involving time-dependent subdifferential operators*, Boll. Un. Mat. Ital. **9** (1995), no. 2, 359–374.

[Pap19]   _____, *On the properties of the solution set of semilinear evolution inclusions*, Nonlinear Anal. TMA **24** (1995), no. 12, 1683–1712.

[Pap20]   _____, *Topological properties of the solution set of integrodifferential inclusions*, Comment. Math. Univ. Carolin. **36** (1995), no. 3, 429–442.

[Pap21]   _____, *On the solution set of nonlinear integrodifferential inclusions in* $\mathbb{R}^N$, Math. Japon. **46** (1997), no. 1, 117–127.

[Pap22]   _____, *On the structure of the solution set of evolution inclusions with time-dependent subdifferentials*, Rend. Sem. Mat. Univ. Padova **97** (1997), 163–186.

[Pap23]   _____, *Topological properties of the solution set of a class of nonlinear evolutions inclusions*, Czechoslovak Math. J. **47** (1997), no. 3, 409–424.

[Pap24]   _____, *On the solution set of evolution inclusions driven by time dependent subdifferentials*, Math. Japon. **37** (1992), 1087–1099.

[PapPa]   N. S. Papageorgiou and F. Papalini, *On the structure of the solution set of evolution inclusions with time-dependent subdifferentials*, Acta Math. Univ. Comenian. (N.S.) **65** (1996), no. 1, 33–51.

[PapSh]   N. S. Papageorgiou and N. Shahzad, *Properties of the solution set of nonlinear evolution inclusions*, Appl. Math. Optim. **36** (1997), no. 1, 1–20.

[Ppl]   F. Papalini, *Properties of the solution set of evolution inclusions*, Nonlinear Anal. **26** (1996), 1279–1292.

[Pr1]   S. Park, *Generalized Leray–Schauder principles for condensing admissible multifunctions*, Ann. Mat. Pura Appl. (4) **172** (1997), 65–85.

[Pr2]   _____, *A unified fixed point theory of multimaps on topological vector spaces*, J. Korean Math. Soc. **35** (1998), 803–829.

[Pat]   S. N. Patnaik, *Fixed points of multiple-valued transformations*, Fund. Math. **65** (1969), 345–349.

[Pav]   N. Pavel, *On dissipative systems*, Boll. U.M.I. **4** (1971), no. 4, 701–707.

[Pea1]   G. Peano, *Sull'integrabilitá delle equazioni differenziali del primo ordine*, Atti della Reale Accad. dell Scienze di Torino **21** (1886), 677–685.

[Pea2]   _____, *Démonstration de l'integrabilite des équations differentielles ordinaires*, Math. Annalen **37** (1890), 182–238.

[PeTo]   A. Pelczynski and H. Torunczyk, *Certain non-locally convex linear metric spaces homeomorphic to Hilbert spaces*, Commentat. Math. **I** (1978), 261–265.

[Pei]   H. O. Peitgen, *On the Lefschetz number for iterates of continuous mappings*, Proc. Amer. Math. Soc. **54** (1976), 441–444.

[Per]   M. P. Pera, *A topological method for solving nonlinear equations in Banach spaces and some related global results on the structure of the solution sets*, Rend. Sem. Mat. Univ. Politec. Torino **41** (1983), 9–30.

[Pet1]   W. V. Petryshyn, *On the solvability of* $x \in Tx + \lambda Fx$ *in quasinormal cones with* T *and* F, *k-set-contractive*, Nonlinear Anal. **5** (1981), no. 5, 585–591.

[Pet2]   _____, *Some results on multiple positive fixed points of multivalued condensing maps*, Contemp. Math. **21** (1983), 171–177.

[Pet3]   _____, *Note on the structure of fixed point sets of 1-set-contractions*, Proc. Amer. Math. Soc. **31** (1972), 189–194.

[Pet4]   _____, *Solvability of various boundary value problems for the equation* $x'' = f(t, x', x'') - y$, Pacific J. Math. **122** (1986), 169–195.

[PetFi]   W. V. Petryshyn and P. M. Fitzpatrick, *A degree theory, fixed-point theorems and mapping theorems for multivalued noncompact mappings*, Trans. Amer. Math. Soc. **194** (1974), 1–25.

[PetYu1]   W. V. Petryshyn and Z. S. Yu, *Solvability of Neumann problems for nonlinear second-order ODEs which need not be solvable for the highest-order derivative*, J. Math. Anal. Appl. **91** (1983), 244–253.

[PetYu2]   ———, *On the solvability of an equation describing the periodic motions of a satellite in its elliptic orbit*, Nonlin. Anal. **9** (1985), 961–975.

[Pia]   G. Pianigiani, *Existence of solutions of ordinary differential equations in Banach spaces*, Bull. Acad. Polon. Math. **23** (1975), 853–857.

[Pit]   H. R. Pitt, *On the Fourier coefficients of almost periodic functions*, J. London Math. Soc. **14** (1939), 143–150.

[Pl1]   S. Plaskacz, *Periodic solutions of differential inclusions on compact subsets of $\mathbb{R}^n$*, J. Math. Anal. Appl. **148** (1990), 202–212.

[Pl2]   ———, *On the solution sets for differential inclusions*, Boll. Un. Mat. Ital. A **6** (1992), 387–394.

[Pli]   A. Pliś, *Measurable orientor fields*, Bull. Polish Acad. Sci. Math. **13** (1965), 565–569.

[Poi]   H. Poincaré, *Mémoire sur les courbes définies par les éqations différentielles*, J. Math. Pures Appl. **3** (1881), no. 7, 375–422; **3** (1882), no. 8, 251–286; **4** (1885), no. 1, 167–244; **4** (1886), no. 2, 151–217; C. R. Acad. Sci. Paris **93** (1881), 951–952; **98** (1884), 287–289.

[Pol]   E. S. Polovinkin, *The properties of continuity and differentiation of solution sets of Lipschitzean differential inclusions*, Modeling, Estimation and Control of Systems with Uncertainty, Birkhäuser, Boston, 1991, pp. 349–360.

[VP]   C. de la Vallée Poussin, *Sur les fonctions presque périodiques de H. Bohr*, Ann. Soc. Sci. Bruxelles **47** (1927), 140–158.

[Po1]   M. J. Powers, *Multi-valued mappings and Lefschetz fixed point theorems*, Proc. Cambridge Philos. Soc. **68** (1970), 619–630.

[Po2]   ———, *Fixed point theorems for non-compact approximative ANR-s*, Fund. Math. **75** (1972), 61–68.

[Po3]   ———, *Lefschetz fixed point theorem for a new class of multi-valued maps*, Pacific J. Math. **42** (1972), 211–220.

[PG]   A. M. Powlotskiĭ and E. A. Gango, *On periodic solutions of differential equations with a multi-valued right-hand side*, Utchen. Zap. Leningrad. Ped. Inst. Im. Gertzena **541** (1972), 145–154. (in Russian)

[P1]   B. Przeradzki, *On a two-point boundary value problem for differential equations on the half-line*, Ann. Polon. Math. **50** (1989), 53–61.

[P2]   ———, *On the solvability of singular BVPs for second-order ordinary differential equations*, Ann. Polon. Math. **50** (1990), 279–289.

[P3]   ———, *The existence of bounded solutions for differential equations in Hilbert spaces*, Ann. Polon. Math. **56** (1992), 103–121.

[Pus]   L. Pust, *Influence of the variable power source on the oscillations of a mechanical system*, Aplikace Matematiky **3** (1958), no. 6, 428–450. (in Russian)

[R1]   I. Rachůnková, *On a Kneser problem for a system of nonlinear ordinary differential equations*, Czech. Math. J. **31** (1981), no. 106, 114–126.

[R2]   ———, *On Kneser problem for differential equations of the 3rd order*, Čas. pěst. mat. **115** (1990), 18–27.

[R3]   ———, *Nonnegative nonincreasing solutions of differential equations of the 3rd order*, Czech. Math. J. **40** (1990), 213–221.

[Rad]   L. Radová, *The Bohr–Neugebauer-type theorems for almost-periodic differential systems*, Preprint (2002).

[RedWa]   R. M. Redheffer and W. Walter, *Flow-invariant sets and differential inequalities in normed spaces*, Appl. Anal. **5** (1975), 149–161.

[Rei] A. Reich, *Präkompakte Gruppen und Fastperiodizität*, Mat. Zeitschr., **116** (1970), 216–234.

[Rss] R. Reissig, *Continua of periodic solutions of the Liénard equation*, Constructive Methods for Nonlinear Boundary Value Problems and Nonlinear Oscillations (J. Albrecht, L. Collatz and K. Kirchgässner, eds.), Birkäuser, Basel, 1979, pp. 126–133.

[Rhe] C. J. Rhee, *Homotopy functors determined by set-valued maps*, Math. Z. **113** (1970), 154–158.

[Rho] B. E. Rhoades, *Some theorems on weakly contractive maps*, Nonlinear Analysis **47** (2001), 2683–2693.

[Ri1] B. Ricceri, *Une propriété topologique de l'ensemble des points fixes d'une contraction multivoque à valeurs convexes*, Atti. Accad. Naz. Lincei Cl. Sci. Fis. Mat. Natur. Rend. Lincei **81** (1987), 283–286.

[Ri2] _____, *Existence theorems for nonlinear problems*, Rend. Accad. Naz. Sci. XL Mem. Mat. **11** (1987), 77–99.

[Ri3] _____, *On the Cauchy problem for the differential equation* $f(t, x, x', \ldots, x^{(k)}) = 0$, Glasgow Math. J. **33** (1991), 343–348.

[Ri4] _____, *On multifunctions with convex graph*, Atti. Accad. Naz. Lincei Rend. Cl. Sci. Fis. Mat. Natur. **77** (1984), 64–70.

[Ri5] _____, *Sur l'approximation des sélections mesurables*, C. R. Acad. Sci. Paris Sér. I Math. **295** (1982), 527–530.

[Ri6] _____, *On the topological dimension of the solution set of a class of nonlinear equations*, C. R. Acad. Sci. Paris, Sér. I Math. **325** (1997), 65–70.

[Ri7] _____, *Covering dimension and nonlinear equations*, RIMS **1031** (1998), Surika-isekikenkyusho-Kokyuroku, Kyoto, 97–100.

[NR] O. Naselli Ricceri, *A-fixed points of multi-valued contractions*, J. Math. Anal. Appl. **135** (1988), 406–418.

[Rob] W. Robertson, *Completions of topological vector spaces*, Proceed. London Math. Soc. **3** (1958), 242–257.

[Roc] R. T. Rockafellar, *Protodifferentiability of set-valued mappings and its applications in optimization*, Ann. Inst. H. Poincare. Anal. Nonlin., Suppl., vol. 6, 1989, pp. 449–482.

[Rox] E. O. Roxin, *Stability in general control systems*, J. Diff. Eqns **1** (1965), 115–150.

[Ryb] K. P. Rybakowski, *Some remarks on periodic solutions of Carathéodory RFDEs via Ważewski principle*, Riv. Mat. Univ. Parma **4** (1982), no. 8, 377–385.

[Rbn1] L. E. Rybiński, *On Carathéodory type selectors*, Fund. Math. **125** (1985), 187–193.

[Rbn2] _____, *A fixed point approach in the study of the solution sets of Lipschitzian functional-differential inclusions*, J. Math. Anal. Appl. **160** (1991), 24–46.

[Rz] B. Rzepecki, *An existence theorem for bounded solutions of differential equations in Banach spaces*, Rend. Sem. Mat. Univ. Padova **73** (1985), 89–94.

[Rz1] T. Rzeżuchowski, *On the set where all the solutions satisfy a differential inclusion*, Qual. Theory Differ. Equations., vol. 2, Amsterdam, 1981, pp. 903–913.

[Rz2] _____, *Scorza-Dragoni type theorem for upper semicontinuous multi-valued functions*, Bull. Polish Acad. Sci. Math. **28** (1980), 61–66.

[Sac] S. K. Sachdev, *Positive solutions of a nonlinear boundary value problem on half-line*, Panamer. Math. J. **1** (1991), no. 2, 27–40.

[SckSe1] R. J. Sacker and G. R. Sell, *Existence of dichotomies and invariant splittings for linear differential systems*, I, J. Diff. Eqns **15** (1974), no. 3, 429–459.

[SckSe2] _____, *Existence of dichotomies and invariant splittings for linear differential systems*, II, J. Diff. Eqns **22** (1976), no. 2, 478–496.

[SckSe3] _____, *Existence of dichotomies and invariant splittings for linear differential systems*, III, J. Diff. Eqns **22** (1976), no. 2, 497–522.

[SckSe4] ———, *A spectral theory for linear differential systems*, J. Diff. Eqns **27** (1978), no. 3, 320–358.

[Sad1] B. N. Sadovskiĭ, *On measures of noncompactness and contracting operators*, Problems in the Mathematical Analysis of Complex Systems, Second Edition, Voronezh, 1968, pp. 89–119. (in Russian)

[Sad2] ———, *Limit-compact and condensing operators*, Uspekhi Mat. Nauk **27** (1972), 81–146 (in Russian); Russian Math. Surveys **27** (1972), 85–155. (Engl. transl.)

[SR1] J. Saint-Raymond, *Points fixes des multiapplications à valeurs convexex*, C. R. Acad. Sci. Paris Sér I Math. **298** (1984), 71–74.

[SR2] ———, *Points fixes des contractions multivoques*, Fixed Point Theory and Appl., vol. 252, Pitman Res. Notes Math., 1991, pp. 359–375.

[SR3] ———, *Multivalued contractions*, Set-Valued Analysis **4** (1994), 559–571.

[Sam1] A. R. Sambucini, *Integrazione per seminorme in spazi localmente convessi*, Riv. Mat. Univ. Parma **5** (1994), no. 3, 321–381.

[Sam2] ———, *Remarks on set valued integrals of multifunctions with nonempty, bounded, closed and convex values*, Commentationes Math **34** (1999), 159–165.

[Sam3] ———, *A survey of multivalued integration*, Atti Sem. Mat. Fis. Univ. Modena **50** (2002), no. 1, 53–63.

[Sav] P. Saveliev, *A Lefschetz-type coincidence theorem*, Preprint (1998).

[Sc] J. Schauder, *Der Fixpunktsatz in Funktionalräumen*, Studia Math. **2** (1930), 171–180.

[Sch1] H. Schirmer, *A Kakutani type coincidence theorem*, Fund. Math. **69** (1970), 219–226.

[Sch2] ———, *A Nielsen number for fixed points and near points of small multifunctions*, Fund. Math. **88** (1977), 145–156.

[Sch3] ———, *An index and a Nielsen number for n-valued multifunctions*, Fund. Math. Soc. **124** (1984), 207–219.

[Sch4] ———, *Fixed points, antipodal points and coincidences of n-acyclicvalued multifunctions*, Proc. Special. Session on Fixed Points, AMS, Toronto, 1982.

[Sch5] ———, *A relative Nielsen number*, Pacific J. Math. **112** (1986), 459–473.

[Sch6] ———, *On the location of fixed points on a pair of spaces*, Topology Appl. **30** (1988), 253–266.

[Sch7] ———, *A survey of relative Nielsen fixed point theory*, Nielsen Theory and Dynamical Systems, Contemp. Math. (C. K. McCord, ed.), vol. 152, Amer. Math. Soc. Colloq. Publ., Providence, R. I., 1993, pp. 291–309.

[Sch8] ———, *A topologist's view of Sharkovskiĭ's theorem*, Houston J. Math. **11** (1985), no. 3, 385–395.

[Schm] R. Schmidt, *Die trigonometrische Approximation für eine Klasse von verallgemeinerten fastperiodischen Funktionen*, Math. Ann. **100** (1928), 333–356.

[SchtWa] K. Schmitt and J. R. Ward, Jr., *Almost periodic solutions of nonlinear second order differential equations*, Results in Math. **21** (1992), 190–199.

[Scho] U. K. Scholz, *The Nielsen fixed point theory for noncompact spaces*, Rocky Mountain J. Math. **4** (1974), 81–87.

[SchWe] R. M. Schori and J. E. West, *The hyperspace of the closed unit interval is a Hilbert cube*, Trans. Am. Math. Soc. **213** (1975), 217–235.

[Scr1] K. W. Schrader, *Existence theorems for second order boundary value problems*, J. Diff. Eqns **5** (1969), 572–584.

[Scr2] ———, *Second and third order boundary value problems*, Proceed. Amer. Math. Soc. **32** (1972), 247–252.

[Scu1] J. D. Schuur, *The existence of proper solutions of a second order ordinary differential equation*, Proceed. Amer. Math. Soc. **17** (1966), 595–597.

[Scu2] ———, *A class of nonlinear ordinary differential equations which inherit linear-like asymptotic behavior*, Nonlin. Anal. TMA **3** (1979), 81–86.

[Sea] Sek Wui Seah, *Bounded solutions of multivalued differential systems*, Houston J. Math. **8** (1982), 587–597.

[Sed]   S. Sedziwy, *Periodic solutions of differential equations in the cylindrical space*, Ann. Polon. Math. **26** (1972), 335–339.

[Se4]   W. Segiet, *Local coincidence index for morphisms*, Bull. Polish Acad. Sci. Math. **30** (1982), 261–267.

[Se5]   _____, *Nonsymmetric Borsuk-Ulam theorem for multivalued mappings*, Bull. Polish Acad. Sci. Math. **32** (1984), 113–119.

[Sei1]   G. Seifert, *On certain solutions of a pendulum-type equation*, Quart. Appl. Math. **11** (1953), no. 1, 127–131.

[Sei2]   _____, *A rotated vector approach to the problem of solvability of solutions of pendulum-type equations*, Ann. Math. Stud. **36** (1956), Princeton Univ. Press, 1–17.

[Sei3]   _____, *Stability conditions for the existence of almost periodic solutions of almost periodic systems*, J. Math. Anal. Appl. **10** (1965), 409–418.

[Sei4]   _____, *Almost periodic solutions for almost periodic systems of ordinary differential equations*, J. Diff. Eqns **2** (1966), 305–319.

[Sel1]   G. R. Sell, *Almost periodic and periodic solutions of difference equations*, Bull. of the Amer. Math. Soc. **72** (1966), no. 2, 261–265.

[Sel2]   _____, *Periodic solutions and asymptotic stability*, J. Diff. Eqns **2** (1966), no. 2, 143–157.

[Sel3]   _____, *The structure of a flow in the vicinity of an almost periodic motion*, J. Diff. Eqns **27** (1978), no. 3, 359–393.

[SerTT]   E. Serra, M. Tarallo and S. Terracini, *On the structure of the solution set of forced pendulum-type equations*, J. Diff. Eqns **131** (1996), 189–208.

[Sgh]   A. Sghir, *On the solution set of second-order delay differential inclusions in Banach spaces*, Ann. Math. Blaise Pascal **7** (2000), 65–79.

[Sh1]   A. N. Sharkovskiǐ, *Coexistence of cycles of a continuous map of a line into itself*, Ukrainian Math. J. **16** (1964), 61–71. (in Russian)

[Sh2]   _____, *A classification of fixed points*, Ukrainian Math. J. **17** (1965), no. 5, 80–95. (in Russian)

[SY]   Zhang Shi-Sheng and Chen Yu-Chin, *Degree theory for multivalued (S) type mappings and fixed point theorems*, Appl. Math. Mech. **11** (1990), no. 5, 441–454.

[Shi]   J. S. Shin, *Kneser type theorems for functional differential equations in a Banach space*, Funk. Ekvacioj **35** (1992), 451–466.

[Shu]   M. A. Shubin, *Almost-periodic functions and pseudo-differential operators*, Russian Math. Surv. **33** (1978), 1–52.

[SeS]   H. W. Siegberg and G. Skordev, *Fixed point index and chain approximations*, Pacific J. Math. **102** (1982), 455–486.

[Sil]   D. B. Silin, *On set-valued differentation and integration*, Set-Valued Anal. **5** (1997), no. 2, 107–146.

[Ski1]   R. Skiba, *On the Lefschetz fixed point theorem for multivalued weighted mappings*, Acta. Univ. Palacki. Olomuc. **40** (2001), 201–214.

[Ski2]   _____, *Topological essentiality for multivalued weighted mappings*, Acta. Univ. Palacki. Olomuc. **41** (2002) (to appear).

[Sk1]   E. G. Skljarenko, *On a theorem of Vietoris and Begle*, Dokl. Akad. Nauk **149** (1963), 264–267.

[Sk2]   _____, *On some applications of sheaf theory in general topology*, Uspekhi Mat. Nauk **19(6)** (1964), 47–70. (in Russian)

[Sko1]   G. Skordev, *Fixed point index for open sets in euclidean spaces*, Fund. Math. **121** (1984), no. 1, 41–58.

[Sko2]   _____, *The multiplicative property of the fixed point index for multivalued maps*, Serdica **15** (1989), no. 2, 160–170.

[Sko3]   _____, *The Lefschetz fixed point theorem*, God. Sofii Sk. Univ. Fak. Mat. i Meh. Mat. **79** (1985), no. 1, 201–213.

[Sko4]   _____, *Fixed points of maps of AANR-s*, Bull. Polish Acad. Sci. Math. **21** (1973), 173–180.

[Smth]   R. A. Smith, *Some applications of Hausdorff dimension inequalities for ordinary differential equations*, Proceed. Royal Soc. Edinb. **104A** (1986), 235–259.

[Sng1]   Z. Song, *Existence of generalized solutions for ordinary differential equations in Banach spaces*, Math. Anal. Appl. **128** (1987), 405–412.

[Sng2]   ———, *The solution set of a differential inclusion on a closed set of a Banach space*, Appl. Math. **23** (1995), 13–23.

[Sos]   W. Sosulski, *Compactness and upper semi continuity of solution set of functional differential equations of hyperbolic type*, Comment. Mat., Prace. Mat. **25** (1985), no. 2, 359–362.

[Spe]   A. Spezamiglio, *Bounded solutions of ordinary differential equations and stability*, Rev. Mat. Statist. **8** (1990), 1–9. (in Portuguese)

[SprBi]   J. S. Spraker and D. C. Biles, *A comparison of the Carathéodory and Filippov solution sets*, J. Math. Anal. Appl. **198** (1996), 571–580.

[Srv]   S. M. Srivastava, *Approximations and approximative selections of upper Carathéodory multifunctions*, Boll. U.M.I. **7** (1994), no. 8-A, 251-262.

[Sr1]   R. Srzednicki, *Periodic and constant solutions via topological principle of Wazewski*, Acta Math. Univ. Iagel. **26** (1987), 183–190.

[Sr2]   ———, *Periodic and bounded solutions in blocks for time-periodic nonautonomous ordinary differential equations*, Nonlin. Anal. TMA **22** (1994), no. 6, 707–737.

[Sr3]   ———, *Generalized Lefschetz theorem and fixed point index formula*, Topology Appl. **81** (1997), 207–224.

[Sr4]   ———, *A generalization of the Lefschetz fixed point theorem and detection of chaos*, Proc. Amer. Math. Soc. (to appear).

[Sr5]   ———, *On periodic solutions inside isolating chains*, J. Diff. Eqns **165** (2000), 42–60.

[Sr6]   ———, *On solutions of two-point boundary value problems inside isolating segments*, Topol. Meth. Nonlin. Anal. **13** (1999), 73–89.

[Stc1]   V. Staicu, *Continuous selections of solution sets to evolution equations*, Proc. Amer. Math. Soc. **113** (1991), 403–413.

[Stc2]   ———, *Qualitative properties of solutions sets to Lipschitzian differential inclusions*, World Sci. Publ. (1993), Singapore, 910–914.

[Stc3]   ———, *On the solution sets to nonconvex differential inclusions of evolution type*, Discrete Contin. Dynam. Systems **2** (1998), 244–252.

[StcWu]   V. Staicu and H. Wu, *Arcwise connectedness of solution sets to Lipschitzean differential inclusions*, Boll. Un. Mat. Ital. A **7** (1991), no. 5, 253–256.

[St1]   S. Staněk, *Bounded solutions of second order functional differential equations*, Acta Univ. Palacki. Olomuc. 100, Math. **30** (1991), 97–105.

[St2]   ———, *On the boundedness and periodicity of solutions of second-order functional differential equations with a parameter*, Czech. Math. J. **42** (1992), 257–270.

[St3]   ———, *On the class of functional boundary-value problems for the equation* $x'' = f(t, x, x', x'', \lambda)$, Ann. Polon. Math. **59** (1994), 225–237.

[St4]   ———, *On an almost periodicity criteria of solutions for systems of nonhomogeneous linear differential equationss with almost periodic coefficients*, Acta Univ. Palacki. Olomuc., Fac. Res. Nat., Math. **28** (1989), no. 94, 27–42.

[SZ]   G. Stefani and P. Zecca, *Properties of convex sets with application to differential theory of multivalued functions*, Nonlinear Analysis T.M.A. **2** (1978), 581–593.

[Stp1]   W. Stepanov, *Sur quelques généralisations des fonctions presque périodiques*, C. R. Acad. Sci. Paris **181** (1925), 90–92.

[Stp2]   ———, *Über einigen Verallgemeinerungen der fastperiodischen Funktionen*, Math. Ann. **95** (1926), 473–498.

[Sto1]   S. Stoinski, *On compactness in variation of almost periodic functions*, Demonstr. Math. **31** (1998), 131–134.

[Sto2]   ———, *On compactness of almost periodic functions in the Lebesgue measure*, Fasc. Math. **30** (1999), 171–175.

[Sto3]   ———, *Almost periodic functions in Lebesgue measure*, Comm. Math. **34** (1994), 189–198.

[Sto4]   ———, *Some remarks on spaces of almost periodic functions*, Fasc. Math. **31** (2001), 105–115.

[Sto5]   ———, L_α - *almost periodic functions*, Demonstr. Math. **28** (1995), 689–696.

[Sto6]   ———, *A note on H-almost periodic functions and S^P-almost periodic functions*, Demonstr. Math. **29** (1996), 557–564.

[Sto7]   ———, $(VC)^{(n)}$-*almost periodic functions*, Demonstr. Math. **32** (1999), 376–383.

[Sto8]   ———, *Real-valued functions almost periodic in variation*, Funct. Approx. Comment. Math. **22** (1993), 141–148.

[Sto9]   ———, *H-almost periodic functions*, Funct. Approx. Comment. Math. **1** (1974), 113–122.

[Sto10]   ———, *Approximation of H-almost periodic functions by means of certain linear operators*, Funct. Approx. Comment. Math. **2** (1976), 233–241.

[Sto11]   ———, *A connection between H-almost periodic functions and almost periodic functions of other types*, Funct. Approx. Comment. Math. **3** (1976), 205–223.

[Str1]   W. L. Strother, *Fixed points, fixed sets and M-retracts*, Bull. Polish Acad. Sci. Math. **22** (1955), 551–556.

[Str2]   ———, *Multi homotopy*, Duke Math. J. **22** (1955), 281–285.

[Su]   A. Suszycki, *Retracts and homotopies for multi-maps*, Fund. Math. **115** (1983), no. 1, 9–26.

[Se1]   V. Šeda, *On an application of the Stone theorem in the theory of differential equations*, Časopis pěst. Mat. **97** (1972), 183–189.

[Se2]   ———, *On a generalization of the Thomas-Fermi equation*, Acta Math. Univ. Comen. **39** (1980), 97–114.

[Se3]   ———, R_δ-*set of solutions to a boundary value problem*, Topol. Meth. Nonlin. Anal. **16** (2000), no. 1, 93–101.

[Se4]   ———, *Fredholm mappings and the generalized boundary value problem*, Diff. Integral Eqns **8**, no. 1995, 19–40.

[Se5]   ———, *Generalized boundary value problems and Fredholm mappings*, Nonlinear Anal. TMA **30** (1997), 1607–1616.

[SK]   V. Šeda and Z. Kubáček, *On the connectedness of the set of fixed points of a compact operator in the Fréchet space $C^m(\langle b, \infty), R^n)$*, Czechosl. Math. J. **42** (1992), no. 117, 577–588.

[Sv1]   M. Švec, *Fixpunktsatz und monotone Lösungen der Differentialgleichung $y^{(n)} + B(x, y, y', \ldots, y^{(n-1)})y = 0$*, Arch. Math. (Brno) **2** (1966), 43–55.

[Sv2]   ———, *Les propriétés asymptotique des solutions d'une équations différentielle nonlinéaire d'ordre n*, Czech. Math. J. **17** (1967), 550–557.

[Sz1]   S. Szufla, *Solutions sets of nonlinear equations*, Bull. Acad. Polon. Math. **21** (1973), 971–976.

[Sz2]   ———, *Sets of fixed points nonlinear mappings in function spaces*, Funkcial. Ekrac. **22** (1979), 121–126.

[Sz3]   ———, *On the existence of bounded solutions of nonlinear differential equations in Banach spaces*, Functiones et Approx. **15** (1986), 117–123.

[Sz4]   ———, *Some remarks on ordinary differential equations in Banach spaces*, Bull. Acad. Polon. Math. **16** (1968), 795–800.

[Sz5]   ———, *Measure of noncompactness and ordinary differential equations in Banach spaces*, Bull Acad. Polon. Sci. **19** (1971), 831–835.

[Sz6]   ———, *Structure of the solutions set of ordinary differential equations in a Banach space*, Bull. Acad. Polon. Sci. **21** (1973), no. 2, 141–144.

[Sz7] ———, *Some properties of the solutions set of ordinary differential equations*, Bull. Acad. Polon. Sci. **22** (1974), no. 7, 675–678.

[Sz8] ———, *On the structure of solutions sets of differential and integral equations in Banach spaces*, Ann. Polon. Math. **34** (1977), 165–177.

[Sz9] ———, *On the equation $\dot{x} = f(t,x)$ in Banach spaces*, Bull. Acad. Polon. Sci. **26** (1978), no. 5, 401–406.

[Sz10] ———, *Kneser's theorem for weak solutions of ordinary differential equations in reflexive Banach spaces*, Bull. Acad. Polon. Sci. **26** (1978), no. 5, 407–413.

[Sz11] ———, *On the existence of solutions of differential equations in Banach spaces*, Bull. Acad. Polon. Sci. **30** (1982), no. 11–12, 507–515.

[Sz12] ———, *On the equation $\dot{x} = f(t,x)$ in locally convex spaces*, Math. Nachr. **118** (1984), 179–185.

[Sz13] ———, *Existence theorems for solutions of integral equations in Banach spaces*, Proc. Conf. Diff. Eqns and Optimal Control (1985), Zielona Góra, 101–107.

[Sz14] ———, *On the application of measure of noncompactness to differential and integral equations in a Banach space*, Fasc. Math. **18** (1988), 5–11.

[Sz15] ———, *Aronszajn type theorems for differential and integral equations in Banach spaces*, Proceed. of the 1st Polish Symp. Nonlin. Anal. (L. Górniewicz et al., eds.), Wyd. Uniw. Łódzkiego, Łódź, 1997, pp. 113–123.

[Sz16] ———, *On the differential equation $x^{(m)} = f(t,x)$ in Banach spaces*, Funkcialaj Ekvacioj **41** (1998), no. 1, 101–105.

[Sz17] ———, *Kneser's theorem for weak solutions of an mth-order ordinary differential equation in Banach spaces*, Nonlin. Anal. **38** (1999), 785–791.

[Sz18] ———, *Solutions sets of non-linear integral equations*, Funkcial. Ekvac. **17** (1974), 67–71.

[Sz19] ———, *On the structure of solution sets of nonlinear equations*, Differential Equations and Optimal Control, Higher College Engrg., Zielona Góra, 1989, pp. 33–39.

[SzSz] S. Szufla and A. Szukała, *An existence theorem for the equation $x^{(m)} = f(t,x)$ in Banach spaces*, Preprint (2001).

[Tad] V. Taddei, *Bound sets for Floquet boundary value problems: the nonsmooth case*, Discr. Cont. Dynam. Syst. **6** (2000), 459–473.

[Tal1] P. Talaga, *The Hukuhara–Kneser property for parabolic systems with nonlinear boundary conditions*, J. Math. Anal. **79** (1981), 461–488.

[Tal2] ———, *The Hukuhara–Kneser property for quasilinear parabolic equations*, Nonlinear Anal. TMA **12** (1988), no. 3, 231–245.

[Tas] U. Tasmetov, *On the connectedness of hyperspaces*, Sov. Math., Dokl. **15** (1974), 502–504.

[Th] R. B. Thompson, *A unified approach to local and global fixed point indices*, Adv. Math. **3** (1969), 1–71.

[Tol1] A. A. Tolstonogov, *On differential inclusions in a Banach space and continuous selectors*, Dokl. Akad. Nauk SSSR **244** (1979), 1088–1092. (in Russian)

[Tol2] ———, *On properties of solutions of differential inclusions in a Banach space*, Dokl. Akad. Nauk SSSR **248** (1979), 42–46. (in Russian)

[Tol3] ———, *On the structure of the solution set for differential inclusions in Banach spaces*, Math. USSR Sbornik **46** (1983), 1–15. (in Russian)

[Tol4] ———, *On the density and "being boundary" for the solution set of a differential inclusion in a Banach space*, Dokl. Akad. Nauk SSSR **261** (1981), 293–296. (in Russian)

[Tol5] ———, *The solution set of a differential inclusion in a Banach space. II*, Sibirsk. Mat. Zh. **25** (1984), 159–173. (in Russian)

[TolChu] A. A. Tolstonogov and P. I. Chugunov, *The solution set of a differential inclusion in a Banach space. I*, Sibirsk. Mat. Zh. **24** (1983), 144–159. (in Russian)

[Tri]   F. Tricomi, *Integrazione di un aequazione differenziale presentatasi in elettrotec-nica*, Ann. R. Sc. Norm. Sup. di Pisa (1933), 1–20.

[Tto]   G. M. Troianiello, *Structure of the solution set for a class of nonlinear parabolic problems*, Nonlinear Parabolic Equations: Qualitative Properties of Solutions, Longman Sci. Tech., Harlow, 1987, pp. 219–225.

[Tua]   H. D. Tuan, *On the continuous dependence on parameter of the solution set of differential inclusions*, Z. Anal. Anwendungen **11** (1992), 215–220.

[U]   S. Umamaheswaran, *Boundary value problems for higher order differential equations*, J. Diff. Eqns **18** (1975), 188–201.

[UV]   S. Umamaheswaran and M. Venkata Rama, *Multipoint focal boundary value problems on infinite intervals*, J. Appl. Math. Stochastic Anal. **5** (1992), no. 3, 283–289.

[Umn]   Ya. I. Umanskiĭ, *On a property of the solution set of differential inclusions in a Banach space*, Differentsial'nye Uravneniya **28** (1992), 1346–1351, 1468. (in Russian)

[Urb]   K. Urbanik, *Fourier analysis in Marcinkiewicz spaces*, Studia Math. **21** (1961), 93–102.

[Urs]   H. D. Ursell, *Parseval's theorem for almost-periodic functions*, Proc. London Math. Soc. **32** (1931), 402–440.

[Vae1]   M. Vaeth, *Coepi maps and generalizations of the Hopf extension theorem* (to appear).

[Vae2]   _____, *Fixed point theorems and fixed-point index for countably condensing maps*, Topol. Methods Nonlinear Anal. **13** (1999), no. 2, 341–363.

[Vae3]   _____, *An axiomatic approach to a coincidence index for noncompact function pairs*, Topol. Methods Nonlinear Anal. **16** (2000), no. 2, 307–338.

[Vae4]   _____, *On the connection of degree theory and 0-epi maps*, J. Math. Anal. Appl. **257** (2001), 321–343.

[Vae5]   _____, *Coincidence points of function pairs based on compactness properties*, Glasgow Math. J. **44** (2002), no. 2, 209–230.

[Vee]   W. A. Veech, *Almost automorphic functions on groups*, Am. J. Math. **87** (1965), 719–751.

[Vej1]   O. Vejvoda, *Note to the paper of L. Pust: Influence of the variable power source on the oscillations of a mechanical system*, Aplikace Matematiky **3** (1958), no. 6, 451–460.

[Vej2]   _____, *On the existence and stability of the periodic solution of the second kind of a cetain mechanical systems*, Czech. Math. J. **9** (1959), no. 84, 390–415.

[Vel]   V. Veliov, *Convergence of the solution set of singularly perturbed differential inclusions*, Nonlinear Anal. **30** (1997), 5505–5514.

[Vid1]   G. Vidossich, *On Peano-phenomenon*, Boll. Un. Mat. Ital. **3** (1970), 33–42.

[Vid2]   _____, *On the structure of solutions set of nonlinear equations*, J. Math. Anal. Appl. **34** (1971), 602–617.

[Vid3]   _____, *A fixed point theorem for function spaces*, J. Math. Anal. Appl. **36** (1971), 581–587.

[Vid4]   _____, *Existence, uniqueness and approximation of fixed points as a generic property*, Bol. Soc. Brasil. Mat. **5** (1974), 17–29.

[Vid5]   _____, *Two remarks on global solutions of ordinary differential equations in the real line*, Proc. Amer. Math. Soc. **55** (1976), 111–115.

[Vie]   L. Vietoris, *Über den höheren Zusammenhang kompakter Räume und eine Klasse von zusammenhangstreuen Abbildungen*, Math. Ann. **97** (1927), 454–472.

[Wat]   P. Waltman, *Asymptotic behavior of solutions of an n-th order differential equation*, Monath. Math. Österr. **69** (1965), no. 5, 427–430.

[Wng]   Z. H. Wang, *Existence of solutions for parabolic type evolution differential inclusions and the property of the solution set*, Appl. Math. Mech. **20** (1999), 314–318.

[Wa1]   J. R. Ward, Jr., *Asymptotic conditions for periodic solutions of ordinary differential equations*, Proceed. Amer. Math. Soc. **81** (1981), 415–420.

[Wa2]   _____, *Averaging, homotopy and bounded solutions of ordinary differential equations*, Diff. Integral Eqns. **3** (1990), 1093–1100.

[Wa3]   _____, *A topological method for bounded solutions of nonautonomous ordinary differential equations*, Trans. Amer. Math. Soc. **333** (1992), no. 2, 709–720.

[Wa4]   _____, *Global continuation for bounded solutions of ordinary differential equations*, Topol. Meth. Nonlin. Anal. **2** (1993), no. 1, 75–90.

[Wa5]   _____, *Homotopy and bounded solutions of ordinary differential equations*, J. Diff. Eqns **107** (1994), 428–445.

[Wa6]   _____, *A global continuation theorem and bifurcation from infinity for infinite dimensional dynamical systems*, Proc. Roy. Soc. Edinburgh Sect. A **126** (1996), 725–738.

[Wa7]   _____, *Bifurcating continua in infinite dimensional dynamical systems and applications to differential equations*, J. Diff. Eqns **125** (1996), 117-132.

[Wa8]   _____, *Bounded and almost periodic solutions of semi-linear parabolic equations*, Rocky Mount. J. Math. **18** (1988).

[Waz1]   T. Ważewski, *Sur un principle topologique pour l'examen de allure asymptotique des intégrales des équations différentieles ordinaires*, Ann. Soc. Polon. Math. **20** (1947), 279–313.

[Waz2]   _____, *Sur l'existence et l'unicité des intégrales des équations différentieles ordinaires au cas de l'espace de Banach*, Bull. Acad. Polon. Math. **8** (1960), 301–305.

[Wei]   Z. K. Wei, *On the existence of unbounded connected branches of solution sets of a class of semilinear operator equations*, Bull. Soc. Math. Belg. Ser. B **38** (1986), 14–30.

[Wen]   S. Wenxian, *Horseshoe motions and subharmonics of the J-J type with small parameters*, Acta Math. Appl. Sinica **4** (1988), no. 4, 345–354.

[Wey]   H. Weyl, *Integralgleichungen und fastperiodische Funktionen*, Math. Ann. **97** (1926), 338–356.

[Wie]   N. Wiener, *On the representation of functions by trigonometric integrals*, Math. Zeitschr. **24** (1925), 575–616.

[Wil]   S. A. Williams, *An index for set-valued maps in infinite-dimensional spaces*, Proc. Amer. Math. Soc. **31** (1972), 557–563.

[Woj]   M. Wojdyslawski, *Retractes absolus et hyperespaces des continus*, Fundam. Math. **32** (1939), 184–192.

[Wo]   Pui- Kei Wong, *Existence and asymptotic behavior of proper solutions of a class of second order nonlinear differential equations*, Pacific J. Math. **13** (1963), 737–760.

[Wng1]   P. Wong, *A note on the local and the extension Nielsen numbers*, Topology Appl. **78** (1992), 207–213.

[Wng2]   _____, *Estimation of Nielsen type numbers for periodic points I, II*, Far East J. Math. Sci. **1** (1993), no. 1, 69–83.

[Wng3]   _____, *Fixed points on pairs of nilmanifolds*, Topology Appl. **62** (1995), 173–179.

[Wjc]   K. Wójcik, *Periodic segments and Nielsen numbers*, Banach Center Publ. **47** (1999), Inst of Math., Polish Acad. Sci., Warsaw, 247–252.

[Wjt]   D. Wójtowicz, *On implicit Darboux problem in Banach spaces*, Bull. Austral. Math. Soc. **56** (1997), 149–156.

[Ye]   X. Ye, *D-function of a minimal set and an extesion of Sharkovskiï's theorem to minimal sets*, Ergod. Th. Dynam. Sys. **12** (1992), 365–376.

[Yor1]   J. A. Yorke, *Extending Liapunov's second method to non-Lipschitz Liapunov functions*, Bull. Amer. Math. Soc. **2** (1968), 322–325.

[Yor2]   _____, *Spaces of solutions*, Lect. Notes Op. Res. Math. Econ., vol. 12, Springer-Verlag, Berlin, 1969, pp. 383–403.

[Yor3] ———, *A continuous differential equation in a Hilbert space without existence*, Funkc. Ekvac. **13** (1970), 19–21.

[Yor4] ———, *Differential inequalities and non-Lipschitz scalar functions*, Math. Systems Theory **4** (1970), 140–151.

[You1] C. Y. You, *A note on periodic points on tori*, Beijing Math. **1** (1995), 224–230.

[You2] ———, *The least number of periodic points on tori*, Adv. Math. (China) **24** (1995), no. 2, 155–160.

[Youj] J. You, *Invariant tori and Lagrange stability of pendulum-type equations*, J. Diff. Eqns **85** (1990), 54–65.

[Zan1] F. Zanolin, *Permanence and positive periodic solutions for Kolmogorov competing species*, Results in Math. **21** (1992), 224-250.

[Zan2] ———, *Continuation theorems for the periodic problem via the translation operator*, Rend. Sem. Mat. Univ. Politec. Torino **54** (1996), 1-23.

[Zan3] ———, *Bound sets, periodic solutions and flow-invariance for ordinary differential equations in $\mathbb{R}^N$: some remarks*, Rend. Ist. Mat. Univ. Trieste **19** (1987), 76-92.

[Zar] S. K. Zaremba, *Sur certaines familles de courbes en relation avec la théorie des équations différentielles*, Rocznik Polskiego Tow. Matemat. **XV** (1936), 83–100.

[ZZ] P. Zecca and P. L. Zezza, *Nonlinear boundary value problems in Banach space for multivalued differential equations on a noncompact interval*, Nonlin. Anal. TMA **3** (1979), 347–352.

[Zel] V. V. Zelikova, *On a topological degree for Vietoris type multimaps in locally convex spaces*, Sb. Stat. Asp. Stud. Mat. Fak. Vgu. (1999), Voronezh, 45–51. (in Russian)

[Zgl1] P. Zgliczyński, *Sharkovskiĭ theorem for multidimensional perturbations of 1-dim maps*, Ergod. Th. Dynam. Sys. **19** (1999), no. 6, 1655–1684.

[Zgl2] ———, *Sharkovskiĭ theorem for multidimensional perturbations of one-dimensional maps, II*, Topol. Math. Nonlin. Anal. **14** (1999), no. 1, 169–182.

[Zha] Chuanyi Zhang, *Pseudo almost periodic solutions of some differential equations*, J. Math. Anal. Appl. **181** (1994), 62–76.

[ZhnCha] L. Zhangju and H. L. Chan, *Periodic perturbations of phase-locking loops*, Ann. Mat. Sinica **7A** (1986), no. 6, 699–704.

[Zha1] W. X. Zhao, *A relative Nielsen number for the complement*, Lecture Notes in Math. **1411** (1989), Springer, Berlin, 189–199.

[Zha2] ———, *Estimation of the number of fixed points on the complement*, Topology Appl. **37** (1990), 257–265.

[Zha3] ———, *Basic relative Nielsen numbers*, Topology, Hawaii, World Scientific, Singapore, 1992, pp. 215–222.

[Zha4] ———, *On minimal fixed point numbers of relative maps*, Topol. Appl. **112** (2001), 229–250.

[ZL] V. V. Zhikov and B. M. Levitan, *The Favard theory*, Uspekhi Matem. Nauk. **32** (1977), 123–171 (in Russian); Russian Math. Surv. **32** (1977), 129–180.

[Zhu] Q. J. Zhu, *On the solution set of differential inclusions in a Banach space*, J. Diff. Eqns **93** (1991), 213–237.

[ZhWu] S. M. Zhu and Y. T. Wu, *Periodic solutions to autonomous functional-differential equations with piecewise constant function*, Acta Sci. Natur. Univ. Sunyatseni **29** (1990), no. 3, 12–19. (in Chinese)

[Zvy] V. G. Zvyagin, *The structure of the solution set of a nonlinear elliptic boundary value problem under fixed boundary conditions*, Topological and Geometric Methods of Analysis, vol. 173, Voronezh. Gos. Univ., Voronezh, 1989, pp. 152–158. (in Russian)

INDEX